Lecture Notes in Computer Science 8349

Commenced Publication in 1973
Founding and Former Series Editors:
Gerhard Goos, Juris Hartmanis, and Jan van Leeuwen

Yehuda Lindell (Ed.)

Theory of Cryptography

11th Theory of Cryptography Conference, TCC 2014
San Diego, CA, USA, February 24-26, 2014
Proceedings

 Springer

Volume Editor

Yehuda Lindell
Bar-Ilan University
Department of Computer Science
Ramat Gan 52900, Israel
E-mail: lindell@biu.ac.il

ISSN 0302-9743   e-ISSN 1611-3349
ISBN 978-3-642-54241-1   e-ISBN 978-3-642-54242-8
DOI 10.1007/978-3-642-54242-8
Springer Heidelberg New York Dordrecht London

Library of Congress Control Number: 2014930318

CR Subject Classification (1998): E.3, D.4.6, K.6.5, F.1.1-2, C.2.0, F.2.1-2, G.2.2, I.1

LNCS Sublibrary: SL 4 – Security and Cryptology

Typesetting: Camera-ready by author, data conversion by Scientific Publishing Services, Chennai, India

Printed on acid-free paper

Springer is part of Springer Science+Business Media (www.springer.com)

Preface

TCC 2014 was held at the University of California San Diego in California, during February 24–26, 2014. TCC 2014 was sponsored by the International Association for Cryptologic Research (IACR). The general chairs of the conference were Mihir Bellare and Daniele Micciancio. I would like to thank them in the name of the TCC community in general, and in the name of all of the participants of TCC 2014 in particular, for their hard work in organizing the conference.

The conference received 90 submissions, of which the Program Committee selected 30 for presentation at the conference. These proceedings consist of the revised versions of the 30 papers. The revisions were not reviewed, and the authors bear full responsibility for the contents of their papers. In addition to the regular paper presentations, TCC 2014 featured a rump session where short presentations of recent results were given, and two invited talks. The invited speakers were Russell Impagliazzo and Silvio Micali, and the Program Committee is very grateful to them for accepting our invitation.

I am greatly indebted to many people who contributed to the success of TCC 2014. First and foremost, I would like to thank all those who submitted their papers to TCC. The success of TCC is due mainly to your work. In addition, I would like to thank the Program Committee for all of their hard work and diligence in reviewing the submissions and choosing the program. A lot of work is involved in this process, and your service to the community is greatly appreciated. I would also like to thank all of the external reviewers who participated in the process and provided in-depth reviews of the papers that they read. Finally, I owe deep thanks to Shai Halevi and Tal Rabin who provided me with valuable advice when I needed it. The TCC Program Committee also used Shai's excellent web-review software, and I thank Shai for writing it and for the support he provided when needed.

This was the 11th Theory of Cryptography Conference, and it was my honor and pleasure to act as the program chair of TCC as it entered its second decade. A quick look at the proceedings herein suffices to appreciate the vibrant and dynamic work being carried out by the TCC community. The proceedings include research on new and exciting topics like obfuscation, as well as basic foundational research on classic topics like zero-knowledge, secure computation, encryption, black-box separations, cryptographic coding theory and more. In addition to the fascinating research presented at TCC, the conference atmosphere is always warm and friendly and is essentially a meeting of friends who come together to study the fundamentals of our field. I thank the entire TCC community for creating this event and for maintaining its unique and special qualities.

February 2014   Yehuda Lindell

TCC 2014
The 11th Theory of Cryptography Conference

University of California San Diego, California, USA
February 24–26, 2014

Sponsored by the *International Association for Cryptologic Research (IACR)*

General Chair

Mihir Bellare   UCSD, USA
Daniele Micciancio   UCSD, USA

Program Chair

Yehuda Lindell   Bar-Ilan University, Israel

Program Committee

Amos Beimel	Ben-Gurion University, Israel
Alexandra Boldyreva	Georgia Tech, USA
Kai-Min Chung	Academia Sinica, Taiwan
Yevgeniy Dodis	New York University, USA
Nelly Fazio	City University of New York, USA
Marc Fischlin	Darmstadt University of Technology, Germany
Jens Groth	University College London, UK
Iftach Haitner	Tel-Aviv University, Israel
Martin Hirt	ETH Zurich, Switzerland
Dennis Hofheinz	Karlsruhe Institute of Technology, Germany
Susan Hohenberger Waters	Johns Hopkins University, USA
Eike Kiltz	Ruhr-Universität Bochum, Germany
Eyal Kushilevitz	Technion – Israel Institute of Technology, Israel
Mohammad Mahmoody	Cornell University, USA
Claudio Orlandi	Aarhus University, Denmark
Christopher J. Peikert	Georgia Tech, USA
Krzysztof Pietrzak	IST, Austria
Mike Rosulek	Oregon State University, USA
Adam Smith	Pennsylvania State University, USA
Salil Vadhan	Harvard University, USA
Vinod Vaikuntanathan	University of Toronto, Canada

External Reviewers

Divesh Aggarwal
Shashank Agrawal
Martin Albrecht
Jacob Alperin-Sheriff
Joel Alwen
Christian Badertscher
Paul Baecher
Abhishek Banerjee
Nir Bitansky
Olivier Blazy
Elette Boyle
Christina Brzuska
Nishanth Chandran
Melissa Chase
Cheng Chen
Alessandro Chiesa
Sherman Chow
Sandro Coretti
Özgür Dagdelen
Ivan Damgård
Grégory Demay
Frederic Dupuis
Serge Fehr
Tore Frederiksen
Georg Fuchsbauer
Felix Günther
Tommaso Gagliardoni
Chaya Ganesh
Sanjam Garg
Peter Gazi
Rosario Gennaro

Sergey Gorbunov
Vipul Goyal
Shai Halevi
Kristiyan Haralambiev
Javier Herranz
Thomas Holenstein
Yuval Ishai
Tibor Jager
Abhishek Jain
Daniel Kraschewski
Sara Krehbiel
Guanfeng Liang
Huijia (Rachel) Lin
Feng-Hao Liu
Zhenming Liu
Adriana Lopez-Alt
Hemanta Maji
Giorgia Azzurra Marson
Daniel Masny
Eric Miles
Payman Mohassel
Antonio Nicolosi
Adam O'Neill
Cristina Onete
Jiaxin Pan
Omer Paneth
Milinda Perera
Manoj Prabhakaran
Carla Rafols
Ananth Raghunathan
Vanishree Rao

Pavel Raykov
Guy Rothblum
Christian Schaffner
Dominique Schröder
Karn Seth
Or Sheffet
Tom Shrimpton
Fang Song
Francois-Xavier
    Standaert
Uri Stemmer
Noah Stephens-
    Davidowitz
Björn Tackmann
Sidharth Telang
Aris Tentes
Stefano Tessaro
Roberto Trifiletti
Daniel Tschudi
Dominique Unruh
Yevgeniy Vahlis
Muthuramakrishnan
    Venkitasubramaniam
Dhinakaran
    Vinayagamurthy
Brent Waters
Daniel Wichs
Scott Yilek
Hong-Sheng Zhou

Invited Talks

Collusion and Privacy in Mechanism Design

Silvio Micali

Laboratory for Computer Science,
MIT, Cambridge, MA 02139
silvio@csail.mit.edu

Abstract. Mechanism design aims at engineering games that, rationally played, yield desired outcomes. In such games, multiple players interact very much as in a cryptographic protocol. But there are some fundamental differences. No player is "good", that is, always follows his prescribed instruction. No player is "malicious", that is, always acts so as to prevent the desired outcome from being achieved. Rather, every player is RATIONAL, that is, always acts so as to maximize HIS OWN utility.

Rational players too, however, have incentives to collude, and value privacy. Thus, privacy and collusion can disrupt the intended course of a game, and ultimately prevent the desired outcome from being achieved. Mechanism design has been only moderately successful in protecting against collusion, and has largely ignored privacy.

I believe that there is an opportunity for cryptographers and game theorists to join forces and produce new mechanisms that are resilient to collusion and privacy issues. I also believe that, to be successful, this effort requires a good deal of modeling and the development of new conceptual frameworks. In sum, there is the promise of a great deal of fun, challenge, and excitement, and I would like to recruit as much talent as possible towards this effort.

As a concrete example of what may be done in this area, I will describe a (quite) resilient mechanism, designed by Jing Chen and I, for achieving a (quite) alternative revenue benchmark in unrestricted combinatorial auctions. In such auctions there are multiple distinct goods for sale, each player privately attributes an arbitrary value to any possible subset of the goods, and the seller has no information about the players valuations. (Traditional mechanisms for unrestricted combinatorial auctions were uniquely "vulnerable" to collusion and privacy.)

Specific versus General Assumptions in Cryptography

Russell Impagliazzo *

CSE Department, UCSD

Abstract. Modern cryptography began with the insight that computational difficulty could limit the ability of an attacker to break encryption or forge signatures. However, it was not for another few years that the required computational difficulty of specific problems on specific distributions for a cryptographic protocol to be secure was made explicit and quantitative. A further advantage of formalizing this connection is that it clarifies the exact properties, both in terms of which aspects should be computationally feasible and which related problems should be computationally intractable, were used to prove security of the protocol. This lays the foundation for proving possibility results in cryptography based on general assumptions, about the existence of types of cryptographically useful tools, rather than based on the difficulty of specific problems. A pattern emerged, where a new cryptographic goal is proposed, an "existence proof" given based on specific assumptions (sometimes untested) is given, then a variety of protocols are given based on different assumptions, and then these protocols are abstracted in terms of more general assumptions that suffice.

This talk will focus on the history of how this pattern emerged, the advantages that proofs of security based on general assumptions gives over protocol design based on specific assumptions, and on both progress and set-backs in basing cryptography on general assumptions.

* Work supported by the Simons Foundation and NSF grant CCF-121351.

Table of Contents

Obfuscation

Applications of Obfuscation

Zero Knowledge

Black-Box Separations

Secure Computation

Coding and Cryptographic Applications

Leakage

Encryption

Hardware-Aided Secure Protocols

Encryption and Signatures

Virtual Black-Box Obfuscation for All Circuits via Generic Graded Encoding

Zvika Brakerski[1] and Guy N. Rothblum[2]

[1] Weizmann Institute of Science
[2] Microsoft Research

Abstract. We present a new general-purpose obfuscator for all polynomial size circuits. The obfuscator uses graded encoding schemes, a generalization of multilinear maps. We prove that the obfuscator exposes no more information than the program's black-box functionality, and achieves *virtual black-box security*, in the generic graded encoded scheme model. This proof is under the Bounded Speedup Hypothesis (BSH, a plausible worst-case complexity-theoretic assumption related to the Exponential Time Hypothesis), in addition to standard cryptographic assumptions. We also prove that it satisfies the notion of *indistinguishability obfuscation* without without relying on BSH (in the same generic model and under standard cryptographic assumptions).

Very recently, Garg et al. (FOCS 2013) used graded encoding schemes to present a candidate obfuscator for indistinguishability obfuscation. They posed the problem of constructing a provably secure indistinguishability obfuscator in the generic graded encoding scheme model. Our obfuscator resolves this problem (indeed, under BSH it achieves the stronger notion of virtual black box security, which is our focus in this work).

Our construction is different from that of Garg et al., but is inspired by it, in particular by their use of permutation branching programs. We obtain our obfuscator by developing techniques used to obfuscate d-CNF formulas (ITCS 2014), and applying them to permutation branching programs. This yields an obfuscator for the complexity class $\mathcal{NC}^1$. We then use homomorphic encryption to obtain an obfuscator for any polynomial-size circuit.

1 Introduction

Code obfuscation is the task of taking a program, and making it "unintelligible" or impossible to reverse engineer, while maintaining its input-output functionality. While this is a foundational question in the theory and practice of cryptography, until recently very few techniques or heuristics were known. Recently, however, several works have leveraged new constructions of cryptographically secure graded encoding schemes (which generalize multilinear maps) [23,20] to propose obfuscators for complex functionalities [10,11] and, in a fascinating recent work of Garg et al [24], even for arbitrary polynomial size circuits.

Y. Lindell (Ed.): TCC 2014, LNCS 8349, pp. 1–25, 2014.

In this work, we propose a new code obfuscator, building on techniques introduced in [10,24,11]. The obfuscator works for any polynomial-time circuit, and its security is analyzed in the idealized generic graded encoding scheme model. We prove that, in this idealized model, the obfuscator achieves the strong "virtual black-box" security notion of Barak et al. [5] (see below). Security in the idealized model relies on a worst-case exponential assumption on the hardness of the $\mathcal{NP}$-complete 3SAT problem (in the flavor of the well known exponential time hypothesis). Our construction relies on *asymmetric graded encoding schemes*, and can be instantiated using the new candidate constructions of Garg, Gentry and Halevi [23], or of Coron, Lepoint and Tibouchi [20].

Obfuscation: Definitions. Intuitively, an obfuscator should generate a new program that preserves the the original program's functionality, but is impossible to reverse engineer. The theoretical study of this problem was initiated by Barak et al. [5]. They formalized a strong simulation-based security requirement of *black box obfuscation*: namely, the obfuscated program should expose nothing more than what can be learned via oracle access to its input-output behavior. We refer to this notion as "black-box" obfuscation, and we use this strong formalization throughout this work.

A weaker notion of obfuscation, known as *indistinguishability* or *best-possible* obfuscation was studied in [5,27]. An indistinguishability obfuscator guarantees that the obfuscations of any two programs (boolean circuits) with identical functionalities are indistinguishable. We note that, unlike the black-box definition of security, indistinguishability obfuscation does not quantify or qualify what information the obfuscation might expose. In particular, the obfuscation might reveal non-black-box information about the functionality. Recently, Sahai and Waters [37] showed that indistinguishability obfuscation suffices for many cryptographic applications, such as transforming private key cryptosystems to public key, and even for constructing deniable encryption schemes.

Prior Work: Negative Results. In their work, [5] proved the *impossibility* of general-purpose black-box obfuscators (i.e. ones that work for any polynomial-time functionality) in the virtual black box model. This impossibility result was extended by Goldwasser and Kalai [26]. Goldwasser and Rothblum [27] showed obstacles to the possibility of achieving indistinguishability obfuscation with information-theoretic security, and to achieving it in the idealized random oracle model.

Looking ahead, we note that the impossibility results of [5,26] *do not extend* to idealized models, such as the random oracle model, the generic group model, and (particularly relevant to our work) the generic graded encoding model.

Prior Work: Positive Results. Positive results on obfuscation focused on specific, simple programs. One program family, which has received extensive attention, is that of "point functions": password checking programs that only accept a single input string, and reject all others. Starting with the work of Canetti [13], several works have shown obfuscators for this family under various assumptions

[17,32,39], as well as extensions [14,8]. Canetti, Rothblum and Varia [18] showed how to obfuscate a function that checks membership in a hyperplane of constant dimension (over a large finite field). Other works showed how to obfuscate cryptographic function classes under different definitions and formalizations. These function classes include checking proximity to a hidden point [21], vote mixing [1], and re-encryption [29]. Several works [13,17,28,29] relaxed the security requirement so that obfuscation only holds for a random choice of a program from the family.

More recently, Brakerski and Rothblum [10] showed that graded encoding schemes could be used to obfuscate richer function families. They constructed a black-box obfuscator for *conjunctions*, the family of functions that test whether a subset of the input bits take on specified values. Building on this result, in a followup work [11], they constructed a black-box obfuscator for d-CNFs and (more generally) conjunctions of $\mathcal{NC}^0$ circuits. These constructions were proved secure in the generic graded encoding model. The conjunction obfuscator was also shown to be secure under falsifiable (see [34]) multilinear DDH-like assumptions, so long as the conjunction is drawn from a family with sufficient entropy.

In recent work, Garg et al. [24] use cryptographic graded encoding schemes to construct a candidate indistinguishability obfuscator (see above) for all polynomial size circuits. This is the first non-trivial candidate in the literature for general-purpose obfuscation. The main differences between our results and theirs are: (*i*) we construct an obfuscator with the stronger security notion of black-box obfuscation (for the same class of functions), and (*ii*) we provide a security proof in the generic graded encoding scheme model. This was posed as a major open question in [24].[1]

Canetti and Vaikuntanathan [19] outline a candidate obfuscator and prove its security in an idealized pseudo-free group model. They also use Barrington's theorem and randomization techniques. The main difference from our work is in the nature of their idealized pseudo-free group model: in particular, we do not know of an instantiation that is conjectured to be secure.

1.1　Our Work: Black-Box Obfuscation for All of $\mathcal{P}$

In this work we construct an obfuscator for any function in $\mathcal{P}$, using cryptographic graded encoding schemes. Our obfuscator can be instantiated using recently proposed candidates [23,20]. The main component of our construction is an obfuscator for the complexity class $\mathcal{NC}^1$, which is then leveraged to an obfuscator for $\mathcal{P}$ using homomorphic encryption. Our main contribution is a proof that the main component is a secure black-box obfuscator in the generic graded encoding scheme model, assuming the *bounded speedup hypothesis* (BSH) [11], a generalization of the exponential time hypothesis. More details follow.

Theorem 1.1. *There exists an obfuscator* PObf *for any circuit in* $\mathcal{P}$, *which is virtual black-box secure in the generic graded encoding scheme model, assuming*

---

[1] [24] provide a proof of security in a more restricted "generic colored matrix model".

the bounded speedup hypothesis, and the existence of homomorphic encryption with an $\mathcal{NC}^1$ decryption circuit.

We also prove that our obfuscator is an indistinguishability obfuscator in the generic graded encoding scheme model (in fact, we show that this is true even for a simplified variant of the construction). This proof does not require the bounded speedup hypothesis.[2]

Theorem 1.2. *There exists an obfuscator* PIndObf *for any circuit in* $\mathcal{P}$, *which is an indistinguishability obfuscator in the generic graded encoding scheme model, assuming the existence of homomorphic encryption with an $\mathcal{NC}^1$ decryption circuit.*

To prove Theorem 1.2, we use an equivalent formulation for the security of indistinguishability obfuscators. This formulation requires the existence of a *computationally unbounded* simulator, which only has *black-box access* to the obfuscated program.

For the remainder of this section, we focus our attention on black-box obfuscation. Our construction proceeds in two steps. As hinted above, the first (and main) step is an obfuscator for $\mathcal{NC}^1$ circuits. Then, in the second step, we use homomorphic encryption [35,25] to obfuscate any polynomial-size circuit. (Which is done by encrypting the input circuit using the homomorphic scheme, and obfuscating a "verified decryption" circuit, as explained in the full version [12].)

Our obfuscator for $\mathcal{NC}^1$ circuits combines ideas from: (*i*) the *d*-CNF obfuscator of [11] and its security proof. In particular, we build on their technique of *randomizing sub-assignments* to prove security in the generic model based on the bounded speedup hypothesis, and we also build on their use of random generators in each ring of the graded encoding scheme. We also use ideas from (*ii*) the indistinguishability obfuscator of [24], in particular their use of Barrington's theorem [6] and randomization techniques for permutation branching programs. We note that the obfuscator of [19] was also based on Barrington's theorem and randomization techniques. See Section 1.2 for an overview of the construction and its proof, Section 3 for the detailed construction, and the full version [12] for the security proof.

The Generic Graded Encoding Scheme Model. We prove that our construction is a black-box obfuscator in the generic graded encoding scheme model. In this model, an adversary must operate independently of group elements' representations. The adversary is given arbitrary strings representing these elements, and can only manipulate them using oracles for addition, subtraction, multilinear operations and more. See Section 2.5 for more details.

The Bounded Speedup Hypothesis. We prove security based on the *Bounded Speedup Hypothesis*, as introduced by [11]. This is a worst-case assumption about

---

[2] Under the BSH, the claim follows immediately from Theorem 1.1, because any virtual black-box obfuscator is also an indistinguishability obfuscator [27].

exponential hardness of 3SAT, a strengthening of the long-standing exponential time hypothesis (ETH) for solving 3SAT [30]. The exponential-time hypothesis states that no sub-exponential time algorithm can resolve satisfiability of 3CNF formulas. Intuitively, the bounded-speedup hypothesis states that no polynomial-time algorithm for resolving satisfiability of 3CNFs can have "super-polynomial speedup" over a brute-force algorithm that tests assignments one-by-one. More formally, there does not exist an ensemble of polynomial-size circuits $\{\mathcal{A}_n\}$, and an ensemble of super-polynomial-size sets of assignments $\{\mathcal{X}_n\}$, such that on input a 3CNF Φ on n-bit inputs, w.h.p. $\mathcal{A}_n$ finds a satisfying assignment for Φ in $\mathcal{X}_n$ if such an assignment exists. We emphasize that this is a worst-case hypothesis, i.e. it only says that for every ensemble $\mathcal{A}$, there *exists* some 3CNF on which $\mathcal{A}$ fails. See Section 2.2 for the formal definition.[3]

Perspective. Barak et al. [5] show that there are function families that are impossible to obfuscate under the black-box security definition. Their results *do not apply to idealized models* such as the random oracle model, the generic group model, and the generic graded encoding model. This is because their adversary needs to be able to execute the obfuscated circuit on parts of its own explicit description. In idealized models, the obfuscated circuit does not have a succinct explicit description, and so these attacks fail. Indeed, our main result, Theorem 1.1, shows that *general-purpose black-box obfuscation is possible in the generic graded encoding model* (under plausible assumptions). Indeed, prior works have shown that virtual black-obfuscation is possible in various idealized models: [5] showed that there *exists* an (arguably contrived) oracle that allows general-purpose obfuscation. [19] proposed a black-box obfuscator in an idealized generic pseudo-free group model, but we do not know an instantiation of this model that is conjectured to be secure. In contrast, Theorem 1.1 provides an (arguably) *natural* idealized model that allows general-purpose black-box obfuscation. It is natural to ask how one should interpret this result in light of the impossibility theorems.

One immediate answer, is that if one implements a graded encoding scheme using opaque secure hardware (essentially implementing the generic model), then the hardware can be used to protect any functionality (under plausible assumptions). The hardware is (arguably) natural, simple, stateless, and independent of the functionality being obfuscated.

Another answer, is that the security proof shows that (under plausible assumptions) our obfuscator is provably resilient to attacks from a rich family: namely, to all attacks that are independent of the encoding scheme's instantiation. While we find this guarantee to be of value, we caution that it should not be over-interpreted. The results of [5] imply that, for any concrete instantiation of the graded encoding scheme, the obfuscation of their unobfuscatable functions *is not secure*. In particular, their result (applied to our construction) provides a

---

[3] We note that if both the adversary and the black-box simulator are allowed to run in quasi-polynomial time, security can be based on the (standard) Exponential-Time Hypothesis.

non-generic attack against any graded encoding candidate. This is similar to the result of [15], showing an attack against any instantiation of a random oracle in a particular protocol. Somewhat differently from their result, however, in our case the primitive in question is altogether impossible in the standard model. In the result of [15], the primitive is not achieved by a specific construction when the idealized model is instantiated with *any* concrete functionality. We find this state of affairs to be of interest, even irrespective of the applications to code obfuscation.

Taking a more optimistic view, the new construction invites us to revisit limits in the negative results: both in the unobfuscatable functionalities, and in the nature of the attacks themselves. It may suggest new relaxations that make obfuscation achievable in standard models, e.g. obfuscating functionalities that inherently do not allow self-execution, or protecting against a class of attackers that cannot execute the obfuscated code on itself.

Finally, the relaxed notion of indistinguishability obfuscation, where only limited hardness and impossibility results are known, remains a promising avenue for future research. Our construction is the first provably secure indistinguishability obfuscator in the generic graded encoding model. It is interesting to explore whether indistinguishability obfuscation can be proved in the standard model under falsifiable assumptions.

Follow-up Work. In recent follow-up work, Barak et al. [4] propose an obfuscator that achieves virtual black-box security in the generic graded encoding scheme model without relying on the BSH. Their construction builds on encoding and randomization techniques introduced in this work.

1.2   Construction Overview

We proceed with an overview of the main step in our construction: an obfuscator for $\mathcal{NC}^1$. We are assuming basic familiarity with graded encoding schemes. The full construction appears in Section 3. Due to space limitations, the proof of its security in the generic model appears in the full version of this manuscript [12].

Permutation Branching Programs. The obfuscator NC^{1}Obf takes as input an $\mathcal{NC}^1$ program, represented as an oblivious width 5 permutation branching program C, as in [24]. Let m denote the depth of C (as is necessary, we allow the obfuscator to expose m or some upper bound thereof). Let $C = \{M_{j,0}, M_{j,1}\}_{j \in [m]}$, where $M_{j,b} \in \{0,1\}^{5 \times 5}$ are matrices, and let $i = \ell(j)$ indicate which variable x_i controls the jth level branch. See Section 2.1 for more background on (oblivious) branching programs.

Graded Encoding Schemes. We begin by recalling the notion of multilinear maps, due to Boneh and Silverberg [9]. Rothblum [36] considered the asymmetric case, where the groups may be different. The obfuscator makes (extensive) use of an asymmetric graded encoding schemes, which are an extension of multilinear maps.

Similarly to [24,19], we assign a group prog_j to each level j of the branching program, and encode the matrices $M_{j,b}$ in the group prog_j.[4] This encoding is done as in [10,11]: we encode each matrix $M_{j,b}$ relative to a unique generator of prog_j (denoted $\rho_{\text{prog}_j,b}$), and also provide an encoding of the generators. In other words, in the j-th group prog_j, we have two pairs:

$$(\rho_{\text{prog}_j,0}, (\rho_{\text{prog}_j,0} \cdot M_{j,0})) \text{ and } (\rho_{\text{prog}_j,0}, (\rho_{\text{prog}_j,0} \cdot M_{j,0}))$$

We note that this encoding, relative to a random generator, is different from what was done in [24], and plays a crucial role in the security proof.

Randomizing the Matrices. As computed above, the encoded pairs clearly hide nothing: $M_{j,b}$ are binary matrices, and so they are completely revealed (via zero-testing). As a first step, we use the $\mathcal{NC}^1$ randomization technique (see [3,31,22]), as was done in [24] (however, unlike [24], we don't need to extend the matrix dimensions beyond 5×5). The idea is to generate a sequence of random matrices $\mathcal{Y}_j$ (over the ring R underlying the encoding scheme), and work with encodings of $\mathcal{N}_{j,b} = \mathcal{Y}_{j-1}^{-1} \cdot M_{j,b} \cdot \mathcal{Y}_j$ instead of the original $M_{j,b}$. This preserves the program's functional behavior, but each matrix, examined in isolation, becomes completely random. In fact, even if we take one matrix out of each pair, the joint distribution of these m matrices is uniformly random (see Section 2.1).

There is an obstacle here, because using the standard graded encoding interface, we can generate a random level 0 encoding of $\mathcal{Y}$, but we cannot derive $\mathcal{Y}^{-1}$ (in fact, this is not possible even for scalars). Indeed, to perform this step, [24] rely on the properties of a specific graded encoding instantiation. We propose a difference solution that works with any graded encoding scheme: instead of $\mathcal{Y}^{-1}$, we use the adjoint matrix $\mathcal{Z} = \text{adj}(\mathcal{Y})$, which is composed of determinants of minors of $\mathcal{Y}$, and is therefore computable given the level 0 encoding of $\mathcal{Y}$. We know that $\mathcal{Y} \cdot \mathcal{Z} = \det(\mathcal{Y}) \cdot I$, which will be sufficient for our purposes.[5] Using the encoding scheme from [10,11], in the j-th group prog_j we encode two pairs:

$$(\rho_{\text{prog}_j,b}, (\rho_{\text{prog}_j,b} \cdot \mathcal{N}_{j,b}))_{b \in \{0,1\}}, \text{ where } \mathcal{N}_{j,b} = \mathcal{Z}_{j-1} \cdot M_{j,b} \cdot \mathcal{Y}_j$$

To efficiently *evaluate* this program, we need an additional group, which we denote by chk. In this group we encode a random generator ρ_{chk}, and the element $(\rho_{\text{chk}} \cdot (\prod_j \det(\mathcal{Y}_j)) \cdot \mathcal{Y}_m[1,1])$. We evaluate the branching program using the graded encoding scheme's zero-test feature, by checking whether:

$$((\rho_{\text{prog}_1,x_{\ell(1)}} \mathcal{N}_{1,x_{\ell(1)}}) \cdots (\rho_{\text{prog}_m,x_{\ell(m)}} \mathcal{N}_{m,x_{\ell(m)}}) \cdot \rho_{\text{chk}}) [1,1] -$$
$$\rho_{\text{prog}_1,x_{\ell(1)}} \cdots \rho_{\text{prog}_m,x_{\ell(m)}} \cdot (\rho_{\text{chk}} \cdot (\prod_j \det(\mathcal{Y}_j)) \cdot \mathcal{Y}_m[1,1]) = 0 \ .$$

---

[4] In the setting of multilinear maps, we typically refer to input groups. The map takes a single element from each input group, and maps them to a target group. In the setting of graded encoding schemes, the groups are replaced with indexed sets. See Section 2.4 for further details.

[5] We use here the fact that all matrices we work with are of constant dimension, and so we can compute the determinants of the minors in polynomial time while using only multilinear operations.

This provides the required functionality, but it does not provide a secure construction.

Enforcing Consistency. An obvious weakness of the above construction is that it does not verify *consistency*. For a variable x_i that appears multiple time in the program, the above scheme does not enforce that the same value will be used at all times. This will be handled, similarly to [24,11], by adding *consistency check variables*. In each group grp that is "associated" with a variable x_i (so far, these only include groups of the form prog_j s.t. $\ell(j) = i$), the obfuscator generates two random variables $\beta_{\mathrm{grp},i,0}$ and $\beta_{\mathrm{grp},i,1}$, and multiplies the relevant variables. Namely, in group prog_j with $\ell(j) = i$, we provide encodings of

$$(\rho_{\mathrm{prog}_j,b}, (\rho_{\mathrm{prog}_j,b} \cdot \beta_{\mathrm{prog}_j,i,b} \cdot \mathcal{N}_{j,b}))_{b \in \{0,1\}}$$

To preserve functionality, we would like to choose the β variables so that the product of all zero-choices and the product of all one-choices are the same (one might even consider imposing a constraint that the product is 1). For clarity of exposition, we prefer the following solution: we use an additional auxiliary group cc_i for every variable x_i, such that

$$\beta_{\mathrm{cc}_i,0} = \beta'_{\mathrm{cc}_i} \cdot \prod_{j:\ell(j)=i} \beta_{\mathrm{prog}_j,1} \ ,$$

and vice versa (and β'_{cc_i} is the same for both cases). This guarantees that the product of all zero-choices, and the product of all one-choices, is the same. We denote this value by γ_i.

To preserve functionality, we multiply the element in the chk group by $\prod_i \gamma_i$. Now in the chk group we have encodings of:

$$(\rho_{\mathrm{chk}}, (\rho_{\mathrm{chk}} \cdot \prod_i \gamma_i \cdot (\prod_j \det(\mathcal{Y}_j)) \cdot \mathcal{Y}_m[1,1]))$$

Intuitively, it seems that this change renders inconsistent assignments useless: if, for some bit i of the input, the β values for i are not all taken according to the same value (0 or 1), then the constraint does not come into play. Therefore, the β values completely randomize these selected values.

One could postulate that the above construction is secure. In fact, we do not know of an explicit generic-model attack on this construction. Still, there are challenges to constructing a simulator. The crux of the difficulty is that an attacker might somehow efficiently produce a multilinear expression that corresponds to the evaluation of *multiple (super-polynomially many) consistent inputs* at the same time (or some function of super-polynomially many inputs: e.g. checking if the circuit accepts all of them simultaneously). This would break the obfuscator's security, since an (efficient) simulator cannot evaluate the function on super-polynomially many inputs.

Indistinguishability Obfuscation via Inefficient Simulation. If we allow a computationally unbounded simulator, then the above is not a problem. We show that the existence of a computationally unbounded black-box simulator implies indistinguishability obfuscation. In fact, the notions are equivalent both in the standard model and in the generic graded encoding scheme model. *Indistinguishability obfuscation* for $\mathcal{NC}^1$ therefore follows, and an indistinguishability obfuscator for $\mathcal{P}$ can be derived using the $\mathcal{NC}^1$ to $\mathcal{P}$ transformation of [24].

We note that the main conceptual difference that allows us to prove indistinguishability obfuscation for our construction, as opposed to [24]'s, is our use of the randomized ρ generators. This allows us, for any multilinear expression computed by the adversary, to isolate the relevant consistent inputs that affect the value of that expression.

Efficient Simulation and Virtual Black-Box Obfuscation. To get efficient black-box simulation (and virtual black-box security), we need to address the above difficulty. To do so, we build on the *randomizing sub-assignments* technique from [11]. Here, we use this technique to *bind the variables together* into triples. This done by adding $\binom{n}{3}$ additional groups, denoted bind$_T$, where $T \in \binom{[n]}{3}$ (i.e. one for each triple of variables). The group bind$_T$ is associated with the triple of variables $\{i_1, i_2, i_3\} \in T$, and contains 8 pairs of encodings:

$$\left(\rho_{\mathsf{bind}_T, b_1 b_2 b_3}, \left(\rho_{\mathsf{bind}_T, b_1 b_2 b_3} \cdot \beta_{\mathsf{bind}_T, i_1, b_1} \cdot \beta_{\mathsf{bind}_T, i_2, b_2} \cdot \beta_{\mathsf{bind}_T, i_3, b_3}\right)\right)_{b_1 b_2 b_3 \in \{0,1\}^3}$$

In evaluating the program on an input x, for each group bind$_T$, the evaluator chooses one of these 8 pairs according to the bits of $x_{|T}$, and uses it in computing the two products that go into the zero test (as above). The aforementioned consistency variables $\beta_{\mathsf{cc}_i, b}$ take the new β's into account, and are accordingly computed as

$$\beta_{\mathsf{cc}_i, 0} = \beta'_{\mathsf{cc}_i} \cdot \prod_{j : \ell(j) = i} \beta_{\mathsf{prog}_j, 1} \cdot \prod_{T : i \in T} \beta_{\mathsf{bind}_T, i, 1} \,,$$

and vice versa. The γ_i values are modified in the same way.

Intuitively, in order to evaluate the program, the adversary now needs not only to consistently choose the value of every single variable, but also to *jointly commit* to the values of each triple, consistently with its choices for the singleton variables. We show that if a polynomial adversary is able to produce an expression that corresponds to a sum of superpolynomially many consistent evaluations, then it can also evaluate a 3SAT formula on superpoynomially many values simultaneously, which contradicts the bounded speedup hypothesis (BSH, see Section 2.2).

A Taste of the Security Proof. The (high-level) intuition behind the security proof is as follows. In the idealized generic graded encoding scheme model, an adversary can only compute (via the encoding scheme) multilinear arithmetic circuits of the items encoded in the obfuscation. Moreover, the expansion of these multilinear circuits into a sum-of-monomials form, will only have one element from each group in each monomial (note that this expansion may be inefficient and the number of monomials can even be exponential). We call this a *cross-linear polynomial*.

The main challenge for simulation is "zero testing" of cross linear polynomials, given their circuit representation:[6] determining whether or not a polynomial f computed by the adversary takes value 0 on the items encoded in the obfuscation. We note that this is where we exploit the generic model—it allows us to reason about what functions the adversary is computing, and to assume that they have the restricted cross-linear form. We also note that zero-testing is a serious challenge, because the simulator does not know the joint distribution of the items encoded in the obfuscation (their joint distribution depends on the branching program C in its entirety). Due to space limitations, the full details on the simulator are deferred to the full version [12].

A cross-linear polynomial f computed by the adversary can be decomposed using monomials that only depend on the ρ variables in the construction outlined above. f must be a sum of such "ρ-monomials", each multiplied with a function of the other variables (the variables derived from the matrices of the branching program and the randomized elements used in the obfuscation). Because of the restricted structure of these ρ-monomials, they each implicitly specify an assignment to every program group prog_j (a bit value for the $\ell(j)$-th bit), every binding group bind_T (a triple of bit values for the input bits in T), and every consistency variable cc_i (a bit value for the i-th bit). We say that the assignment is *full and consistent*, if all of these groups are assigned appropriate values, and the value assigned to each bit i (0 or 1) is the same in throughout all the groups. We show that if a polynomial f computed by the adversary contains even a single such ρ-monomial that is *not full and consistent*, then it will not take 0 value (except with negligible probability over the obfuscator's coins), and thus it can be simulated. Further, if f contains only full and consistent ρ-monomials, the simulator can isolate each of these monomials, discover the associated full and consistent input assignment, and then zero-test f.

The main remaining concern, as hinted at above, is that f *might have superpolynomially many* full and consistent ρ-monomials. Isolating these monomials as described above would then take super-polynomial time (whereas we want a polynomial time black-box simulator). Intuitively, this corresponds to an adversary that can test some condition on the obfuscated circuit's behavior over super-polynomially many inputs (which the black-box simulator cannot do). Let $X \subseteq \{0,1\}^n$ denote the (super-polynomial) set of assignments associated with the above ρ-monomials. We show that given such f (even via black-box access), it is possible to test whether any given 3CNF formula Φ has a satisfying assignment in X. This yields *a worst-case* "super-polynomial speedup" in solving 3SAT. Since the alleged f is computable by a polynomial size arithmetic circuit, we get a contradiction to the Bounded Speedup Hypothesis.

This connection to solving 3SAT is proved by building on the "randomizing sub-assignment" technique from [11] and the groups bind_T. The intuition is as

---

[6] In fact, all we need is black-box access to the polynomial, and the *guarantee* that it is computable by a polynomial-size arithmetic circuit.

follows. The generic adversary implicitly specifies a polynomial-size arithmetic circuit that computes the function f. Recall that f has many full and consistent ρ-monomials, each associated with an input $x \in X \subseteq \{0,1\}^n$. For a given 3CNF formula Φ, we can compute a restriction of f by setting some of the variables $\rho_{\mathsf{bind}_T,i,\vec{v}}$ to 0, in a way that "zeroes out" every ρ-monomial associated with an input $x \in X$ that does not satisfy Φ, which is possible since the binding variables correspond exactly to all possible clauses in a 3CNF formula (see below). We then test to see whether or not the restricted polynomial (which has low degree) is identically 0, i.e. whether any of the ρ-monomials were *not* "zeroed out" by the above restriction. This tells us whether there exists $x \in X$ that satisfies Φ.

All that remains is to compute the restriction claimed above. For every clause in the 3CNF formula Φ, operating on variables $T = \{i_1, i_2, i_3\}$, there is an assignment $\vec{v} \in \{0,1\}^3$ that fails to satisfy the clause. For each such clause, we set the variable $\rho_{\mathsf{bind}_T,\vec{v}}$ to be 0. This effectively "zeroes out" all of the ρ-monomials whose associated assignments do not satisfy Φ (i.e., the ρ-monomials whose assignments simultaneously fail to satisfy *all* clauses in Φ).

The full construction appears in Section 3. The security proof appears in the full version [12], and is omitted from this extended abstract due to space limitations.

2 Preliminaries

For all $n, d \in \mathbb{N}$ we define $\binom{[n]}{d}$ to be the set of lexicographically ordered sets of cardinality d in $[n]$. More formally:

$$\binom{[n]}{d} = \left\{ \langle i_1, \ldots, i_d \rangle \in [n]^d : i_1 < \cdots < i_d \right\}.$$

Note that $\left| \binom{[n]}{d} \right| = \binom{n}{d}$.

For $\vec{x} \in \{0,1\}^n$ and $I = \langle i_1, \ldots, i_d \rangle \in \binom{[n]}{d}$, we let $\vec{x}_{|I} \in \{0,1\}^d$ denote the vector $\langle \vec{x}[i_1], \ldots, \vec{x}[i_d] \rangle$. We often slightly abuse notation when working with $\vec{s} = \vec{x}_{|I}$, and let $\vec{s}[i_j]$ denote the element $x[i_j]$ (rather than the i_jth element in $\vec{s}$).

2.1 Branching Programs and Randomizations

A width-5 length-m permutation branching program C for n-bit inputs is composed of: a sequence of pairs of permutations represented as 0/1 matrices $(M_{j,v} \in \{0,1\}^{5 \times 5})_{j \in [m], v \in \{0,1\}}$, a labelling function $\ell : [m] \to [n]$, an accepting permutation Q_{acc}, and a rejecting permutation Q_{rej} s.t. $Q_{\mathsf{acc}}^T \cdot \vec{e}_1 = \vec{e}_1$ and $Q_{\mathsf{rej}}^T \cdot \vec{e}_1 = \vec{e}_k$ for $k \neq 1$. Without loss of generality, we assume that all permutation branching programs have the same accepting and rejecting permutations.

For an input $\vec{x} \in \{0,1\}$, taking $P = \prod_{j \in [m]} M_{j,\vec{x}[\ell(j)]}$, the program's output is 1 iff $P = Q_{\mathsf{acc}}$, and 0 iff $P = Q_{\mathsf{rej}}$. If P is not equal to either of these permutations, then the output is undefined (this will never be the case in any construction we use).

Barrington's Theorem [6] shows that any function in $\mathcal{NC}^1$, i.e. a function that can be computed by a circuit of depth d can be computed by a permutation branching program of length 4^d. Moreover, the theorem is constructive, and gives an algorithm that efficiently transforms any depth d circuit into a permutation branching program in time $2^{O(d)}$. This program is *oblivious*, in the sense that its labeling function is independent of the circuit C (and depends only on its depth). An immediate implication is that $\mathcal{NC}^1$ circuits, which have depth $d(n) = \log(n)$, can be transformed into polynomial length branching program, in polynomial time.

Theorem 2.1 (Barrington's Theorem [6]).
For any circuit depth d and input size n, there exists a length $m = 4^d$, a labeling function $\ell : [m] \to [n]$, an accepting permutation Q_{acc} and a rejecting permutation Q_{rej}, s.t. the following holds. For any circuit with input size n, depth d and fan-in 2, which computes a function f, there exists a permutation branching program of length m that uses the labeling function $\ell(\cdot)$, has accepting permutation Q_{acc} and rejecting permutation Q_{rej}, and computes the same function f.

The permutation branching program is computable in time $\text{poly}(m)$ given the circuit description.

Randomized Branching Programs. Permutation branching programs are amenable to randomization techniques that have proved very useful in cryptography and complexity theory [3,31,22,2]. The idea is to "randomize" each matrix pair while preserving the program's functional behavior. Specifically, taking p to be a large prime, for $i \in [m]$ multiply the i-th matrix pair (on its right) by a random invertible matrix $\mathcal{Y}_i \in \mathbb{Z}_p^{*5 \times 5}$, and multiply the $(i+1)$-th pair by $\mathcal{Y}_i^{-1}$ (on its left). This gives a new branching program:

$$(\mathcal{N}_{j,v})_{j\in[m],v\in\{0,1\}} : \mathcal{N}_{j,v} = (\mathcal{Y}_{j-1}^{-1} \cdot M_{j,v} \cdot \mathcal{Y}_j)$$

(where we take $\mathcal{Y}_0^{-1}$ to be the identity). The new randomized program preserves functionality in the sense that intermediate matrices cancel out. For an input $\vec{x} \in \{0,1\}^n$, taking $P = \prod_j \mathcal{N}_{j,v}$, the program accepts $\vec{x}$ if $P = (Q_{\text{acc}} \cdot \mathcal{Y}_m)$ (or, equivalently $P[1,1] = \mathcal{Y}_m[1,1]$) and rejects $\vec{x}$ if $P = (Q_{\text{rej}} \cdot \mathcal{Y}_m)$ (or, equivalently, $P[1,1] = \mathcal{Y}_m[k,1]$, for $k \neq 1$). We note that there is a negligible probability of error due to the multiplication by $\mathcal{Y}_m$. In terms of randomization, one can see that for any assignment $y : [m] \to \{0,1\}$, the collection of matrices $(\mathcal{N}_{j,y(j)})_{j\in[m]}$ are uniformly random and independent invertible matrices. We note that this holds even if y is not a "consistent" assignment: for $j, j' \in [m] : \ell(j) = \ell(j') = i$, y can assign different values to j and j' (corresponding to an inconsistent assignment to the i-th bit of $\vec{x}$).

Implementing the randomization idea over graded encoding schemes (see Section 2.4) is not immediate, because we do not know an efficient procedure for computing inverses, and we also do not know how to sample random invertible matrices. To handle these difficulties, we utilize a variant of the above idea (see the discussion in Section 1.2).

Instead of $\mathcal{Y}_j^{-1}$, we use the adjoint matrix $\mathcal{Z}_j = \mathsf{adj}(\mathcal{Y}_j)$, which is composed of determinants of minors of $\mathcal{Y}$, and satisfies $\mathcal{Y} \cdot \mathcal{Z} = \det(\mathcal{Y}) \cdot I$. We take:

$$(\mathcal{N}_{j,v})_{j \in [m], v \in \{0,1\}} : \mathcal{N}_{j,v} = (\mathcal{Z}_{j-1} \cdot M_{j,v} \cdot \mathcal{Y}_j)$$

(where $\mathcal{Z}_0$ is again the identity matrix). Observe that for $\vec{x} \in \{0,1\}^n$, taking $P = \prod_j \mathcal{N}_{j,v}$, the program accepts $\vec{x}$ if $P[1,1] = ((\prod_{j \in [m-1]} \det(\mathcal{Y}_j)) \cdot \mathcal{Y}_m)[1,1]$ and rejects $\vec{x}$ if $P[1,1] = ((\prod_{j \in [m-1]} \det(\mathcal{Y}_j)) \cdot \mathcal{Y}_m)[k,1]$. It is not hard to see that this also preserves the randomization property from above. The only remaining subtlety is that we do not know how to pick a uniformly random invertible matrix (without being able to compute inverses). This is not a serious issue, because for a large enough prime p, we can simply sample uniformly random matrices in $\mathbb{Z}_p^{5 \times 5}$, and their joint distribution will be statistically close to uniformly random *and invertible* matrices.

Lemma 2.2 (Randomized Branching Programs). *For security parameter $\lambda \in \mathbb{N}$, let p_λ be a prime in $[2^\lambda, 2^{\lambda+1}]$. Fix a length-$\ell$ permutation branching program, and let $y : [m] \to \{0,1\}$ be any assignment function. Let $(\mathcal{Y}_j)_{j \in [m]}$ be chosen uniformly at random from $\mathbb{Z}_p^{5 \times 5}$, and for $j \in [m], v \in \{0,1\}$ take $\mathcal{N}_{j,v} = (\mathcal{Z}_{j-1} \cdot M_{j,v} \cdot \mathcal{Y}_j)$. (where $\mathcal{Z}_0$ is the identity matrix).*

Then the joint distribution of $(\mathcal{N}_{j,y(j)})_{j \in [m]}$ is $\mathsf{negl}(\lambda)$-statistically close to uniformly random and independent.

2.2   The Bounded Speedup Hypothesis (BSH)

The *Bounded Speedup Hypothesis* was introduced in [11] as a strengthening of the exponential time hypothesis (ETH). Formally, the hypothesis is as follows.

Definition 2.3 ($\mathcal{X}$-3-SAT Solver). *Consider a family of sets $\mathcal{X} = \{X_n\}_{n \in \mathbb{N}}$ such that $X_n \subseteq \{0,1\}^n$. We say that an algorithm $\mathcal{A}$ is a $\mathcal{X}$-3-SAT solver if it solves the 3-SAT problem, restricted to inputs in $\mathcal{X}$. Namely, given a 3-CNF formula $\Phi : \{0,1\}^n \to \{0,1\}$, $\mathcal{A}$ finds whether there exists $x \in X_n$ such that $\Phi(x) = 1$.*

Assumption 2.4 (Bounded Speedup Hypothesis). *There exists a polynomial $p(\cdot)$, such that for any $\mathcal{X}$-3-SAT solver that runs in time $t(\cdot)$, the family of sets $\mathcal{X}$ is of size at most $p(t(\cdot))$.*

The plausibility of this assumption is discussed in [11, Appendix A], where it is shown that a quasi-polynomial variant of BSH follows from ETH. Further evidence comes from the field of parameterized complexity.

2.3 Obfuscation

Definition 2.5 (Virtual Black-Box Obfuscator [5]).
Let $C = \{C_n\}_{n \in \mathbb{N}}$ be a family of polynomial-size circuits, where C_n is a set of boolean circuits operating on inputs of length n. And let $\mathcal{O}$ be a PPTM algorithm, which takes as input an input length $n \in \mathbb{N}$, a circuit $C \in C_n$, a security parameter $\lambda \in \mathbb{N}$, and outputs a boolean circuit $\mathcal{O}(C)$ (not necessarily in C).
$\mathcal{O}$ is a (black-box) obfuscator for the circuit family C if it satisfies:

1. Preserving Functionality: *For every $n \in \mathbb{N}$, and every $C \in C_n$, and every $\vec{x} \in \{0,1\}^n$, with all but $negl(\lambda)$ probability over the coins of $\mathcal{O}$:*

$$(\mathcal{O}(C, 1^n, \lambda))(\vec{x}) = C(\vec{x})$$

2. Polynomial Slowdown: *For every $n, \lambda \in \mathbb{N}$ and $C \in C$, the circuit $\mathcal{O}(C, 1^n, 1^\lambda)$ is of size at most $poly(|C|, n, \lambda)$.*

3. Virtual Black-Box: *For every (non-uniform) polynomial size adversary $\mathcal{A}$, there exists a (non-uniform) polynomial size simulator $\mathcal{S}$, such that for every $n \in \mathbb{N}$ and for every $C \in C_n$:*

$$\left| \Pr_{\mathcal{O}, \mathcal{A}}[\mathcal{A}(\mathcal{O}(C, 1^n, 1^\lambda)) = 1] - \Pr_{\mathcal{S}}[\mathcal{S}^C(1^{|C|}, 1^n, 1^\lambda) = 1] \right| = negl(\lambda)$$

Remark 2.6. A stronger notion of functionality, which also appears in the literature, requires that with overwhelming probability the obfuscated circuit is correct on *every input simultaneously*. We use the relaxed requirement that for every input (individually) the obfuscated circuit is correct with overwhelming probability (in both cases the probability is only over the obfuscator's coins). We note that our construction can be modified to achieve the stronger functionality property (by using a ring of sufficiently large size and the union bound).

Definition 2.7 (Indistinguishability Obfuscator [5]). *Let C be a circuit family and $\mathcal{O}$ a PPTM as in Definition 2.5. $\mathcal{O}$ is an indistinguishability obfuscator for C if it satisfies the preserving functionality and polynomial slowdown properties of Definition 2.5 with respect to C, but the virtual black-box property is replaced with:*

3. Indistinguishable Obfuscation: *For every (non-uniform) polynomial size adversary $\mathcal{A}$, for every $n \in \mathbb{N}$ and for every $C_1, C_2 \in C_n$ s.t. $|C_1| = |C_2|$ and $C_1 \equiv C_2$:*

$$\left| \Pr_{\mathcal{O}, \mathcal{A}}[\mathcal{A}(\mathcal{O}(C_1, 1^n, 1^\lambda)) = 1] - \Pr_{\mathcal{O}, \mathcal{A}}[\mathcal{A}(\mathcal{O}(C_2, 1^n, 1^\lambda)) = 1] \right| = negl(\lambda)$$

Definition 2.8 (iO2: Indistinguishability Obfuscator, Alternative Formulation). *Let C be a circuit family and $\mathcal{O}$ a PPTM as in Definition 2.5. $\mathcal{O}$ is an indistinguishability obfuscator2 (iO2) for C if it satisfies the preserving functionality and polynomial slowdown properties of Definition 2.5 with respect to C, but the virtual black-box property is replaced with:*

3. Unbounded simulation: *For every (non-uniform) polynomial size adversary $\mathcal{A}$, there exists a* computationally unbounded *simulator $\mathcal{S}$, such that for every $n \in \mathbb{N}$ and for every $C \in \mathcal{C}_n$:*

$$\left| \Pr_{\mathcal{O},\mathcal{A}}[\mathcal{A}(\mathcal{O}(C,1^n,1^\lambda)) = 1] - \Pr_{\mathcal{S}}[\mathcal{S}^C(1^{|C|},1^n,1^\lambda) = 1] \right| = negl(\lambda)$$

Lemma 2.9. *Let $\mathcal{C}$ be a circuit family and $\mathcal{O}$ a PPTM as in Definition 2.5. Then $\mathcal{O}$ is iO if and only if it is iO2.*

Proof. Assume that $\mathcal{O}$ is iO2, let $\mathcal{A}$ be an adversary and let $C_1 \equiv C_2 \in \mathcal{C}_n$ s.t. $|C_1| = |C_2|$. Then by Definition 2.8 there exists a computationally unbounded $\mathcal{S}$ such that

$$\left| \Pr_{\mathcal{O},\mathcal{A}}[\mathcal{A}(\mathcal{O}(C_1,1^n,1^\lambda)) = 1] - \Pr_{\mathcal{S}}[\mathcal{S}^{C_1}(1^{|C_1|},1^n,1^\lambda) = 1] \right| = negl(\lambda)$$

and also

$$\left| \Pr_{\mathcal{O},\mathcal{A}}[\mathcal{A}(\mathcal{O}(C_2,1^n,1^\lambda)) = 1] - \Pr_{\mathcal{S}}[\mathcal{S}^{C_2}(1^{|C_2|},1^n,1^\lambda) = 1] \right| = negl(\lambda) .$$

However, since $C_1 \equiv C_2$, then $|C_1| = |C_2|$ and an oracle to C_1 is identical to an oracle to C_2, and in particular

$$\Pr_{\mathcal{S}}[\mathcal{S}^{C_1}(1^{|C_1|},1^n,1^\lambda) = 1] = \Pr_{\mathcal{S}}[\mathcal{S}^{C_2}(1^{|C_2|},1^n,1^\lambda) = 1] .$$

The triangle inequality immediately implies that

$$\left| \Pr_{\mathcal{O},\mathcal{A}}[\mathcal{A}(\mathcal{O}(C_1,1^n,1^\lambda)) = 1] - \Pr_{\mathcal{O},\mathcal{A}}[\mathcal{A}(\mathcal{O}(C_2,1^n,1^\lambda)) = 1] \right| = negl(\lambda) ,$$

and therefore $\mathcal{O}$ is iO.

In the opposite direction, let $\mathcal{O}$ be iO and let $\mathcal{A}$ be an adversary. We define the following simulator $\mathcal{S}^C(1^{|C|},1^n,1^\lambda)$. The simulator first queries the oracle C on all $x \in \{0,1\}^n$, thus obtaining its truth table. Then it performs exhaustive search over all C' of size $1^{|C|}$ and finds $C' \in \mathcal{C}$ such that $C' \equiv C$ (this can be tested using the truth table). Finally the simulator outputs $\mathcal{A}(\mathcal{O}(C',1^n,1^\lambda))$.

By definition we have that

$$\left| \Pr_{\mathcal{O},\mathcal{A}}[\mathcal{A}(\mathcal{O}(C,1^n,1^\lambda)) = 1] - \Pr_{\mathcal{S}}[\mathcal{S}^C(1^{|C|},1^n,1^\lambda) = 1] \right| =$$

$$\left| \Pr_{\mathcal{O},\mathcal{A}}[\mathcal{A}(\mathcal{O}(C,1^n,1^\lambda)) = 1] - \Pr_{\mathcal{O},\mathcal{A}}[\mathcal{A}(\mathcal{O}(C',1^n,1^\lambda)) = 1] \right| ,$$

and since $C \equiv C'$ and $\mathcal{O}$ is iO, this quantity is negligible and iO2 follows.

2.4   Graded Encoding Schemes

We begin with the definition of a graded encoding scheme, due to Garg, Gentry and Halevi [23]. While their construction is very general, for our purposes a more restricted setting is sufficient as defined below.

Definition 2.10 (τ-Graded Encoding Scheme [23]). *A τ-encoding scheme for an integer $\tau \in \mathbb{N}$ and ring R, is a collection of sets $\mathcal{S} = \{S_{\vec{v}}^{(\alpha)} \subset \{0,1\}^* : \vec{v} \in \{0,1\}^\tau, \alpha \in R\}$ with the following properties:*

1. *For every index $\vec{v} \in \{0,1\}^\tau$, the sets $\{S_{\vec{v}}^{(\alpha)} : \alpha \in R\}$ are disjoint, and so they are a partition of the indexed set $S_{\vec{v}} = \bigcup_{\alpha \in R} S_{\vec{v}}^{(\alpha)}$.*

   In this work, for a 5×5 matrix $\mathcal{Y}$, we use $S_{\vec{v}}^{(\mathcal{Y})}$ to denote the set of 5×5 matrices where for all $i, j \in [5]$, the matrix's $[i,j]$-th entry contains an element in $S_{\vec{v}}^{(\mathcal{Y}[i,j])}$.

2. *There are binary operations "+" and "−" such that for all $\vec{v} \in \{0,1\}^\tau$, $\alpha_1, \alpha_2 \in R$ and for all $u_1 \in S_{\vec{v}}^{(\alpha_1)}$, $u_2 \in S_{\vec{v}}^{(\alpha_2)}$:*

$$u_1 + u_2 \in S_{\vec{v}}^{(\alpha_1 + \alpha_2)} \quad \text{and} \quad u_1 - u_2 \in S_{\vec{v}}^{(\alpha_1 - \alpha_2)} \,,$$

   where $\alpha_1 + \alpha_2$ and $\alpha_1 - \alpha_2$ are addition and subtraction in R.

3. *There is an associative binary operation "×" such that for all $\vec{v}_1, \vec{v}_2 \in \{0,1\}^\tau$ such that $\vec{v}_1 + \vec{v}_2 \in \{0,1\}^\tau$, for all $\alpha_1, \alpha_2 \in R$ and for all $u_1 \in S_{\vec{v}_1}^{(\alpha_1)}$, $u_2 \in S_{\vec{v}_2}^{(\alpha_2)}$, it holds that*

$$u_1 \times u_2 \in S_{\vec{v}_1 + \vec{v}_2}^{(\alpha_1 \cdot \alpha_2)},$$

   where $\alpha_1 \cdot \alpha_2$ is multiplication in R.

In this work, the ring R will always be $\mathbb{Z}_p$ for a prime p.

For the reader who is familiar with [23], we note that the above is the special case of their construction, in which we consider only binary index vectors (in the [23] notation, this corresponds to setting $\kappa = 1$), and we construct our encoding schemes to be *asymmetric* (as will become apparent below when we define our zero-test index $\mathbf{vzt} = \vec{1}$).

Definition 2.11 (Efficient Procedures for τ-Graded Encoding Scheme [23]). *We consider τ-graded encoding schemes (see above) where the following procedures are efficiently computable.*

– *Instance Generation:* InstGen($1^\lambda, 1^\tau$) *outputs the set of parameters params, a description of a τ-Graded Encoding Scheme. (Recall that we only consider Graded Encoding Schemes over the set indices $\{0,1\}^\tau$, with zero testing in the set $S_{\vec{1}}$). In addition, the procedure outputs a subset evparams $\subset$ params that is sufficient for computing addition, multiplication and zero testing[7] (but possibly insufficient for encoding or for randomization).*

– *Ring Sampler:* samp($params$) *outputs a "level zero encoding" $a \in S_0^{(\alpha)}$ for a nearly uniform $\alpha \in_R R$.*

---

[7] The "zero testing" parameter **pzt** defined in [23] is a part of *evparams*.

- *Encode and Re-Randomize:*[8] $\mathsf{encRand}(params, i, a)$ *takes as input an index* $i \in [\tau]$ *and* $a \in S_0^{(\alpha)}$, *and outputs an encoding* $u \in S_{\vec{e}_i}^{(\alpha)}$, *where the distribution of* u *is (statistically close to being) only dependent on* α *and not otherwise dependent of* a.
- *Addition and Negation:* $\mathsf{add}(evparams, u_1, u_2)$ *takes* $u_1 \in S_{\vec{v}}^{(\alpha_1)}, u_2 \in S_{\vec{v}}^{(\alpha_2)}$, *and outputs* $w \in S_{\vec{v}}^{(\alpha_1 + \alpha_2)}$. *(If the two operands are not in the same indexed set, then* add *returns* $\bot$). *We often use the notation* $u_1 + u_2$ *to denote this operation when evparams is clear from the context. Similarly,* $\mathsf{negate}(evparams, u_1) \in S_{\vec{v}}^{(-\alpha_1)}$.
- *Multiplication:* $\mathsf{mult}(evparams, u_1, u_2)$ *takes* $u_1 \in S_{\vec{v}_1}^{(\alpha_1)}, u_2 \in S_{\vec{v}_2}^{(\alpha_2)}$. *If* $\vec{v}_1 + \vec{v}_2 \in \{0, 1\}^\tau$ *(i.e. every coordinate in* $\vec{v}_1 + \vec{v}_2$ *is at most 1), then* mult *outputs* $w \in S_{\vec{v}_1 + \vec{v}_2}^{(\alpha_1 \cdot \alpha_2)}$. *Otherwise,* mult *outputs* $\bot$. *We often use the notation* $u_1 \times u_2$ *to denote this operation when evparams is clear from the context.*
- *Zero Test:* $\mathsf{isZero}(evparams, u)$ *outputs 1 if* $u \in S_{\vec{1}}^{(0)}$, *and 0 otherwise.*

In the [23,20] constructions, encodings are noisy and the noise level increases with addition and multiplication operations, so one has to be careful not to go over a specified noise bound. However, the parameters can be set so as to support $O(\tau)$ *operations, which are sufficient for our purposes. We therefore ignore noise management throughout this manuscript. An additional subtle issue is that with negligible probability the initial noise may be too big. However this can be avoided by adding rejection sampling to* samp *and therefore ignored throughout the manuscript as well.*

As was done in [10,11], our definition deviates from that of [23]. We define two sets of parameters *params* and *evparams*. While the former will be used by the obfuscator in our construction (and therefore will not be revealed to an external adversary), the latter will be used when evaluating an obfuscated program (and thus will be known to an adversary). When instantiating our definition, the guideline is to make *evparams* minimal so as to give the least amount of information to the adversary. In particular, in the known candidates [23,20], *evparams* only needs to contain the zero-test parameter **pzt** (as well as the global modulus).

2.5   The Generic Graded Encoding Scheme Model

We would like to prove the security of our construction against *generic adversaries*. To this end, we will use the *generic graded encoding scheme* model, which was previously used in [10], and is analogous to the *generic group model* (see Shoup [38] and Maurer [33]). In this model, an algorithm/adversary $\mathcal{A}$ can only interact with the graded encoding scheme via oracle calls to the add, mult, and isZero operations from Definition 2.11. Note that, in particular, we only

---

[8] This functionality is not explicitly provided by [23], however it can be obtained by combining their encoding and re-randomization procedures.

allow access to the operations that can be run using *evparams*. To the best of our knowledge, non-generic attacks on known schemes, e.g. [23], require use of *params* and cannot be mounted when only *evparams* is given.

We use $\mathcal{G}$ to denote an oracle that answers adversary calls. The oracle operates as follows: for each index $\vec{v} \in \{0,1\}^\tau$, the elements of the indexed set $S_{\vec{v}} = \bigcup_{\alpha \in R} S_{\vec{v}}^{(\alpha)}$ are arbitrary binary strings. The adversary $\mathcal{A}$ can manipulate these strings using oracle calls (via $\mathcal{G}$) to the graded encoding scheme's functionalities. For example, the adversary can use $\mathcal{G}$ to perform an add call: taking strings $s_1 \in S_{\vec{v}}^{(\alpha_1)}, s_2 \in S_{\vec{v}}^{(\alpha_2)}$, encoding indexed ring elements $(\vec{v}, \alpha_1), (\vec{v}, \alpha_2)$ (respectively), and obtaining a string $s \in S_{\vec{v}}^{(\alpha_1+\alpha_2)}$, encoding the indexed ring element $(\vec{v}, (\alpha_1 + \alpha_2))$.

We say that $\mathcal{A}$ is a generic algorithm (or adversary) for a problem on graded encoding schemes (e.g. for computing a moral equivalent of discreet log), if it can accomplish this task with respect to *any* oracle representing a graded encoding scheme, see below.

In the add example above, there may be many strings/encodings in the set $S_{\vec{v}}^{(\alpha_1+\alpha_2)}$. One immediate question is *which* of these elements should be returned by the call to add. In our abstraction, for each $\vec{v} \in \{0,1\}^\tau$ and $\alpha \in R$, $\mathcal{G}$ always uses a *single unique encoding* of the indexed ring element $(\vec{v}, \alpha)$. I.e. the set $S_{\vec{v}}^\alpha$ is a singleton. Thus, the representation of items in the graded encoding scheme is given by a map $\sigma(\vec{v}, \alpha)$ from $\vec{v} \in \{0,1\}^\tau$ and $\alpha \in R$, to $\{0,1\}^*$. We restrict our attention to the case where this mapping has polynomial blowup.

Remark 2.12 (Unique versus Randomized Representation).

We note that the known candidates of secure graded encoding schemes [23,20] *do not provide unique encodings*: their encodings are probabilistic. Nonetheless, in the generic graded encoding scheme abstraction we find it helpful to restrict our attention to schemes with unique encodings. For the purposes of proving security against generic adversaries, this makes sense: a generic adversary should work for *any* implementation of the oracle $\mathcal{G}$, and in particular also for an implementation that uses unique encodings.

Moreover, our perspective is that unique encodings are more "helpful" to an adversary than randomized encodings: a unique encoding gives the adversary the additional power to "automatically" check whether two encodings are of the same indexed ring element (without consulting the oracle). Thus, we prefer to prove security against generic adversaries even for unique representations.

It is important to note that the set of legal encodings may be very sparse within the set of images of σ, and indeed this is the main setting we will consider when we study the generic model. In this case, the only way for $\mathcal{A}$ to obtain a valid representation of any element in any graded set is via calls to the oracle (except with negligible probability). Finally, we note that if oracle calls contain invalid operators (e.g. the input is not an encoding of an element in any graded set, the inputs to add are not in the same graded set, etc.), then the oracle returns $\perp$.

Random Graded Encoding Scheme Oracle. We focus on a particular randomized oracle: the random generic encoding scheme (GES) oracle $\mathcal{RG}$. $\mathcal{RG}$ operates as follows: for each indexed ring element (with index $\vec{v} \in \{0,1\}^{\tau}$ and ring element $\sigma \in R$), its encoding is of length $\ell = (|\tau| \cdot \log |R| \cdot poly(\lambda))$ (where $|\tau|$ is the bit representation length of τ). The encoding of each indexed ring element is a uniformly random string of length ℓ. In particular, this implies that the only way that $\mathcal{A}$ can obtain valid encodings is by calls to the oracle $\mathcal{RG}$ (except with negligible probability).

The definitions of secure obfuscation in the random GES model are as follows.

Definition 2.13 (Virtual Black-Box in the Random GES Model).
Let $\mathcal{C} = \{\mathcal{C}_n\}_{n \in \mathbb{N}}$ be a family of circuits and $\mathcal{O}$ a PPTM as in Definition 2.5.

A generic algorithm $\mathcal{O}^{\mathcal{RG}}$ is an obfuscator *in the random generic encoding scheme model, if it satisfies the functionality and polynomial slowdown properties of Definition 2.5 with respect to $\mathcal{C}$ and to any GES oracle $\mathcal{RG}$, but the virtual black-box property is replaced with:*

3. Virtual Black-Box in the Random GES Model: *For every (non-uniform) polynomial size generic adversary $\mathcal{A}$, there exists a (non-uniform) generic polynomial size simulator $\mathcal{S}$, such that for every $n \in \mathbb{N}$ and every $C \in \mathcal{C}_n$:*

$$\left| \left(\Pr_{\mathcal{RG}, \mathcal{O}, \mathcal{A}}[\mathcal{A}^{\mathcal{RG}}(\mathcal{O}^{\mathcal{RG}}(C, 1^n, 1^\lambda))] = 1 \right) - \left(\Pr_{\mathcal{S}}[\mathcal{S}^C(1^{|C|}, 1^n, 1^\lambda)] = 1 \right) \right| = negl(\lambda)$$

Remark 2.14. We remark that it makes sense to allow $\mathcal{S}$ to access the oracle $\mathcal{RG}$. However, this is in not really necessary. The reason is that $\mathcal{RG}$ can be implemented in polynomial time (as described below), and therefore $\mathcal{S}$ itself can implement $\mathcal{RG}$.

Definition 2.15 (GiO: Indistinguishability Obfuscator in the Random GES Model). *Let $\mathcal{C} = \{\mathcal{C}_n\}_{n \in \mathbb{N}}$ be a family of circuits and $\mathcal{O}$ a PPTM as in Definition 2.5. A generic algorithm $\mathcal{O}^{\mathcal{RG}}$ is an* indistinguishability obfuscator *in the random generic encoding scheme model, if it satisfies the functionality and polynomial slowdown properties of Definition 2.5 with respect to $\mathcal{C}$ and to any GES oracle $\mathcal{RG}$, but the virtual black-box property is replaced with:*

3. Indistinguishable Obfuscation in the Random GES Model: *For every (non-uniform) polynomial size generic adversary $\mathcal{A}$, for every $n \in \mathbb{N}$ and for every $C_1, C_2 \in \mathcal{C}_n$ s.t. $C_1 \equiv C_2$ and $|C_1| = |C_2|$:*

$$\left| \left(\Pr_{\mathcal{RG}, \mathcal{O}, \mathcal{A}}[\mathcal{A}^{\mathcal{RG}}(\mathcal{O}^{\mathcal{RG}}(C_1, 1^n, 1^\lambda))] = 1 \right) \right.$$
$$\left. - \Pr_{\mathcal{RG}, \mathcal{O}, \mathcal{A}}[\mathcal{A}^{\mathcal{RG}}(\mathcal{O}^{\mathcal{RG}}(C_2, 1^n, 1^\lambda))] = 1 \right| = negl(\lambda)$$

Definition 2.16 (GiO2: Indistinguishability Obfuscator in the Random GES Model, Alternative Formulation).
Let $\mathcal{C} = \{\mathcal{C}_n\}_{n \in \mathbb{N}}$ be a family of circuits and $\mathcal{O}$ a PPTM as in Definition 2.5. A generic algorithm $\mathcal{O}^{\mathcal{RG}}$ is an indistinguishability obfuscator2 *in the random*

generic encoding scheme model *(GiO2)*, *if it satisfies the functionality and polynomial slowdown properties of Definition 2.5 with respect to C and to any GES oracle $\mathcal{RG}$, but the virtual black-box property is replaced with:*

3. Indistinguishable Obfuscation2 in the Random GES Model: *For every (non-uniform) polynomial size* generic *adversary $\mathcal{A}$, there exists a (possibly computationally unbounded) simulator $\mathcal{S}$, such that for every $n \in \mathbb{N}$ and every $C \in \mathcal{C}_n$:*

$$\left| \left(\Pr_{\mathcal{RG}, \mathcal{O}, \mathcal{A}}[\mathcal{A}^{\mathcal{RG}}(\mathcal{O}^{\mathcal{RG}}(C, 1^n, 1^\lambda))] = 1 \right) - \left(\Pr_{\mathcal{S}}[\mathcal{S}^C(1^{|C|}, 1^n, 1^\lambda)] = 1 \right) \right| = negl(\lambda)$$

The following lemma asserts the equivalence between GiO and GiO2 in the generic model. The proof is identical to that of Lemma 2.9 (note that $\mathcal{S}$ does not need access to $\mathcal{RG}$).

Lemma 2.17. *Let C be a circuit family and $\mathcal{O} = \mathcal{O}^{\mathcal{RG}}$ an oracle PPTM as in Definition 2.5. Then $\mathcal{O}$ is GiO if and only if it is GiO2.*

Online Random GES Oracle. In our proof, we will use the property that the oracle $\mathcal{RG}$ can be approximated to within negligible statistical distance by an *online polynomial time process*, which samples the representations on-the-fly. Specifically, the oracle will maintain a table of entries of the form $(\vec{v}, \alpha, \mathsf{label}_{\vec{v}, \alpha})$, where $\mathsf{label}_{\vec{v}, \alpha} \in \{0, 1\}^\ell$ is the representation of $S_{\vec{v}}^{(\alpha)}$ in $\mathcal{RG}$ (the table is initially empty). Every time $\mathcal{RG}$ is called for some functionality, it checks that its operands indeed correspond to an entry in the table, in which case it can retrieve the appropriate $(\vec{v}, \alpha)$ to perform the operation (if the operands are not in the table, $\mathcal{RG}$ returns $\perp$). Whenever $\mathcal{RG}$ needs to return a value $S_{\vec{v}}^{(\alpha)}$, it checks whether $(\vec{v}, \alpha)$ is already in the table, and if so returns the appropriate $\mathsf{label}_{\vec{v}, \alpha}$. Otherwise it samples a new uniform label, and inserts a new entry into the table.

When interacting with an adversary that only makes a polynomial number of calls, the online version of $\mathcal{RG}$ is within negligible statistical distance of the offline version (in fact, the statistical distance is exponentially small in λ) . This is because the only case when the online oracle implementation differs from the offline one is when when the adversary guesses a valid label that it has not seen (in the offline setting). This can only occur with exponentially small probability due to the sparsity of the labels. The running time of the online oracle is polynomial in the number of oracle calls.

3   Obfuscating $\mathcal{NC}^1$

See Section 1.2 for an overview of the construction. We proceed with a formal description of the obfuscator. Functionality follows by construction, virtual black-box security is only stated, and the proof is deferred to the full version [12].

Obfuscator NC^{1}Obf, on input $(1^\lambda, 1^n, C = (\ldots, (M_{j,0}, M_{j,1}), \ldots)_{j \in [m]})$

Input: Security parameter λ; Number of input variables n; Oblivious permutation branching program C (with labeling function ℓ), where $(M_{j,b})_{j\in[m],b\in\{0,1\}}$ are 5×5 permutation matrices. Let $Q_{\mathsf{acc}}, Q_{\mathsf{rej}}$ be the accepting and rejecting permutations for C (see Section 2.1).

Output: Obfuscated program for C.

Execution:

1. **Generate asymmetric encoding scheme.**
   Generate $(params, evparams) \leftarrow \mathsf{InstGen}(1^\lambda, 1^\tau)$, where $\tau = m+n+\binom{n}{3}+1$. Namely, we have τ level-1 groups. As explained above, we denote these groups as follows:

   - Groups $1, \ldots, m$ are related to the execution of the branching program and are denoted $\mathsf{prog}_1, \ldots, \mathsf{prog}_m$.

   - Groups $m+1, \ldots, m+n$ are related to consistency check and are denoted $\mathsf{cc}_1, \ldots, \mathsf{cc}_n$.

   - Groups $m+n+1, \ldots, m+n+\binom{n}{3}$ are used to bind triples of variables and are denoted bind_T, for $T \in \binom{[n]}{3}$.

   - Lastly, the group $m+n+\binom{n}{3}+1$ is the check group and is denoted chk.
   We let $L(i)$ denote the set of groups that are related to the ith variable:

   $$L(i) = \{\mathsf{prog}_j : i = \ell(j)\} \cup \{\mathsf{bind}_T : i \in T\} \cup \{\mathsf{cc}_i\} .$$

   We let L denote the set of all groups: $L = \cup_{i\in[n]} L(i) \cup \mathsf{chk}$.

2. **Generate consistency check variables.**
   For all $i \in [n]$:
   (a) for each $\mathsf{grp} \in (L(i) \setminus \{\mathsf{cc}_i\})$ and $v \in \{0,1\}$: $b_{\mathsf{grp},i,v} \leftarrow \mathsf{samp}(params) \in S_0^{(\beta_{\mathsf{grp},i,v})}$. [9]

   (b) $b'_{\mathsf{cc}_i} \leftarrow \mathsf{samp}(params) \in S_0^{(\beta'_{\mathsf{cc}_i})}$
   for $v \in \{0,1\}$: $b_{\mathsf{cc}_i,i,v} \leftarrow b'_{\mathsf{cc}_i} \times \left(\prod_{\mathsf{grp}\in(L(i)\setminus\{\mathsf{cc}_i\})} b_{\mathsf{grp},i,(1-v)}\right) \in S_0^{(\beta_{\mathsf{cc}_i},v)}$

   (c) $c_i \leftarrow \prod_{\mathsf{grp}\in L(i)} b_{\mathsf{grp},i,0} \in S_0^{(\gamma_i)}$, where

   $$\gamma_i = \prod_{\mathsf{grp}\in L(i)} \beta_{\mathsf{grp},i,0} = \prod_{\mathsf{grp}\in L(i)} \beta_{\mathsf{grp},i,1} .$$

3. **Generate randomizing matrices.**
   For each $j \in [m]$:
   Sample $Y_j \leftarrow \mathsf{samp}(params)^{5\times5} \in S_0^{(\mathcal{Y}_j)}$,
   and compute $Z_j = \mathsf{adj}(Y_j) \in S_0^{(\mathcal{Z}_j)}$, s.t. $\mathcal{Y}_j \cdot \mathcal{Z}_j = \det(\mathcal{Y}_j) \cdot I$
   Sample $Z_0 \leftarrow \mathsf{samp}(params)^{5\times5} \in S_0^{(\mathcal{Z}_0)}$

---

[9] For notational convenience, we drop i when it is uniquely defined by grp. E.g., we use $\beta_{\mathsf{prog}_j,v}$ to refer to $\beta_{\mathsf{prog}_j,\ell(j),v}$.

4. **Encode elements in program groups (prog).**
   For each $j \in [m]$ and $v \in \{0,1\}$:

   $D_{\mathsf{prog}_j,v} \leftarrow b_{\mathsf{prog}_j,v} \cdot (Z_{j-1} \times M_{j,v} \times Y_j) \in S_0^{(\mathcal{D}_{j,v})}$, where $\mathcal{D}_{j,v} = (\beta_{\mathsf{prog}_j,v} \cdot \underbrace{(\mathcal{Z}_{j-1} \cdot \mathcal{M}_{j,v} \cdot \mathcal{Y}_j)}_{\mathcal{N}_{j,v}})$

   $r_{\mathsf{prog}_j,v} \leftarrow \mathsf{samp}(params) \in S_0^{(\rho_{\mathsf{prog}_j,v})}$,

   $K_{\mathsf{prog}_j,v} \leftarrow (r_{\mathsf{prog}_j,v} \cdot D_{j,v}) \in S_0^{(\rho_{\mathsf{prog}_j,v} \cdot \mathcal{D}_{j,v})}$

   $w_{\mathsf{prog}_j,v} \leftarrow \mathsf{encRand}(params, \mathsf{prog}_j, r_{\mathsf{prog}_j,v}) \in S_{\vec{e}_{\mathsf{prog}_j}}^{(\rho_{\mathsf{prog}_j,v})}$,

   $U_{\mathsf{prog}_j,v} \leftarrow \mathsf{encRand}(params, \mathsf{prog}_j, K_{\mathsf{prog}_j,v}) \in S_{\vec{e}_{\mathsf{prog}_j}}^{(\rho_{\mathsf{prog}_j,v} \cdot \mathcal{D}_{j,v})}$

5. **Encode elements in consistency check groups (cc).**
   For each $i \in [n]$ and $v \in \{0,1\}$:

   $d_{\mathsf{cc}_i,v} \leftarrow b_{\mathsf{cc}_i,\vec{v}[i]} \in S_0^{(\delta_{\mathsf{cc}_i,v})}$, where $\delta_{\mathsf{cc}_i,v} = \beta_{\mathsf{cc}_i,v}$.

   $r_{\mathsf{cc}_i,v} \leftarrow \mathsf{samp}(params) \in S_0^{(\rho_{\mathsf{cc}_i,v})}$, $q_{\mathsf{cc}_i,v} \leftarrow (r_{\mathsf{cc}_i,v} \cdot d_{\mathsf{cc}_i,v}) \in S_0^{(\rho_{\mathsf{cc}_i,v} \cdot \delta_{\mathsf{cc}_i,v})}$

   $w_{\mathsf{cc}_i,v} \leftarrow \mathsf{encRand}(params, \mathsf{cc}_i, r_{\mathsf{cc}_i,v}) \in S_{\vec{e}_{\mathsf{cc}_i}}^{(\rho_{\mathsf{cc}_i,v})}$

   $u_{\mathsf{cc}_i,v} \leftarrow \mathsf{encRand}(params, \mathsf{cc}_i, q_{\mathsf{cc}_i,v}) \in S_{\vec{e}_{\mathsf{cc}_i}}^{(\rho_{\mathsf{cc}_i,v} \cdot \delta_{\mathsf{cc}_i,v})}$

6. **Encode elements in binding groups (bind).**
   For each $T \in \binom{[n]}{3}$ and $\vec{v} \in \{0,1\}^3$:

   $d_{\mathsf{bind}_T,\vec{v}} \leftarrow (\prod_{i \in T} b_{\mathsf{bind}_T,i,\vec{v}[i]}) \in S_0^{(\delta_{\mathsf{bind}_T,\vec{v}})}$, where $\delta_{\mathsf{bind}_T,\vec{v}} = \prod_{i \in T} \beta_{\mathsf{bind}_T,i,\vec{v}[i]}$

   $r_{\mathsf{bind}_T,\vec{v}} \leftarrow \mathsf{samp}(params) \in S_0^{(\rho_{\mathsf{bind}_T,\vec{v}})}$,

   $q_{\mathsf{bind}_T,\vec{v}} \leftarrow (r_{\mathsf{bind}_T,\vec{v}} \cdot d_{\mathsf{bind}_T,\vec{v}}) \in S_0^{(\rho_{\mathsf{bind}_T,\vec{v}} \cdot \delta_{\mathsf{bind}_T,\vec{v}})}$

   $w_{\mathsf{bind}_T,\vec{v}} \leftarrow \mathsf{encRand}(params, \mathsf{bind}_T, r_{\mathsf{bind}_T,\vec{v}}) \in S_{\vec{e}_{\mathsf{bind}_T}}^{(\rho_{\mathsf{bind}_T,\vec{v}})}$,

   $u_{\mathsf{bind}_T,\vec{v}} \leftarrow \mathsf{encRand}(params, \mathsf{bind}_T, q_{\mathsf{bind}_T,\vec{v}}) \in S_{\vec{e}_{\mathsf{bind}_T}}^{(\rho_{\mathsf{bind}_T,\vec{v}} \cdot \delta_{\mathsf{bind}_T,\vec{v}})}$

7. **Encode elements in last group (chk).**

   $d_{\mathsf{chk}} \leftarrow ((\prod_{i \in [n]} c_i) \cdot (\prod_{j \in [m-1]} \det(Y_j)) \cdot Z_0 \cdot Y_m)[1,1] \in S_0^{(\delta_{\mathsf{chk}})}$,

   where $\delta_{\mathsf{chk}} = \left((\prod_{i \in [n]} \gamma_i) \cdot (\prod_{j \in [m-1]} \det(\mathcal{Y}_j)) \cdot \mathcal{Z}_0 \cdot \mathcal{Y}_m \right)[1,1]$

   $r_{\mathsf{chk}} \leftarrow \mathsf{samp}(params) \in S_0^{(\rho_{\mathsf{chk}})}$, $q_{\mathsf{chk}} \leftarrow r_{\mathsf{chk}} \cdot d_{\mathsf{chk}} \in S_0^{(\rho_{\mathsf{chk}} \cdot \delta_{\mathsf{chk}})}$

   $w_{\mathsf{chk}} \leftarrow \mathsf{encRand}(params, \mathsf{chk}, r_{\mathsf{chk}}) \in S_{\vec{e}_{\mathsf{chk}}}^{(\rho_{\mathsf{chk}})}$,

   $u_{\mathsf{chk}} \leftarrow \mathsf{encRand}(params, \mathsf{chk}, q_{\mathsf{chk}}) \in S_{\vec{e}_{\mathsf{chk}}}^{(\rho_{\mathsf{chk}} \cdot \delta_{\mathsf{chk}})}$

8. **Output.**
   Output $evparams$ and the obfuscation:

   $$\left(\{(w_{\mathsf{prog}_j,v}, U_{\mathsf{prog}_j,v})\}_{j \in [m], v \in \{0,1\}}, \{(w_{\mathsf{cc}_i,v}, u_{\mathsf{cc}_i,v})\}_{i \in [n], v \in \{0,1\}}, \right.$$

   $$\left. \{(w_{\mathsf{bind}_T,\vec{v}}, u_{\mathsf{bind}_T,\vec{v}})\}_{T \in (\binom{[n]}{3}), \vec{v} \in \{0,1\}^3}, (w_{\mathsf{chk}}, u_{\mathsf{chk}}) \right)$$

Evaluation, on input $x \in \{0,1\}^n$

1. $\mathbf{t} \leftarrow \left(w_{\mathsf{chk}} \cdot \left(\prod_{j \in [m]} U_{\mathsf{prog}_j, x[\ell(j)]}\right) \cdot \left(\prod_{T \in \binom{[n]}{3}} u_{\mathsf{bind}_T, x_{|T}}\right) \cdot \left(\prod_{i \in [n]} u_{\mathsf{cc}_i, x[i]}\right)\right)[1,1]$

2. $\mathbf{t}' \leftarrow \left(u_{\mathsf{chk}} \cdot \left(\prod_{j \in [m]} w_{\mathsf{prog}_j, x[\ell(j)]}\right) \cdot \left(\prod_{T \in \binom{[n]}{3}} w_{\mathsf{bind}_T, x_{|T}}\right) \cdot \left(\prod_{i \in [n]} w_{\mathsf{cc}_i, x[i]}\right)\right)$

3. output the bit: $\mathsf{isZero}(evparams, (\mathbf{t} - \mathbf{t}'))$.

Virtual Black-Box Security. The following theorem states the security properties of our scheme. The proof is deferred to the full version [12].

Theorem 3.1. *Under the bounded speedup hypothesis,* $\mathsf{NC^1Obf}$ *is a virtual black box obfuscator in the random GES model for the class* $\mathcal{NC}^1$.

References

1. Adida, B., Wikström, D.: How to shuffle in public. In: TCC 2007. LNCS, vol. 4392, pp. 555–574. Springer, Heidelberg (2007)
2. Applebaum, B., Ishai, Y., Kushilevitz, E.: Cryptography in nc^0. SIAM J. Comput. 36(4), 845–888 (2006)
3. Babai, L.: Trading group theory for randomness. In: STOC, pp. 421–429 (1985)
4. Barak, B., Garg, S., Kalai, Y.T., Paneth, O., Sahai, A.: Protecting obfuscation against algebraic attacks. Cryptology ePrint Archive, Report 2013/631 (2013), http://eprint.iacr.org/
5. Barak, B., Goldreich, O., Impagliazzo, R., Rudich, S., Sahai, A., Vadhan, S.P., Yang, K.: On the (im)possibility of obfuscating programs. J. ACM 59(2), 6 (2012); Preliminary version in Kilian, J. (ed.): CRYPTO 2001. LNCS, vol. 2139. Springer, Heidelberg (2001)
6. Barrington, D.A.M.: Bounded-width polynomial-size branching programs recognize exactly those languages in NC^1. In: Hartmanis, J. (ed.) STOC, pp. 1–5. ACM (1986); Full version in [7]
7. Barrington, D.A.M.: Bounded-width polynomial-size branching programs recognize exactly those languages in NC^1. J. Comput. Syst. Sci. 38(1), 150–164 (1989)
8. Bitansky, N., Canetti, R.: On strong simulation and composable point obfuscation. In: Rabin, T. (ed.) CRYPTO 2010. LNCS, vol. 6223, pp. 520–537. Springer, Heidelberg (2010)
9. Boneh, D., Silverberg, A.: Applications of multilinear forms to cryptography. IACR Cryptology ePrint Archive 2002, 80 (2002)
10. Brakerski, Z., Rothblum, G.N.: Obfuscating conjunctions. In: Canetti, R., Garay, J.A. (eds.) CRYPTO 2013, Part II. LNCS, vol. 8043, pp. 416–434. Springer, Heidelberg (2013), http://eprint.iacr.org/2013/471
11. Brakerski, Z., Rothblum, G.N.: Black-box obfuscation for d-CNFs. Cryptology ePrint Archive (2013). Extended abstract in ITCS 2014
12. Brakerski, Z., Rothblum, G.N.: Virtual black-box obfuscation for all circuits via generic graded encoding. In: Lindell, Y. (ed.) TCC 2014. LNCS, vol. 8349, pp. 1–25. Springer, Heidelberg (2014); Cryptology ePrint Archive, Report 2013/563 (2013)

13. Canetti, R.: Towards realizing random oracles: Hash functions that hide all partial information. In: Kaliski Jr., B.S. (ed.) CRYPTO 1997. LNCS, vol. 1294, pp. 455–469. Springer, Heidelberg (1997)
14. Canetti, R., Dakdouk, R.R.: Obfuscating point functions with multibit output. In: Smart, N. (ed.) EUROCRYPT 2008. LNCS, vol. 4965, pp. 489–508. Springer, Heidelberg (2008)
15. Canetti, R., Goldreich, O., Halevi, S.: The random oracle methodology, revisited (preliminary version). In: Vitter, J.S. (ed.) STOC, pp. 209–218. ACM (1998); Full version in [16]
16. Canetti, R., Goldreich, O., Halevi, S.: The random oracle methodology, revisited. J. ACM 51(4), 557–594 (2004)
17. Canetti, R., Micciancio, D., Reingold, O.: Perfectly one-way probabilistic hash functions (preliminary version). In: Vitter, J.S. (ed.) STOC, pp. 131–140. ACM (1998)
18. Canetti, R., Rothblum, G.N., Varia, M.: Obfuscation of hyperplane membership. In: Micciancio, D. (ed.) TCC 2010. LNCS, vol. 5978, pp. 72–89. Springer, Heidelberg (2010)
19. Canetti, R., Vaikuntanathan, V.: Obfuscating branching programs using black-box pseudo-free groups. Cryptology ePrint Archive, Report 2013/500 (2013), http://eprint.iacr.org/
20. Coron, J.-S., Lepoint, T., Tibouchi, M.: Practical multilinear maps over the integers. In: Canetti, R., Garay, J.A. (eds.) CRYPTO 2013, Part I. LNCS, vol. 8042, pp. 476–493. Springer, Heidelberg (2013)
21. Dodis, Y., Smith, A.: Correcting errors without leaking partial information. In: Gabow, H.N., Fagin, R. (eds.) STOC, pp. 654–663. ACM (2005)
22. Feige, U., Kilian, J., Naor, M.: A minimal model for secure computation (extended abstract). In: STOC, pp. 554–563 (1994)
23. Garg, S., Gentry, C., Halevi, S.: Candidate multilinear maps from ideal lattices. In: Johansson, T., Nguyen, P.Q. (eds.) EUROCRYPT 2013. LNCS, vol. 7881, pp. 1–17. Springer, Heidelberg (2013)
24. Garg, S., Gentry, C., Halevi, S., Raykova, M., Sahai, A., Waters, B.: Candidate indistinguishability obfuscation and functional encryption for all circuits. Cryptology ePrint Archive, Report 2013/451 (2013); Extended abstract in FOCS 2013
25. Gentry, C.: Fully homomorphic encryption using ideal lattices. In: Mitzenmacher, M. (ed.) STOC, pp. 169–178. ACM (2009)
26. Goldwasser, S., Kalai, Y.T.: On the impossibility of obfuscation with auxiliary input. In: FOCS, pp. 553–562 (2005)
27. Goldwasser, S., Rothblum, G.N.: On best-possible obfuscation. In: Vadhan, S.P. (ed.) TCC 2007. LNCS, vol. 4392, pp. 194–213. Springer, Heidelberg (2007)
28. Hofheinz, D., Malone-Lee, J., Stam, M.: Obfuscation for cryptographic purposes. J. Cryptology 23(1), 121–168 (2010)
29. Hohenberger, S., Rothblum, G.N., Shelat, A., Vaikuntanathan, V.: Securely obfuscating re-encryption. J. Cryptology 24(4), 694–719 (2011)
30. Impagliazzo, R., Paturi, R.: Complexity of k-sat. In: IEEE Conference on Computational Complexity, pp. 237–240. IEEE Computer Society (1999)
31. Kilian, J.: Founding cryptography on oblivious transfer. In: Simon, J. (ed.) STOC, pp. 20–31. ACM (1988)
32. Lynn, B.Y.S., Prabhakaran, M., Sahai, A.: Positive results and techniques for obfuscation. In: Cachin, C., Camenisch, J.L. (eds.) EUROCRYPT 2004. LNCS, vol. 3027, pp. 20–39. Springer, Heidelberg (2004)

33. Maurer, U.: Abstract models of computation in cryptography. In: Smart, N.P. (ed.) Cryptography and Coding 2005. LNCS, vol. 3796, pp. 1–12. Springer, Heidelberg (2005)
34. Naor, M.: On cryptographic assumptions and challenges. In: Boneh, D. (ed.) CRYPTO 2003. LNCS, vol. 2729, pp. 96–109. Springer, Heidelberg (2003)
35. Rivest, R., Adleman, L., Dertouzos, M.: On data banks and privacy homomorphisms. In: Foundations of Secure Computation, pp. 169–177. Academic Press (1978)
36. Rothblum, R.D.: On the circular security of bit-encryption. In: Sahai, A. (ed.) TCC 2013. LNCS, vol. 7785, pp. 579–598. Springer, Heidelberg (2013)
37. Sahai, A., Waters, B.: How to use indistinguishability obfuscation: Deniable encryption, and more. Cryptology ePrint Archive, Report 2013/454 (2013), http://eprint.iacr.org/
38. Shoup, V.: Lower bounds for discrete logarithms and related problems. In: Fumy, W. (ed.) EUROCRYPT 1997. LNCS, vol. 1233, pp. 256–266. Springer, Heidelberg (1997)
39. Wee, H.: On obfuscating point functions. In: Gabow, H.N., Fagin, R. (eds.) STOC, pp. 523–532. ACM (2005)

Obfuscation for Evasive Functions

Boaz Barak[1], Nir Bitansky[2,*], Ran Canetti[2,3,**],
Yael Tauman Kalai[1], Omer Paneth[3,***], and Amit Sahai[4,†]

[1] Microsoft Research
[2] Tel Aviv University
[3] Boston University
[4] UCLA

Abstract. An *evasive circuit family* is a collection of circuits $\mathcal{C}$ such that for every input x, a random circuit from $\mathcal{C}$ outputs 0 on x with overwhelming probability. We provide a combination of definitional, constructive, and impossibility results regarding obfuscation for evasive functions:

1. The (average case variants of the) notions of *virtual black box* obfuscation (Barak et al, CRYPTO '01) and *virtual gray box* obfuscation (Bitansky and Canetti, CRYPTO '10) coincide for evasive function families. We also define the notion of *input-hiding* obfuscation for evasive function families, stipulating that for a random $C \in \mathcal{C}$ it is hard to find, given $\mathcal{O}(C)$, a value outside the preimage of 0. Interestingly, this natural definition, also motivated by applications, is likely not implied by the seemingly stronger notion of average-case virtual black-box obfuscation.

2. If there exist average-case virtual gray box obfuscators for all evasive function families, then there exist (quantitatively weaker) average-case virtual gray obfuscators for *all* function families.

3. There does not exist a *worst-case* virtual black box obfuscator even for evasive circuits, nor is there an average-case virtual gray box obfuscator for evasive *Turing machine* families.

4. Let $\mathcal{C}$ be an evasive circuit family consisting of functions that test if a low-degree polynomial (represented by an efficient arithmetic circuit) evaluates to zero modulo some large prime p. Then under a natural analog of the discrete logarithm assumption in a group supporting multilinear maps, there

* Supported by an IBM Ph.D. Fellowship and the Check Point Institute for Information Security.
** Supported by the Check Point Institute for Information Security, an NSF EAGER grant, and an NSF Algorithmic foundations grant 1218461.
*** Work done while the author was an intern at Microsoft Research New England. Supported by the Simons award for graduate students in theoretical computer science and an NSF Algorithmic foundations grant 1218461.
† Work done in part while visiting Microsoft Research, New England. Research supported in part from a DARPA/ONR PROCEED award, NSF grants 1228984, 1136174, 1118096, and 1065276, a Xerox Faculty Research Award, a Google Faculty Research Award, an equipment grant from Intel, and an Okawa Foundation Research Grant. This material is based upon work supported by the Defense Advanced Research Projects Agency through the U.S. Office of Naval Research under Contract N00014-11-1-0389. The views expressed are those of the author and do not reflect the official policy or position of the Department of Defense, the National Science Foundation, or the U.S. Government.

Y. Lindell (Ed.): TCC 2014, LNCS 8349, pp. 26–51, 2014.

exists an input-hiding obfuscator for C. Under a new *perfectly-hiding multi-linear encoding* assumption, there is an average-case virtual black box obfuscator for the family C.

1 Introduction

The study of *Secure Software Obfuscation* — or, methods to transform a program (say described as a Boolean circuit) into a form that is executable but otherwise completely unintelligible — is a central research direction in cryptography. In this work, we study obfuscation of evasive functions— an *evasive function family* is a collection C_n of Boolean circuits mapping some domain D to $\{0, 1\}$ such that for every $x \in D$ the probability over $C \leftarrow C_n$ that $C(x) = 1$ is negligible. Focusing on evasive functions leads us to new notions of obfuscation, as well as new insights into general-purpose obfuscation.

Why Study Obfuscation of Evasive Functions? To motivate the study of the obfuscation of evasive functions, let us consider the following scenario taken from [13]: Suppose that a large software publisher discovers a new vulnerability in their software that causes the software to behave in undesirable ways on certain (rare) inputs. The publisher then designs a software patch P that tests the input x to see if it falls into the set S of bad inputs, and if so outputs 1 to indicate that the input x should be ignored. If $x \notin S$, the patch outputs 0 to indicate that the software can behave normally. If the publisher releases the patch P "in the clear", an adversary could potentially study the code of P and learn bad inputs $x \in S$ that give rise to the original vulnerability. Since it can take months before a majority of computers install a new patch, this would give the attacker plenty of time to exploit this vulnerability on unpatched systems *even when the set S of bad inputs was evasive to begin with.* If instead, the publisher could obfuscate the patch P before releasing it, then intuitively an attacker would gain no advantage in finding an input $x \in S$ from studying the obfuscated patch. Indeed, assuming that without seeing P the attacker has negligible chance of finding an input $x \in S$, it makes sense to model P as an evasive function.

Aside from the motivating scenario above, evasive functions are natural to study in the context of software obfuscation because they are a natural generalization of the special cases for which obfuscation was shown to exist in the literature such as point functions [8], hyperplanes [10], and conjunctions [6].[1] Indeed, as we shall see, the study of obfuscation of evasive functions turns out to be quite interesting from a theoretical perspective, and sheds light on general obfuscation as well.

What notions of obfuscation makes sense for evasive functions? As the software patch problem illustrates, a very natural property that we would like is *input hiding*: Given an obfuscation of a random circuit $C \leftarrow C_n$, it should be hard for an adversary to find an input x such that $C(x) = 1$. It also makes sense to consider (average case versions of) strong notions of obfuscation, such as "virtual black box" (VBB) obfuscation introduced by Barak, Goldreich, Impagliazzo, Rudich, Sahai, Vadhan and Yang [3],

---

[1] Conjunctions are not necessarily evasive, however the interesting case for obfuscation is large conjunctions which are evasive.

which, roughly speaking, states that any predicate of the original circuit C computed from its obfuscation could be computed with almost the same probability by an efficient simulator having only oracle (i.e., black box) access to the function (see Section 2 for a formal definition). One can also consider the notion of "virtual gray box" (VGB) introduced by Bitansky and Canetti [4], who relaxed the VBB definition to allow the simulator to run in unbounded time (though still with only a polynomial number of queries to its oracle). Another definition proposed in [3], with a recent construction given by [15], is of "indistinguishability obfuscation" (IO). The actual meaning of IO is rather subtle, and we discuss it in Section 1.2.

1.1   Our Results

We provide a combination of definitional, constructive, and impossibility results regarding obfuscation for evasive functions. First, we formalize the notion of input-hiding obfuscation for evasive functions (as sketched above). We give evidence that this notion of input-hiding obfuscation is actually *incomparable* to the standard definition of VBB obfuscation. While it is not surprising that input-hiding obfuscation does not imply VBB obfuscation, it is perhaps more surprising that VBB obfuscation does not imply input-hiding obfuscation (under certain computational assumptions). Intuitively, this is because VBB obfuscation requires only that *predicates* of the circuit being obfuscated are simulatable, whereas input hiding requires that no complete input string x that causes $C(x) = 1$ can be found.

Second, we formalize a notion of *perfect circuit-hiding* obfuscation, which asks that for *every* predicate of the circuit $C \leftarrow \mathcal{C}_n$, the probability that the adversary can guess the predicate (or its complement) given an obfuscation of C is negligibly close to the expected value of the predicate over $C \leftarrow \mathcal{C}_n$. We then show that for any evasive circuit family $\mathcal{C}$, this simple notion of obfuscation is equivalent to both average-case VBB obfuscation and average-case VGB obfuscation. Thus, in particular, we have:

Theorem 1 (Informal). *For every evasive function collection $\mathcal{C}$ and obfuscator $\mathcal{O}$, it holds that $\mathcal{O}$ is an average-case VBB obfuscator for $\mathcal{C}$ if and only if it is an average-case VGB obfuscator for $\mathcal{C}$.*

We also show that evasive functions are at the heart of the VGB definition in the sense that if it is possible to achieve VGB obfuscators for all evasive circuit families then it is possible to achieve a slightly weaker variant of VGB obfuscation for *all* circuits:

Theorem 2 (Informal). *If there exists an average-case VGB obfuscator for every evasive circuit family then there exists a "weak" average-case VGB obfuscator for every (not necessarily evasive) circuit family.*

The notion of "weak" VGB obfuscation allows the simulator to make a slightly super-polynomial number of queries to its oracle. It also allows the obfuscating algorithm to be inefficient. The latter relaxation is not needed if we assume the existence of indistinguishability obfuscators for all circuits, as conjectured in [15].

We then proceed to give new *constructions* of obfuscators for specific natural families of evasive functions. We focus on functions that test if a bounded (polynomial) degree multivariate polynomial, given by an arithmetic circuit, evaluates to zero modulo some large prime p. We provide two constructions that build upon the approximate multilinear map framework developed by Garg, Gentry, and Halevi [14] and continued by Coron, Lepoint, and Tibouchi [12]. We first construct an input-hiding obfuscator whose security is based on a variant of the discrete logarithm assumption in the multilinear setting. We then construct a perfect circuit-hiding obfuscator based on a new assumption, called *perfectly-hiding multilinear encoding*. Very roughly, we assume that given encodings of k and $r \cdot k$, for a random r and k, the value of any predicate of k cannot be efficiently learned.

Theorem 3 (Informal). *Let C be the evasive function family that tests if a bounded (polynomial) degree multivariate polynomial, given by an arithmetic circuit, evaluates to zero modulo some large prime p. Then: (i) Assuming the existence of a group supporting a multilinear map in which the discrete logarithm problem is hard, there exists an input-hiding obfuscator for C, and (ii) Under the perfectly-hiding multilinear encoding assumption, there exists an average-case VBB obfuscator for all log-depth circuits in C.*

These constructions can be combined to obtain a single obfuscator for testing if an input is in the zero-set of a bounded-degree polynomial, that simultaneously achieves input-hiding and average-case VBB obfuscation. We give an informal overview of our construction in Section 3.

Finally, we complement our constructive results by giving two impossibility results regarding the obfuscation of evasive functions. First, we show impossibility with respect to evasive Turing Machines:

Theorem 4 (Informal). *There exists a class of evasive Turing Machines $\mathcal{M}$ such that no input-hiding obfuscator exists with respect to $\mathcal{M}$ and no average-case VGB obfuscator exists with respect to $\mathcal{M}$.*

We also show that there exist classes of evasive *circuits* for which VBB obfuscation is not possible. However, here we only rule out *worst case* obfuscation:

Theorem 5 (Informal). *There exists a class of evasive circuits C such that no worst-case VBB obfuscator exists with respect to C.*

1.2 Alternative Approaches to Obfuscation

We briefly mention two other approaches to general program obfuscation. One is to use the notion of *indistinguishability obfuscation* (IO) [3]. Indeed, in a recent breakthrough result, Garg, Gentry, Halevi, Raykova, Sahai and Waters [15] propose a candidate general obfuscation mechanism and conjecture that it is IO for all circuits. Improved variants of this construction have been proposed in [7, 2]. Roughly speaking, IO implies that an obfuscation of a circuit C hides "as much as possible" about the circuit C in the sense that if $\mathcal{O}$ is an IO obfuscator, and there exists some other obfuscator $\mathcal{O}'$ (with not too long output) that (for instance) achieves input hiding or VBB obfuscation for C, then $\mathcal{O}(C)$ will achieve the same security as well [16]. However, while IO obfuscators have found uses for many cryptographic applications [15, 18, 17], its hard to quantify what security is obtained by directly obfuscating a function with an IO obfuscator since for most classes of functions we do not know what "as much as possible" is. In particular, IO obfuscators are not known to imply properties such as input hiding or VBB/VGB for general circuits. For insance, the "totally unobfuscatable functions" of [3] (which are functions that are hard to learn but whose canonical representation can be recovered from any circuit) can be trivially IO-obfuscated, but this clearly does not give any meaningful security guarantees. Furthermore, we do not know whether IO obfuscators give guarantees such as input hiding or VBB on any families of evasive functions beyond those families that we already know how to VBB-obfuscate. Thus our work here can be seen as complementary to the question of constructing indistinguishability obfuscators. Furthermore, the hardness assumptions made in this work are much simpler than those implied by the IO security of any of the above candidates.

Another approach is to consider obfuscation mechanisms in idealized models where basic group operations are modeled as calls to abstract oracles. Works along this line, using different levels of abstraction, include [15, 11, 7, 2]. It should be stressed however that, as far as we know, proofs of security in any of these idealized models bear no immediate relevance to the security of obfuscation schemes in the plain model.

1.3   Organization of the Paper

In Section 2 we formally define evasive function families and the various notions of obfuscation that apply to them, and show equivalence between several of these notions. Section 3 contains our constructions for obfuscating zero-testing functions for low degree polynomials. In Section 4 we show that obtaining virtual gray box obfuscation for evasive functions implies a weaker variant of VGB for *all* functions. Section 5 contains our impossibility results for worst-case VBB obfuscation of evasive circuits and average-case VGB obfuscation of evasive Turing machines.

2   Evasive Circuit Obfuscation

Let $\mathcal{C} = \{\mathcal{C}_n\}_{n \in \mathbb{N}}$ be a collection of circuits such that every $C \in \mathcal{C}_n$ maps n input bits to a single output bit, and has size $\text{poly}(n)$. We say that $\mathcal{C}$ is *evasive* if on every input, a random circuit in the collection outputs 0 with overwhelming probability:[2]

Definition 2.1 (Evasive Circuit Collection). *A collection of circuits $\mathcal{C}$ is evasive if there exist a negligible function μ such that for every $n \in \mathbb{N}$ and every $x \in \{0,1\}^n$:*

$$\Pr_{C \leftarrow \mathcal{C}_n} [C(x) = 1] \leq \mu(n) \ .$$

An equivalent definition that will be more useful to us states that given oracle access to a random circuit in the collection it is hard to find an input that maps to 1.

---

[2] To avoid confusion, we note that the notion here is unrelated (and quite different) from the notion of evasive relations in [9].

Definition 2.2 (Evasive Circuit Collection - Alternative Definition). *A collection of circuits C is evasive if for every (possibly inefficient) $\mathcal{A}$ and for every polynomial q there exist a negligible function μ such that and for every $n \in \mathbb{N}$:*

$$\Pr_{C \leftarrow \mathcal{C}_n} [C(\mathcal{A}^{C[q(n)]}(1^n)) = 1] \leq \mu(n) \ .$$

Where $C[q(n)]$ denotes an oracle to the circuit C which allows at most $q(n)$ queries.

2.1 Definitions of Evasive Circuit Obfuscation

We start by recalling the syntax and functionality requirement for circuit obfuscation as defined in [3].

Definition 2.3 (Functionality of Circuit Obfuscation). *An obfuscator $\mathcal{O}$ for a circuit collection C is a PPT algorithm such that for all $C \in \mathcal{C}$, $\mathcal{O}(C)$ outputs a description of a circuit that computes the same function as C.*

We suggest two new security notions for obfuscation of evasive collections: perfect circuit-hiding and input-hiding. Both notions are average-case notions, that is, they only guarantee security when the obfuscated circuit is chosen at random from the collection.

The notion of input hiding asserts that given an obfuscation of a random circuit in the collection, it remains hard to find input that evaluates to 1.

Definition 2.4 (Input-Hiding Obfuscation). *An obfuscator $\mathcal{O}$ for a collection of circuits C is input-hiding if for every PPT adversary $\mathcal{A}$ there exist a negligible function μ such that for every $n \in \mathbb{N}$ and for every auxiliary input $z \in \{0,1\}^{\text{poly}(n)}$ to $\mathcal{A}$:*

$$\Pr_{C \leftarrow \mathcal{C}_n} [C(\mathcal{A}(z, \mathcal{O}(C))) = 1] \leq \mu(n) \ ,$$

where the probability is also over the randomness of $\mathcal{O}$.

Our second security notion of perfect circuit-hiding asserts that an obfuscation of a random circuit in the collection does not reveal any partial information about original circuit. We show that for evasive collections, this notion is equivalent to existing definitions of average-case obfuscation such as average-case virtual black-box (VBB) [3], average-case virtual gray-box (VGB) [4], and average-case oracle indistinguishability [8].

Definition 2.5 (Perfect Circuit-Hiding Obfuscation). *Let C be a collection of circuits. An obfuscator $\mathcal{O}$ for a circuit collection C is perfect circuit-hiding if for every PPT adversary $\mathcal{A}$ there exist a negligible function μ such that for every balanced predicate $\mathcal{P}$, every $n \in \mathbb{N}$ and every auxiliary input $z \in \{0,1\}^{\text{poly}(n)}$ to $\mathcal{A}$:*

$$\Pr_{C \leftarrow \mathcal{C}_n} [\mathcal{A}(z, \mathcal{O}(C)) = \mathcal{P}(C)] \leq \frac{1}{2} + \mu(n) \ ,$$

where the probability is also over the randomness of $\mathcal{O}$.

Remark 2.1 (On Definition 2.5). The restriction to the case of balanced predicates is made for simplicity of exposition only. We note that the proof of Theorem 2.1 implies that Definition 2.5 is equivalent to a more general definition that considers all predicates.

Remark 2.2 (On extending the definitions for non-evasive functions). The definitions of input-hiding and perfect circuit-hiding obfuscation are tailored for evasive collections and clearly cannot be achieved for all collections of circuits. For the case of input-hiding, this follows directly from Definition 2.2 of evasive collections. For the case of perfect circuit-hiding, consider the non-evasive collection $\mathcal{C}$ such that for every $C \in \mathcal{C}$, $C(0^n)$ outputs the first bit of C. Clearly, no obfuscator can preserve functionality while hiding the first bit of the circuit.

2.2 On the Relations between the Definitions

An input-hiding obfuscation is not necessarily perfect circuit-hiding since an input-hiding obfuscation might always include, for example, the first bit of the circuit in the output. In the other direction we do not believe that every perfect circuit-hiding obfuscation is also input hiding. Intuitively, the reason is that the perfect circuit-hiding obfuscation only prevents the adversary from learning a predicate of the circuit. Note that there may be many inputs on which the circuit evaluates to 1, and therefore, even if the obfuscation allows the adversary to learn some input that evaluates to 1, it is not clear how to use such an input to learn a predicate of the circuit.

In the full version of this paper [1] we give an example of an obfuscation for some evasive collection that is perfect circuit-hiding but not input-hiding. The example is based on a worst case obfuscation assumption for hyperplanes [10]. Nonetheless, we prove that for evasive collections where every circuit only evaluates to 1 on a *polynomial* number of inputs, every perfect circuit-hiding obfuscation is also input-hiding.

2.3 Perfect Circuit-Hiding Obfuscation Is Equivalent to Existing Notions

We start by recalling the average-case versions of existing security definitions of obfuscation.

Definition 2.6 (Average-Case Virtual Black-Box (VBB) from [3]). *An obfuscator $\mathcal{O}$ for a collection of circuits $\mathcal{C}$ is average-case VBB if for every PPT adversary $\mathcal{A}$ there exists a PPT simulator* Sim *and a negligible function μ such that for every predicate $\mathcal{P}$, every $n \in \mathbb{N}$ and every auxiliary input $z \in \{0,1\}^{\mathrm{poly}(n)}$ to $\mathcal{A}$:*

$$\left| \Pr_{C \leftarrow \mathcal{C}_n} [\mathcal{A}(z, \mathcal{O}(C)) = \mathcal{P}(C)] - \Pr_{C \leftarrow \mathcal{C}_n} [\mathrm{Sim}^C(z, 1^n) = \mathcal{P}(C)] \right| \leq \mu(n) \ ,$$

where the probability is also over the randomness of $\mathcal{O}$ and Sim.

The notion of VGB relaxes VBB by considering a computationally unbounded simulator, however, the number of queries the simulator makes to its oracle is bounded.

Definition 2.7 (Average-Case VGB Obfuscation from [4]). *An obfuscator $\mathcal{O}$ for a collection of circuits $\mathcal{C}$ is average-case VGB if for every PPT adversary $\mathcal{A}$ there exists a negligible function μ, a polynomial q and a (possibly inefficient) simulator Sim such that for every predicate $\mathcal{P}$, every $n \in \mathbb{N}$ and every auxiliary input $z \in \{0,1\}^{\text{poly}(n)}$ to $\mathcal{A}$:*

$$\left| \Pr_{C \leftarrow \mathcal{C}_n} [\mathcal{A}(z, \mathcal{O}(C)) = \mathcal{P}(C)] - \Pr_{C \leftarrow \mathcal{C}_n} [\text{Sim}^{C[q(n)]}(z, 1^n) = \mathcal{P}(C)] \right| \leq \mu(n) \ ,$$

where $C[q(n)]$ denotes an oracle to the circuit C which allows at most $q(n)$ queries and where the probability is also over the randomness of $\mathcal{O}$ and Sim.

The notion of oracle indistinguishability was originally formulated in the context of point functions, and here we give a variation of it for general collections. Similarly to our new notions, this definition is meaningful for evasive collections, but not for arbitrary collections.

Definition 2.8 (Average-Case Oracle-Indistinguishability Obfuscation from [8]). *An obfuscator $\mathcal{O}$ for a collection of circuits $\mathcal{C}$ is average-case oracle indistinguishable if for every PPT adversary $\mathcal{A}$ that outputs one bit, the following ensembles are computationally indistinguishable:*

- $\{(C, \mathcal{A}(z, \mathcal{O}(C))) \mid C \leftarrow \mathcal{C}_n\}_{n \in \mathbb{N}, z \in \{0,1\}^{\text{poly}(n)}}$,
- $\{(C, \mathcal{A}(z, \mathcal{O}(C'))) \mid C, C' \leftarrow \mathcal{C}_n\}_{n \in \mathbb{N}, z \in \{0,1\}^{\text{poly}(n)}}$.

The following theorem showing that, for evasive collections, the above notions are all equivalent to perfect circuit-hiding is proven in the full version of this paper [1]. We note that, for general circuits, average-case VBB and average-case VGB may not be equivalent (follows from [4, Proposition 4.1]). The equivalence of average-case VBB and average-case oracle-indistinguishability was proven for point functions by Canetti [8]. We generalize the claim for all evasive functions.

Theorem 2.1. *Let $\mathcal{O}$ be an obfuscator for an evasive collection $\mathcal{C}$. The following statements are equivalent:*

1. *$\mathcal{O}$ is perfect circuit-hiding (Definition 2.5).*
2. *$\mathcal{O}$ is average-case VBB (Definition 2.6).*
3. *$\mathcal{O}$ is average-case VGB (Definition 2.7).*
4. *$\mathcal{O}$ is average-case oracle-indistinguishability (Definition 2.8).*

Remark 2.3 (On evasive obfuscation with a universal simulator). It follows from the proof of Theorem 2.1 that every obfuscator $\mathcal{O}$ for an evasive collection $\mathcal{C}$ that is average-case VBB-secure (or average-case VGB-secure) can be simulated as follows: given an adversary $\mathcal{A}$, the simulator Sim simply executes $\mathcal{A}$ on a random obfuscation $\mathcal{O}(C')$ of a circuit C' sampled uniformly from the collection $\mathcal{C}$. The simplicity of the above simulator can be demonstrated as follows:

- The simulator is *universal*, that is, the same algorithm Sim can simulate every obfuscator $\mathcal{O}$ and family $\mathcal{C}$ given only black box access to $\mathcal{O}$ and the ability to sample uniformly from $\mathcal{C}$.

– The simulator only makes *black-box* use of the adversary $\mathcal{A}$. This is in contrast to the case of worst-case VBB-security where non-black simulators are required ([8, 20]).
– The simulator does not make any calls to its oracle. This is in contrast to the case of non-evasive function where, for example, the possibility of obfuscating learnable functions can only be demonstrated by a simulator that queries it oracle.

3   Obfuscating Root Sets of Low Degree Polynomials

In this section, we present constructions of input-hiding obfuscation and of perfect circuit-hiding obfuscation for a subclass of evasive collections. We will be able to obfuscate collections that can be expressed as the zero-set of some low-degree polynomial. More concretely, we say that a collection $\mathcal{C}$ is of *low arithmetic degree* if for every n, there is a $\theta(n)$-bit prime p, and a polynomial size low degree arithmetic circuit $U(k, x)$ over $\mathbb{Z}_p$ where $k \in \mathbb{Z}_p^\ell, x \in \mathbb{Z}_p^m$ such that $\mathcal{C}_n = \{C_k\}_{k \in \mathbb{Z}_p^\ell}$ and $C_k(x) = 1$ iff $U(k, x) = 0$.

Note that the Schwartz-Zippel Lemma, together with the fact that $U(k, x)$ is of low degree implies that for every $x \in \mathbb{Z}_p^m$ either

$$\Pr_{k \leftarrow \mathbb{Z}_p^\ell}[C_k(x) = 0] = \mathrm{negl}(n) \quad \text{or,} \quad \Pr_{k \leftarrow \mathbb{Z}_p^\ell}[C_k(x) = 0] = 1 \ .$$

Thus, there exists a single function h such that the collection $\{C_k - h\}_{k \in \mathbb{Z}_p^\ell}$ is evasive, where $h(x) = 1$ iff:

$$\Pr_{k \leftarrow \mathbb{Z}_p^\ell}[C_k(x) = 0] = 1 \ .$$

In other words, a collection of low arithmetic degree is either evasive or it is a "translation" of an evasive collection by a fixed, efficiently computable (in RP) function that is independent of the key. Therefore, we can restrict ourselves WLOG to *evasive* collection of low arithmetic degree.

Both constructions will be based on *graded encoding* as introduced by Garg, Gentry, and Halevi [14]. The high-level idea behind both constructions is as follows. The obfuscation of a circuit C_k will contain some encoding of the elements of k. Using this encoding, an evaluator can homomorphically evaluate the low-degree polynomial $U(k, x)$. Then, the evaluator tests whether the output is an encoding of 0 and learns the value of $C_k(x)$.

The two constructions will encode the key k in two different ways, and their security will be based on two different hardness assumptions. The security of the input-hiding construction will be based on a discrete-logarithm-style assumption on the encoding. The obfuscation will support evasive collections of low arithmetic degree defined by a polynomial size circuit $U(k, x)$ of total degree $\mathrm{poly}(n)$. This class of circuits is equivalent to polynomial size arithmetic circuits of depth $\mathcal{O}(\log^2(n))$ and total degree $\mathrm{poly}(n)$ [19]. The security of the perfect circuit-hiding construction will be based on a new assumption called *the perfectly-hiding multilinear encoding assumption*. The obfuscation will support evasive collections defined by a polynomial size circuit $U(k, x)$ of depth $\mathcal{O}(\log(n))$. We also discuss a stronger variant of this assumption, which like in the input-hiding construction, supports arbitrary arithmetic circuits with total polynomial degree.

3.1 Graded Encoding

We start by defining a variant of the symmetric graded encoding scheme from ([14]), used in our construction, and by specifying hardness assumptions used.

Definition 3.1. *A d-graded encoding system consists of a ring R and a collection of disjoint sets of encodings $\left\{ S_i^{(\alpha)} \mid \alpha \in R, 0 \leq i \leq d \right\}$. The scheme supports the following efficient procedures:*

- *Instance Generation: given security parameter n and the parameter d, $\mathsf{Gen}(1^n, 1^d)$ outputs public parameters pp.*
- *Encoding: given the public parameters pp and $\alpha \in R$, $\mathsf{Enc}(\mathsf{pp}, \alpha)$ outputs $u \in S_1^{(\alpha)}$.*
- *Addition and Negation: given the public parameters pp and two encodings $u_1 \in S_i^{(\alpha_1)}, u_2 \in S_i^{(\alpha_2)}$, $\mathsf{Add}(\mathsf{pp}, u_1, u_2)$ outputs an encoding in $S_i^{(\alpha_1 + \alpha_2)}$, and $\mathsf{Neg}(\mathsf{pp}, u_1)$ outputs an encoding in $S_i^{(-\alpha_1)}$.*
- *Multiplication: given the public parameters pp and two encodings $u_1 \in S_{i_1}^{(\alpha_1)}$, $u_2 \in S_{i_2}^{(\alpha_2)}$ such that $i_1 + i_2 \leq d$, $\mathsf{Mul}(\mathsf{pp}, u_1, u_2)$ outputs an encoding in $S_{i_1 + i_2}^{(\alpha_1 \cdot \alpha_2)}$.*
- *Zero Test: given the public parameters pp and an encodings u, $\mathsf{Zero}(\mathsf{pp}, u)$ outputs 1 if $u \in S_d^{(0)}$ and 0 otherwise.*

The main difference between this formulation and the formulation in [14] is that we assume that there is an efficient procedure for generating level 1 encoding of every ring element. In [14], it is only possible to obliviously generate an encoding of a random element in R, without knowing the underlying encoded element. While we currently do not know how how to use the construction of [14] to instantiate our scheme, we can get a candidate construction with public encoding based on the construction of [12], by publishing encodings of all powers of 2 as part of the public parameters. The known candidate constructions involve noisy encodings. Since in our construction we use at most $O(d)$ operations, we may set the noise parameter to be small enough so that it can be ignored. Therefore, from hereon, we omit the management of noise from the interfaces of the graded encoding.

Our first hardness assumption (used in the construction of input-hiding obfuscation) is that the encoding function is one-way. That is, given the output of $\mathsf{Enc}(\mathsf{pp}, \alpha)$ for randomly generated parameters pp and a random ring element $\alpha \in_R R$, it is hard to find α.

Our second hardness assumption (used in the construction of perfect circuit-hiding obfuscation) is a new assumption called *perfectly-hiding multilinear encoding*. The assumption states that given level 1 encodings of r and $r \cdot k$ for random $r, k \in R$, the value of any one bit predicate of k cannot be efficiently learned. The perfectly-hiding multilinear encoding assumption can be shown to hold in an ideal model where the encoding is generic (such as the generic ideal models described in [7, 2]).

Assumption 3.1 (perfectly-hiding multilinear encoding). *For every PPT adversary $\mathcal{A}$ that outputs one bit, the following ensembles are computationally indistinguishable:*

- $\left\{ \mathsf{pp}, k, \mathcal{A}(\mathsf{Enc}(\mathsf{pp}, r), \mathsf{Enc}(\mathsf{pp}, r \cdot k)) : k, r \leftarrow R, \mathsf{pp} \leftarrow \mathsf{Gen}(1^n, 1^d) \right\}$,
- $\left\{ \mathsf{pp}, k, \mathcal{A}(\mathsf{Enc}(\mathsf{pp}, r), \mathsf{Enc}(\mathsf{pp}, r \cdot k')) : k, k', r \leftarrow R, \mathsf{pp} \leftarrow \mathsf{Gen}(1^n, 1^d) \right\}$.

We note that both of the above assumptions *do not hold* for the candidate construction of [14] (assuming there is an efficient encoding procedure for every ring element). However, they are possibly satisfied by other constructions. In particular, to our knowledge there are no known attacks violating this assumption for the candidate construction of [12].

3.2 Input-Hiding Obfuscation

Let $\mathcal{C}_n$ be an evasive collection defined by the arithmetic circuit $U(k, x)$, $k \in \mathbb{Z}_p^\ell, x \in \mathbb{Z}_p^m$ of degree $d = \mathrm{poly}(n)$. The obfuscator will make use of a d-symmetric graded encoding scheme over the ring $\mathbb{Z}_p$. For every $k \in \mathbb{Z}_p^\ell$ the obfuscation $\mathcal{O}(C_k)$ generates parameters $\mathsf{pp} \leftarrow \mathsf{Gen}(1^n, 1^d)$ for the graded encoding, and outputs a circuit that has the public parameters pp and the encodings $\{\mathsf{Enc}(\mathsf{pp}, k_i)\}$ for $i \in [\ell]$ hardwired into it. The circuit $\mathcal{O}(C_k)$, on input $x \in \mathbb{Z}_p^m$, first generates encodings $\{\mathsf{Enc}(\mathsf{pp}, x_i)\}$ for $i \in [m]$, and uses the evaluation functions of the graded encoding system to compute an encoding $u \in S_d^{(U(k,x))}$. $\mathcal{O}(C_k)$ then uses the zero test to check whether $u \in S_d^{(0)}$. If so it outputs 1 and otherwise it outputs 0.

More concretely, the encoding $u \in S_d^{(U(k,x))}$ is obtained by computing the encoded value for every wire of the arithmetic circuit $U(k, x)$. For every gate in the circuit connecting the wires w_1 and w_2 to w_3, given encodings

$$u_{w_1} \in S_{d_1}^{(\alpha_1)}, u_{w_2} \in S_{d_2}^{(\alpha_2)} \ ,$$

for the values on wires w_1 and w_2, an encoding u^{w_2} of the value on wire w_3 is computed as follows. If the gate is an addition gate we first obtain encodings

$$u'_{w_1} \in S_{\max(d_1, d_2)}^{(\alpha_1)}, u'_{w_2} \in S_{\max(d_1, d_2)}^{(\alpha_2)} \ ,$$

by multiplying either u_{w_1} or u_{w_2} by the appropriate encoding of 1. The encodings u'_{w_1}, u'_{w_2} are added to obtain the encoding $u_{w_3} \in S_{\max(d_1, d_2)}^{(\alpha_1 + \alpha_2)}$. If the gate is a multiplication gate, we multiply the encodings u_{w_1}, u_{w_2} to obtain the encoding $u_{w_3} \in S_{d_1 + d_2}^{(\alpha_1 \cdot \alpha_2)}$. Note that the degree of the encoding computed for every is at most the degree of the polynomial computed by the wire and therefore does not exceed d.

Theorem 3.2. $\mathcal{O}$ *is an input-hiding obfuscator for* $\mathcal{C}$, *assuming* Enc *is one-way.*

Proof. We need to prove that $\mathcal{O}$ satisfies both the functionality requirement and the security requirement. The functionality requirement follows immediately from the guarantees of the graded encoding scheme. Thus, we focus on proving the security requirement. To this end, fix any PPT adversary $\mathcal{A}$, and suppose for the sake of contradiction that for infinitely many values of n,

$$\Pr_{k \leftarrow \mathbb{Z}_p^\ell}[C_k(\mathcal{A}(\mathcal{O}(C_k))) = 1] \geq \frac{1}{\text{poly}(n)} \ . \tag{1}$$

Next we prove that there exists a PPT adversary $\mathcal{A}'$ that brakes the one-wayness of the encoding. The adversary $\mathcal{A}'$ will make use of the the helper procedure Simplify described in the following claim:

Claim 3.3. *There exists an efficient procedure Simplify that, given a multivariate non-trivial polynomial $P : \mathbb{Z}_p^\ell \to \mathbb{Z}_p$ of total degree d, represented as an arithmetic circuit, outputs a set of multivariate polynomials $\{P_j\}_{j \in [\ell]}$ (represented as arithmetic circuits) such that:*

1. *P_j is a multivariate non-trivial polynomial of total degree d.*
2. *For every $r \in \mathbb{Z}_p^\ell$ such that $P(r) = 0$ there exist $j \in [\ell]$ such that $P(r) = 0$ but the univariate polynomial $Q(x) = P(r_1, \ldots, r_{j-1}, x, r_{j+1} \ldots, r_\ell)$ is non-trivial.*

We note that a very similar claim was proven by Bitansky *et al.* [5, Proposition 5.15]. We reprove it here for completeness

Proof (Claim 3.3). Given an arithmetic circuit computing a multivariate polynomial $P : \mathbb{Z}_p^\ell \to \mathbb{Z}_p$ of total degree d such that $P \not\equiv 0$, the procedure Simplify is as follows:

1. Set $P_1 = P$. repeat the following for $j = 1$ to ℓ.
2. Decompose P_j as follows:

$$P_j(k_j, \ldots, k_\ell) = \sum_{i=1}^{d} k_j^i \cdot P_{j,i}(k_{j+1}, \ldots, k_\ell).$$

3. Set P_{j+1} to be the non-zero polynomial $P_{j,i}$ with the minimal i.

Note that decomposing an arithmetic circuit into homogeneous components can be done efficiently. It is left to show that for every $r \in \mathbb{Z}_p^\ell$ if $P(r) = 0$ then there exists $j \in [\ell]$ such that

$$Q(x) = P_j(x, r_{j+1}, \ldots, r_\ell) \not\equiv 0 \ ,$$

and

$$P_j(r_j, r_{j+1}, \ldots, r_\ell) = 0.$$

The proof is by induction on j. The base case is when $j = 1$, for which it holds that:

$$P_1(x_1, \ldots, x_\ell) \not\equiv 0$$
$$P_1(r_1, \ldots, r_\ell) = 0 \ .$$

For any $1 \leq j < \ell$, suppose that:

$$P_j(x_j, \ldots, x_\ell) \not\equiv 0$$
$$P_j(r_j, \ldots, r_\ell) = 0$$
$$P_j(x, r_{j+1}, \ldots, r_\ell) \equiv 0 \ ;$$

then, by the construction of P_{j+1} from P_j,

$$P_{j+1}(x_{j+1}, \ldots, x_\ell) \neq 0$$
$$P_{j+1}(r_{j+1}, \ldots, r_\ell) = 0 .$$

If this inductive process reaches P_ℓ, then it holds that:

$$P_\ell(x_\ell) \neq 0$$
$$P_\ell(r_\ell) = 0 ,$$

which already satisfies the claim.   □

The adversary $\mathcal{A}'$. $\mathcal{A}'$ is given the public parameters pp, and an encoding u of a random element $r \in \mathbb{Z}_p$. $\mathcal{A}'$ is defined as follows:

1. $\mathcal{A}'$ samples a random index $i \in [\ell]$ and a random element $k_j \leftarrow \mathbb{Z}_p$ for every $j \in [\ell] \setminus \{i\}$.
2. $\mathcal{A}'$ generates a random obfuscation $\mathcal{O}(C_k)$ from the encodings $\{\mathsf{Enc}(k_j)\}_{j \in [\ell] \setminus \{i\}}$ and using his input u as the encoding of k_i.
3. $\mathcal{A}'$ executes $\mathcal{A}(\mathcal{O}(C_k))$ and obtains an input x. If $\mathcal{O}(C_k)(x) \neq 1$, $\mathcal{A}'$ aborts.
4. Otherwise, $\mathcal{A}'$ executes the helper procedure Simplify on the polynomial $U(\cdot, x)$ with the values of x fixed and obtains the polynomials $\{P_j\}_{j \in [\ell]}$.
5. $\mathcal{A}'$ constructs the univariate polynomial $Q(x) = P_i(k_1, \ldots, k_{i-1}, x, k_{i+1}, \ldots, k_\ell)$ (the rest of the elements of k are known to $\mathcal{A}'$).
6. If $Q \equiv 0$, $\mathcal{A}'$ aborts, otherwise it outputs a random root of Q.

We show that $\mathcal{A}'$ outputs the correct value of r with noticeable probability. By our assumption on $\mathcal{A}$, $\mathcal{O}(C_k)(x) \neq 1$ with some noticeable probability ϵ. In this case, it follows from the correctness of $\mathcal{O}$ that $U(k, x) = 0$. Let j be the index guaranteed by Claim 3.3. Since the distribution of $k_1, \ldots, k_\ell$ is independent from the choice of i, it follows that conditioned on the event $U(k, x) = 0$, $i = j$ with probability $1/\ell$. In this case by Claim 3.3 it holds that $P_i(k) = 0$ but the univariate polynomial P defined above is not identically zero, which means that r is one of the at most d roots of P. Therefore, $\mathcal{A}'$ will output the correct root with probability at least $\epsilon/(\ell d)$.   □

3.3   Perfect Circuit-Hiding Obfuscation

Let C_n be an evasive collection defined by the arithmetic circuit $U(k, x)$, $k \in \mathbb{Z}_p^\ell, x \in \mathbb{Z}_p^m$ of depth degree $h = O(\log(n))$. The obfuscator will make use of a d-symmetric graded encoding scheme for $d = 2^h$ over the ring $\mathbb{Z}_p$. For every $k \in \mathbb{Z}_p^\ell$, the obfuscation $\mathcal{O}(C_k)$ generates parameters $\mathsf{pp} \leftarrow \mathsf{Gen}(1^n, 1^d)$ for the graded encoding, samples random elements $r_1, \ldots r_\ell \in \mathbb{Z}_p$, and outputs a circuit that has the public parameters pp hardwired into it, and for every for $i \in [\ell]$, has the encodings $\mathsf{Enc}(\mathsf{pp}, r_i)$ and $\mathsf{Enc}(\mathsf{pp}, r_i \cdot k_i)$ hardwired into it.

The circuit $\mathcal{O}(C_k)$, on input $x \in \mathbb{Z}_p^m$, does the following: For $i \in [m]$, it generates the encodings $\mathsf{Enc}(\mathsf{pp}, r_i')$ and $\mathsf{Enc}(\mathsf{pp}, r_i' \cdot x_i)$ where $r_i' = 1$ (Note that when encoding the input we do not need r'_i to be random for security. We only use r'_i the make the encoding of the input x and the key k have the same format). Using the evaluation functions of the graded encoding system, $\mathcal{O}(C_k)$ then computes a pair of encodings $u_0 \in S_d^{(\tilde{r})}$ and $u_1 \in S_d^{(\tilde{r} \cdot U(k,x))}$ for some $\tilde{r} \neq 0$ that is a function of $r_1, \dots, r_\ell$. Finally, it uses the zero test to check whether $u_1 \in S_d^{(0)}$. If so it outputs 1 and otherwise it outputs 0.

We next elaborate on how this encoded output is computed. The circuit $\mathcal{O}(C_k)$ evaluates the arithmetic circuit $U(k, x)$ gate by gate, as follows: Fix any gate in the circuit connecting the wires w_1 and w_2 to w_3. Suppose that for wires w_1 and w_2 we have the pairs of encodings

$$(u_0^{w_1}, u_1^{w_1}) \in S_{d_1}^{(\tilde{r}_1)} \times S_{d_1}^{(\tilde{r}_1 \cdot \alpha_1)}, \quad (u_0^{w_2}, u_1^{w_2}) \in S_{d_2}^{(\tilde{r}_2)} \times S_{d_2}^{(\tilde{r}_2 \cdot \alpha_2)} ,$$

where $\tilde{r}_1, \tilde{r}_2 \neq 0$ (supposedly the value on the wire w_1 is α_1 and the value on wire w_2 is α_2). If the gate is a multiplication gate, one can compute an encoding for wire w_3, by simply computing the pair of encodings:

$$(u_0^{w_3}, u_1^{w_3}) \in S_{d_1+d_2}^{(\tilde{r}_1 \cdot \tilde{r}_2)} \times S_{d_1+d_2}^{(\tilde{r}_1 \cdot \tilde{r}_2 \cdot \alpha_1 \cdot \alpha_2)}.$$

If the gate is an addition gate we compute $u_0^{w_3}$ in the same way. We also compute the encodings:

$$u_1^{w_{3,1}}, u_1^{w_{3,2}} \in S_{d_1+d_2}^{(\tilde{r}_1 \cdot \tilde{r}_2 \cdot \alpha_1)} \times S_{d_1+d_2}^{(\tilde{r}_1 \cdot \tilde{r}_2 \cdot \alpha_2)},$$

which can then be added to get:

$$u_1^{w_3} \in S_{d_1+d_2}^{(\tilde{r}_1 \cdot \tilde{r}_2 \cdot (\alpha_1 + \alpha_2))}.$$

Note that in any case $\tilde{r}_1 \cdot \tilde{r}_2 \neq 0$. Also note that in the evaluation of every level of the circuit U the maximal degree of the encodings at most doubles and therefore it never exceeds $d = 2^h$.

Theorem 3.4. *$\mathcal{O}$ is a perfect circuit-hiding obfuscator for $\mathcal{C}$, assuming the encoding satisfies the perfectly-hiding multilinear encoding assumption.*

Proof. By Theorem 2.1, it suffices to prove that $\mathcal{O}$ is an average-case oracle indistinguishability obfuscator for $\mathcal{C}$. Namely, it suffices to prove that for every PPT adversary $\mathcal{A}$ that outputs a single bit,

$$\{(C_k, \mathcal{A}(\mathcal{O}(C_k))) \mid k \leftarrow \mathbb{Z}_p^\ell\} \approx_c \{(C_k, \mathcal{A}(\mathcal{O}(C_{k'}))) \mid k, k' \leftarrow \mathbb{Z}_p^\ell\} .$$

Suppose for the sake of contradiction there exists a PPT adversary $\mathcal{A}$ (that outputs a single bit), a distinguisher D, and a non-negligible function ϵ, such that

$$\left| \Pr_{k \leftarrow \mathbb{Z}_p^\ell} [D(k, \mathcal{A}(\mathcal{O}(C_k))) = 1] - \Pr_{k, k' \leftarrow \mathbb{Z}_p^\ell} [D(k, \mathcal{A}(\mathcal{O}(C_{k'}))) = 1] \right| \geq \epsilon(n) ,$$

Recall that the obfuscated circuit $\mathcal{O}(C_k)$ consists of the public parameters pp and from the encodings:

$$\{\mathsf{Enc}(\mathsf{pp}, r_i), \mathsf{Enc}(\mathsf{pp}, r_i \cdot k_i)\}_{i \in [\ell]} ,$$

where $r_1, \ldots, r_\ell \leftarrow \mathbb{Z}_p^*$. Since sampling encoding corresponding to random input wires is efficient, the equation above, together with a standard hybrid argument, implies that there exists $i \in [\ell]$, a PPT adversary $\mathcal{A}'$ (that outputs a single bit), a distinguisher D', and a non-negligible function ϵ', such that

$$\left| \begin{array}{l} \Pr_{k_i \leftarrow \mathbb{Z}_p}[D'(\mathsf{pp}, k_i, \mathcal{A}'(\mathsf{Enc}(\mathsf{pp}, r_i), \mathsf{Enc}(\mathsf{pp}, r_i \cdot k_i))) = 1]- \\ \Pr_{k_i, k_i' \leftarrow \mathbb{Z}_p}[D'(\mathsf{pp}, k_i, \mathcal{A}'(\mathsf{Enc}(\mathsf{pp}, r_i), \mathsf{Enc}(\mathsf{pp}, r_i \cdot k_i'))) = 1] \end{array} \right| \geq \epsilon'(n) \; ,$$

contradicting the perfectly-hiding multilinear encoding assumption.   □

Remark 3.1 (On unifying the constructions). Under a stronger variant of the perfectly-hiding multilinear encoding assumption we can directly prove that the input hiding obfuscation construction presented in Section 3.2 is also perfect circuit-hiding. Intuitively, the stronger variant assumes that the the function Enc given a random input k already hides every predicate of k (without adding any additional randomization).

Assumption 3.5 (strong perfectly-hiding multilinear encoding). *For every PPT adversary $\mathcal{A}$ that outputs one bit, the following ensembles are computationally indistinguishable:*

- $\left\{ \mathsf{pp}, k, \mathcal{A}(\mathsf{Enc}(\mathsf{pp}, k)) : k, \leftarrow R, \mathsf{pp} \leftarrow \mathsf{Gen}(1^n, 1^d) \right\}$,
- $\left\{ \mathsf{pp}, k, \mathcal{A}(\mathsf{Enc}(\mathsf{pp}, k')) : k, k', \leftarrow R, \mathsf{pp} \leftarrow \mathsf{Gen}(1^n, 1^d) \right\}$.

Note that this strong perfectly-hiding multilinear encoding assumption cannot hold for a deterministic encoding function (unlike with the perfectly-hiding multilinear encoding assumption). Using the stronger assumption above, we can get perfect circuit-hiding obfuscation for a larger class of functions. Specifically, $U(k, x)$ can be any arithmetic circuit computing a polynomial of degree $\mathrm{poly}(n)$, removing the logarithmic depth restriction.

4   Evasive Function and Virtual Grey Box Obfuscation

We show that average-case VGB obfuscation for all evasive functions implies a slightly weaker form of average-case VGB obfuscation for *all* function.

   We start by giving a slightly more general definition for VGB obfuscation that considers also computationally unbounded obfuscators and allows for the query complexity of the simulator to be super-polynomial. Note that when the obfuscator is unbounded, we need to explicitly require that it has a polynomial slowdown, that is, the that the obfuscated circuit is not too large.

Definition 4.1 (Weak Average-Case VGB Obfuscation). *Let $\mathcal{C} = \{\mathcal{C}_n\}_{n \in \mathbb{N}}$ be a collection of circuits such that every $C \in \mathcal{C}_n$ is a circuit of size $\mathrm{poly}(n)$ that takes n bits as input. A (possibly inefficient) algorithm $\mathcal{O}$ is a weak average-case VGB obfuscator for $\mathcal{C}$ if it satisfies the following requirements:*

- *Functionality: for all $C \in \mathcal{C}$, $\mathcal{O}(C)$ outputs a description of a circuit that computes the same function as C.*

- *Polynomial Slowdown:* There exist a polynomial p such that for every $C \in \mathcal{C}$, $|\mathcal{O}(C)| < p(|C|)$.
- *Security:* For every super-polynomial function $q = q(n)$ and for every PPT adversary $\mathcal{A}$ there exist a (possibly inefficient) simulator Sim and a negligible function μ such that for every predicate $\mathcal{P}$, every $n \in \mathbb{N}$ and every auxiliary input $z \in \{0,1\}^{\text{poly}(n)}$ to $\mathcal{A}$:

$$\left| \Pr_{C \leftarrow \mathcal{C}_n} [\mathcal{A}(z, \mathcal{O}(C)) = \mathcal{P}(C)] - \Pr_{C \leftarrow \mathcal{C}_n} [\text{Sim}^{C[q(n)]}(z, 1^n) = \mathcal{P}(C)] \right| \leq \mu(n) .$$

Where $C[q(n)]$ denotes an oracle to the circuit C which allows at most $q(n)$ queries.

Remark 4.1 (On obfuscation with inefficient obfuscator). The notion of obfuscation with computationally unbounded obfuscators was first considered in [3]. To demonstrate the meaningfulness of this notion we note that assuming the existence of *indistinguishability obfuscation* for a collection $\mathcal{C}$ with an *efffcent* obfuscator, the existence of a (weak) average-case VGB Obfuscation for $\mathcal{C}$ with a computationally unbounded obfuscator already implies the existence of a (weak) average-case VGB Obfuscation for $\mathcal{C}$ with an *efffcent* obfuscator.

Theorem 4.1. *If there exists an average-case VGB obfuscator for every collection of evasive circuits, then there exists a weak average-case VGB obfuscator for every collection of circuits.*

Proof overview of Theorem 4.1. Let $\mathcal{C}$ be a (non-evasive) collection of circuit that we want to VGB obfuscate. We start by showing a computationally unbounded leaning algorithm $\mathcal{L}$ that given oracle access to a circuit $C \in \mathcal{C}$ tries to learn the circuit C. Clearly, If $\mathcal{L}$ can make unbounded number of queries it can learn C exactly. However, if the number of queries $\mathcal{L}$ makes to C is bounded by some super-polynomial function $q(n)$, we show that $\mathcal{L}$ can still learn a circuit $C' \in \mathcal{C}$ that is "close" to C. That is, C and C' only disagree on some negligible fraction of the inputs.

The learning algorithm $\mathcal{L}$ will repeatedly query C on the input that gives maximal information about the circuit C, taking into account the information gathered from all the previous oracle answers. $\mathcal{L}$ stops when it learns C or when there is no query that will give "enough" information about C. In this case, we denote by $K(C)$ the set of all circuits in $\mathcal{C}_n$ that are consistent with all the previous oracle answers. We show that all the circuits in $K(C)$ are close to C, and $\mathcal{L}$ will just output a random circuit $C' \in K(C)$.

The high-level idea behind the construction of the weak average-case VGB obfuscator $\mathcal{O}$ for $\mathcal{C}$ is that given black box access to a random circuit C a weak VGB simulator, that is, an unbounded adversary that can make at most $q(n)$ oracle queries to C, is able to run the learning algorithm $\mathcal{L}$ and learn the set $K(C)$. Therefore, a secure obfuscation $\mathcal{O}(C)$ of C does not need to hide the set $K(C)$. In particular, $\mathcal{O}(C)$ may contain a random circuit $C' \leftarrow K(C)$ in the clear. To satisfy the functionality requirement, the obfuscation cannot contain only C'. Additionally, $\mathcal{O}(C)$ will contain the circuit C_{diff}, where $C_{\text{diff}} = C \oplus C'$ is a circuit that outputs 1 on every input on which C and C' differ. Now, to evaluate C on an input x an evaluator can compute $C(x) = C'(x) \oplus C_{\text{diff}}(x)$. Clearly, $\mathcal{O}(C)$ cannot contain C_{diff} in the clear, since C_{diff} depends on C. instead, $\mathcal{O}(C)$

will obfuscate C_{diff} using the VGB obfuscator for evasive collections. Since C' is a random function in $K(C)$ it only differs from C on a negligible fraction of the inputs, and therefore C_{diff} outputs 1 only on a negligible fraction of the inputs.

Unfortunately, this high-level idea does not quite work. The problem is that since $\mathcal{O}(C)$ contains the circuit C' in the clear, we cannot argue that C_{diff} is taken at random from an evasive collection. In particular, given C it may be easy to find an input where C_{diff} outputs 1. For example, if C outputs 1 only on a single input x, and C' outputs 1 only on a single input x', then C_{diff} will output 1 on both inputs x and x'. Now, given the circuit C' it is easy find the input x' such that $C_{\text{diff}}(x') = 1$ and therefore we do not know how to securely obfuscate C_{diff}.

To fix this problem we do not choose C' to be a random circuit in the set $K(C)$, but instead C' will be a circuit computing the majority function on many random circuits taken from the set $K(C)$. Now we can show that even given the circuit C' it is hard to find an input where C and C' differ, and therefore the obfuscation of C_{diff} is secure.

Proof (Theorem 4.1). Let $\mathcal{O}$ be an average-case VGB secure obfuscator for every collection of evasive circuits. Let $\mathcal{C} = \{\mathcal{C}_n\}_{n \in \mathbb{N}}$ be a (not necessarily evasive) collection of circuits such that every $C \in \mathcal{C}_n$ is a circuit of size $p(n)$, for some polynomial p, that takes n bits as input. Let $q(n)$ be any super-polynomial function. We construct a weak average-case VGB obfuscator $\mathcal{O}'$ for $\mathcal{C}$ where the simulator makes at most $q(n)$ queries to its oracle. $\mathcal{O}'$ will make use of the following learning algorithm $\mathcal{L}$ as a subroutine. The algorithm $\mathcal{L}$ is an inefficient algorithm that has oracle access to C, it queries its oracle at most $q'(n)$ times for $q'(n) \triangleq \frac{q(n)}{(n+1)}$, and outputs a circuit that is "close" to C.

Loosely speaking, algorithm $\mathcal{L}$ starts by setting K to be the set of all circuits in $\mathcal{C}_n$. Then, in each iteration $\mathcal{L}$ reduces the size of K so that, at the end, it contains only circuits that are "close" to C. The number of iterations is at most $q'(n)$ and in each iteration $\mathcal{L}$ makes a single oracle call. However, the computation done by $\mathcal{L}$ in each iteration is inefficient.

Formally, the algorithm $\mathcal{L}$ is defined as follows:

1. Set $K \leftarrow \mathcal{C}_n$.
2. For every $b \in \{0, 1\}$ and every $x \in \{0, 1\}^n$, compute:

$$p_x^b = \Pr_{C' \leftarrow K}[C'(x) = b], \quad p_x = \min(p_x^0, p_x^1) \ .$$

3. Set $x^* = \arg\max_x p_x$.
4. If $p_{x^*} < \frac{p(n)}{q'(n)}$, then return a random $C' \leftarrow K$.
5. Else, query the oracle on x^* and set:

$$K \leftarrow K \cap \{C' | C(x^*) = C'(x^*)\} \ .$$

6. If K contains a single element C, return C.
7. Else, goto Step 2.

We later argue that the output of $\mathcal{L}^C$ is a circuit that is close to C, that is, the circuits only disagree on a negligible fraction of the inputs. Next, we describe a weak average-case VGB obfuscator $\mathcal{O}'$ for $\mathcal{C}$. In the description of $\mathcal{O}'$, we use the following notation:

we denote by $C_1 \oplus C_2$ the circuit that is composed from the circuits C_1 and C_2 where the output wires of these circuits are connected by a XOR gate. Similarly, we we denote by $\mathsf{Majority}_{i \in [n]} \, C_i$ the circuit that is composed from the circuits $C_1, \ldots, C_n$ where the output wires of these circuits are connected by a Majority gate.

Note about notation. For any two circuits C and C', we use $C \equiv C'$ to denote that C and C' are functionally equivalent. That is, the circuits compute the same function, but may be very different as "formal" circuits. We use the notation $C = C'$ when C and C' are not only functionally equivalent, but are also equal as "formal" circuits.

The obfuscator. The obfuscator $\mathcal{O}'$ on input $C \in \mathcal{C}$ operates as follows:

1. For $i \in [n]$, set $C_i \leftarrow \mathcal{L}^C$ where all executions of $\mathcal{L}$ use independent randomness.
2. Construct the circuit $C_{\mathsf{maj}} = \mathsf{Majority}_{i \in [n]} \, C_i$.
3. Construct the circuit $C_{\mathsf{diff}} = C_{\mathsf{maj}} \oplus C$.
4. Construct and output the circuit $C_{\mathsf{out}} = C_{\mathsf{maj}} \oplus \mathcal{O}(C_{\mathsf{diff}})$.

The correctness and polynomial slowdown properties of $\mathcal{O}'$ follow from those of $\mathcal{O}$, and from the fact that the circuits in $\mathcal{C}_n$ are of (approximately) the same size.

To show that $\mathcal{O}'$ is a weak average-case VGB secure, we demonstrate an inefficient simulator Sim. For every PPT adversary $\mathcal{A}$, and for every auxiliary input $z \in \{0, 1\}^{\mathrm{poly}(n)}$ to $\mathcal{A}$, Sim is given z, and oracle access to $C[q(n)]$. Sim acts as follows:

1. For $i \in [n]$, set $C_i \leftarrow \mathcal{L}^{C[q'(n)]}$, where independent randomness is used in different executions of $\mathcal{L}$.
2. Construct the circuit $C'_{\mathsf{maj}} = \mathsf{Majority}_{i \in [n]} \, C_i$.
3. Sample $C' \leftarrow \mathcal{L}^{C[q'(n)]}$ and construct the circuit $C'_{\mathsf{diff}} = C'_{\mathsf{maj}} \oplus C'$.
4. Construct the circuit $C_{\mathsf{sim}} = C'_{\mathsf{maj}} \oplus \mathcal{O}(C'_{\mathsf{diff}})$.
5. Execute $\mathcal{A}(z, C_{\mathsf{sim}})$ and output the result.

Sim invokes the learning algorithm, $\mathcal{L}$, $n+1$ times, and each invocation may include $q'(n) = \frac{q(n)}{(n+1)}$ queries to the oracle. Thus, in total Sim makes at most $q(n)$ oracle calls.

Another note about notation. In the rest of the proof, C denotes a random circuit in $\mathcal{C}_n$ (unless specified otherwise we assume C is distributed uniformly among all circuits in $\mathcal{C}_n$). The random variables C_{maj}, C_{diff} and C_{out} represent the value of the corresponding local variable in a random execution of the obfuscator $\mathcal{O}'$ on input C (this random variable is both over the random choice of C and of the coins used by $\mathcal{O}'$). Similarly, the random variables C'_{maj}, C'_{diff} and C_{sim} represent the value of the corresponding local variable in a random execution of $\mathsf{Sim}^{C[q(n)]}$ (the value of these random variables does not depend on the auxiliary input z passed to Sim). For simplicity of notation, in the rest of the proof, we omit the auxiliary input z from the parameter list of $\mathcal{A}$ and of Sim.

The simulator Sim is valid if for every predicate $\mathcal{P}$:

$$\left| \Pr_{C \leftarrow \mathcal{C}_n} [\mathcal{A}(\mathcal{O}'(C)) = \mathcal{P}(C)] - \Pr_{C \leftarrow \mathcal{C}_n} [\mathsf{Sim}^{C[q(n)]}(1^n) = \mathcal{P}(C)] \right| \leq \mathrm{negl}(n) \ .$$

That is, if:

$$\left| \begin{array}{l} \Pr_{C \leftarrow \mathcal{C}_n}[\mathcal{A}((C_{\mathsf{maj}}, \mathcal{O}(C_{\mathsf{diff}}))) = \mathcal{P}(C)] \\ - \Pr_{C \leftarrow \mathcal{C}_n}[\mathcal{A}((C'_{\mathsf{maj}}, \mathcal{O}(C'_{\mathsf{diff}}))) = \mathcal{P}(C)] \end{array} \right| \leq \mathrm{negl}(n) \ . \tag{2}$$

Before proving Equation (2), we introduce the following notation. For every circuit $C \in \mathcal{C}_n$, let $K(C)$ be the value of the set K when $\mathcal{L}^C$ terminates. Note that once C is fixed $K(C)$ is fully determined; indeed, up to step Step 4, where $\mathcal{L}$ outputs a random circuit from $K(C)$, $\mathcal{L}$ is deterministic. Define $\mathsf{GOOD}(C, C_{\mathsf{maj}})$ to be the event that:

$$C_{\mathsf{maj}} \equiv \underset{C' \in K(C)}{\mathsf{Majority}} C'$$

The following three lemmas imply Equation (2), and thus conclude the proof.

Lemma 4.1. *For every circuit $C \in \mathcal{C}_n$, the variables C_{maj} and C'_{maj} are identically distributed.*

Lemma 4.2.

$$\Pr_{C \leftarrow \mathcal{C}_n}[\mathsf{GOOD}(C, C_{\mathsf{maj}})] \geq 1 - \mathrm{negl}(n) \ .$$

Lemma 4.3. *For every C^*_{maj} in the support of C_{maj}, i.e., such that:*

$$\Pr_{C \leftarrow \mathcal{C}_n}[C_{\mathsf{maj}} = C^*_{\mathsf{maj}}] > 0 \ ,$$

it holds that:

$$\left| \begin{array}{l} \Pr_{C \leftarrow \mathcal{C}_n}\left[\mathcal{A}((C_{\mathsf{maj}}, \mathcal{O}(C_{\mathsf{diff}}))) = \mathcal{P}(C) \left| \begin{array}{l} C_{\mathsf{maj}} = C^*_{\mathsf{maj}} \\ \mathsf{GOOD}(C, C^*_{\mathsf{maj}}) \end{array} \right. \right] \\ - \Pr_{C \leftarrow \mathcal{C}_n}\left[\mathcal{A}((C'_{\mathsf{maj}}, \mathcal{O}(C'_{\mathsf{diff}}))) = \mathcal{P}(C) \left| \begin{array}{l} C'_{\mathsf{maj}} = C^*_{\mathsf{maj}} \\ \mathsf{GOOD}(C, C^*_{\mathsf{maj}}) \end{array} \right. \right] \end{array} \right| \leq \mathrm{negl}(n) \ .$$

Proof (Lemma 4.1). C'_{maj} is computed by Sim in the same way that C_{maj} is computed by $\mathcal{O}'$ except that that Sim limits the learning algorithm $\mathcal{L}$ to make at most $q'(n)$ oracle queries. It is therefore sufficient to show that $\mathcal{L}$ makes at most $q'(n)$ queries. By the choice of x^*, in every execution of $\mathcal{L}$, at Step 5, the size of the set K reduces by a factor of at least $(1 - p^*_x) \geq 1 - \frac{p(n)}{q'(n)}$. Since $|\mathcal{C}_n| \leq 2^{p(n)}$, after $q'(n)$ queries K must contain a single element and will thus $\mathcal{L}$ terminate. □

Proof (Lemma 4.2). Fix C, and denote by $\tilde{C}_{\mathsf{maj}}$ the circuit:

$$\tilde{C}_{\mathsf{maj}} \equiv \underset{C' \in K(C)}{\mathsf{Majority}} C' \ .$$

If $K(C)$ contains a single element (corresponding to Step 6 of $\mathcal{L}$), then GOOD must occur. Else, the stopping condition in Step 4 of $\mathcal{L}$ guarantees that for every $x \in \{0, 1\}^n$, and letting $p^b_x = \Pr_{C'' \leftarrow K(C)}[C''(x) = b]$, it holds that $\min(p^0_x, p^1_x) < \frac{p(n)}{q'(n)}$. That is, almost all circuits in $K(C)$ agree with the majority value $\tilde{C}_{\mathsf{maj}}(x)$. Formally, for every $x \in \{0,1\}^n$,

$$\Pr_{C'' \leftarrow K(C)}[C''(x) \neq \tilde{C}_{\mathsf{maj}}(x)] < \frac{p(n)}{q'(n)} .$$

The event GOOD does not occur if and only if there exist $x \in \{0,1\}^n$ such that at least $n/2$ of the circuits $C_1, \ldots, C_n$ disagree with $\tilde{C}_{\mathsf{maj}}$ on x. Since every C_i is sampled independently from $K(C)$ and since q' is super-polynomial we have that:

$$\Pr_{C \leftarrow C_n}[\neg \mathsf{GOOD}] \leq 2^n \cdot \binom{n}{\frac{n}{2}} \cdot \left(\frac{p(n)}{q'(n)}\right)^{\frac{n}{2}} < \left(\frac{16p(n)}{q'(n)}\right)^{\frac{n}{2}} = \mathsf{negl}(n) .$$

$\square$

Proof (Lemma 4.3). Fix C_{maj}^* such that:

$$\Pr_{C \leftarrow C_n}[C_{\mathsf{maj}} = C_{\mathsf{maj}}^*] > 0 .$$

If $C_{\mathsf{maj}} = C_{\mathsf{maj}}^*$, the circuits have the exact same structure and therefore, C_{maj}^* is of the form $C_{\mathsf{maj}}^* = \mathsf{Majority}_{i \in [n]}\, C_i^*$ for some circuits $C_1^*, \ldots C_n^* \in C_n$. The next claim will be useful for proving the lemma.

Claim 4.2.

1. *For every $C^* \in C_n$ and every $C, C' \in K(C^*)$, the random variable C_{maj} in the execution of $\mathcal{O}(C)$ and the random variable C_{maj} in the execution of $\mathcal{O}(C')$ are identically distributed.*
2. *For every $C_1^*, \ldots, C_n^* \in C_n$, for $C_{\mathsf{maj}}^* = \mathsf{Majority}_{i \in [n]}\, C_i^*$, and for every $C \notin K(C_1^*)$, C_{maj}^* is outside the support of C_{maj} (defined by the execution of $\mathcal{O}'(C)$).*

To prove Claim 4.2, we will use yet another simpler claim:

Claim 4.3. *For every $C^* \in C_n$ and every $C \in K(C^*)$ it holds that $K(C) = K(C^*)$.*

Proof. Fix any $C^* \in C_n$ and any $C \in K(C^*)$. Consider the set of oracle queries made by $\mathcal{L}^C$ and by $\mathcal{L}^{C^*}$ and their answers. If the two query-answer sets are equal then $K(C) = K(C^*)$. Else, both $\mathcal{L}^C$ and $\mathcal{L}^{C^*}$ make some query x^* such that $C(x^*) \neq C^*(x^*)$. However, this contradicts the fact that $C \in K(C^*)$, which means that C agrees with C^* on all the queries performed by $\mathcal{L}^{C^*}$ in the formation of $K(C^*)$.

$\square$

Proof (Claim 4.2). For Part 1, fix any $C^* \in C_n$ and any $C, C' \in K(C^*)$. Claim 4.3 implies that $K(C) = K(C') = K(C^*)$. Since the output of $\mathcal{L}^C$ is just a random element in $K(C)$ if follows that the output of $\mathcal{L}^C$ and the output of $\mathcal{L}^{C'}$ are identically distributed, and therefore the random variable C_{maj} in the execution of $\mathcal{O}'(C)$ and the random variable C_{maj} in the execution of $\mathcal{O}'(C')$ are also identically distributed.

For Part 2, fix $C_1^*, \ldots, C_n^* \in C_n$, and $C \notin K(C_1^*)$, and let $C_{\mathsf{maj}}^* = \mathsf{Majority}_{i \in [n]}\, C_i^*$. If $C_{\mathsf{maj}} = C_{\mathsf{maj}}^*$ (that is, the circuits are formally identical) then it must be that $C_1^* \in$

$K(C)$. Indeed, because the circuits $C_{\mathrm{maj}}, C^*_{\mathrm{maj}}$ are formally equal, C^*_1 equals a circuit $C_1 \in K(C)$ where $C_{\mathrm{maj}} = \mathrm{Majority}_{i \in [n]} C_i$. This, together with Claim 4.3, implies that $K(C) = K(C^*_1)$. Since it is always true that $C \in K(C)$, this also implies that $C \in K(C^*_1)$, contradicting our assumption.

$\square$

We are now ready to prove Lemma 4.3. Recall that $C^*_{\mathrm{maj}} = \mathrm{Majority}_{i \in [n]} C^*_i$. Denote the set $K(C^*_1)$ by K^*. By Claim 4.2 (Part 2), the lemma's statement is equivalent to the following, where we sample C from K^* instead of from C_n:

$$\left| \Pr_{C \leftarrow K^*} \left[\mathcal{A}((C_{\mathrm{maj}}, \mathcal{O}(C_{\mathrm{diff}}))) = \mathcal{P}(C) \middle| \begin{array}{l} C_{\mathrm{maj}} = C^*_{\mathrm{maj}} \\ \mathrm{GOOD}(C, C^*_{\mathrm{maj}}) \end{array} \right] - \Pr_{C \leftarrow K^*} \left[\mathcal{A}((C'_{\mathrm{maj}}, \mathcal{O}(C'_{\mathrm{diff}}))) = \mathcal{P}(C) \middle| \begin{array}{l} C'_{\mathrm{maj}} = C^*_{\mathrm{maj}} \\ \mathrm{GOOD}(C, C^*_{\mathrm{maj}}) \end{array} \right] \right| \leq \mathrm{negl}(n) . \quad (3)$$

Let the adversary $\mathcal{A}'$ be $\mathcal{A}$ with C^*_{maj} hard-coded to it. That is, $\mathcal{A}'(\mathcal{O}(C_{\mathrm{diff}}))$ outputs $\mathcal{A}((C^*_{\mathrm{maj}}, \mathcal{O}(C_{\mathrm{diff}})))$. Now we can rewrite Equation (3) as:

$$\left| \Pr_{C \leftarrow K^*} \left[\mathcal{A}'(\mathcal{O}(C_{\mathrm{diff}})) = \mathcal{P}(C) \middle| \begin{array}{l} C_{\mathrm{maj}} = C^*_{\mathrm{maj}} \\ \mathrm{GOOD}(C, C^*_{\mathrm{maj}}) \end{array} \right] - \Pr_{C \leftarrow K^*} \left[\mathcal{A}'(\mathcal{O}(C'_{\mathrm{diff}})) = \mathcal{P}(C) \middle| \begin{array}{l} C'_{\mathrm{maj}} = C^*_{\mathrm{maj}} \\ \mathrm{GOOD}(C, C^*_{\mathrm{maj}}) \end{array} \right] \right| \leq \mathrm{negl}(n) . \quad (4)$$

Let $\mathcal{P}'$ be a predicate that has C^*_{maj} hardwired into it, and is defined as follows: On inputs of the form $C_{\mathrm{diff}} = C^*_{\mathrm{maj}} \oplus C$ where $\mathrm{GOOD}(C, C^*_{\mathrm{maj}})$ holds, the predicate $\mathcal{P}'(C_{\mathrm{diff}})$ outputs $\mathcal{P}(C)$. On all other inputs the output of $\mathcal{P}'$ is arbitrarily defined to be 0. Now we can rewrite Equation (4) as:

$$\left| \Pr_{C \leftarrow K^*} \left[\mathcal{A}'(\mathcal{O}(C_{\mathrm{diff}})) = \mathcal{P}'(C_{\mathrm{diff}}) \middle| \begin{array}{l} C_{\mathrm{maj}} = C^*_{\mathrm{maj}} \\ \mathrm{GOOD}(C, C^*_{\mathrm{maj}}) \end{array} \right] - \Pr_{C \leftarrow K^*} \left[\mathcal{A}'(\mathcal{O}(C'_{\mathrm{diff}})) = \mathcal{P}'(C_{\mathrm{diff}}) \middle| \begin{array}{l} C'_{\mathrm{maj}} = C^*_{\mathrm{maj}} \\ \mathrm{GOOD}(C, C^*_{\mathrm{maj}}) \end{array} \right] \right| \leq \mathrm{negl}(n) . \quad (5)$$

Recall that $\mathcal{O}'$ sets $C_{\mathrm{diff}} = C_{\mathrm{maj}} \oplus C$. Let $\mathcal{C}_{\mathrm{diff}}$ be the collection:

$$\mathcal{C}_{\mathrm{diff}} = \left\{ C_{\mathrm{diff}} = C_{\mathrm{maj}} \oplus C \middle| \begin{array}{l} C \leftarrow K^* \\ C_{\mathrm{maj}} = C^*_{\mathrm{maj}} \\ \mathrm{GOOD}(C, C^*_{\mathrm{maj}}) \end{array} \right\}_{n \in \mathbb{N}, C^*_{\mathrm{maj}}} .$$

Additionally, recall that Sim sets $C'_{\mathrm{diff}} = C'_{\mathrm{maj}} \oplus C'$ where $C' \leftarrow \mathcal{L}^C$. Let $\mathcal{C}'_{\mathrm{diff}}$ be the collection:

$$\mathcal{C}'_{\mathrm{diff}} = \left\{ C'_{\mathrm{diff}} = C'_{\mathrm{maj}} \oplus C' \middle| \begin{array}{l} C \leftarrow K^* \\ C' \leftarrow \mathcal{L}^C \\ C'_{\mathrm{maj}} = C^*_{\mathrm{maj}} \\ \mathrm{GOOD}(C, C^*_{\mathrm{maj}}) \end{array} \right\}_{n \in \mathbb{N}, C^*_{\mathrm{maj}}} .$$

Now we can rewrite Equation (5) as:

$$\left| \begin{array}{l} \mathrm{Pr}_{C_{\mathsf{diff}} \leftarrow \mathcal{C}_{\mathsf{diff}}} [\mathcal{A}'(\mathcal{O}(C_{\mathsf{diff}})) = \mathcal{P}'(C_{\mathsf{diff}})] \\ - \mathrm{Pr}_{C'_{\mathsf{diff}} \leftarrow \mathcal{C}_{\mathsf{diff}}, C_{\mathsf{diff}} \leftarrow \mathcal{C}'_{\mathsf{diff}}} [\mathcal{A}'(\mathcal{O}(C'_{\mathsf{diff}})) = \mathcal{P}'(C_{\mathsf{diff}})] \end{array} \right| \leq \mathrm{negl}(n) . \tag{6}$$

By the proof of Claim 4.2, for any circuit $C \in K^*$, it holds that $K(C) = K^*$. Noting that C, C' defined in the collections $\mathcal{C}_{\mathsf{diff}}$ and $\mathcal{C}'_{\mathsf{diff}}$ are random circuits in K^*, and thus the two collections are identical. We can now rewrite Equation (6) as:

$$\left| \begin{array}{l} \mathrm{Pr}_{C_{\mathsf{diff}} \leftarrow \mathcal{C}_{\mathsf{diff}}} [\mathcal{A}'(\mathcal{O}(C_{\mathsf{diff}})) = \mathcal{P}'(C_{\mathsf{diff}})] \\ - \mathrm{Pr}_{C'_{\mathsf{diff}}, C_{\mathsf{diff}} \leftarrow \mathcal{C}'_{\mathsf{diff}}} [\mathcal{A}'(\mathcal{O}(C'_{\mathsf{diff}})) = \mathcal{P}'(C_{\mathsf{diff}})] \end{array} \right| \leq \mathrm{negl}(n) . \tag{7}$$

□

To prove equation Equation 7 and conclude the proof of the lemma, we show that the collection $\mathcal{C}_{\mathsf{diff}}$ is evasive. This, together with the fact that $\mathcal{O}$ is average-case VGB secure evasive collections and the proof of Theorem 2.1, imply that Equation 7 holds.

Claim 4.4. *The collection $\mathcal{C}_{\mathsf{diff}}$ is evasive.*

Proof. Let C be the random variable given in the definition of $\mathcal{C}_{\mathsf{diff}}$. If K^* contains a single element (corresponding to Step 6 of $\mathcal{L}$) then $C \equiv C_{\mathsf{maj}}$ and the collection $\mathcal{C}_{\mathsf{diff}}$ contains, in fact, only the all-zero function, and is therefore evasive. Assuming $|K^*| > 1$, the stopping condition in Step 4 of $\mathcal{L}$ guarantees that for every $x \in \{0,1\}^n$, letting $p_x^b = \mathrm{Pr}_{C'' \leftarrow K(C)}[C''(x) = b]$, it holds that for $\min(p_x^0, p_x^1) < \frac{p(n)}{q'(n)}$. This implies:

$$\mathrm{Pr}_{C'' \leftarrow K(C)} \left[C''(x) \neq \underset{\bar{C} \in K(C)}{\mathsf{Majority}} \bar{C}(x) \right] < \frac{p(n)}{q'(n)} .$$

By the proof of Claim 4.2, $K(C) = K^*$ for every $C \in K^*$, and therefore also

$$\mathrm{Pr}_{C'' \leftarrow K^*} [C''(x) \neq \underset{\bar{C} \in K^*}{\mathsf{Majority}} \bar{C}(x)] < \frac{p(n)}{q'(n)} .$$

Plugging-in the definitions of $\mathcal{C}_{\mathsf{diff}}$ and of $\mathsf{GOOD}(C, C^*_{\mathsf{maj}})$, we get that for every $x \in \{0,1\}^n$,

$$\mathrm{Pr}_{C \leftarrow K^*} [C_{\mathsf{diff}}(x) = 1] = \mathrm{Pr}_{C \leftarrow K^*} [C(x) \neq C^*_{\mathsf{maj}}(x)] < \frac{p(n)}{q'(n)} = \frac{p(n)}{n^{\omega(1)}} = \mathrm{negl}(n),$$

which implies that $\mathcal{C}_{\mathsf{diff}}$ is evasive. □

□

5 Impossibility Results

Definitions 2.5 and 2.4 only consider circuit obfuscation with average-case security. In this section we give impossibility results for obfuscating evasive *Turing machines* and for obfuscating evasive circuits with *worst-case* security.

5.1 Impossibility of Turing Machine Obfuscation

Barak et. al. [3] show the impossibility general obfuscation of circuits and Turing machines. We show that the impossibility of Turing machines obfuscation can be extended to the case of evasive functions. Similarly to the result of [3], our negative result applies for VBB obfuscation as well as for weaker notions such as average-case obfuscation (see [3] for more details). In particular, we get an impossibility for the Turing machine versions of Definitions 2.5 and 2.4.

Let $\mathcal{M} = \{\mathcal{M}_n\}_{n \in \mathbb{N}}$ be a collection of Turing machines such that every $M \in \mathcal{M}_n$ has description of size $\text{poly}(n)$ and outputs a bit. We say that $\mathcal{M}$ is *evasive* if given oracle access to a random machine in the collection it is hard to find an input that evaluates to 1.

Definition 5.1 (Evasive Turing Machine Collection). *A collection of Turing machines $\mathcal{M}$ is evasive if there exists a negligible function μ such that for every $n \in \mathbb{N}$ and every $x \in \{0,1\}^*$*

$$\Pr_{M \leftarrow \mathcal{M}_n} [M(x) = 1] \leq \mu(n) .$$

We start by recalling the syntax, functionality, and polynomial slowdown requirements for Turing machine obfuscation as defined in [3]. Then we give security definitions that are the Turing machine versions of Definitions 2.5 and 2.4.

Definition 5.2 (Turing Machine Obfuscation). *An obfuscator $\mathcal{O}$ for $\mathcal{M}$ is a PPT algorithm that satisfies the following requirements:*

- *Functionality: For every $n \in \mathbb{N}$ and every $M \in \mathcal{M}_n$, $\mathcal{O}(M)$ outputs a description of a Turing machine that computes the same function as M.*
- *Polynomial Slowdown: There exists a polynomial p such that for every $M \in \mathcal{M}$ and for every $x \in \{0,1\}^*$ if the running time of $M(x)$ is t, then the running time of $(\mathcal{O}(M))(x)$ is at most $p(t)$.*

Definition 5.3 (Perfect Turing-Machine-Hiding). *An obfuscator $\mathcal{O}$ for a collection of Turing machines $\mathcal{M}$ is perfect circuit-hiding if for every PPT adversary $\mathcal{A}$ there exist a PPT simulator Sim and a negligible function μ such that for every $n \in \mathbb{N}$ and every efficiently computable predicate $\mathcal{P}$:*

$$\left| \Pr_{M \leftarrow \mathcal{M}_n} [\mathcal{A}(\mathcal{O}(M)) = \mathcal{P}(M)] - \Pr_{M \leftarrow \mathcal{M}_n} [\text{Sim}(1^n) = \mathcal{P}(M)] \right| \leq \mu(n) .$$

Definition 5.4 (Input Hiding). *A obfuscator $\mathcal{O}$ for a collection of Turing machines $\mathcal{M}$ is input hiding if there exists a negligible function μ such that for every $n \in \mathbb{N}$ and for every PPT adversary $\mathcal{A}$*

$$\Pr_{M \leftarrow \mathcal{M}_n} [M(\mathcal{A}(\mathcal{O}(M))) = 1] \leq \mu(n) .$$

The impossibility. The impossibility of [3] demonstrates a pair of functions $C_{\alpha,\beta}, D_{\alpha,\beta}$ such that given oracle access to these functions, it is impossible to learn the key (α, β). However, given any efficient implementation of $C_{\alpha,\beta}$ and $D_{\alpha,\beta}$ as a pair of Turing machines, it is possible to learn (α, β). The two functions are then combined into a single function that cannot be obfuscated. The idea is to "embed" the functions $C_{\alpha,\beta}$ and $D_{\alpha,\beta}$ of [3] inside an evasive Turing machine.

For a key $\alpha, \beta \in \{0,1\}^n$ define the machine $C_{\alpha,\beta}$ as follows:

$$C_{\alpha,\beta}(x; i) = \begin{cases} \beta_i & \text{if } x = \alpha \\ 0 & \text{otherwise} \end{cases}$$

The machine $D_{\alpha,\beta}$ takes as input a description of a machine C that is suppose to run in time $p(n)$ and checks whether C computes the same function as $C_{\alpha,\beta}$ on the inputs $\{(\alpha, i)\}_{i \in [n]}$. Namely,

$$D_{\alpha,\beta}(C) = \begin{cases} \beta_1 & \text{if } \forall i \in [n], C(\alpha, i) \text{ outputs } \beta_i \text{ within } p(n) \text{ steps} \\ 0 & \text{otherwise} \end{cases}.$$

The polynomial p is defined to be greater than the running time of $\mathcal{O}(C_{\alpha,\beta})$. Next we define a single machine $M_{\alpha,\beta}$ combining the machines $C_{\alpha,\beta}$ and $D_{\alpha,\beta}$, as follows:

$$M_{\alpha,\beta}(b, z) = \begin{cases} C_{\alpha,\beta}(z) & \text{if } b = 0 \\ D_{\alpha,\beta}(z) & \text{if } b = 1 \end{cases}.$$

It is straightforward to verify that $C_{\alpha,\beta}$ and $D_{\alpha,\beta}$ are evasive, and therefore $M_{\alpha,\beta}$ is also evasive. By construction, an adversary that is given $\mathcal{O}(M_{\alpha,\beta})$ can compute a description of machines M_C and M_D, computing $C_{\alpha,\beta}$ and $D_{\alpha,\beta}$ respectively, where the running time of M_C is at most p. The adversary can therefore execute $M_D(M_C)$ and obtain β_1 with probability 1. Note that a simulator (with no access to $M_{\alpha,\beta}$) can guess β_1 with probability at most $1/2$ and therefore $\mathcal{O}$ is not perfect circuit-hiding (Definition 5.3).

To show that $\mathcal{O}$ is not input hiding (Definition 5.4) we consider an adversary that produces the input $(1, M_C)$ to $M_{\alpha,\beta}$. Since $f_{\alpha,\beta}(1, M_C) = \beta_1$ and β is random, the adversary outputs a preimage of 1 with probability $1/2$.

5.2 Impossibility of Worst-Case Obfuscation

The impossibility of [3] for circuit obfuscation demonstrates a collection of circuits $\mathcal{C}_n = \{C_s\}_{s \in \{0,1\}^n}$ such that given oracle access to C_s for a random seed s it is impossible to learn s. However, given any circuit computing the same function as C_s, an adversary can learn s. In general we do not know how to "embed" $\mathcal{C}$ inside an evasive collection without loosing the above learnability property. However, such embedding is possible when the adversary has some partial knowledge about the seed of the circuit taken from the evasive collection. This type of attack can be used to rule out a worst-case security definition.

We recall the definition of worst-case VBB from [3]. We present an equivalent version of the definition that uses a predicate and resembles Definition 2.6 for average-case VBB. Note that a worst-case version of the input-hiding security definition (Definition 2.4) cannot hold against non-uniform adversaries.

Definition 5.5 (Worst-Case Virtual Black-Box (VBB) from [3]). *An obfuscator $\mathcal{O}$ for a collection of circuits $\mathcal{C}$ is perfect circuit-hiding in the worst-case if for every PPT adversary $\mathcal{A}$ there exists a PPT simulator Sim and a negligible function μ such that for every $n \in \mathbb{N}$, every $C \in \mathcal{C}_n$ and every predicate $\mathcal{P}$:*

$$\left| \Pr[\mathcal{A}(\mathcal{O}(C)) = \mathcal{P}(C)] - \Pr[\mathrm{Sim}^C(1^n) = \mathcal{P}(C)] \right| \le \mu(n) \ .$$

Let $\mathcal{C}_n = \{C_s\}_{s \in \{0,1\}^n}$ be the collection defined by [3]. For $\alpha, s \in \{0,1\}^n$ we define $C'_{\alpha,s}$ as follows:

$$C'_{\alpha,s}(x_1, x_2, i) = \begin{cases} [C_s(x_1)]_i & \text{if } x_2 = \alpha \\ 0 & \text{otherwise} \end{cases} \ .$$

First note that the collection $\mathcal{C}$ is evasive, since for every input (x_1, x_2, i) the probability over a random key (α, s) that $x_2 = \alpha$ is negligible. However, this circuit cannot be VBB obfuscated. There is an adversary that given an obfuscation of $C'_{\alpha,s}$ for $\alpha = 0^n$ and for a random s, can transform this obfuscation into a circuit computing the same function as C_s and thereby learn s. Conversely, every simulator that is given oracle access to $C'_{\alpha,s}$ for $\alpha = 0^n$ and for a random s, cannot learn more than what can be learned with oracle access to C_s, and in particular cannot learn s.

Acknowledgement. We thank Vijay Ganesh for suggesting to us the "software patch" problem.

References

[1] Barak, B., Bitansky, N., Canetti, R., Kalai, Y.T., Paneth, O., Sahai, A.: Obfuscation for evasive functions. Cryptology ePrint Archive, Report 2013/668 (2013), http://eprint.iacr.org/

[2] Barak, B., Garg, S., Kalai, Y.T., Paneth, O., Sahai, A.: Protecting obfuscation against algebraic attacks. Cryptology ePrint Archive, Report 2013/631 (2013), http://eprint.iacr.org/

[3] Barak, B., Goldreich, O., Impagliazzo, R., Rudich, S., Sahai, A., Vadhan, S., Yang, K.: On the (im)possibility of obfuscating programs. In: Kilian, J. (ed.) CRYPTO 2001. LNCS, vol. 2139, pp. 1–18. Springer, Heidelberg (2001)

[4] Bitansky, N., Canetti, R.: On strong simulation and composable point obfuscation. In: Rabin, T. (ed.) CRYPTO 2010. LNCS, vol. 6223, pp. 520–537. Springer, Heidelberg (2010)

[5] Bitansky, N., Chiesa, A., Ishai, Y., Paneth, O., Ostrovsky, R.: Succinct non-interactive arguments via linear interactive proofs. In: Sahai, A. (ed.) TCC 2013. LNCS, vol. 7785, pp. 315–333. Springer, Heidelberg (2013)

[6] Brakerski, Z., Rothblum, G.N.: Obfuscating conjunctions. In: Canetti, R., Garay, J.A. (eds.) CRYPTO 2013, Part II. LNCS, vol. 8043, pp. 416–434. Springer, Heidelberg (2013)

[7] Brakerski, Z., Rothblum, G.N.: Virtual black-box obfuscation for all circuits via generic graded encoding. Cryptology ePrint Archive, Report 2013/563 (2013), http://eprint.iacr.org/

[8] Canetti, R.: Towards realizing random oracles: Hash functions that hide all partial information. In: Kaliski Jr., B.S. (ed.) CRYPTO 1997. LNCS, vol. 1294, pp. 455–469. Springer, Heidelberg (1997)

[9] Canetti, R., Goldreich, O., Halevi, S.: The random oracle methodology, revisited. J. ACM 51(4), 557–594 (2004)

[10] Canetti, R., Rothblum, G.N., Varia, M.: Obfuscation of hyperplane membership. In: Micciancio, D. (ed.) TCC 2010. LNCS, vol. 5978, pp. 72–89. Springer, Heidelberg (2010)

[11] Canetti, R., Vaikuntanathan, V.: Obfuscating branching programs using black-box pseudo-free groups. Cryptology ePrint Archive, Report 2013/500 (2013), http://eprint.iacr.org/

[12] Coron, J.-S., Lepoint, T., Tibouchi, M.: Practical multilinear maps over the integers. In: Canetti, R., Garay, J.A. (eds.) CRYPTO 2013, Part I. LNCS, vol. 8042, pp. 476–493. Springer, Heidelberg (2013)

[13] Ganesh, V., Carbin, M., Rinard, M.C.: Cryptographic path hardening: Hiding vulnerabilities in software through cryptography. CoRR abs/1202.0359 (2012)

[14] Garg, S., Gentry, C., Halevi, S.: Candidate multilinear maps from ideal lattices. In: Johansson, T., Nguyen, P.Q. (eds.) EUROCRYPT 2013. LNCS, vol. 7881, pp. 1–17. Springer, Heidelberg (2013)

[15] Garg, S., Gentry, C., Halevi, S., Raykova, M., Sahai, A., Waters, B.: Candidate indistinguishability obfuscation and functional encryption for all circuits. In: FOCS (2013)

[16] Goldwasser, S., Rothblum, G.N.: On best-possible obfuscation. In: Vadhan, S.P. (ed.) TCC 2007. LNCS, vol. 4392, pp. 194–213. Springer, Heidelberg (2007)

[17] Hohenberger, S., Sahai, A., Waters, B.: Replacing a random oracle: Full domain hash from indistinguishability obfuscation. IACR Cryptology ePrint Archive 2013, 509 (2013)

[18] Sahai, A., Waters, B.: How to use indistinguishability obfuscation: Deniable encryption, and more. IACR Cryptology ePrint Archive 2013, 454 (2013)

[19] Valiant, L.G., Skyum, S., Berkowitz, S., Rackoff, C.: Fast parallel computation of polynomials using few processors. SIAM J. Comput. 12(4), 641–644 (1983)

[20] Wee, H.: On obfuscating point functions. IACR Cryptology ePrint Archive 2005, 1 (2005)

On Extractability Obfuscation

Elette Boyle[1,*], Kai-Min Chung[2], and Rafael Pass[1,**]

[1] Cornell University
ecb227@cornell.edu, rafael@cs.cornell.edu
[2] Academica Sinica
kmchung@iis.sinica.edu.tw

Abstract. We initiate the study of *extractability obfuscation*, a notion first suggested by Barak *et al.* (JACM 2012): An extractability obfuscator $e\mathcal{O}$ for a class of algorithms $\mathcal{M}$ guarantees that if an efficient attacker $\mathcal{A}$ can distinguish between obfuscations $e\mathcal{O}(M_1), e\mathcal{O}(M_2)$ of two algorithms $M_1, M_2 \in \mathcal{M}$, then $\mathcal{A}$ can efficiently recover (given M_1 and M_2) an input on which M_1 and M_2 provide different outputs.

- We rely on the recent candidate virtual black-box obfuscation constructions to provide candidate constructions of extractability obfuscators for NC^1; next, following the blueprint of Garg *et al.* (FOCS 2013), we show how to bootstrap the obfuscator for NC^1 to an obfuscator for all non-uniform polynomial-time Turing machines. In contrast to the construction of Garg *et al.*, which relies on indistinguishability obfuscation for NC^1, our construction enables succinctly obfuscating non-uniform *Turing machines* (as opposed to circuits), without turning running-time into description size.
- We introduce a new notion of *functional witness encryption*, which enables encrypting a message m with respect to an instance x, language L, and function f, such that anyone (and only those) who holds a witness w for $x \in L$ can compute $f(m, w)$ on the message and particular known witness. We show that functional witness encryption is, in fact, equivalent to extractability obfuscation.
- We demonstrate other applications of extractability extraction, including the first construction of fully (adaptive-message) indistinguishability-secure functional encryption for an unbounded number of key queries and unbounded message spaces.
- We finally relate indistinguishability obfuscation and extractability obfuscation and show special cases when indistinguishability obfuscation can be turned into extractability obfuscation.

---

* Supported in part by AFOSR YIP Award FA9550-10-1-0093.
** Pass is supported in part by a Alfred P. Sloan Fellowship, Microsoft New Faculty Fellowship, NSF Award CNS-1217821, NSF CAREER Award CCF-0746990, NSF Award CCF-1214844, AFOSR YIP Award FA9550-10-1-0093, and DARPA and AFRL under contract FA8750-11-2- 0211. The views and conclusions contained in this document are those of the authors and should not be interpreted as representing the official policies, either expressed or implied, of the Defense Advanced Research Projects Agency or the US Government.

Y. Lindell (Ed.): TCC 2014, LNCS 8349, pp. 52–73, 2014.

1   Introduction

Obfuscation. The goal of *program obfuscation* is to "scramble" a computer program, hiding its implementation details (making it hard to "reverse-engineer"), while preserving its functionality (i.e, input/output behavior). A first formal definition of such program obfuscation was provided by Hada [22]: roughly speaking, Hada's definition—let us refer to it as *strongly virtual black-box*—is formalized using the simulation paradigm. It requires that anything an attacker can learn from the obfuscated code, could be simulated using just black-box access to the functionality.[1] Unfortunately, as noted by Hada, only learnable functionalities can satisfy such a strong notion of obfuscation: if the attacker simply outputs the code it is given, the simulator must be able to recover the code by simply querying the functionality and thus the functionality must be learnable.

An in-depth study of program obfuscation was initiated in the seminal work of Barak, Goldreich, Impagliazzo, Rudich, Sahai, Vadhan, and Yang [2]. Their central result shows that even if we consider a more relaxed simulation-based definition of program obfuscation—called *virtual black-box* obfuscation—where the attacker is restricted to simply outputting a single bit, impossibility can still be established (assuming the existence of one-way functions). Their result is even stronger, demonstrating the existence of families of functions such that given black-box access to f_s (for a randomly chosen s), not even a *single* bit of s can be guessed with probability significantly better than $1/2$, but given the code of any program that computes f_s, the entire secret s can be recovered. Thus, even quite weak simulation-based notions of obfuscation are impossible.

Barak *et al.* [2] also suggested an avenue for circumventing these impossibility results:[2] introducing the notions of *indistinguishability* and *"differing-inputs"* *obfuscation*. Roughly speaking, an indistinguishability obfuscator iO for a class of circuits C guarantees that given any two *equivalent* circuits C_1 and C_2 (i.e., whose outputs agree on all inputs) from the class, obfuscations $iO(C_1)$ and $iO(C_2)$ of the circuits are indistinguishable. In a recent breakthrough result, Garg, Gentry, Halevi, Raykova, Sahai, and Waters [14] provide the first candidate construction of indistinguishability obfuscators for all polynomial-size circuits. Additionally, Garg et al [14] and even more recently, the elegant works of Sahai and Waters [29] and Hohenberger, Sahai and Waters [23], demonstrate several beautiful (and surprising) applications of indistinguishability obfuscation.

In this work, we initiate the study of the latter notion of obfuscation— *"differing-inputs"*, or as we call it, *extractability obfuscation*—whose security guarantees are at least as strong as indistinguishability obfuscation, but weaker than virtual black-box obfuscation. We demonstrate candidate constructions of such extractability obfuscators, and new applications.

---

[1] Hada actually considered slight distributional weakening of this definition.

[2] Hada also suggested an approach for circumventing his impossibility result, sticking with a simulation-based definition, but instead restricting to particular classes of attacker. It is, however, not clear (to us) what reasonable classes of attackers are.

Extractability Obfuscation. Roughly speaking, an extractability obfuscator $e\mathcal{O}$ for a class of circuits $\mathcal{C}$ guarantees that if an attacker $\mathcal{A}$ can distinguish between obfuscations $i\mathcal{O}(C_1), i\mathcal{O}(C_2)$ of two circuits $C_1, C_2 \in \mathcal{C}$, then $\mathcal{A}$ can efficiently recover (given C_1 and C_2) a point x on which C_1 and C_2 differ: i.e., $C_1(x) \neq C_2(x)$.[3] Note that if C_1 and C_2 are equivalent circuits, then no such input exists, thus requiring obfuscations of the circuits to be indistinguishable (and so extractability obfuscation implies indistinguishability obfuscation).

We may rely on the candidate obfuscator for NC^1 of Brakerski and Rothblum [9] or Barak *et al.* [1] to obtain extractability obfuscation for the same class. We next demonstrate a bootstrapping theorem, showing how to obtain extractability obfuscation for all polynomial-size circuits. Our transformation follows [14], but incurs a somewhat different analysis.

Theorem 1 (Informal). *Assume the existence of an extractability obfuscator for NC^1 and the existence of a (leveled) fully homomorphic encryption scheme with decryption in NC^1 (implied, e.g., by Learning With Errors). Then there exists an extractability obfuscation for $P/poly$.*

Relying on extractability obfuscation, however, has additional advantages: in particular, it allows us to achieve obfuscation of (non-uniform) *Turing machines.* The size of the obfuscated code preserves a polynomial relation to the size of the original Turing machine. In contrast, existing obfuscator constructions [14,9] can achieve this only by first converting the Turing machine to a circuit, turning running time into size.

To achieve this, we additionally rely on the existence of **P**-certificates in the CRS model—namely, succinct non-interactive arguments for **P**.[4]

Theorem 2 (Informal). *Assume the existence of extractability obfuscation for NC^1, fully homomorphic encryption with decryption in NC^1 and P-certificates (in the CRS model). Then there exists extractability obfuscation for polynomial-size Turing machines.*

On a high level, our construction follows the one from [14] but (1) modifies it to deal with executions of Turing machines (by relying on an oblivious Turing machine), and more importantly (2) compresses "proofs" by using **P**-certificates. We emphasize that this approach does *not* work in the setting of indistinguishability obfuscation. Intuitively, the reason for this is that **P**-certificates of false statements *exist*, but are just hard to find; efficiently extracting such **P**-certificates from a successful adversary is thus crucial (and enabled by the extractability property).

We next explore applications of extractability obfuscation.

---

[3] Pedantically, our notion is a slightly relaxed version of that of [2]; see Section 3.

[4] Such certificates can be either based on knowledge-of-exponent type assumptions [4], or even on *falsifiable* assumptions [12].

Functional Witness Encryption. Consider the following scenario: You wish to encrypt the labels in a (huge) graph (e.g., names of people in a social network) so that no one can recover them, unless there is a clique in the graph of size, say, 100. Then, anyone (and only those) who knows such a clique should be able to recover the labels of the nodes *in the identified clique* (and only these nodes). Can this be done?

The question is very related to the notion of *witness encryption*, recently introduced by Garg, Gentry, Sahai, and Waters [15]. Witness encryption makes it possible to encrypt the graph in such a way that anyone who finds any clique in the graph can recover the *whole* graph; if the graph does not contain any such cliques, the graph remains secret. The stronger notion of extractable witness encryption, introduced by Goldwasser, Kalai, Popa, Vaikuntanathan, and Zeldovich [20], further guarantees that the graph can only be decrypted by someone who actually knowns a clique. However, in contrast to existing notions, here we wish to reveal *only the labels associated with the particular known clique*.

More generally, we put forward the notion of *functional witness encryption (FWE)*. An FWE scheme enables one to encrypt a message m with respect to an NP-language L, instance x and function f, such that anyone who has (and only those who have) a witness w for $x \in L$ can recover $f(m, w)$. In the above example, m contains the labels of the whole graph, w is a clique, and $f(m, w)$ are the labels of all nodes in w. More precisely, our security definition requires that if you can tell apart encryptions of two messages m_0, m_1, then you must know a witness w for $x \in L$ such that $f(m_0, w) \neq f(m_1, w)$.

We observe that general-purpose FWE and extractability obfuscation actually are equivalent (up to a simple transformation).

Theorem 3 (Informal). *There exists a FWE for NP and every polynomial-size function f if and only if there exists an extractability obfuscator for every polynomial-size circuit.*

The idea is very simple: Given an extractability obfuscator $e\mathcal{O}$, an FWE encryption of the message m for the language L, instance x and function f is the obfuscation of the program that on input w outputs $f(m, w)$ if w is a valid witness for $x \in L$. On the other hand, given a general-purpose FWE, to obfuscate a program Π, let f be the universal circuit that on input (Π, y) runs Π on input y, let L be the trivial language where every witness is valid, and output a FWE of the message Π—since every input y is a witness, this makes it possible to evaluate $\Pi(y)$ on every y.

Other Applications. Functional encryption [6,28] enables the release of "targeted" secret keys sk_f that enable a user to recover $f(m)$, and *only* $f(m)$, given an encryption of m. It is known that strong simulation-based notions of security cannot be achieved if users can request an unbounded number of keys. In contrast, Garg *et al.* elegantly showed how to use indistinguishability obfuscation to satisfy an indistinguishability-based notion of functional encryption (roughly, that encryptions of any two messages m_0, m_1 such that $f(m_0) = f(m_1)$ for all the requested secret keys sk_f are indistinguishable). The main construction of Garg et al, however, only achieves *selective-message* security, where the attacker must select the two message m_0, m_1 to distinguish before the experiment begins (and it can request decryption keys sk_f). Garg *et al.* observe that if they make subexponential-time security assumptions, use complexity leveraging, and consider a small (restricted) message space, then they can also achieve adaptive-message security.

We show how to use extractability obfuscation to directly achieve full adaptive-message security for any unbounded size message space (without relying on complexity leveraging).

The idea behind our scheme is as follows. Let the public key of the encryption scheme be the verification key for a signature scheme, and let the master secret key (needed to release secret keys sk_f) be the signing key for the signature scheme. To encrypt a message m, obfuscate the program that on input f and a valid signature on f (with respect to the hardcoded public key) simply computes $f(m)$. The secret key sk_f for a function f is then simply the signature on f. (The high-level idea behind the construction is somewhat similar to the one used in [20], which used witness encryption in combination with signature schemes to obtain simulation-based FE for a *single* function f; in contrast, we here focus on FE for an unbounded number of functions).

Proving that this construction works is somewhat subtle. In fact, to make the proof go through, we here require the signature scheme in use to be of a special *delegatable* kind—namely, we require the use of *functional signatures* [7,3] (which can be constructed based on non-interactive zero-knowledge (NIZK) arguments of knowledge), which make it possible to delegate a signing key sk' that allows one to sign only messages satisfying some predicate. The delegation property is only used in the security reduction and, roughly speaking, makes it possible to simulate key queries without harming security for the messages selected by the attacker.

Theorem 4 (Informal). *Assume the existence of NIZK arguments of knowledge for NP and the existence of extractability obfuscators for polynomial-size circuits. Then there exists an (adaptive-message) indistinguishability-secure functional encryption scheme for arbitrary length messages.*

Another interesting feature of our approach is that it directly enables constructions of Hierarchical Functional Encryption (HiFE) (in analogy with Hierarchical Identity-Based encryption [24]), where the secret keys for functions f can be released in a hierarchical way (the top node can generate keys for subsidiary nodes, those nodes can generate keys for its subsidiaries etc.). To do this, simply modify the encryption algorithm to release the $f(m)$ message in case you provide an appropriate chain of signatures that terminates with a signature on f.

From Indistinguishability Obfuscation to Extractability Obfuscation. A natural question is whether we can obtain extractability obfuscation from indistinguishability obfuscation. We address this question in two different settings: first directly in the context of obfuscation, and second in the language of FWE. (Recall that these two notions are equivalent when dealing with arbitrary circuits and arbitrary functions; however, when considering restricted function classes, there are interesting differences).

– We introduce a weaker form of extractability obfuscation, in which extraction is only required when the two circuits differ on only polynomially many inputs. We demonstrate that any indistinguishability obfuscation in fact implies weak extractability obfuscation.

Theorem 5 (Informal). *Any indistinguishability obfuscator for* P/poly *is also a weak extractability obfuscator for* P/poly.

– Mirroring the definition of indistinguishability obfuscation, we may define a weaker notion of FWE—which we refer to as *indistinguishability FWE* (or iFWE)—which only requires that if $f(m_0, w) = f(m_1, w)$ for *all* witnesses w for $x \in L$, then encryptions of m_0 and m_1 are indistinguishable (in contrast, the stronger notion requires that if you can distinguish between encryptions of m_0 and m_1 you must know a witness on which they differ). It follows that iFWE for languages in NP and functions in P/poly is equivalent to indistinguishability obfuscation for P/poly, up to a simple transformation. We show that if restricting to languages with polynomially many witnesses, it is possible to turn any iFWE to an FWE.

Theorem 6 (Informal). *Assume there exists indistinguishability FWE for every* NP *language with polynomially many witnesses, and the function* f. *Then for every language* L *in* NP *with polynomially many witnesses, there exists an FWE for* L *and* f.

Our proof relies on a local list-decoding algorithm for a large-alphabet Hadamard code due to Goldreich, Rubinfeld and Sudan [19].

Theorems 5 and 6 are incomparable in that Theorem 5 begins with a stronger assumption and yields a stronger conclusion. More precisely, if one begins with iFWE supporting all languages in NP and functions in P/poly, then the equivalence between indistinguishability (respectively, standard) FWE and indistinguishability (resp., extractability) obfuscation, in conjunction with the transformation of Theorem 5, yields a *stronger* outcome in the setting of FWE than Theorem 6: Namely, a form of FWE where (extraction) security holds as long as the function $M(m, w)$ is not "too sensitive" to m: i.e., if for any two messages m_0, m_1 there are only polynomially many witnesses w for which $M(m_0, w) \neq M(m_1, w)$. This captures, for example, functions M that only rarely output nonzero values. Going back to the example of encrypting data m associated with nodes of a social network, we could then allow someone holding clique w to learn whether the nodes in this clique satisfy some chosen rare property (e.g., contains someone with a rare disease, all have the same birthday, etc). Indeed, while there may be many cliques (corresponding to several, even exponentially many, witnesses w), it will be the case that $M(m, w)$ is almost always 0, for all but polynomially many w.

On the other hand, Theorem 6 also provides implications of iFWE for restricted function classes. In particular, Theorem 6 gives a method for transforming indistinguishability FWE for the trivial function $f(m, w) = m$ to FWE for the same function f. It is easy to see that indistinguishability FWE for this particular f is equivalent to the notion of witness encryption [15], and FWE for the same f is equivalent to the notion of extractable witness encryption of [20]. Theorem 6 thus shows how to turn witness encryption to extractable witness encryption for the case of languages with polynomially many witness.

Finally, we leave open whether there are corresponding transformations from indistinguishability obfuscation in the case of many disagreeing inputs, and iFWE in the case of many witnesses. In the latter setting, this is interesting even for the special case of witness encryption (i.e., the function $f(m, w) = m$).

Overview of the Paper. In Section 2, we present definitions and notation for some of the tools used in the paper. In Section 3, we introduce the notion of extractability obfuscation and present a bootstrapping transformation from any extractability obfuscator for NC^1 to one for all poly-time Turing machines. In Section 4, we define functional witness encryption (FWE), and show an equivalence between FWE and extractability obfuscation. In Section 5, we describe an application of extractability obfuscation, in achieving indistinguishability functional encryption with unbounded-size message space. In Section 6, we explore the relationship between indistinguishability and extractability obfuscation, providing transformations from the former to the latter in special cases.

2 Preliminaries

2.1 Fully Homomorphic Encryption

A fully homomorphic encryption scheme $\mathcal{E} = (\mathsf{Gen_{FHE}}, \mathsf{Enc_{FHE}}, \mathsf{Dec_{FHE}}, \mathsf{Eval_{FHE}})$ is a public-key encryption scheme associated with an additional polynomial-time algorithm $\mathsf{Eval_{FHE}}$, which enables computation on encrypted data. Formally, we require $\mathcal{E}$ to have the following correctness property:

Definition 1 (FHE correctness). *There exists a negligible $\nu(k)$ s.t.*

$$\Pr_{\mathsf{pk},\mathsf{sk}\leftarrow\mathsf{Gen}(1^k)} \left[\begin{array}{l} \forall \text{ ciphertexts } c_1, ..., c_n \text{ s.t. } c_i \leftarrow \mathsf{Enc_{pk}}(b_i), \\ \forall \text{ poly-size circuits } f : \{0,1\}^n \to \{0,1\} \\ \mathsf{Dec_{sk}}(\mathsf{Eval_{pk}}(f, c_1, ..., c_n)) = f(b_1, ..., b_n), \end{array} \right] \geq 1 - \nu(k).$$

The size of $c' = \mathsf{Eval_{FHE}}(\mathsf{pk}, \mathsf{Enc_{FHE}}(\mathsf{pk}, m), C)$ must depend polynomially on the security parameter and the length of $C(m)$, but be otherwise independent of the size of the circuit C. For security, we require that $\mathcal{E}$ is semantically secure. We also require that Eval is deterministic, and that the decryption circuit $\mathsf{Dec_{sk}}(\cdot)$ is in NC^1. Most known schemes satisfy these properties. Since the breakthrough of Gentry [17], several fully homomorphic encryption schemes have been constructed with improved efficiency and based on more standard assumptions such as LWE (Learning With Errors) (e.g., [10,8,18,11]), together with a circular security assumption. We refer the reader to these works for more details.

Remark 1 (Homomorphic evaluation of Turing machines). As part of our extractability obfuscation construction for general Turing machines (TM), we require the homomorphic evaluation of an *oblivious* TM with known runtime. Recall that a TM is said to be oblivious if its tape movements are independent of its input. The desired evaluation is done as follows.

Suppose $\hat{x} = (\hat{x}_1, \hat{x}_2, \cdots, \hat{x}_k)$ is an FHE encryption of plaintext message x (where $\hat{x}_\ell$ encrypts the ℓth position of x), $\hat{a} = (\hat{a}_1, \hat{a}_2, \ldots)$ an FHE encryption of the tape values, $\hat{s}$ an FHE ciphertext of the current state, and M an oblivious TM terminating on all inputs within t steps. More specifically, a description of M consists of an initial state s and description of a transition circuit, C_M. In each step $i = 1, \ldots, t$ of evaluation, M accesses some fixed position $\mathsf{pos}_{\mathsf{input}}(i)$ of the input, fixed position $\mathsf{pos}_{\mathsf{tape}}(i)$ of the tape (extending straightforwardly to the multi-tape setting), and the current value of the state, and evaluates C_M on these values.

Homomorphic evaluation of M on the encrypted input $\hat{x}$ then takes place in t steps: In each step i, the transition circuit C_M of M is homomorphically evaluated on the ciphertexts $\hat{x}_{\mathsf{pos}_{\mathsf{input}}}$, $\hat{a}_{\mathsf{pos}_{\mathsf{tape}}}$, and $\hat{s}$, yielding updated values for these ciphertexts. The updated state ciphertext $\hat{s}$ resulting after t steps is the desired output ciphertext. Note that obliviousness of the Turing machine is crucial for this efficient method of homomorphic evaluation, as any input-dependent choices for the head location would only be available to an evaluator in encrypted form.

Overall, homomorphic evaluation of M takes time $O(t(k) \cdot \mathsf{poly}(k))$, and can be described in space $O(|M| \cdot \mathsf{poly}(k))$.

2.2   (Indistinguishability) Functional Encryption

A functional encryption scheme [6,28] enables the release of "targeted" secret keys that enable a user to recover $f(m)$—and only $f(m)$—given an encryption of m. In this work, we will consider the indistinguishability notion of security for functional encryption. Roughly, such a scheme is said to be secure if an adversary who requests and learns secret keys sk_f for a collection of functions f cannot distinguish encryptions of any two messages m_0, m_1 for which $f(m_0) = f(m_1)$ for every requested f. We refer the reader to e.g. [6,28] for a formal definition.

2.3   P-Certificates

P-Certificates are a succinct argument system for **P**. We consider **P** certificates in the CRS model.

For every $c \in \mathbb{N}$, let $L_c = \{(M, x, y) : M(x) = y \text{ within } |x|^c \text{ steps}\}$. Let $T_M(x)$ denote the running time of M on input x.

Definition 2 (P-certificates). *[13] A tuple of probabilistic interactive Turing machines* (CRSGen$_{\mathsf{cert}}$, P_{cert}, V_{cert}) *is a* **P***-certificate system in the CRS model if there exist polynomials* g_P, g_V, ℓ *such that the following hold:*

- **Efficient Verification:** *On input* crs $\leftarrow$ CRSGen(1^k), $c \geq 1$, *and a statement* $q = (M, x, y) \in L_c$, *and* $\pi \in \{0, 1\}^*$, V_{cert} *runs in time at most* $g_V(k + |q|)$.

- **Completeness by a Relatively Efficient Prover:** *For every $c, d \in \mathbb{N}$, there exists a negligible function μ such that for every $k \in \mathbb{N}$ and every $q = (M, x, y) \in L_c$ such that $|q| \leq k^d$,*
  $$\Pr[\text{crs} \leftarrow \text{CRSGen}(1^k); \pi \leftarrow P_{\text{cert}}(\text{crs}, c, q) : V_{\text{cert}}(\text{crs}, c, q, \pi) = 1] \geq 1 - \mu(k).$$
  Furthermore, P_{cert} on input (crs, c, q) outputs a certificate of length $\ell(k)$ in time bounded by $g_P(k + |M| + T_M(x))$.
- **Soundness:** *For every $c \in \mathbb{N}$, and every (not necessarily uniform) PPT P^*, there exists a negligible function μ such that for every $k \in \mathbb{N}$,*
  $$\Pr[\text{crs} \leftarrow \text{CRSGen}(1^k); (q, \pi) \leftarrow P^*(\text{crs}, c) : V_{\text{cert}}(\text{crs}, c, q, \pi) = 1 \wedge q \notin L_c] \leq \mu(k).$$

P-certificates are directly implied by any publicly-verifiable succinct non-interactive argument system (SNARG) for **P**. It was shown by Chung et al. [13] that **P**-certificates can be based on *falsifiable* assumptions [27].

Theorem 7. *Assuming that Micali's CS proof [26] is sound, or assuming the existence of publicly-verifiable fully succinct SNARG system for **P** [4] (which in turn can be based on any publicly-verifiable SNARG [21,25,16,5]), then there exists a **P**-certificate system in the CRS model.*

3 Extractability Obfuscation

We now present and study the notion of *extractability obfuscation*, which is a slight relaxation of "differing-inputs obfuscation" introduced in [2]. Intuitively, such an obfuscator has the property that if a PPT adversary can distinguish between obfuscations of two programs M_0, M_1, then he must "know" an input on which they differ.

Definition 3 (Extractability Obfuscator). *(Variant of [2][5]) A uniform PPT machine $e\mathcal{O}$ is an extractability obfuscator for a class of Turing machines $\{\mathcal{M}_k\}_{k \in \mathbb{N}}$ if the following conditions are satisfied:*
- **Correctness:** *There exists a negligible function $\text{negl}(k)$ such that for every security parameter $k \in \mathbb{N}$, for all $M \in \mathcal{M}_k$, for all inputs x, we have $\Pr[M' \leftarrow e\mathcal{O}(1^k, M) : M'(x) = M(x)] = 1 - \text{negl}(k)$.*
- **Security:** *For every PPT adversary $\mathcal{A}$ and polynomial $p(k)$, there exists a PPT extractor E and polynomial $q(k)$ such that the following holds. For every $k \in \mathbb{N}$, every pair of Turing machines $M_0, M_1 \in \mathcal{M}_k$, and every auxiliary input z,*

$$\Pr\left[\begin{array}{l} b \leftarrow \{0,1\}; \\ M' \leftarrow e\mathcal{O}(1^k, M_b) \end{array} : \mathcal{A}(1^k, M', M_0, M_1, z) = b\right] \geq \frac{1}{2} + \frac{1}{p(k)} \quad (1)$$

$$\implies \quad \Pr\left[w \leftarrow E(1^k, M_0, M_1, z) : M_0(w) \neq M_1(w)\right] \geq \frac{1}{q(k)}. \quad (2)$$

---

[5] Formally, our notion of extractability obfuscation departs from differing-inputs obfuscation of [2] in two ways: First, [2] require the extractor E to extract a differing input for M_0, M_1 given *any* pair of programs M_0', M_1' evaluating equivalent functions. Second, [2] consider also adversaries who distinguish with negligible advantage $\epsilon(k)$, and require that extraction still succeeds in this setting, but within time polynomial in $1/\epsilon$. In contrast, we restrict our attention only to adversaries who succeed with noticeable advantage.

We remark that we can also consider a distributional-variant of the extraction condition, where instead of requiring the condition to hold with respect to every $M_0, M_1 \in \mathcal{M}_k$ and $z \in \{0,1\}^*$, we consider a distribution $\mathcal{D}$ that samples $(M_0, M_1, z) \leftarrow \mathcal{D}$ and requires that for every distribution $\mathcal{D}$, there exists an extractor such that the extraction condition to hold with respect to $\mathcal{D}$. In applications, it often suffices to require the extraction condition to hold with respect to some specific distribution $\mathcal{D}$. We here focus on the above definition for concrete exposition, but our results hold naturally also for the distributional-variant definition.

We contrast this notion with *indistinguishability obfuscation*:

Definition 4 (Indistinguishability Obfuscator). *[2] A uniform PPT machine $i\mathcal{O}$ is an* indistinguishability obfuscator *for a class of circuits $\{\mathcal{C}_k\}$ if $i\mathcal{O}$ satisfies the Correctness and Security properties as in Definition 3 (for circuit class $\{\mathcal{C}_k\}$ and circuits C_0, C_1 in the place of Turing machines), except with Line (2) replaced with the following:*

$$\implies \quad \exists w : C_0(w) \neq C_1(w). \tag{2'}$$

Note that any *extractability* obfuscator is also directly an *indistinguishability* obfuscator, since existence of an efficient extraction algorithm E finding desired distinguishing input w as in (2) in particular implies that such an input exists, as in (2').

Remark 2. Note that in the definition of extractability obfuscation, the extractor is given access to the programs M_0, M_1. One could consider an even stronger notion of obfuscation, in which the extractor is given only black-box access to the two programs. As we show in the full version, however, achieving general-purpose obfuscation satisfying this stronger extractability notion is impossible.

We now present specific definitions of extractability obfuscators for special classes of Turing machines.

Definition 5 (Extractability Obfuscator for NC^1). *A uniform PPT machine $e\mathcal{O}_{NC^1}$ is called an* extractability obfuscator *for NC^1 if for constants $c \in \mathbb{N}$, the following holds: Let $\mathcal{M}_k$ be the class of Turing machines corresponding to the class of circuits of depth at most $c \log k$ and size at most k. Then $e\mathcal{O}(c, \cdot, \cdot)$ is an extractability obfuscator for the class $\{\mathcal{M}_k\}$.*

Definition 6 (Extractability Obfuscator for TM). *A uniform PPT machine $e\mathcal{O}_{TM}$ is called an* extractability obfuscator *for the class TM of polynomial-size Turing machines if it satisfies the following. For each k, let $\mathcal{M}_k$ be the class of Turing machines Π containing a description of a Turing machine M of size bounded by k, such that Π takes two inputs, (t, x), with $|t| = k$, and the output of $\Pi(t, x)$ is defined to be the the output of running the Turing machine $M(x)$ for t steps. Then $e\mathcal{O}_{TM}$ is an extractability obfuscator for $\{\mathcal{M}_k\}$.*

Applying the properties of extractability obfuscation to this class of Turing machines $\{\mathcal{M}_k\}$ implies that for programs $\Pi_0, \Pi_1 \in \mathcal{M}_k$ defined above (corresponding to underlying size-k Turing machines M_0, M_1), efficiently distinguishing between obfuscations of Π_0 and Π_1 implies that one can efficiently extract an input *pair* (t', x') for which $\Pi_0(t', x') \neq \Pi_1(t', x')$. In particular, either $M_0(x') \neq M_1(x')$ or $Runtime(M_0, x') \neq Runtime(M_1, x)$. Thus, if restricting attention to a subclass of $\mathcal{M}_k$ for which each pair of programs satisfies $Runtime(M_0, x) = Runtime(M_1, x)$ for each input x, then "standard" extraction is guaranteed (i.e., such that the extracted input contains x' satisfying $M_0(x') \neq M_1(x')$), while achieving input-specific runtime of the obfuscated program. (Indeed, for an input x of unknown runtime, one simply executes the obfuscated program $\tilde{\Pi}$ sequentially with increasing time bounds $t = k, 2k, 2^2 k, 2^3 k, \cdots$ until a non-$\perp$ output is received). If restricting to a subclass $\mathcal{M}_k$ that has a polynomial runtime bound $t(k)$, then "standard" extraction can be guaranteed by simply defining $Runtime(M, x) = t(k)$ for every $M \in \mathcal{M}_k$ and every input x.

In the sequel, when referring to an extractability obfuscation of a Turing machine M, we will implicitly mean the related program Π_M as above, but will suppress notation of the additional input t.

Definition 7 (Extractability Obfuscator for Bounded-Input TM). *A uniform PPT machine $e\mathcal{O}_{BI}$ is called an extractability obfuscator for bounded-input Turing machines if it satisfies the following. For each k and polynomial $\ell(k)$, let $\mathcal{M}_k^\ell$ be the class of Turing machines Π as in Definition 6, but where the inputs (t, x) of Π are limited by $|t| = k$ and $|x| \leq \ell(k)$. Then there exists a polynomial $p_s(k)$ for which $e\mathcal{O}_{BI}$ is an extractability obfuscator for $\{\mathcal{M}_k^\ell\}$, and for every $k \in \mathbb{N}$, and every $M \in \mathcal{M}_k^\ell$, it holds that the obfuscation $M' \leftarrow e\mathcal{O}_{BI}(1^k, M, \ell(k))$ has size bounded by $p_s(\ell(k), k)$.*

3.1 Extractability Obfuscation for NC^1

In this work, we build upon the existence of any extractability obfuscator for NC^1. In particular, this assumption can be instantiated using the candidate obfuscator for NC^1 given by Brakerski and Rothblum [9] or Barak et al. [1]. These works achieve (stronger) *virtual black-box* security within an idealized model, based on certain assumptions. We refer the reader to [9,1] for more details.

Assumption 8 (NC^1 Extractability Obfuscator). *There exists a secure extractability obfuscator $e\mathcal{O}_{NC^1}$ for NC^1, as in Definition 5*

3.2 Amplifying to General Polynomial-Sized Turing Machines

In this section, we demonstrate how to bootstrap from an extractability obfuscator for NC^1 to one for *all* (bounded-input) Turing machines with a polynomial-sized description, by use of fully homomorphic encryption (FHE), in conjunction with a **P**-certificate system (a succinct argument system for statements in **P**).[6] Our construction provides also two corollaries. Relaxing our assump-

---

[6] **P**-certificates are immediately implied by any succinct non-interactive argument (SNARG) system for NP, but can additionally be based on *falisifiable* assumptions. We refer the reader to Section 2.3 for details.

tions, by using *leveled* FHE, and removing **P**-certificates, we achieve extractability obfuscation for polynomial-size circuits. And strengthening our assumption, replacing **P**-certificates with succinct non-interactive arguments of knowledge (SNARKs), yields extractability obfuscation for all polynomial-size Turing machines. Our construction follows the analogous amplification transformation of Garg et. al [14] in the (weaker) setting of indistinguishability obfuscation.

At a high level, the transformation works as follows. An obfuscation of a Turing machine M consists of two FHE ciphertexts g_1, g_2, each encrypting a description of M under a distinct public key, and an obfuscation of a certain (low-depth) verify-and-decrypt circuit. To evaluate an obfuscation of M on input x, a user will homomorphically evaluate the oblivious[7] universal Turing machine $U(\cdot, x)$ on both ciphertexts g_1 and g_2, and generate a **P**-*certificate* ϕ that the resulting ciphertexts e_1, e_2 were computed correctly. Then, he will provide a low-depth proof π that the certificate ϕ properly verifies (e.g., simply providing the entire circuit evaluation). The collection of e_1, e_2, ϕ, π is then fed into the NC^1-obfuscated program, which checks the proofs, and if valid outputs the decryption of e_1.

Note that the use of computationally sound **P**-certificates enables the size of the obfuscation of M to depend only on the *description size* of M, and not its runtime. This approach is not possible in the setting of *indistinguishability* obfuscation, as certificates of false statements *exist*, but are simply hard to find.

Theorem 9. *There exists a succinct extractability obfuscator eO for bounded-input* TM, *as in Definition 7, assuming the existence of the following tools:*

- eO_{NC^1}: *an extractability obfuscator for the class of circuits* NC^1.
- $\mathcal{E}$ = (Gen$_{FHE}$, Enc$_{FHE}$, Dec$_{FHE}$, Eval$_{FHE}$): *a fully homomorphic encryption scheme with decryption* Dec *in* NC^1.
- (CRSGen$_{cert}$, P_{cert}, V_{cert}): *a* **P**-*certificate system in the CRS model.*

We remark that by replacing the **P**-certificates with succinct non-interactive arguments of knowledge (SNARKs) and additionally using collision resistant hash functions, then the resulting extractability obfuscator is secure for all polynomial-size Turing machines of possibly unbounded input size.

Corollary 1. *Based on any extractability obfuscator for the class of circuits* NC^1, *fully homomorphic encryption, succinct non-interactive arguments of knowledge (SNARKs), there exists an extractability obfuscator for* TM, *as in Definition 6.*

We also observe that by using a *leveled* FHE, and removing the **P**-certificates from the construction, we can still achieve extractability obfuscation for P/poly. Namely, instead of generating a **P**-certificate that the homomorphic evaluation of U_k was performed correctly and then computing a low-depth proof that the

---

[7] A Turing machine is said to be *oblivious* if the tape movements are independent of the input. Without obliviousness, one would be unable to homomorphically evaluate the Turing machine efficiently, as the location of the head of the Turing machine is encrypted.

resulting **P**-certificate properly verifies, simply generate a (large) low-depth proof of correctness of the homomorphic evaluation directly. Further, in the place of FHE, simply sample and utilize keys for a leveled FHE scheme with sufficient levels to support homomorphic evaluation of U_k. The resulting transformation $e\mathcal{O}'$ still satisfies the required correctness and security properties, but no longer achieves succinctness (i.e., the size of the obfuscated Turing machine depends polynomially on its runtime).

Corollary 2. *Based on any extractability obfuscator for the class of circuits NC^1, and leveled fully homomorphic encryption, there exists a (non-succinct) extractability obfuscator for P/poly.*

We refer the reader to the full version of this paper for the full construction and analysis of the bootstrapping procedure and associated corollaries.

4   Functional Witness Encryption

We put forth the notion of *functional witness encryption (FWE)*. An FWE scheme enables one to encrypt a message m with respect to an NP language L, instance x and a function f, such that anyone that has, and *only* those that have, a witness w for $x \in L$ can recover $f(m, w)$. More precisely, our security definition requires that if you can distinguish encryptions of two messages m_0, m_1, then you must know a witness w for $x \in L$ such that $f(m_0, w) \neq f(m_1, w)$.

For example, an FWE scheme would allow one to encrypt the nodes of a large graph in such a way that anybody (and *only those*) who knows a clique in the graph can decrypt the labels *on the corresponding clique*.

Definition 8 (Functional Witness Encryption). *A functional witness encryption scheme for an NP language L (with corresponding witness relation R) and class of Turing machines $\{\mathcal{M}_k\}_{k\in\mathbb{N}}$, consists of the following two polynomial-time algorithms:*

- *$\mathsf{Enc}(1^k, x, m, M)$: On input the security parameter 1^k, an unbounded-length string x, message $m \in MSG$ for some message space MSG, and Turing machine description $M \in \mathcal{M}_k$, Enc outputs a ciphertext c.*
- *$\mathsf{Dec}(c, w)$: On input a ciphertext c and an unbounded-length string w, Dec outputs an evaluation m' or the symbol $\perp$.*

satisfying the following conditions:

Correctness: *There exists a negligible function $\mathsf{negl}(k)$ such that for every security parameter k, for any message $m \in MSG$, for any Turing machine $M \in \mathcal{M}_k$, and for any $x \in L$ such that $R(x, w)$ holds, we have that $\Pr\left[\mathsf{Dec}(\mathsf{Enc}(1^k, x, m, M), w) = M(m, w)\right] = 1 - \mathsf{negl}(k)$.*

Security: *For every PPT adversary $\mathcal{A}$ and polynomials $p(k), \ell(k)$, there exists a PPT extractor E and polynomial $q(k)$ s.t. for every security parameter k, pair of messages $m_0, m_1 \in MSG_k$, Turing machine $M \in \mathcal{M}_k$, string x, and auxiliary input z of length at most $\ell(k)$,*

$$\Pr\left[b \leftarrow \{0,1\}; c \leftarrow \mathsf{Enc}(1^k, x, m_b, M) : \mathcal{A}(1^k, c, z) = b\right] \geq \tfrac{1}{2} + \tfrac{1}{p(k)}$$
$$\Rightarrow \Pr\left[w \leftarrow E(1^k, p(k), x, m_0, m_1, M, z) : M(m_0, w) \neq M(m_1, w)\right] \geq \tfrac{1}{q(k)}.$$

We demonstrate that FWE is, in fact, *equivalent* to extractability obfuscation, up to a simple transformation.

Theorem 10 (Equivalence of FWE and Extractability Obfuscation).
The existence of the following two primitives is equivalent:
1. *Succinct functional witness encryption for NP and P/poly.*
2. *Succinct extractability obfuscation for P/poly.*

Roughly, given an extractability obfuscator $e\mathcal{O}$, an FWE encryption of the message m, for the language L, instance x and function f will be the obfuscation of the program that on input w outputs $f(m, w)$ if w is a valid witness for $x \in L$. On the other hand, given a general-purpose FWE, to obfuscate a program Π, let f be the universal circuit that on input (Π, y) runs Π on input y, let L be the trivial language where every witness is valid, and output a FWE of the message Π. We defer the proof of Theorem 10 to the full version of this paper.

5   Applications to Functional Encryption

We show how to use extractability obfuscation to directly achieve (indistinguishability) functional encryption for unbounded number of key queries and with full adaptive-message security for any unbounded size message space, without relying on complexity leveraging.

The intuition behind our scheme is simple. Let the public key of the FE scheme be the verification key for a signature scheme, and let the master secret key (needed to release secret keys sk_f) be the signing key for the signature scheme. To encrypt a message m, obfuscate the program that on input f and a valid signature on f (given the public key) simply computes $f(m)$. The secret key sk_f for a function f is then simply the signature on f. (The high-level idea behind the construction is somewhat similar to the one used in [20], which uses witness encryption in combination with signature schemes to obtain simulation-based FE for a *single* function f; in contrast, we here focus on FE for an unbounded number of functions).

Proving that this construction works is somewhat subtle. In fact, to make the proof go through, we require the signature scheme in use to be of a special *delegatable* kind—namely, we require the use of *functional signatures* [7,3] (which can be constructed based on non-interactive zero-knowledge arguments of knowledge), which make it possible to delegate a signing key sk' that enables one to sign only messages that satisfy some predicate.[8] The delegation property is only used in the security reduction and, roughly speaking, makes it possible

---

[8] Note that functional signatures were not needed in [20], as they only consider a single key query. In our case, functional signatures are needed to answer "CCA"-type key queries.

to simulate key queries without harming security for the messages selected by the attacker.

We defer the full construction of the functional encryption scheme and proof of security to the full version.

Theorem 11. *Assume the existence of non-interactive zero-knowledge arguments of knowledge (NIZKAoK) for NP and the existence of a extractability obfuscators for P/poly. Then there exists a (fully) indistinguishability-secure functional encryption scheme for arbitrary length messages.*

6   Relating Extractability and Indistinguishability Obfuscation

A natural question is whether we can obtain extractability obfuscation from indistinguishability obfuscation. We address this question in two different settings: first directly in the context of obfuscation, and second in the language of FWE. (Recall that these two notions are equivalent when dealing with arbitrary circuits and arbitrary functions; however, when considering restricted function classes, there are interesting differences).

In Section 6.1, we demonstrate that any indistinguishability obfuscation in fact implies a weak version of extractability obfuscation, in which extraction is only guaranteed when the two circuits differ on only polynomially many inputs. In Section 6.2, we define a weaker notion of FWE mirroring the definition of indistinguishability obfuscation, and provide a transformation from any such indistinguishability FWE to standard FWE for languages with polynomially many witnesses.

The two results are incomparable, in that the former transformation (in Section 6.1) starts with a stronger assumption and yields a stronger result. Indeed, if one begins with indistinguishability FWE for all NP and P/poly, then by the equivalence of FWE and obfuscation, the former transformation yields a stronger outcome in the setting of FWE, guaranteeing indistinguishability of encryptions of messages m_0, m_1 with respect to a function f and NP statement x with potentially exponentially many witnesses, as long as only *polynomially many such witnesses w produce differing outputs $f(m_0, w) \neq f(m_1, w)$*. On the other hand, the FWE transformation (in Section 6.2) also treats the case of restricted function classes. For example, it provides a method for transforming indistinguishability FWE for the trivial function $f(m, w) = m$ to FWE for the same function f. It is easy to see that indistinguishability FWE for this particular f is equivalent to the notion of witness encryption [15], and FWE for the same f is equivalent to the notion of extractable witness encryption of [20]. The transformation in Section 6.2 thus shows how to turn witness encryption to extractable witness encryption for the case of languages with polynomially many witness.

6.1   From Indistinguishability Obfuscation to Extractability Obfuscation for Circuits with Polynomial Differing Inputs

We show that indistinguishability obfuscation directly implies a weak version of extraction obfuscation, where extraction is successful for any pair of circuits C_0, C_1 that vary on polynomially many inputs.

Definition 9 (Weak Extractability Obfuscation). *A uniform transformation $\mathcal{O}$ is a* weak extractability obfuscator *for a class of Turing machines $\mathcal{M} = \{\mathcal{M}_k\}$ if for every PPT adversary $\mathcal{A}$ and polynomial $p(k)$, there exists a PPT algorithm E and polynomials $p_E(k), t_E(k)$ for which the following holds. For every polynomial $d(k)$, for all sufficiently large k, and every pair of circuits $M_0, M_1 \in \mathcal{M}_k$ differing on at most $d(k)$ inputs, and every auxiliary input z,*

$$\Pr\left[b \leftarrow \{0,1\}; \tilde{M} \leftarrow \mathcal{O}(1^k, C_b) : \mathcal{A}(1^k, \tilde{M}, M_0, M_1, z) = b\right] \geq \tfrac{1}{2} + \tfrac{1}{p(k)}$$
$$\implies \Pr\left[x \leftarrow E(1^k, M_0, M_1, z) : M_0(x) \neq M_1(x)\right] \geq \tfrac{1}{p_E(k)},$$

and the runtime of E is $t_E(k, d(k))$.

Theorem 12. *Let $\mathcal{O}$ be an indistinguishability obfuscator for P/poly. Then $\mathcal{O}$ is also a* weak extractability obfuscator *for P/poly.*

Denote by $n = n(k)$ the (polynomial) input length of the circuits in question. At a high level, the extractor E associated with an adversary $\mathcal{A}$ performs a form of binary search over $\{0,1\}^n$ for a desired input by considering a sequence of intermediate circuits lying "in between" C_0 and C_1. The goal is that after n iterations, E will reach a pair of circuits $C^{\mathsf{Left}}, C^{\mathsf{Right}}$ for which: (1) $\mathcal{A}$ can still distinguish between obfuscations $\{\mathcal{O}(C^{\mathsf{Left}})\}$ and $\{\mathcal{O}(C^{\mathsf{Right}})\}$, and yet (2) C^{Left} and C^{Right} are identical on all inputs except a single known x, for which $C^{\mathsf{Left}}(x) = C_0(x)$ and $C^{\mathsf{Right}}(x) = C_1(x)$. Thus, by the indistinguishability security of $\mathcal{O}$, it must be that E has extracted an input x for which $C_0(x) \neq C_1(x)$.

To demonstrate, consider the case where the circuits C_0, C_1 differ on a single unknown input x^*. In the first step, the extractor algorithm E will consider an intermediate circuit C^{Mid} equal to C_0 on the first half of its inputs, and equal to C_1 on the second half of its inputs. Then since $C^{\mathsf{Mid}}(x^*) \in \{C_0(x^*), C_1(x^*)\}$ and all three circuits agree on inputs $x \neq x^*$, it must be that C^{Mid} is *equivalent* to either C_0 or C_1. By the security of the indistinguishability obfuscator, it follows that the obfuscations of such equivalent circuits are indistinguishable. But, if an adversary $\mathcal{A}$ distinguishes between obfuscations of C_0 and C_1 with non-negligible advantage ϵ, then $\mathcal{A}$ must successfully distinguish between obfuscations of C_0 & C^{Mid} or C^{Mid} & C_1. Namely, it must be the case that $\mathcal{A}$'s distinguishing advantage is very small between one of these pairs of distributions (corresponding to the case $C^{\mathsf{Mid}} \equiv C_b$) and is nearly ϵ for the other pair of distributions (corresponding to $C^{\mathsf{Mid}} \not\equiv C_{1-b}$). Thus, by generating samples from these distributions and estimating $\mathcal{A}$'s distinguishing advantages for the two distribution pairs, E can determine whether $C^{\mathsf{Mid}} \equiv C_0$ or $C^{\mathsf{Mid}} \equiv C_1$ and, in turn, has learned whether x^* lies in the first or second half of the input space. This process is then repeated iteratively within a smaller window (i.e., considering a new intermediate circuit lying "in between" C^{Mid} and C_b for which $C^{\mathsf{Mid}} \not\equiv C_b$). In each step, we decrease the input space by a factor of two, until x^* is completely determined.

The picture becomes more complicated, however, when there are several inputs on which C_0 and C_1 disagree. Here the intermediate circuit C^{Mid} need not agree with either C^{Left} or C^{Right} on all inputs. Thus, whereas above $\mathcal{A}$'s distinguishing advantage along one of the two paths was guaranteed to drop no more than a negligible amount, here in each step $\mathcal{A}$'s advantage could split by as much as half. At this rate, after only $\log k$ iterations, $\mathcal{A}$'s advantage will drop below usable levels, and the binary search approach will fail. Indeed, if C_0, C_1 differ on superpolynomially may inputs $d(k) \in k^{\omega(1)}$, there may not even *exist* a pair of adjacent circuits C^{Left} and C^{Right} satisfying the desired properties (1) and (2) described above. (Intuitively, for example, it could be the case that each time one evaluation is changed from $C_0(x)$ to $C_1(x)$, the adversary's probability of outputting 1 increases by the negligible amount $1/d$.)

We show, however, that if there are polynomially many differing inputs $D \subset \{0,1\}^n$ for which $C_0(x) \neq C_1(x)$, then this issue can be overcome. The key insight is that, in any step of the binary search where the adversary's distinguishing advantage may split noticeably among the two possible continuing paths, this step must also split the *set of differing inputs* into two subsets: that is, the number of points d' on which C^{Left} and C^{Right} disagree is equal to the *sum* of the number of points d^L on which C^{Mid} and C^{Left} disagree and the number of points d^R on which C^{Mid} and C^{Right} disagree. Then even though the adversary's distinguishing advantage may split as $\epsilon' = \epsilon^L + \epsilon^R$, for at least one of the two paths $b \in \{L, R\}$, it must be that the ratio of $\epsilon^b/d^b \geq \epsilon'/d'$ is roughly maintained (up to a negligible amount). Since there are only polynomially many total disagreeing inputs $d(k) \in k^{O(1)}$ to start, and assuming $\mathcal{A}$ begins with non-negligible distinguishing advantage, the original ratio ϵ/d at the root node begins as a non-negligible amount. And so we are guaranteed that there exists a path down the tree for which ϵ'/d' (and, in particular, the intermediate distinguishing advantage ϵ') stays above this non-negligible amount ϵ/d. Our extractor E will find this path by simply following *all paths* which maintain distinguishing advantage above this value. By the security of the indistinguishability obfuscation scheme, there will be at most polynomially many such paths (corresponding to those terminating at the inputs $x \in D$), and all other paths in the tree will be pruned.

More specifically, our extractor E runs as follows. At the beginning of execution, it sets a fixed threshold $\mathsf{thresh} = \epsilon/dk$ based on the original (signed) distinguishing advantage ϵ of $\mathcal{A}$ and the number of inputs d on which the circuits differ (note that if $d = d(k)$ is unknown, E will repeat the whole extraction procedure with guesses k, k^2, k^{2^2}, k^{2^3}, etc, for this value). At each step it considers three circuits $C^{\mathsf{Left}}, C^{\mathsf{Mid}}, C^{\mathsf{Right}}$, and estimates $\mathcal{A}$'s (signed) distinguishing advantage between obfuscations of C^{Left} & C^{Mid} and of C^{Mid} & C^{Right}, using repeated sampling with sufficiently low error ($\mathsf{err} = \epsilon/dk^2$). For each pair that yields distinguishing probability above thresh (possibly neither, one, or both pairs), E recurses by repeating this process at a circuit lying between the relevant window. More explicitly, if the left pair yields sufficient distinguishing advantage, then E will repeat the process for the triple of circuits $C^{\text{Left}}, C', C^{\text{Mid}}$ for the circuit C' "halfway between" $C^{\text{Left}}, C^{\text{Mid}}$; analogous for the right pair; if both surpass threshold, E repeats for both; and if neither surpass threshold, then E will not continue down this path of the binary search.

In the full version, we prove that for appropriate choice of threshold, E will only ever visit polynomially many nodes in the binary search tree, and will be guaranteed to find a complete path for which $\mathcal{A}$'s distinguishing advantage maintains above threshold through all n steps down the tree (thus specifying a desired n-bit distinguishing input).

Note that Theorem 12 implies, for example, that for the class of polynomial multipoint locker functions (i.e., functions evaluating to nonzero bit strings at polynomially many hidden points), indistinguishability obfuscation is *equivalent* to extractability obfuscation.

6.2   From Indistinguishability FWE to FWE for Languages with Polynomial Witnesses

We now address this question in the language of FWE.

Mirroring the definition of indistinguishability obfuscation, we define a weaker notion of FWE—which we refer to as *indistinguishability FWE*—which only requires that if $f(m_0, w) = f(m_1, w)$ for *all* witnesses w for $x \in L$, then encryptions of m_0 and m_1 are indistinguishable. Recall that, in contrast, the stronger notion requires that if you can distinguish between encryptions of m_0 and m_1 you must know a witness on which they differ.

Definition 10 (Indistinguishability FWE). *An* indistinguishability functional witness encryption *(iFWE) scheme for an NP language L and class of functions $\mathcal{F} = \{F_k\}$ consists of encryption and decryption algorithms* Enc, Dec *with the same syntax as standard FWE, satisfying the same correctness property, and the following (weaker) security property:*

(Indistinguishability) security: *For every PPT adversary $\mathcal{A}$ and polynomial $\ell(\cdot)$, there exists a negligible function $\nu(\cdot)$ such that for every security parameter k, every function $f \in F_k$, messages $m_0, m_1 \in MSG_k$, string x, and auxiliary information z of length at most $\ell(k)$ for which $f(m_0, w) = f(m_1, w)$ for every witness w of $x \in L$,*

$$\big| \Pr\big[\mathcal{A}(1^k, \text{Enc}(1^k, x, m_0, f), z) = 1\big]$$
$$- \Pr\big[\mathcal{A}(1^k, \text{Enc}(1^k, x, m_1, f), z) = 1\big] \big| < \nu(k).$$

Using the same transformation as in the Extractability Obfuscation-FWE equivalence (see Theorem 10), it can be seen that iFWE for P/poly and NP is directly equivalent to indistinguishability obfuscation for P/poly. We now consider the question of whether we can turn any iFWE into an FWE. We provide an affirmative answer for two restricted cases.

The first result is derived from the transformation from the previous section, combined with the simple extractability obfuscation-to-FWE equivalence transformation (see Theorem 10). Loosely, it says that from iFWE for P/poly, we can obtain a weak form of FWE where (extraction) security holds as long as the function $f(m, w)$ is not "too sensitive" to m: i.e., if for any two messages m_0, m_1 there are only polynomially many witnesses w for which $f(m_0, w) \neq f(m_1, w)$. For example, this captures functions f that rarely output nonzero values. Returning to the example of encrypting data m associated with nodes of a social network, we could allow someone holding clique w to learn whether the nodes in this clique satisfy some chosen rare property (e.g., contains someone with a rare disease, all have the same birthday, etc). Then, while there may be many cliques (corresponding to several, even exponentially many, witnesses w), it will hold that $f(m, w) = 0$ for all but polynomially many w.

As a special case, if the language has only polynomially many witnesses for each statement, then this property holds for any function class.

Definition 11. *We say a class of functions $\mathcal{F} = \{F_k\}$ has t-bounded sensitivity w.r.t. message space MSG and NP language L (with relation R), if for every $f \in F_k$, every $m_0, m_1 \in MSG$, and every $x \in \{0,1\}^*$ there are at most $t(|x|)$ witnesses w s.t. $R(x, w) = 1$ and $f(m_0, w) \neq f(m_1, w)$.*

Corollary 3. *Suppose there exists iFWE for NP and P/poly. Then for any polynomial $t(\cdot)$, there exist FWE schemes for any class of functions $\mathcal{F} = \{F_k\}$, message space MSG, and NP language L, for which $\mathcal{F}$ has t-bounded sensitivity with respect to MSG and L.*

The second result considers iFWE for general function classes (instead of just P/poly), but restricts to NP languages with polynomial witnesses. In the encrypted social network example, this allows basing on a weaker assumption (not requiring the iFWE scheme to support all P/poly), but would restrict to social networks with only polynomially many cliques. The transformation preserves the supported function class: For example, given iFWE for the singleton function class $\{f(m, w) = m\}$ (corresponding to standard witness encryption), one obtains standard FWE for the same class (i.e., *extractable* witness encryption). This result requires a new approach, and makes use of techniques in error-correcting codes.

Definition 12. *Let L be an NP language with corresponding relation R. We say that L has t-bounded witness if for every $x \in \{0,1\}^*$, there are at most $t(|x|)$ distinct witnesses w such that $R(x, w) = 1$.*

Theorem 13. *For every function class $\mathcal{F} = \{F_k\}$ and polynomial $t(\cdot)$, if there exist indistinguishability functional witness encryption schemes for $\mathcal{F}$ and every t-bounded witness NP language, then for every t-bounded witness NP language L (with corresponding relation R), there exists a functional witness encryption schemes for $\mathcal{F}$ and L.*

Proof. Let L be a t-bounded witness NP language with corresponding relation R for some polynomial $t(\cdot)$. Define $q(\cdot)$ such that for every $k \in \mathbb{N}$, $q(k)$ is the smallest prime $\geq 8t(k)$. Assume without loss of generality (by padding) that any witness of any $x \in L$ has length $u(|x|)$ for some polynomial u. To construct a functional witness encryption scheme (Enc, Dec) for L and $\mathcal{F}$, we consider the following NP language L'.

$$L' = \{(x, r, a) : \exists w \in \{0,1\}^{u(|x|)} \text{ s.t. } (R(x,w) = 1) \wedge (r \in \mathbb{F}_{q(|x|)}^{u(|x|)}) \wedge (\langle r, w \rangle = a)\},$$

where $\mathbb{F}_q = \{0, \ldots, q-1\}$ is the prime field of size q and $\langle \cdot, \cdot \rangle$ denotes inner product over $\mathbb{F}_q^u$.

Let (Enc', Dec') be a indistinguishability FWE scheme for L' and $\mathcal{F}$. We construct a FWE scheme (Enc, Dec) for L and $\mathcal{F}$ as follows.

- Enc($1^k, x, m, f$): On input security parameter 1^k, statement $x \in \{0,1\}^*$, message $m \in MSG_k$, and function $f \in \mathcal{F}_k$, Enc generates c as:
  - Let $q = q(|x|)$ and $u = u(|x|)$. Sample $r \leftarrow \mathbb{F}_q^u$ uniformly random.
  - For every $a \in \mathbb{F}_q$, compute $c_a = \text{Enc}'(1^k, (x, q, r, a), m, f)$.
  - Output $c = \{c_a\}_{a \in \mathbb{F}_q}$.
- Dec(c, w): On input a ciphertext $c = \{c_a\}_{a \in \mathbb{F}_q}$ and a witness $w \in \{0,1\}^*$, Dec runs Dec'(c_a, w) for every $a \in \mathbb{F}_q$. If there exists some a such that Dec'(c_a, w) $\neq \perp$, then output the first non-$\perp$ Dec'(c_a, w). Otherwise, output $\perp$.

It is not hard to see that correctness of (Enc', Dec') implies correctness of (Enc, Dec): For every k, x, m, f, w, if w is a witness for $x \in L$, then there exists some $a \in \mathbb{F}_q$ such that w is a witness for $(x, q, r, a) \in L'$, and for the first such a, by the correctness of (Enc', Dec'), Dec'(c_a, w) $= f(m, w)$ with $1 - \text{negl}(k)$ probability, which implies that Dec(Enc($1^k, x, m, f$), w) output $f(m, w)$ with $1 - \text{negl}(k)$ probability as well.

We refer the reader to the full version of this paper for the proof of security of (Enc, Dec). At a high level, we show that if an adversary $\mathcal{A}$ can distinguish Enc($1^k, x, m_0, f$) and Enc($1^k, x, m_1, f$) with a non-negligible advantage, then there is a non-negligible fraction of $r \in \mathbb{F}_q^u$ such that we learn non-trivial information about the value of $\langle r, w \rangle$ for some witness w such that $f(m_0, w) \neq f(m_1, w)$. Note that a linear function $g_w(r) := \langle r, w \rangle$ can be viewed as a q-ary Hadamard code of w. The non-trivial information allows us to obtain a (randomized) function $h(r)$ that agree with $g_w(r)$ on non-negligibly more than $1/q$ fraction of points. We can then apply the local list-decoding algorithm of Goldreich, Rubinfield, and Sudan [19] to recover w.

References

1. Barak, B., Garg, S., Kalai, Y.T., Paneth, O., Sahai, A.: Protecting obfuscation against algebraic attacks. Cryptology ePrint Archive, Report 2013/631 (2013)
2. Barak, B., Goldreich, O., Impagliazzo, R., Rudich, S., Sahai, A., Vadhan, S.P., Yang, K.: On the (im)possibility of obfuscating programs. J. ACM 59(2), 6 (2012)
3. Bellare, M., Fuchsbauer, G.: Policy-based signatures. Cryptology ePrint Archive, Report 2013/413 (2013)

4. Bitansky, N., Canetti, R., Chiesa, A., Tromer, E.: Recursive composition and boot-strapping for snarks and proof-carrying data. In: STOC, pp. 111–120 (2013)
5. Bitansky, N., Chiesa, A., Ishai, Y., Paneth, O., Ostrovsky, R.: Succinct non-interactive arguments via linear interactive proofs. In: Sahai, A. (ed.) TCC 2013. LNCS, vol. 7785, pp. 315–333. Springer, Heidelberg (2013)
6. Boneh, D., Sahai, A., Waters, B.: Functional encryption: a new vision for public-key cryptography. Commun. ACM 55(11), 56–64 (2012)
7. Boyle, E., Goldwasser, S., Ivan, I.: Functional signatures and pseudorandom functions. Cryptology ePrint Archive, Report 2013/401 (2013)
8. Brakerski, Z., Gentry, C., Vaikuntanathan, V.: Fully homomorphic encryption without bootstrapping. Electronic Colloquium on Computational Complexity (ECCC) 18, 111 (2011)
9. Brakerski, Z., Rothblum, G.: Virtual black-box obfuscation for all circuits via generic graded encoding. Cryptology ePrint Archive, Report 2013/563 (2013)
10. Brakerski, Z., Vaikuntanathan, V.: Efficient fully homomorphic encryption from (standard) LWE. In: FOCS, pp. 97–106 (2011)
11. Brakerski, Z., Vaikuntanathan, V.: Lattice-based FHE as secure as PKE. Cryptology ePrint Archive, Report 2013/541 (2013)
12. Canetti, R., Lin, H., Paneth, O.: Public-coin concurrent zero-knowledge in the global hash model. In: Sahai, A. (ed.) TCC 2013. LNCS, vol. 7785, pp. 80–99. Springer, Heidelberg (2013)
13. Chung, K.-M., Lin, H., Pass, R.: Constant-round concurrent zero knowledge from falsifiable assumptions. Cryptology ePrint Archive, Report 2012/563 (2012)
14. Garg, S., Gentry, C., Halevi, S., Raykova, M., Sahai, A., Waters, B.: Candidate indistinguishability obfuscation and functional encryption for all circuits. In: FOCS (2013)
15. Garg, S., Gentry, C., Sahai, A., Waters, B.: Witness encryption and its applications. In: STOC, pp. 467–476 (2013)
16. Gennaro, R., Gentry, C., Parno, B., Raykova, M.: Quadratic span programs and succinct nizks without pcps. In: Johansson, T., Nguyen, P.Q. (eds.) EUROCRYPT 2013. LNCS, vol. 7881, pp. 626–645. Springer, Heidelberg (2013)
17. Gentry, C.: Fully homomorphic encryption using ideal lattices. In: STOC, pp. 169–178 (2009)
18. Gentry, C., Sahai, A., Waters, B.: Homomorphic encryption from learning with errors: Conceptually-simpler, asymptotically-faster, attribute-based. In: Canetti, R., Garay, J.A. (eds.) CRYPTO 2013, Part I. LNCS, vol. 8042, pp. 75–92. Springer, Heidelberg (2013)
19. Goldreich, O., Rubinfeld, R., Sudan, M.: Learning polynomials with queries: The highly noisy case. SIAM J. Discrete Math. 13(4), 535–570 (2000)
20. Goldwasser, S., Kalai, Y.T., Popa, R.A., Vaikuntanathan, V., Zeldovich, N.: How to run turing machines on encrypted data. In: Canetti, R., Garay, J.A. (eds.) CRYPTO 2013, Part II. LNCS, vol. 8043, pp. 536–553. Springer, Heidelberg (2013)
21. Groth, J.: Short pairing-based non-interactive zero-knowledge arguments. In: Abe, M. (ed.) ASIACRYPT 2010. LNCS, vol. 6477, pp. 321–340. Springer, Heidelberg (2010)
22. Hada, S.: Zero-knowledge and code obfuscation. In: Okamoto, T. (ed.) ASIACRYPT 2000. LNCS, vol. 1976, pp. 443–457. Springer, Heidelberg (2000)
23. Hohenberger, S., Sahai, A., Waters, B.: Replacing a random oracle: Full domain hash from indistinguishability obfuscation. Cryptology ePrint Archive, Report 2013/509 (2013), http://eprint.iacr.org/

24. Horwitz, J., Lynn, B.: Toward hierarchical identity-based encryption. In: Knudsen, L.R. (ed.) EUROCRYPT 2002. LNCS, vol. 2332, pp. 466–481. Springer, Heidelberg (2002)
25. Lipmaa, H.: Progression-free sets and sublinear pairing-based non-interactive zero-knowledge arguments. In: Cramer, R. (ed.) TCC 2012. LNCS, vol. 7194, pp. 169–189. Springer, Heidelberg (2012)
26. Micali, S.: Computationally sound proofs. SIAM J. Comput. 30(4), 1253–1298 (2000)
27. Naor, M.: On cryptographic assumptions and challenges. In: Boneh, D. (ed.) CRYPTO 2003. LNCS, vol. 2729, pp. 96–109. Springer, Heidelberg (2003)
28. O'Neill, A.: Definitional issues in functional encryption. Cryptology ePrint Archive, Report 2010/556 (2010), http://eprint.iacr.org/
29. Sahai, A., Waters, B.: How to use indistinguishability obfuscation: Deniable encryption, and more. Cryptology ePrint Archive, Report 2013/454 (2013), http://eprint.iacr.org/

Two-Round Secure MPC
from Indistinguishability Obfuscation*

Sanjam Garg[1], Craig Gentry[1], Shai Halevi[1], and Mariana Raykova[2]

[1] IBM T. J. Watson
[2] SRI International

Abstract. One fundamental complexity measure of an MPC protocol is its *round complexity*. Asharov et al. recently constructed the first three-round protocol for general MPC in the CRS model. Here, we show how to achieve this result with only two rounds. We obtain UC security with abort against static malicious adversaries, and fairness if there is an honest majority. Additionally the communication in our protocol is only proportional to the input and output size of the function being evaluated and independent of its circuit size. Our main tool is indistinguishability obfuscation, for which a candidate construction was recently proposed by Garg et al.

The technical tools that we develop in this work also imply virtual black box obfuscation of a new primitive that we call a *dynamic point function*. This primitive may be of independent interest.

1 Introduction

Secure multiparty computation (MPC) allows a group of mutually distrusting parties to jointly compute a function of their inputs without revealing their inputs to each other. This fundamental notion was introduced in the seminal works of [Yao⁺82, GMW⁺87], who showed that *any* function can be computed securely, even in the presence of malicious parties, provided the fraction of malicious parties is not too high. Since these fundamental feasibility results, much of the work related to MPC has been devoted to improving *efficiency*. There are various ways of measuring the efficiency of a MPC protocol, the most obvious being its computational complexity. In this paper, we focus on minimizing the *communication complexity* of MPC, primarily in terms of the number of *rounds* of interaction needed to complete the MPC protocol, but also in terms of the number of *bits* transmitted between the parties.

---

* The second and third authors were supported by the Intelligence Advanced Research Projects Activity (IARPA) via Department of Interior National Business Center (DoI/NBC) contract number D11PC20202. The U.S. Government is authorized to reproduce and distribute reprints for Governmental purposes notwithstanding any copyright annotation thereon. Disclaimer: The views and conclusions contained herein are those of the authors and should not be interpreted as necessarily representing the official policies or endorsements, either expressed or implied, of IARPA, DoI/NBC, or the U.S. Government.

Y. Lindell (Ed.): TCC 2014, LNCS 8349, pp. 74–94, 2014.

1.1 Our Main Result: Two-Round MPC from Indistinguishability Obfuscation

Our main result is a *compiler* that transforms *any* MPC protocol into a *2-round* protocol in the CRS model. Our compiler is conceptually very simple, and it uses as its main tool *indistinguishability obfuscation* $(i\mathcal{O})$ [BGI$^+$12]. Roughly, in the first round the parties commit to their inputs and randomness, and in the second round each party provides an *obfuscation* of their "next-message" function in the underlying MPC protocol. The parties then separately evaluate the obfuscated next-message functions to obtain the output.

A bit more precisely, our main result is as follows:

Informal Theorem. *Assuming indistinguishability obfuscation, CCA-secure public-key encryption, and statistically-sound noninteractive zero-knowledge, any multiparty function can be computed securely in just two rounds of broadcast.*

We prove that our MPC protocol resists static malicious corruptions in the UC setting [Can$^+$01]. Moreover, the same protocol also achieves fairness if the set of corrupted players is a strict minority. Finally the communication in our protocol can be made to be only proportional to the input and output size of the function being evaluated and independent of its circuit size.

Minimizing round complexity is not just of theoretical interest. Low-interaction secure computation protocols are also applicable in the setting of computing on the web [HLP$^+$11], where a single server coordinates the computation, and parties "log in" at different times without coordination.

1.2 Indistinguishability Obfuscation

Obfuscation was first rigorously defined and studied by Barak et al. [BGI$^+$12]. Most famously, they defined a notion of *virtual black box (VBB)* obfuscation, and proved that this notion is impossible to realize in general – i.e., some functions are VBB unobfuscatable.

Barak et al. also defined a weaker notion of *indistinguishability obfuscation* $(i\mathcal{O})$, which avoids their impossibility results. $i\mathcal{O}$ provides the same functionality guarantees as VBB obfuscation, but a weaker security guarantee. Namely, that for any two circuits C_0, C_1 of similar size that *compute the same function*, it is hard to distinguish an obfuscation of C_0 from an obfuscation of C_1. Barak et al. showed that $i\mathcal{O}$ is *always* realizable, albeit inefficiently: the $i\mathcal{O}$ can simply *canonicalize* the input circuit C by outputting the lexicographically first circuit that computes the same function. More recently, Garg et al. [GGH$^+$13b] proposed an efficient construction of $i\mathcal{O}$ for all circuits, basing security in part on assumptions related to multilinear maps [GGH$^+$13a].

It is clear that $i\mathcal{O}$ is a weaker primitive than VBB obfuscation. In fact, it is not hard to see that we cannot even hope to prove that $i\mathcal{O}$ implies one-way functions: Indeed, if $P = NP$ then one-way functions do not exist but $i\mathcal{O}$ does exist (since the canonicalizing $i\mathcal{O}$ from above can be implemented efficiently). Therefore we do not expect to build many "cryptographically interesting" tools just from $i\mathcal{O}$, but usually need to combine it with other assumptions. (One exception is witness encryption [GGSW+13], which can be constructed from $i\mathcal{O}$ alone.)

It is known that $i\mathcal{O}$ can be combined with one-way functions (OWFs) to construct many powerful primitives such as public-key encryption, identity-based encryption, attribute-based encryption (via witness encryption), as well as NIZKs, CCA encryption, and deniable encryption [SW+12]. However, there are still basic tools that are trivially constructible from VBB obfuscation that we do not know how to construct from $i\mathcal{O}$ and OWFs: for example, collision-resistant hash functions, or compact homomorphic encryption. (Compact homomorphic encryption implies collision-resistant hash functions [IKO+05].) The main challenge in constructing primitives from $i\mathcal{O}$ is that the indistinguishability guarantee holds only in a limited setting: when the two circuits in question are perfectly functionally equivalent.

1.3 Our Techniques

To gain intuition and avoid technical complications, let us begin by considering how we would construct a 2-round protocol if we could use "perfect" VBB obfuscation. For starters, even with VBB obfuscation we still need at least two rounds of interaction, since a 1-round protocol would inherently allow the corrupted parties to repeatedly evaluate the "residual function" associated to the inputs of the honest parties on many different inputs of their choice (e.g., see [HLP+11]).

It thus seems natural to split our 2-round protocol into a commitment round in which all players "fix their inputs," and then an evaluation round where the output is computed. Moreover, it seems natural to use CCA-secure encryption to commit to the inputs and randomness, as this would enable a simulator to extract these values from the corrupted players.

As mentioned above, our idea for the second round is a simple compiler: take any (possibly highly interactive) underlying MPC protocol, and have each party obfuscate their "next-message" function in that protocol, one obfuscation for each round, so that the parties can independently evaluate the obfuscations to obtain the output. Party i's next-message function for round j in the underlying MPC protocol depends on its input x_i and randomness r_i (which are hardcoded in the obfuscations), it takes as input the transcript through round $j-1$, and it produces as output the next broadcast message.

However, there is a complication: unlike the initial interactive protocol, the obfuscations are susceptible to a "reset" attack – i.e., they can be evaluated on multiple inputs. To prevent such attacks, we ensure that the obfuscations can be used for evaluation only on a unique set of values – namely, values consistent with the inputs and randomness that the parties committed to in the first round, and the current transcript of the underlying MPC protocol. To ensure such consistency, naturally we use non-interactive zero-knowledge (NIZK) proofs. Since the NIZKs apply not only to the committed values of the first round, but also to the transcript as it develops in the second round, the obfuscations themselves must output these NIZKs "on the fly". In other words, the obfuscations are now augmented to perform not only the next-message function, but also to prove that their output is consistent. Also, obfuscations in round j of the underlying MPC protocol verify NIZKs associated to obfuscations in previous rounds before providing any output.

If we used VBB obfuscation, we could argue security intuitively as follows. Imagine an augmented version of the underlying MPC protocol, where we prepend a round of commitment to the inputs and randomness, after which the parties (interactively) follow the underlying MPC protocol, except that they provide NIZK proofs that their messages are consistent with their committed inputs and randomness and the developing transcript. It is fairly easy to see that the security of this augmented protocol (with some minor modifications to how the randomness is handled) reduces to the security of the underlying MPC protocol (and the security of the CCA encryption and NIZK proof system). Now, remove the interaction by providing VBB obfuscations of the parties in the second round. These VBB obfuscations "virtually emulate" the parties of the augmented protocol while providing no additional information – in particular, the obfuscations output $\bot$ unless the input conforms exactly to the transcript of the underlying MPC protocol on the committed inputs and randomness; the obfuscations might accept many valid proofs, but since the proofs are statistically sound this gives no more information than one obtains in the augmented protocol.

Instead, we use indistinguishability obfuscation, and while the our protocol is essentially as described above, the proof of security is more subtle. Here, we again make use of the fact that the transcript in the underlying MPC protocol is completely determined by the commitment round, but in a different way. Specifically, there is a step in the proof where we change the obfuscations, so that instead of actually computing the next-message function (with proofs), these values are extracted and simply hardcoded in the obfuscations as the output on any accepting input. We show that these two types of obfuscations are functionally equivalent, and invoke $i\mathcal{O}$ to prove that they are indistinguishable. Once these messages have been "hardcoded" and separated from the computation, we complete the security proof using standard tricks. The most interesting remaining step in the proof is where we replace hardcoded real values with hardcoded simulated values generated by the simulator of the underlying MPC protocol.

1.4   Additional Results

Two-Round MPC with Low Communication. In our basic 2-round MPC protocol, the communication complexity grows polynomially with the circuit size of the function being computed. In Section 3.2, we show how to combine our basic 2-round protocol with *multikey fully homomorphic encryption* [LATV+12] to obtain an MPC that is still only two rounds, but whose communication is basically independent of the circuit size. Roughly speaking, this protocol has a first round where the players encrypt their inputs and evaluate the function under a shared FHE key (and commit to certain values as in our basic protocol), followed by a second round where the players apply the second round of our basic protocol to decrypt the final FHE ciphertext.

Dynamic Point Functions. As a side effect of our technical treatment, we observe that $i\mathcal{O}$ can be used to extend the reach of (some) known VBB obfuscators. For example, we can VBB obfuscate *dynamic point functions*. In this setting, the obfuscation process is partitioned between two parties, the "point owner" Penny and the "function owner" Frank. Penny has a secret string (point) $x \in \{0,1\}^*$, and she publishes a commitment to her point $c_x = \text{com}(x)$. Frank has a function $f : \{0,1\}^* \to \{0,1\}^*$ and knows c_x but not x itself. Frank wants to allow anyone who happens to know x to compute $f(x)$. A dynamic point function obfuscator allows Frank to publish an obfuscated version of the point function

$$F_{f,x}(z) = \begin{cases} f(x) & \text{if } z = x \\ \bot & \text{otherwise.} \end{cases}$$

The security requirement here is that $F_{f,x}$ is obfuscated in the strong VBB sense (and that c_x hides x computationally). We believe that this notion of dynamic point functions is interesting on its own and that it may find future applications.

1.5   Other Related Work

The round complexity of MPC has been studied extensively: both lower and upper bounds, for both the two-party and multiparty cases, in both the semi-honest and malicious settings, in plain, CRS and PKI models. See [AJLA+12, Section 1.3] for a thorough overview of this work.

Here, we specifically highlight the recent work of Asharov et al. [AJLA+12], which achieves 3-round MPC in the CRS model (and 2-round MPC in the PKI model) against static malicious adversaries. They use fully homomorphic encryption (FHE) [RAD+78, Gen+09], but not as a black box. Rather, they construct threshold versions of *particular* FHE schemes – namely, schemes by Brakerski, Gentry and Vaikuntanathan [BV+11, BGV+12] based on the learning with errors (LWE) assumption. (We note that Myers, Sergi and shelat [MSS+11] previously thresholdized a different FHE scheme based on the approximate gcd assumption [vDGHV+10], but their protocol required more rounds.)

In more detail, Asharov et al. observe that these particular LWE-based FHE schemes have a key homomorphic property. Thus, in the first round of their protocol, each party can encrypt its message under its own FHE key, and then the parties can use the key homomorphism to obtain encryptions of the inputs under a shared FHE key. Also, in the last round of their protocol, decryption is a simple one-round process, where decryption of the final ciphertext under the individual keys reveals the decryption under the shared key. In between, the parties use FHE evaluation to compute the encrypted output under the shared key. Unfortunately, they need a third (middle) round for technical reasons: LWE-based FHE schemes typically also have an "evaluation key" – namely, an encryption of a function of the secret key under the public key. They need the extra round to obtain an evaluation key associated to their shared key.

Recently, Gentry, Sahai and Waters [GSW+13] proposed an LWE-based FHE scheme without such an evaluation key. Unfortunately, eliminating the evaluation key in their scheme does not seem to give 2-round MPC based on threshold FHE, since their scheme lacks the key homomorphism property needed by Asharov et al.

We note that our basic two-round protocol does not rely on any *particular* constructions for $i\mathcal{O}$ (or CCA-secure PKE or NIZK proofs), but rather uses these components as black boxes.

Our low-communication two-round protocol uses multikey FHE, but only as a black box. This protocol can be seen as a realization of what Asharov et al. were trying to achieve: a first round where the players encrypt their inputs and evaluate the function under a shared FHE key, followed by a second round where the players decrypt the final FHE ciphertext.

2   Preliminaries

In this section we will start by briefly recalling the definition of different notions essential for our study. We refer the reader to the full version of the paper [GGHR+13] for additional background. The natural security parameter is λ, and all other quantities are implicitly assumed to be functions of λ. We use standard big-O notation to classify the growth of functions. We let $\mathsf{poly}(\lambda)$ denote an unspecified function $f(\lambda) = O(\lambda^c)$ for some constant c. A *negligible* function, denoted generically by $\mathsf{negl}(\lambda)$, is an $f(\lambda)$ such that $f(\lambda) = o(\lambda^{-c})$ for every fixed constant c. We say that a function is *overwhelming* if it is $1 - \mathsf{negl}(\lambda)$.

2.1   Indistinguishability Obfuscators

We will start by recalling the notion of indistinguishability obfuscation $(i\mathcal{O})$ recently realized in [GGH+13b] using candidate multilinear maps[GGH+13a].

Definition 1 (Indistinguishability Obfuscator $(i\mathcal{O})$). *A uniform PPT machine $i\mathcal{O}$ is called an* indistinguishability obfuscator *for a circuit class $\{\mathcal{C}_\lambda\}$ if the following conditions are satisfied:*

- *For all security parameters $\lambda \in \mathbb{N}$, for all $C \in \mathcal{C}_\lambda$, for all inputs x, we have that*

$$\Pr[C'(x) = C(x) : C' \leftarrow i\mathcal{O}(\lambda, C)] = 1$$

- *For any (not necessarily uniform) PPT distinguisher D, there exists a negligible function α such that the following holds: For all security parameters $\lambda \in \mathbb{N}$, for all pairs of circuits $C_0, C_1 \in \mathcal{C}_\lambda$, we have that if $C_0(x) = C_1(x)$ for all inputs x, then*

$$\left| \Pr\left[D(i\mathcal{O}(\lambda, C_0)) = 1 \right] - \Pr\left[D(i\mathcal{O}(\lambda, C_1)) = 1 \right] \right| \le \alpha(\lambda)$$

Definition 2 (Indistinguishability Obfuscator for NC^1).
A uniform PPT machine $i\mathcal{O}$ is called an indistinguishability obfuscator *for NC^1 if for all constants $c \in \mathbb{N}$, the following holds: Let $\mathcal{C}_\lambda$ be the class of circuits of depth at most $c \log \lambda$ and size at most λ. Then $i\mathcal{O}(c, \cdot, \cdot)$ is an indistinguishability obfuscator for the class $\{\mathcal{C}_\lambda\}$.*

Definition 3 (Indistinguishability   Obfuscator   for $P/poly$).   A
uniform PPT machine iO is called an indistinguishability obfuscator *for P/poly*
if the following holds: Let $\mathcal{C}_\lambda$ be the class of circuits of size at most λ. Then iO
is an indistinguishability obfuscator for the class $\{\mathcal{C}_\lambda\}$.

2.2   Semi-honest MPC

We will also use a semi-honest n-party computation protocol π for any functionality f in the stand-alone setting. The existence of such a protocol follows from
the existence of semi-honest 1-out-of-2 oblivious transfer [Yao+82, GMW+87]
protocols. Now we build some notation that we will use in our construction.

Let $\mathcal{P} = \{P_1, P_2, \ldots P_n\}$ be the set of parties participating in a t round protocol π. Without loss of generality, in order to simplify notation, we will assume
that in each round of π, each party broadcasts a single message that depends on
its input and randomness and on the messages that it received from all parties
in all previous rounds. (We note that we can assume this form without loss of
generality, since in our setting we have broadcast channels and CCA-secure encryption, and we only consider security against static corruptions.) We let $m_{i,j}$
denote the message sent by the ith party in the jth round.   We define the function π_i such that $m_{i,j} = \pi_i(x_i, r_i, M_{j-1})$ where $m_{i,j}$ is the jth message generated
by party P_i in protocol π with input x_i, randomness r_i and the series of previous
messages M_{j-1}

$$M_{j-1} = \begin{pmatrix} m_{1,1} & m_{2,1} & \cdots & m_{n,1} \\ m_{1,2} & m_{2,2} & \cdots & m_{n,2} \\ \vdots & & \ddots & \\ m_{1,j-1} & m_{2,j-1} & \cdots & m_{n,j-1} \end{pmatrix}$$

sent by all parties in π.

3   Our Protocol

In this section, we provide our construction of a two-round MPC protocol.

Protocol Π. We start by giving an intuitive description of the protocol. A formal
description appears in Figure 1. The basic idea of our protocol is to start with an
arbitrary round semi-honest protocol π and "squish" it into a two round protocol
using indistinguishability obfuscation. The first round of our protocol helps set
the stage for the "virtual" execution of π via obfuscations that all the parties
provide in the second round.

The common reference string in our construction consists of a CRS σ for a
NIZK Proof system and a public key pk corresponding to a CCA-secure public
key encryption scheme. Next, the protocol proceeds in two rounds as follows:

Protocol Π

Protocol Π uses an Indistinguishability Obfuscator $i\mathcal{O}$, a NIZK proof system (K, P, V), a CCA-secure PKE scheme $(\mathsf{Gen}, \mathsf{Enc}, \mathsf{Dec})$ with perfect correctness and an n-party semi-honest MPC protocol π.

Private Inputs: Party P_i for $i \in [n]$, receives its input x_i.

Common Reference String: Let $\sigma \leftarrow K(1^\lambda)$ and $(pk, \cdot) \leftarrow \mathsf{Gen}(1^\lambda)$ and then output (σ, pk) as the common reference string.

Round 1: Each party P_i proceeds as:
- $c_i = \mathsf{Enc}(i \| x_i)$ and,
- $\forall j \in [n]$, sample randomness $r_{i,j} \in \{0,1\}^\ell$ and generate $d_{i,j} = \mathsf{Enc}(i \| r_{i,j})$. (Here ℓ is the length of the maximum number of random coins needed by any party in π.)

It then sends $Z_i = \{c_i, \{d_{i,j}\}_{j \in [n]}\}$ to every other party.

Round 2: P_i generates:
- For every $j \in [n]$, $j \neq i$ generate $\gamma_{i,j}$ as the NIZK proof under σ for the NP-statement:
$$\left\{ \exists\ \rho_{r_{i,j}} \mid d_{i,j} = \mathsf{Enc}(i \| r_{i,j}; \rho_{r_{i,j}}) \right\}. \tag{1}$$
- A sequence of obfuscations $(i\mathcal{O}_{i,1}, \ldots i\mathcal{O}_{i,t})$ where $i\mathcal{O}_{i,j}$ is the obfuscation of the program $\mathsf{Prog}_{i,j}^{0, x_i, \rho_{x_i}, r_{i,i}, \rho_{r_{i,i}}, \{Z_i\}, 0^{\ell_{i,j}}}$. (Where $\ell_{i,j}$ is output length of the program $\mathsf{Prog}_{i,j}$.)
- It sends $(\{r_{i,j}, \gamma_{i,j}\}_{j \in [n], j \neq i}, \{i\mathcal{O}_{i,j}\}_{j \in [t]})$ to every other party.

Evaluation (MPC in the Head): For each $j \in [t]$ proceed as follows:
- For each $i \in [n]$, evaluate the obfuscation $i\mathcal{O}_{i,j}$ of program $\mathsf{Prog}_{i,j}$ on input $(R, \Gamma, M_{j-1}, \Phi_{j-1})$ where

$$R = \begin{pmatrix} \cdot & r_{2,1} & \cdots & r_{n,1} \\ r_{1,2} & \cdot & \cdots & r_{n,2} \\ \vdots & & \ddots & \\ r_{1,n} & r_{2,n} & \cdots & \cdot \end{pmatrix}, \quad \Gamma = \begin{pmatrix} \cdot & \gamma_{2,1} & \cdots & \gamma_{n,1} \\ \gamma_{1,2} & \cdot & \cdots & \gamma_{n,2} \\ \vdots & & \ddots & \\ \gamma_{1,n} & \gamma_{2,n} & \cdots & \cdot \end{pmatrix}$$

$$M_{j-1} = \begin{pmatrix} m_{1,1} & m_{2,1} & \cdots & m_{n,1} \\ m_{1,2} & m_{2,2} & \cdots & m_{n,2} \\ \vdots & & \ddots & \\ m_{1,j-1} & m_{2,j-1} & \cdots & m_{n,j-1} \end{pmatrix}, \quad \Phi = \begin{pmatrix} \phi_{1,1} & \phi_{2,1} & \cdots & \phi_{n,1} \\ \phi_{1,2} & \phi_{2,2} & \cdots & \phi_{n,2} \\ \vdots & & \ddots & \\ \phi_{1,j-1} & \phi_{2,j-1} & \cdots & \phi_{n,j-1} \end{pmatrix}$$

- And obtain, $m_{1,j}, \ldots, m_{n,j}$ and $\phi_{1,j}, \ldots, \phi_{n,j}$.
  Finally each party P_i outputs $m_{i,t}$.

Fig. 1. Two Round MPC Protocol

$$\mathsf{Prog}_{i,j}^{\mathsf{flag},x_i,\rho_{x_i},r_{i,i},\rho_{r_{i,i}},\{Z_i\},\mathsf{fixedOutput}}$$

Program $\mathsf{Prog}_{i,j}^{\mathsf{flag},x_i,\rho_{x_i},r_{i,i},\rho_{r_{i,i}},\{Z_i\},\mathsf{fixedOutput}}$ takes as input $(R,\varGamma,M_{j-1},\varPhi)$ as defined above and outputs $m_{i,j}$ and $\phi_{i,j}$. Specifically, it proceeds as follows:

- $\forall p,q \in [n]$ such that $p \neq q$ check that $\gamma_{p,q}$ is an accepting proof under σ for the NP-statement:

$$\left\{ \exists\ \rho_{r_{p,q}} \mid d_{p,q} = \mathsf{Enc}(p||r_{p,q};\rho_{r_{p,q}}) \right\}$$

- $\forall p \in [n], q \in [j-1]$ check that $\phi_{p,q}$ is an accepting proof for the NP-statement

$$\left\{ \begin{array}{l} \exists\ (x_p,r_{p,p},\rho_{x_p},\rho_{r_{p,p}})\ | \\ \left(c_p = \mathsf{Enc}(p||x_p;\rho_{x_p}) \bigwedge d_{p,p} = \mathsf{Enc}(p||r_{p,p},\rho_{r_{p,p}}) \bigwedge m_{p,q} = \pi_p(x_p, \oplus_{k\in[n]}r_{k,p}, M_{q-1}) \right) \end{array} \right\}$$

- If the checks above fail, output $\bot$. Otherwise, if flag $= 0$ then output $(\pi_i(x_i, \oplus_{j\in[n]}r_{j,i}, M_{j-1}), \phi_{i,j})$ where $\phi_{i,j}$ is the proof for the NP-statement: (under some fixed randomness)

$$\left\{ \begin{array}{l} \exists\ (x_i,r_{i,i},\rho_{x_i},\rho_{r_{i,i}})\ | \\ \left(c_i = \mathsf{Enc}(i||x_i;\rho_{x_i}) \bigwedge d_{i,i} = \mathsf{Enc}(i||r_{i,i},\rho_{r_{i,i}}) \bigwedge m_{i,j} = \pi_i(x_i, \oplus_{j\in[n]}r_{j,i}, M_{j-1}) \right) \end{array} \right\}$$

Otherwise, output $\mathsf{fixedOutput}$.

Fig. 2. Obfuscated Programs in the Protocol

Round 1: In the first round, the parties "commit" to their inputs and randomness, where the commitments are generated using the CCA-secure encryption scheme. The committed randomness will be used for coin-flipping and thereby obtaining unbiased random coins for all parties. Specifically, every party P_i, proceeds by encrypting its input x_i under the public key pk. Let c_i be the ciphertext. P_i also encrypts randomness $r_{i,j}$ for every $j \in [n]$. Let the ciphertext encrypting $r_{i,j}$ be denoted by $d_{i,j}$. Looking ahead the random coins P_i uses in the execution of π will be $s_i = \oplus_j r_{j,i}$. P_i broadcasts $\{c_i, \{d_{i,j}\}_j\}$ to everyone.

Round 2: In the second round parties will broadcast obfuscations corresponding to the next message function of π allowing for a "virtual emulation" of the interactive protocol π. Every party P_i proceeds as follows:

- P_i reveals the random values $\{r_{i,j}\}_{j\neq i\in[n]}$ and generates proofs $\{\gamma_{i,j}\}_{j\neq i\in[n]}$ that these are indeed the values that are encrypted in the ciphertexts $\{d_{i,j}\}_{j\neq i\in[n]}$.

- Recall that the underlying protocol π is a t round protocol where each party broadcasts one message per round. Each player P_i generates t obfuscations of its next-round function, $(i\mathcal{O}_{i,1}, \ldots, i\mathcal{O}_{i,t})$.
  In more detail, each $i\mathcal{O}_{i,k}$ is an obfuscation of a function $F_{i,k}$ that takes as input the $r_{i,j}$ values sent by all the parties along with the proofs that they are well-formed, and also all the π-messages that were broadcast upto round $k-1$, along with the proof of correct generation of these messages. (These proofs are all with respect to the ciphertexts generated in first round and the revealed $r_{i,j}$ values.) The output of the function

$F_{i,j}$ is the next message of P_i in π, along with a NIZK proof that it was generated correctly.

P_i broadcasts all the values $\{r_{i,j}\}_{j\neq i\in[n]}$, $\{\gamma_{i,j}\}_{j\neq i\in[n]}$, and $\{i\mathcal{O}_{i,k}\}_{k\in[t]}$.

Evaluation: After completion of the second round each party can independently "virtually" evaluate the protocol π using the obfuscations provided by each of the parties and obtain the output.

Theorem 1. *Let f be any deterministic poly-time function with n inputs and single output. Assume the existence of an Indistinguishability Obfuscator $i\mathcal{O}$, a NIZK proof system (K, P, V), a CCA secure PKE scheme $(\mathsf{Gen}, \mathsf{Enc}, \mathsf{Dec})$ with perfect correctness and an n-party semi-honest MPC protocol π. Then the protocol Π presented in Figure 1 UC-securely realizes the ideal functionality $\mathcal{F}_f$ in the $\mathcal{F}_{CRS}$-hybrid model.*

3.1 Correctness and Proof of Security

Correctness. The correctness of our protocol Π in Figure 1 follows from the correctness of the underlying semi-honest MPC protocol and the other primitives used. Next we will argue that all the messages sent in the protocol Π are of polynomial length and can be computed in polynomial time. It is easy to see that all the messages of round 1 are polynomially long. Again it is easy to see that the round 2 messages besides the obfuscations themselves are of polynomial length.

We will now argue that each obfuscation sent in round 2 is also polynomially long. Consider the obfuscation $i\mathcal{O}_{i,j}$, which obfuscates $\mathsf{Prog}_{i,j}$; we need to argue that this program for every i, j is only polynomially long. Observe that this program takes as input $(R, \Gamma, M_{i-1}, \Phi_{j-1})$, where Γ and Φ_{j-1} consist of polynomially many NIZK proofs. This program roughly proceeds by first checking that all the proofs in Γ and Φ_{j-1} are accepting. If the proofs are accepting then Prog outputs $m_{i,j}$ and $\phi_{i,j}$.

Observe that Γ and Φ_{j-1} are proofs of NP-statements each of which is a fixed polynomial in the description of the next message function of the protocol π. Also observe that the time taken to evaluate $m_{i,j}$ and $\phi_{i,j}$ is bounded a fixed polynomial. This allows us to conclude that all the computation done by $\mathsf{Prog}_{i,j}$ can be bounded by a fixed polynomial.

Security. Let $\mathcal{A}$ be a malicious, static adversary that interacts with parties running the protocol Π from Figure 1 in the $\mathcal{F}_{CRS}$-hybrid model. We construct an ideal world adversary $\mathcal{S}$ with access to the ideal functionality $\mathcal{F}_f$, which simulates a real execution of Π with $\mathcal{A}$ such that no environment $\mathcal{Z}$ can distinguish the ideal world experiment with $\mathcal{S}$ and $\mathcal{F}_f$ from a real execution of Π with $\mathcal{A}$.

We now sketch the description of the simulator and the proof of security, restricting ourselves to the stand-alone setting. The fully detailed description of our simulator and the proof of indistinguishability are provided in Appendix A. Those more formal proofs are given for the general setting of UC-security.

Our simulator $\mathcal{S}$ roughly proceeds as follows:

- **Common reference string:** Recall that the common reference string in our construction consists of a CRS σ for a NIZK Proof system and a public key pk corresponding to a CCA secure public key encryption scheme. Our simulator uses the simulator of the NIZK proof system in order to generate the reference string σ. Note that the simulator for NIZK proof system also generates some trapdoor information that can be used to generate simulated NIZK proofs. Our simulator saves that for later use. $\mathcal{S}$ also generates the public key pk along with its secret key sk, which it will later use to decrypt ciphertexts generated by the adversary.
- **Round 1:** Recall that in round 1, honest parties generate ciphertexts corresponding to encryptions of their inputs and various random coins. Our simulator just generates encryptions of the zero-string on behalf of the honest parties. Also $\mathcal{S}$ uses the knowledge of the secret key sk to extract the input and randomness that the adversarial parties encrypt.
- **Round 2:** Recall that in the second round the honest parties are required to "open" some of the randomness values committed to in round 1 along with obfuscations necessary for execution of π.
  $\mathcal{S}$ proceeds by preparing a simulated transcript of the execution of π using the malicious party inputs previously extracted and the output obtained from the ideal functionality, which it needs to force onto the malicious parties. $\mathcal{S}$ opens the randomness on behalf of honest parties such that the randomness of malicious parties becomes consistent with the simulated transcript and generates simulated proofs for the same. The simulator generates the obfuscations on behalf of honest parties by hard-coding the messages as contained in the simulated transcript. The obfuscations also generate proofs proving that the output was generated correctly. Our simulator hard-codes these proofs in the obfuscations as well.

Very roughly, our proof proceeds by first changing all the obfuscations $\mathcal{S}$ generates on behalf of honest parties to output fixed values. The statistical soundness of the NIZK proof system allows us to base security on the weak notion of indistinguishability obfuscation. Once this change has been made, in a sequence of hybrids we change from honest execution of the underlying semi-honest MPC protocol to a the simulated execution. We refer the reader to Appendix A for a complete proof.

3.2   Extensions

Low Communication. Our protocol Π (as described in Figure 1) can be used to UC-securely realize any functionality $\mathcal{F}_f$. However the communication complexity of this protocol grows polynomially in the size of the circuit evaluating function f and the security parameter λ. We would like to remove this restriction and construct a protocol Π' whose communication complexity is independent of the the function being evaluated.

A key ingredient of our construction is *multikey fully homomorphic encryption* [LATV+12]. Intuitively, multikey FHE allows us to evaluate any circuit on

---

Protocol Π'

Let Π be the MPC Protocol from Figure 1.
Let $(\mathsf{Setup}_{MK}, \mathsf{Encrypt}_{MK}, \mathsf{Eval}_{MK}, \mathsf{Decrypt}_{MK})$ be a multikey FHE scheme.
Private Inputs: Party P_i for $i \in [n]$, receives its input x_i.
Common Reference String: Generate the CRS corresponding to Π.

Round 1: P_i proceeds as follows:
- $(pk_i, sk_i) \leftarrow \mathsf{Setup}_{MK}(1^\lambda; \rho_i)$ and generates encryption $c_i :=$ $\mathsf{Encrypt}_{MK}(pk_i, x_i; \varrho_i)$.
- Generates the first round message Z_i of Π playing as P_i with input (x_i, ρ_i, ϱ_i). (Recall that the first message of Π does not depend on the function Π is used to evaluate.)
- Sends3 (pk_i, c_i, Z_i) to all parties.

Round 2: Every party P_i computes $c^* := \mathsf{Eval}_{MK}(C, (c_1, pk_1), \ldots, (c_n, pk_n))$. P_i generates P_i's second round message of Π, where Π computes the following function:
- For every $i \in [n]$, check if $(pk_i, sk_i) \leftarrow \mathsf{Setup}_{MK}(1^\lambda; \rho_i)$ and $c_i := \mathsf{Encrypt}_{MK}(pk_i, x_i; \varrho_i)$.
- If all the checks pass then output $\mathsf{Decrypt}_{MK}(sk_1, \ldots, sk_n, c^*)$ and otherwise output $\perp$.

Evaluation: P_i outputs the output of P_i in Π.

Fig. 3. Two Round MPC Protocol with Low Communication Complexity

ciphertexts that might be encrypted under different public keys. To guarantee semantic security, decryption requires all of the corresponding secret keys. We refer the reader to the full version of the paper [GGHR+13] for more details.

Our protocol Π' works by invoking Π. Recall that Π proceeds in two rounds. Roughly speaking, in the first stage parties commit to their inputs, and in the second round the parties generate obfuscations that allow for "virtual" execution of sub-protocol π on the inputs committed in the first round. Our key observation here is that the function that the sub-protocol π evaluates does not have to be specified until the second round.

We will now give a sketch of our protocol Π'. Every party P_i generates a public key pk_i and a secret key sk_i using the setup algorithm of the multikey FHE scheme. It then encrypts its input x_i under the public key pk_i and obtains ciphertext c_i. It then sends (pk_i, c_i) to everyone along with the first message of Π with input the randomness used in generation of pk_i and c_i. This completes the first round. At this point, all parties can use the values $((pk_1, c_1), \ldots, (pk_n, c_n))$ to obtain an encryption of $f(x_1, \ldots x_n)$, where f is the function that we want to compute. The second round of protocol Π can be used to decrypt this value. A formal description of the protocol appears in Figure 3.

Theorem 2. *Under the same assumptions as in Theorem 1 and assuming the semantic security of the multikey FHE scheme, the protocol Π' presented in Figure 3 UC-securely realizes the ideal functionality $\mathcal{F}_f$ in the $\mathcal{F}_{CRS}$-hybrid model.*

Furthermore the communication complexity of protocol Π' is polynomial in the input lengths of all parties and the security parameter. (It is independent of the size of f.)

Proof. The correctness of the our protocol Π' follows from the correctness of the protocol Π and the correctness of the multikey FHE scheme. Observe that the compactness of the multikey FHE implies that the ciphertext c^* evaluated in Round 2 on the description of Protocol Π (Figure 3) is independent of the size of the function f being evaluated. Also note that no other messages in the protocol depend on the function f. This allows us to conclude that the communication complexity of protocol Π' is independent of the size of f.

We defer the formal description of our simulator and the proof of indistinguishability to the full version of the paper [GGHR+13].   ∎

General Functionality. Our basic MPC protocol as described in Figure 1 only considers deterministic functionalities (See [GGHR+13]) where all the parties receive the same output. We would like to generalize it to handle randomized functionalities and individual outputs (just as in [GGHR+13, AJW+11]). First, the standard transformation from a randomized functionality to a deterministic one (See [Gol+04, Section 7.3]) works for this case as well. In this transformation, instead of computing some randomized function $g(x_1, \ldots x_n; r)$, the parties compute the deterministic function $f((r_1, x_1), \ldots, (r_n, x_n)) \overset{def}{=} g(x_1, \ldots, x_n; \oplus_{i=1}^{n} r_i)$. We note that this computation does not add any additional rounds.

Next, we move to individual outputs. Again, we use a standard transformation (See [LP+09], for example). Given a function $g(x_1, \ldots, x_n) \to (y_1, \ldots, y_n)$, the parties can evaluate the following function which has a single output:

$$f((k_1, x_1), \ldots, (k_n; x_n)) = (g_1(x_1, \ldots, x_n) \oplus k_1 || \ldots || g_n(x_1, \ldots, x_n) \oplus k_n)$$

where $a||b$ denotes a concatenation of a with b, g_i indicates the ith output of g, and k_i is randomly chosen by the ith party. Then, the parties can evaluate f, which is a single output functionality, instead of g. Subsequently every party P_i uses its secret input k_i to recover its own output. The only difference is that f has one additional exclusive-or gate for every circuit-output wire. Again, this transformation does not add any additional rounds of interaction.

Corollary 1. *Let f be any (possibly randomized) poly-time function with n inputs and n outputs. Assume the existence of an Indistinguishability Obfuscator $i\mathcal{O}$, a NIZK proof system (K, P, V), a CCA secure PKE scheme (Gen, Enc, Dec) with perfect correctness and an n-party semi-honest MPC protocol π. Then the protocol Π presented in Figure 1 UC-securely realizes the ideal functionality $\mathcal{F}_f$ in the $\mathcal{F}_{CRS}$-hybrid model.*

Common Random String vs Common Reference String. Our basic MPC protocol as described in Figure 1 uses a common reference string. We can adapt the construction to work in the setting of common random string by assuming the existence of a CCA secure public-key encryption scheme with perfect correctness and pseudorandom public keys and a NIZK scheme [FLS+90]. See [GGHR+13] for details.

Fairness. We note that the same protocol Π can be used to securely and fairly UC-realize the generalized functionality in the setting of honest majority, by using a fair semi-honest MPC protocol for π.

4   Applications

In this section we will discuss additional applications of our results.

4.1   Secure Computation on the Web

In a recent work, Halevi, Lindell and Pinkas [HLP+11] studied secure computation in a client-server model where each client connects to the server once and interacts with it, without any other client necessarily being connected at the same time. They show that, in such a setting, only limited security is achievable. However, among other results, they also point out that if we can get each of the players to connect twice to the server (rather than once), then their protocols can be used for achieving the standard notion of privacy.

One key aspect of the two-pass protocols of Halevi et. al [HLP+11] is that there is a preset order in which the clients must connect to the server. Our protocol Π from Section 3 directly improves on the results in this setting by achieving the same two-pass protocol, but without such a preset order. Also, we achieve this result in the common reference/random string model, while the original protocols of Halevi et. al [HLP+11] required a public key setup.

4.2   Black-Box Obfuscation for More Functions

In this subsection, we generalize the class of circuits that can be obfuscated according to the strong (virtual black box (VBB) notion of obfuscation. This application does not build directly on our protocol for two-round MPC. Rather, the main ideas here are related to ideas (particularly within the security proof) that arose in our MPC construction.

Our Result. Let $\mathcal{C}$ be a class of circuits that we believe to be VBB obfuscatable, e.g., point functions or conjunctions. Roughly speaking, assuming indistinguishability obfuscation, we show that a circuit C can be VBB obfuscated if there exists a circuit C' such that $C' \in \mathcal{C}$ and $C(x) = C'(x)$ for every input x. The non-triviality of the result lies in the fact that it might not be possible to efficiently recover C' from C. We refer the reader to the full version of the paper [GGHR+13] for a formal statement and proof.

Dynamic Point Function Obfuscation. We will now highlight the relevance of the results presented above with an example related to point functions. We know how to VBB obfuscate point functions. Now, consider a setting of three players. Player 1 generates a (perfectly binding) commitment to a value x. Player 2 would like to generate an obfuscation of an arbitrary function f that allows an arbitrary Player 3, if he knows x, to evaluate f on input x alone (and nothing other than x). Our construction above enables such obfuscation. We stress that the challenge here is that Player 2 is not aware of the value x, which is in fact computationally hidden from it.

References

[AJLA+12]   Asharov, G., Jain, A., López-Alt, A., Tromer, E., Vaikuntanathan, V., Wichs, D.: Multiparty computation with low communication, computation and interaction via threshold FHE. In: Pointcheval, D., Johansson, T. (eds.) EUROCRYPT 2012. LNCS, vol. 7237, pp. 483–501. Springer, Heidelberg (2012)

[AJW+11]   Asharov, G., Jain, A., López-Alt, A., Tromer, E., Vaikuntanathan, V., Wichs, D.: Multiparty computation with low communication, computation and interaction via threshold FHE. In: Pointcheval, D., Johansson, T. (eds.) EUROCRYPT 2012. LNCS, vol. 7237, pp. 483–501. Springer, Heidelberg (2012)

[BGI+12]   Barak, B., Goldreich, O., Impagliazzo, R., Rudich, S., Sahai, A., Vadhan, S.P., Yang, K.: On the (im)possibility of obfuscating programs. J. ACM 59(2), 6 (2012)

[BGV+12]   Brakerski, Z., Gentry, C., Vaikuntanathan, V.: (leveled) fully homomorphic encryption without bootstrapping. In: ITCS (2012)

[BV+11]   Brakerski, Z., Vaikuntanathan, V.: Efficient fully homomorphic encryption from (standard) lwe. In: FOCS, pp. 97–106 (2011)

[Can+01]   Canetti, R.: Universally composable security: A new paradigm for cryptographic protocols. In: 42nd Annual Symposium on Foundations of Computer Science, Las Vegas, Nevada, USA, October 14-17, pp. 136–145. IEEE Computer Society Press (2001)

[FLS+90]   Feige, U., Lapidot, D., Shamir, A.: Multiple non-interactive zero knowledge proofs based on a single random string. In: Proceedings of the 31st Annual Symposium on Foundations of Computer Science, vol. 1, pp. 308–317 (1990)

[Gen+09]   Gentry, C.: A fully homomorphic encryption scheme. PhD thesis, Stanford University (2009), http://crypto.stanford.edu/craig

[GGH+13a]   Gentry, C., Sahai, A., Waters, B.: Homomorphic encryption from learning with errors: Conceptually-simpler, asymptotically-faster, attribute-based. In: Canetti, R., Garay, J.A. (eds.) CRYPTO 2013, Part I. LNCS, vol. 8042, pp. 75–92. Springer, Heidelberg (2013)

[GGH+13b]   Garg, S., Gentry, C., Halevi, S., Raykova, M., Sahai, A., Waters, B.: Candidate indistinguishability obfuscation and functional encryption for all circuits. In: FOCS 2013. IEEE (to appear, 2013), http://eprint.iacr.org/2013/451

[GGHR+13]   Garg, S., Gentry, C., Halevi, S., Raykova, M.: Two-round secure mpc
            from indistinguishability obfuscation. Cryptology ePrint Archive, Report
            2013/601 (2013), http://eprint.iacr.org/
[GGSW+13]   Garg, S., Gentry, C., Sahai, A., Waters, B.: Witness encryption and its
            applications. In: STOC (2013)
[GMW+87]   Garg, S., Gentry, C., Sahai, A., Waters, B.: Witness encryption and its
            applications. In: STOC (2013)
[Gol+04]   Goldreich, O.: Foundations of Cryptography: Basic Applications, vol. 2.
            Cambridge University Press, Cambridge (2004)
[GSW+13]   Gentry, C., Sahai, A., Waters, B.: Homomorphic encryption from learn-
            ing with errors: Conceptually-simpler, asymptotically-faster, attribute-
            based. In: Canetti, R., Garay, J.A. (eds.) CRYPTO 2013, Part I. LNCS,
            vol. 8042, pp. 75–92. Springer, Heidelberg (2013)
[HLP+11]   Halevi, S., Lindell, Y., Pinkas, B.: Secure computation on the web: Com-
            puting without simultaneous interaction. In: Rogaway, P. (ed.) CRYPTO
            2011. LNCS, vol. 6841, pp. 132–150. Springer, Heidelberg (2011)
[IKO+05]   Ishai, Y., Kushilevitz, E., Ostrovsky, R.: Sufficient conditions for
            collision-resistant hashing. In: Kilian, J. (ed.) TCC 2005. LNCS,
            vol. 3378, pp. 445–456. Springer, Heidelberg (2005)
[LATV+12]   López-Alt, A., Tromer, E., Vaikuntanathan, V.: On-the-fly multiparty
            computation on the cloud via multikey fully homomorphic encryption.
            In: STOC, pp. 1219–1234 (2012)
[LP+09]   Lindell, Y., Pinkas, B.: A proof of security of Yao's protocol for two-party
            computation. Journal of Cryptology 22(2), 161–188 (2009)
[MSS+11]   Myers, S., Sergi, M., Shelat, A.: Threshold fully homomorphic encryption
            and secure computation. IACR Cryptology ePrint Archive 2011, 454
            (2011)
[RAD+78]   Rivest, R., Adleman, L., Dertouzos, M.L.: On data banks and privacy
            homomorphisms. In: Foundations of Secure Computation, pp. 169–180
            (1978)
[SW+12]   Sahai, A., Waters, B.: How to use indistinguishability obfuscation: Deni-
            able encryption, and more. IACR Cryptology ePrint Archive 2013, 454
            (2013)
[vDGHV+10] van Dijk, M., Gentry, C., Halevi, S., Vaikuntanathan, V.: Fully homo-
            morphic encryption over the integers. In: Gilbert, H. (ed.) EUROCRYPT
            2010. LNCS, vol. 6110, pp. 24–43. Springer, Heidelberg (2010)
[Yao+82]   Yao, A.C.: Protocols for secure computations. In: 23rd Annual Sym-
            posium on Foundations of Computer Science, Chicago, Illinois, Novem-
            ber 3-5, pp. 160–164. IEEE Computer Society Press (1982)

A   Proof of Security of Theorem 1

Let $\mathcal{A}$ be a malicious, static adversary that interacts with parties running the
protocol Π from Figure 1 in the $\mathcal{F}_{CRS}$-hybrid model. We construct an ideal
world adversary $\mathcal{S}$ with access to the ideal functionality $\mathcal{F}_f$, which simulates a
real execution of Π with $\mathcal{A}$ such that no environment $\mathcal{Z}$ can distinguish the ideal
world experiment with $\mathcal{S}$ and $\mathcal{F}_f$ from a real execution of Π with $\mathcal{A}$.

Recall that $\mathcal{S}$ interacts with the ideal functionality $\mathcal{F}_f$ and with the environment $\mathcal{Z}$. The ideal adversary $\mathcal{S}$ starts by invoking a copy of $\mathcal{A}$ and running a

simulated interaction of $\mathcal{A}$ with the environment $\mathcal{Z}$ and the parties running the protocol. Our simulator $\mathcal{S}$ proceeds as follows:

Simulated CRS: The common reference string is chosen by $\mathcal{S}$ in the following manner (recall that $\mathcal{S}$ chooses the CRS for the simulated $\mathcal{A}$ as we are in the $\mathcal{F}_{CRS}$-hybrid model):

1. $\mathcal{S}$ generates $(\sigma, \tau) \leftarrow S_1(1^\lambda)$, the simulated common reference string for the NIZK proof system (K, P, V) with simulator $S = (S_1, S_2)$.
2. $\mathcal{S}$ runs the setup algorithm $\mathsf{Gen}(1^\lambda)$ of the CCA secure encryption scheme and obtains a public key pk and a secret key sk.

$\mathcal{S}$ sets the common reference string to equal (σ, pk) and locally stores (τ, sk). (The secret key sk will be later used to extract inputs of the corrupted parties and the trapdoor τ for the simulated CRS σ will be used to generate simulated proofs.)

Simulating the Communication with $\mathcal{Z}$: Every input value that $\mathcal{S}$ receives from $\mathcal{Z}$ is written on $\mathcal{A}$'s input tape. Similarly, every output value written by $\mathcal{A}$ on its own output tape is directly copied to the output tape of $\mathcal{S}$.

Simulating Actual Protocol Messages in Π: Note that there might be multiple sessions executing concurrently. Let sid be the session identifier for one specific session. We will specify the simulation strategy corresponding to this specific session. The simulator strategy for all other sessions will be the same. Let $\mathcal{P} = \{P_1, \ldots, P_n\}$ be the set of parties participating in the execution of Π corresponding to the session identified by the session identifier sid. Also let $\mathcal{P}^\mathcal{A} \subseteq \mathcal{P}$ be the set of parties corrupted by the adversary $\mathcal{A}$. (Recall that we are in the setting with static corruption.)

In the subsequent exposition we will assume that at least one party is honest. If no party is honest then the simulator does not need to do anything else.

Round 1 Messages $\mathcal{S} \to \mathcal{A}$: In the first round $\mathcal{S}$ must generate messages on behalf of the honest parties, i.e. parties in the set $\mathcal{P} \backslash \mathcal{P}^\mathcal{A}$. For each party $P_i \in \mathcal{P} \backslash \mathcal{P}^\mathcal{A}$ our simulator proceeds as:

1. $c_i = \mathsf{Enc}(i||0^{\ell_{in}})$ and, (recall that ℓ_{in} is the length of inputs of all parties)
2. $\forall j \in [n]$, and generate $d_{i,j} = \mathsf{Enc}(i||0^\ell)$. (Recall that ℓ is the length of the maximum number of random coins needed by any party in π.)

It then sends $Z_i = \{c_i, \{d_{i,j}\}_{j \in [n]}\}$ to $\mathcal{A}$ on behalf of party P_i.

Round 1 Messages $\mathcal{A} \to \mathcal{S}$: Also in the first round the adversary $\mathcal{A}$ generates the messages on behalf of corrupted parties in $\mathcal{P}^\mathcal{A}$. For each party $P_i \in \mathcal{P}^\mathcal{A}$ our simulator proceeds as:

1. Let $Z_i = \{c_i, \{d_{i,j}\}_{j \in [n]}\}$ be the message that $\mathcal{A}$ sends on behalf of P_i. Our simulator $\mathcal{S}$ decrypts the ciphertexts using the secret key sk. In particular $\mathcal{S}$

sets $x'_i = \mathsf{Dec}(sk, c_i)$ and $r'_{i,j} = \mathsf{Dec}(sk, d_{i,j})$. Obtain $x_i \in \{0,1\}^{\ell_{in}}$ such that $x'_i = i||x_i$. If x'_i is not of this form the set $x_i = \bot$. Similarly obtain $r_{i,j}$ from $r'_{i,j}$ for every j setting the value to $\bot$ in case it is not of the right format.

2. $\mathcal{S}$ sends (input, sid, $\mathcal{P}, P_i, x_i$) to $\mathcal{F}_f$ on behalf of the corrupted party P_i. It saves the values $\{r_{i,j}\}_j$ for later use.

Round 2 Messages $\mathcal{S} \to \mathcal{A}$: In the second round $\mathcal{S}$ must generate messages on behalf of the honest parties, i.e. parties in the set $\mathcal{P}\backslash\mathcal{P}^{\mathcal{A}}$. $\mathcal{S}$ proceeds as follows:

- $\mathcal{S}$ obtains the output (output, sid, $\mathcal{P}, y$) from the ideal functionality $\mathcal{F}_f$ and now it needs to force this output onto the adversary $\mathcal{A}$.

- In order to force the output, the simulator $\mathcal{S}$ executes the simulator $\mathcal{S}_\pi$ and obtains a simulated transcript. The simulated transcript specifies the random coins of all the parties in $\mathcal{P}^{\mathcal{A}}$ and the protocol messages. Let s_i denote the random coins of party $P_i \in \mathcal{P}^{\mathcal{A}}$ and let $m_{i,j}$ for $i \in [n]$ and $j \in [t]$ denote the protocol messages. (Semi-honest security of protocol π implies the existence of such a simulator.)

- For each $P_j \in \mathcal{P}^{\mathcal{A}}$ sample $r_{i,j}$ randomly in $\{0,1\}^\ell$ for each $P_i \in \mathcal{P}\backslash\mathcal{P}^{\mathcal{A}}$ subject to the constraint that $\oplus_{i=1}^n r_{i,j} = s_j$.

- For each $P_i \in \mathcal{P}\backslash\mathcal{P}^{\mathcal{A}}$, $\mathcal{S}$ proceeds as follows:

1. For every $j \in [n]$, $j \neq i$ generate $\gamma_{i,j}$ as a simulated NIZK proof under σ for the NP-statement:
$$\{\exists \rho_{r_{i,j}} \mid d_{i,j} = \mathsf{Enc}(i||r_{i,j}; \rho_{r_{i,j}})\}.$$

2. A sequence of obfuscations $(i\mathcal{O}_{i,1}, \ldots i\mathcal{O}_{i,t})$ where $i\mathcal{O}_{i,j}$ is the obfuscation of the program $\mathsf{Prog}_{i,j}^{1,x_i,\rho_{x_i},r_{i,i},\rho_{r_{i,i}},\{Z_i\},\mathsf{fixedOutput}}$, where fixedOutput is the value $(m_{i,j}, \phi_{i,j})$ such that $\phi_{i,j}$ is the simulated proof that $m_{i,j}$ was generated correctly. (Recall that the flag has been set to 1 and this program on accepting inputs always outputs the value fixedOutput.)

3. It sends $(\{r_{i,j}, \gamma_{i,j}\}_{j\in[n], j\neq i}, \{i\mathcal{O}_{i,j}\}_{j\in[t]})$ to $\mathcal{A}$ on behalf of P_i.

Round 2 Messages $\mathcal{A} \to \mathcal{S}$: Also in the second round the adversary $\mathcal{A}$ generates the messages on behalf of corrupted parties $\mathcal{P}^{\mathcal{A}}$. For each party $P_i \in \mathcal{P}\backslash\mathcal{P}^{\mathcal{A}}$ that has obtained "correctly formed" second round messages from all parties in $\mathcal{P}^{\mathcal{A}}$, our simulator sends (generateOutput, sid, $\mathcal{P}, P_i$) to the ideal functionality.

This completes the description of the simulator.

Next we will prove via a sequence of hybrids that no environment $\mathcal{Z}$ can distinguish the ideal world experiment with $\mathcal{S}$ and $\mathcal{F}_f$ (as defined above) from a real execution of Π with $\mathcal{A}$. We will start with the real world execution in which the adversary $\mathcal{A}$ interacts directly with the honest parties holding their inputs and step-by-step make changes till we finally reach the simulator as described above. At each step will argue that the environment cannot distinguish the change except with negligible probability.

- H_1: This hybrid corresponds to the $\mathcal{Z}$ interacting with the real world adversary $\mathcal{A}$ and honest parties that hold their private inputs.

We can restate the above experiment with the simulator as follows. We replace the real world adversary $\mathcal{A}$ with the ideal world adversary $\mathcal{S}$. The ideal adversary $\mathcal{S}$ starts by invoking a copy of $\mathcal{A}$ and running a simulated interaction of $\mathcal{A}$ with the environment $\mathcal{Z}$ and the honest parties. $\mathcal{S}$ forwards the messages that $\mathcal{A}$ generates for it environment directly to $\mathcal{Z}$ and vice versa (as explained in the description of the simulator $\mathcal{S}$). In this hybrid the simulator $\mathcal{S}$ holds the private inputs of the honest parties and generates messages on their behalf using the honest party strategies as specified by Π.

- H_2: In this hybrid we change how the simulator generates the CRS. In particular we will change how $\mathcal{S}$ generates the public key pk of the CCA secure encryption scheme. We will not change the way CRS for the NIZK is generated.

  $\mathcal{S}$ runs the setup algorithm $\mathsf{Gen}(1^\lambda)$ of the CCA secure encryption scheme and obtains a public key pk and a secret key sk. $\mathcal{S}$ will use this public key pk as part of the CRS and use the secret key sk to decrypt the ciphertexts generated by $\mathcal{A}$ on behalf of $\mathcal{P}^{\mathcal{A}}$. In particular for each party $P_i \in \mathcal{P}^{\mathcal{A}}$ our simulator proceeds as:

  - Let $Z_i = \{c_i, \{d_{i,j}\}_{j \in [n]}\}$ be the message that $\mathcal{A}$ sends on behalf of P_i. Our simulator $\mathcal{S}$ decrypts the ciphertexts using the secret key sk. In particular $\mathcal{S}$ sets $x_i' = \mathsf{Dec}(sk, c_i)$ and $r_{i,j}' = \mathsf{Dec}(sk, d_{i,j})$. Obtain $x_i \in \{0,1\}^{\ell_{in}}$ such that $x_i' = i\|x_i$. If x_i' is not of this form the set $x_i = \perp$. Similarly obtain $r_{i,j}$ from $r_{i,j}'$ for every j setting the value to $\perp$ in case it is not of the right format.

Note that in hybrid H_2 the simulator $\mathcal{S}$ additionally uses the secret key sk to extract the inputs of the adversarial parties. Furthermore if at any point in the execution any of the messages of the adversary are inconsistent with the input and randomness extracted but the adversary succeeds in providing an accepting NIZK proof then the simulator aborts, which event we call `Extract Abort`.

The distribution of the CRS, and hence the view of the environment $\mathcal{Z}$, in the two cases is identical. Also note that it follows from the perfect correctness of the encryption scheme and the statistical soundness of the NIZK proof system that the NIZK proofs adversary generates will have to be consistent with the extracted values. In other words over the random choices of the CRS we have that the probability of `Extract Abort` is negligible.

- H_3: In this hybrid we will change how the simulator generates the obfuscations on behalf of honest parties. Roughly speaking we observe that the obfuscations can only be evaluated to output one unique value (consistent with inputs and randomness extracted using sk) and we can just hardcode this value into the obfuscated circuit. More formally in the second round $\mathcal{S}$ generates the messages on behalf of the honest parties, i.e. parties in the set $\mathcal{P} \backslash \mathcal{P}^{\mathcal{A}}$ as follows:

  1. For every P_j, $\mathcal{S}$ obtains $s_j = \oplus_{i=1}^n r_{i,j}$.
  2. $\mathcal{S}$ virtually executes the protocol π with inputs $x_1, \ldots, x_n$ and random coins $s_1, \ldots, s_n$ for the parties $P_1, \ldots P_n$ respectively, and obtains the messages $m_{i,j}$ for all $i \in [n]$ and $j \in [t]$.
  3. For each $P_i \in \mathcal{P} \backslash \mathcal{P}^{\mathcal{A}}$, $\mathcal{S}$ proceeds as follows:

(a) For every $j \in [n]$, $j \neq i$ generate $\gamma_{i,j}$ as a NIZK proof under σ for the NP-statement:

$$\{\exists \ \rho_{r_{i,j}} \mid d_{i,j} = \mathsf{Enc}(i\|r_{i,j}; \rho_{r_{i,j}})\} \ .$$

(b) A sequence of obfuscations $(i\mathcal{O}_{i,1}, \dots i\mathcal{O}_{i,t})$ where $i\mathcal{O}_{i,j}$ is the obfuscation of the program $\mathsf{Prog}_{i,j}^{1,x_i,\rho_{x_i},r_{i,i},\rho_{r_{i,i}},\{Z_i\},\mathsf{fixedOutput}}$, where fixedOutput is the value $(m_{i,j}, \phi_{i,j})$ such that $\phi_{i,j}$ is the proof that $m_{i,j}$ was generated correctly. (Recall that the flag has been set to 1 and this program on all accepting inputs always outputs the value fixedOutput.)

(c) It sends $(\{r_{i,j}, \gamma_{i,j}\}_{j\in[n],j\neq i}, \{i\mathcal{O}_{i,j}\}_{j\in[t]})$ to $\mathcal{A}$ on behalf of P_i.

We will now argue that hybrids H_2 and H_3 and computationally indistinguishable. More formally we will consider a sequence of $t \cdot |\mathcal{P}\backslash\mathcal{P}^{\mathcal{A}}|$ hybrids $H_{3,0,0}, \dots H_{3,|\mathcal{P}\backslash\mathcal{P}^{\mathcal{A}}|,t}$. In hybrid $H_{3,i,j}$ all the obfusctaions by the first $i-1$ honest parties and the first j obfuscations generated by the ith honest party are generated in the modified way as described above. It is easy to see that hybrid $H_{3,0,0}$ is same as hybrid H_2 and hybrid $H_{3,|\mathcal{P}\backslash\mathcal{P}^{\mathcal{A}}|,t}$ is same as hybrid H_3 itself.

We will now argue that the hybrids $H_{3,i,j-1}$ and $H_{3,i,j}$ for $j \in [t]$ are computationally indistinguishable. This implies the above claim, but in order to argue the above claim we first prove the following lemma.

Lemma 1.

$$\Pr \left[\begin{array}{l} \exists \ a,b \ : \\ \mathsf{Prog}_{i,j}^{0,x_i,\rho_{x_i},r_{i,i},\rho_{r_{i,i}},\{Z_i\},0^{\ell_{i,j}}} (a) \neq \mathsf{Prog}_{i,j}^{0,x_i,\rho_{x_i},r_{i,i},\rho_{r_{i,i}},\{Z_i\},0^{\ell_{i,j}}} (b) \\ \wedge \ \mathsf{Prog}_{i,j}^{0,x_i,\rho_{x_i},r_{i,i},\rho_{r_{i,i}},\{Z_i\},0^{\ell_{i,j}}} (a) \neq \bot \\ \wedge \ \mathsf{Prog}_{i,j}^{0,x_i,\rho_{x_i},r_{i,i},\rho_{r_{i,i}},\{Z_i\},0^{\ell_{i,j}}} (b) \neq \bot \end{array} \right] = \mathsf{negl}(\lambda)$$

where the probability is taken over the random choices of the generation of the CRS.

Proof. Recall that program $\mathsf{Prog}_{i,j}^{0,x_i,\rho_{x_i},r_{i,i},\rho_{r_{i,i}},\{Z_i\},0^{\ell_{i,j}}}$ represents the jth message function of the ith party in protocol π. Recall that the input to the program consists of two $(R, \Gamma, M_{j-1}, \Phi_{j-1})$. We will refer to the (R, M_{j-1}) as the *main input part* and the Γ, Φ_{j-1} as the *proof part*.

Observe that since the proofs are always consistent with the extracted inputs and randomness, we have that there is a unique main input part for which adversary can provide valid (or accepting) proof parts. Further note that if the proof part is not accepting then $\mathsf{Prog}_{i,j}$ just outputs $\bot$. In other words if the proof is accepting then the program outputs a fixed value that depends just on the values that are fixed based on $\{Z_i\}$ values. We stress that the output actually does include a NIZK proof as well, however it is not difficult to see that this NIZK proof is also unique as a fixed randomness is used in generation of the proof. $\blacksquare$

Armed with Lemma 1, we can conclude that the programs $\mathsf{Prog}_{i,j}^{0,x_i,\rho_{x_i},r_{i,i},\rho_{r_{i,i}},\{Z_i\},0^{\ell_{i,j}}}$ and $\mathsf{Prog}_{i,j}^{1,x_i,\rho_{x_i},r_{i,i},\rho_{r_{i,i}},\{Z_i\},\mathsf{fixedOutput}}$ are functionally equivalent. Next based on the indistinguisbaility obfuscation property, it is easy to see that the hybrids $H_{3,i,j-1}$ and $H_{3,i,j}$ are computationally indistinguishable.

- H_4: In this hybrid we change how the simulator generates the NIZKs on behalf of honest parties. Formally $\mathcal{S}$ generates the σ using the simulator S_1 of the NIZK proof system and generates all the proofs using the simulator S_2. The argument can be made formal by considering a sequence of hybrids and changing each of the NIZK proofs one at a time.

The indistinguishability between hybrids H_3 and H_4 can be based on the zero-knowledge property of the NIZK proof system.

- H_5: In this hybrid we change how the simulator $\mathcal{S}$ generates the first round messages on behalf of honest parties. In particular $\mathcal{S}$ instead of encrypting inputs and randomness of honest parties just encrypts zero strings of appropriate length.

We could try to base the indistinguishabilty between hybrids H_4 and H_5 on the semantic security of the PKE scheme. However observe that $\mathcal{S}$ at the same time should continue to be able to decrypt the ciphertexts that $\mathcal{A}$ generates on behalf of corrupted parties. Therefore we need to rely on the CCA security of the PKE scheme.

- H_6: In this hybrid instead of generating all the messages $m_{i,j}$ on behalf of honest parties honestly $\mathcal{S}$ uses $\mathcal{S}_\pi$ (the simulator for the underlying MPC protocol) to generated simulated messages.

The indistinguishability between hybrids H_5 and H_6 directly follows for the indistinguishability of honestly generated transcript in the execution of π from the transcript generated by $\mathcal{S}_\pi$.

- H_7: Observe that in hybrid H_6, $\mathcal{S}$ uses inputs of honest parties just in obtaining the output of the computation. It can obtain the same value by sending extracted inputs of the malicious parties to the ideal functionality $\mathcal{F}_f$.

Note that the hybrids H_6 and H_7 are identical. Observe that hybrid H_7 is identical to our simulator, which concludes the proof.

Chosen Ciphertext Security via Point Obfuscation

Takahiro Matsuda and Goichiro Hanaoka

Research Institute for Secure Systems (RISEC),
National Institute of Advanced Industrial Science and Technology (AIST), Japan
{t-matsuda,hanaoka-goichiro}@aist.go.jp

Abstract. In this paper, we show two new constructions of chosen ciphertext secure (CCA secure) public key encryption (PKE) from general assumptions. The key ingredient in our constructions is an obfuscator for point functions with multi-bit output (MBPF obfuscators, for short), that satisfies some (average-case) indistinguishability-based security, which we call AIND security, in the presence of hard-to-invert auxiliary input. Specifically, our first construction is based on a chosen plaintext secure PKE scheme and an MBPF obfuscator satisfying the AIND security in the presence of computationally hard-to-invert auxiliary input. Our second construction is based on a lossy encryption scheme and an MBPF obfuscator satisfying the AIND security in the presence of statistically hard-to-invert auxiliary input. To clarify the relative strength of AIND security, we show the relations among security notions for MBPF obfuscators, and show that AIND security with computationally (resp. statistically) hard-to-invert auxiliary input is implied by the average-case virtual black-box (resp. virtual grey-box) property with the same type of auxiliary input. Finally, we show that a lossy encryption scheme can be constructed from an obfuscator for point functions (point obfuscator) that satisfies re-randomizability and a weak form of composability in the worst-case virtual grey-box sense. This result, combined with our second generic construction and several previous results on point obfuscators and MBPF obfuscators, yields a CCA secure PKE scheme that is constructed *solely* from a re-randomizable and composable point obfuscator. We believe that our results make an interesting bridge that connects CCA secure PKE and program obfuscators, two seemingly isolated but important cryptographic primitives in the area of cryptography.

Keywords: public key encryption, lossy encryption, key encapsulation mechanism, chosen ciphertext security, point obfuscation.

1 Introduction

1.1 Background and Motivation

One of the fundamental research themes in cryptography is to clarify what the minimal assumptions to realize various kinds of cryptographic primitives are, and up to now, a number of relationships among primitives have been investigated and established. Clarifying these relationships gives us a lot of insights for how to construct and/or prove the security of cryptographic primitives, enables us to understand the considered primitives more deeply, and leads to systematizing the research area in cryptography.

Y. Lindell (Ed.): TCC 2014, LNCS 8349, pp. 95–120, 2014.

In this paper, we focus on the constructions of public key encryption (PKE) schemes secure against chosen ciphertext attacks (CCA) [54,29] from general cryptographic assumptions. CCA secure PKE is one of the most important cryptographic primitives that has been intensively studied, due to its resilience against practical attacks such as [10], and its implication to many useful security notions, such as non-malleability [29] and universal composability [18].

The first successful result regarding this line of research is the construction by Dolev, Dwork, and Naor [29] that uses a chosen plaintext secure (CPA secure) PKE scheme and a non-interactive zero-knowledge proof. Since these two primitives can be constructed from (an enhanced variant of) trapdoor permutations (TDP) [35], CCA secure PKE can be constructed solely from TDPs. Canetti, Halevi, and Katz [20] showed that CCA secure PKE can be constructed from an identity-based encryption (IBE). It was later shown that in fact, a weaker primitive called tag-based encryption suffices [45]. Peikert and Waters [53] showed that CCA secure PKE can be constructed from any lossy trapdoor function (TDF), and subsequent works showed that injective TDFs with weaker properties suffice: injective TDFs secure for correlated inputs [55], slightly lossy TDFs [49], adaptive one-way TDFs [46], and adaptive one-way relations [59]. (CPA secure) PKE schemes with additional security/functional properties have also turned out to be useful for constructing CCA secure PKE: Hemenway and Ostrovsky [40] showed that we can construct CCA secure PKE in several ways from homomorphic encryption with appropriate properties. The same authors [41] also showed that CCA secure PKE can be constructed from a lossy encryption scheme [6] if the plaintext space is larger than the randomness space (the results of [40,41] achieve CCA secure PKE via lossy TDFs [53]). Hohenberger, Lewko, and Waters [42] showed that if one has a PKE scheme which satisfies the notion called detectable CCA security, which is somewhere between CCA1 and CCA2 security, then using it one can construct a CCA secure PKE scheme. Myers and Shelat [50] showed how to construct a CCA secure PKE scheme that can encrypt plaintexts with arbitrary length from a CCA secure one with 1-bit plaintext space. Lin and Tessaro [47] showed how to amplify weak CCA security into ordinary one. Very recently, Dachman-Soled [25] constructs CCA secure PKE from PKE satisfying (standard model) plaintext-awareness together with some additional property.

The main purpose of this work is to show that a different kind of cryptographic primitives is also useful for achieving CCA secure PKE. Specifically, we add new recipes for the construction of CCA secure PKE, based on the techniques and results from program obfuscation [3] for the very simple classes of functions, point functions and point functions with multi-bit output. Despite the tremendous efforts, it is not known whether it is possible to construct CCA secure PKE only from CPA secure one (in fact, a partial negative result is known [33]). Clarifying new classes of primitives that serve as building blocks is important for tackling this problem. In particular, it has been shown that there is no black-box construction of IBE and a TDF from (CCA secure) PKE [11,34] and thus to tackle the CPA-to-CCA problem, the attempts to construct IBE or the above TDF-related primitives from a CPA secure PKE scheme seem hopeless (though there is a possibility that some non-black-box construction exists). Our new constructions based on (multi-bit) point obfuscators do not seem to be covered by these negative results, and thus potentially it could serve as a new target for building CCA secure PKE.

1.2   Our Contribution

In this paper, we show two new constructions of CCA secure PKE schemes from general cryptographic assumptions, using the techniques and results from program obfuscation [3]. We actually construct CCA secure key encapsulation mechanisms (KEMs) [24], where a KEM is a "PKE"-part of hybrid encryption that encrypts a random "session-key" for symmetric key encryption (SKE). By combining a CCA secure KEM with a CCA secure SKE scheme, one obtains a full-fledged CCA secure PKE scheme [24]. The key ingredient in our constructions is an obfuscator for point functions with multi-bit output (MBPF obfuscators) [48,19,27,37,21,7], that satisfies a kind of average-case indistinguishability-based security in the presence of "hard-to-invert" auxiliary inputs. The formal definition of this security notion is given in Section 3. For brevity, we call it AIND security.

Our first construction in Section 4.1 is based on a CPA secure PKE scheme and an MBPF obfuscator satisfying the above mentioned AIND security in the presence of computationally hard-to-invert auxiliary input. Our second construction in Section 4.2 is based on a lossy encryption scheme [6] and an MBPO satisfying the above mentioned AIND security in the presence of statistically hard-to-invert auxiliary input. Interestingly, the first and the second constructions are in fact exactly the same, and we show two different security analyses from different assumptions on building blocks. These two constructions add new recipes into the current picture of the constructions of CCA secure PKE schemes/KEMs from general cryptographic assumptions.

In order to clarify where these AIND security definitions for MBPF obfuscators are placed, in Section 5 we show that AIND security with computationally (resp. statistically) hard-to-invert auxiliary inputs is implied by the (average-case) virtual black-box property [3] (resp. virtual grey-box property [7]) in the presence of the same auxiliary inputs. Besides these, we show the relations among several related worst-/average-case virtual black-/grey-box properties under several types of auxiliary inputs, and summarize them in Fig. 2, which we believe is useful for further research on this topic and might be of independent interest.

Finally, in Section 6, we show that a lossy encryption scheme can be constructed from an obfuscator for point functions (point obfuscator) that satisfies re-randomizability [7] and a weak form of composability [48,19,7] in the worst-case virtual grey-box sense. This result, combined with our second generic construction and the results on composable point obfuscators with the virtual grey-box property in [7], shows that a CCA secure PKE scheme can be constructed *solely* from a point obfuscator which is re-randomizable and composable.

We believe that our results make an interesting bridge that connects CCA secure PKE and program obfuscators,[1] two seemingly isolated but important primitives in the area of cryptography, and hope that our results motivate further studies on them.

---

[1] Recently, Sahai and Waters [57] (among others) showed how to construct CCA secure PKE using *indistinguishability obfuscation*. We explain the difference with our results in Section 1.4.

1.3 Overview of Techniques

Our proposed constructions of KEMs are based on the "witness-recovering" technique [53,55,50,42] in which a part of randomness used to generate a ciphertext is somehow embedded into the ciphertext itself, and is later recovered in the decryption process for checking the validity of the ciphertext by re-encryption. What we believe is novel in our constructions is how to implement this mechanism of witness-recovering by using an MBPF obfuscator with an appropriate security property.

Let $\mathcal{I}_{\alpha \to \beta}$ denote an MBPF such that $\mathcal{I}_{\alpha \to \beta}(x) = \beta$ if $x = \alpha$ and $\bot$ otherwise, and let MBPO denotes an MBPF obfuscator which takes an MBPF $\mathcal{I}_{\alpha \to \beta}$ as input, and outputs an obfuscated circuit DL for $\mathcal{I}_{\alpha \to \beta}$. ("DL" stands for "digital locker," the name due to [19].) Let $\Pi = (\mathsf{PKG}, \mathsf{Enc}, \mathsf{Dec})$ be a PKE scheme, where PKG, Enc, and Dec are the key generation, the encryption, and the decryption algorithms of Π, respectively.

Below we give a high level idea behind our main proposed constructions in Section 4 by explaining how the "toy" version of our constructions $\Pi' = (\mathsf{PKG}', \mathsf{Enc}', \mathsf{Dec}')$, constructed using Π and MBPO, is proved CCA1 secure based on the assumptions that Π is CPA secure and that MBPO satisfies the virtual black-box property with respect to dependent auxiliary input [36]. (As mentioned earlier, in this paper we actually construct KEMs rather than PKE schemes, but the intuition for our results are captured by the explanation here.) A public/secret key pair (PK, SK) of Π' is of the form $PK = (pk_1, pk_2)$, $SK = (sk_1, sk_2)$, where each (pk_i, sk_i) is an independently generated key pair by running PKG. To encrypt a plaintext m under PK, Enc' first picks a random string $\alpha \in \{0, 1\}^k$ (where k is the security parameter) and two randomness r_1 and r_2 for Enc, and computes a ciphertext C in the following way:

$$C = (c_1, c_2, \mathsf{DL}) = \Big(\mathsf{Enc}(pk_1, (m\|\alpha); r_1), \mathsf{Enc}(pk_2, (m\|\alpha); r_2), \mathsf{MBPO}(\mathcal{I}_{\alpha \to (r_1\|r_2)}) \Big)$$

where "$\|$" denotes the concatenation of strings, and "$\mathsf{Enc}(pk, m; r)$" means to encrypt the plaintext m under the public key pk using the randomness r. To decrypt C, we first decrypt c_1 by using sk_1 to obtain $(m\|\alpha)$, then run $\mathsf{DL}(\alpha)$ to recover $(r_1\|r_2)$. Finally, m is returned if $c_i = \mathsf{Enc}(pk_i, (m\|\alpha); r_i)$ holds for both $i = 1, 2$, and otherwise we reject C. Here, it should be noted that due to the symmetric roles of pk_1 and pk_2 and the validity check by re-encryption performed in Dec', we can also decrypt C using sk_2, so that the decryption result of C using sk_1 and that using sk_2 always agree.

Now, recall the interface of a CCA1 adversary $\mathcal{A} = (\mathcal{A}_1, \mathcal{A}_2)$, where $\mathcal{A}_1$ and $\mathcal{A}_2$ represent an adversary's algorithm before and after the challenge, respectively. $\mathcal{A}_1$ is firstly given a public key PK, and can start using the decryption oracle $\mathsf{Dec}'(SK, \cdot)$. After that, $\mathcal{A}_1$ terminates with output two plaintexts (m_0, m_1) and some state information st that is passed to $\mathcal{A}_2$. $\mathcal{A}_2$ is given st and the challenge ciphertext $C^* = (c_1^*, c_2^*, \mathsf{DL}^*)$ which is an encryption of m_b (where b is the challenge bit), and outputs a bit as its guess for b.

The key observation is that $\mathcal{A}_2$ can be seen as an adversary for the MBPF obfuscator MBPO, by regarding $(\mathsf{st}, c_1^*, c_2^*)$ as an auxiliary input z about the obfuscated circuit DL^* of the MBPF $\mathcal{I}_{\alpha^* \to (r_1^*\|r_2^*)}$. Then, if MBPO satisfies the virtual black-box property with respect to dependent auxiliary input [36], there exists a simulator $\mathcal{S}$ that takes only $z = (\mathsf{st}, c_1^*, c_2^*)$ as input, has oracle access to $\mathcal{I}_{\alpha^* \to (r_1^*\|r_2^*)}$, and has the property that $\mathcal{A}$'s success probability (in guessing b) is negligibly close to the probability that $\mathcal{S}$ succeeds in guessing b. (For convenience, let us call the latter probability "$\mathcal{S}$'s success probability," although $\mathcal{S}$ is not a CCA1 adversary and thus its task is not to guess a challenge bit.) This means that if $\mathcal{S}$'s success probability is close to $1/2$, then so is $\mathcal{A}$'s success probability, which will prove the CCA1 security of Π'.

To show that $\mathcal{S}$'s success probability is close to $1/2$, we consider the hypothetical experiment for $\mathcal{S}$ in which the auxiliary input z is generated so that decryption queries from $\mathcal{A}_1$ are answered using sk_2, and both c_1^* and c_2^* are an encryption of a fixed value (say, $0^{|m_0|+k}$). Since z contains no information on b and α^*, in this hypothetical experiment $\mathcal{S}$'s success probability is exactly $1/2$ and the probability that $\mathcal{S}$ makes the query α^* (which is chosen randomly) is negligible. Next, we make the experiment closer to the actual $\mathcal{S}$'s experiment, by changing c_1^* into an encryption of $(m_b\|\alpha^*)$. By the CPA security regarding pk_1, $\mathcal{S}$'s success probability as well as the probability of $\mathcal{S}$ making the query α^* is negligibly close to those in the hypothetical experiment. Then, we further modify the previous experiment by changing c_2^* into an encryption of $(m_b\|\alpha^*)$, but this time we use sk_1 for answering $\mathcal{A}_1$'s queries. Notice that this is exactly the actual experiment for $\mathcal{S}$. As mentioned above, switching sk_2 to/from sk_1 for answering $\mathcal{A}_1$'s queries does not affect $\mathcal{A}_1$'s behavior, and thus again by the CPA security regarding pk_2, $\mathcal{S}$'s success probability is negligibly close to $1/2$ and the probability that $\mathcal{S}$ makes the query α^* is negligible. Then, by the virtual black-box property of MBPO with auxiliary input, $\mathcal{A}$'s original success probability is negligibly close to $1/2$, meaning that $\mathcal{A}$ has negligible advantage in breaking the CCA1 security of the scheme Π'.

The above completes a proof sketch of how Π' is proved CCA1 secure. By encrypting a random K, Π' can be used as a CCA1 secure KEM. Our proposed CCA2 secure KEMs are obtained by applying several optimizations and enhancement to this KEM:

- Firstly, we do not need the full virtual black-box property with auxiliary input of [36]. As mentioned earlier, an indistinguishability-based definition in the presence of only "hard-to-invert" auxiliary input is sufficient for a similar argument to work.

- Secondly, we need not include a plaintext into each of c_i. Instead, we pick a randomness $K \in \{0,1\}^k$ used as a plaintext of a KEM, and include this K into the output of the MBPF, i.e now we obfuscate the MBPF $\mathcal{I}_{\alpha \to (r_1\|r_2\|K)}$. (This is the actual our basic construction whose formal description and security proof are given in the full version.)

- Lastly, note that the above construction cannot be proved to be CCA2 secure as it is. In particular, the obfuscated circuit DL could be malleable. To deal with this issue, instead of the Naor-Yung-style double encryption [52], we employ the Dolev-Dwork-Naor-style multiple encryption [29] together with the technique of the "unduplicatable set selection" [56]. Unlike the classical method of using a one-time signature scheme, we implement the technique using a universal one-way hash function (UOWHF) [51], where a hash value of the obfuscated circuit DL is used as a "selector" of the public key components. Another issue is that the second stage adversary $\mathcal{A}_2$ in the CCA2 experiment can also make decryption queries, and thus the above explained idea of replacing $\mathcal{A}_2$ with a simulator $\mathcal{S}$ does not work. However, our indistinguishability-based security definition for MBPF obfuscators enables us to directly work with an original CCA2 adversary, and we can avoid considering how a simulator deal with the queries from $\mathcal{A}_2$. For more details, see Section 4.

1.4 Related Work: Program Obfuscation

Roughly speaking, an obfuscator is an algorithm that takes a program (e.g. Turing machine or circuit) as input, and outputs another program with the same functionality, but otherwise "unintelligible."

After the impossibility of general-purpose program obfuscation satisfying the nowadays standard security notion called *virtual black-box* property shown in the seminal work by Barak et. al. [3], several subsequent works extended the impossibility in various other settings [36,58,38,7]. The other line of research pursues possibilities of obfuscating a specific class of functions. Before 2013, most known positive results were about obfuscation for point functions and their variants, e.g. [48,58,19,22,7]. Relaxing the security requirements to "average-case" in which a program is sampled according to some distribution, several more complex tasks have been shown to be obfuscatable, such as proximity testing [28] and cryptographic tasks such as re-encryption [43].

Since the first candidates of a cryptographic multilinear map have been proposed in 2013 [30,23], the research field of (cryptographic) obfuscation has drastically changed and accelerated. Brakerski and Rothblum [14] showed how to construct an obfuscator for conjunctions from graded encoding schemes [30,23], and the same authors showed a further extension [13]. Most recently, they showed a general-purpose obfuscator satisfying a virtual black-box property in an idealized model called the generic graded encoded scheme model [15]. Barak et al. [2] studied obfuscation for a class of functions called *evasive functions* which in particular includes point functions (with multi-bit output). A series of works [32,57,44,31] (and many other recent works) have shown that a general-purpose obfuscator satisfying a security notion weaker than the virtual black-box property, called *indistinguishability obfuscator*, which seems to be too weak to be useful, is in fact surprisingly powerful and can be used as a building block for constructing a various kinds of cryptographic primitives. Garg et al. [32] constructed the first candidate of general-purpose indistinguishability obfuscation. A security notion stronger than indistinguishability obfuscation, called *differing-inputs obfuscation* (a.k.a. *extractability obfuscation* [12]), has also been shown to be quite powerful and useful [1,12].

Among a number of recent fascinating results, especially relevant to our work is the work by Sahai and Waters [57] who showed (among several other primitives) how to construct CCA secure PKE from an indistinguishability obfuscator (and a one-way function). Although our work and [57] have the common property that both works build CCA secure PKE using techniques and results from obfuscation, our use of obfuscators and that of [57] are quite different: We use an obfuscator for a specific class of functions, point functions and MBPFs, while [57] uses an obfuscator for all polynomial-sized circuits. Furthermore, the indistinguishability-based security notion for MBPF obfuscators used in our main result is about randomly chosen MBPFs, while that used in [57] is for the worst-case choice of circuits (that compute the same functions). We would also like to stress that our work and [57] were done concurrently and independently.

1.5   Paper Organization

The rest of the paper is organized as follows: In Section 2 (and Appendix A) we review the basic notations and definitions of primitives. In Section 3, we introduce the formal definitions of our new indistinguishability-based security notions for MBPF obfuscators. In Section 4, we show our main results: two CCA secure KEMs using a MBPF obfuscator. In Section 5, we investigate relations between our new security notions and other notions for MBPF obfuscators. In Section 6, we show how to construct a lossy encryption scheme from a point obfuscator with re-randomizability and composability. In Section 7, we discuss some issues on the MBPF obfuscators that we use.

2   Preliminaries

Here, we review the basic notation and the definitions for lossy encryption [6] and (cryptographic) obfuscation. The definitions for standard cryptographic primitives that are not given here are given in Appendix A, which include PKE, KEMs, and UOWHFs.

Basic Notation. $\mathbb{N}$ denotes the set of all natural numbers, and if $n \in \mathbb{N}$ then $[n] = \{1, \ldots, n\}$. "$x \leftarrow y$" denotes that x is chosen uniformly at random from y if y is a finite set, x is output from y if y is a function or an algorithm, or y is assigned to x otherwise. If x and y are strings, then "$|x|$" denotes the bit-length of x, and "$x\|y$" denotes the concatenation x and y. "$x \stackrel{?}{=} y$" is the operation that returns 1 if and only if $x = y$. "PPTA" stands for a *probabilistic polynomial time algorithm*. If $\mathcal{A}$ is a probabilistic algorithm then $y \leftarrow \mathcal{A}(x; r)$ denotes that $\mathcal{A}$ computes y as output by taking x as input and using r as randomness. $\mathcal{A}^{\mathcal{O}}$ denotes an algorithm $\mathcal{A}$ with oracle access to $\mathcal{O}$. A function $\epsilon(k) : \mathbb{N} \to [0, 1]$ is said to be *negligible* if for all positive polynomials $p(k)$ and all sufficiently large $k \in \mathbb{N}$, we have $\epsilon(k) < 1/p(k)$. Throughout this paper, we use the character "k" to denote a security parameter.

2.1   Lossy Encryption

Definition 1. *A tuple of PPTAs $\Pi = (\mathsf{PKG}, \mathsf{Enc}, \mathsf{Dec}, \mathsf{LKG})$ is said to be an ϵ-lossy encryption scheme[2] if the following properties are satisfied:*

- *(Syntax)* $(\mathsf{PKG}, \mathsf{Enc}, \mathsf{Dec})$ *constitutes a PKE scheme. The algorithm LKG is called a lossy key generation algorithm, which takes 1^k as input, and outputs a "lossy" public key pk.*
- *(Indistinguishability of ordinary/lossy keys) For all PPTAs $\mathcal{A}$, $\mathsf{Adv}_{\Pi,\mathcal{A}}^{\mathsf{KEY}}(k) := 2 \cdot |\Pr[\mathsf{Expt}_{\Pi,\mathcal{A}}^{\mathsf{KEY}}(k) = 1] - 1/2|$ is negligible, where the experiment $\mathsf{Expt}_{\Pi,\mathcal{A}}^{\mathsf{KEY}}(k)$ is defined as follows:*

$$[(pk_0, sk) \leftarrow \mathsf{PKG}(1^k); \ pk_1 \leftarrow \mathsf{LKG}(1^k); \ b \leftarrow \{0, 1\}; \ b' \leftarrow \mathcal{A}(pk_b);$$
$$\text{Return } (b' \stackrel{?}{=} b)].$$

---

[2] In this paper, we consider the "exact security"-style definition for lossy encryption and CPA secure PKE. This is to quantify the "hardness" of inverting an auxiliary input functions used in the security definitions of MBPF obfuscators. For details, see Section 3.

- (**Statistical lossiness**) *For all computationally unbounded algorithms $\mathcal{A}$ and for all sufficiently large $k \in \mathbb{N}$ it holds that $\text{Adv}_{\Pi,\mathcal{A}}^{\text{LOS-CPA}}(k) := 2 \cdot |\Pr[\text{Expt}_{\Pi,\mathcal{A}}^{\text{LOS-CPA}}(k) = 1] - 1/2| \leq \epsilon(k)$, where the experiment $\text{Expt}_{\Pi,\mathcal{A}}^{\text{LOS-CPA}}(k)$ is defined in the same way as the ordinary CPA experiment $\text{Expt}_{\Pi,\mathcal{A}}^{\text{CPA}}(k)$ except that the public key pk is generated as $pk \leftarrow \text{LKG}(1^k)$. We call ϵ lossiness.*

2.2 Obfuscation for Circuits and Worst-Case Security Definitions

Here, we recall the definition of circuit obfuscations, following the definitions given in [3,48,36,8]. In the following, by $\mathcal{C}$ we denote an ensemble $\{\mathcal{C}_k\}_{k \in \mathbb{N}}$, where $\mathcal{C}_k$ is a collection of circuits whose input length is k and whose size is bounded by some polynomial of k.

Definition 2. *We say that a PPTA Obf is an obfuscator for $\mathcal{C}$ if it satisfies the following:*

- (**Functionality**) *For every $k \in \mathbb{N}$ and every $C \in \mathcal{C}_k$, a circuit output from $\text{Obf}(C)$ computes the same function as C.*
- (**Polynomial blowup**) *There exists a polynomial $p = p(k) > 0$ such that for every $k \in \mathbb{N}$ and every $C \in \mathcal{C}_k$, the size of a circuit output from $\text{Obf}(C)$ is bounded by p.*

Note that Definition 2 is only about the functionality requirements of obfuscators.

Next, we recall the security definitions for "worst-case" choice of circuits.: The *virtual black-box property* is due to Barak et al. [3], the *virtual black-box property with (dependent) auxiliary input* is due to Goldwasser and Kalai [36], and *virtual "grey"-box (with (dependent) auxiliary input)* is due to Bitansky and Canetti [7].

Definition 3. *We say that an obfuscator Obf for $\mathcal{C}$ satisfies:*

- *the worst-case virtual black-box property (WVB security, for short), if for every PPTA $\mathcal{A}$ (adversary) and every positive polynomial $q = q(k)$, there exists a PPTA $\mathcal{S}$ (simulator) such that for all sufficiently large $k \in \mathbb{N}$ and all circuits $C \in \mathcal{C}_k$, it holds that*

$$|\Pr[\mathcal{A}(1^k, \text{Obf}(C)) = 1] - \Pr[\mathcal{S}^C(1^k) = 1]| \leq 1/q,$$

- *the worst-case virtual black-box property w.r.t. auxiliary input (WVB–AI security, for short), if for every PPTA $\mathcal{A}$ and every positive polynomials $q = q(k)$ and $\ell = \ell(k)$, there exists a PPTA $\mathcal{S}$ such that all sufficiently large $k \in \mathbb{N}$, all circuits $C \in \mathcal{C}_k$, and all strings $z \in \{0,1\}^\ell$, it holds that*

$$|\Pr[\mathcal{A}(1^k, z, \text{Obf}(C)) = 1] - \Pr[\mathcal{S}^C(1^k, z) = 1]| \leq 1/q,$$

where the probabilities are over the randomness consumed by Obf, $\mathcal{A}$, and $\mathcal{S}$.

Furthermore, we define the worst-case virtual grey-box property (WVG security), and the worst-case virtual grey-box property w.r.t. auxiliary input (WVG-AI security) of Obf, in the same way as the definitions for the corresponding virtual black-box properties, except that we replace "a PPTA $\mathcal{S}$" in each definition with "a computationally unbounded algorithm $\mathcal{S}$ that makes only polynomially many queries."

Note that in the above definitions, the simulator $\mathcal{S}$ can depend on the polynomial q which represents the "quality of simulation." Wee [58] refers to the simulators of this type as a "weak simulator."

We also define $(t\text{-})$composability of obfuscations [48,19,7,21]. Following [8], we only define the composability in the grey-box (WVG) notion, using a computationally unbounded simulator, which is sufficient for our purpose in this paper.

Definition 4. *([7]) Let $t = t(k) > 0$ be a polynomial. We say that an obfuscator* Obf *for $\mathcal{C}$ satisfies t-composability, if for every PPTA $\mathcal{A}$ and a positive polynomial $q = q(k)$, there exists a computationally unbounded algorithm $\mathcal{S}$ that makes only polynomially many queries, such that for all sufficiently large $k \in \mathbb{N}$ and for all circuits $C_1, \ldots, C_t \in \mathcal{C}_k$, it holds that:*

$$| \Pr[\mathcal{A}(1^k, \mathsf{Obf}(C_1), \ldots, \mathsf{Obf}(C_t)) = 1] - \Pr[\mathcal{S}^{C_1, \ldots, C_t}(1^k) = 1]| \leq 1/q,$$

where the probabilities are over the randomness consumed by Obf, $\mathcal{A}$, *and* $\mathcal{S}$.

Notations for Point Obfuscators and MBPF Obfuscators. Let $\mathcal{X}$ be a finite set, $t \in \mathbb{N}$, $\alpha \in \mathcal{X}$, and $\beta \in \{0,1\}^t$. A *point function* $\mathcal{I}_\alpha$ and a *multi-bit point function* (MBPF) $\mathcal{I}_{\alpha \to \beta}$ are functions defined as follows:

$$\mathcal{I}_\alpha(x) = \begin{cases} \top & \text{if } x = \alpha \\ \bot & \text{otherwise} \end{cases} \quad \text{and} \quad \mathcal{I}_{\alpha \to \beta}(x) = \begin{cases} \beta & \text{if } x = \alpha \\ \bot & \text{otherwise} \end{cases}$$

We refer to α and β as the *point address* and the *point value*, respectively.

In this paper, we will only consider circuits for computing point functions/MBPFs with the properties that (1) the description is given in some canonical form and thus there is a one-to-one correspondence between a point address/value and the circuit for computing the point function/MBPF, and (2) the description of the circuits reveals the point address/value in the clear. Hereafter, we will identify a point function and an MBPF with circuits that compute them (with the above mentioned properties).

For an ensemble $\mathcal{X} = \{\mathcal{X}_k\}_{k \in \mathbb{N}}$, where each $\mathcal{X}_k$ is a set, we denote by $\mathsf{PF}(\mathcal{X})$ the ensemble of point functions $\{\mathcal{I}_\alpha\}_{\alpha \in \mathcal{X}_k}$. Similarly, for $\mathcal{X}$ and a polynomial t, we denote by $\mathsf{MBPF}(\mathcal{X}, t)$ the ensemble MBPFs $\{\mathcal{I}_{\alpha \to \beta}\}_{\alpha \in \mathcal{X}_k, \beta \in \{0,1\}^t}$.

Hereafter, we refer to an obfuscator for point functions as a *point obfuscator* and will denote it by PO. Furthermore, we refer to an obfuscator for MBPFs as an *MBPF obfuscator* and will denote it by MBPO. Moreover, we call an ensemble $\mathcal{X} = \{\mathcal{X}_k\}_{k \in \mathbb{N}}$ a *"domain ensemble" (for point functions and MBPFs)* if (1) for all $k \in \mathbb{N}$, each element of $\mathcal{X}_k$ is k-bit, (2) $|\mathcal{X}_k|$ is superpolynomially large in k (and thus $1/|\mathcal{X}_k|$ is negligible), and (3) we can efficiently sample an element from $\mathcal{X}_k$ uniformly at random.

Concrete Instantiations of a Composable Point Obfuscator and an MBPF Obfuscator. In Appendix B, we recall the concrete construction of a point obfuscator due to the results [17,7], which is originally proposed by Canetti [17] as a perfectly one-way function and is later shown to be t-composable under the t-strong vector decision Diffie-Hellman (t-SVDDH) assumption [7], which is a stronger variant of the decisional Diffie-Hellman (DDH) assumption. There, we also recall the construction of an MBPF obfuscator based on a composable point obfuscator [19,7].

3 New Security Definitions for MBPF Obfuscators

In this section, we introduce and formalize the new security notions for MBPF obfuscators that we call *average-case indistinguishability w.r.t. (computationally/statistically) partially uninvertible auxiliary input*, which will play a central role in our proposed KEMs given in Section 4. This security definition requires that obfuscated circuits of MBPFs hide the point values on average, even in the presence of "dependent" auxiliary inputs [36,27], as long as the auxiliary input has some "hard-to-invert" property.

In the following, we formally define what we mean by "hard-to-invert" auxiliary input in Section 3.1. Then, in Section 3.2, we define the new indistinguishability-based notions.

For notational convenience, in this section, $\mathcal{X}$ will always denote a domain ensemble $\{\mathcal{X}_k\}_{k \in \mathbb{N}}$, and $t = t(k) > 0$ be a polynomial that will be used for MBPF obfuscators for MBPF$(\mathcal{X}, t)$, and do not introduce them in each definition.

3.1 Auxiliary Input Functions and Partial Uninvertibility

For MBPF obfuscators, we will consider the average-case security in the presence of "dependent" auxiliary input [36] that depends on the description of an MBPF $\mathcal{I}_{\alpha \to \beta}$ being obfuscated. We will capture this by a probabilistic function ai that takes as input the point address/value pair $(\alpha, \beta) \in \mathcal{X}_k \times \{0, 1\}^t$. Furthermore, we consider the (average-case) "partial uninvertibility" of the function ai. That is, given z output by ai(α, β) for a randomly chosen (α, β), it is hard to find α. We consider computational and statistical partial uninvertibility.

Definition 5. *Let* $\delta : \mathbb{N} \to [0, 1]$, *and let* ai $: \mathcal{X}_k \times \{0, 1\}^t \to \{0, 1\}^*$ *be a (possibly probabilistic) two-input function. We say that* ai *is a δ-computationally (resp. δ-statistically) partially uninvertible auxiliary input function (δ-cPUAI (resp. δ-sPUAI) function, for short) if (1) it is efficiently computable, and (2) for all PPTAs (resp. computationally unbounded algorithms) $\mathcal{F}$ and for all sufficiently large $k \in \mathbb{N}$, it holds that* $\mathsf{Adv}_{ai,\mathcal{F}}^{P-Inv}(k) := \Pr[\mathsf{Expt}_{ai,\mathcal{F}}^{P-Inv}(k) = 1] - 1/|\mathcal{X}_k| \le \delta(k)$,[3] *where the experiment* $\mathsf{Expt}_{ai,\mathcal{F}}^{P-Inv}(k)$ *is defined as follows:*

$$[\, \alpha \leftarrow \mathcal{X}_k; \ \beta \leftarrow \{0, 1\}^t; \ z \leftarrow ai(\alpha, \beta); \ \alpha' \leftarrow \mathcal{F}(1^k, z); \ \text{Return } (\alpha' \overset{?}{=} \alpha) \,].$$

Furthermore, we say that ai *is ℓ-bounded if the output length of* ai *is bounded by* $\ell = \ell(k)$.

3.2 Average-Case Indistinguishability of Point Values with Auxiliary Input

In our proposed KEM constructions, what we need for an MBPF obfuscator is that it hides the point value "on average," in the presence of auxiliary input that is *simltaneously* dependent on the point address and the point value. This indistinguishability-based definition, formalized below, enables us to avoid using simulator-based security notions, and helps to make the security analyses of our proposed constructions simpler.

---

[3] Here, the subtraction of $1/|\mathcal{X}_k|$ is to offset the trivial success probability by a random guess.

Definition 6. *Let* $\delta : \mathbb{N} \to [0,1]$. *We say that an MBPF obfuscator* MBPO *satisfies average-case indistinguishability w.r.t.* δ-*computationally (resp.* δ-*statistically) partially uninvertible auxiliary input (* AIND-δ-cPUAI *(resp.* AIND-δ-sPUAI*) secure, for short), if for all PPTAs* $\mathcal{A}$ *and all* δ-cPUAI *(resp.* δ-sPUAI*) functions* ai, $\mathsf{Adv}^{\mathrm{AIND-AI}}_{\mathrm{MBPO,ai},\mathcal{A}}(k) := 2 \cdot | \Pr[\mathsf{Expt}^{\mathrm{AIND-AI}}_{\mathrm{MBPO,ai},\mathcal{A}}(k) = 1] - 1/2|$ *is negligible, where the experiment* $\mathsf{Expt}^{\mathrm{AIND-AI}}_{\mathrm{MBPO,ai},\mathcal{A}}(k)$ *is defined as follows:*

$$[\, \alpha \leftarrow \mathcal{X}_k; \; \beta_0, \beta_1 \leftarrow \{0,1\}^t; \; z \leftarrow \mathsf{ai}(\alpha, \beta_0); \; b \leftarrow \{0,1\};$$

$$\mathrm{DL} \leftarrow \mathsf{MBPO}(\mathcal{I}_{\alpha \to \beta_b}); \; b' \leftarrow \mathcal{A}(1^k, z, \mathrm{DL}); \; \text{Return } (b' \overset{?}{=} b) \,].$$

In the experiment, DL stands for a "digital locker" (the name is due to [19]).

The following is a simple fact that in order for the new definitions to be meaningful, δ has to be a negligible function. (The proof is given in the full version.)

Lemma 1. *Let* $\delta : \mathbb{N} \to [0,1]$. *If* δ *is non-negligible, then an MBPF obfuscator cannot be* AIND-δ-sPUAI *secure (and hence it cannot be* AIND-δ-cPUAI *secure, either).*

4   Chosen Ciphertext Security via MBPF Obfuscation

In this section, we show our main results: two constructions of CCA2 secure KEMs. The first and second constructions are given in Sections 4.1 and 4.2, respectively. We also explain several extensions applicable to our proposed constructions in Section 4.3.

4.1   First Construction

Let $\Pi = (\mathsf{PKG}, \mathsf{Enc}, \mathsf{Dec})$ be a PKE scheme with the plaintext space $\{0,1\}^k$, the public key length $\ell_{\mathrm{PK}}(k)$, the randomness length $\ell_{\mathrm{R}}(k)$, and the ciphertext length $\ell_{\mathrm{C}}(k)$ (where the definitions of these are given in Appendix A). We define $t(k) = k \cdot \ell_{\mathrm{R}}(k) + k$ and $t'(k) = k \cdot \ell_{\mathrm{PK}}(k) + k \cdot \ell_{\mathrm{C}}(k) + k$. Let $\mathcal{X} = \{\mathcal{X}_k\}_{k \in \mathbb{N}}$ be a domain ensemble such that each element in $\mathcal{X}_k$ is of length k, and let MBPO be an MBPF obfuscator for MBPF$(\mathcal{X}, t)$. Furthermore, let $\mathcal{H} = (\mathsf{HKG}, \mathsf{H})$ be a UOWHF. Then we construct the proposed KEM $\Gamma = (\mathsf{KKG}, \mathsf{Encap}, \mathsf{Decap})$ as in Fig. 1.

Useful Properties of Γ. To show the CCA2 security of the proposed KEM Γ, it is useful to note the following two simple properties, which are both due to the validity check performed in the last step of Decap (and the correctness of the underlying PKE scheme Π). The first property states that in order to generate a valid ciphertext, an obfuscated circuit DL cannot be copied from other valid ciphertexts.

Lemma 2. *Let* (PK, SK) *be a key pair output by* $\mathsf{KKG}(1^k)$, *and* $C = (c_1, \ldots, c_k, \mathrm{DL})$ *be a ciphertext output by* $\mathsf{Encap}(PK)$. *Then, for any ciphertext* $C' = (c'_1, \ldots, c'_k, \mathrm{DL}')$ *satisfying* $\mathrm{DL}' = \mathrm{DL}$ *and* $(c'_1, \ldots, c'_k) \neq (c_1, \ldots, c_k)$, *it holds that* $\mathsf{Decap}(SK, C') = \bot$.

The second property is the existence of the "alternative" decapsulation algorithm AltDecap. For a k-bit string $h^* = (h^*_1 \| \ldots \| h^*_k) \in \{0,1\}^k$ and a key pair (PK, SK)

KKG(1^k) :	Decap(SK, C) :
$\kappa \leftarrow$ HKG(1^k)	Parse SK as $(\{sk_i^{(j)}\}_{i \in [k], j \in \{0,1\}}, \kappa)$
$(pk_i^{(j)}, sk_i^{(j)}) \leftarrow$ PKG(1^k) for $(i, j) \in [k] \times \{0, 1\}$	Parse C as $(c_1, \ldots, c_k, \text{DL})$
$PK \leftarrow (\{pk_i^{(j)}\}_{i \in [k], j \in \{0,1\}}, \kappa)$	$h \leftarrow$ H$_\kappa$(DL)
$SK \leftarrow (\{sk_i^{(j)}\}_{i \in [k], j \in \{0,1\}}, \kappa)$	View h as $(h_1 \| \ldots \| h_k) \in \{0, 1\}^k$
Return (PK, SK)	$\alpha \leftarrow$ Dec($sk_1^{(h_1)}, c_1$)
	If $\alpha = \perp$ then return $\perp$
Encap(PK) :	$\beta \leftarrow$ DL(α)
Parse PK as $(\{pk_i^{(j)}\}_{i \in [k], j \in \{0,1\}}, \kappa)$	If $\beta = \perp$ then return $\perp$
$\alpha \leftarrow \mathcal{X}_k$; $\beta \leftarrow \{0, 1\}^t$	Parse β as $(r_1, \ldots, r_k, K)$
DL $\leftarrow$ MBPO($\mathcal{I}_{\alpha \rightarrow \beta}$)	$\in (\{0, 1\}^{\ell_R})^k \times \{0, 1\}^k$
$h \leftarrow$ H$_\kappa$(DL)	If $\forall i \in [k]$: Enc($pk_i^{(h_i)}, \alpha; r_i) = c_i$
View h as $(h_1 \| \ldots \| h_k) \in \{0, 1\}^k$	then return K else return $\perp$
Parse β as $(r_1, \ldots, r_k, K) \in (\{0, 1\}^{\ell_R})^k \times \{0, 1\}^k$	
$c_i \leftarrow$ Enc($pk_i^{(h_i)}, \alpha; r_i$) for $i \in [k]$	
$C \leftarrow (c_1, \ldots, c_k, \text{DL})$	
Return (C, K)	

Fig. 1. The proposed CCA2 secure KEM Γ

output by KKG(1^k), where $SK = (\{sk_i^{(j)}\}_{i \in [k], j \in \{0,1\}}, \kappa)$, we define the "alternative" secret key $\widehat{SK}_{h^*}$ associated with h^* by $\widehat{SK}_{h^*} = (h^*, PK, \{sk_i^{(1-h_i^*)}\}_{i \in [k]})$. AltDecap takes an "alternative" secret key $\widehat{SK}_{h^*}$ and a ciphertext $C = (c_1, \ldots, c_k, \text{DL})$ as input, and runs as follows:

AltDecap($\widehat{SK}_{h^*}, C$)**:** First check if H$_\kappa$(DL) $= h^*$, and return $\perp$ if this is the case. Otherwise, let $h =$ H$_\kappa$(DL) and let $\ell \in [k]$ be the smallest index such that $h_\ell = 1 - h_\ell^*$, where h_ℓ is the ℓ-th bit of h. (Note that such ℓ must exist because $h \neq h^*$ in this case.) Run in exactly the same way as Decap(SK, C), except that it executes Dec($sk_\ell^{(1-h_\ell^*)}, c_\ell$) in the fifth step, instead of executing Dec($sk_1^{(h_1)}, c_1$).

Regarding AltDecap, the following lemma holds due to the symmetric role of each of $sk_i^{(j)}$ and the validity check of each c_i by re-encryption performed at the last step.

Lemma 3. *Let $h^* \in \{0, 1\}^k$ be a string, (PK, SK) be a key pair output by* KKG(1^k), *and $\widehat{SK}_{h^*}$ be an alternative secret key defined as above. Then, for any ciphertext $C = (c_1, \ldots, c_k, \text{DL})$ (which could be outside the range of* Encap(PK)) *satisfying* H$_\kappa$(DL) $\neq$ h^*, *it holds that* Decap(SK, C) = AltDecap($\widehat{SK}_{h^*}, C$).

The formal proofs of Lemmas 2 and 3 are given in the full version.

CCA2 *Security of Γ.* The security of Γ is guaranteed by the following theorem. (The formal proof is given in the full version.)

Theorem 1. *Assume that Π is ϵ-CPA secure with negligible ϵ, $\mathcal{H}$ is a UOWHF, and* MBPO *is* AIND-δ-cPUAI *secure with $\delta(k) \geq k\epsilon(k)$. Then, the KEM Γ constructed as in Fig. 1 is* CCA2 *secure.*

Proof Sketch of Theorem 1.   Let $\mathcal{A} = (\mathcal{A}_1, \mathcal{A}_2)$ be any PPTA adversary that attacks the CCA2 security of the KEM Γ. Consider the following sequence of games: (Here, the values with asterisk (*) represent those related to the challenge ciphertext for $\mathcal{A}$.)

Game 1: This is the experiment $\mathsf{Expt}_{\Gamma, \mathcal{A}}^{\mathsf{CCA2}}(k)$ itself. Without loss of generality, we generate the challenge ciphertext $C^* = (c_1^*, \ldots, c_k^*, \mathsf{DL}^*)$ and the challenge session-key K_b^* for $\mathcal{A}$, where b is the challenge bit for $\mathcal{A}$, before running $\mathcal{A}_1$. (Note that this modification does not affect $\mathcal{A}$'s behavior.)

Game 2: Same as Game 1, except that all decapsulation queries $C = (c_1, \ldots, c_k, \mathsf{DL})$ satisfying $\mathsf{DL} = \mathsf{DL}^*$ are answered with $\perp$.

Game 3: Same as Game 2, except that all decapsulation queries $C = (c_1, \ldots, c_k, \mathsf{DL})$ satisfying $\mathsf{H}_\kappa(\mathsf{DL}) = h^* = \mathsf{H}_\kappa(\mathsf{DL}^*)$ are answered with $\perp$.

Game 4: Same as Game 3, except that all decapsulation queries C are answered with $\mathsf{AltDecap}(\widehat{SK}_{h^*}, C)$, where $\widehat{SK}_{h^*}$ is the alternative secret key corresponding to (PK, SK) and $h^* = \mathsf{H}_\kappa(\mathsf{DL}^*) \in \{0,1\}^k$.

Game 5: Same as Game 4, except that DL^* is replaced with an obfuscation of the MBPF $\mathcal{I}_{\alpha^* \to \beta'}$ with an independently chosen random value $\beta' \in \{0,1\}^t$. That is, the step "$\mathsf{DL}^* \leftarrow \mathsf{MBPO}(\mathcal{I}_{\alpha^* \to \beta^*})$" is replaced with the steps "$\beta' \leftarrow \{0,1\}^t$; $\mathsf{DL}^* \leftarrow \mathsf{MBPO}(\mathcal{I}_{\alpha^* \to \beta'})$." (Note that each r_i^* and K_1^* are still generated from β^*.)

For $i \in [5]$, let S_i be the event that $\mathcal{A}$ succeeds in guessing the challenge bit (i.e. $b' = b$ occurs) in Game i. Using the above notation, $\mathcal{A}$'s CCA2 advantage can be calculated as follows:

$$\mathsf{Adv}_{\Gamma, \mathcal{A}}^{\mathsf{CCA2}}(k) = 2 \cdot |\Pr[\mathsf{S}_1] - \frac{1}{2}| \leq 2 \cdot \sum_{i \in [4]} |\Pr[\mathsf{S}_i] - \Pr[\mathsf{S}_{i+1}]| + 2 \cdot |\Pr[\mathsf{S}_5] - \frac{1}{2}|. \quad (1)$$

To complete the proof, it remains to upperbound the right hand side of the above inequality (1).

Firstly, notice that the difference between Game 1 and Game 2 is only in how $\mathcal{A}$'s decapsulation query $C = (c_1, \ldots, c_k, \mathsf{DL})$ satisfying $\mathsf{DL} = \mathsf{DL}^*$ is answered. (It is answered with $\perp$ in Game 2, while it may be answered with some value that is not $\perp$ in Game 1.) However, due to Lemma 2, the only ciphertext C that contains DL^* and can be decapsulated to some value that is not $\perp$ is the challenge ciphertext C^* itself, and $\mathcal{A}_2$ is not allowed to ask it. Furthermore, since DL^* is information-theoretically hidden from $\mathcal{A}_1$'s view, the probability of $\mathcal{A}_1$ making a decapsulation query containing DL^* is negligible. Hence, the oracles behave almost identically in both Game 1 and Game 2, which shows that $|\Pr[\mathsf{S}_1] - \Pr[\mathsf{S}_2]|$ is negligible.

Next, notice that $|\Pr[\mathsf{S}_2] - \Pr[\mathsf{S}_3]|$ can be upperbounded by the probability of $\mathcal{A}$ making a decapsulation query $C = (c_1, \ldots, c_k, \mathsf{DL})$ satisfying $\mathsf{H}_\kappa(\mathsf{DL}) = h^* = \mathsf{H}_\kappa(\mathsf{DL}^*)$ and $\mathsf{DL} \neq \mathsf{DL}^*$ (because Game 2 and Game 3 proceed identically without such a query), but it is easy to see that this probability is negligible due to the security of the UOWHF $\mathcal{H}$.

It is also easy to see that $\Pr[\mathsf{S}_3] = \Pr[\mathsf{S}_4]$, because the behavior of the oracle in Game 3 and that in Game 4, are identical due to Lemma 3.

To show the upperbound of $|\Pr[\mathsf{S}_4] - \Pr[\mathsf{S}_5]|$, we need to use the $\mathtt{AIND}\text{-}\delta\text{-}\mathtt{cPUAI}$ security of MBPO. We therefore first specify the auxiliary input function that we are going to consider. Define the probabilistic function $\mathsf{ai}_\Gamma : \mathcal{X}_k \times \{0,1\}^t \to \{0,1\}^{t'}$ which takes $(\alpha, \beta) \in \mathcal{X}_k \times \{0,1\}^t$ as input, and computes $z = (\{pk_i\}_{i\in[k]}, c_1^*, \ldots, c_k^*, K^*) \in \{0,1\}^{t'}$ in the following way:

$\mathsf{ai}_\Gamma(\alpha, \beta) : [\ (pk_i, sk_i) \leftarrow \mathsf{PKG}(1^k)$ for $i \in [k]$;

$\qquad\qquad$ Parse β as $(r_1^*, \ldots, r_k^*, K^*) \in (\{0,1\}^{\ell_\mathtt{R}})^k \times \{0,1\}^k$;

$\qquad c_i^* \leftarrow \mathsf{Enc}(pk_i, \alpha; r_i^*)$ for $i \in [k]$; Return $z \leftarrow (\{pk_i\}_{i\in[k]}, c_1^*, \ldots, c_k^*, K^*)\ ]$.

Note that ai_Γ is efficiently computable. Furthermore, due to the ϵ-CPA security of the underlying PKE scheme Π and the security of the k-repetition construction Π^k (which is $(k\epsilon)$-CPA secure based on the ϵ-CPA security of Π)[4], it is straightforward to see that ai_Γ is $(k\epsilon)$-computationally partially uninvertible (in particular, in the $\mathtt{P\text{-}Inv}$ experiment regarding ai_Γ, each r_i^* is a uniformly chosen randomness (independently of any other values), and thus we can rely on the CPA security of Π). In the full proof, we will show that there exists a PPTA $\mathcal{B}_o$ such that $\mathsf{Adv}^{\mathtt{AIND\text{-}AI}}_{\mathsf{MBPO},\mathsf{ai}_\Gamma,\mathcal{B}_o}(k) = |\Pr[\mathsf{S}_4] - \Pr[\mathsf{S}_5]|$: $\mathcal{B}_o$ takes an auxiliary input $z = (\{pk_i\}_{i\in[k]}, c_1^*, \ldots, c_k^*, K^*) \leftarrow \mathsf{ai}_\Gamma(\alpha, \beta_0)$ and an obfuscated circuit DL^* which is either computed as $\mathsf{MBPO}(\mathcal{I}_{\alpha\to\beta_0})$ or $\mathsf{MBPO}(\mathcal{I}_{\alpha\to\beta_1})$ as input. $\mathcal{B}_o$ will generate $\mathcal{A}$'s challenge ciphertext C^* based on the auxiliary input z and the obfuscated ciphertext DL^* that it receives, and generates the remaining key materials, which enables $\mathcal{B}_o$ to generate the alternative key $\widehat{SK}_{h^*}$, and thus using AltDecap, $\mathcal{B}_o$ can perfectly simulate the decryption oracle in Game 4 (and Game 5) for $\mathcal{A}$. Here, by regarding α, β_0, and β_1 in $\mathcal{B}_o$'s experiment as α^*, β^*, and β' (in Game 4 and Game 5), respectively, $\mathcal{B}_o$ will simulate the whole of Game 4 or Game 5 perfectly for $\mathcal{A}$ depending on the value of $\mathcal{B}$'s challenge bit, and we can derive $\mathsf{Adv}^{\mathtt{AIND\text{-}AI}}_{\mathsf{MBPO},\mathsf{ai}_\Gamma,\mathcal{B}_o}(k) = |\Pr[\mathsf{S}_4] - \Pr[\mathsf{S}_5]|$. But here, since ai_Γ is a $(k\epsilon)$-cPUAI function and $\delta(k) \geq k\epsilon(k)$, the $\mathtt{AIND}\text{-}\delta\text{-}\mathtt{cPUAI}$ security of MBPO guarantees that $|\Pr[\mathsf{S}_4] - \Pr[\mathsf{S}_5]|$ is negligible.

Finally, observe that in Game 5, the "real" session-key K_1^* is independent of the challenge ciphertext C^* and thus the challenge session-key K_b^* (together with other values available to $\mathcal{A}$ in Game 5) is distributed identically regardless of $\mathcal{A}$'s challenge bit b. This implies $\Pr[\mathsf{S}_5] = 1/2$.

Therefore, the right hand side of the inequality (1) is shown to be negligible, which implies that Γ is CCA2 secure. $\qquad\qquad\qquad\qquad\qquad\qquad\qquad\qquad\qquad\qquad\Box$

4.2 Second Construction

In the first construction shown above, we used an ordinary CPA secure PKE scheme for Π. Now, we consider the construction of the KEM Γ in which Π is replaced with a lossy encryption scheme. Π now has the lossy key generation algorithm LKG, and thus is of the form $\Pi = (\mathsf{PKG}, \mathsf{Enc}, \mathsf{Dec}, \mathsf{LKG})$. (The lossy key generation algorithm

---

[4] Here, by "k-repetition construction" Π^k we mean the PKE scheme in which a public key consists of k independently generated public keys of Π, and a ciphertext consists of k ciphertexts of a same plaintext.

LKG is actually not used in the construction, and is used only in the security proof.) Because of this change, we can now relax the requirement for the MBPF obfuscator MBPO to be secure in the presence of only statistically partially uninvertible auxiliary input. This result is captured by the following theorem. (The formal proof is given in the full version.)

Theorem 2. *Assume* Π *is an* ϵ-lossy encryption scheme with negligible ϵ, $\mathcal{H}$ *is a UOWHF, and* MBPO *is* AIND-δ-sPUAI *secure with* $\delta(k) \geq k\epsilon(k)$. *Then, the KEM* Γ *constructed as in Fig. 1 is* CCA2 *secure.*

Proof Sketch of Theorem 2. The proof proceeds very similarly to that of Theorem 1. The main difference is that we consider an additional game between Game 4 and Game 5 (say, Game 4.5), in which we generate all public keys for $\{pk_i^{(h_i^*)}\}_{i\in[k]}$ by using the lossy key generation algorithm LKG(1^k), instead of PKG(1^k), where h_i^* is the i-th bit of $h^* = \mathsf{H}_\kappa(\mathsf{DL}^*)$. Then the difference between a CCA2 adversary $\mathcal{A}$'s success probability in Game 4 and that in Game 4.5 can be bounded to be negligible by the indistinguishability of keys of the k-repetition lossy encryption scheme Π^k. In particular, the corresponding secret keys $\{sk_i^{(h_i^*)}\}_{i\in[k]}$ are already not used in Game 4, and the reduction algorithm (for distinguishing ordinary/lossy public keys of Π^k) need not use them. Correspondingly to the above, in order to show that the difference between $\mathcal{A}$'s success probability in Game 4.5 and that in Game 5 is negligible, we will use the AIND-δ-sPUAI security of MBPO, with the auxiliary input $\mathsf{ai}'_\Gamma : \mathcal{X}_k \times \{0,1\}^t \to \{0,1\}^{t'}$ that is defined in the same way as ai_Γ used in the proof of Theorem 1 except that the public keys $\{pk_i\}_{i\in[k]}$ are generated by executing the lossy key generation algorithm LKG(1^k). Since the keys $\{pk_i\}_{i\in[k]}$ are generated from LKG, due to ϵ-lossiness of the lossy encryption scheme Π and $(k\epsilon)$-lossiness of the k-repetition construction Π^k (where $(k\epsilon)$-lossiness of Π^k based on ϵ-lossiness of Π can be shown by a standard hybrid argument), we can easily see that ai'_Γ is a $(k\epsilon)$-sPUAI function. The rest of the proof proceeds identically to that of Theorem 1. $\qquad\square$

4.3 Extensions

A-priori Fixed and Bounded-length Auxiliary Input Functions. Note that for both of our proposed constructions, the auxiliary input functions under which the building block MBPF obfuscator MBPO needs to be secure, are dependent only on the building block PKE/lossy encryption scheme Π, which is fixed when Π is fixed. In particular, MBPO is required to satisfy AIND-δ-cPUAI (and AIND-δ-sPUAI) security only for t'-bounded δ-cPUAI (and δ-sPUAI) functions with $t'(k) = k \cdot \ell_{\mathsf{PK}}(k) + k \cdot \ell_{\mathsf{C}}(k) + k$. This a-priori bounded output length for auxiliary input functions might make it easier to achieve AIND-δ-cPUAI (and AIND-δ-sPUAI) secure MBPF obfuscators. We note that a similar observation on the possibility of weakening the requirement regarding auxiliary inputs by bounding the output length is also given in [9].

Using MBPF Obfuscators with Short Point Values. In our constructions, the MBPF obfuscator MBPO needs to obfuscate an MBPF $\mathcal{I}_{\alpha\to\beta}$ whose point value β is relatively long (which consists of k randomness $\{r_i\}_{i\in[k]}$ and a k-bit string K). For our first construction, however, we can shorten the length of a point value of MBPFs that need to be obfuscated by utilizing a pseudorandom generator (PRG). More specifically, let $G : \{0,1\}^k \to \{0,1\}^t$ be a PRG (where $t(k) = k \cdot \ell_{\mathsf{R}}(k) + k$). Then instead of picking $\{r_i\}_{i \in [k]}$ and $K \in \{0,1\}^k$ uniformly at random, these values can be generated from a short seed $s \in \{0,1\}^k$ by $\beta = (r_1 \| \ldots \| r_k \| K) \leftarrow G(s)$, and now we only need to obfuscate $\mathcal{I}_{\alpha \to s}$, instead of $\mathcal{I}_{\alpha \to \beta}$. However, this modification is at the cost of a stronger requirement for AIND-δ-cPUAI security of MBPO. That is, now δ has to be large enough to incorporate the security of the used PRG. Specifically, if the PRG is ϵ_g-secure, then it is required that $\delta \geq k\epsilon + \epsilon_g$ (where a PRG is said to be ϵ-secure if all PPTA adversaries have at most advantage $\epsilon = \epsilon(k)$ in distinguishing a pseudorandom value from a truly random value for all sufficiently large $k \in \mathbb{N}$). We note that this idea of using a PRG does not work for our second construction, because we cannot use a pseudorandom string as a randomness in the encryption algorithm of a lossy encryption scheme. Using a pseudorandomness violates the statistical lossiness property in general.

A Simpler Construction with CCA1 *Security.* We can show that a simpler variant of the proposed construction which employs the Naor-Yung-style double encryption [52] (instead of the Dolev-Dwork-Naor-style multiple encryption), leads a CCA1 secure KEM. This KEM is partly explained in Introduction, and we will show the details in the full version. Interestingly, unlike our CCA2 secure constructions, in the proof of this CCA1 secure variant, we need to use an auxiliary input function that internally runs (a part of) a CCA1 adversary, and thus its output length cannot be a-priori bounded.

5   Relations among Security Notions for MBPF Obfuscators

In this section, we investigate the relations between our new indistinguishability-based security notions for MBPF obfuscators, AIND-δ-cPUAI/sPUAI, and the worst-/average-case virtual black-/grey-box properties in the presence of auxiliary inputs. For the average-case virtual black-/grey-box properties, we consider the auxiliary input functions defined in Section 3.1, and show that our new security notions are implied by the average-case virtual black-/grey-box properties with the same type of auxiliary inputs.

We first formally define the average-case virtual black-/grey-box properties. For notational convenience, for an MBPF obfuscator MBPO, a probabilistic algorithm $\mathcal{M}$ whose output is restricted to be a bit, and a two-input probabilistic function ai : $\mathcal{X}_k \times \{0,1\}^t \to \{0,1\}^*$, we define the following three experiments:

$\mathsf{Expt}^{\mathsf{Real}}_{\mathsf{MBPO},\mathsf{ai},\mathcal{M}}(k)$:	$\mathsf{Expt}^{\mathsf{Sim}}_{\mathsf{ai},\mathcal{M}}(k)$:	$\mathsf{Expt}^{\mathsf{s\text{-}Sim}}_{\mathsf{ai},\mathcal{M}}(k)$:
$\alpha \leftarrow \mathcal{X}_k$	$\alpha \leftarrow \mathcal{X}_k$	$\alpha \leftarrow \mathcal{X}_k$
$\beta \leftarrow \{0,1\}^t$	$\beta \leftarrow \{0,1\}^t$	$\beta \leftarrow \{0,1\}^t$
$z \leftarrow \mathsf{ai}(\alpha,\beta)$	$z \leftarrow \mathsf{ai}(\alpha,\beta)$	$z \leftarrow \mathsf{ai}(\alpha,\beta)$
$\mathsf{DL} \leftarrow \mathsf{MBPO}(\mathcal{I}_{\alpha \to \beta})$	Return $b \leftarrow \mathcal{M}^{\mathcal{I}_{\alpha \to \beta}}(1^k, z)$	Return $b \leftarrow \mathcal{M}(1^k, z)$
Return $b \leftarrow \mathcal{M}(1^k, z, \mathsf{DL})$		

(Note that in $\mathsf{Expt}^{\mathsf{s\text{-}Sim}}_{\mathsf{ai},\mathcal{M}}(k)$, the algorithm $\mathcal{M}$ does not have access to any oracle.)

Definition 7. *We say that an MBPF obfuscator* MBPO *satisfies*

– *the* average-case virtual black-box property w.r.t. δ-computationally (resp. δ-statistically) partially uninvertible auxiliary input *(AVB-δ-cPUAI (resp. AVB-δ-sPUAI) security, for short), if for every PPTA $\mathcal{A}$ and all positive polynomials $q = q(k)$ and $\ell = \ell(k)$, there exists a PPTA $\mathcal{S}$ such that for every ℓ-bounded δ-cPUAI (resp. δ-sPUAI) function* ai *and all sufficiently large $k \in \mathbb{N}$, it holds that*

$$\mathrm{Adv}^{\text{A-MBPO-AI}}_{\text{MBPO,ai},\mathcal{A},\mathcal{S}}(k) := | \Pr[\mathrm{Expt}^{\text{Real}}_{\text{MBPO,ai},\mathcal{A}}(k) = 1] - \Pr[\mathrm{Expt}^{\text{Sim}}_{\text{ai},\mathcal{S}}(k) = 1]| \leq 1/q.$$

– *the* strong average-case virtual black-box property w.r.t. δ-computationally (resp. δ-statistically) partially uninvertible auxiliary input *(SAVB-δ-cPUAI (resp. SAVB-δ-sPUAI) security, for short), if for every PPTA $\mathcal{A}$ and all positive polynomials $q = q(k)$ and $\ell = \ell(k)$, there exists a PPTA $\mathcal{S}$ such that for every ℓ-bounded δ-cPUAI (resp. δ-sPUAI) function* ai *and all sufficiently large $k \in \mathbb{N}$, it holds that*

$$\mathrm{Adv}^{\text{SA-MBPO-AI}}_{\text{MBPO,ai},\mathcal{A},\mathcal{S}}(k) := | \Pr[\mathrm{Expt}^{\text{Real}}_{\text{MBPO,ai},\mathcal{A}}(k) = 1] - \Pr[\mathrm{Expt}^{\text{s-Sim}}_{\text{ai},\mathcal{S}}(k) = 1]| \leq 1/q.$$

Furthermore, we define the (strong) average-case virtual grey-box property w.r.t. δ-computationally (resp. δ-statistically) partially uninvertible auxiliary input *((S)AVG-δ-cPUAI (resp. (S)AVG-δ-sPUAI) security for short) for an MBPF obfuscator* MBPO, *in the same way as the definitions for the corresponding virtual black-box properties, except that we replace "a PPTA $\mathcal{S}$" in each definition with "a computationally unbounded algorithm $\mathcal{S}$ that makes only polynomially many queries."*

Now, we show the relations among security notions, which are summarized in Fig. 2. Most of the relations are obvious. Namely, the virtual black-box properties always imply the virtual grey-box properties for the same class of auxiliary inputs. Furthermore, WVB-AI security implies AVB-δ-cPUAI security for arbitrary (not necessarily negligible) δ, and AVB-δ-cPUAI security implies AVB-δ-sPUAI security because the class of δ-sPUAI functions are smaller than the class of δ-cPUAI functions for the same δ. Moreover, by definition, for both $X \in \{\delta\text{-cPUAI}, \delta\text{-sPUAI}\}$, SAVB-X and SAVG-X imply AVB-X and AVG-X, respectively, because the former notions consider simulators that do not make any oracle queries and thus can also be used as simulator for the latter.

In the following, we show the implications of the non-trivial directions. The following equivalence is due to the result by Bitansky and Canetti [7]. (Note that the following results are only for non-uniform PPTA adversaries, while our default notions in this paper are with respect to uniform PPTA adversaries.)

Lemma 4. *([8, Propositions 8.3 and A.3]) For MBPF obfuscators,* WVB *security for non-uniform PPTA adversaries with non-uniform PPTA simulators,* WVG *security for non-uniform PPTA adversaries, and* WVG-AI *security for PPTA non-uniform adversaries, are equivalent.*

The following is useful for showing the implication to the AIND security notions that we will show later.

Lemma 5. *Let $\delta : \mathbb{N} \to [0,1]$ be a negligible function. For MBPF obfuscators, for both $X \in \{\delta\text{-cPUAI}, \delta\text{-sPUAI}\}$,* AVB-X *security and* SAVB-X *security are equivalent. Furthermore,* AVG-δ-sPUAI *security and* SAVG-δ-sPUAI *security are equivalent.*

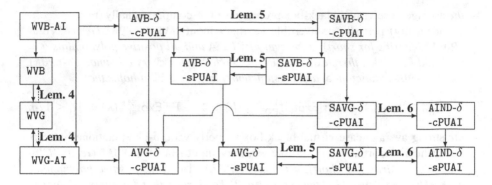

Fig. 2. Relations among security notions for MBPF obfuscators defined in this paper. The arrow "X → Y" indicates that X-security implies Y-security. The dotted arrows indicate the implications that hold only for the non-uniform setting in which an adversary (and a simulator) are non-uniform algorithms. In the figure, δ is a negligible function.

Intuition. For both cPUAI and sPUAI cases, the implication from the latter to the former is trivial by definition. The implications of the opposite directions can be shown because the partial uninvertibility of an auxiliary input function guarantees that a simulator cannot find the point address of the MBPF being obfuscated and thus having oracle access to an MBPF does not give much advantage. The computational uninvertibility and statistical uninvertibility naturally correspond to the uninvertibility of auxiliary input functions against a PPTA simulator and that against a computationally unbounded simulator, respectively.

Finally, the following implications clarify that AIND notions introduced in Section 3.2 are indeed implied by the average-case virtual black-box/grey-box properties.

Lemma 6. *Let* $\delta : \mathbb{N} \to [0, 1]$ *be a negligible function. For both* X $\in$ {δ-cPUAI, δ-sPUAI}, *if an MBPF obfuscator is* SAVG-X *secure, then it is* AIND-X *secure.*

Intuition. This lemma is shown by considering a hybrid experiment in which a (computationally unbounded) simulator $\mathcal{S}$ (due to SAVG-δ-cPUAI/sPUAI security) is given only an auxiliary input ai(α, β) (for randomly chosen (α, β)) as input, and outputs a bit.; By the SAVG-δ-cPUAI/sPUAI security, for both cases $b \in \{0, 1\}$, the probability that an adversary (attacking the AIND-δ-cPUAI/sPUAI security) on input ai(α, β_0) and MBPO($\mathcal{I}_{\alpha \to \beta_b}$) (for randomly chosen α, β_0, β_1) outputs 1 can be shown to be negligibly close to the probability that the simulator $\mathcal{S}$ outputs 1 in the hybrid experiment, which proves the lemma.

6   Lossy Encryption from Re-randomizable Point Obfuscation

In this section, we show that a re-randomizable point obfuscator yields a lossy encryption scheme. We first recall the definition of re-randomizability [7].

PKG(1^k) :	Enc(pk, m) :	Dec(sk, c) :
$\alpha_0 \leftarrow \mathcal{X}_k$	Parse pk as $(\widehat{\mathsf{P}}_0, \widehat{\mathsf{P}}_1)$	Parse c as $(\mathsf{P}_1, \ldots, \mathsf{P}_t)$
$\alpha_1 \leftarrow \mathcal{X}_k \backslash \{\alpha_0\}$	View m as	For $i \in [t]$:
$\widehat{\mathsf{P}}_i \leftarrow \mathsf{PO}(\mathcal{I}_{\alpha_i})$ for $i \in \{0, 1\}$	$(m_1 \| \ldots \| m_t) \in \{0, 1\}^t$	$m_i \leftarrow \begin{cases} 0 & \text{if } \mathsf{P}_i(sk) = \top \\ 1 & \text{otherwise} \end{cases}$
$pk \leftarrow (\widehat{\mathsf{P}}_0, \widehat{\mathsf{P}}_1); \quad sk \leftarrow \alpha_0$	$\mathsf{P}_i \leftarrow \mathsf{ReRand}(\widehat{\mathsf{P}}_{m_i})$	
Return (pk, sk)	for $i \in [t]$	End For
LKG(1^k) :	Return $c \leftarrow (\mathsf{P}_1, \ldots, \mathsf{P}_t)$	Return $m \leftarrow (m_1 \| \ldots \| m_t)$
$\alpha \leftarrow \mathcal{X}_k$		
$\widehat{\mathsf{P}}_i \leftarrow \mathsf{PO}(\mathcal{I}_\alpha)$ for $i \in \{0, 1\}$		
Return $pk \leftarrow (\widehat{\mathsf{P}}_0, \widehat{\mathsf{P}}_1)$		

Fig. 3. Lossy encryption from a re-randomizable point obfuscator

Definition 8. *([7]) Let $\mathcal{X} = \{\mathcal{X}_k\}_{k \in \mathbb{N}}$ be a domain ensemble and let PO be a point obfuscator for $\mathsf{PF}(\mathcal{X})$ whose randomness space is $\{0, 1\}^{\ell(k)}$. We say that PO is re-randomizable if there exists a PPTA ReRand (called the re-randomization algorithm) such that for all $k \in \mathbb{N}$, all $\alpha \in \mathcal{X}_k$, and for all $r \in \{0, 1\}^\ell$, the distribution of* $\mathsf{ReRand}(\mathsf{PO}(\mathcal{I}_\alpha; r))$ *and the distribution of* $\mathsf{PO}(\mathcal{I}_\alpha)$ *are identical.*

We note that the point obfuscator based on the perfect one-way hash function by Canneti [17] is re-randomizable. (We review the construction in Appendix B.)

Now, we formally describe our proposed lossy encryption scheme. Let $\mathcal{X} = \{\mathcal{X}_k\}_{k \in \mathbb{N}}$ be a domain ensemble, and let PO be a re-randomizable point obfuscator for $\mathsf{PF}(\mathcal{X})$ with the re-randomization algorithm ReRand, and let $t = t(k) > 0$ be a polynomial. Then we construct a lossy encryption scheme $\Pi = (\mathsf{PKG}, \mathsf{Enc}, \mathsf{Dec}, \mathsf{LKG})$ whose plaintext space is $\{0, 1\}^t$ as in Fig. 3.

Our construction is inspired partly by the construction of a PKE scheme from a re-randomizable point obfuscator due to Bitansky and Canetti [7], and partly by the construction of lossy encryption from a re-randomizable encryption scheme due to Hemenway et al. [39]. The following theorem guarantees that Π constructed as above is indeed a lossy encryption scheme. (The formal proof is given in the full version.)

Theorem 3. *If PO is re-randomizable and 2-composable, then Π constructed as in Fig. 3 is a 0-lossy encryption scheme.*

Intuition. Theorem 3 is shown by using the equivalence of t-composability and t-distributional indistinguishability for coordinate-wise well-spread (CWS) distributions, established by Bitansky and Canetti [8]. The latter property roughly states that if $(\alpha_1, \ldots, \alpha_t)$ are chosen from a distribution so that each α_i has high min-entropy (but α_i's could be arbitrarily correlated), $(\mathsf{PO}(\alpha_1), \ldots, \mathsf{PO}(\alpha_t))$ is computationally indistinguishable from $(\mathsf{PO}(u_1), \ldots, \mathsf{PO}(u_t))$ where each u_i is chosen uniformly at random (the formal definition appears in the full version). This property can be used to show the indistinguishability of keys, which is easy to see due to the design of PKG and LKG. Moreover, note that a lossy key consists of a pair of obfuscated circuits of point functions with a same point address. Therefore, due to the re-randomizability, an encryption of any plaintext have identical distribution, which implies 0-statistical lossiness.

CCA2 *Secure PKE/KEM Based Solely on Re-randomizable, Composable Point Obfuscators.* Recall that when considering non-uniform PPTA adversaries, WVB security (with non-uniform PPTA simulators), WVG security, and WVG-AI security for MBPF obfuscators are equivalent (see Lemma 4). Therefore, the WVG secure MBPF obfuscator for t-bit point values due to [19,7] based on a $(t + 1)$-composable point obfuscator can be used as an AIND-δ-sPUAI secure MBPF obfuscator (with any negligible δ). Note that if we denote by ℓ the length of the randomness used by ReRand, then the randomness length ℓ_R of the lossy encryption scheme Π for the k-bit plaintext space is $\ell_R(k) = k \cdot \ell(k)$. Combining these results with our second generic construction, we obtain the following.

Theorem 4. *Assume there exists a point obfuscator which is (1) re-randomizable where* ReRand *uses $\ell(k)$-bit randomness, and (2) $(k^2 \cdot \ell(k) + k + 1)$-composable for non-uniform PPTA adversaries. Then there exists a* CCA2 *secure PKE scheme/KEM.*

7 Discussion

On Replacing MBPF Obfuscators with SKE. As has been clarified in several previous works [19,27,37,21], there is a strong connection between MBPF obfuscators and SKE schemes. More specifically, an MBPF obfuscator can always be used as a SKE scheme. In order for the opposite direction to be true, among other things regarding security, it is necessary that a SKE scheme has the property called the *unique-key* property [27,37,21]. Therefore, a variant of our KEM Γ in Section 4 in which an MBPF obfuscator is replaced with a SKE scheme that has the unique-key property and satisfies the security that we call AIND-δ-cPUAI (and AIND-δ-sPUAI) security (which is defined similarly to that for MBPF obfuscator), can also be proved CCA2 secure.

Since the unique-key property is not satisfied by SKE schemes in general, it may be the case that a SKE scheme is in general a weaker primitive than an MBPF obfuscator, and is potentially easier to achieve. (Although a generic transformation of a SKE scheme into one that has this property was proposed in [21], we could not figure out whether this transformation preserves AIND-δ-cPUAI security and AIND-δ-sPUAI security.) Motivated by this observation, in the full version we will show another variant of the proposed KEM based on a SKE scheme without the unique-key property.

On the Difficulty of Achieving AIND-δ-cPUAI *Security.* We have shown that AIND-δ-sPUAI security is implied by the virtual grey-box properties (see Fig. 2), and thus by the results established by [19,7] we can construct an AIND-δ-sPUAI secure MBPF obfuscator (or SKE) from any composable point obfuscator. Unfortunately, however, we could not come up with a natural assumption that is sufficient to realize an AIND-δ-cPUAI secure MBPF obfuscator, and we would like to leave it as an interesting open problem. In the full version, we will show that constructing it is at least as difficult as constructing a SKE scheme which is one-time chosen plaintext secure in the presence of computationally hard-to-invert leakage where leakage occurs only from a key. There, we will also show that the MBPF obfuscator by Lynn et al. [48] can be shown to be AIND-δ-cPUAI secure for any negligible δ. This at least suggests that it can be achieved under a strong assumption. We conjecture that the MBPF obfuscator by Lynn et al. can be shown to be AIND-δ-cPUAI secure for any negligible δ if we instantiate the random oracle as a family of hash functions satisfying (some version of) UCE security that is recently introduced by Bellare et al. [5].

We see that the difficulty of achieving AIND-δ-cPUAI security is that it allows a leakage from a random point address/value pair (α, β) (or a key/message pair in the case of SKE) that could be arbitrarily correlated, as long as partial uninvertibility is satisfied. This definition allows β to be (a part of) the source of the hardness of the partial uninvertibility. For example, we could consider an auxiliary input function ai(α, β) that returns an encryption of the "plaintext" α under the "key" β, using some SKE scheme, which will be a δ-cPUAI function under a reasonable assumption on the SKE scheme. This is quite different from a usual indistinguishability-based security definition (e.g. CPA security of a SKE scheme) in which a point value (or a message in SKE) is chosen by an adversary, and thus cannot be a source of hardness. This is one of the reasons why we cannot straightforwardly use the existing results on MBPF obfuscators/SKE [27,21] (or a stronger primitive of PKE secure under hard-to-invert leakage [26]). We notice that the formulation of AIND-δ-cPUAI security looks close to the security definition for deterministic encryption in the hard-to-invert auxiliary input setting [16], which considers a leakage from a plaintext (as opposed to a key). This setting is in some sense a "dual" of the settings that consider leakage only from a key. We also notice the similarity to the notion called security under *chosen distribution attacks* [4] that considers the security under a correlated leakage from a message and randomness simultaneously (this is a security notion for PKE but can be considered for SKE as well), but this does not consider a leakage from a key or leakage with computational uninvertibility. It would be worth clarifying further whether it is possible to leverage techniques from these various kinds of "leakage resilient" cryptography for achieving AIND-δ-cPUAI/sPUAI secure MBPF obfuscators/SKE schemes.

Acknowledgement. The authors would like thank the members of the study group "Shin-Akarui-Angou-Benkyou-Kai" and the anonymous reviewers for their invaluable comments and suggestions.

References

1. Ananth, P., Boneh, D., Garg, S., Sahai, A., Zhandry, M.: Differing-inputs obfuscation and Applications (2013), http://eprint.iacr.org/2013/689
2. Barak, B., Bitansky, N., Canetti, R., Kalai, Y.T., Paneth, O., Sahai, A.: Obfuscation for evasive functions. In: Lindell, Y. (ed.) TCC 2014. LNCS, vol. 8349, pp. 26–51. Springer, Heidelberg (2014), http://eprint.iacr.org/2013/668
3. Barak, B., Goldreich, O., Impagliazzo, R., Rudich, S., Sahai, A., Vadhan, S., Yang, K.: On the (im)possibility of obfuscating programs. In: Kilian, J. (ed.) CRYPTO 2001. LNCS, vol. 2139, pp. 1–18. Springer, Heidelberg (2001)
4. Bellare, M., Brakerski, Z., Naor, M., Ristenpart, T., Segev, G., Shacham, H., Yilek, S.: Hedged public-key encryption: How to protect against bad randomness. In: Matsui, M. (ed.) ASIACRYPT 2009. LNCS, vol. 5912, pp. 232–249. Springer, Heidelberg (2009)
5. Bellare, M., Hoang, V.T., Keelveedhi, S.: Instantiating random oracles via UCEs. In: Canetti, R., Garay, J.A. (eds.) CRYPTO 2013, Part II. LNCS, vol. 8043, pp. 398–415. Springer, Heidelberg (2013)

6. Bellare, M., Hofheinz, D., Yilek, S.: Possibility and impossibility results for encryption and commitment secure under selective opening. In: Joux, A. (ed.) EUROCRYPT 2009. LNCS, vol. 5479, pp. 1–35. Springer, Heidelberg (2009)
7. Bitansky, N., Canetti, R.: On strong simulation and composable point obfuscation (2010); Full version of [8], http://eprint.iacr.org/2010/414
8. Bitansky, N., Canetti, R.: On strong simulation and composable point obfuscation. In: Rabin, T. (ed.) CRYPTO 2010. LNCS, vol. 6223, pp. 520–537. Springer, Heidelberg (2010)
9. Bitansky, N., Paneth, O.: Point obfuscation and 3-round zero-knowledge. In: Cramer, R. (ed.) TCC 2012. LNCS, vol. 7194, pp. 189–207. Springer, Heidelberg (2012)
10. Bleichenbacher, D.: Chosen ciphertext attacks against protocols based on the RSA encryption standard PKCS #1. In: Krawczyk, H. (ed.) CRYPTO 1998. LNCS, vol. 1462, pp. 1–12. Springer, Heidelberg (1998)
11. Boneh, D., Papakonstantinou, P.A., Rackoff, C., Vahlis, Y., Waters, B.: On the impossibility of basing identity based encryption on trapdoor permutations. In: FOCS 2008, pp. 283–292 (2008)
12. Boyle, E., Chung, K.-M., Pass, R.: On extractability obfuscation. In: Lindell, Y. (ed.) TCC 2014. LNCS, vol. 8349, pp. 52–73. Springer, Heidelberg (2014), http://eprint.iacr.org/2013/650
13. Brakerski, Z., Rothblum, G.N.: Black-box obfuscation for d-CNFs. To appear in ITCS 2014 (2013), http://eprint.iacr.org/2013/557
14. Brakerski, Z., Rothblum, G.N.: Obfuscating conjunctions. In: Canetti, R., Garay, J.A. (eds.) CRYPTO 2013, Part II. LNCS, vol. 8043, pp. 416–434. Springer, Heidelberg (2013)
15. Brakerski, Z., Rothblum, G.N.: Virtual black-box obfuscation for all circuits via generic graded encoding. In: Lindell, Y. (ed.) TCC 2014. LNCS, vol. 8349, pp. 1–25. Springer, Heidelberg (2014), http://eprint.iacr.org/2013/563
16. Brakerski, Z., Segev, G.: Better security for deterministic public-key encryption: The auxiliary-input setting. In: Rogaway, P. (ed.) CRYPTO 2011. LNCS, vol. 6841, pp. 543–560. Springer, Heidelberg (2011)
17. Canetti, R.: Towards realizing random oracles: Hash functions that hide all partial information. In: Kaliski Jr., B.S. (ed.) CRYPTO 1997. LNCS, vol. 1294, pp. 455–469. Springer, Heidelberg (1997)
18. Canetti, R.: Universally composable security: A new paradigm for cryptographic protocols. In: FOCS 2001, pp. 136–145 (2001)
19. Canetti, R., Dakdouk, R.R.: Obfuscating point functions with multibit output. In: Smart, N. (ed.) EUROCRYPT 2008. LNCS, vol. 4965, pp. 489–508. Springer, Heidelberg (2008)
20. Canetti, R., Halevi, S., Katz, J.: Chosen-ciphertext security from identity-based encryption. In: Cachin, C., Camenisch, J.L. (eds.) EUROCRYPT 2004. LNCS, vol. 3027, pp. 207–222. Springer, Heidelberg (2004)
21. Canetti, R., Tauman Kalai, Y., Varia, M., Wichs, D.: On symmetric encryption and point obfuscation. In: Micciancio, D. (ed.) TCC 2010. LNCS, vol. 5978, pp. 52–71. Springer, Heidelberg (2010)
22. Canetti, R., Rothblum, G.N., Varia, M.: Obfuscation of hyperplane membership. In: Micciancio, D. (ed.) TCC 2010. LNCS, vol. 5978, pp. 72–89. Springer, Heidelberg (2010)
23. Coron, J.-S., Lepoint, T., Tibouchi, M.: Practical multilinear maps over the integers. In: Canetti, R., Garay, J.A. (eds.) CRYPTO 2013, Part I. LNCS, vol. 8042, pp. 476–493. Springer, Heidelberg (2013)
24. Cramer, R., Shoup, V.: Design and analysis of practical public-key encryption schemes secure against adaptive chosen ciphertext attack. SIAM J. Computing 33(1), 167–226 (2003)
25. Dachman-Soled, D.: A black-box construction of a CCA2 encryption scheme from a plaintext aware encryption scheme (2013), http://eprint.iacr.org/2013/680

26. Dodis, Y., Goldwasser, S., Tauman Kalai, Y., Peikert, C., Vaikuntanathan, V.: Public-key encryption with auxiliary inputs. In: Micciancio, D. (ed.) TCC 2010. LNCS, vol. 5978, pp. 361–381. Springer, Heidelberg (2010)
27. Dodis, Y., Kalai, Y.T., Lovett, S.: On cryptography with auxiliary input. In: STOC 2009, pp. 621–630 (2009)
28. Dodis, Y., Smith, A.: Correcting errors without leaking partial information. In: STOC 2005, pp. 654–663 (2005)
29. Dolev, D., Dwork, C., Naor, M.: Non-malleable cryptography. In: STOC 1991, pp. 542–552 (1991)
30. Garg, S., Gentry, C., Halevi, S.: Candidate multilinear maps from ideal lattices. In: Johansson, T., Nguyen, P.Q. (eds.) EUROCRYPT 2013. LNCS, vol. 7881, pp. 1–17. Springer, Heidelberg (2013)
31. Garg, S., Gentry, C., Halevi, S., Raykova, M.: Two-round secure MPC from indistinguishability obfuscation. In: Lindell, Y. (ed.) TCC 2014. LNCS, vol. 8349, pp. 74–94. Springer, Heidelberg (2014), http://eprint.iacr.org/2013/601
32. Garg, S., Gentry, C., Halevi, S., Raykova, M., Sahai, A., Waters, B.: Candidate indistinguishability obfuscation and functional encryption for all curcuits. To appear in FOCS (2013), http://eprint.iacr.org/2013/451
33. Gertner, Y., Malkin, T., Myers, S.: Towards a separation of semantic and CCA security for public key encryption. In: Vadhan, S.P. (ed.) TCC 2007. LNCS, vol. 4392, pp. 434–455. Springer, Heidelberg (2007)
34. Gertner, Y., Malkin, T., Reingold, O.: On the impossibility of basing trapdoor functions on trapdoor predicates. In: FOCS 2001, pp. 126–135 (2001)
35. Goldreich, O., Rothblum, R.D.: Enhancements of trapdoor permutations. J. of Cryptology 26(3), 484–512 (2013)
36. Goldwasser, S., Kalai, Y.T.: On the impossibility of obfuscation with auxiliary input. In: FOCS 2005, pp. 553–562 (2005)
37. Goldwasser, S., Kalai, Y.T., Peikart, C., Vaikuntanathan, V.: Robustness of the learning with errors assumption. In: ICS 2010, pp. 230–240 (2010)
38. Goldwasser, S., Rothblum, G.N.: On best-possible obfuscation. In: Vadhan, S.P. (ed.) TCC 2007. LNCS, vol. 4392, pp. 194–213. Springer, Heidelberg (2007)
39. Hemenway, B., Libert, B., Ostrovsky, R., Vergnaud, D.: Lossy encryption: Constructions from general assumptions and efficient selective opening chosen ciphertext security. In: Lee, D.H., Wang, X. (eds.) ASIACRYPT 2011. LNCS, vol. 7073, pp. 70–88. Springer, Heidelberg (2011)
40. Hemenway, B., Ostrovsky, R.: On homomorphic encryption and chosen-ciphertext security. In: Fischlin, M., Buchmann, J., Manulis, M. (eds.) PKC 2012. LNCS, vol. 7293, pp. 52–65. Springer, Heidelberg (2012)
41. Hemenway, B., Ostrovsky, R.: Building lossy trapdoor functions from lossy encryption. In: Sako, K., Sarkar, P. (eds.) ASIACRYPT 2013, Part II. LNCS, vol. 8270, pp. 241–260. Springer, Heidelberg (2013)
42. Hohenberger, S., Lewko, A., Waters, B.: Detecting dangerous queries: A new approach for chosen ciphertext security. In: Pointcheval, D., Johansson, T. (eds.) EUROCRYPT 2012. LNCS, vol. 7237, pp. 663–681. Springer, Heidelberg (2012)
43. Hohenberger, S., Rothblum, G.N., Shelat, A., Vaikuntanathan, V.: Securely obfuscating re-encryption. In: Vadhan, S.P. (ed.) TCC 2007. LNCS, vol. 4392, pp. 233–252. Springer, Heidelberg (2007)
44. Hohenberger, S., Sahai, A., Waters, B.: Replacing a random oracle: Full domain hash from indistinguishability obfuscation (2013), http://eprint.iacr.org/2013/509
45. Kiltz, E.: Chosen-ciphertext security from tag-based encryption. In: Halevi, S., Rabin, T. (eds.) TCC 2006. LNCS, vol. 3876, pp. 581–600. Springer, Heidelberg (2006)

46. Kiltz, E., Mohassel, P., O'Neill, A.: Adaptive trapdoor functions and chosen-ciphertext security. In: Gilbert, H. (ed.) EUROCRYPT 2010. LNCS, vol. 6110, pp. 673–692. Springer, Heidelberg (2010)
47. Lin, H., Tessaro, S.: Amplification of chosen-ciphertext security. In: Johansson, T., Nguyen, P.Q. (eds.) EUROCRYPT 2013. LNCS, vol. 7881, pp. 503–519. Springer, Heidelberg (2013)
48. Lynn, B.Y.S., Prabhakaran, M., Sahai, A.: Positive results and techniques for obfuscation. In: Cachin, C., Camenisch, J.L. (eds.) EUROCRYPT 2004. LNCS, vol. 3027, pp. 20–39. Springer, Heidelberg (2004)
49. Mol, P., Yilek, S.: Chosen-ciphertext security from slightly lossy trapdoor functions. In: Nguyen, P.Q., Pointcheval, D. (eds.) PKC 2010. LNCS, vol. 6056, pp. 296–311. Springer, Heidelberg (2010)
50. Myers, S., Shelat, A.: Bit encryption is complete. In: FOCS 2009, pp. 607–616 (2009)
51. Naor, M., Yung, M.: Universal one-way hash functions and their cryptographic applications. In: STOC 1989, pp. 33–43 (1989)
52. Naor, M., Yung, M.: Public-key cryptosystems provably secure against chosen ciphertext attacks. In: STOC 1990, pp. 427–437 (1990)
53. Peikert, C., Waters, B.: Lossy trapdoor functions and their applications. In: STOC 2008, pp. 187–196 (2008)
54. Rackoff, C., Simon, D.R.: Non-interactive zero-knowledge proof of knowledge and chosen ciphertext attack. In: Feigenbaum, J. (ed.) CRYPTO 1991. LNCS, vol. 576, pp. 433–444. Springer, Heidelberg (1992)
55. Rosen, A., Segev, G.: Chosen-ciphertext security via correlated products. In: Reingold, O. (ed.) TCC 2009. LNCS, vol. 5444, pp. 419–436. Springer, Heidelberg (2009)
56. Sahai, A.: Non-malleable non-interactive zero knowledge and adaptive chosen-ciphertext security. In: FOCS 1999, pp. 543–553 (1999)
57. Sahai, A., Waters, B.: How to use indistinguishability obfuscation: Deniable encryption, and more (2013), http://eprint.iacr.org/2013/454
58. Wee, H.: On obfuscating point functions. In: STOC 2005, pp. 523–532 (2005)
59. Wee, H.: Efficient chosen-ciphertext security via extractable hash proofs. In: Rabin, T. (ed.) CRYPTO 2010. LNCS, vol. 6223, pp. 314–332. Springer, Heidelberg (2010)

A   Basic Cryptographic Primitives

Public Key Encryption. A public key encryption (PKE) scheme Π consists of the three PPTAs $(\mathsf{PKG}, \mathsf{Enc}, \mathsf{Dec})$ with the following interface:

Key Generation:	**Encryption:**	**Decryption:**
$(pk, sk) \leftarrow \mathsf{PKG}(1^k)$	$c \leftarrow \mathsf{Enc}(pk, m)$	$m \text{ (or } \bot) \leftarrow \mathsf{Dec}(sk, c)$

where Dec is a deterministic algorithm, (pk, sk) is a public/secret key pair, and c is a ciphertext of a plaintext m under pk. We require for all $k \in \mathbb{N}$, all (pk, sk) output by $\mathsf{PKG}(1^k)$, and all m, it holds that $\mathsf{Dec}(sk, \mathsf{Enc}(pk, m)) = m$.

We define the *"public key length"* $\ell_{\mathsf{PK}}(k)$ as the length of pk output by $\mathsf{PKG}(1^k)$. Moreover, if Enc can encrypt k-bit plaintexts (for security parameter k), we define the *"randomness length"* $\ell_{\mathsf{R}}(k)$ and the *"ciphertext length"* $\ell_{\mathsf{C}}(k)$, respectively, as the length of randomness used by Enc and the length of ciphertexts output from Enc.

We say that a PKE scheme Π is ϵ-CPA secure if for all PPTAs $\mathcal{A} = (\mathcal{A}_1, \mathcal{A}_2)$ and for all sufficiently large $k \in \mathbb{N}$, it holds that $\mathsf{Adv}_{\Pi,\mathcal{A}}^{\mathsf{CPA}}(k) := 2 \cdot |\Pr[\mathsf{Expt}_{\Pi,\mathcal{A}}^{\mathsf{CPA}}(k) = 1] - 1/2| \leq \epsilon(k)$, where the experiment $\mathsf{Expt}_{\Pi,\mathcal{A}}^{\mathsf{CPA}}(k)$ is defined as in Fig. 4 (left). In the experiment, it is required that $|m_0| = |m_1|$.

$\mathrm{Expt}_{\Pi,\mathcal{A}}^{\mathrm{CPA}}(k):$	$\mathrm{Expt}_{\Gamma,\mathcal{A}}^{\mathrm{CCA2}}(k):$	$\mathrm{Expt}_{\mathcal{H},\mathcal{A}}^{\mathrm{UOW}}(k):$
$(pk, sk) \leftarrow \mathsf{PKG}(1^k)$	$(pk, sk) \leftarrow \mathsf{KKG}(1^k)$	$(m, \mathsf{st}) \leftarrow \mathcal{A}_1(1^k)$
$(m_0, m_1, \mathsf{st}) \leftarrow \mathcal{A}_1(pk)$	$\mathsf{st} \leftarrow \mathcal{A}_1^{\mathsf{Decap}(sk,\cdot)}(pk)$	$\kappa \leftarrow \mathsf{HKG}(1^k)$
$b \leftarrow \{0,1\}$	$(c^*, K_1^*) \leftarrow \mathsf{Encap}(pk)$	$m' \leftarrow \mathcal{A}_2(\mathsf{st}, \kappa)$
$c^* \leftarrow \mathsf{Enc}(pk, m_b)$	$K_0^* \leftarrow \{0,1\}^k;\ b \leftarrow \{0,1\}$	If $\mathsf{H}_\kappa(m') = \mathsf{H}_\kappa(m) \wedge m' \neq m$
$b' \leftarrow \mathcal{A}_2(\mathsf{st}, c^*)$	$b' \leftarrow \mathcal{A}_2^{\mathsf{Decap}(sk,\cdot)}(\mathsf{st}, c^*, K_b^*)$	then return 1 else return 0
Return $(b' \overset{?}{=} b)$	Return $(b' \overset{?}{=} b)$	

Fig. 4. The CPA security experiment for a PKE scheme Π (left), the CCA2 security experiment for a KEM Γ (center), and the security experiment for a UOWHF $\mathcal{H}$ (right)

Key Encapsulation Mechanism. A key encapsulation mechanism (KEM) Γ consists of the three PPTAs (KKG, Encap, Decap) with the following interface:

Key Generation:	**Encapsulation:**	**Decapsulation:**
$(pk, sk) \leftarrow \mathsf{KKG}(1^k)$	$(c, K) \leftarrow \mathsf{Encap}(pk)$	K (or $\perp$) $\leftarrow \mathsf{Decap}(sk, c)$

where Decap is a deterministic algorithm, (pk, sk) is a public/secret key pair, and c is a ciphertext of a session-key $K \in \{0,1\}^k$ under pk. We require for all $k \in \mathbb{N}$, all (pk, sk) output by $\mathsf{KKG}(1^k)$, and all $(c, K) \leftarrow \mathsf{Encap}(pk)$, it holds that $\mathsf{Decap}(sk, c) = K$.

We say that a KEM Γ is CCA2 secure if for all PPTAs $\mathcal{A} = (\mathcal{A}_1, \mathcal{A}_2)$, $\mathsf{Adv}_{\Gamma,\mathcal{A}}^{\mathrm{CCA2}}(k) := 2 \cdot |\Pr[\mathsf{Expt}_{\Gamma,\mathcal{A}}^{\mathrm{CCA2}}(k) = 1] - 1/2|$ is negligible, where the experiment $\mathsf{Expt}_{\Gamma,\mathcal{A}}^{\mathrm{CCA2}}(k)$ is defined as in Fig. 4 (center). In the experiment, $\mathcal{A}_2$ is not allowed to query c^*.

Universal One-Way Hash Function. We say that a pair of PPTAs $\mathcal{H} = (\mathsf{HKG}, \mathsf{H})$ is a universal one-way hash function (UOWHF) if the following two properties are satisfied: (1) On input 1^k, HKG outputs a hash-key κ. For any hash-key κ output from $\mathsf{HKG}(1^k)$, H defines an (efficiently computable) function of the form $\mathsf{H}_\kappa : \{0,1\}^* \to \{0,1\}^k$. (2) For all PPTAs $\mathcal{A} = (\mathcal{A}_1, \mathcal{A}_2)$, $\mathsf{Adv}_{\mathcal{H},\mathcal{A}}^{\mathrm{UOW}}(k) := \Pr[\mathsf{Expt}_{\mathcal{H},\mathcal{A}}^{\mathrm{UOW}}(k) = 1]$ is negligible, where the experiment is defined as in Fig. 4 (right).

B   Concrete Instantiations of Point/MBPF Obfuscators

Composable Point Obfuscator. Here we recall the point obfuscator due to Canetti [17] (which was originally introduced as a perfectly one-way hash function). Let $\mathbb{G}$ be a cyclic group with prime order p (where the size of p is determined by the security parameter k). Then, consider the following point obfuscator PO for $\mathrm{PF}(\mathbb{Z}_p)$:

$\mathsf{PO}(\mathcal{I}_\alpha):$ (where $\alpha \in \mathbb{Z}_p$) Pick a group element $r \leftarrow \mathbb{G}$ uniformly at random, and outputs the circuit $\mathcal{C}_{r,r^\alpha}(\cdot) : \mathbb{Z}_p \to \{\top, \perp\}$, where $\mathcal{C}_{A,B}$ is the circuit which takes $x \in \mathbb{Z}_p$ as input, and outputs $\top$ if $A^x = B$ and otherwise outputs $\perp$.

Bitansky and Canetti [7] showed that the above point obfuscator is t-composable, under a strong variant of the decisional Diffie-Hellman (DDH) assumption, called the t-strong vector DDH (t-SVDDH) assumption (see [7] for a formal definition).

$\mathsf{MBPO}(\mathcal{I}_{\alpha \to \beta})$:	$\mathcal{C}_{\mathsf{P}_0,\ldots,\mathsf{P}_t}(x)$:
$\quad \mathsf{P}_0 \leftarrow \mathsf{PO}(\mathcal{I}_\alpha)$ $\quad$ View β as $(\beta_1 \| \ldots \| \beta_t) \in \{0,1\}^t$ $\quad \alpha' \leftarrow \mathcal{X}_k \backslash \{\alpha\}$ $\quad$ For $i \in [t]$: $\qquad \mathsf{P}_i \leftarrow \begin{cases} \mathsf{PO}(\mathcal{I}_\alpha) & \text{if } \beta_i = 1 \\ \mathsf{PO}(\mathcal{I}_{\alpha'}) & \text{otherwise} \end{cases}$ $\quad$ End For $\quad$ Return $\mathsf{DL} \leftarrow \mathcal{C}_{\mathsf{P}_0,\ldots,\mathsf{P}_t}$.	$\quad$ If $\mathsf{P}_0(x) = \perp$ then return $\perp$ $\quad$ For $i \in [t]$: $\qquad \beta_i \leftarrow \begin{cases} 1 & \text{if } \mathsf{P}_i(x) = \top \\ 0 & \text{otherwise} \end{cases}$ $\quad$ End For $\quad$ Return $\beta \leftarrow (\beta_1 \| \ldots \| \beta_t)$.

Fig. 5. The construction of an MBPF obfuscator MBPO from a composable point obfuscator PO [19,7]. MBPO takes an MBPF $\mathcal{I}_{\alpha \to \beta}$ as input, and returns a circuit $\mathsf{DL} = \mathcal{C}_{\mathsf{P}_0,\ldots,\mathsf{P}_t}$ that is described in the right column.

We remark that as mentioned in [7], the point obfuscator based on the t-SVDDH assumption described here satisfies the re-randomizability in the sense of Definition 8. Specifically, we can just re-randomize two group elements in an obfuscated circuit output from PO without changing the point address.

WVG *Secure MBPF Obfuscator from Composable Point Obfuscator.* We recall the construction of an MBPF obfuscator based on a composable point obfuscator, due to Canetti and Dakdouk [19] and Bitansky and Canetti [7]. Let PO be a point obfuscator for $\mathsf{PF}(\mathcal{X})$ and let $t = t(k) > 0$ be a polynomial. Then an MBPF obfuscator MBPO for $\mathsf{MBPF}(\mathcal{X}, t)$ is constructed as in Fig. 5.

Based on the result of [19], Bitansky and Canetti [7] showed that if PO is $(t+1)$-composable, then the MBPF obfuscator MBPF constructed as in Fig. 5 is a WVG secure. By instantiating this conversion with the above mentioned point obfuscator, we obtain a WVG secure t-bit-output MBPF obfuscator under the $(t+1)$-SVDDH assumption.

Probabilistically Checkable Proofs of Proximity with Zero-Knowledge*

Yuval Ishai and Mor Weiss

Department of Computer Science, Technion, Haifa
{yuvali,morw}@cs.technion.ac.il

Abstract. A probabilistically Checkable Proof (PCP) allows a randomized verifier, with oracle access to a purported proof, to probabilistically verify an input statement of the form "$x \in L$" by querying only few bits of the proof. A *PCP of proximity* (PCPP) has the additional feature of allowing the verifier to query only few bits of the *input x*, where if the input is accepted then the verifier is guaranteed that (with high probability) the input is *close* to some $x' \in L$.

Motivated by their usefulness for sublinear-communication cryptography, we initiate the study of a natural zero-knowledge variant of PCPP (ZKPCPP), where the view of any verifier making a bounded number of queries can be efficiently simulated by making the same number of queries to *the input oracle alone*. This new notion provides a useful extension of the standard notion of zero-knowledge PCPs. We obtain two types of results.

- **Constructions.** We obtain the first constructions of query-efficient ZKPCPPs via a general transformation which combines standard query-efficient PCPPs with protocols for secure multiparty computation. As a byproduct, our construction provides a conceptually simpler alternative to a previous construction of honest-verifier zero-knowledge PCPs due to Dwork et al. (Crypto '92).
- **Applications.** We motivate the notion of ZKPCPPs by applying it towards sublinear-communication implementations of commit-and-prove functionalities. Concretely, we present the first sublinear-communication commit-and-prove protocols which make a *black-box* use of a collision-resistant hash function, and the first such multiparty protocols which offer *information-theoretic* security in the presence of an honest majority.

1 Introduction

In this work we initiate the study of probabilistically checkable proofs of proximity with a zero-knowledge property, and use such proofs to design efficient cryptographic protocols. Before describing our main results, we first give a short overview of probabilistic proof systems.

* Research supported by the European Union's Tenth Framework Programme (FP10/ 2010-2016) under grant agreement no. 259426 ERC-CaC. The first author was additionally supported by ISF grant 1361/10 and BSF grants 2008411 and 2012366.

Y. Lindell (Ed.): TCC 2014, LNCS 8349, pp. 121–145, 2014.

Probabilistically Checkable Proof (PCP) systems [1,2] are proof systems that allow an efficient randomized verifier, with oracle access to a purported proof, to probabilistically verify claims such as "$x \in L$" (for some input x and an NP-language L) by probing only few bits of the proof. The verifier accepts the proof of a true claim with probability 1 (the *completeness* property), and rejects false claims with high probability (the probability that the verifier accepts a false claim is called *the soundness error*). The celebrated PCP theorem [1,2,11] asserts that any NP language admits a PCP system with soundness error $1/2$ in which the verifier reads only a *constant* number of bits from the proof. The soundness error can be reduced to $2^{-\sigma}$ by running the verifier σ times.

Probabilistically Checkable Proofs of Proximity (PCPPs), also known as assignment testers, are proof systems that allow probabilistic verification of claims by probing few bits *of the input* and a purported proof. Needless to say, the verifier of such a system cannot generally be expected to distinguish inputs in the language from inputs that are not in the language, but rather it should accept every $x \in L$ with probability 1 and reject (with high probability) every input that is "far" from all $x' \in L$. First introduced in [7,12,11] as building blocks for the construction of more efficient PCPs, there are currently known PCPP systems for NP with parameters comparable to those of the best known PCP systems [8,23].

A seemingly unrelated concept is that of zero-knowledge (ZK) proofs [15], namely proofs that carry no extra knowledge other than being convincing. Combining the advantages of ZK proofs and PCPs, a *zero-knowledge PCP* (ZKPCP) is defined similarly to a traditional PCP, with the additional guarantee that the view of any (possibly malicious) verifier can be efficiently simulated up to a small statistical distance. ZKPCPs were first constructed by Kilian et al. [21], building on a previous weaker "honest-verifier" notion of ZKPCPs implicitly constructed by Dwork et al. [13]. More concretely, the work of [21] combines the weaker variant of ZKPCPs from [13] with an unconditionally secure oracle-based commitment primitive called a "locking scheme" to obtain ZKPCPs for NP that guarantee statistical zero-knowledge against *query-bounded* malicious verifiers, namely ones who are limited to asking at most $p(|x|)$ queries for some *fixed* polynomial p. A simpler construction of locking schemes was recently given in [17].

ZKPCPPs. In this work we put forward and study the new notion of *zero-knowledge PCPP* (or ZKPCPP for short), which extends the previous notion of ZKPCP and makes it more useful. A ZKPCPP is a PCPP with a probabilistic choice of proof and the additional guarantee that the view of any (possibly-malicious) verifier, making at most q^* queries to his input and his proof oracle, can be efficiently simulated by making the same number of queries *to the input alone*. ZKPCPPs are a natural extension of ZKPCPs (indeed, the existence of a ZKPCPP system implies the existence of a ZKPCP system with related parameters) and interesting objects on their own. As we explain next, they are also motivated by cryptographic applications that involve sublinear-communication zero-knowledge arguments on distributed or committed data.

To give the flavor of these applications, suppose that a database owner (prover) commits to a large sensitive database D by robustly secret-sharing it among a large number of potentially unreliable servers. At a later point in time, a user (verifier) may want to learn the answer to a query $q(D)$ on the committed database. ZKPCPPs provide the natural tool for efficiently verifying that the answer a provided by the prover is indeed consistent with the committed database, namely that $a = q(D)$, without revealing any additional information about the database to the verifier and a colluding set of servers. Concretely, the prover distributes between the servers a ZKPCPP asserting that the shares of D (the input) are indeed valid shares of some database D' such that $q(D') = a$. The verifier, by probing only few entries in the input and the proof string, is convinced that the shares held by the servers are indeed *close* to being consistent with valid shares of some database D' such that $q(D') = a$. If not "too many" servers are corrupted, the robustness of the underlying secret-sharing scheme guarantees that $D' = D$. (Unlike the ZKPCPP model, the answers provided by malicious servers may depend on the identity of the verifier's queries; this difficulty can be overcome by ensuring that with sufficiently high probability the verifier's queries are answered by honest servers.) The above approach can also be used for verifiable updates of a secret distributed database, where a ZKP-PCPP is used to convince a verifier that the shares of the updated version of the database are consistent with the original shares with respect to the update relation.

A similar idea can be used to get a sublinear-communication implementation of a "commit-and-prove" functionality in the two-party setting. Here the prover first succinctly commits, using a Merkle tree, to the shares of D. To later prove that $q(D) = a$, the prover again uses a Merkle tree to succinctly commit to a ZKPCPP asserting that the values it committed to in the previous phase are valid shares of some database D' such that $q(D') = a$. This gives the first sublinear-communication implementations of commit-and-prove which only make a *black-box* use of a collision-resistant hash function. (See Section 5.2 for a non-black-box alternative using standard sublinear arguments.)

1.1 Summary of Results

We introduce the notion of ZKPCPPs and construct *query-efficient* ZKPCPPs for any NP language L. More precisely, given an input $x \in L$, a corresponding witness w, and a zero-knowledge parameter q^*, the prover can efficiently generate a proof string π of length $\mathrm{poly}(|x|, q^*)$ which is statistical zero-knowledge against (possibly malicious) verifiers that make at most q^* queries to (x, π); by making only a polylogarithmic number of queries (in $|x|$ and q^*), an honest verifier can get convinced that x is at most δ-far from L, except with negligible soundness error, where δ can be any positive constant (or even inverse polylogarithmic). We then present applications of this construction to sublinear commit-and-prove protocols in both the two-party and the multiparty setting, as discussed above.

1.2   Techniques

Our main ZKPCPP construction is obtained by combining an (efficient) PCPP system without zero-knowledge with a protocol for secure multiparty computation (MPC), inheriting the efficiency from the PCPP component and the zero-knowledge from the MPC component. The transformation has two parts. The first consists of a *general* transformation from a PCPP and a secure MPC protocol to a PCPP system with zero-knowledge against semi-honest verifiers (HVZKPCPP, for *honest-verifier* ZKPCPP). The transformation can also be applied to PCPs, yielding an HVZKPCP comparable to that of [13] that is conceptually simpler. (Thus, our construction also simplifies the ZKPCP of Kilian et al. [21] which uses the HVZKPCP of [13] as a building block.)

The second part strengthens the zero-knowledge property to hold against *arbitrary* (query-bounded) malicious verifiers, by forcing the queries of any such verifier to be distributed (almost) as the queries of the honest verifier of the HVZKPCPP system. This part follows the approach of [21] of using a locking scheme. Concretely, we use the combinatorial construction of locking schemes from [17], except that to achieve negligible soundness error and negligible simulation error simultaneously we need to apply a natural amplification technique for reducing the error of the previous construction.

Organization. We first give the necessary preliminaries in Section 2. In Section 3 we describe the construction of a ZKPCPP from MPC protocols, and in Section 4 we state and prove our result regarding the existence of efficient ZKPCPP systems for NP. We describe our cryptographic applications in Section 5. Due to space limitations, we defer several definitions, constructions, and proofs, as well as the discussion regarding amplification of locking schemes, to the full version.

2   Preliminaries

We consider efficient probabilistic proof system for NP relations. (We refer to a relation rather than a language because we require the prover to be computationally efficient given an NP-witness.) Recall that an NP relation $\mathcal{R}$ is a polynomial-time recognizable binary relation which is *polynomially bounded* in the sense that there is a polynomial p such that if $(x, w) \in \mathcal{R}$ then $|w| \leq p(|x|)$. We refer to x as an *input* and to w as a *witness*.

We denote by $L_{\mathcal{R}}$ the NP language corresponding to $\mathcal{R}$, namely $L_{\mathcal{R}} = \{x : \exists w, (x, w) \in \mathcal{R}\}$. We say that x is δ-*far* from $L_{\mathcal{R}}$ (for some $\delta \in [0, 1]$) if the relative hamming distance between x and every $x' \in L_{\mathcal{R}}$ of the same length is more than δ.

We say that two distribution ensembles X_n, Y_n are computationally (resp. statistically) indistinguishable if every computationally bounded (respectively, computationally unbounded) distinguisher achieves only a negligible advantage in distinguishing a sample drawn according to X_n from a sample drawn according to Y_n.

A *probabilistic proof system* (P, V) for $\mathcal{R}$ consists of a PPT prover P, that on input (x, w) outputs a proof π, and a PPT verifier V that given input $1^{|x|}$ and oracle access to x (the *input oracle*) and π (the *proof oracle*) outputs either accept or reject. Intuitively, P tries to convince V of the claim "$x \in L_{\mathcal{R}}$" using w such that $(x, w) \in \mathcal{R}$. All the probabilistic proof systems studied in this work will have perfect completeness (i.e. V accepts true claims with probability 1). The system has soundness error ϵ if every input $x \notin L_{\mathcal{R}}$ is accepted by V with probability at most ϵ, *regardless* of the proof oracle. Our systems are sometimes parameterized by a statistical security parameter σ and a zero-knowledge parameter q^*. These parameters are given as additional inputs to both P and V.

In the following sections, we define several classes consisting of NP relations that have probabilistic proof systems with additional properties, and discuss the containment relations between these classes. (We associate each class of relations with the corresponding class of proof systems.) Every containment stated in this paper follows from *constructive transformations*, namely if we claim that $\text{Class}_1 \subseteq \text{Class}_2$, then the proof also shows an efficient transformation from a pair $(P, V) \in \text{Class}_1$ to a pair $(P', V') \in \text{Class}_2$.

PCPs. A standard probabilistically checkable proof (PCP) is a probabilistic proof system (P, V) in which V can query his input oracle x freely (that is, his queries to x are not counted towards the query complexity). We write $\mathcal{R} \in \text{PCP}_{\Sigma}[r, q, \epsilon, \ell]$ if $\mathcal{R}$ admits a PCP with verifier randomness complexity r, query complexity q, soundness error ϵ, and a proof π of length ℓ over the alphabet Σ.

PCPPs. Intuitively, the verifier of a PCPP system tries to validate the claim "$x \in L_{\mathcal{R}}$", while reading only *few* bits of x. Of course, V cannot generally be expected to distinguish the case that $x \in L_{\mathcal{R}}$ from the case that $x \notin L_{\mathcal{R}}$ but is very "close" to it. Instead, V is only expected to reject when x is "far" from $L_{\mathcal{R}}$. Concretely, A probabilistically checkable proof of proximity (PCPP) system with soundness error $0 \leq \epsilon < 1$ and proximity parameter $0 \leq \delta < 1$ is a probabilistic proof system (P, V) for which the following holds. For every x that is δ-far from $L_{\mathcal{R}}$, V accepts x with probability at most ϵ, *regardless* of his proof oracle. In this case we write $\mathcal{R} \in \text{PCPP}_{\Sigma}[r, q, \delta, \epsilon, \ell]$ where r, q, ℓ are as above. We refer to a PCPP system with proximity parameter $\delta = 0$ (i.e., in which soundness holds for every $x \notin L_{\mathcal{R}}$) as an *exact* PCPP.

We also consider systems with the following notion of *strong soundness*, where *every* $x \notin L_{\mathcal{R}}$ is rejected with probability that is proportional to its distance from $L_{\mathcal{R}}$. That is, there exists a function $\epsilon_S = \epsilon_S(\delta, |x|) : [0, 1] \times \mathbb{N} \to [0, 1]$ such that for every $\delta \in [0, 1]$, *every* x that is δ-far from $L_{\mathcal{R}}$ is accepted by V with probability at most $\epsilon_S(\delta, |x|)$. Such PCPPs are called *strong PCPPs*, see [8,22]. A strong PCPP system has *rejection ratio* β if every x that is δ-far from $L_{\mathcal{R}}$ is rejected with probability at least $\beta\delta$.

ZKPCPPs and ZKPCPs. We are interested in PCPs and PCPPs that reveal (almost) no information to verifiers who do not make "too many" queries. Intuitively, a probabilistic proof system is q^*-zero-knowledge if whatever a (possibly malicious) verifier learns by making $q' \leq q^*$ queries to x, π can be simulated by making q' queries *to x alone*. In particular, zero-knowledge of a PCP system implies the witness is entirely hidden (the queries of the verifier to the input oracle are not counted towards the query complexity, so the simulator can query all of x, and consequently only the witness is hidden), while in a zero-knowledge PCPP system not only is the witness hidden, but so is most of x. Notice that the prover in a zero-knowledge proof system must be *probabilistic* (while the prover in standard proof systems for NP can be deterministic).

More formally, let (P, V) be a probabilistic proof system, and let V^* be a (possibly malicious) *q-bounded* verifier (namely a verifier that never makes more than q queries). We compare the real-life interaction between P and V^* with an *ideal-world* interaction, in which a simulator Sim with oracle access to V^* interacts with a trusted third party (TTP) that knows only x. Denote the distribution ensemble describing the view of V^* with oracles x, π (where π was honestly generated by P on input x, w) by $\mathsf{View}_{V^* x, \pi}$, and let q_{V^*} denote the total number of queries V^* sent to the input and proof oracles. Let $\mathsf{Real}_{V^*, P}(x, w) = (\mathsf{View}_{V^* x, \pi}, q_{V^*})$. Similarly, for an ideal-world simulator Sim let $\mathsf{Sim}(x)$ denote the distribution ensemble describing the output of Sim (after making his queries to x), and let q_S denote the number of queries Sim made. We define $\mathsf{Ideal}_{\mathsf{Sim}}(x) = (\mathsf{Sim}(x), q_S)$.

We say that (P, V) is (ϵ, q^*)-zero knowledge with respect to $\mathcal{R}$ (for some $\epsilon \in [0, 1]$ and $q^* \in \mathbb{N}$) if for every real-life q^*-bounded verifier V^* there exists an ideal-world simulator Sim such that for every $(x, w) \in \mathcal{R}$, we have $\mathsf{Real}_{V^*, P}(x, w) \approx^\epsilon \mathsf{Ideal}_{\mathsf{Sim}}(x)$, where $\approx^\epsilon$ denotes statistical distance of ϵ. If (P, V) is $(0, q^*)$-zero-knowledge we say that it has *perfect q^*-zero-knowledge*, and write $\mathsf{Real}_{V^*, P}(x, w) \equiv \mathsf{Ideal}_{\mathsf{Sim}}(x)$. We may choose ϵ, q^* to be functions of a security parameter σ and the input size $|x|$. By default, we will make the stronger requirement that there exist a single, PPT *black-box* simulator S such that for every q^*-bounded V^*, the simulator $\mathsf{Sim} = \mathsf{S}^{V^*}$ satisfies the above requirement. Moreover, S can only interact with V^* in a straight-line fashion (i.e., it cannot rewind V^*). The latter straight-line simulation requirement is useful for one of our motivating applications.

Remark 1. The above notion of zero-knowledge requires that the number of input bits read by the simulator be the same as the *total* number of bits read by the verifier. One may consider stronger notions which require that the number of input bits read by the simulator coincide with the number of input bits read by the verifier, or even that the *same* input bits are read by the verifier and the simulator. The latter is captured by letting Real and Ideal, instead of including the number of queries V^* and Sim made (respectively), include the *specific indices* V^*, Sim queried in x. Our constructions do not satisfy these stronger notions.

Notation 1. *If a system* $(P, V) \in \mathsf{PCPP}_\Sigma[r, q, \delta, \epsilon_S, \ell]$ *for relation* $\mathcal{R}$ *guarantees* (ϵ_{ZK}, q^*)-*zero-knowledge, we say that* (P, V) *is a* q^*-*zero-knowledge PCPP and write* $(P, V) \in \mathsf{ZKPCPP}_\Sigma[r, q, \epsilon_{ZK}, \delta, \epsilon_S, \ell]$. *Similarly, if* $(P, V) \in \mathsf{PCP}_\Sigma[r, q, \epsilon_S, \ell]$ *for relation* $\mathcal{R}$ *guarantees* (ϵ_{ZK}, q^*)-*zero-knowledge, we write* $(P, V) \in \mathsf{ZKPCP}_\Sigma[r, q, \epsilon_{ZK}, \epsilon_S, \ell]$.

We also consider the following *honest-verifier* variant of zero-knowledge which is used as a simpler building block and is also of independent interest. We say that (P, V) has *honest-verifier zero-knowledge* (HVZK) with statistical distance $\epsilon \in [0, 1]$ if for the honest verifier V there exists a PPT simulator Sim such that the previous zero-knowledge requirement holds, namely for every pair $(x, w) \in \mathcal{R}$, $\text{Real}_{V,P}(x, w) \approx^\epsilon \text{Ideal}_{\text{Sim}}(x)$.

Secure MPC. We follow the terminology and notation of [16]. Let $P_1, ..., P_n$ be n parties, where every party P_i holds a private input z_i (we allow z_i to be empty, which we denote by $z_i = \lambda$). We consider protocols for securely realizing an n-party functionality g, that maps the tuple of inputs $(z_1, ..., z_n)$ to an output in $\{0, 1\}$. All parties are expected to have the same output. The *view* of a party P_i, denoted V_i, includes his private input z_i and a random input r_i, together with all the messages that P_i received during the protocol execution. (The messages P_i sends during the execution, as well as his local output, are determined by this view.) A pair V_i, V_j of views are *consistent with respect to z_i, z_j and Π*, if the outgoing messages (from P_i to P_j) implicit in V_i in an execution of Π on inputs z_i, z_j, are identical to the incoming messages (from P_i to P_j) reported in V_j, and vice versa. Consistency between a view and one of its incident communication channels is defined similarly.

We consider the execution of the protocol in the presence of an adversary $\mathcal{A}$ who may corrupt up to t parties. A *semi-honest* adversary can only passively corrupt parties (i.e., it does not modify their behavior but can learn their entire view), whereas a *malicious* adversary can arbitrarily modify the behavior of corrupted parties. A *static* adversary is restricted to pick the set of corrupted parties in advance, whereas an *adaptive* adversaries may pick them one by one, choosing the next party to corrupt based on its view so far.

A protocol Π realizes a deterministic n-party functionality $g(z_1, ..., z_n)$ with *perfect correctness* if for all inputs $z_1, ..., z_n$, when no party is corrupted, all parties output $g(z_1, ..., z_n)$. For a security threshold $1 \le t \le n$, we say that Π realizes g with *perfect t-privacy* if for every semi-honest adversary $\mathcal{A}$ corrupting a set $T \subseteq [n], |T| \le t$ of parties there exists a simulator Sim that can perfectly simulate the view of $\mathcal{A}$ given only the inputs of corrupted parties and the output. One can naturally define a variant of privacy that applies to adaptive adversaries. (In the adaptive case, we require the existence of a PPT black-box straight-line simulator.) We say that Π realizes g with *perfect T-robustness* (for some subset $T \subseteq [n]$) if for every *malicious* adversary $\mathcal{A}$ corrupting the parties in T, and for every tuple $z_{\bar{T}}$ of inputs of uncorrupted parties, the following holds. If g evaluates to 0 on *all* choices of inputs z consistent with $z_{\bar{T}}$, then all uncorrupted parties are guaranteed to output 0.[1] This property is implied by the standard simulation-based notion of security against malicious adversaries.

---

[1] Notice that we only define robustness for the case that g evaluates to 0, which suffices for our purposes since we only consider functions g representing relations $\mathcal{R}$. More specifically, robustness is used to construct *sound* proofs systems, where the corrupted party is the party holding the witness (and the bits of the input are partitioned between the honest parties). As soundness concerns the case $x \notin L_{\mathcal{R}}$,

Locking schemes [21,17]. Informally, a locking scheme allows a sender S to commit some secret to a receiver R, such that given a key the receiver can "open" the lock and retrieve the secret, whereas without the key this is almost impossible (for a query-bounded receiver). More formally, a locking scheme (S, R) for message space $\mathcal{W}$ consists of a sender S and a receiver R that interact in two phases: *Commitment*, during which S sends a locking oracle L_w to R, thus committing to some $w \in \mathcal{W}$; and *Decommitment*, in which S decommits w by sending R a key K_w that "opens" the lock. The requirements from the locking scheme are as follows. First, for every honestly-generated pair (L_w, K_w), R with key K_w and oracle access to L_w outputs w at the end of the decommitment phase with probability 1 (this is called *perfect completeness*). Second, the scheme is *hiding*, namely without knowing the key, R learns nothing about w, even if he probes many bit coordinates of the lock. Thirdly, we require *binding*, i.e. every (possibly ill-formed) lock commits the sender to *some* value w'. (See [17] for formal definitions.)

3   From MPC Protocols to (Inefficient) EZKPCPPs

We show a general connection between secure MPC protocols and (exact) ZKPCPPs. More specifically, given an NP-relation $\mathcal{R}$, we define the *characteristic function* $g_{\mathcal{R}_m} : \{0,1\}^* \times \{0,1\}^m \to \{0,1\}$ of $\mathcal{R}_m = \{(x, w) \in \mathcal{R} : |x| = m\}$ (or simply g, when $\mathcal{R}, m$ are clear from the context) as follows. $g_{\mathcal{R}_m}(w, x_1, ..., x_m) = 1$ if and only if $(x_1 \circ ... \circ x_m, w) \in \mathcal{R}_m$. Following techniques of Ishai et al. [16], we transform a protocol Π securely realizing $g_{\mathcal{R}_m}$ into an EZKPCPP system for $\mathcal{R}_m$, with perfect zero-knowledge against malicious (query-bounded) verifiers. Concretely, for any $t = t(m)$, if the underlying n-party protocol is t-private (for some $n = n(m, t)$), then the system has perfect zero-knowledge against t-bounded verifiers.

Construction 2 (EZKPCPP from MPC.). *The system is parameterized by a length parameter $m \in \mathbb{N}$, a zero-knowledge parameter $t = t(m)$, and employs an n-party protocol Π realizing $g_{\mathcal{R}_m}$ with perfect adaptive t-privacy and perfect static 1-robustness.[2] We assume without loss of generality that the bits $x_1, ..., x_m$ are given as input to $P_1, ..., P_m$.*

Prover algorithm. On input $(x, w), 1^t$ the prover P_E emulates "in his head" a random execution of Π on inputs $(w, x_1, ..., x_m)$. Let $\mathsf{V}_0, ..., \mathsf{V}_n$ be the views of

---

i.e., $(x, w^*) \notin \mathcal{R}$ for every "witness" w^*, then g evaluates to 0 on *every* input of the party holding the witness.

[2] We could get the same results using secure protocols in the *semi-honest* model, by sharing the witness w between $t + 1$ parties (similar to the construction of zero-knowledge protocols from MPC of [16]). However, this solution requires a larger number of parties than in our solution. We prefer to rely on robust protocols, since it suffices to have $\{P_0\}$-robustness, and such protocols can be instantiated by more efficient protocols in the semi-honest model (e.g., [5]).

$P_0, ..., P_n$ in this execution, and for every $0 \leq i < j \leq n$, let $\mathsf{Ch}_{i,j}$ describe the messages sent over the communication channel between i, j during the execution. P_E outputs the proof π consisting of the concatenation of the views $\mathsf{V}_1, ..., \mathsf{V}_n$ and the communication channels $\mathsf{Ch}_{i,j}$ for $0 \leq i < j \leq n$, where every view and communication channel constitutes a symbol of the proof. (Notice that the proof does not include the view V_0, since V_0 reveals the witness w.)

Verifier algorithm. The verifier V_E with input $1^t, m$ and oracles x, π flips a random coin to decide which test to perform. If the outcome was 0, V_E picks a random view $\mathsf{V}_i, i \in_R [m]$, and verifies that the input of P_i in the protocol execution was x_i (this ensures the protocol execution is consistent with x). If the outcome was 1, V_E picks $i \in_R [n]$ and $j \in_R \{0, 1, ..., n\}, i \neq j$ and verifies that V_i is consistent with $\mathsf{Ch}_{i,j}$ (this ensures the emulated execution is consistent). In both cases V_E verifies that the output of the protocol (implicit in V_i) was 1.

Lemma 1 (From MPC to EZKPCPPs). For any NP-relation $\mathcal{R} = \mathcal{R}(x, w)$, Construction 2 is a perfectly t-zero-knowledge exact-PCPP for $\mathcal{R}$, with soundness error $\left(1 - \Omega\left(\frac{1}{n^2}\right)\right)$, where the honest verifier makes only 2 oracle queries.

Proof (sketch). Set some $m \in \mathbb{N}$ and let $\Pi = \Pi_m$. The perfect completeness follows from the perfect completeness of Π. As for soundness, if $x \notin L_\mathcal{R}$ then a malicious prover has three possible courses of action. First, he can emulate an execution of Π on some $x' \neq x$, which is detected by the verifier with probability at least $\frac{1}{2m} \geq \frac{1}{n^2}$. Second, he can provide a proof in which some view V_i is inconsistent with some incident communication channel (either $\mathsf{Ch}_{j,i}, 0 \leq j < i$ or $\mathsf{Ch}_{i,j}, 1 \leq i < j \leq n$), which V_E detects with probability at least $\frac{1}{2n(n+1)}$. Thirdly, P_E can generate a proof in which every view V_i is consistent with all incident communication channels (with respect to Π, x). In this case, it can be shown that there exists an execution of Π on x, in which all parties (except, possibly, P_0) are honest, such that the view of P_i in the execution is V_i, and the messages exchanged between P_i, P_j are according to $\mathsf{Ch}_{i,j}$. Therefore, the P_0-robustness of Π guarantees that the output implicit in $\mathsf{V}_1, ..., \mathsf{V}_n$ is 0, so V_E rejects (with probability 1). We note that the soundness error can be reduces by repetition ($\lceil \frac{t}{2} \rceil$ iterations can be performed while preserving zero-knowledge). The t-zero-knowledge follows from the privacy of Π. Indeed, for every $i, j \in [n]$, the communication channel $\mathsf{Ch}_{i,j}$ can be reconstructed from V_i (and from V_j). Therefore, the answers to the queries of every (possibly malicious) verifier V^* can be simulated given the views of (a specific subset of) t parties $P_{i_1}, ..., P_{i_t}$. By the privacy of Π, these views can be perfectly simulated given $x_{i_1}, ..., x_{i_t}$. Therefore, the view of V^* can be simulated with only t TTP-queries. $\square$

Notice that if we only require *honest-verifier* zero-knowledge, then it suffices for Π to be 1-private. (P_E, V_E) is *weakly-sound* in the sense that its soundness error is large. (As noted above, the error can be reduced by repetition, but this increases the query complexity and again requires Π to be private against larger coalitions of parties.)

We note that Construction 2 is inefficient in the sense that the alphabet size may be exponential in m, t, since it contains symbols for all possible views and communication channels in Π. This inefficiency will not pose a problem in later constructions. Indeed, the construction of honest-verifier ZKPCPPs uses EZKPCPPs only for *constant sized* claims, and the construction of a ZKPCPP (with zero-knowledge against malicious verifiers) uses EZKPCPPs for claims of size $O(\sigma)$, where σ denotes the security parameter of the ZKPCPP. Moreover, we will only use EZKPCPPs for relations in P.

Basing Construction 2 on efficient multiparty protocols that can withstand a constant fraction of corrupted parties, we obtain the following result.

Corollary 1 (EZKPCPPs for NP). *Let $\mathcal{R} = \mathcal{R}(x, w)$ be an NP-relation. Then for every zero-knowledge parameter $t = t(|x|)$, $\mathcal{R} \in$ EZKPCPP$_\Sigma [r, q, \epsilon_{ZK}, \delta, \epsilon_S, \ell]$, where $\Sigma = 2^{\mathrm{poly}(t,|x|)}$, $r = O(\log t + \log |x|)$, $q = 2$, $\epsilon_{ZK} = 0$, $\epsilon_S = 1 - \frac{1}{\mathrm{poly}(t,|x|)}$, $\ell = \mathrm{poly}(t, |x|)$, and the EZKPCPP system is t-zero-knowledge. Furthermore, $\mathcal{R}$ has an EZKPCPP system over the binary alphabet with $q = 3$ (and $r, \epsilon_{ZK}, \epsilon_S, \ell$ are as above).*

The existence of the EZKPCPP over a large alphabet follows from Contruction 2. The natural approach towards reducing the alphabet size, is to define the proofs over Σ and represent every view and communication channel using several symbols, and have the verifier read all the bits corresponding to the symbol he wishes to query. However, this solution does not preserve zero-knowledge. Indeed, it increases the query complexity of the honest verifier, and consequently a malicious (even query bounded) verifier may query many *parts* of views, thus potentially breaking the privacy of the underlying protocol, and consequently the zero-knowledge of the system.

Proof (sketch). The existence of an EZKPCPP system over a large alphabet follows from Lemma 1, when Construction 2, based on an efficient multiparty protocol (e.g., the protocols of [5]).

The EZKPCPP over a binary alphabet, denoted $(P_{\mathrm{bin}}, V_{\mathrm{bin}})$, is obtained using techniques of Dwork et al. [13]. The general idea is to represent a proof generated by P_E over the binary alphabet, but avoid increasing the query complexity of the honest verifier by having V_{bin} *probabilistically* check that the oracles satisfy the decision circuit of V_E. More specifically, P_{bin} on input (x, w) uses P_E to generate a proof π_E. Then, for every random string r of V_E, P_{bin} writes down the assignment A_r to the inner gates of the verification circuit of V_E (i.e., the circuit V_E uses when he has randomness r). P_{bin} outputs the proof π_E, concatenated with the assignments A_r for all random strings r of V_E. (The proof should actually include, for every r, a proof that A_r is consistent with the verification circuit of V_E and the bits of x, π_E that V_E queries. We refer the reader to [20] or the full version for additional details. We note that these "proofs" have length $O(A_r)$ so they can be ignored when analyzing the efficiency properties of the system.) To verify that $x \in L_{\mathcal{R}}$, V_{bin} picks a random string r for V_E, and checks that x, π_E, A_r satisfy a random gate in the verification circuit of V_E.

Notice that V_E reads only poly $(|x|, t)$ bits from his oracles (i.e., the symbols he reads can be represented using poly $(|x|, t)$ bits), and his verification circuit has size poly $(|x|, t)$ (since V_E is efficient in the number of bits he reads). Therefore, the randomness complexity increases by only $O(\log t + \log|x|)$ (V_{bin} needs to pick a random gate in the verification circuit of V_E), and the proof increases by a factor of poly $(|x|, t)$ (there are poly $(|x|, t)$ random strings for V_E, and every random string corresponds to a circuit of size poly $(|x|, t)$). Moreover, the soundness error degrades only by a factor of $\frac{1}{\mathrm{poly}(|x|, t)}$, since in every verification circuit of V_E which x, π_E do not satisfy, at least one gate (out of poly $(|x|, t)$) is not satisfied. Regarding zero-knowledge, notice that every t-bounded verifier algorithm V_{bin}^* in the modified system induces a verifier algorithm V^* in the original system, whose queries correspond to the queries V_E makes in t independent invocations. Therefore, the view of V^* can be simulated given the views $V_{i_1}, ..., V_{i_t}$ which V^* queries, and these views can be simulated given the inputs of $P_{i_1}, ..., P_{i_t}$ in Π. As the view of V_{bin}^* can be reconstructed from the view of V^*, this implies the system is t-zero-knowledge.   □

Remark 2 (Strong HVZK). Both of the EZKPCPP systems mentioned above have a *stronger* honest-verifier zero-knowledge guarantee, which is formalized next. For an integer parameter c and a soundness parameter ϵ, we say a proof system has (ϵ, c)-strong honest-verifier zero-knowledge, if there exists a straight-line simulator Sim such that the following holds for every $c' \leq c$ and every $(x,) \in \mathcal{R}$. Sim interacts with $V^{c'}$ ($V^{c'}$ denotes c' random and independent invocations of the honest verifier V) without rewinding the verifier. During the simulation, Sim makes only c' TTP queries, and generates a view which is statistically close (up to distance ϵ) to the real-world view of $V^{c'}$, when $V^{c'}$ has oracle access to x and a random honestly-generated proof for x. Both our EZKPCPP systems have *perfect* t-strong honest-verifier zero-knowledge (where t is the zero-knowledge parameter), i.e., the simulated view is indistinguishable from the real world view. (This stronger zero-knowledge feature will be used in Section 4.1 to construct an HVZKPCPP with similar properties, which in turn will be used to construct a ZKPCPP in Section 4.2.)

4   From Efficient PCPPs to Efficient ZKPCPPs

We show a general transformation from PCPPs to ZKPCPPs, and construct an efficient ZKPCPP system for any NP-relation $\mathcal{R}$. (Using the same methods one can transform a PCP into a ZKPCP.) First, we use proof-composition techniques to transform a PCPP into an HVZKPCPP, using an EZKPCPP as the inner proof system. Then, we show a transformation from an HVZKPCPP and a locking scheme, into a ZKPCPP that guarantees zero-knowledge against *malicious* query-bounded verifiers. Finally, by applying the first transformation to an efficient PCPP, and the second to an efficient locking scheme and to the HVZKPCPP obtained through the first transformation, we get an efficient ZKPCPP.

4.1 From PCPPs to HVZKPCPPs

In this section we present the general transformation from PCPPs to HVZKPCPPs. We first describe a basic transformation (with weak parameters), and then improve it. The high-level idea is to use proof composition (see, e.g., [7,12,11]). In the context of PCPs, proof composition is used to reduce the query complexity of a PCP verifier: instead of making q queries and applying some predicate to the oracle answers, the verifier delegates the verification task to an "inner verifier", who probabilistically checks the oracle satisfies the decision circuit of the outer verifier (the query complexity is reduced since the inner verifier makes *less* queries than the original verifier). Intuitively, as the verifier of the composed system emulates the verification procedure of the *inner* verifier, then the composed system should have a zero-knowledge guarantee if the inner system does (even when the outer system has *no* zero-knowledge guarantee). The advantage in using composition in this case is similar to the advantage achieved by composition in standard PCP constructions: the inner system may be *very* inefficient, but the composed system is efficient if the outer system is.

More specifically, let $\mathcal{R}$ be a relation, and let $(P_{\text{out}}, V_{\text{out}})$ be a PCPP for $\mathcal{R}$ with soundness error ϵ and proximity parameter δ, where V_{out} makes q oracle queries and uses r random bits. Then every random string rand of V_{out} corresponds to a set of q queries, and a predicate $\varphi_{\text{rand}} : \{0,1\}^q \rightarrow \{0,1\}$ describing the decision of V_{out}. Denote the vector of the 2^r predicates corresponding to *all* random strings of V_{out} by $(\varphi_1, ..., \varphi_{2^r})$, then the following holds. If $x \in L_{\mathcal{R}}$ and π was honestly-generated by P_{out} for x, then (x, π) satisfies φ_i for every $1 \leq i \leq 2^r$, and if x is δ-far from $L_{\mathcal{R}}$ then for *any* "proof" π^*, (x, π^*) satisfies at most an ϵ-fraction of $\varphi_1, ..., \varphi_{2^r}$. In standard proof-composition constructions, the prover concatenates π with proofs $\pi_{\text{in}}^1, ..., \pi_{\text{in}}^{2^r}$, where π_{in}^i should convince the inner verifier that (x, π) satisfies φ_i. The verifier then runs the outer verifier to generate $rand$ and φ_{rand}, and the inner verifier to check that (x, π) satisfies φ_{rand}. However, the verification procedure of the inner verifier may query π, and is consequently *not* zero-knowledge (since π may reveal additional information about x). Therefore, we need to use proof composition, together with a form of *secret-sharing* which guarantees that π also remains hidden. Concretely, we replace every proof bit π_i (i.e., every predicate variable corresponding to a proof bit) with a set of bits $\{\pi_{i,j}\}$ (i.e., with a set of *new* predicate variables) such that π_i is reconstructable given all the new bits $\pi_{i,j}$, but (any) subset of the new bits $\{\pi_{i,j}\}$ reveals *no* information about π_i. We refer to a predicate obtained thorough this "secret-sharing" transformation as a *private* predicate, since a partial assignment to few predicate variables reveals no information about the proof π.

More specifically, given a predicate $\varphi : \{0,1\}^q \rightarrow \{0,1\}$ over variables $v_1, ..., v_q$, we partition its variables to a set V_{inp} of *input variables* (i.e., variables corresponding to bits of the input oracle) and a set V_{pf} of *proof variables*. The k-*private form of* φ (for any $k \in \mathbb{N}$), denoted $\varphi(k)$, is obtained from φ by replacing every proof variable $v_i \in V_{\text{pf}}$ with the exclusive-or of $k + 1$ new variables $y_{i,1}, ..., y_{i,k+1}$ (i.e., every appearance of v_i is replaced with $\overset{k+1}{\underset{j=1}{\oplus}} y_{i,j}$). The private-predicates relation $\mathcal{R}_{\text{priv}}$ consists of all pairs of private predicates and satisfying assignments for them, i.e.,

$$\mathcal{R}_{\text{priv}} = \{((w, \varphi(k)), \lambda) : w \text{ satisfies } \varphi(k)\}.$$

We now describe the basic transformation from PCPPs to (weakly-sound) HVZKPCPPs.

Construction 3 (HVZKPCPPs from PCPPs.). *The basic HVZKPCPP system, denoted (P_B, V_B), will be the composition of a PCPP system $(P_{\text{out}}, V_{\text{out}})$ for $\mathcal{R}$ as the outer PCPP system, and the inner EZKPCPP system $(P_{\text{in}}, V_{\text{in}})$ for the relation $\mathcal{R}_{\text{priv}}$. The system is parameterized by $d \in \mathbb{N}$ which determines the zero-knowledge requirement from the inner system. (Without loss of generality, $d \geq 3$.)*

Prover algorithm. On input $1^d, x, w$ such that $(x, w) \in \mathcal{R}$, P_B:

- *Generates the verification predicates $\varphi_1, ..., \varphi_m$ of V_{out} (for $m := 2^r$, where r denotes the length of the randomness of V_{out}), and a proof $\pi \in P_{\text{out}}(x, w)$.*
- *Generates the d-private form $\varphi_i(d)$ of every predicate φ_i, and replaces every proof variable $v_j \in V_{\text{pf}}$ with the exclusive-or of $d + 1$ new variables $y_{j,1}, ..., y_{j,d+1}$, such that $\overset{d+1}{\underset{k=1}{\oplus}} \pi(d)_{y_{j,k}} = \pi_{v_j}$. (As (x, π) is interpreted as an assignment to the predicates $\varphi_1, ..., \varphi_m$, this transforms (x, π) into an assignment to the private predicates. We denote this partial assignment to $\varphi_1(d) \wedge ... \wedge \varphi_m(d)$ by $\pi(d)$.)[3]*
- *"Proves" that $(x, \pi(d))$ satisfies the private predicates. Concretely, let $(x, \pi(d))_i$ denote the restriction of $(x, \pi(d))$ to the variables of the private predicate $\varphi_i(d)$. Then P_B generates a proof $\pi_{\text{in}}^i \in P_{\text{in}}(1^d, ((x, \pi(d))_i, \varphi_i(d)), \lambda)$ for the claim $(((x, \pi(d))_i, \varphi_i(d)), \lambda) \in \mathcal{R}_{\text{priv}}$.*
- *Outputs the proof $\pi_B = \pi_{\text{in}}^1 \circ ... \circ \pi_{\text{in}}^m \circ \pi(d)$.*

Verifier algorithm. V_B on input $1^d, |x|$ and given oracle access to x and a proof $\pi_B = \pi_{\text{in}}^1 \circ ... \circ \pi_{\text{in}}^m \circ \pi(d)$, picks an $i \in_R [m]$, uses V_{out} to generate the predicate φ_i, and transforms it into the d-private predicate $\varphi_i(d)$. Then, V runs V_{in} to check that $(x, \pi(d))_i$ satisfies $\varphi_i(d)$ ($(x, \pi(d))_i$ is used as the input oracle, and π_{in}^i as the proof oracle, of V_{in}).

Lemma 2. *Let $\mathcal{R} \in \text{PCPP}[r, q, \delta, \epsilon_{\text{out}}, \ell]$ with the PCPP system $(P_{\text{out}}, V_{\text{out}})$. Let $(P_{\text{in}}, V_{\text{in}})$ be a perfectly d-zero-knowledge EZKPCPP system for $\mathcal{R}_{\text{priv}}$ with soundness error $\epsilon_{\text{in}}(\ell, d)$ (where ϵ_{in} is non-decreasing and ℓ is the length of the input to the EZKPCPP system), in which the honest verifier makes $q_{\text{in}} \leq d$ queries. Then Construction 3, based on $(P_{\text{out}}, V_{\text{out}})$ and $(P_{\text{in}}, V_{\text{in}})$, is a PCPP*

---

[3] We say that π is a *partial* assignment to the predicates, since some of the variables are assigned values by the input x.

system for $\mathcal{R}$ with perfect completeness, perfect honest-verifier zero-knowledge, and soundness error $\epsilon_{\mathrm{out}} \left(1 - \epsilon_{\mathrm{in}} \left(\ell \left(d+1\right), d\right)\right) + \epsilon_{\mathrm{in}} \left(\ell \left(d+1\right), d\right)$. Moreover, V_B makes only q_{in} queries, and the prover generates proofs of length $O_{q,d} \left(\ell + 2^r\right)$.

Proof (sketch). The completeness follows directly from the completeness of the underlying proof systems. As for zero-knowledge, the zero-knowledge of the inner EZKPCPP system guarantees there exists a simulator $\mathsf{Sim}_{\mathrm{in}}$ that can perfectly simulate the view of the honest verifier V_{in} (since V_{in} is d-query bounded), given oracle access to the input oracle of V_{in} (through the TTP). Notice that the "input oracle" of V_{in} is of the form $(x, \pi\,(d))_i$ for some $i \in [m]$, i.e., $\mathsf{Sim}_{\mathrm{in}}$ may query bits of $\pi\,(d)$. However, in a *random* proof $\pi_B \in_R P_B \left(1^d, x, w\right)$, $\pi\,(d)$ is a *random* sharing of π (that is, every set of bits $\pi_{j,1}, \ldots, \pi_{j,d+1}$ that correspond to a bit π_j of π, is random such that $\pi_{j,1} \oplus \ldots \oplus \pi_{j,d+1} = \pi_j$). Therefore, Sim can simulate the view of V_B by running $\mathsf{Sim}_{\mathrm{in}}$, and answering his TTP queries with random bits. These bits are distributed as the answers $\mathsf{Sim}_{\mathrm{in}}$ would have been given by its TTP, so it suffices to prove indistinguishability conditioned on the "input" oracle $\pi\,(d)$. In this case, indistinguishability follows from the zero-knowledge of $(P_{\mathrm{in}}, V_{\mathrm{in}})$.

Regarding soundness, if x is δ-far from $L_{\mathcal{R}}$ then the soundness of $(P_{\mathrm{out}}, V_{\mathrm{out}})$ guarantees that for every "proof" π^* at most an ϵ_{out}-fraction of the predicates $\varphi_1, \ldots, \varphi_m$ are satisfied by (x, π^*). Consequently, for every "proof" $\pi^*\,(d)$, $(x, \pi^*\,(d))$ satisfies at most an ϵ_{out}-fraction of the private predicates $\varphi_1\,(d), \ldots, \varphi_m\,(d)$. If V_B chooses to verify a predicate $\varphi_i\,(d)$ that is *not* satisfied by $x \circ \pi^*\,(d)$, then the soundness of $(P_{\mathrm{in}}, V_{\mathrm{in}})$ guarantees that he accepts with probability at most $\epsilon_{\mathrm{in}} \left(q\left(d+1\right), d\right)$. (Indeed, every predicate contains at most q proof variables, so V_{in} has input of length at most $q\left(d+1\right)$, and ϵ_{in} is non-decreasing.) □

It is clear from Lemma 2 that the soundness error degrades through this transformation (since the soundness error of the composed system depends not only on the soundness error of the outer PCPP system, but also on that of the inner EZKPCPP system). Therefore, our next goal is to reduce the soundness error.

Reducing the soundness error. The main idea is to have the verifier repeat the verification procedure of V_B. However, we must change the ZKPCPP itself since repetition does not necessarily preserve zero-knowledge. (That is, if the verifier simply repeats the verification procedure, then his queries may exceed the upper bound for which zero-knowledge is guaranteed.) Intuitively, the prover can generate several independent copies of *basic proofs* (i.e., a proof generated by P_B), and the verifier can repeat the basic verification scheme, using a "fresh" proof in every iteration. This "assumption" that the verifier uses every proof at most once, is the reason the system guarantees only honest-verifier zero-knowledge. Indeed, we increase the query complexity of the verifier without increasing the zero-knowledge guarantee of the basic system (since increasing the zero-knowledge parameter will also increase the soundness error). Therefore, a malicious verifier can potentially break the zero-knowledge by using the same proof in several iterations.

Construction 4 (HVZKPCPP). *The modified HVZKPCPP system (P_H, V_H) uses the the system (P_B, V_B) as a building block, and is parameterized by l, the number of basic proofs in a proof generated by P_H; t, the number of runs (of V_B) that V_H emulates; and d, to be passed on to the underlying HVZKPCPP system. We assume without loss of generality that $l \geq t$.*

Prover algorithm. P_H *on input $1^l, 1^d$ and $(x, w) \in \mathcal{R}$, uses P_B to generate l independent (basic) proofs $\pi_B^1, ..., \pi_B^l$ for the claim $(x, w) \in \mathcal{R}$, and outputs the proof $\pi_H = \pi_B^1 \circ ... \circ \pi_B^l$.*

Verifier algorithm. V_H *on input $1^t, l, 1^d, |x|$, and given access to oracles x, π_H, picks at random t different basic proofs $\pi_B^1, ..., \pi_B^t$, and for every $1 \leq i \leq t$, runs V_B with parameter d and oracles x, π_B^i (all emulations of V_B are performed in parallel). V_H accepts if V_B accepted in all t iterations, otherwise he rejects.*

Theorem 5 (HVZKPCPPs from PCPPs). *Let σ be a security parameter. Then for any $q \in \mathbb{N}$, $\epsilon_S = \epsilon_S(\sigma, |x|)$, $\delta = \delta(\sigma, |x|)$, $r = r(\sigma, |x|)$ and $\ell = \ell(\sigma, |x|)$,*

$$\text{PCPP}\left[r, q, \delta, \frac{1}{2}, \ell\right] \subseteq \text{HVZKPCPP}\left[r', q', \epsilon'_{\text{ZK}} = 0, \delta' = \delta, \epsilon_S, \ell'\right]$$

where $r' = O_q\left(r \cdot \text{poly} \log \frac{1}{\epsilon_S}\right)$, $q' = O_q\left(\log \frac{1}{\epsilon_S}\right)$ and $\ell' = O_q\left((\ell + 2^r) \log \frac{1}{\epsilon_S}\right)$.

Proof (sketch). We take $d = O(1)$ and $t = l = O_q\left(\log \frac{1}{\epsilon_S}\right)$ in Construction 4, which increase the query complexity and proof length (of the basic system) by a factor of $\log \frac{1}{\epsilon_S}$, and the randomness complexity by a factor of poly $\log \frac{1}{\epsilon_S}$ (since in every iteration the verifier needs to pick a new basic proof to use). Completeness follows from the completeness of the basic HVZKPCPP system. Regarding soundness, the soundness error of (P_B, V_B) is $\frac{1}{2}(1 + \epsilon_{\text{in}})$ (where $\epsilon_{\text{in}} < 1$ is the soundness error of the EZKPCPP, and depends only on q since d is constant). As V_H emulates t *independent* runs of V_B, and accepts only if all iterations succeed, then V_H accepts an x that is δ-far from $L_{\mathcal{R}}$ with probability at most $\left(\frac{1}{2}(1 + \epsilon_{\text{in}})\right)^t = \epsilon_S$ (for an appropriate choice of the constant defining t). As for zero-knowledge, every emulation of V_B can be perfectly simulated (by some simulator Sim_B) while making at most $d = O(1)$ TTP queries, and as the emulations are independent (and use independent basic proof), a simulator Sim for V_H can run Sim_B t independent times, and forward the TTP queries of Sim_B to his own TTP. $\square$

Remark 3 (Strong HVZK). The strong honest-verifier zero-knowledge feature of Construction 2 (see Section 3, Remark 2) implies that both the HVZKPCPP systems described in this section also guarantee $(\epsilon_{\text{ZK}}, q^*)$-strong honest-verifier zero-knowledge, as defined in Remark 2. More specifically, to get (ϵ_{ZK}, q^*)-strong HVZK it suffices to take $l = \text{poly}(q^*)$ in Construction 4, and use the EZKPCPP (over a binary alphabet) of Section 3 with zero-knowledge parameter $d = O\left(\log \frac{1}{\epsilon_{ZK}}\right)$. Consequently, the proof length increases by a factor of $\text{poly}\left(q^*, \log \frac{1}{\epsilon_{ZK}}\right)$, the randomness complexity by a factor of $\text{poly}\left(\log q^*, \log \log \frac{1}{\epsilon_{ZK}}\right)$, and the query complexity by a factor of $\text{poly}\log \frac{1}{\epsilon_{ZK}}$. Moreover, if the original PCPP has strong soundness then so does the HVZKPCPP. (To get soundness error ϵ_S on inputs that are δ-far from the relation for some $\delta \in (0, 1)$, the proof length, query complexity and randomness complexity increase by a factor of $O\left(\frac{1}{\delta}\right)$.)

It is evident from Theorem 5 that Construction 4 inherits many of its properties from the underlying PCPP system, so efficient PCPPs yield efficient HVZKPCPPs. More specifically, we can use the following PCPP due to Dinur [11], to obtain an efficient HVZKPCPP.

Theorem 6 (PCPP, implicit in [11]). *Let $\mathcal{R} = \mathcal{R}(x, w) \subseteq \text{DTIME}(t(n))$, then $\mathcal{R}$ has a strong PCPP system (P, V) with constant rejection ratio, such that V on inputs of length n tosses $O(\log t(n))$ coins and reads $O(1)$ bits from his oracles.*

Plugging the PCPP system of Theorem 6 into Theorem 5, we get the following result.

Corollary 2 (Efficient HVZKPCPP). *Let ϵ be a soundness parameter and let δ be a proximity parameter. Then every relation $\mathcal{R} = \mathcal{R}(x, w) \in \text{DTIME}(t(n))$ has an HVZKPCPP system (P_H, V_H) with perfect completeness, perfect honest-verifier zero-knowledge, and soundness error ϵ with proximity parameter δ. On input x, P_H generates a proof of size $\text{poly}\left(t(|x|), \log \frac{1}{\epsilon}, \frac{1}{\delta}\right)$ and V_H makes $O\left(\frac{1}{\delta} \log \frac{1}{\epsilon}\right)$ queries.*

4.2   From HVZKPCPPs and Locking Schemes to ZKPCPPs

In this section we construct a ZKPCPP with zero-knowledge against *arbitrary* query-bounded verifiers, from a locking scheme and an HVZKPCPP with strong honest-verifier zero-knowledge (see Remark 2 for a discussion of this zero-knowledge property). We first give a high-level description of the transformation. For $q^* \in \mathbb{N}$, let (P_H, V_H) be an HVZKPCPP with q^*-strong honest-verifier zero-knowledge (e.g., the system of Construction 4, see Remark 3). Intuitively, all we need to do to achieve zero-knowledge against *arbitrary* (q^*-bounded) verifiers is to force the queries of every (possibly malicious) verifier to be distributed as the queries in q^* random and independent invocations of V_H. Following techniques of Kilian et al. [21], we achieve this by employing a locking scheme. Hiding a few technical details, the high-level idea is as follows. The proof consists of three sections: the PCPP section in which the prover P locks (using the locking scheme) every bit of the HVZKPCPP; the PERM section which contains a locked permutation of the random strings of the honest verifier V_H (namely, P picks a random permutation τ over the space of random strings of V_H, and for every possible random string r of V_H, P lock the image $\tau(r)$ in the PERM section); and the MIX section, where the location indexed by $\tau(r)$ contains r and the collection of keys for the locks holding the HVZKPCPP bits V_H (with randomness r) queries. To verify the proof, V picks a random string r', queries $\text{MIX}_{r'}$, and retrieves some (other) random string r and a set of keys, which he uses to unlock the corresponding locks. Then, V verifies that the lock PERM_r holds the string r' and that V_H would accept (if he was given the HVZKPCPP bits locked in the PCPP section of the proof).

Applying this transformation to an efficient HVZKPCPP and an efficient locking scheme, we get the following result (full details are deferred to the full version).

Theorem 7 (Efficient ZKPCPP). *Let ϵ be a soundness parameter, let δ be a proximity parameter and let $q^* \in \mathbb{N}$. Then every relation $\mathcal{R}(x, w) \in$ DTIME$(t(n))$ has a ZKPCPP system (P, V) with soundness error ϵ, proximity parameter δ, and straight-line (ϵ, q^*)-zero-knowledge. P on input x generates proofs of length* poly$\left(t(|x|), q^*, \log \frac{1}{\epsilon}, \frac{1}{\delta}\right)$ *and V on input $|x|$ makes* poly$\left(\log t(|x|), \log q^*, \log \frac{1}{\epsilon}, \frac{1}{\delta}\right)$ *queries.*

(P, V) inherits its properties from those of the HVZKPCPP and the locking scheme combined. More specifically, perfect completeness follows from the perfect completeness of both building blocks. As for soundness, the binding of the locking scheme guarantees the proof oracle V uses to emulate V_H is consistent with *some* proof oracle for V_H, and therefore (by the soundness of (P_H, V_H)) if x is far from $L_{\mathcal{R}}$ then V_H rejects (with high probability). As for zero-knowledge, the hiding of the locking scheme guarantees that by probing the locks, V learns almost nothing about the values locked within them. Therefore, V can only "hope" to gather some information by retrieving the keys and using them to open the locks (i.e. by reading MIX entries and then the corresponding PCPP entries). However, in this case the random permutation τ guarantees that his queries to π_H are distributed as in random and independent emulations of V_H, so the oracle-answers to his queries can be simulated (by the strong honest-verifier zero-knowledge of (P_H, V_H)).

THE ADAPTIVITY OF THE HONEST VERIFIER. Unlike our HVZKPCPP systems (Section 4.1), the verifier in Theorem 7 is inherently adaptive. Indeed, to decommit the locks the verifier must first retrieve the corresponding keys from the appropriate MIX entry, and therefore cannot make his queries non-adaptively. However, all iterations of the verification procedure may be executed in parallel (i.e. all MIX-queries are asked simultaneously, all locks are then simultaneously unlocked etc.), giving a verifier with adaptivity $k_{\text{lock}} + 1$, where k_{lock} is the adaptivity of the locking scheme receiver.

5   Cryptographic Applications

In this section we describe several applications of ZKPCPPs. Concretely, we construct two-party and multiparty protocols that allow a dealer to commit to a secret and prove (with sublinear communication) that it satisfies an NP predicate.

5.1   Certifiable VSS

Motivated by applications that require verification with no information leakage, we focus on reducing the communication complexity of verifying the shares in a verifiable secret sharing (VSS) protocol [10,14,5,9,6]. Roughly speaking, VSS allows a dealer D to distribute a secret s among n servers in a way that prevents a coalition of up to t servers from learning or modifying the secret, while on the other hand guaranteeing unique reconstruction, even if D and up to t servers can collude. We study a *certifiable* variant of VSS (which we call cVSS) which is similar to traditional VSS, except that it provides the additional guarantee that the secret satisfies some NP predicate. Similar to [18], to achieve sublinear verification we consider networks that include an additional receiver entity R who eventually receives the secret, and may assist in the verification. We now provide more details about the model we consider.

We assume that the participating parties can interact over a synchronous network of secure point-to-point channels. The parties also have access to a *broadcast* channel, where a message sent over this channel is received by all other parties. When measuring communication complexity, we count a message sent over a broadcast channel only once towards the total communication. Alternatively, our protocols can be implemented with similar communication complexity using a public bulletin board, where every time a message is written to or read from the board is counted towards the communication complexity.

The security of protocols is defined by considering their execution in the presence of a malicious, static adversary, who may corrupt and control a subset of the parties. The adversary is capable of *rushing*, namely sending his messages only after receiving all messages sent by honest parties in the same round.

A cVSS protocol for an NP-relation $\mathcal{R}$ consists of three phases. In the *sharing* phase, the dealer D is give input $(x, w) \in \mathcal{R}$ and sends a message to each server. In the *verification* phase, the receiver R can freely interact with the servers, possibly using a broadcast channel. Finally, in the *reconstruction* phase, each server sends a single message to R, and R reconstructs the secret. We note that R and the servers are given $1^{|x|}$ as input.

A protocol as above is said to be (t, ϵ)-secure if it satisfies the following requirements:

- *Correctness.* For every adversary $\mathcal{A}$ corrupting t out of n servers and for every $(x, w) \in \mathcal{R}$, in the end of the reconstruction phase R outputs x, except with at most ϵ probability.

- *Secrecy*. For every adversary $\mathcal{A}$ corrupting R and t servers there exists a PPT simulator Sim such that for every $(x, w) \in \mathcal{R}$, $\text{View}_{\mathcal{A}}(x, w) \approx^{\epsilon} \text{Sim}(|x|)$, where $\text{View}_{\mathcal{A}}(x, w)$ denotes the view of $\mathcal{A}$ during the sharing and verification phases.
- *Commitment*. For every adversary $\mathcal{A}$ corrupting R and t servers and for every $(x, w) \in \mathcal{R}$, the following holds except with at most ϵ failure probability over the randomness of the sharing and verification phases. In the end of the verification phase, either R outputs $\perp$, or there is a unique secret x^* (determined by the messages exchanged up to this point), such that $x^* \in L_{\mathcal{R}}$, $|x^*| = |x|$, and R will output x^* regardless of the messages sent by the adversary during the reconstruction phase.

We note that traditional VSS is stronger than our certifiable VSS in that the verification phase does not involve the receiver R. Thus, when there are multiple receivers, traditional VSS can guarantee that the same secret x^* is reconstructed by all receivers. However, traditional VSS protocols do not guarantee that the reconstructed secret possess any specific properties, as guaranteed by certifiable VSS, and also do not achieve sublinear verification. (We note that certifiable VSS can be implemented using general MPC protocols, but the communication complexity required to verify the shares will not be sublinear.)

CERTIFIABLE VSS FROM ZKPCPPs. The protocol uses a ZKPCPP system (P, V), and a robust secret sharing scheme. (A robust secret sharing scheme maps a secret x into a vector $(s_1, ..., s_m)$ of shares such that "few" shares reveal no information on x, but x can be reconstructed from the shares even if "few" of them are replaced with incorrect values.) We note that for the protocol to be efficient, P, V should be efficient, as well as the sharing and reconstruction algorithms of the secret sharing scheme.

In the sharing phase, the dealer D secret shares $x \in L_{\mathcal{R}}$ using the secret sharing scheme, generates a ZKPCPP for the claim 'the secret shares are "close" to a vector of "legal" secret shares and $x \in L_{\mathcal{R}}$', and partitions the shares and the proof between the servers. In the verification phase, the receiver R verifies that the shares D distributed are close to a sharing of some $x' \in L_{\mathcal{R}}$ by emulating a the verifier V, where R broadcasts the queries of V and the servers answer. (The use of broadcast prevents R from contacting too many servers, which would violate the secrecy requirement.) If the verification fails, R outputs $\perp$ and ignores further messages. For reconstruction, the servers holding the secret shares send them to R, who reconstructs the secret x.

This description is in fact an over-simplification of the actual protocol. Indeed, the verification procedure of the ZKPCPP cannot be used as-is since in the context of VSS, verification is executed in the presence of an adversary that can determine the answers of the corrupted servers *after seeing the queries of the verifier*, while ZKPCPPs guarantee completeness and soundness when the verification is performed *with oracles* (in particular, the oracle answers are *independent* of the queries). Intuitively, to restrict the influence the adversary has on the verification procedure, it suffices to guarantee that symbols held by corrupted servers are queried with low probability, which can be done as follows. The dealer distributes *several copies* of the secret shares and the proof, and the value of every specific symbol V queries is determined by the majority vote over the corresponding symbols in several *randomly selected* copies. (This can be thought of as applying a sort of "error correction" to the symbols of the secret shares and the proof.) In addition, the verification procedure is repeated several times, and the verification phase passes (i.e., R does not abort) only if V accepted in most of the iterations. Further details are differed to the full version.

The secrecy property follows from the secrecy of the secret sharing scheme and from the straight-line zero-knowledge property of (P, V). Indeed, the zero knowledge implies R learns only few shares, and the secrecy of the secret sharing scheme guarantees that these shares reveal no information on x. (Straight-line zero-knowledge is required to guarantee that the view of the adversary can be simulated.) The "error correction" applied to the secret shares and the proof guarantees that with high probability, corrupted servers are queried only in few of the emulations of V. Therefore, in most emulations we can think of the verification as being performed *with oracles*, which is useful both for correctness and for binding. Indeed, for correctness, if D is honest then with high probability V accepts in most iterations (by the completeness of the ZKPCPP), and the robustness of the secret sharing guarantees that R will reconstruct $x^* = x \in L_{\mathcal{R}}$ in the end of the reconstruction phase, even if t servers are corrupted. As for binding, a corrupted D has 2 possible courses of actions. First, if he distributes a shares vector that is far from every "legal" shares-vector, or close to a shares vector of some $x^* \notin L_{\mathcal{R}}$, then the soundness of the ZKPCPP implies that V rejects in most of the emulations (since corrupted servers are queried only in few of these emulations), so R outputs $\bot$ with high probability. Second, if he distributes a shares vector that is close to a legal shares vector of some $x^* \in L_{\mathcal{R}}$, then either R outputs $\bot$ at the end of the verification phase, or x^* will be reconstructed (due to the robustness of the secret sharing). Thus, we obtain the following result.

Theorem 8 (verification-efficient certifiable VSS). *Let $\mathcal{R} = \mathcal{R}(x, w)$ be an NP-relation. Then for every corruption threshold $t \in \mathbb{N}$ and every soundness parameter ϵ, there exists a (t, ϵ)-secure cVSS protocol for $\mathcal{R}$, with $n = \text{poly}\left(|x|, t, \log \frac{1}{\epsilon}\right)$ servers, total communication complexity $\text{poly}\left(|x|, t, \log \frac{1}{\epsilon}\right)$, and a verification phase that uses $\text{poly}\left(\log |x|, \log t, \log \frac{1}{\epsilon}\right)$ bits of communication.*

Our certified VSS protocol has non-interactive single-round sharing and reconstruction phases, and a 6-round verification phase. During the sharing phase D sends a single bit to each server, and during the reconstruction phase every server sends a single bit to R. The servers are deterministic and the communication complexity of every server (throughout the protocol) is $O(1)$. Moreover, there is no direct communication between the servers.

5.2   Two-Party Commit-and-Prove

We use ZKPCPPs to construct a "2-party analog" of cVSS, or alternatively, a "certifiable" generalization of a commitment scheme, which we call *Commit-and-Prove*. A commitment scheme is a two-phase protocol between a sender S and a receiver R. In the first phase, called the *commit* phase, the server on input x freely interacts with R (who has input $1^{|x|}$). The messages exchanged between S, R during the phase are called the *commitment*. In the second phase, called the *reveal* pahse, S sends x, together with a decommitment string dec to R, and R decides whether to accept of reject x, based on dec and the commitment.

A commitment scheme should have the following properties. First, it should be *hiding*, in the sense that a (possibly malicious) receiver interacting with the honest sender learns nothing about the secret x during the commit phase. Second, it should be *binding*, namely there exists no efficient malicious sender that, after the interaction with R during the commit phase, can find distinct x, x' of the same length, and two decommitment strings dec, dec', such that R would have accepted x, x' with decommitment dec, dec', respectively.

A commit-and-prove protocol is *certifiable* in the sense that S not only commits to x, but also proves it satisfies some predicate. (The relation between commitment schemes and commit-and-prove protocols is similar to the relation between VSS and cVSS.) Specifically, it is similar to a commitment scheme, but at the end of the reveal phase R either outputs x and $x \in L_{\mathcal{R}}$ (for some relation $\mathcal{R}$), or aborts. As R, S are both efficient algorithms, the sender cannot generally be expected to find on its own a "witness" to the fact that x satisfies the predicate (think, for example, of an NP predicate). Therefore, S is given a witness w (in addition to the input x).

We say a commit-and-prove protocol for a relation $\mathcal{R}$ is *secure* if it satisfies the following requirements:

- *Correctness.* For every $(x, w) \in \mathcal{R}$, if S, R are honest then R outputs x at the end of the reveal phase.
- *Binding.* Every efficient (possibly malicious) sender algorithm S^* wins the following game with only negligible probability. First, S^* interacts with R in the commit phase, with common input 1^n. Then, S^* outputs two pairs $(x, \mathsf{dec}), (x', \mathsf{dec'})$ such that $|x| = |x'| = n$. S^* wins if R would have accepted x, x' given the decommitments dec, dec' (respectively), and in addition either $x \neq x'$ or $x \notin L_{\mathcal{R}}$.
- *Hiding.* There exists a PPT oracle machine Sim such that for every (possibly malicious) PPT receiver algorithm R^* and for every sender input $(x, w) \in \mathcal{R}$, $\mathsf{Sim}^{R^*}(|x|)$ is computationally indistinguishable from the view of R^* during the commitment phase, when interacting with $S(x, w)$.
- *Zero-knowledge after reveal.* There exists a PPT oracle machine Sim such that for every (possibly malicious) PPT receiver algorithm R^* and for every sender input $(x, w) \in \mathcal{R}$, $\mathsf{Sim}^{R^*}(x)$ is computationally indistinguishable from the view of R^* during the *entire* interaction with $S(x, w)$.

Similar to standard commitments, one can also consider stronger variants in which the binding or the hiding property is statistical. Our construction in fact satisfies the statistical variant of hiding and zero-knowledge after reveal.

Using techniques similar to those employed by [17] to construct sublinear ZK arguments, we construct a commit-and-prove protocol with *polylogarithmic* communication during the commit phase, and the protocol makes a *black-box* use of an exponentially-hard collision-resistant hash function. (By relaxing the communication requirements such that the communication during commit is sublinear, instead of polylogarithmic, the protocol can be based on a super-polynomially hard hash function.)

COMMIT-AND-PROVE FROM ZKPCPPs. As in the cVSS protocol described in Section 5.1, the protocol is based on a robust secret sharing scheme, and an *HVZKPCPP* system (P, V), e.g., the HVZKPCPP system of Construction 3. (Notice that *honest-verifier* zero-knowledge suffices in this case, since the sender can refuse to answer queries the honest ZKPCPP verifier would not make.) In addition, the protocol employs a family $\mathcal{H}$ of collision resistant hash functions. In the commit phase, R chooses a function $h \in \mathcal{H}$ and sends (the index of) h to S. S secret-shares $x \in L_\mathcal{R}$ into shares $s_1, ..., s_n$ and, using P, generates a proof π for the claim that the secret shares are "close" to the shares of some $x^* \in L_\mathcal{R}$. Next, S commits to π using a computationally-binding and statistically-hiding commitment scheme Com_h,[4] and "compresses" the commitments, using a "Merkle Hash Tree" [19]. (That is, the commitments are compressed by repeatedly applying the hash function h to pairs of adjacent strings, where every application of h shrinks the input. Thus, a "tree" of hash values is generated, and its root is used as the compressed commitment.) S commits to $(s_1, ..., s_n)$ in a similar manner. Then, S sends the compressed commitments C_π (of π) and C_x (of $(s_1, ..., s_n)$) to R, and R verifies the commitments as follows. R picks a randomness r for V and sends it to S. S determines the set Q of queries V, with randomness r, would have made, and answers every query $q \in Q$ by decommitting the corresponding bit of $\pi, s_1, ..., s_n$ (using the reveal phase of Com_h) and sending the pre-images of all the hash values computed along the path in the Merkle hash tree leading from that bit to the root. R verifies that the values on the paths are consistent with C_π, C_x and h, that V makes the queries Q when using randomness r, and that V would accept given these oracle answers. In the reveal phase, S sends R the entire hash tree used to compress the commitments to $s_1, ..., s_n$, together with the random strings used to generates the commitments (through Com_h) of $s_1, ..., s_n$. R verifies that the commitments and the Merkle tree are consistent with $s_1, ..., s_n$, and if so reconstructs x from the shares, and outputs it.

The properties of the protocol follow from a combination of the properties of the HVZKPCPP, the secret sharing scheme, and the collision-resistent hash function. More specifically, hiding follows from the secrecy of the secret sharing scheme and from the zero-knowledge property of (P, V), and holds even if

---

[4] Such a scheme can be constructed from a collision-resistant hash function, with no additional assumptions. Moreover, if the hash function has exponential hardness then the resultant commitments can be polylogarithmic (in the length of the input).

the commitments to $s_1, ..., s_n, \pi$ are not compressed (the compression is required to "save" on communication during commit). Indeed, by the zero-knowledge of (P, V) even a malicious R^* learns only few shares, which reveal no information on x. Zero-knowledge after reveal follows in a similar manner, since a simulator for the entire view of a (possibly malicious) receiver R^* receives x, and can therefore emulate a simulation of a malicious verifier in the underlying HVZKPCPP system. As for binding, the collision-resistance of h and the binding of Com_h guarantee that except with negligible probability, S is committed to some shares vector $(s_1^*, ..., s_n^*)$. (If C_x is inconsistent with all possible Merkle hash tree commitments to all vectors $(s_1^*, ..., s_n^*)$ then R necessarily aborts during the reveal phase.) Therefore, if $(s_1^*, ..., s_n^*)$ is far from every "legal" shares-vector, or close to a shares vector of some $x^* \notin L_{\mathcal{R}}$, then R detects this during the commit phase with high probability (even when interacting with a malicious sender S^*). Otherwise, $(s_1^*, ..., s_n^*)$ is "close" to a legal shares-vector of some $x^* \in L_{\mathcal{R}}$, meaning the only value a (possibly malicious) S^* can successfully decommit during the reveal phase, is x^*.

We note that the protocol as described above achieves a constant error, which can be reduced (while preserving hiding and zero-knowledge after reveal) by sequential repetition of the commit phase (further details are differed to the full version). Consequently, we obtain the following result.

Theorem 9 (Sublinear Commit-and-Prove). *Let $\mathcal{H}$ be any family of exponentially-hard collision-resistant hash functions. Then there exists a computationally-binding and statistically-hiding Commit-and-Prove protocol with negligible soundness error and polylogarithmic communication complexity during the Commit phase. Moreover, the protocol makes only black-box use of $\mathcal{H}$.*

We note that if $\mathcal{H}$ only satisfies the usual notion of super-polynomial hardness, the communication complexity (during commit) of the resulting Commit-and-Prove protocol can be $O(n^{\epsilon})$, for an arbitrarily small $\epsilon > 0$.

A NON-BLACK-BOX ALTERNATIVE. We have shown how to apply ZKPCPPs towards obtaining sublinear-communication commit-and-prove protocols that make a black-box use of any collision-resistant hash function. Settling for a non-black-box use of the hash function, one could avoid the use of ZKPCPPs by combining a sublinear commitment Com with sublinear zero-knowledge arguments of knowledge [3,4]. Concretely, during the commit phase S first commits to x using Com, and then proves to R that he knows a witness w and randomness r such that (x, r) are consistent with the transcript of Com and $(x, w) \in \mathcal{R}$. For the reveal phase, S sends (x, r) to R. Both of the above primitives can be based on a collision-resistant hash function. However, the commit phase of the protocol is inherently non-black-box because the sublinear argument applies to an NP-relation which depends on the hash function (since Com depends on the hash function).

References

1. Arora, S., Lund, C., Motwani, R., Sudan, M., Szegedy, M.: Proof verification and the hardness of approximation problems. Electronic Colloquium on Computational Complexity (ECCC) 5(8) (1998)
2. Arora, S., Safra, S.: Probabilistic checking of proofs: A new characterization of NP. J. ACM 45(1), 70–122 (1998)
3. Barak, B.: How to go beyond the black-box simulation barrier. In: FOCS, pp. 106–115 (2001)
4. Barak, B., Goldreich, O.: Universal arguments and their applications. SIAM J. Comput. 38(5), 1661–1694 (2008)
5. Ben-Or, M., Goldwasser, S., Wigderson, A.: Completeness theorems for non-cryptographic fault-tolerant distributed computation (extended abstract). In: STOC, pp. 1–10 (1988)
6. Ben-Or, M., Rabin, T.: Verifiable secret sharing and multiparty protocols with honest majority. In: Proceedings of the 21st Annual ACM Symposium on Theory of Computing (STOC), Seattle, Washigton, USA, May 14-17, pp. 73–85. ACM (1989)
7. Ben-Sasson, E., Goldreich, O., Harsha, P., Sudan, M., Vadhan, S.P.: Robust PCPs of proximity, shorter PCPs, and applications to coding. SIAM J. Comput. 36(4), 889–974 (2006)
8. Ben-Sasson, E., Sudan, M.: Short PCPs with polylog query complexity. SIAM J. Comput. 38(2), 551–607 (2008)
9. Chaum, D., Crépeau, C., Damgård, I.: Multiparty unconditionally secure protocols (extended abstract). In: STOC, pp. 11–19 (1988)
10. Chor, B., Goldwasser, S., Micali, S., Awerbuch, B.: Verifiable secret sharing and achieving simultaneity in the presence of faults (extended abstract). In: FOCS, pp. 383–395 (1985)
11. Dinur, I.: The PCP theorem by gap amplification. In: STOC, pp. 241–250 (2006)
12. Dinur, I., Reingold, O.: Assignment testers: Towards a combinatorial proof of the PCP-theorem. In: FOCS, pp. 155–164 (2004)
13. Dwork, C., Feige, U., Kilian, J., Naor, M., Safra, S.: Low communication 2-prover zero-knowledge proofs for NP. In: Brickell, E.F. (ed.) Advances in Cryptology - CRYPTO 1992. LNCS, vol. 740, pp. 215–227. Springer, Heidelberg (1993)
14. Feldman, P.: A practical scheme for non-interactive verifiable secret sharing. In: FOCS, pp. 427–437. IEEE Computer Society (1987)
15. Goldwasser, S., Micali, S., Rackoff, C.: The knowledge complexity of interactive proof systems. SIAM J. Comput. 18(1), 186–208 (1989)
16. Ishai, Y., Kushilevitz, E., Ostrovsky, R., Sahai, A.: Zero-knowledge from secure multiparty computation. In: STOC, pp. 21–30 (2007)
17. Ishai, Y., Mahmoody, M., Sahai, A.: On efficient zero-knowledge PCPs. In: Cramer, R. (ed.) TCC 2012. LNCS, vol. 7194, pp. 151–168. Springer, Heidelberg (2012)
18. Ishai, Y., Sahai, A., Viderman, M., Weiss, M.: Zero knowledge LTCs and their applications. In: APPROX-RANDOM, pp. 607–622 (2013)
19. Kilian, J.: Uses of randomness in algorithms and protocols. MIT Press (1990)

20. Kilian, J., Naor, M.: On the complexity of statistical reasoning. In: Proceedings of the Third Israel Symposium on the Theory of Computing and Systems, pp. 209–217. IEEE (1995)
21. Kilian, J., Petrank, E., Tardos, G.: Probabilistically checkable proofs with zero knowledge. In: STOC, pp. 496–505 (1997)
22. Meir, O.: Combinatorial construction of locally testable codes. SIAM J. Comput. 39(2), 491–544 (2009)
23. Mie, T.: Short PCPPs verifiable in polylogarithmic time with O(1) queries. Ann. Math. Artif. Intell. 56(3-4), 313–338 (2009)

Achieving Constant Round Leakage-Resilient Zero-Knowledge[*]

Omkant Pandey

University of Illinois at Urbana-Champaign
omkant@uiuc.edu

Abstract. Recently there has been a huge emphasis on constructing cryptographic protocols that maintain their security guarantees even in the presence of side channel attacks. Such attacks exploit the physical characteristics of a cryptographic device to learn useful information about the internal state of the device. Designing protocols that deliver meaningful security even in the presence of such leakage attacks is a challenging task.

The recent work of Garg, Jain, and Sahai formulates a meaningful notion of zero-knowledge in presence of leakage; and provides a construction which satisfies a weaker variant of this notion called $(1 + \epsilon)$-leakage-resilient-zero-knowledge, for every constant $\epsilon > 0$. In this weaker variant, roughly speaking, if the verifier learns ℓ bits of leakage during the interaction, then the simulator is allowed to access $(1 + \epsilon) \cdot \ell$ bits of leakage. The round complexity of their protocol is $\lceil \frac{n}{\epsilon} \rceil$.

In this work, we present the first construction of leakage-resilient zero-knowledge satisfying the ideal requirement of $\epsilon = 0$. While our focus is on a feasibility result for $\epsilon = 0$, our construction also enjoys a constant number of rounds. At the heart of our construction is a new "public-coin preamble" which allows the simulator to recover arbitrary information from a (cheating) verifier in a "straight line." We use non-black-box simulation techniques to accomplish this goal.

1 Introduction

The concept of zero-knowledge interactive proofs, originating in the seminal work of Goldwasser, Micali, and Rackoff [39], is a fundamental concept in theoretical cryptography. Informally speaking, a zero-knowledge proof allows a prover P to prove an assertion x to a verifier V such that V learns "nothing more" beyond the validity of x. The proof is an interactive and randomized process. To formulate "nothing more," the definition of zero-knowledge requires that for every malicious V^* attempting to lean more from the proof, there exists a polynomial time simulator S which on input *only* x, simulates a "real looking" interaction for V^*.

In formulating the zero-knowledge requirement, it is assumed that the prover P is able to keep its internal state — the witness and the random coins —

---

[*] IACR Eprint Archive Report 2012/362.

Y. Lindell (Ed.): TCC 2014, LNCS 8349, pp. 146–166, 2014.

perfectly hidden from the verifier V^*. It has been observed, however, that this assumption may not hold in many settings where an adversary has the ability to perform *side channel attacks*. These attacks enable the adversary to learn useful information about the internal state of a cryptographic device (see e.g., [48,6,68,59] and the references therein). In presence of such attacks, standard cryptographic primitives often fail to deliver any meaningful notion of security. As a matter of fact, even *formulating* a meaningful security notion under such attacks—as is the case with leakage-resilient zero-knowledge—can be a challenging task.

To deliver meaningful security in the presence side channel attacks, many recent works consider stronger adversarial models in which the device implementing the honest algorithm *leaks* information about its internal state to the adversary. The goal of these works is then to construct cryptographic primitives that are "resilient" to such leakage. Leakage resilient constructions for many basic cryptographic tasks are now known [29,3,64,26,4,5,57,47,15,25,24,50,30,49,14,2].

Leakage-resilient zero-knowledge. Very recently Garg, Jain, and Sahai [33] initiated an investigation of leakage-resilient zero-knowledge (LRZK). Their notion considers a cheating verifier V^* who can learn an arbitrary amount of leakage on the internal state of the honest prover, including the witness. This is formulated by allowing V^* to make leakage queries $F_1, F_2, \ldots$ throughout the execution of the protocol. Then the definition of LRZK, roughly speaking, captures the intuition that no such V^* learns anything *beyond the validity of the assertion* and *the leakage.*

The actual formulation of this intuition is slightly more involved. Observe that during the simulation, S will need to answer leakage queries of V^*, which may contain information about the witness to V^*. Simulator S cannot answer such queries without having access to the witness. The definition of [33] therefore provides S with access to a *leakage oracle* which holds a witness to x. The oracle, $\mathcal{L}_w^n(\cdot)$, is parameterized by the witness w and $n = |x|$; on input a function F expressed as a boolean circuit, it returns $F(w)$. To ensure that S can answer leakage requests of V^*, the simulator is also allowed to query $\mathcal{L}_w^n$ on leakage functions of its choice. Of course, providing S with uncontrolled access to the witness will render the notion meaningless.[1] Therefore, to ensure that the notion delivers meaningful security, the LRZK definition requires the following restriction on the length of bits that S can read from $\mathcal{L}_w^n$. Suppose that $S^{\mathcal{L}_w^n}$ outputs a simulated view v for V^*. Denote by $\ell_S(v)$ the number of bits S reads from $\mathcal{L}_w^n$ in generating this particular view v. Denote by $\ell_{V^*}(v)$ the total length of the leakage answers that S provides to V^* (which are already included in v, and can be different from answers received by S). Then, it is required that:

$$\ell_S(v) \leq \ell_{V^*}(v). \tag{1}$$

---

[1] S can simply access the entire witness, and then simulate.

More precisely, in [33], a slightly more general notion of $(1 + \epsilon)$-LRZK is defined in which the above condition is relaxed to:

$$\ell_S(v) \le (1 + \epsilon) \cdot \ell_{V^*}(v),$$

where $\epsilon > 0$ is a constant. In addition, [33] also present a protocol of $\lceil \frac{n}{\epsilon} \rceil$ rounds, which achieves $(1 + \epsilon)$-LRZK for every a-priori fixed constant $\epsilon > 0$. Since $\epsilon > 0$, the resulting notion is weaker than the one required by equation 1. Nevertheless, [33] show that despite this relaxation, $(1 + \epsilon)$-LRZK still delivers meaningful security. By applying this notion in the context of cryptography based on hardware-tokens, [33] were able to weaken the requirements of tamper-proofness on the hardware tokens.

Our main result. Although a protocol with $\epsilon > 0$ is still useful certain contexts, it is significantly weaker than the ideal requirement of $\epsilon = 0$—both qualitatively and philosophically. Qualitatively, a constant $\epsilon > 0$ allows the simulator to learn *strictly more* information about the internal secret than the actual leakage allows! Qualitatively, it means that a protocol proven to be $(1 + \epsilon)$-LRZK "secure" may actually expose additional parts of the internal secret than the actual leakage. Furthermore, even in situations where $(1 + \epsilon)$-LRZK is sufficient, protocol of [33] requires a large round complexity, which continues to increase as we lower the value of ϵ.

Philosophically, an $\epsilon > 0$ essentially defies the very nature of simulation-based security. In particular, as argued above, since it allows S to learn strictly more than what the verifier does, it is not "zero" knowledge, but only an "ϵ-approximation" of it, and closer in spirit to super-polynomial time simulation [61,67,65]. Furthermore, this is not merely a philosophical issue—$(1 + \epsilon)$-LRZK can be particularly problematic in protocol composition [16,17]. For example, using such a simulator in place of a cheating party may result in learning more "outputs" than allowed.

In this work, we present the first construction of an LRZK protocol satisfying $\epsilon = 0$. Although our main goal is to obtain a feasibility result, our protocol also enjoys a *constant* number of rounds. Our protocol uses standard cryptographic tools. However, it requires some of them – particularly, oblivious transfer – to have an *invertible sampling* property [20,42]. To the best of our knowledge, instantiations satisfying this property are known only based on the decisional Diffie-Hellman assumption (DDH) [23]. We leave constructing an LRZK proof system based on general assumption as an interesting open question.

Theorem 1 (Main Result). *Suppose that the decision Diffie-Hellman assumption holds. Then, there exists a constant-round leakage-resilient zero-knowledge proof system for all languages in $\mathcal{NP}$.*

We remark that the low round-complexity is usually a desirable protocol feature [37,7,69,66,70]. In the context of side channel attacks, however, it can be a particularly attractive one to have. This is because a protocol with high round complexity may require the device to maintain state for more rounds, and therefore may give an adversary more opportunities to launch side-channel attacks.

1.1   An Overview of Our Approach

Let us start by recalling the main difficulty in constructing an LRZK protocol. Recall that a zero-knowledge simulator S "cheats" in the execution of the protocol so that it can produce a convincing view. When dealing with leakage, not only the simulator needs to continue executing the protocol, but it also needs to "explain its actions" so far by maintaining a state consistent with an honest prover.

To be able to simultaneously perform these two actions, the GJS simulator does the following. It combines the following two different but well-known methods of "cheating." The first method, due to Goldreich-Kahan [37], requires the verifier to commit its challenge ch; the second method, due to Feige-Shamir [31], requires the prover to use equivocal[2] commitments. The GJS simulator uses these methods *together*. It uses ch to perform its main simulation (by using [37] strategy), and uses the trapdoor of equivocal commitments, denoted t_1, to "explain its actions" so far. We call (t_1, ch) the double trapdoor.

The GJS simulator "rewinds" the verifier to obtain the two trapdoors before it actually enters the main proof stage. By using a precise rewinding strategy [53], GJS achieves $(1 + \epsilon)$-LRZK. However, since rewinding strategy is crucial to their simulation, this approach by itself seems insufficient for achieving LRZK.

A fundamentally different simulation strategy, in which the simulator uses the program of the malicious verifier V^*, was presented in Barak's work [7]. This method does not need to "rewind" the verifier to produce its output. Our first idea there is to somehow use Barak's simulation strategy along with the use of equivocal commitments as in [31]. Unfortunately, this does not work since the trapdoor t_1 for equivocation has to be somehow recovered and only then any other simulation strategy (such as knowing the challenge ch) can be used.

We therefore modify this approach so that we can use Barak's method to recover arbitrary information from the verifier during the simulation. For the purpose of this discussion, let us assume that Barak's technique provides a way for P and V to interactively generate a statement σ for some $\mathcal{NP}$-relation $\mathbf{R}_{sim}$ so that no cheating prover P^* can prove $\sigma \in \mathbf{L}_{sim}$, but a simulator S holding the program of the cheating verifier V^* will always have a witness ω such that $\mathbf{R}_{sim}(\sigma, \omega) = 1$. At this point, let us just assume that the verifier does not ask leakage queries.

Then, we need to design a two party protocol for the following task. The first party P holds a private input ω, the second party V holds an arbitrary private message m, the common input to both parties is σ. The protocol allows P to learn m if and only if $\mathbf{R}_{sim}(\sigma, \omega) = 1$, nothing otherwise; V learns nothing. This is similar in spirit to the *conditional disclosure* primitive of [34], except that here the condition is an arbitrary $\mathcal{NP}$-relation $\mathbf{R}_{sim}(\sigma, \omega) = 1$. Constructing such protocols for $\mathcal{NP}$-conditions has not been studied, since they follow from work on secure two-party computation [71,35,54]. Clearly, we cannot directly use

---

[2] These are commitments which, given appropriate trapdoor information, can be opened to both 0 or 1.

secure two-party computation since their security-guarantee is often *simulation-based*—which is essentially what our protocol is trying to achieve in the first place.

Our next observation is that we do not really require the strong simulation-based guarantee. We only need to construct a conditional disclosure protocol for a very specific $\mathcal{NP}$-relation. We construct such a protocol based on Yao's garbled circuit technique. We show that if we use properly chosen OT protocols (constructed in [1,56]) — then we get a conditional disclosure protocol. In addition, the protocol ensures that the messages of P are pseudorandom (more precisely, invertible samplable [20,42]). As a result, the protocol maintains its security claims even in the presence of leakage. This is a two-round protocol, and a crucial ingredient in achieving leakage resilience.

Armed with this new tool, simulation now seems straightforward: use the conditional disclosure protocol to recover both (t_1, ch) and then use the GJS-simulator. While this idea works, there is a difficulty in proving the *soundness* of this protocol. Recall that in Barak's protocol, one must find collisions in the hash function h to prove that no cheating P^* can succeed in learning a witness to statement σ. Typically, this is achieved by applying "rewinding techniques" to extract knowledge P^*. However, ensuring this typical requires the simulator to demonstrate "knowledge"—which is difficult to "explain" later when leakage queries are asked by the cheating prover.

To resolve this difficulty, we need to ensure that *extraction* can be performed from a party *without requiring it to maintain knowledge.*[3] We ensure this by using a variant of the commitment protocol of Barak and Lindell [11]. We use this protocol to extract useful information directly from Barak's preamble [7], without requiring the honest party to maintain knowledge explicitly.

Recall that we work in the model of [33]. In this model the verifier is allowed to ask arbitrary leakage queries $F_1, F_2, \ldots$ on prover's state. The state of the prover at any given round only consists of its witness and the randomness up to that round. In particular, the randomness of future rounds is determined only at the beginning of those round. Observe that all ingredients described by us so far actually require the prover to send only random strings. Therefore, it is easy to asnwer the leakage queries up to this point in the simulation. By the time simulator enters the main body, it recovers (t_1, ch) and use them to answer leakage queries as in [33].

1.2   Related Work

Relevant to our work are the works on zero-knowledge proofs in other more complex attack models such as man-in-middle attacks [27], concurrent attacks [28], resettable attacks [19,8], and so on. Also relevant to our work are different variants of non-black-box simulation used in the literature [7,9,62,63,22] as well as efficient and universal arguments [46,52,10,43].

---

[3] Indeed, there is a difference between the two, see discussion in [11].

The explicit study of leakage-resilient cryptography was started by Dziembowski and Pietrzak [29]. Related study on protecting devices appears in the works of Ishai, Prabhakaran, Sahai, and Wagner [45,44]. After these works a long line of research has focussed on constructing primitives resilient to leakage including public-key encryption and signatures [3,64,26,4,5,57,47,15,25,24,49,14,50], devices [30,2], and very recently interactive protocols [33,12].

Also relevant to our work are the works on adaptive security [18] and invertible sampling [20,42]. Adaptively secure protocols and leakage-resilience in interactive protocols were shown to be tightly connected in the work of Bitansky, Canetti, and Halevi [12].

2 Notation and Definitions

Notation. For a randomized algorithm A we write $A(x; r)$ the process of evaluating A on input x with random coins r. We write $A(x)$ the process of sampling a uniform r and then evaluating $A(x; r)$. We define $A(x, y; r)$ and $A(x, y)$ analogously. The set of natural numbers is represented by $\mathbb{N}$. Unless specified otherwise, $n \in \mathbb{N}$ represents a security parameter available as an implicit input when necessary. All inputs are assumed to be of length at most polynomial in n. We assume familiarity with standard concepts such as interactive Turing machines (ITM), computational indistinguishability, commitment schemes, $\mathcal{NP}$-languages, witness relations and so on (see [36]).

For two randomized ITMs A and B, we denote by $[A(x, y) \leftrightarrow B(x, z)]$ the interactive computation between A and B, with A's inputs (x, y) and B's inputs (x, z), and uniform randomness; and $[A(x, y; r_A) \leftrightarrow B(x, z; r_B)]$ when we wish to specify randomness. We denote by $\text{VIEW}_B[A(x, y) \leftrightarrow B(x, z)]$ and $\text{OUT}_B[A(x, y) \leftrightarrow B(x, z)]$ the view and output of B in this computation; VIEW_A, OUT_A are defined analogously. Finally, $\text{TRANS}[A(x, y) \leftrightarrow B(x, z)]$ denotes the public transcript of the interaction $[A(x, y) \leftrightarrow B(x, z)]$.

For two probability distributions D_1 and D_2, we write $D_1 \overset{c}{\equiv} D_2$ to mean that D_1 and D_2 are computationally indistinguishable.

Definition 1 (Interactive Proofs). *A pair of probabilistic polynomial time interactive Turing machines $\langle P, V \rangle$ is called an interactive proof system for a language $L \in \mathcal{NP}$ with witness relation R if the following two conditions with respect to some negligible function $\text{negl}(\cdot)$:*

– *Completeness: for every $x \in L$, and every witness w such that $R(x, w) = 1$,*

$$\Pr\left[\text{OUT}_V[P(x, w) \leftrightarrow V(x)] = 1\right] \geq 1 - \text{negl}(|x|).$$

– *Soundness: for every $x \notin L$, every interactive Turing machine P^*, and every $y \in \{0, 1\}^*$,*

$$\Pr\left[\text{OUT}_V[P^*(x, y) \leftrightarrow V(x)] = 1\right] \leq \text{negl}(|x|).$$

If the soundness condition holds only against polynomial time machines P^*, $\langle P, V \rangle$ is called an argument system. We will only need/construct argument systems in this work.

Leakage attack. Machine P and V are modeled as randomized ITM which interact in rounds. It is assumed that the the random coins used by a party in any particular round are determined only at the beginning of that round. Denote by state a variable initialized to prover's private input w. At the end beginning of each round i, P flips coins r_i to be used for that round, and updates state := state $\| r_i$. A leakage query on prover's state in round i corresponds to verifier sending a function F_i (represented as a polynomial-sized circuit), to which the prover responds with $F_i(\text{state})$. The verifier is allowed to any number of arbitrary leakage queries throughout the interaction. A malicious verifier who obtains leakage under this setting is said to be launching a *leakage attack.*

To formulate zero-knowledge under a leakage attack, we consider a PPT machine S called the simulator, which receives access to an oracle $\mathcal{L}_w^n(\cdot)$. $\mathcal{L}_w^n(\cdot)$ is called the leakage oracle, and parametrized by the witness w and the security parameter n. A query to the leakage oracle consists of an efficiently computable function F, to which the oracle responds with $F(w)$. The leakage-resilient zero-knowledge is defined by requiring that the output of S be computationally indistinguishable from the real view; in addition the length of all bits read by S from $\mathcal{L}_w^n$ in producing a particular view v is at most the length of leakage answers contained in the v.

For $x \in L$, w such that $R(x, w) = 1$, $z \in \{0, 1\}^*$, and randomness $r \in \{0, 1\}^*$ defining the output $v = S^{\mathcal{L}_w^n(\cdot)}(x, z; r)$, we let the function $\ell_S(v, r)$ denote the number of bits that S receives from $\mathcal{L}_w^n(\cdot)$ in generating view v with randomness r. Further, we let the function $\ell_{V^*}(v)$ denote the total length of leakage answers that V^* receives in the output v. By convention, randomness r will be included in the notation only when we need to be explicit about it.

Definition 2 (Leakage-resilient Zero-knowledge). *We say that an interactive proof system $\langle P, V \rangle$ for a language $L \in \mathcal{NP}$ with a witness relation R, is leakage-resilient zero-knowledge if for every probabilistic polynomial time machine V^* launching a leakage attack on P, there exists a probabilistic polynomial time machine S such that the following two conditions hold:*

1. *For every $x \in L$, every w such that $R(x, w) = 1$, and every $z \in \{0, 1\}^*$, distributions $\text{VIEW}_{V^*}[P(x, w) \leftrightarrow V^*(x, z)]$ and $S^{\mathcal{L}_w^n(\cdot)}(x, z)$ are computationally indistinguishable.*
2. *For every $x \in L$, every w such that $R(x, w) = 1$, every $z \in \{0, 1\}^*$, and every sufficiently long $r \in \{0, 1\}^*$ defining the output $v = S^{\mathcal{L}_w^n(\cdot)}(x, z; r)$, it holds that $\ell_S(v, r) \leq \ell_{V^*}(v)$.*

The definition of standard zero-knowledge is obtained by removing condition 2, and enforcing that no leakage queries are allowed to any machine.

3   Cryptographic Tools

We start by recalling some standard cryptographic tools and two-party protocols. Looking ahead, we will require that our protocols satisfy the following important property. For a specific party (chosen depending upon the protocol), all messages sent by this party be pseudorandom strings. In some cases where this is not possible, it will be sufficient if the messages are pseudorandom elements of group (e.g., a prime-order subgroup of $\mathbb{Z}_p^*$ for a (safe) prime p of length n).[4] We will provide necessary details when appropriate.

Statistically-Binding Commitments. We use Naor's scheme [55], based on a pseudorandom generator (prg). Recall that in this scheme, first the receiver sends a random string τ of length $3n$; to commit to bit b, the sender selects a uniform seed s of length n and sends y such that if $b = 0$ then $y = \mathsf{prg}(s)$, otherwise $y = \tau \oplus \mathsf{prg}(s)$. This scheme is statistically binding; in addition, *sender's message is pseudorandom.* A string can be committed by committing bitwise, and it suffices to use same τ for all the bits. We write $\mathsf{sbcom}_\tau(m; s)$ to represent sender's string, when receiver's first message is τ.

Statistically-Hiding Commitments. We use a statistically hiding commitment scheme as well. We require the receiver of this scheme to be *public coin.* Such schemes are known, including a two-round string commitment scheme, based on collision-resistant hash functions (crhf) [58,41,21]. We write $\mathsf{shcom}_\rho(m; s)$ to denote sender's commitment string, when receiver's first message is ρ. Without loss of generality, $|\rho| = n$.

Zero-Knowledge Proofs. We use a statistical zero-knowledge argument-of-knowledge (szkaok) protocol for proving $\mathcal{NP}$-statements. We require the verifier of this protocol to be *public coin.* Such protocols are known to exist; including a constant-round protocol based on crhf [7,10,63], and a $\omega(1)$-round protocol based on statistically-hiding commitments [38,13].

We choose the constant-round protocol of Pass and Rosen, denoted Π_{PR}, as our candidate szkaok. Let S_{PR} denote the corresponding simulator for Π_{PR}. We remark that S_{PR} is a "straight-line" simulator, with strict polynomial running time.

3.1 Oblivious Transfer

We will use a *two-round* oblivious transfer protocol $\mathsf{OT} := \langle S_{\mathrm{OT}}, R_{\mathrm{OT}} \rangle$. For the choice bit b of the receiver, we denote by $\{R_{\mathrm{OT}}(1^n, b)\}_{n \in \mathbb{N}}$ the message sent by R_{OT} on input $(1^n, b)$.

Let p, q be primes such that $p = 2q + 1$ and $|p| = n$. Then, we require the OT protocol to satisfy the following requirement. There exists a randomized PPT algorithm R_{OT}^{pub} such that for every $n \in \mathbb{N}$, every $b \in \{0,1\}$, and every safe prime $p = 2q + 1$, the following two conditions hold:

---

[4] This will be sufficient since public sampling from such a group admits *invertible sampling* [20,42]. However, it is more convenient to directly work with the assumption that algorithms can receive random elements in such a group as part of their random tape.

1. $R_{OT}(1^n, 0) \overset{c}{\equiv} R_{OT}^{pub}(1^n, p)$
2. The output of $R_{OT}^{pub}(1^n, p)$ consists of components $\{\alpha_i\}_{i=1}^{\text{poly}(n)}$ such that every α_i is a uniform and independent element in an order q subgroup of $\mathbb{Z}_p^*$.

We can formulate the second requirement by simply requiring the output to contain independent and random bits. The difficulty is that we do not know any OT protocol that would satisfy such a requirement. We therefore choose the above formulation. Note that without loss of generality, we can assume that uniform and independent elements can be provided as part of the random tape.[5] We will call algorithm R_{OT}^{pub} the "fake" receiver algorithm.

Concrete Instantiation: The existence of R_{OT}^{pub} is extremely crucial for our construction. Unfortunately, no OT protocol satisfying this requirement are known to exist based on *general* assumptions. However, two round OT protocols of [56,1] based on the DDH assumption, do satisfy both of our requirements. For concreteness, we fix the Naor-Pinkas oblivious transfer (protocol 4.1 in [56]) as our choice, and denote it by OT_{NP}. Algorithm R_{OT}^{pub} in this protocol consists of sending random and independent elements in order q subgroup of $\mathbb{Z}_p^*$. When multiple secrets must be exchanged we simply repeat this protocol in parallel.

Security of OT. In terms of security, the protocols in [56,1] are secure against malicious adversaries. However, they do not satisfy the usual simulation based (i.e., "ideal/real") security. Instead, they satisfy the following (informally stated) security notions:

1. *Indistinguishability for receiver:* it ensures that $\{R_{OT}(1^n, 0)\}_{n \in \mathbb{N}} \overset{c}{\equiv} \{R_{OT}(1^n, 1)\}_{n \in \mathbb{N}}$, where $\{R_{OT}(1^n, b)\}_{n \in \mathbb{N}}$ denotes the message sent by honest receiver on input $(1^n, b)$.
2. *Statistical secrecy for sender:* it ensures either $\{S_{OT}(1^n, m_0, m_1, q)\}_{n \in \mathbb{N}} \overset{s}{\equiv} \{S_{OT}(1^n, m_0, m')\}_{n \in \mathbb{N}}$ or $\{S_{OT}(1^n, m_0, m_1, q)\}_{n \in \mathbb{N}} \overset{s}{\equiv} \{S_{OT}(1^n, m', m_1)\}_{n \in \mathbb{N}}$, where m' is an arbitrary message and $S_{OT}(1^n, m_0, m_1, q)$ denotes the message sent by the honest sender on input $(1^n, m_0, m_1)$ when receiver's message is q.

This type of security notion is sufficient for our purpose. A formal description, following [40], is given in the full version of this work [60].

3.2 Extractable Commitments

We will need a perfectly-binding commitment scheme which satisfies the following two properties. First, if a cheating committer C^* successfully completes the protocol, then there exists an extractor algorithm E which outputs the value

---

[5] This assumption is easily removed by requiring an invertible sampling algorithm for R_{OT}^{pub}, which are known to exist. Also, the two-round requirement is not essential and can be relaxed.

committed by C^* during the commit stage. Second, there exists a public-coin algorithm C_{pub} such that no cheating receiver can tell if it is interacting with C_{pub} or the honest committing algorithm C. Algorithm C_{pub} is essentially the "fake" committing algorithm for C (much like the fake receiver R_{OT}^p define above). Let us first define these properties.

Commit-with-Extract. We will actually need a slightly property than mere extraction, called *commit-with-extract* [11,51]. Informally, commit-with-extract requires that for every cheating C^*, there exists an extractor E which first simulates the view of the cheating *committer* in an execution with honest receiver; further, if the view is accepting then it also outputs the value committed to in this view. Our specific use requires that the quality of simulation be *statistical*.

Definition 3 (Commit-with-extract [11]). *Let $n \in \mathbb{N}$ be the security parameter. A perfectly-binding commitment scheme $\Pi_{com} := \langle C, R \rangle$ is a commit-with-extract scheme if the following holds: there exists a strict PPT commitment-extractor E such that for every PPT committer C^*, for every $m \in \{0,1\}^n$, every (advice) $z \in \{0,1\}^*$ and every $r \in \{0,1\}^*$, upon input (C^*, m, z, r), machine E outputs a pair, denoted $(E_1(C^*, m, z, r), E_2(C^*, m, z, r))$, such that the following conditions hold:*

1. $E_1(C^*, m, z, r) \overset{s}{\equiv} \text{VIEW}_{C^*}[C^*(m, z; r) \leftrightarrow R()]$
2. $\Pr[E_2(C^*, m, z, r) = \text{value}(E_1(C^*, m, z, r))] \geq 1 - \text{negl}(n)$

where $\text{value}(\cdot)$ is a deterministic function which outputs either the unique value committed to in the view $E_1(C^*, m, z, r)$, or $\bot$ if no such value exists.

We say that a perfectly binding commitment scheme Π_{com} admits *public decommitment* if there exists a deterministic polynomial time algorithm D_{com} which on input the *public transcript* of interaction $\widehat{m}$, and the decommitment information d, outputs the *unique* value m committed in $\widehat{m}$. If there is no such value, the algorithm outputs $\bot$. For perfectly binding commitment schemes, the function value is well defined on the public transcripts as well. Therefore, we can write $D_{com}(d, \widehat{m}) = \text{value}(\widehat{m})$.

We now specify our "fake" public-coin sender requirement. Since we are working with DDH based construction, we will use a safe prime $p = 2q + 1$ of length n, (as used in R_{OT}^{pub}).

Let $n \in \mathbb{N}$ be the security parameter. We say that a perfectly binding commitment scheme $\Pi_{com} := \langle C, R \rangle$ has a *fake public-coin sender* if there exists an algorithm C_{pub} such that for every malicious PPT R^*, every $m \in \{0,1\}^n$, every safe prime p of length n, every advice $z \in \{0,1\}^*$, the following two conditions hold:

1. $\text{VIEW}_{R^*}[C(m) \leftrightarrow R^*(z)] \overset{c}{\equiv} \text{VIEW}_{R^*}[C_{pub}(p) \leftrightarrow R^*(z)]$
2. The output of $C_{pub}(p)$ consists of components $\{\alpha_i\}_{i=1}^{\text{poly}(n)}$ such that for every i: α_i is a uniform and independent element either in $\{0, 1\}$ or in an order q subgroup of $\mathbb{Z}_p^*$.

Concrete Instantiation. Unfortunately, no commitment protocol satisfying these requirements is known. The central reason behind this is that the fake public-coin sender C_{pub} requirement interferes with the commit-with-extract requirement. In [11], Barak and Lindell constructed a commitment protocol with the goal of *strict polynomial time* extraction. We observe that somewhat surprisingly, with some very minor changes, this protocol actually satisfies all our requirements. In particular, this commitment scheme is a *commit-with-extract* scheme, has a *fake public-coin sender*, and admits *public decommitment*. However, as with the OT protocol, this change requires us to use ElGamal [32] and hence DDH (instead of a general trapdoor permutation). For completeness, we present the protocol of [11] and explain the required modifications in the full version of this work [60].

Important Notation. For concreteness, fix $\Pi_{com} := \langle C, R \rangle$ to be a specific commitment protocol satisfying all three conditions above, and let D_{com} denote it's public decommitment algorithm. Let $\mathbf{L}_{com} := \{(m, \widehat{m}) : \exists d \text{ s.t. } D_{com}(\widehat{m}, d) = m\}$. That is, $\mathbf{L}_{com}$ is an $\mathcal{NP}$-language containing statements $(m, \widehat{m})$ such that $\widehat{m}$ is a commitment-transcript for value m. Let $\mathbf{R}_{com}$ be the corresponding $\mathcal{NP}$-relation so that $\mathbf{R}_{com}((m, \widehat{m}), d) = 1$ if $D_{com}(\widehat{m}, d) = m$ and 0 otherwise.

3.3   Barak's Preamble

In this section, we will recall Barak's non-black-box simulation method. In addition, we will make a slight change to this protocol which requires us to reprove some of the claims. We start by recalling Barak's relation for the complexity class $\mathbf{NTIME}(n^{\log \log(n)})$.

Barak's Relation. Let $n \in \mathbb{N}$ be the security parameter, and $\{\mathcal{H}_n\}_n$ be a family of crhf, $h : \{0,1\}^* \to \{0,1\}^n$. Since we are using Naor's commitment scheme, we will have an extra string τ for the commitment scheme sbcom. Barak's relation, $\mathbf{R}_\mathrm{B}$ takes as input an instance of the form $\langle h, \tau, c, r \rangle \in \{0,1\}^n \times \{0,1\}^{3n} \times \{0,1\}^{3n^2} \times \{0,1\}^{n+n^2}$ and a witness of the form $\langle M, y, s \rangle \in \{0,1\}^* \times \{0,1\}^* \times \{0,1\}^{\mathrm{poly}(n)}$.

> **Relation:** $\mathbf{R}_\mathrm{B}(\langle h, \tau, c, r \rangle, \langle M, y, s \rangle) = 1$ if and only if:
> 1. $|y| \leq |r| - n$.
> 2. $c = \mathsf{sbcom}_\tau(h(M); s)$.
> 3. $M(c, y) = r$ within $n^{\log \log n}$ steps.

Let $\mathbf{L}_\mathrm{B}$ be the language corresponding to $\mathbf{R}_\mathrm{B}$. We use this more complex version involving y, since it will allow us to successfully simulate even in the presence of leakage queries, which a cheating verifier obtains during the protocol execution.[6]

---

[6] This relation is identical to the one used for constructing bounded concurrent zero-knowledge in constant rounds in [7].

Universal Arguments and Statement Generation. Universal arguments (UARG) are four-round public-coin interactive argument systems [46,52,10], which can be used to prove statements in $\mathbf{L_B}$. Let $\langle P_{\mathrm{UA}}, V_{\mathrm{UA}} \rangle$ be such a system. We will denote the four rounds of this UARG by $\langle \alpha, \beta, \gamma, \delta \rangle$. Consider the following protocol between a party P_B and a party V_B.

> **Protocol** GenStat: Let $\{\mathcal{H}_n\}_n$ be a family of crhf functions.
> 1. V_B sends $h \leftarrow \mathcal{H}_n$ and $\tau \leftarrow \{0,1\}^{3n}$
> 2. P_B sends $c \leftarrow \{0,1\}^{3n^2}$
> 3. V_B sends $r \leftarrow \{0,1\}^{n+n^2}$.

Note that that length of r is $n^2 + n$ which allows y to be of length at most n^2. Length of c is $3n^2$ since it is supposed to be a commitment to n bits. We have the following lemma:[7]

Composed Protocol $\langle P^\otimes, V^\otimes \rangle$. We define this for convenience. The composed protocol is simply the GenStat protocol followed by an universal argument that the transcript $\sigma := \langle h, \tau, c, r \rangle$ is in $\mathbf{R_B}$. More precisely, strategy $P^\otimes := P_B \odot P_{\mathrm{UA}}$ is the composed prover, and $V^\otimes := V_B \odot V_{\mathrm{UA}}$ is the composed verifier, where $A \odot B$ denotes the process of running ITM A first, and then continuing ITM B from then onwards.[8] The following lemma states that the composed verifier $V^\otimes$ almost always rejects in an interaction with any PPT prover (i.e., it always rejects that $\sigma \in \mathbf{L_B}$).

Lemma 1 ([7]). *Suppose that $\{\mathcal{H}_n\}_n$ is a family of crhf functions. There exists a negligible function* negl *such that for every* PPT *strategy P^*, every $z \in \{0,1\}^*$, every $r \in \{0,1\}^*$, and every sufficiently large n,*

$$\Pr\left[\mathrm{OUT}_{V^\otimes}[P^*(z;r) \leftrightarrow V^\otimes()] \right] \leq \mathrm{negl}(n)$$

where the probability is taken over the randomness of $V^\otimes$.

The "Encrypted" Version. In Barak's protocol, an "encrypted" version of the above protocol is used in which the honest prover sends commitments to its UARG-messages (instead of the messages themselves). This is possible to do since the verifier is public coin.

We will use our commit-with-extract scheme $\Pi_{com} := \langle C, R \rangle$ for this purpose.[9] Recall that for Π_{com}, there exists a fake public-coin sender algorithm C_{pub} whose

---

[7] The version of Barak's relation that we use is actually a somewhat simplified form of the relation given in [10], which results only in a reduction to hash functions that are crhf against circuits of size $n^{\log n}$. By using the more complex version of [10], we get a reduction to standard crhf, without affecting any of our claims.

[8] A and B do not share states and run with their own independent inputs.

[9] Recall that Π_{com} is perfectly-binding commitment scheme which satisfies the commit-with-extract notion. In addition, the protocol has a public decommitment algorithm D_{com}, an associated $\mathcal{NP}$-relation $\mathbf{R}_{com}$, and $\mathcal{NP}$-language $\mathbf{L}_{com}$, and a fake public-coin sender algorithm. See section 3.2.

execution is indistinguishable from that of C. During the commitment phase, our prover algorithm will follow instructions of C_{pub}; the verifier will continue to use the normal receiver strategy R.

"Encrypted" preamble. $\langle \widehat{P}_\mathrm{B}, \widehat{V}_\mathrm{B} \rangle$: Let $\{\mathcal{H}_n\}_n$ be a family of crhf functions.

1. $\widehat{P}_\mathrm{B}$ and $\widehat{V}_\mathrm{B}$ run the GENSTAT protocol.
   Let $\langle h, \tau, c, r \rangle$ denote the resulting statement.
2. $\widehat{P}_\mathrm{B}$ and $\widehat{V}_\mathrm{B}$ execute UARG for the statement $\langle h, \tau, c, r \rangle$.
   (a) $\widehat{V}_\mathrm{B}$ sends α, obtained from V_{UA}.
   (b) $\widehat{P}_\mathrm{B}$ runs C_{pub}, and $\widehat{V}_\mathrm{B}$ runs R;
      Let $\widehat{\beta}$ be the commitment transcript.
   (c) $\widehat{V}_\mathrm{B}$ sends γ, obtained from V_{UA}.
   (d) $\widehat{P}_\mathrm{B}$ runs C_{pub}, and $\widehat{V}_\mathrm{B}$ runs R;
      Let $\widehat{\delta}$ be the commitment transcript.

The full transcript of the preamble is $\langle h, \tau, c, r, \alpha, \widehat{\beta}, \gamma, \widehat{\delta} \rangle$.

Since the prover messages are committed, we cannot make a claim along the lines of lemma 1. Therefore, we define the following $\mathcal{NP}$-relation $\mathbf{R}_{sim}$ and claim that it is a "hard" relation. This relation simply tests that there exist valid de-commitments (d_1, d_2) for strings $\widehat{\beta}, \widehat{\delta}$ so that the transcript is accepted by the UARG verifier.

Relation: $\mathbf{R}_{sim}(\langle h, \tau, c, r, \alpha, \widehat{\beta}, \gamma, \widehat{\delta} \rangle, \langle \beta, d_1, \delta, d_2 \rangle) = 1$ if and only if:
1. $\mathbf{R}_{com}(\langle \beta, \widehat{\beta} \rangle, d_1) = 1$.
2. $\mathbf{R}_{com}(\langle \delta, \widehat{\delta} \rangle, d_2) = 1$.
3. $V_{\mathrm{UA}}(h, \tau, c, r, \alpha, \beta, \gamma, \delta) = 1$.

The language corresponding to relation $\mathbf{R}_{sim}$ is denoted by $\mathbf{L}_{sim}$. Also note that $\widehat{P}_\mathrm{B}$ sends either random strings of uniform elements in a prime order group of $\mathbb{Z}_p^*$. The proof of the following lemma appears in the full version of this work [60].

Lemma 2. *Suppose that $\{\mathcal{H}_n\}_n$ is a family of* crhf *functions. There exists a negligible function* negl *such that for every* PPT *strategy P^*, every $z \in \{0,1\}^*$, every $r \in \{0,1\}^*$, and every sufficiently large n,*

$$\Pr\left[\sigma \leftarrow \mathrm{TRANS}[P^*(z; r) \leftrightarrow \widehat{V}_\mathrm{B}()]; \sigma \in \mathbf{L}_{sim} \right] \leq \mathrm{negl}(n)$$

where the probability is taken over the randomness of $\widehat{V}_\mathrm{B}$.

4   Conditional Disclosure via Garbled Circuits

Yao's garbled circuit method [72] allows two parties to compute any arbitrary function f of their inputs in a "secure" manner. Without loss of generality, let $f : \{0,1\}^n \times \{0,1\}^n \to \{0,1\}^n$.

The Method. The garbled circuit method specifies two polynomial time algorithms (Garble, Eval). Algorithm Garble is randomized; on input 1^n and the description of a circuit (that computes) f, it outputs a triplet $(\mathcal{C}, \text{key}_0, \text{key}_1)$. $\mathcal{C}$, which consists of a set of tables containing encrypted values, is called the *garbled circuit*; and $\text{key}_0 = \{(k_{0,i}^0, k_{0,i}^1)\}_{i=1}^n$ and $\text{key}_1 = \{(k_{1,i}^0), k_{1,i}^1\}_{i=1}^n$ are called the keys. Let $a = a_1, \ldots, a_n$ and $b = b_1, \ldots, b_n$ be binary strings. Algorithm Eval, on input $(\mathcal{C}, K_a, K_b)$ outputs a value $v \in \{0,1\}^n$ such that if $K_a = \{k_{0,i}^{a_i}\}$ and $K_b = \{k_{1,i}^{b_i}\}$ then $v = f(a,b)$.

For an $\mathcal{NP}$-relation R, and $\sigma \in \{0,1\}^*$, let $f_{\sigma,R}$ be the following function.

> **Function $f_{\sigma,R}(\omega, m)$:**
> If $R(\sigma, \omega) = 1$, output m; otherwise output $0^{|m|}$.

That is, $f_{\sigma,R}$ discloses m if and only if ω is a valid witness for the statement σ. We will use the garbled circuit method for such functions $f_{\sigma,R}$. Jumping ahead, we will use $f_{\sigma,\mathbf{R}_{sim}}$ for the $\mathcal{NP}$-relation $\mathbf{R}_{sim}$ described in section 3.3.

Conditional disclosure via garbled-circuits. In the two party setting, one party prepares the garbled circuit $\mathcal{C}$ and sends the keys K_b corresponding to her input b to the other party. An OT protocol is used by the first party to receive keys K_a for her input a, so that it can execute the evaluation algorithm. This allows the receiver of the garbled circuit (and OT) to learn $f(a,b)$ but "nothing more". In addition, receiver's input remains secure due to OT-security for receiver.

Looking forward, we will require our protocol so that it will admit a "fake" receiver algorithm. Therefore, we will use the Naor-Pinkas OT protocol, denoted OT_{NP} (see section 3.1). For a technical reason, our protocol starts by first executing steps of OT_{NP}, and then executes the garbled circuit step. Note that the first step involves n parallel executions of OT, one for each input bit. The resulting two-round conditional disclosure protocol, Π_{cd}, is as follows.

Protocol Π_{cd} for computing $f_{\sigma,R}(\omega, m)$: The protocol consists of two parties, a receiver R_{cd} and a sender S_{cd}. R_{cd}'s private input is bit string $\omega = \omega_1, \ldots, \omega_n$, and S_{cd}'s private input is bit string $m = m_1, \ldots, m_n$. The common input to the parties is the description of the function $f_{\sigma,R}$ as a circuit (equivalently, just σ).
  1. R_{cd} computes $v = (v_1, \ldots, v_n)$, where v_i is the first message of OT_{NP} using the input ω_i and fresh randomness for $i \in [n]$. It then sends v.
  2. S_{cd} prepares a garbled circuit for the function $f_{\sigma,R}$: $(\mathcal{C}_{\sigma,R}, \text{key}_0, \text{key}_1) \leftarrow$ Garble$(f_{\sigma,R})$. Next, S_{cd} prepares $v' = (v'_1, \ldots, v'_n)$ where v'_i is the second message of OT_{NP} computed using $(k_{0,i}^0, k_{0,i}^1)$ as sender's input and v_i as receiver's first message. Here the keys $(k_{0,i}^0, k_{0,i}^1)$ are the ith component of key_0. Finally, let K_m denote the keys taken from key_1 corresponding to m. S_{cd} sends $(\mathcal{C}_{\sigma,R}, v', K_m)$.

Recall that OT_{NP} is a two-round protocol, it provides statistical secrecy for the sender, and has a fake public-coin receiver. Also recall that OT_{NP} does not satisfy the the standard simulation-based security. As a result, we cannot directly use known results about the security of Yao's protocol. Nevertheless, we can make weaker indistinguishability-style claims which suffice for our purpose. First notice that the OT-security for receiver, intuitively guarantees indistinguishability for the input of R_{cd}. For the sender, we can prove the following claim, whose proof appears in the full version of this work [60].

Lemma 3 (Security for sender). *Let* $L \in \mathcal{NP}$ *with witness relation* R *and* $\sigma \in \{0,1\}^*$. *For the security parameter* n, *let* $S_{cd}(1^n, f_{\sigma,R}, m, q)$ *represent the response of the honest sender (of protocol* Π_{cd}*), with input* $(f_{\sigma,R}, m)$ *when receiver's first message is* q. *Then, for every pair of distinct messages* (m, m'), *every* $q \in \{0,1\}^*$ *(from a possibly malicious* PPT *receiver), and every* $\sigma \notin L$, *it holds that*

$$S_{cd}(1^n, f_{\sigma,R}, m, q) \overset{c}{\equiv} S_{cd}(^n, f_{\sigma,R}, m', q).$$

5   A Constant Round Protocol

In this section we will present our constant round protocol. The protocol will use the dual simulation idea, introduced in [33], as an important tool. To simplify the exposition and the proofs, we isolate a part of the protocol from [33], and present it as a separate building block.[10]

Shortened GJS Protocol $\langle P_{\text{GJS}}, V_{\text{GJS}} \rangle$. The common input is an n vertex graph G in the form of an adjacency matrix, and prover's auxiliary input is a Hamiltonian cycle H in G. The protocol proceeds in following three steps.

1. Commitment stage:
   (a) P_{GJS} sends a random string ρ.
   (b) V_{GJS} sends $\widehat{t_1} = \text{shcom}_\rho(t_1; s_1)$ and $\widehat{ch} = \text{shcom}_\rho(ch; s_2)$,
       where $t_1 \leftarrow \{0,1\}^{3n^4}$, $ch \in \{0,1\}^n$, and $s_1, s_2 \leftarrow \{0,1\}^{\text{poly}(n)}$.
2. Coin flipping stage:
   (a) P_{GJS} sends a random string t_2.
   (b) V_{GJS} opens $\widehat{t_1}$ by sending (t_1, s_1).
       Let $\mathbf{t} = t_1 \oplus t_2$.
3. Blum Hamiltonicity protocol:
   (a) Let $\mathbf{t} = \mathbf{t}_1, \ldots, \mathbf{t}_{n^3}$ so that $|\mathbf{t}_i| = 3n$ for $i \in [n^3]$.
       Prover chooses n random permutations $\pi_1, \ldots, \pi_n$ and sets
$G_i = \pi_i(G)$
       for each $i \in [n]$. It then commits to each bit b_j in G_i using
$\text{sbcom}_{\mathbf{t}_{i \times j}}$.

---

[10] The only difference is that the challenge-response slots in the [33] protocol have been removed. As a result, many other parameters of their protocol become irrelevant, and also do not appear in this protocol. This does not affect the soundness of the protocol.

　　　　　(b) Verifier opens to $\widehat{ch}$ by sending (ch, s_2).
　　　　　　(c) Let $ch = ch_1, \ldots, ch_n$. For every $i \in [n]$, if $ch_i = 0$ then
prover opens

　　　　　　　　each edge in G_i and reveals π_i; else, it opens edges of the
cycle in G_i.

The following lemma has been shown in [33].

Lemma 4 ([33]). *Protocol* $\langle P_{\text{GJS}}, V_{\text{GJS}} \rangle$ *is a sound interactive argument system for all of* $\mathcal{NP}$.

5.1 Our Protocol

We are now ready to present our protocol $\langle P, V \rangle$. The protocol starts with an execution of the "encrypted" preamble protocol (see section 3.3); this is followed by the first i.e., commitment, stage of the GJS protocol. Before completing the GJS protocol, verifier executes the garbled-circuit protocol Π_{cd} for $f_{\sigma, \mathbf{R}_{sim}}$ and a specific m (described shortly), and proves using an szkaok that this step was performed honestly. This will enable the simulator to extract useful information in m. Finally, the rest of the GJS protocol is executed to complete the proof. The full description of the protocol is given in figure 1. It is easy to see that our protocol has constant rounds. The completeness of the protocol follows directly from the completeness of $\langle P_{\text{GJS}}, V_{\text{GJS}} \rangle$. In next two sections, we prove the soundness and zero-knowledge of this protocol. Note that the the prover is actually "public coin" *up until the final step*.

Proving Leakage-resilient Zero-Knowledge. Due to space constraints, the proof of security of this protocol—theorem 1—appears in the full version of this work [60]. At a high level, we use Barak's non-black-box simulation idea along with GJS simulation. Let V^* be an arbitrary PPT verifier whose program is given as an input to the simulator S. There are four main ideas:

1. First, the simulation uses V^*'s code to execute the preamble in such a way, that at the end of the preamble, $\sigma \in \mathbf{L}_{sim}$. In addition, the simulator will also have a witness ω so that $\mathbf{R}_{sim}(\sigma, \omega) = 1$. The properties of the components used in the preamble (in particular the use of fake sampling algorithms that are public coin) guarantee that simulator's actions in the preamble are indistinguishable from a real execution with an honest prover. In addition, it is easy to answer leakage queries since the messages exchanged so far represent the entire random-tape of the prover at this point. This allows the simulator to answer leakage queries by simply appending these messages to the state, and sending an appropriate query to the leakage oracle.

2. Next, the simulator will use ω in the garbled circuit step to obtain keys K_ω. Once again, since the first message of OT_{NP} provides indistinguishability for receiver's input, this step does not affect the simulation. Further, since P is public coin in this step as well, the simulator can continue to answer leakage queries as before.

Protocol $\langle P, V \rangle$. The common input consists of 1^n, and an n vertex graph G in the form of its adjacency matrix. Prover's private input is a Hamiltonian cycle H in G.

1. **"Encrypted" preamble:** $P \Rightarrow V$
   P and V run Barak's encrypted preamble. P runs the public-coin strategy $\widehat{P}_{\text{B}}$, and V runs strategy $\widehat{V}_{\text{B}}$. Let the transcript be $\sigma := \langle h, \tau, c, r, \alpha, \widehat{\beta}, \gamma, \delta \rangle$.

2. **Commitment step:** $V \Rightarrow P$
   P and V run the first, i.e. commitment, step of $\langle P_{\text{GJS}}, V_{\text{GJS}} \rangle$.
   (a) P sends a random string ρ
   (b) V sends $\widehat{t_1} = \text{shcom}_\rho(t_1; s_1)$ and $\widehat{ch} = \text{shcom}_\rho(ch; s_2)$, where $t_1 \leftarrow \{0,1\}^{3n^4}$,
   $ch \leftarrow \{0,1\}^n$, and $s_1, s_2 \leftarrow \{0,1\}^{\text{poly}(n)}$; let $m := (t_1, s_1, ch, s_2)$.

3. **Garbled-circuit step:** $V \Rightarrow P$
   P and V run the *two-round* garbled circuit protocol, Π_{cd}, for the function $f_{\sigma, \mathbf{R}_{sim}}$. V acts as the sender with private input m.
   (a) P runs the fake receiver, $v_1 \leftarrow R^{pub}_{\text{OT}}(1^n, p)$ for a random safe prime p; sends v_1.
   (b) V sends $(\mathcal{C}, v_2, K_m) \leftarrow S_{cd}(f_{\sigma, R}, m, v_1; s_3)$, using fresh coins s_3.

4. **Proof of correctness:** $V \Rightarrow P$
   V proves to P using public-coin szkaok Π_{PR} the knowledge of s_3 and $m = (t_1, s_1, ch, s_2)$ so that:
   (a) $\widehat{t_1} = \text{shcom}_\rho(t_1; s_1)$,
   (b) $\widehat{ch} = \text{shcom}_\rho(ch; s_2)$,
   (c) $S_{cd}(f_{\sigma, R}, m, v_1; s_3) = (\mathcal{C}, v_2, K_m)$.

5. **Final step:** $P \Rightarrow V$
   P and V complete all remaining five rounds of $\langle P_{\text{GJS}}, V_{\text{GJS}} \rangle$. P uses H as the witness.

Fig. 1. Our Constant Round LRZK Protocol

3. Having obtained K_ω along with $\mathcal{C}, K_m$ in the garbled circuit step, the simulator can evaluate the $\mathcal{C}$ and learn $f_{\sigma, \mathbf{R}_{sim}}(\omega, m)$ to learn m. By the soundness of szkaok of the next step, it is guaranteed that m contains valid openings (t_1, s_1, ch, s_2) for $\widehat{t_1}$ and $\widehat{ch}$.

4. Finally, observe that (t_1, ch) is precisely the information needed by the GJS simulation method to successfully simulate the last step, while answering leakage queries properly. Briefly, ch is the challenge for Blum's protocol, and a first message can be created by the simulator to successfully answer V^*'s challenge in the last message. At the same time, since t_1 is known prior to the coin-flipping stage of the GJS protocol (see section 5), the simulator will have the ability to equivocate in Naor's commitment scheme, allowing it to successfully answer leakage queries.

An important point to note is that if V^* asks more than n^2 bits of leakage after receiving c and before sending r (see GENSTAT), the simulator will not be able to ensure that $\sigma \in \mathbf{L}_{sim}$. However, if this happens, the simulator can simply ask for the entire witness H from the leakage oracle since the length of leakage is more than the witness size. The simulator can then continue to run like the honest prover and output a view. See full proof in the full version of this work [60].

References

1. Aiello, W., Ishai, Y., Reingold, O.: Priced oblivious transfer: How to sell digital goods. In: EUROCRYPT 2001. LNCS, vol. 2045, pp. 119–135. Springer, Heidelberg (2001)
2. Ajtai, M.: Secure computation with information leaking to an adversary. In: STOC, pp. 715–724 (2011)
3. Akavia, A., Goldwasser, S., Vaikuntanathan, V.: Simultaneous hardcore bits and cryptography against memory attacks. In: Reingold, O. (ed.) TCC 2009. LNCS, vol. 5444, pp. 474–495. Springer, Heidelberg (2009)
4. Alwen, J., Dodis, Y., Wichs, D.: Leakage-resilient public-key cryptography in the bounded-retrieval model. In: Halevi, S. (ed.) CRYPTO 2009. LNCS, vol. 5677, pp. 36–54. Springer, Heidelberg (2009)
5. Alwen, J., Dodis, Y., Wichs, D.: Leakage-resilient public-key cryptography in the bounded-retrieval model. In: Halevi, S. (ed.) CRYPTO 2009. LNCS, vol. 5677, pp. 36–54. Springer, Heidelberg (2009)
6. Anderson, R.J., Kuhn, M.G.: Low cost attacks on tamper resistant devices. In: Security Protocols Workshop, pp. 125–136 (1997)
7. Barak, B.: How to go beyond the black-box simulation barrier. In: FOCS, pp. 106–115 (2001)
8. Barak, B., Goldreich, O., Goldwasser, S., Lindell, Y.: Resettably-sound zero-knowledge and its applications. In: FOCS 2001, pp. 116–125 (2001)
9. Barak, B.: Constant-round coin-tossing with a man in the middle or realizing the shared random string model. In: FOCS (2002)
10. Barak, B., Goldreich, O.: Universal arguments and their applications. In: Annual IEEE Conference on Computational Complexity (CCC), vol. 17 (2002); Preliminary full version available as Cryptology ePrint Archive, Report 2001/105
11. Barak, B., Lindell, Y.: Strict polynomial-time in simulation and extraction. SIAM Journal on Computing 33(4), 783–818 (2004); Extended abstract appeared in STOC 2002
12. Bitansky, N., Canetti, R., Halevi, S.: Leakage-tolerant interactive protocols. In: Cramer, R. (ed.) TCC 2012. LNCS, vol. 7194, pp. 266–284. Springer, Heidelberg (2012)
13. Blum, M.: How to prove a theorem so no one else can claim it. In: Proceedings of the International Congress of Mathematicians, pp. 1444–1451 (1987)
14. Boyle, E., Segev, G., Wichs, D.: Fully leakage-resilient signatures. In: Paterson, K.G. (ed.) EUROCRYPT 2011. LNCS, vol. 6632, pp. 89–108. Springer, Heidelberg (2011)
15. Brakerski, Z., Kalai, Y.T., Katz, J., Vaikuntanathan, V.: Overcoming the hole in the bucket: Public-key cryptography resilient to continual memory leakage. In: FOCS, pp. 501–510 (2010)

16. Canetti, R.: Security and composition of multiparty cryptographic protocols. Journal of Cryptology: The Journal of the International Association for Cryptologic Research 13(1), 143–202 (2000)
17. Canetti, R.: Universally composable security: A new paradigm for cryptographic protocols. In: Werner, B. (ed.) Proc. 42nd FOCS, pp. 136–147 (2001); Preliminary full version available as Cryptology ePrint Archive Report 2000/067
18. Canetti, R., Feige, U., Goldreich, O., Naor, M.: Adaptively secure multi-party computation. In: STOC, pp. 639–648 (1996)
19. Canetti, R., Goldreich, O., Goldwasser, S., Micali, S.: Resettable zero-knowledge. In: Proc. 32th STOC, pp. 235–244 (2000)
20. Damgård, I., Nielsen, J.B.: Improved non-committing encryption schemes based on a general complexity assumption. In: Bellare, M. (ed.) CRYPTO 2000. LNCS, vol. 1880, pp. 432–450. Springer, Heidelberg (2000)
21. Damgård, I., Pedersen, T.P., Pfitzmann, B.: On the existence of statistically hiding bit commitment schemes and fail-stop signatures. J. Cryptology 10(3), 163–194 (1997)
22. Deng, Y., Goyal, V., Sahai, A.: Resolving the simultaneous resettability conjecture and a new non-black-box simulation strategy. In: FOCS (2009)
23. Diffie, W., Hellman, M.E.: New directions in cryptography. IEEE Transactions on Information Theory 22(6), 644–654 (1976)
24. Dodis, Y., Haralambiev, K., López-Alt, A., Wichs, D.: Cryptography against continuous memory attacks. In: FOCS, pp. 511–520 (2010)
25. Dodis, Y., Haralambiev, K., López-Alt, A., Wichs, D.: Efficient public-key cryptography in the presence of key leakage. In: Abe, M. (ed.) ASIACRYPT 2010. LNCS, vol. 6477, pp. 613–631. Springer, Heidelberg (2010)
26. Dodis, Y., Kalai, Y.T., Lovett, S.: On cryptography with auxiliary input. In: STOC, pp. 621–630 (2009)
27. Dolev, D., Dwork, C., Naor, M.: Non-malleable cryptography (extended abstract). In: STOC, pp. 542–552 (1991)
28. Dwork, C., Naor, M., Sahai, A.: Concurrent zero knowledge. In: Proc. 30th STOC, pp. 409–418 (1998)
29. Dziembowski, S., Pietrzak, K.: Leakage-resilient cryptography. In: FOCS, pp. 293–302 (2008)
30. Faust, S., Rabin, T., Reyzin, L., Tromer, E., Vaikuntanathan, V.: Protecting circuits from leakage: the computationally-bounded and noisy cases. In: Gilbert, H. (ed.) EUROCRYPT 2010. LNCS, vol. 6110, pp. 135–156. Springer, Heidelberg (2010)
31. Feige, U., Shamir, A.: Zero knowledge proofs of knowledge in two rounds. In: Brassard, G. (ed.) Advances in Cryptology - CRYPTO 1989. LNCS, vol. 435, pp. 526–544. Springer, Heidelberg (1990)
32. ElGamal, T.: A public key cryptosystem and a signature scheme based on discrete logarithms. In: Blakely, G.R., Chaum, D. (eds.) Advances in Cryptology - CRYPTO 1984. LNCS, vol. 196, pp. 10–18. Springer, Heidelberg (1985)
33. Garg, S., Jain, A., Sahai, A.: Leakage resilient zero knowledge. In: Rogaway, P. (ed.) CRYPTO 2011. LNCS, vol. 6841, pp. 297–315. Springer, Heidelberg (2011), http://www.cs.ucla.edu/~abhishek/papers/lrzk.pdf
34. Gertner, Y., Ishai, Y., Kushilevitz, E., Malkin, T.: Protecting data privacy in private information retrieval schemes. J. Comput. Syst. Sci. 60(3), 592–629 (2000)
35. Goldreich, O., Micali, S., Wigderson, A.: How to play ANY mental game. In: ACM (ed.) Proc. 19th STOC, pp. 218–229 (1987); For more details see ([36], ch. 7)

36. Goldreich, O.: Foundations of Cryptography: Basic Applications. Cambridge University Press (2004)
37. Goldreich, O., Kahan, A.: How to construct constant-round zero-knowledge proof systems for NP. Journal of Cryptology 9(3), 167–189 (1996)
38. Goldreich, O., Micali, S., Wigderson, A.: Proofs that yield nothing but their validity or all languages in NP have zero-knowledge proof systems. Journal of the ACM 38(3), 691–729 (1991); Preliminary version in FOCS 1986
39. Goldwasser, S., Micali, S., Rackoff, C.: The knowledge complexity of interactive proof-systems. In: Proc. 17th STOC, pp. 291–304. ACM, Providence (1985)
40. Halevi, S., Kalai, Y.T.: Smooth projective hashing and two-message oblivious transfer. J. Cryptology 25(1), 158–193 (2012)
41. Halevi, S., Micali, S.: Practical and provably-secure commitment schemes from collision-free hashing. In: Koblitz, N. (ed.) CRYPTO 1996. LNCS, vol. 1109, pp. 201–215. Springer, Heidelberg (1996)
42. Ishai, Y., Kumarasubramanian, A., Orlandi, C., Sahai, A.: On invertible sampling and adaptive security. In: Abe, M. (ed.) ASIACRYPT 2010. LNCS, vol. 6477, pp. 466–482. Springer, Heidelberg (2010)
43. Ishai, Y., Kushilevitz, E., Ostrovsky, R.: Efficient arguments without short pcps. In: IEEE Conference on Computational Complexity, pp. 278–291 (2007)
44. Ishai, Y., Prabhakaran, M., Sahai, A., Wagner, D.: Private circuits ii: Keeping secrets in tamperable circuits. In: Vaudenay, S. (ed.) EUROCRYPT 2006. LNCS, vol. 4004, pp. 308–327. Springer, Heidelberg (2006)
45. Ishai, Y., Sahai, A., Wagner, D.: Private circuits: Securing hardware against probing attacks. In: Boneh, D. (ed.) CRYPTO 2003. LNCS, vol. 2729, pp. 463–481. Springer, Heidelberg (2003)
46. Kilian, J.: A note on efficient zero-knowledge proofs and arguments (extended abstract). In: Proc. 24th STOC, pp. 723–732 (1992)
47. Kiltz, E., Pietrzak, K.: Leakage resilient elgamal encryption. In: Abe, M. (ed.) ASIACRYPT 2010. LNCS, vol. 6477, pp. 595–612. Springer, Heidelberg (2010)
48. Kocher, P.C.: Timing attacks on implementations of diffie-hellman, rsa, dss, and other systems. In: Koblitz, N. (ed.) CRYPTO 1996. LNCS, vol. 1109, pp. 104–113. Springer, Heidelberg (1996)
49. Lewko, A., Rouselakis, Y., Waters, B.: Achieving leakage resilience through dual system encryption. In: Ishai, Y. (ed.) TCC 2011. LNCS, vol. 6597, pp. 70–88. Springer, Heidelberg (2011)
50. Lewko, A.B., Waters, B.: On the insecurity of parallel repetition for leakage resilience. In: FOCS, pp. 521–530 (2010)
51. Lindell, Y.: Parallel coin-tossing and constant-round secure two-party computation. In: Kilian, J. (ed.) CRYPTO 2001. LNCS, vol. 2139, pp. 171–189. Springer, Heidelberg (2001)
52. Micali, S.: CS proofs. In: Proc. 35th FOCS, pp. 436–453 (1994)
53. Micali, S., Pass, R.: Local zero knowledge. In: Kleinberg, J.M. (ed.) STOC, pp. 306–315. ACM (2006)
54. Micali, S., Rogaway, P.: Secure computation. In: Feigenbaum, J. (ed.) CRYPTO 1991. LNCS, vol. 576, pp. 392–404. Springer, Heidelberg (1992)
55. Naor, M.: Bit commitment using pseudo-randomness (extended abstract). In: Brassard, G. (ed.) CRYPTO 1989. LNCS, vol. 435, pp. 128–136. Springer, Heidelberg (1990)
56. Naor, M., Pinkas, B.: Efficient oblivious transfer protocols. In: SODA, pp. 448–457 (2001)

57. Naor, M., Segev, G.: Public-key cryptosystems resilient to key leakage. In: Halevi, S. (ed.) CRYPTO 2009. LNCS, vol. 5677, pp. 18–35. Springer, Heidelberg (2009)
58. Naor, M., Yung, M.: Universal one-way hash functions and their cryptographic applications. In: Proc. 21st STOC, pp. 33–43 (1989)
59. Osvik, D.A., Shamir, A., Tromer, E.: Cache attacks and countermeasures: The case of aes. In: Pointcheval, D. (ed.) CT-RSA 2006. LNCS, vol. 3860, pp. 1–20. Springer, Heidelberg (2006)
60. Pandey, O.: Achieving constant round leakage-resilient zero-knowledge. IACR Cryptology ePrint Archive (2012), http://eprint.iacr.org/2012/362.pdf
61. Pass, R.: Simulation in quasi-polynomial time, and its application to protocol composition. In: Biham, E. (ed.) EUROCRYPT 2003. LNCS, vol. 2656, pp. 160–176. Springer, Heidelberg (2003)
62. Pass, R.: Bounded-concurrent secure multi-party computation with a dishonest majority. In: Proc. 36th STOC, pp. 232–241 (2004)
63. Pass, R., Rosen, A.: New and improved constructions of non-malleable cryptographic protocols. In: STOC (2005)
64. Pietrzak, K.: A leakage-resilient mode of operation. In: Joux, A. (ed.) EUROCRYPT 2009. LNCS, vol. 5479, pp. 462–482. Springer, Heidelberg (2009)
65. Prabhakaran, M.: New Notions of Security. PhD thesis, Department of Computer Science, Princeton University, Princeton, NJ, USA (2005)
66. Prabhakaran, M., Rosen, A., Sahai, A.: Concurrent zero knowledge with logarithmic round-complexity. In: FOCS (2002)
67. Prabhakaran, M., Sahai, A.: New notions of security: achieving universal composability without trusted setup. In: STOC, pp. 242–251 (2004)
68. Quisquater, J.-J., Samyde, D.: Electromagnetic analysis (ema): Measures and counter-measures for smart cards. In: Attali, I., Jensen, T. (eds.) E-smart 2001. LNCS, vol. 2140, pp. 200–210. Springer, Heidelberg (2001)
69. Rosen, A.: A note on the round-complexity of concurrent zero-knowledge. In: Bellare, M. (ed.) CRYPTO 2000. LNCS, vol. 1880, pp. 451–468. Springer, Heidelberg (2000)
70. Rosen, A.: A note on constant-round zero-knowledge proofs for NP. In: Naor, M. (ed.) TCC 2004. LNCS, vol. 2951, pp. 191–202. Springer, Heidelberg (2004)
71. Yao, A.C.: Theory and applications of trapdoor functions. In: Proc. 23rd FOCS, pp. 80–91 (1982)
72. Yao, A.C.-C.: How to generate and exchange secrets. In: Proc. 27th FOCS, pp. 162–167 (1986)

Statistical Concurrent Non-malleable Zero Knowledge

Claudio Orlandi[1]*, Rafail Ostrovsky[23], Vanishree Rao[2],
Amit Sahai[2], and Ivan Visconti[4]

[1] Department of Computer Science, Aarhus University, Denmark
orlandi@cs.au.dk
[2] Department of Computer Science, UCLA, USA
[3] Department of Mathematics, UCLA, USA
{rafail,vanishri,sahai}@cs.ucla.edu
[4] Dipartimento di Informatica, University of Salerno, Italy
visconti@unisa.it

Abstract. The notion of Zero Knowledge introduced by Goldwasser, Micali and Rackoff in STOC 1985 is fundamental in Cryptography. Motivated by conceptual and practical reasons, this notion has been explored under stronger definitions. We will consider the following two main strengthened notions.

Statistical Zero Knowledge: here the zero-knowledge property will last forever, even in case in future the adversary will have unlimited power.

Concurrent Non-Malleable Zero Knowledge: here the zero-knowledge property is combined with non-transferability and the adversary fails in mounting a concurrent man-in-the-middle attack aiming at transferring zero-knowledge proofs/arguments.

Besides the well-known importance of both notions, it is still unknown whether one can design a zero-knowledge protocol that satisfies both notions simultaneously.

In this work we shed light on this question in a very strong sense. We show a *statistical concurrent non-malleable* zero-knowledge argument system for $\mathcal{NP}$ with a *black-box* simulator-extractor.

1 Introduction

The notion of zero knowledge, first introduced in [10], is one of the most pivotal cryptographic constructs. Depending on both natural and real-world attack scenarios, zero knowledge has been studied considering different conceptual flavors and practical applications.

---

* Work done while visiting UCLA.

Y. Lindell (Ed.): TCC 2014, LNCS 8349, pp. 167–191, 2014.
© International Association for Cryptologic Research 2014

Zero Knowledge and Man-in-the-Middle Attacks. In distributed settings such as the Internet, an adversary that controls the network can play concurrently as a verifier in some proofs[1] and as a prover in the other proofs. The goal of the adversary is to exploit the proofs it receives from the provers to then generate new proofs for the verifiers. The original notion of zero knowledge does not prevent such attacks since it assumes the adversarial verifier to only play as a verifier and only in sequential sessions.

The need of providing non-transferable proofs secure against such man-in-the-middle (MiM, for short) attacks was first studied by Dolev, Dwork and Naor in [7]. In [1], Barak, Prabhakaran and Sahai achieved for the first time such a strong form of zero knowledge, referred to as concurrent non-malleable zero knowledge (CNMZK, for short) is possible in the plain model. They provide a poly(λ)-round construction, for λ being the security parameter, based on one-way functions, and a $O(\log(\lambda))$-round construction based on collision-resistant hash functions. More recent results focused on achieving round efficiency with a mild setup [23], computationally efficient constructions [22], security with adaptive inputs [16].

Zero Knowledge and Forward Security. The zero-knowledge property says that the view of the adversarial verifier does not help her in gaining any useful information. This means that it does not include information that can be exploited by a PPT machine. However, even though the execution of a zero-knowledge protocol can be based on the current hardness of some complexity assumptions, it is quite risky to rely on the assumed resilience of such assumptions against more powerful machines of the future. What is zero knowledge in a transcript produced today could not be zero knowledge in the eyes of a distinguisher that will read the transcript in 2040.

It is therefore appealing to provide some forward security flavor so that whatever is zero knowledge today will be zero knowledge forever. Statistical zero knowledge [2,25,21,9,20,12,19] is the notion that satisfies this requirement. It has been achieved in constant rounds using collision-resistant hash functions [14], and even under the sole assumption that one-way functions exist requiring more rounds [13].

Unfortunately, all the known constructions for CNMZK protocols strongly rely on the computational indistinguishability of the output of the simulator. Techniques so far used to design protocols that are then proved to be CNMZK require the protocol to fix a witness in a commitment, that therefore must be statistically binding and thus only computationally hiding. There is therefore no hope to prove those protocol to be statistical zero knowledge. Moreover it does not seem that minor changes can establish the statistical zero knowledge property still allowing to prove CNMZK.

---

[1] While in our general discussion, we often refer to zero-knowledge proofs, we will finally need to resort to only arguments since our goal is to achieve statistical zero-knowledge property.

The Open Problem. Given the above state-of-the-art a natural question is the following: *is it possible to design an argument system that combines the best of both worlds, namely, a statistical concurrent non-malleable zero-knowledge argument system?*

1.1   Our Contribution

In this work, we provide the first statistical concurrent non-malleable zero-knowledge argument system. Our construction is an argument of knowledge (AoK, for short) and has a black-box simulator-extractor producing a statistically indistinguishable distribution.

As mentioned earlier, Barak et al. [1] presented the first CNMZKAoK protocol; we will refer to their work here as BPS. However, their construction had an inherent limitation that the simulation can only be computational, the reason being the following. In their protocol, the prover needs to commit to a valid witness via a statistically binding non-malleable commitment scheme. The commitment scheme being statistically binding is extremely crucial in their proof of security. This implies that when the simulator cheats and commits to a non-witness, the simulated view can only be computationally indistinguishable and not statistically so.

In this work, we overcome this shortcoming with the following idea. We take the BPS argument as a starting point and modify it. Firstly, we work on the root of the problem – the non-malleable commitment. We replace it with a special kind of a commitment scheme called '*mixed non-malleable commitment*' scheme. The notion of mixed commitment was first introduced by Damgård and Nielsen [6]. Our mixed non-malleable commitment is parameterized by a string that if sampled with uniform distribution makes the scheme statistically hiding and computationally binding. Instead, when it is taken from another (computationally indistinguishable) distribution it is a statistically binding, computationally hiding, and non-malleable. We will construct such a scheme by using as distributions non-DDH and DDH tuples.

The next idea would be to append the (modified) BPS argument to a coin-flipping phase in which the prover and the verifier generate a random string. Thus, in the real-world the above mixed commitment is statistically hiding. This thus enables us to prove statistical simulatability of our protocol. Furthermore, in order to also achieve extractability of witnesses for the arguments given by the adversary, we switch to a hybrid which biases the coin-flipping outcome to a random DDH tuple. Typically, a coin-flipping protocol would involve the verifier committing to its share of randomness, the prover sending its share of randomness in the clear, and finally, the verifier opening the commitment. However, in order to enable the simulator to bias the outcome, instead of the verifier opening the commitment to its share of randomness, it gives only the committed value in the clear and presents an AoK for the randomness used. This argument is again played by using the BPS AoK, since we would need concurrent non-malleability here.

In order to simplify our proofs, we rely on the Robust Extraction Lemma of Goyal et al. [11] that generalizes concurrent extractability of the PRS preamble (or concurrently extractable commitments – CECom, for short) [24] in the following sense. Consider an adversary who sends multiple CECom commitments interleaving them arbitrarily and also interacts with an external party B in an arbitrary protocol. Then, [11] shows how to perform concurrent extraction of the CECom commitments without rewinding the external party B. The extractor designed by them is called the 'robust simulator'.

Technical Challenges. While we will encounter multiple technical challenges, which will be clear as we go ahead, we point out the core technical challenge here and the way we will solve it.

One of the main technical challenges is when we prove witness extractability of our protocol. Namely, in our hybrid argument, we will encounter two consecutive hybrids H_a and H_b, wherein a coin-flipping phase of a particular right hand session is 'intact' in H_a, but is biased in H_b. This results in the mixed commitment changing from statistically hiding to statistically binding. In order to finally be able to argue that the extracted values are indeed valid witnesses, we will need to argue for the hybrid H_b that the value committed in this commitment is a valid witness. Herein, we will need to reduce our claim to computational binding of a CECom commitment in the protocol. Thus, the requirement in this reduction would be that no extraction performed should rewind the external CECom sender. Even the Robust Extraction Lemma will not be helpful here as the Lemma requires that the external protocol have round complexity strictly less than the round complexity of CECom commitments (on which the robust simulator performs extraction) and the external protocol in this case is a CECom commitment itself. The condition for the Lemma thus cannot be met. We get around this difficulty through a carefully designed sequence of hybrid arguments. A similar difficulty arises in the proof of statistical simulatability of our protocol. Here again, we rely on a carefully designed sequence of hybrids.

The second main technical challenge, still of the same flavor as the first one above, is in the proof of witness extractability. Here, we encounter a pair of hybrids: in the former hybrid, we would have a few CECom commitments of the right session being extracted by the robust simulator; in the latter hybrid, the modification introduced would be to change the value committed in a (statistically hiding) CECom commitment of a left session from a valid witness to a zero-string. Here again, we will not be able to argue a reduction to the hiding property of the CECom commitment of the left session in question, just by relying on the Robust Extraction Lemma. Here, we instead present a more detailed hybrid argument. Namely, in the CECom commitment, we change the committed value one sub-commitment at a time [24]. Since every sub-commitment in the standard CECom commitment of [24] ranges over just three rounds, we are now still able to apply the Robust Extraction Lemma.

2   Background

We assume familiarity with interactive Turing machines, denoted ITM. Given a pair of ITMs, A and B, we denote by $\langle A(x), B(y) \rangle(z)$ the random variable representing the (local) output of B, on common input z and private input y, when interacting with A with private input x, when the random tape of each machine is uniformly and independently chosen. In addition, we denote $\text{view}_B^A(x, z)$ to be the random variable representing the content of the random tape of B together with the messages received by B from A during the interaction on common input x and auxiliary input z to B.

If $\mathcal{D}_1$ and $\mathcal{D}_2$ are two distributions, then we denote that they are statistically close by $\mathcal{D}_1 \approx_s \mathcal{D}_2$; we denote that they are computationally indistinguishable by $\mathcal{D}_1 \approx_c \mathcal{D}_2$; and we denote that they are identical by $\mathcal{D}_1 \equiv \mathcal{D}_2$.

Definition 1 (Pseudorandom Language). *An NP-language $L \subseteq \{0,1\}^*$ is said to be a pseudorandom language if the following holds. For $\lambda \in \mathbb{N}$, let $\mathcal{D}_\lambda$ be a uniform distribution over $L \cap \{0,1\}^\lambda$. Then, for every distinguisher $\mathcal{D}$ running in time polynomial in λ, there exists a negligible function $\mathsf{negl}(\cdot)$ such that $\mathcal{D}$ can distinguish between $\mathcal{D}_\lambda$ and U_λ with probability at most $\mathsf{negl}(\lambda)$.*

We assume familiarity with notions like witness relation, interactive argument systems, and statistical witness-indistinguishable argument of knowledge (sWIAoK).

The verifier's view of an interaction consists of the common input x, followed by its random tape and the sequence of prover messages the verifier receives during the interaction. We denote by $\text{view}_{\mathcal{V}^*}^{\mathcal{P}}(x, z)$ a random variable describing $\mathcal{V}^*(z)$'s view of the interaction with $\mathcal{P}$ on common input x.

We will use various forms of commitment schemes. We will denote by SB, SH, CB, CH the usual properties that can be enjoyed by classic commitment schemes, namely: statistical binding, statistical hiding, computational binding and computational hiding.

Statistical Concurrent Non-malleable Zero Knowledge. The definition of statistical CNMZK is taken almost verbatim from [1] except for the additional requirement on the simulation being statistical. Let $\langle \mathcal{P}, \mathcal{V} \rangle$ be an interactive proof for an NP-language L with witness relation R_L, and let λ be the security parameter. Consider a man-in-the-middle adversary $\mathcal{M}$ that participates in m_L "left interactions" and m_R "right interactions" described as follows. In the left interactions, the adversary $\mathcal{M}$ interacts with $\mathcal{P}_1, \dots, \mathcal{P}_{m_L}$, where each $\mathcal{P}_i$ is an honest prover and proves the statement $x_i \in L$. In the right interactions, the adversary proves the validity of statements $\overline{x}_1, \dots, \overline{x}_{m_R}$. Prior to the interactions, both $\mathcal{P}_1, \dots, \mathcal{P}_{m_L}$ receive $(x_1, w_1), \dots, (x_{m_L}, w_{m_L})$, respectively, where for all i, $(x_i, w_i) \in R_L$. The adversary $\mathcal{M}$ receives $x_1, \dots, x_{m_L}$ and the auxiliary input z, which in particular might contain a-priori information about $(x_1, w_1), \dots, (x_{m_L}, w_{m_L})$. On the other hand, the statements proved in the right interactions $\overline{x}_1, \dots, \overline{x}_{m_R}$ are chosen by $\mathcal{M}$. Let $\text{view}_{\mathcal{M}}(x_1, \dots, x_{m_L}, z)$ denote a random variable that describes the view of $\mathcal{M}$ in the above experiment. Loosely speaking, an interactive argument is statistical concurrent non-malleable zero-knowledge (sCNMZK) if for every man-in-the-middle adversary $\mathcal{M}$, there exists a probabilistic polynomial time machine (called the simulator-extractor) that can *statistically* simulate both the left and the right interactions for $\mathcal{M}$, while outputting a witness for every statement proved by the adversary in the right interactions.

Definition 2 ((Black-Box) Statistical Concurrent Non-Malleable Zero Knowledge Argument of Knowledge). *An interactive protocol $\langle \mathcal{P}, \mathcal{V} \rangle$ is said to be a* (Black-Box) Statistical Concurrent Non-Malleable Zero Knowledge *(sCNMZK) argument of knowledge for membership in an NP language L with witness relation R_L, if the following hold:*

1. *$\langle \mathcal{P}, \mathcal{V} \rangle$ is an interactive argument system;*
2. *For every m_L and m_R that are polynomial in λ, for every PPT adversary $\mathcal{M}$ launching a concurrent non-malleable attack (i.e., $\mathcal{M}$ interacts with honest provers $\mathcal{P}_1, \ldots, \mathcal{P}_{m_L}$ in "left sessions" and honest verifiers $\mathcal{V}_1, \ldots, \mathcal{V}_{m_R}$ in "right sessions"), there exists an expected polynomial time simulator-extractor $\mathcal{SE}$ such that for every set of "left inputs" $x_1, \ldots, x_{m_L}$ we have $\mathcal{SE}(x_1, \ldots, x_{m_L}) = (\text{view}, \overline{w}_1, \ldots, \overline{w}_{m_R})$ such that:*
   - *view is the simulated joint view of $\mathcal{M}$ and $\mathcal{V}_1, \ldots, \mathcal{V}_{m_R}$. Further, for any set of witnesses $(w_1, \ldots, w_{m_L})$ defining the provers $\mathcal{P}_1, \ldots, \mathcal{P}_{m_L}$, the view view is distributed statistically indistinguishable from the view of $\mathcal{M}$, denoted $\text{view}_{\mathcal{M}}(x_1, \ldots, x_{m_L}, z)$, in a real execution;*
   - *In the view view, let trans_ℓ denote the transcript of ℓ-th left execution, and $\overline{\text{trans}}_t$ that of t-th right execution, $\ell \in [m_L]$, $t \in [m_R]$. If $\overline{x}_t$ is the common input in $\overline{\text{trans}}_t$, $\overline{\text{trans}}_t \neq \text{trans}_\ell$ (for all ℓ) and $\mathcal{V}_t$ accepts, then $R_L(\overline{x}_t, \overline{w}_t) = 1$ except with probability negligible in λ.*
   The probability is taken over the random coins of $\mathcal{SE}$. Further, the protocol is black-box sCNMZK, *if $\mathcal{SE}$ is a universal simulator that uses $\mathcal{M}$ only as an oracle, i.e., $\mathcal{SE} = \mathcal{SE}^{\mathcal{M}}$.*

We remark here that the statistical indistinguishability is considered only against computationally unbounded distinguishers, and not against unbounded man-in-the-middle adversaries. This restriction is inherent to the definition since we require statistical zero-knowledge and thus cannot simultaneously ask for soundness against unbounded provers.

Extractable Commitment Schemes.

Definition 3 (Extractable Commitment Schemes). *An extractable commitment scheme $\langle \text{Sender}, \text{Receiver} \rangle$ is a commitment scheme such that given oracle access to any PPT malicious sender Sender^*, committing to a string, there exists an expected PPT extractor E that outputs a pair (τ, σ^*) such that the following properties hold:*

Simulatability. The simulated view τ is identically distributed to the view of Sender^* (when interacting with an honest Receiver) in the commitment phase.

Extractability. the probability that τ is accepting and σ^* correspond to $\perp$ is at most $1/2$. Moreover if $\sigma^* \neq \perp$ then the probability that Sender opens τ to a value different than σ^* is negligible.*

Lemma 1. *[15] $\mathsf{Com}_{\mathsf{nm}}$ is an extractable commitment scheme.*

As shown in [15], $\mathsf{Com}_{\mathsf{nm}}$ is an extractable commitment scheme. This is in fact the core property of the scheme that is relied upon in proving its non-malleability in [8,15].

Extractable Mixed Robust Non-malleable Commitments w.r.t. 1-Round Protocols. In our protocol we make use of a special kind of commitment scheme, that we call a *extractable mixed robust non-malleable commitment scheme*. These are basically the mixed commitment schemes introduced by Damgård and Nielsen [6] that are also non-malleable (or robust) not only w.r.t. themselves but also w.r.t. 1-round protocols and also extractable.

We shall first discuss how we get mixed non-malleable commitments, and then at the end, we shall discuss how we also get mixed non-malleable commitments that are also robust w.r.t. 1-round protocols.

Intuitively, a mixed non-malleable commitment scheme is a commitment scheme that is parameterized by a string srs in such a way that if srs is from some specific distribution, then commitment scheme is SH, and if srs is from another specific indistinguishable distribution, then the scheme is non-malleable. We require that both the distributions be efficiently samplable. When srs is randomly sampled (from the dominion over which both the distributions are defined), we would require that srs is such that with all but negligible probability the scheme is SH. We denote such a scheme by $\mathsf{NMMXCom}_{\mathsf{srs}}$. More formally:

Definition 4 (Mixed Non-Malleable Commitments). *A commitment scheme is said to be a mixed non-malleable commitment scheme if it is parameterized by a string srs and if there exist two efficiently samplable distributions $\mathcal{D}_1$, $\mathcal{D}_2$, such that, $\mathcal{D}_1 \approx_c \mathcal{D}_2$, and if srs $\leftarrow \mathcal{D}_1$ then the commitment scheme is SH and if srs $\leftarrow \mathcal{D}_2$ then the commitment scheme is non-malleable. Furthermore, $|\mathrm{Supp}(\mathcal{D}_2)|/|\mathrm{Supp}(\mathcal{D}_1)| = \mathsf{negl}(\lambda)$.*

Below, we show how to construct such a scheme. At a high level, we achieve this by using a *mixed commitment scheme* which, roughly speaking, is a commitment scheme parameterized by a string srs in such a way that if srs is from some specific efficiently samplable distribution, then commitment scheme is SH, and if srs is from another specific indistinguishable efficiently samplable distribution, then the scheme is SB. We denote such a scheme by $\mathsf{MXCom}_{\mathsf{srs}}$. More formally:

Definition 5 (Mixed Commitments). *A commitment scheme is said to be a mixed commitment scheme if it is parameterized by a string srs and if there exist two efficiently samplable distributions $\mathcal{D}_1$, $\mathcal{D}_2$, such that, $\mathcal{D}_1 \approx_c \mathcal{D}_2$, and if srs $\leftarrow \mathcal{D}_1$ then the commitment scheme is SH and if srs $\leftarrow \mathcal{D}_2$ then the commitment scheme is SB. Furthermore, $|\mathrm{Supp}(\mathcal{D}_2)|/|\mathrm{Supp}(\mathcal{D}_1)| = \mathsf{negl}(\lambda)$.*

In [6], Damgård and Nielsen gave two constructions of mixed commitment schemes, one based on one based on the Paillier cryptosystem and the other based on the Okamoto-Uchiyama cryptosystem. For concreteness, we provide a construction below based on Σ-protocols and that builds on previous ideas presented in [5,3,4].

Constructing Mixed Commitments. Let us first describe how to construct a mixed commitment scheme. The idea is to have $\mathcal{D}_1$ be uniform over $\{0,1\}^{\mathrm{poly}(\lambda)}$ and $\mathcal{D}_2$ be uniform over a pseudorandom language L (as per Definition 1) with a Σ-protocol (i.e., public-coin 3-round special-sound special honest-verifier zero-knowledge proof system). Then, to commit to a value β, sender would first run the simulator of the Σ-protocol for the statement that $\mathsf{srs} \in L$ such that the simulated proof has β as the challenge; let (α, β, γ) be the simulated proof. Then the commitment would just be α. The opening would be γ.

Observe that if $\mathsf{srs} \notin L$, then for any β there is only one accepting (α, β, γ), making the scheme parameterized by this srs to be SB. Furthermore, with srs sampled uniformly at random from $\{0,1\}^* \setminus L$, we will also be able to argue that the resulting scheme is CH. On the other hand, if $\mathsf{srs} \in L$, then, for every α (in its valid domain as defined by the Σ-protocol), there exists γ' for every β' such that $(\alpha, \beta', \gamma')$ is an accepting transcript. This implies that there exists an opening of α to any β'. This makes the scheme SH. Furthermore, with srs sampled uniformly at random from L, it shall hold for any PPT machine that it can only run the simulator and it is infeasible for the machine to open α to *also* any $\beta' \neq \beta$ (with some γ' as an opening), assuming special-soundness of the Σ-protocol (Otherwise, one could extract the witness from $(\alpha, \beta, \gamma, \beta', \gamma')$). This makes the system only computationally binding. In detail:

Mixed Commitment from Σ-protocol. Let R_L be a hard relation for a pseudo-random language L i.e., $L = \{\mathsf{srs} \in \{0,1\}^{\lambda} | \exists w : R_L(\mathsf{srs}, w) = 1\}$ and $L \approx_c U_{\lambda}$. Consider a Σ-protocol for the above language L. The special honest-verifier zero-knowledge property of the Σ-protocol implies existence of a simulator S that on input the instance srs, a string β and a randomness r, outputs a pair (α, γ) such that $(\mathsf{srs}, \alpha, \beta, \gamma)$ is computationally indistinguishable from a transcript $(\mathsf{srs}, \alpha, \beta, \gamma)$ played by the honest prover when receiving β as challenge.

The commitment scheme played by sender C and receiver R that we need goes as follows.

Shared Random String: A random string $\mathsf{srs} \in \{0,1\}^{\lambda}$ is given as a common input to both the parties;

Commitment Phase: We denote the commitment function by $\mathsf{MXCom}_{\mathsf{srs}}(\cdot; \cdot)$ and to commit to a string $\beta \in \{0,1\}^{\lambda}$:
1. C runs the Σ-protocol simulator $S(\mathsf{srs}, \beta, r)$ to obtain (α, γ);
2. C sends α to R;

Decommitment Phase: To open α to β:
1. C sends (β, γ) to R;
2. R accepts if $(\mathsf{srs}, \alpha, \beta, \gamma)$ is an accepting transcript for the Σ-protocol.

If srs $\in L$, then the commitment is computationally binding (since, with two openings one gets two accepting conversations for the same α, and from the special-soundness property of the Σ-protocol one can extract the witness) and statistically hiding (which is directly implied by perfect completeness of the Σ-protocol; i.e., for any α output as the first message by the simulator – for any β as the challenge – for every β', given the witness, one can efficiently compute a final message γ' such that the verifier accepts). If srs $\notin L$ the commitment is statistically binding (since, for any α, there exists at most one β that makes R accept the decommitment, as there is no witness for srs $\in L$ and two accepting transcripts (α, β, γ), $(\alpha, \beta', \gamma')$ with $\beta \neq \beta'$ implies a witness owing to the special-soundness property of the Σ-protocol) and computationally hiding (since, if on input α, one can guess β efficiently, then this can be used to decide whether or not srs $\in L$, a contradiction).

While there are many instantiations for L, we shall work with the following simple one. Define $L = \{(g_1, g_2, g_3, g_4) \in \mathbb{G}^4 | \exists a, b : a \neq b \wedge g_1^a = g_2 \wedge g_3^b = g_4\}$ with $\mathbb{G}$ being a prime order group, where DDH is believed to be hard. That is, L is the language of non-DDH triplets. Note that in this case if srs is chosen uniformly at random from $\mathbb{G}^4$ the commitment is statistically hiding with overwhelming probability (most strings are not DDH triplets).

Relaxing the Assumption. Another example for L is the following language: let (G, E, D) be a *dense* cryptosystem (i.e., valid public keys and ciphertexts can be easily extracted from random strings). The language L is:

$$L = \{(pk_0, pk_1, c_0, c_1) | \exists r_0, r_1, m_0, m_1, s_0, s_1 : m_0 \neq m_1, (pk_0, sk_0) \leftarrow G(1^k, r_0),$$

$$c_0 = E_{pk_0}(m_0, s_0), (pk_1, sk_1) \leftarrow G(1^k, r_1), c_1 = E_{pk_1}(m_1, s_1))\}.$$

Also in this case most strings are in the language, while the simulator can choose a string not in the language (i.e., with $m_0 = m_1$).

Moreover, we can plug this mixed commitment MXCom in a zero-knowledge protocol in the SRS model NMMXCom, so that when srs is a random DDH triple, the zero-knowledge protocol is a proof (i.e., statistically sound) and computational zero-knowledge, while when the srs is a random non-DDH triple then the zero-knowledge protocol is statistical zero-knowledge (and computationally sound). For eg., an implementation of Blum's protocol by using MXCom as commitment scheme when the prover commits to the permuted adjacency matrices gives us a computational zero-knowledge proof-of-knowledge (ZKPoK, for short) if srs of the MXCom commitment used is a random DDH tuple and a statistical zero-knowledge argument-of-knowledge (ZKAoK, for short) if the srs is a random non-DDH tuple.

Constructing Mixed Non-malleable Commitments. As mentioned earlier, we show how to construct a mixed non-malleable commitment scheme by using a mixed commitment scheme. For concreteness, we shall work with the mixed commitment scheme MXCom described earlier. To thus recall, by the construction of MXCom, our mixed non-malleable commitment scheme will be non-malleable when srs is a random DDH tuple and, is statistically hiding and computationally binding when srs is a random non-DDH tuple.

Our scheme NMMXCom$_{srs}$ is described as follows. At a high level, our approach is to slightly modify the DDN non-malleable commitment scheme in [8]. In fact, we shall describe our modification by considering the concurrent non-malleable commitment scheme that appears in [15] (whose analysis of non-malleability is similar to that of the DDN commitment and is simpler). The protocol in [15] is in fact non-malleable w.r.t. any arbitrary protocols of logarithmic round-complexity, a property that is called $\log(\lambda)$-robust non-malleability. This is one of the properties which will be of a crucial use to us and we shall elaborate on this property shortly. In fact, we only need 1-robust non-malleability. The scheme of [15] is described below.

Common Input : An identifier ID $\in \{0,1\}^L$, where $L = \text{poly}(\lambda)$. Define
$\ell := \log(L) + 1$.

Input for Sender : A string $V \in \{0,1\}^\lambda$.

    Sender ← Receiver: Sender chooses $V_1, V_2, \ldots, V_L \leftarrow \{0,1\}^\lambda$ such that
    $V_1 \oplus V_2 \oplus \ldots \oplus V_L = V$. For each $i \in [L]$, run Stage 1 and Stage 2 in
    parallel with $v := V_i$ and id $= (i, \text{ID}_i)$, where ID_i is the i-th bit of ID.

Stage 1 :

    Sender ← Receiver: Receiver samples $x \leftarrow \{0,1\}^\lambda$, computes $y = f(x)$,
    and sends s to Sender. Sender aborts if y is not in the range of f.

    Sender → Receiver: Sender chooses randomness $\leftarrow \{0,1\}^\lambda$ and sends
    $c = \text{Com}_{sb}(v; \text{randomness})$.

Stage 2 :

    Sender → Receiver: 4ℓ special-sound $\mathcal{WI}$ proofs of the statement:
    either there exists values v, randomness such that $c = \text{Com}_{sb}(v; \text{randomness})$
    or there exists a value x such that $y = f(x)$
    with 4ℓ $\mathcal{WI}$ proofs in the following schedule:
    For $j = 1$ to ℓ do: Execute $\text{design}_{\text{id}_j}$ followed by $\text{design}_{1-\text{id}_j}$.

Fig. 1. $O(\log(\lambda))$-round Non-Malleable Commitment of [15]

At a high level, the protocol of the sender who wishes to commit to some value v proceeds as follows. To catch the core of the intuition, we describe here a simplified version of the protocol while ignoring the currently unnecessary details (such as parallel repetitions, etc.); later in the formal description, we shall present the original protocol of [15]. The sender proceeds as follows. In the first stage, upon receiving an output of a one-way function from the receiver, commit to v using a statistically binding commitment scheme Com_{sb}. In the second stage, engage in $\log(\lambda)$ (special-sound) $\mathcal{WI}$ proofs of knowledge of either the value committed to using Com_{sb} or of a pre-image of the one-way function output sent by the receiver. (The number of $\mathcal{WI}$ proofs is logarithmic in the length of the identities of the senders; hence, it is considered to be $\log(\lambda)$ in general). We note here that a special-sound $\mathcal{WI}$ proof can be instantiated by using Blum's Hamiltonicity protocol, wherein the commitment sent by the $\mathcal{WI}$ prover in this protocol is SB.

Now to construct the mixed non-malleable commitment, the idea is to replace the SB commitment Com_{sb} of the first stage and the SB commitment within the Blum's Hamiltonicity protocol (where both the commitments are given by the sender to the receiver) with the mixed commitment MXCom_{srs}. We shall analyze the properties of the resulting commitment scheme, denoted by $\mathsf{NMMXCom}_{srs}$, below.

Recall that if srs is a random DDH tuple, then MXCom_{srs} is SB and CH. Under this case, the resulting scheme would have the properties identical to the original scheme of [18]; namely it is SB, CH, and non-malleable. On the other hand, if srs is a random non-DDH tuple, then MXCom_{srs} is SH and CB. This would render the the resulting scheme to be SH (owing to the SH property of the commitment scheme in the first phase and witness-indistinguishability of the Hamiltonicity protocol that is instantiated with SH commitment) and CB (owing to the computational binding property of the commitment scheme in the first phase; this is due to the fact that decommitment of the scheme in [15] is simply an opening of the commitment of the first phase). In fact, if srs is a random string, then it is a non-DDH tuple with all but negligible probability. Hence, we also have that when srs is a random string, MXCom_{srs} is SH and CB with all but negligible probability. For future reference, we shall bookmark this into the following proposition.

Proposition 1. *If* srs *is a uniform DDH tuple, then* MXCom_{srs} *is SB, CH, and non-malleable. If* srs *is a uniform random string, then* MXCom_{srs} *is SH and CB.*

Robustness w.r.t. 1-Round Protocols of the Mixed Non-Malleable Commitments. Recall that we modified the [15] non-malleable commitment scheme that is robust w.r.t. 1-round protocols to get mixed non-malleable commitment scheme. It turns out that the modified scheme still retains robust w.r.t. 1-round protocols. Here, we only give a high-level description of the reason behind this fact as this can be easily verified. The reason is that robustness of the non-malleable commitment scheme in Figure 1 is proved in [15] by relying only upon the structure (the 'designs', in particular) of the commitment scheme in Figure 1. In particular, this proof does not rely upon the specifics of the underlying commitment scheme. Now recall that the only modification we introduced in the robust non-malleable commitment scheme of [15] to get a mixed non-malleable commitment scheme is the following. Instead of using any underlying commitment scheme, we used a mixed commitment scheme. Thus, the scheme continues to be non-malleable commitment scheme robust w.r.t. 1-round protocols even when the underlying commitment schemes are mixed commitments.

Non-malleability of $\mathsf{NMMXCom}_{srs}$ *w.r.t.* Com_{nm}. Another property of $\mathsf{NMMXCom}_{srs}$ that we need is the following. Let Com_{nm} be the NMCom commitment robust w.r.t. 1-round protocol. We shall argue below that $\mathsf{NMMXCom}_{srs}$ is non-malleable w.r.t. Com_{nm}.

Proposition 2. *The non-malleable commitment* $\mathsf{NMMXCom_{srs}}$ *is robust w.r.t. the non-malleable commitment* $\mathsf{Com_{nm}}$.

Proof Sketch.   Essentially, the proof is exactly the same as the proof of non-malleability of the non-malleable commitment scheme of [15] presented in Figure 1. We argue this here next. Consider a MiM adversary against non-malleability of $\mathsf{NMMXCom_{srs}}$ that executes a $\mathsf{Com_{nm}}$ session on the left by playing the role of the receiver and a $\mathsf{NMMXCom_{srs}}$ session on the right by playing the role of a sender. The key technique in proving non-malleability in [8,18,15] is to show that, immaterial of the way a MiM adversary interleaves the left and right commitments, there exists at least one $\mathcal{WI}$ proof (within some design) on the right session such that it is 'safe' to rewind the MiM adversary for this proof; by 'safe', we mean that rewinding the MiM adversary at this point can be done without rewinding the external sender on the left. (Recall that to rewind a $\mathcal{WI}$ proof is to rewind to the point between the first and the second message of the proof). To then understand what $\mathcal{WI}$ proof qualifies to be safe to rewind, we begin by giving a high level idea of when a proof *does not* qualify to be safe. Consider any $\mathcal{WI}$ proof $(\alpha_r, \beta_r, \gamma_r)$ on the right. If it is trying to use and 'maul' some $\mathcal{WI}$ proof $(\alpha_l, \beta_l, \gamma_l)$ on the left, then the right proof is positioned in time with respect to the left one as shown in Figure 2. Observe that rewinding such a proof on the right with a new challenge may make the MiM adversary send a new challenge for the left proof too asking for a new response which tantamounts to rewinding the sender on the left. [8,18,15] provide a characterization for the $\mathcal{WI}$ proofs on the right that qualify as safe for being rewound; however, the details of this characterization itself will not be important to us; the core argument in proving non-malleability in [8,18,15] is an argument that, immaterial of the way a MiM adversary interleaves the left and right commitments, there exists a $\mathcal{WI}$ proof on the right that is safe to rewind. This is so owing to the fact that the adversary can use only one proof on the left for every proof on the right and to the fact that there are exactly the same number of proofs on the left and the right. This would imply that if the left and the right identities are distinct (at least at one bit position), then at proofs corresponding to this bit position, design_0 on the left 'matches up' with design_1 on the right, depicted in Figure 2. With a closer look at this interleaving, it can be easily derived that at least one of the $\mathcal{WI}$ proofs within this design_1 on the right is safe to be rewound.

We first observe that the only way $\mathsf{NMMXCom_{srs}}$ differs from $\mathsf{Com_{nm}}$ in Figure 1 is that a specific kind of commitment, namely, a mixed commitment is used to instantiate the underlying commitments used in building $\mathsf{Com_{nm}}$ in Figure 1. Next, we observe that non-malleability of the commitment scheme $\mathsf{NMMXCom_{srs}}$ is mainly due to the structure (or designs) of the $\mathcal{WI}$ proofs, and the same arguments on interleaving and safety of rewinding would hold even if the left commitment is under an $\mathsf{Com_{nm}}$ session.   □

We remark that in fact the non-malleable commitments $\mathsf{NMMXCom_{srs}}$ and $\mathsf{Com_{nm}}$ are robust w.r.t. each other by the same arguments as above. However, it suffices for us that $\mathsf{NMMXCom_{srs}}$ is robust w.r.t. $\mathsf{Com_{nm}}$.

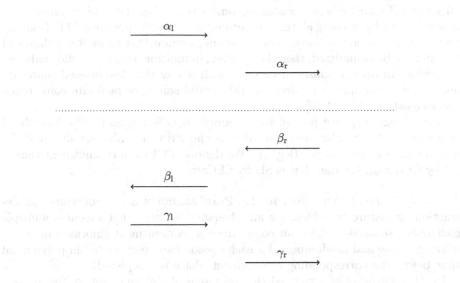

Fig. 2. Prefix (until the dotted line) that is not a safe point

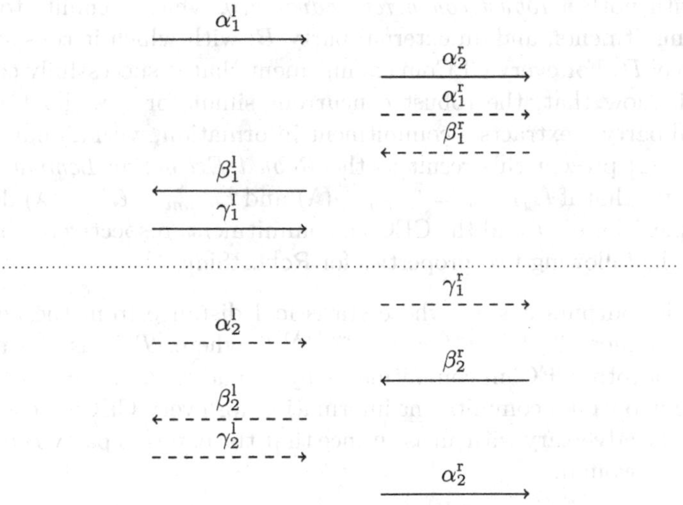

Fig. 3. A design$_0$ matches up with design$_1$

Concurrently Extractable Commitment Schemes. Concurrently extractable commitment (CECom) schemes consist of committing using the PRS preamble, and decommitting by opening all the commitments within the preamble [24]. Roughly speaking, the preamble consists of the sender committing to multiple shares of the value to be committed; then the receiver, in multiple rounds, would challenge the sender to open a subset of them in such a way that the opened shares do not reveal the committed value, but this would somehow facilitate consistency checks as shown in [24,20].

A challenge-response pair in the preamble is called a 'slot'. [20] formalized concurrent extractability and showed that the PRS preamble satisfies it if the number of slots therein is $\omega(\log(\lambda))$. We denote a CECom commitment that is SB by CECom_{sb}, the one that is SH by CECom_{sh}.

Robust Concurrent Extraction. In [24], Prabhakaran et al. demonstrated an extraction procedure by which, for an adversary Sender* that executes multiple concurrent sessions of CECom commitments, commitment information (commitment value and randomness) for each session can be extracted in polynomial time before the corresponding commitment phase is completed.

In [11], Goyal et al. extended the technique of [24] and showed how to perform efficient extractions of CECom commitments when an adversary Sender*, besides concurrently performing CECom commitments, also interacts with an 'external' party B in some arbitrary protocol Π. This setting now additionally requires that the extraction procedure rewinds the adversary Sender* in a way that B does not get rewound in the process. This is achieved in [11] by building a *robust concurrent simulator* (or just 'robust simulator') RobustSim that interacts with both a *robust concurrent adversary*, which commits to multiple CECom commitments, and an external party B, with which it runs some arbitrary protocol Π. For every CECom commitment that is successfully completed, Goyal et al. show that, the robust concurrent simulator – without rewinding the external party – extracts a commitment information, with all but negligible probability. [11] present this result as the *Robust Extraction Lemma* which informally states that if $\ell_{external} = \ell_{external}(\lambda)$ and $\ell_{cecom} = \ell_{cecom}(\lambda)$ denote the round complexities of Π and the CECom commitment, respectively, the Lemma guarantees the following two properties for RobustSim:

- RobustSim outputs a view whose statistical distance from the adversary's view is at most $2^{-(\ell_{cecom}-\ell_{external}\cdot\log(T(\lambda)))}$, where, $T(\lambda)$ is the maximum number of total CECom commitments by the adversary.
- RobustSim outputs commitment information for every CECom commitment sent by the adversary with an assurance that the external party B of protocol Π is not rewound.

3   Statistical Concurrent Non-malleable Zero-Knowledge

We start by giving an intuition behind the design of our protocol. In [1], Barak et al. gave a construction of a computational CNMZK argument of knowledge.

The simulation for this protocol was restricted to be only computational due to the following reason. In their protocol, one of the messages sent by the prover is a non-malleable commitment to a valid witness. Since the non-malleable commitment is SB, and the simulator, unlike an honest prover, does not use a valid witness in this non-malleable commitment, the simulated view was only computationally indistinguishable from the real-world view of a MiM adversary. It will be quite relevant for us to note that the non-malleable commitment being SB was crucially used in the proof of concurrent non-malleability of their protocol, therefore it is not possible to replace the above commitment scheme with a statistically hiding non-malleable commitment. More specifically, the proof would begin with the real-world view and through a series of hybrids would move towards the simulated view. In some certain hybrid along the way there would be introduced PRS rewindings to facilitate simulation. Given such a hybrid that performs PRS rewindings, it would be difficult to establish that one can extract a value out of the non-malleable commitment and that the extracted value is a valid-witness. The difficulty here is in ensuring that the PRS rewindings would not interfere with the non-malleable commitment on which the NMCom extractor is run. The idea in their proof instead was to first prove for the real-world view itself that the value committed in the NMCom commitment is a valid witness, and then make transitions to hybrids by introducing PRS rewindings. The point to be noted here is that it was crucial in their proof that the non-malleable commitment is a *statistically binding* commitment, so that they could put forth arguments on the values committed in it. With this, since introducing PRS rewindings would only bias the distribution of the view output by at most a negligible amount, their proof boiled down to proving that the value committed in the NMCom commitment does not adversely change as we move across various hybrids. Now, since we began with a hybrid where the values committed were valid witnesses, the values committed in the NMCom commitments after the PRS rewindings too are valid witnesses by non-malleability (and in particular statistical binding) of the commitment scheme.

Our idea begins from noticing that statistical binding of the NMCom commitment is crucial in proving extractability of valid witnesses and not important in simulating the view of the adversary. So the core idea is to somehow ensure that when we prove the indistinguishability of the simulation, the commitment scheme is statistically hiding. Instead, when we need to argue that the distribution of the extracted message does not change, then the commitment should be statistically binding. With this being the crux of our idea, the way we shall execute it is via what we call 'mixed non-malleable commitments'. Intuitively, a mixed non-malleable commitment scheme is associated with two efficiently samplable, computationally indistinguishable distributions, and every commitment is parameterized by some string. Furthermore, one of the distributions is such that if the string is uniformly sampled from this distribution then the commitment is SH and CB; on the other hand, a commitment that is parameterized by a string that is uniformly sampled from the other distribution is SB and CH. Given such a commitment scheme, our protocol basically is an instantiation of the BPS protocol except that the NMCom commitment in the BPS protocol is replaced by a mixed non-malleable commitment. Also, the string that parameterizes this commitment computed jointly by both the prover and the verifier is the outcome of a coin-flipping protocol. Namely, in our mixed non-malleable commitment scheme, the distribution on the parameter that produces a SH, CB commitment is the uniform distribution. Hence, the parameter generated via the coin-flipping protocol is SH and CB, as required. The BPS protocol forms the **Main BPS Phase** and the coin-flipping protocol is run in the **Coin-flipping Phase** of our protocol.

A traditional coin-flipping protocol would involve the verifier committing to a random string in the first round, followed by the prover sending another random string in the clear in the second round, the verifier opening the commitment in the third round, and finally having the prover's and the verifier's strings XOR-ed as the outcome of the coin-flipping protocol. However, now that we would also like to be able to cheat and bias the outcome to another (computationally indistinguishable) distribution (so that the mixed non-malleable commitment would then be SB), we modify the third round. Namely, instead of the third round being the verifier opening the commitment by giving both the committed value and the randomness used, the verifier would only give the committed value and then give an argument that there exists a randomness that would explain the commitment to this value. However, we won't be able to work with just any argument since we are in the concurrent setting. Furthermore, we also would like to ensure that when our simulator cheats in the argument to bias the coin-flipping outcome, the MiM adversary will not get any undue advantage. Thus, the argument that we use here is a CNMZK argument. In particular, we use the BPS argument itself. This argument forms the **BPS^{CFP} Phase** in our protocol.

Furthermore, towards simplifying our proof, we introduce the following slight modification of the BPS protocol in the 'Main BPS Phase'. In the original BPS protocol, the commitment in which the prover commits the valid witness to is an NMCom commitment; on the other hand, in the 'Main BPS Phase', besides sending the NMCom commitment to the witness, the prover also sends a concurrently extractable (CECom) commitment to the same witness. The simplification we achieve by adding the CECom commitment is that even the extraction of the witnesses (by the simulator-extractor) can be performed just like an extraction on any other CECom commitments in the protocol. Since, for simulation, we anyway need to employ certain techniques for the extraction from the other CE-Com commitments, we are now able to recycle the same techniques for witness extractions too, thus letting our focus stay on the other crucial subtleties (which we shall see as we get to the proofs of security).

We will now give a formal description of the protocol.

3.1   Our sCNMZKAoK Protocol $\langle \mathcal{P}, \mathcal{V} \rangle$

Ingredients.

1. Let CECom_{sh} and CECom_{sb} be SH and SB concurrently-extractable commitment scheme, respectively. Let each of them be of k_{cecom}-slots, where

$k_{\text{cecom}} \in \omega(\log \lambda)$. Let the sender's randomness space for these commitments be RandSpace_{cecom}.

2. Let Com_{sh} be a SH commitment scheme. Let k_{sh} be its round-complexity, where k_{sh} is a constant.
3. Let sWIAoK be a statistical WIAoK protocol. Let k_{swiaok} be its round-complexity, where k_{swiaok} is a constant.
4. Let $\text{NMMXCom}_{(.)}$ be our mixed non-malleable commitment scheme. Recall that it satisfies extractability and is robust w.r.t. 1-round protocols. Let k_{nmmxcom} be its round-complexity, where k_{nmmxcom} is $O(\log(\lambda))$.
5. Let Com_{nm} be the non-malleable commitment scheme (described in Fig. 1). Recall that it satisfies extractability and is robust w.r.t. 1-round protocols. Let k_{nmcom} be its round-complexity.

In summary, the round complexities of the sub-protocols in our protocol are as follows: $k_{\text{cecom}} \in \omega(\log \lambda)$, k_{swiaok}, k_{sh} are constants, and $k_{\text{nmcom}}, k_{\text{nmmxcom}} \in O(\log(\lambda))$.

Coin-Flipping Phase (CFP).

cfp_1 $(\mathcal{V} \to \mathcal{P})$: Sample $r_V \leftarrow \{0,1\}^\lambda$, rand $\leftarrow \text{RandSpace}_{cecom}$ and commit to r_V using CECom_{sh} and randomness rand.
cfp_2 $(\mathcal{P} \to \mathcal{V})$: Sample $r_P \leftarrow \{0,1\}^\lambda$ and send r_P.
cfp_3 $(\mathcal{V} \to \mathcal{P})$: Send r_V.

BPS^{CFP} Phase.

$\text{bps}^{\text{cfp}}_1$ $(\mathcal{P} \to \mathcal{V})$: Sample $\alpha \leftarrow \{0,1\}^\lambda$ and commit to α using CECom_{sb}.
$\text{bps}^{\text{cfp}}_2$ $(\mathcal{V} \to \mathcal{P})$: Commit to 0^λ using Com_{sh} and argue knowledge of a commitment information (i.e., a commitment value and randomness) using sWIAoK.
$\text{bps}^{\text{cfp}}_3$ $(\mathcal{P} \to \mathcal{V})$: Open the commitment of Step $\text{bps}^{\text{cfp}}_1$ to α.
$\text{bps}^{\text{cfp}}_4$ $(\mathcal{V} \to \mathcal{P})$: Commit to rand (used as commitment randomness in Step cfp_1) using the NMCom commitment Com_{nm}. In the rest of the paper, we shall refer to rand as the *sub-witness*.
$\text{bps}^{\text{cfp}}_5$ $(\mathcal{V} \to \mathcal{P})$: Send sWIAoK to argue knowledge of either rand or r_{comsh} such that:
   1. the value committed to by $\mathcal{V}$ with the NMCom commitment at Step $\text{bps}^{\text{cfp}}_4$ is rand and rand explains the CECom commitment at Step cfp_1 to r_V.
   2. Randomness r_{comsh} explains Com_{sh} at Step $\text{bps}^{\text{cfp}}_2$ being committed to α.

Let $\text{srs} = r_P \oplus r_V$.

Main BPS Phase.

bps_1 $(\mathcal{V} \to \mathcal{P})$: Sample $\sigma \leftarrow \{0,1\}^\lambda$ and commit to it using CECom_{sb}.
bps_2 $(\mathcal{P} \to \mathcal{V})$: Commit to 0^λ using Com_{sh} and argue knowledge of a commitment information (i.e., a commitment value and randomness) using sWIAoK.

bps_3 $(\mathcal{V} \to \mathcal{P})$: Open the commitment of Step bps_1 to σ.

bps_4 $(\mathcal{P} \to \mathcal{V})$: Commit to the witness w using mixed commitment $\mathsf{NMMXCom}_{\mathsf{srs}}$.

bps_{4+} $(\mathcal{P} \to \mathcal{V})$: Commit to the witness w using $\mathsf{CECom}_{sh}{}^2$.

bps_5 $(\mathcal{P} \to \mathcal{V})$: Send sWIAoK to argue knowledge of either w, r_{nm}, r_{cecom} or r'_{comsh} such that:

1. r_{nm} and r_{cecom} explain the $\mathsf{NMMXCom}_{\mathsf{srs}}$ commitment of Step bps_4 and the CECom commitment of Step bps_{4+} to w, respectively, and w is such that $R_L(x, w) = 1$,

2. Randomness r'_{comsh} explains Com_{sh} at Step bps_2 being committed to σ.

3.2   Proofs of Security

In this section, we prove that our proposed protocol $\langle \mathcal{P}, \mathcal{V} \rangle$ is a statistical concurrent non-malleable zero-knowledge argument of knowledge. In other words, we show that there exists a simulator-extractor $\mathcal{SE}$ that, for every concurrent MiM adversary $\mathcal{M}$, outputs a view view that is statistically indistinguishable from the view $\mathrm{view}_{\mathcal{M}}(x_1, \ldots, x_{m_L}, z)$ of $\mathcal{M}$ in a real execution, and also outputs valid witnesses $\bar{y}_1, \ldots, \bar{y}_{m_R}$ for all accepting right sessions.

Our Simulator-Extractor. The Simulator-Extractor $\mathcal{SE}$ runs RobustSim which is the robust concurrent simulator for a robust concurrent attack. The adversary of the robust concurrent attack is a procedure I that we describe below. $\mathcal{SE}$ will then output the output of $\mathsf{RobustSim}^I(z)$. Recall that RobustSim runs a given adversary that mounts a robust concurrent attack by committing to multiple CECom commitments, where the adversary also interacts with an external party B in an arbitrary external protocol. RobustSim then is guaranteed to extract commitment information from every CECom commitment sent by the adversary before the completion of its commitment phase, in such a way that the external party B does not get rewound.

Procedure $I(z)$. I incorporates the MiM adversary $\mathcal{M}$, initiates an execution, and simulates its view as follows. Let the m_L left sessions be ordered with some arbitrary ordering. Let the m_R right sessions be ordered as follows: Consider any two right sessions, the i-th and the j-th; $i \leq j$ if and only if the CECom_{sb} commitment at Step bps_1 of the i-th session begins earlier to the CECom_{sb} commitment at Step bps_1 of the j-th session.

For every right session: Run the code of the verifier except isolate CECom_{sh} at Step bps_{4+} and relay it to external receiver. Let value y'_t be received from the outside (RobustSim) at the end of the CECom_{sh} commitment.

---

[2] In order to make the difference from the BPS protocol more easily noticeable, the five steps here that are common to the BPS protocol are numbered in sequence from bps_1 through bps_5, while this 'extra' step is given a distinctive notation, bps_{4+}.

For every left session: When $\mathcal{M}$ initiates an ℓ-th new session on the left, I proceeds as follows.

- Run the coin-flipping phase and the $\mathsf{BPS}^{\mathsf{CFP}}$ phase honestly. Let srs be the outcome.
- Isolate CECom_{sb} at Step bps_1 and relay it to an external receiver. Let σ' be the value received from the outside ($\mathsf{RobustSim}$) at the end of the CECom_{sb} commitment.
- Then commit to σ' using Com_{sh} at Step bps_2; also, use the same extracted value as the witness in executing the sWIAoK of Step bps_2.
- In Step bps_3, let $\mathcal{M}$ opens its CECom_{sb} (of Step bps_1) to σ. Abort if $\sigma \neq \sigma'$.
- Commit to 0^λ using the mixed non-malleable commitment $\mathsf{NMMXCom}_{\mathsf{srs}}$ in Step bps_4.
- Commit to 0^λ using the CECom_{sh} commitment in Step bps_{4+}.
- Use σ' committed to in Step bps_2 as the witness in executing sWIAoK of Step bps_5.

When $\mathcal{M}$ halts, I outputs the view of $\mathcal{M}$ together with $y'_1, \ldots, y'_{m_R}$, and halts.

Statistical simulation. We shall prove that the view output by $\mathcal{SE}$ is distributed statistically close to the real-world view of the MiM adversary $\mathcal{M}$.

Theorem 1. *For every PPT adversary* $\mathcal{M}$, $\{\mathsf{view}_{\mathcal{M}}(x_1, \ldots, x_{m_L})\}_{x_1, \ldots, x_{m_L} \in L} \approx_s \{\mathsf{view}\}_{x_1, \ldots, x_{m_L} \in L}$.

We only provide an intuition to the proof here below. Full proof appears in the full version of the paper.

Proof Sketch. To prove the indistinguishability, we first take note of the ways in which the view generated by the simulator differs from the real-world view of the MiM adversary. Basically, the differences are that: for left sessions, the simulator does not use valid witnesses but tries to get 'fake' witnesses via the robust simulator; and for the right sessions, the simulator tries to extract witnesses via the robust simulator. While we know that using the robust simulator can incur at most negligible distance, what still remains to be shown is that the simulator using fake-witnesses for the left sessions also creates at most negligible distance from the real-view. For this, we simply rely on the statistical properties of the sub-protocols in which the simulator uses different values; namely, we rely upon SH of Com_{sh} of Step bps_2, sWI property of sWIAoK of Step bps_2, SH of the mixed non-malleable commitment of Step bps_4, and sWI of sWIAoK of Step bps_5– the steps at which the simulator uses different values in left sessions. Except for SH of the mixed non-malleable commitment of Step bps_4, all the above properties are already guaranteed by the corresponding primitives themselves; however, on the other hand, to ensure that the mixed non-malleable commitment – parameterized by srs which is the outcome of the coin-flipping protocol – is SH, we need to ensure that srs is uniformly random with all but negligible probability. Before we proceed, we thus prove that in the real-world view srs is uniform in every left session with all but negligible probability.

Claim. In the real-world view $\text{view}_{\mathcal{M}}(x_1, \ldots, x_{m_L})$, for every left session, srs is uniformly random with all but negligible probability.

Proof Sketch.   We begin by outlining the structure of the proof.

1. First, we show that, there exists a PPT algorithm that can extract a value r'_V from CECom_{sh} of Step cfp_1 of every left session *before* Step cfp_2 of that session is reached. Thus, since r_P is sent to the adversary after r'_V is extracted, r'_V is independent of r_P, and since r_P is uniformly random, $r_P \oplus r'_V$ is also uniformly random with all but negligible probability.
2. Then, we show that, in every left session, with all but negligible probability, $r'_V = r_V$, where, r_V is the value sent by $\mathcal{M}$ in Step cfp_3.

The above items together imply that $\text{srs} = r_P \oplus r_V$ is uniformly random, with all but negligible probability.

We prove the first step above by relying upon the Robust Extraction Lemma. Basically, the PPT algorithm (mentioned in the first step above) just emulates honest provers and honest verifiers to $\mathcal{M}$ except that it relays the CECom_{sh} of Step cfp_1 of every left session to RobustSim for extraction. We establish the second step as follows. Recall that a commitment information for r'_V of CECom_{sh} of Step cfp_1 in question is extractable as shown for the first step. Furthermore, from the witness-extractability of the BPS protocol in BPS^{CFP} phase, we can extract a witness – that we call sub-witness – for r_V being committed in the same CECom_{sh} commitment. Thus, if $r_V \neq r'_V$, we break CB of CECom_{sh}.

However, the proof is still not complete. The reason is for an implicit assumption in proving the second step above that the BPS argument given by the adversary in BPS^{CFP} phase of the left session is sound. To prove this, we establish the following claim.

Sub-Claim 1. *In the real world view, if BPS^{CFP} phase of the ℓ-th left session is accepted by the prover $\mathcal{P}_\ell$, then the value committed to by $\mathcal{M}$ in Com_{nm} at Step $\text{bps}^{\text{cfp}}_4$ of the ℓ-th left session is a valid sub-witness.*

Proof Sketch.   Intuitively, Com_{nm} at Step $\text{bps}^{\text{cfp}}_4$ of the ℓ-th left session contains a valid sub-witness owing to

computational hiding of CECom_{sb} – to argue that $\mathcal{M}$ does not learn α, committed to by the prover in CECom_{sb}, and use it in its commitment Com_{sh} and sWIAoK at Step $\text{bps}^{\text{cfp}}_2$,

knowledge-soundness of sWIAoK in Step $\text{bps}^{\text{cfp}}_2$– to extract knowledge of commitment information (i.e., commitment value and randomness) for Com_{sh} in Step $\text{bps}^{\text{cfp}}_2$ and to verify that the extracted value will not be α,

knowledge-soundness of sWIAoK in Step $\text{bps}^{\text{cfp}}_5$– to argue that either the value committed to in Com_{nm} at Step $\text{bps}^{\text{cfp}}_4$ is a valid sub-witness or to argue knowledge of a commitment information for Com_{sh} in Step $\text{bps}^{\text{cfp}}_2$ with commitment value as α,

and finally, computational binding of Com_{sh} at Step $\text{bps}^{\text{cfp}}_2$ to show the knowledge extracted is not α as a commitment value.

We prove each of the above steps by carefully designing interfaces that launch robust concurrent attacks and by crucially relying upon the Robust Extraction Lemma for extraction of commitment information out of these interfaces.   □

With this, we continue with a hybrid argument by moving from the real-world view to the simulated view. This is facilitated by the already established facts that the messages where the simulator deviates in its behavior from the real-world are statistically hiding (in some sense).   □

Witness Extractability. We shall prove that the values $y'_1, \ldots, y'_{m_R}$ extracted by the simulator-extractor $\mathcal{SE}$ are valid witnesses for the statements of the corresponding right sessions.

Theorem 2. *For every PPT adversary $\mathcal{M}$, the output of the simulator $\mathcal{SE}(x_1, \ldots, x_{m_L}, z) = (\text{view}, \overline{y}_1, \ldots, \overline{y}_{m_R})$ is such that, $\forall i \in [m_R], (\overline{x}_i, \overline{y}_i) \in R_L$.*

We discuss some of the core technical difficulties of the proof together with a high-level proof structure. Full proof appears in the full version of the paper

Proof Sketch.   Recall that in our protocol, the prover commits to a valid witness in NMMXCom$_{\text{srs}}$ at Step bps$_4$ and also commits to the same valid witness in CECom$_{sh}$ at Step bps$_{4+}$ (accompanied by a sWIAoK later in Step bps$_5$ for correctness of behavior). Note that both of these commitments are extractable. However, we cannot in a straight-forward manner employ the proof techniques of [1] or [17] to prove that the values extracted from these commitments by the simulator are indeed valid witnesses.

We begin by pointing out the reason why we are not able to simply make use of the proofs of [1] or [17]. In both [1] and [17], the prover commits to the witness with a non-malleable commitment. Thus, the commitment is *statistically binding*. Their proofs essentially proceed in the following manner: First, prove that the values committed to in the non-malleable commitments are valid witnesses. Secondly, move to a hybrid where extractions are performed to extract 'trapdoors' for cheating in the left sessions and to extract witnesses of the right sessions. Although cheating by the simulator on the left sessions may adversely change the values committed by $\mathcal{M}$ in the commitments of the right sessions, one can argue that the values committed to in the commitments of the right sessions are still valid witnesses owing to non-malleability of the commitment schemes.

Indeed, the statistically binding NMCom commitments are the reason why the protocols of [1] and [17] are not statistical CNMZK, but only computationally so. Our approach, to recall, is to use a mixed NMCom commitment which is parameterized by a string that is output of the coin-flipping phase that precedes the main argument phase. Thus, in the real-world, as proven earlier for Theorem 1, the parameter is a uniform random string rendering the mixed NMCom commitment to be SH. (Recall that the commitment being SH was crucial in proving statistical simulation in Theorem 1). Thus, it is not clear how to solely rely on the proof techniques of [1,17] for our proof.

Our proof technique instead is as follows. We begin with the real-world experiment where the outcome of the coin-flipping protocol is a uniform random string and thus the commitment scheme at Step bps_4 is a SH commitment. Then we start moving towards the hybrid which cheats in right sessions by biasing the outcome of the coin-flipping protocol to a uniform DDH tuple. The technical challenge will be the following. Fix any right session. Let H_a and H_b be the two hybrids in our hybrid sequence such that, the commitment at Step bps_4 in H_a is SH while the same commitment is SB in H_b (due to cheating in the coin-flipping protocol). Here, we need to establish that in H_b, the committed value in the commitment at Step bps_4 is a valid witness. We establish this through a careful design of hybrids and their sequence. We expand on our techniques and the whole high-level structure of the proof here below. We shall discuss the further multiple technical difficulties in the full proof in the full version of the paper.

We begin with a hybrid that is identical to the real-world view. Then we gradually modify the behavior of the hybrid for the right sessions towards biasing the coin-flipping protocol outcome to a random DDH tuple (from a uniform random string). Here, we will also prove that the values committed to by the MiM adversary in the mixed commitment at Step bps_4 is a valid witness (note that, with the outcome of coin-flipping being a random DDH tuple, this commitment scheme is now SB, thus allowing us to put forth arguments on the values committed in it). Next, we further move to hybrids which also behave differently in the left sessions by using 'trapdoors' (or fake-witnesses) extracted from the adversary itself (instead of valid witnesses). Here, we argue that such deviation in the hybrids' behavior for the left sessions does not adversely change the values committed to in the mixed NMCom commitments of the right sessions. Finally, we thereby reach a hybrid that behaves the same as our simulator-extractor, thus proving that the values extracted by $\mathcal{SE}$ are indeed valid witnesses.

Observe that it is easy to prove indistinguishability of hybrids as we change hybrids' behavior for the left sessions. The reason is that the left sessions will still have the outcome of coin-flipping to be uniformly random and thus the corresponding mixed commitment is SH. Thus, hybrids using fake-witnesses instead of the real ones will only introduce negligible statistical distance. However, the challenging part would be to argue indistinguishability of hybrids as they deviate in their behavior on the right sessions. We expand on the difficulty and our techniques briefly here below.

In order for hybrids to start cheating in coin-flipping phases of the right sessions, it is crucial that the hybrids are ordered carefully. Note that, we cannot at once move to a hybrid which changes the outcome of the coin-flipping phase due to soundness of the BPS protocol in $\mathsf{BPS}^{\mathsf{CFP}}$ phase. Thus, we first *simulate* this BPS protocol. We do so by extracting a trapdoor from the adversary in a way similar to [1]. Then, the next hybrid would be 'free' to bias the coin-flipping outcome to a random DDH tuple. However, note that this change is not statistically indistinguishable but only computationally so. Hence, this may adversely change the values committed to in the NMCom commitments in the protocol. However, with a careful sequence of arguments, we will be able to obtain a reduction to robustness w.r.t. 1-round protocols. Here it will be crucial to ensure that the other rewindings performed by the hybrids would not rewind the external NMCom receiver of the reduction.

Let us now consider the first hybrid that biases the coin-flipping outcome of the i-th right session. By this hybrid, we will already have biased coin-flipping outcomes of the first $i-1$ sessions. We thus need to make sure that this biasing will also not adversely change the values committed to in the mixed NMCom commitments at Step bps_4 of the first $i-1$ right sessions. Here again we rely on w.r.t. 1-round protocols for these NMCom commitments too.

A major technical difficulty would be the following. Fix any right session. Consider the first hybrid that biases the coin-flipping outcome of this session. Note that the previous hybrid had coin-flipping outcome to be a random string and thus the mixed commitment at Step bps_4 of the right session here to be SH. But in the current hybrid, due to the bias, the commitment scheme is SB. Here we need to argue that the committed value is a valid witness. As shown in the full proof, this would entail proving computational binding of a CECom_{sh} commitment. Here, we are no longer able to rely only upon the Robust Extraction Lemma to ensure us of successful extractions for the following reason. In Robust Extraction Lemma, it is essential that the external protocol whose party is not supposed to be rewound is such that its round complexity is strictly less than the number of slots of the CECom commitments extracted from. However, in the current case, the external protocol itself is a CECom commitment and hence this condition can not be met. We get around this difficulty again with a careful sequencing of hybrid arguments.

Furthermore, the above technical difficulty arises at another juncture in the proof of witness extractability. Namely, we encounter a hybrid where coin-flippings of all right sessions are biased, and in the subsequent hybrid we start changing the values committed in CECom_{sh} commitments of the left sessions. Here, we are still able to rely on the robustness of the concurrent extraction as follows. Although one cannot use the Robust Extraction Lemma for a reduction to statistical hiding of the entire left CECom_{sh} commitment, we can consider intermediate hybrids where, at a time, only one sub-commitment of the CECom_{sh} commitment is changed. Thus, we are still able to use robustness of the concurrent extraction since the sub-protocol in question is only of three rounds (as per the standard CECom commitment of [24]).

Then, once we ensure that the commitments at Step bps_4 of right sessions contain valid witnesses, we proceed to argue that the values extracted from the CECom_{sh} commitments are are valid witnesses with the following argument. We, along the way, show that the adversary cannot have a trapdoor, namely, r'_{comsh} that explains Com_{sh} at Step bps_2 being committed to σ. This implies that, for every right session, the witness that is extractable from the sWIAoK argument at Step bps_5 of is an opening of the CECom_{sh} commitment (together with the opening of the $\mathsf{NMMXCom}_{srs}$ commitment of Step bps_4) to a valid witness.

With this, we finally are at a hybrid that extracts valid witnesses from the right sessions. Furthermore, this hybrid is identical to our simulator-extractor, thus proving witness extractability of our protocol $\langle \mathcal{P}, \mathcal{V} \rangle$.   □

Acknowledgments. C. Orlandi is in part supported by the Danish Council for Independent Research (DFF).

R. Ostrovsky and V. Rao are in part supported by NSF grants 09165174, 1065276, 1118126 and 1136174, OKAWA Foundation Research Award, IBM Faculty Research Award, Xerox Faculty Research Award, B. John Garrick Foundation Award, Teradata Research Award, and Lockheed-Martin Corporation Research Award. This material is based upon work supported by the Defense Advanced Research Projects Agency through the U.S. Office of Naval Research under Contract N00014 − 11 − 1 − 0392. The views expressed are those of the author and do not reflect the official policy or position of the Department of Defense or the U.S. Government.

V. Rao and A. Sahai are in part supported by a DARPA/ONR PROCEED award, NSFgrants 1228984, 1136174, 1118096, and 1065276, a Xerox Faculty Research Award, a Google Faculty Research Award, an equipment grant from Intel, and an Okawa Foundation Research Grant. This material is based upon work supported by the Defense Advanced Research Projects Agency through the U.S. Office of Naval Research under Contract N00014-11- 1-0389. The views expressed are those of the author and do not reflect the official policy or position of the Department of Defense, the National Science Foundation, or the U.S. Government.

I. Visconti is in part supported by MIUR Project PRIN "GenData 2020".

References

1. Barak, B., Prabhakaran, M., Sahai, A.: Concurrent non-malleable zero knowledge. In: FOCS, p. 345 (2006); full version available on eprint arhive
2. Bellare, M., Micali, S., Ostrovsky, R.: The (true) complexity of statistical zero knowledge. In: STOC, pp. 494–502 (1990)
3. Catalano, D., Visconti, I.: Hybrid trapdoor commitments and their applications. In: Caires, L., Italiano, G.F., Monteiro, L., Palamidessi, C., Yung, M. (eds.) ICALP 2005. LNCS, vol. 3580, pp. 298–310. Springer, Heidelberg (2005)
4. Catalano, D., Visconti, I.: Hybrid commitments and their applications to zero-knowledge proof systems. Theor. Comput. Sci. 374(1-3), 229–260 (2007)
5. Damgård, I., Groth, J.: Non-interactive and reusable non-malleable commitment schemes. In: Proceedings of the 35th Annual ACM Symposium on Theory of Computing, San Diego, CA, USA, June 9-11, pp. 426–437. ACM (2003)
6. Damgård, I., Nielsen, J.B.: Perfect hiding and perfect binding universally composable commitment schemes with constant expansion factor. In: Yung, M. (ed.) CRYPTO 2002. LNCS, vol. 2442, pp. 581–596. Springer, Heidelberg (2002)
7. Dolev, D., Dwork, C., Naor, M.: Non-malleable cryptography (extended abstract). In: STOC, pp. 542–552 (1991)
8. Dolev, D., Dwork, C., Naor, M.: Nonmalleable cryptography. SIAM J. Comput. 30(2), 391–437 (2000); preliminary version in STOC 1991

9. Goldreich, O., Sahai, A., Vadhan, S.P.: Honest-verifier statistical zero-knowledge equals general statistical zero-knowledge. In: STOC, pp. 399–408 (1998)
10. Goldwasser, S., Micali, S., Rackoff, C.: The knowledge complexity of interactive proof-systems. In: Proc. 17th STOC, pp. 291–304 (1985)
11. Goyal, V., Lin, H., Pandey, O., Pass, R., Sahai, A.: Round-efficient concurrently composable secure computation via a robust extraction lemma. IACR Cryptology ePrint Archive 2012, 652 (2012)
12. Goyal, V., Moriarty, R., Ostrovsky, R., Sahai, A.: Concurrent statistical zero-knowledge arguments for np from one way functions. In: Kurosawa, K. (ed.) ASIACRYPT 2007. LNCS, vol. 4833, pp. 444–459. Springer, Heidelberg (2007)
13. Haitner, I., Nguyen, M.H., Ong, S.J., Reingold, O., Vadhan, S.P.: Statistically hiding commitments and statistical zero-knowledge arguments from any one-way function. SIAM J. Comput. 39(3), 1153–1218 (2009)
14. Halevi, S., Micali, S.: Practical and provably-secure commitment schemes from collision-free hashing. In: Koblitz, N. (ed.) CRYPTO 1996. LNCS, vol. 1109, pp. 201–215. Springer, Heidelberg (1996)
15. Lin, H., Pass, R.: Non-malleability amplification. In: STOC, pp. 189–198 (2009)
16. Lin, H., Pass, R.: Concurrent non-malleable zero knowledge with adaptive inputs. In: Ishai, Y. (ed.) TCC 2011. LNCS, vol. 6597, pp. 274–292. Springer, Heidelberg (2011)
17. Lin, H., Pass, R., Tseng, W.-L.D., Venkitasubramaniam, M.: Concurrent non-malleable zero knowledge proofs. In: Rabin, T. (ed.) CRYPTO 2010. LNCS, vol. 6223, pp. 429–446. Springer, Heidelberg (2010)
18. Lin, H., Pass, R., Venkitasubramaniam, M.: Concurrent non-malleable commitments from any one-way function. In: Canetti, R. (ed.) TCC 2008. LNCS, vol. 4948, pp. 571–588. Springer, Heidelberg (2008)
19. Mahmoody, M., Xiao, D.: Languages with efficient zero-knowledge pcps are in szk. In: Sahai, A. (ed.) TCC 2013. LNCS, vol. 7785, pp. 297–314. Springer, Heidelberg (2013)
20. Micciancio, D., Ong, S.J., Sahai, A., Vadhan, S.: Concurrent zero knowledge without complexity assumptions. In: Halevi, S., Rabin, T. (eds.) TCC 2006. LNCS, vol. 3876, pp. 1–20. Springer, Heidelberg (2006)
21. Okamoto, T.: On relationships between statistical zero-knowledge proofs. J. Comput. Syst. Sci. 60(1), 47–108 (2000)
22. Ostrovsky, R., Pandey, O., Visconti, I.: Efficiency preserving transformations for concurrent non-malleable zero knowledge. In: Micciancio, D. (ed.) TCC 2010. LNCS, vol. 5978, pp. 535–552. Springer, Heidelberg (2010)
23. Ostrovsky, R., Persiano, G., Visconti, I.: Constant-round concurrent non-malleable zero knowledge in the bare public-key model. In: Aceto, L., Damgård, I., Goldberg, L.A., Halldórsson, M.M., Ingólfsdóttir, A., Walukiewicz, I. (eds.) ICALP 2008, Part II. LNCS, vol. 5126, pp. 548–559. Springer, Heidelberg (2008)
24. Prabhakaran, M., Rosen, A., Sahai, A.: Concurrent zero knowledge with logarithmic round-complexity. In: Proc. 43rd FOCS (2002)
25. Sahai, A., Vadhan, S.P.: A complete problem for statistical zero knowledge. J. ACM 50(2), 196–249 (2003)

4-Round Resettably-Sound Zero Knowledge

Kai-Min Chung[1], Rafail Ostrovsky[2], Rafael Pass[3],
Muthuramakrishnan Venkitasubramaniam[4], and Ivan Visconti[5]

[1] Academia Sinica, Taiwan
kmchung@iis.sinica.edu.tw
[2] UCLA, Los Angeles, CA, USA
rafail@cs.ucla.edu
[3] Cornell University, Ithaca, NY 14850, USA
chung@cs.cornell.edu
[4] University of Rochester, Rochester, NY 14627, USA
muthuv@cs.rochester.edu
[5] University of Salerno, Italy
visconti@unisa.it

Abstract. While 4-round constructions of zero-knowledge arguments are known based on the existence of one-way functions, constuctions of *resettably-sound* zero-knowledge arguments require either stronger assumptions (the existence of a fully-homomorphic encryption scheme), or more communication rounds. We close this gap by demonstrating a 4-round resettably-sound zero-knowledge argument for NP based on the existence of one-way functions.

1 Introduction

Zero-knowledge (ZK) interactive protocols [18] are paradoxical constructs that allow one player (called the Prover) to convince another player (called the Verifier) of the validity of a mathematical statement $x \in L$, while providing *zero additional knowledge* to the Verifier. We are here interested in a stronger notion of zero-knowledge arguments known as *resettably-sound zero-knowledge*. This notion, first introduced by Barak, Goldwasser, Goldreich and Lindell (BGGL)[2], additionally requires the soundness property to hold even if the malicious prover is allowed to "reset" and "restart" the verifier. This model is particularly relevant for cryptographic protocols being executed on embedded devices, such as smart cards. BGGL provided a construction of a resettably-sound zero-knowledge argument for NP based on the existence of collision-resistant hash-functions. More recently, Bitansky and Paneth [5] presented a resettably-sound zero-knowledge argument based on the existence of an *oblivious transfer (OT) protocol*. Finally, Chung, Pass and Seth (CPS) [10] show how to construct such protocol based on the minimal assumption of one-way functions (OWFs).[1]

---

[1] As shown by Ostrovsky and Wigderson, one-way functions are also "essentially" necessary for non-trivial zero-knowledge [25]. In [9] one-way functions have been shown to suffice also when resettable zero knowledge is desired.

Y. Lindell (Ed.): TCC 2014, LNCS 8349, pp. 192–216, 2014.
© International Association for Cryptologic Research 2014

Our focus here is on the *round-complexity* of resettably-sound zero-knowledge arguments. All the above protocols only require a constant number of rounds; but what is the exact round-complexity? The original BGGL protocol requires 8 rounds and collision-resistant hash functions (CRHs); an implementation in 6 rounds of the BGGL construction has been shown in [24]. More recently, Bitansky and Paneth in [6] improved the round complexity of resettably-sound zero knowledge to 4 rounds but additionally requiring the existence of a fully homomorphic encryption (FHE) schemes [13,8]. Additionally they showed a 6-round protocol based on trapdoor permutations. In contrast, for "plain" (i.e., not resettably-sound) zero-knowledge, Bellare, Jakobsson and Yung [4] show how to obtain a 4-round zero-knowledge argument for NP based on the existence of the existence of one-way functions. This leaves open the question of whether round-efficient (namely 4-round) resettably-sound arguments can be based on weaker assumptions than FHE.

1.1 Our Results

We close the gap between resettably-sound and "plain" zero-knowledge arguments, demonstrating a 4-round resettably sound zero-knowledge argument (of knowledge) based solely on the existence of OWFs.

Theorem 1 (Informal). *Assume the existence of one-way functions. Then there exists a 4-round resettably-sound zero-knowledge argument of knowledge for every language in NP.*

Our starting point is the constant-round resettably-sound zero-knowledge argument for NP due to CPS. Our central contribution is a method for "collapsing" rounds in this protocol. A key feature of the CPS protocol is that, although the protocol consist of many rounds, the honest prover actually just sends commitments to 0 in all but *two* of these rounds. These "commitment to 0" preamble messages are only used by the simulator; roughly speaking, the simulator uses these message to come up with a "fake witness" that it can use in the remaining part of the protocol. On a very high-level, we show that all these preamble messages can be run in parallel, if appropriately adjusting the remaining two messages. An initial observation is that if we simply run all the preamble rounds in parallel—in a single "preamble slots"—then both completeness and zero-knowledge will still hold; the problem is that soundness no longer holds. In fact, soundness of the CPS protocol relies on the fact that the preamble messages are executed in sequence. Our key-idea for dealing with this issue is to have the verifier additionally provide a *signature* on the message-response pair for the "preamble" slot, and we now modify the "fake witness" part of the protocol to be a *chain of signatures* of the preamble messages *in the right order*. Soundness is now restored, and zero-knowledge simulation can be re-established by having the simulator *rewind* the preamble slot to get a signed sequence of messages in the right order.

1.2   Techniques

To explain our techniques in more detail, let us first recall Barak's non-black-box zero knowledge protocol on which BGGL is based, and then recall how CPS modify this protocol to only rely on OWF. We finally explain how to "collapse" rounds in this protocol.

Barak's Protocol and the BGGL Transformation. Recall that Barak's protocol relies on the existence of a family of collision-resistant hash function $h : \{0,1\}^* \to \{0,1\}^n$; note that any such family of collision-resistant hash functions can be implemented from a family of collision-resistant hash functions mapping n-bit string into $n/2$-bit strings using *tree hashing* [21]. Roughly speaking, in Barak's protocol, on common input 1^n and $x \in \{0,1\}^{\mathrm{poly}(n)}$, the Prover P and Verifier V, proceed in two stages. In Stage 1, V starts by selecting a function h from a family of collision-resistant hash function and sends it to P; P next sends a commitment $c = \mathsf{Com}(0^n)$ of length n, and finally, V next sends a "challenge" $r \in \{0,1\}^{2n}$; we refer to this as the "commit-challenge" round. In Stage 2, P shows (using a witness indistinguishable argument of knowledge) that either x is true, or that c is a commitment to a "hash" (using h) of a program M (i.e., $c = \mathsf{Com}(h(M))$ such that $M(c) = r$.

Roughly speaking, soundness follows from the fact that even if a malicious prover P^* tries to commit to (the hash of) some program M (instead of committing to 0^n), with high probability, the string r sent by V will be different from $M(c)$ (since r is chosen independently of c). To prove ZK, consider the non-black-box simulator S that commits to a hash of the code of the malicious verifier V^*; note that, by definition, it thus holds that $M(c) = r$, and the simulator can use c as a "fake" witness in the final proof. To formalize this approach, the witness indistinguishable argument in Stage 2 must actually be a witness indistinguishable *universal argument* (WIUARG) [22,1] since the statement that c is a commitment to a program M of *arbitrary* polynomial-size, and that proving $M(c) = r$ within some *arbitrary* polynomial time, is not in $\mathcal{NP}$. WIUARGs are known based on the existence of CRH and those protocols are constant-round public-coin; as a result, the whole protocol is constant-round and public-coin.

Finally, BGGL show that any constant-round public-coin zero-knowledge argument of knowledge can be transformed into a resettable-sound zero-knowledge argument, by simply having the verifier generate its (random) message by applying a pseudorandom function to the current partial transcript.[2]

The CPS Protocol. We now turn to recall the ideas from CPS for removing the use of CRHs in Barak's protocol. Note that hash functions are needed in two locations in Barak's protocol. First, since there is no *a-priori* polynomial upper-bound of the length of the code of V^*, we require a simulator to commit to the

---

[2] Strictly speaking, Barak's protocol is not a argument of knowledge, but rather a "weak" argument of knowledge (see [1,2] for more details), but the transformation of [2] applies also to such protocol.

hash of the code of V^*. Secondly, since there is no *a-priori* polynomial upper-bound on the running-time of V^*, we require the use of universal arguments (and such constructions are only known based on the existence of collision-resistant hash functions).

The main idea of CPS is to notice that digital signature schemes—which can be constructed based on one-way functions—share many of the desirable properties of CRHs, and to show how to appropriately instantiate (a variant of) Barak's protocol using signature schemes instead of using CRHs. More precisely, CPS show that by relying on strong fixed-length signature schemes, which can be constructed based on one-way functions, one can construct *signature tree* analogous to the tree hashing that could be used to compress arbitrary length messages into a signature of length n and satisfies an analogue collision-resistance property. A strong fixed-length signature scheme allows signing messages of arbitrary polynomial-length (e.g length $2n$) using a length n signature and satisfies that no polynomial time attacker can obtain a *new* signature even for messages that it has seen a signature on [14].

CPS then show how to replace tree hashing by signature trees by appropriately modifying Barak's protocol. Firstly, CPS adds a signature slot at the beginning of the protocol. More precisely, in an initial stage of the protocol, the verifier generates a signature key-pair SK, VK and sends only the verification key VK to the prover. Next, in a "signature slot", the prover sends a commitment c of some message to the verifier, and the verifier computes and returns a valid signature σ of c (using SK). This is used by the simulator to construct a signature tree through rewinding the (malicious) verifier as a fake witness for WIUARG in an analogous way as before. Note that the commitment is used to hide the message to be signed from the malicious verifier, and as such, the signature tree is constituted by signatures of commitments of signatures...etc—this is referred to as a *Sig-com tree*. On the other hand, soundness follows in a similar way to Barak's protocol by relying on the fact that Sig-com tree satisfy a strong "collision-resistance" property—namely, no attacker getting the VK can find collisions, even given access to a signing oracle.

Secondly, CPS use a variant of Barak's protocol due to Pass and Rosen [26], which relies on a special-purpose WIUARG, in which the honest prover never needs to perform any hashing.[3] More precisely, the WIUARG consist two phases: a first phase where the honest provery simply sends commitments to 0^n, and a second phase where it proves that either $x \in L$ or the messages it committed to consistutes a valid UARG proving that the prover knows a fake witness.

While this protocol is not public-coin, CPS nevertheless shows that it suffices to apply the PRF transformation of BGGL to just the the public-coin part of the protocol to obtain a resettably soundness protocol; recall that the only part of the protocol that is not public-coin is the "signature slot" and, thus, intuitively, the only "advantages" a resetting prover gets is that it may rewind the signature slot, and thus get an arbitrary polynomial number of signatures on messages of its choice. But, as noted above, signature trees are collision-resistant even with

---

[3] In fact, an early version of Barak's protocol also had this property.

respect to an attacker that gets an arbitrary polynomial number of queries to a signing oracle and thus resettable-soundness follows in exactly the same way as the (non-resetting) soundness property.

Formalizing this intuition, however, is subtle. CPS first introduce an "oracle-aided" model where both players have access to a signing oracle, and construct a public-coin zero knowledge argument of knowledge in this model. Then the transformation of [2] is applied to this protocol to obtain an oracle-aided resettably-sound zero-knowledge argument of knowledge. CPS then show a general transformation for turning the protocol into a "fixed-input" resettably-sound zero-knowledge argument (of knowledge) in the "plain" model (i.e. without any oracle); fixed-input resettable-soundness means that resettable soundness is only required to hold with respect to a single fixed input. Finally, CPS show another general transformation that turns any fixed-input resettable soundness *argument of knowledge* into "full-fledged" resettable sound argument (or knowledge). Combining all these steps leads to constant-round resettably-sound zero-knowledge argument of knowledge for $\mathcal{NP}$ based on one-way functions.

Collapsing Rounds for the CPS Protocol. We are now ready to explain our method for collapsing rounds in the CPS protocol. Note that, although the CPS protocol consists of many rounds, the honest prover actually just sends commitments to 0, in all but the final two rounds, where the prover shows that it either has a "fake witness" or that $x \in L$. More precisely, in the final "proof phase" of the protocol (where the prover only sends two messages), the prover shows the verifier that either $x \in L$ or that the "committed UARG" transcript is accepting. The key idea is to modify the protocol to let the prover show in the "proof phase" that either $x \in L$ or it *knows* a "commit-challenge" pair (c, r) and a committed UARG transcript showing that the commit-challenge pair was successful. This, alone, clearly does not work: soundness no longer hold if the prover can come up with its own "invented transcript". Inspired by the work of Lin and Pass [20], we instead require the prover to show that it knows a transcript—that has been signed, message-by-message, by the verifier through a "signature-chain". A similar approach was used also in [11,19]. Once we have done this change, we can simply remove all messages in the preamble phase (where the honest prover commits to 0) and just replace them with a signature slot. More precisely, we modify the CPS protocol in the following way:

- We start by running *two* signature slots in parallel: the first one is used for the signature-trees as in the original CPS protocol; the second one is used for the "signature-chain".
- In parallel with the signature slots, we start running the modified "proof phase" where the prover is requested to (using a WI argument of knowledge) prove that either $x \in L$ or it knows a "successful" transcript for the preamble phase that has been signed, message-by-message, in the right sequence using the second signature key.

Intuitively, simulation can be performed similarly to CPS, except that instead of simply providing the UARG messages in the protocol, the simulator rewinds the signature slot to get an appropriately signed transcript of the UARG protocol. (Proving this is a bit delicate since the CPS simulator is already providing its own rewindings, so we need to be careful to ensure that the composition of these rewindings does not blow up the expected running-time.)

The key challenge, however, is proving resettable-soundness of the resulting protocol. On a very high-level, we shows that how to transform any resetting attacker to a "stand-alone" (i.e., non-resetting) attacker for oracle-aided CPS protocol (recall that the CPS protocol was first constructed in an oracle-model where the prover and verifier have access to signature oracles, and then the oracle-aided protocol was transformed into a protocol in the "plain" model by adding the the signature slots).[4] Roughly speaking, we show how to extract out the implicit transcript messages from any successful resetting prover and we can then use these messages in the (oracle-aided) CPS protocol. This is not entirely trivial, since in the CPS protocol these messages need to be provided one-by-one, whereas we can only extract out a full transcript. Our key technical contribution consist of showing how to appropriately rewind the resetting attacker to make it provide accepting transcript that are consistent with a current partial transcript of the CPS protocol. We here rely on the properties of signature-chains, and the fact the the protocol only has a constant number of rounds.

2   Definitions

We now give definitions for interactive proof/argument systems with all variants that are useful in this work.

Definition 1 (interactive proofs [17]). *A proof system for the language L, is a pair of interactive Turing machines (P, V) running on common input x such that:*

- *Efficiency: P and V are PPT.*
- *Completeness: There exists a negligible function $\nu(\cdot)$ such that for every pair (x, w) such that $R_L(x, w) = 1$,*

$$\text{Prob}[\ \langle P(w), V \rangle(x) = 1\ ] \geq 1 - \nu(|x|).$$

- *Soundness: For every $x \notin L$ and for every interactive Turing machine P^* there exists a negligible function $\nu(\cdot)$ such that*

$$\text{Prob}[\ \langle P^*, V \rangle(x) = 1\ ] < \nu(|x|).$$

In the above definition we can relax the soundness requirement by considering P^* as PPT. In this case, we say that (P, V) is an argument system.

We denote by $\texttt{view}_{V^*(x,z)}^{P(w)}$ the view (i.e., its private coins and the received messages) of V^* during an interaction with $P(w)$ on common input x and auxiliary input z.

---

[4] This is a slight oversimplification; we actually need to slightly modify the oracle-aided CPS protocol. See Section 3 for more details.

Definition 2 (zero-knowledge arguments [17]). *Let (P, V) be an interactive argument system for a language L. We say that (P, V) is zero knowledge (ZK) if, for any probabilistic polynomial-time adversary V^* receiving an auxiliary input z, there exists a probabilistic polynomial-time algorithm S_{V^*} such for all pairs $(x, w) \in R_L$ the ensembles $\{\text{view}_{V^*(x,z)}^{P(w)}\}$ and $\{S_{V^*}(x, z)\}$ are computationally indistinguishable.*

Arguments of knowledge are arguments where there additionally exists an expected PPT *extractor* that can extract a witness from any successful prover, and this is a stronger notion of soundness. We will give now a definition that is slightly weaker than the standard definition of [3] but is useful for our constructions.

Note, also, that in the following definition, the extractor is given non-black box access to the prover. This is an essential property for our techniques.

Definition 3 (arguments of knowledge [2]). *Let R be a binary relation. We say that a probabilistic, polynomial-time interactive machine V is a knowledge verifier for the relation R with negligible knowledge error if the following two conditions hold:*

- *Non-triviality: There exists a probabilistic polynomial-time interactive machine P such that for every $(x, w) \in R$, all possible interactions of V with P on common input x, where P has auxiliary input w, are accepting, except with negligible probability.*
- *Validity (or knowledge soundness) with negligible error: There exists a probabilistic polynomial-time machine K such that for every probabilistic polynomial-time machine P^*, every polynomial $p(\cdot)$ and all sufficiently large x's,*
  $$Pr[w \leftarrow K(desc(P^*), x) \wedge R_L(x, w) = 1] > Pr[\langle P^*, V \rangle(x) = accept] - \frac{1}{p(|x|)}$$
  where $\langle P^*, V \rangle(x)$ denotes V's output after interacting with P^* upon common input x and $desc(P^*)$ denotes the description of P^*'s strategy.

Further, (P, V) is an argument of knowledge for relation R.

Definition 4 (witness indistinguishability [12]). *Let L be a language in $\mathcal{NP}$ and R_L be the corresponding relation. An interactive argument (P, V) for L is witness indistinguishable (WI) if for every verifier V^*, every pair (w_0, w_1) such that $(x, w_0) \in R_L$ and $(x, w_1) \in R_L$ and every auxiliary input z, the following ensembles are computationally indistinguishable:*

$$\{\text{view}_{V^*(x,z)}^{P(w_0)}\} \quad and \quad \{\text{view}_{V^*(x,z)}^{P(w_1)}\}.$$

2.1 Resettably-Sound Proofs

A polynomial-time relation R is a relation for which it is possible to verify in time polynomial in $|x|$ whether $R(x, w) = 1$. Let us consider an $\mathcal{NP}$-language L and denote by R_L the corresponding polynomial-time relation such that $x \in L$

if and only if there exists w such that $R_L(x, w) = 1$. We will call such a w a *valid witness for* $x \in L$. A *negligible* function $\nu(k)$ is a non-negative function such that for any constant $c < 0$ and for all sufficiently large k, $\nu(k) < k^c$. We will denote by $\text{Prob}_r[\,X\,]$ the probability of an event X over coins r. The abbreviation "PPT" stands for probabilistic polynomial time. We will use the standard notion of computational indistinguishability [16].

Let us recall the definition of resettable soundness due to [2].

Definition 5 (resettably-sound arguments [2]). *A resetting attack of a cheating prover P^* on a resettable verifier V is defined by the following two-step random process, indexed by a security parameter k.*

1. *Uniformly select and fix $t = \texttt{poly}(k)$ random-tapes, denoted $r_1, \ldots, r_t$, for V, resulting in deterministic strategies $V^{(j)}(x) = V_{x,r_j}$ defined by $V_{x,r_j}(\alpha) = V(x, r_j, \alpha)$,[5] where $x \in \{0,1\}^k$ and $j \in [t]$. Each $V^{(j)}(x)$ is called an incarnation of V.*
2. *On input 1^k, machine P^* is allowed to initiate $\texttt{poly}(k)$-many interactions with the $V^{(j)}(x)$'s. The activity of P^* proceeds in rounds. In each round P^* chooses $x \in \{0,1\}^k$ and $j \in [t]$, thus defining $V^{(j)}(x)$, and conducts a complete session with it.*

Let (P, V) be an interactive argument for a language L. We say that (P, V) is a resettably-sound argument for L if the following condition holds:

- *Resettable-soundness: For every polynomial-size resetting attack, the probability that in some session the corresponding $V^{(j)}(x)$ has accepted and $x \notin L$ is negligible.*

We will also consider a slight weakening of the notion of resettable soundness, where the statement to be proven is fixed, and the verifier uses a single random tape (that is, the prover cannot start many independent instances of the verifier).

Definition 6 (fixed-input resettably-sound arguments [27]). *An interactive argument (P, V) for a $\mathcal{NP}$ language L with witness relation R_L is fixed-input resettably-sound if it satisfies the following property: For all non-uniform polynomial-time adversarial resetting prover P^*, there exists a negligible function $\mu(\cdot)$ such that for every all $x \notin L$,*

$$\Pr[R \leftarrow \{0,1\}^\infty; (P^{*V_R(x)}, V_R)(x) = 1] \leq \mu(|x|)$$

The following theorem was proved in [10]

Claim 1. *Let (P, V) be a fixed-input resettably sound zero-knowledge (resp. witness indistinguishable) argument of knowledge for a language $L \in \mathcal{NP}$. Then there exists a protocol (P', V') that is a (full-fledged) resettably-sound zero-knowledge (resp. witness indistinguishable) argument of knowledge for L.*

As a result, in the sequel, we only focus on proving fixed-input resettable-soundness.

---

[5] Here, $V(x, r, \alpha)$ denotes the message sent by the strategy V on common input x, random-tape r, after seeing the message-sequence α.

2.2   Commitment Schemes

We now give a definition for a commitment scheme. For readability we will use "for all m" to mean any possible message m of length polynomial in the security parameter.

Definition 7. (Gen, Com, Ver) *is a* commitment scheme *if:*

- **efficiency:** Gen, Com *and* Ver *are polynomial-time algorithms;*
- **completeness:** *for all m it holds that* $\Pr[h_{com} \leftarrow \mathsf{Gen}(1^n); (\mathrm{COM}, \mathrm{dec}) \leftarrow \mathsf{Com}(h_{com}, m) : \mathsf{Ver}(h_{com}, \mathrm{COM}, \mathrm{dec}, m) = 1] = 1;$
- **binding:** *for any polynomial-time algorithm* committer* *there is a negligible function ν such that for all sufficiently large k it holds that:*
  $\Pr[h_{com} \leftarrow \mathsf{Gen}(1^n); (\mathrm{COM}, m_0, m_1, \mathrm{dec}_0, \mathrm{dec}_1) \leftarrow$ committer*$(h_{com}) :$
  $m_0 \neq m_1$ and $\mathsf{Ver}(h_{com}, \mathrm{COM}, \mathrm{dec}_0, m_0) = \mathsf{Ver}(h_{com}, \mathrm{COM}, \mathrm{dec}_1, m_1) = 1] \leq \nu(k);$
- **hiding:** *for any algorithm polynomial-time* receiver* *there is a negligible function ν such that for all m_0, m_1 where $|m_0| = |m_1|$ and all sufficiently large k it holds that*

$$\Pr\left[(h_{com}, \mathrm{aux}) \leftarrow \mathtt{receiver}(1^n); b \leftarrow \{0,1\}; (\mathrm{COM}, \mathrm{dec}) \leftarrow \mathsf{Com}(h_{com}, m_b)\right.$$
$$\left. : b \leftarrow \mathtt{receiver}^*(\mathrm{COM}, \mathrm{aux})\right] \leq \frac{1}{2} + \nu(n)$$

When h_{com} is clear from context, we often say "m, dec is a valid opening for COM" to mean that $\mathsf{Ver}(h_{com}, \mathrm{COM}, \mathrm{dec}, m) = 1$.

Collision-resistant hash functions. We will use hash functions as defined below.

Definition 8. *Let $\mathcal{H} = \{h_\alpha\}$ be an efficiently sampleable hash function ensemble where $h_\alpha : \{0,1\}^* \to \{0,1\}^\alpha$. We say that $\mathcal{H}$ is* collision-resistant against polynomial size circuits *if for every (non-uniform) polynomial-size circuit family $\{A_n\}_{n \in N}$, for all positive constants c, and all sufficiently large k, it holds that*

$$\mathrm{Prob}[\, \alpha \xrightarrow{R} \{0,1\}^k : A_n(\alpha) = (x, x') \wedge h_\alpha(x) = h_\alpha(x') \,] < n^{-c}.$$

2.3   Signature Trees

Constructions of universal arguments (defined later) rely on Merkle-trees and collision-resistant hash-functions to be able to commit to a program of arbitrary polynomial length where no apriori-bound is known. In [10], they construct an analog to Merkle-trees, called signature trees, while relying only on one-way functions. Below, we recall definitions from [10]. Some of the text in this section, is copied verbatim from [10]

Definition 9 (Strong Signatures). *A strong, length-ℓ, signature scheme* SIG *is a triple* (Gen, Sign, Ver) *of* PPT *algorithms, such that*

1. *for all* $n \in \mathbf{N}, m \in \{0,1\}^*$,

$$\Pr[(\text{SK}, \text{VK}) \leftarrow \text{Gen}(1^n); \sigma \leftarrow \text{Sign}_{\text{SK}}(m); \text{Ver}_{\text{VK}}(m, \sigma) = 1 \wedge |\sigma| = \ell(n)] = 1$$

2. *for every non-uniform* PPT *adversary* A, *there exists a negligible function* $\mu(\cdot)$ *such that*

$$\Pr\Big[(\text{SK}, \text{VK}) \leftarrow \text{Gen}(1^n), (m, \sigma) \leftarrow A^{\text{Sign}_{\text{SK}}(\cdot)}(1^n) :$$

$$\text{Ver}_{\text{VK}}(m, \sigma) = 1 \wedge (m, \sigma) \notin L] \leq \mu(n),$$

where L *denotes the list of query-answer pairs of* A*'s queries to its oracle.*

Strong, length-ℓ, deterministic signature schemes with $\ell(n) = n$ are known based on the existence of OWFs; see [23,28,14] for further details.

Definition 10 (Signature Trees). *Let* SIG $=$ (Gen, Sign, Ver) *be a strong, length-n signature scheme. Let* (SK, VK) *be a key-pair of* SIG, *and* s *be a string of length 2^d. A signature tree of the string s w.r.t.* (SK, VK) *is a complete binary tree of depth d, defined as follows.*

- *A leaf l_γ indexed by $\gamma \in \{0,1\}^d$ is set as the bit at position γ in s.*
- *An internal node l_γ indexed by $\gamma \in \bigcup_{i=0}^{d-1} \{0,1\}^i$ satisfies that* $\text{Ver}_{\text{VK}}((l_{\gamma 0}, l_{\gamma 1}), l_\gamma) = 1.$

To *verify* whether a Γ is a valid signature-tree of a string s w.r.t. the signature scheme SIG and the key-pair (SK, VK) knowledge of the secret key SK is not needed. However, to *create* a signature-tree for a string s, the secret key SK is needed.

Definition 11 (Signature Path). *A signature path w.r.t.* SIG, VK *and a root l_λ for a bit b at leaf $\gamma \in \{0,1\}^d$ is a vector $\boldsymbol{\rho} = ((l_0, l_1), ((l_{\gamma_{\leq 1}0}, l_{\gamma_{\leq 1}1}), \dots (l_{\gamma_{\leq d-1}0}, l_{\gamma_{\leq d-1}1}))$ such that for every $i \in \{0, \dots, d-1\}$, $\text{Ver}_{\text{VK}}((l_{\gamma_{\leq i}0}, l_{\gamma_{\leq i}1}), l_{\gamma_{\leq i}}) = 1.$*
Let $\text{PATH}^{\text{SIG}}(\boldsymbol{\rho}, b, \gamma, l_\lambda, \text{VK}) = 1$ *if ρ is a signature path w.r.t.* SIG, VK, l_λ *for b at γ.*

2.4 Sig-Com Schemes

Definition 12 (Sig-Com Schemes). *Let* SIG $=$ (Gen, Sign, Ver) *be a strong, length-n, signature scheme, and let* COM *be a non-interactive commitment schemes. Define* SIG' $=$ (Gen', Sign', Ver') *to be a triple of* PPT *machines defined as follows:*

- $\mathsf{Gen}' = \mathsf{Gen}$.
- $\mathsf{Sign}'_{\mathrm{SK}}(m)$: *compute a commitment* $c = \mathrm{COM}(m; \tau)$ *using a uniformly selected* τ, *and let* $\sigma = \mathsf{Sign}_{\mathrm{SK}}(c)$; *output* (σ, τ)
- $\mathsf{Ver}'_{\mathrm{VK}}(m, \sigma, \tau)$: *Output* 1 *iff* $\mathsf{Ver}_{\mathrm{VK}}(\mathrm{COM}(m, \tau), \sigma) = 1$.

We call SIG' the Sig-Com Scheme *corresponding to* SIG *and* COM.

Definition 13 (Sig-Com Trees). *Let* $\mathsf{SIG} = (\mathsf{Gen}, \mathsf{Sign}, \mathbf{SHVer}_{h_{com}})$ *be a strong, length-n signature scheme, let* COM *be a non-interactive commitment and let* $\mathsf{SIG}' = (\mathsf{Gen}', \mathsf{Sign}', \mathbf{SHVer}'_{h_{com}})$ *be the sig-com scheme corresponding to* SIG *and* COM. *Let* $(\mathrm{SK}, \mathrm{VK})$ *be a key-pair of* SIG', *and s be a string of length* 2^d. *A* signature tree *of the string s w.r.t.* $(\mathrm{SK}, \mathrm{VK})$ *is a complete binary tree of depth d, defined as follows.*

- *A leaf l_γ indexed by $\gamma \in \{0, 1\}^d$ is set as the bit at position γ in s.*
- *An internal node l_γ indexed by $\gamma \in \bigcup_{i=0}^{d-1}\{0, 1\}^i$ satisfies that there exists some τ_γ such that* $\mathsf{Ver}'_{\mathrm{VK}}((l_{\gamma 0}, l_{\gamma 1}), l_\gamma, \tau_\gamma) = 1$.

Definition 14 (Sig-Com Path). *Let* $\mathsf{SIG}' = (\mathsf{Gen}', \mathsf{Sign}', \mathsf{Ver}')$ *be a sig-com scheme. A sig-com path w.r.t.* SIG', VK *and a root l_λ for a bit b at leaf $\gamma \in \{0, 1\}^d$ is a vector* $\rho = ((l_0, l_1, \tau_\lambda), ((l_{\gamma \leq 1}0, l_{\gamma \leq 1}1, \tau_{\gamma \leq 1}), \ldots, (l_{\gamma \leq d-1}0, l_{\gamma \leq d-1}0, \tau_{\gamma \leq d-1})$ *such that for every* $i \in \{0, \ldots, d-1\}$, $\mathsf{Ver}'_{\mathrm{VK}}((l_{\gamma \leq i}0, l_{\gamma \leq i}1), l_{\gamma \leq i}, \tau_{\gamma \leq i})) = 1$. *Let* $\mathsf{PATH}^{\mathsf{SIG}'}(\rho, b, \gamma, l_\lambda, \mathrm{VK}) = 1$ *if ρ is a signature path w.r.t.* SIG', VK, l_λ *for b at γ.*

2.5 Oracle-Aided Zero Knowledge Protocols

In this section we recall definitions of oracle-aided protocols from [10].

Let $\mathcal{O}$ be a probabilistic algorithm that on input a security parameter n, outputs a polynomial-length (in n) public-parameter pp, as well as the description of an oracle O. The oracle-aided execution of an interactive protocol with common input x between a prover P with auxiliary input y and a verifier V consist of first generating $\mathsf{pp}, O \leftarrow \mathcal{O}(1^{|x|})$ and then letting $P^O(x, y, \mathsf{pp})$ interact with $V(x, \mathsf{pp})$.

Definition 15 (Oracle-aided Interactive Arguments). *A pair of oracle algorithms (P, V) is an $\mathcal{O}$-oracle aided argument for a $\mathcal{NP}$ language L with witness relation R_L if it satisfies the following properties:*

- *Completeness: There exists a negligible function $\mu(\cdot)$, such that for all $x \in L$, if $w \in R_L(x)$,*

$$\Pr[\mathsf{pp}, O \leftarrow \mathcal{O}(1^{|x|}); (P^O(w), V)(x, \mathsf{pp}) = 1] = 1 - \mu(|x|)$$

- *Soundness: For all non-uniform polynomial-time adversarial prover P^*, there exists a negligible function $\mu(\cdot)$ such that for every all $x \notin L$,*

$$\Pr[\mathsf{pp}, O \leftarrow \mathcal{O}(1^{|x|}); (P^{*O}, V)(x, \mathsf{pp}) = 1] \leq \mu(|x|)$$

Additionally, if the following condition holds, (P, V) is an $\mathcal{O}$-oracle aided argument of knowledge:

- *Argument of knowledge: There exists a expected* PPT *algorithm E such that for every polynomial-size P^*, there exists a negligible function $\mu(\cdot)$ such that for every x,*

$$\Pr[\mathsf{pp}, O \leftarrow \mathcal{O}(1^{|x|}); w \leftarrow E^{P^{*O}(x,\mathsf{pp})}(x, \mathsf{pp}); w \in R_L(x)]$$
$$\geq \Pr[\mathsf{pp}, O \leftarrow \mathcal{O}(1^{|x|}); (P^{*O}, V)(x, \mathsf{pp}) = 1] - \mu(|x|)$$

Definition 16 (Oracle-aided Resettably-sound Interactive Arguments). *An $\mathcal{O}$-oracle aided resetting attack of a cheating prover P^* on a resettable verifier V is defined by the following three-step random process, indexed by a security parameter n.*

1. *An initial setup where a public parameter and an oracle are generated:* $\mathsf{pp}, O \leftarrow \mathcal{O}(1^n)$. *$P^*$ is given* pp *and oracle access to O.*
2. *Uniformly select and fix $t = poly(n)$ random-tapes, denoted $r_1, \ldots, r_t$, for V, resulting in deterministic strategies $V^{(j)}(x) = V_{x, r_j}$ defined by $V_{x, r_j}(\alpha) = V(x, r_j, \alpha)$, where $x \in \{0, 1\}^n$ and $j \in [t]$. Each $V^{(j)}(x)$ is called an incarnation of V.*
3. *On input 1^n, machine P^* is allowed to initiate $poly(n)$-many interactions with the $V^{(j)}(x)$'s. The activity of P^* proceeds in rounds. In each round P^* chooses $x \in \{0, 1\}^n$ and $j \in [t]$, thus defining $V^{(j)}(x)$, and conducts a complete session with it.*

Let (P, V) be an $\mathcal{O}$-oracle aided interactive argument for a language L. We say that (P, V) is an $\mathcal{O}$-oracle aided resettably-sound argument for L if the following condition holds:

- *$\mathcal{O}$-oracle aided resettable soundness: For every polynomial-size resetting attack, the probability that in some session the corresponding $V^{(j)}(x)$ has accepted and $x \notin L$ is negligible.*

Oracle-Aided Universal Arguments. Universal arguments (introduced in [1] and closely related to CS-proofs [22]) are used in order to provide "efficient" proofs to statements of the form $y = (M, x, t)$, where y is considered to be a true statement if M is a non-deterministic machine that accepts x within t steps. The corresponding language and witness relation are denoted $L_\mathcal{U}$ and $\mathbf{R}_\mathcal{U}$ respectively, where the pair $((M, x, t), w)$ is in $\mathbf{R}_\mathcal{U}$ if M (viewed here as a two-input deterministic machine) accepts the pair (x, w) within t steps. Notice that every language in $\mathcal{NP}$ is linear time reducible to $L_\mathcal{U}$. Thus, a proof system for $L_\mathcal{U}$ allows us to handle all $\mathcal{NP}$-statements. In fact, a proof system for $L_\mathcal{U}$ enables us to handle languages that are beyond $\mathcal{NP}$, as the language $L_\mathcal{U}$ is $\mathcal{NE}$-complete (hence the name universal arguments).[6]

---

[6] Furthermore, every language in $\mathcal{NEXP}$ is polynomial-time (but not linear-time) reducible to $L_\mathcal{U}$.

Definition 17 (Oracle-aided Universal Argument). *An oracle-aided protocol (P, V) is called an $\mathcal{O}$-oracle-aided* universal argument *system if it satisfies the following properties:*

- Efficient verification: *There exists a polynomial p such that for any $y = (M, x, t)$, and for any pp, O generated by $\mathcal{O}$, the total time spent by the (probabilistic) verifier strategy V, on common input y, pp, is at most $p(|y| + |pp|)$. In particular, all messages exchanged in the protocol have length smaller than $p(|y| + |pp|)$.*
- Completeness with a relatively efficient oracle-aided prover: *For every $(y = (M, x, t), w)$ in $\mathbf{R}_{\mathcal{U}}$,*

$$\Pr[pp, O \leftarrow \mathcal{O}(1^{|y|}); (P^O(w), V)(y, pp) = 1] = 1.$$

Furthermore, there exists a polynomial q such that the total time spent by $P^O(w)$, on common input $y = (M, x, t)$, pp, is at most $q(T_M(x, w) + |pp|) \leq q(t + |pp|)$, where $T_M(x, w)$ denotes the running time of M on input (x, w).

- Weak proof of knowledge for adaptively chosen statements: *For every polynomial p there exists a polynomial p' and a probabilistic polynomial-time oracle machine E such that the following holds: for every non-uniform polynomial-time oracle algorithm P^*, if*

$$\Pr[pp, O \leftarrow \mathcal{O}(1^n); R \leftarrow \{0, 1\}^{\infty}; y \leftarrow P_R^{*O}(pp):$$

$$(P_R^{*O}(pp), V(y, pp)) = 1] > 1/p(n)$$

then

$$\Pr[pp, O \leftarrow \mathcal{O}(1^n); R, r \leftarrow \{0, 1\}^{\infty}; y \leftarrow P_R^{*O}(pp): \exists w = w_1, \ldots w_t \in \mathbf{R}_{\mathcal{U}}(y)$$

$$s.t. \; \forall i \in [t], E_r^{P_R^{*O}}(pp, y, i) = w_i] > \frac{1}{p'(n)}$$

where $\mathbf{R}_{\mathcal{U}}(y) \overset{\text{def}}{=} \{w : (y, w) \in \mathbf{R}_{\mathcal{U}}\}$.

Let SIG' be a canonical sig-com scheme with $\mathsf{SIG} = (\mathsf{Gen}, \mathsf{Sign}, \mathsf{Ver})$ and COM being its underlying signature scheme and commitment scheme.

Definition 18 (Signature Oracle). *Given $\mathsf{SIG} = (\mathsf{Gen}, \mathsf{Sign}, \mathsf{Ver})$ a signature scheme , we define a signature oracle $\mathcal{O}^{\mathsf{SIG}}$ as follows: On input a security parameter n, $\mathcal{O}^{\mathsf{SIG}}(1^n)$ generates $(\mathrm{VK}, \mathrm{SK}) \leftarrow \mathsf{Gen}(1^n)$ and lets $pp = \mathrm{VK}$ and $O(m) = \mathsf{Sign}_{\mathrm{SK}}(m)$ for every $m \in \{0, 1\}^{\mathtt{poly}(n)}$.*

Definition 19 (Valid Sig-com Oracle). *An oracle $\mathcal{O}'$ is a valid (SIG', ℓ) oracle if there is a negligible $\mu(\cdot)$ such that for every $n \in N$, the following holds with probability $1 - \mu(n)$ over $pp, O \leftarrow \mathcal{O}'(1^n)$: for every $m \in \{0, 1\}^{\ell(n)}$, $O(m)$ returns (σ, τ) such that $\mathsf{Ver}'_{\mathrm{VK}}(m, \sigma, \tau) = 1$ with probability at least $1 - \mu(n)$.*

Definition 20. *An $\mathcal{O}^{\mathsf{SIG}}$-aided universal arg. (P, V) has (SIG', ℓ)-completeness if there exists a prover P' such that the completeness condition holds for (P', V) when the oracle $\mathcal{O}^{\mathsf{SIG}}$ is replaced by any valid (SIG', ℓ) oracle $\mathcal{O}'$.*

The following theorem was proved in [10] (relying on Barak and Goldreich [1])

Theorem 2. *Let* SIG' *be a canonical sig-com scheme with* SIG *and* COM *being its underlying signature scheme and commitment scheme. Then there exists a* (SIG', ℓ)-*complete* $\mathcal{O}^{\text{SIG}}$-*aided universal argument with* $\ell(n) = 2n$.

3   A Variant of the Signature Oracle-Aided ZK Protocol from CPS

In this section, we provide a formal protocol description and theorem statement for a slight variant of the CPS protocol in a signature oracle-aided model. We will show in the next section how to collapse rounds of this protocol, and prove resettable soundness of the collapsed protocol by reducing the resetting attacker to a stand-alone (i.e., non-resetting) adversary that breaks soundness of this protocol.

Common Input: An instance x of a language $L \in \mathcal{NP}$ with witness relation $\mathbf{R}_L$.
Auxiliary input to P: A witness w such that $(x, w) \in \mathbf{R}_L$.
Primitives Used: A canonical sig-com scheme SIG' with SIG and COM as the underlying signature and commitment schemes, and a (SIG', ℓ)-complete $\mathcal{O}^{\text{SIG}}$-aided universal argument $(P_{\text{UA}}, V_{\text{UA}})$ with $\ell(n) = 2n$.
Set Up: Run $(\text{pp}, O) \leftarrow \mathcal{O}^{\text{SIG}}(1^n)$, add pp to common input for P and V. Furthermore, allow P oracle access to O.
Stage One (Trapdoor—Commit-Challenge):
  P_1: Send $c_0 = \text{COM}(0^{2n}, \tau_0)$ to V with uniform τ_0
  V_1: Send $r \xleftarrow{\$} \{0, 1\}^{4n}$ to P
Stage Two (Encrypted Universal Argument):
  P_2: Send $c_1 = \text{COM}(0^{2n}, \tau_1)$ for uniformly selected τ_1
  V_3: Send r', uniformly chosen random tape for V_{UA}
  P_3: Send $c_2 = \text{COM}(0^k, \tau_2)$ for uniformly selected τ_2, where k is the length of P_{UA}'s second message.
Stage Three: (Main Proof)
  $P \Leftrightarrow V$: A WI-AOK $\langle P_{\text{WI}}, V_{\text{WI}} \rangle$ proving the OR of the following statements:
    1. $\exists\, w \in \{0, 1\}^{\text{poly}(|x|)}$ s.t. $(x, w) \in \mathbf{R}_L$.
    2. $\exists\, \langle p_1, p_2, \tau_1, \tau_2 \rangle$ s.t. $((\langle c_0, r, c_1, c_2, r', \text{pp} \rangle, \langle p_1, p_2, \tau_1, \tau_2 \rangle) \in \mathbf{R}_{L_2}$ (defined in Fig. 2).

Fig. 1. $\mathcal{O}^{\text{SIG}}$-aided ZK Argument of Knowledge

We refer the readers to Section 1.2 for the ideas and intuition behind the CPS protocol. A formal description of the protocol can be found in Figure 1 and 2, where we make a slight modification to the language proved in the UA where we require the committed program either output the string r when fed a commitment to its own description or output r as the second component of

4-tuple output when fed by a string of length shorter than r. This modification is inconsequential to the soundness property of the protocol, but will be useful for us to prove soundness of the collapsed protocol in the next section. The following theorem follows by [10].

Theorem 3. *Assume the existence of one-way functions. The protocol defined in Figure 1 and 2 is a signature oracle-aided zero-knowledge argument of knowledge for NP.*

Relation 1: Let SIG' a sig-com scheme, with underlying signature scheme SIG and commitment scheme COM. Let ECC be a binary error-correcting code with constant min-distance and efficient encoding algorithm. We say that $\langle c_0, r, \mathsf{pp}\rangle \in L_1$ if $\exists \langle \tau_0, d, l_\lambda, C, \{\boldsymbol{\rho}_i\}_{i\in[2^d]}, y\rangle$ such that

- $c_0 = \mathsf{COM}((d, l_\lambda), \tau_0)$
- (d, l_λ) are the depth and root of a sig-com tree for C w.r.t. pp
- Each $\boldsymbol{\rho}_i$ is a valid sig-com path for leaf i of this sig-com tree. That is, $\mathsf{PATH}^{\mathsf{SIG}'}(\boldsymbol{\rho}_i, C_i, i, l_\lambda, \mathsf{pp}) = 1$ for each i.
- $C = \mathsf{ECC}(\Pi)$ for some circuit Π
- $\Pi(c_0) = r$ or $|y| < 2n$ and $\Pi(y) = (s_1, r, s_2, s_3)$ for some strings s_1, s_2, s_3 and appropriate encoding of the 4-tuple.

We let $\mathbf{R}_{L_1}$ be the witness relation corresponding to L_1.

Relation 2: Let L_1 be described as above, with respect to SIG' and ECC. Let $(P_{\mathsf{UA}}, V_{\mathsf{UA}})$ be a (SIG', ℓ)-complete $\mathcal{O}^{\mathsf{SIG}}$-aided universal argument with $\ell(n) = 2n$. We say that $\langle c_0, r, c_1, c_2, r', \mathsf{pp}\rangle \in L_2$ if $\exists \langle p_1, p_2, \tau_1, \tau_2\rangle$ such that

- $c_1 = \mathsf{COM}(p_1, \tau_1)$, $c_2 = \mathsf{COM}(p_2, \tau_2)$.
- (p_1, r', p_2) constitutes an accepting $(P_{\mathsf{UA}}, V_{\mathsf{UA}})$ transcript for $\langle c_0, r, \mathsf{pp}\rangle \in L_1$.

We let $\mathbf{R}_{L_2}$ be the witness relation corresponding to L_2.

Fig. 2. Relations used in the $\mathcal{O}^{\mathsf{SIG}}$-aided ZK protocol in Fig. 1

4   4-Round Resettably-Sound Zero Knowledge

We are now ready to describe our 4-round protocol. Our protocol relies on Blum's 4-round Hamiltonicity WI-AOK, $\langle P_{\mathsf{WI}}, V_{\mathsf{WI}}\rangle$ [7]. Our protocol is obtained by first constructing a "basic" protocol where the verifier uses "fresh" randomness in each round, and then applying the BGGL transformation to this protocol (i.e., having the verifier pick its randomness by applying a PRF to the current transcript). The "basic" protocol proceeds as follows.

1. The verifier V picks two signature key pairs (vk, sk) and (vk', sk') using $\mathsf{Gen}(1^k)$. V also generates the first message BH_1 for the WI AoK.

The language considered for the WI argument of knowledge is identical to one used in the protocol presented in the previous section, i.e. $(x, vk, vk') \in L^*$ iff

(a) $\exists\, w \in \{0,1\}^{\text{poly}(|x|)}$ s.t. $(x, w) \in \mathbf{R}_L$.

(b) $\exists\, \langle c_0, r, c_1', r', p_1, p_2, \tau_1, \tau_2, \sigma_1, \sigma_2 \rangle$ s.t.
$(\langle vk, vk' \rangle, \langle c_0, r, c_1, r', p_1, p_2, \tau_1, \sigma_1, \sigma_2 \rangle) \in \mathbf{R}_{L_3}$ (defined in Fig. 4).

V sends vk, vk', BH_1 to the prover P.

2. P responds with a commitment c to the all 0's string of length k and the second message for the WI AoK, BH_2.

3. V sends $r, \sigma, \sigma', \text{BH}_3$ to the prover where $r \leftarrow \{0,1\}^{3k}$, σ and σ' are signatures of messages $c|r$ and c under keys sk and sk' respectively and BH_3 is the third message of WI AoK .

4. P finally sends BH_4, the fourth message of the WI AoK. The verifier accepts if the transcript $(\text{BH}_1, \text{BH}_2, \text{BH}_3, \text{BH}_4)$ is accepting for $(x, h, vk) \in L^*$.

We finally modify the basic protocol by having the verifier first pick a random seed s for a PRF f and then, at each round, generating the randomness it needs by applying the f_s to the current transcript.

A formal description of the protocol is presented in Figure 3.

Theorem 4. *Assume the existence of OWFs, then protocol in Fig. 3 is a 4-round resettably sound zero knowledge argument of knowledge.*

Proof. We prove completeness and resettable-soundness of the protocol. As proved in [10], it suffices to prove fixed-input resettable-soundness.

Completeness. Completeness of $\langle P, V \rangle$ follows directly from the completeness of the WI-AOK protocol.

Common Input: An instance x of a language $L \in \mathcal{NP}$ with witness relation $\mathbf{R}_L$.
Auxiliary input to P: A witness w such that $(x, w) \in \mathbf{R}_L$.

$1 : V \to P$: Send BH_1, vk, vk' where $(vk, sk) \leftarrow \text{Gen}(1^n)$ and $(vk', sk') \leftarrow \text{Gen}(1^n)$.

$2 : P \to V$: Send $\text{BH}_2, c = \text{COM}(0^{2n}, \tau)$ for a randomly chosen τ.

$3 : V \to P$: Send $\text{BH}_3, r, \sigma, \sigma'$ where $r \leftarrow \{0,1\}^{3n}$, $\sigma \leftarrow \text{SIGN}(sk, c|r)$ and $\sigma' \leftarrow \text{SIGN}(sk', c)$.

$4 : P \to V$: Send BH_4.

We finally modify the above protocol by having the verifier first pick a random seed s for a PRF f and then, at each round, generating the randomness it needs by applying the f_s to the current transcript.

Fig. 3. Our 4-round rsZK Argument of Knowledge $\pi = (P, V)$

Relation 3: Let L_1 be as described in Fig 2, with respect to SIG$'$ and ECC. Let $(P_{\mathsf{UA}}, V_{\mathsf{UA}})$ be a (SIG$'$, ℓ)-complete $\mathcal{O}^{\mathsf{SIG}}$-aided universal argument with $\ell(n) = 2n$. We say that $\langle vk, vk' \rangle \in \mathbf{R}_{L_3}$, if $\exists \langle c_0, r, c_1, r', p_1, p_2, \tau_1, \sigma_1, \sigma_2 \rangle) \in \mathbf{R}_{L_3}$ such that

- $\mathsf{Ver}_{vk}(c_0|r, \sigma_1) = 1$, $c_1 = \mathsf{COM}(p_1|\sigma_1, \tau_1)$ and $\mathsf{Ver}_{vk}(c_1|r', \sigma_2) = 1$.
- (p_1, r', p_2) constitutes an accepting $(P_{\mathsf{UA}}, V_{\mathsf{UA}})$ transcript for $\langle c_0, r, vk' \rangle \in L_1$.

We let $\mathbf{R}_{L_3}$ be the witness relation corresponding to L_3.

Fig. 4. Relations used in the protocol in Fig. 3

Soundness. To prove the fixed-input resettable-soundness of $\langle P, V \rangle$, we show how to convert a malicious prover P^* for $\langle P, V \rangle$ into an *oracle-aided* malicious prover B that violates the *stand-alone* soundness of $\langle P_{\mathsf{zk}}, V_{\mathsf{zk}} \rangle$ (from the previous section).

First, we consider the experiment $\mathrm{HYB}_1^A(n, z)$ where we run an adversary A on input (n, z) by supplying the messages of an honest verifier, with the exception that the verifier challenges, i.e. r and BH_3 in the third message are chosen uniformly at random even in the rewindings instead of applying the PRF. Upon completion, we run the extractor of the WI AoK in a random session to obtain witness w. If this witness is not a real witness, output the transcript along with w. Otherwise output $\perp$.

From the pseudo-randomness of $\mathcal{F}$, we know that if P^* convinces an honest verifier of a false statement with non-negligible probability in the original experiment, then it will succeed in proving a false statement with non-negligible probability in HYB_1 as well. Since there are only polynomially many sessions, $\mathrm{HYB}_1^{P^*}(n, z)$ outputs the second (or fake) witness with non-negligible probability.

More precisely, for a statement $(x, vk, vk') \in L^*$ the fake witness contains $\langle c_0, r, c_1', r', p_1, p_2, \tau_1, \sigma_1, \sigma_2 \rangle$. From the unforgeability of the signature scheme under verifier key vk, it follows that, if P^* proves using the fake witness, then P^* must have obtained σ_1, σ_2 by querying the verifier with the appropriate commitment as part of the second message of the protocol. Let $\mathcal{J}_1$ (and $\mathcal{J}_2$) be the random variable representing the message index where the commitment c_0 and the corresponding signature σ_1 (resp., c_1' and σ_2) were sent in the experiment $\mathrm{HYB}_1^{P^*}(n, z)$. We also denote by $\mathcal{J}_3$ the message index where P^* sends (the same) BH_2 of the convincing session. We set each of them to $\perp$ if no such session exists. From the unforgeability of the signature scheme and the binding property of the commitment, we have the following claims.

Claim 2. *For every adversary A there exists a negligible function $\nu_1()$ such that for all $n \in \mathbf{N}$, $z \in \{0, 1\}^*$, the probability that the output of $\mathrm{HYB}_1^A(n, z)$ is not $\perp$ and any of $\mathcal{J}_1, \mathcal{J}_2$ or $\mathcal{J}_3$ is $\perp$ is at most $\nu_1(n)$.*

Claim 3. *For every adversary A there exists a negligible function $\nu_2()$ such that for all $n \in \mathbf{N}, z \in \{0,1\}^*$, the probability that the output of $\mathrm{HYB}_1^A(n,z)$ is not $\perp$, $\mathcal{J}_1, \mathcal{J}_2, \mathcal{J}_3 \neq \perp$ and $\mathcal{J}_1 \geq \mathcal{J}_2$ or $\mathcal{J}_2 > \mathcal{J}_3$ is at most $\nu_2(n)$.*

Before proving Claims 2 and 3, we prove soundness using these claims. Consider $B^O(1^n, \mathsf{pp})$ that internally incorporates P^* and begins an internal emulation by supplying the verifier messages internally and proceeds as follows:

1. B picks three integers i_1, i_2, i_3 at random such that $i_1 < i_2 < i_3$.
2. B selects keys $(vk, sk) \leftarrow \mathsf{Gen}(1^n)$. It then internally feeds P^* with $(\mathrm{BH}_1, vk, \mathsf{pp})$ where BH_1 is the first message of the WI-AOK proving language L^*. To generate the third message as the verifier, B^* first queries the oracle with the commitment c received in the second message of that session and obtains σ'. Then it generates a random string r and obtains a signature for $c|r, \sigma$ under key sk. B then feeds P^* with $\mathrm{BH}_3, r, \sigma, \sigma'$ where BH_3 is honestly generated. In this manner B continues with the emulation internally.
3. B continues the emulation until the partial transcript has i_1 messages. If this is not a second message of any session, it halts. Otherwise, it takes the commitment c as part of this message and forwards it to $V_{\mathbf{zk}}$ as the first message in the external execution. Upon receiving the challenge r from the external verifier, it forwards that challenge internally as part of the third message corresponding to the same session; it generates σ, σ' as before. It then continues the emulation until the partial transcript has i_2 messages. If this is not a second message of any session, it halts. Otherwise, let β be the partial transcript and α be its session number.
4. Next, it continues the emulation from β until the partial transcript has totally i_3 messages. If the last message is not the third message of session α it halts. Otherwise, let the partial transcript be $(\beta :: \beta_1)$ (where :: denotes concatenating transcripts). Now, it runs two random continuations from i_3 to completion and extracts a witness use in the WI-AOK using the special-sound property. Let the two transcripts be $(\beta :: \beta_1 :: \beta_{11})$ and $(\beta :: \beta_1 :: \beta_{12})$. If it fails to extract a fake witness internally then it halts. If it obtains a fake witness but the witness does not contain c, r from the previous step it halts. Otherwise, it takes p_1 from the witness and sends $\mathrm{COM}(p_1, \tau_1)$ where τ_1 is randomly chosen externally to $V_{\mathbf{zk}}$.
5. Upon receiving the challenge r' from $V_{\mathbf{zk}}$, B internally rewinds P^* to the prefix β. B starts a new continuation from this point and feeds r' as part of the third message in the current session. B then continues the internal emulation until the partial transcript $(\beta :: \beta_2)$ has i_3 messages. Once again B extracts the witness in the WI-AOK by emulating two random continuations to completion from $(\beta :: \beta_2)$, say $(\beta :: \beta_2 :: \beta_{21})$ and $(\beta :: \beta_2 :: \beta_{22})$. If c, r, p_1, r' are not part of the witness B aborts. Otherwise it takes p_2 from the witness and sends $\mathrm{COM}(p_2, \tau_2)$ where τ_2 is randomly chosen externally to $V_{\mathbf{zk}}$.
6. B stops the internal emulation and proceeds to complete the external execution with $V_{\mathbf{zk}}$ by using $(p_1, p_2, \tau_1, \tau_2)$ as the witness for the proof phase.

It follows from the soundness of the WI AOK and the way $\mathbf{R}_{L_3}$ is defined, that if B succeeds in extracting the fake witness that contains the appropriate previous messages, then, except with negligible probability, B succeeds in convincing V_{zk} in the external execution. It suffices to argue that B is able to achieve this with non-negligible probability. Recall that P^* succeeds in convincing a false statement to V with non-negligible probability, say $\frac{1}{p(n)}$.

By Claims 2 and 3, it holds for sufficiently large n that with probability at least $\frac{1}{p(n)} - \nu_1(n) - \nu_2(n)$ that P^* cheats and $\mathcal{J}_1, \mathcal{J}_2, \mathcal{J}_3 \neq \bot$ and $\mathcal{J}_1 < \mathcal{J}_2 < \mathcal{J}_3$ in $\mathrm{HYB}_1^{P^*}(n, z)$. Since there are only polynomially many sessions we can further assume that there exists a polynomial $p_1(n)$ and functions $i_1(), i_2(), i_3()$ such that, for sufficiently large n, with probability $\frac{1}{p_1(n)}$ over the experiment $\mathrm{HYB}_1^{P^*}(n, z)$, it holds that $\mathcal{J}_1 = i_1(n)$, $\mathcal{J}_2 = i_2(n)$ and $\mathcal{J}_3 = i_3(n)$. For a complete transcript β of an interaction with P^*, we say event $\mathsf{WO}(\beta)$ occurs if $\mathcal{J}_1(\beta) = i_1(n)$, $\mathcal{J}_2(\beta) = i_2(n)$ and $\mathcal{J}_3(\beta) = i_3(n)$ (for *well-ordered*).

We now analyze the success probability of B. We do this by analyzing the probability that B succeeds in each of the steps iteratively.

Event E_1: We say E_1 holds if $i_1 = i_1(n), i_2 = i_2(n)$ and $i_3 = i_3(n)$. Since there are only polynomially many sessions, this happens with polynomial probability, say $\frac{1}{p_2(n)}$.

Event E_2: We say that E_2 holds for a partial transcript β, i.e. $\mathsf{E}_2(\beta) = 1$, if β is of length i_2 and WO holds in random continuation from β with probability $\frac{1}{2p_1(n)}$. Since WO holds with probability $\frac{1}{p_1(n)}$, using an averaging argument, we can conclude that with probability at least $\frac{1}{2p_1(n)}$ over partial transcripts of length i_2, WO holds in a random continuation with probability at least $\frac{1}{2p_1(n)}$. So conditioned on E_1, $\mathsf{E}_2(\beta)$ holds with probability $\frac{1}{2p_1(n)}$ over β.

Event E_3: We say that E_3 holds for a partial transcript β, i.e. $\mathsf{E}_3(\beta) = 1$, if β is of length i_3 and WO holds in random continuation from β with probability $\frac{1}{4p_1(n)}$. We estimate the probability E_3 holds conditioned on E_2 and E_1. If E_1 and E_2 holds for transcript β, we know a random continuation from β yields a transcript where WO holds with probability at least $\frac{1}{2p_1(n)}$. So using another averaging argument, we get that, $\Pr[\mathsf{E}_3(\beta :: \beta_1)|\mathsf{E}_2(\beta) \wedge \mathsf{E}_1] \geq \frac{1}{4p_1(n)}$

B succeeds if it extracts the correct witness in Steps 4 and 5. More precisely, B will succeed except with negligible probability, if WO holds in all of $(\beta :: \beta_1 :: \beta_{11})$, $(\beta :: \beta_1 :: \beta_{12})$, $(\beta :: \beta_1 :: \beta_{21})$ and $(\beta :: \beta_1 :: \beta_{21})$ as the witness will be correct and the special-sound extractor will succeed. This probability can be written as

$$\Pr[B \text{ succeeds}] = \Pr\left[\mathsf{WO}(\beta :: \beta_1 :: \beta_{11}) \wedge \mathsf{WO}(\beta :: \beta_1 :: \beta_{12})\right.$$
$$\left. \wedge \mathsf{WO}(\beta :: \beta_1 :: \beta_{21}) \wedge \mathsf{WO}(\beta :: \beta_1 :: \beta_{22})\right] - 2\nu(n) \quad (1)$$

where $\nu(\cdot)$ is the probability that the special-sound extractor fails. From the description of the events, it holds that

$$\Pr[\mathsf{WO}(\beta :: \beta_1 :: \beta_{11})|\mathsf{E}_3(\beta :: \beta_1) \wedge \mathsf{E}_1] \geq \frac{1}{4p_1(n)}$$

$$\Pr[\mathsf{E}_3(\beta :: \beta_1)|\mathsf{E}_2(\beta) \wedge \mathsf{E}_1] \geq \frac{1}{4p_1(n)}$$

$$\Pr[\mathsf{E}_2(\beta)|\mathsf{E}_1] \geq \frac{1}{2p_1(n)}$$

$$\Pr[\mathsf{E}_1] \geq \frac{1}{p_2(n)}$$

And similar bounds hold for the other transcripts as well. Therefore, simplifying Equation 1, we get that

$$\Pr[B \text{ succeeds}] \geq \frac{1}{p_2(n)}\frac{1}{2p_1(n)}\left(\frac{1}{4p_1(n)}\right)^2\left(\frac{1}{4p_1(n)}\right)^4 - 2\nu(n)$$

which is non-negligible

We remark that the transformation works only for a constant-round protocol since B makes a guess for each round (i.e., i_1, i_2 and i_3) each correct only with polynomial probability.

It only remains to prove Claims 2 and 3. This on a high-level will follow from the binding property of the commitment and the unforgeability of the signature scheme.

Proof of Claim 2. Since the output of HYB$_1$ is not $\perp$, it immediately follows that $\mathcal{J}_3 \neq \perp$. We now show that P^* must have obtained the signature σ_1, σ_2 obtained from the witness by sending the commitment and receiving the corresponding random string with the signature in some session. Suppose not, then we can violate the unforgeability of the signature scheme by constructing an adversary C that receives a verification key vk as input conducts the HYB$_1$ experiment by supplying vk to P^* and forwarding all signing queries to the signing oracle. Finally upon extracting a fake witness, C simply outputs either $(c_0|r, \sigma_1)$ or $(c_1'|r', \sigma_2)$ which ever is valid.

Proof of Claim 3. Using the preceding argument, we can conclude that the signatures must be obtained before P^* convinces the verifier in some session, i.e. $\mathcal{J}_1 < \mathcal{J}_3$ and $\mathcal{J}_2 < \mathcal{J}_3$.[7] It only remains to argue that $\mathcal{J}_3 > \mathcal{J}_1 > \mathcal{J}_2$ does not happen. Assume for contradiction that with non-negligible probability $\mathcal{J}_1, \mathcal{J}_2, \mathcal{J}_3 \neq \perp$ and $\mathcal{J}_3 > \mathcal{J}_1 > \mathcal{J}_2$. This means that P^* was able to commit to a signature σ_1 as part of $p_1|\sigma_1$ in session $\mathcal{J}_2$ before it obtained the signature σ_1 from the verifier in session $\mathcal{J}_1$. We construct an adversary C that violates the collision-resistance property of the signature scheme.

C on input (n, vk) and oracle access to a signing oracle $\mathsf{Sign}_{sk}()$ first selects i_1, i_2 and i_3 at random. Then it internally incorporates $P^*(n, z)$ and begins an

---

[7] Consider C that proceeds as in Claim 2, but stops at a random session, extracts the witness and outputs the signature obtained from the witness.

internal emulation of an execution of P^* as follows. It forwards the verification-key vk internally to P^* as part of the first message and generates all the verifier messages honestly except the signatures corresponding to vk which it obtains by feeding the corresponding message to the signing oracle. C then runs the emulation until the partial transcript, say β, has i_2 messages. If this is not the second message of a session, C halts. Otherwise, it spawns two random continuation from β until the partial transcripts, say $(\beta :: \beta_1)$ and $(\beta :: \beta_2)$ of both threads has i_3 messages. If in either of the thread the current message is not the second message of a session C halts. Otherwise, it runs two random continuations from both $(\beta :: \beta_1)$ and $(\beta :: \beta_2)$ to obtain $(\beta :: \beta_1 :: \beta_{11})$, $(\beta :: \beta_1 :: \beta_{12})$, $(\beta :: \beta_1 :: \beta_{21})$ and $(\beta :: \beta_1 :: \beta_{21})$ and run the special-sound extractor of the WI-AOK protocol to obtain two witnesses. If the extractor succeeds in extracting a fake witness from both these sessions and σ_1 is the same in both these witnesses, then the message signed will be different with high-probability. This is because the message being signed has a random string r of length $O(n)$ and for two threads to have the same challenge is exponentially small, say $\nu_1(n)$. Therefore, by the soundness of the WI-AOK protocol we have two different messages with the same signature. C outputs them as a collision.

To argue that C succeeds with non-negligible probability we proceed exactly as in the previous argument. We know that with non-negligible probability, there exists $i_1(n), i_2(n), i_3(n)$ such that $\mathcal{J}_1 = i_1(n), \mathcal{J}_2 = i_2(n), \mathcal{J}_3 = i_3(n)$ and $\mathcal{J}_2 > \mathcal{J}_1 > \mathcal{J}_3$ with probability $\frac{1}{p_1(n)}$. Lets call this event WO as before. Define events $\mathsf{E}_1, \mathsf{E}_2$ and E_3 exactly as before. Following the same approach we can conclude that C succeeds with probability at least

$$\frac{1}{p_2(n)} \frac{1}{2p_1(n)} \left(\frac{1}{4p_1(n)}\right)^2 \left(\frac{1}{4p_1(n)}\right)^4 - 2\nu(n) - \nu_1(n)$$

which is non-negligible and thus we arrive at a contradiction.

Argument of Knowledge Since the $\mathcal{O}^{\mathsf{SIG}}$-oracle aided $\langle P_{\mathsf{zk}}, V_{\mathsf{zk}} \rangle$ protocol is also a argument of knowledge, from the proof of soundness, it holds that our 4-round protocol is also an argument of knowledge.

Zero Knowledge. Before we describe the simulator, we need the following definition of a valid SIG''-oracle similar to Definition 19.

Definition 21 (Valid SIG'' Oracle). *An oracle $\mathcal{O}''$ is a valid (SIG'', ℓ) oracle if there is a negligible $\mu(\cdot)$ such that for every $n \in N$, the following holds with probability $1 - \mu(n)$ over $\mathsf{pp}, O \leftarrow \mathcal{O}''(1^n)$: for every $m \in \{0,1\}^{\ell(n)}, O(m)$ returns $(\mathsf{BH}_2, c, r, \sigma, \tau)$ such that $\mathsf{Ver}_{vk}(c|r, \sigma) = 1, c = \mathrm{COM}(m, \tau)$ and r is the second string in the tuple output by V^* when fed BH_2, c with probability at least $1 - \mu(n)$.*

Consider some malicious (w.l.o.g. deterministic) verifier $\tilde{V}^*$ for (P, V) of size $T_{\tilde{V}^*}$. We remark that while the simulator for the resettably-sound ZK protocol in [10] had one signing slot, here we have a slot that serves as a signing slot for two different keys sk and sk'. We use two signing keys for simplicity. We use two keys for simplicity. We construct a simulator S for $\tilde{V}^*$ that starts simulating $(P, \tilde{V}^*)$ until it receives BH_1 and two verification keys vk, vk'. Let V^* be the "residual" verifier after the first message is sent. It then proceeds as follows.

1. S prepares a valid $(\mathsf{SIG}', 2n)$ oracle $\mathcal{O}'$ and $(\mathsf{SIG}'', 2n)$ oracle $\mathcal{O}''$ by rewinding V^* and using the second and third message of the protocol as a Signing Slot for both sk and sk'. This step is essentially the same as what the simulator does in the protocol presented in [10] which in turn is inspired by Goldreich-Kahan [15]),

2. S will convince V^* in the WI-AOK using the second witness. Towards this, S will first use oracle $\mathcal{O}'$ to produce a Sig-com tree for $C = \mathsf{ECC}(\Pi)$ where $\Pi = V^*$. Let d and l_λ be the depth and root of the Sig-com tree. Using the oracle $\mathcal{O}''$, S obtains (c_0, r, σ_1, τ) where $(c_0, r, vk') \in \mathbf{R}_{L_1}$ and $\mathsf{Ver}_{vk}(c_0|r, \sigma_1) = 1$.

3. S then generates the first prover message p_1 using the witness for $(c_0, r, vk') \in \mathbf{R}_{L_1}$. Using the oracle $\mathcal{O}''$ again, S generates $c_1, r', \sigma_2, \tau_1$ such that $c_1 = \mathsf{COM}(p_1|\sigma_1, \tau_1)$ and $\mathsf{Ver}_{vk}(c_1|r', \sigma_2) = 1$. S now generates the second prover message p_2 for the UA using r' as the challenge message for the UA.

4. Finally, S rewinds V^* to the top and completes the interaction with V^* by using $\langle c_0, r, c_1, r', p_1, p_2, \tau_1, \sigma_1, \sigma_2 \rangle$ as the second witness in the WI-AOK.

The correctness of S follows essentially using the same proof as in [10]. First, we argue that S can prepare valid oracles for both the keys. Given valid oracles, S obtains a valid second witness for the WI-AOK. It then runs V^* in a straight-line manner by generating messages for the WI-AOK protocol using the second witness and all the other messages as the honest prover. Indistinguishability of the output of the simulator follows directly from the witness-indistinguishability property of the WI-AOK protocol. It only remains to argue that S can prepare valid $\mathcal{O}^{\mathsf{SIG}'}$ and $\mathcal{O}^{\mathsf{SIG}''}$ oracles. We remark that the approach we take is similar to [10], with the exception that the preamble phase of the oracle preparation is executed only once for both oracles. First S executes the following preamble.

– S sends c, BH_2 to V^* where $c = \mathsf{COM}(0^{2n}; \tau)$ with uniform τ and BH_2 is a random dummy second message of the Blum-Hamiltonicity protocol[8], and then receives $\mathsf{BH}_3, r, \sigma, \sigma'$ from V^*. If σ is not a valid signature of $c|r$ under verification vk or σ' is not a valid signature of c under vk', then the simulation halts immediately and outputs the transcript up to that point.

– S repetitively queries V^* with fresh commitments $\mathsf{COM}(0^{2n}; \tau)$ at the Signing Slot along with dummy BH_2 messages until it collects $2n$ valid signatures. Let t be the number of queries $\tilde{S}$ makes.

Preparing $\mathcal{O}^{\mathsf{SIG}''}$ Oracle: Define $\mathcal{O}''$ that outputs $\mathsf{pp} = vk$, and an oracle O that on input a message $m \in \{0, 1\}^{2n}$, proceeds as follows: O repetitively queries V^* at the Signing Slot with fresh commitments $c_m = \mathsf{COM}(m; \tau)$ with dummy BH_2 messages for at most t times. If V^* ever replies $\mathsf{BH}_3, r, \sigma, \sigma'$ where $\mathsf{Ver}_{vk}(c_m|r, \sigma) = 1$, then O outputs $(\mathsf{BH}_2, c_m, r, \sigma, \tau)$. Otherwise, O returns $\bot$.

---

[8] Recall that, in the second message of the Blum-Hamiltonicity protocol, the prover sends a set of commitments. Hence, to generate a dummy message, the simulator can simply commit to the all 0's string.

Preparing $\mathcal{O}^{\mathsf{SIG}'}$ *Oracle:* Define $\mathcal{O}'$ that outputs $\mathsf{pp} = vk'$, and an oracle O that on input a message $m \in \{0,1\}^{2n}$, proceeds as follows: O repetitively queries V^* at the Signing Slot with fresh commitments $\mathrm{COM}(m;\tau)$ for at most t times. If V^* ever replies a valid signature σ' for $\mathrm{COM}(m,\tau)$, then O outputs (σ',τ). Otherwise, O returns $\perp$.

We now analyze the running time. If $t \geq 2^{n/2}$, then S aborts. To analyze this part of the simulator S, we introduce some notation. Let $p(m)$ be the probability that V^* on query BH_2, c_m where BH_2 is the specific second message of the Blum-Hamiltonicity protocol and $c_m = \mathrm{COM}(m,\tau)$ of $m \in \{0,1\}^{2n}$ a random commitment returns a valid signature of $c_m|r$ under sk where r is part of V^*'s output when fed BH_2, c_m and a valid signature of c_m under sk'. Let $p = p(0^{2n})$.

We first show that S runs in expected polynomial time. To start, note that S aborts at the end of the Signature Slot with probability $1 - p$, and in this case, S runs in polynomial time. With probability p, S continues to invoke a strictly polynomial-time simulator S for the residual V^*, which has size bounded by $T_{\tilde{V}^*}$. Thus, S runs in some $T = \mathtt{poly}(T_{\tilde{V}^*})$ time and makes at most T queries to both its oracles, which in turn runs in time $t \cdot \mathtt{poly}(\mathtt{n})$ to answer each query. Also note that S runs in time at most 2^n, since S aborts when $t \geq 2^{n/2}$. Now, we claim that $t \leq 10n/p$ with probability at least $1 - 2^{-n}$, and thus the expected running time of S is at most

$$(1 - p) \cdot \mathtt{poly}(\mathtt{n}) + \mathtt{p} \cdot \mathtt{T} \cdot (10\mathtt{n}/\mathtt{p}) \cdot \mathtt{poly}(\mathtt{n}) + 2^{-\mathtt{n}} \cdot 2^{\mathtt{n}} \leq \mathtt{poly}(T_{\tilde{V}^*}, \mathtt{n}).$$

To see that $t \leq 10n/p$ with overwhelming probability, let $X_1, \ldots, X_{10n/p}$ be i.i.d. indicator variables on the event that V^* returns a valid signature for the message 0^{2n}, and note that $t \leq 10n/p$ implies $\sum_i X_i \leq 2n$, which by a standard Chernoff bound, can only happen with probability at most 2^{-n}.

Finally, we argue indistinguishability. First, the computational hiding property of COM implies that there exists some negligible $\nu(\cdot)$ such that $|p(m) - p| \leq \nu(n)$ for every $m \in \{0,1\}^{2n}$. Now we consider two cases. If $p \leq 2\nu$, then the indistinguishability trivially holds since the interaction aborts at the end of the Signature Slot (in this case, the view is perfectly simulated) with all but negligible probability. On the other hand, if $p \geq 2\nu$, we show that $\mathcal{O}''$ generated by S is a valid $(\mathsf{SIG}'', 2n)$ oracle for SIG'' and $\mathcal{O}'$ generated by S is a valid $(\mathsf{SIG}', 2n)$ oracle for SIG' with overwhelming probability, and thus the indistinguishability of S follows by the indistinguishability of S.

To see that $\mathcal{O}''$ is a valid $(\mathsf{SIG}'', 2n)$ oracle for SIG'' with overwhelming probability, note again by a Chernoff bound that $n/p \leq t \leq 2^{n/2}$ with probability at least $1 - 2^{-\Omega(n)}$. In this case, for every $m \in \{0,1\}^{2n}$, $p(m) \geq p - \nu \geq p/2$ implies that $t \geq n/2p(m)$, and thus $O(m)$ learns a valid signature of $\mathrm{COM}(m;\tau)$ from V^* with probability at least $1 - 2^{-\Omega(n)}$. A similar argument establishes that $\mathcal{O}'$ is a valid $(\mathsf{SIG}', 2n)$ oracle for SIG' with overwhelming probability. This concludes the proof of correctness.

Acknowledgment. Ostrovsky's research is supported in part by NSF grants CCF-0916574; IIS-1065276; CCF-1016540; CNS-1118126; CNS-1136174;

US-Israel BSF grant 2008411, OKAWA Foundation Research Award, IBM Faculty Research Award, Xerox Faculty Research Award, B. John Garrick Foundation Award, Teradata Research Award, Lockheed-Martin Corporation Research Award, Defense Advanced Research Projects Agency through the U.S. Office of Naval Research under Contract N00014-11-1-0392.

Pass is supported in part by a Alfred P. Sloan Fellowship, Microsoft New Faculty Fellowship, NSF Award CNS-1217821, NSF CAREER Award CCF-0746990, NSF Award CCF-1214844, AFOSR YIP Award FA9550-10-1-0093, and DARPA and AFRL under contract FA8750-11-2- 0211.

Chung is supported by NSF Award CNS-1217821, NSF Award CCF-1214844 and Pass' Sloan Fellowship.

Visconti's research is supported in part by the MIUR Project PRIN "GenData 2020".

The views and conclusions contained in this document are those of the authors and should not be interpreted as representing the official policies or positions, either expressed or implied, of the Department of Defense, the Defense Advanced Research Projects Agency or the U.S. Government.

References

1. Barak, B., Goldreich, O.: Universal arguments and their applications. In: Computational Complexity, pp. 162–171 (2002)
2. Barak, B., Goldreich, O., Goldwasser, S., Lindell, Y.: Resettably-sound zero-knowledge and its applications. In: FOCS 2002, pp. 116–125 (2001)
3. Bellare, M., Goldreich, O.: On defining proofs of knowledge. In: Brickell, E.F. (ed.) CRYPTO 1992. LNCS, vol. 740, pp. 390–420. Springer, Heidelberg (1993)
4. Bellare, M., Jakobsson, M., Yung, M.: Round-optimal zero-knowledge arguments based on any one-way function. In: Fumy, W. (ed.) EUROCRYPT 1997. LNCS, vol. 1233, pp. 280–305. Springer, Heidelberg (1997)
5. Bitansky, N., Paneth, O.: From the impossibility of obfuscation to a new non-black-box simulation technique. In: FOCS (2012)
6. Bitansky, N., Paneth, O.: On the impossibility of approximate obfuscation and applications to resettable cryptography. In: STOC (2013)
7. Blum, M.: How to prove a theorem so no one else can claim it. In: Proc. of the International Congress of Mathematicians, pp. 1444–1451 (1986)
8. Brakerski, Z., Gentry, C., Vaikuntanathan, V.: (leveled) fully homomorphic encryption without bootstrapping. In: ITCS, pp. 309–325. ACM (2012)
9. Chung, K.M., Ostrovsky, R., Pass, R., Visconti, I.: Simultaneous resettability from one-way functions. In: 54th Annual IEEE Symposium on Foundations of Computer Science, FOCS 2013, pp. 60–69. IEEE Computer Society (2013)
10. Chung, K.M., Pass, R., Seth, K.: Non-black-box simulation from one-way functions and applications to resettable security. In: STOC. ACM (2013)
11. Di Crescenzo, G., Persiano, G., Visconti, I.: Improved setup assumptions for 3-round resettable zero knowledge. In: Lee, P.J. (ed.) ASIACRYPT 2004. LNCS, vol. 3329, pp. 530–544. Springer, Heidelberg (2004)
12. Feige, U., Shamir, A.: Witness indistinguishable and witness hiding protocols. In: STOC 1990, pp. 416–426 (1990)

13. Gentry, C.: Fully homomorphic encryption using ideal lattices. In: STOC, pp. 169–178. ACM (2009)
14. Goldreich, O.: Foundations of Cryptography — Basic Tools. Cambridge University Press (2001)
15. Goldreich, O., Kahan, A.: How to construct constant-round zero-knowledge proof systems for NP. Journal of Cryptology 9(3), 167–190 (1996)
16. Goldwasser, S., Micali, S.: Probabilistic encryption. J. Comput. Syst. Sci. 28(2), 270–299 (1984)
17. Goldwasser, S., Micali, S., Rackoff, C.: The knowledge complexity of interactive proof-systems. In: STOC 1985, pp. 291–304. ACM (1985), http://doi.acm.org/10.1145/22145.22178
18. Goldwasser, S., Micali, S., Rackoff, C.: The knowledge complexity of interactive proof systems. SIAM J. Comput. 18(1), 186–208 (1989)
19. Goyal, V., Jain, A., Ostrovsky, R., Richelson, S., Visconti, I.: Constant-round concurrent zero knowledge in the bounded player model. In: Sako, K., Sarkar, P. (eds.) ASIACRYPT 2013, Part I. LNCS, vol. 8269, pp. 21–40. Springer, Heidelberg (2013)
20. Lin, H., Pass, R.: Constant-round non-malleable commitments from any one-way function. In: STOC, pp. 705–714 (2011)
21. Merkle, R.C.: A digital signature based on a conventional encryption function. In: Pomerance, C. (ed.) CRYPTO 1987. LNCS, vol. 293, pp. 369–378. Springer, Heidelberg (1988)
22. Micali, S.: Computationally sound proofs. SIAM Journal on Computing 30(4), 1253–1298 (2000)
23. Naor, M., Yung, M.: Universal one-way hash functions and their cryptographic applications. In: STOC 1989, pp. 33–43 (1989)
24. Ostrovsky, R., Visconti, I.: Simultaneous resettability from collision resistance. Electronic Colloquium on Computational Complexity (ECCC) 19, 164 (2012)
25. Ostrovsky, R., Wigderson, A.: One-way fuctions are essential for non-trivial zero-knowledge. In: ISTCS, pp. 3–17 (1993)
26. Pass, R., Rosen, A.: New and improved constructions of non-malleable cryptographic protocols. In: STOC 2005, pp. 533–542 (2005)
27. Pass, R., Tseng, W.L.D., Wikström, D.: On the composition of public-coin zero-knowledge protocols. SIAM J. Comput. 40(6), 1529–1553 (2011)
28. Rompel, J.: One-way functions are necessary and sufficient for secure signatures (1990)

Can Optimally-Fair Coin Tossing
Be Based on One-Way Functions?

Dana Dachman-Soled[1], Mohammad Mahmoody[2], and Tal Malkin[3]

[1] University of Maryland
danadach@ece.umd.edu
[2] University of Virginia
mohammad@cs.virginia.edu
[3] Columbia University and Bar-Ilan University
tal@cs.columbia.edu

Abstract. Coin tossing is a basic cryptographic task that allows two distrustful parties to obtain an unbiased random bit in a way that neither party can bias the output by deviating from the protocol or halting the execution. Cleve [STOC'86] showed that in any r round coin tossing protocol one of the parties can bias the output by $\Omega(1/r)$ through a "fail-stop" attack; namely, they simply execute the protocol honestly and halt at some chosen point. In addition, relying on an earlier work of Blum [COMPCON'82], Cleve presented an r-round protocol based on one-way functions that was resilient to bias at most $O(1/\sqrt{r})$. Cleve's work left open whether "'optimally-fair'" coin tossing (i.e. achieving bias $O(1/r)$ in r rounds) is possible. Recently Moran, Naor, and Segev [TCC'09] showed how to construct optimally-fair coin tossing based on oblivious transfer, however, it was left open to find the *minimal* assumptions necessary for optimally-fair coin tossing. The work of Dachman-Soled et al. [TCC'11] took a step toward answering this question by showing that any black-box construction of optimally-fair coin tossing based on a one-way functions with n-bit input and output needs $\Omega(n/\log n)$ rounds.

In this work we take another step towards understanding the complexity of optimally-fair coin-tossing by showing that this task (with an arbitrary number of rounds) cannot be based on one-way functions in a black-box way, as long as the protocol is "'oblivious'" to the implementation of the one-way function. Namely, we consider a natural class of black-box constructions based on one-way functions, called *function oblivious*, in which the output of the protocol does not depend on the specific implementation of the one-way function and only depends on the randomness of the parties. Other than being a natural notion on its own, the known coin tossing protocols of Blum and Cleve (both based on one-way functions) are indeed function oblivious. Thus, we believe our lower bound for function-oblivious constructions is a meaningful step towards resolving the fundamental open question of the complexity of optimally-fair coin tossing.

Keywords: Coin-Tossing, One-Way Functions, Black-Box Separations.

1 Introduction

In this work, we address the fundamental problem of secure, two-party coin-tossing, where two mutually distrustful parties wish to generate a common random bit. A secure

Y. Lindell (Ed.): TCC 2014, LNCS 8349, pp. 217–239, 2014.

coin-tossing scheme has the following complementary properties: (1) Security—even if one of the parties deviates arbitrarily from the protocol, the output bit of the honest party should be almost completely unbiased (namely be equal to 1 with probability that is at most negligibly far from $1/2$) and (2) Correctness—when both parties follow the protocol they are guaranteed to output the same random bit. Unfortunately, a classic result by Cleve [Cle86] shows that even if the party which deviates from the protocol misbehaves only by choosing whether or not to abort early (this is known as a fail-stop adversary), then secure coin-tossing cannot be achieved. In particular, Cleve proved that for any coin tossing protocol running for $\hat{r}$ rounds, there exists an efficient fail-stop adversary that can bias the output bit of the honest party by at least $\Omega(1/\hat{r})$.

It turns out that in a weaker model we can, in fact, construct secure coin-tossing protocols. An early result by Blum [Blu82] uses one-way functions (OWF) to construct a *weak* coin tossing protocol where no party can increase the probability of 0 or 1 by more than negligible by deviating from the protocol or halting. In a weak coin tossing protocol whenever a party aborts, the other party is not required output anything. Weak coin tossing can be useful for scenarios where Alice and Bob each have a preferred outcome in mind (e.g., Alice wants 0 and Bob wants 1) simply because if any party aborts the other one can take their desired outcome as the output. However, note that the weak coin tossing definition doesn't preclude the possibility that a malicious party can cause the output to always be either 0 or abort; indeed this is the case for Blum's protocol (where a malicious party can discover first the emerging output and then choose whether to abort or continue). In contrast, in a *strong* coin tossing protocol (which is the focus of our work), the protocol always requires an output. A strong coin tossing protocol with bias at most δ is one where each honest party always announces an output (even if the other party aborted), and yet no malicious party can bias the honest party's output (in any direction) by more than δ. The weak coin tossing protocol of Blum was used as a building block by Cleve [Cle86] to construct a *strong* coin tossing protocol that (for any polynomial $\hat{r}$) runs for $\hat{r}$ rounds and for which no efficient adversary can bias the output bit by more than $O(1/\sqrt{\hat{r}})$. In our work, whenever not explicitly mentioned, we are referring to *strong* coin tossing protocols.

The question of closing the gap between this best known upper bound ($O(1/\sqrt{\hat{r}})$ based on OWF) and lower bound ($\Omega(1/\hat{r})$ regardless of any assumption) remained unresolved for more than two decades. A few years ago, the gap was closed by Moran et al. [MNS09] who constructed a protocol for coin tossing whose bias matches the lower-bound of [Cle86]. Specifically, for any $\hat{r}$ they constructed an $O(\hat{r})$-round protocol with the property that no efficient adversary can bias the output by more than $O(1/\hat{r})$. Thus, they demonstrated that the $O(1/\hat{r})$ lower-bound is tight. We call a protocol which achieves bias $O(1/\hat{r})$ *optimally-fair*, because no protocol can achieve asymptotically-lower bias. The protocol of [MNS09], however, uses general secure computation and thus requires the strong assumption that protocols for oblivious transfer exist. In contrast, the coin tossing protocol of Blum [Blu82] and the protocol of [Cle86] achieving bias of $O(1/\sqrt{\hat{r}})$ can be constructed from any one-way function, and in fact, rely only on the existence of a commitment scheme. This leads us to our main question:

Can optimally-fair coin tossing be based on one-way functions?

This question was already asked by Moran et al. [MNS09] as a challenging open problem. Indeed, the question of whether one-way functions suffice for optimally-fair coin-tossing seems to be a difficult problem and remains open, despite much effort. A partial answer to the main question above was presented in the work of Dachman-Soled et al. [DSLMM11]. Informally, they show that if C is a black-box construction of optimally-fair coin tossing based on one-way functions with input and output length n, then the number of rounds of interaction in C is at least $\Omega(n/\log n)$. Thus, they rule out such black-box constructions with a "small" number of rounds. However, their results say nothing about constructions with a higher number of rounds. For example, they do not rule out the possibility of constructing coin-tossing protocols from one-way functions of input size n, which have $\hat{r} = n^3$ number of rounds and for which no efficient adversary can bias the output by more than $1/n^3 = 1/\hat{r}$.

In this work, we make an important step towards answering our main question. In particular, we manage to remove the limitation on the round complexity in the impossibility result of [DSLMM11]. Indeed, we consider protocols with an arbitrary polynomial number of rounds: $\hat{r} = \text{poly}(n)$. However, we introduce another limitation: our impossibility results only rule out protocols which posses the following property.

Definition 1 (Function-Obliviousness:). *A coin-tossing protocol* $C^f = \langle A^f, B^f \rangle$ *based on one-way functions is called* function-oblivious *if the outcome of the coin tossing protocol* $\langle A^f(r_A), B^f(r_B) \rangle$, *when both parties are honest, depends only on the random tapes* r_A, r_B *of the two parties and not on the choice of one-way function* f.

Function-obliviousness captures the intuition that the one-way function f is being used only to achieve *security* for the coin-tossing protocol but does not affect *correctness*. In this work, we rule out (fully) black-box constructions of optimally-fair coin-tossing protocols which are function-oblivious from one-way functions.

Theorem 1 (Main Theorem, Informal). *There is no (fully) black-box and function-oblivious* construction of optimally-fair coin-tossing protocols from one-way functions.

Our result is incomparable to that of [DSLMM11]: we restrict ourselves to function-oblivious protocols but handle protocols with *arbitrary* polynomial number of rounds.

We believe that function-obliviousness is a natural assumption on coin-tossing protocols. Indeed, the known one-way-function based coin tossing protocols of Blum [Blu82] and Cleve [Cle86], as well as any other protocols based only on commitment schemes, are function-oblivious. The notion of function-obliviousness as defined in Definition 1 can be directly generalized to other pairs of primitives as well, and so understanding the limits of oblivious black-box constructions could be considered as a first step towards understanding the full power of black-box constructions. Thus, introducing the notion of oblivious black-box constructions, as a natural form of black-box constructions, is a conceptual contribution of our work. On a technical level, to deal with function-oblivious protocols, we introduce several new techniques which were not needed/applicable in the case of black-box $O(n/\log n)$-round protocols. These techniques also may be of independent interest and indicate that we are making progress on a fundamental question by considering this class of protocols. Thus, we believe that our partial negative result is meaningful and improves our understanding of the relative complexity of one-way functions and optimally-fair coin-tossing protocols.

An important remaining open question is whether function-obliviousness is necessary for our result, or we can completely rule out any black-box construction of optimally-fair coin tossing from one-way functions. Our results together with those of [DSLMM11] indicate that if any such construction exists, it must have many $(\omega(n/\log n))$ rounds, and must use the one-way function in a novel way, not only for commitment but to determine the coin toss outcome even when both parties are honest.

1.1 Black-Box Separations

One of the main goals of modern cryptography has been to identify the minimal assumptions necessary to construct secure cryptographic primitives. For example, [Yao82, GM84, Rom90, HILL99, GGM86, LR88, IL89, NY89, Nao91] have shown that private key encryption, pseudorandom generators, pseudorandom functions and permutations, bit commitment, and digital signatures exist if and only if one-way functions exist. On the other hand, some cryptographic primitives such as public key encryption, oblivious transfer, and key agreement are not known to be equivalent to one way functions. Thus, it is natural to ask whether the existence of one-way functions implies these primitives. However, it seems unclear how to formalize such a question; since it is widely believed that both one-way functions and public key encryption exist, this would imply in a trivial logical sense that the existence of one-way functions implies the existence of public key encryption. Thus, we can only hope to rule out restricted types of constructions that are commonly used to prove implications in cryptography. Impagliazzo and Rudich [IR89] were the first to develop a framework and techniques to rule out the existence of an important class of reductions between primitives known as black-box reductions. Intuitively, this is a reduction where the primitive is treated as an oracle or a "black-box". There are actually several flavors of black-box reductions (fully black-box, semi black-box and weakly black-box [RTV04]). In our work, we only deal with fully black-box reduction, and so we will focus on this notion here.

Informally, a fully black-box reduction from a primitive $\mathcal{Q}$ to a primitive $\mathcal{P}$ is a pair of oracle PPT machines $(G; S)$ such that the following two properties hold:

Correctness: For every implementation f of primitive $\mathcal{P}$, $g = G^f$ implements $\mathcal{Q}$.
Security: For every implementation f of primitive $\mathcal{P}$, and every adversary $\mathcal{A}$, if $\mathcal{A}$ breaks G^f (as an implementation of $\mathcal{Q}$) then $S^{\mathcal{A};f}$ breaks f.

We remark that an implementation of a primitive is any specific scheme that meets the syntactical requirements of that primitive (e.g., an implementation of a public-key encryption scheme provides samplability of key pairs, encryption with the public-key, and decryption with the private key). Correctness thus states that when G is given oracle access to any valid implementation of $\mathcal{P}$, the result is a valid implementation of $\mathcal{Q}$. Furthermore, security states that any adversary breaking G^f yields an adversary breaking f. The reduction here is *fully* black-box [RTV04] in the sense that the adversary S breaking f uses $\mathcal{A}$ in a black-box manner.

Separation from One-Way Functions. A common technique to separate a cryptographic primitive $\mathcal{P}$ from one-way functions is to show that any implementation of $\mathcal{P}$ in the random oracle can be broken by an attacker that asks "a few" (more specifically $2^{o(n)}$) queries to the random oracle (e.g. see [BM07] or [DSLMM11]). The reason, roughly speaking, is that if a $2^{o(n)}$ attacker Adv exists, the security reduction could turn Adv into a $2^{o(n)}$-query attack to invert the random oracle, which is not possible [IR89, GT00].[1] We will also take this approach in this work.

1.2 Related Work

Cleve and Impagliazzo [CI93] showed that the bias $O(1/\sqrt{r})$ is optimal when the attacker is *computationally unbounded*, and so their result does not resolve our main question.[2] However, using the result of [CI93], Dachman-Soled et al. [DSLMM11] gave a partial answer to the question of whether optimally-fair coin-tossing can be constructed in a black-box manner from one-way functions. As mentioned previously, they showed that if C is a construction of optimally-fair coin tossing based on one-way functions with input and output length n, then the number of rounds of interaction in C is at least $\Omega(n/\log n)$. More specifically, [DSLMM11] shows how to "compile out" the random oracle from the coin-tossing protocol by asking $(\operatorname{poly}(n))^{\hat{r}}$ oracle queries by the parties where $\hat{r}$ is the number of rounds of the protocol. Note that whenever $\hat{r} = o(n/\log n)$, the number of queries asked by the parties will be $(\operatorname{poly}(n))^{\hat{r}} = 2^{o(n)}$, and using the result of [CI93] for the no-oracle protocols one obtains a $2^{o(n)}$-query attacker for one of the parties A or B in the with-oracle protocol which, as explained above, leads to a black-box separation. Unfortunately, the techniques of [DSLMM11] do not seem to extend to the case where $\hat{r} = \omega(n/\log n)$ since in this case, the adversarial strategy will require $(\operatorname{poly}(n))^{\omega(n/\log n)} = 2^{\omega(n)}$ number of queries which is in fact enough to successfully invert the random oracle and does not lead to contradiction.

We also mention two other works, which deal with seemingly unrelated problems to ours, but leverage similar techniques. The first work of Haitner et al. [HOZ13] considers the question of constructing protocols for semi-honest no-input two-party computation in the random oracle model. They show that any semi-honest no-input two-party functionality which can be realized in the random oracle model, is trivial in the sense that, essentially, it can also be realized in the information-theoretic semi-honest setting with no random oracle. Note, however, that our coin-tossing in the semi-honest setting is trivial and our setting deals with malicious adversaries, so the result of [HOZ13] does not apply to our case. Mahmoody et al. [MMP13] consider semi-honest, deterministic functionalities with polynomial-sized domains and show that any such functionality which can be realized in the random oracle model is "trivial" in the same sense as above. As in our work, both of the above works utilize the "Eve" algorithm of [BM09] and rely on its specific properties (as described in Lemma 1). Moreover, some of the techniques in the work of [MMP13], where one of the players resamples a fake view and proceeds to compute using this fake view, are similar to our techniques. See the Technical Overview Section (Section 1.3) for additional details.

---

[1] Note that this technique works even if the attacker is not efficient.

[2] Although, as we will see, we use the results and approach of [CI93] as a starting point.

1.3 Technical Overview

We consider two-party coin tossing protocols $C = \langle A, B \rangle$ with $\hat{r}$ rounds (i.e., $2\hat{r}$ messages). Let C denote the outcome of coin tossing protocol $C = \langle A, B \rangle$ and let T_j denote the transcript of the protocol immediately after message j is sent. Moreover, since we consider the setting where one party may abort early, we denote by C_j the output of the other party, when one party aborts *before* sending the j-th message.

Cleve and Impagliazzo [CI93] showed that for any coin tossing protocol we have that with probability at least $1/5$ over the choice of random tapes of the parties, there is some point in the execution of the protocol that $|\mathbf{E}[C \mid T_j] - \mathbf{E}[C \mid T_{j+1}]| \geq \Omega(1/\sqrt{\hat{r}})$. Moreover, in [CI93], it was observed that if for some x, y, we have that an x-fraction of the executions of the coin-tossing protocol with uniformly chosen random tapes for the two parties reach a point where $|\mathbf{E}[C \mid T_j] - \mathbf{E}[C_j \mid T_j]| \geq y$ then the party who sends the mesasge j has a strategy for biasing the output towards either 0 or 1 by $\Omega(x \cdot y)$ by aborting before sending the j-th message when the above event occurs. Thus, the fact that with probability $1/5$ there is some point in the execution such that $|\mathbf{E}[C|T_j] - \mathbf{E}[C|T_{j+1}]| \geq \Omega(1/\sqrt{\hat{r}})$ immediately implies a strategy for either A or B for imposing bias $\Omega(1/\sqrt{\hat{r}})$.

In our work, we extend the [CI93] observation in two ways. First, we allow a party to condition not just on the current transcript, but on its entire view and abort before sending message j when $|\mathbf{E}[C \mid V_{A,j}] - \mathbf{E}[C_j \mid V_{A,j}]| \geq y$, where A is the party sending the j-th message and $V_{A,j}$ denotes her partial view right before sending the j-th message. Additionally, we allow a party to abort immediately *after* sending a message. More specifically, we allow a party to abort immediately after sending message j when $|\mathbf{E}[C \mid T_j] - \mathbf{E}[C_{j+2} \mid T_j]| \geq y$. Although this is technically equivalent to waiting to get the $(j+1)$-st message and aborting immediately after, it will be conceptually helpful to think of the party as aborting immediately after sending the j-th message. In what follows, we refer to the above described strategies as "[CI93]-type strategies". We do not consider our extensions of the [CI93] strategies as our main technical contribution, but we consider tham as useful tools for our proofs.

In our setting, we would like to apply the result of [CI93] in the random oracle model (see Section 1.1). The reason is that, it is well-known that, roughly speaking, (even inefficient) attacks in the random oracle model imply black-box separations from one-way functions. One might might correctly say that it is in fact *impossible* to break a coin-tossing protocol in the random oracle model through a fail-stop attack, simply because the parties can trivially use oracle's answer to a fixed query as their output. However, recall that: (1) our goal is to obtain black-box separations from one-way functions and the mentioned trivial protocol does not work when the random oracle is substituted with an actual one-way function, and (2) we are in fact focusing on function-oblivious protocols that prevent using the random oracle for obtaining the output.

Unfortunately, a straightforward implementation of [CI93] in the random oracle model (where expectations are taken also over the choice of oracle) fails due to the fact that in order for the [CI93] techniques to go through, it must be the case that $\mathbf{E}[C_j \mid T_{j-1}] = \mathbf{E}[C_j \mid T_j]$ (or at least that $|\mathbf{E}[C_j \mid T_{j-1}] - \mathbf{E}[C_j \mid T_j]| = o(1/\sqrt{\hat{r}})$) in all rounds. However, due to dependencies between parties' views created by the random oracle (in addition to the dependencies created by the transcript) it may in fact be the case that $|\mathbf{E}[C_j \mid T_{j-1}] - \mathbf{E}[C_j \mid T_j]| = \Omega(1/\sqrt{\hat{r}})$ in the random oracle model. I.e., the distribution over views of B till end of round j, denoted by $V_{B,j}$, conditioned on T_{j-1} may be very far from the the distribution over $V_{B,j}$ conditioned on T_j. This is due to the fact oracle answers received by A during the computation of T_j can affect the distribution over views of B even though A sends the j'th message.

A natural approach to solving the above problem, would be to leverage the results of [IR89, BM09] on finding so-called "intersection queries" in 2-party protocols. Intersection queries are queries made by both parties A, B during an execution of a protocol in the random oracle model. Intuitively, it is these intersection qeries that cause dependencies between the views of A and B. Moreover, in [IR89, BM09], it was shown that an eavesdropping adversary "Eve" can with high probability find all these intersection queries made by A and B, while making only a polynomial number of queries to the random oracle. Intuitively, one could hope that by running the "Eve" protocol of [IR89, BM09] after each pass of the protocol (which we call running the Eve protocol alongside the main protocol) these intersection queries could be found before they are made, thus eliminating dependencies between the views of A and B. It turns out that there is a subtle problem here: In order for [CI93] techniques to go through, we must prevent intersection queries even between Eve queries made alongside the j-th message (sent by A) and private queries that were made previously by B. Unfortunately, this property is not guaranteed by the Eve algorithm of [IR89, BM09]. However, [DSLMM11] showed that the Eve algorithm can be modified (becoming far less efficient) to guarantee that the above does not occur. Using such (inefficient) Eve [DSLMM11] still managed to rule out optimally-fair coin-tossing protocols with $O(n/\log n)$ number of rounds, where n is the input-output length of the one-way function. In this work, we shall find a different approach that allows us to deal with an arbitrary polynomial number of rounds.

Our Approach. In the following, $\mathcal{D}$ denotes the distribution over views of A, B running the coin-tossing protocol $\mathcal{C}$ with uniformly random coins. Additionally, for joint random variables X, Y, we denote by $X \mid Y$ the distribution over X drawn from $\mathcal{D}$, conditioned on Y. M_i denotes the i-th message of the coin-tossing protocol and T_i denotes the transcript which includes both the messages $M_1, \ldots, M_i$ of the protocol $\mathcal{C}$ as well as Eve$_i$, the information "Eve" has learned by making her queries alongside in the first i messages. Finally, the partial view of a party after the i-th message is sent, denoted by $V_{A,i}$ or $V_{B,i}$ includes its random tape r_A or r_B, its queries to the oracle and the responses, as well as the transcript $M_1, \ldots, M_i$ of the $\mathcal{C}$ protocol.

We consider the "middle value": $MV = \mathbf{E}_{V_{B,j}, \text{Eve}_j \mid T_{j+1}}[\mathbf{E}[C \mid V_{B,j}, \text{Eve}_j]]$, where A sends the $(j + 1)$'st message of the protocol. We shall clarify that for brevity, in our notation above Eve$_j$ is consistent with T_{j+1} from which we are sampling $V_{B,j}$ (even though this his not explicitly mentioned). Intuitively, this means that we sample views of B, $V_{B,j}$, conditioned on the transcript at the $j + 1$-st pass (which A knows before B) and look at the expectation of the outcome of the coin-toss conditioned on these views, $V_{B,j}$. Then we take the expectation over these expected values. Here, we give some intuition as to why MV is significant for our analysis. Observe that B, given its real view can of course compute $\mathbf{E}[C \mid V_{B,j}, \text{Eve}_j] = \mathbf{E}[C \mid V_{B,j}, T_j]$, which is the expected value of the outcome of the coin-toss, C, from B's point-of-view. Note that A cannot compute this value since conditioned on its view, it does not know the real $V_{B,j}$. Before computing its message in the $j + 1$-st pass, A would only be able to compute $E_{V_{B,j}, Eve_j | T_j}[E[C \mid V_{B,j}, Eve_j]]$, the expectation when views of B, $V_{B,j}$ are sampled conditioned on only T_j, which is equivalent to $E[C|T_j]$. However, after computing T_{j+1}, A gets more information about B's view and thus can get a better estimate of B's real expected value by sampling views of B, $V_{B,j}$ conditioned on T_{j+1}. It is this advantage which we leverage in the final strategy in order to allow A to impose bias.

In the following we give some more details of our approach. First, using a similar argument to that of [CI93] shows that one of the following cases occurs:

1. With probability at least $1/20$ there is some point in the execution such that $E[C|T_j] - MV \geq \Omega(1/\sqrt{r})$.
2. With probability at least $1/20$ there is some point in the execution such that $MV - E[C|T_{j+1}] \geq \Omega(1/\sqrt{r})$.
3. With probability at least $1/20$ there is some point in the execution such that $MV - E[C|T_j] \geq \Omega(1/\sqrt{r})$.
4. With probability at least $1/20$ there is some point in the execution such that $E[C|T_{j+1}] - MV \geq \Omega(1/\sqrt{r})$.

For each of the above cases, we need to come up with corresponding strategies that allow A or B to impose bias on the final outcome. It turns out that Cases 1 and 2 give rise to adversarial strategies for biasing towards 0, while Cases 3 and 4 will give rise to adversarial strategies for biasing towards 1. In the following, we give some intuition for the analysis of cases 1 and 2; cases 3 and 4 are entirely analogous.

It is not difficult to see (details can be found in Section 4) that if Case 1 occurs, then one of the following will occur:

- With probability $\Omega(1/\hat{r}^{1/4})$ there is a (first) point where $E[C \mid T_j] - E[C_{j+2} \mid T_j] \geq \Omega(1/\sqrt{\hat{r}})$.
- With probability $\Omega(1/\hat{r}^{1/2})$ there is a (first) point where $E[C \mid T_j] - E[C_{j+2} \mid T_j] \geq \Omega(1/\hat{r}^{1/4})$.
- With probability $\Omega(1/\hat{r}^{1/4})$ there is a (first) point where $E[C \mid V_{B,j}, Eve_j] - E[C_{j+2} \mid V_{B,j}, Eve_j] \geq \Omega(1/\sqrt{\hat{r}})$.
- With probability $\Omega(1/\hat{r}^{1/4})$ there is a (first) point where $E[C \mid V_{B,j}, Eve_j] - E[C_{j+2} \mid V_{B,j}, Eve_j] \geq \Omega(1/\sqrt{\hat{r}})$.

By directly using [CI93]-type strategies we can impose bias of $\Omega(1/\hat{r}^{3/4})$ when any of the above items occurs. On the other hand, if Case 2 occurs then we have either:

- with probability at least $1/40$ there is a (first) point where $E[C_{j+1} \mid T_{j+1}] - E[C \mid T_{j+1}] \geq \Omega(1/\sqrt{\hat{r}})$, or:
- with probability at least $1/40$ there is a (first) point where $E_{V_{B,j}, Eve_j | T_{j+1}}[E[C \mid V_{B,j}, Eve_j]] - E[C_{j+1} \mid T_{j+1}] \geq \Omega(1/\sqrt{\hat{r}})$.

Again, if the first item above occurs, we can impose bias of $\Omega(1/\sqrt{\hat{r}})$ using [CI93]-type strategies. However, in order to utilize the second item above, which we refer to as Case (2b), to impose $\omega(1/\hat{r})$ bias, we need quite a bit of additional work. More specifically, we show that in order to leverage Case (2b), it is sufficient to present a way to simulate a fake transcripts T_{j+1}, which we denote by T'_{j+1} such that:

- Real transcripts T_{j+1} and fake transcripts T'_{j+1} are distributed nearly identically.
- The expected value of outcomes conditioned on views of B sampled w.r.t. the real transcript $T_{j+1} = t_{j+1}$ is nearly the same as the expected value of outcomes conditioned on views of B sampled w.r.t. $T'_{j+1} = t_{j+1}$. Formally, we have that: $\mathbf{E}_{V_{B,j},\mathsf{Eve}_j|T_{j+1}}[\mathbf{E}[C \mid V_{B,j}, \mathsf{Eve}_j]] \approx \mathbf{E}_{V_{B,j},\mathsf{Eve}_j|T'_{j+1}}[\mathbf{E}[C \mid V_{B,j}, \mathsf{Eve}_j]]$.
- T'_{j+1} reveals almost no information about the real $V_{A,j+1}$.

In what follows, we give some intuition as to how the simulated T'_{j+1} is constructed. We note that some of the techniques we use here are similar to those used by [MMP13]. In order to construct T'_{j+1}, we critically use independence of the views of A and B (once the Eve queries have been made). We sample a fake view for A, $V'_{A,j+1}$, conditioned only on T_j and use it to compute a fake next message M'_{j+1}. Then we run the Eve algorithm (pretending that M'_{j+1} is the real $j+1$-st message) carefully choosing which queries to answer w.r.t. the real oracle and which queries to "lie" about. The main idea (although the actual algorithm is slightly more complicated) is the following: All queries made by Eve that are in $V'_{A,j+1}$ are answered according to $V'_{A,j+1}$, all queries made by Eve that are in the real $V_{A,j+1}$ and not in $V'_{A,j+1}$ are answered uniformly at random. All remaining queries are asked to the oracle and the response from the oracle is returned. Now, intuitively, items (1) and (2) above hold since by independence, it is highly likely that all "modified" Eve queries (i.e. queries that appear in $V_{A,j+1}$ or $V'_{A,j+1}$) *do not* intersect with the real $V_{B,j}$. For item (3), recall that T'_{j+1} is computed by "ignoring" the real $V_{A,j+1}$, sampling a new $V'_{A,j+1}$ and continuing with the computation as though $V'_{A,j+1}$ were the real view. Intuitively, T'_{j+1} is close to independent of $V_{A,j+1}$ (conditioned on T_j) and so knowledge of T'_{j+1} does not give additional information on $V_{A,j+1}$ beyond what is already given by T_j.

Properties (1) and (2) are used to argue that if with high probability there is a first point where $\mathbf{E}_{V_{B,j},\mathsf{Eve}_j|T_{j+1}}[\mathbf{E}[C \mid V_{B,j}, \mathsf{Eve}_j]] - \mathbf{E}[C_{j+1} \mid T_{j+1}]$ is large (as occurs in Case (2b)) then with high probability there is a first point where $\mathbf{E}_{V_{B,j},\mathsf{Eve}_j|T'_{j+1}}[\mathbf{E}[C \mid V_{B,j}, \mathsf{Eve}_j]] - \mathbf{E}[C_{j+1} \mid T'_{j+1}]$ is large (see Claim 1 for the precise statement).

Property (3) is used to argue that $\mathbf{E}_{V_{B,j},\mathsf{Eve}_j|T'_{j+1}}[\mathbf{E}[C \mid V_{B,j}, \mathsf{Eve}_j]]$ is close to $\mathbf{E}[C \mid T'_{j+1}]$ (see Claim 2 for the precise statement). To give some intuition into why this holds, note that $\mathbf{E}[C \mid T'_{j+1}]$ can be re-written as: $\mathbf{E}_{V_{A,j+1},V_{B,j}|T'_{j+1}}[\mathbf{E}[C \mid V_{A,j+1}, V_{B,j}]]$. Now, the quantity $\mathbf{E}_{V_{B,j},\mathsf{Eve}_j|T'_{j+1}}[\mathbf{E}[C \mid V_{B,j}, \mathsf{Eve}_j]]$ is nearly the same, except views of B, $V_{B,j}$ are sampled conditioned on T'_{j+1}, but views of A, $V_{A,j+1}$, are sampled conditioned only on $(V_{B,j}, \mathsf{Eve}_j)$ (which in particular includes T_j). Intuitively, this reflects the fact that T'_{j+1} does not provide additional information about the real view of A, $V_{A,j+1}$ over what is contained in T_j. However, the fact that T'_{j+1} does not leak additional information on $V_{A,j+1}$, is not sufficient to argue that $\mathbf{E}_{V_{B,j},\mathsf{Eve}_j|T'_{j+1}}[\mathbf{E}[C \mid V_{B,j}, \mathsf{Eve}_j]]$ is close to $\mathbf{E}[C \mid T'_{j+1}]$. This is because T'_{j+1} still contains additional Eve queries which, although they do not provide additional information about $V_{A,j+1}$, do provide additional overall information about the oracle. Thus, in order for item (3) to hold, we need the additional "function-obliviousness" property (see Property 1) which guarantees that the outcome C of the coin-toss does not depend on the oracle, but only on the random tapes of the two parties. We note that this is the only place in the proof where the "fuction-obliviousness" property is used. Thus, as long as we can sample partial views of A and B, $V_{A,j+1}, V_{B,j}$ according to the correct distribution, we can compute the expected value of the coin toss $C(V_{A,j+1}, V_{B,j}) = C(r_A, r_B)$.

When the above are combined, we get that with high probability there is a first point where $\mathbf{E}[C \mid T'_{j+1}] - \mathbf{E}[C_{j+1} \mid T'_{j+1}]$ is large, which means that an adversary can impose bias by adopting a [CI93]-type strategy.

Unfortunately, the actual argument is somewhat more complicated than what is described above, because once the adversarial party A' playing the role of A has computed the simulated transcript T'_{j+1} and the associated information (which we denote by K_{j+1}), A' cannot just throw away the additional information in K_{j+1} and start afresh when computing expectations in the $j + 3$-rd pass. This is because just *the fact that A' has not aborted* itself gives information that might impact the expected value of the coin toss. Thus, A' cannot decide to abort by conditioning only on T'_{j+3}, but must additionally condition on its extra knowledge K_{j+1}, which it obtained in the previous round, when deciding whether or not to abort. Therefore, all the information in the K_j variables must be used when A' computes subsequent expectations. Moreover, when $V'_{A,j+1}$ is sampled (in order to compute T'_{j+1}), it must be consistent not only with T_j but also with the additional knowledge K_j collected thus far. See Section 4.1 for the precise description and analysis of the final adversarial strategy.

2   Preliminaries

Definition 2 (Black-Box Coin Tossing from One-Way Functions). *For (interactive) oracle algorithms* A, B *we call* $C = \langle A, B \rangle$ *a black-box construction of coin tossing with bias at most* δ *based on one-way functions with input/output length* n, *if the following properties hold:*

- *The parties* A *and* B *get access to private randomness* r_A, r_B *and common input* 1^n *and run in time* poly(n) *and interact for* $\hat{r}(n) = $ poly(n) *number of rounds. The transcript of their interaction determines an output* a. *Also, if during the protocol,* A *(resp.* B*) receives the special message* $\perp$ *(denoting that the other party has stopped playing in the protocol) then* A *(resp.* B*) outputs a bit* a *(resp* b*) on their own which will be the output of the protocol.*
- **Completeness:** *For any function* f, *if* A *and* B *are given oracle access to* f *and execute the protocol honestly, then the output is an unbiased random bit.*
- **Security:** *There is an oracle algorithm* S *running in polynomial time over its input length with the following property. Given any adversary* $\mathcal{A}$ *(playing on behalf of* A *or* B*) that achieves bias* $\delta(n)$ *over common input* 1^n *w.r.t a function* f, $S^{f,\mathcal{A}}(1^n, 1^{1/\delta(n)})$ *breaks the security of* f *as a one-way function.*

2.1   The Eavesdropper Algorithm Eve

In this section, we recall the Eve algorithm, first introduced by Impagliazzo and Rudich [IR89] in the context of separating one-way function and key agreement. The Eve algorithm of [IR89] was later improved by Barak and Mahmoody [BM13]. In our work, we will use the Eve algorithm of [BM13] in a black-box manner. Thus, we do not describe the algorithm itself, and simply state the properties we will need from the Eve algorithm of [BM13] in the following lemma.

Lemma 1 (Implied by Theorem 4.2 in [BM13]). *Let* $C = \langle A, B \rangle$ *be an oracle protocol in which the parties* A, B *ask at most* m *queries each from the oracle* O. *Then there is an Eve algorithm who only gets to see the public messages and asks her own oracle queries after each message is sent and on input parameter* $\epsilon < 1/100$:

- poly(m/ϵ)**-Efficiency:** *Eve is deterministic and, over the randomness of the oracle and* A *and* B*'s private randomness, the expected number of Eve queries from the oracle* O *is at most* $(10m/\epsilon)^{10}$.
- $(1 - \epsilon)$**-Security:** *Let* $T_i = M_1, \ldots, M_i \| \text{Eve}_i$ *be the transcript of messages sent between* A *and* B *so far, including the the additional information that Eve has learned till the* end *of the* i*'th pass. Let* $(V_A, V_B) \mid T_i$ *be the* joint *distribution over the views* (V_A, V_B) *of* A *and* B *only conditioned on* T_i. *By* $V_A \mid T_i$ *and* $V_B \mid T_i$ *we refer to the projections of* $\mathcal{D}(T_i)$ *over its first or second components. Then, with probability at least* $1 - \epsilon$ *over the randomness of* A, B, *and the random oracle* O, *the following holds at* all *moments during the protocol when Eve is done with her learning phase in that round:*

  1. The statistical distance between $V_A \mid T_i \times V_B \mid T_i$ *and* $\mathcal{D}(T_i)$ *is at most* ϵ. *Namely:* $\Delta(V_A \mid T_i \times V_B \mid T_i, (V_A, V_B) \mid T_i) \le \epsilon$.
  2. For every oracle query $q \notin \text{Eve}_i$ *it holds that* $\Pr_{(V_A, V_B) \mid T_i}[q \in Q_{V_A} \cup Q_{V_B}] \le \epsilon$.

In the following, we will run the Eve algorithm with input parameter $\epsilon = \frac{1}{3m\hat{r}^4}$.

For simplicity of the notation and when it is clear from the context, in the following, for probabilities and expected values taken over $(V_A, V_B) \sim \mathcal{D}$, instead of writing $\mathbf{E}_{(V_A, V_B) \sim \mathcal{D}}$ or $\Pr_{(V_A, V_B) \sim \mathcal{D}}$, we simply write $\mathbf{E}$ and $\Pr$.

We consider coin-tossing protocols C, where the Eve algorithm is run alongside the protocol and Eve queries are made immediately after every message M_j is sent. We denote by Eve_j the set of queries made by the Eve algorithm up to and including the queries made immediately after the j-th message is sent. We denote by T_j, the transcript of the protocol with the Eve queries made alongside. Thus, $T_j = M_1, \ldots, M_j \| \text{Eve}_j$.

3 Types of Coin Tossing Protocols We Consider

Consider a coin-tossing protocol $C = \langle A, B \rangle$ with $\hat{r} = \hat{r}(k) = \text{poly}(k)$ rounds and $2\hat{r}$ passes. For $1 \le w \le \hat{r}$, let C_{2w-1} denote the output of party B in the case that A aborts before sending the $2w - 1$-st message. Similarly, For $1 \le w \le \hat{r}$, let C_{2w} denote the output of party A in the case that B aborts before sending the $2w$-th message.

Let $V_{A,j}$ (resp. $V_{B,j}$) denote the partial view of A (resp. B) up to and including pass j. In particular, $V_{A,j}$ consists of the transcript M_j thus far as well as the random tape r_A of A and the queries and responses, $Q_{V_{A,j}}$, that have been made by A thus far. $V_{B,j}$ and $Q_{V_{B,j}}$ are defined analogously.

We consider the distribution $\mathcal{D}$ to be the distribution over pairs of complete views $(V_{A,2\hat{r}}, V_{B,2\hat{r}})$ (also denoted simply by V_A, V_B) generated by a run of C with a random oracle. More specifically, a draw from $\mathcal{D}$ is obtained as follows:

- Draw $O \sim \Upsilon$, $r_A, r_B \leftarrow \{0,1\}^{p(n)}$, for some polynomial $p(\cdot)$ and execute $C = \langle A, B \rangle$ with O, r_A, r_B.
- Output the views (V_A, V_B) resulting from the execution of $C = \langle A, B \rangle$ above.

We prove our result for so-called *instant* constructions as defined in [DSLMM11]. Instant constructions are coin-tossing protocols where for $1 \leq w \leq \hat{r}$, A (resp. B) computes the value C_{2w} (resp. C_{2w+1}) before sending message M_{2w-1} (resp. M_{2w}). Thus, in case a party (say A) aborts before sending message M_{2w+1}, then B can simply output its precomputed value C_{2w+1} which depends only on B's view at the point right after B computed message M_{2w} without making any additional oracle queries. It is not hard to see that the restriction of instant constructions can be removed as was shown by [DSLMM11]. This is a subtle argument relying on the fact that our ultimate goal is to rule out separations from one-way functions and not random oracles (since in the random oracle model coin tossing is trivial). In the following we sketch the argument of [DSLMM11] on why assuming the protocol to be instant is w.l.o.g.

Instant vs. General Protocols. Dealing with non-instant protocols can be done exactly as it was done in [DSLMM11], so in this work we focus on instant protocols and leave the full discussions on dealing with non-instant protocols for the full version of the paper. However, here we give a sketch of how this can be done. Firstly, note that any general coin tossing protocol using an oracle can be made "almost instant" without losing the security as follows. Whenever a party A (or B) wants to send a message M_i, they also go ahead and ask any oracle query that they would need to ask in case the other party halts the execution of the protocol and not sent M_{i+1}. This way, the protocol becomes almost instant because the only time that the instant property might be violated is when the first message is aborted by Alice in which case, Bob might still need to query the oracle to decide the output. However, as shown in [DSLMM11], it is always possible to "fix" a "small" set S of queries of the random oracle in a way that (1) Bob does not ask any query to decide the output if he gets aborted in the first message, and (2) the protocol remains as secure. Roughly speaking, the set S is determined (and its answers are fixed) as follows. The set S contains any query q that has a "non-negligible" chance of being asked by Bob in case of not receiving the first message. It is easy to show that $|S| \leq \mathrm{poly}(n)$, and by sampling (and fixing) the answer of the queries in S, Bob will not need to ask any oracle queries in case of getting aborted in the first round. Finally, observe that a partially-fixed random oracle is still one-way and so one can apply the argument of our work for the instant protocols to the final instant protocol.

We consider coin-tossing protocols that are so-called "function oblivious." As defined in Definition 1, these are coin-tossing protocols such that the outcome of protocol $C^f = \langle A^f, B^f \rangle$ when both parties are honest depends only on the random tapes r_A, r_B of the two parties and not on the choice of one-way function f. We denote by $C(r_A, r_B)$ the output of protocol C when run with random tapes r_A, r_B. When the settings of r_A, r_B are clear from context, we denote the output of the protocol by C.

We are now ready to state our main theorem.

Theorem 2 (Main Theorem, Formal). *There is no (fully) black-box construction of an $\hat{r} = \hat{r}(n)$-round, function-oblivious coin-tossing protocol $C^f = \langle A^f, B^f \rangle$ with bias $o(1/\hat{r}^{3/4})$ from one-way functions.*

4 Proof of the Main Theorem

Towards proving Theorem 2, we begin with the following fact, which follows straightforwardly from [CI93].

Fact 1. *Let* C *be a coin-tossing protocol and let* $\{Y_1, \ldots, Y_{2\hat{r}}\}$ *be a set of random variables, where* Y_j *is associated with some state of protocol* C *immediately after the* j-*th message* (M_j, Eve_j) *has been computed.*

- *For* $1 \leq w \leq \hat{r}$, *set* $j = 2w - 2$ *and define the indicator variable* $I_{\mathsf{Val}^A_{j+1}}$ *in the following way:* $I_{\mathsf{Val}^A_{j+1}} = 1$ *if* $|\mathbf{E}[C|Y_{j+1}] - \mathbf{E}[C_{j+1}|Y_{j+1}]| \geq \beta$ *and for* $1 \leq \ell \leq w$, $I_{\mathsf{Val}^A_{2\ell-1}} = 0$. *Otherwise* $I_{\mathsf{Val}^A_{j+1}} = 0$.
- *For* $1 \leq w \leq \hat{r}$, *set* $j = 2w$ *and define the indicator variable* $I_{\mathsf{Val}^A_{j+1}}$ *in the following way:* $I_{\mathsf{Val}^B_j} = 1$ *if* $|\mathbf{E}[C|Y_j] - \mathbf{E}[C_{j+2}|Y_j]| \geq \beta$ *and for* $1 \leq \ell \leq w$, $I_{\mathsf{Val}^B_{2\ell}} = 0$. *Otherwise* $I_{\mathsf{Val}^B_j} = 0$.

If for some (α, β)

$$\sum_{w=1}^{\hat{r}} \Pr[I_{\mathsf{Val}^A_{2w-1}} = 1] \geq \alpha$$

then player A *has a fail-stop strategy for imposing bias* $\pm 1/2 \cdot \alpha \cdot \beta$ *on* C *by aborting before sending message* M_{2j-1} *either when* $\mathbf{E}[C|Y_{2w-1}] - \mathbf{E}[C_{2w-1}|Y_{2w-1}] \geq \beta$ *or when* $\mathbf{E}[C|Y_{2w-1}] - \mathbf{E}[C_{2w-1}|Y_{2w-1}] \leq -\beta$. *An analogous claim holds for player* B.
If for some (α, β),

$$\sum_{w=1}^{\hat{r}} \Pr[I_{\mathsf{Val}^B_{2w}} = 1] \geq \alpha$$

then player B *has a fail-stop strategy for imposing bias* $\pm 1/2 \cdot \alpha \cdot \beta$ *on* C *by aborting after sending message* M_{2j} *either when* $\mathbf{E}[C|Y_{2w}] - \mathbf{E}[C_{2w+2}|Y_{2w}] \geq \beta$ *or when* $\mathbf{E}[C|Y_{2w}] - \mathbf{E}[C_{2w+2}|Y_{2w}] \leq -\beta$. *An analogous claim holds for player* A.

The following fact is implicit in [CI93]:

Fact 2. *With prob. at least* $1/5$ *over choice of random tapes and oracle there is some point in the execution such that* $|\mathbf{E}[C|T_j] - \mathbf{E}[C|T_{j+1}]| \geq \Omega(1/\sqrt{r})$.

Let us choose the quantity

$$MV = \mathbf{E}_{V_{B,j}, Eve_j | T_{j+1}}[\mathbf{E}[C|V_{B,j}, Eve_j]]$$

as the "middle value."

Thus, it must be the case that either with prob. at least $1/10$ there is some point in the execution such that $|\mathbf{E}[C|T_j] - MV| \geq \Omega(1/\sqrt{r})$. OR with prob. at least $1/10$ there is some point in the execution such that $|MV - \mathbf{E}[C|T_{j+1}]| \geq \Omega(1/\sqrt{r})$.

In particular, there are four possible cases.

1. With probability $1/20$ there is some point s.t. $\mathbf{E}[C|T_j] - \mathrm{MV} \geq \Omega(1/\sqrt{r})$.
2. With probability $1/20$ there is some point s.t. $\mathrm{MV} - \mathbf{E}[C|T_{j+1}] \geq \Omega(1/\sqrt{r})$.
3. With probability $1/20$ there is some point s.t. $\mathrm{MV} - \mathbf{E}[C|T_j] \geq \Omega(1/\sqrt{r})$.
4. With probability $1/20$ there is some point s.t. $\mathbf{E}[C|T_{j+1}] - \mathrm{MV} \geq \Omega(1/\sqrt{r})$.

Note that Cases 1 and 2 will give rise to adversarial strategies for biasing towards 0, while Cases 3 and 4 will give rise to adversarial strategies for biasing towards 1. In the following, we analyze only cases 1 and 2; cases 3 and 4 are entirely analogous.

Lemma 2. *Assume Case 1 occurs with probability at least $1/20$, then there is a strategy that biases the output by $\Omega(1/\sqrt{r})$.*

Proof. Assume that Case (1) occurs. Then this means that with prob. at least $1/20$ there is some point in the execution such that

$$\mathbf{E}[C \mid T_j] - \mathrm{MV} = \mathbf{E}[C \mid T_j] - \mathbf{E}_{\mathsf{V}_{\mathsf{B},j},\mathsf{Eve}_j|T_{j+1}}[\mathbf{E}[C \mid \mathsf{V}_{\mathsf{B},j}, \mathsf{Eve}_j]] \geq \Omega(1/\sqrt{r}).$$

Fix T_{j+1} such that Case (1) occurs. Note that T_{j+1} completely defines T_j and so the quantity above can be calculated for every valid T_{j+1}. Now, for each such T_{j+1} we must have that one of the following two subcases occurs:

(1a) $\mathrm{Pr}_{\mathsf{V}_{\mathsf{B},j},\mathsf{Eve}_j|T_{j+1}}[\mathbf{E}[C \mid \mathsf{V}_{\mathsf{B},j}, \mathsf{Eve}_j] - \mathbf{E}[C \mid T_j] \geq \Omega(1/\hat{r}^{1/4})] \geq \Omega(1/\sqrt{\hat{r}})$ OR
(1b) $\mathrm{Pr}_{\mathsf{V}_{\mathsf{B},j},\mathsf{Eve}_j|T_{j+1}}[\mathbf{E}[C \mid \mathsf{V}_{\mathsf{B},j}, \mathsf{Eve}_j] - \mathbf{E}[C \mid T_j] \geq \Omega(1/\sqrt{r})] \geq \Omega(1/\hat{r}^{1/4})$.

To see this, assume towards contradiction that neither item (1a) nor item (1b) occur. Then this means that when $\mathsf{V}_{\mathsf{B},j}$ is sampled conditioned on T_{j+1}, we have that the contribution from $\mathsf{V}_{\mathsf{B},j}$ such that $\mathbf{E}[C \mid \mathsf{V}_{\mathsf{B},j}, \mathsf{Eve}_j] - \mathbf{E}[C \mid T_j] \geq \Omega(1/\sqrt{r})$ and $\mathbf{E}[C \mid \mathsf{V}_{\mathsf{B},j}, \mathsf{Eve}_j] - \mathbf{E}[C \mid T_j] \leq \Omega(1/\hat{r}^{1/4})$ is at most $o(1/\hat{r}^{1/4}) \cdot 1/\hat{r}^{1/4} = o(1/\sqrt{r})$. Additionally, the contribution from $\mathsf{V}_{\mathsf{B},j}$ such that $\mathbf{E}[C \mid \mathsf{V}_{\mathsf{B},j}, \mathsf{Eve}_j] - \mathbf{E}[C \mid T_j] \geq \Omega(1/\hat{r}^{1/4})$ is at most $o(1/\sqrt{r}) \cdot 1 = o(1/\sqrt{r})$. This is a contradiction to Case 1 occurring.

Now, if item (1a) occurs then this means that either with probability $\Omega(1/\hat{r}^{1/4})$ we have that $\mathbf{E}[C \mid T_j] - \mathbf{E}[C_{j+2} \mid T_j] \geq \Omega(1/\sqrt{\hat{r}})$ occurs OR that with probability $\Omega(1/\hat{r}^{1/4})$ we have that $\mathbf{E}[C \mid \mathsf{V}_{\mathsf{B},j}, \mathsf{Eve}_j] - \mathbf{E}[C \mid T_j] \geq \Omega(1/\sqrt{\hat{r}})$ AND $\mathbf{E}[C \mid T_j] - \mathbf{E}[C_{j+2} \mid T_j] \leq o(1/\sqrt{\hat{r}})$.

In the first case, B can employ the following strategy:

Abort immediately **after** sending message M_j if:

$$\mathbf{E}[C|T_j] - \mathbf{E}[C_{j+2}|T_j] \geq \Omega(1/\sqrt{\hat{r}})$$

By Fact 1 the strategy above imposes bias of at least $\Omega(1/\hat{r}^{3/4})$ towards 0.

In the second case, note that since C_{j+2} is a function of only $\mathsf{V}_{\mathsf{A},j+1}$, we have by the properties of the Eve algorithm given in Lemma 1 we have that with probability $1 - O(1/\hat{r}^2)$, we have that $|\mathbf{E}[C_{j+2} \mid T_j] - E[C_{j+2} \mid \mathsf{V}_{\mathsf{B},j}, \mathsf{Eve}_j]| \leq O(1/\hat{r}^2)$. Thus, in this case, we have that with probability $\Omega(1/\hat{r}^{1/4})$, $\mathbf{E}[C \mid \mathsf{V}_{\mathsf{B},j}, \mathsf{Eve}_j] - \mathbf{E}[C_{j+2} \mid \mathsf{V}_{\mathsf{B},j}, \mathsf{Eve}_j] \geq \Omega(1/\sqrt{\hat{r}})$ and in this case, B can employ the following strategy:

Abort immediately **after** sending message M_j if:

$$\mathbf{E}[C \mid V_{B,j}, Eve_j] - \mathbf{E}[C_{j+2} \mid V_{B,j}, Eve_j] \geq \Omega(1/\sqrt{\hat{r}})$$

Thus by Fact 1 the above strategy imposes bias of at least $\Omega(1/\hat{r}^{3/4})$ towards 0. The analysis for item (1b) is entirely analogous.

Lemma 3. *Assume Case 2 occurs with porbability at least 1/20, then there is a strategy that biases the output by $\Omega(1/\hat{r}^{3/4})$.*

Proof. Case (2) implies that one of the following occurs with prob. at least $1/40$:

(2a)
$$\mathbf{E}[C_{j+1} \mid T_{j+1}] - \mathbf{E}[C \mid T_{j+1}] \geq \Omega(1/\sqrt{\hat{r}})$$

(2b)
$$\mathsf{MV} - \mathbf{E}[C_{j+1} \mid T_{j+1}] = \mathbf{E}_{V_{B,j}, Eve_j \mid T_{j+1}}[\mathbf{E}[C \mid V_{B,j}, Eve_j]] - \mathbf{E}[C_{j+1} \mid T_{j+1}] \geq \Omega(1/\sqrt{\hat{r}})$$

Note that in Case (2a), A can employ the following strategy:

Abort **before** sending message M_{j+1} if:

$$\mathbf{E}[C \mid T_{j+1}] - \mathbf{E}[C_{j+1} \mid T_{j+1}] \geq \Omega(1/\sqrt{\hat{r}})$$

and thus, by Fact 1 imposes bias $\Omega(1/\sqrt{\hat{r}})$ towards 0. Thus, to complete the lemma, we need to show that if Case (2b) occurs with probability at least $1/40$ then there is a strategy for imposing bias of $\Omega(1/\hat{r}^{3/4})$ on the outcome.

Since this case becomes more complicated, we devote the following section to show how to deal with Case (2b).

4.1 Analysis for Case (2b)

The Protocol C'. The modified protocol C' will execute the regular C protocol with Eve queries made alongside. B behaves as in the original protocol. A' behaves as A in the original protocol and additionally computes extra state information K_i and related values in each round i.

For each pass $0 \leq j \leq 2\hat{r} - 1$, we consider the distribution $\mathcal{D}_{extend,j+1}$, which is a distribution over a tuple consisting of partial views $V_{A,j+1}, V_{B,j+1}$, transcripts (with Eve queries alongside) T_{j+1}, and additional knowledge K_{j+1} generated by a random execution of C' with random oracle $\mathcal{O}$.

More specifically, a draw from $\mathcal{D}_{extend,j+1}$ is obtained as follows:

- Draw $O \sim \Upsilon$, $r_{A'}, r_B \leftarrow \{0,1\}^{p'(n)}$, for some polynomial $p'(\cdot)$ and execute $C' = \langle A', B \rangle$ with $O, r_{A'}, r_B$.
- Output a tuple consisting of the views $V_{A,j+1}, V_{B,j+1}$, transcript T_{j+1}, and additional state information K_{j+1} resulting from the execution of $C' = \langle A', B \rangle$ above.

We are now ready to describe how K_{j+1} is computed: For $j = 0$, the variable K_0 is set to empty. For each round $1 \leq w \leq \hat{r}$, set $j = 2(w - 1)$. A' computes the state information K_{j+1} in the following way:

- Sample a random partial view $V'_{A,j+1}$ from $\mathcal{D}_{extend,j+1}(T_j, K_j)$. Recall that this denotes the distribution $\mathcal{D}_{extend}$, conditioned on the current transcript with Eve queries, T_j, and the additional state information K_j. Note that $V'_{A,j+1}$ includes the next message, M'_{j+1}.
- We run a modified version of the Eve algorithm, called the Eve' algorithm. For each pass ℓ, let $Q_{\mathsf{Eve}'_\ell}$ denote the set of queries and responses made by Eve' in the j-th pass. In the $j + 1$-st pass do the following: Run the Eve algorithm at pass $j + 1$ conditioned on $T_j \| M'_{j+1}$ (i.e. as if M'_{j+1} is the real next message). Answer oracle queries made by Eve' in the following way[3]:
  - If the query q appears in $V'_{A,j+1}$, answer according to $V'_{A,j+1}$ (without querying the oracle).
  - Otherwise, if for some $i \leq j$, the query q appears in $Q_{\mathsf{Eve}'_i} \setminus Q_{V'_{A,i}}$, respond according to the value listed in $Q_{\mathsf{Eve}'_i}$ (without querying the oracle).
  - Otherwise, if a query q appears in $V_{A,j}$, sample and return a uniformly random string (without querying the oracle).
  - Otherwise, query the oracle and return the oracle's response.
- We denote by T'_{j+1} the *fake transcript* generated. More specifically, $T'_{j+1} = T_j \| M'_{j+1} \| Q_{\mathsf{Eve}'_{j+1}}$.
- Set K_{j+1} and K_{j+2} to be K_j with the variables $V'_{A,j+1}, Q_{\mathsf{Eve}'_{j+1}}$ appended.

Intuitively, the point of the protocol C' is that it allows a malicious A to sample fake transcripts T'_{j+1}, which, conditioned on T_j, K_j, are distributed (almost) identically to real transcripts T_{j+1}, but reveal (almost) no additional information about the real $V_{A,j+1}$, beyond what was revealed by T_j, K_j. In particular, a '"fake"' view $V'_{A,j+1}$, independent of the real $V_{A,j+1}$, is sampled and a fake next message M'_{j+1} is computed. Now, when we run the Eve' algorithm, ideally we would like to answer all oracle queries q appearing in $Q_{V'_{A,j+1}}$ dishonestly according to $V'_{A,j+1}$ and all other queries honestly according to the real oracle. However, there is a sublte issue here: Queries in the real $Q_{V_{A,j+1}}$ may be '"incorrectly"' distributed if they are answered according to the real oracle. In particular, queries which appear in $Q_{\mathsf{Eve}'_i} \setminus Q_{V'_{A,i}}$ and do not appear in $V'_{A,j+1}$, must be answered according to the value listed there (regardless of whether they are in $V_{A,j+1}$). Queries which do not appear in $Q_{V'_{A,j+1}}$ and do not appear in $Q_{\mathsf{Eve}'_i} \setminus Q_{V'_{A,i}}$, but do appear in $Q_{V_{A,j+1}}$ are answered uniformly at random.

We are now ready to describe the final adversarial strategy:

- Set $f(\hat{r}) = 1/\sqrt{\hat{r}}$ or $f(\hat{r}) = 1/\hat{r}^{1/4}$.
- Play the role of A' in an execution C', while interacting with an honest B.
- Abort immediately **before** sending message M_{j+1} if:

$$\mathbf{E}[C | T'_{j+1}, K_j] - \mathbf{E}[C_{j+1} | T'_{j+1}, K_j] = \Omega(f(\hat{r}))$$

---

[3] We assume that Eve' never re-queries a query that is already contained in Eve_j

Fact 1 implies that all we need to show is that the event occurs "frequently." More specifically, we prove the following lemma, which is sufficient for completing the proof of Case (2b).

Lemma 4. *If Case (2b) occurs with probability* $1/40$ *then either:*

- *With probability* $\Omega(1/\hat{r}^{1/2})$ *over executions of* C' *and choice of oracle* $\mathcal{O}$ *there is a first message* j *where*

$$\mathbf{E}[C|T'_{j+1}, K_j] - \mathbf{E}[C_{j+1}|T'_{j+1}, K_j] = \Omega(1/\hat{r}^{1/4})$$

- *With probability* $\Omega(1/\hat{r}^{1/4})$ *over executions of* C' *and choice of oracle* $\mathcal{O}$ *there is a first message* j *where*

$$\mathbf{E}[C|T'_{j+1}, K_j] - \mathbf{E}[C_{j+1}|T'_{j+1}, K_j] = \Omega(1/\hat{r}^{1/2})$$

Before proving Lemma 4, we introduce the following notation. For $f(\hat{r}) = 1/\sqrt{\hat{r}}$ or $f(\hat{r}) = 1/\hat{r}^{1/4}$ and for every $j = 0, 1 \leq j \leq 2\hat{r}$, we define the indicator random variables $I^f_{EV_j}$ and $I^f_{EV'_j}$ which are set before the j-th message is sent during an execution of C'. For $j = 0$, $I^f_{EV_0} = 0$ and $I^f_{EV_0} = 0$. For $j \geq 1$, $I^f_{EV'_{j+1}}$, $I^f_{EV'_{j+2}}$ are set to 1 if:

- $I^f_{EV'_j} = 1$ OR
- $\mathbf{E}_{V_{B,j}|T'_{j+1}=t'_{j+1}, K_j=k_j}[\mathbf{E}[C|V_{B,j}, K_j = k_j]] - \mathbf{E}_{V_{B,j}|T'_{j+1}=t'_{j+1}, K_j=k_j}[C_{j+1}(V_{B,j})] = \Omega(f(\hat{r}))$.

For $j \geq 1$, $I^f_{EV_{j+1}}$, $I^f_{EV_{j+2}}$ are set to 1 if:

- $I^f_{EV_j} = 1$ OR
- $\mathbf{E}_{V_{B,j}|T'_{j+1}=t_{j+1}, K_j=k_j}[\mathbf{E}[C|V_{B,j}, K_j = k_j]] - \mathbf{E}_{V_{B,j}|T'_{j+1}=t_{j+1}, K_j=k_j}[C_{j+1}(V_{B,j})] = \Omega(f(\hat{r}))$.

Note that in the last expression, we condition on $T'_{j+1} = t_{j+1}$. This means that $T_{j+1} = t_{j+1}$ is sampled via a run of the protocol C'. Then, the expectation above is computed using this same value of t_{j+1}, but conditioning on the variable T'_{j+1} being equal to this value.

Let the event B_j be the event that upon a draw from $\mathcal{D}_{extend,j+1}$ there is a query q such that $q \in Q_{V_{B,j}} \cap (Q_{V'_{A,j+1}} \cup Q_{V_{A,j+1}})$ and $q \notin Eve_j$. Note that by Lemma 1, for each j, the probability that B_j occurs is at most $\frac{3m}{3m\hat{r}^4} = O(1/\hat{r}^4)$. Let $\mathcal{D}^{Good_j}_{extend,j+1}$ denote the distribution $\mathcal{D}_{extend,j+1}$, conditioned on $\overline{B}_j$.

By $\mathcal{D}_{extend}$ we denote the distribution $\mathcal{D}_{extend,2\hat{r}}$. Additionally, for joint random variables X, Y, we denote by $X \mid Y$ the distribution over X drawn from $\mathcal{D}_{extend}$, conditioned on Y.

We first consider three important properties of the C' protocol which which will help us prove the lemma:

Property 1 (T_{j+1} and T'_{j+1} are close). The two distributions

$$\mathcal{D}^{\mathsf{Good}_j}_{\mathsf{T}'_{j+1}}(\mathsf{T}_j = t_j, \mathsf{K}_j = k_j) \qquad\qquad \mathcal{D}^{\mathsf{Good}_j}_{\mathsf{T}_{j+1}}(\mathsf{T}_j = t_j, \mathsf{K}_j = k_j)$$

are identical.

Since B_j occurs with probability at most $O(1/\hat{r}^4)$, Property 1 immediately implies the following: With probability $1 - O(1/\hat{r}^2)$ over $\mathsf{T}_j = t_j, \mathsf{K}_j = k_j$ drawn from $\mathcal{D}_{extend}$, we have that

$$\mathcal{D}_{\mathsf{T}'_{j+1}}(\mathsf{T}_j = t_j, \mathsf{K}_j = k_j) \qquad\qquad \mathcal{D}_{\mathsf{T}_{j+1}}(\mathsf{T}_j = t_j, \mathsf{K}_j = k_j)$$

are $O(1/\hat{r}^2)$-close.

Property 2 (V_B conditioned on T_{j+1} or T'_{j+1} are close). For every $1 \le j \le 2\hat{r}$, the two distributions

$$\mathcal{D}^{\mathsf{Good}_j}_{\mathsf{V}_{B,j}}(\mathsf{T}'_{j+1} = t_{j+1}, \mathsf{K}_j = k_j) \qquad\qquad \mathcal{D}^{\mathsf{Good}_j}_{\mathsf{V}_{B,j}}(\mathsf{T}_{j+1} = t_{j+1}, \mathsf{K}_j = k_j)$$

are identical.

Since B_j occurs with probability at most $O(1/\hat{r}^4)$, Property 2 immediately implies the following: For every $1 \le j \le 2\hat{r}$, we have that with probability $1 - O(1/\hat{r}^2)$ over draws of $T_{j+1} = t_{j+1}$ and $K_j = k_j$ from $\mathcal{D}_{extend}$, the statistical distance between the following:

$$\mathcal{D}_{\mathsf{V}_{B,j}}(\mathsf{T}'_{j+1} = t_{j+1}, \mathsf{K}_j = k_j) \qquad\qquad \mathcal{D}_{\mathsf{V}_{B,j}}(\mathsf{T}_{j+1} = t_{j+1}, \mathsf{K}_j = k_j)$$

is at most $O(1/\hat{r}^2)$.

Property 3 (T'_{j+1} does not reveal much information about $\mathsf{V}_{A,j}$). With probability $1 - O(1/\hat{r}^2)$ over $\mathsf{T}_j = t_j, \mathsf{T}'_{j+1} = t_j \| m'_{j+1}, eve'_{j+1}, \mathsf{K}_j = k_j$ drawn from $\mathcal{D}_{extend}$, we have that

$$\mathcal{D}_{\mathsf{V}_{A,j+1}}(\mathsf{T}'_{j+1} = t'_{j+1}, \mathsf{K}_j = k_j) \qquad\qquad \mathcal{D}_{\mathsf{V}_{A,j+1}}(\mathsf{T}_j = t_j, \mathsf{K}_j = k_j)$$

are $O(1/\hat{r}^2)$-close.

We defer the proofs of Properties 1, 2, 3 to the full version and now complete the proof of Lemma 4 via the following claims and facts:

Claim 1. *If Case (2b) occurs with probability $1/40$ then either:*

- *With probability $\Omega(1/\hat{r}^{1/2})$ there is a first point where*

$$\mathbf{E}_{\mathsf{V}_{B,j},\mathsf{Eve}_j|\mathsf{T}'_{j+1},\mathsf{K}_j}[E[C|\mathsf{V}_{B,j},\mathsf{Eve}_j,\mathsf{K}_j]] - \mathbf{E}_{\mathsf{V}_{B,j}|\mathsf{T}'_{j+1},\mathsf{K}_j}[C_{j+1}(\mathsf{V}_{B,j})] = \Omega(1/\hat{r}^{1/4}).$$

- *With probability $\Omega(1/\hat{r}^{1/4})$ there is a first point where*

$$\mathbf{E}_{\mathsf{V}_{B,j},\mathsf{Eve}_j|\mathsf{T}'_{j+1},\mathsf{K}_j}[E[C|\mathsf{V}_{B,j},\mathsf{Eve}_j,\mathsf{K}_j]] - \mathbf{E}_{\mathsf{V}_{B,j}|\mathsf{T}'_{j+1},\mathsf{K}_j}[C_{j+1}(\mathsf{V}_{B,j})] = \Omega(1/\hat{r}^{1/2}).$$

Claim 2. *With probability* $1 - O(1/\hat{r}^2)$, *we have that*

$$|\mathbf{E}[C|T'_{j+1}, K_j] - \mathbf{E}_{V_{B,j}, \text{Eve}_j|T'_{j+1}, K_j}[E[C|V_{B,j}, \text{Eve}_j, K_j]]| = O(1/\hat{r}^2).$$

Fact 3. *We have the following equivalence:*

$$\mathbf{E}[C_{j+1}|T'_{j+1}, K_j]] = \mathbf{E}_{V_{B,j}|T'_{j+1}, K_j}[C_{j+1}(V_{B,j})]$$

The above immediately imply Lemma 4. We now proceed to prove Claims 1 and 2

Proof. (Claim 1) The hypothesis of Claim 1 and Markov's inequality imply that one of the following must occur:

- With probability $\Omega(1/\hat{r}^{1/2})$ there is a first point where

$$\mathbf{E}_{V_{B,j}, T_j|T_{j+1}, K_j}[\mathbf{E}[C|V_{B,j}, T_j, K_j]] - \mathbf{E}_{V_{B,j}|T_{j+1}, K_j}[C_{j+1}(V_{B,j})] \geq \Omega(1/\hat{r}^{1/4})$$

- With probability $\Omega(1/\hat{r}^{1/4})$ there is a first point where

$$\mathbf{E}_{V_{B,j}, T_j|T_{j+1}, K_j}[\mathbf{E}[C|V_{B,j}, T_j, K_j]] - \mathbf{E}_{V_{B,j}|T_{j+1}, K_j}[C_{j+1}(V_{B,j})] \geq \Omega(1/\hat{r}^{1/2})$$

Let us assume that the first case above occurs. The analysis for the remaining case is entirely analogous. Now, by Claim 2 we have that with probability $1 - O(1/\hat{r}^2)$, over $T_{j+1} = t_{j+1}, K_j = k_j$ drawn from $\mathcal{D}_{extend}$ the distributions $V_{B,j} \mid T_{j+1} = t_{j+1}, K_j = k_j$ and $V_{B,j} \mid T'_{j+1} = t_{j+1}, K_j = k_j$ are $O(1/\hat{r}^2)$-close.

Thus, we have that with probability $\Omega(1/\hat{r}^{1/2})$ over $T_{j+1} = t_{j+1}, K_j = k_j$ drawn from $\mathcal{D}_{extend}$

$$\mathbf{E}_{V_{B,j}, T_j|T_{j+1}=t_{j+1}, K_j}[\mathbf{E}[C|V_{B,j}, T_j, K_j]] - \mathbf{E}_{V_{B,j}|T'_{j+1}=t_{j+1}, K_j}[C_{j+1}(V_{B,j})] = \Omega(1/\hat{r}^{1/4}). \tag{4.1}$$

By definition of $I^{f=1/\hat{r}^{1/4}}_{EV_{j+1}}$ we have by (4.1) that $\Pr[I^{f=1/\hat{r}^{1/4}}_{EV_{j+1}} = 1$ for some $1 \leq j \leq 2\hat{r}] = \Omega(1/\hat{r}^{1/2})$. In the following, we will use this fact to show that $\Pr[I^{f=1/\hat{r}^{1/4}}_{EV'_{j+1}} = 1$ for some $1 \leq j \leq 2\hat{r}] = \Omega(1/\hat{r}^{1/2})$ as well. This will immediately imply the Claim. First, for $1 \leq w \leq \hat{r}$, where $j = 2w - 2$, we define

$$v_{2w-1} = \Pr[I^{f=1/\hat{r}^{1/4}}_{EV_{j+1}} = 1 \wedge I^{f=1/\hat{r}^{1/4}}_{EV_j} = 0 \wedge I^{f=1/\hat{r}^{1/4}}_{EV'_j} = 0]$$

$$y_{2w-1} = \Pr[I^{f=1/\hat{r}^{1/4}}_{EV'_{j+1}} = 1 \wedge I^{f=1/\hat{r}^{1/4}}_{EV_j} = 0 \wedge I^{f=1/\hat{r}^{1/4}}_{EV'_j} = 0].$$

Now, we have by Claim 1 that for every $1 \leq w \leq \hat{r}$, $j = 2w - 2$, one of the following occurs:

- $\Pr[I^{f=1/\hat{r}^{1/4}}_{EV_j} = 0 \wedge I^{f=1/\hat{r}^{1/4}}_{EV'_j} = 0] = O(1/\hat{r}^2)$
- The distributions $T_{j+1}, K_j \mid I^{f=1/\hat{r}^{1/4}}_{EV_j} = 0, I^{f=1/\hat{r}^{1/4}}_{EV'_j} = 0$ and $T'_{j+1}, K_j \mid I^{f=1/\hat{r}^{1/4}}_{EV_j} = 0, I^{f=1/\hat{r}^{1/4}}_{EV'_j} = 0$ are at most $O(1/\hat{r}^2)$-far.

We show that in both cases, we must have that

$$v_{2w-1} = y_{2w-1} \pm O(1/\hat{r}^2). \tag{4.2}$$

In the first case, we clearly must have that $0 \le v_{2w-1}, y_{2w-1} \le O(1/\hat{r}^2)$. In the second case, we bound the difference between v_{2w-1}, y_{2w-1} in the following way:

$$
\begin{aligned}
v_{2w-1} &= \Pr[I^{f=1/\hat{r}^{1/4}}_{\mathsf{EV}_{j+1}} = 1 \wedge I^{f=1/\hat{r}^{1/4}}_{\mathsf{EV}_j} = 0 \wedge I^{f=1/\hat{r}^{1/4}}_{\mathsf{EV}'_j} = 0] \\
&= \Pr[I^{f=1/\hat{r}^{1/4}}_{\mathsf{EV}_{j+1}} = 1 \mid I^{f=1/\hat{r}^{1/4}}_{\mathsf{EV}_j} = 0, I^{f=1/\hat{r}^{1/4}}_{\mathsf{EV}'_j} = 0] \cdot \Pr[I^{f=1/\hat{r}^{1/4}}_{\mathsf{EV}_j} = 0 \wedge I^{f=1/\hat{r}^{1/4}}_{\mathsf{EV}'_j} = 0] \\
&= \left(\Pr[I^{f=1/\hat{r}^{1/4}}_{\mathsf{EV}'_{j+1}} = 1 \mid I^{f=1/\hat{r}^{1/4}}_{\mathsf{EV}_j} = 0, I^{f=1/\hat{r}^{1/4}}_{\mathsf{EV}'_j} = 0] \pm O(1/\hat{r}^2) \right) \\
&\quad \cdot \Pr[I^{f=1/\hat{r}^{1/4}}_{\mathsf{EV}_j} = 0 \wedge I^{f=1/\hat{r}^{1/4}}_{\mathsf{EV}'_j} = 0] \\
&= \Pr[I^{f=1/\hat{r}^{1/4}}_{\mathsf{EV}'_{j+1}} = 1 \wedge I^{f=1/\hat{r}^{1/4}}_{\mathsf{EV}_j} = 0, I^{f=1/\hat{r}^{1/4}}_{\mathsf{EV}'_j} = 0] \pm O(1/\hat{r}^2) \\
&= y_{2w-1} \pm O(1/\hat{r}^2),
\end{aligned}
$$

where the third equality follows since $I^{f=1/\hat{r}^{1/4}}_{\mathsf{EV}_{j+1}}, I^{f=1/\hat{r}^{1/4}}_{\mathsf{EV}'_{j+1}}$ are completely determined by $\mathsf{T}_{j+1}, \mathsf{K}_j$ and $\mathsf{T}'_{j+1}, \mathsf{K}_j$, respectively.

Now, using the definition of $I^{f=1/\hat{r}^{1/4}}_{\mathsf{EV}_{j+1}}$ and (4.2) above, we have that

$$\Pr[I^{f=1/\hat{r}^{1/4}}_{\mathsf{EV}_{j+1}} = 1 \text{ for some } j = 2w - 2, 1 \le w \le \hat{r}] \le \sum_{w=1}^{\hat{r}} v_{2w-1} + y_{2w-1}$$

$$\le O(1/\hat{r}) + 2 \sum_{w=1}^{\hat{r}} y_{2w-1}.$$

Moreover, (4.1) implies that

$$\Omega(1/\hat{r}^{1/2}) = \Pr[I^{f=1/\hat{r}^{1/4}}_{\mathsf{EV}_{j+1}} = 1 \text{ for some } j = 2w - 2, 1 \le w \le \hat{r}]$$

$$\le O(1/\hat{r}) + 2 \sum_{w=1}^{\hat{r}} y_{2w-1}.$$

Thus, it must be the case that

$$\sum_{w=1}^{\hat{r}} y_{2w-1} = \sum_{w=1}^{\hat{r}} \Pr[I^{f=1/\hat{r}^{1/4}}_{\mathsf{EV}'_{2w-1}} = 1 \wedge I^{f=1/\hat{r}^{1/4}}_{\mathsf{EV}_{2w-2}} = 0 \wedge I^{f=1/\hat{r}^{1/4}}_{\mathsf{EV}'_{2w-2}} = 0]$$

$$= \Omega(1/\hat{r}^{1/2}).$$

Finally, by definition, this implies that with probability $\Omega(1/\hat{r}^{1/2})$ over $\mathcal{D}_{extend}$ we have some j such that $I^{f=1/\hat{r}^{1/4}}_{\mathsf{EV}_j} = 0, I^{f=1/\hat{r}^{1/4}}_{\mathsf{EV}'_j} = 0$ and

$$\mathbf{E}_{\mathsf{V}_{\mathsf{B},j}, \mathsf{Eve}_j \mid \mathsf{T}'_{j+1}, \mathsf{K}_j}[E[C \mid \mathsf{V}_{\mathsf{B},j}, \mathsf{Eve}_j, \mathsf{K}_j]] - \mathbf{E}_{\mathsf{V}_{\mathsf{B},j} \mid \mathsf{T}'_{j+1}, \mathsf{K}_j}[C_{j+1}(\mathsf{V}_{\mathsf{B},j})] = \Omega(1/\hat{r}^{1/4}).$$

and so the claim is proved.

Proof. (Claim 2) Towards proving the claim, note that by Lemma 1 we have that with probability $1 - O(1/\hat{r}^4)$ for every j

$$V_{A,j}, V_{B,j} \mid T_j \qquad\qquad V_{A,j} \mid T_j \times V_{B,j} \mid T_j$$

are $O(1/\hat{r}^4)$-close.

Thus, by applying Markov's inequality, we have that with probability $1 - O(1/\hat{r}^2)$, for every $1 \leq j \leq 2\hat{r}$,

$$V_{A,j}, V_{B,j} \mid T'_{j+1}, K_j \qquad\qquad V_{A,j} \mid T'_{j+1}, K_j \times V_{B,j} \mid T'_{j+1}, K_j$$

are $O(1/\hat{r}^2)$-close.

Now, by applying Property 3 we have that for every $1 \leq j \leq 2\hat{r}$ with probability $1 - O(1/\hat{r}^2)$, over draws of $T_j = t_j$, $T'_{j+1} = t_j \| m'_{j+1}, eve'_{j+1}, K_j = k_j$,

$$V_{A,j} \mid T'_{j+1}, K_j \times V_{B,j} \mid T'_{j+1}, K_j \qquad\qquad V_{A,j} \mid T_j, K_j \times V_{B,j} \mid T'_{j+1}, K_j$$

are $O(1/\hat{r}^2)$-close.

By combining the above, we have that for every j, with probability $1 - O(1/\hat{r}^2)$, over draws of $T_j = t_j$, $T'_{j+1} = t_j \| m'_{j+1}, eve'_{j+1}, K_j = k_j$,

$$V_{A,j}, V_{B,j} \mid T'_{j+1}, K_j, \qquad\qquad V_{A,j} \mid T_j, K_j \times V_{B,j} \mid T'_{j+1}, K_j \qquad (4.3)$$

are $O(1/\hat{r}^2)$-close.

Now, let us consider the expression $\mathbf{E}_{V_{B,j}, T_j \mid T'_{j+1}, K_j}[\mathbf{E}[C \mid V_{B,j}, T_j, K_j]]$ and the expression $\mathbf{E}[C \mid T'_{j+1}, K_j]$. If we expand notation, we have that:

$$\mathbf{E}_{V_{B,j}, T_j \mid T'_{j+1}, K_j}[\mathbf{E}[C \mid V_{B,j}, T_j, K_j]] = \mathbf{E}_{V_{B,j}, T_j \mid T'_{j+1}, K_j}[\mathbf{E}_{V_{A,2\hat{r}}, V_{B,2\hat{r}} \mid T_j, K_j, V_{B,j}}[C(V_{A,2\hat{r}}, V_{B,2\hat{r}})]].$$

and that

$$\mathbf{E}[C \mid T'_{j+1}, K_j] = \mathbf{E}_{V_{A,2\hat{r}}, V_{B,2\hat{r}} \mid T'_{j+1}, K_j}[C(View_{A,2\hat{r}}, V_{B,2\hat{r}})].$$

Due to the function-obliviousness property (see Property 1), we have that $C(V_A, V_B)$ depends only on the random tapes r_A, r_B of A, B, which are contained in the partial views $V_{A,j}, V_{B,j}$, and so

$$\mathbf{E}_{V_{B,j}, T_j \mid T'_{j+1}, K_j}[\mathbf{E}_{V_A, V_B \mid T_j, K_j, V_{B,j}}[C(r_A, r_B)]] = \mathbf{E}_{V_{B,j}, T_j \mid T'_{j+1}, K_j}[\mathbf{E}_{V_{A,j} \mid T_j, K_j, V_{B,j}}[C(r_A, r_B)]]$$

and that

$$\mathbf{E}[C \mid T'_{j+1}, K_j] = \mathbf{E}_{V_{A,j}, V_{B,j} \mid T'_{j+1}, K_j}[C(r_A, r_B)].$$

Next, we have by Lemma 1 and Markov's inequality, that with probability $1 - O(1/\hat{r}^2)$,

$$V_{A,j} \mid V_{B,j}, T_j, K_j \qquad\qquad V_{A,j} \mid T_j, K_j$$

are $O(1/\hat{r}^2)$-close.

Thus, we have that with probability $1 - O(1/\hat{r}^2)$:

$$\left| \mathbf{E}_{\mathsf{V}_{\mathsf{B},j},\mathsf{T}_j | \mathsf{T}'_{j+1},\mathsf{K}_j} [\mathbf{E}[C \mid \mathsf{V}_{\mathsf{B},j},\mathsf{T}_j,\mathsf{K}_j]] - \mathbf{E}_{\mathsf{V}_{\mathsf{B},j},\mathsf{T}_j | \mathsf{T}'_{j+1},\mathsf{K}_j} [\mathbf{E}_{\mathsf{V}_{\mathsf{A},j} | \mathsf{T}_j,\mathsf{K}_j} [C(\mathsf{V}_{\mathsf{A},j},\mathsf{V}_{\mathsf{B},j})]] \right| = O(1/\hat{r}^2).$$

Equivalently, we have that with probability $1 - O(1/\hat{r}^2)$ over draws of $\mathsf{T}_j = t_j$, $\mathsf{T}'_{j+1} = t_j \| m'_{j+1}, eve'_{j+1}, \mathsf{K}_j = k_j$:

$$\left| \mathbf{E}_{\mathsf{V}_{\mathsf{B},j},\mathsf{T}_j | \mathsf{T}'_{j+1},\mathsf{K}_j} [\mathbf{E}[C \mid \mathsf{V}_{\mathsf{B},j},\mathsf{T}_j,\mathsf{K}_j]] - \mathbf{E}_{\mathsf{V}_{\mathsf{A},j} | \mathsf{T}_j,\mathsf{K}_j \times \mathsf{V}_{\mathsf{B},j} | \mathsf{T}'_{j+1},\mathsf{K}_j} [C] \right| = O(1/\hat{r}^2).$$

Finally, by applying (4.3), and since $\mathsf{V}_{\mathsf{B},j},\mathsf{T}_j$ and $\mathsf{V}_{\mathsf{B},j},\mathsf{Eve}_j$ contain the same information, we have that with all but $O(1/\hat{r}^2)$ probability,

$$\left| \mathbf{E}_{\mathsf{V}_{\mathsf{B},j},\mathsf{Eve}_j | \mathsf{T}'_{j+1},\mathsf{K}_j} [\mathbf{E}[C | \mathsf{V}_{\mathsf{B},j},\mathsf{Eve}_j,\mathsf{K}_j]] - \mathbf{E}_{\mathsf{V}_{\mathsf{A},j},\mathsf{V}_{\mathsf{B},j} | \mathsf{T}'_{j+1},\mathsf{K}_j)} [C(\mathsf{V}_{\mathsf{A},j},\mathsf{V}_{\mathsf{B},j})] \right| = O(1/\hat{r}^2).$$

Equivalently, we have that with all but $O(1/\hat{r}^2)$ probability,

$$\left| \mathbf{E}[C | \mathsf{T}'_{j+1},\mathsf{K}_j] - \mathbf{E}_{\mathsf{V}_{\mathsf{B},j},\mathsf{Eve}_j | \mathsf{T}'_{j+1},\mathsf{K}_j} [\mathbf{E}[C | \mathsf{V}_{\mathsf{B},j},\mathsf{Eve}_j,\mathsf{K}_j]] \right| = O(1/\hat{r}^2),$$

and so the claim is proved.

References

[Blu82]   Blum, M.: Coin flipping by telephone - a protocol for solving impossible problems. In: COMPCON, pp. 133–137 (1982)

[BM07]   Barak, B., Mahmoody, M.: Lower bounds on signatures from symmetric primitives. In: FOCS: IEEE Symposium on Foundations of Computer Science, FOCS (2007)

[BM09]   Barak, B., Mahmoody-Ghidary, M.: Merkle puzzles are optimal–an $o(n^2)$-query attack on key exchange from a random oracle. In: Halevi, S. (ed.) CRYPTO 2009. LNCS, vol. 5677, pp. 374–390. Springer, Heidelberg (2009)

[BM13]   Barak, B., Mahmoody, M.: Merkle's key agreement protocol is optimal - an $O(n^2)$-query attack on any key exchange from random oracles (2013), http://www.cs.cornell.edu/~mohammad/files/papers/MerkleFull.pdf

[CI93]   Cleve, R., Impagliazzo, R.: Martingales, collective coin flipping and discrete control processes (1993) (unpublished)

[Cle86]   Cleve, R.: Limits on the security of coin flips when half the processors are faulty (extended abstract). In: STOC, pp. 364–369 (1986)

[DSLMM11] Dachman-Soled, D., Lindell, Y., Mahmoody, M., Malkin, T.: On the black-box complexity of optimally-fair coin tossing. In: Ishai, Y. (ed.) TCC 2011. LNCS, vol. 6597, pp. 450–467. Springer, Heidelberg (2011)

[GGM86]   Goldreich, O., Goldwasser, S., Micali, S.: How to construct random functions. J. ACM 33(4), 792–807 (1986)

[GM84]   Goldwasser, S., Micali, S.: Probabilistic encryption. J. Comput. Syst. Sci. 28(2), 270–299 (1984)

[GT00]   Gennaro, R., Trevisan, L.: Lower bounds on the efficiency of generic cryptographic constructions. In: FOCS, pp. 305–313 (2000)

[HILL99]   Håstad, J., Impagliazzo, R., Levin, L.A., Luby, M.: A pseudorandom generator from any one-way function. SIAM J. Comput. 28(4), 1364–1396 (1999)

[HOZ13]   Haitner, I., Omri, E., Zarosim, H.: Limits on the usefulness of random oracles. In: Sahai, A. (ed.) TCC 2013. LNCS, vol. 7785, pp. 437–456. Springer, Heidelberg (2013)

[IL89]   Impagliazzo, R., Luby, M.: One-way functions are essential for complexity based cryptography (extended abstract). In: FOCS, pp. 230–235 (1989)

[IR89]   Impagliazzo, R., Rudich, S.: Limits on the provable consequences of one-way permutations. In: STOC, pp. 44–61 (1989)

[LR88]   Luby, M., Rackoff, C.: How to construct pseudorandom permutations from pseudorandom functions. SIAM J. Comput. 17(2), 373–386 (1988)

[MMP13]   Mahmoody, M., Maji, H.K., Prabhakaran, M.: Limits of random oracles in secure computation. To Appear in: Innovations in Theoretical Computer Science, ITCS (2013)

[MNS09]   Moran, T., Naor, M., Segev, G.: An optimally fair coin toss. In: Reingold, O. (ed.) TCC 2009. LNCS, vol. 5444, pp. 1–18. Springer, Heidelberg (2009)

[Nao91]   Naor, M.: Bit commitment using pseudorandomness. J. Cryptology 4(2), 151–158 (1991)

[NY89]   Naor, M., Yung, M.: Universal one-way hash functions and their cryptographic applications. In: STOC, pp. 33–43 (1989)

[Rom90]   Rompel, J.: One-way functions are necessary and sufficient for secure signatures. In: STOC, pp. 387–394 (1990)

[RTV04]   Reingold, O., Trevisan, L., Vadhan, S.: Notions of reducibility between cryptographic primitives. In: Naor, M. (ed.) TCC 2004. LNCS, vol. 2951, pp. 1–20. Springer, Heidelberg (2004)

[Yao82]   Yao, A.C.-C.: Theory and applications of trapdoor functions. In: FOCS, pp. 80–91 (1982)

On the Power of Public-Key Encryption
in Secure Computation

Mohammad Mahmoody[1], Hemanta K. Maji[2], and Manoj Prabhakaran[3]

[1] University of Virginia*
mohammad@cs.virginia.edu
[2] University of California, Los Angeles**
hmaji@cs.ucla.edu
[3] University of Illinois, Urbana-Champaign***
mmp@cs.uiuc.edu

Abstract. We qualitatively separate semi-honest secure computation of non-trivial secure-function evaluation (SFE) functionalities from existence of key-agreement protocols. Technically, we show the existence of an oracle (namely, PKE-oracle) relative to which key-agreement protocols exist; but it is useless for semi-honest secure realization of symmetric 2-party (deterministic finite) SFE functionalities, i.e. any SFE which can be securely performed relative to this oracle can also be securely performed in the plain model.

Our main result has following consequences.

- There exists an oracle which is useful for some 3-party deterministic SFE; but useless for semi-honest secure realization of any general 2-party (deterministic finite) SFE.
- With respect to semi-honest, standalone or UC security, existence of key-agreement protocols (if used in black-box manner) is only as useful as the commitment-hybrid for general 2-party (deterministic finite) SFE functionalities.

This work advances (and conceptually simplifies) several state-of-the-art techniques in the field of black-box separations:

1. We introduce a general *common-information learning* algorithm (CIL) which extends the "eavesdropper" in prior work [1,2,3], to protocols whose message can depend on information gathered by the CIL so far.
2. With the help of this CIL, we show that in a secure 2-party protocol using an idealized PKE oracle, surprisingly, decryption queries are useless.
3. The idealized PKE oracle with its decryption facility removed can be modeled as a collection of *image-testable random-oracles*. We extend the analysis approaches of prior work on random oracle [1,2,4,5,3] to apply to this class of oracles. This shows that these oracles are useless for semi-honest 2-party SFE (as well as for key-agreement).

These information theoretic impossibility results can be naturally extended to yield black-box separation results (cf. [6]).

* Research done while at Cornell and supported in part by NSF Awards CNS-1217821 and CCF-0746990, AFOSR Award FA9550-10-1-0093, and DARPA and AFRL under contract FA8750-11-2- 0211.
** Supported by NSF CI Postdoctoral Fellowship.
*** Research supported in part by NSF grants 1228856 and 0747027.

Y. Lindell (Ed.): TCC 2014, LNCS 8349, pp. 240–264, 2014.

1 Introduction

Public-key encryption (PKE) is an important security primitive in a system involving *more than two parties*. In this work, we ask if PKE could be useful for protecting two mutually distrusting parties against each other, *if there is no other party involved*. More specifically, we ask if the existence of PKE can facilitate 2-party secure function evaluation (SFE). Informally, our main result in this work shows the following:

> *The existence of PKE (as a computational complexity assumption, when used in a black-box manner) is useless for semi-honest secure evaluation of any finite, deterministic 2-party function.*

Here, a complexity assumption being "useless" for a task means that the task can be realized using that assumption alone (in a black-box manner) if and only if it can be realized unconditionally (i.e., information-theoretically).[1] As is typical in this line of research, our focus is on deterministic functions whose domain-size is finite. (However, all our results extend to the case when the domain-size grows *polynomially* in the security parameter; our proofs (as well as the results we build on) do not extend to exponentially growing domain-sizes, though.) Technically, we show an "oracle-separation" result, by presenting a randomized oracle which enables PKE in the information-theoretic setting, but does not enable SFE for any 2-party function for which SFE was impossible without the oracle. Then, using standard techniques, this information theoretic impossibility result is translated into the above black-box separation result [6]. While the above statement refers to semi-honest security, as we shall shortly see, a similar statement holds for security against active corruption, as well.

It is instructive to view our result in the context of "cryptographic complexity" theory [7]: with every (finite, deterministic) multi-party function f, one can associate a computational intractability assumption that there exists a secure computation protocol for f that is secure against semi-honest corruption.[2] Two assumptions are considered distinct unless they can be black-box reduced to each other. Then, the above result implies that secure key agreement (i.e., the interactive analog of PKE) does not belong to the universe of assumptions associated with 2-party functions. However, it is not hard to see that there are 3-party functions f such that a semi-honest secure protocol for f (in the broadcast channel model) is equivalent to a key agreement protocol.[3] Thus we obtain the following important conclusion:

---

[1] The task here refers to 2-party SFE in the "plain" model. We do not rule out the possibility that PKE is useful for 2-party SFE in a "hybrid" model, where the parties have access to a trusted third party.

[2] This is the simplest form of assumptions associated with functionalities in [7], where a more general framework is presented.

[3] As an example, consider the 3-party function $f(x, y, z) = x \oplus y$. A semi-honest secure protocol π for f over a broadcast channel can be black-box converted to a key-agreement protocol between Alice and Bob, where, say, Alice plays the role of the first party in π with the key as its input, and Bob plays the role of the second and third parties with random inputs. Conversely, a key-agreement protocol can be used as a black-box in a semi-honest secure protocol for f, in which the first party sends its input to the second party encrypted using a key that the two of them generate using the key-agreement protocol.

> *The set of computational complexity assumptions associated (in the above sense)*
> *with 3-party functions is strictly larger than the set associated with 2-party*
> *functions.*

This answers an open question posed in [7], but raises many more questions. In particular, we ask if the same conclusion holds if we consider $(n + 1)$-party functions and n-party functions, for every $n > 2$.

Another consequence of our main result is its implications for SFE secure against *active* corruption. Following a related work in [5], using characterizations of functions that have SFE protocols secure against semi-honest and active corruptions [8,9,10,11], we obtain the following corollary of our main result.

> *The existence of PKE (as a black-box assumption) is exactly as useful as a*
> *commitment functionality (given as a trusted third party) for secure evaluation*
> *of any finite, deterministic 2-party function. This holds for semi-honest security,*
> *standalone active security and UC-security.*

Note that for semi-honest security, the commitment functionality is not useful at all (since semi-honest parties can commit using a trivial protocol), and this agrees with the original statement. The interesting part of the corollary is the statement about active (standalone or UC) security. Commitment is a "minicrypt" functionality that can be implemented using one-way functions (in the standalone setting) or random oracles. PKE, on the other hand, is not a minicrypt primitive [1]. Yet, in the context of guaranteeing security for *two* mutually distrusting parties, computing a (finite, deterministic) function, without involving a trusted third party, PKE is no more useful than the commitment functionality.

In the rest of this section, we state our main results more formally, and present an overview of the techniques. But first we briefly mention some of the related work.

1.1 Related Work

Impagliazzo and Rudich [1] showed that random oracles are not useful against a computationally unbounded adversary for the task of secure key agreement. This analysis was recently simplified and sharpened in [2,3]. Haitner, Omri, and Zarosim [12,3] show that random oracles are essentially useless in any *inputless protocol*.[4]

Following [1] many other black-box separation results have appeared (e.g. [13,14,15,16,17]). In particular, Gertner et. al [18] insightfully asked the question of comparing oblivious-transfer (OT) and key agreement (KA) and showed that OT is strictly more complex (in the sense of [1]). Another trend of results has been to prove lower-bounds on the efficiency of the implementation in black-box constructions (e.g. [19,20,21,22,23,2,2]). A complementary approach has been to find black-box reductions when they exist (e.g. [24,25,26,27,28]). Also, results in the black-box separation framework of [1,6] have immediate consequences for computational

---

[4] Ideally, a result similar to that of [3] should be proven in our setting of secure function evaluation too, where parties do have private inputs, as it would extend to randomized functions as well. While quite plausible, such a result remains elusive.

complexity theory. Indeed, as mentioned above, separations in this framework can be interpreted as new worlds in Impagliazzo's universe [29].

Our work relies heavily on [5], where a similar result was proven for one-way functions instead of PKE. While we cannot use the result of [5] (which we strictly improve upon) in a black-box manner, we do manage to exploit the modularity of the proof there and avoid duplicating any significant parts of the proof.

1.2 Our Contribution

For brevity, in the following we shall refer to "2-party deterministic SFE functions with polynomially large domains" simply as SFE functions. Also, we consider security against semi-honest adversaries in the information theoretic setting, unless otherwise specified (as in Corollary 1).

Our main result establishes that there exists an oracle which facilitates key-agreement while being useless to 2-party SFE.

Theorem 1. *There exists an oracle* $\mathbb{PKE}$ *such that, the following hold:*

- *There is a secure key-agreement protocol (or equivalently, a semi-honest secure 3-party XOR protocol) using* $\mathbb{PKE}$.
- *A general 2-party deterministic function* f, *with a polynomially large domain, has a semi-honest secure protocol against computationally unbounded adversaries using* $\mathbb{PKE}$ *if and only if* f *has a perfectly semi-honest secure protocol in the plain model.*

As discussed below, this proof breaks into two parts — a compiler that shows that the decryption queries can be omitted, and a proof that the oracle without the decryption queries is not useful for SFE. For proving the latter statement, we heavily rely on a recent result from [5] for random oracles; however, this proof is modular, involving a "frontier analytic" argument, which uses a few well-defined properties regarding the oracles. Our contribution in this is to prove these properties for a more sophisticated oracle class (namely, family of image-testable random oracles), rather than random oracles themselves.

As in [5], Theorem 1 translates to a black-box separation of the primitive PKE from non-trivial SFE. Also, it yields the following corollary, that against active corruption, our $\mathbb{PKE}$ oracle is only as useful as the commitment-hybrid model, as far as secure protocols for 2-party SFE is concerned.

Corollary 1. *There exists an oracle* $\mathbb{PKE}$ *such that, the following hold:*

- *There is a secure key-agreement protocol (or equivalently, a semi-honest secure 3-party XOR protocol) using* $\mathbb{PKE}$.
- *A general 2-party deterministic function* f, *with a polynomially large domain, has a statistically semi-honest, standalone or UC-secure protocol relative to* $\mathbb{PKE}$ *if and only if* f *has a perfectly, resp., semi-honest, standalone or UC-secure protocol in the commitment-hybrid.*

Apart from there results, and their implications to the complexity of 2-party and 3-party functions, we make important technical contributions in this work. As described below, our "common-information learner" is simpler than that in prior work. This also helps us handle a more involved oracle class used to model PKE. Another module in our proof is a compiler that shows that the decryption facility in PKE is not needed in a (semi-honest secure) protocol that uses PKE, even if the PKE is implemented using an idealized oracle.

1.3   Technical Overview

The main result we need to prove (from which our final results follow, using arguments in [5]) is that there is an oracle class $\mathbb{PKE}_\kappa$ relative to which secure public-key encryption (i.e., 2-round key agreement) protocol exists, but there is no secure protocol for any non-trivial SFE function relative to it.

The oracle class $\mathbb{PKE}_\kappa$ is a collection of following correlated oracles:

- Gen($\cdot$): It is a (length-tripling injective) random oracle which maps secret keys sk to respective public keys pk.
- Enc($\cdot, \cdot$): For each public key pk, it is an independently chosen (length tripling injective) random oracle which maps messages m to cipher texts c.
- Dec($\cdot, \cdot$): Given a valid secret key sk and a valid cipher text c it outputs m such that message m was encrypted using public key $pk = \mathsf{Gen}(sk)$.
- Additionally, it provides test oracles Test which output whether a public key pk is a valid public key or not; and whether a cipher text c has been created using a public key pk or not.

Note that without the Test oracle, this oracle class can be used to semi-honest securely perform OT; hence, all 2-party SFE will be trivial relative to it (see discussion in [18,30]). The main technical contribution of this paper is the negative result which shows that the above mentioned oracle class $\mathbb{PKE}_\kappa$ is useless for 2-party SFE against semi-honest adversaries.

This is shown in two steps:

1. First, we show that the decryption oracle Dec($\cdot, \cdot$) is not useful against semi-honest adversaries. That is, given a (purported) semi-honest secure protocol ρ for a 2-party SFE f we compile it into another semi-honest secure protocol Π (with identical round complexity, albeit polynomially more query complexity) which has slightly worse security but performs no decryption-queries.
2. Finally, we observe that the oracle class "$\mathbb{PKE}_\kappa$ minus the decryption oracle" is identical to image-testable random-oracles. And we extend the negative result of [5] to claim that this oracle class is useless for 2-party SFE.

The key component in both these steps is the *Common Information Learner* algorithm, relative to image-testable random oracle class. But we begin by introducing image-testable random oracles.

Image-testable Random-oracle Class. It is a pair of correlated oracles (R, T), where R is a (length-tripling injective) random oracle and test oracle T which outputs whether a point in range has a pre-image or not. We consider *keyed-version* of these oracle, where for each key in an exponentially large key-space $\mathbb{K}$ we have an independent copy of image-testable random oracle.

Note that the answer to an image test query can slightly change the distribution of exponentially many other queries; namely, when we know that y is not in the image of R, the answer to any query x for R will not be uniformly distributed (because it cannot be y). However, since the number of tested images are polynomial-size and the number of possible queries to R are exponentially large, this will affect the distribution of the answers by R only negligibly. Also, because of the expansion of the random oracle R, the fraction of the image-size of R is negligibly small relative to the range of R. So an algorithm, with polynomially-bounded query complexity, who queries the test oracle T has negligible chance of getting a positive answer (i.e. an image) without actually calling R. We emphasize that our whole analysis is conditioned on this event (i.e. accidentally discovering y in the image of the oracle) not taking place; and this requires careful accounting of events because it holds only for (polynomially) bounded query algorithms.

Common Information Learner. The common information learner is a procedure that can see the transcript of an oracle-based protocol between Alice and Bob, and by making a polynomial number of publicly computable queries to the oracle, derives sufficient information such that conditioned on this information, the views of Alice and Bob are almost independent of each other. Our common information learner is similar in spirit to those in [1,2,5,3] but is different and more general in several ways:

- **Handling Image-Testable Oracles.** Our common information learner applies to the case when the oracle is not just a random oracle, but an image-testable random oracle family.[5]
- **Interaction between Learner and the System.** It is important for the first part of our proof (i.e. compiling out the decryption queries) that the common information learner *interacts with the protocol execution itself.* That is, at each round the information gathered by the common information learner is used by the parties in subsequent rounds. We require the common information learner to still make only a polynomial number or oracle queries while ensuring that conditioned on the information it gathers, the views of the two parties remain almost independent. In showing that the common information learner is still efficient, we show a more general result in terms of an interactive process between an oracle system (the Alice-Bob system, in our case) and a learner, both with access to an arbitrary oracle (possibly correlated with the local random tapes of Alice and Bob).
- **Simpler Description of the Learner.** The common information learner in our work has a simpler description than that in [1,2,5]. Our learner queries the random oracle with queries that are most likely to be queried by Alice or Bob in a protocol execution. The learner in [2,5] is similar, but uses probabilities not based on the actual

---

[5] The work of [3] also handles a larger set of oracles than random oracles (called *simple* oracle), but that class is not known to include image-testable oracles as special case [31].

protocol, but a variant of it; this makes the description of their common information learner more complicated, and somewhat complicates the proofs of the query
efficiency of the learner.[6]

Showing that Image-Testable Random Oracles are Useless for SFE. In [5] it was shown
that random oracles are useless for SFE. This proof is modular in that there are four
specific results that depended on the nature of the oracle and the common information
learner. The rest of the proof uses a "frontier analytic" argument that is agnostic to the
oracle and the common information learner. Thus, in this work, to extend the result of
[5] to a family of image-testable random oracles, we need only ensure that these four
properties continue to hold. The four properties are as follows:

1. Alice's message conditioned on the view of the CIL is almost independent of Bob's
   input, see Section 6.1 item 1.
2. Safety holds with high probability, see Section 6.1 item 2.
3. A strong independence property of Alice's and Bob's views conditioned on that of
   the CIL, see Section 6.1 item 3.
4. Finally, local samplability and oblivious rerandomizability of image-testable random oracles which permit simulation of alternate views, see Section 6.

Compiling Out the Decryption Queries. The main idea behind compiling out the decryption queries is that if Alice has created a ciphertext by encrypting a message using
a public-key that was created by Bob, and she realizes that there is at least a small (but
significant) probability that Bob would be querying the decryption oracle on this ciphertext (since he has the secret key), then she would preemptively send the actual message
to him. We need to ensure two competing requirements on the compiler:

1. **Security.** Note that with some probability Alice might send this message even if
   Bob was not about to query the decryption oracle. To argue that this is secure, we
   need to argue that a curious Bob *could have* called the decryption oracle at this
   point, for the same ciphertext.
2. **Completeness.** We need to ensure that in the compiled protocol, Bob will never
   have to call the decryption oracle, as Alice would have sent him the required decryptions ahead of time.

For security, firstly we need to ensure that Alice chooses the set of encryptions to be
revealed based *only* on the common information that Alice and Bob have. This ensures
that Bob can sample a view for himself from the same distribution used by Alice to
compute somewhat likely decryption queries, and obtain the ciphertext and secret-key
from the decryption query made in this view. The one complication that arises here is
the possibility that the secret-key in the sampled view is not the same as the secret-key
in the actual execution. To rule this out, we rely on the independence of the views of
the parties conditioned on the common information. This, combined with the fact that
it is unlikely for a valid public-key to be discovered by the system without either party

---

[6] [1] uses an indirect mechanism to find the heavy queries, and reasoning about their common
information learner is significantly more involved.

having actually called the key-generation oracle using the corresponding secret-key, we can show that it is unlikely for a sampled view to have a secret-key different from the actual one.

For completeness of the compiler, we again rely on the common information learner to ensure that if Alice uses the distribution based on common information to compute which decryption queries are likely, then it is indeed unlikely for Bob to make a decryption query that is considered unlikely by Alice.

1.4 Overview of the Paper

The full version of the paper is available at [32]. In Section 2 we formally define all the relevant oracle classes. Section 3 introduces relevant definitions and notations for this paper. The efficiency of an algorithm which performs heavy-queries is argued in Section 4.1. This is directly used to provide an independence learner for protocols where parties do not have private inputs in Section 4.2. In Section 5 we show that for 2-party deterministic SFE Decryption queries in $\mathbb{PKE}_\kappa$ are useless. Next, in Section 6, we extend Lemma 2 to protocols where parties have private inputs. Finally, we prove our main result (Theorem 1) in Section 7.

2 Oracle Classes

General class of oracles shall be represented by $\mathbb{O}$. We are interested in three main classes of oracles, each parameterized by the security parameter κ.

2.1 Image-Testable Random Oracle Class

The set $\mathbb{O}_\kappa$ consists of the all possible pairs of correlated oracles $O \equiv (R, T)$ of the form:

1. $R : \{0,1\}^\kappa \mapsto \{0,1\}^{3\kappa}$ is a function, and
2. $T : \{0,1\}^{3\kappa} \mapsto \{0,1\}$ is defined by: $T(\beta) = 1$ if there exists $\alpha \in \{0,1\}^\kappa$ such that $R(\alpha) = \beta$; otherwise $T(\beta) = 0$.

This class of oracles is known as *image-testable random oracle* class. Based on the length of the query string we can uniquely determine whether it is a query to R or T oracle. We follow a notational convention. Queries to R oracle shall be denoted by α and its corresponding answer shall be denoted by β.

2.2 Keyed Version of Image-Testable Random Oracle Class

Given a class $\mathbb{K}$ of keys,[7] consider the following oracle $O^{(\mathbb{K})}$: For every $k \in \mathbb{K}$, let $O^{(k)} \in \mathbb{O}_\kappa$. Given a query $\langle k, q \rangle$, where $k \in \mathbb{K}$ and q is the query to an oracle in $\mathbb{O}_\kappa$, answer it with $O^{(k)}(q)$. Let $\mathbb{O}_\kappa^{(\mathbb{K})}$ be the set of all possible oracles $O^{(\mathbb{K})}$. This class of oracle $\mathbb{O}_\kappa^{(\mathbb{K})}$ is called *keyed-version of image-testable random oracle* class.

---

[7] Note that the description of the keys in $\mathbb{K}$ is poly(κ); so the size of the set $\mathbb{K}$ could possibly be exponential in κ.

2.3   Public-Key Encryption Oracle

We shall use a "PKE-enabling" oracle similar to the one used in [18]. With access to this oracle, a semantically secure public-key encryption scheme can be readily constructed,[8] yet we shall show that it is useless for SFE. This oracle, which we will call $\mathbb{PKE}_\kappa$ (or simply $\mathbb{PKE}$, when κ is understood), is a collection of the oracles (Gen, Enc, Test$_1$, Test$_2$, Dec) defined as follows:

- Gen: It is a length-tripling random oracle from the set of inputs $\{0,1\}^\kappa$ to $\{0,1\}^{3\kappa}$. It takes as input a secret key sk and provides a public-key pk corresponding to it, i.e. Gen$(sk) = pk$.
- Enc: This is an "encryption" oracle. It can be defined as a collection of length-tripling random oracles, keyed by strings in $\{0,1\}^{3\kappa}$. For each key $pk \in \{0,1\}^{3\kappa}$, the oracle implements a random function from $\{0,1\}^\kappa$ to $\{0,1\}^{3\kappa}$. When queried with a (possibly invalid) public key pk, and a message $m \in \{0,1\}^\kappa$, this oracle provides the corresponding cipher text $c \in \{0,1\}^{3\kappa}$ for it, i.e. Enc$(pk, m) = c$.
- Test$_1$: It is a test function which tests the validity of a public key, i.e. given a public-key pk, it outputs 1 if and only if there exists a secret key sk such that Gen$(sk) = pk$.
- Test$_2$: It is a test function which tests the validity of a public key and cipher text pair, i.e. given a public-key pk and cipher text c, it outputs 1 if and only if there exists m such that Enc$(pk, m) = c$.
- Dec: This is the decryption oracle, from $\{0,1\}^\kappa \times \{0,1\}^{3\kappa}$ to $\{0,1\}^\kappa \cup \{\bot\}$, which takes a secret-key, cipher-text pair (sk, c) and returns the lexicographically smallest m such that Enc(Gen$(sk), m) = c$. If no such m exists, it outputs $\bot$.

We note that the encryption oracle produces cipher texts for public keys pk irrespective of whether there exists sk satisfying Gen$(sk) = pk$. This is crucial because we want to key set $\mathbb{K}$ to be defined independent of the Gen oracle.

$\mathbb{PKE}_\kappa$ *Without* Dec. We note that if we remove the oracle Dec, the above oracle is exactly the same as the image-testable random oracle $\mathbb{O}_\kappa^{(\mathbb{K})}$, with $\mathbb{K} = \{0,1\}^{3\kappa} \cup \{\bot\}$. Here we identify the various queries to $\mathbb{PKE}_\kappa$ with queries to $\mathbb{O}_\kappa^{(\mathbb{K})}$ as follows: Gen(sk) corresponds to the query $\langle \bot, sk \rangle$, Enc(pk, m) corresponds to $\langle pk, m \rangle$, Test$_1(pk)$ corresponds to $\langle \bot, pk \rangle$ and Test$_2(pk, c)$ corresponds to $\langle pk, c \rangle$.

3   Preliminaries

We say $a = b \pm c$ if $|a - b| \leq c$. We shall use the convention that a random variable shall be represented by a bold face, for example $\mathbf{X}$; and a corresponding value of the random variable without bold face, i.e. X in this case. We say that two distributions $\mathcal{D}_1$ and $\mathcal{D}_2$ are ε-close to each other if $\Delta(\mathcal{D}_1, \mathcal{D}_2) \leq \varepsilon$.

---

[8] To encrypt a message of length, say, $\kappa/2$, a random string of length $\kappa/2$ is appended to it, and passed to the "encryption" oracle, along with the public-key.

Two-party Secure Function Evaluation. Alice and Bob have inputs $x \in \mathcal{X}$ and $y \in \mathcal{Y}$ and are interested in evaluating $f(x, y)$ securely, where f is a deterministic function with output space $\mathcal{Z}$.

Protocols and Augmented Protocols. We shall consider two-party protocols π between Alice and Bob relative to an oracle class. Alice and Bob may or may not have private inputs for the parties. An augmentation of the protocol with a third party Eve, represented as π^+, is a three party protocol where parties have access to a broadcast channel and speak in following order: Alice, Eve, Bob, Eve, and so on. In every round one party speaks and then Eve speaks.

Views of Parties. We shall always consider Eve who have no private view; her complete view is public. Such Eve shall be referred to as *public query-strategy.* Transcript message sent by Eve in a round is her sequence of oracle query-answer pairs performed in that round. The oracle query-answer sets of Alice, Bob and Eve are represented by P_A, P_B and P_E, respectively. The transcript is represented by m (note that m only contains messages from Alice and Bob). View of Eve is $V_E = (m, P_E)$; view of Alice is $V_A = (x, r_A, m, P_A, P_E)$ (where x is Alice's private input and r_A is her local random tape; in input-less protocols x is not present) and view of Bob is $V_B = (y, r_B, m, P_B, P_E)$.
  If i is odd then Alice performs local query-answers $P_{A,i}$ and sends the message m_i in that round followed by Eve message $P_{E,i}$. If i is even then Bob sends the message m_i, followed by Eve message $P_{E,i}$. View of Alice up to round i is represented by $V_A^{(i)} = (x, r_A, m^{(i)}, P_A^{(i)}, P_E^{(i)})$, where $m^{(i)} = m_1 \ldots m_i$; and $P_A^{(i)}$ and $P_E^{(i)}$ are similarly defined.

Relative to Oracle Class $\mathbb{O}_\kappa^{(\mathbb{K})}$. Our sample space is distribution over complete Alice-Bob joint views when: $r_A \xleftarrow{\$} \mathbf{U}$, $r_B \xleftarrow{\$} \mathbf{U}$ and $O \xleftarrow{\$} \mathbb{O}_\kappa^{(\mathbb{K})}$.

Definition 1 (Canonical). *A canonical sequence of query-answer pairs is a sequence of query-answer pairs such that an R-query of form $\langle k, \alpha \rangle$ is immediately followed by a T-query of form $\langle k, \beta \rangle$, where the query $\langle k, \alpha \rangle$ was answered by β.*

Definition 2 (Normal Form for Protocols). *A three party protocol between Alice, Bob and (public query strategy) Eve is in normal form, if:*

1. *In every round Alice or Bob sends a message; followed by a sequence of query-answer pairs from Eve. We allow Alice and Bob to base their messages on prior messages broadcast by Eve.*
2. *In rounds $i = 1, 3, \ldots$ Alice sends the message m_i; and in $i = 2, 4, \ldots$ Bob sends a message.*
3. *In every round i after Alice/Bob has sent the message m_i, Eve broadcasts $P_{E,i}$.*
4. *Alice, Bob and Eve always perform canonical queries.*

4 Common Information Learner

In this section we shall introduce a *Heavy-query Performer* algorithm (see Fig. 2). Using this heavy querier, we shall *augment* any two-party protocol with a third party algorithm. Relative to the oracle class $\mathbb{O}_\kappa^{(\mathbb{K})}$ we show that the distribution of Alice-Bob joint views is (nearly) independent of each other conditioned on the transcript of the augmented protocol. Thus, the third party is aptly called an *common information learner* (see Eve_π in Fig. 3).

4.1 Heavy-Query Performer

In this section we shall introduce a *Heavy-query Performer* algorithm. Let $\mathbb{O}$ be a finite class of oracles with finite domain D. Our experiment is instantiated by an oracle system Σ and a deterministic "Heavy-query Performer" $\mathcal{H}$ (with implicit parameter σ, see Fig. 2).

The oracle system Σ takes a random tape as input which has finite length. Let $\mathbb{S}$ be the set of pairs of random tape r for Σ and oracle $O \in \mathbb{O}$. The system Σ could possibly be computationally unbounded; but its round complexity is finite.

Consider the experiment in Fig. 1.

---

1. Let $\mathcal{D}_\mathbb{S}$ be a distribution over $\mathbb{S}$ such that $\mathsf{Supp}(\mathcal{D}_\mathbb{S}) = \mathbb{S}$. Sample $(r, O) \sim \mathcal{D}_\mathbb{S}$.
2. Start an interactive protocol between $\Sigma^O(r)$, i.e. the oracle system Σ with access to oracle O and local random tape r, and the heavy-query performer $\mathcal{H}$.

Fig. 1. Protocol between Oracle system Σ and the Heavy-query Performer $\mathcal{H}$

We emphasize that the heavy-query performer $\mathcal{H}$ never performs a query unless its answer is uncertain. If the answer to the query q^* in uncertain, we say that the answer to this query has (max) entropy. Let $\mathcal{Q}_\Sigma \left(\langle \Sigma^O(r), \mathcal{H} \rangle \right)$ represent the query-answer set of the oracle system Σ when its local random tape is r, has oracle access to O and is interacting with the heavy-query performer $\mathcal{H}$. Similarly, $\mathcal{Q}_\mathcal{H} \left(\langle \Sigma^O(r), \mathcal{H} \rangle \right)$ represents the query-answer set of the heavy-query performer $\mathcal{H}$ which were actually performed to the oracle when Σ has local random tape r and has oracle access to O. Note that $\mathcal{Q}_\Sigma \left(\langle \Sigma^O(r), \mathcal{H} \rangle \right)$ and $\mathcal{Q}_\mathcal{H} \left(\langle \Sigma^O(r), \mathcal{H} \rangle \right)$ could possibly be correlated to each other.

Efficiency of the Heavy-query Performer. We argue that the expected query complexity of the heavy-query performer cannot be significantly larger than the query complexity of the system Σ itself:

Lemma 1 (Efficiency of Heavy-query Performer). *Let $\mathcal{D}_\mathbb{S}$ be a joint distribution over the space $\mathbb{S}$ as defined above. For every (randomized) oracle system Σ, the expected query complexity of the heavy-query performer $\mathcal{H}$ (presented an Fig. 2) is at most $\frac{1}{\sigma}$*

After every message sent by the oracle system Σ, perform the following step:

- Repeatedly call Heavy-Query-Finder to obtain a query-answer pair (q^*, a^*); and add the query-answer pair (q^*, a^*) to the transcript T. Until it reports that there are no more heavy queries left.

Heavy-Query-Finder: Let T be the transcript between the oracle system Σ and heavy-query performer $\mathcal{H}$. The messages added by Σ are represented by T_Σ and the set of query-answer pairs added by $\mathcal{H}$ are represented by $T_\mathcal{H}$. It has an implicit parameter σ, which is used to ascertain whether a query is heavy or not.

1. For every $q \in D \setminus T_\mathcal{H}$, compute the probability that Σ performs the query q when $(\tilde{r}, \tilde{O}) \sim \mathcal{D}_\mathbb{S}$ conditioned on transcript T.
2. If there is no query q with probability $\geq \sigma$ then report that there are no more heavy queries left and quit. Otherwise, let q^* be the lexicographically smallest such query.
3. If the answer to q^* is uncertain (given the transcript T) then query O at q^* and obtain the answer a^*. Otherwise, let a^* be the fixed answer to q^*.
4. Return (q^*, a^*).

Fig. 2. Heavy-Query-Performer $\mathcal{H}$

---

times the expected query complexity of the oracle system Σ in the experiment shown in Fig. 1. Formally,

$$\mathop{\mathbb{E}}_{(r,O)\sim\mathcal{D}_\mathbb{S}} \left[\left\| \mathcal{Q}_\mathcal{H}\left(\langle \Sigma^O(r), \mathcal{H} \rangle \right) \right\| \right] \leq \frac{\mathbb{E}_{(r,O)\sim\mathcal{D}_\mathbb{S}} \left[\left\| \mathcal{Q}_\Sigma\left(\langle \Sigma^O(r), \mathcal{H} \rangle \right) \right\| \right]}{\sigma}$$

In particular, the probability that $\mathcal{H}$ asks more than $\frac{\mathbb{E}_{(r,O)\sim\mathcal{D}_\mathbb{S}}\left[\left\| \mathcal{Q}_\Sigma(\langle \Sigma^O(r), \mathcal{H} \rangle) \right\| \right]}{\sigma^2}$ queries is at most σ.

The proof is provided in the full version of the paper [32]. We mention some highlights of the current proof. The proof is significantly simpler and is more general than the ones presented in [2,3] because our learner is directly working with heavy queries rather than concluding the heaviness of the queries being asked by the learner. Also note that in our setting the oracles might be correlated with local random tape of the system Σ; and the future messages of the oracle system Σ could, possibly, depend on prior messages of $\mathcal{H}$. We also note that the same proof also works in the setting where Σ cannot read the transcript T[9] but $\mathcal{H}$ also considers *queries performed in the future* by Σ while computing the set of heavy-queries.[10] We emphasize that it is possible that the future messages of the oracle system Σ could possible depend on the prior messages sent by the heavy-query performer $\mathcal{H}$. This property is inherited by Lemma 2, which (in turn) is crucially used by Theorem 2.

---

[9] More specifically, it cannot read $T_\mathcal{H}$; note that Σ already knows the part T_Σ generated by Σ itself.

[10] Note that if Σ can also read from T then the distribution of future queries is not well defined. But if Σ cannot read T, then future queries are well defined after (r, O) is instantiated.

Specific to Image-testable Random-oracles. Relative to the oracle class $\mathbb{O}_\kappa^{(\mathbb{K})}$, we can make an assumption that after performing a R-query $\langle k, \alpha \rangle$ and receiving β as answer, it immediately performs the next query as $\langle k, \beta \rangle$. Note that this query has no entropy (because this query will surely be answered 1); and, hence, need not be performed to the oracle.

4.2   Common Information Learner for Input-Less Protocols

In this section we shall consider two-party protocols where parties have access to an oracle $O \in \mathbb{O}_\kappa^{(\mathbb{K})}$. For a two-party input-less protocol π, we augment it with the following eavesdropper strategy, referred as Eve_π, to obtain π^+:

---

1. Interpret the two-party oracle protocol π as the oracle system Σ in Fig. 1. Messages produced by Alice or Bob in round i is interpreted as the message of Σ.
2. Define $\mathbb{O} = \mathbb{O}_\kappa^{(\mathbb{K})}$ and $\mathcal{D}_\mathbb{S}$ as the uniform distribution over $\mathbb{O}$ and the space of local random tapes of Alice and Bob.
3. Let Eve_π be the heavy-query performer algorithm in Fig. 2 instantiated with a suitably small parameter σ.

Fig. 3. Eavesdropper strategy to augment an input-less protocol π

Note that the query-complexity of Eve_π is $\text{poly}(\kappa)$ with $1 - 1/\text{poly}(\kappa)$ probability, if σ is set to $1/\text{poly}(\kappa)$ and the query complexity of the parties in π is (at most) $\text{poly}(\kappa)$ (due to Lemma 1).

Lemma 2 (Common Information Learner for Input-less Protocols). *Let π be an input-less protocol in normal form between Alice and Bob relative to $\mathbb{O}_\kappa^{(\mathbb{K})}$, and Eve_π be as defined in Fig. 3. Let the distributions $\mathbf{V}_{AB}^{(i)}$ and $\mathbf{V}_{A \times B}^{(i)}$ for each round i of π be as follows:*

1. *$\mathbf{V}_{AB}^{(i)} = (\mathbf{V}_A^{(i)}, \mathbf{V}_B^{(i)})$;*
2. *Distribution $\mathbf{V}_{A \times B}^{(i)}$ defined as: Sample $V_E^{(i)} \sim \mathbf{V}_E^{(i)}$ and output $(V_A, V_B) \sim (\mathbf{V}_A^{(i)} | V_E^{(i)}) \times (\mathbf{V}_B^{(i)} | V_E^{(i)})$.*

For every $\varepsilon = 1/\text{poly}(\kappa)$, there exists a choice of Eve_π's parameter $\sigma = 1/\text{poly}(\kappa)$ such that, for every i,

$$\Delta \left(\mathbf{V}_{AB}^{(i)}, \mathbf{V}_{A \times B}^{(i)} \right) \le \varepsilon.$$

Below, we sketch the ideas behind proving this lemma. Interested reader may refer to the full version of this paper [32] for the proof.

The Case of Random Oracles. First we consider the case of random oracles without image-testing. This case was already analyzed in [1,2], but it will be helpful to rephrase this proof, so that we can extend it to the case when image-testing is present. At the beginning of the execution, the views of Alice and Bob are indeed independent of each other. As the execution progresses, at each round, we introduce a "tidy" distribution over (V_A, V_B, V_E), which has the following properties: a tidy distribution is obtained simply by restricting the support of the real execution to "good" tuples. Below, (P_A, P_B, P_E) stand for the query-answer sets of (V_A, V_B, V_E).

Definition 3 (Good). *Three query-answer sets P_A, P_B and P_E are called good, represented by* Good(P_A, P_B, P_E), *if* Consistent$(P_A \cup P_B \cup P_E)$ *and* $P_A \cap P_B \subseteq P_E$.

This has the consequence that a tidy distribution is identical to a "conditional product distribution" – i.e., a distribution which, when conditioned on each Eve view in its support, is a product distribution – when restricted to the same support as the tidy distribution.

When the execution evolves for one step (an Alice or Bob round), we start with the tidy distribution at that step, but will end up with a distribution that is not tidy. This distribution is again tidied up to obtain a new tidy distribution.

Then we argue the following:

1. Claim: At any point, the tidy distribution is close to a "conditional product distribution" – i.e., a distribution which, when conditioned on each Eve view in its support, is close to a conditional product distribution.

This closeness property is maintained inductively. Indeed, during an Eve round, it is easy to see that the distance from the conditional product distribution can only decrease. In an Alice or Bob step, we bound the *additional* distance from a conditional product distribution using the fact that, since Eve had just finished its step before the beginning of the current step, every query not in Eve's view was of low probability for either party ("lightness" guarantee). A lightness threshold parameter for Eve controls this additional distance.

2. Claim: After each Alice or Bob step, the statistical difference incurred in modifying the resulting distribution to become a tidy distribution is small.

Note that in an Alice or Bob step, even tough we start from a tidy distribution, after that step, tuples that are not good can indeed be introduced. But their probability mass can be bounded by the fact that the tidy distribution was close to a conditional product distribution.

Thus at each step, the statistical difference from the actual execution incurred by tidying up can be bounded, as well as the distance of the tidy distribution from a conditional product distribution. By choosing the lightness threshold parameter for Eve to be sufficiently small, after a polynomial number of steps, we obtain that the distribution of (V_A, V_B, V_E) in the actual execution is close to a tidy distribution, which is in turn close to a conditional product distribution.

We remark that, the "lightness" guarantee in [2] was ensured directly for the tidy distribution. However, it is enough to ensure that the lightness holds for the original distribution, since the tidy distribution is obtained by restricting the support of the actual distribution (without changing the relative probabilities within the support). This allows for a more modular description of the Eve's dropper's algorithm, independent of the definition of the tidy distributions. This turns out to be helpful when we move to the setting of image-testable random oracles, where the tidy distributions are much more complicated.

The Case of Image-testable Random Oracles. To adapt the above argument to accommodate test queries, we need to change several elements from above. Firstly, we replace the notion of good tuples, with a more refined notion of "nice" tuples, which takes into account the presence of *positive* test queries. (As it turns out, negative test queries by themselves have a negligible effect in the probability of individual views.)

Given a query-answer set P relative to $\mathbb{O}_\kappa^{(\mathbb{K})}$, we say that a query $q = \langle k, \beta \rangle \in \mathbb{K} \times \{0,1\}^{3\kappa}$ is *unexplained* if $(q,1) \in P$ (i.e. $T(q) = 1$) but there is no $q' = \langle k, \alpha \rangle \in \mathbb{K} \times \{0,1\}^\kappa$ such that $(q,\beta) \in P$ (i.e. $R(q') = \beta$). We define $\mathsf{T_1Guess}(P)$ as the total number of unexplained queries in P.

Definition 4 (Typical and Nice Views). *A query-answer set P relative to $\mathbb{O}_\kappa^{(\mathbb{K})}$ is typical, represented by* $\mathsf{Typical}(P)$*, if* $\mathsf{T_1Guess}(P|_k) = 0$*, for every $k \in \mathbb{K}$.*

Alice, Bob and Eve views in a normal protocol are called nice*, represented as* $\mathsf{Nice}(V_A, V_B, V_E)$ *if:*

1. $\mathsf{Consistent}(P_A, P_B, P_E)$, $\mathsf{Good}(P_A, P_B, P_E)$, *and*
2. $\mathsf{Typical}(P_A \cup P_B \cup P_E)$, $\mathsf{Typical}(P_A \setminus P_E)$ *and* $\mathsf{Typical}(P_B \setminus P_E)$.

Apart from replacing goodness with niceness, the tidy distributions we use are different in a few other important ways. Firstly, a tidy distribution's support would typically not contain all the nice tuples in the actual execution; we remove certain kinds of nice tuples too from the support, to ensure that test queries do not lead to increased distance from a conditional product distribution. Secondly, to ensure that a tidy distribution is identical to a conditional product distribution, when the latter is restricted to the supported of the former, we let it be different from the actual distribution restricted to the same support. The definition of niceness however, ensures that this difference is at most a $1 \pm \mathsf{negl}$ factor point-wise. (The $1 \pm \mathsf{negl}$ factor corresponds to the *negative* test queries in the actual distribution, which are ignored in defining the probabilities in a tidy distribution.)

More formally, let $\mathcal{A}_i$, $\mathcal{B}_i$ and $\mathcal{E}_i$ denote the set of possible views of Alice, Bob and Eve respectively, after i steps of the augmented protocol execution (where each "round" consists of an Alice or Bob step, and an Eve step). To specify a tidy distribution Γ over the views after i steps of the augmented protocol we need to specify two sets $\mathcal{S}_A^{(i)} \subseteq \mathcal{A}_i \times \mathcal{E}_i$ and $\mathcal{S}_B^{(i)} \subseteq \mathcal{B}_i \times \mathcal{E}_i$. Then the distribution is defined as follows:

$$\Gamma(V_A, V_B, V_E) = \begin{cases} Z \cdot \gamma(V_A, V_E) \cdot \gamma(V_B, V_E) & \text{if } (V_A, V_E) \in \mathcal{S}_A^{(i)}, (V_B, V_E) \in \mathcal{S}_B^{(i)}, \\ & \text{and } \mathsf{Nice}(V_A, V_B, V_E) \\ 0 & \text{otherwise} \end{cases}$$

Here Z is a normalization factor, and $\gamma(V_A, V_E) = 2^{-r} N^{-3w}$ where r is the length of the random tape in V_A and w is the number of random oracle queries (not including test queries) in $P_A \setminus P_E$. Note that by restricting to $\mathsf{Nice}(V_A, V_B, V_E)$, the positive test queries are taken into account by the definition of Γ, but the number of negative test queries in the views are not accounted for. But the probability of (V_A, V_B, V_E) in an actual distribution of any protocol, when restricted to the support of Γ, can be shown to be $\Gamma(V_A, V_B, V_E)(1 \pm \text{negl})$.

Another major difference in our proof is the tidying up operation itself. Unlike in the random oracle case, we need to introduce a tidying up step even during the Eve round. In fact, this tidying up is done per query that Eve makes. Before each fresh query that Eve makes, we tidy up the distribution to ensure that at most one of Alice or Bob could have made that query previously. Further, after an Eve test query that is answered positively for which Eve does not have an explanation (i.e., none of the random oracle queries that Eve has made so far returned the image being tested), we remove the possibility that neither party has an explanation. (By tidying up before this query, we would have already required that *at most one party* had made that query previously; the current tidying up ensures that, exactly one party has an explanation for this query.)

Though this tidying up is carried per query that Eve makes, we ensure that the entire statistical difference incurred by the tidying up process during one round of Eve's execution is bounded in terms of the distance to the conditional product distribution at the start of this round. Indeed, this latter distance can only decrease through out Eve's round.

The tidying up procedure when Alice or Bob makes a query is similar to that in the case of the random oracle setting. It ensures that the tidied up distribution is close to a conditional product distribution, and the *additional* distance can be bounded in terms of the lightness threshold parameter for Eve, as before.

With these modifications, the resulting proof follows the outline mentioned above. At each round we tidy up the distribution over (V_A, V_B, V_E), by incurring a statistical difference related to the distance between the previous tidy distribution and a conditional product distribution. In turn, we bound the increase in the latter distance (during Alice's and Bob's rounds) in terms of the lightness guarantee by Eve.

As a direct consequence of Lemma 2, we can conclude that:

Corollary 2. *There is no key-agreement protocol relative to* $\mathbb{O}_\kappa^{(\mathbb{K})}$, *for any key set* $\mathbb{K}$.

5   Compiling out Decryption Queries

In this section we show that a family of PKE-enabling oracles is only as useful as a family of image testable random oracles, for semi-honest SFE. Combined with the result that this image testable random oracle family is useless for SFE, we derive the main result in this paper, that PKE is useless for semi-honest SFE.

As pointed out by [18], care must be taken in modeling such an oracle so that it does not allow oblivious transfer. In our case, we need to separate it from not just oblivious transfer but any non-trivial SFE.

In our proof we shall use the oracle $\mathbb{PKE}_\kappa$ defined in Section 2.3. This oracle facilitates public-key encryption (by padding messages with say $\kappa/2$ random bits before calling Enc), and hence key agreement. But, as mentioned before, by omitting the Dec oracle, the collection (Gen, Enc, Test$_1$, Test$_2$) becomes an image-testable random oracle family $\mathbb{O}^{(\mathbb{K})}$. As we will see in Section 6, an image-testable random oracle is not useful for SFE or key agreement. The challenge is to show that even given the decryption oracle Dec, which does help with key-agreement, the oracle remains useless for SFE. [18] addressed this question for the special case of oblivious transfer, relying on properties that are absent from weaker (yet non-trivial) SFE functionalities. Our approach is to instead show that the decryption facility is completely useless in SFE, by giving a carefully compiled protocol whereby the parties help each other in finding decryptions of ciphertexts without accessing Dec oracle, while retaining the security against honest-but-curious adversaries. We show the following.

Theorem 2. *Suppose Π is an N-round 2-party protocol with input domain $\mathcal{X} \times \mathcal{Y}$, that uses the oracle $\mathbb{PKE}_\kappa$. Then for any polynomial poly, there is an N-round protocol Π^* using the oracle $\mathbb{O}_\kappa^{(\mathbb{K})}$ that is as secure as Π against semi-honest adversaries, up to a security error of $|\mathcal{X}||\mathcal{Y}|/\mathsf{poly}(\kappa)$.*

Below we present the compiler used to prove this theorem, and sketch why it works. The full proof appears in the full version of the paper [32].

The Idea Behind the Compiler. For ease of presentation, we assume here that the oracles $\mathsf{Gen}(\cdot)$ and $\mathsf{Enc}(\cdot, \cdot)$ are injective (which is true, except with negligible probability, because they are length tripling random oracles). Conceptually the compiler is simple: each party keeps track of the ciphertexts that it *created* that the other party becomes "capable of" decrypting and sends the message in the ciphertext across at the right time. This will avoid the need for calling the decryption oracle. But we need to also argue that the compilation preserves security: if the original protocol was a secure protocol for some functionality, then so is the compiled protocol. To ensure this, a party, say Bob, should reveal the message in an encryption it created only if there is a high probability that Alice (or a curious adversary with access to Alice) *can* obtain that message by decryption. Further, the fact that Bob found out that Alice could decrypt a ciphertext should not compromise Bob's security. This requires that just based on common information between the two parties it should be possible to accurately determine which ciphertexts each party can possibly decrypt. This is complicated by the fact that the protocol can have idiosyncratic ways of transferring ciphertexts and public and private keys between the parties, and even if a party *could* carry out a decryption, it may choose to not extract the ciphertext or private key implicit in its view. By using the common information learner for image testable random oracles, it becomes possible to

Definition of the Compiler. Given a 2-party protocol Π, with input domains $\mathcal{X} \times \mathcal{Y}$, we define the compiled protocol Π^* below. Π has access to $\mathbb{PKE}_\kappa$, where as Π^* will have access to the interface of $\mathbb{PKE}_\kappa$ consisting only of $(\mathsf{Gen}, \mathsf{Enc}, \mathsf{Test}_1, \mathsf{Test}_2)$ (or equivalently, to $\mathbb{O}_\kappa^{(\mathbb{K})}$ as described in Section 2). For convenience, we require a normal form for Π that before making a decryption query $\mathsf{Dec}(sk, c)$ a party should make queries $\mathsf{Gen}(sk)$, $\mathsf{Test}_1(pk)$ and $\mathsf{Test}_2(pk, c)$ where pk was what was returned by $\mathsf{Gen}(sk)$.

We define Π^* in terms of a 3-party protocol involving $\mathsf{Alice}_0, \mathsf{Bob}_0, \mathsf{Eve}$, over a broadcast channel. In the following we will define Alice_0 and Bob_0; this then defines an (inputless) system Σ which consists of them interacting with each other internally, while interacting with an external party; in Σ, the inputs to Alice_0 and Bob_0 are picked uniformly at random. Then, Eve is defined to be $\mathcal{H}$ for the system Σ, as defined in Fig. 3: after each message from Σ (i.e., from Alice or Bob), Eve responds with a set of publicly computable queries to the oracle. Finally, Π^* is defined as follows: Alice runs Alice_0 and Eve internally, and Bob runs Bob_0 and Eve.[11]

So to complete the description of the compiled protocol, it remains to define the programs Alice_0 and Bob_0. We will define Alice_0; Bob_0 is defined symmetrically.

Alice_0 internally maintains the state of an execution of Alice's program in Π (denoted by Alice_Π). In addition, Alice_0 maintains a list L_A of entries of the form (m, pk, c), one for every call $\text{Enc}(pk, m) = c$ that Alice_Π has made so far, along with such triples from the (secondary) messages from Bob_0.

Corresponding to a single message m_i from Alice in Π, Alice_0 will send out two messages — a primary message m_i and a secondary message $C_{A,i}$ (with an intermediate message from Eve) — as follows. (For the sake of brevity we ignore the boundary cases $i = 1$ and $i - 1$ being the last message in the protocol; they are handled in a natural way.)

The list $L_{A,}$, before receiving the $i - 1^{\text{st}}$ message, is denoted by $L_{A,i-2}$.

- On receiving m_{i-1} and $C_{B,i-1}$ from Bob_0 (and the corresponding messages from Eve), first Alice sets $L_{A,i-1} := L_{A,i-2} \cup C_{B,i-1}$ (where $C_{B,i-1}$ is parsed as a set of entries of the form (m, pk, c)).
- Then Alice_0 passes on m_{i-1} to Alice_Π, and Alice_Π is executed. During this execution Alice_Π is given direct access to $(\text{Gen}, \text{Enc}, \text{Test1}, \text{Test2})$; but for every query of the form $\text{Dec}(sk, c)$ from Alice_Π, Alice_0 obtains $pk = \text{Enc}(sk)$ and checks if any entry in $L_{A,i-1}$ is of the form (m, pk, c) for some m. If it is, Alice_0 will respond to this query with m. Otherwise Alice_0 responds with $\perp$. At the end of this computation, the message output by Alice_Π is sent out as m_i.
  Also Alice updates the list $L_{A,i-1}$ (which was defined above as $L_{A,i-2} \cup C_{B,i-1}$) to $L_{A,i}$ by including in it a tuple (m, pk, c) for each encryption query $\text{Enc}(pk, m) = c$ that Alice_Π made during the above execution.
- Next it reads a message from Eve. Let $T^{(i)}$ denote the entire transcript at this point (including messages sent by Alice_0, Bob_0 and Eve). Based on this transcript Alice_0 computes a set $D_B^{T^{(i)}}$ of ciphertexts that Bob is "highly likely" to be able to decrypt in the next round, but has not encrypted itself,[a] and then creates a message $C_{A,i}$ that would help Bob decrypt all of them without querying the decryption oracle. The algorithm Assist_A used for this is detailed below in Fig. 5. Alice finishes her turn by sending out $C_{A,i}$.

---

[a] The threshold δ used in defining $D_B^{T^{(i)}}$ by itself does not make it *highly likely* for the honest Bob to be able to decrypt a ciphertext. However, as we shall see, this will be sufficient for a *curious* Bob to be able to decrypt with high probability.

Fig. 4. Definition of Alice_0 procedure

---

[11] Note that Eve follows a deterministic public-query strategy, and can be run by both parties. Alternately, in Π^*, one party alone could have run Eve. But letting both parties run Eve will allow us to preserve the number of rounds exactly, when consecutive messages from the same party are combined into a single message.

For each possible view V_B of Bob$_\Pi$ at the point $T^{(i)}$ is generated let,

$d_B(V_B) := \{(pk, c)|\exists sk \text{ s.t. } [\text{Gen}(sk) = pk], [\text{Test}_1(pk) = 1], [\text{Test}_2(pk, c) = 1] \in V_B$
$\quad\quad\quad \text{and } \not\exists m \text{ s.t. } [\text{Enc}(pk, m) = c] \in V_B\}.$

We define the set

$$D_B^{T^{(i)}} := \{(pk, c)|\Pr[(pk, c) \in d_B(V_B)|T^{(i)}] > \delta\} \tag{1}$$

where the probability is over a view V_B for Bob$_\Pi$ sampled conditioned on $T^{(i)}$, in the interaction between Σ (i.e., Alice$_0$ and Bob$_0$ with a random input pair) and Eve.[a] The threshold δ which will be set to an appropriately small quantity (larger than, but polynomially related to, σ associated with Eve).

The message $C_{A,i}$ is a set computed as follows: for each $(pk, c) \in D_B^{T^{(i)}}$, if there is an m such that $\text{Enc}(pk, m) = c$ appears in $L_{A,i}$, then the triple (m, pk, c) is added to $C_{A,i}$. If for any (pk, c), if there is no such m, then the entire protocol is aborted.

---

[a] Even though we define Alice$_0$ in terms of a probability that is in terms of the behavior of a system involving Alice$_0$, we point out that this probability is well-defined. This is because the probability computed in this round refers only to the behavior of the system up till this round. Also Eve, which is also part of the system generating $T^{(i)}$, depends only on the prior messages from Alice$_0$.

Fig. 5. Procedure Assist$_A$ for computing $C_{A,i}$

Security of the Compiled Protocol. To formally argue the security of the compiled protocol we must show an honest-but-curious *simulator* with access to either party in an execution of Π, which can simulate the view of an honest-but-curious adversary in Π^*. Here we do allow a small (polynomially related to σ), but possibly non-negligible simulation error. We give a detailed analysis of such a simulation in the full version of the paper. Below we sketch some of the important arguments.

Firstly, it must be the case that the probability of Alice$_0$ aborting in Π^* while computing a secondary message $C_{A,i}$, is small. Suppose, with probability p Alice fails to find an encryption for some $(pk, c) \in D_B^{T^{(i)}}$. Then, by the independence property guaranteed by Lemma 2, with about probability δp this Alice execution takes place in conjunction with a Bob view V_B such that $(pk, c) \in d_B(V_B)$. This would mean that with close to probability δp we get an execution of the original protocol Π in which (pk, c) is present in the parties' views, but neither Alice nor Bob created this ciphertext. This probability must then be negligible.

The more interesting part is to show that it is safe to reveal an encrypted message, when there is only a small (but inverse polynomial) probability that the other party would have decrypted it. For concreteness, consider when the honest-but-curious adversary has access to Alice. In the execution of Π^* it sees the messages $C_{B,i}$ that are sent by Bob (assuming Bob does not abort). These contain the messages for each (pk, c) pair in D_A^T where T is the common information so far. So we need to show that the simulator would be able to extract all these messages as well. Consider a $(pk, c) \in D_A^T$. If Alice's view contains an Enc query that generates c, or a Gen query that generates pk, then the simulator can use this to extract the encrypted message. Otherwise it samples a view A' for Alice consistent with T, but conditioned on $(pk, c) \in d_A(A')$ (such A' must exist since $(pk, c) \in D_A^T$). Then A' does contain a secret key sk' for pk that Alice will use to decrypt the ciphertext.

However, note that the view of the oracle in A' need not be consistent with the given oracle. Thus it may not appear meaningful to use sk' as a secret key. But intuitively, if it is the case that with significant probability Alice did not generate pk herself, then it must have been generated by Bob, and then the only way Alice could have carried out the decryption is by extracting Bob's secret key from the common information. Thus this secret key is fixed by the common information. Further, by sampling an Alice view in which a secret key for pk occurs, this secret key must, with high probability agree with the unique secret key implicit in the common information. Formalizing this intuition heavily relies on the independence characterization: otherwise the common information need not fix the secret key, even if it fixes the public key.

In the full version of this paper we give a detailed proof of security of Π^*, by defining a complete simulation, and using a coupled execution to analyze how good the simulation is. We show that the compiled protocol is as secure as the original protocol up to a security error of $O|\mathcal{X}||\mathcal{Y}|(N(\sigma/\delta + \delta)) = O(1/\text{poly})$ by choosing appropriate parameters, assuming $|\mathcal{X}||\mathcal{Y}|$ is polynomial.

The proof relies on Lemma 2. It shows that even when the protocol allows the parties to use the information from the common information learner, it holds that the views of the two parties (in an inputless version of the protocol considered in the proof) are nearly independent of each other's, conditioned on the common information gathered by Eve.

6 Limits of Image-Testable Random Oracles

Applying the compiler from the above section, we can convert a protocol using the $\mathbb{PKE}$ oracle to one using an image-testable random oracle $\mathbb{O}^{(\mathbb{K})}$. Then, to complete the proof of Theorem 1 it will suffice to prove the following result, which asserts that no protocol ρ using $\mathbb{O}^{(\mathbb{K})}$ can be a secure realization of f, if f is semi-honest non-trivial.

Lemma 3. *Suppose ρ is a $1 - \lambda(\kappa)$ semi-honest secure protocol (with round complexity N) for 2-party finite semi-honest non-trivial f relative to oracle class $\mathbb{O}_\kappa^{(\mathbb{K})}$, for any key set $\mathbb{K}$. There exists $\Lambda = 1/\text{poly}(\cdot)$ such that, for infinitely many κ, we have $\lambda(\kappa) > \Lambda(N, \kappa)$.*

The proof of this lemma follows the proof of a similar result of [5], which considered the class of random oracles instead of image-testable random-oracles. The proof in [5] uses a detailed frontier analysis, in which the following two properties of the random oracles (informally stated here) were used, which we extend to the case of image-testable random oracles.

1. Local Samplability: We need Bob to sample hypothetical Bob view V_B', based on his actual view V_B, but without exactly knowing what view V_A Alice has. A crucial

step in this is to sample a new query-answer set P'_B which is consistent with P_E; and this sampling has to be independent of the exact query-answer set P_A of Alice.

2. Oblivious Re-randomizability: Once Bob has sampled a hypothetical Bob view, it needs to simulate the view further (for just one round, before the next message from Alice arrives). This simulation includes answering further queries to the (hypothetical) oracle. A crucial step in this is to answer these new queries with answers which are consistent with Alice's query-answer set P_A, but are otherwise independent of P_B. That is, in simulating answers to further oracle queries, Bob should *rerandomize* the part of the oracle which is consistent with the actual Bob query-answer pairs P_B.

Local samplability is a direct consequence of Lemma 2, which characterized the views in the actual execution of the prototocl as close to a conditional product distribution. For proving the oblivious rerandomization property, we need to specify the rerandomization procedure. Such a procedure for the case of random oracles was provided in [5]. In Fig. 6 we extend this to the case of image-testable random oracles.

Suppose Alice has private query-answer sequence P_A, Eve has P_E and Bob has P_B. Assume that Bob has been provided with P'_B; and Typical$(P_A \cup P \cup P_E)$ and Good(P_A, P, P_E) hold, for $P \in \{P_B, P'_B\}$.

Let D be the set of R-queries in P_B which are not already included in $\mathcal{Q}(P_E \cup P'_B)$. We re-emphasize that the queries in $\mathcal{Q}(P_B) \cap \mathcal{Q}(P'_B)$ outside $\mathcal{Q}(P_E)$ could possible by inconsistently answered.

Initialize a global set $R_{\text{local}} = \emptyset$.

Query-Answering (q) :

1. If q is answered in $P_E \cup P'_B$ use that answer.
2. If $q = \langle k, \alpha \rangle$ is a new R-query and $q \in D$, answer with $a \xleftarrow{\$} \{0,1\}^{3\kappa}$. Add $\langle k, a \rangle$ to R_{local}.
3. If q is a T-query which is already in R_{local} then answer 1.
4. Otherwise (i.e. if the conditions above are not met) forward the query to the actual oracle and obtain the answer a.

Fig. 6. Bob's algorithm to answer future queries using re-randomization (oblivious to P_A)

We need to show that the result of the simulated execution using the rerandomized oracle is close to that of an actual execution with the hypothetical view V'_B that was sampled. Formally, the analysis requires the following "safety" property to hold with high probability, when Bob samples V'_B at a point when the views are (V_A, V_B, V_E).

Definition 5 (Safety). *For Alice view V_A, pair of Bob views (V_B, V'_B) and Eve view V_E, we define the following predicate:*

$$\mathsf{Safety}(V_A, (V_B, V'_B), V_E) := \mathsf{Nice}(V_A, V_B, V_E) \land \mathsf{Nice}(V_A, V'_B, V_E).$$

In the full version we show that at all rounds of the protocol, the safety condition is satisfied with high probability. The proof, again, is a consequence of the fact that the actual distributions are close to tidy distributions, and the tidy distributions are close to conditional product distributions.

6.1 Extending to Protocols with Inputs

The final ingredient we need to extend the proof in [5] to the case of image-testable random oracles is to extend our common information learner to protocols with private inputs for Alice and Bob. As was done in [5], such a common information learner can be easily reduced to one for inputless protocols, *as long as the domain size of the inputs is polynomial in the security parameter.*[12] For this we transform the given protocol to an inputless protocol by randomly sampling inputs for Alice and Bob. We augment the protocol ρ where parties have private inputs with an eavesdropper strategy as guaranteed by Lemma 2 when we assume that parties have picked their input uniformly at random. This augmented protocol is referred to as ρ^+.

By choosing the threshold parameter σ of the eavesdropper to be suitably small, we can ensure the following strong independence properties:

1. Suppose i is an even round in the augmented protocol ρ^+. If $x \in \mathcal{X}$ and $y, y' \in \mathcal{Y}$ are likely inputs at $V_E^{(i)}$ (transcript of the augmented protocol), then the message sent by Alice is nearly independent of Bob's private input being y or y'.
2. Suppose i is an even round in the augmented protocol ρ^+. If $x \in \mathcal{X}$ and $y, y' \in \mathcal{Y}$ are likely inputs at $V_E^{(i)}$, then sample a Alice-Bob joint view $(V_A^{(i+1)}, V_B^{(i)})$ just after Alice has sent the message m_{i+1}. Conditioned on the transcript $V_E^{(i)}$, message sent by Alice m_{i+1} and Bob input being y', sample a new Bob view $V'^{(i)}_B$. With high probability: $\mathsf{Safety}(V_A^{(i+1)}, (V_B^{(i)}, V'^{(i)}_B), V_E^{(i)})$ holds, i.e. $\mathsf{Nice}(V_A^{(i+1)}, V_B^{(i)}, V_E^{(i)})$ and $\mathsf{Nice}(V_A^{(i+1)}, V'^{(i)}_B, V_E^{(i)})$.
3. For an even round i, and likely inputs $x \in \mathcal{X}$ and $y, y' \in \mathcal{Y}$ the distribution of $(V_A^{(i+1)}, V_B^{(i)} V'^{(i)}_B)$ is close to a product distribution where each component is independently sampled conditioned on $(V_E^{(i)}, m_{i+1})$.

Analogous conditions hold when i is odd. For a formal version of this result, refer to the full version of the paper [32].

Given the above results, the frontier analysis of [5] can be carried out for image-testable random oracles.

7 Putting Things Together

Now we show how to complete the proof of Theorem 1. Suppose f is a 2-party finite semi-honest non-trivial SFE. Assume that there exists a $1 - \mathsf{negl}(\kappa)$ secure protocol ρ relative to oracle class $\mathbb{PKE}_\kappa$, with round complexity N. By Theorem 2, we construct a

---

[12] Note that we depend on the input domains being of polynomial size also for applying the decomposability characterization of [8,10].

$1 - \lambda^*(\kappa)$ secure protocol ρ^* relative to oracle class $\mathbb{O}_\kappa^{(\mathbb{K})}$, where λ^* could be arbitrarily small $1/\text{poly}$ and $\mathbb{K} = \{0,1\}^{2\kappa} \cup \{\perp\}$.

Now, if we choose λ^* in Lemma 3 to be sufficiently small so that $\lambda^*(\kappa) < \Lambda(N, \kappa)$, ρ^* contradicts Lemma 3 and hence also the assumption that ρ is a $(1 - \mathsf{negl}(\kappa))$ secure protocol for semi-honest non-trivial f relative to $\mathbb{PKE}_\kappa$. Note that this result crucially relies on the fact that Theorem 2 preserves round-complexity and the simulation error exhibited in Lemma 3 is function of only round complexity (and independent of the query complexity).

8   Conclusions and Open Problems

As mentioned in the introduction, our result can be set in the larger context of the "cryptographic complexity" theory of [7]: with every (finite, deterministic) multi-party function f, one can associate a computational intractability assumption that there exists a secure computation protocol for f that is secure against semi-honest corruption. The main result of this work shows that the set of such assumptions associated with 3-party functions is strictly larger than the set associated with 2-party functions. However, *we do not characterize this set either for the 3-party case or for the 2-party case.*

It remains a major open problem in this area to understand what all computational intractability assumptions could be associated with multi-party functions. For the 3-party case, this question is far less understood than that for 2-party functions. Intuitively, there are many more "modes of secrecy" when more than two parties are involved, and these modes will be associated with a finer gradation of intractability assumptions. Our result could be seen as a first step in understanding such a finer gradation. It raises the question whether there are further modes of secrecy for larger number of parties, and if they always lead to "new" complexity assumptions.

Stepping further back, the bigger picture involves randomized and reactive functionalities, various different notions of security, and "hybrid models" (i.e., instead of considering each multi-party function f and a secure protocol for it in plain model, we can consider a pair of functions (f, g) and consider a secure protocol for f given ideal access to g). The cryptographic complexity questions of such functions remain wide open.

References

1. Impagliazzo, R., Rudich, S.: Limits on the provable consequences of one-way permutations. In: Johnson, D.S. (ed.) STOC, pp. 44–61. ACM (1989)
2. Barak, B., Mahmoody-Ghidary, M.: Merkle puzzles are optimal - an $O(n^2)$-query attack on any key exchange from a random oracle. In: Halevi, S. (ed.) CRYPTO 2009. LNCS, vol. 5677, pp. 374–390. Springer, Heidelberg (2009)
3. Haitner, I., Omri, E., Zarosim, H.: Limits on the usefulness of random oracles. In: Sahai, A. (ed.) TCC 2013. LNCS, vol. 7785, pp. 437–456. Springer, Heidelberg (2013)
4. Dachman-Soled, D., Lindell, Y., Mahmoody, M., Malkin, T.: On black-box complexity of optimally-fair coin-tossing. In: Ishai, Y. (ed.) TCC 2011. LNCS, vol. 6597, pp. 450–467. Springer, Heidelberg (2011)

5. Mahmoody, M., Maji, H.K., Prabhakaran, M.: Limits of random oracles in secure computation. CoRR **abs/1205.3554** (2012); To appear in ITCS 2014
6. Reingold, O., Trevisan, L., Vadhan, S.: Notions of reducibility between cryptographic primitives. In: Naor, M. (ed.) TCC 2004. LNCS, vol. 2951, pp. 1–20. Springer, Heidelberg (2004)
7. Maji, H.K., Prabhakaran, M., Rosulek, M.: Cryptographic complexity classes and computational intractability assumptions. In: Yao, A.C.C. (ed.) ICS, pp. 266–289. Tsinghua University Press (2010)
8. Kushilevitz, E.: Privacy and communication complexity. In: [33], pp. 416–421.
9. Beaver, D.: Perfect privacy for two-party protocols. In: Feigenbaum, J., Merritt, M. (eds.) Proceedings of DIMACS Workshop on Distributed Computing and Cryptography, vol. 2, pp. 65–77. American Mathematical Society (1989)
10. Maji, H.K., Prabhakaran, M., Rosulek, M.: Complexity of multi-party computation problems: The case of 2-party symmetric secure function evaluation. In: Reingold, O. (ed.) TCC 2009. LNCS, vol. 5444, pp. 256–273. Springer, Heidelberg (2009)
11. Maji, H.K., Prabhakaran, M., Rosulek, M.: A unified characterization of completeness and triviality for secure function evaluation. In: Galbraith, S., Nandi, M. (eds.) INDOCRYPT 2012. LNCS, vol. 7668, pp. 40–59. Springer, Heidelberg (2012)
12. Haitner, I., Omri, E., Zarosim, H.: On the power of random oracles. Electronic Colloquium on Computational Complexity (ECCC) 19, 129 (2012)
13. Simon, D.R.: Finding collisions on a one-way street: Can secure hash functions be based on general assumptions? In: Nyberg, K. (ed.) EUROCRYPT 1998. LNCS, vol. 1403, pp. 334–345. Springer, Heidelberg (1998)
14. Gertner, Y., Malkin, T., Reingold, O.: On the impossibility of basing trapdoor functions on trapdoor predicates. In: FOCS, pp. 126–135 (2001)
15. Boneh, D., Papakonstantinou, P.A., Rackoff, C., Vahlis, Y., Waters, B.: On the impossibility of basing identity based encryption on trapdoor permutations. In: FOCS, pp. 283–292. IEEE Computer Society (2008)
16. Katz, J., Schröder, D., Yerukhimovich, A.: Impossibility of blind signatures from one-way permutations. In: [34], pp. 615–629
17. Matsuda, T., Matsuura, K.: On black-box separations among injective one-way functions. In: [34], pp. 597–614
18. Gertner, Y., Kannan, S., Malkin, T., Reingold, O., Viswanathan, M.: The relationship between public key encryption and oblivious transfer. In: FOCS, pp. 325–335. IEEE Computer Society (2000)
19. Kim, J.H., Simon, D.R., Tetali, P.: Limits on the efficiency of one-way permutation-based hash functions. In: FOCS, pp. 535–542 (1999)
20. Gennaro, R., Gertner, Y., Katz, J., Trevisan, L.: Bounds on the efficiency of generic cryptographic constructions. SIAM J. Comput. 35(1), 217–246 (2005)
21. Lin, H., Trevisan, L., Wee, H.: On hardness amplification of one-way functions. In: Kilian, J. (ed.) TCC 2005. LNCS, vol. 3378, pp. 34–49. Springer, Heidelberg (2005)
22. Haitner, I., Hoch, J.J., Reingold, O., Segev, G.: Finding collisions in interactive protocols - a tight lower bound on the round complexity of statistically-hiding commitments. In: FOCS, pp. 669–679. IEEE Computer Society (2007)
23. Barak, B., Mahmoody, M.: Lower bounds on signatures from symmetric primitives. In: FOCS: IEEE Symposium on Foundations of Computer Science, FOCS (2007)
24. Impagliazzo, R., Luby, M.: One-way functions are essential for complexity based cryptography (extended abstract). In: [33], pp. 230–235
25. Ostrovsky, R.: One-way functions, hard on average problems, and statistical zero-knowledge proofs. In: Structure in Complexity Theory Conference, pp. 133–138 (1991)

26. Ostrovsky, R., Wigderson, A.: One-way functions are essential for non-trivial zero-knowledge. Technical Report TR-93-073, International Computer Science Institute, Preliminary version in Proc. 2nd Israeli Symp. on Theory of Computing and Systems, 1993, pp. 3–17, Berkeley, CA (November 1993)
27. Haitner, I.: Semi-honest to malicious oblivious transfer - the black-box way. In: Canetti, R. (ed.) TCC 2008. LNCS, vol. 4948, pp. 412–426. Springer, Heidelberg (2008)
28. Haitner, I., Nguyen, M.H., Ong, S.J., Reingold, O., Vadhan, S.P.: Statistically hiding commitments and statistical zero-knowledge arguments from any one-way function. SIAM J. Comput. 39(3), 1153–1218 (2009)
29. Impagliazzo, R.: A personal view of average-case complexity. In: Structure in Complexity Theory Conference, pp. 134–147 (1995)
30. Lindell, Y., Omri, E., Zarosim, H.: Completeness for symmetric two-party functionalities - revisited. In: Wang, X., Sako, K. (eds.) ASIACRYPT 2012. LNCS, vol. 7658, pp. 116–133. Springer, Heidelberg (2012)
31. Haitner, I.: Personal communication (January 21, 2013)
32. Mahmoody, M., Maji, H.K., Prabhakaran, M.: On the power of public-key encryption in secure computation. Electronic Colloquium on Computational Complexity (ECCC) 20, 137 (2013)
33. 30th Annual Symposium on Foundations of Computer Science, Research Triangle Park, North Carolina, USA, 30 October-1 November (1989); In: FOCS. IEEE (1989)
34. Ishai, Y. (ed.): TCC 2011. LNCS, vol. 6597. Springer, Heidelberg (2011)

On the Impossibility of Basing Public-Coin One-Way Permutations on Trapdoor Permutations

Takahiro Matsuda

Research Institute for Secure Systems (RISEC),
National Institute of Advanced Industrial Science and Technology (AIST), Japan
t-matsuda@aist.go.jp

Abstract. One of the fundamental research themes in cryptography is to clarify what the minimal assumptions to realize various kinds of cryptographic primitives are, and up to now, a number of relationships among primitives have been investigated and established. Among others, it has been suggested (and sometimes explicitly claimed) that a family of one-way trapdoor permutations (TDP) is sufficient for constructing almost all the basic primitives/protocols in both "public-key" and "private-key" cryptography. In this paper, however, we show strong evidence that this is not the case for the constructions of a one-way permutation (OWP), one of the most fundamental primitives in private cryptography. Specifically, we show that there is no black-box construction of a OWP from a TDP, even if the TDP is *ideally secure*, where, roughly speaking, ideal security of a TDP corresponds to security satisfied by random permutations and thus captures major security notions of TDPs such as one-wayness, claw-freeness, security under correlated inputs, etc. Our negative result might at first sound unexpected because both OWP and (ideally secure) TDP are primitives that implement a "permutation" that is "one-way". However, our result exploits the fact that a TDP is a "secret-coin" family of permutations whose permutations become available only after some sort of key generation is performed, while a OWP is a publicly computable function which does not have such key generation process.

Keywords: black-box separation, trapdoor permutation, one-way permutation, family of one-way permutations.

1 Introduction

1.1 Background and Motivation

One of the fundamental research themes in cryptography is to clarify what the minimal assumptions to realize various kinds of cryptographic primitives are, and up to now, a number of relationships among primitives have been investigated and established. Clarifying these relationships gives us a lot of insights for how to construct and/or prove the security of cryptographic primitives, enables us to understand the considered primitives more deeply, and leads to systematizing the research area in cryptography.

In this paper, we focus on two central cryptographic primitives, a family of trapdoor permutations (TDP) and a one-way permutation (OWP). Among others, it has been suggested, and sometimes explicitly claimed (see, e.g. [9]), that a TDP is sufficient for constructing (almost) all basic primitives/protocols in both "public-key" and

Y. Lindell (Ed.): TCC 2014, LNCS 8349, pp. 265–290, 2014.

"private-key" cryptography. In particular, it has been shown that a TDP can be used for constructing a family of one-way trapdoor functions, public-key encryption schemes, key agreement protocols, private information retrieval, oblivious transfer, etc. Moreover, it has also been shown that a OWP is sufficient to construct most of private-key cryptographic primitives/protocols including symmetric key encryption schemes, message authentication codes, digital signature schemes [37], pseudorandom generators/functions/permutations [7,47,16,32], bit commitment schemes [35], etc. (Some of them later turned out to be possible to construct from any one-way function, e.g. a pseudorandom generator from any one-way function [22].) These primitives can also be constructed from a TDP as well.

Somewhat surprisingly, however, the following simple but fundamental question has not been answered yet: *"Can we construct a OWP from a TDP?"* The main motivation of this paper is to clarify the answer to this question, in order to fully establish the relationships among these very basic and important primitives. One might think that the answer is trivially yes (and that this is obvious), because a TDP is trivially a family of one-way permutations if we keep trapdoors secret. However, we show strong evidence that the answer to the above question is *no* by showing that there is no *black-box construction* of a OWP from a TDP. Roughly, a black-box construction of a target primitive P from a building block primitive Q requires that the construction of P treats an instance of Q as a black-box (i.e. treats as an oracle) and furthermore that the reduction algorithm for the security proof treats an adversary that breaks the security of the construction of P (and the instance of Q) as a black-box. (The impossibility of the opposite direction, i.e. constructing a TDP from a OWP in a black-box way, is due to [25].)

Actually, to tackle the above question, we have to be careful about the difference between a "single" one-way permutation and a "family" of one-way permutations (one-way permutation family, OWPF).[1] Our black-box separation result mentioned above separates a "single" one-way permutation from a TDP. Furthermore, for OWPFs, we have to be also careful about the difference between the *public-coin* case and the *secret-coin* case. Informally, a OWPF is said to be *public-coin* if the randomness for choosing a permutation from the family can be revealed together with the description of the permutation. On the other hand, a OWPF is said to be *secret-coin* if the security (one-wayness) is not guaranteed if the randomness is revealed. (The distinction between public-coin primitives and secret-coin primitives is studied by Hsiao and Reyzin [24] for the case of collision-resistant hash function families.) With these categorizations, it is straightforward to see that any one-way TDP can always be seen as a secret-coin OWPF by regarding an evaluation-key (public-key) output from a key generation algorithm of the TDP as an index specifying a permutation in the family. However, the same OWPF derived from a TDP is *not* secure as a public-coin OWPF, because the randomness for choosing the evaluation-key (public-key) cannot be revealed: If revealed, then anyone can compute the corresponding trapdoor, which makes the permutation invertible. Furthermore, it is also straightforward to see that a single OWP is a special type of a public-coin OWPF (by implementing the permutations in the family with the given

---

[1] In order not to mix up with the difference between single function and function family of one-way permutations, when we just write "OWP", we always mean it is a "single" one-way permutation (i.e. not a family), and when we mean a family of OWPs, we write "OWPF".

single OWP). Here, what is not at all trivial is whether we can construct a public-coin OWPFs from a TDP in general. We also partially answer to this question in the negative.

1.2 Our Contribution

In this paper, we show that there is no black-box construction of a OWP from a TDP, even if the TDP is *ideally secure* [11,29], where, roughly speaking, ideal security of a TDP corresponds to the security satisfied by random permutations (see Section 2.3 for the formal definition), and thus captures major security notions for a TDP such as one-wayness, claw-freeness [19], security under correlated inputs [42], etc. Therefore, our impossibility result rules out the black-box constructions of a OWP from TDP satisfying these security notions, and is strictly stronger than the result by Chang et al. [9] who showed the black-box separation of a OWP from a family of injective trapdoor functions. Our impossibility result might at first sound unexpected because both OWP and (one-way) TDP are primitives that implement a "permutation" that is "one-way". However, our result is established by exploiting the essential difference between a family of functions and a single function, that a TDP is a "secret-coin" family of permutations whose permutations become available only after some sort of key generation is performed, while a OWP is a publicly computable function which does not have such key generation process. (We explain the overview of the proof in Section 1.3.)

The type of black-box constructions that our main result rules out is called a *fully-black-box* construction in the taxonomy of Reingold et al. [41]. (The formal definition for a fully-black-box construction of a OWP from an ideal TDP is given in Section 3.) In fact, our result can be easily strengthened to rule out a *semi*-black-box construction, which is a less restrictive type than fully-black-box one, using the technique called "embedding" by Reingold et al. [41]. (We discuss this extension in Section 4.) Although the absence of (fully- and semi-)black-box constructions of a OWP from an ideal TDP does not necessarily mean that constructing a OWP from an ideal TDP is generally impossible, it should be emphasized that most of the known primitive-to-primitive constructions are fully-black-box, and thus the impossibility of black-box constructions is considered as a very strong evidence that "natural" and "efficient" constructions are impossible.

Our result also sheds light on the difference between "public-coin" and "secret-coin" OWPFs (their formal definitions can be found in Section 2.2). Whether a primitive remains secure in the sense of public-coin is usually related to whether we need some kind of trusted setup in a cryptographic protocol such as multi-party computation. Hsiao and Reyzin [24] conjectured that there is no (fully-)black-box construction of a public-coin OWPF from a secret-coin one. We partially answer to this conjecture: Specifically, we show that there is no black-box construction of a public-coin OWPF that satisfies a special property called *canonical domain sampling* (the formal definition is given in Section 2.2) from an ideal TDP (and especially from a secret-coin OWPF). This result is obtained as a corollary of our main result above by combining it with the result by Goldreich et al. [17] who showed that a OWP can be constructed, in a black-box manner, from a public-coin OWPF with the canonical domain sampling property. (See Section 4 for more details.) We note that the techniques we use to prove the black-box separation of a public-coin OWPF from a secret-coin one (and the black-box separation of a OWP from an ideal TDP) are different from those used by Hsiao and Reyzin in [24] (in fact, we use a part of the results in [24]).

Why Studying OWP vs. TDP? Historically, OWP and (public-coin/secret-coin) OWPF have much more often been treated as assumptions rather than as target primitives that are constructed from other primitives, and thus one may wonder why we should care the (im)possibility of constructing a OWP from TDP (or from other primitives).

Our opinion is that firstly, OWP, OWPF, and TDP are very basic primitives, and thus clarifying any of their properties as well as relations is important, and we believe that our results contribute to correctly understanding and firmly establishing relationships among these basic cryptographic primitives. Specifically, our results suggest that there is no simple hierarchy of black-box constructions even among very basic cryptographic primitives. Our results also clarify explicitly that there is a real difference among single function, public-coin and secret-coin families of functions in the case of permutations, which should be contrasted with the case of "functions" because the existence of a single one-way function is equivalent to the existence of a family of one-way functions (regardless of whether the family is secret-coin or public-coin). Furthermore, our results also show that it is not always the case that "public-key"-type primitives are stronger than "non-public-key"-type primitives (at least in the case of permutations). This should be again contrasted with the case of "functions", where there is a (trivial) black-box construction of a one-way function from basically all known "public-key"-type primitives (because key generation algorithms typically have to be a one-way function), but there does not exist a black-box construction for the opposite direction [25].

Secondly, there might actually be a cryptographic primitive that can be constructed from a OWP, but not from a TDP. One of such candidates may be a public-coin point obfuscation (an obfuscator for a point function) [1,45]. Wee [45] showed that a point obfuscator can be constructed from a (very strong) OWP, while his point obfuscator does not seem to be proved secure if we replace the OWP in his construction with a permutation from a TDP together with its public-key (at least the "public-coin" property will be lost unless we assume some additional property for the TDP). We believe that there are much more (natural) examples of this sort, and that it is interesting to seek for such examples. (In particular, the difference between public-coin and secret-coin primitives will stand out more in the context of interactive protocols.)

1.3   Technical Overview

The main result of our paper builds on the results and techniques from several previous work [43,26,15,24,9,30,23], and our technical contribution lies in coming up with an appropriate combination of these results/techniques for achieving our purpose of separating OWP from (ideal) TDP.

We will use the "two oracle separation" paradigm [15,24] (which is an extension of the one oracle separation [25,41]) to show that there is no fully-black-box construction of a OWP from an ideal TDP. That is, we will use two oracles (more precisely, a random instance picked from all possible instances of oracles): the first oracle models a "building block" primitive (TDP in our case) and the second oracle is the "breaking" oracle that is useful for breaking all candidates of a target primitive (OWP in our case) but useless for breaking the security of the building block oracle. As the "building block" oracle, we use a random instance of a *TDP oracle* $\mathcal{T}$ that consists of suboracles $(\mathcal{G}, \mathcal{E}, \mathcal{D})$ that essentially constitutes a (random) TDP, namely, $\mathcal{G}$ is the key generation, $\mathcal{E}$ is the evaluation of permutations, and $\mathcal{D}$ is the inversion of permutations. As the "breaking" oracle, we use the PSPACE oracle that has often been used in the literature of black-box separations, e.g. [25,15,9], mainly in order to guarantee that any computational hardness comes only from the building block oracle. If we pick $\mathcal{T}$ randomly, then $\mathcal{T}$ can be shown to be "ideally secure" even against computationally unbounded adversary that makes only polynomially many queries to $\mathcal{T}$. Since such adversary can simulate the PSPACE oracle by itself, it follows that an "ideally secure" TDP exists relative to $\mathcal{T}$ and PSPACE.

The difficult part of the proof is to show that any permutation $\mathsf{P}^{\mathcal{T}}$ is inverted, and thus a OWP does not exist relative to $\mathcal{T}$ and PSPACE. Here, we note that the evaluation-key space of $\mathcal{T}$ cannot be *dense* [20] (i.e. an inverse-polynomial fraction of strings are in the range of $\mathcal{G}$), because in this case, an evaluation-key ek of permutations in $\mathcal{E}$ could be picked without using $\mathcal{G}$, and thus implementing a permutation $\mathsf{P}^{\mathcal{T}}$ by the permutation (in $\mathcal{E}$) made available by this picked ek might lead to a OWP (even in the presence of the PSPACE oracle). To prevent this, we make the range of $\mathcal{G}$ sparse, and make $\mathcal{E}$ useless unless it is invoked with an honestly generated evaluation-key that is generated by making a query to $\mathcal{G}$. This guarantees that when calculating the permutation $\mathsf{P}^{\mathcal{T}}$, permutations in $\mathcal{E}$ become available only after making a query to $\mathcal{G}$ and obtaining an evaluation-key ek, *together with the corresponding trapdoor td*. Put differently, from the viewpoint of an entity computing the permutation $\mathsf{P}^{\mathcal{T}}$, every permutation in $\mathcal{E}$ associated with ek that becomes available during the computation of $\mathsf{P}^{\mathcal{T}}$ can be seen as an *invertible permutation*, because the entity must have known td corresponding to ek. This observation leads to the idea of simulating the TDP oracle $\mathcal{T}$ in $\mathsf{P}^{\mathcal{T}}$ with a *block cipher* oracle, which is a family of invertible permutations. More specifically, we introduce a new oracle $\mathcal{B}$, which we call *block cipher* oracle that models an ideally secure block cipher, and show that for any permutation $\mathsf{P}^{\mathcal{T}}$, there is another permutation $\widehat{\mathsf{P}}^{\mathcal{B}}$ such that inverting $\widehat{\mathsf{P}}^{\mathcal{B}}$ is as hard as inverting $\mathsf{P}^{\mathcal{T}}$. The idea and the technique of simulating a TDP oracle $\mathcal{T}$ (used in a constructed primitive) with a block cipher oracle is previously used by Lindell and Zarosim [30] who showed the black-box separation of an adaptively secure oblivious transfer protocol from a TDP. Furthermore, by using the result by Holenstein et al. [23] who showed that a random invertible permutation is simulatable by the fourteen-round Feistel-network construction of a permutation [32] in which each round function is an independent random function,[2] we can simulate the block cipher oracle $\mathcal{B}$ in the permutation $\widehat{\mathsf{P}}^{\mathcal{B}}$ with another oracle $\mathcal{R}$ (which we call *round function oracle*) that consists only of random functions (not permutations). More specifically, we show that for any permutation $\widehat{\mathsf{P}}^{\mathcal{B}}$, there is another permutation $\widetilde{\mathsf{P}}^{\mathcal{R}}$ such that inverting $\widetilde{\mathsf{P}}^{\mathcal{R}}$ is as hard as inverting $\widehat{\mathsf{P}}^{\mathcal{B}}$. Finally, using the previous results by Rudich [43], Kahn et al. [26], and Chang et al. [9] on the black-box separations of

---

[2] More precisely, [23] shows that the fourteen-round Feistel-network is *indifferentialble* [34] from an (invertible) random permutation. The statement that a constant-round Feistel-network was sufficient was originally suggested by Coron et al. [10]. However, it was pointed out in [23] that the original proof in [10] for six rounds had a gap and was not completed.

a OWP from random (injective) functions, we can show that there is a good inverter (which uses the PSPACE oracle) for any permutation $\widetilde{P}^{\mathcal{R}}$.[3] Then, this inverter can be used to invert not only $\widetilde{P}^{\mathcal{R}}$ but also $P^{\mathcal{T}}$, and thus any permutation $P^{\mathcal{T}}$ is inverted using the PSPACE oracle.

It is already known that a OWP is black-box separated from a one-way function (OWF) [43,26] and that there is a black-box construction of a pseudorandom permutation, which is a standard security notion of a block cipher, from a OWF [22,16,32]. Therefore, one might wonder that if we give up the "ideal security" of a TDP and just consider one-way TDPs, then we may be able to conclude that there is no black-box construction of a OWP from a one-way TDP, as soon as we reduce a TDP-based permutation $P^{\mathcal{T}}$ to a block-cipher-based permutation $\widehat{P}^{\mathcal{B}}$. However, that a OWP is separated from a OWF in a black-box manner does not immediately mean that our block-cipher-based permutation $\widehat{P}^{\mathcal{B}}$ cannot be proved one-way, because our block-cipher oracle $\mathcal{B}$ contains random permutations which may help $\widehat{P}^{\mathcal{B}}$ to be one-way (with some clever use of permutations in $\mathcal{B}$). This is the main reason why we further reduce the block-cipher-based permutation $\widehat{P}^{\mathcal{B}}$ to a random function-based permutation $\widetilde{P}^{\mathcal{R}}$ by using the result of [23], so that random permutations in the oracle $\mathcal{B}$ do not help achieving a OWP any better than random "functions" in the oracle $\mathcal{R}$ do.

1.4 Related Work

Up to now, a number of black-box separations among various kinds of primitives have been established. For an excellent survey of the literature and the techniques of black-box separations, we refer the reader to [48]. Here, we review black-box separations related to OWPs and TDPs.

Regarding the black-box separations of a OWP from other primitives, it is known that it is separated from one-way functions [43,26], from injective trapdoor functions and a private information retrieval protocols [9], and from length-increasing injective one-way functions (even if they are just 1-bit-increasing) [33].

On the other hand, recently, several black-box separation results have shown the limitations of a (one-way) TDP as a base primitive for constructing and/or proving the security of several "highly functional" cryptographic primitives or basic primitives with special functional/security properties. Those include the impossibility of constructing identity-based encryption [8], a wide class of predicate encryption [27], lossy trapdoor functions [42], trapdoor functions secure under correlated inputs [44], encryption schemes secure under key-dependent inputs [21], adaptively secure oblivious transfer protocols [30], non-interactive or perfectly binding commitment schemes secure under selective-opening attacks [2], verifiable random functions [12], a natural class of three-move blind signature schemes [13], succinct non-interactive argument systems [14], constant-round sequentially witness-hiding special-sound protocols for unique witness

---

[3] We note that a random function (which is length preserving) is indistinguishable from a random permutation for any (even computationally unbounded) algorithm that can make only polynomially many queries to the random function (even in the presence of the PSPACE oracle), but this fact does not mean that we can construct a OWP from a random function in a black-box way (in fact, it is not possible [43,26,9,33]).

relations [39], and many of the cryptographic primitives that admit the so-called simulatable attacks [46]. We note that in fact, the results of [21,2,13,14,39,46] rule out the possibility of constructions (and/or, security proofs) of the target primitives based not only on one-way TDP but also on much broader class of primitives or assumptions, such as all falsifiable assumptions [36].

Black-box separations for a particular construction that uses a TDP as a building block are also known. The unforgeability of the FDH signature scheme [4] cannot be based on an ideal TDP, if the TDP is treated as a black-box [11]. [6] shows a similar result for the PSS signature scheme, and [29] shows the impossibility of basing chosen ciphertext security of padding-based encryption schemes which include many known TDP-based encryption schemes such as the OAEP encryption scheme [3], on the (ideal) security of the building block TDP.

1.5 Paper Organization

The rest of this paper is organized as follows. In Section 2 we review some basic definitions and terminology. In Section 3, we show our main result on the black-box separation of a OWP from an ideal TDP, and we discuss further results, and the possibility of more general separation results in Section 4.

2 Preliminaries

In this section, we review the basic notation and the definitions of primitives.

Basic Notation. $\mathbb{N}$ denotes the set of natural numbers. For $n \in \mathbb{N}$, we define $[n] = \{1, \ldots, n\}$. If x and y are strings, then "$|x|$" denotes the bit-length of x, and "$(x||y)$" denotes a concatenation of x and y. "$x \leftarrow y$" denotes an assignment of y to x. If S is a set then "$|S|$" denotes its size, and "$x \leftarrow_R S$" denotes that x is chosen uniformly at random from S. "PPTA" denotes *probabilistic polynomial time algorithm*. If $\mathcal{A}$ is a probabilistic algorithm, then "$z \leftarrow_R \mathcal{A}(x, y, \ldots)$" means that $\mathcal{A}$ takes $x, y, \ldots$ as input and outputs z, and "$z \leftarrow \mathcal{A}(x, y, \ldots; r)$" means that $\mathcal{A}$ takes $x, y, \ldots$ as input, uses r as an internal randomness, and outputs z. For an oracle algorithm $\mathcal{A}^{\mathcal{O}}$, we say that $\mathcal{A}^{\mathcal{O}}$ has query complexity q if $\mathcal{A}$ makes queries to the oracle $\mathcal{O}$ at most q times. "Perm_n" denotes the set of all permutations over $\{0,1\}^n$. If f is a function and D is its domain, then we define $\mathrm{Range}(f) = \{f(x)|x \in D\}$.

A function $f : \mathbb{N} \to [0,1]$ is said to be *negligible* if $f(k) < 1/p(k)$ for all positive polynomials $p(k)$ and all sufficiently large $k \in \mathbb{N}$, and a function $g : \mathbb{N} \to [0,1]$ is said to be *overwhelming* if the function $f(k) = 1 - g(k)$ is negligible.

2.1 One-Way Permutations

Typically, security of a OWP is defined so that the security parameter k is its input length. However, since later we consider constructions of a OWP from another primitive, it will be convenient to consider the security parameter and the input length of the constructed permutation separately, so that the one-wayness advantage of an adversary and the input length of the constructed permutation are a function of the security parameter of the building block. Moreover, it is also convenient to identify a (one-way) permutation with a PPTA that computes it. Therefore, we take these approaches for the definition of a OWP.

Let $\ell = \ell(k)$ be a positive polynomial and P be a PPTA such that P is a permutation over $\{0,1\}^\ell$. We say that a PPTA P is a *one-way permutation (OWP) for length ℓ* if the following advantage function $\mathsf{Adv}^{\mathsf{OWP}}_{\mathsf{P},\mathcal{A},\ell}(k)$ is negligible for any PPTA adversary $\mathcal{A}$ (we assume that P is also given 1^k but omit to write it for simplicity):

$$\mathsf{Adv}^{\mathsf{OWP}}_{\mathsf{P},\mathcal{A},\ell}(k) = \Pr[x^* \leftarrow_\mathsf{R} \{0,1\}^\ell; y^* \leftarrow \mathsf{P}(x^*); x' \leftarrow_\mathsf{R} \mathcal{A}(1^k, y^*) : x' = x^*].$$

2.2 One-Way Permutation Families

A family of permutations (permutation family) PF consists of the following three PP-TAs (Gen, Eval, Samp): Gen is the probabilistic evaluation-key generation algorithm which takes 1^k as input and outputs an evaluation-key ek. (An evaluation-key is also called an index.) Eval is the deterministic evaluation algorithm which takes ek and an element $x \in D_{ek}$ as input, and outputs $y \in D_{ek}$, where D_{ek} is the domain of Eval$(ek, \cdot)$ that is determined by ek. Samp is the probabilistic sampling algorithm which takes ek as input, and outputs a (random) element $x \in D_{ek}$. As a correctness requirement, we require that for all $k \in \mathbb{N}$ and all $ek \leftarrow_\mathsf{R} \mathsf{Gen}(1^k)$, (i) Samp$(ek)$ is a uniform distribution over D_{ek}, and (ii) Eval$(ek, \cdot)$ is a permutation over D_{ek}.

We say that PF $=$ (Gen, Eval, Samp) is a *one-way permutation family (OWPF)* if the following advantage function $\mathsf{Adv}^{\mathsf{OWPF}}_{\mathsf{PF},\mathcal{A}}(k)$ is negligible for any PPTA adversary $\mathcal{A}$:

$$\mathsf{Adv}^{\mathsf{OWPF}}_{\mathsf{PF},\mathcal{A}}(k) = \Pr[ek \leftarrow_\mathsf{R} \mathsf{Gen}(1^k); x^* \leftarrow_\mathsf{R} \mathsf{Samp}(ek); y^* \leftarrow \mathsf{Eval}(ek, x^*);$$
$$x' \leftarrow_\mathsf{R} \mathcal{A}(ek, y^*) : x' = x^*].$$

If a permutation family PF remains one-way even when $\mathcal{A}$ is given the randomness r that is used to generate $ek = \mathsf{Gen}(1^k; r)$, then we call PF a *public-coin*[4] OWPF, and in order to distinguish it from an ordinary one, we call an ordinary OWPF a *secret-coin* OWPF.

Canonical Domain Sampling Property. We say that a OWPF PF has the *canonical domain sampling* property [17] if the following two conditions are satisfied:

1. **(Recognizable domain)** There exists a PPTA which, on input ek and x, tells if $x \in D_{ek}$ or not.
2. **(Dense domain)** There exist a polynomial time computable function $\ell = \ell(k)$ and a positive polynomial $p = p(k)$ so that $D_{ek} \subseteq \{0,1\}^\ell$ and $|D_{ek}| > 2^\ell/p$.

Goldreich et al. [17] showed that a OWP can be constructed in a black-box manner from a public-coin OWPF with the above property, and we briefly review their construction. Given a public-coin OWPF (Gen, Eval, Samp) with the canonical domain

---

[4] Goldreich et al. [17] called this property *"augmented one-wayness."* Here we use the name due to Hsiao and Reyzin [24].

sampling property, where $\text{Gen}(1^k)$ uses a $\lambda = \lambda(k)$-bit randomness, we construct a single permutation P for length $\lambda + \ell$ that works as follows: On input $(r_g \| z)$ such that $|r_g| = \lambda$ and $|z| = \ell$, P first calculates $ek \leftarrow \text{Gen}(1^k; r_g)$, and then outputs $(r_g \| \text{Eval}(ek, z))$ if $z \in D_{ek}$ or $(r_g \| z)$ otherwise. This P is indeed a permutation, and can be shown to be weakly one-way. Then, this weak one-wayness can be amplified by a standard technique (e.g. [47]) to obtain a OWP (with ordinary one-wayness).

2.3 Trapdoor Permutations

A family of trapdoor permutations (TDP) is a special class of secret-coin permutation family (Gen, Eval, Samp) with the following additional properties: (1) The algorithm Gen is a deterministic polynomial-time algorithm that takes 1^k and a trapdoor $td \in \{0,1\}^k$ as input, and outputs a corresponding evaluation-key ek.[5] (This process is denoted by "$ek \leftarrow \text{Gen}(1^k, td)$".) (2) There is a deterministic *inversion* algorithm Inv which takes $td \in \{0,1\}^k$ and an element $y \in D_{ek}$ as input (where $ek = \text{Gen}(1^k, td)$), and outputs $x \in D_{ek}$ such that $\text{Eval}(ek, x) = y$.

Hard Games and Ideal Security. In this paper, we consider "ideal security" of a TDP, following [11,29]. Roughly, ideal security of a TDP corresponds to security satisfied by random permutations.

Let G be a PPTA (called a challenger) that can exchange messages with another algorithm (called an adversary) $\mathcal{A}$ by a shared communication tape. We say that G defines a game regarding random permutations if both G and $\mathcal{A}$ have access to t independent random permutations $\pi_1, \ldots, \pi_t$ over $\{0,1\}^k$, where $t = t(k)$ is a polynomial determined by G, G interacts with $\mathcal{A}$, and finally outputs a decision bit d. This process is denoted by "$d \leftarrow_{\text{R}} \text{Expt}_{\text{RP}, \mathcal{A}^{\pi_1(\cdot), \ldots, \pi_t(\cdot)}}^{\mathsf{G}^{\pi_1(\cdot), \ldots, \pi_t(\cdot)}}(k)$." (Here, "RP" stands for "random permutations.") We say that the adversary $\mathcal{A}$ wins the game G if $d = 1$.

Informally, an oracle PPTA G defines a δ-hard game regarding random permutations, where $0 \leq \delta < 1$, if no oracle algorithm $\mathcal{A}$ can win the game G regarding random permutations with probability significantly better than δ. Typically, $\delta = 0$ for "search games" (e.g. one-wayness experiment) or $\delta = 1/2$ for "distinguishing games" (e.g. security experiment for a pseudorandom generator). We define the advantage of an adversary $\mathcal{A}$ in a game G as follows:

$$\text{Adv}_{\text{RP}, \mathcal{A}}^{\mathsf{G}}(k) = \Pr[\pi_1, \ldots, \pi_t \leftarrow_{\text{R}} \text{Perm}_k; d \leftarrow_{\text{R}} \text{Expt}_{\text{RP}, \mathcal{A}^{\pi_1(\cdot), \ldots, \pi_t(\cdot)}}^{\mathsf{G}^{\pi_1(\cdot), \ldots, \pi_t(\cdot)}}(k) : d = 1].$$

Then, we define the δ-hardness of the game G as follows.

Definition 1. *We say that a game* G *is δ-hard (for some $0 \leq \delta \leq 1$) for adversaries with polynomial query complexity if for any (even computationally unbounded) algorithm $\mathcal{A}$ whose query complexity is at most polynomial, there is a negligible function $\mu(k)$ such that $\text{Adv}_{\text{RP}, \mathcal{A}}^{\mathsf{G}}(k) - \delta \leq \mu(k)$. We call "$\delta(\mathsf{G})$" the hardness of the game* G *and is the smallest value such that* G *is δ-hard for adversaries with polynomial query complexity.*

---

[5] It is usual to define the Gen algorithm as a probabilistic algorithm so that it takes 1^k as input, and outputs a pair (ek, td). However, in terms of existence, a TDP with such definition is equivalent to one defined in this paper, because without loss of generality we can identify the randomness r for generating $(ek, td) \leftarrow \text{Gen}(1^k; r)$ with the trapdoor of a TDP.

We stress that unlike [11,29], our definition of the hardness $\delta(\mathsf{G})$ of a game G regarding random permutations is with respect to *computationally unbounded* adversaries, and the restriction on an adversary is only on its query complexity, rather than its running time. Though this requirement for hard games is stronger than the ones used in [11,29] (and thus potentially harder to achieve), most security games that are δ-hard for all PPTAs remain δ-hard for computationally unbounded adversaries with polynomial query complexity. Examples include one-wayness, claw-freeness [19], and security under $t(k)$-correlated inputs [42] for any predetermined polynomial $t(k)$. See also [29, Table 1] for other types of security games that can be captured by δ-hard games. We note that, since G does not have access to inversions of permutations, our definition of hard games does not capture adaptive one-wayness [38,28].

A game for a TDP is then defined by replacing the random permutations in a δ-hard game with instantiations of permutations in the TDP. More specifically, we define the advantage of an adversary $\mathcal{A}$ in a game G for a TDP TDP $=$ (Gen, Eval, Samp, Inv) as follows:

$$\mathsf{Adv}^{\mathsf{G}}_{\mathsf{TDP},\mathcal{A}}(k) = \Pr\left[\begin{array}{c} td_1, \ldots, td_t \leftarrow_{\mathsf{R}} \{0,1\}^k; ek_i \leftarrow \mathsf{Gen}(1^k, td_i) \text{ for } i \in [t] \\ d \leftarrow_{\mathsf{R}} \mathsf{Expt}^{\mathsf{G}^{\mathsf{Eval}(ek_1,\cdot),\ldots,\mathsf{Eval}(ek_t,\cdot)}}_{\mathsf{TDP},\mathcal{A}(ek_1,\ldots,ek_t)}(k) \end{array} : d = 1 \right]$$

Note that in the above experiment, the interface of G is exactly the same as that of a game defined for random permutations. However, the interface of $\mathcal{A}$ is changed. Unlike the games regarding random permutations, we do not provide $\mathcal{A}$ with oracle access to $\mathsf{Eval}(ek_i, \cdot)$'s because it gets evaluation keys $\{ek_i\}$ and thus can compute each $\mathsf{Eval}(ek_i, \cdot)$ by itself.

Definition 2. *We say that* TDP *is secure for game* G *if for all PPTAs* $\mathcal{A}$*, there is a negligible function* $\mu(k)$ *such that* $\mathsf{Adv}^{\mathsf{G}}_{\mathsf{TDP},\mathcal{A}}(k) - \delta(\mathsf{G}) \le \mu(k)$*. Furthermore, we say that* TDP *is an* ideal TDP *if it is secure for all games.*

Note that the definition of the hard games for a TDP considers only PPTA adversaries, although the hardness $\delta(\mathsf{G})$ is defined with respect to (computationally unbounded) adversaries with polynomial query complexity.

It has been observed in [11] that ideal security is too strong to be satisfied by TDPs implemented by PPTAs. However, we will show the *impossibility* of constructing a OWP from an ideal TDP in a black-box manner, and thus ruling out a black-box construction from a TDP with such strong security makes our result *stronger*.

3   Black-Box Separation of OWP from Ideal TDP

In this section, we show our main result: there is no black-box construction of a OWP from an ideal TDP.

We first recall the formal definition of the type of black-box constructions that we will rule out, which is called a *fully*-black-box construction (reduction) in the taxonomy of Reingold et al. [41]. (The definition can be easily adapted to other primitives.)

Definition 3. *We say that there exists a fully-black-box construction of a OWP from an ideal TDP, if there exist a positive polynomial* $\ell = \ell(k)$*, an oracle PPTA* P *(called* "construction"), and an oracle PPTA $\mathcal{R}$ (called "reduction") such that for all tuples of algorithms $\mathsf{TDP} = (\mathsf{Gen}, \mathsf{Eval}, \mathsf{Samp}, \mathsf{Inv})$ that implement a TDP with security parameter k and all algorithms $\mathcal{A}$ (where each algorithm in TDP and $\mathcal{A}$ are of arbitrary complexity) the following two conditions hold:

(Correctness): $\mathsf{P}^{\mathsf{TDP}}$ is a permutation over $\{0,1\}^\ell$.
(Security): If $\mathsf{Adv}^{\mathsf{OWP}}_{\mathsf{P}^{\mathsf{TDP}}, \mathcal{A}, \ell}(k)$ is non-negligible, then so is $\mathsf{Adv}^{\mathsf{G}}_{\mathsf{TDP}, \mathcal{R}^{\mathcal{A}}, \mathsf{TDP}}(k) - \delta(\mathsf{G})$ for some game G.

The main result in this paper is the following.

Theorem 1. *There is no fully-black-box construction of a OWP from an ideal TDP.*

Recall that the security games for most of the security notions of a TDP, such as (ordinary) one-wayness, security under $t(k)$-correlated-inputs [42] for any predetermined polynomial $t = t(k)$, and claw-freeness [19], can be captured by the δ-hard games. Since "a (fully-)black-construction of a primitive from another primitive" is a transitive relation, we obtain the following as a corollary of Theorem 1.

Corollary 1. *There is no fully-black-box construction of a OWP from a one-way TDP[6], a TDP secure under t-correlated-input for any predetermined polynomial t, or a claw-free TDP.*

To prove Theorem 1, we will use the following "two oracle separation" technique [15,24] (which is an extension from the "one oracle separation" by [25,41]). Specifically, to prove Theorem 1, it is sufficient to show the following lemma.

Lemma 1. *(adapted from [15,24].) Let* PSPACE *be an oracle for a* PSPACE-*complete problem. Assume there exist a set* $\mathbb{O}$ *of oracles and a tuple of oracle PPTAs* $\mathsf{TDP} = (\mathsf{Gen}, \mathsf{Eval}, \mathsf{Samp}, \mathsf{Inv})$ *that satisfy the following three conditions:*

(1): $\mathsf{TDP}^{\mathcal{O}} = (\mathsf{Gen}^{\mathcal{O}}, \mathsf{Eval}^{\mathcal{O}}, \mathsf{Samp}^{\mathcal{O}}, \mathsf{Inv}^{\mathcal{O}})$ *is correct as a TDP for all* $\mathcal{O} \in \mathbb{O}$.
(2): For any game G *and for any oracle PPTA* $\mathcal{A}$, $\mathbf{E}_{\mathcal{O} \leftarrow_{\mathsf{R}} \mathbb{O}}[\mathsf{Adv}^{\mathsf{G}}_{\mathsf{TDP}^{\mathcal{O}}, \mathcal{A}^{\mathcal{O}}, \mathsf{PSPACE}}(k)] - \delta(\mathsf{G})$ *is negligible.*
(3): For any positive polynomial $\ell = \ell(k)$ *and for any oracle PPTA* P, *if* $\mathsf{P}^{\mathcal{O}}$ *is a permutation over* $\{0,1\}^\ell$ *for all* $\mathcal{O} \in \mathbb{O}$, *then there exists an oracle PPTA* $\mathcal{A}$ *such that* $\mathbf{E}_{\mathcal{O} \leftarrow_{\mathsf{R}} \mathbb{O}}[\mathsf{Adv}^{\mathsf{OWP}}_{\mathsf{P}^{\mathcal{O}}, \mathcal{A}^{\mathcal{O}}, \mathsf{PSPACE}, \ell}(k)]$ *is overwhelming.*

Then, there is no fully-black-box construction of a OWP from an ideal TDP.

In order to use Lemma 1 for showing our main result, we define the set $\mathbb{T}$ of "TDP" oracles $\mathcal{T}$ below, which will be used as $\mathbb{O}$ in the above lemma. Next, in Section 3.1, we show Lemmas 2 and 3 which guarantee that there is a tuple of oracle PPTAs $\mathsf{TDP} = (\mathsf{Gen}, \mathsf{Eval}, \mathsf{Samp}, \mathsf{Inv})$ such that $\mathbb{T}$ and TDP satisfy the conditions (1) and (2) of the above lemma, respectively. Then, in Section 3.2, we show Lemma 4 which guarantees that the set $\mathbb{T}$ satisfies the condition (3) of the above lemma. Theorem 1 follows by combining these lemmas.

---

[6] Actually, permutations in our TDP have a trivial domain $\{0,1\}^k$ and thus the TDP satisfies *doubly enhanced one-wayness* [18]. Furthermore, given a $2k$-bit string ek, whether $\mathcal{E}(ek, \cdot)$ defines a permutation can also be checked easily by checking the result of $\mathcal{E}(ek, 0^k)$, and thus it also satisfies the *certified* property [5]. Thus, our result also rules out constructions from a one-way TDP with these properties.

TDP Oracle $\mathcal{T}$. The *TDP oracle* $\mathcal{T}$ models an ideally secure TDP whose evaluation-key space is sparse. Formally, a TDP oracle $\mathcal{T}$ consists of the following three suboracles $(\mathcal{G}, \mathcal{E}, \mathcal{D})$:

- $\mathcal{G} : \{0,1\}^k \to \{0,1\}^{2k}$: (Corresponding to the key generation for the TDP) This is an injective function that takes $td \in \{0,1\}^k$ as input, and returns $ek \in \{0,1\}^{2k}$.
- $\mathcal{E} : \{0,1\}^{2k} \times \{0,1\}^k \to \{0,1\}^k \cup \{\perp\}$: (Corresponding to evaluation) For every $ek \in$ Range$(\mathcal{G})$, $\mathcal{E}(ek, \cdot)$ is a permutation over $\{0,1\}^k$, and for every $ek \notin$ Range$(\mathcal{G})$ and every $\alpha \in \{0,1\}^k$, $\mathcal{E}(ek, \alpha) = \perp$.
- $\mathcal{D} : \{0,1\}^k \times \{0,1\}^k \to \{0,1\}^k$: (Corresponding to inversion) This function takes $td \in \{0,1\}^k$ and $\beta \in \{0,1\}^k$ as input, and returns $\alpha \in \{0,1\}^k$ such that $\mathcal{E}(\mathcal{G}(td), \alpha) = \beta$.

We denote by $\mathbb{T}$ the set consisting of all possible TDP oracles $\mathcal{T}$ that satisfy the above syntax.

3.1   Ideal Trapdoor Permutation Based on $\mathcal{T}$

Here, we show that there exists an ideal TDP that uses a TDP oracle $\mathcal{T} = (\mathcal{G}, \mathcal{E}, \mathcal{D}) \in \mathbb{T}$. Consider the following tuple $\mathsf{TDP}^{\mathcal{T}} = (\mathsf{Gen}^{\mathcal{T}}, \mathsf{Eval}^{\mathcal{T}}, \mathsf{Samp}^{\mathcal{T}}, \mathsf{Inv}^{\mathcal{T}})$ of oracle PPTAs, which are constructed straightforwardly from $\mathcal{T}$:

- $\mathsf{Gen}^{\mathcal{T}}(1^k, td)$: Compute $ek \leftarrow \mathcal{G}(td)$ and output the evaluation-key ek.
- $\mathsf{Eval}^{\mathcal{T}}(ek, x)$: Compute $y \leftarrow \mathcal{E}(ek, x)$ and output y. (We define the domain D_{ek} of $\mathsf{Eval}^{\mathcal{T}}(ek, \cdot)$ to be $\{0,1\}^k$ for all $ek \in$ Range$(\mathcal{G})$.)
- $\mathsf{Samp}^{\mathcal{T}}(ek)$: Pick $x \in \{0,1\}^k$ uniformly at random, and output x. (Note that this algorithm does not use $\mathcal{T}$ at all.)
- $\mathsf{Inv}^{\mathcal{T}}(td, y)$: Compute $x \leftarrow \mathcal{D}(td, y)$ and output x.

Regarding $\mathsf{TDP}^{\mathcal{T}}$ described above, the following two lemmas can be shown:

Lemma 2. *For any $\mathcal{T} \in \mathbb{T}$, $\mathsf{TDP}^{\mathcal{T}}$ is correct as a TDP.*

Lemma 3. *For all games G and any oracle PPTA adversary $\mathcal{A}$, there exists a negligible function $\mu(k)$ such that $\mathbf{E}_{\mathcal{T} \leftarrow_{\mathrm{R}} \mathbb{T}}[\mathsf{Adv}^{\mathsf{G}}_{\mathsf{TDP}^{\mathcal{T}}, \mathcal{A}^{\mathcal{T}}, \mathrm{PSPACE}}(k)] - \delta(\mathsf{G}) \leq \mu(k)$.*

Lemma 2 is immediate from the definition of the TDP oracle $\mathcal{T}$. The formal proof of Lemma 3 is given in the full version (but we will give a proof sketch below). Note that if we pick $\mathcal{T} = (\mathcal{G}, \mathcal{E}, \mathcal{D})$ uniformly from $\mathbb{T}$, then $\mathcal{G}$ is a random injective function that is length-doubling, and every permutation $\mathcal{E}(ek, \cdot)$ with $ek \in$ Range$(\mathcal{G})$ is an independent random permutation. Kiltz and Pietrzak [29] showed that a similar construction of a TDP oracle whose "key generation oracle" is also a random permutation is ideally secure even against computationally unbounded adversary that makes only polynomially many queries. Our proof of Lemma 3 is similar to theirs.

Proof Sketch of Lemma 3. Fix an arbitrary δ-hard game G, and let $t = t(k)$ be a polynomial implicitly determined by G. Fix also an arbitrary PPTA adversary $\mathcal{A}$.

The expectation (over the choice of $\mathcal{T}$) of the advantage of the adversary $\mathcal{A}$ attacking $\mathsf{TDP}^{\mathcal{T}} = (\mathsf{Gen}^{\mathcal{T}}, \mathsf{Eval}^{\mathcal{T}}, \mathsf{Samp}^{\mathcal{T}}, \mathsf{Inv}^{\mathcal{T}})$ in the game G (in the presence of the PSPACE oracle) can be written as follows:

$$\mathop{\mathbf{E}}_{\mathcal{T} \leftarrow_{\mathrm{R}} \mathbb{T}} \left[\mathsf{Adv}^{\mathsf{G}}_{\mathsf{TDP}^{\mathcal{T}}, \mathcal{A}^{\mathcal{T}}, \mathsf{PSPACE}}(k) \right]$$

$$= \mathop{\mathbf{E}}_{\mathcal{T} \leftarrow_{\mathrm{R}} \mathbb{T}} \left[\Pr \left[\begin{array}{l} td_1^*, \ldots, td_t^* \leftarrow_{\mathrm{R}} \{0,1\}^k; \ ek_i^* \leftarrow \mathsf{Gen}^{\mathcal{T}}(1^k, td_i^*) \text{ for } i \in [t]; \\ d \leftarrow_{\mathrm{R}} \mathsf{Expt}^{\mathsf{G}^{\mathsf{Eval}^{\mathcal{T}}(ek_1^*, \cdot), \ldots, \mathsf{Eval}^{\mathcal{T}}(ek_t^*, \cdot)}}_{\mathsf{TDP}^{\mathcal{T}}, \mathcal{A}^{\mathcal{T}}, \mathsf{PSPACE}(ek_1^*, \ldots, ek_t^*)}(k) \end{array} : d = 1 \right] \right]$$

$$= \Pr \left[\begin{array}{l} \mathcal{T} \leftarrow_{\mathrm{R}} \mathbb{T}; \ td_1^*, \ldots, td_t^* \leftarrow_{\mathrm{R}} \{0,1\}^k; \ ek_i^* \leftarrow \mathcal{G}(td_i^*) \text{ for } i \in [t]; \\ d \leftarrow_{\mathrm{R}} \mathsf{Expt}^{\mathsf{G}^{\mathcal{E}(ek_1^*, \cdot), \ldots, \mathcal{E}(ek_t^*, \cdot)}}_{\mathsf{TDP}^{\mathbb{T}}, \mathcal{A}^{\mathbb{T}}, \mathsf{PSPACE}(ek_1^*, \ldots, ek_t^*)}(k) \end{array} : d = 1 \right].$$

Let us denote by $\widetilde{\mathsf{Expt}}^{\mathsf{G}}_{\mathsf{TDP}^{\mathbb{T}}, \mathcal{A}^{\mathbb{T}}, \mathsf{PSPACE}}(k)$ the experiment in the probability in the last equation.

Now, consider the following two games.

Game 1: This is the ordinary δ-hard game G for $\mathsf{TDP}^{\mathcal{T}}$, in which sampling of the oracle $\mathcal{T}$ from $\mathbb{T}$ is also taken into account, i.e. $\widetilde{\mathsf{Expt}}^{\mathsf{G}}_{\mathsf{TDP}^{\mathbb{T}}, \mathcal{A}^{\mathbb{T}}, \mathsf{PSPACE}}(k)$.

Game 2: Same as Game 1, except that $\mathcal{A}$'s queries of the following types are answered with $\bot$: (i) a $\mathcal{G}$-query td_i^* for some $i \in [t]$, and (ii) a $\mathcal{D}$-query $(td_i^*, *)$ for some $i \in [t]$.

For $i \in \{1, 2\}$, let Succ_i be the event that $\mathcal{A}$ wins (i.e. $d = 1$ occurs) in Game i. By definition we have $\mathbf{E}_{\mathcal{T} \leftarrow_{\mathrm{R}} \mathbb{T}}[\mathsf{Adv}^{\mathsf{G}}_{\mathsf{TDP}^{\mathcal{T}}, \mathcal{A}^{\mathcal{T}}, \mathsf{PSPACE}}(k)] = \Pr[\mathsf{Succ}_1]$. Furthermore, we have

$$\mathop{\mathbf{E}}_{\mathcal{T} \leftarrow_{\mathrm{R}} \mathbb{T}}[\mathsf{Adv}^{\mathsf{G}}_{\mathsf{TDP}^{\mathcal{T}}, \mathcal{A}^{\mathcal{T}}, \mathsf{PSPACE}}(k)] - \delta(\mathsf{G}) = \Pr[\mathsf{Succ}_1] - \delta(\mathsf{G})$$

$$\leq |\Pr[\mathsf{Succ}_1] - \Pr[\mathsf{Succ}_2]| + \Pr[\mathsf{Succ}_2] - \delta(\mathsf{G}). \quad (1)$$

In the full version, we will show how to upperbound each term in the right hand side of the inequality (1), which will prove Lemma 3. Below we explain the sketches for how to show these.

$|\Pr[\mathsf{Succ}_1] - \Pr[\mathsf{Succ}_2]|$ can be shown to be negligible, because the adversary $\mathcal{A}$, who can make only polynomially many queries, cannot tell the difference between Game 1 and Game 2 (except with negligible probability). More specifically, Game 1 and Game 2 differ only in the response to $\mathcal{A}$'s $\mathcal{G}$-queries and $\mathcal{D}$-queries that contain the preimages $\{td_i^*\}_{i \in [t]}$ of the evaluation keys $\{ek_i^*\}_{i \in [t]}$, and thus in order for $\mathcal{A}$ to distinguish these games, $\mathcal{A}$ has to find one of $\{td_i^*\}_{i \in [t]}$. However, intuitively, finding any of the preimages $\{td_i^*\}_{i \in [t]}$ is hard because the TDP oracle $\mathcal{T}$ is chosen randomly and especially the function $\mathcal{G}$ is a random injective function, and we will formally show that this intuition works.

$\Pr[\mathsf{Succ}_2] - \delta(\mathsf{G})$ can be shown to be negligible, roughly because Game 2 can be perfectly simulated by another computationally unbounded adversary $\mathcal{S}$ with polynomial query complexity that interacts with the PPTA (challenger) G for *random permutations* (not for the TDP $\mathsf{TDP}^{\mathcal{T}}$), in such a way that $\mathsf{Adv}^{\mathsf{G}}_{\mathsf{RP}, \mathcal{S}}(k) = \Pr[\mathsf{Succ}_2]$. But by the assumption that G is a δ-hard game, $\mathsf{Adv}^{\mathsf{G}}_{\mathsf{RP}, \mathcal{S}}(k) - \delta(\mathsf{G}) = \Pr[\mathsf{Succ}_2] - \delta(\mathsf{G})$ is negligible.

This completes the proof sketch of Lemma 3. $\qquad \square$

3.2 Breaking Any Candidate of One-Way Permutation Based on $\mathcal{T}$

Here, we show that any candidate of a OWP $\mathsf{P}^{\mathcal{T}}$ based on a TDP oracle $\mathcal{T} \in \mathbb{T}$ can be broken by some oracle PPTA almost perfectly (using the PSPACE oracle). Specifically, this subsection is devoted to proving the following lemma.

Lemma 4. *Let $\ell = \ell(k)$ be a positive polynomial and P be an oracle PPTA such that $\mathsf{P}^{\mathcal{T}}$ is a permutation over $\{0,1\}^\ell$ for all $\mathcal{T} \in \mathbb{T}$. Then there exists an oracle PPTA $\mathcal{A}$ such that $\mathbf{E}_{\mathcal{T} \leftarrow_{\mathrm{R}} \mathbb{T}}[\mathsf{Adv}^{\mathsf{OWP}}_{\mathsf{P}^{\mathcal{T}}, \mathcal{A}^{\mathcal{T}, \mathsf{PSPACE}}, \ell}(k)]$ is overwhelming.*

To prove Lemma 4, we need some further notations, two other oracles than $\mathcal{T}$, and several intermediate lemmas. Thus, we first introduce them, and in the last of this subsection show the proof of Lemma 4. The intuitive explanation on how the above lemma is proved can be found in Section 1.3.

Further Notations. For notational convenience, we introduce two notations. Let $\mathbb{O}$ be a set of oracles $\mathcal{O}$, $\ell = \ell(k)$ be a positive polynomial, and P and $\mathcal{A}$ be oracle PP-TAs. If $\mathsf{P}^{\mathcal{O}}$ is a permutation over $\{0,1\}^\ell$ for all oracles $\mathcal{O} \in \mathbb{O}$, then we denote by $\widetilde{\mathsf{Expt}}^{\mathsf{OWP}}_{\mathsf{P}^{\mathcal{O}}, \mathcal{A}^{\mathcal{O}, \mathsf{PSPACE}}, \ell}(k)$ the following experiment:

$$[\, \mathcal{O} \leftarrow_{\mathrm{R}} \mathbb{O};\ x^* \leftarrow_{\mathrm{R}} \{0,1\}^\ell;\ y^* \leftarrow \mathsf{P}^{\mathcal{O}}(x^*);\ x' \leftarrow_{\mathrm{R}} \mathcal{A}^{\mathcal{O}, \mathsf{PSPACE}}(1^k, y^*) \,].$$

Note that $\widetilde{\mathsf{Expt}}^{\mathsf{OWP}}_{\mathsf{P}^{\mathcal{O}}, \mathcal{A}^{\mathcal{O}, \mathsf{PSPACE}}, \ell}(k)$ includes sampling an oracle $\mathcal{O}$ from $\mathbb{O}$.

Then, we define $\widetilde{\mathsf{Adv}}^{\mathsf{OWP}}_{\mathsf{P}^{\mathcal{O}}, \mathcal{A}^{\mathcal{O}, \mathsf{PSPACE}}, \ell}(k) := \mathbf{E}_{\mathcal{O} \leftarrow_{\mathrm{R}} \mathbb{O}}[\, \mathsf{Adv}^{\mathsf{OWP}}_{\mathsf{P}^{\mathcal{O}}, \mathcal{A}^{\mathcal{O}, \mathsf{PSPACE}}, \ell}(k) \,]$, i.e.,

$$\widetilde{\mathsf{Adv}}^{\mathsf{OWP}}_{\mathsf{P}^{\mathcal{O}}, \mathcal{A}^{\mathcal{O}, \mathsf{PSPACE}}, \ell}(k)$$
$$= \Pr[\mathcal{O} \leftarrow_{\mathrm{R}} \mathbb{O}; x^* \leftarrow_{\mathrm{R}} \{0,1\}^\ell; y^* \leftarrow \mathsf{P}^{\mathcal{O}}(x^*); x' \leftarrow_{\mathrm{R}} \mathcal{A}^{\mathcal{O}, \mathsf{PSPACE}}(1^k, y^*) : x' = x^*].$$

(Our goal in this subsection is to show that $\widetilde{\mathsf{Adv}}^{\mathsf{OWP}}_{\mathsf{P}^{\mathcal{T}}, \mathcal{A}^{\mathcal{T}, \mathsf{PSPACE}}, \ell}(k)$ is overwhelming.)

Block Cipher Oracle $\mathcal{B}$. Here we introduce a "block cipher" oracle $\mathcal{B}$ which models an ideally secure block cipher (or, keyed invertible permutation) whose key space is sparse. Formally, a block cipher oracle $\mathcal{B}$ consists of the following three suboracles $(\widehat{\mathcal{G}}, \mathcal{P}, \mathcal{P}^{-1})$:

- $\widehat{\mathcal{G}} : \{0,1\}^k \to \{0,1\}^{2k}$: (Corresponding to the key generation for the block cipher) This is an injective function that takes $td \in \{0,1\}^k$ as input, and returns $ek \in \{0,1\}^{2k}$.
- $\mathcal{P} : \{0,1\}^{2k} \times \{0,1\}^k \to \{0,1\}^k \cup \{\bot\}$: (Corresponding to encryption) For every $ek \in \mathsf{Range}(\widehat{\mathcal{G}})$, $\mathcal{P}(ek, \cdot)$ is a permutation over $\{0,1\}^k$, and for every $ek \notin \mathsf{Range}(\widehat{\mathcal{G}})$ and every $\alpha \in \{0,1\}^k$, $\mathcal{P}(ek, \alpha) = \bot$.
- $\mathcal{P}^{-1} : \{0,1\}^{2k} \times \{0,1\}^k \to \{0,1\}^k \cup \{\bot\}$: (Corresponding to decryption) For every $ek \in \mathsf{Range}(\widehat{\mathcal{G}})$, $\mathcal{P}^{-1}(ek, \cdot)$ is the inversion of $\mathcal{P}(ek, \cdot)$, and for every $ek \notin \mathsf{Range}(\widehat{\mathcal{G}})$ and every $\beta \in \{0,1\}^k$, $\mathcal{P}^{-1}(ek, \beta) = \bot$.

We denote by $\mathbb{B}$ the set consisting of all possible block cipher oracles $\mathcal{B}$ that satisfy the above syntax.

Relationship between T and B. We will use the following simple fact shown by Lindell and Zarosim [30].

Lemma 5. *([30]) Let ϕ be the mapping that maps a block cipher oracle $B = (\widehat{G}, P, P^{-1}) \in \mathbb{B}$ to a tuple of oracles $\phi(B) = (G, \mathcal{E}, D)$, where the suboracles G, $\mathcal{E}$, and D are defined in the following way: For all $td \in \{0,1\}^k$, $ek \in \{0,1\}^{2k}$, $\alpha \in \{0,1\}^k$ and $\beta \in \{0,1\}^k$, we let*

$$G(td) := \widehat{G}(td), \quad \mathcal{E}(ek, \alpha) := P(ek, \alpha), \quad \text{and} \quad D(td, \beta) := P^{-1}(\widehat{G}(td), \beta).$$

Then, ϕ is a bijection from $\mathbb{B}$ to $\mathbb{T}$.

Round Function Oracle $\mathcal{R}$. Here, we introduce a "round function" oracle $\mathcal{R}$ which models a set of "round functions" in the Feistel-network construction of permutations [32] (whose evaluation key space is sparse). Formally, a round function oracle $\mathcal{R}$ consists of the following two suboracles $(\widetilde{G}, \mathcal{F})$:

$\widetilde{G} : \{0,1\}^k \to \{0,1\}^{2k}$: (Corresponding to the key generation for each round function) This is an injective function that takes $td \in \{0,1\}^k$ as input, and returns $ek \in \{0,1\}^{2k}$.

$\mathcal{F} : [14] \times \{0,1\}^{2k} \times \{0,1\}^{k/2} \to \{0,1\}^{k/2} \cup \{\bot\}$: (Corresponding to the round functions in the Feistel-network). For every index $i \in [14]$ and $ek \in \text{Range}(\widetilde{G})$, $\mathcal{F}(i, ek, \cdot)$ is a function from $k/2$ bit to $k/2$ bit, and for every $ek \notin \text{Range}(\widetilde{G})$ and every $(i, \gamma) \in [14] \times \{0,1\}^{k/2}$, $\mathcal{F}(i, ek, \gamma) = \bot$.

We denote by $\mathbb{R}$ the set consisting of all possible round function oracles $\mathcal{R}$ that satisfy the above syntax.

Relationship between B and $\mathcal{R}$. Holenstein et al. [23] showed that the random oracle model and the ideal cipher model are equivalent. (The statement itself was posed by Coron et al. [10].) More concretely, they proved that a random invertible permutation can be simulated by the fourteen-round Feistel-network construction of a permutation in which each round function is an independent random function. (Technically, this means that the latter is *indifferentiable* [34] from the former.) Based on their result, we can also construct oracle PPTAs C and S such that $(C^{\mathcal{R}}, \mathcal{R})$ and (B, S^B) are indistinguishable.

More formally, consider the following PPTA C that, given access to $\mathcal{R} = (\widetilde{G}, \mathcal{F}) \in \mathbb{R}$, tries to simulate a block cipher oracle $C^{\mathcal{R}} = (\widehat{G}, P, P^{-1})$ as follows:

$\widehat{G}(\cdot)$: Define $\widehat{G}(\cdot) = \widetilde{G}(\cdot)$.

$P(\cdot, \cdot)$: On input $(ek, \alpha) \in \{0,1\}^{2k} \times \{0,1\}^k$, check if $ek \in \text{Range}(\widetilde{G})$ by making an $\mathcal{F}$-query $(1, ek, 0^{k/2})$. If the answer from $\mathcal{F}$ is $\bot$ (meaning $ek \notin \text{Range}(\widetilde{G})$), then return $\bot$. Otherwise, regard α as $\alpha = (L_0 \| R_0)$ so that $|L_0| = |R_0| = k/2$. Then, for each $i \in [14]$, compute $L_i \leftarrow R_{i-1}$ and $R_i \leftarrow \mathcal{F}(i, ek, R_{i-1}) \oplus L_{i-1}$, and finally output $\beta \leftarrow (L_{14} \| R_{14})$.

$P^{-1}(\cdot, \cdot)$: On input $(ek, \beta) \in \{0,1\}^{2k} \times \{0,1\}^k$, check if $ek \in \text{Range}(\widetilde{G})$ as above. If $ek \notin \text{Range}(\widetilde{G})$, then return $\bot$. Otherwise, compute and output the inversion of $P(ek, \cdot)$ using $\mathcal{F}$.

Constructed as above, it is guaranteed that $C^{\mathcal{R}} \in \mathbb{B}$ for all $\mathcal{R} \in \mathbb{R}$, because the Feistel-network construction yields a permutation no matter what round functions are used. Moreover, the result in [23] yields the following.

Lemma 6. *(follows from [23].) Let* C *be the oracle PPTA as above. Then, for any polynomial* $q = q(k)$*, there exists an oracle PPTA* S *such that for all (computationally unbounded) oracle algorithms* $\mathcal{D}$ *making at most* q *queries, the following difference is negligible:*

$$\left| \Pr_{\mathcal{R} \leftarrow_{\mathrm{R}} \mathbb{R}}[\mathcal{D}^{C^{\mathcal{R}}, \mathcal{R}}(1^k) = 1] - \Pr_{\mathcal{B} \leftarrow_{\mathrm{R}} \mathbb{B}}[\mathcal{D}^{\mathcal{B}, S^{\mathcal{B}}}(1^k) = 1] \right|.$$

TDP Oracle $\mathcal{T}$ *Can Be Simulated.* Here, we show that if there exists a TDP-based permutation $P^{\mathcal{T}}$, then so does a "random function"-based permutation $\widetilde{P}^{\mathcal{R}}$ such that inverting $\widetilde{P}^{\mathcal{R}}$ is as hard as inverting $P^{\mathcal{T}}$. Furthermore, the latter is true even in the presence of PSPACE oracle.

Lemma 7. *Let* $\ell = \ell(k)$ *be a positive polynomial and* P *be an oracle PPTA such that* $P^{\mathcal{T}}$ *is a permutation over* $\{0,1\}^\ell$ *for all* $\mathcal{T} \in \mathbb{T}$*. Then, there exists another oracle PPTA* $\widetilde{P}$ *that satisfies the following two properties: (1) For all* $\mathcal{R} \in \mathbb{R}$*,* $\widetilde{P}^{\mathcal{R}} \in \mathrm{Perm}_\ell$*. (2) For any oracle PPTA* $\widetilde{\mathcal{A}}$*, there exist another oracle PPTA* $\mathcal{A}$ *and a negligible function* $\mu(k)$ *such that* $\widetilde{\mathrm{Adv}}^{\mathrm{OWP}}_{P^{\mathrm{T}}, \mathcal{A}^{\mathrm{T}}, \mathrm{PSPACE}, \ell}(k) \geq \widetilde{\mathrm{Adv}}^{\mathrm{OWP}}_{\widetilde{P}^{\mathrm{R}}, \widetilde{\mathcal{A}}^{\mathrm{R}}, \mathrm{PSPACE}, \ell}(k) - \mu(k).$

Proof of Lemma 7. (The intuitive explanation can be found in Section 1.3.) Let ℓ and P be as stated in the lemma. First, define the "intermediate" oracle PPTA $\widehat{P}$ by $\widehat{P}^{\mathcal{B}}(\cdot) = P^{\phi(\mathcal{B})}(\cdot)$, where ϕ is the bijection from $\mathbb{B}$ to $\mathbb{T}$ due to Lemma 5. This construction of $\widehat{P}$ also guarantees that $P^{\mathcal{T}}(\cdot) = \widehat{P}^{\phi^{-1}(\mathcal{T})}(\cdot)$ where ϕ^{-1} is the inversion function of ϕ (i.e. ϕ^{-1} is also a bijection from $\mathbb{T}$ to $\mathbb{B}$). Next, define the oracle PPTA $\widetilde{P}$ by $\widetilde{P}^{\mathcal{R}}(\cdot) = \widehat{P}^{C^{\mathcal{R}}}(\cdot)$, where C is the oracle PPTA due to Lemma 6. Then, since $P^{\mathcal{T}} \in \mathrm{Perm}_\ell$ for all $\mathcal{T} \in \mathbb{T}$, we have $\widehat{P}^{\mathcal{B}} \in \mathrm{Perm}_\ell$ for all $\mathcal{B} \in \mathbb{B}$. This in turn guarantees that $\widetilde{P}^{\mathcal{R}} \in \mathrm{Perm}_\ell$ for all $\mathcal{R} \in \mathbb{R}$, because $C^{\mathcal{R}} \in \mathbb{B}$ for all $\mathcal{R} \in \mathbb{R}$. Therefore, $\widetilde{P}$ satisfies the property (1).

Next, we show that $\widetilde{P}$ satisfies the property (2). Let $\widetilde{\mathcal{A}}$ be an arbitrary oracle PPTA adversary that runs in the experiment $\widetilde{\mathrm{Expt}}^{\mathrm{OWP}}_{\widetilde{P}^{\mathrm{R}}, \widetilde{\mathcal{A}}^{\mathrm{R}}, \mathrm{PSPACE}, \ell}(k)$ and makes in total $q = q(k)$ oracle queries. Note that since $\widetilde{\mathcal{A}}$ is a PPTA, q is a polynomial. Let S be the simulator corresponding to the polynomial q, which is guaranteed to exist by Lemma 6, and define an oracle PPTA $\widehat{\mathcal{A}}^{(\cdot),(\cdot)}$ (which expects to have access to an oracle $\mathcal{B} \in \mathbb{B}$ and the PSPACE oracle) by $\widetilde{\mathcal{A}}^{S^{(\cdot)},(\cdot)}$. That is, given access to any $\mathcal{B} \in \mathbb{B}$ and the PSPACE oracle, $\widehat{\mathcal{A}}^{\mathcal{B}, \mathrm{PSPACE}}$ and $\widetilde{\mathcal{A}}^{S^{\mathcal{B}}, \mathrm{PSPACE}}$ behave identically. Since both $\widetilde{\mathcal{A}}$ and S are oracle PPTAs, $\widehat{\mathcal{A}}$ is also an oracle PPTA and thus makes at most polynomially many queries.

Then, consider the following sequence of games.

Game 1. This is the ordinary experiment $\widetilde{\mathrm{Expt}}^{\mathrm{OWP}}_{\widetilde{P}^{\mathrm{R}}, \widetilde{\mathcal{A}}^{\mathrm{R}}, \mathrm{PSPACE}, \ell}(k)$ that $\widetilde{\mathcal{A}}$ runs in. That is:

$[\mathcal{R} \leftarrow_{\mathrm{R}} \mathbb{R}; \; x^* \leftarrow_{\mathrm{R}} \{0,1\}^\ell; \; y^* \leftarrow \widetilde{P}^{\mathcal{R}}(x^*); \; x' \leftarrow_{\mathrm{R}} \widetilde{\mathcal{A}}^{\mathcal{R}, \mathrm{PSPACE}}(1^k, y^*)].$

Game 2. This game is defined as follows:

$[\mathcal{B} \leftarrow_{\mathrm{R}} \mathbb{B}; \; x^* \leftarrow_{\mathrm{R}} \{0,1\}^\ell; \; y^* \leftarrow \widehat{P}^{\mathcal{B}}(x^*); \; x' \leftarrow_{\mathrm{R}} \widehat{\mathcal{A}}^{\mathcal{B}, \mathrm{PSPACE}}(1^k, y^*)].$

Game 3. This game is defined as follows:

$$[\mathcal{T} \leftarrow_R \mathbb{T}; \ x^* \leftarrow_R \{0,1\}^\ell; \ y^* \leftarrow \mathsf{P}^\mathcal{T}(x^*); \ x' \leftarrow_R \widehat{\mathcal{A}}^{\phi^{-1}(\mathcal{T}),\mathsf{PSPACE}}(1^k, y^*)].$$

Game 4. Same as Game 3, except that when $\widehat{\mathcal{A}}$ makes a $\mathcal{P}$-query (ek, α) or a $\mathcal{P}^{-1}$-query (ek, β) such that ek is not an answer to some of $\widehat{\mathcal{A}}$'s previous $\widehat{\mathcal{G}}$-queries, the query is answered with $\bot$.

For $i \in [4]$, let Succ_i be the event that $x' = x^*$ occurs in Game i. Then we have

$$\widetilde{\mathsf{Adv}}_{\widetilde{\mathsf{PR}},\widetilde{\mathcal{A}}^{\mathbb{R},\mathsf{PSPACE}},\ell}(k) = \Pr[\mathsf{Succ}_1] \leq \sum_{i \in [3]} |\Pr[\mathsf{Succ}_i] - \Pr[\mathsf{Succ}_{i+1}]| + \Pr[\mathsf{Succ}_4]. \quad (2)$$

To complete the proof, we upperbound each term in the above inequality.

Claim 1. $|\Pr[\mathsf{Succ}_1] - \Pr[\mathsf{Succ}_2]|$ *is negligible.*

Proof of Claim 1. We show that we can construct a *computationally unbounded* oracle algorithm (distinguisher) $\mathcal{D}$ that, using $\widehat{\mathsf{P}}$ and $\widetilde{\mathcal{A}}$ as its subroutines, makes at most q queries, and satisfies

$$|\Pr_{\mathcal{R} \leftarrow_R \mathbb{R}}[\mathcal{D}^{\mathsf{C}^{\mathcal{R}},\mathcal{R}}(1^k) = 1] - \Pr_{\mathcal{B} \leftarrow_R \mathbb{B}}[\mathcal{D}^{\mathcal{B},\mathsf{S}^{\mathcal{B}}}(1^k) = 1]| = |\Pr[\mathsf{Succ}_1] - \Pr[\mathsf{Succ}_2]|. \quad (3)$$

$\mathcal{D}$ is given access to two oracles $(\mathcal{O}_1, \mathcal{O}_2)$, which is either $(\mathsf{C}^{\mathcal{R}}, \mathcal{R})$ or $(\mathcal{B}, \mathsf{S}^{\mathcal{B}})$, and runs as follows:

$\mathcal{D}^{\mathcal{O}_1,\mathcal{O}_2}(1^k)$: $\mathcal{D}$ picks $x^* \leftarrow_R \{0,1\}^\ell$, computes $y^* \leftarrow \widehat{\mathsf{P}}^{\mathcal{O}_1}(x^*)$, and then simulates $\widetilde{\mathcal{A}}^{\mathcal{O}_2,\mathsf{PSPACE}}(1^k, y^*)$. Note that $\mathcal{D}$ is computationally unbounded, and thus can simulate the PSPACE oracle perfectly for $\widetilde{\mathcal{A}}$.

When $\widetilde{\mathcal{A}}$ terminates with output x', $\mathcal{D}$ checks whether $x' = x^*$. If this is the case, then $\mathcal{D}$ outputs 1, otherwise outputs 0, and terminates.

The above completes the description of $\mathcal{D}$. Note that the number of queries that $\mathcal{D}$ makes is at most the number of queries made by $\widetilde{\mathcal{A}}$, and thus is at most q.

Now, consider the case when $(\mathcal{O}_1, \mathcal{O}_2) = (\mathsf{C}^{\mathcal{R}}, \mathcal{R})$. Then it is clear that $\mathcal{D}$ simulates Game 1 perfectly for $\widetilde{\mathcal{A}}$. In particular, in this case we have $\widehat{\mathsf{P}}^{\mathcal{O}_1}(x^*) = \widehat{\mathsf{P}}^{\mathsf{C}^{\mathcal{R}}}(x^*) = \widetilde{\mathsf{P}}^{\mathcal{R}}(x^*)$, and $\widetilde{\mathcal{A}}$ is given access to $\mathcal{O}_2 = \mathcal{R}$ and PSPACE as in Game 1. Under this situation, the probability that $\mathcal{D}$ outputs 1 is exactly the same as the probability that $\widetilde{\mathcal{A}}$ succeeds in outputting the preimage x^* under $\widetilde{\mathsf{P}}^{\mathcal{R}}$ in Game 1, i.e. $\Pr_{\mathcal{R} \leftarrow_R \mathbb{R}}[\mathcal{D}^{\mathsf{C}^{\mathcal{R}},\mathcal{R}}(1^k) = 1] = \Pr[\mathsf{Succ}_1]$.

Next, consider the case when $(\mathcal{O}_1, \mathcal{O}_2) = (\mathcal{B}, \mathsf{S}^{\mathcal{B}})$. Recall that we defined $\widehat{\mathcal{A}}^{\mathcal{B},\mathsf{PSPACE}}$ by $\widetilde{\mathcal{A}}^{\mathsf{S}^{\mathcal{B}},\mathsf{PSPACE}}$, and thus $\widetilde{\mathcal{A}}^{\mathcal{O}_2,\mathsf{PSPACE}} = \widetilde{\mathcal{A}}^{\mathsf{S}^{\mathcal{B}},\mathsf{PSPACE}} = \widehat{\mathcal{A}}^{\mathcal{B},\mathsf{PSPACE}}$. Recall also that $\mathcal{D}$ can simulate PSPACE perfectly by its computationally unbounded power. Therefore, in this case $\mathcal{D}$ perfectly simulates Game 2 for $\widehat{\mathcal{A}}$. In particular, in this case we have $\widehat{\mathsf{P}}^{\mathcal{O}_1}(x^*) = \widehat{\mathsf{P}}^{\mathcal{B}}(x^*)$, and $\widehat{\mathcal{A}}$'s oracle queries are perfectly answered as in Game 2, using $\mathcal{O}_1 = \mathcal{B}$ and $\mathcal{D}$'s computationally unbounded power. Therefore the probability that $\mathcal{D}$ outputs 1 is exactly the same as the probability that $\widehat{\mathcal{A}}$ outputs x^* in Game 2, i.e. $\Pr_{\mathcal{B} \leftarrow_R \mathbb{B}}[\mathcal{D}^{\mathcal{B},\mathsf{S}^{\mathcal{B}}}(1^k) = 1] = \Pr[\mathsf{Succ}_2]$.

In summary, our distinguisher $\mathcal{D}$ makes in total q queries and satisfies the equation (3). Thus, Lemma 6 guarantees that $|\Pr[\mathsf{Succ}_1] - \Pr[\mathsf{Succ}_2]|$ is upperbounded to be negligible. This completes the proof of Claim 1. □

Claim 2. $\Pr[\mathsf{Succ}_2] = \Pr[\mathsf{Succ}_3]$.

Proof of Claim 2. Recall that due to Lemma 5, ϕ (and thus ϕ^{-1}) is a bijection between $\mathbb{B}$ and $\mathbb{T}$. Therefore, the uniform distribution over $\mathbb{B}$ is equivalent to the distribution of $\phi^{-1}(\mathcal{T})$ when $\mathcal{T} \leftarrow_{\mathsf{R}} \mathbb{T}$. Moreover, $\mathsf{P}^{\mathcal{T}}(\cdot) = \hat{\mathsf{P}}^{\phi^{-1}(\mathcal{T})}(\cdot)$ for all $\mathcal{T} \in \mathbb{T}$ by definition. These imply that from $\widehat{\mathcal{A}}$'s view point, all values in Game 2 and those in Game 3 are distributed identically, and thus $\Pr[\mathsf{Succ}_2] = \Pr[\mathsf{Succ}_3]$. This completes the proof of Claim 2. □

Claim 3. $|\Pr[\mathsf{Succ}_3] - \Pr[\mathsf{Succ}_4]|$ *is negligible.*

Proof Sketch of Claim 3. For $i \in \{3, 4\}$, let Find_i be the event that in Game i, $\widehat{\mathcal{A}}$ makes at least one $\mathcal{P}$- or $\mathcal{P}^{-1}$-query such that ek is not an answer to some of previous $\widehat{\mathcal{A}}$'s $\mathcal{G}$-queries and $ek \in \mathsf{Range}(\mathcal{G})$. Note that Game 3 and Game 4 proceed identically until Find_3 or Find_4 occurs in the corresponding games. Therefore, we have

$$|\Pr[\mathsf{Succ}_3] - \Pr[\mathsf{Succ}_4]| \leq \Pr[\mathsf{Find}_3] = \Pr[\mathsf{Find}_4].$$

Hence, to prove the claim it is sufficient to bound $\Pr[\mathsf{Find}_4]$.

Recall that in Game 4 (and in Game 3) the oracle $\mathcal{T} \in \mathbb{T}$ is picked uniformly, and thus $\mathcal{G}$ oracle is a random injective function which is length-doubling. Therefore, the probability that Find_4 occurs is exactly the same as the probability that an oracle algorithm with polynomial query complexity, which is given access to a random length-doubling injective function and the corresponding "membership" function for its range (this membership function tells if a given value is in the range of the injective function), finds a "fresh" element that is not obtained by actually making a query to the function but belongs to its range. However, it is easy to prove that such a probability is negligible (as long as the query complexity of the algorithm is at most polynomial), and this in turn bounds $\Pr[\mathsf{Find}_4]$ to be negligible. (The formal proof is provided in the full version.) This completes the proof sketch of Claim 3. □

Claim 4. *There exists an oracle PPTA $\mathcal{A}$ such that* $\Pr[\mathsf{Succ}_4] = \widetilde{\mathsf{Adv}}^{\mathsf{OWP}}_{\mathsf{PT},\mathcal{A}^{\mathsf{T},\mathsf{PSPACE}},\ell}(k)$.

Proof of Claim 4. Using the oracle PPTA $\widehat{\mathcal{A}}$ as a building block, we construct an oracle PPTA $\mathcal{A}$ that runs in $\widetilde{\mathsf{Expt}}^{\mathsf{OWP}}_{\mathsf{PT},\mathcal{A}^{\mathsf{T},\mathsf{PSPACE}},\ell}(k)$: $\mathcal{A}$ is given $(1^k, y^*)$ as input, where $y^* = \mathsf{P}^{\mathcal{T}}(x^*)$ for a randomly chosen $x^* \in \{0,1\}^\ell$ and $\mathcal{T} \in \mathbb{T}$, given access to $\mathcal{T}$ and PSPACE, and runs as follows:

$\mathcal{A}^{\mathcal{T},\mathsf{PSPACE}}(1^k, y^*)$: $\mathcal{A}$ generates an empty list L used to store "known" $\mathcal{G}$-query/answer pairs, and then runs $\widehat{\mathcal{A}}(1^k, y^*)$.

   $\mathcal{A}$ responds to the queries from $\widehat{\mathcal{A}}$ as follows:
   - For a $\mathcal{G}$-query td, $\mathcal{A}$ forwards it to $\mathcal{G}$, receives ek from $\mathcal{G}$, and returns this ek to $\widehat{\mathcal{A}}$. $\mathcal{A}$ also stores the pair (td, ek) into the list L.

- For a $\mathcal{P}$-query (ek, α), if there is no entry of the form $(*, ek)$ in L, then $\mathcal{A}$ responds with $\perp$. Otherwise, $\mathcal{A}$ makes a $\mathcal{E}$-query (ek, α), receives β from $\mathcal{E}$, and finally returns this β to $\widehat{\mathcal{A}}$.
- For a $\mathcal{P}^{-1}$-query (ek, β), if there is no entry of the form $(*, ek)$ in L, then $\mathcal{A}$ responds with $\perp$. Otherwise, $\mathcal{A}$ retrieves td that corresponds to ek from L, makes a $\mathcal{D}$-query (td, β), receives α from $\mathcal{D}$, and finally returns this α to $\widehat{\mathcal{A}}$.
- For a PSPACE-query, $\mathcal{A}$ answers to it by using $\mathcal{A}$'s own PSPACE oracle.

When $\widehat{\mathcal{A}}$ terminates with output x', $\mathcal{A}$ also terminates with output this x'.

It is easy to see that $\mathcal{A}$ perfectly simulates Game 4 for $\widehat{\mathcal{A}}$ in which the oracles given access to $\widehat{\mathcal{A}}$ are $\phi^{-1}(\mathcal{T})$ (that works as specified in Game 4) and PSPACE. Under this situation, when $\widehat{\mathcal{A}}$ succeeds in outputting the value x^* such that $\widehat{\mathsf{P}}^{\phi^{-1}(\mathcal{T})}(x^*) = y^*$, since $\mathsf{P}^{\mathcal{T}}(\cdot) = \widehat{\mathsf{P}}^{\phi^{-1}(\mathcal{T})}(\cdot)$ for all $\mathcal{T} \in \mathbb{T}$ by definition, $\mathcal{A}$ also succeeds in outputting the preimage under $\mathsf{P}^{\mathcal{T}}$. Therefore, we have $\mathsf{Adv}_{\mathsf{P}^{\mathcal{T}}, \mathcal{A}^{\mathcal{T}}, \mathsf{PSPACE}}^{\mathsf{OWP}}(k) = \Pr[\mathsf{Succ}_4]$. This completes the proof of Claim 4.   □

Claims 1 to 4 imply that for any oracle PPTA $\widetilde{\mathcal{A}}$, there exist an oracle PPTA $\mathcal{A}$ and a negligible function $\mu(k)$ such that $\mathsf{Adv}_{\mathsf{P}^{\mathcal{T}}, \mathcal{A}^{\mathcal{T}}, \mathsf{PSPACE}, \ell}^{\mathsf{OWP}}(k) \geq \mathsf{Adv}_{\widetilde{\mathsf{P}}^{\mathcal{R}}, \widetilde{\mathcal{A}}^{\mathcal{R}}, \mathsf{PSPACE}, \ell}^{\mathsf{OWP}}(k) - \mu(k)$, and thus the property (2) is satisfied as well. This completes the proof of Lemma 7.   □

"Mimicking" Algorithm N and Good Inverter Q for N. The combination of the results by Rudich [43] and Kahn et al. [26] shows that any permutation which has oracle access to a set of random functions can be inverted using the PSPACE oracle. On the other hand, Lemma 7 shows that for any TDP-based permutation $\mathsf{P}^{\mathcal{T}}$, there is another "random function"-based permutation $\widetilde{\mathsf{P}}^{\mathcal{R}}$ such that if $\widetilde{\mathsf{P}}^{\mathcal{R}}$ can be inverted using the PSPACE oracle, then so can be $\mathsf{P}^{\mathcal{T}}$. Here, it seems that by combining the results [43,26] and Lemma 7 we can invert the "random function"-based permutation $\widetilde{\mathsf{P}}^{\mathcal{R}}$ using the PSPACE oracle. However, there is a subtle issue here: The suboracle $\mathcal{F}$ in a round function oracle $\mathcal{R}$ is not a pure random function, even if $\mathcal{R}$ is sampled randomly from the set $\mathbb{R}$. Specifically, $\mathcal{F}$ returns an "invalid" symbol $\perp$ for some inputs, and thus we cannot directly use the results [43,26].

For convenience, let us refer to a query to the suboracle $\mathcal{F}$ in a round function oracle $\mathcal{R} \in \mathbb{R}$ as *invalid* if the answer to the query is $\perp$, and an oracle algorithm N that expects to access to an oracle $\mathcal{R} \in \mathbb{R}$ as *legal* if $\mathsf{N}^{\mathcal{R}}$ never makes an invalid query for all $\mathcal{R} \in \mathbb{R}$ and for all inputs.

To resolve the subtlety on invalid queries, we will use the approach by Chang et al. [9]: we show two lemmas that enable us to finally show that a TDP-based permutation can be inverted almost perfectly. The first lemma below (Lemma 8) roughly states that for a permutation $\widetilde{\mathsf{P}}^{\mathcal{R}}$ based on a round function oracle $\mathcal{R}$, there is a "mimicking" algorithm $\mathsf{N}^{\mathcal{R}}$ which is legal and, for most inputs, computes almost the same result as $\widetilde{\mathsf{P}}^{\mathcal{R}}$ for most oracles $\mathcal{R} \in \mathbb{R}$.

Lemma 8. *Let $\ell = \ell(k) > 0$ be a polynomial and $\widetilde{\mathsf{P}}$ be an oracle PPTA such that $\widetilde{\mathsf{P}}^{\mathcal{R}}$ is a permutation over $\{0, 1\}^{\ell}$ for all $\mathcal{R} \in \mathbb{R}$. Then, there exists an oracle PPTA N (that expects to access to an oracle from $\mathbb{R}$) with the following properties: (i) N is legal, and (ii) For sufficiently large k's, for at least $1 - 2 \cdot 2^{-k/6}$ fraction of strings $y \in \{0, 1\}^{\ell}$, $(\mathsf{N}^{\mathcal{R}})^{-1}(y) = (\widetilde{\mathsf{P}}^{\mathcal{R}})^{-1}(y)$ holds for at least $1 - 2^{-k/3}$ fraction of oracles $\mathcal{R} \in \mathbb{R}$.*

The formal proof proceeds closely to that of [9, Claim 3 and Lemma 3], and is given in the full version. We give a proof sketch.

Proof Sketch of Lemma 8. Let ℓ and $\widetilde{P}$ be as stated in the lemma. Using $\widetilde{P}$ as a subroutine, we construct the oracle PPTA N that satisfies the properties (i) and (ii). N takes a string $x \in \{0,1\}^{\ell}$ as input, has access to an oracle $\mathcal{R} \in \mathbb{R}$, and runs as follows:

$N^{\mathcal{R}}(x)$: N firstly generates an empty list L into which "known" evaluation-keys $ek \in$ Range$(\widetilde{\mathcal{G}})$ will be stored, and then runs $\widetilde{P}(x)$. N responds to queries from $\widetilde{P}$ as follows:

- When $\widetilde{P}$ makes a $\widetilde{\mathcal{G}}$-query $td \in \{0,1\}^{k}$, N forwards it to $\widetilde{\mathcal{G}}$, receives a result ek from $\widetilde{\mathcal{G}}$, and returns this ek to $\widetilde{P}$. N also stores ek into the list L.
- When $\widetilde{P}$ makes a $\mathcal{F}$-query $(i, ek, \gamma) \in [14] \times \{0,1\}^{2k} \times \{0,1\}^{k/2}$, N responds with $\bot$ if $ek \notin L$. Otherwise, N forwards (i, ek, γ) to $\mathcal{F}$, receives an answer $\delta \in \{0,1\}^{k/2}$ from $\mathcal{F}$, and returns δ to $\widetilde{P}$.

When $\widetilde{P}$ terminates with output y, N also terminates with output y.

The above completes the description of N. Note that N is legal, because N's $\mathcal{F}$-queries always satisfy $ek \in$ Range$(\widetilde{\mathcal{G}})$. Hence, the property (i) is satisfied.

To show that the above N satisfies the property (ii), we will show the following two claims that together imply what we want (the formal proofs are given in the full version), and hence enable us to complete the proof of Lemma 8:

Claim 5. *For any string* $x \in \{0,1\}^{\ell}$, $\Pr_{\mathcal{R} \leftarrow_{\mathbb{R}} \mathbb{R}}[N^{\mathcal{R}}(x) \neq \widetilde{P}^{\mathcal{R}}(x)] \leq 2^{-k/2}$ *holds for sufficiently large* k's.

Claim 6. *For sufficiently large* k's, *the following holds. There are at least* $1 - 2 \cdot 2^{-k/6}$ *fraction of strings* $y \in \{0,1\}^{\ell}$ *such that* $(N^{\mathcal{R}})^{-1}(y) = (\widetilde{P}^{\mathcal{R}})^{-1}(y)$ *holds for at least* $1 - 2^{-k/3}$ *fraction of oracles* $\mathcal{R} \in \mathbb{R}$.

Claim 5 can be shown in a similar manner to the negligible upperbound of $\Pr[\mathsf{Find}_4]$ in the proof of Claim 3. Specifically, it is clear from the description of N that for any input $x \in \{0,1\}^{\ell}$, the output of N and that of $\widetilde{P}$ agree unless $\widetilde{P}$ makes a $\mathcal{F}$-query $(*, ek, *)$ such that ek is not an answer to $\widetilde{P}$'s previous $\widetilde{\mathcal{G}}$-queries. Therefore, "$N^{\mathcal{R}}(x) \neq \widetilde{P}^{\mathcal{R}}(x)$" must mean that $\widetilde{P}$ makes such a $\mathcal{F}$-query. However, if $\mathcal{R}$ is chosen uniformly, $\widetilde{\mathcal{G}}$ is a random length-doubling injective function, and thus the probability of $\widetilde{P}$ finding a "fresh" element that belongs to Range$(\widetilde{\mathcal{G}})$ is exponentially small. (Here, $\mathcal{F}$ works as the "membership" oracle regarding the range of $\widetilde{\mathcal{G}}$, but it does not help much.)

For showing Claim 6, consider the Boolean matrix $M = (M_{(y,\mathcal{R})})$ whose rows are indexed by $y \in \{0,1\}^{\ell}$ and whose columns are indexed by $\mathcal{R} \in \mathbb{R}$, so that $M_{(y,\mathcal{R})} = 1$ if and only if $(N^{\mathcal{R}})^{-1}(y) \neq (\widetilde{P}^{\mathcal{R}})^{-1}(y)$. By Claim 5, we know that for sufficiently large k's, we have that for each $x \in \{0,1\}^{\ell}$, $N^{\mathcal{R}}(x) \neq \widetilde{P}^{\mathcal{R}}(x)$ holds for at most $2^{-k/2}$ fraction of oracles $\mathcal{R} \in \mathbb{R}$. Since any such pair $(x, \mathcal{R})$ contributes at most two 1's to the matrix M (namely, to the entries $M_{(N^{\mathcal{R}}(x),\mathcal{R})}$ and $M_{(\widetilde{P}^{\mathcal{R}}(x),\mathcal{R})}$), the total fraction of 1's in M is at most $2 \cdot 2^{-k/2}$. That is, $\Pr_{y \leftarrow_{\mathbb{R}} \{0,1\}^{\ell}, \mathcal{R} \leftarrow_{\mathbb{R}} \mathbb{R}}[M_{(y,\mathcal{R})} = 1] \leq 2 \cdot 2^{-k/2}$. Then, a simple counting argument yields Claim 6.

This completes the proof sketch of Lemma 8.   □

We note that even if $\widetilde{\mathsf{P}}^{\mathcal{R}}$ is a permutation, $\mathsf{N}^{\mathcal{R}}$ in Lemma 8 is not guaranteed to be a permutation (although $\mathsf{N}^{\mathcal{R}}$ is very close to a permutation), and this is the main reason why we cannot directly use the results from [43,26]. A similar situation was encountered in [9] where the authors could not directly apply the results from [43,26] to show the separation of a OWP from a trapdoor function.

Fortunately, we can use the next lemma, which is implied by the one shown and used in [9, Section 3.2] (which is in turn based on [43,26]). The following lemma roughly says that if most of the images under a legal oracle algorithm $\mathsf{N}^{\mathcal{R}}$ have a unique preimage, (and in particular these properties are satisfied by the algorithm $\mathsf{N}^{\mathcal{R}}$ in Lemma 8), then there is an oracle algorithm $\mathsf{Q}^{\mathcal{R},\mathsf{PSPACE}}$ that can invert $\mathsf{N}^{\mathcal{R}}$ almost always, using the PSPACE oracle.

Lemma 9. *(follows from [9, Lemma 4].) Let $\ell = \ell(k)$ be a positive polynomial. There exists a constant $\lambda > 0$ such that for every legal oracle PPTA $\mathsf{N}^{(\cdot)} : \{0,1\}^\ell \to \{0,1\}^\ell$ (that expects to access to an oracle from $\mathbb{R}$), there is another oracle PPTA Q with the following property: For any $\epsilon < \lambda$ and any $y \in \{0,1\}^\ell$, if the size of the set $(\mathsf{N}^{\mathcal{R}})^{-1}(y) = \{x \in \{0,1\}^\ell | \mathsf{N}^{\mathcal{R}}(x) = y\}$ is one for $1 - \epsilon$ fraction of oracles $\mathcal{R} \in \mathbb{R}$, then $\mathsf{Q}^{\mathcal{R},\mathsf{PSPACE}}(1^k, y) = (\mathsf{N}^{\mathcal{R}})^{-1}(y)$ holds for $1 - \sqrt{\epsilon}$ fraction of oracles $\mathcal{R} \in \mathbb{R}$.*

Inverting Any Permutation Based on $\mathcal{T}$: Proof of Lemma 4. Now, we are ready to prove Lemma 4. Let ℓ and P be as stated in the lemma. By lemma 7, for this P, there is an oracle PPTA $\widetilde{\mathsf{P}}$ such that $\widetilde{\mathsf{P}}^{\mathcal{R}} \in \mathrm{Perm}_\ell$ for all $\mathcal{R} \in \mathbb{R}$. Then, Lemma 8 tells us that for this $\widetilde{\mathsf{P}}$, there exists an oracle PPTA N that satisfies the properties (i) and (ii). Since $\widetilde{\mathsf{P}}^{\mathcal{R}}$ is a permutation for all $\mathcal{R} \in \mathbb{R}$, the size of the set $(\widetilde{\mathsf{P}}^{\mathcal{R}})^{-1}(y) = \{x \in \{0,1\}^\ell | \widetilde{\mathsf{P}}^{\mathcal{R}}(x) = y\}$ is one for all $y \in \{0,1\}^\ell$ and all $\mathcal{R} \in \mathbb{R}$. Thus, if $(\mathsf{N}^{\mathcal{R}})^{-1}(y) = (\widetilde{\mathsf{P}}^{\mathcal{R}})^{-1}(y)$, the size of the set $(\mathsf{N}^{\mathcal{R}})^{-1}(y) = \{x \in \{0,1\}^\ell | \mathsf{N}^{\mathcal{R}}(x) = y\}$ must also be one. By the property (ii) of N in Lemma 8, for at least $1 - 2 \cdot 2^{-k/6}$ fraction of strings $y \in \{0,1\}^\ell$, the size of the set $(\mathsf{N}^{\mathcal{R}})^{-1}(y) = \{x \in \{0,1\}^\ell | \mathsf{N}^{\mathcal{R}}(x) = y\}$ is one for at least $1 - 2^{-k/3}$ fraction of oracles $\mathcal{R} \in \mathbb{R}$.

Set $\epsilon' = 2^{-k/3}$. For any constant $\lambda > 0$, $\epsilon' < \lambda$ holds for all sufficiently large k's, and thus this ϵ' can be used as the ϵ in Lemma 9. Call $y \in \{0,1\}^\ell$ *good* if $(\mathsf{N}^{\mathcal{R}})^{-1}(y) = (\widetilde{\mathsf{P}}^{\mathcal{R}})^{-1}(y)$ holds for $1 - 2^{-k/3}$ fraction of oracles $\mathcal{R} \in \mathbb{R}$. By definition, if y is good, then it is guaranteed that the size of the set $(\mathsf{N}^{\mathcal{R}})^{-1}(y) = \{x \in \{0,1\}^\ell | \mathsf{N}^{\mathcal{R}}(x) = y\}$ is one for at least $1 - \epsilon' = 1 - 2^{-k/3}$ fraction of oracles $\mathcal{R} \in \mathbb{R}$, and it is also guaranteed that $\mathrm{Pr}_{y \leftarrow_{\mathrm{R}} \{0,1\}^\ell}[y \text{ is good}] \geq 1 - 2 \cdot 2^{-k/6}$ holds. Furthermore, by using N and ϵ', Lemma 9 implies that there is an oracle PPTA Q such that for sufficiently large k's and for all good y's, $\mathsf{Q}^{\mathcal{R},\mathsf{PSPACE}}(y) = (\mathsf{N}^{\mathcal{R}})^{-1}(y)$ holds for $1 - \sqrt{\epsilon'}$ fraction of oracles $\mathcal{R} \in \mathbb{R}$. Recall that for $y \in \{0,1\}^\ell$ and $\mathcal{R} \in \mathbb{R}$ such that $(\mathsf{N}^{\mathcal{R}})^{-1}(y) = (\widetilde{\mathsf{P}}^{\mathcal{R}})^{-1}(y)$ and $\mathsf{Q}^{\mathcal{R},\mathsf{PSPACE}}(1^k, y) = (\mathsf{N}^{\mathcal{R}})^{-1}(y) = x$, it holds that $\widetilde{\mathsf{P}}^{\mathcal{R}}(x) = y$, i.e. Q succeeds in calculating the preimage x of y under the permutation $\widetilde{\mathsf{P}}^{\mathcal{R}}$. Therefore, considering sufficiently large k's, we have

$$\mathrm{Pr}[\mathcal{R} \leftarrow_{\mathrm{R}} \mathbb{R}; x \leftarrow_{\mathrm{R}} \mathsf{Q}^{\mathcal{R},\mathsf{PSPACE}}(1^k, y) : \mathsf{N}^{\mathcal{R}}(x) = \widetilde{\mathsf{P}}^{\mathcal{R}}(x) = y | y \text{ is good}]$$

$$\geq 1 - \sqrt{\epsilon'} = 1 - 2^{-k/6}.$$

Now, define an oracle PPTA adversary $\widetilde{\mathcal{A}}$, which runs in $\widetilde{\mathsf{Expt}}_{\widetilde{\mathsf{PR}},\widetilde{\mathcal{A}}^{\mathsf{R}},\mathsf{PSPACE},\ell}^{\mathsf{OWP}}(k)$, by $\widetilde{\mathcal{A}}^{\mathcal{R},\mathsf{PSPACE}}(1^k, y^*) = \mathsf{Q}^{\mathcal{R},\mathsf{PSPACE}}(1^k, y^*)$. Since x^* is chosen uniformly from $\{0,1\}^\ell$ in $\widetilde{\mathsf{Expt}}_{\widetilde{\mathsf{PR}},\widetilde{\mathcal{A}}^{\mathsf{R}},\mathsf{PSPACE},\ell}^{\mathsf{OWP}}(k)$ and $\widetilde{\mathsf{P}}^{\mathcal{R}}$ is a permutation, $y^* = \widetilde{\mathsf{P}}^{\mathcal{R}}(x^*)$ is distributed uniformly over $\{0,1\}^\ell$. Therefore, for sufficiently large k's, we have:

$$\widetilde{\mathsf{Adv}}_{\widetilde{\mathsf{PR}},\widetilde{\mathcal{A}}^{\mathsf{R}},\mathsf{PSPACE},\ell}^{\mathsf{OWP}}(k)$$

$$= \Pr[\mathcal{R} \leftarrow_{\mathsf{R}} \mathbb{R}; x^* \leftarrow_{\mathsf{R}} \{0,1\}^\ell; y^* \leftarrow \widetilde{\mathsf{P}}^{\mathcal{R}}(x^*); x' \leftarrow_{\mathsf{R}} \widetilde{\mathcal{A}}^{\mathcal{R},\mathsf{PSPACE}}(1^k, y^*) : x' = x^*]$$

$$\geq \Pr[\mathcal{R} \leftarrow_{\mathsf{R}} \mathbb{R}; y^* \leftarrow_{\mathsf{R}} \{0,1\}^\ell; x' \leftarrow_{\mathsf{R}} \mathsf{Q}^{\mathcal{R},\mathsf{PSPACE}}(1^k, y^*) : \mathsf{N}^{\mathcal{R}}(x') = \widetilde{\mathsf{P}}^{\mathcal{R}}(x') = y^*]$$

$$\geq \Pr[\mathcal{R} \leftarrow_{\mathsf{R}} \mathbb{R}; x' \leftarrow_{\mathsf{R}} \mathsf{Q}^{\mathcal{R},\mathsf{PSPACE}}(1^k, y^*) : \mathsf{N}^{\mathcal{R}}(x') = \widetilde{\mathsf{P}}^{\mathcal{R}}(x') = y^* | y^* \text{ is good}]$$

$$\times \Pr_{y^* \leftarrow \{0,1\}^\ell}[y^* \text{ is good}]$$

$$\geq (1 - 2^{-k/6}) \cdot (1 - 2 \cdot 2^{-k/6}) \geq 1 - 3 \cdot 2^{-k/6}.$$

Finally, by the property (2) of P in Lemma 7, for this $\widetilde{\mathcal{A}}$, there exist an oracle PPTA adversary $\mathcal{A}$, which runs in $\widetilde{\mathsf{Expt}}_{\mathsf{PT},\mathcal{A}^{\mathsf{T}},\mathsf{PSPACE},\ell}^{\mathsf{OWP}}(k)$, and a negligible function $\mu(k)$ such that for sufficiently large k's:

$$\widetilde{\mathsf{Adv}}_{\mathsf{PT},\mathcal{A}^{\mathsf{T}},\mathsf{PSPACE}}^{\mathsf{OWP}}(k) \geq \widetilde{\mathsf{Adv}}_{\widetilde{\mathsf{PR}},\widetilde{\mathcal{A}}^{\mathsf{R}},\mathsf{PSPACE},\ell}^{\mathsf{OWP}}(k) - \mu(k) \geq 1 - 3 \cdot 2^{-k/6} - \mu(k).$$

What we have shown thus far is that there exists an oracle PPTA $\mathcal{A}$ such that $\mathbf{E}_{\mathcal{T} \leftarrow_{\mathsf{R}} \mathbb{T}}[\mathsf{Adv}_{\mathsf{PT},\mathcal{A}^{\mathsf{T}},\mathsf{PSPACE},\ell}^{\mathsf{OWP}}(k)] = \widetilde{\mathsf{Adv}}_{\mathsf{PT},\mathcal{A}^{\mathsf{T}},\mathsf{PSPACE}}^{\mathsf{OWP}}(k)$ is overwhelming. The above can be shown for all positive polynomials $\ell(k)$ and any PPTA P such that $\mathsf{P}^{\mathcal{T}} \in \mathsf{Perm}_\ell$ for all $\mathcal{T} \in \mathbb{T}$. This completes the proof of Lemma 4. $\qquad\square$

4   Towards More General Separations

Broader Class of Permutations and Permutation Families. As in the previous black-box separation results of a OWP from other basic primitives [43,26,9,33], our separation results rule out a black-box construction of a OWP which is *defined over strings* (i.e. the domain is a set of strings of a fixed length determined by the security parameter). However, we can consider a more general form of a permutation whose domain is not just a set of strings but an arbitrary set D, and which has a corresponding sampling algorithm Samp to sample an element from the domain D (although such formulation of a OWP is not standard). Furthermore, as a more natural and closely related primitive to a OWP, we can also consider a public-coin OWPF.

Therefore, a natural question regarding our result will be: "*Can our impossibility result be extended to also rule out a black-box construction of a OWP with such general form of domain or of a public-coin OWPF?*"

We note that previously to our work, Hsiao and Reyzin [24] conjectured that there is no black-box construction of a public-coin OWPF from a secret-coin OWPF. We can partially answer to the above question in the positive due to the result by Goldreich et al. [17], who showed that there is a (fully-)black-box construction of a OWP from a public-coin OWPF with the *canonical domain sampling* property (see Section 2.2 for the definition and a brief review of the construction of [17]). This result, combined with Theorem 1, yields the following corollary.

Corollary 2. *There is no fully-black-box construction of a public-coin OWPF with canonical domain sampling from an ideal TDP.*

It seems to us that if we consider another restricted type of a constructed public-coin permutation family such that the sampling algorithm Samp of the constructed permutation family does not use the algorithms of a building block TDP, then we can rule out a black-box construction of such public-coin OWPF from an ideal TDP, with essentially the same approach used to show Theorem 1 (although we have not formally checked this). This is because if Samp of the constructed public-coin permutation family does not depend on the TDP used as a building block, then whenever we use a same evaluation key ek, the domain D_{ek} of a permutation Eval($ek, \cdot$) remains the same, and thus slight modifications of Lemmas 7 to 9 seem to work accordingly.

Other than these observations, so far we do not know how to rule out the possibility of constructing a public-coin OWPF from a TDP (or even from an ordinary secret-coin OWPF) in general, and thus we would like to leave it as an interesting open problem. Goldreich et al. [17] showed that under the standard RSA assumption or a discrete logarithm assumption in the integer group $\mathbb{Z}_p^*$ (with some appropriate condition on p), we can construct a public-coin OWPF with the canonical domain sampling property, and hence a OWP. However, they noted that how to construct a OWP or a public-coin OWPF under the standard factoring assumption is still open. Tackling the above open problem of clarifying whether there exists a black-box construction of a public-coin OWPF from a secret-coin OWPF will also contribute to this problem: If it turns out to be possible (which we think is unlikely), then we can use the Rabin TDP [40] as a building block to construct a public-coin OWPF, while if it is not possible, one has to essentially use some specific algebraic property to build a public-coin OWPF under the factoring assumption.

Stronger Separation. So far, all our results are impossibility of a fully-black-box construction, which is the most restrictive type of black-box constructions. With a slight modification, however, our separation results can be strengthened to show that there is no *semi*-black-box construction (in the taxonomy of Reingold et al. [41]) of a OWP (and a public-coin OWPF with canonical domain sampling) from an ideal TDP. Specifically, to show such a result, we need to show a "single" oracle which simultaneously implements an ideal TDP and PSPACE. However, our TDP oracle $\mathcal{T}$ can be easily modified to such an oracle by using the "embedding" technique due to Reingold et al. [41]. We discuss more details in the full version.

Acknowledgement. The author would like to thank Jacob Schuldt for many insightful comments and discussions on several drafts of this paper. The author would also like to thank the members of the study group "Shin-Akarui-Angou-Benkyou-Kai" and the anonymous reviewers for their helpful comments and suggestions.

References

1. Barak, B., Goldreich, O., Impagliazzo, R., Rudich, S., Sahai, A., Vadhan, S., Yang, K.: On the (im)possibility of obfuscating programs. In: Kilian, J. (ed.) CRYPTO 2001. LNCS, vol. 2139, pp. 1–18. Springer, Heidelberg (2001)
2. Bellare, M., Hofheinz, D., Yilek, S.: Possibility and impossibility results for encryption and commitment secure under selective opening. In: Joux, A. (ed.) EUROCRYPT 2009. LNCS, vol. 5479, pp. 1–35. Springer, Heidelberg (2009)
3. Bellare, M., Rogaway, P.: Optimal asymmetric encryption. In: De Santis, A. (ed.) EURO-CRYPT 1994. LNCS, vol. 950, pp. 92–111. Springer, Heidelberg (1995)
4. Bellare, M., Rogaway, P.: The exact security of digital signatures – how to sign with RSA and Rabin. In: Maurer, U. (ed.) EUROCRYPT 1996. LNCS, vol. 1070, pp. 399–416. Springer, Heidelberg (1996)
5. Bellare, M., Yung, M.: Certifying permutations: Noninteractive zero-knowledge based on any trapdoor permutation. J. of Cryptology 9(3), 149–166 (1996)
6. Bhattacharyya, R., Mandal, A.: On the impossibility of instantiating PSS in the standard model. In: Catalano, D., Fazio, N., Gennaro, R., Nicolosi, A. (eds.) PKC 2011. LNCS, vol. 6571, pp. 351–368. Springer, Heidelberg (2011)
7. Blum, M., Micali, S.: How to generate cryptographically strong sequences of pseudo-random bits. SIAM J. Computing 13(4), 850–864 (1984)
8. Boneh, D., Papakonstantinou, P.A., Rackoff, C., Vahlis, Y., Waters, B.: On the impossibility of basing identity based encryption on trapdoor permutations. In: FOCS 2008, pp. 283–292 (2008)
9. Chang, Y.-C., Hsiao, C.-Y., Lu, C.-J.: The impossibility of basing one-way permutations on central cryptographic primitives. J. of Cryptology 19(1), 97–114 (2006)
10. Coron, J.-S., Patarin, J., Seurin, Y.: The random oracle model and the ideal cipher model are equivalent. In: Wagner, D. (ed.) CRYPTO 2008. LNCS, vol. 5157, pp. 1–20. Springer, Heidelberg (2008)
11. Dodis, Y., Oliveira, R., Pietrzak, K.: On the generic insecurity of the full domain hash. In: Shoup, V. (ed.) CRYPTO 2005. LNCS, vol. 3621, pp. 449–466. Springer, Heidelberg (2005)
12. Fiore, D., Schröder, D.: Uniqueness is a different story: Impossibility of verifiable random functions from trapdoor permutations. In: Cramer, R. (ed.) TCC 2012. LNCS, vol. 7194, pp. 636–653. Springer, Heidelberg (2012)
13. Fischlin, M., Schröder, D.: On the impossibility of three-move blind signature schemes. In: Gilbert, H. (ed.) EUROCRYPT 2010. LNCS, vol. 6110, pp. 197–215. Springer, Heidelberg (2010)
14. Gentry, C., Wichs, D.: Separating succinct non-interactive arguments from all falsifiable assumptions. In: STOC 2011, pp. 99–108 (2011)
15. Gertner, Y., Malkin, T., Reingold, O.: On the impossibility of basing trapdoor functions on trapdoor predicates. In: FOCS 2001, pp. 126–135 (2001)
16. Goldreich, O., Goldwasser, S., Micali, S.: How to construct random functions. J. ACM 33(4), 792–807 (1986)
17. Goldreich, O., Levin, L.A., Nisan, N.: On constructing 1-1 one-way functions. In: Goldreich, O. (ed.) Studies in Complexity and Cryptography. LNCS, vol. 6650, pp. 13–25. Springer, Heidelberg (2011)
18. Goldreich, O., Rothblum, R.D.: Enhancements of trapdoor permutations. J. of Cryptology 26(3), 484–512 (2013)
19. Goldwasser, S., Micali, S., Rivest, R.: A digital signature schemes secure against adaptive chosen-message attacks. SIAM J. Computing 17(2), 281–308 (1988)

20. Haitner, I.: Implementing oblivious transfer using collection of dense trapdoor permutations. In: Naor, M. (ed.) TCC 2004. LNCS, vol. 2951, pp. 394–409. Springer, Heidelberg (2004)
21. Haitner, I., Holenstein, T.: On the (Im)Possibility of key dependent encryption. In: Reingold, O. (ed.) TCC 2009. LNCS, vol. 5444, pp. 202–219. Springer, Heidelberg (2009)
22. Håstad, J., Impagliazzo, R., Levin, L., Luby, M.: Construction of a pseudorandom generator from any one-way function. SIAM J. Computing 28(4), 1364–1396 (1999)
23. Holenstein, T., Künzler, R., Tessaro, S.: The equivalence of the random oracle model and the ideal cipher model, revisited. In: STOC 2011, pp. 89–98 (2011)
24. Hsiao, C.-Y., Reyzin, L.: Finding collisions on a public road, or do secure hash functions need secret coins? In: Franklin, M. (ed.) CRYPTO 2004. LNCS, vol. 3152, pp. 92–105. Springer, Heidelberg (2004)
25. Impagliazzo, R., Rudich, S.: Limits on the provable consequences of one-way permutations. In: STOC 1989, pp. 44–61 (1989)
26. Kahn, J., Saks, M., Smyth, C.: A dual version of Reimer's inequality and a proof of Rudich's conjecture. In: CoCo 2000, pp. 98–103 (2000)
27. Katz, J., Yerukhimovich, A.: On black-box constructions of predicate encryption from trapdoor permutations. In: Matsui, M. (ed.) ASIACRYPT 2009. LNCS, vol. 5912, pp. 197–213. Springer, Heidelberg (2009)
28. Kiltz, E., Mohassel, P., O'Neill, A.: Adaptive trapdoor functions and chosen-ciphertext security. In: Gilbert, H. (ed.) EUROCRYPT 2010. LNCS, vol. 6110, pp. 673–692. Springer, Heidelberg (2010)
29. Kiltz, E., Pietrzak, K.: On the security of padding-based encryption schemes - or - why we cannot prove OAEP secure in the standard model. In: Joux, A. (ed.) EUROCRYPT 2009. LNCS, vol. 5479, pp. 389–406. Springer, Heidelberg (2009)
30. Lindell, Y., Zarosim, H.: Adaptive zero-knowledge proofs and adaptively secure oblivious transfer. In: Full version of [13] (2009), http://u.cs.biu.ac.il/~zarosih/papers/adaptive-full version.pdf
31. Lindell, Y., Zarosim, H.: Adaptive zero-knowledge proofs and adaptively secure oblivious transfer. In: Reingold, O. (ed.) TCC 2009. LNCS, vol. 5444, pp. 183–201. Springer, Heidelberg (2009)
32. Luby, M., Rackoff, C.: How to construct pseudorandom permutations from pseudorandom functions. SIAM J. Computing 17(2), 373–386 (1988)
33. Matsuda, T., Matsuura, K.: On black-box separations among injective one-way functions. In: Ishai, Y. (ed.) TCC 2011. LNCS, vol. 6597, pp. 597–614. Springer, Heidelberg (2011)
34. Maurer, U., Renner, R., Holenstein, C.: Indifferentiability, impossibility results on reductions and applications to the random oracle methodology. In: Naor, M. (ed.) TCC 2004. LNCS, vol. 2951, pp. 21–39. Springer, Heidelberg (2004)
35. Naor, M.: Bit commitment using pseudorandomness. J. of Cryptology 4(2), 151–158 (1991)
36. Naor, M.: On cryptographic assumptions and challenges. In: Boneh, D. (ed.) CRYPTO 2003. LNCS, vol. 2729, pp. 96–109. Springer, Heidelberg (2003)
37. Naor, M., Yung, M.: Universal one-way hash functions and their cryptographic applications. In: STOC 1989, pp. 33–43 (1989)
38. Pandey, O., Pass, R., Vaikuntanathan, V.: Adaptive one-way functions and applications. In: Wagner, D. (ed.) CRYPTO 2008. LNCS, vol. 5157, pp. 57–74. Springer, Heidelberg (2008)
39. Pass, R.: Limits of provable security from standard assumptions. In: STOC 2011, pp. 109–118 (2011)
40. Rabin, M.O.: Digitalized signatures as intractable as factorization. Technical Report MIT/LCS/TR-212, MIT Laboratory for Computer Science (January 1979)
41. Reingold, O., Trevisan, L., Vadhan, S.: Notions of reducibility between cryptographic primitives. In: Naor, M. (ed.) TCC 2004. LNCS, vol. 2951, pp. 1–20. Springer, Heidelberg (2004)

42. Rosen, A., Segev, G.: Chosen-ciphertext security via correlated products. In: Reingold, O. (ed.) TCC 2009. LNCS, vol. 5444, pp. 419–436. Springer, Heidelberg (2009)
43. Rudich, S.: Limits on the provable consequences of one-way functions, PhD thesis, University of California at Berkeley (1988)
44. Vahlis, Y.: Two is a crowd? A black-box separation of one-wayness and security under correlated inputs. In: Micciancio, D. (ed.) TCC 2010. LNCS, vol. 5978, pp. 165–182. Springer, Heidelberg (2010)
45. Wee, H.: On obfuscating point functions. In: STOC 2005, pp. 523–532 (2005)
46. Wichs, D.: Barriers in cryptography with weak, correlated and leaky sources. In: Proc of ITCS 2013, pp. 111–126 (2013)
47. Yao, A.C.-C.: Theory and application of trapdoor functions. In: FOCS 1982, pp. 80–91 (1982)
48. Yerukhimovich, A.: A study of separation in cryptography: New results and new models, PhD thesis, the University of Maryland (2011),
    http://www.cs.umd.edu/~arkady/thesis/thesis.pdf

Towards Characterizing Complete Fairness in Secure Two-Party Computation*

Gilad Asharov

Department of Computer Science,
Bar-Ilan University, Israel
asharog@cs.biu.ac.il

Abstract. The well known impossibility result of Cleve (STOC 1986) implies that in general it is impossible to securely compute a function with *complete fairness* without an honest majority. Since then, the accepted belief has been that *nothing* non-trivial can be computed with complete fairness in the two party setting. The surprising work of Gordon, Hazay, Katz and Lindell (STOC 2008) shows that this belief is false, and that there exist *some* non-trivial (deterministic, finite-domain) boolean functions that can be computed fairly. This raises the fundamental question of characterizing complete fairness in secure two-party computation.

In this work we show that not only that some or few functions can be computed fairly, but rather an *enormous amount* of functions can be computed with complete fairness. In fact, *almost all* boolean functions with distinct domain sizes can be computed with complete fairness (for instance, more than 99.999% of the boolean functions with domain sizes 31×30). The class of functions that is shown to be possible includes also rather involved and highly non-trivial tasks, such as set-membership, evaluation of a private (Boolean) function and private matchmaking.

In addition, we demonstrate that fairness is not restricted to the class of symmetric boolean functions where both parties get the same output, which is the only known feasibility result. Specifically, we show that fairness is also possible for asymmetric boolean functions where the output of the parties is not necessarily the same. Moreover, we consider the class of functions with *non-binary* output, and show that fairness is possible *for any finite range*.

The constructions are based on the protocol of Gordon et. al, and the analysis uses tools from convex geometry.

Keywords: Complete fairness, secure two-party computation, foundations, malicious adversaries.

---

* This research was supported by the European Research Council under the European Union's Seventh Framework Programme (FP/2007-2013) / ERC Grant Agreement n. 239868 and by the ISRAEL SCIENCE FOUNDATION (grant No. 189/11).

Y. Lindell (Ed.): TCC 2014, LNCS 8349, pp. 291–316, 2014.

1   Introduction

In the setting of secure multiparty computation, some mutually distrusting parties wish to compute some function of their inputs in the presence of adversarial behavior. The security requirements of such a computation are that nothing is learned from the protocol other than the output (privacy), that the outputs are distributed according to the prescribed functionality (correctness) and that the parties cannot choose their inputs as a function of the others' inputs (independence of inputs). Another important security property is that of *fairness*, which intuitively means that the adversary learns the output if and only if, the honest parties learn their output.

In the multiparty case, where a majority of the parties are honest, it is possible to compute any functionality while guaranteeing all the security properties mentioned above [14,6,8,25,13]. In the multiparty case when honest majority is not guaranteed, including the important case of the two-party settings where one may be corrupted, it is possible to compute any function while satisfying *all* security properties mentioned above *except* for fairness [29,14,13]. The deficiency of fairness is not just an imperfection of theses constructions, but rather a result of inherent limitation. The well-known impossibility result of Cleve [9] shows that there exist functions that cannot be computed by two parties with complete fairness, and thus, fairness cannot be achieved *in general*. Specifically, Cleve showed that the coin-tossing functionality, where two parties toss an unbiased fair coin, cannot be computed with complete fairness. This implies that any function that can be used to toss a fair coin (like, for instance, the boolean XOR function) cannot be computed fairly as well.

Since Cleve's result, the accepted belief has been that *only trivial functions*[1] can be computed with complete fairness. This belief is based on a solid and substantiate intuition: In any protocol computing any interesting function, the parties move from a state of no knowledge about the output to full knowledge about it. Protocols proceed in rounds and the parties cannot exchange information simultaneously, therefore, apparently, there must be a point in the execution where one party knows more about the output than the other party. Aborting at that round yields the unfair situation where one party can guess better the output, and learn the output alone.

Our understanding regarding fairness has been changed recently by the surprising work of Gordon, Hazay, Katz and Lindell [17]. This work shows that there *exist* some non-trivial (deterministic, finite-domain) boolean functions that can be computed in the malicious settings with *complete fairness*, and re-opens the research on this subject. The fact that *some* functions can be computed fairly, while some other were proven to be impossible to compute fairly, raises the following fundamental question:

Which functions can be computed with complete fairness?

---

[1] In our context, the term "trivial functions" refers to constant functions, functions that depend on only one party's input and functions where only one party receives output. It is easy to see that these functions can be computed fairly.

Recently, [3] provided a full characterization for the class of functions that imply fair coin-tossing and thus are ruled out by Cleve's impossibility. This extends our knowledge on what functions cannot be computed with complete fairness. However, there have been no other works that further our understanding regarding which (boolean) functions can be computed fairly, and the class of functions for which [17] shows possibility are the only known possible functions. There is therefore a large class of functions for which we have no idea as to whether or not they can be securely computed with complete fairness.

To elaborate further, the work of [17] show that any function that does not contain an embedded XOR (i.e., inputs x_1, x_2, y_1, y_2 such that $f(x_1, y_1) = f(x_2, y_2) \neq f(x_1, y_2) = f(x_2, y_1)$) can be computed fairly. Examples of functions without an embedded XOR include the boolean OR / AND functions and the greater-than function. Given the fact that Cleve's impossibility result rules out completely fair computation of boolean XOR, a natural conjuncture is that any function that does contain an embedded XOR is impossible to compute fairly. However, the work shows that this conclusion is incorrect. Namely, it considers a *specific* function that does contain an embedded XOR, and constructs a proto-col that securely computes this function with complete fairness. Furthermore, it presents a generalization of this protocol that may potentially compute a large class of functions. It also shows how to construct a (rather involved) set of equations for a given function, that indicates whether the function can be computed fairly using this protocol.

These results are ground-breaking and completely change our perception re-garding fairness. The fact that *something* non-trivial can be computed fairly is very surprising, it contradicts the aforementioned natural intuition and com-mon belief and raises many interesting questions. For instance, are there many functions that can be computed fairly, or only a few? Which functions can be computed fairly? Which functions can be computed using the generalized GHKL protocol? What property distinguishes these functions from the functions that are impossible to compute fairly? Furthermore, the protocol of GHKL is espe-cially designed for deterministic symmetric boolean functions with finite domain, where both parties receive the same output. Is fairness possible in any other class of functions, over larger ranges, or for asymmetric functions? Overall, our understanding of what can be computed fairly is very vague.

1.1 Our Work

In this paper, we study the fundamental question of characterizing which func-tions can be computed with complete fairness. We show that *any* function that defines a full-dimensional geometric object, *can be computed with complete fair-ness*. That is, we present a simple property on the truth table of the function, and show that every function that satisfies this property, the function can be computed fairly. This extends our knowledge of what can be computed fairly, and is an important step towards a full characterization for fairness.

Our results deepen out understanding of fairness and show that many more functions can be computed fairly than what has been thought previously. Using results of combinatorics, we show that a random function with distinct domain sizes (i.e., functions $f : X \times Y \to \{0,1\}$ where $|X| \neq |Y|$) defines a full-dimensional geometric object with overwhelming probability. Therefore, surprisingly, *almost all* functions with distinct domain sizes can be computed with complete fairness.

Although only one bit of information is revealed by output, the class of boolean functions that define full-dimensional geometric object is very rich, and includes fortune of interesting and non-trivial tasks. For instance, the task of *set-membership*, where P_1 holds some set $S \subseteq \Omega$, P_2 holds an element $x \in \Omega$, and the parties wish to find (privately) whether $x \in S$, is a part of this class. Other examples are tasks like *private matchmaking* and *secure evaluation of a private (boolean) function*, where the latter task is very general and can be applied in many practical situations. Unexpectedly, it turns out that all of these tasks can be computed with complete fairness.

In addition to the above, we provide an additional property that indicates that a function *cannot* be computed using the protocol of GHKL. This property is almost always satisfied in the case where $|X| = |Y|$. Thus, at least at the intuitive level, almost all functions with $|X| \neq |Y|$ can be computed fairly, whereas almost all functions with $|X| = |Y|$ cannot be computed using the protocol of GHKL. This negative result does not rule out the possibility of these functions using some other protocols, however, it shows that the only known possibility result does not apply to this class of functions. Combining this result with [3] (i.e., characterization of coin-tossing), there exists a large class of functions for which the only known possibility result does not apply, the only known impossibility result does not apply either, and so fairness for this set of functions is left as an interesting open problem.

Furthermore, we also consider larger families of functions rather than the symmetric boolean functions with finite domain, and show that fairness is also possible in these classes. We consider the class of asymmetric functions where the parties do not necessarily get the same output, as well as the class of functions with non-binary outputs. This is the first time that fairness is shown to be possible in both families of functions, and it shows that fairness can be achieved in a much larger and wider class of functions than previously known.

Intuition. We present some intuition before proceeding to our results in more detail. The most important and acute point is to understand what distinguishes functions that can be computed fairly from functions that cannot. Towards this goal, let us reconsider the impossibility result of Cleve. This result shows that fair coin-tossing is impossible by constructing concrete adversaries that *bias* and *influence* the output of the honest party in any protocol implementing coin-tossing. We believe that such adversaries can be constructed for any protocol computing any function, and not specific to coin-tossing. In any protocol, one party can better predict the outcome than the other, and abort the execution if it is not satisfied with the result. Consequently, it has a concrete ability to *influence* the output of the honest party by aborting prematurely. Of course, a fair protocol should limit and decrease this ability to the least possible, but in general, this phenomenon cannot be totally eliminated and cannot be prevented.

So if this is the case, how do fair protocols exist? The answer to this question does not lie in the real execution but rather in the ideal process: *the simulator can simulate this influence in the ideal execution.* In some sense, for some functions, the simulator has the ability to significantly influence the output of the honest party in the ideal execution and therefore the bias in the real execution is not considered a breach of security. This is due to the fact that in the malicious setting the simulator has an ability that is crucial in the context of fairness: it can *choose* what input it sends to the trusted party. Indeed, the protocol of GHKL uses this switching-input ability in the simulation, and as pointed out by [3], once we take off this advantage from the simulator – every function that contains an embedded XOR cannot be computed fairly, and fairness is almost always impossible.

Therefore, the algebraic structure of the function plays an essential role in the question of whether a function can be computed fairly or not. This is because this structure reflects the "power" and the "freedom" that the simulator has in the ideal world and how it can influence the output of the honest party. The question of whether a function can be computed fairly is related to the amount of "power" the simulator has in the ideal execution. Intuitively, the more freedom that the simulator has, it is more likely that the function can be computed fairly.

A Concrete Example. We demonstrate this "power of the simulator" on two functions. The first is the XOR function, which is impossible to compute by a simple implication of Cleve's result. The second is the specific function for which GHKL has proved to be possible (which we call "the GHKL function"). The truth tables of the functions are as follows:

(a)

	y_1	y_2
x_1	0	1
x_2	1	0

(b)

	y_1	y_2
x_1	0	1
x_2	1	0
x_3	1	1

Fig. 1. (a) The XOR function – impossible, (b) The GHKL function – possible

What is the freedom of the simulator in each case? Consider the case where P_1 is corrupted (that is, we can assume that P_1 is the first to receive an output, and thus it is "harder" to simulate). In the XOR function, let p be the probability that the simulator sends the input x_1 to the trusted party, and let $(1-p)$ be the probability that it sends x_2. Therefore, the output of P_2 in the ideal execution can be represented as $(q_1, q_2) = p \cdot (0, 1) + (1-p) \cdot (1, 0) = (1-p, p)$, which means that if P_2 inputs y_1, then it receives 1 with probability $1-p$, and if it uses input y_2, then it receives 1 with probability p. We call this vector "*the output distribution vector*" for P_2, and the set of all possible output distribution vectors reflects the freedom that the simulator has in the ideal execution. In the XOR function, this set is simply $\{(1-p, p) \mid 0 \le p \le 1\}$, which gives the simulator one degree of freedom. Any increment of the probability in the first coordinate, must be balanced with an equivalent decrement in the second coordinate, and vice versa.

On the other hand, consider the case of the GHKL function. Assume that the simulator chooses x_1 with probability p_1, x_2 with probability p_2 and x_3 with probability $1-p_1-p_2$. Then, all the output vector distributions are of the form:

$$(q_1, q_2) = p_1 \cdot (0, 1) + p_2 \cdot (1, 0) + (1 - p_1 - p_2) \cdot (1, 1) = (1 - p_1, 1 - p_2) .$$

This gives the simulator two degrees of freedom, which is significantly more power.

Geometrically, we can refer to the rows of the truth table as points in $\mathbb{R}^2$, and so in the XOR function we have the two points $(0, 1)$ and $(1, 0)$. All the output distribution vectors are of the form $p \cdot (0, 1) + (1-p) \cdot (1, 0)$ which is exactly the line segment between these two points (geometric object of dimension 1). In the GHKL function, all the output distribution vectors are the triangle between the points $(0, 1), (1, 0)$ and $(1, 1)$, which is a geometric object of dimension 2 (a full dimensional object in $\mathbb{R}^2$).

The difference between these two geometric objects already gives a perception for the reason why the XOR function is impossible to compute, whereas the GHKL function is possible, as the simulator has significantly more options in the latter case. However, we provide an additional refinement. At least in the intuitive level, fix some output distribution vector of the honest party (q_1, q_2). Assume that there exists a real-world adversary that succeeds to bias the output and obtain output distribution vector (q_1', q_2') that is at most ϵ-far from (q_1, q_2). In the case of the XOR function, this results in points that are not on the line, and therefore this adversary cannot be simulated. On the contrary, in case of the GHKL function, these points are still in the triangle, and therefore this adversary can be simulated.

In Figure 2, we show the geometric objects defined by the XOR and the GHKL functions. The centers of the circuits are the output distribution of honest executions, and the circuits represent the possible biases in the real execution. In (a) there exist small biases that are invalid points, whereas in (b) all small biases are valid points that can be simulated.

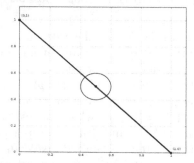
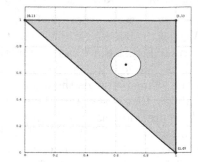

(a) The potential output distribution vectors of the XOR function: a line segment between $(0, 1)$ and $(1, 0)$

(b) The potential output distribution vectors of the GHKL function: the triangle between $(0, 1)$, $(1, 0)$ and $(1, 1)$

Fig. 2. The geometric objects defined by the XOR (a) and the GHKL (b) functions

1.2 Our Results

For a given function $f : \{x_1, \ldots, x_\ell\} \times \{y_1, \ldots, y_m\} \to \{0, 1\}$, we consider its geometric representation as ℓ points over $\mathbb{R}^m$, where the jth coordinate of the ith point is simply $f(x_i, y_j)$. We then prove that any function that its geometric representation is of full dimension *can be computed with complete fairness*. We prove the following theorem:

Theorem 1.1 (informal). *Let $f : X \times Y \to \{0, 1\}$ be a function. Under suitable cryptographic assumptions, if the geometric object defined by f is of full-dimension, then the function can be computed with complete fairness.*

For the proof, we simply use the extended GHKL protocol. Moreover, the proof uses tools from convex geometry. We find the connection between the problem of fairness and convex geometry very appealing.

On the other hand, we show that if the function is not full dimensional, and satisfies some additional requirements (that are almost always satisfied in functions with $|X| = |Y|$), then the function cannot be computed using the protocol of [17].

We then proceed to the class of asymmetric functions where the parties do not necessarily get the same output, and the class of non-binary output. Interestingly, the GHKL protocol can be extended to these classes of functions. We show:

Theorem 1.2 (informal). *Under suitable cryptographic assumptions,*

1. *There exists a large class of asymmetric boolean functions that can be computed with complete fairness.*
2. *For any finite range Σ, there exists a large class of functions $f : X \times Y \to \Sigma$ that can be computed with complete-fairness.*

For the non-binary case, we provide a general criteria that holds only for functions for which $|X| > (|\Sigma| - 1) \cdot |Y|$, that is, when the ratio between the domain sizes is greater than $|\Sigma| - 1$. This, together with the results in the binary case, may refer to an interesting relationship between the size of the domains and possibility of fairness. This is the first time that a fair protocol is constructed for both non-binary output, and asymmetric boolean functions. This shows that fairness is not restricted to a very specific and particular type of functions, but rather a property that under certain circumstances can be achieved. Moreover, it shows the power that is concealed in the GHKL protocol alone.

Related Work. Several other impossibility results regarding fairness, rather than the result of Cleve, have been published [12,1]. However, it seems that only Cleve's impossibility can be reduced into the family of boolean functions with finite domain. The work of [3] identifies which function imply fair coin-tossing and are ruled out by the impossibility result of Cleve. Interestingly, the class of functions that imply fair coin-tossing shares a similar (but yet distinct) algebraic structure with the class of functions that we show that cannot be computed using the GHKL protocol. We link between the two criterions in the body of our work.

For decades fairness was believed to be impossible, and so researchers have simply resigned themselves to being unable to achieve this goal. Therefore, a huge amount of works consider several relaxations like gradual release, partial fairness and rational adversaries ([10,15,5,19,4,21] to state a few. See [16] for a survey of fairness in secure computation).

Open Problems. Our work is an important step towards a full characterization of fairness of finite domain functions. The main open question is to finalize this characterization. In addition, it seems appealing to generalize our results to functions with infinite domains (domains with sizes that depend on the security parameter). Finally, in the non-binary case, we have a positive result only when the ratio between the domain sizes is greater than $|\Sigma| - 1$. A natural question is whether fairness be achieved in any other case, or for any other ratio.

2   Definitions and Preliminaries

We assume that the reader is familiar with the definitions of secure computation, and with the ideal-real paradigm. We distinguish between security-with-abort, for which the adversary may receive output while the honest party does not (security without fairness), and security with fairness, where all parties receive output (this is similar to security with respect to honest majority as in [7], although we do not have honest majority). In the following, we present the necessary notations, and we cover the mathematical background that is needed for our results.

Notations. We let κ denote the security parameter. We use standard O notation, and let poly denote a polynomial function. A function $\mu(\cdot)$ is *negligible* if for every positive polynomial poly$(\cdot)$ and all sufficiently large κ's it holds that $\mu(\kappa) < 1/\text{poly}(\kappa)$. In most of the paper, we consider binary deterministic functions over a finite domain; i.e., functions $f : X \times Y \to \{0,1\}$ where $X, Y \subset \{0,1\}^*$ are finite sets. Throughout the paper, we denote $X = \{x_1, \ldots, x_\ell\}$ and $Y = \{y_1, \ldots, y_m\}$, for constants $\ell, m \in \mathbb{N}$. Let M_f be the $\ell \times m$ matrix that represents the function, i.e., a matrix whose entry position (i,j) is $f(x_i, y_j)$. For $1 \leq i \leq \ell$, let X_i denote the ith row of M_f, and for $1 \leq j \leq m$ let Y_j denote the jth column of M_f. A vector $\mathbf{p} = (p_1, \ldots, p_\ell)$ is a *probability vector* if $p_i \geq 0$ for every $1 \leq i \leq \ell$ and $\sum_{i=1}^{\ell} p_i = 1$. As a convention, we use **bold**-case letters to represent a vector (e.g., $\mathbf{p}$, $\mathbf{q}$), and sometimes we use upper-case letters (e.g., X_i, as above). All vectors will be assumed to be row vectors. We denote by $\mathbf{1}_k$ (resp. $\mathbf{0}_k$) the all one (resp. all zero) vector of size k. We work in the Euclidian space $\mathbb{R}^m$, and use the Euclidian norm $||x|| = \sqrt{\langle x, x \rangle}$ and the distance function as $d(x,y) = ||x - y||$.

2.1   Mathematical Background

Our characterization is based on the geometric representation of the function f. In the following, we provide the necessary mathematical background, and link it to the context of cryptography whenever possible. Most of the following Mathematical definitions are taken from [26,20].

Output Vector Distribution and Convex Combination. We now analyze the "power of the simulator" in the ideal execution. The following is an inherent property of the concrete function and the ideal execution, and is correct for any protocol computing the function. Let $\mathcal{A}$ be an adversary that corrupts the party P_1, and assume that the simulator $\mathcal{S}$ chooses its input according to some distribution $\mathbf{p} = (p_1, \ldots, p_\ell)$. That is, the simulator sends an input x_i with probability p_i, for $1 \le i \le \ell$. Then, the length m vector $\mathbf{q} = (q_{y_1}, \ldots, q_{y_m}) \stackrel{\text{def}}{=} \mathbf{p} \cdot M_f$ represents the *output distribution vector* of the honest party P_2. That is, in case the input of P_2 is y_j for some $1 \le j \le m$, then it gets 1 with probability q_{y_j}.

Convex Combination. The output distribution vector is in fact a convex combination of the rows $\{X_1, \ldots, X_\ell\}$ of the matrix M_f. That is, when the simulator uses $\mathbf{p}$, the output vector distribution of P_2 is:

$$\mathbf{p} \cdot M_f = (p_1, \ldots, p_\ell) \cdot M_f = p_1 \cdot X_1 + \ldots + p_\ell \cdot X_\ell \; .$$

A convex combination of points $X_1, \ldots, X_\ell$ in $\mathbb{R}^m$ is a linear combination of the points, where all the coefficients (i.e., $(p_1, \ldots, p_\ell)$) are non-negative and sum up to 1.

Convex Hull. The set of all possible output distributions vectors that the simulator can produce in the ideal execution is:

$$\{\mathbf{p} \cdot M_f \mid \mathbf{p} \text{ is a probability vector}\} \; .$$

In particular, this set reflects the "freedom" that the simulator has in the ideal execution. This set is in fact, the *convex hull* of the row vectors $X_1, \ldots, X_\ell$, and is denoted as $\mathbf{conv}(\{X_1, \ldots, X_\ell\})$. That is, for a set $S = \{X_1, \ldots, X_\ell\}$, $\mathbf{conv}(S) = \left\{ \sum_{i=1}^{\ell} p_i \cdot X_i \mid 0 \le p_i \le 1, \sum_{i=1}^{m} p_i = 1 \right\}$. The convex-hull of a set of points is a convex set, which means that for every $X, Y \in \mathbf{conv}(S)$, the line segment between X and Y also lies in $\mathbf{conv}(S)$, that is, for every $X, Y \in \mathbf{conv}(S)$ and for every $0 \le \lambda \le 1$, it holds that $\lambda \cdot X + (1 - \lambda) \cdot Y \in \mathbf{conv}(S)$.

Geometrically, the convex-hull of two (distinct) points in $\mathbb{R}^2$, is the line-segment that connects them. The convex-hull of three points in $\mathbb{R}^2$ may be a line (in case all the points lie on a single line), or a triangle (in case where all the points are collinear). The convex-hull of 4 points may be a line, a triangle, or a parallelogram. In general, the convex-hull of k points in $\mathbb{R}^2$ may define a convex polygon of at most k vertices. In $\mathbb{R}^3$, the convex-hull of k points can be either a line, a triangle, a tetrahedron, a parallelepiped, etc.

Affine-Hull and Affine Independence. A subset B of $\mathbb{R}^m$ is an affine subspace if $\lambda \cdot \mathbf{a} + \mu \cdot \mathbf{b} \in B$ for every $\mathbf{a}, \mathbf{b} \in B$ and $\lambda, \mu \in \mathbb{R}$ such that $\lambda + \mu = 1$. For a set of points $S = \{X_1, \ldots, X_\ell\}$, its affine hull is defined as: $\mathbf{aff}(S) = \left\{ \sum_{i=1}^{\ell} \lambda_i \cdot X_i \mid \sum_{i=1}^{\ell} \lambda_i = 1 \right\}$, which is similar to convex hull, but without the additional requirement for non-negative coefficients. The set of points $X_1, \ldots, X_\ell$ in $\mathbb{R}^m$ is affinely independent if $\sum_{i=1}^{\ell} \lambda_i X_i = \mathbf{0}_m$ holds with $\sum_{i=1}^{\ell} \lambda_i = 0$ only if $\lambda_1 = \ldots = \lambda_\ell = 0$. In particular, it means that one of the points is in the affine hull of the other points. It is easy to see that the set of points $\{X_1, \ldots, X_\ell\}$ is affinely independent if and only if the set $\{X_2 - X_1, \ldots, X_\ell - X_1\}$ is a linearly independent set. As a result, any $m + 2$ points in $\mathbb{R}^m$ are affine dependent, since any $m + 1$ points in $\mathbb{R}^m$ are linearly dependent. In addition, it is easy to see that the points $\{X_1, \ldots, X_\ell\}$ over $\mathbb{R}^m$ is affinely independent if and only if the set of points $\{(X_1, 1), \ldots, (X_\ell, 1)\}$ over $\mathbb{R}^{m+1}$ is linearly independent.

If the set $S = \{X_1, \ldots, X_\ell\}$ over $\mathbb{R}^m$ is affinely independent, then $\mathbf{aff}(S)$ has dimension $\ell - 1$, and we write $\dim(\mathbf{aff}(S)) = \ell - 1$. In this case, S is the affine basis for $\mathbf{aff}(S)$. Note that an affine basis for an m-dimensional affine space has $m + 1$ elements.

Linear Hyperplane. A linear hyperplane in $\mathbb{R}^m$ is a $(m-1)$-dimensional affine-subspace of $\mathbb{R}^m$. The linear hyperplane can be defined as all the points $X = (x_1, \ldots, x_m)$ which are the solutions of a linear equation:

$$a_1 x_1 + \ldots a_m x_m = b ,$$

for some constants $\mathbf{a} = (a_1, \ldots, a_m) \in \mathbb{R}^m$ and $b \in \mathbb{R}$. We denote this hyperplane by:

$$\mathcal{H}(\mathbf{a}, b) \overset{\text{def}}{=} \{X \in \mathbb{R}^m \mid \langle X, \mathbf{a} \rangle = b\} .$$

Throughout the paper, for short, we will use the term hyperplane instead of linear hyperplane. It is easy to see that indeed this is an affine-subspace. In $\mathbb{R}^1$, an hyperplane is a single point, in $\mathbb{R}^2$ it is a line, in $\mathbb{R}^3$ it is a plane and so on. We remark that for any m affinely independent points in $\mathbb{R}^m$ there exists a *unique* hyperplane that contains all of them (and infinitely many in case they are not affinely independent). This is a simple generalization of the fact that for any distinct 2 points there exists a single line that passes through them, for any 3 (collinear) points there exists a single plane that contains all of them and etc.

Convex Polytopes. Geometrically, a full dimensional convex polytope in $\mathbb{R}^m$ is the convex-hull of a finite set S where $\dim(\mathbf{aff}(S)) = m$. Polytopes are familiar objects: in $\mathbb{R}^2$ we get *convex polygons* (a triangle, a parallelogram etc.). In $\mathbb{R}^3$ we get *convex polyhedra* (a tetrahedron, a parallelepiped etc.). Convex polytopes play an important role in solutions of linear programming.

In addition, a special case of polytope is simplex. If the set S is affinely independent of cardinality $m + 1$, then $\mathbf{conv}(S)$ is an m-dimensional *simplex* (or, m-simplex). For $m = 2$, this is simply a triangle, whereas in $m = 3$ we get a tetrahedron. A simplex in $\mathbb{R}^m$ consists of $m + 1$ *facets*, which are themselves simplices of lower dimensions. For instance, a tetrahedron (which is a 3-simplex) consists of 4 facets, which are themselves triangles (2-simplex).

3   The Protocol of Gordon, Hazay, Katz and Lindell [17]

In the following, we give a high level overview of the protocol of [17]. We also present its simulation strategy, and the set of equations that indicates whether a given function can be computed with this protocol, which is the important part for our discussion.

The Protocol. Assume the existence of an online dealer (a reactive functionality that can be replaced using standard secure computation that is secure-with-abort). The parties invoke this online-dealer and send it their respective inputs

$(x, y) \in X \times Y$. The online dealer computes values $a_1, \ldots, a_R$ and $b_1, \ldots, b_R$ (we will see later how they are defined). In round i the dealer sends party P_1 the value a_i and afterward it sends b_i to P_2. At each point of the execution, each party can abort the online-dealer, preventing the other party from receiving its value at that round. In such a case, the other party is instructed to halt and output the last value it has received from the dealer. For instance, if P_1 aborts at round i after it learns a_i and prevents from P_2 to learn b_i, P_2 halts and outputs b_{i-1}.

The values $(a_1, \ldots, a_R), (b_1, \ldots, b_R)$ are generated by the dealer in the following way: The dealer first chooses a round i^* according to some geometric distribution with parameter α. In each round $i < i^*$, the parties receive bits (a_i, b_i), that depend on their respective inputs solely and uncorrelated to the input of the other party. In particular, for party P_1 the dealer computes $a_i = f(x, \hat{y})$ for some random $\hat{y}$, and for P_2 it computes $b_i = f(\hat{x}, y)$ for some random $\hat{x}$. For every round $i \geq i^*$, the parties receive the correct output $a_i = b_i = f(x, y)$. In case one of the party initially aborts (i.e., does not invoke the online-dealer in the first round and the parties do not receive a_1, b_1), each party can locally compute initial outputs a_0, b_0 similarly to the way the values a_i, b_i are computed by the online-dealer for $i < i^*$. Note that if we set $R = \alpha^{-1} \cdot \omega(\ln \kappa)$, then $i^* < R$ with overwhelming probability, and so correctness holds.

Security. Since P_2 is the second to receive an output, it is easy to simulate an adversary that corrupts P_2. If the adversary aborts before i^*, then it has not obtained any information about the input of P_1. If the adversary aborts at or after i^*, then in the real execution the honest party P_1 already receives the correct output $f(x, y)$, and fairness is obtained. Therefore, the protocol is secure with respect to corrupted P_2, *for any function f*.

The case of corrupted P_1 is more delicate, and defines some requirements from f. Intuitively, if the adversary aborts before i^*, then the outputs of both parties are uncorrelated, and no one gets any advantage. If the adversary aborts after i^*, then both parties receive the correct output and fairness is obtained. The worst case, then, occurs when P_1 aborts exactly in iteration i^*, as P_1 has then learned the correct value of $f(x, y)$ while P_2 has not. Since the simulator has to give P_1 the true output if it aborts at i^*, it sends the trusted party the *true* input x_i in round i^*. As a result, P_2 in the ideal execution learns the correct output $f(x, y)$ at round i^*, unlike the real execution where it outputs a random value $f(\hat{x}, y)$. [17] overcomes this problem in a very elegant way: in order to balance this advantage of the honest party in the ideal execution in case the adversary aborts at i^*, the simulator chooses a random value $\hat{x}$ *different* from the way it is chosen in the real execution in case the adversary abort *before i^** (that is, according to a different distribution than the one the dealer uses in the real execution). The calculations show that overall, the output distribution of the honest party is distributed identically in the real and ideal executions. This balancing is possible only sometimes, and depends on the actual function f that is being evaluated.

In more detail, in the real execution the dealer before i^* chooses b_i as $f(\hat{x}, y)$, where $\hat{x}$ is chosen according to some distribution X_{real}. In the ideal execution, in case the adversary sends x to the simulated online-dealer, aborts at round $i < i^*$ upon viewing some a_i, the simulator chooses the input $\tilde{x}$ it sends to the trusted party according to distribution X_{ideal}^{x,a_i}. Then, define $Q^{x,a_i} = X_{ideal}^{x,a_i} \cdot M_f$, the output distribution vector of the honest party P_2 in this case. In fact, the protocol and the simulation define the output distribution vectors Q^{x,a_i}, and simulation is possible only if the corresponding X_{ideal}^{x,a_i} distribution exists, which depends on the function f being computed. Due to lack of space, we now show the definitions of the desired output distribution vectors Q^{x,a_i} without getting into the calculations for why these are defined like that. We refer the reader to [17] or the full version of this paper [2] to see how the protocol defines these requirements.

The Output Distributions Vectors $Q^{x,a}$. Let $f : \{x_1, \ldots, x_\ell\} \times \{y_1, \ldots, y_m\} \to \{0, 1\}$. Fix X_{real}, and let U_Y denote the uniform distribution over Y. For every $x \in X$, denote by p_x the probability that $a_i = 1$ before i^*. Similarly, for every $y_j \in Y$, let p_{y_j} denote the probability $b_i = 1$ before i^*. That is: $p_x \stackrel{\text{def}}{=} \Pr_{\hat{y} \leftarrow U_Y}[f(x, \hat{y}) = 1]$, and $p_{y_j} \stackrel{\text{def}}{=} \Pr_{\hat{x} \leftarrow X_{real}}[f(\hat{x}, y_j) = 1]$. For every $x \in X$, $a \in \{0, 1\}$, define the row vectors $Q^{x,a} = (q_{y_1}^{x,a}, \ldots, q_{y_m}^{x,a})$ indexed by $y_j \in Y$ as follows:

$$
q_{y_j}^{x,0} \stackrel{\text{def}}{=} \begin{cases} p_{y_j} & \text{if } f(x, y_j) = 1 \\ p_{y_j} + \frac{\alpha \cdot p_{y_j}}{(1-\alpha)\cdot(1-p_x)} & \text{if } f(x, y_j) = 0 \end{cases}
$$

$$
q_{y_j}^{x,1} \stackrel{\text{def}}{=} \begin{cases} p_{y_j} + \frac{\alpha \cdot (p_{y_j}-1)}{(1-\alpha)\cdot p_x} & \text{if } f(x, y_j) = 1 \\ p_{y_j} & \text{if } f(x, y_j) = 0 \end{cases} \tag{1}
$$

In case for every $x \in X, a \in \{0, 1\}$ there exists a probability vector $X_{ideal}^{x,a}$ such that $X_{ideal}^{x,a} \cdot M_f = Q^{x,a}$, then the simulator succeeds to simulate the protocol. We therefore have the following theorem:

Theorem 3.1. *Let $f : \{x_1, \ldots, x_\ell\} \times \{y_1, \ldots, y_m\} \to \{0, 1\}$ and let M_f be as above. If there exist probability vector X_{real} and a parameter $0 < \alpha < 1$ (where $\alpha^{-1} \in O(\text{poly}(\kappa))$), such that for every $x \in X$, $a \in \{0, 1\}$, there exists a probability vector $X_{ideal}^{x,a}$ for which:*

$$
X_{ideal}^{x,a} \cdot M_f = Q^{x,a} ,
$$

then the protocol securely computes f with complete fairness.

An alternative formulation of the above, is to require that for every x, a, the points $Q^{x,a}$ are in $\mathbf{conv}(\{X_1, \ldots, X_\ell\})$, where X_i is the ith row of M_f. Moreover, observe that in order to decide whether a function can be computed using the protocol, there are 2ℓ linear systems that should be satisfied, with m constraints each, and with $2\ell^2$ variables overall. This criterion depends heavily on some parameters of the protocols (like p_x, p_{y_j}) rather than properties of the function. We are interested in a simpler and easier way to validate criteria.

4 Our Criteria

4.1 Possibility of Full-Dimensional Functions

In this section, we show that any function that defines a full-dimensional geometric object, can be computed using the protocol of [17]. A full dimensional function is defined as follows:

Definition 4.1 (full-dimensional function). *Let $f : X \times Y \to \{0,1\}$ be a function, and let $X_1, \ldots, X_\ell$ be the ℓ rows of M_f over $\mathbb{R}^m$. We say that f is a full-dimensional function if* $\dim(\mathbf{aff}(\{X_1, \ldots, X_\ell\})) = m$.

Recall that for a set of points $S = \{X_1, \ldots, X_\ell\} \in \mathbb{R}^m$, if $\dim(\mathbf{aff}(S)) = m$ then the convex-hull of the points defines a full-dimensional convex polytope. Thus, intuitively, the simulator has enough power to simulate the protocol. Recall that a basis for an affine space of dimension m has cardinality $m+1$, and thus we must have that $\ell > m$. Therefore, we assume without loss of generality that $\ell > m$ (and consider the transposed function $f^T : \{y_1, \ldots, y_m\} \times \{x_1, \ldots, x_\ell\} \to \{0,1\}$, defined as $f^T(y, x) = f(x, y)$, otherwise). Overall, our property inherently holds only if $\ell \neq m$.

Alternative Representation. Before we prove that any full-dimensional function can be computed fairly, we give a different representation for this definition. This strengthens our understanding of this property, and is also related to the balanced property defined in [3] (we will elaborate more about these two criterions in Subsection 4.3). The proof for the following claim appears in the full version [2]:

Claim 4.2. *Let $f : \{x_1, \ldots, x_\ell\} \times \{y_1, \ldots, y_m\} \to \{0,1\}$ be a function, let M_f be as above and let $S = \{X_1, \ldots, X_\ell\}$ be the rows of M_f (ℓ points in $\mathbb{R}^m$). The following are equivalent:*

1. The function is right-unbalanced with respect to arbitrary vectors.
   That is, for every non-zero $\mathbf{q} \in \mathbb{R}^m$ and any $\delta \in \mathbb{R}$ it holds that: $M_f \cdot \mathbf{q}^T \neq \delta \cdot \mathbf{1}_\ell$.
2. The rows of the matrix do not lie on the same hyperplane.
   That is, for every non-zero $\mathbf{q} \in \mathbb{R}^m$ and any $\delta \in \mathbb{R}$, there exists a point X_i such that $X_i \notin \mathcal{H}(\mathbf{q}, \delta)$. Alteratively, $\mathbf{conv}(\{X_1, \ldots, X_\ell\}) \nsubseteq \mathcal{H}(\mathbf{q}, \delta)$.
3. The function is full-dimensional.
   There exists a subset of $\{X_1, \ldots, X_\ell\}$ of cardinality $m + 1$, that is affinely independent. Thus, $\dim(\mathbf{aff}(\{X_1, \ldots, X_\ell\})) = m$.

From Alternative 1, checking whether a function is full-dimensional can be done efficiently. Giving that $\ell > m$, all we have to do is to verify that the only possible solution $\mathbf{q}$ for the linear system $M_f \cdot \mathbf{q}^T = \mathbf{0}_\ell^T$ is the trivial one (i.e., $\mathbf{q} = \mathbf{0}$), and that there is no solution $\mathbf{q}$ for the linear system $M_f \cdot \mathbf{q}^T = \mathbf{1}_\ell^T$. This implies that the function is unbalanced for every $\delta \in \mathbb{R}$.

The Proof of Possibility. We now show that any function that is full dimensional can be computed with complete fairness, using the protocol of [17]. The proof for this Theorem is geometrical. Recall that by Theorem 3.1, we need to show that there exists a solution for some set of equations. In our proof here, we show that such a solution exists without solving the equations explicitly. We show that all the points $Q^{x,a}$ that the simulator needs (by Theorem 3.1) are in the convex-hull of the rows $\{X_1, \ldots, X_\ell\}$, and therefore there exist probability vectors $X^{x,a}_{ideal}$ as required. We show this in two steps. First, we show that all the points are very "close" to some point $\mathbf{c}$, and therefore, all the points are inside the Euclidian ball centered at $\mathbf{c}$ for some small radius ϵ (defined as $B(\mathbf{c}, \epsilon) \stackrel{\text{def}}{=} \{Z \in \mathbb{R}^m \mid d(Z, \mathbf{c}) \leq \epsilon\}$). Second, we show that this whole ball is embedded inside the convex-polytope that is defined by the rows of the function, which implies that all the points $Q^{x,a}$ are in the convex-hull and simulation is possible.

In more detail, fix some distribution X_{real} for which the point $\mathbf{c} = (p_{y_1}, \ldots, p_{y_m})$ $= X_{real} \cdot M_f$ is inside the convex-hull of the matrix. Then, we observe that by adjusting α, all the points $Q^{x,a}$ that we need are very "close" to this point $\mathbf{c}$. This is because each coordinate $q^{x,a}_{y_j}$ is exactly p_{y_j} plus some term that is multiplied by $\alpha/(1 - \alpha)$, and therefore we can control its distance from p_{y_j} (see Eq. (1)). In particular, if we choose $\alpha = 1/\ln \kappa$, then for all sufficiently large κ's the distance between $Q^{x,a}$ and $\mathbf{c}$ is smaller than any constant. Still, for $\alpha = 1/\ln \kappa$, the number of rounds of the protocol is $R = \alpha^{-1} \cdot \omega(\ln \kappa) = \ln \kappa \cdot \omega(\ln \kappa)$, and thus asymptotically remains unchanged.

All the points $Q^{x,a}$ are close to the point $\mathbf{c}$. This implies that they all lie in the m-dimensional Euclidian ball of some constant radius $\epsilon > 0$ centered at $\mathbf{c}$. Moreover, since the function is of full-dimension, the convex-hull of the function defines a full-dimensional convex polytope, and therefore this ball is embedded in this polytope. We prove this by showing that the center of the ball $\mathbf{c}$ is "far" from each facet of the polytope, using the separation theorems of closed convex sets. As a result, all the points that are "close" to $\mathbf{c}$ (i.e., our ball) are still "far" from each facet of the polytope, and thus they are inside it. As an illustration, consider again the case of the GHKL function in Figure 2 (in Section 1). We conclude that all the points that the simulator needs are in the convex-hull of the function, and therefore the protocol can be simulated.

Before we proceed to the full proof formally, we give an additional definition and an important Claim. For a set $F \subseteq \mathbb{R}^m$ and a point $\mathbf{p} \in \mathbb{R}^m$, we define the distance between $\mathbf{p}$ and F to be the minimal distance between $\mathbf{p}$ and a point in F, that is: $d(\mathbf{p}, F) = \min\{d(\mathbf{p}, \mathbf{f}) \mid \mathbf{f} \in F\}$. The following claim shows that if a point is not on a closed convex set, then there exists a constant distance between the point and the convex set. We use this claim to show that the point $\mathbf{c}$ is far enough from each one of the facets of the polytope (and therefore the ball centered in $\mathbf{c}$ is in the convex). The proof for this claim is a simple implication of the separation theorems for convex sets, see [26]. We have:

Claim 4.3. *Let $\mathcal{C}$ be a closed convex subset of $\mathbb{R}^m$, and let $\mathbf{a} \in \mathbb{R}^m$ such that $\mathbf{a} \notin \mathcal{C}$. Then, there exists a constant $\epsilon > 0$ such that $d(\mathbf{a}, \mathcal{C}) > \epsilon$ (that is, for every $Z \in \mathcal{C}$ it holds that $d(\mathbf{a}, Z) > \epsilon$).*

We now ready for our main theorem of this section:

Theorem 4.4. *Let $f : \{x_1, \ldots, x_\ell\} \times \{y_1, \ldots, y_m\} \to \{0, 1\}$ be a boolean function. If f is of full-dimension, then f can be computed with complete fairness.*

Proof: Since f is full-dimensional, there exists a subset of $m + 1$ rows that are affinely independent. Let $S' = \{X_1, \ldots, X_{m+1}\}$ be this subset of rows. We now locate $\mathbf{c}$ to be inside the simplex that is defined by S', by choosing X_{real} to be the uniform distribution over S' (i.e., the ith position of X_{real} is 0 if $X_i \notin S'$, and $1/(m+1)$ if $X_i \in S'$). We then let $\mathbf{c} = (p_{y_1}, \ldots, p_{y_m}) = X_{real} \cdot M_f$. Finally, we set $\alpha = 1/\ln \kappa$. We consider the GHKL protocol with the above parameters, and consider the set of points $\{Q^{x,a}\}_{x \in X, a \in \{0,1\}}$. The next claim shows that all these points are close to $\mathbf{c}$, and in the m-dimensional ball $B(\mathbf{c}, \epsilon)$ for some small $\epsilon > 0$. That is:

Claim 4.5. *For every constant $\epsilon > 0$, for every $x \in X, a \in \{0, 1\}$, and for all sufficiently large κ's it holds that:*

$$Q^{x,a} \in B(\mathbf{c}, \epsilon)$$

Proof: Fix ϵ. Since $\alpha = 1/\ln \kappa$, for every constant $\delta > 0$ and for all sufficiently large κ's it holds that: $\alpha/(1 - \alpha) < \delta$. We show that for every x, a, it holds that $d(Q^{x,a}, \mathbf{c}) \leq \epsilon$, and thus $Q^{x,a} \in B(\mathbf{c}, \epsilon)$.

Recall the definition of $Q^{x,0}$ as in Eq. (1): If $f(x, y_j) = 1$ then $q_{y_j}^0 = p_{y_j}$ and thus $|p_{y_j} - q_{y_j}^0| = 0$. In case $f(x, y_j) = 1$, for $\delta = \epsilon(1-p_x)/\sqrt{m}$ and for all sufficiently large κ's it holds that:

$$\left| p_{y_j} - q_{y_j}^{x,0} \right| = \left| p_{y_j} - p_{y_j} - \frac{\alpha}{1 - \alpha} \cdot \frac{p_{y_j}}{(1 - p_x)} \right| \leq \frac{\alpha}{1 - \alpha} \cdot \frac{1}{(1 - p_x)} \leq \frac{\delta}{(1 - p_x)} = \frac{\epsilon}{\sqrt{m}} \ .$$

Therefore, for all sufficiently large κ's, $\left| p_{y_j} - q_{y_j}^{x,0} \right| \leq \epsilon/\sqrt{m}$ irrespectively to whether $f(x, y_j)$ is 1 or 0. In a similar way, for all sufficiently large κ's it holds that: $\left| p_{y_j} - q_{y_j}^{x,1} \right| \leq \epsilon/\sqrt{m}$. Overall, for every $x \in X$, $a \in \{0, 1\}$ we have that the distance between the points $Q^{x,a}$ and $\mathbf{c}$ is:

$$d(Q^{x,a}, \mathbf{c}) = \sqrt{\sum_{j=1}^{m} \left(q_{y_j}^{x,b} - p_{y_j} \right)^2} \leq \sqrt{\sum_{j=1}^{m} \left(\frac{\epsilon}{\sqrt{m}} \right)^2} \leq \epsilon$$

and therefore $Q^{x,a} \in B(\mathbf{c}, \epsilon)$. ∎

We now show that this ball is embedded inside the simplex of S'. That is:

Claim 4.6. *There exists a constant $\epsilon > 0$ for which $B(\mathbf{c}, \epsilon) \subset \mathbf{conv}(S')$.*

Proof: Since $S' = \{X_1, \ldots, X_{m+1}\}$ is affinely independent set of cardinality $m + 1$, $\mathbf{conv}(S')$ is a simplex. Recall that $\mathbf{c}$ is a point in the simplex (since it assigns 0 to any row that is not in S'), and so $\mathbf{c} \in \mathbf{conv}(S')$. We now show that for every *facet* of the simplex, there exists a constant distance between the point $\mathbf{c}$ and the facet. Therefore, there exists a small ball around $\mathbf{c}$ that is "far" from each facet of the simplex, and inside the simplex.

For every $1 \leq i \leq m + 1$, the ith facet of the simplex is the set $F_i = \mathbf{conv}(S' \setminus \{X_i\})$, i.e., the convex set of the vertices of the simplex without the vertex X_i. We now show that $\mathbf{c} \notin F_i$, and therefore, using Claim 4.3, $\mathbf{c}$ is ϵ-far from F_i, for some small $\epsilon > 0$.

In order to show that $\mathbf{c} \notin F_i$, we show that $\mathbf{c} \notin \mathcal{H}(\mathbf{q}, \delta)$, where $\mathcal{H}(\mathbf{q}, \delta)$ is an hyperplane that contains F_i. That is, let $\mathcal{H}(\mathbf{q}, \delta)$ be the unique hyperplane that contains all the points $S' \setminus \{X_i\}$ (these are m affinely independent points and therefore there is a unique hyperplane that contains all of them). Recall that $X_i \notin \mathcal{H}(\mathbf{q}, \delta)$ (otherwise, S' is affinely dependent). Observe that $F_i = \mathbf{conv}(S' \setminus \{X_i\}) \subset \mathcal{H}(\mathbf{q}, \delta)$, since each point X_i is in the hyperplane, and the hyperplane is an affine set. We now show that since $X_i \notin \mathcal{H}(\mathbf{q}, \delta)$, then $\mathbf{c} \notin \mathcal{H}(\mathbf{q}, \delta)$ and therefore $\mathbf{c} \notin F_i$.

Assume by contradiction that $\mathbf{c} \in \mathcal{H}(\mathbf{q}, \delta)$. We can write:

$$\delta = \langle \mathbf{c}, \mathbf{q} \rangle = \Big\langle \sum_{j=1}^{m+1} \frac{1}{m+1} \cdot X_j, \mathbf{q} \Big\rangle = \frac{1}{m+1} \langle X_i, \mathbf{q} \rangle + \frac{1}{m+1} \sum_{j \neq i} \langle X_j, \mathbf{q} \rangle$$

$$= \frac{1}{m+1} \langle X_i, \mathbf{q} \rangle + \frac{m}{m+1} \cdot \delta$$

and so, $\langle X_i, \mathbf{q} \rangle = \delta$, which implies that $X_i \in \mathcal{H}(\mathbf{q}, \delta)$ in contradiction.

Since $\mathbf{c} \notin F_i$, and since F_i is a closed[2] convex, we can apply Claim 4.3 to get the existence of a constant $\epsilon_i > 0$ such that $d(\mathbf{c}, F_i) > \epsilon_i$.

Now, let $F_1, \ldots, F_{m+1}$ be the facets of the simplex. We get the existence of $\epsilon_1, \ldots, \epsilon_{m+1}$ for each facet as above. Let $\epsilon = \min\{\epsilon_1, \ldots, \epsilon_{m+1}\}/2$, and so for every i, we have: $d(\mathbf{c}, F_i) > 2\epsilon$.

Consider the ball $B(\mathbf{c}, \epsilon)$. We show that any point in this ball is of distance at least ϵ from each facet F_i. Formally, for every $\mathbf{b} \in B(\mathbf{c}, \epsilon)$, for every facet F_i it holds that: $d(\mathbf{b}, F_i) > \epsilon$. This can be easily derived from the triangle inequality, where for every $\mathbf{b} \in B(\mathbf{c}, \epsilon/2)$:

$$d(\mathbf{c}, \mathbf{b}) + d(\mathbf{b}, F_i) \geq d(\mathbf{c}, F_i) > 2\epsilon ,$$

and so $d(\mathbf{b}, F_i) > \epsilon$ since $d(\mathbf{b}, \mathbf{c}) \leq \epsilon$.

Overall, all the points $\mathbf{b} \in B(\mathbf{c}, \epsilon)$ are of distance at least ϵ from each facet of the simplex, and inside the simplex. This shows that $B(\mathbf{c}, \epsilon) \subset \mathbf{conv}(S')$. ∎

For conclusion, there exists a constant $\epsilon > 0$ for which $B(\mathbf{c}, \epsilon) \subset \mathbf{conv}(S') \subseteq \mathbf{conv}(\{X_1, \ldots, X_\ell\})$. Moreover, for all $x \in X, a \in \{0, 1\}$ and for all sufficiently large κ's, it holds that $Q^{x,a} \in B(\mathbf{c}, \epsilon)$. Therefore, the requirements of Theorem 3.1 are satisfied, and the protocol securely computes f with complete fairness. ∎

---

[2] The convex-hull of a finite set S of vectors in $\mathbb{R}^m$ is a compact set, and therefore is closed (See [26, Theorem 15.4]).

On the Number of Full-Dimensional Functions. We count the number of functions that are full dimensional. Recall that a function with $|X| = |Y|$ cannot be full-dimensional, and we consider only functions where $|X| \neq |Y|$. Interestingly, the probability that a random function with distinct domain sizes is full-dimensional tends to 1 when $|X|, |Y|$ grow. Thus, almost always, a random function with distinct domain sizes can be computed with complete fairness(!). The answer for the frequency of full-dimensional functions within the class of boolean functions with distinct sizes relates to a beautiful problem in combinatorics and linear algebra, that has received careful attention: Estimating the probability that a random boolean matrix of size $m \times m$ is singular. Denote this probability by P_m. The answer for our question is simply $1 - P_m$, and is even larger when the difference between $|X|$ and $|Y|$ increases (see Claim 4.7 below).

The value of P_m is conjectured to be $(1/2 + o(1))^m$, and recent results [23,22,28] are getting closer to this conjuncture, by showing that $P_m \leq (1/\sqrt{2} + o(1))^m$, which is roughly the probability to have two identical or compliments rows or columns. Since our results hold only for the case of *finite* domain, it is remarkable to address that P_m is small already for very small dimensions m. For instance, $P_{10} < 0.29$, $P_{15} < 0.047$ and $P_{30} < 1.6 \cdot 10^{-6}$ (and so $> 99.999\%$ of the 31×30 functions can be computed fairly). See more experimental results in [27]. The following Claim is based on [30, Corollary 14]:

Claim 4.7. *With a probability that tends to 1 when $|X|, |Y|$ grow, a random function with $|X| \neq |Y|$ is full-dimensional.*

Proof: An alternative question for the first item is the following: What is the probability that the convex-hull of $m + 1$ (or even more) random 0/1-points in $\mathbb{R}^m$ is an m-dimensional simplex?

Recall that P_m denotes the probability that a random m vectors of size m are linearly dependent. Then, the probability for our first question is simply $1 - P_m$. This is because with very high probability our $m + 1$ points will be distinct, we can choose the first point X_1 arbitrarily, and the rest of the points $S = \{X_2, \ldots, X_{m+1}\}$ uniformly at random. With probability $1 - P_m$, the set S is linearly independent, and so it linearly spans X_1. It is easy to see that this implies that $\{X_2 - X_1, \ldots, X_{m+1} - X_1\}$ is a linearly independent set, and thus $\{X_1, \ldots, X_{m+1}\}$ is affinely-independent set. Overall, a random set $\{X_1, \ldots, X_{m+1}\}$ is affinely independent with probability $1 - P_m$. ∎

4.2   Functions That Are Not Full-Dimensional

A Negative Result. We now consider the case where the functions are not full-dimensional. This includes the limited number of functions for which $|X| \neq |Y|$, and *all* functions with $|X| = |Y|$. In particular, for a function that is not full-dimensional, all the rows of the function lie in some hyperplane (a $(m - 1)$-dimensional subspace of $\mathbb{R}^m$), and all the columns of the matrix lie in a different hyperplane (in $\mathbb{R}^\ell$). We show that under some additional requirements, the protocol of [17] cannot be simulated for any choice of parameters, with respect to the specific simulation strategy defined in the proof of Theorem 3.1. We have the following Theorem:

Theorem 4.8. *Let* $f, M_f, \{X_1, \ldots, X_\ell\}$ *be as above, and let* $\{Y_1, \ldots, Y_m\}$ *be the columns of* M_f. *Assume that the function is not full-dimensional, that is, there exist non-zero* $\mathbf{p} \in \mathbb{R}^\ell$, $\mathbf{q} \in \mathbb{R}^m$ *and some* $\delta_1, \delta_2 \in \mathbb{R}$ *such that:*

$$X_1, \ldots, X_\ell \in \mathcal{H}(\mathbf{q}, \delta_2) \quad \text{and} \quad Y_1, \ldots, Y_m \in \mathcal{H}(\mathbf{p}, \delta_1) .$$

Assume that in addition, $\mathbf{0}_\ell, \mathbf{1}_\ell \notin \mathcal{H}(\mathbf{p}, \delta_1)$ *and* $\mathbf{0}_m, \mathbf{1}_m \notin H(\mathbf{q}, \delta_2)$. *Then, the function* f *cannot be computed using the GHKL protocol, for any choice of parameters* (α, X_{real}), *with respect to the specific simulation strategy used in Theorem 3.1.*

Proof: We first consider the protocol where P_1 plays the party that inputs $x \in X$ and P_2 inputs $y \in Y$ (that is, P_2 is the second to receive output, exactly as GHKL protocol is described in Section 3). Fix any X_{real}, α, and let $\mathbf{c} = (p_{y_1}, \ldots, p_{y_m}) = X_{real} \cdot M_f$. First, observe that $\mathbf{conv}(\{X_1, \ldots, X_\ell\}) \subseteq \mathcal{H}(\mathbf{q}, \delta_2)$, since for any point $Z \in \mathbf{conv}(\{X_1, \ldots, X_\ell\})$, we can represent Z as $\mathbf{a} \cdot M_f$ for some probability vector $\mathbf{a}$. Then, we have that $\langle Z, \mathbf{q} \rangle = \langle \mathbf{a} \cdot M_f, \mathbf{q} \rangle = \mathbf{a} \cdot \delta_2 \cdot \mathbf{1}_\ell = \delta_2$ and so $Z \in \mathcal{H}(\mathbf{q}, \delta_2)$. Now, assume by contradiction that the set of equations is satisfied. This implies that $Q^{x,a} \in \mathcal{H}(\mathbf{q}, \delta_2)$ for every $x \in X$, $a \in \{0, 1\}$, since $Q^{x,a} \in \mathbf{conv}(\{X_1, \ldots, X_\ell\}) \subseteq \mathcal{H}(\mathbf{q}, \delta_2)$.

Let $\circ$ denote the entrywise product over $\mathbb{R}^m$, that is for $Z = (z_1, \ldots, z_m)$, $W = (w_1, \ldots, w_m)$, the point $Z \circ W$ is defined as $(z_1 \cdot w_1, \ldots, z_m \cdot w_m)$. Recall that $\mathbf{c} = (p_{y_1}, \ldots, p_{y_m})$. We claim that for every X_i, the point $\mathbf{c} \circ X_i$ is also in the hyperplane $\mathcal{H}(\mathbf{q}, \delta_2)$. This trivially holds if $X_i = \mathbf{1}_m$. Otherwise, recall the definition of $Q^{x_i, 0}$ (Eq. (1)):

$$q_{y_j}^{x_i, 0} \overset{\text{def}}{=} \begin{cases} p_{y_j} & \text{if } f(x_i, y_j) = 1 \\ p_{y_j} + \frac{\alpha \cdot p_{y_j}}{(1-\alpha) \cdot (1-p_{x_i})} & \text{if } f(x_i, y_j) = 0 \end{cases} ,$$

Since $X_i \neq \mathbf{1}_m$, it holds that $p_{x_i} \neq 1$. Let $\gamma = \frac{\alpha}{(1-\alpha) \cdot (1-p_{x_i})}$. We can write $Q^{x,0}$ as follows:

$$Q^{x,0} = (1 + \gamma) \cdot \mathbf{c} - \gamma \cdot (\mathbf{c} \circ X_i) .$$

Since for every i, the point $Q^{x_i, 0}$ is in the hyperplane $\mathcal{H}(\mathbf{q}, \delta_2)$, we have:

$$\delta_2 = \langle Q^{x,0}, \mathbf{q} \rangle = \langle (1 + \gamma) \cdot \mathbf{c} - \gamma \cdot (\mathbf{c} \circ X_i), \mathbf{q} \rangle = (1 + \gamma) \cdot \langle \mathbf{c}, \mathbf{q} \rangle - \gamma \cdot \langle \mathbf{c} \circ X_i, \mathbf{q} \rangle$$
$$= (1 + \gamma) \cdot \delta_2 - \gamma \cdot \langle \mathbf{c} \circ X_i, \mathbf{q} \rangle$$

and thus, $\langle \mathbf{c} \circ X_i, \mathbf{q} \rangle = \delta_2$ which implies that $\mathbf{c} \circ X_i \in \mathcal{H}(\mathbf{q}, \delta_2)$.

We conclude that all the points $(\mathbf{c} \circ X_1), \ldots, (\mathbf{c} \circ X_\ell)$ are in the hyperplane $\mathcal{H}(\mathbf{q}, \delta_2)$. Since all the points $Y_1, \ldots, Y_m$ are in $\mathcal{H}(\mathbf{p}, \delta_1)$, it holds that $\mathbf{p} \cdot M_f = \delta_1 \cdot \mathbf{1}_m$. Thus, $\sum_{i=1}^\ell p_i \cdot X_i = \delta_1 \cdot \mathbf{1}_m$, which implies that:

$$\sum_{i=1}^\ell p_i \cdot \delta_2 = \sum_{i=1}^\ell p_i \cdot \left\langle \mathbf{c} \circ X_i, \mathbf{q} \right\rangle = \left\langle \sum_{i=1}^\ell p_i \cdot (\mathbf{c} \circ X_i), \mathbf{q} \right\rangle = \left\langle \mathbf{c} \circ \left(\sum_{i=1}^\ell p_i \cdot X_i \right), \mathbf{q} \right\rangle$$
$$= \langle \mathbf{c} \circ (\delta_1 \cdot \mathbf{1}_m), \mathbf{q} \rangle = \delta_1 \cdot \langle \mathbf{c}, \mathbf{q} \rangle = \delta_1 \cdot \delta_2$$

and thus it must hold that either $\sum_{i=1}^{\ell} p_i = \delta_1$ or $\delta_2 = 0$, which implies that $1 \in \mathcal{H}(\mathbf{p}, \delta_1)$ or $0 \in \mathcal{H}(\mathbf{q}, \delta_2)$, in contradiction to the additional requirements.

The above shows that the protocol does not hold when the P_1 party is the first to receive output. We can change the roles and let P_2 to be the first to receive an output (that is, we can use the protocol to compute f^T). In such a case, we will get that it must hold that $\sum_{i=1}^{m} q_i = \delta_2$ or $\delta_1 = 0$, again, in contradiction to the assumptions that $1 \notin \mathcal{H}(\mathbf{q}, \delta_2)$ and $0 \notin \mathcal{H}(\mathbf{p}, \delta_1)$. ∎

This negative result does not rule out the possibility of these functions using some other protocol. However, it rules out the only known possibility result that we have in fairness. Moreover, incorporating this with the characterization of coin-tossing [3], there exists a large set of functions for which the only possibility result does not hold, and the only impossibility result does not hold either. Moreover, this class of functions shares similar (but yet distinct) algebraic structure with the class of functions that imply fair coin-tossing. See more in Subsection 4.3.

Our theorem does not hold in cases where either $\mathbf{0}_\ell \in \mathcal{H}(\mathbf{p}, \delta_1)$ or $\mathbf{1}_\ell \in \mathcal{H}(\mathbf{p}, \delta_1)$ (likewise, for $\mathcal{H}(\mathbf{q}, \delta_2)$). These two requirements are in some sense equivalent. This is because the alphabet is not significant, and we can switch between the two symbols 0 and 1. Thus, if for some function f the hyperplane $\mathcal{H}(\mathbf{p}, \delta_1)$ passes through the origin $\mathbf{0}$, the corresponding hyperplane for the function $\bar{f}(x, y) = 1 - f(x, y)$ passes through $\mathbf{1}$ and vice versa. Feasibility of fairness for f and $\bar{f}$ is equivalent.

On the Number of Functions That Satisfy the Additional Requirements. We now count on the number of functions with $|X| = |Y|$ that satisfy these additional requirements, that is, define hyperplanes that do not pass through the origin $\mathbf{0}$ and the point $\mathbf{1}$. As we have seen in Theorem 4.8, these functions cannot be computed with complete fairness using the protocol of [17]. As we will see, only negligible amount of functions with $|X| = |Y|$ do not satisfy these additional requirements. Thus, our characterization of [17] is almost tight: Almost all functions with $|X| \neq |Y|$ can be computed fairly, whereas almost all functions with $|X| = |Y|$ cannot be computed using the protocol of [17]. We have the following Claim:

Claim 4.9. *With a probability that tends to 0 when $|X|, |Y|$ grow, a random function with $|X| = |Y|$ define hyperplanes that pass through the points $\mathbf{0}$ or $\mathbf{1}$.*

Proof: Let $m = |X| = |Y|$. Recall that P_m denotes the probability that a random m vectors of size m are linearly dependent. Moreover, by Claim 4.7, the probability that a random set $\{X_1, \ldots, X_{m+1}\}$ is affinely independent with probability $1 - P_m$, even when one of the points is chosen arbitrarily.

Thus, with probability P_m, the set $\{X_1, \ldots, X_m, \mathbf{1}\}$ where $X_1, \ldots, X_m$ are chosen at random is affinely dependent. In this case, the hyperplane defined by $\{X_1, \ldots, X_m\}$ contains the point $\mathbf{1}$. Similarly, the set $\{X_1, \ldots, X_m, \mathbf{0}\}$ is affinely dependent with the same probability P_m. Overall, using union-bound, the probability that the hyperplane of random points $X_1, \ldots, X_m$ contains the points $\mathbf{1}$ or

0 is negligible. From similar arguments, the probability that the hyperplane that is defined by the columns of the matrix contains either **1** or **0** is also negligible.

∎

Functions with Monochromatic Input. We consider a limited case where the above requirements do not satisfy, that is, functions that are not full-dimensional but define hyperplanes that pass through **0** or **1**. For this set of functions, the negative result does not apply. We now show that for some subset in this class, fairness is possible. Our result here does not cover all functions in this subclass.

Assume that a function contains a "monochromatic input", that is, one party has an input that causes the same output irrespectively to the input of the other party. For instance, P_2 has input y_j such that for every $x \in X$: $f(x, y_j) = 1$. In this case, the point $\mathbf{1}_\ell$ is one of the columns of the matrix, and therefore, the hyperplane $\mathcal{H}(\mathbf{p}, \delta_1)$ must pass through it. We show that in this case we can ignore this input and consider the "projected" $m \times (m-1)$ function f' where we remove the input y_j. This latter function may now be full-dimensional, and the existence of a protocol for f' implies the existence of a protocol for f. Intuitively, this is because when P_2 uses y_j, the real-world adversary P_1 cannot bias its output since it is always 1. We have:

Claim 4.10. Let $f : X \times Y \to \{0, 1\}$, and assume that M_f contains the all-one (resp. all-zero) column. That is, there exists $y \in Y$ such that for every $\hat{x} \in X$, $f(\hat{x}, y) = 1$ (resp. $f(\hat{x}, y) = 0$).

If the function $f' : X \times Y' \to \{0, 1\}$, where $Y' = Y \setminus \{y\}$ is full-dimensional, then f can be computed with complete-fairness.

Proof: Assume that the function contains the all one column, and that it is obtained by input y_m (i.e., the mth column is the all-one column). Let $X_1, \ldots, X_m$ be the rows of M_f, and let X_i' be the rows over $\mathbb{R}^{m-1}$ without the last coordinate, that is, $X_i = (X_i', 1)$. Consider the "projected" function $f' : \{x_1, \ldots, x_m\} \times \{y_1, \ldots, y_{m-1}\} \to \{0, 1\}$ be defined as $f'(x, y) = f(x, y)$, for every x, y in the range (we just remove y_m from the possible inputs of P_2). The rows of $M_{f'}$ are $X_1', \ldots, X_m'$.

Now, since f' is of full-dimensional, the function f' can be computed using the GHKL protocol. Let $X_{ideal}^{x,a}$ be the solutions for equations of Theorem 3.1 for the function f'. It can be easily verified that $X_{ideal}^{x,a}$ are the solutions for equations for the f function as well, since for every x, a, the first $m-1$ coordinates of $Q^{x,a}$ are the same as f', and the last coordinate of $Q^{x,a}$ is always 1. For $Q^{x,0}$ it holds immediately, for $Q^{x,1}$ observe that $p_{y_m} = 1$ no matter what X_{real} is, and thus $p_{y_j} + \frac{\alpha \cdot (p_{y_j} - 1)}{(1-\alpha) \cdot p_x} = 1 + 0 = 1$). Therefore, $X_{ideal}^{x,a}$ are the solutions for f as well, and Theorem 3.1 follows for f as well.   ∎

The above implies an interesting and easy to verify criterion:

Proposition 4.11. Let $f : \{x_1, \ldots, x_m\} \times \{y_1, \ldots, y_m\} \to \{0, 1\}$ be a function. Assume that f contains the all-one column, and that M_f is of full rank. Then, the function f can be computed with complete fairness.

Proof: Let $X_1, \ldots, X_m$ be the rows of M_f, and assume that the all-one column is the last one (i.e., input y_m). Consider the points $X'_1, \ldots, X'_m$ in $\mathbb{R}^{m-1}$, where for every i, $X_i = (X'_i, 1)$ (i.e., X'_i is the first $m-1$ coordinates of X_i). Since M_f is of full-rank, the rows $X_1, \ldots, X_m$ are linearly independent, which implies that m points $X'_1, \ldots, X'_m$ in $\mathbb{R}^{m-1}$ are affinely independent. We therefore can apply Claim 4.10 and fairness in f is possible.   ■

Finally, from simple symmetric properties, almost *always* a random matrix that contains the all one row / vector is of full rank, in the sense that we have seen in Claims 4.7 and 4.9. Therefore, almost always a random function that contains a monochromatic input can be computed with complete fairness.

4.3 Conclusion: Symmetric Boolean Functions with Finite Domain

We summarize all the known results in complete fairness for symmetric boolean functions with finite domain, and we link our results to the balanced property of [3].

Characterization of Coin-tossing [3]. The work of Asharov, Lindell and Rabin [3] considers the task of coin-tossing, which was shown to be impossible to compute fairly [9]. The work provides a simple property that indicates whether a function implies fair coin-tossing or not. If the function satisfies the property, then the function implies fair coin tossing, in the sense that a fair protocol for the function implies the existence of a fair protocol for coin-tossing, and therefore it cannot be computed fairly by Cleve's impossibility. On the other hand, if a function f does not satisfy the property, then for any protocol for coin-tossing in the f-hybrid model there exists an (inefficient) adversary that biases the output of the honest party. Thus, the function does not imply fair coin-tossing, and may potentially be computed with complete fairness. The results hold also for the case where the parties have an ideal access to Oblivious Transfer [24,11]. The property that [3] has defined is as follows:

Definition 4.12 (strictly-balanced property [3]). *Let* $f : \{x_1, \ldots, x_\ell\} \times \{y_1, \ldots, y_m\} \to \{0,1\}$ *be a function. We say that the function is* balanced with respect to probability vectors *if there exist probability vectors* $\mathbf{p} = (p_1, \ldots, p_\ell)$, $\mathbf{q} = (q_1, \ldots, q_m)$ *and a constant* $0 < \delta < 1$ *such that:*

$$\mathbf{p} \cdot M_f = \delta \cdot \mathbf{1}_m \qquad \text{and} \qquad M_f \cdot \mathbf{q}^T = \delta \cdot \mathbf{1}_\ell^T .$$

Intuitively, if such probability vectors exist, then in a single execution of the function f, party P_1 can choose its input according to distribution $\mathbf{p}$ which fixes the output distribution vector of P_2 to be $\delta \cdot \mathbf{1}_m$. This means that no matter what input (malicious) P_2 uses, the output is 1 with probability δ. Likewise, honest P_2 can choose its input according to distribution $\mathbf{q}$, and malicious P_1 cannot bias the result. We therefore obtain a fair coin-tossing protocol. On the other hand, [3] shows that if the function does not satisfy the condition above, then there always exists a party that can bias the result of any coin-tossing protocol that can be constructed using f.

The Characterization. A full-dimensional function is an important special case of this unbalanced property, as was pointed out in Claim 4.2. Combining the above characterization of [3] with ours, we get the following Theorem:

Theorem 4.13. *Let* $f : \{x_1, \ldots, x_\ell\} \times \{y_1, \ldots, y_m\} \to \{0, 1\}$, *and let* M_f *be the corresponding matrix representing* f *as above. Then:*

1. **Balanced with respect to probability vectors [3]:**
   If there exist probability vectors $\mathbf{p} = (p_1, \ldots, p_\ell), \mathbf{q} = (q_1, \ldots, q_m)$ *and a constant* $0 < \delta < 1$ *such that:*

$$\mathbf{p} \cdot M_f = \delta \cdot \mathbf{1}_m \qquad \text{and} \qquad M_f \cdot \mathbf{q}^T = \delta \cdot \mathbf{1}_\ell^T .$$

   Then, the function f *implies fair coin-tossing, and is impossible to compute fairly.*

2. **Balanced with respect to arbitrary vectors, but not balanced with respect to probability vectors:**
   If there exist two non-zero vectors $\mathbf{p} = (p_1, \ldots, p_\ell) \in \mathbb{R}^\ell$, $\mathbf{q} = (q_1, \ldots, q_m) \in \mathbb{R}^m$, $\delta_1, \delta_2 \in \mathbb{R}$, *such that:*

$$\mathbf{p} \cdot M_f = \delta_1 \cdot \mathbf{1}_m \qquad \text{and} \qquad M_f \cdot \mathbf{q}^T = \delta_2 \cdot \mathbf{1}_\ell^T$$

   then we say that the function is balanced with respect to arbitrary vectors. Then, the function does not (information-theoretically) imply fair-coin tossing [3]. Moreover:

   (a) *If* δ_1 *and* δ_2 *are non-zero,* $\sum_{i=1}^\ell p_i \neq \delta_1$ *and* $\sum_{i=1}^m q_i \neq \delta_2$, *then the function* f *cannot be computed using the GHKL protocol (Theorem 4.8).*

   (b) *Otherwise: this case is left not characterized. For a subset of this subclass, we show possibility (Proposition 4.10).*

3. **Unbalanced with respect to arbitrary vectors:**
   If for every non-zero $\mathbf{p} = (p_1, \ldots, p_\ell) \in \mathbb{R}^\ell$ *and any* $\delta_1 \in \mathbb{R}$ *it holds that:* $\mathbf{p} \cdot M_f \neq \delta_1 \cdot \mathbf{1}_m$, **OR** *for every non-zero* $\mathbf{q} = (q_1, \ldots, q_m) \in \mathbb{R}^m$ *and any* $\delta_2 \in \mathbb{R}$ *it holds that:* $M_f \cdot \mathbf{q}^T \neq \delta_2 \cdot \mathbf{1}_\ell^T$, *then* f *can be computed with complete fairness (Theorem 4.4).*

We remark that in general, if $|X| \neq |Y|$ then almost always a random function is in subclass 3. Moreover, if $|X| = |Y|$, only negligible amount of functions are in subclass 2b, and thus only negligible amount of functions are left not characterized.

If a function is balanced with respect to arbitrary vectors (i.e., the vector may contain negative values), then all the rows of the function lie in the hyperplane $\mathcal{H}(\mathbf{q}, \delta_2)$, and all the columns lie in the hyperplane $\mathcal{H}(\mathbf{p}, \delta_1)$. Observe that $\delta_1 = 0$ if and only if $\mathcal{H}(\mathbf{p}, \delta_1)$ passes through the origin, and $\sum_{i=1}^\ell p_i = \delta_1$ if and only if $\mathcal{H}(\mathbf{p}, \delta_1)$ passes through the all one point $\mathbf{1}$. Thus, the requirements of subclass 2a are a different formalization of the requirements of Theorem 4.8. Likewise, the requirements of subclass 3 are a different formalization of Theorem 4.4, as was proven in Claim 4.2.

5 Extensions: Asymmetric Functions and Non-binary Outputs

5.1 Asymmetric Functions

We now move to a richer class of functions, and consider asymmetric boolean functions where the parties do not necessarily get the same output. We consider functions $f(x,y) = (f_1(x,y), f_2(x,y))$, where each f_i, $i \in \{1,2\}$ is defined as: f_i : $\{x_1, \ldots, x_\ell\} \times \{y_1, \ldots, y_m\} \rightarrow \{0,1\}$. Interestingly, our result here shows that if the function f_2 is of full-dimensional, then f can be computed fairly, irrespectively to the function f_1. This is because simulating P_1 is more challenging (because it is the first to receive an output) and the simulator needs to assume the rich description of f_2 in order to be able to bias the output of the honest party P_2. On the other hand, since P_2 is the second to receive an output, simulating P_2 is easy and the simulator does not need to bias the output of P_1, thus, nothing is assumed about f_1.

In the full version of this paper [2], we revise the protocol of [17] to deal with this class of functions. This is done in a straightforward way, where the online dealer computes at each round the value a_i according to the function f_1, and b_i according to f_2. We then derive a set of equations, similarly to Eq. (1) and obtain an analogue theorem to Theorem 3.1. We then show the following Corollary:

Corollary 5.1. *Let $f : \{x_1, \ldots, x_\ell\} \times \{y_1, \ldots, y_m\} \rightarrow \{0,1\} \times \{0,1\}$, where $f = (f_1, f_2)$. If f_2 is a full-dimensional function, then f can be computed with complete fairness.*

5.2 Functions with Non-binary Output

Until now, all the known possibility results in fairness deal with the case of binary output. We now extend the results to the case of non-binary output. Let $\Sigma = \{\sigma_1, \ldots, \sigma_k\}$ be an alphabet for some finite $k > 0$, and consider functions $f : \{x_1, \ldots, x_\ell\} \times \{y_1, \ldots, y_m\} \rightarrow \Sigma$.

The protocol is exactly the GHKL protocol presented in Section 3, where here a_i, b_i are elements in Σ and not just bits. However, the analysis for this case is more involved. For instance, in the binary case for every input $y_j \in Y$, we considered the parameter p_{y_j}, the probability that P_2 receives 1 in each round before i^* when its input is y_j. In the non-binary case, we have to define an equivalent parameter $p_{y_j}(\sigma)$ for *any* symbol σ in the alphabet Σ (i.e., the probability that P_2 receives σ in each round before i^* when its input is y_j). This makes things harder, and in order to obtain fairness, several requirements should be satisfied simultaneously for every $\sigma \in \Sigma$.

In order to see this, fix some X_{real}. For any symbol $\sigma \in \Sigma$, and for every $y_j \in Y$, let $p_{y_j}(\sigma)$ denote the probability that b_i is σ (when $i \leq i^*$). That is:

$$p_{y_j}(\sigma) \stackrel{\text{def}}{=} \Pr_{\hat{x} \leftarrow X_{real}} [f(\hat{x}, y_j) = \sigma] .$$

Observe that $\sum_{\sigma \in \Sigma} p_{y_j}(\sigma) = 1$. For every $\sigma \in \Sigma$, we want to represent the vector $(p_{y_1}(\sigma), \ldots, p_{y_m}(\sigma))$ as a function of X_{real} and M_f, as we did in the binary case (where there we just had: $(p_{y_1}, \ldots, p_{y_m}) = X_{real} \cdot M_f$). However, here M_f does not represent exactly what we want, and the multiplication $X_{real} \cdot M_f$ gives the "expected output distribution vector" and not exactly what we want. Instead, for any $\sigma \in \Sigma$ we define the binary matrix M_f^σ as follows:

$$M_f^\sigma(i, j) = \begin{cases} 1 & \text{if } f(x_i, y_j) = \sigma \\ 0 & \text{otherwise} \end{cases}.$$

Now, we can represent the vector $(p_{y_1}(\sigma), \ldots, p_{y_m}(\sigma))$ as $X_{real} \cdot M_f^\sigma$. However, here a *single* vector X_{real} determines the values of $|\Sigma|$ vectors, one for each $\sigma \in \Sigma$. Therefore, we overall get $|\Sigma|$ systems of equations, one for each symbol in the alphabet. In [2] we show that it is enough to consider only $|\Sigma| - 1$ systems since our probabilities sum-up to 1 (i.e., $\sum_{\sigma \in \Sigma} p_{y_j}(\sigma) = 1$), and provide the sets of equations that guarantees fairness. In the following, we provide a corollary of our result which provides a simpler criterion.

Given a function $f : X \times Y \to \Sigma$, let $\rho \in \Sigma$ be arbitrarily, and define $\Sigma_\rho = \Sigma \setminus \{\rho\}$. Define the *boolean* function $f' : X \times Y^{\Sigma_\rho} \to \{0, 1\}$, where $Y^{\Sigma_\rho} = \{y_j^\sigma \mid y_j \in Y, \sigma \in \Sigma_\rho\}$, as follows:

$$f'(x, y_j^\sigma) = \begin{cases} 1 & \text{if } f(x, y_j) = \sigma \\ 0 & \text{otherwise} \end{cases}$$

Observe that $|Y^{\Sigma_\rho}| = (|\Sigma| - 1) \cdot |Y|$. We show that if the boolean function f' is full-dimensional, then the function f can be computed with complete-fairness. Observe that this property can be satisfied only when $|X|/|Y| > |\Sigma| - 1$.

An Example. We give an example for a non-binary function that can be computed with complete-fairness. We consider trinary alphabet $\Sigma = \{0, 1, 2\}$, and thus we consider a function of dimensions 5×2. We provide the trinary function f and the function f' that it reduced to. Since the binary function f' is a full-dimensional function in $\mathbb{R}^4$, it can be computed fairly, and thus the trinary function f can be computed fairly as well. We have:

f	y_1	y_2		f'	y_1^1	y_2^1	y_1^2	y_2^2
x_1	0	1		x_1	0	1	0	0
x_2	1	0		x_2	1	0	0	0
x_3	1	1	$\implies$	x_3	1	1	0	0
x_4	2	0		x_4	0	0	1	0
x_5	1	2		x_5	1	0	0	1

Acknowledgement. The author gratefully thanks Eran Omri, Ran Cohen, Carmit Hazay, Yehuda Lindell and Tal Rabin for very helpful discussions and helpful comments.

References

1. Agrawal, S., Prabhakaran, M.: On fair exchange, fair coins and fair sampling. In: Canetti, R., Garay, J.A. (eds.) CRYPTO 2013, Part I. LNCS, vol. 8042, pp. 259–276. Springer, Heidelberg (2013)
2. Asharov, G.: Towards characterizing complete fairness in secure two-party computation. IACR Cryptology ePrint Archive (to appear)
3. Asharov, G., Lindell, Y., Rabin, T.: A full characterization of functions that imply fair coin tossing and ramifications to fairness. In: Sahai, A. (ed.) TCC 2013. LNCS, vol. 7785, pp. 243–262. Springer, Heidelberg (2013)
4. Beimel, A., Lindell, Y., Omri, E., Orlov, I.: $1/p$-secure multiparty computation without honest majority and the best of both worlds. In: Rogaway, P. (ed.) CRYPTO 2011. LNCS, vol. 6841, pp. 277–296. Springer, Heidelberg (2011)
5. Ben-Or, M., Goldreich, O., Micali, S., Rivest, R.L.: A fair protocol for signing contracts. IEEE Transactions on Information Theory 36(1), 40–46 (1990)
6. Ben-Or, M., Goldwasser, S., Wigderson, A.: Completeness theorems for non-cryptographic fault-tolerant distributed computation (extended abstract). In: STOC, pp. 1–10 (1988)
7. Canetti, R.: Security and composition of multiparty cryptographic protocols. J. Cryptology 13(1), 143–202 (2000)
8. Chaum, D., Crépeau, C., Damgård, I.: Multiparty unconditionally secure protocols (extended abstract). In: STOC, pp. 11–19 (1988)
9. Cleve, R.: Limits on the security of coin flips when half the processors are faulty (extended abstract). In: STOC, pp. 364–369 (1986)
10. Cleve, R.: Controlled gradual disclosure schemes for random bits and their applications. In: Brassard, G. (ed.) CRYPTO 1989. LNCS, vol. 435, pp. 573–588. Springer, Heidelberg (1990)
11. Even, S., Goldreich, O., Lempel, A.: A randomized protocol for signing contracts. In: CRYPTO, pp. 205–210 (1982)
12. Even, S., Yacobi, Y.: Relations among public key signature schemes. Technical Report #175, Technion Israel Institute of Technology, Computer Science Department (1980), http://www.cs.technion.ac.il/users/wwwb/cgi-bin/trinfo.cgi/1980/CS/CS0175
13. Goldreich, O.: The Foundations of Cryptography - Basic Applications, vol. 2. Cambridge University Press (2004)
14. Goldreich, O., Micali, S., Wigderson, A.: How to play any mental game or a completeness theorem for protocols with honest majority. In: STOC, pp. 218–229 (1987)
15. Goldwasser, S., Levin, L.: Fair computation of general functions in presence of immoral majority. In: Menezes, A.J., Vanstone, S.A. (eds.) CRYPTO 1990. LNCS, vol. 537, pp. 77–93. Springer, Heidelberg (1991)
16. Gordon, S.D.: On fairness in secure computation. PhD thesis, University of Maryland (2010)
17. Gordon, S.D., Hazay, C., Katz, J., Lindell, Y.: Complete fairness in secure two-party computation. In: STOC, pp. 413–422 (2008); Extended full version available on: http://eprint.iacr.org/2008/303. Journal version: [18]
18. Gordon, S.D., Hazay, C., Katz, J., Lindell, Y.: Complete fairness in secure twoparty computation. J. ACM 58(6), 24 (2011)
19. Gordon, S.D., Katz, J.: Partial fairness in secure two-party computation. In: Gilbert, H. (ed.) EUROCRYPT 2010. LNCS, vol. 6110, pp. 157–176. Springer, Heidelberg (2010)

20. Grünbaum, B.: Convex Polytopes. In: Kaibel, V., Klee, V., Ziegler, G. (eds.) Graduate Texts in Mathematics, 2nd edn. Springer (May 2003)
21. Halpern, J.Y., Teague, V.: Rational secret sharing and multiparty computation: extended abstract. In: STOC, pp. 623–632 (2004)
22. Kahn, J., Komlòs, J., Szemerèdi, E.: On the probability that a random ±1-matrix is singular. Journal of Amer. Math. Soc. 8, 223–240 (1995)
23. Komlòs, J.: On the determinant of $(0, 1)$ matrices. Studia Sci. Math. Hungar 2, 7–21 (1967)
24. Rabin, M.O.: How to exchange secrets with oblivious transfer. Technical Report TR-81, Aiken Computation Lab, Harvard University (1981)
25. Rabin, T., Ben-Or, M.: Verifiable secret sharing and multiparty protocols with honest majority (extended abstract). In: STOC, pp. 73–85 (1989)
26. Roman, S.: Advanced Linear Algebra, 3rd edn. Graduate Texts in Mathematics, vol. 135, p. xviii. Springer, New York (2008)
27. Voigt, T., Ziegler, G.M.: Singular 0/1-matrices, and the hyperplanes spanned by random 0/1-vectors. Combinatorics, Probability and Computing 15(3), 463–471 (2006)
28. Wood, P.J.: On the probability that a discrete complex random matrix is singular. PhD thesis, Rutgers University, New Brunswick, NJ, USA, AAI3379178 (2009)
29. Yao, A.C.C.: How to generate and exchange secrets (extended abstract). In: FOCS, pp. 162–167 (1986)
30. Ziegler, G.M.: Lectures on 0/1-polytopes. Polytopes: Combinatorics and Computation, Birkhauser, Basel. DMV Seminar, vol. 29, pp. 1–40 (2000)

On the Cryptographic Complexity
of the Worst Functions*

Amos Beimel[1], Yuval Ishai[2], Ranjit Kumaresan[2], and Eyal Kushilevitz[2]

[1] Dept. of Computer Science, Ben Gurion University of the Negev, Be'er Sheva, Israel
amos.beimel@gmail.com
[2] Department of Computer Science, Technion, Haifa, Israel
{yuvali,ranjit,eyalk}@cs.technion.ac.il

Abstract. We study the complexity of realizing the "worst" functions in several standard models of information-theoretic cryptography. In particular, for the case of security against passive adversaries, we obtain the following main results.

- **OT complexity of secure two-party computation.** Every function $f : [N] \times [N] \to \{0,1\}$ can be securely evaluated using $\tilde{O}(N^{2/3})$ invocations of an oblivious transfer oracle. A similar result holds for securely sampling a uniform pair of outputs from a set $S \subseteq [N] \times [N]$.
- **Correlated randomness complexity of secure two-party computation.** Every function $f : [N] \times [N] \to \{0,1\}$ can be securely evaluated using $2^{\tilde{O}(\sqrt{\log N})}$ bits of correlated randomness.
- **Communication complexity of private simultaneous messages.** Every function $f : [N] \times [N] \to \{0,1\}$ can be securely evaluated in the non-interactive model of Feige, Kilian, and Naor (STOC 1994) with messages of length $O(\sqrt{N})$.
- **Share complexity of forbidden graph access structures.** For every graph G on N nodes, there is a secret-sharing scheme for N parties in which each pair of parties can reconstruct the secret if and only if the corresponding nodes in G are connected, and where each party gets a share of size $\tilde{O}(\sqrt{N})$.

The worst-case complexity of the best previous solutions was $\Omega(N)$ for the first three problems and $\Omega(N/\log N)$ for the last one. The above results are obtained by applying general transformations to variants of private information retrieval (PIR) protocols from the literature, where different flavors of PIR are required for different applications.

1 Introduction

How bad are the worst functions? For most standard complexity measures of boolean functions, the answer to this question is well known. For instance, the circuit complexity of the worst function $f : [N] \to \{0,1\}$ is $\Theta(N/\log N)$ [53,49] and

---

* Research by the first three authors received funding from the European Union's Tenth Framework Programme (FP10/2010-2016) under grant agreement no. 259426 ERC-CaC. The fourth author was supported by ISF grant 1361/10 and BSF grant 2008411.

Y. Lindell (Ed.): TCC 2014, LNCS 8349, pp. 317–342, 2014.

the two-party communication complexity of the worst function $f : [N] \times [N] \to \{0, 1\}$ is $\Theta(\log N)$ in every standard model of communication complexity [48].[1] In sharp contrast, this question is wide open for most natural complexity measures in information-theoretic cryptography that involve *communication* or *randomness* rather than computation. Standard counting techniques or information inequalities only yield very weak lower bounds, whereas the best known upper bounds are typically linear in the size of the input domain (and exponential in the bit-length of the inputs).

The only exceptions to this state of affairs are in the context of secure multi-party computation where it is known that, when a big majority of honest parties is guaranteed, the communication and randomness complexity can always be made sublinear in the input domain size [5,40] (see Section 1.2 for discussion of these and other related works). However, no similar results were known for secure computation with no honest majority and, in particular, in the two-party case.

In the present work we study the complexity of the worst-case functions in several standard models for information-theoretic secure two-party computation, along with a related problem in the area of secret sharing.

We restrict our attention to security against *passive* (aka semi-honest) adversaries. We will usually also restrict the attention to deterministic two-party functionalities captured by boolean functions $f : [N] \times [N] \to \{0, 1\}$, where the output is learned by both parties.[2] In the following, the term "secure" will refer by default to perfect security in the context of positive results and to statistical security in the case of negative results. In this setting, we consider the following questions.

OT Complexity. The first model we consider is secure two-party computation in the *OT-hybrid model*, namely in a model where an ideal oracle implementing 1-out-of-2 oblivious transfer [52,29] (of bits) is available. Secure computation in this model is motivated by the possibility of realizing OT using noisy communication channels [22], the equivalence between OT and a large class of complete functionalities [45,46], and the possibility of efficiently precomputing [4] and (in the computational setting) extending [3,37] OTs. See [43] for additional motivating discussion.

Viewing OT as an "atomic currency" for secure two-party computation, it is natural to study the minimal number of OT calls required for securely computing a given two-party functionality f. We refer to this quantity as the *OT complexity* of f. Special cases of this question were studied in several previous works (e.g., [25,3,56]), and a more systematic study was conducted in [12,51].

---

[1] Here and in the following, we let $[N]$ denote the set $\{1, 2, \ldots, N\}$ and naturally identify an input $x \in [N]$ with a $\lceil \log_2 N \rceil$-bit binary representation.

[2] Using standard reductions (cf. [31]), our results can be extended to general (possibly randomized or even reactive) functionalities that may deliver different outputs to the two parties. While some of our results can also be extended to the case of k-party secure computation, we focus here on the two-party case for simplicity.

The GMW protocol [32,33,31] shows that the OT complexity of any function f is at most twice the size of the smallest boolean circuit computing f. For most functions $f : [N] \times [N] \to \{0,1\}$, this only gives an upper bound of $O(N^2 / \log N)$ on the OT complexity.[3]

A simpler and better upper bound can be obtained by using 1-out-of-N OT (denoted $\binom{N}{1}$-OT). Concretely, the first party P_1, on input x_1, prepares a truth-table of the function $f_{x_1}(x_2)$ obtained by restricting f to its own input, and using $\binom{N}{1}$-OT lets the second party P_2 select the entry of this table indexed by x_2. Since $\binom{N}{1}$-OT can be reduced to $N - 1$ instances of standard OT [17], we get an upper bound of $N - 1$ on the OT complexity of the worst-case f. This raises the following question:

Question 1. What is the OT complexity of the worst function $f : [N] \times [N] \to \{0,1\}$? In particular, can every such f be securely realized using $o(N)$ OTs?

Given the existence of constant-rate reductions between OT and any finite complete functionality [35,42], the answer to Question 1 remains the same, up to a constant multiplicative factor, even if the OT oracle is replaced by a different complete functionality, such as binary symmetric channel. In particular, the OT complexity of f is asymptotically the same as the "AND complexity" of f considered in [12].

We will also be interested in a sampling variant of Question 1, where the goal is to securely sample from some probability distribution over output pairs from $[N] \times [N]$ using a minimal number of OTs. This captures the rate of securely reducing complex correlations to simple ones, a question which was recently studied in [51].

Correlated Randomness Complexity. The second model we consider is that of secure two-party computation with an arbitrary source of correlated randomness. That is, during an offline phase, which takes place before the inputs are known, the two parties are given a pair of random strings (r_1, r_2) drawn from some fixed joint distribution, where r_i is known only to P_i. During the online phase, once the inputs (x_1, x_2) are known, the parties can use their correlated random inputs, possibly together with independent secret coins, to securely evaluate f. This model can be viewed as a relaxation of the OT-hybrid model discussed above, since each OT call is easy to realize given correlated randomness corresponding to a random instance of OT [4]. The model is motivated by the possibility of generating the correlated randomness using semi-trusted servers or a (computationally) secure interactive protocol, thus capturing the goal of minimizing the online complexity of secure computation via offline preprocessing. See [14,41,24] for additional discussion.

General correlations have several known advantages over OT correlations in the context of secure computation. Most relevant to our work is a result from [41],

---

[3] The GMW protocol can handle XOR and NOT gates for free, but it is not clear if this can be used to significantly lower the complexity of the worst-case functions. A negative result in a restricted computation model is given in [20].

showing that for any $f : [N] \times [N] \to \{0,1\}$ there is a source of correlated randomness (r_1, r_2) given which f can be realized using only $O(\log N)$ bits of communication. However, the *correlated randomness complexity* of this protocol, namely the length of the random strings r_1, r_2, is $O(N^2)$. Minimizing the correlated randomness complexity is desirable because the correlated randomness needs to be communicated and stored until the online phase begins. The simple OT complexity upper bound discussed above also implies an $O(N)$ upper bound on the correlated randomness complexity of the worst functions. No better bound is known. This raises the following question:

Question 2. What is the correlated randomness complexity of the worst function $f : [N] \times [N] \to \{0,1\}$? In particular, can every such f be securely realized using $o(N)$ bits of correlated randomness?

Communication Complexity of Private Simultaneous Messages Protocols. Feige, Kilian, and Naor [30] considered the following non-interactive model for secure two-party computation. The two parties simultaneously send messages to an external referee, where the message of party P_i depends on its input x_i and a common source of randomness r that is kept secret from the referee. From the two messages it receives, the referee should be able to recover $f(x_1, x_2)$ but learn no additional information about x_1, x_2. Following [38], we refer to such a protocol as a *private simultaneous messages* (PSM) protocol for f. A PSM protocol for f can be alternatively viewed as a special type of randomized encoding of f [39,1], where the output of f is encoded by the output of a randomized function $\hat{f}((x_1, x_2); r)$ such that $\hat{f}$ can be written as $\hat{f}((x_1, x_2); r) = (\hat{f}_1(x_1; r), \hat{f}_2(x_2; r))$. This is referred to as a "2-decomposable" encoding in [36].

It was shown in [30] that every $f : [N] \times [N] \to \{0,1\}$ admits a PSM protocol with $O(N)$ bits of communication. While better protocols are known for functions that have small formulas or branching programs [30,38], this still remains the best known upper bound on the communication complexity of the worst-case functions, or even most functions, in this model. We thus ask:

Question 3. What is the PSM communication complexity of the worst function $f : [N] \times [N] \to \{0,1\}$? In particular, does every such f admit a PSM protocol which uses $o(N)$ bits of communication?

Share Complexity of Forbidden Graph Access Structures. A longstanding open question in information-theoretic cryptography is whether every (monotone) access structures can be realized by a secret-sharing scheme in which the share size of each party is polynomial in the number of parties. Here we consider a "scaled down" version of this question, where the access structure only specifies, for each *pair* of parties, whether this pair should be able to reconstruct the secret from its joint shares or learn nothing about the secret.[4] This type of graph-based

---

[4] In contrast to the more standard notion of graph-based access structures, we make no explicit requirement on bigger or smaller sets of parties. However, one can easily enforce the requirement that every single party learns nothing about the secret and every set of 3 parties can reconstruct the secret.

access structures was considered in [54] under the name "forbidden graph" access structures.

A simple way of realizing such an access structure is by independently sharing the secret between each authorized pair. For most graphs, this solution distributes a share of length $\Omega(N)$ to each party. This can be improved by using covers by complete bipartite graphs implying that every graph access structure can be realized by a scheme in which the share size of each party is $O(N/\log N)$ [18,16,28]. This raises the following question:

Question 4. What is share length required for realizing the worst graphs G? In particular, can every forbidden graph access structure on N nodes be realized by a secret-sharing scheme in which each party receives $o(N/\log N)$ bits?

1.1 Our Results

For each of the above questions, we obtain an improved upper bound. Our upper bounds are obtained by applying general transformations to variants of information-theoretic private information retrieval (PIR) protocols from the literature (see Section 1.2), where different flavors of PIR are required for different applications. At a high level, our results exploit new connections between 2-server PIR and OT complexity, between 3-server PIR and correlated randomness complexity, and between a special "decomposable" variant of 3-server PIR and PSM complexity. The secret sharing result is obtained by applying a general transformation to the PSM result, in the spirit of a transformation implicit in [9]. More concretely, we obtain the following main results.

OT Complexity of Secure Two-Party Computation. We show that every function $f : [N] \times [N] \to \{0,1\}$ can be securely evaluated using $\widetilde{O}(N^{2/3})$ invocations of an oblivious transfer oracle. In fact, the total communication complexity and randomness complexity of the protocol are also bounded by $\widetilde{O}(N^{2/3})$. We also obtain a similar result for securely sampling a uniform pair of outputs from a set $S \subseteq [N] \times [N]$. More generally and precisely, for any joint probability distribution (U, V) over $[N] \times [N]$ and any $\epsilon > 0$, we obtain an ϵ-secure protocol for sampling correlated outputs from (U, V) using $N^{2/3} \cdot \text{poly}(\log N, \log 1/\epsilon)$ OTs. This can be viewed as a nontrivial secure reduction of complex correlations (or "channels") to simple ones. These results apply the 2-server PIR protocol from [21]. See full version for more details.

Correlated Randomness Complexity of Secure Two-Party Computation. We show that every function $f : [N] \times [N] \to \{0,1\}$ can be securely evaluated using $2^{\widetilde{O}(\sqrt{\log N})}$ bits of correlated randomness. In fact, the same bound holds also for the total randomness complexity of the protocol (counting private independent coins as well) and also for the *communication* complexity of the protocol. This result applies the 3-server PIR protocol of [27]. It was previously observed in [30,41] that secure two-party computation with correlated randomness gives rise to a 3-server PIR protocol. Here we show a connection in the other direction.

Communication Complexity of Private Simultaneous Messages. We show that every function $f : [N] \times [N] \to \{0, 1\}$ can be realized by a PSM protocol with messages of length $O(\sqrt{N})$. The construction is based on a special "decomposable" variant of 3-server PIR which we realize by modifying a PIR protocol from [21]. In the hope of improving our $O(\sqrt{N})$ upper bound, we reduce the problem of decomposable 3-server PIR to a combinatorial question of obtaining a decomposable variant of matching vector sets [58,26]. See full version for more details. We leave open the existence of decomposable matching vector sets with good parameters.

In the terminology of randomized encoding of functions, the above result shows that every $f : [N] \times [N] \to \{0, 1\}$ admits a 2-decomposable randomized encoding of length $O(\sqrt{N})$. It is instructive to note that whereas previous PSM protocols from [30,38] employ a *universal* decoder (i.e., referee algorithm), which does not depend on the function f other than on a size parameter, the decoder in our construction strongly depends on f. It follows by a simple counting argument that this is inherent.

Share Complexity of Forbidden Graph Access Structures. We show that for every graph G with N nodes, the corresponding forbidden graph access structure can be realized by a secret-sharing scheme in which each party gets a share of size $\tilde{O}(\sqrt{N})$. This result is obtained by applying a general transformation to our new PSM protocols. Curiously, while our secret-sharing scheme is not linear, a simple generalization of a result of Mintz [50] implies a lower bound of $\Omega(\sqrt{N})$ on the share complexity of any *linear* scheme realizing the worst forbidden graph access structure. This extends a previous lower bound from [7] that applies to the stricter notion of graph-based access structures. The existence of *linear* secret-sharing schemes meeting this lower bound is left open.

1.2 Related Work

Prior to our work, the only previous context in which sublinear communication was known is that of secure multiparty computation in the presence of an honest majority. While the complexity of standard protocols [13,19] grows linearly with the circuit size, it is possible to do much better when there is a sufficiently large majority of honest parties. Beaver et al. [5] have shown that when only $\log n$ parties are corrupted, any function $f : \{0, 1\}^n \to \{0, 1\}$ can be securely evaluated using only poly(n) bits of communication and randomness, namely the complexity is polylogarithmic in the input domain size. Their technique makes an ad-hoc use of locally random reductions, which are in turn related to the problem of information-theoretic *private information retrieval* (PIR) [21]. A k-server PIR protocol allows a client to retrieve an arbitrary bit D_i from a database $D \in \{0, 1\}^N$, which is held by k servers, while hiding the selection i from each individual server. The main optimization goal for PIR protocols is their communication complexity, which is required to be sublinear in N.

Ishai and Kushilevitz [40] present a general method for transforming communication-efficient PIR protocols into communication-efficient secure multiparty protocols in which the number of parties is independent of the total input length n. In contrast to our constructions, which require the underlying PIR protocols to satisfy additional computational and structural requirements, the transformation from [40] is completely general. On the down side, it does not apply in the two-party case and it requires (information theoretic) PIR protocols with polylogarithmic communication, which are not known to exist for a constant number of servers k.

Beimel and Malkin [12] put forward the general goal of studying the minimal number of OTs/ANDs required for securely realizing a given two-party functionality f, observe that this quantity can be smaller in some cases than the circuit size of f, and obtain several connections between this question and communication complexity. These connections are mainly useful for proving lower bounds that are logarithmic in the domain size N or upper bounds for specific functions that have low communication complexity. More results in this direction are given in [44]. Prabhakaran and Prabhakaran [51] put forward the question of characterizing the rate of secure reductions between *sampling* functionalities, and strengthen previous negative results from [56] on the rate of secure reductions between different OT correlations. None of the above results give nontrivial upper bounds for the worst (or most) functions f. Winkler and Wulschlegger [56] prove an $\Omega(\log N)$ lower bound on the correlated randomness complexity of secure two-party computation. Except for very few functions, this lower bound is very far from the best known upper bounds even when considering the results of this work.

The complexity of secret sharing for graph-based access structures was extensively studied in a setting where the edges of the graph represent the *only* minimal authorized sets, that is, any set of parties that does not contain an edge should learn nothing about the secret. The notion of forbidden graph access structures we study, originally introduced in [54], can be viewed as a natural "promise version" of this question, where one is only concerned about sets of size 2. It is known that every graph access structure can be realized by a (linear) scheme in which the share size of each party is $O(N/\log N)$ [18,16,28]. The best lower bound for the total share size required to realize a graph access structure by a general secret-sharing scheme is $\Omega(N \log N)$ [55,15,23]. The best lower bound for total share size required to realize a graph access structure by a linear secret-sharing scheme is $\Omega(N^{3/2})$ [7]. The problem of secret sharing for dense graphs was studied in [8]. Additional references on secret sharing of graph access structures can be found in [8].

2   Preliminaries

2.1   Models and Definitions

Notation. Let $[n]$ denote the set $\{1, 2, \ldots, n\}$. Let $\mathcal{F}_N$ denote the set of all boolean functions from $[N] \times [N]$ to $\{0,1\}$. We will interpret $f \in \mathcal{F}_N$ as a 2-party function from $[N] \times [N]$ to $\{0,1\}$. For an algorithm $\mathcal{B}$, let $\tau(\mathcal{B})$ denote the size (measured as number of AND gates) of a boolean circuit, over the basis $\{\wedge, \oplus, \neg\}$, that represents $\mathcal{B}$.

Computational Model. Since our results refer to *perfect* security, we incorporate perfect uniform sampling of $[m]$, for an arbitrary positive integer m, into the computational model as an atomic computation step.

Protocols. A k-party protocol can be formally defined by a *next message function*. This function on input (i, x_i, j, m) specifies a k-tuple of messages sent by party P_i in round j, when x_i is its input and m describes all the messages P_i received in previous rounds. The next message function may also instruct P_i to terminate the protocol, in which case it also specifies the output of P_i.

Protocols with Preprocessing. In the *preprocessing model*, the specification of a protocol also includes a joint distribution $\mathcal{D}$ over $R_1 \times R_2 \ldots \times R_k$, where the R_i's are finite randomness domains. This distribution is used for sampling correlated random inputs $(r_1, \ldots, r_k)$ that the parties receive before the beginning of the protocol (in particular, the preprocessing is independent of the inputs). The next message function, in this case, may also depend on the private random input r_i received by P_i from $\mathcal{D}$. We assume that for every possible choice of inputs and random inputs, all parties eventually terminate.

OT Correlations and the OT-Hybrid Model. We will be interested in the special case of the 2-party setting when the correlated random inputs (X, Y) given to the two parties are random OT correlations, corresponding to a random instance of oblivious transfer, in which the receiver obtains one of two bits held by the sender. That is, $X = (X_0, X_1)$ is uniformly random over $\{0,1\}^2$ and $Y = (b, X_b)$ for a random bit b. We refer to a model in which the correlated randomness given to the parties consists of random OT correlations, as the *OT preprocessing model*. Alternatively, we may consider a setting where (each pair of) parties have access to an ideal (bit) OT functionality that receives from one of the parties, designated as the sender, a pair of bits (x_0, x_1), and a choice bit b from the other party, designated as the receiver, and sends back to the receiver the value x_b. We call this model the *OT-hybrid model*.

Security Definition. We use the standard ideal-world/real-world simulation paradigm. We restrict our attention mainly to the case of semi-honest (passive) corruptions. (In Appendix B, we show how to extend some of our results to the malicious setting.) Using the standard terminology of secure computation, the preprocessing model can be thought of as a *hybrid model* where the parties have a one-time access to an ideal randomized functionality $\mathcal{D}$ (with no inputs) providing them with correlated, private random inputs r_i. For lack of space, we omit the full security definitions (see, e.g., [41, App. A] adapted to the semi-honest setting).

2.2   Private Information Retrieval

The following is a somewhat non-standard view of PIR protocols, where the index is thought of as a pointer into a two-dimensional table, which in turn is thought of as a two-argument function.

Definition 1 (Private Information Retrieval). *Let $\mathcal{F}_N$ be the set of all boolean functions $f : [N] \times [N] \to \{0, 1\}$. A k-server private information retrieval (PIR) scheme $\mathcal{P} = (\mathcal{Q}, \mathcal{A}, \mathcal{R})$ for $\mathcal{F}_N$ is composed of three algorithms: a randomized query algorithm $\mathcal{Q}$, an answering algorithm $\mathcal{A}$, and a reconstruction algorithm $\mathcal{R}$. At the beginning of the protocol, the client has an input $x \in [N] \times [N]$ and each server has an identical input f representing a function in $\mathcal{F}_N$. Using its private randomness $r \in \{0, 1\}^{\gamma(N)}$, the client computes a tuple of k queries $(q_1, \ldots, q_k) = \mathcal{Q}(x, r)$, where $q_i \in \{0, 1\}^{\alpha(N)}$, for all $i \in [k]$. The client then sends the query q_j to server $\mathcal{S}_j$, for every $j \in [k]$. Each server $\mathcal{S}_j$ responds with an answer $a_j = \mathcal{A}(j, q_j, f)$, with $a_j \in \{0, 1\}^{\beta(N)}$. Finally, the client computes the value $f(x)$ by applying the reconstruction algorithm $\mathcal{R}(x, r, a_1, \ldots, a_k)$. We ask for the following correctness and privacy requirements:*

Correctness. *The client always outputs the correct value of $f(x)$. Formally, for every function $f \in \mathcal{F}_N$, every input $x \in [N] \times [N]$, and every random string r, if $(q_1, \ldots, q_k) = \mathcal{Q}(x, r)$ and $a_j = \mathcal{A}(j, q_j, f)$, for $j \in [k]$, then $\mathcal{R}(x, r, a_1, \ldots, a_k) = f(x)$.*

Client's Privacy. *Each server learns no information about x. Formally, for every two inputs $x, x' \in [N] \times [N]$, every $j \in [k]$, and every query q, the server $\mathcal{S}_j$ cannot know if the query q was generated with input x or with input x'; that is, $\Pr[\mathcal{Q}_j(x, r) = q] = \Pr[\mathcal{Q}_j(x', r) = q]$, where $\mathcal{Q}_j$ denotes the jth query in the k-tuple that $\mathcal{Q}$ outputs and the probability is taken over a uniform choice of the random string r.*

The communication complexity *of a protocol $\mathcal{P}$ is the total number of bits communicated between the client and the k servers (i.e., $\sum_j (|q_j| + |a_j|) = k(\alpha(N) + \beta(N))$).*

Every function $f \in \mathcal{F}_N$ is represented by an N^2-bit string $y = (y_{1,1}, \ldots, y_{N,N})$, where $f(i, j) = y_{i,j}$. The string y is also called a database, and we think of the client as querying a bit $y_{i,j}$ from the database.

Observe that the query received by each server is independent of the client's input x. In particular, this holds for the first query q_1, which therefore, may be thought of as depending only on the private randomness, say r, of the client, and not on the client input x. That is, we may assume that the query generation algorithm $\mathcal{Q}$ is expressed as the combination of two algorithms $\mathcal{Q}_1, \mathcal{Q}_{-1}$ and we assume that the client, with private randomness r, computes a tuple of k queries $(q_1, \ldots, q_k)$ as $q_1 = \mathcal{Q}_1(r)$, and $q_2, \ldots, q_k = \mathcal{Q}_{-1}(x, r)$.

2.3 Private Simultaneous Messages

The Private Simultaneous Messages (PSM) model was introduced by [30] as a minimal model for secure computation. It allows k players $P_1, \ldots, P_k$ with access to shared randomness, to send a single message each to a referee Ref, so that the referee learns the value of a function $f(x_1, \ldots, x_k)$ (where x_i is the private input of P_i) but nothing else. It is formally defined as follows:

Definition 2 (Private Simultaneous Messages). *Let* $X_1, \ldots, X_k, Z$ *be finite domains, and let* $X = X_1 \times \cdots \times X_k$. *A private simultaneous messages (PSM) protocol* $\mathcal{P}$, *computing a* k-*argument function* $f : X \to Z$, *consists of:*

- *A finite domain* R *of shared random inputs, and* k *finite message domains* $M_1, \ldots, M_k$.
- *Message computation function* $\mu_1, \ldots, \mu_k$, *where* $\mu_i : X_i \times R \to M_i$.
- *A reconstruction function* $g : M_1 \times \cdots \times M_k \to Z$.

Let $\mu(x, r)$ *denote the* k-*tuple of messages* $(\mu_1(x_1, r), \ldots, \mu_k(x_k, r))$. *We say that the protocol* $\mathcal{P}$ *is* correct *(with respect to* f*), if for every input* $x \in X$ *and every random input* $r \in R$, $g(\mu(x, r)) = f(x)$. *We say that the protocol* $\mathcal{P}$ *is* private *(with respect to* f*), if the distribution of* $\mu(x, r)$, *where* r *is a uniformly random element of* R, *depends only on* $f(x)$. *That is, for every two inputs* $x, x' \in X$ *such that* $f(x) = f(x')$, *the random variables* $\mu(x, r)$ *and* $\mu(x', r)$ *(over a uniform choice of* $r \in R$*) are identically distributed.* $\mathcal{P}$ *is a PSM protocol computing* f *if it is both correct and private.*

The communication complexity of the PSM protocol $\mathcal{P}$ *is naturally defined as* $\sum_{i=1}^{n} \log |M_i|$. *The randomness complexity of the PSM protocol* $\mathcal{P}$ *is defined as* $\log |R|$.

3 Our Results

Secure Computation in the OT-hybrid Model. We show a connection between secure computation in the (bit) OT-hybrid model and 2-server PIR. More formally, we show:

Theorem 1. *Let* $\mathcal{P} = (\mathcal{Q}, \mathcal{A}, \mathcal{R})$ *be a 2-server PIR scheme for* $\mathcal{F}_N$ *as described in Definition 1. Then, for any 2-party functionality* $f : [N] \times [N] \to \{0,1\}$, *there is a protocol* π *which realizes* f *in the (bit) OT-hybrid model, and has the following features:*

- π *is perfectly secure against semi-honest parties;*
- *The total communication complexity, and in particular the number of calls to the OT oracle, is* $O(\tau(\mathcal{Q}) + \tau(\mathcal{R}))$.

Plugging in parameters from the best known 2-server PIR protocol [21] in Theorem 1, we obtain:

Corollary 1. *For any 2-party functionality $f : [N] \times [N] \to \{0,1\}$, there is a protocol π that realizes f in the (bit) OT-hybrid model; this protocol is perfectly secure against semi-honest parties, and has total communication complexity (including communication with the OT oracle) $\widetilde{O}(N^{2/3})$.*

Prior to our work, the best upper bound on the communication complexity of an information-theoretically secure protocol in the OT-hybrid model for evaluating arbitrary functions $f : [N] \times [N] \to \{0,1\}$ was $\Omega(N)$. This can, for instance, be achieved by formulating the secure evaluation of $f : [N] \times [N] \to \{0,1\}$ as a 1-out-of-N OT problem between the two parties, where party P_1 participates as sender with inputs $\{f(x_1, i)\}_{i \in [N]}$ and party P_2 participates as receiver with input x_2. An instance of 1-out-of-N OT can be obtained information theoretically from $O(N)$ instances of 1-out-of-2 OT by means of standard reductions [17].

Secure Computation in the Preprocessing Model. Since OTs can be precomputed [4], the protocol implied by Theorem 1 yields a perfectly secure semi-honest protocol in the OT-preprocessing model where the communication complexity of the protocol and number of OTs required are both $O(\tau(\mathcal{Q}) + \tau(\mathcal{R}))$.

Our next theorem shows that it is possible to obtain much better communication complexity in a setting where we are not restricted to using precomputed OT correlations alone. We show this by demonstrating a connection between secure computation in the preprocessing model and 3-server PIR. More formally,

Theorem 2. *Let $\mathcal{P} = (\mathcal{Q}, \mathcal{A}, \mathcal{R})$ be a 3-server PIR scheme for $\mathcal{F}_N$ as described in Definition 1. Then, for any 2-party functionality $f : [N] \times [N] \to \{0,1\}$, there is a protocol π that realizes f in the preprocessing model, and has the following features:*

- *π is perfectly secure against semi-honest parties;*
- *The total communication complexity is $O(\tau(\mathcal{Q}) + \tau(\mathcal{R}))$;*
- *The total correlated randomness complexity is $O(\tau(\mathcal{Q}) + \tau(\mathcal{R}))$.*

Remark 1. We point out that a transformation in the other direction (i.e., constructing 3-server PIR protocols from protocols in the preprocessing model) was shown in [41]. In more detail, they show that a semi-honest secure protocol in the preprocessing model for $f : [N] \times [N] \to \{0,1\}$ with correlated randomness complexity $s(N)$ implies the existence of a 3-server, interactive PIR protocol, with communication complexity $s(\widehat{N}^{1/2}) + O(\log \widehat{N})$, where $\widehat{N}$ is the size of the database held by the servers. Taken together with our Theorem 2, this shows a two-way connection between the communication complexity of 3-server PIR protocols and the correlated randomness complexity of protocols in the preprocessing model.

Plugging in parameters from the best known 3-server PIR protocols [27,11] in Theorem 2, we obtain:

Corollary 2. *For any 2-party functionality* $f : [N] \times [N] \to \{0,1\}$, *there is a protocol* π *that realizes* f *in the preprocessing model; this protocol is perfectly secure against semi-honest parties, and has total communication complexity and correlated randomness complexity* $2^{\tilde{O}(\sqrt{\log N})}$.

While we mainly focus here on efficiency of 2-party secure computation, we show how to construct protocols in the multiparty setting, and also for the setting with honest majority in Appendix A. We summarize our results on t-private k-party semihonest secure computation in Table 1. In Appendix B we show how to extend our results on secure computation to the malicious setting.

Table 1. Summary of upper bounds on different complexity measures of t-private k-party semihonest secure computation of the worst function $f : [N]^k \to \{0,1\}$

Complexity measure	(t,k)	This work	Reference
OT complexity in the OT-hybrid model	$(1,2)$	$O(N^{2/3})$	Cor. 1
	$(t, k \leq 2t)$	$N^{k/\lfloor 2k-1/t \rfloor} \cdot \mathrm{poly}(k)$	Cor. 5
Correlated randomness complexity in the preprocessing model	$(1,2)$	$2^{\tilde{O}(\sqrt{\log N})}$	Cor. 2
	$(t > 1, k \leq 2t)$	$N^{k/\lfloor 2k+1/t \rfloor} \cdot \mathrm{poly}(k)$	Cor. 4
Communication complexity in the plain model	$(t, 2t < k < 3^t)$	$N^{k/\lfloor 2k-1/t \rfloor} \cdot \mathrm{poly}(k)$	Cor. 4
	$(t, k \geq 3^t)$	$2^{\tilde{O}(\sqrt{k \log N})} \cdot \mathrm{poly}(k)$	Cor. 5

Private Simultaneous Messages (PSM) Model. We obtain the following upper bound for 2-party protocols in the PSM model.

Theorem 3. *For any 2-party functionality* $f : [N] \times [N] \to \{0,1\}$, *there is a PSM protocol* π *that realizes* f, *and has the following features:*

- π *is perfectly secure against semi-honest parties;*
- *The total communication complexity and the randomness complexity are* $O(N^{1/2})$.

This improves upon the best known upper bound of $O(N)$ on the communication and randomness complexity of PSM protocols [30].

Secret Sharing for Forbidden Graph Access Structures. Consider a graph $G = (V, E)$. We are interested in the following graph access structure $\mathcal{A}^G$ in which the parties correspond to the vertices of the graph and (1) every vertex set of size three or more is authorized, and (2) every pair of vertices that is not connected by an edge in E is authorized. Such an access structure is called a *forbidden graph* access structure [54] since pairs of vertices connected by an edge in G are forbidden from reconstructing the secret. We obtain the following upper bound on the share size for a secret-sharing scheme realizing $\mathcal{A}^G$, for all G.

Theorem 4. *Let $G = (V, E)$ be a graph with $|V| = N$, and let $\mathcal{A}^G$ be the corresponding access structure. Then, there exists a perfect secret-sharing scheme realizing $\mathcal{A}^G$ with total share size $O(N^{3/2} \log N)$.*

4   Secure Computation in the OT-Hybrid Model

In this section, we construct a 2-party secure computation protocol realizing $f : [N] \times [N] \to \{0, 1\}$ in the (bit) OT-hybrid model from a 2-server PIR protocol $\mathcal{P} = (\mathcal{Q}, \mathcal{A}, \mathcal{R})$. The resulting protocol has communication complexity $\widetilde{O}(N^{2/3})$ and makes $\widetilde{O}(N^{2/3})$ calls to the ideal OT functionality, improving over prior work whose worst-case complexity (both in terms of communication and calls to the ideal OT functionality) was $\Omega(N)$ [17,25].

Let $\mathcal{P} = (\mathcal{Q}, \mathcal{A}, \mathcal{R})$ be a 2-server PIR protocol. Let the truth table of the function $f : [N] \times [N] \to \{0, 1\}$ that we are interested in, serve as the database (of length N^2). The high level idea behind our protocol is that the two parties P_1 and P_2, with their respective inputs x_1, x_2, securely emulate a virtual client with input $x = x_1 \| x_2$, and two virtual servers holding as database the truth table of f, in the PIR protocol $\mathcal{P}$. In more detail, parties P_1 and P_2, with inputs $x_1 \in [N], r^{(1)} \in \{0, 1\}^{\gamma(N)}$ and $x_2 \in [N], r^{(2)} \in \{0, 1\}^{\gamma(N)}$ respectively, emulate a PIR client by securely evaluating the query generation algorithm $\mathcal{Q}$ on input $x = x_1 \| x_2 \in [N^2]$ and randomness $r = r^{(1)} \oplus r^{(2)}$, such that party P_1 obtains query q_1 and party P_2 obtains query q_2. Then, using the PIR queries as their respective inputs, the parties locally emulate the PIR servers by running the PIR answer generation algorithm $\mathcal{A}$ and obtaining PIR answers a_1 and a_2, respectively. Finally, using the answers a_1, a_2, the inputs x_1, x_2, and the randomness $r^{(1)}, r^{(2)}$, parties P_1 and P_2 once again participate in a secure computation protocol to securely evaluate the PIR reconstruction algorithm $\mathcal{R}$ to obtain the final output z. The protocol is described in Figure 1. It is easy to see that the communication complexity as well as the number of calls to the ideal OT functionality is $O(\tau(\mathcal{Q}) + \tau(\mathcal{R}))$, that is, the complexity is proportional to the circuit size of the query and reconstruction algorithms. For a detailed proof, see full version.

Intuitively, the protocol is private because (1) each individual PIR query does not leak any information about the query location and the reconstruction algorithms outputs nothing but the desired bit (both follow from the definition of PIR schemes); and (2) emulation of the algorithms run by the PIR client is done via secure computation protocols.

Instantiating the protocol in Figure 1 with the 2-server PIR protocol of Chor et al. [21] yields a perfectly secure protocol in the OT-hybrid model whose communication complexity is $\widetilde{O}(N^{2/3})$ and which makes $\widetilde{O}(N^{2/3})$ calls to the ideal OT functionality. This proves Corollary 1.

Preliminaries: Let $\mathcal{P} = (\mathcal{Q}, \mathcal{A}, \mathcal{R})$ be a 2-server PIR protocol where servers hold as database the truth table of a function $f : [N] \times [N] \to \{0,1\}$. Parties P_1, P_2 have inputs $x_1, x_2 \in [N]$ respectively. At the end of the protocol, both parties learn $z = f(x_1, x_2)$.

Protocol:

1. P_1, P_2 choose uniformly random $r^{(1)}, r^{(2)} \in \{0,1\}^{\gamma(N)}$, respectively (where $\gamma(N)$ is the size of the randomness required by algorithm $\mathcal{Q}$). Let $\widetilde{\mathcal{Q}}$ denote an algorithm that takes as input $(x_1, r^{(1)}), (x_2, r^{(2)})$ and runs algorithm $\mathcal{Q}(x_1 \| x_2, r^{(1)} \oplus r^{(2)})$. Party P_1 with inputs $(x_1, r^{(1)})$ and P_2 with inputs $(x_2, r^{(2)})$ run a 2-party semi-honest secure GMW protocol in the OT-hybrid model to evaluate circuit $C(\widetilde{\mathcal{Q}})$. Let q_1, q_2 denote their respective outputs.
2. P_1 and P_2 locally compute $a_1 = \mathcal{A}(1, q_1, f)$ and $a_2 = \mathcal{A}(2, q_2, f)$ respectively.
3. Let $\widetilde{\mathcal{R}}$ denote an algorithm that takes as input $(a_1, x_1, r^{(1)}), (a_2, x_2, r^{(2)})$ and runs algorithm $\mathcal{R}(x_1 \| x_2, r^{(1)} \oplus r^{(2)}, a_1, a_2)$. Party P_1 with inputs $(a_1, x_1, r^{(1)})$ and P_2 with inputs $(a_2, x_2, r^{(2)})$ run a 2-party semi-honest secure GMW protocol in the OT-hybrid model to evaluate circuit $C(\widetilde{\mathcal{R}})$, where z denotes their common output. Both parties output z and terminate the protocol.

Fig. 1. A perfectly secure protocol in the OT-hybrid model

5   Secure Computation in the Preprocessing Model

In this section, we construct a 2-party secure computation protocol realizing $f : [N] \times [N] \to \{0,1\}$ in the preprocessing model from a 3-server PIR protocol $\mathcal{P} = (\mathcal{Q}, \mathcal{A}, \mathcal{R})$. The resulting protocol will have communication and correlated randomness complexity $2^{\widetilde{O}(\sqrt{\log N})}$ improving over prior work whose worst-case complexity was $\Omega(N)$ [17,25]. Note that we manage to emulate a protocol with 3 servers and one client by a protocol with 2 parties.

Let $\mathcal{P} = (\mathcal{Q}, \mathcal{A}, \mathcal{R})$ be a 3-server PIR protocol. We assume that the database represents the truth table of the function $f : [N] \times [N] \to \{0,1\}$ that we are interested in. The high level idea behind our protocol is that the two parties P_1 and P_2 with their respective inputs x_1, x_2 securely emulate a virtual client with input $x = x_1 \| x_2$, and two of the three virtual servers, say $\mathcal{S}_2$ and $\mathcal{S}_3$, holding as database the truth table of f, in the PIR protocol $\mathcal{P}$. The key observation is that server $\mathcal{S}_1$'s inputs and outputs can be precomputed and shared between P_1 and P_2 as preprocessed input. This is possible because $\mathcal{S}_1$'s input, namely the PIR query q_1, is distributed independently of the client's input, and thus can be computed beforehand. Similarly, a_1, the answer of $\mathcal{S}_1$, is completely determined by q_1 and the truth table of the function f, and thus can be precomputed as well. Thus, the preprocessed input along with the emulation done by P_1 and P_2 allow them to securely emulate all PIR algorithms $\mathcal{Q}$, $\mathcal{A}$, $\mathcal{R}$ of the 3-server PIR protocol $\mathcal{P}$. We provide a more detailed description of the protocol below.

Parties P_1 and P_2, are provided as preprocessed input, values $(r^{(1)}, a_1^{(1)})$ and $(r^{(2)}, a_1^{(2)})$ respectively along with sufficient OT correlations (whose use we will see later). The values $r^{(1)}$ and $r^{(2)}$ together determine the randomness used in PIR query generation algorithm $\mathcal{Q}$ as $r = r^{(1)} \oplus r^{(2)}$. Given randomness r, the first server's query q_1 (resp. answer a_1) is completely determined as $\mathcal{Q}_1(r)$ (resp. $\mathcal{A}(1, q_1, f)$). The values $a_1^{(1)}$ and $a_1^{(2)}$ together form a random additive sharing of a_1.

In the online phase, when parties obtain their respective inputs x_1 and x_2, they proceed to emulate the PIR client by securely evaluating the query generation algorithm on input $x = x_1 \| x_2 \in [N^2]$ and randomness $r = r^{(1)} \oplus r^{(2)}$, such that party P_1 obtains query q_2 and party P_2 obtains query q_3. Then, using the PIR queries as their respective inputs, the parties locally emulate the PIR servers by running the PIR answer generation algorithm $\mathcal{A}$ and obtain PIR answers a_2 and a_3 respectively. Recall that a random sharing of answer a_1 is already provided to the parties as preprocessed input. Using this random sharing of answer a_1, the locally computed answers a_2, a_3, the inputs x_1, x_2, and the randomness $r = r^{(1)} \oplus r^{(2)}$, parties P_1 and P_2 once again participate in a secure computation protocol to securely evaluate the PIR reconstruction algorithm $\mathcal{R}$ to obtain the final output z. It is easy to see that the communication and correlated randomness complexity of the protocol equals $O(\tau(\mathcal{Q}) + \tau(\mathcal{R}))$.

Intuitively, the protocol is private because (1) each party knows at most one PIR query, and (2) each individual PIR query does not leak any information about the query location (follows from the definition of PIR properties), and (3) emulation of the algorithms run by the PIR client is done via secure computation protocols. Instantiating the protocol described above with the best known 3-server PIR protocol [58,27,11] we obtain a perfectly secure protocol in the preprocessing model whose communication and correlated randomness complexity is $2^{\tilde{O}(\sqrt{\log N})}$. The details are deferred to the full version.

6   Private Simultaneous Messages

In this section, we provide a new framework for constructing PSM protocols (cf. Definition 2). Our proposed framework is based on a new variant of PIR protocols that we call *decomposable* PIR protocols. We define decomposable PIR in Section 6.1. We construct a 2-party PSM protocol using 3-server decomposable PIR protocols in Section 6.2, and we present a concrete decomposable 3-server PIR protocol in Section 6.3. The PSM protocol of Section 6.2, instantiated with this concrete decomposable 3-server PIR protocol, has communication (and randomness) complexity $O(N^{1/2})$, for all $f : [N] \times [N] \to \{0, 1\}$.

6.1   Decomposable PIR Schemes

A k-server decomposable PIR protocol allows a client with input $x = (x_1, \ldots, x_{k-1}) \in [N]^{k-1}$ to query k servers, each holding a copy of a database of size N^{k-1} and retrieve the contents of the database at index x while offering (possibly relaxed) privacy guarantees to the client. Loosely speaking, decomposable PIR protocols differ from standard PIR protocols (cf. Definition 1) in two ways: (1) the query generation and reconstruction algorithms can be decomposed into "simpler" algorithms that depend only on parts of the entire input. (2) We change the privacy requirement and require that the query of server $\mathcal{S}_k$ together with some information about the answers of the first $k-1$ servers does not disclose information about the input of the client. We note that the privacy of the first $k-1$ queries follows from the decomposability of the query generation algorithm. We provide the formal definition below.

Definition 3 (Decomposable PIR). *Let $\mathcal{F}_{N,k-1}$ be the set of all boolean functions $f : [N]^{k-1} \to \{0,1\}$. A k-server decomposable PIR protocol $\mathcal{P} = (\mathcal{Q}, \mathcal{A}, \mathcal{R})$ for $\mathcal{F}_{N,k-1}$ consists of three algorithms: a randomized query algorithm $\mathcal{Q}$, an answering algorithm $\mathcal{A}$, and a reconstruction algorithm $\mathcal{R}$. The client has an input $x = (x_1, \ldots, x_{k-1}) \in [N]^{k-1}$ (i.e., x is from the input domain of $\mathcal{F}_{N,k-1}$) and each server has an identical input f representing a function in $\mathcal{F}_{N,k-1}$. Using its private randomness $r \in \{0,1\}^{\gamma(N)}$, the client computes a tuple of k queries $(q_1, \ldots, q_k) = \mathcal{Q}(x,r)$, where each $q_i \in \{0,1\}^{\alpha(N)}$. The client then sends the query q_j to server $\mathcal{S}_j$, for every $j \in [k]$. Each server $\mathcal{S}_j$ responds with an answer $a_j = \mathcal{A}(j, q_j, f)$, with $a_j \in \{0,1\}^{\beta(N)}$. Finally, the client computes the value $f(x)$ by applying the reconstruction algorithm $\mathcal{R}(x, r, a_1, \ldots, a_k)$. The query generation algorithm $\mathcal{Q}$ and the reconstruction algorithm $\mathcal{R}$ satisfy the following "decomposability" properties.*

Decomposable Query Generation. *The randomized query generation algorithm $\mathcal{Q}$ can be decomposed into k algorithms $\mathcal{Q}_1, \ldots, \mathcal{Q}_{k-1}, \mathcal{Q}_k = (\mathcal{Q}_k^1, \ldots, \mathcal{Q}_k^{k-1})$, such that for every input $x = (x_1, \ldots, x_{k-1}) \in [N]^{k-1}$, and for every random string $r \in \{0,1\}^{\gamma(N)}$, the queries $(q_1, \ldots, q_k) = \mathcal{Q}(x,r)$ are computed by the client as $q_j = \mathcal{Q}_j(x_j, r)$ for $j \in [k-1]$, and $q_k = (q_k^1, \ldots, q_k^{k-1}) = (\mathcal{Q}_k^1(x_1, r), \ldots, \mathcal{Q}_k^{k-1}(x_{k-1}, r))$.*

Decomposable Reconstruction. *There exists algorithms $\mathcal{R}', \mathcal{R}''$ such that for every input $x = (x_1, \ldots, x_{k-1}) \in [N]^{k-1}$, and for every random string $r \in \{0,1\}^{\gamma(N)}$, if $(q_1, \ldots, q_k) = \mathcal{Q}(x,r)$, and $a_j = \mathcal{A}(j, q_j, f)$ for $j \in [k]$, then the output of the reconstruction algorithm $\mathcal{R}(x, r, a_1, \ldots, a_k)$ equals $\mathcal{R}''(a_k, \mathcal{R}'(x, r, a_1, \ldots, a_{k-1}))$.*

We ask for the following correctness and privacy requirements:

Correctness. *The client always outputs the correct value of $f(x)$. Formally, for every function $f \in \mathcal{F}_{N,k-1}$, every input $x \in [N]^{k-1}$, and every random string r, if $(q_1, \ldots, q_k) = \mathcal{Q}(x,r)$ and $a_j = \mathcal{A}(j, q_j, f)$, for $j \in [k]$, then $\mathcal{R}(x, r, a_1, \ldots, a_k) = f(x)$.*

Privacy. *We require that q_k, the query of $\mathcal{S}_k$, and $\mathcal{R}'(x, r, a_1, \ldots, a_{k-1})$ do not disclose information not implied by $f(x)$. Formally, for every $f \in \mathcal{F}_{N,k-1}$, for every two inputs $x, x' \in [N]^{k-1}$ such that $f(x) = f(x')$, and every values q, b, letting $a_j = \mathcal{A}(j, \mathcal{Q}_j(x, r), f)$ and $a'_j = \mathcal{A}(j, \mathcal{Q}_j(x', r), f)$ for $j \in [k-1]$, and $q_k = \mathcal{Q}_k(x, r)$, $q'_k = \mathcal{Q}_k(x', r)$*

$$\Pr_r[q_k = q \wedge \mathcal{R}'(x, r, a_1, \ldots, a_{k-1}) = b] = \Pr_r[q'_k = q \wedge \mathcal{R}'(x', r, a'_1, \ldots, a'_{k-1}) = b],$$

where the probability is taken over a uniform choice of the random string r.

As usual, the communication complexity of such a protocol $\mathcal{P}$ is the total number of bits communicated between the client and the k servers (i.e., $\sum_j(|q_j| + |a_j|) = k(\alpha(N) + \beta(N)))$.

6.2   From 3-Server Decomposable PIR to 2-Party PSM

Given a function $f : [N] \times [N] \to \{0, 1\}$, we construct a 2-party PSM protocol for f using a 3-Server Decomposable PIR protocol. We give an informal description of the protocol. The shared randomness of the two parties is composed of two strings, one string for the decomposable PIR protocol and one for a PSM protocol π for computing $\mathcal{R}'$. (We remark that $\mathcal{R}'$ is "simpler" than f, and consequently existing PSM protocols (e.g., [38,47]) can realize $\mathcal{R}'$ very efficiently.) In the protocol, P_1, holding x_1 and f, computes the query q_1 and its part of the query of server $\mathcal{S}_3$, namely q_3^1 (party P_1 can compute these queries by the decomposability of the query generation). P_1 also computes a_1. Similarly, P_2, holding x_2 and f, computes q_2, its part of the query of server $\mathcal{S}_3$, namely q_3^2, and a_2. Parties P_1 and P_2 send q_3^1 and q_3^2 to the referee, who uses this information and f to compute a_3. Furthermore, P_1 and P_2 execute a PSM protocol that enables the referee to compute $z' = \mathcal{R}'((x_1, x_2), r, a_1, a_2)$. The referee reconstructs $f(x)$ by computing $\mathcal{R}''(a_3, z')$, where a_3 is the answer computed by the referee for query $q_3 = (q_3^1, q_3^2)$.

The correctness of the protocol described above follows immediately from the definition of decomposable PIR. Furthermore, the information that the referee gets is q_3 and the messages of a PSM protocol computing $\mathcal{R}'$. By the privacy of the PSM protocol, the referee only learns the output of $\mathcal{R}'$ from this PSM protocol. Thus, the referee only learns q_3 and the output of $\mathcal{R}'$; by the privacy requirement of the decomposable PIR protocol the referee learns only $f(x)$. We summarize the properties of our PSM protocol in the following lemma.

Lemma 1. *Let $\mathcal{P}$ be a 3-server decomposable PIR protocol where the query length is $\alpha(N)$ and the randomness complexity is $\gamma(N)$. Furthermore, assume that $\mathcal{R}'$ can be computed by a 2-party PSM protocol π' with communication complexity $\alpha'(N)$ and randomness complexity $\gamma'(N)$. Then, every function $f \in \mathcal{F}_N$ can be computed by a 2-party PSM protocol π with communication complexity $\alpha(N) + \alpha'(N)$ and randomness complexity $\gamma(N) + \gamma'(N)$.*

6.3   A 3-Server Decomposable PIR Protocol

In this section, we show how to construct a decomposable 3-server PIR protocol. Our construction is inspired by the cubes approach of [21]. We start with a high level description of this approach, specifically for the case of 4-dimensional cubes and of its adaptation to the decomposable case. In the following, for set S and element i, let $S \oplus \{i\}$ denote the set $S \backslash \{i\}$ if $i \in S$, and $S \cup \{i\}$ otherwise.

The starting point of the CGKS [21] cubes approach (restricted here to dimension 4) is viewing the n-bit database as a 4-dimensional cube (i.e., $[n^{1/4}]^4$). Correspondingly, the index that the client wishes to retrieve is viewed as a 4-tuple $i = (i_1, \ldots, i_4)$. The protocol starts by the client choosing a random subset for each dimension, i.e. $S_1, \ldots, S_4 \subseteq_R [n^{1/4}]$. It then creates 16 queries of the form $(T_1, \ldots, T_4)$ where each T_j is either S_j itself or $S_j \oplus \{i_j\}$ (we often use vectors in $\{0,1\}^4$ to describe these 16 combinations; e.g., 0000 refers to the query $(S_1, \ldots, S_4)$ while 1111 refers to the query $(S_1 \oplus \{x_1\}, \ldots, S_4 \oplus \{x_4\})$). If there were 16 servers available, the client could send each query $(T_1, \ldots, T_4)$ to a different server ($4 \cdot n^{1/4}$ bits to each), who would reply with a single bit which is the XOR of all bits in the sub-cube $T_1 \otimes \ldots \otimes T_4$. The observation made in [21] is that each element of the cube appears in an even number of those 16 sub-cubes, and the only exception is the entry $i = (i_1, \ldots, i_4)$ that appears exactly once. Hence, taking the XOR of the 16 answer bits, all elements of the cube are canceled out except for the desired element in position i.

The next observation of the cubes approach is that a server who got a query $(T_1, \ldots, T_4)$ can provide a longer answer (but still of length $O(n^{1/4})$ bits) from which the answers to some of the other queries can be derived (and, hence, the corresponding servers in the initial solution can be eliminated). Specifically, it can provide also the answers to the queries $(T_1 \oplus \{\ell\}, T_2, T_3, T_4)$, for all possible values $\ell \in [n^{1/4}]$. One of these is the bit corresponding to $\ell = i_1$ which is the desired answer for another one of the 16 queries; and, clearly, the same can be repeated in each of the 4 dimensions. Stated in the terminology of 4-bit strings, a server that gets the query represented by some $b \in \{0,1\}^4$ can reply with $O(n^{1/4})$ bits from which the answer to the 5 queries of hamming distance at most one from b can be obtained; further, it can be seen that 4 servers that will answer the queries corresponding to $\{1100, 0011, 1000, 0111\}$ provide all the information needed to answer the 16 queries in the initial solution (this corresponds also to the notion of "covering codes" from the coding theory literature).

Next, we informally describe how to turn the above ideas into a decomposable 3-server PIR protocol. We still view the database as 4-dimensional cube and the client is still interested in obtaining the answers to the same 16 queries. Moreover, we are allowed to use only 3 servers for this. However, the requirements of decomposable PIR give us some freedom that we did not have before; specifically, we allow the answer of the first server to depend on $x_1 = (i_1, i_2)$ and the answer of the second server to depend on $x_2 = (i_3, i_4)$. The query to the third server should still give no information about i. Specifically, we will give the first server the basic sets $S_1, \ldots, S_4$ along with the values i_1, i_2. This server can easily compute the answer to all 4 queries of the form $(T_1, \ldots, T_4)$ with T_1 being either S_1 or $S_1 \oplus \{i_1\}$ and T_2 being either S_2 or $S_2 \oplus \{i_2\}$ (in vectors notation, those correspond to the queries 0000,0100,1000,1100). Moreover, using the idea described above, even though the first server does not know the value of i_3 it can provide $O(n^{1/4})$-bit answer corresponding to all choices of i_3 from which the client can select the right ones (in vectors notation, those corresponding to the queries 0010,0110,1010,1110). Similarly it can provide $O(n^{1/4})$-bit answer corresponding to all choices of i_4 from which the client can select the right ones (in vectors notation, those corresponding to the queries 0001,0101,1001,1101). The query to the second server consists of $S_1, \ldots, S_4$ along with the values i_3, i_4. In a similar way, this server provides an answer of $O(n^{1/4})$ bits that can be used to answer the queries 0000,0010,0001,0011 directly and 1000,1010,1001,1011, 0100,0110,0101,0111 by enumerating all values of i_1 and then all values of i_2 (some queries are answered by both servers; this small overhead can be easily saved – see full version). So, based on a_1, a_2, the only query that remained unanswered is the 1111 query. For this, the client asks the third server the query $(S_1 \oplus \{i_1\}, \ldots, S_4 \oplus \{i_4\})$ (which is independent of i) and gets the missing bit, denoted a_3, back. Finally, note that the reconstruction has the desired "decomposable" form: the client output can be obtained by processing the answers of the first two servers to get the sum v of the first 15 queries (this is the desired $\mathcal{R}'$) and then adding a_3 to it. Moreover, the pair (q_3, v) gives no information on i beyond the output: q_3 is independent of i (it is just a random sub-cube), and v is just the exclusive-or of the output and a_3 (which depends only on q_3 and hence independent of i).

7 Secret Sharing

We present a generic transformation from any 2-party PSM protocol to secret-sharing schemes for forbidden graph access structures, and then use the results from Section 6 to obtain efficient secret-sharing schemes for these access structures. Specifically, we obtain N-party secret-sharing schemes for forbidden graph access structures whose total share size is $O(N^{3/2})$. The best previous constructions for these access structures had total share size $O(N^2 / \log N)$ [18,16,28].

In Section 7.1, we demonstrate our transformation from PSM protocols to secret-sharing schemes for forbidden graph access structures, for the simple case when the graph is bipartite. For lack of space, our generalized construction is presented in the full version. We start by formally defining forbidden graph access structures.

Definition 4. *Let $G = (V, E)$ be an arbitrary graph. A forbidden graph access structure, denoted $\mathcal{A}^G$, is an access structure where the parties are the vertices in V and the only unauthorized sets are singletons (i.e., sets containing a single vertex in V), and sets of size 2 corresponding to edges on G (i.e., sets $\{x, y\}$ with $(x, y) \in E$).*

7.1 Secret Sharing Schemes for Forbidden Bipartite Graph Access Structures

We first show how to realize forbidden graph access structures $\mathcal{A}^G$, where the graph G is bipartite.

Definition 5. *Let $G = (L, R, E)$ be a bipartite graph, where $|L| = |R| = N$. We label the vertices in L by $1, 2, \ldots, N$, and similarly, vertices in R by $1, 2, \ldots, N$. We associate the bipartite graph $G = (L, R, E)$ with a boolean function $f_G : [N] \times [N] \to \{0, 1\}$, where $f(x, y)$ equals 0 iff there exists an edge between vertex $x \in L$ and vertex $y \in R$.*

Lemma 2. *Let $G = (L, R, E)$ be a bipartite graph where $|L| = |R| = N$ and $f_G : [N] \times [N] \to \{0, 1\}$ be the function associated with G. Let $\mathcal{P}$ be a PSM protocol for computing f_G with communication complexity $c_{\mathcal{P}}(N)$. Then, there exists a secret sharing realizing $\mathcal{A}^G$ with domain of secrets $\{0, 1\}$ and total share size $O(N \cdot c_{\mathcal{P}}(N))$.*

Proof. In a forbidden bipartite graph access structure the sets that can reconstruct the secret are: (1) All sets of 3 or more parties, (2) all pairs of parties that correspond to vertices from the same "side" of the graph (L or R), and (3) all pairs of parties that correspond to vertices from different sides of the graph and are not connected by an edge.

We construct a secret-sharing scheme for $\mathcal{A}^G$ by dealing with the three types of authorized sets. First, the dealer shares the secret with Shamir's 3-out-of-$2N$ threshold secret-sharing scheme among the $2N$ parties of the access structure. Next, the dealer independently shares the secret with Shamir's 2-out-of-N threshold secret-sharing scheme among the parties in L, and independently among the parties in R.

The interesting case is how to share the secret for sets $\{x, y\}$ such that $x \in L, y \in R$, and $(x, y) \notin E$. Let μ_1, μ_2 represent the message computation functions of the PSM protocol $\mathcal{P}$ (as defined in Definition 2). To share a secret $s \in \{0, 1\}$, the dealer chooses the randomness r, required for $\mathcal{P}$. Then, depending on the value of s, it distributes the shares to the parties as follows:

- If $s = 0$, then the dealer chooses arbitrary $x_0, y_0 \in [N]$ such that $f_G(x_0, y_0) = 0$, and gives the share $m_x = \mu_1(x_0, r)$ to each party $x \in L$, and the share $m_y = \mu_2(y_0, r)$ to each party $y \in R$.
- Else, if $s = 1$, then the dealer gives the share $m_x = \mu_1(x, r)$ to each party $x \in L$, and the share $m_y = \mu_2(y, r)$ to each party $y \in R$.

Any two parties $x \in L$ and $y \in R$ that are not connected by an edge in G reconstruct the secret by returning the output of the PSM reconstruction function $s' = g(m_x, m_y)$ (cf. Definition 2). Correctness of this reconstruction for $(x, y) \notin E$ follows from the correctness of the PSM protocol $\mathcal{P}$. Specifically, (1) when $s = 0$, the parties x and y reconstruct $f(x_0, y_0) = 0 = s$, and (2) when $s = 1$, the parties x and y reconstruct $f_G(x, y) = 1 = s$.

For the privacy, consider a pair of parties x, y such that $x \in L, y \in R$, and $(x, y) \in E$. When $s = 0$, these parties hold shares $\mu_1(x_0, r)$ and $\mu_2(y_0, r)$ respectively. When $s = 1$, these parties hold shares $\mu_1(x, r)$ and $\mu_2(y, r)$ respectively. Since $f_G(x, y) = f_G(x_0, y_0) = 0$, the shares do not reveal any information about s (by the privacy of the PSM protocol).   □

Using the PSM protocols described in Theorem 3 in Lemma 2, we get the following corollary.

Corollary 3. *Let* $G = (L, R, E)$ *be a bipartite graph where* $|L| = |R| = N$. *There exists a secret sharing realizing* $\mathcal{A}^G$ *with domain of secrets* $\{0, 1\}$ *and total share size* $O(N^{3/2})$.

In the full version, we show how to construct secret-sharing schemes realizing $\mathcal{A}^G$ for general graphs, using the secret-sharing scheme for forbidden bipartite graph access structures.

References

1. Applebaum, B., Ishai, Y., Kushilevitz, E.: Computationally private randomizing polynomials and their applications. In: CCC, pp. 260–274 (2005)
2. Barkol, O., Ishai, Y., Weinreb, E.: On locally decodable codes, self-correctable codes, and t-private PIR. In: Charikar, M., Jansen, K., Reingold, O., Rolim, J.D.P. (eds.) APPROX and RANDOM 2007. LNCS, vol. 4627, pp. 311–325. Springer, Heidelberg (2007)
3. Beaver, D.: Correlated pseudorandomness and the complexity of private computations. In: STOC, pp. 479–488 (1996)
4. Beaver, D.: Precomputing oblivious transfer. In: Coppersmith, D. (ed.) CRYPTO 1995. LNCS, vol. 963, pp. 97–109. Springer, Heidelberg (1995)
5. Beaver, D., Feigenbaum, J., Kilian, J., Rogaway, P.: Security with low communication overhead. In: Menezes, A.J., Vanstone, S.A. (eds.) CRYPTO 1990. LNCS, vol. 537, pp. 62–76. Springer, Heidelberg (1991)
6. Beaver, D., Feigenbaum, J., Kilian, J., Rogaway, P.: Locally random reductions: Improvements and applications. Journal of Cryptology 10(1), 17–36 (1997)
7. Beimel, A., Gal, A., Paterson, M.: Lower bounds on monotone span programs. In: FOCS, pp. 674–681 (1995)
8. Beimel, A., Farràs, O., Mintz, Y.: Secret sharing for very dense graphs. In: Safavi-Naini, R., Canetti, R. (eds.) CRYPTO 2012. LNCS, vol. 7417, pp. 144–161. Springer, Heidelberg (2012)
9. Beimel, A., Ishai, Y.: On the power of nonlinear secret sharing. In: CCC, pp. 188–202 (2001)
10. Beimel, A., Ishai, Y., Kushilevitz, E.: General constructions for information-theoretic private information retrieval. Jour. Comput. Syst. & Sci. 71(2), 213–247 (2005)
11. Beimel, A., Ishai, Y., Kushilevitz, E., Orlov, I.: Share conversion and private information retrieval. In: CCC, pp. 258–268 (2012)
12. Beimel, A., Malkin, T.: A quantitative approach to reductions in secure computation. In: Naor, M. (ed.) TCC 2004. LNCS, vol. 2951, pp. 238–257. Springer, Heidelberg (2004)

13. Ben-Or, M., Goldwasser, S., Wigderson, A.: Completeness theorems for noncryptographic fault-tolerant distributed computations. In: STOC, pp. 1–10 (1988)
14. Bendlin, R., Damgrard, I., Orlandi, C., Zakarias, S.: Semi-homomorphic encryption and multiparty computation. In: Paterson, K.G. (ed.) EUROCRYPT 2011. LNCS, vol. 6632, pp. 169–188. Springer, Heidelberg (2011)
15. Blundo, C., De Santis, A., de Simone, R., Vaccaro, U.: Tight bounds on information rate of secret sharing schemes. In: Designs, Codes and Cryptography, pp. 107–122 (1997)
16. Blundo, C., De Santis, A., Gargano, L., Vaccaro, U.: On information rate of secret sharing schemes. Theoretical Computer Science, 283–306 (1996)
17. Brassard, G., Crepeau, C., Robert, J.-M.: Information theoretic reduction among disclosure problems. In: FOCS, pp. 168–173 (1986)
18. Bublitz, S.: Decomposition of graphs and monotone formula size of homogeneous functions. In: Acta Informatica, pp. 689–696 (1986)
19. Chaum, D., Crépeau, C., Damgard, I.: Multiparty unconditionally secure protocols. In: STOC, pp. 11–19 (1988)
20. Chen, X., Kayal, N., Wigderson, A.: Partial derivatives in arithmetic complexity and beyond. In: FSTTCS, pp. 1–138 (2011)
21. Chor, B., Goldreich, O., Kushilevitz, E., Sudan, M.: Private information retrieval. In: FOCS, pp. 41–50 (1995)
22. Crepeau, C., Kilian, J.: Achieving oblivious transfer using weakened security assumptions (extended abstract). In: FOCS, pp. 42–52 (1988)
23. Csirmaz, L.: Secret sharing schemes on graphs. In: ePrint 2005/059 (2005)
24. Damgrard, I., Zakarias, S.: Constant-overhead secure computation of boolean circuits using preprocessing. In: Sahai, A. (ed.) TCC 2013. LNCS, vol. 7785, pp. 621–641. Springer, Heidelberg (2013)
25. Dodis, Y., Micali, S.: Parallel reducibility for information-theoretically secure computation. In: Bellare, M. (ed.) CRYPTO 2000. LNCS, vol. 1880, pp. 74–92. Springer, Heidelberg (2000)
26. Dvir, Z., Gopalan, P., Yekhanin, S.: Matching vector codes. In: FOCS, pp. 705–714 (2010)
27. Efremenko, K.: 3-query locally decodable codes of subexponential length. In: STOC, pp. 39–44 (2009)
28. Erdos, P., Pyber, L.: Covering a graph by complete bipartite graphs. In: Discrete Mathematics, pp. 249–251 (1997)
29. Even, S., Goldreich, O., Lempel, A.: A randomized protocol for signing contracts. In: Crypto, pp. 205–210 (1983)
30. Feige, U., Kilian, J., Naor, M.: A minimal model for secure computation (extended abstract). In: STOC, pp. 554–563 (1994)
31. Goldreich, O.: Foundations of cryptography - basic applications, vol. 2 (2004)
32. Goldreich, O., Micali, S., Wigderson, A.: How to play any mental game, or a completeness theorem for protocols with honest majority. In: STOC, pp. 218–229 (1987)
33. Goldreich, O., Vainish, R.: How to solve any protocol problem - an efficiency improvement. In: Pomerance, C. (ed.) CRYPTO 1987. LNCS, vol. 293, pp. 73–86. Springer, Heidelberg (1988)
34. Harnik, D., Ishai, Y., Kushilevitz, E.: How many oblivious transfers are needed for secure multiparty computation? In: Menezes, A. (ed.) CRYPTO 2007. LNCS, vol. 4622, pp. 284–302. Springer, Heidelberg (2007)
35. Harnik, D., Ishai, Y., Kushilevitz, E., Nielsen, J.B.: OT-combiners via secure computation. In: Canetti, R. (ed.) TCC 2008. LNCS, vol. 4948, pp. 393–411. Springer, Heidelberg (2008)

36. Ishai, Y.: Randomization techniques for secure computation. In: Prabhakaran, M., Sahai, A. (eds.) Secure Multi-Party Computation (2013)
37. Ishai, Y., Kilian, J., Nissim, K., Petrank, E.: Extending oblivious transfers efficiently. In: Boneh, D. (ed.) CRYPTO 2003. LNCS, vol. 2729, pp. 145–161. Springer, Heidelberg (2003)
38. Ishai, Y., Kushilevitz, E.: Private simultaneous messages protocols with applications. In: ISTCS, pp. 174–184 (1997)
39. Ishai, Y., Kushilevitz, E.: Randomizing polynomials: A new representation with applications to round-efficient secure computation. In: FOCS, pp. 294–304 (2000)
40. Ishai, Y., Kushilevitz, E.: On the hardness of information-theoretic multiparty computation. In: Cachin, C., Camenisch, J.L. (eds.) EUROCRYPT 2004. LNCS, vol. 3027, pp. 439–455. Springer, Heidelberg (2004)
41. Ishai, Y., Kushilevitz, E., Meldgaard, S., Orlandi, C., Paskin-Cherniavsky, A.: On the power of correlated randomness in secure computation. In: Sahai, A. (ed.) TCC 2013. LNCS, vol. 7785, pp. 600–620. Springer, Heidelberg (2013)
42. Ishai, Y., Kushilevitz, E., Ostrovsky, R., Sahai, A.: Extracting correlations. In: FOCS, pp. 261–270 (2009)
43. Ishai, Y., Prabhakaran, M., Sahai, A.: Founding cryptography on oblivious transfer - efficiently. In: Wagner, D. (ed.) CRYPTO 2008. LNCS, vol. 5157, pp. 572–591. Springer, Heidelberg (2008)
44. Kaplan, M., Kerenidis, I., Laplante, S., Roland, J.: Non-local box complexity and secure function evaluation. In: FSTTCS, pp. 239–250 (2009)
45. Kilian, J.: Founding cryptography on oblivious transfer. In: STOC, pp. 20–31 (1988)
46. Kilian, J.: More general completeness theorems for secure two-party computation. In: STOC, pp. 316–324 (2000)
47. Kolesnikov, V.: Gate evaluation secret sharing and secure one-round two-party computation. In: Roy, B. (ed.) ASIACRYPT 2005. LNCS, vol. 3788, pp. 136–155. Springer, Heidelberg (2005)
48. Kushilevitz, E., Nisan, N.: Communication complexity (1997)
49. Lupanov, O.: A method of circuit synthesis. In: Izvesitya VUZ, Radiofizika, pp. 120–140 (1958)
50. Mintz, Y.: Information ratios of graph secret-sharing schemes. Master's thesis, Ben Gurion University, Israel (2012)
51. Prabhakaran, V., Prabhakaran, M.: Assisted common information with an application to secure two-party sampling. In: arxiv:1206.1282v1 (2012)
52. Rabin, M.O.: How to exchange secrets with oblivious transfer. In: Technical Report TR-81, Aiken Computation Lab, Harvard University (1981)
53. Shannon, C.: The synthesis of two-terminal switching circuits. Bell System Technical Journal, 59–98 (1949)
54. Sun, H., Shieh, S.: Secret sharing in graph-based prohibited structures. In: INFOCOM, pp. 718–724 (1997)
55. van Dijk, M.: On the information rate of perfect secret sharing schemes. In: Designs, Codes and Cryptography, pp. 143–169 (1995)
56. Winkler, S., Wullschleger, J.: On the efficiency of classical and quantum oblivious transfer reductions. In: Rabin, T. (ed.) CRYPTO 2010. LNCS, vol. 6223, pp. 707–723. Springer, Heidelberg (2010)
57. Woodruff, D., Yekhanin, S.: A geometric approach to information-theoretic private information retrieval. SIAM J. Comp. 37(4), 1046–1056 (2007)
58. Yekhanin, S.: Towards 3-query locally decodable codes of subexponential length. In: STOC, pp. 266–274 (2007)

A Multiparty Secure Computation

Notation. Let $\mathcal{F}_N^k$ denote the set of all boolean functions from $[N]^k$ to $\{0,1\}$. We will interpret $f \in \mathcal{F}_N^k$ as a k-party function. Also, we consider t-private k-server PIR for $\mathcal{F}_N^k$, a natural generalization of 1-private k-server PIR for $\mathcal{F}_N$ defined in Section 2.

The following theorems summarize the connections between t-private k-server PIR, and multiparty secure computation in the plain model, OT-hybrid model, and the preprocessing model. The protocols implied by the theorems are straightforward extensions of the ideas behind the protocols of Sections 4 and 5.

Theorem 5. *Let* $\mathcal{P} = (\mathcal{Q}, \mathcal{A}, \mathcal{R})$ *be a t-private k-server PIR scheme for* $\mathcal{F}_N^k$. *Then, for any k-party functionality* $f : [N]^k \to \{0,1\}$, *the following hold:*

- *There is a perfectly secure k-party protocol π that realizes f in the* plain model, *tolerates $t < k/2$ passively corrupt parties, and has communication complexity* $O(k^2 \cdot (\tau(\mathcal{Q}) + \tau(\mathcal{R})))$.
- *There is a perfectly secure k-party protocol π that realizes f in the* OT-hybrid model, *tolerates $t < k$ passively corrupt parties, and has communication complexity* $O(k^2 \cdot (\tau(\mathcal{Q}) + \tau(\mathcal{R})))$.

Theorem 6. *Let* $\mathcal{P} = (\mathcal{Q}, \mathcal{A}, \mathcal{R})$ *be a t-private $(k+1)$-server PIR scheme for* $\mathcal{F}_N^k$. *Then, for any k-party functionality* $f : [N]^k \to \{0,1\}$, *there is a perfectly secure k-party protocol π that realizes f in the* preprocessing *model, and tolerates $t < k$ passively corrupt parties, and has correlated randomness complexity (and communication complexity)* $O(k^2 \cdot (\tau(\mathcal{Q}) + \tau(\mathcal{R})))$.

Plugging in parameters from the best known t-private k-server (resp. $(k+1)$-server) PIR protocols [10,57] in Theorem 5 (resp. Theorem 6), we obtain the following corollary.

Corollary 4. *Let* $f : [N]^k \to \{0,1\}$ *be any k-party functionality. Then,*

- *There is a perfectly secure k-party protocol π that realizes f in the* plain model, *tolerates $t < k/2$ passively corrupt parties, and has communication complexity* $N^{k/\lfloor 2k-1/t \rfloor} \cdot \mathrm{poly}(k)$.
- *There is a perfectly secure k-party protocol π that realizes f in the* OT-hybrid model, *tolerates $t < k$ passively corrupt parties, and has communication complexity* $N^{k/\lfloor 2k-1/t \rfloor} \cdot \mathrm{poly}(k)$.
- *There is a perfectly secure k-party protocol π that realizes f in the* preprocessing model, *tolerates $t < k$ passively corrupt parties, and has correlated randomness complexity (and communication complexity)* $N^{k/\lfloor 2k+1/t \rfloor} \cdot \mathrm{poly}(k)$.

We point out that for the specific case of $t = k - 1$ our protocol in the OT-hybrid model has communication complexity $N^{k/2} \cdot \mathrm{poly}(k)$ which improves over prior work which had complexity $N^{k-1} \cdot \mathrm{poly}(k)$ [34].

For the case of honest majority, it is possible to obtain better results for t-private k-party computation when $k \geq 3^t$ via the best known t-private 3^t-server PIR protocols obtained by boosting (via [2]) the PIR protocols of [58,27,11].

Corollary 5. *For any $t \geq 0$, and for any $k \geq 3^t$-party functionality $f : [N]^k \rightarrow \{0,1\}$, there is a protocol π that realizes f in the plain model, and has the following features:*

- *π is perfectly secure, and tolerates t passively corrupt parties;*
- *The total communication complexity is $2^{\tilde{O}(\sqrt{k \log N})} \cdot \mathrm{poly}(k)$.*

B   Extension to the Malicious Setting

In this section, we show how to compile our semihonest secure protocols for secure computation in the OT-hybrid/preprocessing/plain model in to malicious secure protocols for secure computation in the respective models. The high level idea is to use the IPS compiler [43], which is parameterized by an outer malicious secure protocol (that helps computing the target function) and an inner semihonest secure protocol (for simulating the next message function of the outer protocol). The main challenge is in implementing the compiler while somewhat preserving the complexity of the underlying semihonest secure protocol.

To this end, the outer protocol that we employ is inspired by the instance hiding scheme of Beaver et al. [6]. If f represents the target function that we need to realize, then we set the target function of the outer protocol, say g to be, for parameter m, an m-variate degree-d polynomial over $\mathbb{F}$ obtained by *arithmetizing* f. To evaluate a function g, our outer protocol will use k parties (where k depends on the size of the input domain N), that evaluate g on *shares* of the actual input. Note that (1) the actual parties need to distribute shares computed from the joint input of both parties to the k virtual parties, and (2) each of the k virtual parties compute their next message which is the evaluation of g on the share they received. The share computation step depends only on the length of g's input, and the number of virtual parties. To evaluate g, the actual parties first interpret g as a boolean function g^* (with multi-bit output), and then use our semihonest secure protocol multiple times to evaluate each output bit of g^*. In other words, our semihonest secure protocol acts as the IPS compiler's inner protocol. The final output is obtained as in the scheme of [6] via polynomial interpolation, which in our compiled protocol will be performed using a secure computation protocol.

We summarize the discussion by stating the final theorems that we obtain, and defer the proofs to the full version.

Theorem 7. *Let σ be a statistical security parameter. For all $\epsilon > 0$, and for any 2-party functionality $f : [N] \times [N] \rightarrow \{0,1\}$, there is a protocol π that realizes f in the OT-hybrid model; this protocol is statistically secure against malicious parties, and has total communication complexity (including communication with the OT oracle) $\tilde{O}(N^{\frac{2}{3}+\epsilon}) + \mathrm{poly}(\sigma, \log N, 1/\epsilon)$.*

Theorem 8. *Let σ be a statistical security parameter. For any 2-party functionality $f : [N] \times [N] \to \{0,1\}$, there is a protocol π that realizes f in the preprocessing model; this protocol is statistically secure against malicious parties, and has total communication complexity and correlated randomness complexity $2^{\widetilde{O}(\sqrt{\log N})} + \mathrm{poly}(\sigma, \log N)$.*

Theorem 9. *Let σ be a statistical security parameter. For any 3-party functionality $f : [N] \times [N] \times [N] \to \{0,1\}$, there is a protocol π that realizes f in the plain model; this protocol is statistically secure against a single malicious party, and has total communication complexity $2^{\widetilde{O}(\sqrt{\log N})} + \mathrm{poly}(\sigma, \log N)$.*

Constant-Round Black-Box Construction of Composable Multi-Party Computation Protocol

Susumu Kiyoshima[1], Yoshifumi Manabe[2], and Tatsuaki Okamoto[1]

[1] NTT Secure Platform Laboratories, Japan
{kiyoshima.susumu,okamoto.tatsuaki}@lab.ntt.co.jp
[2] Kogakuin University, Japan
manabe@cc.kogakuin.ac.jp

Abstract. We present the first general MPC protocol that satisfies the following: (1) the construction is black-box, (2) the protocol is universally composable in the plain model, and (3) the number of rounds is constant. The security of our protocol is proven in angel-based UC security under the assumption of the existence of one-way functions that are secure against sub-exponential-time adversaries and constant-round semi-honest oblivious transfer protocols that are secure against quasi-polynomial-time adversaries. We obtain the MPC protocol by constructing a constant-round CCA-secure commitment scheme in a black-box way under the assumption of the existence of one-way functions that are secure against sub-exponential-time adversaries. To justify the use of such a sub-exponential hardness assumption in obtaining our constant-round CCA-secure commitment scheme, we show that if black-box reductions are used, there does not exist any constant-round CCA-secure commitment scheme under any falsifiable polynomial-time hardness assumptions.

1 Introduction

Protocols for *secure multi-party computation* (MPC) enable mutually distrustful parties to compute a functionality without compromising the correctness of the outputs and the privacy of the inputs. In the seminal work of Goldreich et al. [14], a general MPC protocol was constructed in a model with malicious adversaries and a dishonest majority.[1] (By "a general MPC protocol," we mean a protocol that can be used to securely compute any functionality.)

Black-box Constructions. A construction of a protocol is *black-box* if it uses the underlying cryptographic primitives only in a black-box way (that is, only through their input/output interfaces). In contrast, if a construction uses the codes of the underlying primitives, it is *non-black-box*.

Obtaining black-box constructions is an important step toward obtaining practical MPC protocols. This is because black-box constructions are typically more efficient than non-black-box ones. (Typical non-black-box constructions,

---

[1] In the following, we consider only such a model.

Y. Lindell (Ed.): TCC 2014, LNCS 8349, pp. 343–367, 2014.

such as that of [14], use the codes of the primitives to compute NP reductions in general zero-knowledge proofs. Thus, they should be viewed as feasibility results.) Black-box constructions are also theoretically interesting, since understanding whether non-black-box use of primitives is necessary for a cryptographic task is of great theoretical interest.

Recently, a series of works showed black-box constructions of general MPC protocols. Ishai et al. [20] showed the first construction of a general MPC protocol that uses the underlying low-level primitives in a black-box way. Combined with the subsequent work of Haitner [18], their work showed a black-box construction of a general MPC protocol based on a semi-honest oblivious transfer protocol [19]. Subsequently, Wee [37] showed an $O(\log^* n)$-round protocol under polynomial-time hardness assumptions and a constant-round protocol under sub-exponential-time hardness assumptions, and Goyal [15] showed a constant-round protocol under polynomial-time hardness assumptions.

The security of these black-box protocols is considered in the *stand-alone setting*. That is, the protocols of [15, 20, 37] are secure in the setting where only a single instance of the protocol is executed at a time.

Composable Security. The *concurrent setting*, in which many instances of protocols are executed concurrently in an arbitrary schedule, is a more general and realistic setting than the stand-alone one. In the concurrent setting, an adversary can perform a coordinated attack in which he chooses his messages in an instance based on the executions of the other instances.

As a strong and realistic security notion in the concurrent setting, Canetti [2] proposed *universally composable (UC) security*. The main advantage of UC security is *composability*, which guarantees that when we compose many UC-secure protocols, we can prove the security of the resultant protocol using the security of its components. Thus, UC security enables us to construct protocols in a modular way. Composability also guarantees that a protocol remains secure even when it is concurrently executed with any other protocols in any schedule. Canetti et al. [8] constructed a UC-secure general MPC protocol in the *common reference string (CRS) model* (i.e., in a model in which all parties are given a common public string that is chosen by a trusted third party).

UC security, however, turned out to be too strong to achieve in the *plain model* (i.e., in a model without any trusted setup except for authenticated communication channels). That is, we cannot construct UC-secure general MPC protocols in the plain model [3, 6].

To achieve composable security in the plain model, Prabhakaran and Sahai [36] proposed a variant of UC security called *angel-based UC security*. Roughly speaking, angel-based UC security is the same as UC security except that the adversary and the simulator have access to an additional entity—the *angel*—that allows some judicious use of super-polynomial-time resources. It was proven that, like UC security, angel-based UC security guarantees composability. Furthermore, as argued in [36], angel-based UC security guarantees meaningful security in many cases. (For example, angel-based UC security implies *super-polynomial-time simulation (SPS) security* [1, 12, 29, 31]. In SPS security, we allow the simulator to run in super-polynomial time. Thus, SPS security guarantees that whatever an adversary can do in the real world can also be done in the ideal world in super-polynomial time.) Then, Prabhakaran and Sahai [36] presented a general MPC protocol that satisfies this security notion in the plain model, based on new (unstudied and non-standard) assumptions. Subsequently, Malkin et al. [25] constructed another general MPC protocol that satisfies this security notion in the plain model based on new number-theoretic assumption. In [1], Barak and Sahai remarked that their protocol (which is SPS secure under subexponential-time hardness assumptions) can be shown to be secure in angel-based UC security.

Recently, Canetti et al. constructed a polynomial-round general MPC protocol in angel-based UC security based on a standard assumption (the existence of enhanced trapdoor permutations). Subsequently, Lin [21] and Goyal et al. [17] reduced the round complexity to $\widetilde{O}(\log n)$ under the same assumption. They also proposed constant-round protocols, where the security is based on a super-polynomial-time hardness assumption (the existence of enhanced trapdoor permutations that are secure against quasi-polynomial-time adversaries). These constructions, however, use the underlying primitives in a non-black-box way.

Black-Box Constructions of Composable Protocols. Lin and Pass [23] showed the first black-box construction of a general MPC protocol that guarantees composable security in the plain model. The security of their protocol is proven under angel-based UC security, and based on the minimum assumption of the existence of semi-honest oblivious transfer (OT) protocols.

The round complexity of their protocol is $O(n^\epsilon)$, where $\epsilon > 0$ is an arbitrary constant. In contrast, for non-black-box constructions of composable protocols, we have constant-round protocols in the plain model (under non-standard assumptions or super-polynomial-time hardness assumptions) [17,21,25,36]. Thus, a natural question is the following.

> *Does there exist a constant-round black-box construction of a general MPC protocol that guarantees composability in the plain model (possibly under super-polynomial-time hardness assumptions)?*

1.1   Our Result

In this paper, we answer the above question affirmatively.

Theorem (Informal). *Assume the existence of one-way functions that are secure against sub-exponential-time adversaries and constant-round semi-honest oblivious transfer protocols that are secure against quasi-polynomial-time adversaries. Then, there exists a constant-round black-box construction of a general MPC protocol that satisfies angel-based UC security in the plain model.*

The formal statement of this theorem is given in Section 7.

CCA-Secure Commitment Schemes. We prove the above theorem by constructing a constant-round *CCA-secure commitment scheme* [7,23] in a black-box way. Once we obtain a CCA-secure commitment scheme, we can construct a general MPC protocol in essentially the same way as Lin and Pass do in [23].

Roughly speaking, a CCA-secure commitment scheme is a tag-based commitment scheme (i.e., a commitment scheme that takes an n-bit string, or *tag*, as an additional input) such that the committed value of a commitment with tag id remains hidden even if the receiver has access to a super-polynomial-time oracle—the *committed-value oracle*—that returns the committed value of any commitment with tag id$'$ $\neq$ id. Lin and Pass [23] showed an $O(n^\epsilon)$-round black-box construction of a CCA-secure commitment scheme for arbitrary $\epsilon > 0$ by assuming the minimum assumption of the existence of one-way functions.

Our main technical result is the following.

Theorem (Informal). *Assume the existence of one-way functions that are secure against sub-exponential-time adversaries. Then, there exists a constant-round black-box construction of a CCA-secure commitment scheme.*

The formal statement of this theorem is given in Section 7.

To obtain our CCA-secure commitment scheme, we use the idea of *non-malleability amplification* that was used in previous works on concurrent non-malleable (NM) commitment schemes [22,34]. That is, we construct a CCA commitment scheme in the following steps.

Step 1. We say that a commitment scheme is *one-one CCA secure* if it is CCA secure with respect to restricted classes of adversaries that receive only a single answer from the oracle. Then, we construct a constant-round one-one CCA-secure commitment for tags of length $O(\log \log \log n)$.

Step 2. We construct a transformation from the commitment scheme constructed in Step 1 to a CCA-secure commitment for tags of length $O(n)$ with a constant additive increase in round complexity. Toward this end, we construct the following two transformations:

- A transformation from any one-one CCA-secure commitment scheme for tags of length $t(n)$ to a CCA-secure commitment scheme for tags of length $t(n)$ with a constant additive increase in round complexity
- A transformation from any CCA-secure commitment scheme for tags of length $t(n)$ to a one-one CCA-secure commitment scheme for tags of length $2^{t(n)-1}$ with no increase in round complexity

(The latter transformation is essentially the same as the "DDN $\log n$ trick" [11,24].) By repeatedly composing these two transformations, we obtain the desired transformation.

On the Use of Super-Polynomial-Time Hardness Assumption. Although the round complexity of our CCA-secure commitment scheme is constant, it relies on a super-polynomial-time hardness assumption. (Recall that the $O(n^\epsilon)$-round CCA-secure commitment scheme of [23] relies on a polynomial-time hardness assumption.)

We show that the use of such a strong assumption is *inevitable*, as long as the security of a constant-round CCA-secure commitment scheme is proven under *falsifiable assumptions* [13, 28] by using a *black-box reduction*. Roughly speaking, a falsifiable assumption is an assumption that is modeled as an interactive game between a challenger and an adversary such that the challenger can decide whether the adversary won the game in polynomial time. Then, we say that *the CCA security of a commitment scheme $\langle C, R \rangle$ is proven under a falsifiable assumption by using a black-box reduction* if the CCA security of $\langle C, R \rangle$ is proven by constructing a PPT Turing machine $\mathcal{R}$ such that for any adversary $\mathcal{A}$ that breaks the CCA security of $\langle C, R \rangle$, $\mathcal{R}$ can break the assumption by using $\mathcal{A}$ only in a black-box way. Then, we show the following theorem.

Theorem (Informal). *Let $\langle C, R \rangle$ be any constant-round commitment scheme. Then, the CCA security of $\langle C, R \rangle$ cannot be proven by using black-box reductions under any falsifiable polynomial-time hardness assumption.*

(Due to lack of space, we defer the formal statement of this theorem and its proof to the full version. Roughly speaking, we obtain this theorem by using techniques of the negative result on concurrent zero-knowledge protocols [5].) Since all standard cryptographic assumptions are falsifiable, this theorem says that if we want to construct a constant-round CCA-secure commitment scheme based on standard assumptions, we must use either super-polynomial-time hardness assumptions (as this paper does) or non-black-box reductions.[2]

We note that this negative result holds *even for non-black-box constructions*. That is, we cannot construct constant-round CCA-secure commitment schemes even when we use primitives in a non-black-box way, as long as we use black-box reductions and polynomial-time hardness assumptions.

2 Overview of the Protocols

In this section, we give overviews of our main technical results: a one-one CCA-secure commitment scheme for short tags and a transformation from one-one CCA security to CCA security.

2.1 One-One CCA-Security for Short Tags

We obtain our one-one CCA-secure commitment scheme by observing that the non-black-box construction of a NM commitment scheme of [34] is one-one CCA secure and converting it into a black-box one.

First, we recall the scheme of [34].[3] The starting point of the scheme is "two-slot message length" technique [30]. The basic idea of the technique is to let the receiver sequentially send two challenges—one "long" and one "short"—where

---

[2] We note that, although very recently Goyal [16] showed how to use non-black-box techniques in the fully concurrent setting, Goyal's technique requires polynomially many rounds.

[3] In the following, some of the text is taken from [34].

the length of the challenges are determined by the tag of the commitment. The protocol is designed so that the response to a shorter challenge does not help a man-in-the-middle adversary to provide a response to a longer challenge. A key conceptual insight of [34] is to rely on the complexity leveraging technique [4] to construct these challenges: For one-way functions with sub-exponential hardness, an oracle for inverting challenges of length $n^{o(1)}$ (the "short" challenge) does not help invert random challenges of length n (the "long" challenge), since we can simulate such an oracle by brute force in time $2^{n^{o(1)}}$.

More precisely, the scheme of [34] is as follows. Let $d = O(\log\log n)$ be the number of tags, and let $n^{\omega(1)} = T_0(n) \ll T_1(n) \ll \cdots \ll T_{d+2}(n)$ be a hierarchy of running times. Then, to commit to $v \in \{0,1\}^n$ with tag id $\in \{0,1,\ldots,d-1\}$, the committer C does the following with the receiver R.

1. C commits to v by using a statistically binding commitment Com that is hiding against $T_{d+1}(n)$-time adversaries but is completely broken in time $T_{d+2}(n)$.
2. (Slot 1) C proves knowledge of v by using a zero-knowledge argument of knowledge that is computationally sound against $T_{\mathsf{id}+1}(n)$-time adversaries and can be simulated in straight line in time $o(T_{\mathsf{id}+2}(n))$, where the simulated view is indistinguishable from the real one in time $T_{d+2}(n)$.
3. (Slot 2) C proves knowledge of v by using a zero-knowledge argument of knowledge that is computationally sound against $T_{d-\mathsf{id}}(n)$-time adversaries and can be simulated in straight line in time $o(T_{d-\mathsf{id}+1}(n))$, where the simulated view is indistinguishable from the real one in time $T_{d+2}(n)$.

We can show that the scheme of [34] is one-one CCA secure as follows (by using essentially the same proof as the proof of its non-malleability). Recall that a commitment scheme is one-one CCA secure if it is hiding against adversaries that give a single query to the committed-value oracle $\mathcal{O}$. Let id be the tag used in the *left session* (a commitment from the committer to the adversary $\mathcal{A}$) and $\widetilde{\mathsf{id}}$ be the tag used in the *right session* (a commitment from $\mathcal{A}$ to $\mathcal{O}$). Then, let us consider a hybrid experiment in which the proofs in the second and third steps are replaced with the straight-line simulations in the left session. Since the running time of $\mathcal{O}$ is at most $T_{d+2}(n)$, the zero-knowledge property guarantees that the view of $\mathcal{A}$ in the hybrid experiment is indistinguishable from that of $\mathcal{A}$ in the real experiment even when $\mathcal{A}$ interacts with $\mathcal{O}$. Furthermore, in the right session of the hybrid experiment, the soundness of the zero-knowledge argument still holds either in the second step or in the third step. This follows from the following reasons. For simplicity, let us consider a synchronized adversary.[4] Then, since the simulation of the second step takes at most time $o(T_{\mathsf{id}+2}(n))$ and the soundness of the second step holds against $T_{\widetilde{\mathsf{id}}+1}(n)$-time adversaries, the soundness of the second step holds if id $< \widetilde{\mathsf{id}}$; similarly, the soundness of the third step holds if id $> \widetilde{\mathsf{id}}$. In the hybrid experiment, therefore, the committed value v can be extracted by using the knowledge extractor either in the

---

[4] An synchronized adversary sends the i-th round message to $\mathcal{O}$ immediately after receiving the i-th round messages from the committer, and vise verse.

second step or in the third step, and thus the committed value oracle $\mathcal{O}$ can be simulated in time $o(\max(T_{\text{id}+2}(n), T_{d-\text{id}+1}(n))) \cdot \mathsf{poly}(n) \ll T_{d+1}(n)$. Then, from the hiding property of Com in the first step, the view of $\mathcal{A}$ in the hybrid experiment is computationally independent of the value v. Thus, one-one CCA security follows.

To convert the scheme of [34] into a black-box protocol, we use a black-box trapdoor commitment scheme TrapCom of [33]. We observe that TrapCom has similar properties to the zero-knowledge argument used in the scheme of [34]: TrapCom is extractable and a TrapCom commitment can be simulated in straight line in super-polynomial time. Then, we modify the scheme of [34] and let the committer commit to v instead of proving the knowledge of v. To ensure the "soundness," that is, to ensure that the committed value of TrapCom is v, we use the cut-and-choose technique and Shamir's secret sharing scheme in a similar manner to previous works on black-box protocols [9, 10, 23, 37]. That is, we let the committer commit to Shamir's secret sharing $s = (s_1, \ldots, s_{10n})$ of value v in all steps, let the receiver choose a random subset $\Gamma \subset [10n]$ of size n, and let the committer reveal s_j and decommit the corresponding commitments for every $j \in \Gamma$. The resultant scheme uses the underlying primitives only in a black-box way, and can be proven to be one-one CCA secure from a similar argument to the scheme of [34]. (We note that the actual scheme is a little more complicated. For details, see Section 4.) We note that Lin and Pass [23] also use TrapCom to convert a non-black-box protocol into a black-box one. Unlike them, who mainly use the fact that TrapCom is extractable and is secure against selective opening attacks, we also use the fact that TrapCom commitments are straight-line simulatable.

2.2 CCA Security from One-One CCA Security

We give an overview of the transformation from any one-one CCA-secure commitment scheme to a CCA-secure commitment scheme. Let $n^{\omega(1)} = T_0(n) \ll T_1(n) \ll T_2(n) \ll T_3(n)$ be a hierarchy of running times. Then, we construct a CCA-secure commitment scheme CCACom_0 that is secure against $T_0(n)$-time adversaries from a one-one CCA-secure commitment scheme $\mathsf{CCACom}_3^{1:1}$ that is secure against $T_3(n)$-time adversaries. Let Com_1 be a 2-round statistically binding commitment scheme that is secure against $T_1(n)$-time adversaries but is completely broken in time $o(T_2(n))$, and CECom_2 be a constant-round commitment scheme that is hiding against $T_2(n)$-time adversaries and is concurrently extractable by rewinding the committer $\mathsf{poly}(n^{\log n})$ times [26, 32]. Then, to commit to value v, the committer C does the following with the receiver R.

1. R commits to a random subset $\Gamma \subset [10n]$ of size n by using $\mathsf{CCACom}_3^{1:1}$.
2. C computes an $(n+1)$-out-of-$10n$ Shamir's secret sharing $s = (s_1, \ldots, s_{10n})$ of value v and commits to s_j for each $j \in [10n]$ in parallel by using Com_1.
3. C commits to s_j for each $j \in [10n]$ in parallel by using CECom_2.
4. R decommits the commitment of the first step and reveal Γ.
5. For each $j \in \Gamma$, C decommits the Com_1 and CECom_2 commitments whose committed values are s_j.

The committed value of CCACom_0 is determined by the committed values of Com_1. Thus, the running time of $\mathcal{O}$ is at most $o(T_2(n)) \cdot \mathsf{poly}(n) \ll T_2(n)$.

To prove the CCA security of the scheme, we consider a series of hybrid experiments.

In the first hybrid, in the left interaction the committed value Γ of $\mathsf{CCACom}_3^{1:1}$ is extracted by brute force and the committed value of CECom_2 is switched from s_j to 0 for every $j \notin \Gamma$. Note that, during the CECom_2 commitments of the left, the combined running time of $\mathcal{A}$ and $\mathcal{O}$ is at most $T_2(n)$. Thus, from the hiding property of CECom_2, the view of $\mathcal{A}$ in the first hybrid is indistinguishable from that of $\mathcal{A}$ in the honest experiment.

The second hybrid is the same as the first one except for the following: in every right session of which the second step ends after the start of the second step of the left session, the committed values of the CECom commitments are extracted; then, the answer of $\mathcal{O}$ are computed from the extracted values (instead of the committed values of Com_1). We note that, since the second hybrid differs from the first one only in how the answers of $\mathcal{O}$ are computed, to show the indistinguishability it suffices to show that in the first hybrid the committed values of CECom_2 agree with those of Com_1 in "most" indexes in every right session. We first note that if we ignore the messages that $\mathcal{A}$ receives in the left session, we can prove that the committed values of CECom_2 agree with those of Com_1 in most indexes by using the property of the cut-and-choose technique. In the hybrid, however, $\mathcal{A}$ receives messages in the left session, in which Γ is extracted by brute force and the committed values of CECom_2 disagree with those of Com_1 in 90% of indexes. Thus, $\mathcal{A}$ may be able to use the messages in the left to break the hiding property of $\mathsf{CCACom}^{1:1}$ in the right. (Note that, if $\mathcal{A}$ can break the hiding property of $\mathsf{CCACom}^{1:1}$, we cannot use the property of the cut-and-choose technique.) We show that $\mathcal{A}$ cannot break the hiding property of $\mathsf{CCACom}^{1:1}$ even with the messages of the left session. A key is that given Γ, the left session can be simulated in polynomial time. Hence, one-one CCA security of $\mathsf{CCACom}^{1:1}$ guarantees that the messages of the left session are useless for breaking the hiding property of $\mathsf{CCACom}^{1:1}$. Thus, even with messages of the left session, the cut-and-choose guarantees that the committed values of CECom_2 agree with those of Com_1 in most indexes. The view of $\mathcal{A}$ in the second hybrid is therefore indistinguishable from that of $\mathcal{A}$ in the first one.

The third hybrid is the same as the first one except that in the left session, the committed value of Com_1 is switched from s_j to 0 for every $j \notin \Gamma$. Note that during the Com_1 commitments of the left, the combined running time of $\mathcal{A}$ and $\mathcal{O}$ is at most $T_0(n) \cdot \mathsf{poly}(n^{\log n}) \ll T_1(n)$. This is because

- for every right session in which $\mathcal{A}$ completes the second step before the start of the second step of the left session, the answer of $\mathcal{O}$ (i.e., the committed value of CCACom_0) can be computed before the start of Com_1 commitments of the left session, and

- for every right session in which $\mathcal{A}$ completes the second step after the start of the second step of the left session, the answer of $\mathcal{O}$ is computed by extracting the committed values of CECom_2, which requires rewinding $\mathcal{A}$ at most $\mathsf{poly}(n^{\log n})$ times.

Thus, from the hiding property of Com_1, the view of $\mathcal{A}$ in the third hybrid is indistinguishable from that of $\mathcal{A}$ in the second one.

Note that, since s is $(n+1)$-out-of-$10n$ secret sharing, $\mathcal{A}$ receives no information of v in the third hybrid. Thus, the view of $\mathcal{A}$ in the third hybrid is independent of v, and thus the CCA security follows.

3 Preliminaries

In this section, we explain the assumptions and the definitions that we use in this paper.

3.1 Assumptions

For our CCA-secure commitment scheme, we use a one-way function f that is secure against 2^{n^ϵ}-time adversaries, where $\epsilon < 1$ is a positive constant. Without loss of generality, we assume that f can be inverted in time 2^n. Let $T_i(n) \stackrel{\text{def}}{=} 2^{(\log n)^{(2/\epsilon)^{10i+1}}}$ for $i \in \mathbb{N}$. Then, by setting the security parameter of f to $\ell_i(n) = (\log n)^{(2/\epsilon)^{10i+2}}$, we obtain a one-way function f_i that is secure against $T_i(n)$-time adversaries but can be inverted in time less than $T_{i+0.5}(n)$. We note that when $i \leq O(\log\log n)$, we have $\ell_i(n) \leq \mathsf{poly}(n)$.

For our composable MPC protocol, we additionally use semi-honest oblivious transfer protocols that are secure against $2^{\mathsf{poly}(\log n)}$-time adversaries.

3.2 Shamir's Secret Sharing Scheme

In this paper, we use Shamir's $(n+1)$-out-of-$10n$ secret sharing scheme. For any positive real number $x \leq 1$ and any $s = (s_1, \ldots, s_{10n})$ and $s' = (s'_1, \ldots, s'_{10n})$, we say that s and s' are x-close if $|\{i \mid s_i = s'_i\}| \geq x \cdot 10n$. We note that Shamir's secret sharing is a codeword of Reed-Solomon code with minimum relative distance 0.9. Thus, for any $x > 0.55$ and any s that is x-close to a valid codeword w, we can compute w from s.

3.3 Commitment Schemes

Recall that commitment schemes are two-party protocols between the committer C and the receiver R. A transcript of the commit phase is *accepted* if R does not abort in the commit phase. A transcript of the commit phase is *valid* if there exists a valid decommitment of this transcript. We use a 2-round statistically binding commitment scheme Com based on one-way functions [27].

Strong computational binding property. We say that a commitment scheme $\langle C, R \rangle$ satisfies a *strong computational binding property* if for any PPT committer C^* interacting with the honest receiver R, the probability that C^* generates a commitment that has more than one committed value is negligible.[5]

3.4   Extractable Commitments

Roughly speaking, a commitment scheme is *extractable* if there exists an expected polynomial-time oracle machine (or *extractor*) E such that for any committer C^*, E^{C^*} extracts the value that C^* commits to whenever the commitment is valid. We note that when the commitment is invalid, E may output a garbage value. (This is called *over-extraction.*)

There exists a 4-round extractable commitment scheme ExtCom based on one-way functions [33]. The commit phase of ExtCom consists of three stages— commit, challenge, and reply—and given two accepted transcripts that have the same commit message but have different challenge messages, we can extract the committed value. Thus, we can extract the committed value by rewinding the committer and obtaining two such transcripts. In the following, we use *slot* to denote a pair of the challenge and reply messages in ExtCom. As shown in [33], ExtCom is in fact *parallel extractable*. Thus, even when a committer commits to many values in parallel, we can extract all committed values.

3.5   Concurrently Extractable Commitments

Roughly speaking, a commitment scheme is *concurrently extractable* if there exists an expected polynomial-time extractor E such that for any committer C^* that concurrently commits to many values, E^{C^*} extracts the committed value of each commitment immediately after C^* generates each commitment.

Micciancio et al. [26] showed a concurrently extractable commitment CECom, which consists of r executions of ExtCom, where r is a parameter (see Figure 1). Note that CECom has r sequential slots. Then, by using the rewinding strategy of [35], the committed values of CECom are concurrently extractable when $r = \omega(\log n)$.

Concurrently $T(n)$-Extractable Commitments

For any function $T(n)$, we consider a relaxed notion of concurrent extractability called *concurrent $T(n)$-extractability*, which is the same as concurrent extractability except that the expected running time of the extractor is $T(n)$.

By using the rewinding strategy of [32], we can show that CECom is concurrently $\mathsf{poly}(n^{\log n})$-extractable when $r \geq 3$. Note that in the stand-alone setting, we can extract the committed value of CECom by rewinding any single slot.

---

[5] The standard computational binding property guarantees only that for any PPT committer C^*, the commitment that C^* generates cannot be decommitted to more than one value *in polynomial time*. Thus, this commitment may have more than one committed value.

Commit Phase. The committer C and the receiver R receive common inputs 1^n and parameter r. To commit to $v \in \{0,1\}^n$, the committer C does the following.

Step 1. C and R execute commit stage of ExtCom r times in parallel.

Step $2i$ $(i \in [r])$. R sends the challenge message of ExtCom for the i-th session.

Step $2i+1$ $(i \in [r])$. C sends the reply message of ExtCom for the i-th session.

Decommit Phase. C sends v to R and decommits all the ExtCom commitments in the commit phase.

Fig. 1. Concurrently extractable commitment CECom [26]

3.6 Trapdoor Commitments

Roughly speaking, *trapdoor commitments* are ones such that there exists a simulator that can generate a simulated commitment and can later decommit it to any value.

Pass and Wee [33] showed that the black-box protocol TrapCom in Figure 2 is a trapdoor bit commitment scheme. In fact, given the receiver's challenge e in advance, we can generate a simulated commitment and decommit it to both 0 and 1 in a straight-line manner (i.e., without rewinding the receiver) as follows. To generate a simulated commitment, the simulator internally simulates an interaction between C and $\mathcal{R}^*$ honestly except that in Step 2, the simulator chooses random $\gamma \in \{0,1\}$ and lets each v_i be a matrix such that the e_i-th row of v_i is (η_i, η_i) and the $(1-e_i)$-th row of v_i is $(\gamma \oplus \eta_i, (1-\gamma) \oplus \eta_i)$. To decommit the simulated commitment to $\sigma \in \{0,1\}$, the simulator decommits all the commitments in the $(\sigma \oplus \gamma)$-th column of each v_i.

From the extractability of ExtCom, we can show that TrapCom is extractable. In addition, by using the hiding property of Com, we can show that TrapCom satisfies the strong computational binding property. (Roughly speaking, if C^* generates a commitment that has more than one committed value, we can compute the committed value e of Com by extracting $v_1, \ldots, v_n$.)

Pass and Wee [33] showed that by running TrapCom in parallel, we obtain a black-box trapdoor commitment PTrapCom for multiple bits. PTrapCom also satisfies the strong computational binding property and extractability.

3.7 CCA-Secure Commitments

We recall the definition of CCA security and κ-robustness [7, 23]. *Tag-based commitment schemes* are ones such that both the committer and the receiver receive a string, or *tag*, as an additional input.

Commit Phase. To commit to $\sigma \in \{0,1\}$ on common input 1^n, the committer C does the following with the receiver R:

Step 1. R chooses a random n-bit string $e = (e_1, \ldots, e_n)$ and commits to e by using Com.

Step 2. For each $i \in [n]$, the committer C chooses a random $\eta_i \in \{0,1\}$ and sets

$$v_i := \begin{pmatrix} v_i^{00} & v_i^{01} \\ v_i^{10} & v_i^{11} \end{pmatrix} = \begin{pmatrix} \eta_i & \eta_i \\ \sigma \oplus \eta_i & \sigma \oplus \eta_i \end{pmatrix} .$$

Then, for each $i \in [n]$, $\alpha \in \{0,1\}$, and $\beta \in \{0,1\}$ in parallel, C commits to $v_i^{\alpha\beta}$ by using ExtCom; let $(v_i^{\alpha\beta}, d_i^{\alpha\beta})$ be the corresponding decommitment.

Step 3. R decommits the Step 1 commitment to e.

Step 4. For each $i \in [n]$, C sends $(v_i^{e_i 0}, d_i^{e_i 0})$ and $(v_i^{e_i 1}, d_i^{e_i 1})$ to R. Then, R checks whether these are valid decommitments and whether $v_i^{e_i 0} = v_i^{e_i 1}$.

Decommit Phase. C sends σ and random $\gamma \in \{0,1\}$ to R. In addition, for every $i \in [n]$, C sends $(v_i^{0\gamma}, d_i^{0\gamma})$ and $(v_i^{1\gamma}, d_i^{1\gamma})$ to R. Then, R checks whether $(v_i^{0\gamma}, d_i^{0\gamma})$ and $(v_i^{1\gamma}, d_i^{1\gamma})$ are valid decommitments for every $i \in [n]$ and whether $v_0^{0\gamma} \oplus v_0^{1\gamma} = \cdots = v_n^{0\gamma} \oplus v_n^{1\gamma} = \sigma$.

Fig. 2. Black-box trapdoor bit commitment TrapCom

CCA Security (w.r.t. the Committed-value Oracle). Roughly speaking, a tag-based commitment scheme $\langle C, R \rangle$ is *CCA-secure* if the hiding property of $\langle C, R \rangle$ holds even against adversary $\mathcal{A}$ that interacts with the *committed-value oracle* during the interaction with the committer. The committed-value oracle $\mathcal{O}$ interacts with $\mathcal{A}$ as an honest receiver in many concurrent sessions of the commit phase of $\langle C, R \rangle$ using tags chosen adaptively by $\mathcal{A}$. At the end of each session, if the commitment of this session is invalid or has multiple committed values, $\mathcal{O}$ returns $\perp$ to $\mathcal{A}$. Otherwise, $\mathcal{O}$ returns the unique committed value to $\mathcal{A}$.

More precisely, let us consider the following probabilistic experiment $\text{IND}_b(\langle C, R \rangle, \mathcal{A}, n, z)$ for each $b \in \{0,1\}$. On input 1^n and auxiliary input z, adversary $\mathcal{A}^{\mathcal{O}}$ adaptively chooses a pair of challenge values $v_0, v_1 \in \{0,1\}^n$ and an n-bit tag id $\in \{0,1\}^n$. Then, $\mathcal{A}^{\mathcal{O}}$ receives a commitment to v_b with tag id, and $\mathcal{A}$ outputs y. The output of the experiment is $\perp$ if during the experiment, $\mathcal{A}$ sends $\mathcal{O}$ any commitment using tag id. Otherwise, the output of the experiment is y. Let $\text{IND}_b(\langle C, R \rangle, \mathcal{A}, n, z)$ denote the output of experiment $\text{IND}_b(\langle C, R \rangle, \mathcal{A}, n, z)$.

Then, the CCA security of $\langle C, R \rangle$ is defined as follows.

Definition 1. *Let $\langle C, R \rangle$ be a tag-based commitment scheme and $\mathcal{O}$ be the committed-value oracle of $\langle C, R \rangle$. Then, $\langle C, R \rangle$ is CCA-secure (w.r.t the* committed-value oracle) *if for any* PPT *adversary* $\mathcal{A}$, *the following are computationally indistinguishable:*

- $\{\mathsf{IND}_0(\langle C, R \rangle, \mathcal{A}, n, z)\}_{n \in \mathbb{N}, z \in \{0,1\}^*}$
- $\{\mathsf{IND}_1(\langle C, R \rangle, \mathcal{A}, n, z)\}_{n \in \mathbb{N}, z \in \{0,1\}^*}$

If the length of the tags chosen by $\mathcal{A}$ *is* $t(n)$ *instead of* n, $\langle C, R \rangle$ *is* CCA-*secure for tags of length* $t(n)$. ◇

We also consider a relaxed notion of CCA security called *one-one CCA security*. In the definition of one-one CCA security, we consider adversaries that interact with $\mathcal{O}$ only in a single session of the commit phase.

In the following, we use *left session* to denote the session of the commit phase between the committer and $\mathcal{A}$, and use *right sessions* to denote the sessions between $\mathcal{A}$ and $\mathcal{O}$.

κ-robustness (w.r.t. the Committed-value Oracle). Roughly speaking, a tag-based commitment scheme is *κ-robust* if for any adversary $\mathcal{A}$ and any ITM B, the joint output of a κ-round interaction between $\mathcal{A}^{\mathcal{O}}$ and B can be simulated without $\mathcal{O}$ by a PPT simulator. Thus, the κ-robustness guarantees that the committed-value oracle is useless in attacking any κ-round protocol.

Formally, let $\langle C, R \rangle$ be a tag-based commitment scheme and $\mathcal{O}$ be the committed-value oracle of $\langle C, R \rangle$. For any constant $\kappa \in \mathbb{N}$, we say that $\langle C, R \rangle$ is κ-robust (w.r.t. the committed value oracle) if there exists a PPT oracle machine (or simulator) S such that for any PPT adversary $\mathcal{A}$ and any κ-round PPT ITM B, the following are computationally indistinguishable:

- $\{\mathsf{output}_{B, \mathcal{A}^{\mathcal{O}}}[\langle B(y), \mathcal{A}^{\mathcal{O}}(z) \rangle (1^n, x)]\}_{n \in \mathbb{N}, x, y, z \in \{0,1\}^n}$
- $\{\mathsf{output}_{B, S^{\mathcal{A}}}[\langle B(y), S^{\mathcal{A}}(z) \rangle (1^n, x)]\}_{n \in \mathbb{N}, x, y, z \in \{0,1\}^n}$

Here, for any ITM A and B, we use $\mathsf{output}_{A, B}[\langle A(y), B(z) \rangle (x)]$ to denote the joint output of A and B in an interaction between them on inputs x, y to A and x, z to B respectively.

We also consider a relaxed notion of κ-robustness called *κ-PQT-robustness*. In the definition of κ-PQT-robustness, we allow the simulator to run in quasi-polynomial time.

4 One-One CCA Security for Short Tags

In this section, we construct a one-one CCA-secure commitment for tags of length $O(\log \log \log n)$. (Due to lack of space, the full proof is deferred to the full version.) Since the length of the tags is $O(\log \log \log n)$, we can view each tag as a value in $\{0, 1 \ldots, d - 1 = O(\log \log n)\}$.

4.1   Building Blocks

Let $T_i(n) \stackrel{\text{def}}{=} 2^{(\log n)^{(2/\epsilon)^{10i+1}}}$ for $i \in \mathbb{N}$. Then, for constants $a, b \in \mathbb{N}$, $\mathsf{PTrapCom}_a^b$ is a commitment scheme such that

- the hiding property holds against any $T_a(n)$-time adversary but is completely broken in time $T_{a+0.5}(n)$,
- the strong computational binding property holds against any $T_b(n)$-time adversary, and
- there exists a $T_{b+0.5}(n)$-time straight-line simulator (of the trapdoor property) such that the simulated commitment is indistinguishable from the actual commitment in time $T_a(n)$. (This holds even when $T_{b+0.5}(n) \gg T_a(n)$.)

We can construct $\mathsf{PTrapCom}_a^b$ by appropriately setting the security parameters of Com and ExtCom in PTrapCom.

$\mathsf{PCETrapCom}_a^b$ is the same as $\mathsf{PTrapCom}_a^b$ except that we use CECom in Step 2 instead of ExtCom.

4.2   One-One CCA Security for Tags of Length $O(\log \log \log n)$

Lemma 1. *Let $\epsilon < 1$ be a positive constant, and for any $i \in \mathbb{N}$, let $T_i(n) \stackrel{\text{def}}{=} 2^{(\log n)^{(2/\epsilon)^{10i+1}}}$. Assume the existence of one-way functions that are secure against 2^{n^ϵ}-time adversaries. Then, for any $i \in \mathbb{N}$, there exists a constant-round commitment scheme $\mathsf{CCACom}_i^{1:1}$ that satisfies the following for any $T_i(n)$-time adversary.*

- *Strong computational binding property, and*
- *One-one CCA security for tags of length $O(\log \log \log n)$.*

Furthermore, $\mathsf{CCACom}_i^{1:1}$ uses the underlying one-way function only in a black-box way.

Proof. $\mathsf{CCACom}_i^{1:1}$ is shown in Figure 3. The binding property follows from that of $\mathsf{PTrapCom}_{i+d+1}^{i+d+1}$. Thus, it remains to show that $\mathsf{CCACom}_i^{1:1}$ is one-one CCA secure for tags of length $O(\log \log \log n)$.

To show that $\mathsf{CCACom}_1^{1:1}$ is one-one CCA secure, we show that for any $T_i(n)$-time adversary $\mathcal{A}$ that interacts with $\mathcal{O}$ only in a single session, the following are computationally indistinguishable:

- $\{\mathsf{IND}_0(\mathsf{CCACom}_i^{1:1}, \mathcal{A}, n, z)\}_{n \in \mathbb{N}, z \in \{0,1\}^*}$
- $\{\mathsf{IND}_1(\mathsf{CCACom}_i^{1:1}, \mathcal{A}, n, z)\}_{n \in \mathbb{N}, z \in \{0,1\}^*}$

At the end of the right session, the committed-value oracle $\mathcal{O}$ does the following. First, $\mathcal{O}$ computes the committed values $s = (s_1, \ldots, s_{10n})$ of the Stage 1 commitments by brute force. (If the committed value of the j-th commitment is not uniquely determined, s_j is defined to be $\perp$.) Then, $\mathcal{O}$ checks whether the following conditions hold: (1) s is 0.9-close to a valid codeword $w = (w_1, \ldots, w_{10n})$ and (2) for every $j \in \Gamma$ (where Γ is the subset that $\mathcal{O}$ sends to $\mathcal{A}$ in Stage 4),

Commit Phase. The committer C and the receiver R receive common inputs 1^n and id $\in \{0, 1, \ldots, d-1 = O(\log \log n)\}$. To commit to $v \in \{0,1\}^n$, the committer C does the following with the receiver R.

Stage 1. C computes an $(n + 1)$-out-of-$10n$ Shamir's secret sharing $s = (s_1, \ldots, s_{10n})$ of value v. Then, for each $j \in [10n]$ in parallel, C commits to s_j by using $\mathsf{PTrapCom}_{i+d+1}^{i+d+1}$. Let (s_j, d_j) be the decommitment of the j-th commitment.

Stage 2. For each $j \in [10n]$ in parallel, C commits to (s_j, d_j) by using $\mathsf{PCETrapCom}_{i+d+2}^{i+\mathsf{id}+1}$. Here, the number of slots in $\mathsf{PCETrapCom}_{i+d+2}^{i+\mathsf{id}+1}$ is $\max(3, r + 1)$, where r is the round complexity of $\mathsf{PTrapCom}_{i+d+1}^{i+d+1}$ in Stage 1.

Stage 3. For each $j \in [10n]$ in parallel, C commits to (s_j, d_j) by using $\mathsf{PCETrapCom}_{i+d+2}^{i+d-\mathsf{id}}$. Here, the number of slots in $\mathsf{PCETrapCom}_{i+d+2}^{i+d-\mathsf{id}}$ is $\max(3, r + 1)$.

Stage 4. R sends a random subset $\Gamma \subseteq [10n]$ of size n to C.

Stage 5. For each $j \in \Gamma$, C decommits the j-th Stage 2 commitment and the j-th Stage 3 commitment to (s_j, d_j). Then, R checks whether (s_j, d_j) is a valid decommitment of the j-th Stage 1 commitment.

Decommit Phase. C sends v, $s = (s_1, \ldots, s_{10n})$, and $d = (d_1, \ldots, d_{10n})$ to R. Then, R checks whether (s_j, d_j) is a valid decommitment of the j-th Stage 1 commitment for every $j \in [10n]$. Furthermore, R checks whether (1) s is 0.9-close to a valid codeword $w = (w_1, \ldots, w_{10n})$ and (2) for each $j \in \Gamma$, w_j is equal to the share that was revealed in Stage 5. Finally, R checks whether w is a codeword corresponding to v.

Fig. 3. One-one CCA-secure commitment $\mathsf{CCACom}_i^{1:1}$

w_j is equal to the share that was revealed in Stage 5. If both conditions hold, $\mathcal{O}$ recovers v from w and returns v to $\mathcal{A}$. Otherwise, $\mathcal{O}$ returns $v := \bot$ to $\mathcal{A}$. We note that the running time of $\mathcal{O}$ is at most $\mathrm{poly}(n) \cdot T_{i+d+1.5}(n)$.

To show the indistinguishability, we consider hybrid experiments $G_a^b(n, z)$ for $a \in \{0, 1, 2, 3\}$ and $b \in \{0, 1\}$.

Hybrid $G_0^b(n, z)$ is the same as experiment $\mathrm{IND}_b(\mathsf{CCACom}_i^{1:1}, \mathcal{A}, n, z)$.

Hybrid $G_1^b(n, z)$ is the same as $G_0^b(n, z)$ except for the following:

- In Stage 2 (resp., Stage 3) on the left, the left committer simulates the $10n$ commitments of $\mathsf{PCETrapCom}_{i+d+2}^{i+\mathsf{id}+1}$ (resp., $\mathsf{PCETrapCom}_{i+d+2}^{i+d-\mathsf{id}}$) by using the straight-line simulator.
- In Stage 5 on the left, for each $j \in \Gamma$, the left committer decommits the simulated commitment of $\mathsf{PCETrapCom}_{i+d+2}^{i+\mathsf{id}+1}$ (resp., $\mathsf{PCETrapCom}_{i+d+2}^{i+d-\mathsf{id}}$) to (s_j, d_j) by using the simulator.

We note that the running time of $G_1^b(n, z)$ is at most $\mathrm{poly}(n) \cdot T_{i+d+1.5}(n)$.

Hybrid $G_2^b(n, z)$ is the same as $G_1^b(n, z)$ except for the following:

- Let $\widetilde{\mathsf{id}}$ be the tag of the right session. In Stage 2 (resp., Stage 3) of the right session, the committed values of the $\mathsf{PCETrapCom}_{i+d+2}^{i+\widetilde{\mathsf{id}}+1}$ (resp., $\mathsf{PCETrapCom}_{i+d+2}^{i+d-\widetilde{\mathsf{id}}}$) commitments are extracted *without rewinding Stage 1 on the left* by using the technique of [7,22]. (That is, in Step 2 of each $\mathsf{PCETrapCom}_{i+d+2}^{i+\widetilde{\mathsf{id}}+1}$ (resp. $\mathsf{PCETrapCom}_{i+d+2}^{i+d-\widetilde{\mathsf{id}}}$) commitment, the committed values of CECom are extracted by rewinding a single slot that does not contain any Stage 1 messages of the left session. Such a slot must exist, since the number of slots in CECom is $\max(3, r+1)$.) Then, $\widehat{\boldsymbol{s}} = (\widehat{s}_1, \dots, \widehat{s}_{10n})$ is defined as follows: if there exists $a \in \{2, 3\}$ such that the extracted value $(\widehat{s}_j^{(a)}, \widehat{d}_j^{(a)})$ of the j-th commitment in Stage a is a valid decommitment of the j-th commitment in Stage 1, let $\widehat{s}_j \overset{\text{def}}{=} \widehat{s}_j^{(a)}$ (if both $(\widehat{s}_j^{(2)}, \widehat{d}_j^{(2)})$ and $(\widehat{s}_j^{(3)}, \widehat{d}_j^{(3)})$ are valid decommitments but $\widehat{s}_j^{(2)} \neq \widehat{s}_j^{(3)}$, let $\widehat{s}_j \overset{\text{def}}{=} \bot$); otherwise, let $\widehat{s}_j \overset{\text{def}}{=} \bot$.
- At the end of the right session, $\mathcal{O}$ checks whether the following conditions hold: (1) $\widehat{\boldsymbol{s}}$ is *0.8-close* to a valid codeword $\widehat{\boldsymbol{w}} = (\widehat{w}_1, \dots, \widehat{w}_{10n})$ and (2) for every $j \in \Gamma$, $\widehat{w}_j$ is equal to the share that was revealed in Stage 5. If both conditions hold, $\mathcal{O}$ recovers $\widehat{v}$ from $\widehat{\boldsymbol{w}}$ and returns $\widehat{v}$ to $\mathcal{A}$. Otherwise, $\mathcal{O}$ returns $\widehat{v} := \bot$ to $\mathcal{A}$. We note that $\mathcal{O}$ does not extract the committed values of the Stage 1 commitments.

We note that the expected running time of $G_2^b(n, z)$ is $\mathsf{poly}(n) \cdot T_{i+d+0.5}(n)$. **Hybrid** $G_3^b(n, z)$ is the same as $G_2^b(n, z)$ except that on the left, the Stage 1 commitments are simulated by the straight-line simulator of $\mathsf{PTrapCom}_{i+d+1}^{i+d+1}$.

Since $\mathcal{A}$ receives no information about $\{s_j\}_{j \notin \Gamma}$ in $G_3^0(n, z)$ and $G_3^1(n, z)$, the output of $G_3^0(n, z)$ and that of $G_3^1(n, z)$ are identically distributed. Then, we consider the following claims. In what follows, we use $\mathsf{G}_i^b(n, z)$ to denote the output of experiment $G_i^b(n, z)$.

Claim 1. *For each $b \in \{0, 1\}$, $\{\mathsf{G}_0^b(n, z)\}_{n \in \mathbb{N}, z \in \{0,1\}^*}$ and $\{\mathsf{G}_1^b(n, z)\}_{n \in \mathbb{N}, z \in \{0,1\}^*}$ are computationally indistinguishable.*

Claim 2. *For each $b \in \{0, 1\}$, $\{\mathsf{G}_1^b(n, z)\}_{n \in \mathbb{N}, z \in \{0,1\}^*}$ and $\{\mathsf{G}_2^b(n, z)\}_{n \in \mathbb{N}, z \in \{0,1\}^*}$ are statistically indistinguishable.*

Claim 3. *For each $b \in \{0, 1\}$, $\{\mathsf{G}_2^b(n, z)\}_{n \in \mathbb{N}, z \in \{0,1\}^*}$ and $\{\mathsf{G}_3^b(n, z)\}_{n \in \mathbb{N}, z \in \{0,1\}^*}$ are computationally indistinguishable.*

The lemma follows from these claims. $\square$

Proof (of Claim 1). $G_1^b(n, z)$ differs from $G_0^b(n, z)$ only in that the Stage 2 commitments and the Stage 3 commitments on the left are simulated by the simulator of $\mathsf{PCETrapCom}_{i+d+2}^{i+\widetilde{\mathsf{id}}+1}$ and that of $\mathsf{PCETrapCom}_{i+d+2}^{i+d-\widetilde{\mathsf{id}}}$. Then, since the running time of $G_0^b(n, z)$ and that of $G_1^b(n, z)$ are at most $\mathsf{poly}(n) \cdot T_{i+d+1.5}(n) \ll T_{i+d+2}(n)$, the claim follows from the trapdoor property of $\mathsf{PCETrapCom}_{i+d+2}^{i+\widetilde{\mathsf{id}}+1}$ and that of $\mathsf{PCETrapCom}_{i+d+2}^{i+d-\widetilde{\mathsf{id}}}$. $\square$

Next, we consider Claim 2. Note that $G_2^b(n, z)$ differs from $G_1^b(n, z)$ in that $\mathcal{O}$ computes the committed value of the right session from the extracted values of the Stage 2 commitments and those of the Stage 3 commitments instead of from those of Stage 1 commitments. We prove Claim 2 by showing that in the right session of $G_2^b(n, z)$, the value $\widehat{v}$ that $\mathcal{O}$ computes is the same as the value v that $\mathcal{O}$ computes in $G_1^b(n, z)$. Toward this end, we first show that in the right session of $G_1^b(n, z)$, the strong computational binding property holds in Stage 1 and either in Stage 2 or in Stage 3. (Note that from the property of the cut-and-choose technique, this implies that the committed values of either the Stage 2 commitments or the Sage 3 commitments are 0.9-close to those of the Stage 1 commitments except with negligible probability.) Let us say that $\mathcal{A}$ *cheats* in Stage 1 if at least one of $10n$ PTrapCom commitments in Stage 1 on the right has more than one committed value. We define cheating in Stage 2 and cheating in Stage 3 similarly. Then, we prove two subclaims.

Subclaim 1. *In $G_1^b(n, z)$, the probability that $\mathcal{A}$ cheats in Stage 1 is negligible.*

Proof (sketch). This subclaim follows directly from the strong computational binding property of $\mathsf{PTrapCom}_{i+d+1}^{i+d+1}$, since the running time of $G_1^b(n, z)$ is at most $\mathsf{poly}(n) \cdot T_{i+d+0.5}(n) \ll T_{i+d+1}(n)$ when $\mathcal{A}$ completes Stage 1 on the right. $\square$

Subclaim 2. *In $G_1^b(n, z)$, the probability that $\mathcal{A}$ cheats in Stage 2 and Stage 3 simultaneously is negligible.*

Proof (sketch). To prove this subclaim, we need to show that even though the left committer "cheats," $\mathcal{A}$ cannot use the messages received on the left to cheat on the right. This can be proven by following the proof of the scheme of [34]. Roughly speaking, we show that there always exists $a^* \in \{2, 3\}$ such that during Stage a^* on the right, the left session can be simulated in "short" time (i.e., the left session can be simulated without breaking the strong computational binding property of PCETrapCom in Stage a^*). A little more precisely, we show the following. Recall that the commitment of PCETrapCom can be simulated in polynomial time if we know the committed value of the Step 1 commitment of PCETrapCom. Then, we show that in the left session, either this committed value can be extracted in "short" time (during Stage a^* of the right session) or it can be extracted before $\mathcal{A}$ starts Stage a^* on the right (and thus can be considered as an auxiliary input). Once we show that $\mathcal{A}$ cannot use the messages received on the left to cheat on the right, the subclaim follows from the strong computational binding property of PCETrapCom on the right. $\square$

Now, we are ready to prove Claim 2.

Proof (sketch of Claim 2). As noted above, we prove Claim 2 by showing that in the right session, the value computed by $\mathcal{O}$ in $G_2^b(n, z)$ is equal to the value computed by $\mathcal{O}$ in $G_1^b(n, z)$. From Subclaim 2, there exists $a \in \{2, 3\}$ such that the committed values of the Stage a commitments are uniquely determined. Then, since the committed values of the Stage 1 commitments and those of

Stage a commitments are uniquely determined before Γ is chosen, the committed values of the Stage 1 commitments and those of Stage a commitments are 0.9-close except with negligible probability. Then, since we have carefully defined the behavior of $\mathcal{O}$ in $G_2^b(n, z)$ (in particular, since $\mathcal{O}$ checks whether the share is 0.8-close to a valid codeword in $G_2^b(n, z)$), we can show that the value computed by $\mathcal{O}$ from the extracted values of Stage 2 and 3 is the same as the one computed from the committed values of Stage 1 in a similar manner to the previous works on black-box constructions [9, 10, 23, 37]. □

Finally, we prove Claim 3.

Proof (of Claim 3). $G_3^b(n, z)$ differs from $G_2^b(n, z)$ only in that on the left, the Stage 1 commitments and their decommitments are generated by the simulator of $\mathsf{PTrapCom}_{i+d+1}^{i+d+1}$. Then, since the running time of $G_2^b(n, z)$ and that of $G_3^b(n, z)$ are at most $\mathsf{poly}(n) \cdot T_{i+d+0.5}(n) \ll T_{i+d+1}(n)$ except for Stage 1 on the left, and since Stage 1 on the left is not rewound in $G_2^b(n, z)$ and in $G_3^b(n, z)$, the claim follows from the trapdoor property of $\mathsf{PTrapCom}_{i+d+1}^{i+d+1}$. □

5   CCA Security from One-One CCA Security

In this section, we show a transformation from any one-one CCA-secure commitment scheme to a CCA-secure commitment scheme. To use this transformation to obtain a general MPC protocol, we also show that the resultant CCA-secure commitment satisfies κ-PQT-robustness for any $\kappa \in \mathbb{N}$. (Due to lack of space, the full proof is deferred to the full version.)

Lemma 2. *Let $\epsilon < 1$ be a positive constant, and assume the existence of one-way functions that are secure against 2^{n^ϵ}-time adversaries. Let $r(\cdot)$ and $t(\cdot)$ be arbitrary functions, let $T_i(n) \stackrel{\text{def}}{=} 2^{(\log n)^{(2/\epsilon)^{10i+1}}}$ for any $i \in \mathbb{N}$, and let $\mathsf{CCACom}_{i+3}^{1:1}$ be an $r(n)$-round commitment scheme that satisfies the following for any $T_{i+3}(n)$-time adversary.*

- *Strong computational binding property, and*
- *One-one CCA security for tags of length $t(n)$.*

Then, for any $\kappa \in \mathbb{N}$, there exists an $(r(n) + O(1))$-round commitment scheme CCACom_i that satisfies the following for any $T_i(n)$-time adversary.

- *Statistical binding property,*
- *CCA security for tags of length $t(n)$, and*
- *κ-PQT-robustness.*

If $\mathsf{CCACom}_{i+3}^{1:1}$ uses the underlying one-way function only in a black-box way, then CCACom_i uses the underlying one-way function only in a black-box way.

In the proof of Lemma 2, we use the following building blocks, which we can obtain by appropriately setting the security parameters of known protocols [26, 27, 32].

Commit Phase. The committer C and the receiver R receive common inputs 1^n and id $\in \{0,1\}^{t(n)}$. To commit to $v \in \{0,1\}^n$, the committer C does the following with the receiver R.

Stage 1. R chooses a random subset $\Gamma \subseteq [10n]$ of size n. Then, R commits to Γ by using $\mathsf{CCACom}_{i+3}^{1:1}$ with tag id.

Stage 2. C computes an $(n+1)$-out-of-$10n$ Shamir's secret sharing $s = (s_1, \ldots, s_{10n})$ of value v. Then, for each $j \in [10n]$ in parallel, C commits to s_j by using Com_{i+1}.

Stage 3. For each $j \in [10n]$ in parallel, C commits to s_j by using CECom_{i+2}.

Stage 4. R decommits the Stage 1 commitment to Γ.

Stage 5. For every $j \in [10n]$, let the j-th *column* denote the j-th commitment in Stage 2 and the j-th one in Stage 3 (that is, the commitments whose committed value is s_j). Then, for each $j \in \Gamma$, C decommits the commitments of the j-th column to s_j.

Decommit Phase. C sends v to R and decommits the Stage 2 commitments to s. Then, R checks whether all of these decommitments are valid. Furthermore, R checks whether (1) s is 0.9-close to a valid codeword $w = (w_1, \ldots, w_{10n})$ and (2) for every $j \in \Gamma$, w_j is equal to the share that was revealed in Stage 5. Finally, R checks whether w is a codeword corresponding to v.

Fig. 4. CCA-secure commitment CCACom_i

- A 2-round statistically binding commitment Com_{i+1} that is secure against $T_{i+1}(n)$-time adversaries but is completely broken in time $T_{i+1.5}(n)$.
- A constant-round concurrently $\mathrm{poly}(n^{\log n})$-extractable commitment CECom_{i+2} that is secure against $T_{i+2}(n)$-time adversaries but is completely broken in time $T_{i+2.5}(n)$. The number of slots in CECom_{i+2} is $\kappa + 3$.

We note that both Com_{i+1} and CECom_{i+2} use the underlying one-way function in a black-box way.

Proof (of Lemma 2). CCACom_i is shown in Figure 4. The statistical binding property of CCACom_i follows from that of Com_{i+1}. Then, we consider the following propositions.

Proposition 1. *For any $T_i(n)$-time adversary, CCACom_i is CCA secure for tags of length $t(n)$.*

Proposition 2. *For any $T_i(n)$-time adversary, CCACom_i is κ-PQT-robust.*

The lemma follows from these propositions. $\qquad\square$

Below, we prove Proposition 1. The proof of Proposition 2 is given in the full version. (Proposition 2 can be proven by extending the proof of Proposition 1.)

Proof (of Proposition 1). We show that for any $T_i(n)$-time adversary $\mathcal{A}$, the following are computationally indistinguishable:

- $\{\mathsf{IND}_0(\mathsf{CCACom}_i, \mathcal{A}, n, z)\}_{n \in \mathbb{N}, z \in \{0,1\}^*}$
- $\{\mathsf{IND}_1(\mathsf{CCACom}_i, \mathcal{A}, n, z)\}_{n \in \mathbb{N}, z \in \{0,1\}^*}$

Note that $\mathcal{O}$ does the following in each right session. First, $\mathcal{O}$ extracts the committed values $\boldsymbol{s} = (s_1, \ldots, s_{10n})$ of the Stage 2 commitments by brute force. (If the committed value of the j-th commitment is not uniquely determined, s_j is defined to be $\perp$.) Then, at the end of the session, $\mathcal{O}$ checks whether the following conditions hold: (1) $\boldsymbol{s}$ is 0.9-close to a valid codeword $\boldsymbol{w} = (w_1, \ldots, w_{10n})$, and (2) for every $j \in \Gamma$ (where Γ is the value that $\mathcal{O}$ sends to $\mathcal{A}$ in Stage 4), w_j is equal to the share that was revealed in Stage 5. If both conditions hold, $\mathcal{O}$ recovers v from $\boldsymbol{w}$ and returns v to A. Otherwise, $\mathcal{O}$ returns $v := \perp$ to $\mathcal{A}$. We note that the running time of $\mathcal{O}$ is at most $\mathsf{poly}(n) \cdot T_{i+1.5}(n)$.

To show the indistinguishability, we consider hybrid experiments $H_a^b(n, z)$ for $a \in \{0, 1, 2, 3\}$ and $b \in \{0, 1\}$.

Hybrid $H_0^b(n, z)$ is the same as experiment $\mathsf{IND}_b(\mathsf{CCACom}_i, \mathcal{A}, n, z)$.
Hybrid $H_1^b(n, z)$ is the same as $H_0^b(n, z)$ except for the following:
- In Stage 1 of the left session, the committed value Γ is extracted by brute force. If the commitment is invalid or has multiple committed values, Γ is defined to be a random subset.[6]
- In Stage 3 of the left session, the left committer commits to 0 instead of s_j for each $j \notin \Gamma$.

The running time of $H_1^b(n, z)$ is at most $\mathsf{poly}(n) \cdot T_{i+1.5}(n)$ except for the brute-force extraction of the Stage 1 commitment on the left.
Hybrid $H_2^b(n, z)$ is the same as $H_1^b(n, z)$ except for the following:
- In every right session of which Stage 2 ends after $\mathcal{A}$ starts Stage 2 on the left, the committed values of the Stage 3 commitments are extracted by using the concurrent $\mathsf{poly}(n^{\log n})$-extractability of CECom_{i+2}. Let $\widehat{\boldsymbol{s}} = (\widehat{s}_1, \ldots, \widehat{s}_{10n})$ be the extracted values, where $\widehat{s}_j$ is defined to be $\perp$ if the extraction of the j-th commitment fails.
- At the end of each right session in which $\widehat{\boldsymbol{s}} = (\widehat{s}_1, \ldots, \widehat{s}_{10n})$ is extracted, $\mathcal{O}$ does the following. First, $\mathcal{O}$ checks whether the following conditions hold: (1) $\widehat{\boldsymbol{s}}$ is *0.8-close* to a valid codeword $\widehat{\boldsymbol{w}} = (\widehat{w}_1, \ldots, \widehat{w}_{10n})$ and (2) for every $j \in \widehat{\Gamma}$ (where $\widehat{\Gamma}$ is the value that $\mathcal{O}$ sends to $\mathcal{A}$ in this session), $\widehat{w}_j$ is equal to the share that was revealed in Stage 5. If both conditions hold, $\mathcal{O}$ recovers $\widehat{v}$ from $\widehat{\boldsymbol{w}}$ and returns $\widehat{v}$ to $\mathcal{A}$. Otherwise, $\mathcal{O}$ returns $\widehat{v} := \perp$ to $\mathcal{A}$. We note that $\mathcal{O}$ does not extract the committed values of the Stage 2 commitments in such right sessions.

The expected running time of $H_2^b(n, z)$ is at most $\mathsf{poly}(n^{\log n}) \cdot T_i(n)$ after the start of Stage 2 on the left.

---

[6] Since the running time of $\mathcal{A}$ and $\mathcal{O}$ is at most $\mathsf{poly}(n) \cdot T_{i+1.5}(n) \ll T_{i+2}(n)$, the strong computational binding property of $\mathsf{CCACom}_{i+3}^{1:1}$ guarantees that the Stage 1 commitment has at most one committed value except with negligible probability.

Hybrid $H_3^b(n, z)$ is the same as $H_2^b(n, z)$ except that in Stage 2 on the left, the left committer commits to 0 instead of s_j for each $j \notin \Gamma$.

Since $\mathcal{A}$ receives no information about $\{s_j\}_{j \notin \Gamma}$ on the left in $H_3^0(n, z)$ and $H_3^1(n, z)$, and since s is $(n+1)$-out-of-$10n$ secret sharing, the output of $H_3^0(n, z)$ and that of $H_3^1(n, z)$ are identically distributed. Then, we consider the following claims. In what follows, we use $\mathsf{H}_i^b(n, z)$ to denote the output of experiment $H_i^b(n, z)$.

Claim 4. *For each* $b \in \{0, 1\}$, $\{\mathsf{H}_0^b(n, z)\}_{n \in \mathbb{N}, z \in \{0,1\}^*}$ *and* $\{\mathsf{H}_1^b(n, z)\}_{n \in \mathbb{N}, z \in \{0,1\}^*}$ *are computationally indistinguishable.*

Claim 5. *For each* $b \in \{0, 1\}$, $\{\mathsf{H}_1^b(n, z)\}_{n \in \mathbb{N}, z \in \{0,1\}^*}$ *and* $\{\mathsf{H}_2^b(n, z)\}_{n \in \mathbb{N}, z \in \{0,1\}^*}$ *are statistically indistinguishable.*

Claim 6. *For each* $b \in \{0, 1\}$, $\{\mathsf{H}_2^b(n, z)\}_{n \in \mathbb{N}, z \in \{0,1\}^*}$ *and* $\{\mathsf{H}_3^b(n, z)\}_{n \in \mathbb{N}, z \in \{0,1\}^*}$ *are computationally indistinguishable.*

The proposition follows from these claims. □

Proof (sketch of Claim 4). The view of $\mathcal{A}$ in $H_0^b(n, z)$ and that of $\mathcal{A}$ in $H_1^b(n, z)$ differ only in the committed values of CECom_{i+2} on the left. In addition, the running time of $H_0^b(n, z)$ and that of $H_1^b(n, z)$ are $\mathsf{poly}(n) \cdot T_{i+1.5}(n) \ll T_{i+2}(n)$ (except for the brute force extraction of the Stage 1 commitment on the left in $H_1^b(n, z)$). Thus, by considering Γ as non-uniform advice, we can prove indistinguishability from the hiding property of CECom_{i+2}. □

Next, we consider Claim 5. As in Section 4.2, we first show that in every right session of $H_1^b(n, z)$, the committed values of the Stage 2 commitments and those of Stage 3 commitments are 0.9-close. Formally, for any right session, let $s^{(2)} = (s_1^{(2)}, \ldots, s_{10n}^{(2)})$ be the committed values of the Stage 2 commitments (if the committed value of the j-th commitment is not uniquely determined, $s_j^{(2)}$ is defined to be $\bot$) and let $s^{(3)} = (s_1^{(3)}, \ldots, s_{10n}^{(3)})$ be the committed values of the Stage 3 commitments. Then, for every $j \in [10n]$, we say that the j-th column of this session is *bad* if $s_j^{(2)} = \bot$, $s_j^{(3)} = \bot$, or $s_j^{(2)} \neq s_j^{(3)}$. In addition, we say that $\mathcal{A}$ *cheats* in this session if the session is accepted and the number of bad columns is at least n. Then, we prove the following subclaim.

Subclaim 3. *In any right session of* $H_1^b(n, z)$, $\mathcal{A}$ *cheats with at most negligible probability.*

Proof (sketch). At first sight, it seems that we can prove this subclaim by simply using the hiding property of $\mathsf{CCACom}_{i+3}^{1:1}$ and the property of cut-and-choose technique (i.e., it seems that, since the committed value Γ of the Stage 1 commitment on the right is hidden from $\mathcal{A}$, the probability that there are at least n bad columns but the session is accepted is negligible). However, $\mathcal{A}$ interacts with the left committer as well as with $\mathcal{O}$, and the left committer "cheats" in the left session (i.e., on the left, the committed values of the Stage 2 commitments and those of the Stage 3 commitments are not 0.9-close). Thus, $\mathcal{A}$ may be able to cheat in a right session by using the messages received on the left. A key to prove this subclaim is that the left session can be simulated by using the committed-value oracle of $\mathsf{CCACom}_{i+3}^{1:1}$ (i.e., if we know the committed value Γ of the Stage 1 commitment on the left, we can simulate the later stages in polynomial time). Thus, the one-one CCA security of $\mathsf{CCACom}_{i+3}^{1:1}$ guarantees that $\mathcal{A}$ cannot break the hiding property of $\mathsf{CCACom}_{i+3}^{1:1}$ even with the messages of the left session. We can therefore use the cut-and-choose technique to prove the subclaim.   □

Given Subclaim 3, we can prove Claim 5 in a similar manner to Claim 2 in Section 4.2.

Finally, we prove Claim 6.

Proof (sketch of Claim 6). $H_2^b(n,z)$ and $H_3^b(n,z)$ differ only in the committed values of Com_{i+1}. Since the running time of $H_2^b(n,z)$ and that of $H_3^b(n,z)$ are $\mathsf{poly}(n^{\log n}) \cdot T_i(n) \ll T_{i+1}(n)$ after the start of Stage 2 on the left, we can prove Claim 6 from the hiding property of Com_{i+1} (by considering Γ of the left session and the answers of $\mathcal{O}$ for some right sessions as non-uniform advice). Here, we use the fact that Com_{i+1} is a *2-round* commitment scheme. This fact enables us to rewind $\mathcal{A}$ in the right sessions of $H_2^b(n,z)$ without breaking the hiding property of Com_{i+1}.   □

6 One-One CCA Security for Long Tags from CCA Security for Short Tags

In this section, we consider a transformation from any CCA-secure commitment scheme for tags of length $t(n)$ to a one-one CCA-secure commitment scheme for tags of length $2^{t(n)-1}$. The transformation are essentially the same as those in [24], which shows a transformation from any concurrent NM commitment scheme for short tags to a NM commitment scheme for long tags.

Lemma 3. *Let $\epsilon < 1$ be a positive constant, and assume the existence of one-way functions that are secure against 2^{n^ϵ}-time adversaries. Let $r(\cdot)$ and $t(\cdot)$ be arbitrary functions such that $t(n) \leq O(\log n)$, let $T_i(n) \stackrel{\text{def}}{=} 2^{(\log n)^{(2/\epsilon)^{10i+1}}}$ for $i \in \mathbb{N}$, and let CCACom_{i+1} be an $r(n)$-round commitment scheme that satisfies the following for any $T_{i+1}(n)$-time adversary.*

- *Statistical binding property, and*
- *CCA security for tags of length $t(n)$.*

Then, there exists an $r(n)$-round commitment scheme $\mathsf{CCACom}_i^{1:1}$ that satisfies the following for any $T_i(n)$-time adversary.

- *Statistical binding property, and*
- *One-one CCA security for tags of length $2^{t(n)-1}$.*

If CCACom_{i+1} *uses the underlying one-way function only in a black-box way, then* $\mathsf{CCACom}_i^{1:1}$ *uses the underlying one-way function only in a black-box way.*

Due to lack of space, the proof is deferred to the full version.

7 Constant-Round Black-Box Composable Protocol

In this section, we show a constant-round black-box construction of a general MPC protocol that satisfies angel-based UC security. Roughly speaking, the framework of angel-based UC security (called $\mathcal{H}$-EUC framework) is the same as the UC framework except that both the adversary and the environment in the real and the ideal worlds have access to a super-polynomial-time angel $\mathcal{H}$.

To construct our protocol, we use the following theorem, which we obtain by combining Lemmas 1, 2, and 3.

Theorem 1. *Assume the existence of one-way functions that are secure against sub-exponential-time adversaries. Then, for any constant* $\kappa \in \mathbb{N}$, *there exists a constant-round commitment scheme that is CCA secure and* κ-*PQT-robust. This commitment scheme uses the underlying one-way functions only in a black-box way.*

We additionally use the following results of [7] and [23].

Let $\langle C, R \rangle$ be any $r_{cca}(n)$-round commitment scheme that is CCA secure and κ-robust for any constant κ, $\langle S, R \rangle$ be any $r_{ot}(n)$-round semi-honest OT protocol, and $\mathcal{H}$ be an angel that breaks $\langle C, R \rangle$ essentially in the same way as the committed-value oracle of $\langle C, R \rangle$ does. Then, Lin and Pass [23] showed that there exists a black-box $O(\max(r_{ot}(n), r_{cca}(n)))$-round protocol that securely realizes the ideal OT functionality $\mathcal{F}_{OT}$ in the $\mathcal{H}$-EUC framework. By using essentially the same security proof as that of [23], we can show that even when $\langle C, R \rangle$ is CCA secure and only κ-PQT-robust for a sufficiently large κ, the protocol of [23] is still secure if $\langle S, R \rangle$ is secure against any PQT adversary.[7] Thus, we have the following theorem from [23].

Theorem 2. *Assume the existence of an* $r_{cca}(n)$-*round commitment scheme* $\langle C, R \rangle$ *that is CCA secure and* κ-*PQT-robust for a sufficiently large* κ, *and assume the existence of an* $r_{ot}(n)$-*round semi-honest oblivious transfer protocol* $\langle S, R \rangle$ *that is secure against any PQT adversary. Then, there exists an* $O(\max(r_{cca}(n), r_{ot}(n)))$-*round protocol that* $\mathcal{H}$-*EUC-realizes* $\mathcal{F}_{OT}$. *This protocol uses* $\langle C, R \rangle$ *and* $\langle S, R \rangle$ *only in a black-box way.*

In [7], Canetti et al. showed the following.

Theorem 3 ([7]). *For every well-formed functionality* $\mathcal{F}$, *there exists a constant-round* $\mathcal{F}_{OT}$-*hybrid protocol that* $\mathcal{H}$-*EUC-realizes* $\mathcal{F}$.

Then, by combining Theorems 1, 2, and 3, we obtain the following theorem.

---

[7] This is because κ-PQT-robustness guarantees that the committed-value oracle is useless in attacking any κ-round protocol *if the protocol is* PQT-*secure.*

Theorem 4. *Assume the existence of one-way functions that are secure against sub-exponential-time adversaries and constant-round semi-honest oblivious transfer protocols that are secure against quasi-polynomial-time adversaries. Then, there exists an angel* $\mathcal{H}$ *such that for every well-formed functionality* $\mathcal{F}$, *there exists a constant-round protocol that* $\mathcal{H}$-EUC-*realizes* $\mathcal{F}$. *This protocol uses the underlying one-way functions and oblivious transfer protocols only in a black-box way.*

References

1. Barak, B., Sahai, A.: How to play almost any mental game over the net—concurrent composition via super-polynomial simulation. In: FOCS, pp. 543–552 (2005)
2. Canetti, R.: Universally composable security: A new paradigm for cryptographic protocols. In: FOCS, pp. 136–145 (2001)
3. Canetti, R., Fischlin, M.: Universally composable commitments. In: Kilian, J. (ed.) CRYPTO 2001. LNCS, vol. 2139, pp. 19–40. Springer, Heidelberg (2001)
4. Canetti, R., Goldreich, O., Goldwasser, S., Micali, S.: Resettable zero-knowledge. In: STOC, pp. 235–244 (2000)
5. Canetti, R., Kilian, J., Petrank, E., Rosen, A.: Black-box concurrent zero-knowledge requires (almost) logarithmically many rounds. SIAM J. Comput. 32(1), 1–47 (2002)
6. Canetti, R., Kushilevitz, E., Lindell, Y.: On the limitations of universally composable two-party computation without set-up assumptions. In: Biham, E. (ed.) EUROCRYPT 2003. LNCS, vol. 2656, pp. 68–86. Springer, Heidelberg (2003)
7. Canetti, R., Lin, H., Pass, R.: Adaptive hardness and composable security in the plain model from standard assumptions. In: FOCS, pp. 541–550 (2010)
8. Canetti, R., Lindell, Y., Ostrovsky, R., Sahai, A.: Universally composable two-party and multi-party secure computation. In: STOC, pp. 494–503 (2002)
9. Choi, S.G., Dachman-Soled, D., Malkin, T., Wee, H.: Black-box construction of a non-malleable encryption scheme from any semantically secure one. In: Canetti, R. (ed.) TCC 2008. LNCS, vol. 4948, pp. 427–444. Springer, Heidelberg (2008)
10. Choi, S.G., Dachman-Soled, D., Malkin, T., Wee, H.: Simple, black-box constructions of adaptively secure protocols. In: Reingold, O. (ed.) TCC 2009. LNCS, vol. 5444, pp. 387–402. Springer, Heidelberg (2009)
11. Dolev, D., Dwork, C., Naor, M.: Nonmalleable cryptography. SIAM J. Comput. 30(2), 391–437 (2000)
12. Garg, S., Goyal, V., Jain, A., Sahai, A.: Concurrently secure computation in constant rounds. In: Pointcheval, D., Johansson, T. (eds.) EUROCRYPT 2012. LNCS, vol. 7237, pp. 99–116. Springer, Heidelberg (2012)
13. Gentry, C., Wichs, D.: Separating succinct non-interactive arguments from all falsifiable assumptions. In: STOC, pp. 99–108 (2011)
14. Goldreich, O., Micali, S., Wigderson, A.: How to play any mental game or a completeness theorem for protocols with honest majority. In: STOC. pp. 218–229 (1987)
15. Goyal, V.: Constant round non-malleable protocols using one way functions. In: STOC, pp. 695–704 (2011)
16. Goyal, V.: Non-black-box simulation in the fully concurrent setting. In: STOC, pp. 221–230 (2013)
17. Goyal, V., Lin, H., Pandey, O., Pass, R., Sahai, A.: Round-efficient concurrently composable secure computation via a robust extraction lemma. Cryptology ePrint Archive, Report 2012/652 (2012), http://eprint.iacr.org/

18. Haitner, I.: Semi-honest to malicious oblivious transfer - the black-box way. In: Canetti, R. (ed.) TCC 2008. LNCS, vol. 4948, pp. 412–426. Springer, Heidelberg (2008)
19. Haitner, I., Ishai, Y., Kushilevitz, E., Lindell, Y., Petrank, E.: Black-box constructions of protocols for secure computation. SIAM J. Comput. 40(2), 225–266 (2011)
20. Ishai, Y., Kushilevitz, E., Lindell, Y., Petrank, E.: Black-box constructions for secure computation. In: STOC, pp. 99–108 (2006)
21. Lin, H.: Concurrent Security. Ph.D. thesis, Cornell University (2011)
22. Lin, H., Pass, R.: Non-malleability amplification. In: STOC, pp. 189–198 (2009)
23. Lin, H., Pass, R.: Black-box constructions of composable protocols without setup. In: Safavi-Naini, R., Canetti, R. (eds.) CRYPTO 2012. LNCS, vol. 7417, pp. 461–478. Springer, Heidelberg (2012)
24. Lin, H., Pass, R., Venkitasubramaniam, M.: Concurrent non-malleable commitments from any one-way function. In: Canetti, R. (ed.) TCC 2008. LNCS, vol. 4948, pp. 571–588. Springer, Heidelberg (2008)
25. Malkin, T., Moriarty, R., Yakovenko, N.: Generalized environmental security from number theoretic assumptions. In: Halevi, S., Rabin, T. (eds.) TCC 2006. LNCS, vol. 3876, pp. 343–359. Springer, Heidelberg (2006)
26. Micciancio, D., Ong, S.J., Sahai, A., Vadhan, S.: Concurrent zero knowledge without complexity assumptions. In: Halevi, S., Rabin, T. (eds.) TCC 2006. LNCS, vol. 3876, pp. 1–20. Springer, Heidelberg (2006)
27. Naor, M.: Bit commitment using pseudorandomness. J. Cryptology 4(2), 151–158 (1991)
28. Naor, M.: On cryptographic assumptions and challenges. In: Boneh, D. (ed.) CRYPTO 2003. LNCS, vol. 2729, pp. 96–109. Springer, Heidelberg (2003)
29. Pass, R.: Simulation in quasi-polynomial time, and its application to protocol composition. In: Biham, E. (ed.) EUROCRYPT 2003. LNCS, vol. 2656, pp. 160–176. Springer, Heidelberg (2003)
30. Pass, R.: Bounded-concurrent secure multi-party computation with a dishonest majority. In: STOC, pp. 232–241 (2004)
31. Pass, R., Lin, H., Venkitasubramaniam, M.: A unified framework for UC from only OT. In: Wang, X., Sako, K. (eds.) ASIACRYPT 2012. LNCS, vol. 7658, pp. 699–717. Springer, Heidelberg (2012)
32. Pass, R., Venkitasubramaniam, M.: On constant-round concurrent zero-knowledge. In: Canetti, R. (ed.) TCC 2008. LNCS, vol. 4948, pp. 553–570. Springer, Heidelberg (2008)
33. Pass, R., Wee, H.: Black-box constructions of two-party protocols from one-way functions. In: Reingold, O. (ed.) TCC 2009. LNCS, vol. 5444, pp. 403–418. Springer, Heidelberg (2009)
34. Pass, R., Wee, H.: Constant-round non-malleable commitments from sub-exponential one-way functions. In: Gilbert, H. (ed.) EUROCRYPT 2010. LNCS, vol. 6110, pp. 638–655. Springer, Heidelberg (2010)
35. Prabhakaran, M., Rosen, A., Sahai, A.: Concurrent zero knowledge with logarithmic round-complexity. In: FOCS, pp. 366–375 (2002)
36. Prabhakaran, M., Sahai, A.: New notions of security: Achieving aniversal composability without trusted setup. In: STOC, pp. 242–251 (2004)
37. Wee, H.: Black-box, round-efficient secure computation via non-malleability amplification. In: FOCS, pp. 531–540 (2010)

One-Sided Adaptively Secure Two-Party Computation

Carmit Hazay[1] and Arpita Patra[2]

[1] Faculty of Engineering, Bar-Ilan University, Israel
`carmit.hazay@biu.ac.il`
[2] Dept. of Computer Science, University of Bristol, UK
`arpita.patra@bristol.ac.uk`

Abstract. Adaptive security is a strong security notion that captures additional security threats that are not addressed by static corruptions. For instance, it captures real-world scenarios where "hackers" actively break into computers, possibly while they are executing secure protocols. Studying this setting is interesting from both theoretical and practical points of view. A primary building block in designing adaptively secure protocols is a non-committing encryption (NCE) that implements secure communication channels in the presence of adaptive corruptions. Current constructions require a number of public key operations that grows linearly with the length of the message. Furthermore, general two-party protocols require a number of NCE calls that is linear in the circuit size.

In this paper we study the two-party setting in which at most one of the parties is adaptively corrupted, which we believe is the right security notion in the two-party setting. We study the feasibility of (**1**) NCE with constant number of public key operations for large message spaces (**2**) Oblivious transfer with constant number of public key operations for large input spaces of the sender, and (**3**) constant round secure computation protocols with a number of NCE calls, and an overall number of public key operations, that are independent of the circuit size. Our study demonstrates that such primitives indeed exist in the presence of single corruptions, while this is not known for fully adaptive security.

Keywords: Adaptively Secure Computation, Non-Committing Encryption, Oblivious Transfer.

1 Introduction

1.1 Background

Secure two-party computation. In the setting of secure two-party computation, two parties with private inputs wish to jointly compute some function of their inputs while preserving certain security properties like privacy, correctness and more. In this setting, security is formalized by viewing a protocol execution as if the computation is executed in an ideal setting where the parties send inputs to a trusted party that performs the computation and returns its result (also known by simulation-based security). Starting with the work of [36,22], it is by now well known that (in various settings) any polynomial-time function can be compiled into a secure function evaluation protocol with practical complexity; see [30,16,33] for a few recent works. The security proofs of these constructions assume that a party is statically corrupted. Meaning, corruptions take place at

Y. Lindell (Ed.): TCC 2014, LNCS 8349, pp. 368–393, 2014.

the outset of the protocol execution and the identities of the corrupted parties are fixed throughout the computation. Adaptive security is a stronger notion where corruptions takes place *at any point* during the course of the protocol execution. That is, upon corruption the adversary sees the internal data of the corrupted party which includes its input, randomness and the incoming messages. This notion is much stronger than static security due to the fact that the adversary may choose at any point which party to corrupt, even after the protocol is completed! It therefore models real world threats more accurately than the static corruption model.

Typically, when dealing with adaptive corruptions we distinguish between corruptions with erasures and without erasures. In the former case honest parties are trusted to erase data if are instructed to do so by the protocol, whereas in the latter case no such assumption is made. This assumption is often problematic since it relies on the willingness of the honest parties to carry out this instruction without the ability to verify its execution. In settings where the parties are distrustful it may not be a good idea to base security on such an assumption. In addition, it is generally unrealistic to trust parties to fully erase data since this may depend on the operating system. Nevertheless, assuming that there are no erasures comes with a price since the complexity of adaptively secure protocols without erasures is much higher than the analogue complexity of protocols that rely on erasures. In this paper we do not rely on erasures.

Adaptive Security. It is known by now that security against adaptive attacks captures important real-world concerns that are not addressed by static corruptions. For instance, such attacks capture scenarios where "hackers" actively break into computers, possibly while they are running secure protocols, or when the adversary learns from the communication which parties are worth to corrupt more than others. This later issue can be demonstrated by the following example. Consider a protocol where some party (denoted by the dealer) shares a secret among a public set of $\sqrt{n}$ parties, picked at random from a larger set of n parties. This scheme is insecure in the adaptive model if the adversary corrupts $\sqrt{n}$ parties since it can always corrupt the particular set of parties that share the secret. In the static setting the adversary corrupts the exact same set of parties that share the secret with a negligible probability in n.

Other difficulties also arise when proving security. Consider the following protocol for transferring a message: A receiver picks a public key and sends it to a sender that uses it to encrypt its message. Then, security in the static model is simple and relies on the semantic security of the underlying encryption scheme. However, this protocol is insecure in the adaptive model since standard semantically secure encryption binds the receiver to a single message (meaning, given the public key, a ciphertext can only be decrypted into a single value). Thus, upon corrupting the receiver at the end of the protocol execution it would not be possible to "explain" the simulated ciphertext with respect to the real message. This implies that adaptive security is much harder to achieve.

Adaptively Secure Two-Party Computation. In the two-party setting there are scenarios where the system is comprised from two devices communicating between themselves without being part of a bigger system. For instance, consider a scenario where two devices share an access to an encrypted database that contains highly sensitive data (like passwords). Moreover, the devices communicate via secure computation but do not communicate with other devices due to high risk of breaking into the database. Thus, attacking one of the devices does not disclose any useful information about the content of the database, while attacking both devices is a much harder task. It is reasonable to assume that the devices are not necessarily statically corrupted since they are protected by other means, while attackers may constantly try to break into these devices (even while running secure computation).

In 2011, RSA secureID authentication products were breached by hackers that leveraged the stolen information from RSA in order to attack the U.S. defense contractor Lockheed Martin. The attackers targeted SecurID data as part of a broader scheme to steal defense secrets and related intellectual property. Distributing the SecureID secret keys between two devices potentially enables to defend against such an attack since in order to access these keys the attackers need to *adaptively* corrupt both devices, which is less likely to occur. Many other applications face similar threats when attempt to securely protect their databases.

We therefore focus on a security notion that seems the most appropriate in this context. In this paper, we study secure two-party computation with single adaptive corruptions in the non-erasure model where at most one party is adaptively corrupted. To distinguish this notion from fully adaptive security, where both parties may get corrupted, we denote it by *one-sided* adaptive security. Our goal in this work is to make progress in the study of the efficiency of two-party protocols with one-sided security. Our measure of efficiency is the number of public key encryption (PKE) operations. Loosely speaking, our primitives are parameterized by a public key encryption scheme for which we count the number of key generation/encryption/decryption operations. More concretely, these operations are captured by the number of exponentiations in several important groups (e.g., groups where the DDH assumption is hard and composite order groups where the assumptions DCR and QR are hard), and further considered in prior works such as [19]. Finally, our proofs are given in the universal composable (UC) setting [6] with a common reference string (CRS) setup. The reductions of our non-committing encryption and oblivious transfer with one-sided security are tight. The reductions of our general two-party protocols are tighter than in prior works since we do not need to encrypt the entire communication using non-committing encryption; see more details below. All our theorems *are not* known to hold in the fully adaptive setting.

1.2   Our Results

One-sided NCE with Constant Overhead. A non-committing encryption (NCE) scheme [8] implements secure channels in the presence of adaptive corruptions and is an important building block in designing adaptively secure protocols. In [13], Damgård and Nielsen presented a theoretical improvement in the one-sided setting by designing an NCE under strictly weaker assumptions than simulatable public key encryption scheme (the assumption for fully adaptive NCE). Nevertheless, all known one-sided [8,13] and fully adaptive NCE constructions [13,11] require $\mathcal{O}(1)$ PKE operations for each transmitted bit. It was unknown whether this bound can be reduced for one-sided NCEs and even matched with the overhead of standard PKEs.

We suggest a new approach for designing NCEs secure against one-sided adaptive attacks. Our protocols are built on two cryptographic building blocks that are non-committing with respect to a single party. We denote these by NCE for the sender and NCE for the receiver. *Non-committing for the receiver* (NCER) implies that one can efficiently generate a secret key that decrypts a simulated ciphertext into any plaintext. Whereas *non-committing for the sender* (NCES) implies that one can efficiently generate randomness for any plaintext for proving that a ciphertext, encrypted under a fake key, encrypts this plaintext. A core building block in our one-sided construction is (a variant) of the following protocol, in which the receiver generates two sets of public/secret keys; one pair of keys for each public key system, and sends these public keys to the sender. Next, the sender partitions its message into two shares and encrypts the distinct shares under the distinct public keys. Finally, the receiver decrypts the ciphertexts and reconstructs the message. Both NCES and NCER are semantically secure PKEs and they are as efficient as standard PKEs. Informally, we prove that,

Theorem 1. (Informal) *Assume the existence of NCER and NCES with constant number of PKE operations for message space $\{0,1\}^q$ and simulatable PKE. Then there exists a one-sided NCE with constant number of PKE operations for message space $\{0,1\}^q$, where $q = O(n)$ and n is the security parameter.*

Importantly, the security of this protocol only works if the simulator knows the identity of the corrupted party since fake public keys and ciphertexts cannot be explained as valid ones. We resolve this issue by slightly modifying this protocol using somewhat NCE [19] in order to encrypt only three bits. Namely, we use somewhat NCE to encrypt the *choice* of having fake/valid keys and ciphertexts (which only requires a single non-committing bit per choice). This enables the simulator to "explain" fake keys/ciphertext as valid and vice versa using only a constant number of asymmetric operations. In this work we consider two implementations of NCER and NCES. For polynomial-size message spaces the implementations are secure under the DDH assumption, whereas for exponential-size message spaces security holds under the DCR assumption. The NCER implementations are taken from [25,9]. NCES was further discussed in [17] and realized under the DDH assumption in [5] using the closely related notion of lossy encryption.[1] In this paper we realize NCES under the DCR assumption.

One-sided Oblivious Transfer with Constant Overhead. We use our one-sided NCEs to implement 1-out-of-2 oblivious transfer (OT) between a sender and a receiver. We consider a generic framework that abstracts the statically secure OT of [34] that is based on a dual-mode PKE primitive, while encrypting only a small portion of the communication using our one-sided NCE. Our construction requires a constant number of PKE operations for an input space $\{0,1\}^q$ of the sender, where $q = O(n)$. This is significantly better than the fully adaptively secure OT of [19] (currently the most efficient fully adaptive construction), that requires $\mathcal{O}(q)$ such operations. We prove that:

Theorem 2. (Informal) *Assume the existence of one-sided NCE with constant number of PKE operations for message space $\{0,1\}^q$ and dual-mode PKE. Then there exists a*

---

[1] This notion differs from NCES by not requiring an efficient opening algorithm that enables to equivocate the ciphertext's randomness. We further observe that the notion of NCES is also similar to mixed commitments [14].

one-sided OT with constant number of PKE operations for sender's input space $\{0,1\}^q$, where $q = O(n)$ and n is the security parameter.

We build our one-sided OT based on the PVW protocol as follows. (**1**) First, we require that the sender sends its ciphertexts via a one-sided non-committing channel (based on our previous result, this only inflates the overhead by a constant). (**2**) We fix the common parameters of the dual-mode PKE in a single mode (instead of alternating between two modes as in the [19] protocol). To ensure correctness, we employ a special type of ZK PoK which uses a novel technique; see below for more details. Finally, we discuss two instantiations based on the DDH and QR assumptions.

Constant Round One-Sided Secure Computation. Theoretically, it is well known that any statically secure protocol can be transformed into a one-sided adaptively secure protocol by encrypting the *entire* communication using NCE. This approach, adopted by [26], implies that the number of PKE operations grows linearly with the circuit size times a computational security parameter.[2] A different approach in the OT-hybrid model was taken in [24] and achieved a similar overhead as well.

In this work we demonstrate the feasibility of designing generic constant round protocols based on Yao's garbled circuit technique with one-sided security, tolerating semi-honest and malicious attacks. Our main observation implies that one-sided security can be obtained even if only the keys corresponding to the inputs and output wires are communicated via a one-sided adaptively secure channel. This implies that the bulk of communication is transmitted as in the static setting. Using our one-sided secure primitives we obtain protocols that outperform the constant round one-sided constructions of [26,24] and all known generic fully adaptively secure two-party protocols. Our proofs take a different simulation approach, circumventing the difficulties arise due to the simulation technique from [28] that builds a fake circuit (which cannot be applied in the adaptive setting). Specifically, we prove that

Theorem 3. (Informal) *Under the assumptions of achieving statically secure two-party computation and one-sided OT with constant number of PKE operations for sender's input space $\{0,1\}^q$, where $q = O(n)$ and n is the security parameter, there exists a constant round one-sided semi-honest adaptively secure two-party protocol that requires $\mathcal{O}(|C|)$ private key operations and $\mathcal{O}(|\text{input}| + |\text{output}|)$ public key operations.*

In order to obtain one-sided security against malicious attacks we adapt the cut-and-choose based protocol introduced in [30]. The idea of the cut-and-choose technique is to ask one party to send s garbled circuits and later open half of them by the choice of the other party. This ensures that with very high probability the majority of the unopened circuits are valid. Proving security in the one-sided setting requires dealing with new subtleties and requires a modified cut-and-choose OT protocol, since [30] defines the public parameters of their cut-and-choose OT protocol in a way that precludes the equivocation of the receiver's input. Our result in the malicious setting follows.

---

[2] We note that this statement is valid regarding protocols that do not employ fully homomorphic encryptions (FHE). To this end, we only consider protocols that do not take the FHE approach. As a side note, it was recently observed in [27] that adaptive security is impossible for FHE satisfying compactness.

Theorem 4. (Informal.) *Under the assumptions of achieving static security in [30], one-sided cut-and-choose OT with constant number of PKE operations for sender's input space $\{0,1\}^q$, where $q = O(n)$ and n is the security parameter, and simulatable PKE, there exists a* constant round *one-sided malicious adaptively secure two-party protocol that requires $\mathcal{O}(s \cdot |C|)$ private key operations and $\mathcal{O}(s \cdot (|input| + |output|))$ public key operations where s is a statistical parameter that determines the cut-and-choose soundness error.*

This asymptotic efficiency is significantly better than in prior protocols [26,24].

Witness Equivocal UC ZK PoK for Compound Statements. As a side result, we demonstrate a technique for efficiently generating statically secure UC ZK PoK for known Σ-protocols. Our protocols use a new approach where the prover commits to an additional transcript which enables to extract the witness with a constant overhead.

We further focus on compound statements (where the statement is comprised of substatements for which the prover only knows a subset of the witnesses), and denote a UC ZK PoK by *witness equivocal* if the simulator knows the witnesses for *all* substatements but not which subset is given to the real prover. We extend our proofs for this notion to the adaptive setting as well. In particular, the simulator must be able to convince an adaptive adversary that it does *not know* a different subset of witnesses. This notion is weaker than the typical one-sided security notion (that requires simulation without the knowledge of any witness), but is still meaningful in designing one-sided secure protocols. In this work, we build witness equivocal UC ZK PoKs for a class of fundamental compound Σ-protocols, without relying on NCE. Our protocols are round efficient and achieve a negligible soundness error. Finally, they are proven secure in the UC framework [6].

To conclude, our results may imply that one-sided security is strictly easier to achieve than fully adaptive security, and for some applications this is indeed the right notion to consider. We leave open the feasibility of constant round one-sided secure protocols in the multi-party setting. Currently, it is not clear how to extend our techniques beyond the two-party setting (such as within the [4] protocol), and achieve secure constructions with a number of PKE operations that does not depend on the circuit size.

1.3 Prior Work

We describe prior work on NCE, adaptively secure OT and two-party computation.

Non-committing Encryption. One-sided NCE was introduced in [8] which demonstrated feasibility of the primitive under the RSA assumption. Next, NCE was studied in [13,11]. The construction of [13] requires constant rounds on the average and is based on simulatable PKE, whereas [11] presents an improved expected two rounds NCE based on a weaker primitive. [13] further presented a one-sided NCE based on a weakened simulatable PKE notion. The computational overhead of these constructions is $\mathcal{O}(1)$ PKE operations for each transmitted bit. An exception is the somewhat NCE introduced in [19] (see Section 2.5 for more details). This primitive enables to send arbitrarily long messages at the cost of $\log \ell$ PKE operations, where ℓ is the equivocality parameter that determines the number of messages the simulator needs to explain.

This construction improves over NCEs for sufficiently small ℓ's. Finally, in [32] Nielsen proved that adaptively secure non-interactive encryption scheme must have a decryption key that is at least as long as the transmitted message.

Adaptively Secure Oblivious Transfer. [1,10] designed semi-honest adaptively secure OT (using NCE) and then compiled it into the malicious setting using generic ZK proofs. More recently, in a weaker model that assumes erasures, Lindell [29] used the method of [35] to design an efficient transformation from any static OT to a semi-honest composable adaptively secure OT. Another recent work by Garay et al. [19] presented a UC adaptively secure OT, building on the static OT of [34] and somewhat NCE. This paper introduces an OT protocol with security under a weaker *semi-adaptive* notion, that is then compiled into a fully adaptively secure OT by encrypting the transcript of the protocol using somewhat NCE.[3] Finally, [12] presented an improved compiler for a UC adaptively secure OT in the malicious setting (using NCE as well).

Adaptively Secure Two-Party Computation. In the non-erasure model, adaptively secure computation has been extensively studied [10,15,7,26,24,29,11,12,20]. Starting with the work of [10], it is known by now how to compute any well-formed two-party functionality in the adaptive settings. The followup work of [15] showed how to use a threshold encryption to achieve UC adaptive security but requires honest majority. A generic compiler from static to adaptive security was shown in [7] (yet without considering post-execution corruptions). Then the work by Katz and Ostrovsky [26] studied the round complexity in the one-sided setting. Their protocol is the first round efficient construction, yet it takes the naive approach of encrypting the entire communication using NCE. Moreover, the work of [24] provided a UC adaptively secure protocol given an adaptively secure OT. Their compiler generates one-sided schemes that either require a number of adaptively secure OTs that is proportional to the circuit's size, or a number of rounds that is proportional to the depth of the circuit. Finally, a recent work by Garg and Sahai [20] shows adaptively secure constant round protocols tolerating $n-1$ out of n corrupted parties using a non-black box simulation approach. Their approach uses the OT hybrid compiler of [24].

In the erasure model, one of the earliest works by Beaver and Haber [3] showed an efficient generic transformation from adaptively secure protocols with ideally secure communication channels, to adaptively secure protocols with standard (authenticated) communication channels. A more recent work by Lindell [29] presents an efficient semi-honest constant round two-party protocol with adaptive security.

2   Preliminaries

We denote the security parameter by n. A function $\mu(\cdot)$ is *negligible* if for every polynomial $p(\cdot)$ there exists a value N such that for all $n > N$ it holds that $\mu(n) < \frac{1}{p(n)}$. We denote by $a \leftarrow A$ the random sampling of element a from a set A and write PPT for probabilistic polynomial-time. We denote the message spaces of our schemes and the message space of the sender in our OT protocols by $\{0,1\}^q$ for $q = O(n)$.

---

[3] We stress that the semi-adaptive notion is incomparable to the one-sided notion since the former assumes that either one party is statically corrupted or none of the parties get corrupted.

Definition 5 (Computational indistinguishability). *Let* $X = \{X_n(a)\}_{n\in\mathbb{N}, a\in\{0,1\}^*}$ *and* $Y = \{Y_n(a)\}_{n\in\mathbb{N}, a\in\{0,1\}^*}$ *be distribution ensembles. We say that X and Y are computationally indistinguishable, denoted $X \approx_c Y$, if for every family $\{\mathcal{C}_n\}_{n\in\mathbb{N}}$ of polynomial-size circuits, there exists a negligible function $\mu(\cdot)$ such that for all $a \in \{0,1\}^*$, $| \Pr[\mathcal{C}_n(X_n(a)) = 1] - \Pr[\mathcal{C}_n(Y_n(a)) = 1]| < \mu(n)$.*

We denote a PKE by three algorithms $\Pi = (\mathsf{Gen}, \mathsf{Enc}, \mathsf{Dec})$. We say that a protocol π *realizes functionality* $\mathcal{F}$ *with t PKE operations* (relative to Π) if the number of calls π makes to either one of $(\mathsf{Gen}, \mathsf{Enc}, \mathsf{Dec})$ is at most t. Importantly, this definition is not robust in the sense that one might define an encryption algorithm Enc' that consists of encrypting n times in parallel using Enc. In this work we do not abuse this definition and achieve a single basic operation relative to algorithms $(\mathsf{Gen}, \mathsf{Enc}, \mathsf{Dec})$, which are implemented by $O(1)$ group exponentiations in various group descriptions.

2.1 Simulatable Public Key Encryption

A simulatable public key encryption scheme is a semantically secure PKE with four additional algorithms. I.e., an oblivious public key generator $\widetilde{\mathsf{Gen}}$ and a corresponding key faking algorithm $\widetilde{\mathsf{Gen}}^{-1}$, and an oblivious ciphertext generator $\widetilde{\mathsf{Enc}}$ and a corresponding ciphertext faking algorithm $\widetilde{\mathsf{Enc}}^{-1}$. Intuitively, the key faking algorithm is used to explain a legitimately generated public key as an obliviously generated public key. Similarly, the ciphertext faking algorithm is used to explain a legitimately generated ciphertext as an obliviously generated one.

Definition 6 (Simulatable PKE [13]). *A Simulatable PKE is a tuple of algorithms* $(\mathsf{Gen}, \mathsf{Enc}, \mathsf{Dec}, \widetilde{\mathsf{Gen}}, \widetilde{\mathsf{Gen}}^{-1}, \widetilde{\mathsf{Enc}}, \widetilde{\mathsf{Enc}}^{-1})$ *that satisfy the following properties:*

- **Semantic Security.** $(\mathsf{Gen}, \mathsf{Enc}, \mathsf{Dec})$ *is a semantically secure encryption scheme.*
- **Oblivious public key generation.** *Consider the experiment* $(\mathrm{PK}, \mathrm{SK}) \leftarrow \mathsf{Gen}(1^n)$, $r \leftarrow \widetilde{\mathsf{Gen}}^{-1}(\mathrm{PK})$ *and* $\mathrm{PK}' \leftarrow \widetilde{\mathsf{Gen}}(r')$. *Then,* $(r, \mathrm{PK}) \approx_c (r', \mathrm{PK}')$.
- **Oblivious ciphertext generation.** *For any message m in the appropriate domain, consider the experiment* $(\mathrm{PK}, \mathrm{SK}) \leftarrow \mathsf{Gen}(1^n)$, $c_1 \leftarrow \widetilde{\mathsf{Enc}}_{\mathrm{PK}}(r_1)$, $c_2 \leftarrow \mathsf{Enc}_{pk}(m; r_2)$, $r_1' \leftarrow \widetilde{\mathsf{Enc}}^{-1}(c_2)$. *Then* $(\mathrm{PK}, r_1, c_1) \approx_c (\mathrm{PK}, r_1', c_2)$.

The El Gamal PKE [18] is one example for simulatable PKE.

2.2 Dual-Mode PKE

A dual-mode PKE Π_{DUAL} is specified by the algorithms $(\mathsf{Setup}, \mathsf{dGen}, \mathsf{dEnc}, \mathsf{dDec}, \mathsf{FindBranch}, \mathsf{TrapKeyGen})$ described below.

- Setup is the system parameters generator algorithm. Given a security parameter n and a mode $\mu \in \{0,1\}$, the algorithm outputs (CRS, t). The CRS is a common string for the remaining algorithms, and t is a trapdoor value that is given to either $\mathsf{FindBranch}$ or $\mathsf{TrapKeyGen}$, depends on the mode. The setup algorithms for messy and decryption modes are denoted by $\mathsf{SetupMessy}$ and $\mathsf{SetupDecryption}$, respectively; namely $\mathsf{SetupMessy} := \mathsf{Setup}(1^n, 0)$ and $\mathsf{SetupDecryption} := \mathsf{Setup}(1^n, 1)$.

– dGen is the key generation algorithm that takes a bit α and the CRS as input. If $\alpha = 0$, then it generates left public and secret key pair. Otherwise, it creates right public and secret key pair.
– dEnc is the encryption algorithm that takes a bit β, a public key PK and a message m as input. If $\beta = 0$, then it creates the left encryption of m, else it creates the right encryption.
– dDec decrypts a message given a ciphertext and a secret key SK.
– FindBranch finds whether a given public key (in messy mode) is left key or right key given the messy mode trapdoor t.
– TrapKeyGen generates a public key and two secret keys using the decryption mode trapdoor t such that both left encryption as well as the right encryption using the public key can be decrypted using the secret keys.

Definition 7 (Dual-mode PKE). *A dual-mode PKE is a tuple of algorithms described above that satisfy the following properties:*

1. **Completeness.** *For every mode $\mu \in \{0, 1\}$, every $(CRS, t) \leftarrow$ Setup$(1^n, \mu)$, every $\alpha \in \{0, 1\}$, every $(PK, SK) \leftarrow$ dGen(α), and every $m \in \{0, 1\}^\ell$, decryption is correct when the public key type matches the encryption type, i.e., dDec$_{SK}$(dEnc$_{PK}(m, \alpha)) = m$.*
2. **Indistinguishability of modes.** *The CRS generated by* SetupMessy *and* SetupDecryption *are computationally indistinguishable, i.e.,* SetupMessy$(1^n) \approx_c$ SetupDecryption(1^n).
3. **Trapdoor extraction of key type (messy mode).** *For every $(CRS, t) \leftarrow$* SetupMessy(1^n) *and every (possibly malformed)* PK, FindBranch(t, PK) *outputs the public key type $\alpha \in \{0, 1\}$. Encryption at branch $1 - \alpha$ is then message-lossy; namely, for every $m_0, m_1 \in \{0, 1\}^\ell$, dEnc$_{PK}(m_0, 1 - \alpha) \approx_s$ dEnc$_{PK}(m_1, 1 - \alpha)$.*
4. **Trapdoor generation of keys decrypt both branches (decryption mode).** *For every $(CRS, t) \leftarrow$* SetupDecryption(1^n), TrapKeyGen(t) *outputs (PK, SK_0, SK_1) such that for every α, $(PK, SK_\alpha) \approx_c$ dGen(α).*

2.3 NCE for the Receiver

An NCE for the receiver is a semantically secure PKE with an additional property that enables generating a secret key that decrypts a *simulated* (i.e., fake) ciphertext into any plaintext. Specifically, the scheme operates in two modes. The "real mode" enables to encrypt and decrypt as in the standard definition of PKE. The "simulated mode" enables to generate simulated ciphertexts that are computationally indistinguishable from real ciphertexts. Moreover, using a special trapdoor one can produce a secret key that decrypts a fake ciphertext into any plaintext. Intuitively, this implies that simulated ciphertexts are generated in a lossy mode where the plaintext is not well defined given the ciphertext and the public key. This leaves enough entropy for the secret key to be sampled in a way that determines the desired plaintext. Formally,

Definition 8 (NCE for the receiver (NCER)). *An NCE for the receiver encryption scheme is a tuple of algorithms* (Gen, Enc, Enc*, Dec, Equivocate) *specified as follows:*

- Gen, Enc, Dec *are as specified for public key encryption scheme.*
- Enc*, *given the public key* PK *output a ciphertext* c^* *and a trapdoor* t_{c^*}.
- Equivocate, *given the secret key* SK, *trapdoor* t_{c^*} *and a plaintext* m, *output* SK* *such that* $m \leftarrow \mathrm{Dec}_{\mathrm{SK}^*}(c^*)$.

Definition 9 (Secure NCER). *An NCE for the receiver* $\Pi_{\mathrm{NCR}} = (\mathrm{Gen}, \mathrm{Enc}, \mathrm{Dec}, \mathrm{Enc}^*,$ Equivocate) *is secure if it satisfies the following properties:*

- Gen, Enc, Dec *are as specified in the standard semantically secure encryption scheme.*
- *The following* ciphertext indistinguishability *holds for any plaintext* m: $(\mathrm{PK}, \mathrm{SK}^*, c^*, m)$ *and* $(\mathrm{PK}, \mathrm{SK}, c, m)$ *are computationally indistinguishable, for* $(\mathrm{PK}, \mathrm{SK}) \leftarrow \mathrm{Gen}(1^n)$, $(c^*, t_{c^*}) \leftarrow \mathrm{Enc}^*(\mathrm{PK})$, SK* $\leftarrow$ Equivocate$(\mathrm{SK}, c^*, t_{c^*}, m)$ *and* $c \leftarrow \mathrm{Enc}_{\mathrm{PK}}(m)$.

A review of two implementations of NCER under the DDH [25,9] and DCR [9] assumptions is found in our full version [23].

2.4   NCE for the Sender

NCE for the sender is a semantically secure PKE with an additional property that enables generating a fake public key, such that any ciphertext encrypted under this key can be viewed as the encryption of any message together with the matched randomness. Specifically, the scheme operates in two modes. The "real mode" that enables to encrypt and decrypt as in standard PKEs and the "simulated mode" that enables to generate simulated public keys and an additional trapdoor, such that the keys in the two modes are computationally indistinguishable. In addition, given this trapdoor and a ciphertext generated using the simulated public key, one can produce randomness that is consistent with any plaintext. We continue with a formal definition.

Definition 10 (NCE for the sender (NCES)). *An NCE for the sender encryption scheme is a tuple of algorithms* (Gen, Gen*, Enc, Dec, Equivocate) *specified as follows:*

- Gen, Enc, Dec *are as specified for public key encryption scheme.*
- Gen* *generates public key* PK* *and a trapdoor* t_{PK^*}.
- Equivocate, *given a ciphertext* c^* *computed using* PK*, *a trapdoor* t_{PK^*} *and a plaintext* m, *output* r *such that* $c^* \leftarrow \mathrm{Enc}(m, r)$.

Definition 11 (Secure NCES). *An NCE for the sender* $\Pi_{\mathrm{NCES}} = (\mathrm{Gen}, \mathrm{Gen}^*, \mathrm{Enc}, \mathrm{Dec},$ Equivocate) *is secure if it satisfies the following properties:*

- Gen, Enc, Dec *are as specified in the standard semantically secure encryption scheme.*
- *The following* public key indistinguishability *holds for any plaintext* m: $(\mathrm{PK}^*, r^*, m, c^*)$ *and* (PK, r, m, c) *are computationally indistinguishable, for* $(\mathrm{PK}^*, t_{\mathrm{PK}^*}) \leftarrow \mathrm{Gen}^*(1^n)$, $c^* \leftarrow \mathrm{Enc}_{\mathrm{PK}^*}(m', r')$, $r^* \leftarrow$ Equivocate$(c^*, t_{\mathrm{PK}^*}, m)$ *and* $c \leftarrow \mathrm{Enc}_{\mathrm{PK}}(m, r)$.

A review of the DDH based implementation from [5] and a new DCR based implementation is found in our full version [23].

2.5   Somewhat Non-committing Encryption [19]

The idea of somewhat NCE is to exploit the fact that it is often unnecessary for the simulator to explain a fake ciphertext for *any* plaintext. Instead, in many scenarios it suffices to explain a fake ciphertext with respect to a small set of size ℓ determined in advance (where ℓ might be as small as 2). Therefore there are two parameters that are considered here: a plaintext of bit length l and an equivocality parameter ℓ which is the number of plaintexts that the simulator needs to explain a ciphertext for (namely, the non-committed domain size). Note that for NCE $\ell = 2^l$. Somewhat NCE typically improves over fully NCE whenever ℓ is very small but the plaintext length is still large, say $O(n)$ where n is the security parameter.

3   One-Sided Adaptively Secure NCE

In this section we design one-sided NCE, building on NCE for the sender (NCES) and NCE for the receiver (NCER). The idea of our protocol is to have the receiver create two public/secret key pairs where the first pair is for NCES and the second pair is for NCER. The receiver sends the public keys and the sender encrypts two shares of its message m, each share with a different key. Upon receiving the ciphertexts the receiver recovers the message by decrypting the ciphertexts. Therefore, equivocality of the sender's input can be achieved if the public key of the NCES is fake, whereas, equivocality of the receiver's input can be achieved if the ciphertext of the NCER is fake. Nevertheless, this idea only works if the simulator is aware of the identity of the corrupted party prior to the protocol execution in order to decide whether the keys or the ciphertexts should be explained as valid upon corruption (since it cannot explain fake keys/ciphertext as valid). We resolve this problem using somewhat NCE in order to commit to *the choice* of having fake/valid keys and ciphertexts. Specifically, it enables the simulator to "explain" fake keys/ciphertext as valid and vice versa using only a constant number of asymmetric operations, as each such non-committing bit requires an equivocation space of size 2.

Formally, denote by $\mathcal{F}_{\mathrm{SC}}(m, -) \mapsto (-, m)$ the secure message transfer functionality, and let $\Pi_{\mathrm{NCES}} = (\mathsf{Gen}, \mathsf{Gen}^*, \mathsf{Enc}, \mathsf{Dec}, \mathsf{Equivocate})$ and $\Pi_{\mathrm{NCER}} = (\mathsf{Gen}, \mathsf{Enc}, \mathsf{Enc}^*,$ $\mathsf{Dec}, \mathsf{Equivocate})$ denote secure NCES and NCER for a message space $\{0,1\}^q$. Consider the following one-sided protocol for $\mathcal{F}_{\mathrm{SC}}$.

Protocol 1 (One-sided NCE (Π_{OSC}))

- **Inputs:** *Sender SEN is given input message $m \in \{0,1\}^q$. Both parties are given security parameter 1^n.*
- **The Protocol:**
    1. **Message from the receiver.** *REC invokes $\mathsf{Gen}(1^n)$ of Π_{NCES} and Π_{NCER} and obtains two public/secret key pairs $(\mathrm{PK}_0, \mathrm{SK}_0)$ and $(\mathrm{PK}_1, \mathrm{SK}_1)$, respectively. REC sends PK_1 on clear and PK_0 using somewhat NCE with equivocality parameter $\ell = 2$.*
    2. **Message from the sender.** *Upon receiving PK_0 and PK_1, SEN creates two shares of m, m_0 and m_1, such that $m = m_0 \oplus m_1$. It then encrypts each m_i using PK_i, creating ciphertext c_i, and sends c_0 and c_1 using two instances of somewhat NCE with equivocality parameter $\ell = 2$.*

3. **Output.** *Upon receiving* c_0, c_1, REC *decrypts* c_i *using* SK_i *and outputs the bitwise XOR of the decrypted plaintexts.*

Note that the message space of our one-sided NCE is equivalent to the message space of the NCES/NCER schemes, where q can be as large as n. Therefore, our protocol transmits q-bits messages using a *constant number of PKE operations*, as opposed to NCEs that require $O(q)$ such operations. We provide two instantiations for the above protocol. One for polynomial-size message spaces using DDH based NCES and NCER, and another for exponential-size message spaces using DCR based NCES and NCER. We conclude with the following theorem and the complete proof.

Theorem 12. *Assume the existence of NCER and NCES with constant number of PKE operations for message space* $\{0, 1\}^q$ *and simulatable PKE. Then Protocol 1 UC realizes* $\mathcal{F}_{SC}$ *in the presence of one-sided adaptive malicious adversaries with constant number of PKE operations for message space* $\{0, 1\}^q$, *where* $q = O(n)$ *and* n *is the security parameter.*

Proof: Let ADV be a malicious probabilistic polynomial-time adversary attacking Protocol 1 by adaptively corrupting one of the parties. We construct an adversary SIM for the ideal functionality $\mathcal{F}_{SC}$ such that no environment ENV distinguishes with a non-negligible probability whether it is interacting with ADV in the real setting or with SIM in the ideal setting. We recall that SIM interacts with the ideal functionality $\mathcal{F}_{SC}$ and the environment ENV. We refer to the interaction of SIM with $\mathcal{F}_{SC}$ and ENV as the external interaction. The interaction of SIM with the simulated ADV is the internal interaction. We explain the strategy of the simulation for all corruption cases.

Simulating the communication with ENV. Every input value received by the simulator from ENV is written on ADV's input tape. Likewise, every output value written by ADV on its output tape is copied to the simulator's output tape (to be read by its environment ENV).

SEN is corrupted at the onset of the protocol. SIM begins by activating ADV and emulates the honest receiver by sending to ADV, PK_0 using the somewhat NCE and PK_1 in clear. Upon receiving two ciphertexts c_0 and c_1 from ADV, SIM extracts m by computing $Dec_{SK_0}(c_0) \oplus Dec_{SK_1}(c_1)$. SIM externally forwards m to the ideal functionality $\mathcal{F}_{SC}$.

REC is corrupted at the onset of the protocol. SIM begins by activating ADV and obtains REC's output m from $\mathcal{F}_{SC}$. SIM invokes ADV and receives PK_0 from ADV via the somewhat NCE and PK_1 in clear. Next, SIM completes the execution playing the role of the honest sender on input m. Note that it does not make a difference whether REC generates invalid public keys since SIM knows m and thus perfectly emulates the receiver's view.

If none of the parties is corrupted as above, SIM emulates the receiver's message as follows. It creates public/secret key pair (PK_1, SK_1) for Π_{NCER} and sends the public key in clear. It then creates a valid public/secret key pair (PK_0, SK_0) and a fake public key with a trapdoor $(PK_0^*, t_{PK_0^*})$ for Π_{NCES} (using Gen and Gen*, respectively). SIM sends (PK_0, PK_0^*) using somewhat NCE. Namely, the simulator does not send the valid PK_0 as the honest receiver would do, rather it encrypts both valid and invalid keys within the somewhat NCE.

SEN is corrupted after Step 1 is concluded. Since no message was sent yet on behalf of the sender, SIM completes the simulation playing the role of the honest sender using m.

REC is corrupted after Step 1 is concluded. Upon receiving m, SIM explains the receiver's internal state which is independent of the message m so far. Specifically, it reveals the randomness for generating PK_0, SK_0 and PK_1, SK_1 and presents the randomness for the valid key PK_0 being the message sent by the somewhat NCE. SIM plays the role of the honest sender with input m as the message.

If none of the above corruption cases occur, SIM emulates the sender's message as follows. It first chooses two random shares m'_0, m'_1 and generates a pair of ciphertexts (c_0, c_0^*) for Π_{NCES} that encrypts m'_0 using PK_0 and PK_0^*. It then generates a pair of ciphertexts (c_1, c_1^*) for Π_{NCER} such that c_1 is a valid encryption of m'_1 using the public key PK_1, and c_1^* is a fake ciphertext generated using Enc^* and PK_1. SIM sends (c_0, c_0^*) and (c_1^*, c_1) via two instances of somewhat NCE.

SEN is corrupted after Step 2 is concluded. Upon receiving a corruption instruction from ENV, SIM corrupts the ideal SEN and obtains SEN's input m. It then explains the sender's internal state as follows. It explains PK_0^* for being the public key sent by the receiver using the somewhat NCE. Furthermore, it presents the randomness for c_0^* and c_1 being the ciphertexts sent via the somewhat NCE. Finally, it computes $r'' \leftarrow \text{Equivocate}_{PK_0^*}(t_{PK_0^*}, m'_0, r, m''_0)$ for m''_0 such that $m = m''_0 \oplus m'_1$ and r the randomness used to encrypt m'_0, and presents r'' as the randomness used to generate c_0^* that encrypts m''_0. The randomness used for generating c_1 is revealed honestly.

REC is corrupted after Step 2 is concluded. Upon receiving a corruption instruction from ENV, SIM corrupts the ideal REC and obtains REC's output m. It then explains the receiver's internal state as follows. It presents the randomness for PK_0 for being the public key sent via the somewhat NCE and presents the randomness for generating (PK_0, SK_0). It then explains c_0 and c_1^* for being sent via the somewhat NCE. Finally, it generates a secret key SK_1^* so that $m''_1 \leftarrow \text{Dec}_{SK_1^*}(c_1^*)$ and $m''_1 \oplus m'_0 = m$. That is, it explains (PK_1, SK_1^*) as the other pair of keys generated by the receiver.

We now show that for every corruption case described above, there is not any polynomial-time ENV that distinguishes with a non-negligible probability the real execution with ADV and the simulated execution with SIM.

SEN/REC is corrupted at the onset of the protocol. In these corruption cases there is no difference between the real execution and the simulated execution and the views are statistically indistinguishable.

SEN/REC is corrupted after Step 1 is concluded. In these cases the only difference between the real and simulated executions is with respect to the somewhat NCE that delivers the public key of NCES. Specifically, in the real execution it always delivers a valid public key while in the simulated execution it delivers a fake key. Indistinguishability follows from the security of the somewhat NCE.

SEN/REC is corrupted after Step 2 is concluded. Here the adversary sees in the
simulation either a fake public key or a fake ciphertext. Indistinguishability follows
from the security of Π_{NCES} and Π_{NCER} and the security of somewhat NCE.

∎

4 One-Sided Adaptively Secure OT

A common approach to design an adaptive OT [2,10] is by having the receiver generate two public keys $(\text{PK}_0, \text{PK}_1)$ such that it only knows the secret key associated
with PK_σ. The sender then encrypts x_0, x_1 under these respective keys so that the receiver decrypts the σth ciphertext. The security of this protocol in the adaptive setting
holds if the underlying encryption scheme is an *augmented non-committing encryption
scheme* [10]. In this section we follow the approach from [19] and build one-sided OT
based on the static OT from [34], which is based on a primitive called *dual-mode PKE*.

The PVW OT. Dual-mode PKE is a semantically secure encryption scheme that is initialized with system parameters of two types. For each type one can generate two types
of public/secret key pair, labeled by the left key pair and the right key pair. Similarly,
the encryption algorithm generates a left or a right ciphertext. Moreover, if the key label matches the ciphertext label (i.e., a left ciphertext is generated under the left public
key), then the ciphertext can be correctly decrypted. (A formal definition of dual-mode
PKE is given in Section 2.2.) This primitive was introduced in [34] which demonstrates
its usefulness in designing efficient statically secure OTs under various assumptions.
First, the receiver generates a left key if $\sigma = 0$, and a right key otherwise. In response,
the sender generates a left ciphertext for x_0 and a right ciphertext for x_1. The receiver
then decrypts the σth ciphertext.

The security of dual-mode PKE relies on the two indistinguishable modes of generating the system parameters: *messy* and *decryption* mode. In a messy mode the system
parameters are generated together with a messy trapdoor. Using this trapdoor, any public key (even malformed ones) can be labeled as a left or as a right key. Moreover,
when the key type does not match the ciphertext type, the ciphertext becomes statistically independent of the plaintext. The messy mode is used to ensure security when the
receiver is corrupted since it allows to extract the receiver's input bit while hiding the
sender's other input. On the other hand, the system parameters in a decryption mode are
generated together with a decryption trapdoor that can be used to decrypt both left and
right ciphertexts. Moreover, left public keys are statistically indistinguishable from right
keys. The decryption mode is used to ensure security when the sender is corrupted since
the decryption trapdoor enables to extract the sender's inputs while statistically hiding
the receiver's input. [34] instantiated dual-mode PKE and their generic OT construction
based on various assumptions, such as DDH, QR and lattice-based assumptions.

Our Construction. We build our one-sided OT based on the PVW protocol considering
the following modifications. (**1**) First, we require that the sender sends its ciphertexts
using one-sided NCE (see Section 3). (**2**) We fix the system parameters in a decryption

mode, which immediately implies extractability of the sender's input and equivocality of the receiver's input. We further achieve equivocality of the sender's input using our one-sided NCE. In order to ensure extractability of the receiver's input we employ a special type of ZK PoK. Namely, this proof exploits the fact that the simulator knows both witnesses for the proof yet it does not know which witness will be used by the real receiver, since this choice depends on σ. Specifically, it allows the simulator to use both witnesses and later convince the adversary that it indeed used a particular witness. In addition, it enables to extract σ since the real receiver does not know both witnesses. We denote these proofs for compound statements by witness equivocal and refer to Section 6 for more details.

Our construction is one-sided UC secure in the presence of malicious adversaries, and uses a number of non-committed bits that is *independent* of the sender's input size or the overall communication complexity. We formally denote the dual-mode PKE of [34] by $\Pi_{\mathrm{DUAL}} = (\mathsf{SetupMessy}, \mathsf{SetupDecryption}, \mathsf{dGen}, \mathsf{dEnc}, \mathsf{dDec}, \mathsf{FindBranch}, \mathsf{TrapKeyGen})$ and describe our construction in the $(\mathcal{F}_{\mathrm{SC}}, \mathcal{F}_{\mathrm{ZKPoK}}^{\mathcal{R}_{\mathrm{LR}}})$-hybrid model, where $\mathcal{F}_{\mathrm{SC}}$ is instantiated with one-sided NCE. Furthermore, the latter functionality is required to ensure the correctness of the public key and is defined for a compound statement that is comprised from the following two relations,

$$\mathcal{R}_{\mathrm{LEFT}} = \{(\mathrm{PK}, r_0) \mid (\mathrm{PK}, \mathrm{SK}) \leftarrow \mathsf{dGen}(\mathrm{CRS}, 0; r_0)\},$$

where CRS are the system parameters. Similarly, we define $\mathcal{R}_{\mathrm{RIGHT}}$ for the right keys. Specifically, $\mathcal{F}_{\mathrm{ZKPoK}}^{\mathcal{R}_{\mathrm{LR}}}$ receives a public key PK and randomness r_σ for $\sigma \in \{0,1\}$ and returns Accept if either $\sigma = 0$ and $\mathrm{PK} = \mathsf{dGen}(\mathrm{CRS}, 0; r_0)$, or $\sigma = 1$ and $\mathrm{PK} = \mathsf{dGen}(\mathrm{CRS}, 1; r_1)$ holds. Security is proven by implementing this functionality using a witness equivocal ZK PoK that allows the simulator to equivocate the witness during the simulation (i.e., explaining the proof transcript with respect to either r_0 or r_1). We consider two instantiations of dual-mode PKE (based on the DDH and QR assumptions). For each implementation we design a concrete ZK PoK, proving that the prover knows r_σ with respect to $\sigma \in \{0,1\}$; see details below.

We define our OT protocol as follows,

Protocol 2 (One-sided OT (Π_{OT}))

- **Inputs:** *Sender SEN has $x_0, x_1 \in \{0,1\}^q$ and receiver REC has $\sigma \in \{0,1\}$.*
- **CRS:** *CRS such that $(\mathrm{CRS}, t) \leftarrow \mathsf{SetupDecryption}$.*
- **The Protocol:**
  1. *REC sends SEN PK, where $(\mathrm{PK}, \mathrm{SK}) \leftarrow \mathsf{dGen}(\mathrm{CRS}, \sigma; r_\sigma)$. REC calls $\mathcal{F}_{\mathrm{ZKPoK}}^{\mathcal{R}_{\mathrm{LR}}}$ with (PK, r_σ).*
  2. *Upon receiving Accept from $\mathcal{F}_{\mathrm{ZKPoK}}^{\mathcal{R}_{\mathrm{LR}}}$ and PK from REC, SEN generates $c_0 \leftarrow \mathsf{dEnc}_{\mathrm{PK}}(x_0, 0)$ and $c_1 \leftarrow \mathsf{dEnc}_{\mathrm{PK}}(x_1, 1)$. SEN calls $\mathcal{F}_{\mathrm{SC}}$ twice with inputs c_0 and c_1, respectively.*
  3. *Upon receiving (c_0, c_1), REC outputs $\mathsf{dDec}_{\mathrm{SK}}(c_\sigma)$.*

Theorem 13. *Assume the existence of one-sided NCE with constant number of PKE operations for message space $\{0,1\}^q$ and dual-mode PKE. Then Protocol 2 UC realizes $\mathcal{F}_{\mathrm{OT}}$ in the $(\mathcal{F}_{\mathrm{SC}}, \mathcal{F}_{\mathrm{ZKPoK}}^{\mathcal{R}_{\mathrm{LR}}})$-hybrid model in the presence of one-sided adaptive malicious adversaries with constant number of PKE operations for sender's input space $\{0,1\}^q$, where $q = O(n)$ and n is the security parameter.*

Proof: Let ADV be a probabilistic polynomial-time malicious adversary attacking Protocol2 by adaptively corrupting one of the parties. We construct an adversary SIM for the ideal functionality $\mathcal{F}_{OT}$ such that no environment ENV distinguishes with a non-negligible probability whether it is interacting with ADV in the real setting or with SIM in the ideal setting. We recall that SIM interacts with the ideal functionality $\mathcal{F}_{OT}$ and the environment ENV. We refer to the interaction of SIM with $\mathcal{F}_{OT}$ and ENV as the external interaction. The interaction of SIM with the simulated ADV is the internal interaction. Upon computing $(CRS, t) \leftarrow \mathsf{SetupDecryption}(1^n)$, SIM proceeds as follows:

Simulating the communication with ENV. Every input value received by the simulator from ENV is written on ADV's input tape. Likewise, every output value written by ADV on its output tape is copied to the simulator's output tape (to be read by its environment ENV).

SEN is corrupted at the outset of the protocol. SIM begins by activating ADV and emulates the receiver by running $(PK, SK_0, SK_1) \leftarrow \mathsf{TrapKeyGen}(t)$. It then sends PK and an `Accept` message to ADV on behalf of $\mathcal{F}_{ZKP_0K}^{\mathcal{R}_{LR}}$. Whenever ADV returns c_0, c_1 via $\mathcal{F}_{SC}$, SIM extracts SEN's inputs x_0, x_1 by invoking $\mathsf{dDec}_{SK_0}(c_0)$ and $\mathsf{dDec}_{SK_1}(c_1)$ as in static case. It then sends x_0, x_1 to $\mathcal{F}_{OT}$ and completes the execution playing the role of the receiver using an arbitrary σ.

Note that, in contrast to the hybrid execution where the receiver uses its real input σ to dGen in order to create public/secret keys pair, the simulator does not know σ and thus creates the keys using TrapKeyGen. Nevertheless, when the CRS is set in a decryption mode the left public key is statistically indistinguishable from right public key. Furthermore, the keys (PK, SK_i) that are generated by TrapKeyGen are statistically close to the keys generated by dGen with input bit i. This implies that the hybrid and simulated executions are statistically close.

REC is corrupted at the outset of the protocol. SIM begins by activating ADV and receives its public key PK and a witness r_σ on behalf of $\mathcal{F}_{ZKP_0K}^{\mathcal{R}_{LR}}$. Given r_σ, SIM checks if PK is the left or the right key and use it to extract the receiver's input σ. It then sends σ to $\mathcal{F}_{OT}$, receiving back x_σ. Finally, SIM computes the sender's message using x_σ and an arbitrary $x_{1-\sigma}$.

Unlike in the hybrid execution, the simulator uses an arbitrary $x_{1-\sigma}$ instead of the real $x_{1-\sigma}$. However, a decryption mode implies computational privacy of $x_{1-\sigma}$. Therefore, the hybrid view is also computationally indistinguishable from the simulated view as in the static setting proven in [34].

If none of the above corruption cases occur SIM invokes $(PK, SK_0, SK_1) \leftarrow \mathsf{TrapKeyGen}(t)$ and sends PK to the sender. Note that the simulator knows a witness r_0 such that $PK = \mathsf{dGen}(CRS, 0; r_0)$ and a witness r_1 such that $PK = \mathsf{dGen}(CRS, 1; r_1)$.

SEN is corrupted between Steps 1 and 2. SIM trivially explains the the sender's internal state since SEN did not compute any message so far. The simulator completes the simulation by playing the role of REC using arbitrary σ as in the case when the sender is corrupted at the outset of the execution.

Indistinguishability for this case follows similarly to the prior corruption case when SEN is corrupted at the outset of the execution.

REC is corrupted between Steps 1 and 2. Upon corrupting the receiver SIM obtains σ, x_σ from $\mathcal{F}_{\text{OT}}$ and explains the receiver's internal state as follows. It first explains r_σ as the witness given to $\mathcal{F}_{\text{ZKPoK}}^{\mathcal{R}_{\text{LR}}}$ and PK as the outcome of $\text{dGen}(\text{CRS}, \sigma; r_\sigma)$. The simulator completes the simulation playing the role of the honest sender with x_σ and an arbitrary $x_{1-\sigma}$.

Indistinguishability for this case in the hybrid setting follows similarly to the prior corruption case, since the only difference in the simulation is relative to the witness equivocality proof which only makes a difference in the real execution.

If none of the above corruption cases occur then SIM chooses two arbitrary inputs x_0', x_1' for the sender and encrypts them using the dual-mode encryption. Denote these ciphertexts by c_0', c_1'. SIM pretends sending these ciphertexts using $\mathcal{F}_{\text{SC}}$.

SEN is corrupted after Step 2. Upon corrupting the sender, SIM obtains (x_0, x_1) from $\mathcal{F}_{\text{OT}}$. It then explains the sender's internal state as follows. It first computes c_0, c_1 that encrypts x_0 and x_1 respectively. It then explains c_0 and c_1 as being sent using $\mathcal{F}_{\text{SC}}$.

In the hybrid setting indistinguishability follows as in the prior corruption case of the sender, since the simulator emulates the sender's message via the one-sided non-committing channel. In the real execution, security is reduced to the security of the one-sided encryption scheme implementation.

REC is corrupted after Step 2. Upon corrupting the receiver, SIM obtains REC's input and output (σ, x_σ) from $\mathcal{F}_{\text{OT}}$. It then explains the receiver's internal state as follows. It first explains r_σ as the witness given to $\mathcal{F}_{\text{ZKPoK}}^{\mathcal{R}_{\text{LR}}}$ and PK as the outcome of $\text{dGen}(\text{CRS}, \sigma; r_\sigma)$. Finally, it explains the output of $\mathcal{F}_{\text{SC}}$ as c_σ so that c_σ is indeed a valid encryption of x_σ.

Indistinguishability follows similarly to the prior corruption case of the receiver since the second message is computed by the sender which is not corrupted. ∎

Concrete Instantiations. In the DDH-based instantiation the CRS is a Diffie-Hellman tuple (g_0, g_1, h_0, h_1) and the trapdoor is $\log_{g_0} g_1$. Moreover, the concrete ZK PoK functionality is $\mathcal{F}_{\text{ZKPoK}}^{\mathcal{R}_{\text{DL}},\text{OR}}$ which is invoked with the statement and witness $\big(((g_0 h_0, g_\sigma^r h_\sigma^r), (g_1 h_1, g_\sigma^r h_\sigma^r)), r\big)$, such that PK $= (g_\sigma^r, h_\sigma^r)$, SK $= r$ and $r \leftarrow \mathbb{Z}_p$.

In the QR-based instantiation the CRS is a quadratic residue y and the trapdoor is s such that $y = s^2 \bmod N$ and N is an RSA composite. The concrete ZK PoK functionality is $\mathcal{F}_{\text{ZKPoK}}^{\mathcal{R}_{\text{QR}},\text{OR}}$ which is invoked with the statement and witness $\big((y \cdot \text{PK}, \text{PK}), r\big)$, such that PK $= r^2/y^\sigma$, SK $= r$ and $r \leftarrow \mathbb{Z}_N^*$.

5   Constant Round One-Sided Adaptively Secure Computation

In the following section we demonstrate the feasibility of one-sided adaptively secure two-party protocols in the presence of semi-honest and malicious adversaries. Our constructions are constant round and UC secure and use a number of non-committed bits that is independent of the circuit size, thus reduce the number of PKE operations so that it only depends on the input and output lengths. A high-level overview of Yao's garbled circuit construction $G(C)$ for a circuit C is found in the full version.

5.1 One-Sided Secure Computation for Semi-honest Adversaries

Our first construction adapts the semi-honest two-party protocol [36,28] into the one-sided adaptive setting at a cost of $\mathcal{O}(|C|)$ private key operations and $\mathcal{O}(|\text{input}| + |\text{output}|)$ public key operations. Using our one-sided secure primitives we obtain efficient protocols that outperform the constant round one-sided constructions of [26,24] and all known fully adaptively secure two-party protocols. Namely, we show that one-sided security can be obtained by only communicating the keys corresponding to the input/output wires via a non-committing channel. This implies that the number of PKE operations *does not* depend on the garbled circuit size as in prior work.

Informally, the input keys that correspond to P_0's input are transferred to P_1 using somewhat NCE with equivocation parameter $\ell = 2$, whereas P_1's input keys are sent using one-sided OT. Next, the entire garbled circuit (without the output decryption table) is sent to P_1 using a standard communication channel. P_1 evaluates the garbled circuit and finds the keys for the output wires. The parties then run a one-sided bit OT for each output key where P_1 plays the role of the receiver, and learns the output bit that corresponds to its output wire. Finally, P_1 sends P_0 the output using one-sided NCE. We note that obtaining the output via one-sided OT is crucial to our proof since it enables us to circumvent the difficulties arise when implementing the simulation technique from [28] that uses a fake circuit. To carry out these OTs successfully we require that the keys associated with a output wire have distinct most significant bits that are fixed independently of the bits they correspond to. For simplicity we only consider deterministic and same-output functionalities. This can be further generalized using the reductions specified in [21]. The formal description of our one-sided semi-honest protocol Π_f^{SH} is given below in the $\mathcal{F}_{\text{OT}}$-hybrid model.

Protocol 3 (One-sided adaptively secure semi-honest Yao (Π_f^{SH}))

- **Inputs:** P_0 has $x_0 \in \{0,1\}^n$ and P_1 has $x_1 \in \{0,1\}^n$. Let $x_0 = x_0^1, \ldots, x_0^n$ and $x_1 = x_1^1, \ldots, x_1^n$.
- **Auxiliary Input:** *A boolean circuit C such that for every $x_0, x_1 \in \{0,1\}^n$, $C(x,y) = f(x,y)$ where $f : \{0,1\}^n \times \{0,1\}^n \to \{0,1\}^n$. Furthermore, we assume that C is such that if a circuit-output wire leaves some gate, then the gate has no other wires leading from it into other gates (i.e. no circuit-output wire is also a gate-output wire). Likewise, a circuit-input wire that is also a circuit-output wire enters no gates.*
- **The Protocol:**
    1. **Setup and garbling circuit computation.** P_0 constructs garbled circuit $G(C)$ subject to the constraint that the keys corresponding to each circuit-output wire have a distinct most significant bit.
    2. **Transferring the garbled circuit and input keys to P_1.** Let (k_i^0, k_i^1) be the key pair corresponding to the circuit-input wire that takes the ith bit of x_0 and let (k_{n+i}^0, k_{n+i}^1) be the key pair corresponding to the circuit-input wire that takes the ith bit of x_1. Then,
        (a) *For all $i \in [1, \ldots, n]$, P_0 sends $k_i^{x_0^i}$ using an instance of somewhat NCE with $\ell = 2$.*
        (b) *For all $i \in [1, \ldots, n]$, P_0 and P_1 call $\mathcal{F}_{\text{OT}}$ with input (k_{n+i}^0, k_{n+i}^1) and x_1^i, respectively. Let $k_{n+i}^{x_1^i}$ denotes P_1's ith output.*
        (c) *P_0 sends $G(C)$ without the output decryption table to P_1.*

3. **Circuit evaluation and interactive output computation.** P_1 *evaluates* $G(C)$ *on the above input keys and obtains the keys that correspond to* $f(x_0, x_1)$ *in the circuit-output wires. Let* (k_{2n+i}^0, k_{2n+i}^1) *be the key pair corresponding to the ith circuit-output wire with distinct most significant bits. Also assume* P_1 *obtains key* k_{2n+i}^α *corresponding to the ith circuit-output wire of* $G(C)$. *Then,*

   (a) *For all* $i \in [1, \ldots, n]$, P_0 *and* P_1 *call* $\mathcal{F}_{OT}$ *in which* P_0's *input equals* $(0, 1)$ *if the most significant bit of* k_{2n+i}^0 *is 0, and* $(1, 0)$ *otherwise.* P_1's *input is the most significant bit of* k_{2n+i}^α.

   (b) P_1 *computes* $f(x_0, x_1)$ *by concatenating the bits received from the above n calls.*

4. **Output communication.** P_1 *sends* y *using an instance of one-sided NCE.*

Theorem 14 (One-sided semi-honest). *Let* f *be a deterministic same-output functionality and assume that the encryption scheme for garbling has indistinguishable encryptions under chosen plaintext attacks, and an elusive and efficiently verifiable range. Furthermore, assume that* $\mathcal{F}_{OT}$ *is realized in the presence of one-sided semi-honest adversaries with constant number of PKE operations for sender's input space* $\{0, 1\}^q$, *where* $q = O(n)$ *and* n *is the security parameter. Then Protocol 3 UC realizes* $\mathcal{F}_f$ *in the presence of one-sided semi-honest adversaries at a cost of* $\mathcal{O}(|C|)$ *private key operations and* $\mathcal{O}(|input| + |output|)$ *public key operations.*

We note that the ideal OT calls in Step 2 can be realized using string one-sided OTs, whereas the OT calls in Step 3 can be replaced with bit one-sided OTs. The complete proof can be found in our full version [23].

5.2   Security against Malicious Adversaries

Next, we modify Π_f^{SH} and adapt the cut-and-choose OT protocol introduced in [30] in order to achieve security against malicious adversaries. The idea of the cut-and-choose technique is to ask P_0 to send s garbled circuits and later open half of them (aka, *check circuits*) by the choice of P_1. This ensures that with very high probability the majority of the unopened circuits (aka, *evaluation circuits*) are valid. The cut-and-choose OT primitive of [30] allows P_1 to choose a secret random subset $\mathcal{J}$ of size $s/2$ for which it learns both keys for each input wire that corresponds to the check circuits, and the keys associated with its input with respect to the evaluation circuits.

In order to ensure that P_0 hands P_1 consistent input keys for all the circuits, the [30] protocol ensures that the keys associated with P_0's input are obtained via a Diffie-Hellman pseudorandom synthesizer [31]. Namely, P_0 chooses $g^{a_1^0}, g^{a_1^1}, \ldots, g^{a_n^0}, g^{a_n^1}$ and $g^{c_1}, \ldots, g^{c_s}$, where n is the input/output length, s is the cut-and-choose parameter and g is a generator of a prime order group $\mathbb{G}$. So that the pair of keys associated with the ith input of P_0 in the jth circuit is $(g^{a_i^0 c_j}, g^{a_i^1 c_j})$.[4] Given values $\{g^{a_i^0}, g^{a_i^1}, g^{c_j}\}$ and any subset of keys associated with P_0's input, the remaining keys associated with its input are pseudorandom by the DDH assumption. Furthermore, when the keys are prepared this way P_0 can efficiently prove that it used the same input for all circuits.

---

[4] The actual key pair used in the circuit garbling is derived from $(g^{a_i^0 c_j}, g^{a_i^1 c_j})$ using an extractor. A universal hash function is used in [30] for this purpose, where the seeds for the function are picked by P_0 before it knows $\mathcal{J}$.

P_1 then evaluates the evaluation circuits and takes the majority value for the final output. In Section 5.2 we demonstrate how to adapt the cut-and-choose OT protocol into the one-sided setting using the building blocks introduced in this paper. This requires dealing with new subtleties regarding the system parameters and the ZK proofs. Formally,

Theorem 15 (One-sided malicious). *Let f be a deterministic same-output functionality and assume that the encryption scheme for garbling has indistinguishable encryptions under chosen plaintext attacks, an elusive and efficiently verifiable range, and that the DDH and DCR assumptions are hard in the respective groups. Then Protocol Π_f^{MAL} UC realizes $\mathcal{F}_f$ in the presence of one-sided malicious adversaries at a cost of $\mathcal{O}(s \cdot |C|)$ private key operations and $\mathcal{O}(s \cdot (|\mathsf{input}| + |\mathsf{output}|))$ public key operations where s is a statistical parameter that determines the cut-and-choose soundness error.*

Specifically, the concrete DCR assumption implies cut-and-choose OT with constant number of PKE operations for sender's input space $\{0,1\}^q$, where $q = O(n)$ and n is the security parameter.

One-sided Single Choice Cut-and-Choose OT. We describe next the single choice cut-and-choose OT functionality $\mathcal{F}_{\mathrm{CCOT}}$ from [30] and present a protocol that implements this functionality with UC one-sided malicious security. We then briefly describe our batch single choice cut-and-choose OT construction using a single choice cut-and-choose OT, which is used as a building block in our two-party protocol. Formally, $\mathcal{F}_{\mathrm{CCOT}}$ is defined as follows

1. **Inputs:**
   (a) SEN inputs a vector of pairs $\{(x_0^j, x_1^j)\}_{j=1}^s$.
   (b) REC inputs a bit σ and a set of indices $\mathcal{J} \subset [s]$ of size exactly $s/2$.
2. **Output:** If $\mathcal{J}$ is not of size $s/2$, then SEN and REC receive $\perp$ as output. Otherwise,
   (a) For all $j \in \mathcal{J}$, REC obtains the pair (x_0^j, x_1^j).
   (b) For all $j \notin \mathcal{J}$, REC obtains x_σ^j.

This functionality is implemented in [30] by invoking the DDH based [34] OT s times, where the receiver generates the system parameters in a decryption mode for $s/2$ indices corresponding to $\mathcal{J}$ and the remaining system parameters are generated in a messy mode. The decryption mode trapdoor enables the receiver to learn both sender's inputs for the instances corresponding to $\mathcal{J}$. This idea is coupled with two proofs that are run by the receiver: (**i**) a ZK PoK for proving that half of the system parameters set is in a messy mode which essentially boils down to a ZK PoK realizing functionality $\mathcal{F}_{\mathrm{ZKPoK}}^{\mathcal{R}_{\mathrm{DDH}}, \mathrm{COMP}(s, s/2)}$ (namely, the statement is a set of s tuples and the prover proves the knowledge of $s/2$ Diffie-Hellman tuples within this set). (**ii**) A ZK PoK to ensure that the same input bit σ has been used for all s instances which boils down to a ZK proof realizing functionality $\mathcal{F}_{\mathrm{ZKPoK}}^{\mathcal{R}_{\mathrm{DDH}}, \mathrm{OR}(s)}$ (namely, the statement contains two sets of tuples, each of size s, for which the prover proves that one of the sets contains DH tuples).

Our first step towards making the [30] construction one-sided adaptively secure is to invoke our one-sided OT scheme s times with all system parameters in a decryption mode. Notably, we cannot use the messy mode for the $s/2$ instances not in $\mathcal{J}$ as in the static settings since that would preclude the equivocation of the receiver's bit. Similarly to [30], our constructions have two phases; a *setup phase* and a *transfer phase*. In the setup phase, the receiver generates the system parameters in a decryption mode for the $s/2$ OTs corresponding to indices in $\mathcal{J}$, while the remaining system parameters are generated in the same mode but in a way that does not allow REC to learn the trapdoor. This is obtained by fixing two random generators g_0, g_1, so that the receiver sets the first component of every CRS from the system parameters to be g_0. Moreover, the second component in positions $j \notin \mathcal{J}$ is a power of g_1, else this element is a power of g_0. Note that REC does not know $\log_{g_0} g_1$ which is the decryption mode trapdoor for $j \notin \mathcal{J}$. To ensure correctness, REC proves that it knows the discrete logarithm of the second element with respect to g_1 of at least $s/2$ pairs. This is achieved using a witness equivocal proof for functionality $\mathcal{F}_{\text{ZKPoK}}^{\mathcal{R}_{\text{DL,COMP}(s,s/2)}}$.

In the transfer phase, the receiver uses these system parameters to create a public/secret key pair for each OT execution, for keys not in the set $\mathcal{J}$. For the rest of the OT executions the receiver invokes the TrapKeyGen algorithm of the dual-mode PKE and obtains a public key and two secret keys that enable it to decrypt both of the sender's inputs. In order to ensure that the receiver uses the same input bit σ for all OTs the receiver proves its behavior using a proof for functionality $\mathcal{F}_{\text{ZKPoK}}^{\mathcal{R}_{\text{DDH,OR}(s)}}$. To ensure one-sided security, the proof if further witness equivocal (see Section 6). Finally, we prove the equivocality of the sender's input and the receiver's output based on our one-sided NCE.

Formally, denote by $\Pi_{\text{DUAL}} = (\text{SetupMessy, SetupDecryption, dGen, dEnc, dDec,}$ FindBranch, TrapKeyGen$)$ the DDH based dual-mode PKE of [34]. We present our one-sided QT Π_{CCOT} in the $(\mathcal{F}_{\text{SC}}, \mathcal{F}_{\text{ZKPoK}}^{\mathcal{R}_{\text{DL,COMP}(s,s/2)}}, \mathcal{F}_{\text{ZKPoK}}^{\mathcal{R}_{\text{DDH,OR}(s)}})$-hybrid model.

Protocol 4 (One-sided adaptive single choice cut-and-choose OT (Π_{CCOT}))

- **Inputs:** SEN *inputs a vector of pairs* $\{(x_0^i, x_1^i)\}_{i=1}^s$ *and* REC *inputs a bit* σ *and a set of indices* $\mathcal{J} \subset [s]$ *of size exactly* $s/2$.
- **Auxiliary Inputs:** *Both parties hold a security parameter* 1^n *and* $\mathbb{G}, p$, *where* $\mathbb{G}$ *is an efficient representation of a group of order* p *and* p *is of length* n.
- **CRS:** *The CRS consists of a pair of random group elements* g_0, g_1 *from* $\mathbb{G}$.
- **Setup phase:**
    1. REC *chooses a random* $x_j \in \mathbb{Z}_p$ *and sets* $g_1^j = g_0^{x_j}$ *for all* $j \in \mathcal{J}$ *and* $g_1^j = g_1^{x_j}$ *otherwise.*

       For all j, REC *chooses a random* $y_j \in \mathbb{Z}_p$ *and sets* $\text{CRS}_j = (g_0, g_1^j, h_0^j = (g_0)^{y_j}, h_1^j = (g_1^j)^{y_j})$. *It then sends* $\{\text{CRS}_j\}_{j=1}^s$ *to* SEN.
       Furthermore, for all $j \in \mathcal{J}$, REC *stores the decryption mode trapdoor* $t_j = x_j$.
    2. REC *calls* $\mathcal{F}_{\text{ZKPoK}}^{\mathcal{R}_{\text{DL,COMP}(s,s/2)}}$ *with* $(\{g_1, g_1^j\}_{j=1}^s, \{x_j\}_{j \in \mathcal{J}})$ *to prove the knowledge of the discrete logarithms of* $s/2$ *values within the second element in* $\{\text{CRS}_j\}_j$ *and with respect to* g_1.

- **Transfer phase (repeated in parallel for all j):**
    1. *For all* $j \notin \mathcal{J}$, REC *computes* $(\text{PK}_j, \text{SK}_j) = ((g_j, h_j), r_j) \leftarrow \text{dGen}(\text{CRS}_j, \sigma)$.
       For all $j \in \mathcal{J}$, REC *computes* $(\text{PK}_j, \text{SK}_j^0, \text{SK}_j^1) = ((g_j, h_j), r_j, r_j/t_j) \leftarrow$ TrapKeyGen(CRS_j, t_j).
       Finally, REC *sends the set* $\{\text{PK}_j\}_{j=1}^s$ *and stores the secret keys.*

2. REC calls $\mathcal{F}_{\mathrm{ZKPoK}}^{\mathcal{R}_{\mathrm{DDH}},\mathrm{OR}(s)}$ with input $((\{(g_0, h_0^j, g_j, h_j)\}_{j=1}^s, \{(g_1^j, h_1^j, g_j, h_j)\}_{j=1}^s),$ $\{r_j\}_{j=1}^s)$ to prove that all the tuples in one of the sets $\{(g_0, h_0^j, g_j, h_j)\}_{j=1}^s$ or $\{(g_1^j, h_1^j, g_j, h_j)\}_{j=1}^s$ are DH tuples.

3. For all j, SEN generates $c_0^j \leftarrow \mathsf{dEnc}_{\mathrm{PK}_j}(x_0^j, 0)$ and $c_1^j \leftarrow \mathsf{dEnc}_{\mathrm{PK}_j}(x_1^j, 1)$. Let $c_0^j = (c_{00}^j, c_{01}^j)$ and $c_1^j = (c_{10}^j, c_{11}^j)$. SEN calls $\mathcal{F}_{\mathrm{SC}}$ with c_{01}^j and c_{11}^j.

- **Output:** *Upon receiving* (c_{01}^j, c_{11}^j) *from* $\mathcal{F}_{\mathrm{SC}}$,

  1. REC outputs $x_\sigma^j \leftarrow \mathsf{dDec}_{\mathrm{SK}_j}(c_\sigma^j)$ for all $j \notin \mathcal{J}$.

  2. REC outputs $(x_0^j, x_1^j) \leftarrow (\mathsf{dDec}_{\mathrm{SK}_j^0}(c_0^j), \mathsf{dDec}_{\mathrm{SK}_j^1}(c_1^j))$ for all $j \in \mathcal{J}$.

Theorem 16. *Assume that the DDH assumption is hard in* $\mathbb{G}$. *Then Protocol 4 UC realizes* $\mathcal{F}_{\mathrm{CCOT}}$ *in the* $(\mathcal{F}_{\mathrm{SC}}, \mathcal{F}_{\mathrm{ZKPoK}}^{\mathcal{R}_{\mathrm{DL}},\mathrm{COMP}(s, s/2)}, \mathcal{F}_{\mathrm{ZKPoK}}^{\mathcal{R}_{\mathrm{DDH}},\mathrm{OR}(s)})$-*hybrid model in the presence of one-sided malicious adversaries.*

The complete proof can be found in our full version [23].

Malicious One-Sided Adaptively Secure Two-Party Computation. First, we remark that the single choice cut-and-choose protocol is executed for every input bit of P_1 in the main two-party computation protocol, but with respect to *the same* set $\mathcal{J}$. In order to ensure that the same $\mathcal{J}$ is indeed used the parties engage in a *batch single choice cut-and-choose OT* where a single setup phase is run first, followed by n parallel invocations of the transfer phase. We note that CRS and the set $\mathcal{J}$ are fixed in the setup phase and remain the same for all n parallel invocations of the transfer phase. We denote the batch functionality by $\mathcal{F}_{\mathrm{CCOT}}^{\mathrm{BATCH}}$ and the protocol by $\Pi_{\mathrm{CCOT}}^{\mathrm{BATCH}}$.

We are now ready to describe the steps of our generic protocol Π_f^{MAL} computing any functionality f on inputs x_0 and x_1. We continue with a high-level overview of [30] adapted to the one-sided setting.

Step 1. P_0 constructs s copies of Yao's garbled circuit for computing the function f. All wires keys are picked at random. Keys that are associated with P_0's input wires are picked as follows. P_0 picks n pairs of random values $((a_1^0, a_1^1), \ldots, (a_n^0, a_n^1))$ and $(c_1, \ldots, c_s)$ and sets the keys associated with the ith input wire of the jth circuit as the pair $(g^{a_i^0 c_j}, g^{a_i^1 c_j})$. These values constitute commitments to all $2ns$ keys of P_0.[5] This set of keys forms a pseudorandom synthesizer [31], implying that if some subset of the keys is revealed then the remaining keys are still pseudorandom. We also require that each pair of keys that is associated with a circuit output wire differs within the most significant bit.

Step 2. The parties call $\mathcal{F}_{\mathrm{CCOT}}^{\mathrm{BATCH}}$ where P_0 inputs the key pairs associated with P_1's input and P_1 inputs its input x_1 and a random subset $\mathcal{J} \subset [s]$ of size $s/2$. P_1 receives from $\mathcal{F}_{\mathrm{CCOT}}^{\mathrm{BATCH}}$ the keys that are associated with its input wires for the $s/2$ circuits indexed by $\mathcal{J}$ (denoted the check circuits). In addition, it receives the keys corresponding to its input for the remaining circuits (denoted the evaluation circuits).

Step 3. P_0 sends P_1 s copies of the garbled circuit (except for the output tables) and the values $((g^{a_1^0}, g^{a_1^1}), \ldots, (g^{a_n^0}, g^{a_n^1}), (g^{c_1}, \ldots, g^{c_s}))$ which are the commitments to

---

[5] Recall that the actual symmetric keys of the ith input within the jth circuit are derived from $(g^{a_i^0 c_j}, g^{a_i^1 c_j})$ using randomness extractor such as a universal hash function.

the input keys on the wires associated with P_0's input. At this point P_0 is committed to all the keys associated with the s circuits.

Step 4. P_1 reveals $\mathcal{J}$ and proves that it used this subset in the cut-and-choose batch OT protocol by sending the keys that are associated with P_1's first input bit in each check circuit. Note that P_1 knows the keys corresponding to both bits only for the check circuits.

Step 5. In order to let P_1 know the keys for the input wires of P_0 within the check circuits, P_0 sends c_j for $j \in \mathcal{J}$. P_1 computes the key pair $(g^{a_i^0 c_j}, g^{a_i^1 c_j})$.

Step 6. P_1 verifies the validity of the check circuits using all the keys associated with their input wires. This ensures that the evaluation circuits are correct with high probability.

Step 7. To complete the evaluation phase P_1 is given the keys for the input wires of P_0. P_0 must be forced to give the keys that are associated with the same input for all circuits. Specifically, the following code is executed for all input bits of P_0:

1. For every evaluation circuit C_j, P_0 sends $y_{i,j} = g^{a_i^{x_0^i} c_j}$ using an instance of somewhat NCE with $\ell = 2$, where x_0^i is the ith input bit of P_0.

2. P_0 then proves that $a_i^{x_0^i}$ is in common for all keys associated with the ith input bit, which is reduced to showing that either the set $\{(g, g^{a_i^{x_0^i}}, g^{c_j}, y_{i,j})\}_{j=1}^s$ or the set $\{(g, g^{a_i^{1-x_0^i}}, g^{c_j}, y_{i,j})\}_{j=1}^s$ is comprised of DH tuples. Notably, it is sufficient to use a single UC ZK proof for the simpler relation $\mathcal{R}_{\text{DDH,OR}}$ since the above statement can be compressed into a compound statement of two DH tuples as follows: P_0 first chooses s random values $\gamma_1, \ldots, \gamma_s \in \mathbb{Z}_p$ and sends them to P_1. Both parties compute $\tilde{g} = \prod_{j=1}^s (g^{c_j})^{\gamma_j}$, $\tilde{y} = \prod_{j=1}^s (y_{i,j})^{\gamma_j}$, of which P_0 proves that either $(g, g^{a_i^{x_0^i}}, \tilde{g}, \tilde{y})$ or $(g, g^{a_i^{1-x_0^i}}, \tilde{g}, \tilde{y})$ is a DH tuple.

Step 8. Upon receiving Accept from $\mathcal{F}_{\text{ZKPoK}}^{\mathcal{R}_{\text{DDH,OR}}}$, P_1 completes the evaluation of the circuits. Namely, for every $i \in [1, \ldots, ns]$ P_0 and P_1 call $\mathcal{F}_{\text{OT}}$ in which P_0's input equals $(0, 1)$ if the most significant bit of the output wire key is associated with 0, and $(1, 0)$ otherwise. Moreover, P_1's input is the most significant bit of its output key. P_1 concatenates the bits obtained from these OTs and sets the majority of these values as the output y.

Step 9. P_1 sends y using an instance of one-sided NCE.

To ensure the one-sided security of Π_f^{MAL} we realize the functionalities used in the protocol as follows: **(1)** $\mathcal{F}_{\text{CCOT}}^{\text{BATCH}}$ is realized in **Step 2** using our one-sided batch single choice cut-and-choose OT. This implies the equivocation of P_1's input. **(2)** The statement of $\mathcal{F}_{\text{ZKPoK}}^{\mathcal{R}_{\text{DDH,OR}}}$ is transferred in **Step 7.1** via a somewhat NCE with $\ell = 2$. To obtain a witness equivocal proof for functionality $\mathcal{F}_{\text{ZKPoK}}^{\mathcal{R}_{\text{DDH,OR}}}$ (invoked in **Step 7.2**), it is sufficient to employ a standard static proof realizing this ZK functionality where the prover sends the third message of the proof using a somewhat NCE with $\ell = 2$ (this is due to the fact that we anyway send the statement using a somewhat NCE). Specifically, a statically secure proof is sufficient whenever both the statement and the third message of the (Σ-protocol) proof can be equivocated. This implies the equivocation of P_0's input. **(3)** Finally, in **Step 8** the $\mathcal{F}_{\text{OT}}$ calls are realized using one-sided bit OT. This implies output equivocation.

6 Efficient Statically Secure and Witness Equivocal UC ZK PoKs

This section includes two results that are given in details in the full version. First, we discuss a technique for generating efficient statically secure UC ZK PoK for various Σ-protocols. Our protocols take a new approach where the prover commits to an additional transcript which, in turn, enables witness extraction without using rewinding. Our instantiations imply UC ZK PoK constructions that incur constant overhead and achieve negligible soundness error. Briefly, the prover is instructed to send two responses to a pair of distinct challenges. The first response is sent on clear and publicly verified as specified in the protocol, whereas the second response is encrypted using a homomorphic PKE and its validity is carried out by a UC ZK proof of consistency.

Next, we show how to generate efficient witness equivocal UC ZK PoK for various *compound Σ-protocols*. The additional feature that witness equivocal UC ZK PoK offers over statically secure UC ZK PoK is that it allows the simulator to equivocate the simulated proof upon corrupting the prover. Interestingly, we build witness equivocal UC ZK PoKs for a class of fundamental compound Σ-protocols without relying on NCE. Our approach yields witness equivocal UC ZK PoK only for compound statements where the simulator knows the witnesses for all sub-statements (but not the real witness). This notion is weaker than the notion of one-sided UC ZK PoK where the simulator is required to simulate the proof obliviously of the witness, and later prove consistency with respect to the real witness. The rest of the details can be found in [23].

References

1. Beaver, D.: Plug and play encryption. In: Kaliski Jr., B.S. (ed.) CRYPTO 1997. LNCS, vol. 1294, pp. 75–89. Springer, Heidelberg (1997)
2. Beaver, D.: Adaptively secure oblivious transfer. In: Ohta, K., Pei, D. (eds.) ASIACRYPT 1998. LNCS, vol. 1514, pp. 300–314. Springer, Heidelberg (1998)
3. Beaver, D., Haber, S.: Cryptographic protocols provably secure against dynamic adversaries. In: Rueppel, R.A. (ed.) EUROCRYPT 1992. LNCS, vol. 658, pp. 307–323. Springer, Heidelberg (1993)
4. Beaver, D., Micali, S., Rogaway, P.: The round complexity of secure protocols (extended abstract). In: STOC, pp. 503–513 (1990)
5. Bellare, M., Hofheinz, D., Yilek, S.: Possibility and impossibility results for encryption and commitment secure under selective opening. In: Joux, A. (ed.) EUROCRYPT 2009. LNCS, vol. 5479, pp. 1–35. Springer, Heidelberg (2009)
6. Canetti, R.: Universally composable security: A new paradigm for cryptographic protocols. In: FOCS, pp. 136–145 (2001)
7. Canetti, R., Damgård, I., Dziembowski, S., Ishai, Y., Malkin, T.: Adaptive versus non-adaptive security of multi-party protocols. J. Cryptology 17(3), 153–207 (2004)
8. Canetti, R., Feige, U., Goldreich, O., Naor, M.: Adaptively secure multi-party computation. In: STOC, pp. 639–648 (1996)
9. Canetti, R., Halevi, S., Katz, J.: Adaptively-secure, non-interactive public-key encryption. In: Kilian, J. (ed.) TCC 2005. LNCS, vol. 3378, pp. 150–168. Springer, Heidelberg (2005)
10. Canetti, R., Lindell, Y., Ostrovsky, R., Sahai, A.: Universally composable two-party and multi-party secure computation. In: STOC (2002)

11. Choi, S.G., Dachman-Soled, D., Malkin, T., Wee, H.: Improved non-committing encryption with applications to adaptively secure protocols. In: Matsui, M. (ed.) ASIACRYPT 2009. LNCS, vol. 5912, pp. 287–302. Springer, Heidelberg (2009)

12. Choi, S.G., Dachman-Soled, D., Malkin, T., Wee, H.: Simple, black-box constructions of adaptively secure protocols. In: Reingold, O. (ed.) TCC 2009. LNCS, vol. 5444, pp. 387–402. Springer, Heidelberg (2009)

13. Damgård, I., Nielsen, J.: Improved non-committing encryption schemes based on a general complexity assumption. In: Bellare, M. (ed.) CRYPTO 2000. LNCS, vol. 1880, pp. 432–450. Springer, Heidelberg (2000)

14. Damgård, I., Nielsen, J.B.: Perfect hiding and perfect binding universally composable commitment schemes with constant expansion factor. In: Yung, M. (ed.) CRYPTO 2002. LNCS, vol. 2442, pp. 581–596. Springer, Heidelberg (2002)

15. Damgård, I., Nielsen, J.B.: Universally composable efficient multiparty computation from threshold homomorphic encryption. In: Boneh, D. (ed.) CRYPTO 2003. LNCS, vol. 2729, pp. 247–264. Springer, Heidelberg (2003)

16. Damgård, I., Pastro, V., Smart, N., Zakarias, S.: Multiparty computation from somewhat homomorphic encryption. In: Safavi-Naini, R., Canetti, R. (eds.) CRYPTO 2012. LNCS, vol. 7417, pp. 643–662. Springer, Heidelberg (2012)

17. Fehr, S., Hofheinz, D., Kiltz, E., Wee, H.: Encryption schemes secure against chosen-ciphertext selective opening attacks. In: Gilbert, H. (ed.) EUROCRYPT 2010. LNCS, vol. 6110, pp. 381–402. Springer, Heidelberg (2010)

18. El Gamal, T.: A public key cryptosystem and a signature scheme based on discrete logarithms. IEEE Transactions on Information Theory 31(4), 469–472 (1985)

19. Garay, J.A., Wichs, D., Zhou, H.-S.: Somewhat non-committing encryption and efficient adaptively secure oblivious transfer. In: Halevi, S. (ed.) CRYPTO 2009. LNCS, vol. 5677, pp. 505–523. Springer, Heidelberg (2009)

20. Garg, S., Sahai, A.: Adaptively secure multi-party computation with dishonest majority. In: Safavi-Naini, R., Canetti, R. (eds.) CRYPTO 2012. LNCS, vol. 7417, pp. 105–123. Springer, Heidelberg (2012)

21. Goldreich, O.: Foundations of Cryptography: Basic Applications. Cambridge University Press (2004)

22. Goldreich, O., Micali, S., Wigderson, A.: How to play any mental game or a completeness theorem for protocols with honest majority. In: STOC, pp. 218–229 (1987)

23. Hazay, C., Patra, A.: One-sided adaptively secure two-party computation. IACR Cryptology ePrint Archive 2013, 593 (2013)

24. Ishai, Y., Prabhakaran, M., Sahai, A.: Founding cryptography on oblivious transfer - efficiently. In: Wagner, D. (ed.) CRYPTO 2008. LNCS, vol. 5157, pp. 572–591. Springer, Heidelberg (2008)

25. Jarecki, S., Lysyanskaya, A.: Adaptively secure threshold cryptography: Introducing concurrency, removing erasures. In: Preneel, B. (ed.) EUROCRYPT 2000. LNCS, vol. 1807, pp. 221–242. Springer, Heidelberg (2000)

26. Katz, J., Ostrovsky, R.: Round-optimal secure two-party computation. In: Franklin, M. (ed.) CRYPTO 2004. LNCS, vol. 3152, pp. 335–354. Springer, Heidelberg (2004)

27. Katz, J., Thiruvengadam, A., Zhou, H.-S.: Feasibility and infeasibility of adaptively secure fully homomorphic encryption. In: Kurosawa, K., Hanaoka, G. (eds.) PKC 2013. LNCS, vol. 7778, pp. 14–31. Springer, Heidelberg (2013)

28. Lindell, Y., Pinkas, B.: A proof of security of yaos protocol for two-party computation. Journal of Cryptology 22(2), 161–188 (2009)

29. Lindell, A.Y.: Adaptively secure two-party computation with erasures. In: Fischlin, M. (ed.) CT-RSA 2009. LNCS, vol. 5473, pp. 117–132. Springer, Heidelberg (2009)

30. Lindell, Y., Pinkas, B.: Secure two-party computation via cut-and-choose oblivious transfer. J. Cryptology 25(4), 680–722 (2012)
31. Naor, M., Reingold, O.: Synthesizers and their application to the parallel construction of psuedo-random functions. In: FOCS, pp. 170–181 (1995)
32. Nielsen, J.B.: Separating random oracle proofs from complexity theoretic proofs: The non-committing encryption case. In: Yung, M. (ed.) CRYPTO 2002. LNCS, vol. 2442, pp. 111–126. Springer, Heidelberg (2002)
33. Nielsen, J.B., Nordholt, P.S., Orlandi, C., Burra, S.S.: A new approach to practical active-secure two-party computation. In: Safavi-Naini, R., Canetti, R. (eds.) CRYPTO 2012. LNCS, vol. 7417, pp. 681–700. Springer, Heidelberg (2012)
34. Peikert, C., Vaikuntanathan, V., Waters, B.: A framework for efficient and composable oblivious transfer. In: Wagner, D. (ed.) CRYPTO 2008. LNCS, vol. 5157, pp. 554–571. Springer, Heidelberg (2008)
35. Wolf, S., Wullschleger, J.: Oblivious transfer is symmetric. In: Vaudenay, S. (ed.) EUROCRYPT 2006. LNCS, vol. 4004, pp. 222–232. Springer, Heidelberg (2006)
36. Yao, A.C.-C.: Protocols for secure computations (extended abstract). In: FOCS, pp. 160–164 (1982)

Multi-linear Secret-Sharing Schemes

Amos Beimel[1,*], Aner Ben-Efraim[1,2,*], Carles Padró[3], and Ilya Tyomkin[2,**]

[1] Dept. of Computer Science, Ben Gurion University of the Negev, Be'er Sheva, Israel
[2] Dept. of Mathematics, Ben Gurion University of the Negev, Be'er Sheva, Israel
[3] Nanyang Technological University, Singapore

Abstract. Multi-linear secret-sharing schemes are the most common secret-sharing schemes. In these schemes the secret is composed of some field elements and the sharing is done by applying some fixed linear mapping on the field elements of the secret and some randomly chosen field elements. If the secret contains one field element, then the scheme is called linear. The importance of multi-linear schemes is that they provide a simple non-interactive mechanism for computing shares of linear combinations of previously shared secrets. Thus, they can be easily used in cryptographic protocols.

In this work we study the power of multi-linear secret-sharing schemes. On one hand, we prove that ideal multi-linear secret-sharing schemes in which the secret is composed of p field elements are more powerful than schemes in which the secret is composed of less than p field elements (for every prime p). On the other hand, we prove super-polynomial lower bounds on the share size in multi-linear secret-sharing schemes. Previously, such lower bounds were known only for linear schemes.

Keywords: Ideal secret-sharing schemes, multi-linear matroids, Dowling geometries.

1 Introduction

Consider a scenario where a user holds some secret information and wants to store it on some servers such that only some predefined sets of servers (i.e., trusted sets) can reconstruct this information. Secret-sharing schemes enable such storage, where the dealer – the user holding the secret – computes some strings, called shares, and privately gives one share to each server. In the sequence we will refer to the servers as the parties and to the collection of sets of parties that can reconstruct the secret as an access structure. Secret-sharing schemes are an important cryptographic primitive and they are used nowadays as a basic tool in many cryptographic protocols, e.g., [2,9,10,12,27,18,41,34,37].

In this work we study the most useful construction of secret-sharing schemes, namely, multi-linear secret-sharing schemes. In these schemes the secret is a

---

* Partially supported by ISF grants 938/09, 544/13 and by the Frankel Center for Computer Science.
** Partially supported by ISF grants No. 1018/11.

Y. Lindell (Ed.): TCC 2014, LNCS 8349, pp. 394–418, 2014.

sequence of elements from some finite field, and each share is a linear combination of these elements and some random elements from the field. If the secret contains exactly one element of the field, then the scheme is called linear. Linear and multi-linear secret-sharing schemes are very useful as they provide a simple non-interactive mechanism for computing shares of linear combinations of previously shared secrets.

We prove two results on the power of multi-linear secret-sharing schemes. Our first results shows advantages of multi-linear secret-sharing schemes compared to linear schemes, that is, we prove that ideal schemes in which the secret contains p elements of the field are more efficient than schemes in which the secret contains less than p field elements (for every prime p). Our second results proves super polynomial lower bounds on the size of shares in multi-linear secret-sharing schemes.

Previous Results. Threshold secret-sharing schemes, where all sets of parties whose size is at least some threshold, were introduced by Shamir [33] and Blakley [5]. Secret-sharing schemes for general access structures were introduced and constructed by Ito et al. [19]. Better constructions were introduced by Benaloh and Leichter [3]. Linear secret-sharing schemes were presented by Brickel [7] for the case that each share is one field element and by Krachmer and Wigderson [20] for the case that each share can contain more than one field element. Karchmer and Wigderson's motivation was studying a complexity model called span programs; in particular, they proved that monotone span programs are equivalent to linear secret-sharing schemes. It is important to note that all previously mentioned constructions of secret-sharing schemes are linear. Multi-linear secret-sharing schemes were studied by [4,13], who gave the conditions when a multi-linear scheme realizes an access structure. Construction of multi-linear secret-sharing schemes were given by, e.g., [36,6,39,38].

To explain why linear secret-sharing schemes are useful, we describe the basic idea in using secret-sharing schemes in protocols, starting from [2]. In such protocols the parties share their inputs among the other parties, and, thereafter, the shares of different secrets are "combined" to produce shares of some function of the original secrets. For example, the parties hold shares of two secrets a and b, and they want to compute shares of $a + b$ (without reconstructing the original secrets). If the schemes are multi-linear, the two secrets a and b are shared using the same multi-linear scheme, and each party sums the shares of the two secrets, then the resulting shares are of the secret $a + b$.

In any secret-sharing scheme, the size of the share of each party is at least the size of the secret [21]. An ideal secret-sharing scheme is a scheme in which the size of the share of each party is exactly the size of the secret. For example, Shamir's scheme [33] is ideal. Brickell [7] considered ideal schemes and constructed ideal schemes for some access structures, e.g., for hierarchical access structures. Brickell and Davenport [8] showed an interesting connection between ideal access structures and matroids, that is, (1) If an access structure is ideal then it is induced by a matroid, (2) If an access structure is induced by a representable matroid, then the access structure is ideal. Following this work, many works have studied ideal access structures and matroids, e.g. [32,35,25,24]. In particular, if an access structure is induced by a multi-linear representable matroid, then it is ideal [35].

Simonis and Ashikhmin [35] considered the access structure induced by the Non-Pappus matroid. They construct an ideal multi-linear secret-sharing scheme realizing this access structure, where the secret contains two field elements, and they prove (using known results about matroids) that there is no ideal linear secret-sharing realizing this access structure (that is, in any linear secret-sharing realizing this access structure at least one share must contain more than one field element). Pendavingh and van Zwam [29] (implicitly) provided another example of an access structure that can be realized by an ideal multi-linear secret-sharing scheme, where the secret contains two field elements, but cannot be realized by an ideal linear secret-sharing scheme. Their example is the access structure induced by the rank-3 Dowling matroid of the quaternion group. Note that the rank-3 Dowling matroid [15,14] can be defined with an arbitrary group (see Definition 2.9); in this paper we will use it with properly chosen groups.

For a scheme to be efficient and useful, the size of the shares should be small (i.e., polynomial in the number of parties). The best known schemes for general access structures, e.g., [19,3,20,13], are highly inefficient, that is, for most access structures the size of shares is $2^{O(n)}$ times the size of the secret, where n is the number of parties in the access structure. The best lower bound known on the total share size for an access structure is $\Omega(n^2/\log n)$ times the size of the secret [11]. Thus, there exists a large gap between the known upper and lower bounds. Bridging this gap is one of the most important questions in the study of secret-sharing schemes. In contrast to general secret-sharing schemes, super-polynomial lower bounds are known for linear secret-sharing schemes. That is, there exist explicit access structures such that the total share size of any linear secret-sharing scheme realizing them is $n^{\Omega(\log n)}$ times the size of the secret [1,16,17].

Our Results and Techniques. The simplest way to construct a multi-linear secret-sharing scheme, where the secret is composed of k field elements, is to share each field element independently using a linear secret-sharing scheme. This results in a multi-linear scheme whose information ratio (the ratio between the length of the shares and the length of the secret) is the same as the information ratio of the linear scheme. The question is if one can construct multi-linear secret-sharing schemes whose information ratio is better than linear schemes. Our first result gives a positive answer to this question. Our second result implies that in certain cases the answer is no – we show that the lower bound of [17] for linear secret-sharing schemes holds also for multi-linear secret-sharing schemes.

Our first results shows advantages of multi-linear secret-sharing schemes compared to linear schemes. For every prime $p > 2$, we show that there is an access structure such that: (1) It has an ideal multi-linear secret-sharing scheme in which the secret is composed of p field elements. (2) It does not have an ideal multi-linear secret-sharing scheme in which the secret is composed of k field elements, for every $k < p$. In other words, we prove that schemes in which the secret is composed of p field elements are more efficient than schemes in which the secret is composed of less than p field elements. Previously, this was known only for $p = 2$.

To prove this result we consider the access structures induced by rank-3 Dowling matroids of various groups. By known results, it suffices to study when these matroids are k-linearly representable. We study this question and show that it can be answered using tools from representation theory. The important step in our proof is showing that the Dowling matroid of a group G is k-linearly representable if and only if the group G has a fixed-point free representation of dimension k (see Section 2.5 for definition of these terms). To complete our proof, we show that for every p there is a group G_p that has a fixed-point free representation of dimension p and does not have a fixed-point free representation of dimension $k < p$.

Our second results is super polynomial lower bounds on the size of shares in multi-linear secret-sharing schemes. Prior to our work, such lower bounds were known only for linear secret-sharing schemes. As proving super polynomial lower bounds for general secret-sharing schemes is a major open question, any extension of the lower bounds to a broader class of schemes is important. Specifically, as the class of multi-linear secret-sharing schemes is the class that is useful for applications, it is interesting to prove lower bounds for this class. We show that the method of Gál and Pudlák [17] for proving lower bounds for linear secret-sharing schemes applies also to multi-linear secret-sharing schemes. As a result, we get that there exist access structures such that the total share size of any *multi-linear* secret-sharing scheme realizing them is $n^{\Omega(\log n)}$ times the size of the secret (even when the secret contains any number of field elements).

2 Preliminaries

Notations. We will frequently use block matrices throughout this paper. To differentiate these block matrices, they will be inside square brackets, or in bold letters (e.g. $\mathbf{A} = \begin{bmatrix} A & B \\ C & D \end{bmatrix}$, where A, B, C, D are matrices). In all the proofs and examples, except in the proof of Theorem 4.5, all blocks are of size $k \times k$. For a matrix A, we denote the ith column of A by A_i. We denote fields by $\mathbb{F}$ or $\mathbb{E}$ (general fields), $\mathbb{C}$ (complex numbers), $\widetilde{\mathbb{F}}$ (algebraic closure of $\mathbb{F}$), and $\mathbb{F}_{p^m}$ (the unique field with p^m elements). We denote the integers by $\mathbb{Z}$ and the non-negative integers by $\mathbb{N}$.

2.1 Secret-Sharing Schemes

A secret-sharing scheme is, informally, an algorithm in which a dealer distributes a secret to a set of parties in such that only authorized subsets of parties can reconstruct the secret, while unauthorized subsets cannot learn anything about the secret. We next define secret-sharing schemes, starting with some notations.

Definition 2.1. *Let $\{p_1, \ldots, p_n\}$ be a set of parties. A collection $\mathcal{A} \subseteq 2^{\{p_1,\ldots,p_n\}}$ is* monotone *if $B \in \mathcal{A}$ and $B \subseteq C$ imply that $C \in \mathcal{A}$. An* access structure *is a monotone collection $\mathcal{A} \subseteq 2^{\{p_1,\ldots,p_n\}}$ of non-empty subsets of $\{p_1, \ldots, p_n\}$. Sets in $\mathcal{A}$ are called* authorized, *and sets not in $\mathcal{A}$ are called* unauthorized.

Definition 2.2 (secret-sharing). *A secret-sharing scheme Σ with domain of secrets S is a pair $\Sigma = \langle \Pi, \mu \rangle$, where μ is a probability distribution on some finite set R called the set of random strings and Π is a mapping from $S \times R$ to a set of n-tuples $K_1 \times K_2 \times \cdots \times K_n$, where K_j is called the* domain of shares *of p_j. A dealer distributes a secret $s \in S$ according to Σ by first sampling a random string $r \in R$ according to μ, and applying the mapping Π on s and r, that is, computing a vector of shares $\Pi(s,r) = (s_1, \ldots, s_n)$, and privately communicating each share s_j to party p_j. For a set $A \subseteq \{p_1, \ldots, p_n\}$, we denote $\Pi_A(s,r)$ as the restriction of $\Pi(s,r)$ to its A-entries.*

Correctness. *The secret s can be reconstructed by any authorized set of parties. That is, for any set $B \in \mathcal{A}$ (where $B = \{p_{i_1}, \ldots, p_{i_{|B|}}\}$), there exists a reconstruction function $\mathrm{RECON}_B : K_{i_1} \times \ldots \times K_{i_{|B|}} \to S$ such that for every $s \in S$,*

$$\Pr[\, \mathrm{RECON}_B(\Pi_B(s,r)) = s \,] = 1. \tag{1}$$

Perfect Privacy. *Every unauthorized set cannot learn anything about the secret (in the information theoretic sense) from their shares. Formally, for any set $T \notin \mathcal{A}$, for every two secrets $a, b \in S$, and for every possible vector of shares $\langle s_j \rangle_{p_j \in T}$:*

$$\Pr[\, \Pi_T(a,r) = \langle s_j \rangle_{p_j \in T} \,] = \Pr[\, \Pi_T(b,r) = \langle s_j \rangle_{p_j \in T} \,]. \tag{2}$$

The *information ratio* of a secret-sharing scheme is $\frac{\max_{1 \le j \le n} \log |K_j|}{\log |S|}$, where S is the domain of secrets and K_j is the domain of shares of p_j.

In every secret-sharing scheme, the information ratio is at least 1 [21]. Ideal secret-sharing schemes are those where the information ratio is exactly 1, which means that the size of the domain of the shares is exactly the size of the domain of the secret.

Multi-linear secret-sharing schemes are schemes in which the computation of the shares is a linear mapping. More formally, in a multi-linear secret-sharing scheme over a finite field $\mathbb{F}$, the secret is a vector of elements of the field. To share a secret $s \in \mathbb{F}^k$, the dealer first chooses a random vector $r \in \mathbb{F}^m$ with uniform distribution (for some integer m). Each share is a vector over the field such that each coordinate of this vector is some fixed linear combination of the coordinates of the secret s and the coordinates of the random string r.

2.2   Matroids

Matroids are combinatorial objects that can be defined in many equivalent ways. To make things simple, we will stick to one definition based on rank function.

Definition 2.3. *A matroid M is an ordered pair (E, r) with E a finite set (usually $E = \{1, ..., n\}$) called the ground set and a rank function $r \colon 2^E \to \mathbb{N}$ satisfying the following conditions, called the matroid axioms:*

1. *$r(\emptyset) = 0$,*
2. *If $X \subseteq E$ and $x \in E$, then $r(X) \leq r(X \cup \{x\}) \leq r(X) + 1$,*
3. *If $X \subseteq E$ and $x, y \in E$ such that $r(X \cup \{x\}) = r(X \cup \{y\}) = r(X)$ then $r(X \cup \{x\} \cup \{y\}) = r(X)$.*

A set $X \subseteq E$ is independent if $r(X) = |X|$, otherwise X is dependent. The rank of the matroid is defined $r(M) := r(E)$. A base of M is an independent set $X \subseteq E$ such that $r(X) = r(M)$. The set of bases of a matroid uniquely identifies the matroid. A circuit is a minimal dependent set. The set of all circuits of a matroid also uniquely identifies the matroid. Throughout this paper we will assume that every set $X \subseteq E$ of size 2 is independent (called simple matroids *or* geometries *in the literature).*

The simplest example of a matroid is the size of a group, i.e., let $E = \{1, ..., n\}$ and $r(X) = |X|$. The 3 axioms are trivially verified. In this matroid, all sets are independent. Matroids originated from trying to generalize axioms in graph theory and linear algebra.

Example 2.4. Let $E = \{v_1, ..., v_n\}$ be a set of vectors over some field $\mathbb{F}$. For $X \subseteq E$ let $r(X) = \dim(\text{span}(X))$. By linear algebra, the 3 matroid axioms hold. Furthermore, we can look at the matrix A, in which the ith column is the vector v_i. In this case, $r(X)$ is the rank of the submatrix containing the columns of the vectors in X. Matroids that arise in this manner are called linearly representable (over $\mathbb{F}$). This can also be generalized as follows:

Definition 2.5. *Let $M = (E = \{1, ..., n\}, r)$ be a matroid and $\mathbb{F}$ a field. A k-linear representation of M over $\mathbb{F}$ is a matrix A with $k \cdot n$ columns $A_1, ..., A_{k \cdot n}$ such that the rank of every set $X = \{i_1, ..., i_j\} \subseteq E$ satisfies*

$$r(X) = \frac{\dim(\text{span}(U_{i_1} \cup \cdots \cup U_{i_j}))}{k},$$

where $U_\ell = \{A_{(\ell-1) \cdot k+1}, A_{(\ell-1) \cdot k+2}, ..., A_{\ell \cdot k}\}$ for $1 \leq \ell \leq n$. If such a representation of M exists then M is k-linearly representable. One-linearly represetable matroids are called linearly representable. *A matroid is* multi-linearly representable *if it is k-linearly representable for some $k \in \mathbb{N}$.*

An example of a multi-linear representation is given in Example 2.6.

Matroids of rank 3 can be expressed by a geometric representation on a plane as follows – the bases are the sets of 3 points that are not on a single line. For a diagram on the plane to represent a matroid it must satisfy the following condition: Every 2 distinct points lie on a single line. Since every 2 points lie on a line, usually only lines that pass through at least 3 points are drawn. See [28, Chapter 1.5] for more details and the more general statement.

Example 2.6. Let A and $\mathbf{B}$ be the following matrix and block matrix:

$$
\begin{array}{cc}
1\,2\,3\ g_1'\ g_1''\ g_1''' & \qquad 1\ 2\ 3\ \ g_1'\ g_1''\ g_1''' \\
A = \begin{pmatrix} 1 & 0 & 0 & -1 & 0 & 1 \\ 0 & 1 & 0 & 1 & -1 & 0 \\ 0 & 0 & 1 & 0 & 1 & -1 \end{pmatrix}, & \mathbf{B} = \begin{bmatrix} I_k & 0 & 0 & -I_k & 0 & I_k \\ 0 & I_k & 0 & I_k & -I_k & 0 \\ 0 & 0 & I_k & 0 & I_k & -I_k \end{bmatrix}.
\end{array}
$$

For any field $\mathbb{F}$, the matrix A (resp. the block matrix $\mathbf{B}$) is a linear (k-linear) representation of the matroid with 6 points whose geometric representation is Figure 1 (a). For example, the columns labelled by $1, 2, g_1''$ are independent. Therefore, they do not lie on the same line in Figure 1 (a). On the other hand, the columns labelled by $1, 2, g_1'$ are dependent, thus, they lie on a line.

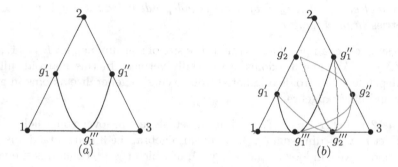

Fig. 1. Geometric Representation of the matroids $Q_3(\{1\})$ and $Q_3(\mathbb{Z}_2)$

Definition 2.7. *Let M be a matroid and $\mathbb{F}$ a field. We say that that M is k-minimally representable over $\mathbb{F}$ if there is a k-linear representation of M over $\mathbb{F}$, but for every $j < k$ there is no j-linear representation of M over $\mathbb{F}$. We will say that M is k-minimally representable if it is k-minimally representable over some field $\mathbb{F}$, but not j-linearly representable over any field for $j < k$.*

Example 2.8. The Non-Pappus matroid (cf. [28, Example 1.5.15, page 39]) whose geometric representation appears in Figure 2 is not linearly representable over any field [28, Proposition 6.1.10], but has a 2-linear representation over $\mathbb{F}_3$ [35]. Therefore, the Non-Pappus matroid is 2-minimally representable.

Our primary focus in the first part of the paper will be the multi-linear representability of the rank-3 Dowling Matroids. These matroids were presented by Dowling [15,14]. We will show that for every prime p there is a Dowling Matroid which is p-minimally representable, and furthermore, over a relatively small field. The Dowling Matroid is defined as follows:

Definition 2.9. *Let $G = \{1_G = g_1, g_2, \ldots, g_n\}$ be a finite group. The rank-3 Dowling Matroid of G, denoted $Q_3(G)$, is a matroid of rank 3 on the set $E = \{1, 2, 3, g_1', \ldots, g_n', g_1'', \ldots, g_n'', g_1''', \ldots, g_n'''\}$. That is, for every element*

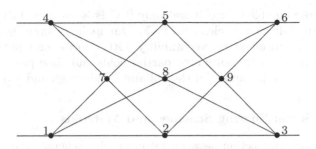

Fig. 2. The Non-Pappus matroid

$g_i \in G$, there are 3 elements in the ground set of the matroid $g_i', g_i'', g_i''' \in E$ and there are 3 additional ground set elements $1, 2, 3$ not related to the group. Every subset of 3 elements not in $C_1 \cup C_2 \cup C_3 \cup C_4$ is a base of the matroid, where,

$$C_1 = \{\{1, 2, g_i'\} | 1 \leq i \leq n\} \cup \{\{1, g_i', g_j'\} | 1 \leq i < j \leq n\} \cup \{\{2, g_i', g_j'\} | 1 \leq i < j \leq n\},$$
$$C_2 = \{\{2, 3, g_i''\} | 1 \leq i \leq n\} \cup \{\{2, g_i'', g_j''\} | 1 \leq i < j \leq n\} \cup \{\{3, g_i'', g_j''\} | 1 \leq i < j \leq n\},$$
$$C_3 = \{\{1, 3, g_i'''\} | 1 \leq i \leq n\} \cup \{\{1, g_i''', g_j'''\} | 1 \leq i < j \leq n\} \cup \{\{3, g_i''', g_j'''\} | 1 \leq i < j \leq n\},$$
$$C_4 = \{\{g_i', g_j'', g_\ell'''\} | g_j \cdot g_i \cdot g_\ell = 1\}.$$

Alternatively, it can be defined by the geometric representation appearing in Figure 3, with additional lines that go through points g_i', g_j'', g_ℓ''' if and only if $g_j \cdot g_i \cdot g_\ell = 1_G$ (e.g., there is always a line that goes through g_1', g_1'', g_1''' since $g_1 = 1_G$ and $1_G \cdot 1_G \cdot 1_G = 1_G$).[1]

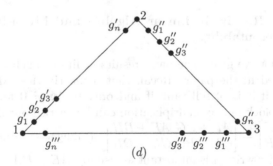

Fig. 3. The Rank-3 Dowling matroid with the lines corresponding to sets $\{g_i', g_j'', g_\ell'''\}$ such that $g_j \cdot g_i \cdot g_\ell = 1_G$ missing

We note that the matroid in Example 2.6 is the Dowling matroid of the trivial group. Figure 1 (b) is a geometric representation of the Dowling matroid of the group $\mathbb{Z}_2$, the unique group with 2 elements. Dowling [15,14] showed that $Q_3(G)$

---

[1] In the literature, the matroid is sometimes defined a bit differently, e.g., a line goes through g_i', g_j'', g_ℓ''' if and only if $(g_j)^{-1} \cdot (g_i)^{-1} \cdot g_\ell = 1_G$. This is just a different naming of the ground set elements.

is linearly representable over $\mathbb{F}$ if and only if G is isomorphic to a subgroup of $\mathbb{F}^*$, the group of invertible elements in $\mathbb{F}$. Our main theorem generalizes this statement for multi-linear representability. Other forms of representability of $Q_3(G)$, namely representability over partial fields and skew partial fields, have been studied by Semple and Whittle [30] and Pendavingh and Van Zwam [29].

2.3   Ideal Secret-Sharing Schemes and Matroids

There is a strong connection between secret-sharing schemes and matroids. Every matroid with ground set $E = \{p_0, p_1, \ldots, p_n\}$ induces an access structure $\mathcal{A}$ with n parties $E' = \{p_1, \ldots, p_n\}$ by the rule $\forall A \subseteq E', A \in \mathcal{A}$ if and only if $r(A \cup \{p_0\}) = r(A)$. The access structure $\mathcal{A}$ is also known as the matroid port. In a sense, we think of p_0 as the dealer. Brickell and Davenport [8] showed that all access structures admitting ideal secret-sharing schemes are induced by matroids. However, not all access structures induced by matroids are ideal [32][25]. The class of matroids inducing ideal access structures are called secret-sharing matroids and also almost affinely representable, and discussed in [35]. Every multi-linearly representable matroid is a secret-sharing matroid. It is still open whether this inclusion is proper. There is also a strong connection between ideal multi-linear secret-sharing schemes and multi-linearly representable matroids [20,13].

Proposition 2.10. *The class of access structures induced by multi-linearly representable matroids is exactly the access structures admitting an ideal multi-linear secret-sharing scheme.*

2.4   Basic Results in Linear Algebra and Multi-linear Representability

In this section we give some basic results in linear algebra and matroid theory that are used in the paper. Recall that a matrix $A \in M_{n \times n}(\mathbb{F})$ is invertible if and only if it is of full rank if and only if $A\boldsymbol{v} \neq \boldsymbol{0}$ for every $\boldsymbol{v} \neq \boldsymbol{0}$. Also recall that block matrix multiplication can be carried out in block fashion, e.g., $\begin{bmatrix} A & B \\ C & D \end{bmatrix} \cdot \begin{bmatrix} E & F \\ G & H \end{bmatrix} = \begin{bmatrix} AE + BG & AF + BH \\ CE + BG & CF + DH \end{bmatrix}$, as long as the dimensions match (note that the order written is important as usually $AE \neq EA$, etc.).

Proposition 2.11. *Let A, B, C be $k \times k$ matrices then*

$$(a) \ \operatorname{rank} \begin{bmatrix} -I_k & 0 & C \\ A & -I_k & 0 \\ 0 & B & -I_k \end{bmatrix} = 2k + \operatorname{rank}(BAC - I).$$

$$(b) \ \operatorname{rank} \begin{bmatrix} -I_k & -I_k \\ A & B \\ 0 & 0 \end{bmatrix} = k + \operatorname{rank}(B - A).$$

Proof. Multiplying by invertible matrices does not change the rank of a matrix. Therefore,

$$
\operatorname{rank}\begin{bmatrix} -I_k & 0 & C \\ A & -I_k & 0 \\ 0 & B & -I_k \end{bmatrix} = \operatorname{rank}\left(\begin{bmatrix} -I_k & 0 & C \\ A & -I_k & 0 \\ 0 & B & -I_k \end{bmatrix} \cdot \begin{bmatrix} I_k & 0 & C \\ 0 & I_k & A\cdot C \\ 0 & 0 & I_k \end{bmatrix} \right)
$$

$$
= \operatorname{rank}\begin{bmatrix} -I_k & 0 & 0 \\ A & -I_k & 0 \\ 0 & B & BAC - I_k \end{bmatrix} = 2k + \operatorname{rank}(BAC - I_k).
$$

and

$$
\operatorname{rank}\begin{bmatrix} -I_k & -I_k \\ A & B \\ 0 & 0 \end{bmatrix} = \operatorname{rank}\left(\begin{bmatrix} -I_k & -I_k \\ A & B \\ 0 & 0 \end{bmatrix} \cdot \begin{bmatrix} I_k & -I_k \\ 0 & I_k \end{bmatrix} \right)
$$

$$
= \operatorname{rank}\begin{bmatrix} -I_k & 0 \\ A & B - A \\ 0 & 0 \end{bmatrix} = k + \operatorname{rank}(B - A).
$$

Proposition 2.12. *Let* $\mathbf{B} := \begin{bmatrix} B_{1,1} & \dots & B_{1,n} \\ \vdots & \ddots & \vdots \\ B_{m,1} & \dots & B_{m,n} \end{bmatrix}$ *be a k-linear representation of a matroid* M, *with* $B_{i,j}$ *being* $k \times k$ *block matrices, and let* G *be any invertible* $k \times k$ *matrix. Then:*

a) *For every* $1 \le i \le n$ *then* $\begin{bmatrix} B_{1,1} & \dots & B_{1,j}\cdot G & \dots & B_{1,n} \\ \vdots & \ddots & \vdots & \ddots & \vdots \\ B_{m,1} & \dots & B_{m,j}\cdot G & \dots & B_{m,n} \end{bmatrix}$ *is a k-linear repre-*

sentation of M.

b) *For every* $1 \le i \le m$ *then* $\begin{bmatrix} B_{1,1} & \dots & B_{1,n} \\ \vdots & \ddots & \vdots \\ G\cdot B_{i,1} & \dots & G\cdot B_{i,n} \\ \vdots & \ddots & \vdots \\ B_{m,1} & \dots & B_{m,n} \end{bmatrix}$ *is a k-linear representation*

of M.

c) *If* $\{1,\dots,m\}$ *is a base of* M *then there exists a matrix of the form*
$\begin{bmatrix} I_k & \dots & 0 & B'_{1,m+1} & \dots & B'_{1,n} \\ \vdots & \ddots & \vdots & & \ddots & \vdots \\ 0 & \dots & I_k & B'_{m,m+1} & \dots & B'_{m,n} \end{bmatrix}$ *that is also a k-linear representation of* M.

Proof. a) Since G is invertible, it is immediate from basic linear algebra that

$$\text{rank}\begin{bmatrix} B_{1,i_1} & \cdots & B_{1,i_\ell} & \cdots & B_{1,i_s} \\ \vdots & \ddots & \vdots & \ddots & \vdots \\ B_{m,i_1} & \cdots & B_{1,i_\ell} & \cdots & B_{m,i_s} \end{bmatrix}$$

$$= \text{rank}\left(\begin{bmatrix} B_{1,i_1} & \cdots & B_{1,i_\ell} & \cdots & B_{1,i_s} \\ \vdots & \ddots & \vdots & \ddots & \vdots \\ B_{m,i_1} & \cdots & B_{1,i_\ell} & \cdots & B_{m,i_s} \end{bmatrix} \cdot \begin{bmatrix} I_k & \cdots & 0 & \cdots & 0 \\ \vdots & \ddots & \vdots & \ddots & \vdots \\ 0 & \cdots & G & \cdots & 0 \\ \vdots & \ddots & \vdots & \ddots & \vdots \\ 0 & \cdots & 0 & \cdots & I_k \end{bmatrix} \right)$$

$$= \text{rank}\begin{bmatrix} B_{1,i_1} & \cdots & B_{1,i_\ell} \cdot G & \cdots & B_{1,i_s} \\ \vdots & \ddots & \vdots & \ddots & \vdots \\ B_{m,i_1} & \cdots & B_{1,i_\ell} \cdot G & \cdots & B_{m,i_s} \end{bmatrix},$$

for any submatrix (with $j = i_\ell$), which is exactly what we need to prove.

b) Similarly, for any submatrix,

$$\text{rank}\begin{bmatrix} B_{1,1} & \cdots & B_{1,n} \\ \vdots & \ddots & \vdots \\ B_{i,1} & \cdots & B_{i,n} \\ \vdots & \ddots & \vdots \\ B_{m,1} & \cdots & B_{m,n} \end{bmatrix} = \text{rank}\left(\begin{bmatrix} I_k & \cdots & 0 & \cdots & 0 \\ \vdots & \ddots & \vdots & \ddots & \vdots \\ 0 & \cdots & G & \cdots & 0 \\ \vdots & \ddots & \vdots & \ddots & \vdots \\ 0 & \cdots & 0 & \cdots & I_k \end{bmatrix} \cdot \begin{bmatrix} B_{1,1} & \cdots & B_{1,n} \\ \vdots & \ddots & \vdots \\ B_{i,1} & \cdots & B_{i,n} \\ \vdots & \ddots & \vdots \\ B_{m,1} & \cdots & B_{m,n} \end{bmatrix} \right)$$

$$= \text{rank}\begin{bmatrix} B_{1,1} & \cdots & B_{1,n} \\ \vdots & \ddots & \vdots \\ G \cdot B_{i,1} & \cdots & G \cdot B_{i,n} \\ \vdots & \ddots & \vdots \\ B_{m,1} & \cdots & B_{m,n} \end{bmatrix}.$$

c) Since $\{1,\ldots,m\}$ is a base of M then the columns $c_1,\ldots,c_{m\cdot k}$ of $\mathbf{B}$ are a basis of the column space of $\mathbf{B}$ (which is, therefore, $\mathbb{F}^{k\cdot m}$). Therefore, there is an invertible linear transformation T such that $\forall 1 \le i \le mk, T(c_i) = e_i$. Since T is invertible $\dim(\text{span}\{T(c_{i_1}),\ldots,T(c_{i_j})\}) = \dim(\text{span}\{c_{i_1},\ldots,c_{i_j}\})$ for any set of columns $\{c_{i_1},\ldots,c_{i_j}\}$, which implies that by applying T to all the columns of $\mathbf{B}$ we get that

$$\begin{bmatrix} I_k & 0 & \cdots & 0 & \vdots & \cdots & \vdots \\ 0 & I_k & \cdots & 0 & T(c_{km+1}) & \cdots & T(c_{kn}) \\ \vdots & \vdots & \ddots & \vdots & \vdots & & \vdots \\ 0 & 0 & \cdots & I_k & \cdot & & \cdot \end{bmatrix}$$

is a k-linear representation of M

We will call operations Proposition 2.12(a) and 2.12(b) column and row *block-scaling* respectively.

2.5 Fixed-Point Free Representations

A standard tool in studying groups is representation theory. Our result relies heavily on theorems from this extensively researched field of mathematics. We will only give the necessary definitions and state the result. We then sketch the main ideas of the proof of this result. The complete proof, which requires much more representation theory, will appear in the full version.

Definition 2.13. *Let G be a finite group and $\mathbb{F}$ a field. A* representation *of G is a group homomorphism $\rho : G \to \mathrm{GL}_n(\mathbb{F})$ (the group of $n \times n$ invertible matrices). The* dimension *or* degree *of a representation is n. A representation is called* faithful *if it is injective. A representation $\rho : G \to \mathrm{GL}_n(\mathbb{F})$ is* fixed-point free *if for every $1 \neq g \in G$ the field element 1 is not an eigenvalue of $\rho(g)$, i.e., $\rho(g) \cdot v \neq v$ for every $g \neq 1$ and for every $v \neq 0$. A* fixed-point free group *is one which has a fixed-point free representation.*

We note that not all representations of a fixed-point free group G are fixed-point free, even if the representation is faithful. For example, cyclic groups are fixed-point free, but also admit non fixed-point free representations:

Example 2.14. Let $G = \mathbb{Z}_m$ be the additive group with m elements. Denote $\zeta = e^{\frac{2\pi i}{m}}$. If $\rho : G \to \mathrm{GL}_2(\mathbb{C})$ is defined by $\rho(k) = \begin{pmatrix} \zeta^k & 0 \\ 0 & 1 \end{pmatrix}$ then ρ is faithful (because $i \neq k \Rightarrow \rho(i) \neq \rho(k)$) but not fixed-point free because $\begin{pmatrix} \zeta^k & 0 \\ 0 & 1 \end{pmatrix} \begin{pmatrix} 0 \\ 1 \end{pmatrix} = \begin{pmatrix} 0 \\ 1 \end{pmatrix}$ (and this should only happen for $k = 0$). However, if we define $\rho(k) = \begin{pmatrix} \zeta^k & 0 \\ 0 & \zeta^k \end{pmatrix}$ then ρ is fixed-point free, because if $k \neq 0$ then 1 is not an eigenvalue of $\begin{pmatrix} \zeta^k & 0 \\ 0 & \zeta^k \end{pmatrix}$. We note that the group $\mathbb{Z}_m$ also has a fixed-point free representation of dimension 1, by $\rho(k) = (e^{\frac{2k\pi i}{m}})$.

Fixed-point free groups have been completely classified by the works of Burnside and later Vincent [40] and Zassenhaus [44]. The classification can be found, for example, in [42]. For our purposes we will require only the following result, easily achieved from the classification:

Proposition 2.15. *For every prime $p > 2$, there exist a prime $q > p$ and a group G_p of order p^2q such that:*

1. *G_p has a fixed-point free representation of dimension p over the field $\mathbb{F}_{2^{pq}}$, i.e., the field of characteristic 2 with 2^{pq} elements.*
2. *The group G_p does not admit a fixed-point representation of dimension less than p over any field.*

Moreover, there exists such q with $q = O(p^{5.18})$, so the field $\mathbb{F}_{2^{pq}}$ has $2^{O(p^{6.18})}$ elements.

Our proof uses the construction of semidirect product of groups. The definition can be found in most group theory books. See for example [26]. It also requires some classical theorems from representation theory, which can be found, for example, in [31].

The complete proof of Proposition 2.15 will be given in the full version. We now sketch the main ideas of the proof.

Proof Sketch. Let $p > 2$ be a prime number. From Linnik's Theorem [22,23], there exists a prime q such that $q = np + 1$ for some $n \in \mathbb{N}$, and q is polynomially bounded by p. The state of the art improvement, by Xylouris [43], shows that $q = O(p^{5.18})$.

From the fact that $q = np + 1$, it can be deduced that there exists a non-trivial semidirect product $G_p = \mathbb{Z}_q \rtimes \mathbb{Z}_{p^2}$, with the action of $\mathbb{Z}_p$ (we give a brief explanation of the construction of this group, and why this group works, in Appendix A). We then show, both directly and using the classification of fixed-point free groups, that the group G_p is fixed-point free, and, thus, has a fixed-point free representation.

Then, using classical theorems from representation theory, we show that a fixed-point free representation of G_p is of dimension at least p, and that there indeed exists a fixed-point free representation of dimension p. In particular, we show directly that there exists such a representation over the field $\mathbb{F}_{2^{pq}}$, which has $2^{pq} = 2^{O(p^{6.18})}$ elements.

3   Main Theorem and Result

In this section, we prove that there is an access structure that has an ideal p-linear secret-sharing scheme and does not have an ideal k-linear secret-sharing scheme for every $k < p$. As explained in Section 2.3, it suffices to prove that there is a matroid that is p-minimally representable. We prove this result for the Dowling matroid, for an appropriate group G. We next state our main theorem.

Theorem 3.1. *For a finite group G, the matroid $Q_3(G)$ is k-linearly representable over a field $\mathbb{F}$ if and only if there is a fixed-point free representation $\rho : G \to \mathrm{GL}_k(\mathbb{F})$.*

The main contribution of the theorem is the new connection between multi-linear representation of the Dowling matroid over G to the existence of a fixed-point free representation of the group G. The theorem transfers the problem of multi-linear representablity of $Q_3(G)$ to finding fixed-point free representations of G. Since fixed-point free groups and representations have been completely classified, it gives a complete answer to this problem.

To discuss the representations of $Q_3(G)$, we define the following block matrix $\mathbf{A}_\rho$. In Lemma 3.2, we will prove that if $Q_3(G)$ is multi-linearly representable, then $\mathbf{A}_\rho$ is a multi-linear representation of $Q_3(G)$ for some representation ρ of G. Then we prove in Lemmas 3.3 and 3.4 that $\mathbf{A}_\rho$ represents $Q_3(G)$ if and only if ρ is fixed-point free.

For a finite group $G = \{1 = g_1, g_2, \ldots, g_n\}$, a field $\mathbb{F}$, and a faithful representation $\rho : G \to GL_k(\mathbb{F})$ we denote by $\mathbf{A}_\rho$ the following block matrix, which contains $3k$ rows and $3(n+1)k$ columns.

$$\mathbf{A}_\rho := \begin{bmatrix} I_k & 0 & 0 & -I_k & \cdots & -I_k & 0 & \cdots & 0 & \rho(g_1) & \cdots & \rho(g_n) \\ 0 & I_k & 0 & \rho(g_1) & \cdots & \rho(g_n) & -I_k & \cdots & -I_k & 0 & \cdots & 0 \\ 0 & 0 & I_k & 0 & \cdots & 0 & \rho(g_1) & \cdots & \rho(g_n) & -I_k & \cdots & -I_k \end{bmatrix}.$$

Lemma 3.2. *If $M = Q_3(G)$ is k-linearly representable over $\mathbb{F}$, then there exists a faithful representation $\rho : G \to GL_k(\mathbb{F})$ such that $\mathbf{A}_\rho$ is a k-linear representation of M.*

Proof. The technique we use to prove this lemma is a standard one (e.g., see the proofs of [28, Proposition 6.4.8, Lemma 6.8.5, Theorem 6.10.10] and [29, Lemma 3.35]). We generalize this technique to multi-linear representations by looking at the representation matrix as a block matrix and using Proposition 2.12. We repeatedly use the fact that for any multi-linear representation of M, if $X \subseteq E$ and $r(X) = n$ then the rank of the relevant sub-matrix of the representation (i.e., deleting the columns of elements not in X) is $n \cdot k$.

Suppose that

$$\mathbf{B} := \begin{bmatrix} B_{1,1} & B_{1,2} & B_{1,3} & B_{1,g_1'} & \cdots & B_{1,g_n'} & B_{1,g_1''} & \cdots & B_{1,g_n''} & B_{1,g_1'''} & \cdots & B_{1,g_n'''} \\ B_{2,1} & B_{2,2} & B_{2,3} & B_{2,g_1'} & \cdots & B_{2,g_n'} & B_{2,g_1''} & \cdots & B_{2,g_n''} & B_{2,g_1'''} & \cdots & B_{2,g_n'''} \\ B_{3,1} & B_{3,2} & B_{3,3} & B_{3,g_1'} & \cdots & B_{3,g_n'} & B_{3,g_1''} & \cdots & B_{3,g_n''} & B_{3,g_1'''} & \cdots & B_{3,g_n'''} \end{bmatrix}$$

is a k-linear representation of M. Then $r(\{1, 2, 3\}) = 3 = r(M)$ so $\mathbf{B}_1, \ldots, \mathbf{B}_{3k}$ span the columns of $\mathbf{B}$. By changing the basis of the column space of $\mathbf{B}$ (see Proposition 2.12(c)) there exists a block matrix $\mathbf{C}$ of the form

$$\mathbf{C} := \begin{bmatrix} I_k & 0 & 0 & C_{1,g_1'} & \cdots & C_{1,g_n'} & C_{1,g_1''} & \cdots & C_{1,g_n''} & C_{1,g_1'''} & \cdots & C_{1,g_n'''} \\ 0 & I_k & 0 & C_{2,g_1'} & \cdots & C_{1,g_n'} & C_{2,g_1''} & \cdots & C_{2,g_n''} & C_{2,g_1'''} & \cdots & C_{2,g_n'''} \\ 0 & 0 & I_k & C_{3,g_1'} & \cdots & C_{1,g_n'} & C_{3,g_1''} & \cdots & C_{3,g_n''} & C_{3,g_1'''} & \cdots & C_{3,g_n'''} \end{bmatrix}$$

that is a k-linear representation of M. As $\forall g \in G, r(\{1, 2, g'\}) = 2$, we have that

$$\mathrm{rank} \begin{bmatrix} I_k & 0 & C_{1,g'} \\ 0 & I_k & C_{2,g'} \\ 0 & 0 & C_{3,g'} \end{bmatrix} = 2k.$$

Thus, $C_{3,g'} = 0$. Also $r(\{1, g'\}) = 2$, so

$$\mathrm{rank} \begin{bmatrix} I_k & C_{1,g'} \\ 0 & C_{2,g'} \\ 0 & C_{3,g'} \end{bmatrix} = 2k,$$

therefore, $C_{2,g'}$ is invertible (it has to be of full rank since $C_{3,g'} = 0$). Since $r(\{2, g'\}) = 2$, by the same argument $C_{1,g'}$ is also invertible. Similarly $\forall g \in G$, $C_{1,g''} = 0$, and $C_{3,g''}, C_{2,g''}$ are invertible, and $C_{2,g'''} = 0$, and $C_{1,g'''}, C_{3,g'''}$ are invertible.

We now apply column block-scaling (Proposition 2.12(a)) on the columns of $g'_1, \ldots, g'_n$ by $-(C_{1,g'_1})^{-1}, \ldots, -(C_{1,g'_n})^{-1}$ respectively to get that

$$
\begin{bmatrix}
I_k & 0 & 0 & C_{1,g'_1}(-(C_{1,g'_1})^{-1}) & \cdots & C_{1,g'_n}(-(C_{1,g'_n})^{-1}) & 0 & \cdots & 0 & C_{1,g'''_1} & \cdots & C_{1,g'''_n} \\
0 & I_k & 0 & C_{2,g'_1}(-(C_{1,g'_1})^{-1}) & \cdots & C_{2,g'_n}(-(C_{1,g'_n})^{-1}) & C_{2,g''_1} & \cdots & C_{2,g''_n} & 0 & \cdots & 0 \\
0 & 0 & I_k & 0 & \cdots & 0 & C_{3,g''_1} & \cdots & C_{3,g''_n} & C_{3,g'''_1} & \cdots & C_{3,g'''_n}
\end{bmatrix}
$$

$$
=
\begin{bmatrix}
I_k & 0 & 0 & -I_k & \cdots & -I_k & 0 & \cdots & 0 & C_{1,g'''_1} & \cdots & C_{1,g'''_n} \\
0 & I_k & 0 & C'_{2,g'_1} & \cdots & C'_{2,g'_n} & C_{2,g''_1} & \cdots & C_{2,g''_n} & 0 & \cdots & 0 \\
0 & 0 & I_k & 0 & \cdots & 0 & C_{3,g''_1} & \cdots & C_{3,g''_n} & C_{3,g'''_1} & \cdots & C_{3,g'''_n}
\end{bmatrix}.
$$

is a k-linear representation of M. Now by row block-scaling (Propostion 2.12(b)) on the second row by $(C'_{2,g'_1})^{-1}$ we get that

$$
\begin{bmatrix}
I_k & 0 & 0 & -I_k & \cdots & -I_k & 0 & \cdots & 0 & C_{1,g'''_1} & \cdots & C_{1,g'''_n} \\
0 & (C'_{2,g'_1})^{-1} & 0 & I_k & \cdots & C'_{2,g'_n}(C'_{2,g'_1})^{-1} & C_{2,g''_1}(C'_{2,g'_1})^{-1} & \cdots & C_{2,g''_n}(C'_{2,g'_1})^{-1} & 0 & \cdots & 0 \\
0 & 0 & I_k & 0 & \cdots & 0 & C_{3,g''_1} & \cdots & C_{3,g''_n} & C_{3,g'''_1} & \cdots & C_{3,g'''_n}
\end{bmatrix}
$$

is a k-linear representation of M. We continue in the same fashion by block-scaling on the columns of $g''_1, \ldots, g''_n$, then row block-scaling on the third row, then column block-scaling of columns $g'''_1, \ldots, g'''_n$, and finally column block scaling of columns $2, 3$ to get that

$$
\mathbf{D} :=
\begin{bmatrix}
I_k & 0 & 0 & -I_k & -I_k & \cdots & -I_k & 0 & 0 & \cdots & 0 & D_{1,g'''_1} & D_{1,g'''_2} & \cdots & D_{1,g'''_n} \\
0 & I_k & 0 & I_k & D_{2,g'_2} & \cdots & D_{2,g'_n} & -I_k & -I_k & \cdots & -I_k & 0 & 0 & \cdots & 0 \\
0 & 0 & I_k & 0 & 0 & \cdots & 0 & I_k & D_{3,g''_2} & \cdots & D_{3,g''_n} & -I_k & -I_k & \cdots & -I_k
\end{bmatrix}
$$

is a k-linear representation of M.

We next use the fact that $\mathbf{D}$ is a multi-linear representation of $Q_3(G)$ to prove that blocks in different parts of the representation are equal, e.g., $D_{3,g''} = D_{2,g'}$. Since $r(\{g'_1, g''_1, g'''_1\}) = 2$, we have that

$$
\mathrm{rank}
\begin{bmatrix}
-I_k & 0 & D_{1,g'''_1} \\
I_k & -I_k & 0 \\
0 & I_k & -I_k
\end{bmatrix}
= 2k,
$$

and this forces $D_{1,g'''_1} = I_k$. For j, ℓ such that $g_j = g_\ell^{-1}$ (thus, $g_j \cdot g_1 \cdot g_\ell = 1$), we have that $r(\{g'_1, g''_j, g'''_\ell\}) = 2$. So,

$$
\mathrm{rank}
\begin{bmatrix}
-I_k & 0 & D_{1,g'''_\ell} \\
I_k & -I_k & 0 \\
0 & D_{3,g''_j} & -I_k
\end{bmatrix}
= 2k.
$$

By Proposition 2.11(a) we get that $\mathrm{rank}(D_{3,g''_j} \cdot D_{1,g'''_\ell} - I_k) = 0$ so $D_{3,g''_j} = (D_{1,g'''_\ell})^{-1}$. By symmetric arguments, $D_{1,g'''_j} = (D_{2,g'_\ell})^{-1}$ and $D_{2,g'_j} = (D_{3,g''_\ell})^{-1}$.

Therefore,

$$\forall g \in G, D_{3,g''} = D_{2,g'} = D_{1,g'''}. \tag{3}$$

Now let $\rho : G \to \mathrm{GL}_k(\mathbb{F})$ be the map $\rho(g) = D_{2,g'}$. We see that $\rho(1) = I$ (because $D_{2,g'_1} = I$). By Proposition 2.11(a)

$$\mathrm{rank} \begin{bmatrix} -I_k & 0 & D_{1,g'''_\ell} \\ D_{2,g'_i} & -I_k & 0 \\ 0 & D_{3,g''_j} & -I_k \end{bmatrix} = 2k + \mathrm{rank}(D_{3,g''_j} D_{2,g'_i} D_{1,g'''_\ell} - I)$$

$$= 2k + \mathrm{rank}(\rho(g_j) \cdot \rho(g_i) \cdot \rho(g_\ell) - I). \tag{4}$$

By the matroid rank, it is equal to $2k$ if $g_j \cdot g_i \cdot g_\ell = 1$ and $3k$ otherwise, thus,

$$\forall g_i, g_j, g_\ell \in G, g_j \cdot g_i \cdot g_\ell = 1 \Leftrightarrow \rho(g_j) \cdot \rho(g_i) \cdot \rho(g_\ell) = I. \tag{5}$$

We now use (5) to show that ρ is an injective group homomorphism, which completes the proof:

For every $g \in G$, since $1 \cdot g^{-1} \cdot g = 1$, we have $I = \rho(1) \cdot \rho(g^{-1}) \cdot \rho(g) = I \cdot \rho(g^{-1}) \cdot \rho(g)$, forcing $\rho(g)^{-1} = \rho(g^{-1})$.

Therefore, $\forall g, h \in G$, as $g \cdot h \cdot (gh)^{-1} = 1$, we have $I = \rho(g) \cdot \rho(h) \cdot \rho((gh)^{-1}) = \rho(g) \cdot \rho(h) \cdot \rho(gh)^{-1}$. Thus, $\rho(gh) = \rho(g) \cdot \rho(h)$. This proves that ρ is a group homomorphism.

For injectivity, if $g \neq h$ then $g \cdot h^{-1} \cdot 1 \neq 1$, which implies that $\rho(g) \cdot \rho(h)^{-1} \cdot \rho(1) \neq I$, so $\rho(g) \neq \rho(h)$.

Lemma 3.3. *Let $\rho : G \to \mathrm{GL}_k(\mathbb{F})$ be a faithful representation. If $\mathbf{A}_\rho$ is a k-linear representation of $Q_3(G)$ then ρ is fixed-point free.*

Proof. Since $\mathbf{A}_\rho$ is a k-linear representation of $Q_3(G)$, for every $g \neq 1_G$ we have that $r(\{g'_1, g'\}) = 2$. So

$$\mathrm{rank} \begin{bmatrix} -I_k & -I_k \\ I_k & \rho(g) \\ 0 & 0 \end{bmatrix} = 2k. \tag{6}$$

By Proposition 2.11(b) we have that

$$\mathrm{rank} \begin{bmatrix} -I_k & -I_k \\ I_k & \rho(g) \\ 0 & 0 \end{bmatrix} = k + \mathrm{rank}(\rho(g) - I_k). \tag{7}$$

By combining (6) and (7), $\mathrm{rank}(\rho(g) - I_k) = k$. This implies that $\rho(g) - I_k$ is invertible, so $\forall v \neq 0, (\rho(g) - I_k)v \neq 0$, therefore, $\forall v \neq 0, \rho(g)v \neq v$, which means that 1 is not an eigenvalue of $\rho(g)$. So, ρ is fixed-point free, as desired.

Lemma 3.4. *If $\rho : G \to \mathrm{GL}_k(\mathbb{F})$ is a fixed-point free representation, then $\mathbf{A}_\rho$ is a k-linear representation of $Q_3(G)$.*

Proof. To prove that $\mathbf{A}_\rho$ is a k-linear representation of M, we need to verify that $\forall X \subset E$, if $r(X) = n$ then the rank of the relevant sub-matrix of $\mathbf{A}_\rho$ (i.e., deleting the columns of elements not in X) is nk. Ranks of most sub-matrices are trivially verified, e.g.,

$$\text{rank} \begin{bmatrix} I_k & 0 & -I_k \\ 0 & I_k & 0 \\ 0 & 0 & \rho(g) \end{bmatrix} = \text{rank} \begin{bmatrix} I_k & -I_k & -I_k \\ 0 & \rho(g_i) & 0 \\ 0 & 0 & \rho(g_j) \end{bmatrix} = 3k, \text{rank} \begin{bmatrix} I_k & -I_k \\ 0 & \rho(g) \\ 0 & 0 \end{bmatrix} = 2k.$$

(Note that $\forall g \in G$, the matrix $\rho(g)$ is invertible, and, therefore, of rank k). So it is necessary and sufficient to ensure that the following 2 requirements hold:

1. For every two distinct elements $g_i \neq g_j$

$$\text{rank} \begin{bmatrix} -I_k & -I_k \\ \rho(g_i) & \rho(g_j) \\ 0 & 0 \end{bmatrix} = 2k, \tag{8}$$

2. For all $g_i, g_j, g_\ell \in G$ (not necessarily distinct)

$$\text{rank} \begin{bmatrix} -I_k & 0 & \rho(g_\ell) \\ \rho(g_i) & -I_k & 0 \\ 0 & \rho(g_j) & -I_k \end{bmatrix} = r(\{g_i', g_j'', g_\ell'''\}) = \begin{cases} 2k & \text{if } g_j \cdot g_i \cdot g_\ell = 1, \\ 3k & \text{otherwise.} \end{cases} \tag{9}$$

Ranks of all other relevant sub-matrices follow from similar arguments.

We first show that Equation (8) holds. By Proposition 2.11(*b*)

$$\text{rank} \begin{bmatrix} -I_k & -I_k \\ \rho(g_i) & \rho(g_j) \\ 0 & 0 \end{bmatrix} = k + \text{rank}(\rho(g_j) - \rho(g_i)), \tag{10}$$

so in order to show that Equation (8) holds, we need to verify that for every two distinct group elements g_i, g_j $\text{rank}(\rho(g_i) - \rho(g_j)) = k$. Since ρ is fixed-point free and $g_i^{-1} g_j \neq 1$, for every $v \neq 0, v \neq \rho(g_i^{-1} g_j)v = (\rho(g_i)^{-1}\rho(g_j))v$, so $\forall v \neq 0, \rho(g_i)v \neq \rho(g_j)v$, thus, $\forall v \neq 0, (\rho(g_i) - \rho(g_j))v \neq 0$, which implies that $\rho(g_i) - \rho(g_j)$ is invertible and, therefore, of rank k, so (8) holds.

We next show that Equation (9) holds. By Proposition 2.11(*a*) and the definition of a homomorphism,

$$\text{rank} \begin{bmatrix} -I_k & 0 & \rho(g_\ell) \\ \rho(g_i) & -I_k & 0 \\ 0 & \rho(g_j) & -I_k \end{bmatrix} = 2k + \text{rank}(\rho(g_j \cdot g_i \cdot g_\ell) - I_k). \tag{11}$$

So, to prove that (9) holds, we need to show that

$$\text{rank}(\rho(g_j \cdot g_i \cdot g_\ell) - I_k) = \begin{cases} 0 & \text{if } g_j \cdot g_i \cdot g_\ell = 1 \\ k & \text{otherwise.} \end{cases}$$

By arguments similar to the above

1. If $g_j \cdot g_i \cdot g_\ell \neq 1$ then $\text{rank}(\rho(g_j \cdot g_i \cdot g_\ell) - I_k) = k$, as ρ is fixed-point free.
2. If $g_j \cdot g_i \cdot g_\ell = 1$ then $\text{rank}(\rho(g_j \cdot g_i \cdot g_\ell) - I_k) = 0$. (This in fact true for any representation because $\rho(g_j \cdot g_i \cdot g_\ell) = \rho(1) = I_k$.)

*Proof (**Proof of Theorem 3.1**).* Combining the lemmas we get Theorem 3.1: If G has a fixed-point free representation ρ of dimension k, then by Lemma 3.4, the block matrix $\mathbf{A}_\rho$ is a k-linear representation of $Q_3(G)$, and, in particular, $Q_3(G)$ has a k-linear representation. On the other hand, if $Q_3(G)$ is k-linearly representable then, by Lemma 3.2, it has a faithful representation ρ of dimension k such that $\mathbf{A}_\rho$ is a k-linear representation of $Q_3(G)$, so, by Lemma 3.3, ρ is fixed-point free.

We combine Theorem 3.1 with Proposition 2.15 to get our desired result:

Corollary 3.5. *For every prime $p > 2$ there is a matroid that is p-minimally representable. Moreover, the matroid has $\text{poly}(p)$ ground points and this representation exists over a finite field with $2^{O(p^{6.18})}$ elements.*

Proof. Let q and G_p be as in Proposition 2.15. By Theorem 3.1 and Proposition 2.15, over the field $\mathbb{F}_{2^{pq}}$, the matroid $Q_3(G_p)$, which has $3p^2q + 3$ elements in the ground set, is p-linearly representable. Furthermore, over any field, the matroid $Q_3(G_p)$ is not j-linearly representable for any $j < p$. So, $Q_3(G_p)$ is p-minimally representable. By Proposition 2.15, if we chose the appropriate q, then the field $\mathbb{F}_{2^{pq}}$ has $2^{O(p^{6.18})}$ elements.

We next rephrase the result in secret-sharing terms.

Corollary 3.6. *For every prime p, there exists an access structure with $\text{poly}(p)$ parties, which has an ideal p-linear secret-sharing scheme with secrets of length $\text{poly}(n)$, but has no ideal k-linear secret-sharing scheme for every $k < p$.*

Since the matroid has $3p^2q + 3$ elements in the ground set, the corresponding access structure has $3p^2q + 2$ parties. Therefore, for every prime p, the smallest access structure of this type has $O(p^{7.18})$ parties. Also note that the schemes is over a field with $2^{\text{poly}(p)}$ elements, so every share can be represented by $\text{poly}(p)$ bits.

4   Lower Bounds for Multi-linear Secret-Sharing Schemes

The best known lower bounds for linear secret-sharing schemes is $n^{\Omega(\log n)}$ [1,16,17]. By modification of the claims in [17], we show that these lower bounds hold also for multi-linear secret-sharing schemes. Thus, even using multi-linear schemes one cannot construct efficient schemes for general access structures.

We will use the following alternative definition of multi-linear secret sharing schemes, proven to be equivalent in [13] (following [7,20]).

Definition 4.1 (Multi-Target Monotone Span Program). *A multi-target monotone span program is a quadruple* $\mathcal{M} = (\mathbb{F}, M, \rho, X)$, *where* $\mathbb{F}$ *is a finite field,* M *is an* $a \times b$ *matrix over* $\mathbb{F}$, *the function* $\rho : \{1, \ldots, a\} \to \{p_1, \ldots, p_n\}$ *labels each row of* M *by a party, and* X *is a set of* k *independent vectors in* $\mathbb{F}^b$ *such that for every* $A \subseteq \{p_1, \ldots, p_n\}$ *either*

- *The rows of the sub-matrix obtained by restricting* M *to the rows labeled by parties in* A, *denoted* M_A, *span every vector in* X. *In this case, we say that* $\mathcal{M}$ *accepts* A, *or,*
- *The rows of* M_A *span no non-zero vector in the linear space spanned by* X. *In this case, we say that* $\mathcal{M}$ *rejects* B.

We say that $\mathcal{M}$ *accepts an access structure* $\mathcal{A}$ *if* $\mathcal{M}$ *accepts a set* B *if* $B \in \mathcal{A}$, *and rejects every set* $B \notin \mathcal{A}$. *The size of a multi-target monotone span program is* a/k, *where* a *is the number of rows in the matrix and* k *is the number of vectors in the set* X.

Note that not every labeled matrix is a multi-target span program. For example, if $k > 1$ and for some set A, the rows in M_A span exactly one vector in X, then this is not a multi-target span program. By [13] a multi-linear secret-sharing scheme realizing an access structure $\mathcal{A}$ with total share size a exists if and only if there exists a multi-target monotone span program accepting $\mathcal{A}$ that has a rows. In particular, if there exists a multi-target monotone span program accepting $\mathcal{A}$ with a_j rows labeled by p_j for $1 \leq j \leq n$ and k vectors in the set X, then the exists a multi-linear secret-sharing scheme realizing $\mathcal{A}$ with information ratio $\max_{1 \leq j \leq n} a_j/k$. In ideal multi-linear secret-sharing schemes $a_j = k$ for every j.

Assume, w.l.o.g., that $X = \{e_1, \ldots, e_k\}$. We make 2 observations regarding multi-target monotone span program.

Observation 4.2. *If* $B \in \mathcal{A}$ *and* $N = M_B$ *then the rows of* N *span* X, *thus* $\forall 0 < s < k$ *there exists some vector* $\boldsymbol{v}_s$ *such that* $\boldsymbol{e}_s = \boldsymbol{v}_s N$.

Observation 4.3. *If* $T \notin \mathcal{A}$ *then for every* $s \in \{1, \ldots, k\}$ *there exists a vector* $\boldsymbol{w}_s \in \mathbb{F}^b$ *such that the following hold: (1)* $M_T \boldsymbol{w}_s = 0$, *(2)* $\forall i \neq s, \boldsymbol{e}_i \cdot \boldsymbol{w}_s = 0$, *and (3)* $\boldsymbol{e}_s \cdot \boldsymbol{w}_s = 1$ *(that is, the coordinate* s *in* $\boldsymbol{w}_s$ *is 1).*

Proof. If $T \notin \mathcal{A}$, then the rows of M_T do not span any of the vectors in X. Let $M_{T,X}$ be the matrix containing the rows of M_T and additional rows $\boldsymbol{e}_1, \ldots, \boldsymbol{e}_k$ and $M_{T,X\backslash\{s\}}$ the same matrix with the row $\boldsymbol{e}_s$ deleted. By simple linear algebra, for every $1 \leq s \leq k$, we have that $\operatorname{rank} M_{T,X} > \operatorname{rank} M_{T,X\backslash\{s\}}$, which implies that $|\operatorname{kernel} M_{T,X}| < |\operatorname{kernel} M_{T,X\backslash\{s\}}|$, and so there is some vector $\boldsymbol{w}_s \in \mathbb{F}^b$ such that $\boldsymbol{e}_s \cdot \boldsymbol{w}_s = 1$ and $M_{T,X\backslash\{s\}} \boldsymbol{w}_s = \mathbf{0}$ (so evidently $M_T \boldsymbol{w}_s = 0$ and $\forall i \neq s, \boldsymbol{e}_i \cdot \boldsymbol{w}_s = 0$).

We next quote the definition of a collection with unique intersection from [17]. Such collections are used in [17] to prove lower bounds for monotone span programs; we show that the same lower bound holds for multi-target monotone span programs.

Definition 4.4. *Let $\mathcal{A}$ be a monotone access structure, with $\mathcal{B} = \{B_1, \ldots, B_\ell\}$ the collection of minimal authorized sets in $\mathcal{A}$. Let $\mathcal{C} = \{(C_{1,0}, C_{1,1}), (C_{2,0}, C_{2,1}), \ldots, (C_{t,0}, C_{t,1})\}$ be a collection of pairs of sets of parties. We say that $\mathcal{C}$ satisfies the* unique intersection *property for $\mathcal{A}$ if*

1. *For every $1 \leq j \leq t$, $\{p_1, \ldots, p_n\} \setminus (C_{j,0} \cup C_{j,1}) \notin \mathcal{A}$.*
2. *For every $1 \leq i \leq \ell$ and every $1 \leq j \leq t$, exactly one of the following conditions hold (1) $B_i \cap C_{j,0} \neq \emptyset$, (2) $B_i \cap C_{j,1} \neq \emptyset$.*

Note that if $B \in \mathcal{A}$ and $\{p_1, \ldots, p_n\} \setminus C \notin \mathcal{A}$, then $B \cap C \neq \emptyset$ (otherwise, $B \subseteq \{p_1, \ldots, p_n\} \setminus C$, contradicting the monotonicity of $\mathcal{A}$). Thus, Condition (2) in Definition 4.4 requires that B_i intersects at most one of the sets $C_{j,0}, C_{j,1}$.

Theorem 4.5. *Let $\mathcal{C}$ be a collection satisfying the unique intersection property for $\mathcal{A}$. Define a matrix D of size $\ell \times t$, with $D_{i,j} = 0$ if $B_i \cap C_{i,0} \neq \emptyset$ and $D_{i,j} = 1$ if $B_i \cap C_{i,1} \neq \emptyset$. Then, the size of every multi-target monotone span program accepting $\mathcal{A}$ is at least $\mathrm{rank}_{\mathbb{F}}(D)$.*

Proof. Let $\mathcal{M} = (\mathbb{F}, M, \rho, X = \{e_1, \ldots, e_k\})$ be a multi-target monotone span program accepting $\mathcal{A}$, and denote the number of rows of M by m. For every $1 \leq i \leq \ell$ since $B_i \in \mathcal{A}$ the rows of M labeled by the parties of B_i span X. By Observation 4.2, for every $1 \leq r \leq k$, there exists $v_{i,r}$ such that $v_{i,r}M = e_r$ and the non-zero coordinates of $v_{i,r}$ are only in rows labeled by B_i.

Fix $1 \leq j \leq t$ and let $T_j = \{p_1, \ldots, p_n\} \setminus (C_{j,0} \cup C_{j,1})$. Since $T_j \notin \mathcal{A}$, by Observation 4.3, for every $1 \leq s \leq k$ there exists a vector $w_{j,s}$ such that $M_{T_j} w_{j,s} = 0$, $e_s \cdot w_{j,s} = 1$ and $\forall r \neq s, e_r \cdot w_{j,s} = 0$. Let $y_{j,s} := M w_{j,s}$ and define $z_{j,s}$ to be the column vector achieved from $y_{j,s}$ by replacing all coordinates in $y_{j,s}$ labeled by parties in $C_{j,0}$ with zero. The only non-zero coordinates in $z_{j,s}$ are in coordinates labeled by $C_{j,1}$.

Define L as the matrix with rows $v_{1,1}, \ldots, v_{\ell,1}, v_{1,2}, \ldots, v_{\ell,2}, \ldots, v_{\ell,k}$ and R the matrix with columns $z_{1,1}, \ldots, z_{\ell,1}, z_{1,2}, \ldots, z_{\ell,2}, \ldots, z_{\ell,k}$. Note that by definition the rows of L are of length m, so L has m columns, thus, $\mathrm{rank}(L) \leq m$.

Let $\mathbf{D} = LR$. We next prove that $\mathbf{D}$ is a block matrix of the form:

$$\mathbf{D} = \begin{bmatrix} D & 0 & \cdots & 0 \\ 0 & D & \cdots & 0 \\ \vdots & \vdots & \ddots & \vdots \\ 0 & 0 & \cdots & D \end{bmatrix}, \tag{12}$$

where D is the matrix defined in the Theorem. We need to show that $v_{i,r} \cdot z_{s,j} = 0$ if $r \neq s$ (off the diagonal matrix block) and $v_{i,r} \cdot z_{s,j} = D_{i,j}$ if $r = s$.

- If $B_i \cap C_{j,0} \neq \emptyset$, $D_{i,j} = 0$. Furthermore, $B_i \cap C_j, 1 = \emptyset$, thus, $v_{i,r}$ and $z_{s,j}$ do not share non-zero coordinates and $v_{i,r} \cdot z_{s,j} = 0$. In particular, if $r = s$ then $v_{i,r} \cdot z_{r,j} = 0 = D_{i,j}$, and if $r \neq s$ then $v_{i,r} \cdot z_{s,j} = 0$ as desired.
- If $B_i \cap C_{j,1} \neq \emptyset$, then $D_{i,j} = 1$, $B_i \cap C_{j,0} = \emptyset$, and all coordinates in $v_{i,r}$ labeled by $C_{j,0}$ are zero, thus,

$$v_{i,r} \cdot z_{s,j} = v_{i,r} \cdot y_{s,j} = v_{i,r} M w_{s,j} = e_r \cdot w_{s,j} = \begin{cases} 0 & r \neq s \\ 1 & r = s \end{cases}.$$

In particular, if $r = s$ then $\boldsymbol{v}_{i,r} \cdot \boldsymbol{z}_{r,j} = 1 = D_{i,j}$ and if $r \neq s$ then $\boldsymbol{v}_{i,r} \cdot \boldsymbol{z}_{s,j} = 0$.

So $\mathrm{rank}_{\mathbb{F}}(\mathbf{D}) = k \cdot \mathrm{rank}_{\mathbb{F}}(D)$, and since $\mathcal{M}$ is a k-linear representation, its size is $\frac{m}{k} \geq \frac{\mathrm{rank}_{\mathbb{F}}(L)}{k} \geq \frac{\mathrm{rank}_{\mathbb{F}}(\mathbf{D})}{k} = \mathrm{rank}_{\mathbb{F}}(D)$.

By [17], for every n there is an access structure $\mathcal{A}$ with n parties, for which there exists a collection $\mathcal{C}$ satisfying the unique intersection property, such that $\mathrm{rank}_{\mathbb{F}}(D) \geq n^{\Omega(\log n)}$ (where D is as defined in Theorem 4.5). So by Theorem 4.5,

Corollary 4.6. *For every n, there exists an access structure $\mathcal{N}_n$ with n parties such that every multi-target monotone span program over any field accepting it has size $n^{\Omega(\log n)}$.*

As multi-target monotone span program are equivalent to multi-linear secret-sharing schemes [20], the same lower bound applies to multi-linear secret-sharing schemes.

Corollary 4.7. *For every n, there exists an access structure $\mathcal{N}_n$ with n parties such that the information ratio of every multi-linear secret-sharing scheme realizing it is $n^{\Omega(\log n)}$.*

References

1. Babai, L., Gál, A., Wigderson, A.: Superpolynomial lower bounds for monotone span programs. Combinatorica 19(3), 301–319 (1999)
2. Ben-Or, M., Goldwasser, S., Wigderson, A.: Completeness theorems for noncryptographic fault-tolerant distributed computations. In: Proc. of the 20th ACM Symp. on the Theory of Computing, pp. 1–10 (1988)
3. Benaloh, J., Leichter, J.: Generalized secret sharing and monotone functions. In: Goldwasser, S. (ed.) CRYPTO 1988. LNCS, vol. 403, pp. 27–35. Springer, Heidelberg (1990)
4. Bertilsson, M., Ingemarsson, I.: A construction of practical secret sharing schemes using linear block codes. In: Zheng, Y., Seberry, J. (eds.) AUSCRYPT 1992. LNCS, vol. 718, pp. 67–79. Springer, Heidelberg (1993)
5. Blakley, G.R.: Safeguarding cryptographic keys. In: Merwin, R.E., Zanca, J.T., Smith, M. (eds.) Proc. of the 1979 AFIPS National Computer Conference. AFIPS Conference proceedings, vol. 48, pp. 313–317. AFIPS Press (1979)
6. Blundo, C., De Santis, A., Stinson, D.R., Vaccaro, U.: Graph decompositions and secret sharing schemes. J. Cryptology 8(1), 39–64 (1995)
7. Brickell, E.F.: Some ideal secret sharing schemes. Journal of Combin. Math. and Combin. Comput. 6, 105–113 (1989)
8. Brickell, E.F., Davenport, D.M.: On the classification of ideal secret sharing schemes. J. of Cryptology 4(73), 123–134 (1991)
9. Chaum, D., Crépeau, C., Damgård, I.: Multiparty unconditionally secure protocols. In: Proc. of the 20th ACM Symp. on the Theory of Computing, pp. 11–19 (1988)

10. Cramer, R., Damgård, I., Maurer, U.: General secure multi-party computation from any linear secret-sharing scheme. In: Preneel, B. (ed.) EUROCRYPT 2000. LNCS, vol. 1807, pp. 316–334. Springer, Heidelberg (2000)
11. Csirmaz, L.: The size of a share must be large. J. of Cryptology 10(4), 223–231 (1997)
12. Desmedt, Y., Frankel, Y.: Shared generation of authenticators and signatures. In: Feigenbaum, J. (ed.) CRYPTO 1991. LNCS, vol. 576, pp. 457–469. Springer, Heidelberg (1992)
13. van Dijk, M.: A linear construction of secret sharing schemes. Designs, Codes and Cryptography 12(2), 161–201 (1997)
14. Dowling, T.A.: A class of geometric lattices based on finite groups. J. Comb. Theory, Ser. B 14(1), 61–86 (1973)
15. Dowling, T.A.: A q-analog of the partition lattice. A Survey of Combinatorial Theory, 101–115 (1973)
16. Gál, A.: A characterization of span program size and improved lower bounds for monotone span programs. Computational Complexity 10(4), 277–296 (2001)
17. Gál, A., Pudlák, P.: A note on monotone complexity and the rank of matrices. Inform. Process. Lett. 87, 321–326 (2003)
18. Goyal, V., Pandey, O., Sahai, A., Waters, B.: Attribute-based encryption for fine-grained access control of encrypted data. In: Proc. of the 13th ACM Conference on Computer and Communications Security, pp. 89–98 (2006)
19. Ito, M., Saito, A., Nishizeki, T.: Secret sharing schemes realizing general access structure. In: Proc. of the IEEE Global Telecommunication Conf., Globecom 1987, pp. 99–102 (1987); Journal version: Multiple assignment scheme for sharing secret. J. of Cryptology 6(1), 15–20 (1993)
20. Karchmer, M., Wigderson, A.: On span programs. In: Proc. of the 8th IEEE Structure in Complexity Theory, pp. 102–111 (1993)
21. Karnin, E.D., Greene, J.W., Hellman, M.E.: On secret sharing systems. IEEE Trans. on Information Theory 29(1), 35–41 (1983)
22. Linnik, Y.V.: On the least prime in an arithmetic progression I. the basic theorem. Rec. Math (Mat. Sbornik) N.S. 15(57), 139–178 (1944)
23. Linnik, Y.V.: On the least prime in an arithmetic progression II. the deuring-heilbronn phenomenon. Rec. Math (Mat. Sbornik) N.S. 15(57), 347–368 (1944)
24. Martí-Farré, J., Padró, C.: On secret sharing schemes, matroids and polymatroids. Journal of Mathematical Cryptology 4(2), 95–120 (2010)
25. Matúš, F.: Matroid representations by partitions. Discrete Mathematics 203, 169–194 (1999)
26. Milne, J.S.: Group theory, v3.12 (2012), http://www.jmilne.org/math/
27. Naor, M., Wool, A.: Access control and signatures via quorum secret sharing. In: 3rd ACM Conf. on Computer and Communications Security, pp. 157–167 (1996)
28. Oxley, J.G.: Matroid Theory, 2nd edn. Oxford University Press (2011)
29. Pendavingh, R.A., van Zwam, S.H.M.: Skew partial fields, multilinear representations of matroids, and a matrix tree theorem. Advances in Applied Mathematics 50(1), 201–227 (2013)
30. Semple, C., Whittle, G.: Partial fields and matroid representation. Advances in Applied Mathematics 17(2), 184–208 (1996)
31. Serre, J.-P.: Linear Representations of Finite Groups. Springer (1977)
32. Seymour, P.D.: On secret-sharing matroids. J. of Combinatorial Theory, Series B 56, 69–73 (1992)

33. Shamir, A.: How to share a secret. Communications of the ACM 22, 612–613 (1979)
34. Shankar, B., Srinathan, K., Rangan, C.P.: Alternative protocols for generalized oblivious transfer. In: Rao, S., Chatterjee, M., Jayanti, P., Murthy, C.S.R., Saha, S.K. (eds.) ICDCN 2008. LNCS, vol. 4904, pp. 304–309. Springer, Heidelberg (2008)
35. Simonis, J., Ashikhmin, A.: Almost affine codes. Designs, Codes and Cryptography 14(2), 179–197 (1998)
36. Stinson, D.R.: Decomposition construction for secret sharing schemes. IEEE Trans. on Information Theory 40(1), 118–125 (1994)
37. Tassa, T.: Generalized oblivious transfer by secret sharing. Des. Codes Cryptography 58(1), 11–21 (2011)
38. van Dijk, M., Jackson, W.-A., Martin, K.M.: A general decomposition construction for incomplete secret sharing schemes. Des. Codes Cryptography 15(3), 301–321 (1998)
39. van Dijk, M., Kevenaar, T.A.M., Schrijen, G.J., Tuyls, P.: Improved constructions of secret sharing schemes by applying (lambda, omega)-decompositions. Inform. Process. Lett. 99(4), 154–157 (2006)
40. Vincent, G.: Les groupes lineaires finis sans point fixes. Commentarii Mathematici Helvetici 20, 117–171 (1947)
41. Waters, B.: Ciphertext-policy attribute-based encryption: An expressive, efficient, and provably secure realization. In: Catalano, D., Fazio, N., Gennaro, R., Nicolosi, A. (eds.) PKC 2011. LNCS, vol. 6571, pp. 53–70. Springer, Heidelberg (2011)
42. Wolf, J.A.: Spaces of Constant Curvature, 5th edn. Publish or Perish, Inc. (1984)
43. Xylouris, T.: On the least prime in an arithmetic progression and estimates for the zeros of Dirichlet L-functions. Acta Arith. 150(1), 65–91 (2011)
44. Zassenhaus, H.: Uber endliche faskorper. Abhandlungen aus dem Mathematischen Seminar der Hamburgischen Universitat 11, 187–220 (1935)

A   The Construction of the Group G_p

In this section we briefly explain the construction of the group G_p (for any prime p), which appears in Proposition 2.15. We then give a partial explanation of why any fixed-point free representation of G_p is of dimension at least p. The complete proofs, and more details on the construction, will be given in the full version.

Semidirect products. We assume some familiarity with group basics, such as group homomorphisms and automorphisms. Let N be a group. Recall that the set of all automorphisms of N, denoted $Aut(N)$, is also a group, with group operation being composition, and the identity element being the identity map. We now recall the definition of an action of a group H on a group N.

Definition A.1. *Let H and N be two groups. By an action of H on N, denoted $H \curvearrowright N$, we mean a group homomorphism $\phi\colon H \to Aut(N)$.*

To simplify the notation, if no confusion is possible, we use shorter notation $x^g := (\phi(g))(x)$. Since ϕ is a homomorphism, the identity of G is mapped to the identity automorphism.

Example A.2. For any pair of groups H and N, there always exists the *trivial action* $\tau\colon H \to Aut(N)$, which maps every element of H to the identity automorphism. A non-trivial action, however, does not always exist, and depends on the choice of H and N.

Example A.3. Let $H = \mathbb{Z}_2$ and $N = \mathbb{Z}_3$. To avoid confusion we denote $N = \{0, 1, 2\}$ and $H = \{f_0, f_1\}$. Then H acts on N by $\phi(f_1)(1) = 2$. We note that this completely identifies the action because f_1 and 1 are generators of H and N respectively. Thus, for example, $f_1(2) = f_1(1 + 1) = f_1(1) + f_1(1) = 2 + 2 = 1$ and $f_0(1) = f_1 \circ f_1(1) = f_1(2) = 1$. So it remains to verify only that this is well defined, which is a very small task.

Lemma A.4. *Let $\psi\colon G_1 \to G_2$ be a group homomorphism, and $\phi\colon G_2 \curvearrowright N$ a group action. Then ψ induces a group action $\psi^*(\phi)\colon G_1 \curvearrowright N$, given by composition $(\psi^*(\phi))(x) := \phi(\psi(x))$. Furthermore, if ψ is surjective and the action ϕ is non-trivial then so is $\psi^*(\phi)$.*

Proof. Follows easily from the definitions.

The following proposition is well known.

Proposition A.5. *For any prime q, the group of automorphisms of $\mathbb{Z}_q$ is isomorphic to the group $\mathbb{Z}_{q-1}$.*

This allows us to build a non-trivial action of $\mathbb{Z}_p$ on $\mathbb{Z}_q$, if p, q are primes such that $q \equiv 1 \mod p$.

Proposition A.6. *Let $p, q \in \mathbb{N}$ be two primes such that $q \equiv 1 \mod p$. Then $\mathbb{Z}_p$ admits a non-trivial action on $\mathbb{Z}_q$.*

Proof. From Proposition A.5 $Aut(\mathbb{Z}_q) \simeq \mathbb{Z}_{q-1}$. Thus, it suffices to construct a non-trivial homomorphism $\phi\colon \mathbb{Z}_p \to \mathbb{Z}_{q-1}$. Let $n \in \mathbb{N}$ be such that $q - 1 = np$, and set $\phi(x) := nx \mod q$. Then ϕ is a non-trivial homomorphism.

Corollary A.7. *Let p, q be as in Proposition A.6. Then $\mathbb{Z}_{p^2}$ admits a non-trivial action on $\mathbb{Z}_q$.*

Proof. We have a natural surjective homomorphism $\psi\colon \mathbb{Z}_{p^2} \to \mathbb{Z}_p$ given by $\psi(x) := x \mod p$. Thus, by Proposition A.6 and Lemma A.4, $\psi^*(\phi)$ is a non-trivial action of $\mathbb{Z}_{p^2}$ on $\mathbb{Z}_q$.

There may exist other non-trivial actions of $\mathbb{Z}_{p^2}$ on $\mathbb{Z}_q$. However, from now on when we mention *the* action of $\mathbb{Z}_{p^2}$ on $\mathbb{Z}_q$, we mean that we have fixed an isomorphism $Aut(\mathbb{Z}_q) \simeq \mathbb{Z}_{q-1}$ and we refer to the non-trivial action $\psi^*(\phi)$ constructed in the proof of Corollary A.7.

Definition A.8. *Let H be a group acting on another group N, and $\phi\colon H \to Aut(N)$ the action. The* semidirect product, *denoted $N \rtimes_\phi H$, is the set $N \times H = \{(n, h) | n \in N, h \in H\}$ equipped with the following operation*

$$(n_1, h_1) \cdot (n_2, h_2) := (n_1 \cdot n_2^{h_1}, h_1 \cdot h_2). \tag{13}$$

We leave to the reader to verify that (13) indeed defines a group-law. We will often omit ϕ in the notation of the semidirect product, and write simply $N \rtimes G$. When the action of G on N is not trivial we will say that the semidirect product is *non-trivial*. An attractive property of non-trivial semidirect products is that they are not abelian, even if H and N are.

Lemma A.9. *If $N \rtimes G$ is a non-trivial semidirect product then it is not abelian.*

Proof. Since G acts non-trivially, there exist $g \in G$ and $h \in N$ such that $h^g \neq h$. Therefore $(e_N, g) \cdot (h, e_G) = (h^g, g) \neq (h, g) = (h, e_G) \cdot (e_N, g)$.

Proposition A.10. *Let p and q be prime integers satisfying $q \equiv 1 \mod p$. Then there exists a non-trivial semidirect product $G_p = \mathbb{Z}_q \rtimes \mathbb{Z}_{p^2}$. The group has $p^2 \cdot q$ elements.*

Proof. Follows immediately from the definitions and Corollary A.7.

Suitability of G_p. We now explain why the above construction works for us. Since a full proof requires quite a few pages of background in representation theory, we will only give a brief overview and refrain from proving the following claims, which rely on some classical theorems in representation theory. But first we state Linnik's theorem:

Theorem A.11 (Linnik's Theorem). *There exists constants c, L such that for any pair of co-prime integers a and d, with $1 \leq a < d$, the smallest prime of the form $a + nd$ ($n \geq 1$) is smaller than cd^L.*

Linnik didn't give an explicit bound on L, but later works have shown that L is in fact very small. The current state of the art is $L \leq 5.18$ due to Xylouris [43].

Corollary A.12. *For every prime p, there exists a prime q, with $q = O(p^{5.18})$, for which a non-trivial semidirect product $\mathbb{Z}_q \rtimes \mathbb{Z}_{p^2}$ exists.*

Now fix a prime p and a prime q such that $q \equiv 1 \mod p$, and let $G_p = \mathbb{Z}_q \rtimes \mathbb{Z}_{p^2}$ be the non-trivial semidirect product explained above.

Lemma A.13. *The group G_p is solvable and every proper subgroup of G_p is cyclic. Thus, from the classification of solvable fixed-point free groups (see for example [42, Theorem 6.1.11]), the group G_p admits a fixed-point free representation.*

Lemma A.14. *The group G_p is not abelian. This implies that G_p does not have fixed-point free representations of dimension 1.*

Lemma A.15. *The group G_p does not have any fixed-point free representations over fields of characteristic p, q.*

Lemma A.16. *Over fields of characteristic different from p, q, the dimension of the smallest fixed-point representation of G_p divides the order of G_p. Thus, since the dimension cannot be 1, G_p has a fixed-point free representation of dimension $\geq p$.*

The completion of Proposition 2.15 (i.e., bounding the size of the field) is done by explicitly building a fixed-point free representation of G_p of dimension p over the field $\mathbb{F}_{2^{pq}}$.

Broadcast Amplification

Martin Hirt, Ueli Maurer, and Pavel Raykov

ETH Zurich, Switzerland
{hirt,maurer,raykovp}@inf.ethz.ch

Abstract. A d-broadcast primitive is a communication primitive that allows a sender to send a value from a domain of size d to a set of parties. A broadcast protocol emulates the d-broadcast primitive using only point-to-point channels, even if some of the parties cheat, in the sense that all correct recipients agree on the same value v (consistency), and if the sender is correct, then v is the value sent by the sender (validity). A celebrated result by Pease, Shostak and Lamport states that such a broadcast protocol exists if and only if $t < n/3$, where n denotes the total number of parties and t denotes the upper bound on the number of cheaters.

This paper is concerned with broadcast protocols for any number of cheaters ($t < n$), which can be possible only if, in addition to point-to-point channels, another primitive is available. Broadcast amplification is the problem of achieving d-broadcast when d'-broadcast can be used once, for $d' < d$. Let $\phi_n(d)$ denote the minimal such d' for domain size d.

We show that for $n = 3$ parties, broadcast for any domain size is possible if only a single 3-broadcast is available, and broadcast of a single bit ($d' = 2$) is not sufficient, i.e., $\phi_3(d) = 3$ for any $d \geq 3$. In contrast, for $n > 3$ no broadcast amplification is possible, i.e., $\phi_n(d) = d$ for any d.

However, if other parties than the sender can also broadcast some short messages, then broadcast amplification is possible for *any* n. Let $\phi_n^*(d)$ denote the minimal d' such that d-broadcast can be constructed from primitives d'_1-broadcast,..., d'_k-broadcast, where $d' = \prod_i d'_i$ (i.e., $\log d' = \sum_i \log d'_i$). Note that $\phi_n^*(d) \leq \phi_n(d)$. We show that broadcasting $8n \log n$ bits in total suffices, independently of d, and that at least $n - 2$ parties, including the sender, must broadcast at least one bit. Hence $\min(\log d, n - 2) \leq \log \phi_n^*(d) \leq 8n \log n$.

1 Introduction

1.1 Byzantine Broadcast

We consider a set $\mathcal{P} = \{P_1, \ldots, P_n\}$ of n parties connected by authenticated synchronous point-to-point channels.[1] The broadcast problem (also known as

---

[1] Synchronous means that the parties work in synchronous rounds such that the messages are guaranteed to be delivered within the same round in which they were sent. If the sending party inputs no message (or a message outside the agreed domain) to the channel, then the receiving party gets a default output.

Y. Lindell (Ed.): TCC 2014, LNCS 8349, pp. 419–439, 2014.

the Byzantine generals problem) is defined as follows [PSL80]: A specific party, the sender, wants to distribute a message to the other parties in such a way that all correct parties obtain the same message, even if up to t of the parties cheat (also called Byzantine) and deviate arbitrarily from the prescribed protocol. We assume that P_1 is the sender and $\mathcal{R} = \mathcal{P} \setminus \{P_1\}$ is the set of recipients. Formally:

Definition 1. *A protocol for the set $\mathcal{P} = \{P_1, \ldots, P_n\}$ of parties, where P_1 has an input value $v \in \mathcal{D}$ and each party in $\mathcal{R}$ outputs a value in $\mathcal{D}$, is called a broadcast protocol for domain $\mathcal{D}$ if the following conditions are satisfied:*[2]

CONSISTENCY: *All correct parties $P_i \in \mathcal{R}$ output the same value $v \in \mathcal{D}$.*
VALIDITY: *If the sender P_1 is correct, then v is the input value of P_1.*
TERMINATION: *Every correct party in $\mathcal{P}$ terminates.*

A broadcast protocol can be understood as emulating a so-called broadcast primitive (or channel), i.e., an ideal communication primitive where P_1 inputs a value which is output to all other parties. Broadcast is one of the most fundamental primitives in distributed computing. It is used as building block in various protocols like voting, bidding, collective contract signing, secure multiparty computation, etc.

A celebrated result by Pease, Shostak and Lamport states that for any non-trivial $\mathcal{D}$ (i.e., $|\mathcal{D}| \geq 2$), a broadcast protocol exists if and only if the upper bound t on the number of cheaters satisfies $t < n/3$ [PSL80, BGP92, CW92].

1.2 Broadcast Amplification

This paper is concerned with broadcast protocols for any number of cheaters ($t < n$), which can be possible only if, in addition to point-to-point channels, another primitive is available.[3] We consider perfect security, which means that the cheating probability is zero.

The perhaps most natural choice of such an additional primitive is the availability of some broadcast primitives for smaller domain sizes. Let d-broadcast be a broadcast primitive (or broadcast channel) for message domain size d for a specific sender.

We assume that in addition to point-to-point channels, the parties have access to a system called BBB which provides a broadcast primitive as a black-box. If invoked for sender P_i and domain $\mathcal{D}'$, BBB takes input $v \in \mathcal{D}'$ from P_i and outputs v to all parties in $\mathcal{P}$ (except P_i).

In this setting, the first and most natural question that arises is: Can a sender broadcast a message with domain size d by using point-to-point communication and broadcasting only a *single* message with domain size $d' < d$?

---

[2] The domain can without essential loss of generality be assumed to be $\mathcal{D} = [d]$, where here and below we define $[k] = \{1, \ldots, k\}$.

[3] One type of primitive considered previously is a so-called trusted set-up [DS83, PW96]. In such a model, perfect security is not achievable, but statistical or cryptographic security is.

Definition 2. *Let $\phi_n(d)$ denote the minimal d' such that d-broadcast can be constructed from d'-broadcast.*

Trivially, $\phi_n(d) \leq d$, as d-broadcast can be constructed directly from d-broadcast.

The most natural generalization of the above question is the following:[4] If any party can broadcast short messages, what is the minimal total number of bits that need to be broadcast to construct an ℓ-bit broadcast? More precisely, since we consider arbitrary alphabet sizes (not just powers of 2), the question is to determine the quantity $\phi_n^*(d)$ defined below.

Definition 3. *Let $\phi_n^*(d)$ denote the minimal d' such that d-broadcast can be constructed from the k primitives d_1'-broadcast, $\ldots$, d_k'-broadcast, where $d' = \prod_i d_i'$.*

Note that $\log d' = \sum_i \log d_i'$ is the total number of bits of information[5] broadcast using BBB. It is therefore often natural to state results for the quantity $\log \phi_n^*(d)$. It is obvious that $\phi_n^*(d) \leq \phi_n(d)$.

A protocol that amplifies the domain of a broadcast, in the sense of the above two definitions, is called a *broadcast-amplification protocol*. A broadcast amplification protocol for domain size d can be used to replace a call to a d-broadcast primitive within another protocol. Hence broadcast amplification protocols can be constructed recursively.

One can call $\phi_n(d)$ and $\phi_n^*(d)$ the *intrinsic broadcast complexity* of domain size d, in the single-sender and in the general multi-sender model, respectively.[6]

The goals of this paper are twofold. First, we study feasibility results, i.e., what is possible in principle. Therefore while studying the quantities $\phi_n(d)$ and $\phi_n^*(d)$ we do not make any restriction on the use of point-to-point channels (In fact, our protocols which are optimized for the BBB usage communicate exponential number of messages over point-to-point channels and are built for succinctness of the proof, not for communication complexity.) Second, based on the obtained bounds for $\phi_n(d)$ and $\phi_n^*(d)$ we search for protocols which are both efficient in terms of the BBB and point-to-point channels usage.

1.3   Contributions of This Paper

This paper introduces the concept of broadcast amplification and proves a number of results, both feasibility results in terms of protocols as well as infeasibility results in terms of impossibility proofs.

We first study the first question mentioned above, namely the setting where the sender uses a single broadcast primitive of smaller domain. For the case of

---

[4] More refined versions of this question exist but will not be considered.

[5] Not necessarily exactly the number of actual bits.

[6] One could also consider a single-sender multi-shot model, i.e., the model where the sender can broadcast with BBB multiple times. Later we give a protocol for the single-sender setting which requires only a single call to BBB and is optimal even in the multi-shot model.

three parties ($n = 3$), the smallest non-trivial case, we show the quite surprising result that broadcast for any domain size d is possible if only a single 3-broadcast ($d' = 3$) is available. Moreover broadcast of a single bit ($d' = 2$) is not sufficient. In other words, $\phi_3(d) = 3$ for any $d \geq 3$.

In contrast, for $n > 3$ no broadcast amplification is possible, i.e., $\phi_n(d) = d$ for any d.

If not only the sender, but also other parties can broadcast some short messages, then (strong) broadcast amplification is possible for *any* n. We show that broadcasting $8n \log n$ bits of information in total suffices, independently of d, i.e., $\log \phi_n^*(d) \leq 8n \log n$. On the negative side, we show that at least $n - 2$ parties must broadcast at least one bit, i.e., $\min(\log d, n - 2) \leq \log \phi_n^*(d)$.

The protocol that uses $8n \log n$ bits to broadcast a value of domain size d communicates exponentially many messages over point-to-point channels. We give an optimized version of this protocol which communicates a polynomial number of messages over point-to-point channels but needs to broadcast $\mathcal{O}(n^2 \log \log d)$ bits with BBB.

1.4   Related Work

All known protocols for efficient multi-valued broadcast [TC84, FH06, LV11, Pat11] can be interpreted as broadcast-amplification protocols, as they actually employ an underlying broadcast scheme for short messages (besides the point-to-point channels). These protocols tolerate only $t < n/3$ or $t < n/2$, where the underlying broadcast itself is realizable with a normal broadcast protocol (hence the given broadcast channels are not needed at all).

Another approach for broadcast amplification can be derived from existing signature-based broadcast protocols [DS83, PW96]. One can use the available black-box broadcast to generate an appropriate setup (e.g., a PKI) and then use the corresponding protocol over point-to-point channels to broadcast the ℓ bit message. Thus we obtain broadcast-amplification protocols for $t < n$ with cryptographic and statistical security that require all parties to broadcast $\text{Poly}(n) \log \ell$ bits in total for the construction of an ℓ-bit broadcast.

Fitzi and Maurer considered amplification of the broadcast recipient set [FM00]. That is, they showed that with the access to local broadcast among every k parties one can construct broadcast among n parties iff $t < \frac{k-1}{k+1} n$ [FM00, CFF+05].

Another related line of research is the amplification of other primitives, like OT extension [Bea96, IKNP03] or coin-toss extension [HMQU06].

In [HR13] the authors give a protocol for 3 players allowing to broadcast message of any length by broadcasting 10 bits only is given. In our notation this shows that $\log \phi_3(d) \leq 10$ for all d.

Broadcast amplification is an example of the construction of a consistency primitive from another consistency primitive as defined in [Mau04].

2   Broadcast-Amplification Model

A broadcast-amplification protocol consists of the programs $\pi_1, \ldots, \pi_n$ that the players $P_1, \ldots, P_n$ use. Each program π_i is a randomized algorithm (which takes an input from domain $\mathcal{D}$ in case of the sender's program π_1) and produces an output. The program π_i has $n - 1$ interfaces to point-to-point channels to communicate with the other programs and additional interfaces to access BBB.

We now describe how the programs interact with BBB. First, we extend the notion of d-broadcast given in Section 1.2. Let (r, P_i, d)-broadcast be a broadcast channel available in round r which allows P_i to broadcast one single value from a domain of size d among the parties. We assume that each program π_j has an interface to each (r, P_i, d)-broadcast channel. Whenever we say that the parties broadcast with BBB, we mean that they actually access the corresponding broadcast channel by explicitly giving input/asking for an output on that channel's interface.

The protocol must ensure that the correct parties agree on which (r, P_i, d)-broadcast channels to invoke, that is, on r, P_i and d.[7] We say that a (r, P_i, d)-broadcast channel is *used* if the correct parties access it, i.e., in round r correct parties expect an output provided by P_i of a domain of the size d (in case of a correct P_i, he provides the corresponding input). Note that which channels are used by the protocol may not be necessarily fixed a priori and may depend on the execution. We say that a broadcast-amplification protocol has a *static* BBB usage pattern if the broadcast channels used are fixed beforehand. As opposed to the static case, protocols with a *dynamic* BBB usage pattern allow to broadcast with BBB adaptively to the execution, where of course still agreement on which broadcast channels to use is required among the correct parties.

Depending on which channels are used we distinguish the following models.

Definition 4. *The **single-sender** model allows for protocols where only (r, P_1, d)-broadcast channels are used, i.e., only P_1 broadcasts with BBB (If only one channel is used then such a single-sender model is called* single-shot; *otherwise, it is called* multi-shot.) *The **multi-sender** model does not put any limitations on the broadcast channels used.*

The costs d' of BBB usage of a broadcast-amplification protocol with a static BBB pattern is defined to be $\prod_i d_i$, where d_i's are the domain sizes of the broadcast channels used. The protocols with a dynamic BBB usage pattern have costs d' to be computed as the maximum of $\prod_i d_i$ among all possible executions.

We say that a broadcast-amplification protocol is *non-trivial* if its costs d' is strictly smaller than the size of the broadcast value domain $d = |\mathcal{D}|$, i.e., $d' < d$.

---

[7] This requirement stems from the observation that the broadcast channel may be implemented via a different protocol and hence in order to employ it all correct parties must start its execution together while agreeing on the broadcasting party and the domain of the broadcast value. Note that without this requirement, the BBB could be abused to reach agreement on "hidden" information, e.g., one could broadcast an ℓ-bit message v with using BBB only for a single bit (in round v).

3   Single-Sender Model

In this section we consider a single-sender model, that is, only the sender is allowed to use the BBB oracle. First, we completely investigate the situation for $n = 3$, that is, we show that 3-broadcast is enough to simulate any d-broadcast while 2-broadcast is not. On the negative side, we prove that for any $n > 3$ perfectly secure broadcast amplification is not possible, showing that $n = 3$ is a peculiar case in the context of broadcast-amplification protocols.

3.1   Broadcast Amplification for 3 Parties

We construct a broadcast-amplification protocol for three parties that allows the sender to broadcast a value v from domain $\mathcal{D}$ of size d, where the sender uses BBB to broadcast one value from a domain $\mathcal{D}'$ of size $d' = 3$. For ease of presentation, we assume that $\mathcal{D} = [d]$ and $\mathcal{D}' = [3]$.

The protocol works recursively. For $d = 3$, v is broadcast directly via BBB. For $d \geq 4$, the sender transmits v to both recipients, who then exchange the received values and forward the exchanged values back to the sender. Finally, the sender broadcasts a hint h from domain $[d-1]$, which allows each recipient to decide which of the values he holds is the right one. Broadcasting the hint is realized via recursion.[8]

The crucial trick in this protocol is the computation of the hint h. Very generically, this computation is expressed as a special function which takes as input three values (the original value v and the two values sent back to the sender) and outputs h. Given the hint h, the recipients decide on the value received from the sender if it is consistent with h. Otherwise, if the other recipient's value (as received in the exchange phase) is consistent with h, then that value is taken. Otherwise, some default value (say $\perp$) is taken.

More formally, denote the value of the sender by v; the values received by the recipients P_2 and P_3 by v_2 and v_3, respectively; the values received by the recipients in the exchange phase by v_{32} and v_{23}, respectively; and the values sent back to the sender by v_{321} and v_{231}, respectively. The function producing the hint is denoted with g_d and maps triples of values from $[d] \times [d] \times [d]$ into the hint domain $[d-1]$. Then the sender computes the hint $h = g_d(v, v_{321}, v_{231})$ and broadcasts it. Recipient P_2 outputs v_2 if $h = g_d(v_2, v_{32}, \widetilde{v_{231}})$ for some $\widetilde{v_{231}} \in [d]$. Otherwise, P_2 outputs v_{32} if $h = g_d(v_{32}, \widetilde{v_{321}}, v_2)$ for some $\widetilde{v_{321}} \in [d]$. Otherwise, P_2 outputs $\perp$. P_3 decides analogously. Clearly, this protocol guarantees validity. Consistency is achieved as long as

$$\forall v_2, v_3, \widetilde{v_{231}}, \widetilde{v_{321}} \in [d] : v_2 \neq v_3 \Rightarrow g_d(v_2, v_3, \widetilde{v_{231}}) \neq g_d(v_3, \widetilde{v_{321}}, v_2). \quad (1)$$

For $d \geq 4$, the function $g_d(x, y, z)$ can be constructed as follows: For $x \leq d-1$, let $g_d(x, y, z) = x$ (for any y, z). For $x = d$, let $g_d(x, y, z) = \min([d-1] \setminus \{y, z\})$. One can easily verify that g_d satisfies (1).

---

[8] As we see later, the recursion can be made much more efficient with the help of so-called identifying predicates. We focus on the feasibility results and hence do not optimize the protocols.

Protocol $\texttt{AmplifyBC}_3(d, v)$
1. If $d = 3$ then broadcast v using the BBB.
2. Otherwise:
   2.1 P_1 sends v to P_2 and P_3. Denote the values received with v_2 and v_3, respectively.
   2.2 P_2 sends v_2 to P_3 and P_3 sends v_3 to P_2. Denote the values received by P_2 and P_3 with v_{32} and v_{23}, respectively.
   2.3 P_2 sends v_{32} to P_1 and P_3 sends v_{23} to P_1. Denote the values received by v_{321} and v_{231}, respectively.
   2.4 P_1 computes $h = g_d(v, v_{321}, v_{231})$. Parties invoke $\texttt{AmplifyBC}_3(d-1, h)$.
   2.5 P_2: If there exists $\widetilde{v_{231}}$ such that $h = g_d(v_2, v_{32}, \widetilde{v_{231}})$ decide on v_2. Else if there exists $\widetilde{v_{321}}$ such that $h = g_d(v_{32}, \widetilde{v_{321}}, v_2)$ decide on v_{32}. Otherwise decide on $\perp$.
   2.6 P_3: If there exists $\widetilde{v_{321}}$ such that $h = g_d(v_3, \widetilde{v_{321}}, v_{23})$ decide on v_3. Else if there exists $\widetilde{v_{231}}$ such that $h = g_d(v_{23}, v_3, \widetilde{v_{231}})$ decide on v_{23}. Otherwise decide on $\perp$.

Lemma 1. *The protocol $\texttt{AmplifyBC}_3$ achieves broadcast. The sender P_1 broadcasts one value from domain* [3] *via BBB.*

Proof. We prove by induction that the broadcast properties are satisfied. For $d = 3$, broadcast is achieved by assumption of BBB. Now consider $d \geq 4$:

VALIDITY: If the sender is correct, then P_2 and P_3 receive $h = g_d(v, v_{321}, v_{231})$ as output from the recursive call to $\texttt{AmplifyBC}_3$. As $h = g_d(v_2, v_{32}, \widetilde{v_{231}})$ for $\widetilde{v_{231}} = v_{231}$, a correct P_2 decides on $v_2 = v$. Analogously, $h = g_d(v_3, \widetilde{v_{321}}, v_{23})$ for $\widetilde{v_{321}} = v_{321}$, a correct P_3 decides on $v_3 = v$.

CONSISTENCY: This property is non-trivial only if both P_2 and P_3 are correct, hence $v_{23} = v_2$ and $v_{32} = v_3$. Due to the Consistency property of the recursive call to $\texttt{AmplifyBC}_3$ both P_2 and P_3 receive the same hint h. If $v_2 = v_3$, then by inspection of the protocol both parties decide on the same value (namely on v_2 if $h \in \{g_d(v_2, v_2, \cdot), g_d(v_2, \cdot, v_2)\}$ and on $\perp$ otherwise). If $v_2 \neq v_3$, then (1) implies that if P_2 decides on v_2 (i.e., $h = g_d(v_2, v_{32}, \widetilde{v_{231}})$), then P_3 does not decide on v_3 (i.e., $h \neq g_d(v_3, \widetilde{v_{321}}, v_{23})$), but decides on $v_{23} = v_2$ (i.e., $h = g_d(v_{23}, v_3, \widetilde{v_{231}})$). Analogously, if P_3 decides on v_3, then P_2 decides on v_3 as well.

TERMINATION: Follows by inspection.

$\square$

3.2 Generic Structure of Impossibility Proofs

The given lower-bounds proofs employ a standard indistinguishability argument that is used to prove that certain security goals cannot be achieved by any protocol in the Byzantine environment [PSL80]. Such a proof goes by contradiction, i.e., by assuming that the security goals can be satisfied by means of some protocol $(\pi_1, \ldots, \pi_n)$. Then the programs π_i are used to build a *configuration* with

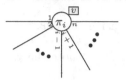

Fig. 1. Drawing of a program π_i. It has $n-1$ interfaces to bilateral channels with other players $\mathcal{P} \setminus \{P_i\}$ labeled accordingly. The program π_i is given v as input.

contradictory behavior. The configuration consists of multiple copies of π_i connected with bilateral channels and given admissible inputs. A pictorial drawing of a program in such a configuration is shown in Figure 1. When describing a configuration we will often use such a drawing accompanied with a textual description. If in the drawing an interface to a bilateral channel is not depicted then it is connected to a "null" device which simulates the program sending no messages. The interfaces to BBB are never drawn. Once the configuration is built, one simultaneously starts all the programs in the configuration and analyzes the outputs produced by the programs locally. By arguing that the view of some programs π_i and π_j in the configuration is indistinguishable from their view when run by the corresponding players P_i and P_j (while the adversary corrupts the remaining players in $\mathcal{P} \setminus \{P_i, P_j\}$) one deduces consistency conditions on the outputs by π_i and π_j that lead to a contradiction.

The main novelty of the proofs presented in this paper is that we consider an extended communication model where in addition to bilateral channels players are given access to BBB. While following the path described above, one needs to additionally define the BBB behavior in the configuration.

In the following impossibility proofs we assume that the BBB usage pattern is static. (In the full version of the paper we show how to adapt the impossibility proofs given to include protocols with a dynamic BBB usage pattern.) Furthermore, the lower bounds are given only for perfectly-secure protocols, i.e., those that fail with probability 0.

3.3   Lower Bounds in the Single-Sender Model

Lemma 2. *There is no perfectly-secure protocol among 3 parties achieving broadcast amplification for domain $\mathcal{D}$ with $|\mathcal{D}| \geq 3$ by broadcasting only 1 bit via BBB.*

Proof. Assume towards a contradiction that there is such a protocol (π_1, π_2, π_3). Without loss of generality, assume that $\mathcal{D} = [d]$ for some $d \geq 3$.

We consider the following configuration: For $i = 1, 2, 3$ and $j = 1, 2, 3$ let π_i^j be an instance of π_i. For $j = 1, 2, 3$ let π_1^j be given input j. We construct the configuration by connecting programs π_i^j as shown in Figure 2. Now we execute the programs. Whenever any program π_1^j broadcasts a bit with BBB it is given to programs π_2^j and π_3^j.

Since there are 3 programs $\pi_1^1, \pi_1^2, \pi_1^3$ broadcasting 1 bit only, there exist two of them π_1^i and π_1^j broadcasting the same bit. Without loss of generality, assume

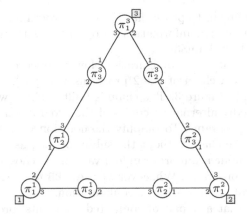

Fig. 2. The configuration for $n = 3$ to show the impossibility of broadcast amplification with broadcasting 1 bit only via BBB

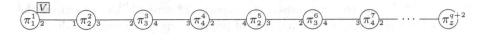

Fig. 3. The configuration for $n = 4$ to show the impossibility of non-trivial broadcast amplification with only the sender broadcasting

that π_1^1 and π_1^2 broadcast the same bit. The configuration can be interpreted in three different ways, which lead to contradicting requirements on the outputs of the programs. (i) P_1 holds input 1 and executes π_1^1, P_3 executes π_3^1, and P_2 is corrupted and executes the remaining programs in the configuration. Due to the validity property, π_3^1 must output 1. (ii) P_1 holds input 2 and executes π_1^2, P_2 executes π_2^2, and P_3 is corrupted and executes the remaining programs in the configuration. Due to the validity property, π_2^2 must output 2. (iii) P_3 executes π_3^1, P_2 executes π_2^2, and P_1 is corrupted and executes the remaining programs in the configuration. Due to the consistency property, π_3^1 and π_2^2 must output the same value. These three requirements cannot be satisfied simultaneously, hence whatever output the programs make, the protocol (π_1, π_2, π_3) is not a perfectly-secure broadcast-amplification protocol. □

Lemma 3. *There is no perfectly-secure protocol among $n \geq 4$ parties achieving non-trivial broadcast amplification in the single-sender multi-shot model.*

Proof. We first prove the lemma for $n = 4$, then reduce the case of arbitrary $n > 4$ to $n = 4$.

(Case n = 4). Assume towards a contradiction that there exists a perfectly-secure protocol $(\pi_1, \pi_2, \pi_3, \pi_4)$ achieving non-trivial broadcast-amplification in the single-sender model in q rounds (for some $q \in \mathbb{N}$). On the highest level our proof consists of three steps. (i) we define a configuration. (ii) we show that all programs in the configuration must output the same value v. (iii) we use an information flow argument to prove that there is a program in the configuration that does not have enough information to output v with probability 1 (this argument is inspired by [Lam83]).

(i) We consider the following configuration: Let π_i^j denote an instance of the program π_i. Consider a chain of $q + 2$ programs $\pi_1^1, \pi_2^2, \pi_3^3, \pi_4^4, \pi_2^5, \pi_3^6, \ldots, \pi_z^{q+2}$ connected as shown in Figure 3. The chain is built starting with a program π_1^1 and then by repeatedly alternating copies of the programs π_2, π_3 and π_4 until the chain has $q + 2$ programs. To simplify the notation we will sometimes refer to the programs in the chain without the subscript, i.e., as to $\pi^1, \pi^2, \ldots, \pi^{q+2}$. Let π_1^1 be given as input a uniform random variable V chosen from domain $\mathcal{D}$. Now we execute the programs. Whenever π_1^1 uses BBB to broadcast some x, the value x is given to all programs in the configuration.

(ii) First, we prove that any pair of connected recipients' programs (π_i^a, π_j^{a+1}) ($a \geq 2$) in the chain output the same value. One can view the configuration as the player P_i running the program π_i^a and P_j running π_j^{a+1} while the adversary corrupting $\{P_1, P_2, P_3, P_4\} \setminus \{P_i, P_j\}$ is simulating the programs $\pi^1, \ldots, \pi^{a-1}$ and $\pi^{a+2}, \ldots, \pi^{q+2}$. Due to the consistency property, π_i^a and π_j^{a+1} must output the same value. Since every connected pair of the recipients' programs in the chain outputs the same value, then the programs $\pi^2, \ldots, \pi^{a+1}$ in the configuration output the same value. Moreover, the configuration can be viewed as P_1 executing π_1^1, P_2 executing π_2^2 while the adversary who corrupts $\{P_3, P_4\}$ is simulating the remaining chain. Due to the validity property, π_2^2 must output V. Finally, each recipient's program $\pi^2, \ldots, \pi^{q+2}$ in the chain outputs V.

(iii) Let S_i^r be a random variable denoting the state of the program π^i in the chain after r rounds of the protocol execution. By state we understand the input that the program has, the set of all messages that the program received up to the rth round over point-to-point channels and on the BBB's interface together with the random coins it has used. Let B^r be a random variable denoting the list of the values that have been broadcast with BBB up to the rth round.

After r rounds only programs $\pi^1, \pi^2, \ldots, \pi^{r+1}$ can receive full information about V. The remaining programs in the chain $\pi^{r+2}, \pi^{r+3}, \ldots, \pi^{q+2}$ can receive only the information that was distributed with BBB, i.e., the information contained in B^r. That is, one can verify by induction that for any r and for all $i \geq r + 2$ holds $I(V; S_i^r | B^r) = 0$. Hence, for the last program in the chain π^{q+2} after q rounds of computation it holds that $I(V; S_{q+2}^q | B^q) = 0$ and hence $I(V; S_{q+2}^q) \leq H(B^q)$. Because we assumed that the protocol achieves non-trivial broadcast-amplification we have that $H(B^q) < H(V)$. Combining these facts we get that $I(V; S_{q+2}^q) < H(V)$. Hence, the last program π^{q+2} cannot output V with probability one, a contradiction.

(Case n > 4). Assume towards a contradiction that there is a protocol $(\pi_1, \pi_2, \pi_3, \ldots, \pi_n)$ allowing to do broadcast amplification in the single-sender model. One particular strategy of the adversary is to corrupt parties $P_5, \ldots, P_n$ and make them not execute their corresponding programs $\pi_5, \ldots, \pi_n$. Still, the remaining protocol $(\pi_1, \pi_2, \pi_3, \pi_4)$ must achieve broadcast, which contradicts the first case.   □

3.4   Summary

Theorem 1. *If* $n = 3$ *then* $\forall d \geq 3$ $\phi_3(d) = 3$; *otherwise, if* $n > 3$ *then* $\forall d$ $\phi_n(d) = d$.

The first statement follows from combining Lemma 1 and Lemma 2. The second statement follows from Lemma 3.

4   Multi-sender Model

As we have seen in the previous section in the single-sender model no broadcast-amplification is achievable for $n \geq 4$. In this section we consider a generalization of this model by allowing recipients to broadcast with BBB as well. In such a model we show that broadcast-amplification is achievable for any n. Moreover, we prove that in order to achieve a non-trivial broadcast-amplification for arbitrary n essentially all recipients must broadcast with BBB.

4.1   Broadcast Amplification for n Parties

In this section we present a broadcast-amplification protocol for n parties, where the parties broadcast with BBB at most $8n \log n$ bits in total. We first introduce the notion of identifying predicates and give an efficient construction of them. Then we present a protocol for graded broadcast, which achieves only a relaxed variant of broadcast, but only requires the sender to use BBB. Finally, we give the main broadcast-amplification protocol, which uses graded broadcast and BBB (by each party) to achieve broadcast.

While the presented protocol is very efficient in terms of the BBB usage (it broadcasts via BBB only $8n \log n$ bits to achieve broadcast of any ℓ bits), it communicates exponentially many messages over authenticated channels. We then show how to optimize this protocol such that it communicates only a polynomial (in n) number of messages at the expense of a higher BBB usage.

Identifying Predicates. An identifying predicate allows to identify a specific element v from some small subset $S \subseteq \mathcal{D}$, where $\mathcal{D}$ is a potentially large domain. To our knowledge, this concept has been firstly introduced in [HR13].

Definition 5. *A c-identifying predicate for domain $\mathcal{D}$ is a family of functions $Q_{k \in \mathcal{K}} : \mathcal{D} \to \{0, 1\}$ such that for any $S \subseteq \mathcal{D}$ with $|S| \leq c$ and any value $v \in S$ there exists a key $k \in \mathcal{K}$ with $Q_k(v) = 1$ and $Q_k(v') = 0$ for all $v' \in S \setminus \{v\}$. We say that such v is* uniquely identified *by Q_k in S.*

Note that any identifying predicates Q_k achieve monotonicity in the following sense:

Lemma 4. *If v is uniquely identified by Q_k in S, then v in uniquely identified in any $S' \subseteq S$ with $v \in S'$.*

The goal of constructing an identifying predicate family is to have $|\mathcal{K}|$ as small as possible given c and $|\mathcal{D}|$. We give a construction of a c-identifying predicate with domain $\mathcal{D}$ below.

Polynomial-Based Identifying Predicate Construction. Let $\ell = \log|\mathcal{D}|$. For $\kappa \in$ $\mathbb{N}$, let any value $v \in \mathcal{D}$ be interpreted as a polynomial f_v over $\mathrm{GF}(2^\kappa)$ of degree at most $\lfloor \ell/\kappa \rfloor$. We find a point $x \in \mathrm{GF}(2^\kappa)$ such that $f_v(x)$ is different from all other values $f_{v'}(x)$ for $v' \in S \setminus \{v\}$. For such a point x to always exist we need that the total number of points in the field is larger than the number of points in which f_v may coincide with other polynomials $f_{v'}$, i.e., $2^\kappa > (c-1)\lfloor \ell/\kappa \rfloor$. To satisfy this condition, it is enough to choose $\kappa := \lceil \log(c\ell) \rceil$. The key for the identifying predicate is defined as $k = (x, f_v(x))$, which is encoded using $2\lceil \log(c\ell) \rceil$ bits.[9] The predicate is defined as follows:

$$Q_{(x,y)}(v) = \begin{cases} 1, & \text{if } f_v(x) = y; \\ 0, & \text{otherwise.} \end{cases}$$

Lemma 5. *The polynomial-based construction gives a c-identifying predicate Q with domain $\mathcal{D}$ and key space $\mathcal{K}_c^{\mathcal{D}} = \{0,1\}^{2\lceil \log(c \log |\mathcal{D}|) \rceil}$.*

Graded Broadcast. Graded broadcast (a.k.a. gradecast) was introduced by Feldman and Micali [FM88]. It allows to broadcast a value among the set of recipients but with weaker consistency guarantees. In addition to the value v_i each recipient P_i also outputs a grade g_i describing the level of agreement reached by the players. In this paper we extend the original gradecast definition [FM88] with a more flexible grading system:

Definition 6. *A protocol achieves graded broadcast if it allows the sender P_1 to distribute a value v among parties $\mathcal{R}$ with every party P_i outputting a value v_i with a grade $g_i \in [n]$ such that:*

VALIDITY: *If the sender P_1 is correct, then every correct $P_i \in \mathcal{R}$ outputs $(v_i, g_i) = (v, 1)$.*

GRADED CONSISTENCY: *If a correct $P_i \in \mathcal{R}$ outputs (v_i, g_i) with $g_i < n$, then every correct $P_j \in \mathcal{R}$ outputs (v_j, g_j) with $v_j = v_i$ and $g_j \le g_i + 1$.*

TERMINATION: *Every correct party in $\mathcal{P}$ terminates.*

Intuitively, the grade can be understood as the consistency level achieved. The "strongest" grade $g_i = 1$ means that from the point of view of P_i, the sender "looks correct". Grade $g_i = 2$ means that P_i actually knows that the sender is incorrect; however, there might be an honest P_j for whom the sender looks correct. Grade $g_i = 3$ means that P_i knows that the sender is incorrect and every honest P_j knows so, too; however, there might be an honest P_k who does not know that every honest P_j knows that the sender is incorrect. And so on till the "weakest" grade $g_i = n$.

The protocol proceeds as follows: The sender sends the value v he wants to broadcast to all parties, who then exchange the received value(s) during $2n$

---

[9] Such a point x can be efficiently found by random sampling elements in $\mathrm{GF}(2^\kappa)$. Indeed, for $\kappa = \lceil \log(c\,\ell) \rceil$ more than half of the elements in $\mathrm{GF}(2^\kappa)$ are points where f_v is different from all other $f_{v'}$.

rounds. That is, in every round each party sends the set of values received so far to every other party. In this way each recipient P_i forms a growing sequence of sets $M_i^1 \subseteq M_i^2 \subseteq \cdots \subseteq M_i^{2n}$ (the set M_i^r represents the set of all messages received by P_i up to the round r). Finally, the sender distributes a hint consisting of the key k for an identifying predicate Q_k that should identify v among the values that the recipients hold. Then each recipient P_i computes his grade g_i to be the smallest number in $[n]$ such that both $M_i^{g_i}$ and $M_i^{2n-g_i}$ contain a uniquely identified message. There could be only one value v_i uniquely identified in both sets since $M_i^{g_i} \subseteq M_i^{2n-g_i}$. Then P_i outputs v_i with the grade g_i. Clearly, if the sender is correct, then each correct recipient outputs $g_i = 1$. Otherwise, since for every pair P_i, P_j of correct recipients it holds that $M_i^{g_i} \subseteq M_j^{g_i+1}$ and $M_j^{2n-(g_i+1)} \subseteq M_i^{2n-g_i}$ we have $g_j \le g_i + 1$.

Let us detail the step when sender distributes his hint k. While it can be done directly with the help of BBB (which would lead to a less efficient construction), we let the parties to invoke gradecast recursively for the distribution of k. Once each player P_i outputs a key k_i with a grade g_i' he uses k_i as a hint. Then the final grade is computed by P_i as the maximum of two grades g_i and g_i', i.e., it is computed as the "weakest" grade among the two.

Protocol GradedBC($P_1, \mathcal{D}, v$)

1. If $|\mathcal{D}| \le |\mathcal{K}_{n^{2n}}^{\mathcal{D}}|$ then P_1 broadcasts v using BBB, and every $P_i \in \mathcal{R}$ outputs $(v, 1)$.
2. Otherwise:
   2.1 Sender P_1: Set $M_1^0 := \{v\}$. $\forall P_i \in \mathcal{R}$: Set $M_i^0 := \emptyset$.
   2.2 For $r = 0, \ldots, 2n - 1$:
   $\forall P_i \in \mathcal{P}$: Send M_i^r (of size at most n^r) to all $P_j \in \mathcal{P}$, P_j denotes the union of the received sets with M_j^{r+1}, i.e., $M_j^{r+1} = \bigcup_i M_i^r$.
   2.3 Sender P_1: Choose a key k for the n^{2n}-identifying predicate Q with domain $\mathcal{D}$, the set of values M_1^{2n} and the value v.
   2.4 Players $\mathcal{P}$ invoke GradedBC($P_1, \mathcal{K}_{n^{2n}}^{\mathcal{D}}, k$) recursively. Let (k_i, g_i') denote the output of $P_i \in \mathcal{R}$.
   2.5 $\forall P_i \in \mathcal{R}$: Let g be the smallest number in $[n]$ such that there exists u which is uniquely identified by Q_{k_i} in M_i^g and in M_i^{2n-g}. Output $(v_i, g_i) = (u, \max(g, g_i'))$. If such g does not exist output $(v_i, g_i) = (\perp, n)$;

Lemma 6. *The protocol GradedBC achieves graded broadcast while requiring only the sender to use BBB to broadcast one value of at most $\lceil 7n \log n \rceil$ bits.*

Proof. We prove by induction that graded broadcast is achieved. For $|\mathcal{D}| \le |\mathcal{K}_{n^{2n}}^{\mathcal{D}}|$, graded broadcast is achieved by assumption of BBB. For $|\mathcal{D}| > |\mathcal{K}_{n^{2n}}^{\mathcal{D}}|$:

VALIDITY: If the sender is correct then he selects a key k for the n^{2n}-identifying predicate Q_k such that only his value v is identified by Q_k in M_1^{2n}. All correct players get $(k, 1)$ as output from the recursive call to GradedBC (due to the Validity property of the recursive GradedBC). Since for every correct player P_i

it holds that $M_i^1 \subseteq M_i^{2n-1} \subseteq M_1^{2n}$ and $v \in M_i^1$ this implies that v is uniquely identified in M_i^1 and in M_i^{2n-1}. Hence P_i computes $v_i = v$ and $g_i = 1$.

GRADED CONSISTENCY: Let P_i denote a correct recipient outputting the smallest grade g_i. If $g_i = n$ then Graded Consistency holds trivially. Now assume that $g_i < n$, and hence $g_i' < n$. Consider any other correct recipient P_j. Due to the Graded Consistency property of the recursive GradedBC, the fact that $g_i' < n$ implies that P_i and P_j have the same keys k_i and k_j which we denote with k. Observe that $M_i^{g_i} \subseteq M_j^{g_i+1} \subseteq M_j^{2n-(g_i+1)} \subseteq M_i^{2n-g_i}$. The value v_i is uniquely identified by Q_k in both $M_i^{g_i}$ and $M_i^{2n-g_i}$, hence v_i is uniquely identified in both $M_j^{g_i+1}$ and $M_j^{2n-(g_i+1)}$. Hence the grade $g_j \in \{g_i, g_i + 1\}$. If $g_j = g_i + 1$ then $v_j = v_i$. If $g_j = g_i$ then, since $M_j^{g_i} \subseteq M_j^{g_i+1}$ and v_i is uniquely identified in $M_j^{g_i+1}$, the only value that can be uniquely identified by Q_k in $M_j^{g_i}$ is v_i. This implies that $v_j = v_i$.

TERMINATION: Follows by inspection.

It remains to prove the stated usage complexity of BBB. Note that BBB is only used at the deepest recursion level. We denote the logarithm of broadcast domain size at the rth recursive level to be ℓ_r. We have that $\ell_0 = \log|\mathcal{D}|$ and ℓ_{i+1} is defined recursively to be $2\lceil \log(n^{2n}\ell_i) \rceil$. It can be verified that $\ell_{i+1} < \ell_i$ for any $\ell_i > 7n \log n$. Hence, the sender P_1 broadcasts with BBB at most $\lceil 7n \log n \rceil$ bits. □

Main Protocol. The broadcast-amplification protocol first invokes graded broadcast. Then, each party broadcasts his grade (using BBB), and decides depending on the grades broadcast whether to use the output of graded broadcast or to use some default value (say $\perp$) as output.

The core idea of the protocol lies in the analysis of the grades broadcast. Denote the set of all grades by $G = \{g_i\}_i$. As $|\mathcal{R}| = n - 1$, there exists a grade $g \in [n]$ with $g \notin G$. Consider the smallest grade g_i of an honest party P_i. If $g_i > g$, then clearly the grade g_j of each honest party P_j is $g_j > g$. On the other hand, if $g_i < g$, then by the definition of graded broadcast, the grade g_j of any honest party P_j is $g_j \leq g_i + 1$, hence $g_j < g$. In other words, either the grades of all honest parties are below g, or the grades of all honest parties are above g. In the former case, every honest party P_i has $g_i < n$ and hence all values v_i are equal (and are a valid output of broadcast). In the latter case, no honest party P_i has grade $g_i = 1$, hence the recipients can output some default value $\perp$.

Protocol $\text{AmplifyBC}_n(P_1, \mathcal{D}, v)$

1. Players $\mathcal{P}$ invoke $\text{GradedBC}(P_1, \mathcal{D}, v)$, let (g_i, v_i) denote the output of P_i.
2. $\forall P_i \in \mathcal{R}$: Broadcast g_i using BBB. Let G denote the set of all g_i broadcast.
3. $\forall P_i \in \mathcal{R}$: Let $g = \min([n] \setminus G)$. If $g_i < g$, then decide on v_i, otherwise decide on $\perp$.

Lemma 7. *The protocol $\mathtt{AmplifyBC}_n$ achieves broadcast and requires the sender to broadcast with BBB one value of at most $\lceil 7n \log n \rceil$ bits and each of the recipients to broadcast one value from domain $[n]$. In total at most $8n \log n$ bits need to be broadcast via BBB.*

Proof. We show that each of the broadcast properties are satisfied:

VALIDITY: If the sender is correct then all correct parties get $(v, 1)$ as an output from GradedBC and decide on v.

CONSISTENCY: Let $g = \min([n] \setminus G)$, and let P_i denote a correct recipient outputting the smallest grade g_i. If $g_i < g$, then clearly $g_i < n$, and all honest parties P_j hold the same value $v_j = v_i$ and grade $g_j \leq g_i + 1$. As $g_j \neq g$, it follows $g_j < g$. Hence, every honest party P_j outputs $v_j = v_i$. On the other hand, if $g_i > g$, then every honest party P_j holds $g_j > g$ and outputs $\bot$.

TERMINATION: Follows by inspection.

It remains to prove the stated usage complexity of BBB. The protocol $\mathtt{AmplifyBC}_n$ requires the sender to broadcast one value of at most $\lceil 7n \log n \rceil$ bits during the GradedBC invocation (cf. Lemma 6). Furthermore, each recipient broadcasts the grade (of domain $[n]$) using BBB. This sums up to $8n \log n$ bits overall. □

Efficient Protocol. The main disadvantage of the protocol $\mathtt{AmplifyBC}_n$ is that the underlying gradecast protocol GradedBC requires exponential message communication. Here we briefly sketch how one can achieve polynomial communication complexity in GradedBC at the cost of higher BBB usage. The main idea of the optimized protocol GradedBC$^+$ is to allow recipients to use BBB such that they can filter out messages from the sets M_i^r. Roughly speaking, if a recipient holds a set of messages M_i^r then he broadcasts a "challenge" forcing the sender in his response to invalidate at least all but one values in M_i^r. After each of the recipients has his set M_i^r filtered, recipients continue exchanging sets consisting of at most one element. The detailed description of this protocol and its analysis is given in Appendix A.

Lemma 8. *The protocol $\mathtt{AmplifyBC}_n$ with the underlying gradecast implementation by GradedBC$^+$ allows to broadcast an ℓ-bit message while broadcasting $\mathcal{O}(n^2 \log \ell)$ bits with BBB and communicating $\mathcal{O}(n^3 \ell)$ bits over point-to-point channels.*

4.2 Lower Bounds in the Multi-sender Model

Based on the approach presented in Section 3.2 we investigate the lower bounds on the broadcast-amplification protocols in the multi-sender model. As it was shown for the single-sender model there is no broadcast-amplification possible when only the sender uses BBB for $n \geq 4$. We extend this result by showing that the sender and at least all but 2 recipients are required to broadcast some information via BBB to achieve non-trivial broadcast-amplification.

Fig. 4. The configuration for $n = 3$ to show that the sender must use BBB to broadcast at least one bit

Lemma 9. *Every perfectly-secure broadcast-amplification protocol for domain $\mathcal{D}$ requires the sender P_1 to broadcast at least 1 bit via BBB.*

Proof. We first prove the theorem for $n = 3$, then reduce the case of arbitrary $n > 3$ to $n = 3$.

(Case n = 3) Assume towards a contradiction that there is a protocol (π_1, π_2, π_3) allowing the parties P_1, P_2, P_3 to do broadcast amplification, where the sender does not broadcast with BBB. We consider the following configuration: Let π_1^u and π_1^v denote two instances of the program π_1, where π_1^u is given input u and π_1^v is given input v for $u, v \in \mathcal{D}$ and $u \neq v$. We connect programs π_1^u, π_2, π_3 and π_1^v with bilateral channels as shown in Figure 4. Now we execute the programs. Whenever π_2 or π_3 use BBB to broadcast some x, the value x is given to all programs.

The configuration can be interpreted in three different ways, which lead to contradicting requirements on the outputs of the programs. (i) P_1 holds input u and executes π_1^u, P_2 executes π_2, and P_3 is corrupted and executes π_3 and π_1^v. Due to the validity property, π_2 must output u. (ii) P_1 holds input v and executes π_1^v, P_3 executes π_3, and P_2 is corrupted and executes π_2 and π_1^u. Due to the validity property, π_3 must output v. (iii) P_2 executes π_2, P_3 executes π_3, and P_1 is corrupted and executes π_1^u and π_1^v. Due to the consistency property, π_2 and π_3 must output the same value. These three requirements cannot be satisfied simultaneously, hence whatever output the programs make, the protocol (π_1, π_2, π_3) is not a perfectly-secure broadcast-amplification protocol.

(Case n > 3) Assume towards a contradiction that there is a protocol $(\pi_1, \pi_2, \pi_3, \dots, \pi_n)$ allowing to do broadcast amplification where the sender does not broadcast with BBB. One particular strategy of the adversary is to corrupt parties $P_4, \dots, P_n$ and make them not execute their corresponding programs $\pi_4, \dots, \pi_n$. Still, the remaining protocol (π_1, π_2, π_3) must achieve broadcast, which contradicts the first case.   □

Lemma 10. *Every perfectly-secure non-trivial broadcast-amplification protocol requires that at least all but 2 of the recipients broadcast at least 1 bit with BBB.*

Proof. Assume towards a contradiction that there is a protocol $(\pi_1, \pi_2, \pi_3, \dots, \pi_n)$ allowing to do non-trivial broadcast amplification with three recipients' programs not broadcasting with BBB. Without loss of generality, assume that these programs are π_2, π_3, π_4.[10] One particular strategy of the adversary is to corrupt parties $P_5, \ldots, P_n$ and make them not execute their corresponding programs $\pi_5, \ldots, \pi_n$. The programs $\pi_1, \pi_2, \pi_3, \pi_4$ of the remaining honest players can then put the values sent and broadcast by the corrupted parties to some default value (say $\perp$). The remaining protocol $(\pi_1, \pi_2, \pi_3, \pi_4)$ achieves non-trivial broadcast amplification, which contradicts Lemma 3. □

4.3   Summary

The following theorem summarizes results obtained in this section (The proof of this theorem follows from Lemmas 7, 9 and 10.)

Theorem 2. *For all n, d we have $8n \log n \geq \log \phi_n^*(d) \geq \min(\log d, n - 2)$.*[11]

Additionally, we give an efficient protocol that allows to broadcast an ℓ-bit value while broadcasting $\mathcal{O}(n^2 \log \ell)$ bits with BBB and communicating $\mathcal{O}(n^3 \ell)$ bits over point-to-point channels.

5   Conclusions

Broadcast amplification is the task of achieving d-broadcast given point-to-point channels and access to a d'-broadcast primitive, for $d' < d$. The existence of such a broadcast-amplification protocol means in a certain sense that d-broadcast and d'-broadcast are equivalent (respectively that d'-broadcast is "as good as" d-broadcast).

It is well known that perfectly-secure broadcast cannot be constructed from point-to-point channels when the number of cheaters is not limited. In this paper, we have shown that:

- For three parties, 3-broadcast and d-broadcast are equivalent for any $d \geq 3$. However, 2-broadcast and 3-broadcast are not equivalent.
- For an arbitrary number of parties, $(8n \log n)$-bit broadcast and ℓ-bit broadcast are equivalent for any $\ell \geq 8n \log n$. However, for $n \geq 4$ parties, $(n-3)$-bit broadcast and ℓ-bit broadcast are not equivalent for large enough ℓ.

In summary, for three parties, we have given a complete picture of equivalence of broadcast primitives for different domains, under the assumption that point-to-point channels are freely available. For $n \geq 4$ parties, we have proved a lower bound and an upper bound on the broadcast primitive necessary for broadcasting arbitrary messages, namely $\Omega(n)$ and $\mathcal{O}(n \log n)$ bits, respectively.

---

[10] Such not broadcasting programs are fixed because we considered protocols with static BBB usage pattern.

[11] The last inequality combines the facts that any non-trivial broadcast amplification protocol broadcasts at least $n - 2$ bits, whereas the trivial protocol always uses $\log d$ bits.

References

[Bea96]   Beaver, D.: Correlated pseudorandomness and the complexity of private computations. In: Proceedings of the Twenty-Eighth Annual ACM Symposium on Theory of Computing (STOC 1996), pp. 479–488. ACM (1996)

[BGP92]   Berman, P., Garay, J.A., Perry, K.J.: Bit optimal distributed consensus. In: Computer Science Research, pp. 313–322. Plenum Publishing Corporation, New York (1992)

[CFF+05]   Considine, J., Fitzi, M., Franklin, M., Levin, L.A., Maurer, U., Metcalf, D.: Byzantine agreement given partial broadcast. Journal of Cryptology 18(3), 191–217 (2005)

[CW92]   Coan, B.A., Welch, J.L.: Modular construction of a byzantine agreement protocol with optimal message bit complexity. Information and Computation 97, 61–85 (1992)

[DS83]   Dolev, D., Strong, H.R.: Authenticated algorithms for Byzantine agreement. SIAM Journal on Computing 12(4), 656–666 (1983)

[FH06]   Fitzi, M., Hirt, M.: Optimally efficient multi-valued Byzantine agreement. In: Proceedings of the 26th Annual ACM Symposium on Principles of Distributed Computing, PODC, 2006, pp. 163–168. ACM, New York (2006)

[FM88]   Feldman, P., Micali, S.: Optimal algorithms for byzantine agreement. In: Proceedings of the Twentieth Annual ACM Symposium on Theory of Computing, STOC 1988, pp. 148–161. ACM, New York (1988)

[FM00]   Fitzi, M., Maurer, U.: From partial consistency to global broadcast. In: Yao, F. (ed.) Proc. 32nd ACM Symposium on Theory of Computing — STOC 2000, pp. 494–503. ACM (May 2000)

[HMQU06]   Hofheinz, D., Müller-Quade, J., Unruh, D.: On the (Im-)Possibility of extending coin toss. In: Vaudenay, S. (ed.) EUROCRYPT 2006. LNCS, vol. 4004, pp. 504–521. Springer, Heidelberg (2006)

[HR13]   Hirt, M., Raykov, P.: On the complexity of broadcast setup. In: Fomin, F.V., Freivalds, R., Kwiatkowska, M., Peleg, D. (eds.) ICALP 2013, Part I. LNCS, vol. 7965, pp. 552–563. Springer, Heidelberg (2013)

[IKNP03]   Ishai, Y., Kilian, J., Nissim, K., Petrank, E.: Extending oblivious transfers efficiently. In: Boneh, D. (ed.) CRYPTO 2003. LNCS, vol. 2729, pp. 145–161. Springer, Heidelberg (2003)

[Lam83]   Lamport, L.: The weak byzantine generals problem. J. ACM 30(3), 668–676 (1983)

[LV11]   Liang, G., Vaidya, N.: Error-free multi-valued consensus with Byzantine failures. In: Proceedings of the 30th Annual ACM Symposium on Principles of Distributed Computing, PODC 2011, pp. 11–20. ACM, New York (2011)

[Mau04]   Maurer, U.: Towards a theory of consistency primitives. In: Guerraoui, R. (ed.) DISC 2004. LNCS, vol. 3274, pp. 379–389. Springer, Heidelberg (2004)

[Pat11]   Patra, A.: Error-free multi-valued broadcast and Byzantine agreement with optimal communication complexity. In: Fernàndez Anta, A., Lipari, G., Roy, M. (eds.) OPODIS 2011. LNCS, vol. 7109, pp. 34–49. Springer, Heidelberg (2011)

[PSL80]   Pease, M.C., Shostak, R.E., Lamport, L.: Reaching agreement in the presence of faults. Journal of the ACM 27(2), 228–234 (1980)

[PW96]   Pfitzmann, B., Waidner, M.: Information-theoretic pseudosignatures and Byzantine agreement for t ≥ n/3. Technical report, IBM Research (1996)

[TC84]   Turpin, R., Coan, B.A.: Extending binary Byzantine agreement to multi-valued Byzantine agreement. Information Processing Letters 18(2), 73–76 (1984)

A   A Broadcast-Amplification Protocol with Polynomial Number of Messages

In this section we present an optimized implementation $\texttt{GradedBC}^+$ of the graded broadcast protocol $\texttt{GradedBC}$. We first introduce the notion of resolution functions (which is closely related to identifying predicates) and give an efficient construction of them. Then we describe the optimized protocol $\texttt{GradedBC}^+$ that communicates polynomially many messages over point-to-point channels. After substituting $\texttt{GradedBC}$ with $\texttt{GradedBC}^+$ in $\texttt{AmplifyBC}_n$ we get a broadcast-amplification protocol with the following properties.

Lemma 8. *The protocol $\texttt{AmplifyBC}_n$ with the underlying gradecast implementation by $\texttt{GradedBC}^+$ allows to broadcast an ℓ-bit message while broadcasting $\mathcal{O}(n^2 \log \ell)$ bits with BBB and communicating $\mathcal{O}(n^3\ell)$ bits over point-to-point channels.*

A.1   Resolution Functions

An identifying predicate allows to identify a specific element v from some set S of potentially a large domain $\mathcal{D}$. Resolution functions extend this notion by providing a collision-free way of choosing one of the values in S while explicitly not choosing the others.

Definition 7. *A c-resolution function for domain $\mathcal{D}$ and range $\mathcal{Y}$ is a family of functions $F_{k \in \mathcal{K}} : \mathcal{D} \rightarrow \mathcal{Y}$ such that for any $S \subseteq \mathcal{D}$ with $|S| \leq c$ there exists a key $k \in \mathcal{K}$ with $F_k(v) \neq F_k(v')$ for any $v \neq v'$ from S. Such a key k is said to resolve the set S.*

We say that v is *identified* by a pair (k, y) in S if $F_k(v) = y$ (trivially, only one value can be identified if k resolves the set S). The goal of constructing such a function F is to have $|\mathcal{K}|$ and $|\mathcal{Y}|$ as small as possible given c and $|\mathcal{D}|$. We give a construction of a c-resolution function with domain $\mathcal{D}$ below.

Polynomial-Based Resolution Function Construction. This construction is very similar to the polynomial-based construction for identifying predicates presented before. Let $\ell = \log |\mathcal{D}|$. Consider any set $S \subseteq \mathcal{D}$ with $|S| \leq c$. For $\kappa \in \mathbb{N}$, let any value $v \in \mathcal{D}$ be interpreted as a polynomial f_v over $GF(2^\kappa)$ of degree at most $\lfloor \ell/\kappa \rfloor$. We find a point $x \in GF(2^\kappa)$ such that $f_v(x) \neq f_{v'}(x)$ for any two $v \neq v' \in S$. For such a point x to always exist we need that the total number of points in the field is larger than the number of points in which any

f_v may coincide with other polynomials $f_{v'}$, i.e., $2^\kappa > \frac{c(c-1)}{2}\lfloor \ell/\kappa \rfloor$. To satisfy this condition, it is enough to choose $\kappa := \lceil \log(c^2\ell) \rceil$. The resolution function is defined as $F_x(v) = f_v(x)$ with the key space and range space being $\mathrm{GF}(2^\kappa)$. So, the key and the value of the resolution function can be encoded using $\lceil \log(c^2\ell) \rceil$ bits.

Lemma 11. *The polynomial-based construction gives a c-resolution function F for domain $\mathcal{D}$ with key and range spaces $\mathcal{K}_c^\mathcal{D} = \mathcal{Y}_c^\mathcal{D} = \{0,1\}^{\lceil \log(c^2 \log |\mathcal{D}|) \rceil}$.*

A.2   The Optimized Graded Broadcast Protocol

The protocol proceeds as follows: The sender sends the value v he wants to broadcast among all recipients $\mathcal{R}$, who then exchange the received value(s) during $n-1$ rounds. In each round r each party P_i sends the value it currently holds (denoted with v_i^r) to every other party and forms the set of at most n received values. Then each party broadcasts a key k_i for an n-resolution function which resolves the set of the values received. In the end of the round the sender broadcasts the values $y_1, \ldots, y_n$ of the resolution function for the keys $k_1, \ldots, k_n$ so that each of the recipients keeps at most one value identified by all (k_i, y_i). Finally, each recipient P_i decides on the grade g_i to be the first "stable" round starting from which the value he holds remain unchanged, i.e., $v_i^{g_i} = v_i^{g_i+1} = \cdots = v_i^n$.

Protocol $\mathtt{GradedBC^+}(P_1, \mathcal{D}, v)$:

1. Sender P_1: Send v to every $P_i \in \mathcal{R}$.
   $\forall P_i \in \mathcal{R}$: Denote the message received from the sender by v_i^1.
r. In each step $r = 2, \ldots, n$, execute the following sub-steps:
   $r.1$ $\forall P_i \in \mathcal{R}$: Send the value v_i^{r-1} to all $P_j \in \mathcal{R}$, P_j denotes the set of the received values with M_j^r.
   $r.2$ $\forall P_i \in \mathcal{R}$: Choose a key k_i^r for an n-resolution function F with domain $\mathcal{D}$, that resolves the set of values S_i^r. Broadcast the key k_i^r using the BBB.
   $r.3$ Sender P_1: Broadcast a list of values $(F_{k_2^r}(v), \ldots, F_{k_n^r}(v))$ using the BBB. Denote the list broadcast with $(y_2^r, \ldots, y_n^r)$.
   $r.4$ $\forall P_i \in \mathcal{R}$: Select v_i^r to be some $u \in M_i^r$ such that u is identified by (k_j^r, y_j^r) for all j; set v_i^r to $\perp$ if no such u exists.
$n{+}1.\forall P_i \in \mathcal{R}$: Compute g_i to be the smallest step r such that $v_i^r = v_i^{r+1} = \cdots = v_i^n$. Output (v_i^n, g_i).

Lemma 12. *The protocol $\mathtt{GradedBC^+}$ achieves graded broadcast while requiring $\mathcal{O}(n^2 \log \ell)$ bits to be broadcast with BBB and communicating $\mathcal{O}(n^3 \ell)$ bits over point-to-point channels.*

Proof. We show that each of the graded broadcast properties is satisfied:

VALIDITY: If the sender is correct then for any key k he broadcasts $y = F_k(v)$ such that only his value v is chosen by correct recipients at every iteration. Hence each correct P_i computes $v_i = v$ and $g_i = 1$.

GRADED CONSISTENCY: Let P_i denote a correct recipient outputting the smallest grade g_i. If $g_i = n$ then Graded Consistency holds trivially. Now assume that $g_i < n$. Consider any other correct recipient P_j. Observe that $v_i^{g_i} \in M_j^{g_i+1}$. Since P_i kept the value $v_i^{g_i}$ till the round n it implies that $v_i^{g_i}$ is identified in S_i^r by all pairs (k_a^r, y_a^r) for all a and $r \geq g_i$. Hence, $v_i^{g_i}$ is identified in $S_j^{g_i+1}, S_j^{g_i+2}, \ldots, S_j^n$ by all pairs (k_a^r, y_a^r) for all a and $r \geq g_i+1$. Moreover, since P_j chose the keys k_j^r for a resolution function faithfully, only a unique value can be identified in $S_j^{g_i+1}, S_j^{g_i+2}, \ldots, S_j^n$. Since only a unique value can be identified then P_j outputs $v_j = v_i^{g_i}$ with the grade $g_j \leq g_i + 1$.

TERMINATION: Follows by inspection.

It remains to prove the stated usage complexity of BBB. At each step $r = 2, \ldots, n$ of the protocol $\mathtt{GradedBC^+}$, every recipient P_i broadcasts a key k_i^r for a family of n-resolution functions and the sender broadcasts a list of n values of the function. If the polynomial-based construction of the resolution function is used then each key and a value of function consists of $\lceil \log(n^2\ell) \rceil$ bits. Since in total the protocol works in $n - 1$ rounds we broadcast $2(n - 1)^2 \lceil \log(n^2\ell) \rceil$ bits with BBB. This expression can be rewritten as $\mathcal{O}(n^2(\log n + \log \ell))$. We can assume that $\ell > n$, since for $\ell \leq n$ it is easier to run the trivial algorithm that broadcasts a message bit by bit. Summing up the analysis above, we have that the total number of the BBB invocations during the protocol run is $\mathcal{O}(n^2 \log \ell)$.

The communication costs of $\mathtt{GradedBC^+}$ over authenticated channels consist of distributing $n - 1$ ℓ-bit messages during Step 1 and exchanging of $(n-1)^2(n-1)$ ℓ-bit messages during Steps $2, \ldots, n$. Hence, the total number of bits that need to be communicated is $\mathcal{O}(n^3\ell)$.   □

Non-malleable Coding against Bit-Wise and Split-State Tampering*

Mahdi Cheraghchi[1,**] and Venkatesan Guruswami[2,***]

[1] CSAIL, Massachusetts Institute of Technology
mahdi@csail.mit.edu
[2] Computer Science Department, Carnegie Mellon University
guruswami@cmu.edu

Abstract. Non-malleable coding, introduced by Dziembowski, Pietrzak and Wichs (ICS 2010), aims for protecting the integrity of information against tampering attacks in situations where error-detection is impossible. Intuitively, information encoded by a non-malleable code either decodes to the original message or, in presence of any tampering, to an unrelated message. Non-malleable coding is possible against any class of adversaries of bounded size. In particular, Dziembowski et al. show that such codes exist and may achieve positive rates for any class of tampering functions of size at most $2^{2^{\alpha n}}$, for any constant $\alpha \in [0, 1)$. However, this result is existential and has thus attracted a great deal of subsequent research on explicit constructions of non-malleable codes against natural classes of adversaries.

In this work, we consider constructions of coding schemes against two well-studied classes of tampering functions; namely, bit-wise tampering functions (where the adversary tampers each bit of the encoding independently) and the much more general class of split-state adversaries (where two independent adversaries arbitrarily tamper each half of the encoded sequence). We obtain the following results for these models.

1. For bit-tampering adversaries, we obtain explicit and efficiently encodable and decodable non-malleable codes of length n achieving rate $1 - o(1)$ and error (also known as "exact security") $\exp(-\tilde{\Omega}(n^{1/7}))$. Alternatively, it is possible to improve the error to $\exp(-\tilde{\Omega}(n))$ at the cost of making the construction Monte Carlo with success probability $1 - \exp(-\Omega(n))$ (while still allowing a compact description of the code). Previously, the best known construction of bit-tampering coding schemes was due to Dziembowski et al. (ICS 2010), which is a Monte Carlo construction achieving rate close to .1887.

---

* A draft of the full version of this paper appears in [6].
** Research supported in part by the Swiss National Science Foundation research grant PA00P2-141980, V. Guruswami's Packard Fellowship, and MSR-CMU Center for Computational Thinking.
*** Research supported in part by the National Science Foundation under Grant No. CCF-0963975. Any opinions, findings, and conclusions or recommendations expressed in this material are those of the author(s) and do not necessarily reflect the views of the National Science Foundation.

2. We initiate the study of *seedless non-malleable extractors* as a natural variation of the notion of non-malleable extractors introduced by Dodis and Wichs (STOC 2009). We show that construction of non-malleable codes for the split-state model reduces to construction of non-malleable two-source extractors. We prove a general result on existence of seedless non-malleable extractors, which implies that codes obtained from our reduction can achieve rates arbitrarily close to $1/5$ and exponentially small error. In a separate recent work, the authors show that the optimal rate in this model is $1/2$. Currently, the best known explicit construction of split-state coding schemes is due to Aggarwal, Dodis and Lovett (ECCC TR13-081) which only achieves vanishing (polynomially small) rate.

Keywords: coding theory, cryptography, error detection, information theory, randomness extractors, tamper-resilient storage.

1   Introduction

Non-malleable codes were introduced by Dziembowski, Pietrzak, and Wichs [12] as a relaxation of the classical notions of error-detection and error-correction. Informally, a code is non-malleable if the decoding a corrupted codeword either recovers the original message, or a completely unrelated message. Non-malleable coding is a natural concept that addresses the basic question of storing messages securely on devices that may be subject to tampering, and they provide an elegant solution to the problem of protecting the integrity of data and the functionalities implemented on them against "tampering attacks" [12]. This is part of a general recent trend in theoretical cryptography to design cryptographic schemes that guarantee security even if implemented on devices that may be subject to physical tampering. The notion of non-malleable coding is inspired by the influential theme of non-malleable encryption in cryptography which guarantees the intractability of tampering the ciphertext of a message into the ciphertext encoding a related message.

The definition of non-malleable codes captures the requirement that if some adversary (with full knowledge of the code) tampers the codeword $\mathsf{Enc}(s)$ encoding a message s, corrupting it to $f(\mathsf{Enc}(s))$, he cannot control the relationship between s and the message the corrupted codeword $f(\mathsf{Enc}(s))$ encodes. For this definition to be feasible, we have to restrict the allowed tampering functions f (otherwise, the tampering function can decode the codeword to compute the original message s, flip the last bit of s to obtain a related message $\tilde{s}$, and then re-encode $\tilde{s}$), and in most interesting cases also allow the encoding to be randomized. Formally, a (binary) non-malleable code against a family of tampering functions $\mathcal{F}$ each mapping $\{0,1\}^n$ to $\{0,1\}^n$, consists of a randomized encoding function $\mathsf{Enc} : \{0,1\}^k \to \{0,1\}^n$ and a deterministic decoding function $\mathsf{Dec} : \{0,1\}^n \to \{0,1\}^k \cup \{\bot\}$ (where $\bot$ denotes error-detection) which satisfy $\mathsf{Dec}(\mathsf{Enc}(s)) = s$ always, and the following non-malleability property with error ϵ: For every message $s \in \{0,1\}^k$ and every function $f \in \mathcal{F}$, the distribution of $\mathsf{Dec}(f(\mathsf{Enc}(s))$ is ϵ-close to a distribution $\mathcal{D}_f$ that depends only on f and is independent of s (ignoring the issue that f may have too many fixed points).

If some code enables error-detection against some family $\mathcal{F}$, for example if $\mathcal{F}$ is the family of functions that flips between 1 and t bits and the code has minimum distance more than t, then the code is also non-malleable (by taking $\mathcal{D}_f$ to be supported entirely on $\perp$ for all f). Error-detection is also possible against the family of "additive errors," namely $\mathcal{F}_{\mathsf{add}} = \{f_\Delta \mid \Delta \in \{0,1\}^n\}$ where $f_\Delta(x) := x + \Delta$ (the addition being bit-wise XOR). Cramer et al. [8] constructed "Algebraic Manipulation Detection" (AMD) codes of rate approaching 1 such that offset by an arbitrary $\Delta \neq 0$ will be detected with high probability, thus giving a construction of non-malleable codes against $\mathcal{F}_{\mathsf{add}}$.

The notion of non-malleable coding becomes more interesting for families against which error-detection is not possible. A simple example of such a class consists of all constant functions $f_c(x) := c$ for $c \in \{0,1\}^n$. Since the adversary can map all inputs to a valid codeword c^*, one cannot in general detect tampering in this situation. However, non-malleability is trivial to achieve in this case as the output distribution of a constant function is trivially independent of the message (so the rate 1 code with identity encoding function is itself non-malleable).

The original work [12] showed that non-malleable codes of positive rate exist against *every* not-too-large family $\mathcal{F}$ of tampering functions, specifically with $|\mathcal{F}| \leqslant 2^{2^{\alpha n}}$ for some constant $\alpha < 1$. In a companion paper [5], we proved that in fact one can achieve a rate approaching $1 - \alpha$ against such families, and this is best possible in that there are families of size $\approx 2^{2^{\alpha n}}$ for which non-malleable coding is not possible with rate exceeding $1 - \alpha$. (The latter is true both for random families as well as natural families such as functions that only tamper the first αn bits of the codeword.)

1.1   Our Results

This work is focused on two natural families of tampering functions that have been studied in the literature.

Bit-Tampering Functions. The first class consists of *bit-tampering functions* f in which the different bits of the codewords are tampered independently (i.e., each bit is either flipped, set to $0/1$, or left unchanged, independent of other bits); formally $f(x) = (f_1(x_1), f_2(x_2), \ldots, f_n(x_n))$, where $f_1, \ldots, f_n \colon \{0,1\} \to \{0,1\}$. As this family is "small" (of size 4^n), by the above general results, it admits non-malleable codes with positive rate, in fact rate approaching 1 by our recent result [5].

Dziembowski et al. [12] gave a Monte Carlo construction of a non-malleable code against this family; i.e., they gave an efficient randomized algorithm to produce the code along with efficient encoding and decoding functions such that w.h.p the encoder/decoder pair ensures non-malleability against all bit-tampering functions. The rate of their construction is, however, close to .1887 and thus falls short of the "capacity" (best possible rate) for this family of tampering functions, which we now know equals 1.

Our main result in this work is the following:

Theorem 1. *For all integers $n \geqslant 1$, there is an explicit (deterministic) construction, with efficient encoding/decoding procedures, of a non-malleable code against bit-tampering functions that achieves rate $1 - o(1)$ and error at most $\exp(-n^{\Omega(1)})$.*

If we seek error that is $\exp(-\tilde{\Omega}(n))$, we can guarantee that with an efficient Monte Carlo construction of the code that succeeds with probability $1 - \exp(-\Omega(n))$.

The basic idea in the above construction (described in detail in Section 4.1) is to use a concatenation scheme with an outer code of rate close to 1 that has large relative distance and large dual relative distance, and as (constant-sized) inner codes the non-malleable codes guaranteed by the existential result (which may be deterministically found by brute-force if desired). This is inspired by the classical constructions of concatenated codes [13,16]. The outer code provides resilience against tampering functions that globally fix too many bits or alter too few. For other tampering functions, in order to prevent the tampering function from locally freezing many entire inner blocks (to possibly wrong inner codewords), the symbols of the concatenated codeword are permuted by a *pseudorandom permutation*[1]. The seed for the permutation is itself included as the initial portion of the final codeword, after encoding by a non-malleable code (of possibly low rate). This protects the seed and ensures that any tampering of the seed portion results in the decoded permutation being essentially independent of the actual permutation, which then results in many inner blocks being error-detected (decoded to $\perp$) with noticeable probability each. The final decoder outputs $\perp$ if any inner block is decoded to $\perp$, an event which happens with essentially exponentially small probability in n with a careful choice of the parameters. Though the above scheme uses non-malleable codes in two places to construct the final non-malleable code, there is no circularity as the codes for the inner blocks are of constant size, and the code protecting the seed can have very low rate (even sub-constant) as the seed can be made much smaller than the message length.

The structure of our construction bears some high level similarity to the optimal rate code construction for correcting a bounded number of additive errors in [15]. The exact details though are quite different; in particular, the crux in the analysis of [15] was ensuring that the decoder can recover the seed correctly, and towards this end the seed's encoding was distributed at random locations of the final codeword. Recovering the seed is both impossible and not needed in our context here.

Split-State Adversaries. Bit-tampering functions act on different bits independently. A much more general class of tampering functions considered in

---

[1] Throughout the paper, by pseudorandom permutation we mean t-wise independent permutation (as in Definition 8) for an appropriate choice of t. This should not be confused with cryptographic pseudorandom permutations, which are not used in this work.

the literature [12,11,1] is the so-called *split-state model*. Here the function f : $\{0,1\}^n \to \{0,1\}^n$ must act on each half of the codeword independently (assuming n is even), but can act arbitrarily within each half. Formally, $f(x) = (f_1(x_1), f_2(x_2))$ for some functions $f_1, f_2 : \{0,1\}^{n/2} \to \{0,1\}^{n/2}$ where x_1, x_2 consist of the first $n/2$ and last $n/2$ bits of x. This represents a fairly general and useful class of adversaries which are relevant for example when the codeword is stored on two physically separate devices, and while each device may be tampered arbitrarily, the attacker of each device does not have access to contents stored on the other device.

The capacity of non-malleable coding in the split-state model equals 1/2, as established in our recent work [5]. A natural question therefore is to construct *efficient* non-malleable codes of rate approaching 1/2 in the split-state model (the results in [12] and [5] are existential, and the codes do not admit polynomial size representation or polynomial time encoding/decoding). This remains a challenging open question, and in fact constructing a code of positive rate itself seems rather difficult. A code that encodes one-bit messages is already non-trivial, and such a code was constructed in [11] by making a connection to two-source extractors with sufficiently strong parameters and then instantiating the extractor with a construction based on the inner product function over a finite field. We stress that this connection to two-source extractor only applies to encoding one-bit messages, and does not appear to generalize to longer messages.

Recently, Aggarwal, Dodis, and Lovett [1] solved the central open problem left in [11] — they construct a non-malleable code in the split-state model that works for arbitrary message length, by bringing to bear elegant techniques from additive combinatorics on the problem. The rate of their code is polynomially small: k-bit messages are encoded into codewords with $n \approx k^7$ bits.

In the second part of this paper (Section 5), we study the problem of non-malleable coding in the split-state model. We do not offer any explicit constructions, and the polynomially small rate achieved in [1] remains the best known. Our contribution here is more conceptual. We define the notion of non-malleable two-source extractors, generalizing the influential concept of non-malleable extractors introduced by Dodis and Wichs [10]. A non-malleable extractor is a regular seeded extractor Ext whose output $\text{Ext}(X, S)$ on a weak-random source X and uniform random seed S remains uniform even if one knows the value $\text{Ext}(X, f(S))$ for a related seed $f(S)$ where f is a tampering function with no fixed points. In a two-source non-malleable extractor we allow both sources to be weak and independently tampered, and we further extend the definition to allow the functions to have fixed points in view of our application to non-malleable codes. We prove, however, that for construction of two-source non-malleable extractors, it suffices to only consider tampering functions that have no fixed points, at cost of a minor loss in the parameters.

We show that given a two-source non-malleable extractor NMExt with exponentially small error in the output length, one can build a non-malleable code in the split-state model by setting the extractor function NMExt to be the decoding function (the encoding of s then picks a pre-image in $\text{NMExt}^{-1}(s)$).

This identifies a possibly natural avenue to construct improved non-malleable codes against split-state adversaries by constructing non-malleable two-source extractors, which seems like an interesting goal in itself. Towards confirming that this approach has the potential to lead to good non-malleable codes, we prove a fairly general existence theorem for seedless non-malleable extractors, by essentially observing that the ideas from the proof of existence of seeded non-malleable extractors in [10] can be applied in a much more general setting. Instantiating this result with split-state tampering functions, we show the existence of non-malleable two-source extractors with parameters that are strong enough to imply non-malleable codes of rate arbitrarily close to $1/5$ in the split-state model.

Explicit construction of (ordinary) two-source extractors and closely-related objects is a well-studied problem in the literature and an abundance of explicit constructions for this problem is known[2] (see, e.g., [2,3,7,17,20,21]). The problem becomes increasingly challenging, however, (and remains open to date) when the entropy rate of the two sources may be noticeably below $1/2$. Fortunately, we show that for construction of constant-rate non-malleable codes in the split-state model, it suffices to have two-source non-malleable extractors for source entropy rate .99 and with some output length $\Omega(n)$ (against tampering functions with no fixed points). Thus the infamous "$1/2$ entropy rate barrier" on two-source extractors does not concern our particular application.

Furthermore, we note that for seeded non-malleable extractors (which is a relatively recent notion) there are already a few exciting explicit constructions [9,14,19][3]. The closest construction to our application is [9] which is in fact a two-source non-malleable extractor when the adversary may tamper with either of the two sources (but not simultaneously both). Moreover, the coding scheme defined by this extractor (which is the character-sum extractor of Chor and Goldreich [7]) naturally allows for an efficient encoder and decoder. Nevertheless, it appears challenging to extend known constructions of seeded non-malleable extractors to the case when both inputs can be tampered. We leave explicit constructions of non-malleable two-source extractors, even with sub-optimal parameters, as an interesting open problem for future work.

2 Preliminaries

2.1 Notation

We use $\mathcal{U}_n$ for the uniform distribution on $\{0,1\}^n$ and U_n for the random variable sampled from $\mathcal{U}_n$ and independently of any existing randomness. For a random variable X, we denote by $\mathscr{D}(X)$ the probability distribution that X is sampled from. Generally, we will use calligraphic symbols (such as $\mathcal{X}$) for probability distributions and the corresponding capital letters (such as X) for related random

---

[2] Several of these constructions are structured enough to easily allow for efficient sampling of a uniform pre-image from $\mathsf{Ext}^{-1}(s)$.

[3] [19] also establishes a connection between seeded non-malleable extractors and ordinary two-source extractors.

variables. We use $X \sim \mathcal{X}$ to denote that the random variable X is drawn from the distribution $\mathcal{X}$. Two distributions $\mathcal{X}$ and $\mathcal{Y}$ being ϵ-close in statistical distance is denoted by $\mathcal{X} \approx_\epsilon \mathcal{Y}$. We will use $(\mathcal{X}, \mathcal{Y})$ for the product distribution with the two coordinates independently sampled from $\mathcal{X}$ and $\mathcal{Y}$. All unsubscripted logarithms are taken to the base 2. Support of a discrete random variable X is denoted by $\mathsf{supp}(X)$. A distribution is said to be *flat* if it is uniform on its support. We use $\tilde{O}(\cdot)$ and $\tilde{\Omega}(\cdot)$ to denote asymptotic estimates that hide poly-logarithmic factors in the involved parameter.

2.2 Definitions

In this section, we review the formal definition of non-malleable codes as introduced in [12]. First, we recall the notion of *coding schemes*.

Definition 2 (Coding schemes). A pair of functions $\mathsf{Enc}\colon \{0,1\}^k \to \{0,1\}^n$ and $\mathsf{Dec}\colon \{0,1\}^n \to \{0,1\}^k \cup \{\bot\}$ where $k \leqslant n$ is said to be a coding scheme with block length n and message length k if the following conditions hold.

1. The encoder Enc is a randomized function; i.e., at each call it receives a uniformly random sequence of coin flips that the output may depend on. This random input is usually omitted from the notation and taken to be implicit. Thus for any $s \in \{0,1\}^k$, $\mathsf{Enc}(s)$ is a random variable over $\{0,1\}^n$. The decoder Dec is; however, deterministic.
2. For every $s \in \{0,1\}^k$, we have $\mathsf{Dec}(\mathsf{Enc}(s)) = s$ with probability 1.

The *rate* of the coding scheme is the ratio k/n. A coding scheme is said to have relative distance δ (or minimum distance δn), for some $\delta \in [0,1)$, if for every $s \in \{0,1\}^k$ the following holds. Let $X := \mathsf{Enc}(s)$. Then, for any $\Delta \in \{0,1\}^n$ of Hamming weight at most δn, $\mathsf{Dec}(X + \Delta) = \bot$ with probability 1.   □

Before defining non-malleable coding schemes, we find it convenient to define the following notation.

Definition 3. For a finite set Γ, the function $\mathsf{copy}\colon (\Gamma \cup \{\underline{\mathsf{same}}\}) \times \Gamma \to \Gamma$ is defined as follows:

$$\mathsf{copy}(x,y) := \begin{cases} x & x \neq \underline{\mathsf{same}}, \\ y & x = \underline{\mathsf{same}}. \end{cases}$$ □

The notion of non-malleable coding schemes from [12] can now be rephrased as follows.

Definition 4 (Non-malleability). A coding scheme $(\mathsf{Enc}, \mathsf{Dec})$ with message length k and block length n is said to be non-malleable with error ϵ (also called *exact security*) with respect to a family $\mathcal{F}$ of tampering functions acting on $\{0,1\}^n$ (i.e., each $f \in \mathcal{F}$ maps $\{0,1\}^n$ to $\{0,1\}^n$) if for every $f \in \mathcal{F}$ there is a distribution $\mathcal{D}_f$ over $\{0,1\}^k \cup \{\bot, \underline{\text{same}}\}$ such that the following holds for all $s \in \{0,1\}^k$. Define the random variable $S := \text{Dec}(f(\text{Enc}(s)))$, and let S' be independently sampled from $\mathcal{D}_f$. Then, $\mathscr{D}(S) \approx_\epsilon \mathscr{D}(\text{copy}(S', s))$. □

Dziembowski et al. [12] also consider the following stronger variation of non-malleable codes, and show that strong non-malleable codes imply regular non-malleable codes as in Definition 4.

Definition 5 (Strong non-malleability). A pair of functions as in Definition 4 is said to be a *strong* non-malleable coding scheme with error ϵ with respect to a family $\mathcal{F}$ of tampering functions acting on $\{0,1\}^n$ if the following holds. For any message $s \in \{0,1\}^k$, let $E_s := \text{Enc}(s)$, consider the random variable

$$D_s := \begin{cases} \underline{\text{same}} & \text{if } f(E_s) = E_s, \\ \text{Dec}(f(E_s)) & \text{otherwise,} \end{cases}$$

and let $\mathcal{D}_{f,s} := \mathscr{D}(D_s)$. It must be the case that for every pair of distinct messages $s_1, s_2 \in \{0,1\}^k$, $\mathcal{D}_{f,s_1} \approx_\epsilon \mathcal{D}_{f,s_2}$. □

Remark 1 (Efficiency of sampling $\mathcal{D}_f$). The original definition of non-malleable codes in [12] also requires the distribution $\mathcal{D}_f$ to be efficiently samplable given oracle access to the tampering function f. It should be noted; however, that for any non-malleable coding scheme equipped with an efficient encoder and decoder, it can be shown that the following is a valid and efficiently samplable choice for the distribution $\mathcal{D}_f$ (possibly incurring a constant factor increase in the error parameter): "Let $S \sim \mathcal{U}_k$, and $X := f(\text{Enc}(S))$. If $\text{Dec}(X) = S$, output $\underline{\text{same}}$. Otherwise, output $\text{Dec}(X)$."

Definition 6 (Sub-cube). A sub-cube over $\{0,1\}^n$ is a set $S \subseteq \{0,1\}^n$ such that for some $T = \{t_1, \ldots, t_\ell\} \subseteq [n]$ and $w = (w_1, \ldots, w_\ell) \in \{0,1\}^\ell$, $S = \{(x_1, \ldots, x_n) \in \{0,1\}^n : x_{t_1} = w_1, \ldots, x_{t_\ell} = w_\ell\}$. The ℓ coordinates in T are said to be *frozen* and the remaining $n - \ell$ are said to be *random*.

Throughout the paper, we use the following notions of limited independence.

Definition 7 (Limited independence of bit strings). A distribution $\mathcal{D}$ over $\{0,1\}^n$ is said to be ℓ-*wise* δ-*dependent* for an integer $\ell > 0$ and parameter $\delta \in [0,1)$ if the marginal distribution of $\mathcal{D}$ restricted to any subset $T \subseteq [n]$ of the coordinate positions where $|T| \leqslant \ell$ is δ-close to $\mathcal{U}_{|T|}$. When $\delta = 0$, the distribution is ℓ-wise independent.

Definition 8 (Limited independence of permutations). The distribution of a random permutation $\Pi : [n] \to [n]$ is said to be ℓ-*wise* δ-*dependent* for an integer $\ell > 0$ and parameter $\delta \in [0,1)$ if for every $T \subseteq [n]$ such that $|T| \leqslant \ell$, the marginal distribution of the sequence $(\Pi(t) : t \in T)$ is δ-close to that of $(\bar{\Pi}(t) : t \in T)$, where $\bar{\Pi} : [n] \to [n]$ is a uniformly random permutation.

We will use the following notion of *Linear Error-Correcting Secret Sharing Schemes* (LECSS) as formalized by Dziembowski et al. [12] for their construction of non-malleable coding schemes against bit-tampering adversaries.

Definition 9 (LECSS). [12] A coding scheme (Enc, Dec) of block length n and message length k is a (d, t)-*Linear Error-Correcting Secret Sharing Scheme* (LECSS), for integer parameters $d, t \in [n]$ if

1. The minimum distance of the coding scheme is at least d,
2. For every message $s \in \{0,1\}^k$, the distribution of $\mathsf{Enc}(s) \in \{0,1\}^n$ is t-wise independent (as in Definition 7).
3. For every $w, w' \in \{0,1\}^n$ such that $\mathsf{Dec}(w) \neq \perp$ and $\mathsf{Dec}(w') \neq \perp$, we have $\mathsf{Dec}(w + w') = \mathsf{Dec}(w) + \mathsf{Dec}(w')$, where we use bit-wise addition over $\mathbb{F}_2$.

3 Existence of Optimal Bit-Tampering Coding Schemes

In this section, we recall the probabilistic construction of non-malleable codes introduced in [5]. This construction, depicted as Construction 1, is defined with respect to an integer parameter $t > 0$ and a *distance parameter* $\delta \in [0, 1)$.

- *Given:* Integer parameters $0 < k \leqslant n$ and integer $t > 0$ such that $t2^k \leqslant 2^n$, and a distance parameter $\delta \geqslant 0$.
- *Output:* A pair of functions $\mathsf{Enc} \colon \{0,1\}^k \to \{0,1\}^n$ and $\mathsf{Dec} \colon \{0,1\}^n \to \{0,1\}^k$, where Enc may also use a uniformly random seed which is hidden from that notation, but Dec is deterministic.
- *Construction:*
    1. Let $\mathcal{N} := \{0,1\}^n$.
    2. For each $s \in \{0,1\}^k$, in an arbitrary order,
        - Let $E(s) := \emptyset$.
        - For $i \in \{1, \ldots, t\}$:
            (a) Pick a uniformly random vector $w \in \mathcal{N}$.
            (b) Add w to $E(s)$.
            (c) Let $\Gamma(w)$ be the Hamming ball of radius δn centered at w. Remove $\Gamma(w)$ from $\mathcal{N}$ (note that when $\delta = 0$, we have $\Gamma(w) = \{w\}$).
    3. Given $s \in \{0,1\}^k$, $\mathsf{Enc}(s)$ outputs an element of $E(s)$ uniformly at random.
    4. Given $w \in \{0,1\}^n$, $\mathsf{Dec}(s)$ outputs the unique s such that $w \in E(s)$, or $\perp$ if no such s exists.

Construction 1. Probabilistic construction of non-malleable codes in [5]

Non-malleability of the construction (for an appropriate choice of the parameters) against any bounded-size family of adversaries, and in particular bit-tampering adversaries, follows from [5]. We derive additional properties of the construction that are needed for the explicit construction of Section 4. In particular, we state the following result which is proved in the final version of the paper.

Lemma 10. *Let $\alpha > 0$ be any parameter. Then, there is an $n_0 = O(\log^2(1/\alpha)/\alpha)$ such that for any $n \geqslant n_0$, Construction 1 can be set up so that with probability $1 - 3\exp(-n)$ over the randomness of the construction, the resulting coding scheme (Enc, Dec) satisfies the following properties:*

1. *(Rate)* Rate of the code is at least $1 - \alpha$.
2. *(Non-malleability)* The code is non-malleable against bit-tampering adversaries with error $\exp(-\Omega(\alpha n))$.
3. *(Cube property)* For any sub-cube $S \subseteq \{0,1\}^n$ of size at least 2, and $U_S \in \{0,1\}^n$ taken uniformly at random from S, $\Pr_{U_S}[\mathsf{Dec}(U_S = \perp)] \geqslant 1/2$.
4. *(Bounded independence)* For any message $s \in \{0,1\}^k$, the distribution of $\mathsf{Enc}(s)$ is $\exp(-\Omega(\alpha n))$-close to an $\Omega(\alpha n)$-wise independent distribution with uniform entries.
5. *(Error detection[4])* Let $f \colon \{0,1\}^n \to \{0,1\}^n$ be any bit-tampering adversary that is neither the identity function nor a constant function. Then, for every $s \in \{0,1\}^k$, $\Pr[\mathsf{Dec}(f(\mathsf{Enc}(s))) = \perp] \geqslant 1/3$, where the probability is taken over the randomness of the encoder.

4   Explicit Construction of Optimal Bit-Tampering Coding Schemes

In this section, we describe an explicit construction of codes achieving rate close to 1 that are non-malleable against bit-tampering adversaries. Throughout this section, we use N to denote the block length of the final code.

4.1   The Construction

At a high level, we combine the following tools in our construction: 1) an inner code C_0 (with encoder Enc_0) of constant length satisfying the properties of Lemma 10; 2) an existing non-malleable code construction C_1 (with encoder Enc_1) against bit-tampering achieving a possibly low (even sub-constant) rate; 3) a linear error-correcting secret sharing scheme (LECSS) C_2 (with encoder Enc_2); 4) an explicit function Perm that, given a uniformly random seed, outputs a pseudorandom permutation (as in Definition 8) on a domain of size close to N. Figure 1 depicts how various components are put together to form the final code construction.

At the outer layer, LECSS is used to pre-code the message. The resulting string is then divided into blocks, where each block is subsequently encoded by the inner encoder Enc_0. For a "typical" adversary that flips or freezes a prescribed fraction of the bits, we expect many of the inner blocks to be sufficiently tampered so that many of the inner blocks detect an error when the corresponding inner decoder is called. However, this ideal situation cannot necessarily be achieved if the fraction of global errors is too small, or if too many bits are frozen by the adversary (in particular, the adversary may freeze all but few of the blocks to valid inner codewords). In this case, we rely on distance and bounded independence properties of LECSS to ensure that the outer decoder, given the tampered information, either detects an error or produces a distribution that is independent of the source message.

---

[4] This property is a corollary of non-malleability, cube property and bounded independence.

A problem with the above approach is that the adversary knows the location of various blocks, and may carefully design a tampering scheme that, for example, freezes a large fraction of the blocks to valid inner codewords and leaves the rest of the blocks intact. To handle adversarial strategies of this type, we permute the final codeword using the pseudorandom permutation generated by Perm, and include the seed in the final codeword. Doing this has the effect of randomizing the action of the adversary, but on the other hand creates the problem of protecting the seed against tampering. In order to solve this problem, we use the sub-optimal code $\mathcal{C}_1$ to encode the seed and prove in the analysis that non-malleability of the code $\mathcal{C}_1$ can be used to make the above intuitions work.

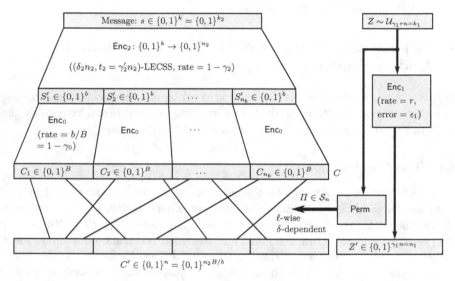

Fig. 1. Schematic description of the encoder Enc from our explicit construction

The Building Blocks. In the construction, we use the following building blocks, with some of the parameters to be determined later in the analysis.

1. An inner coding scheme $\mathcal{C}_0 = (\mathsf{Enc}_0, \mathsf{Dec}_0)$ with rate $1 - \gamma_0$ (for an arbitrarily small parameter $\gamma_0 > 0$), some block length B, and message length $b = (1 - \gamma_0)B$. We assume that $\mathcal{C}_0$ is an instantiation of Construction 1 and satisfies the properties promised by Lemma 10.
2. A coding scheme $\mathcal{C}_1 = (\mathsf{Enc}_1, \mathsf{Dec}_1)$ with rate $r > 0$ (where r can in general be sub-constant), block length $n_1 := \gamma_1 n$ (where n is defined later), and message length $k_1 := \gamma_1 r n$, that is non-malleable against bit-tampering adversaries with error ϵ_1. Without loss of generality, assume that Dec_1 never outputs $\bot$ (otherwise, identify $\bot$ with an arbitrary fixed message; e.g., 0^k). The non-malleable code $\mathcal{C}_1$ need not be strong.

3. A linear error-correcting secret sharing (LECSS) scheme $C_2 = (\mathsf{Enc}_2, \mathsf{Dec}_2)$ (as in Definition 9) with message length $k_2 := k$, rate $1 - \gamma_2$ (for an arbitrarily small parameter $\gamma_2 > 0$) and block length n_2. We assume that C_2 is a $(\delta_2 n_2, t_2 := \gamma_2' n_2)$-linear error-correcting secret sharing scheme (where $\delta_2 > 0$ and $\gamma_2' > 0$ are constants defined by the choice of γ_2). Since b is a constant, without loss of generality assume that b divides n_2, and let $n_b := n_2/b$ and $n := n_2 B/b$.

4. A polynomial-time computable mapping $\mathsf{Perm} \colon \{0,1\}^{k_1} \to S_n$, where S_n denotes the set of permutations on $[n]$. We assume that $\mathsf{Perm}(U_{k_1})$ is an ℓ-wise δ-dependent permutation (as in Definition 8, for parameters ℓ and δ. In fact, it is possible to achieve $\delta \leqslant \exp(-\ell)$ and $\ell = \lceil \gamma_1 rn/\log n \rceil$ for some constant $\gamma > 0$. Namely, we may use the following result due to Kaplan, Naor and Reingold [18]:

Theorem 11. *[18] For every integers $n, k_1 > 0$, there is an explicit function $\mathsf{Perm} \colon \{0,1\}^{k_1} \to S_n$ computable in worst-case polynomial-time (in k_1 and n) such that $\mathsf{Perm}(U_{k_1})$ is an ℓ-wise δ-dependent permutation, where $\ell = \lceil k_1/\log n \rceil$ and $\delta \leqslant \exp(-\ell)$.*

The Encoder. Let $s \in \{0,1\}^k$ be the message that we wish to encode. The encoder generates the encoded message $\mathsf{Enc}(s)$ according to the following procedure.

1. Let $Z \sim U_{k_1}$ and sample a random permutation $\Pi \colon [n] \to [n]$ by letting $\Pi := \mathsf{Perm}(Z)$. Let $Z' := \mathsf{Enc}_1(Z) \in \{0,1\}^{\gamma_1 n}$.
2. Let $S' = \mathsf{Enc}_2(s) \in \{0,1\}^{n_2}$ be the encoding of s using the LECSS code C_2.
3. Partition S' into blocks $S'_1, \ldots, S'_{n_b}$, each of length b, and encode each block independently using C_0 so as to obtain a string $C = (C_1, \ldots, C_{n_b}) \in \{0,1\}^n$.
4. Let $C' := \Pi(C)$ be the string C after its n coordinates are permuted by Π.
5. Output $\mathsf{Enc}(s) := (Z', C') \in \{0,1\}^N$, where $N := (1 + \gamma_1)n$, as the encoding of s.

A schematic description of the encoder summarizing the involved parameters is depicted in Figure 1.

The Decoder. We define the decoder $\mathsf{Dec}(\bar{Z}', \bar{C}')$ as follows:

1. Compute $\bar{Z} := \mathsf{Dec}_1(\bar{Z}')$.
2. Compute the permutation $\bar{\Pi} \colon [n] \to [n]$ defined by $\bar{\Pi} := \mathsf{Perm}(\bar{Z})$.
3. Let $\bar{C} \in \{0,1\}^n$ be the permuted version of $\bar{C}'$ according to $\bar{\Pi}^{-1}$.
4. Partition $\bar{C}$ into n_1/b blocks $\bar{C}_1, \ldots, \bar{C}_{n_b}$ of size B each (consistent to the way that the encoder does the partitioning of $\bar{C}$).
5. Call the inner code decoder on each block, namely, for each $i \in [n_b]$ compute $\bar{S}'_i := \mathsf{Dec}_0(\bar{C}_i)$. If $\bar{S}'_i = \perp$ for any i, output $\perp$ and return.
6. Let $\bar{S}' = (\bar{S}'_1, \ldots, \bar{S}'_{n_b}) \in \{0,1\}^{n_2}$. Compute $\bar{S} := \mathsf{Dec}_2(\bar{S}')$, where $\bar{S} = \perp$ if $\bar{S}'$ is not a codeword of C_2. Output $\bar{S}$.

Remark 2. As in the classical variation of concatenated codes of Forney [13] due to Justesen [16], the encoder described above can enumerate a *family* of inner codes instead of one fixed code in order to eliminate the exhaustive search for a good inner code C_0. In particular, one can consider all possible realizations of Construction 1 for the chosen parameters and use each obtained inner code to encode one of the n_b inner blocks. If the fraction of good inner codes (i.e., those satisfying the properties listed in Lemma 10) is small enough (e.g., $1/n^{\Omega(1)}$), our analysis still applies.

In the following theorem, we prove that the above construction is indeed a coding scheme that is non-malleable against bit-tampering adversaries with rate arbitrarily close to 1. Proof of the theorem appears in Section 4.3.

Theorem 12. *For every $\gamma_0 > 0$, there is a $\gamma_0' = \gamma_0^{O(1)}$ and $N_0 = O(1/\gamma_0^{O(1)})$ such that for every integer $N \geqslant N_0$, the following holds[5]. The pair $(\mathsf{Enc}, \mathsf{Dec})$ defined in Section 4.1 can be set up to be a strong non-malleable coding scheme against bit-tampering adversaries, achieving block length N, rate at least $1 - \gamma_0$ and error $\epsilon \leqslant \epsilon_1 + 3 \exp\left(-\Omega\left(\frac{\gamma_0' r N}{\log^3 N}\right)\right)$, where r and ϵ_1 are respectively the rate and the error of the assumed non-malleable coding scheme C_1.* □

4.2   Instantiations

We present two possible choices for the non-malleable code C_1 based on existing constructions. The first construction, due to Dziembowski et al. [12], is a Monte Carlo result that is summarized below.

Theorem 13. *[12, Theorem 4.2] For every integer $n > 0$, there is an efficient coding scheme C_1 of block length n, rate at least .18, that is non-malleable against bit-tampering adversaries achieving error $\epsilon = \exp(-\Omega(n))$. Moreover, there is an efficient randomized algorithm that, given n, outputs a description of such a code with probability at least $1 - \exp(-\Omega(n))$.*

More recently, Aggarwal et al. [1] construct an *explicit* coding scheme which is non-malleable against the much more general class of split-state adversaries. However, this construction achieves inferior guarantees than the one above in terms of the rate and error. Below we rephrase this result restricted to bit-tampering adversaries.

Theorem 14. *[1, implied by Theorem 5] For every integer $k > 0$ and $\epsilon > 0$, there is an efficient and explicit[6] coding scheme C_1 of message length k that is non-malleable against bit-tampering adversaries achieving error at most ϵ.*

---

[5] We can extend the construction to arbitrary block lengths N by standard padding techniques and observing that the set of block lengths for which construction of Figure 1 is defined is dense enough to allow padding without affecting the rate.

[6] To be precise, explicitness is guaranteed assuming that a large prime $p = \exp(\tilde{\Omega}(k + \log(1/\epsilon)))$ is available.

Moreover, the block length n of the coding scheme satisfies $n = \tilde{O}((k+\log(1/\epsilon))^7)$. By choosing $\epsilon := \exp(-k)$, we see that we can have $\epsilon = \exp(-\tilde{\Omega}(n^{1/7}))$ while the rate r of the code satisfies $r = \tilde{\Omega}(n^{-6/7})$.

By instantiating Theorem 12 with the Monte Carlo construction of Theorem 13, we arrive at the following corollary.

Corollary 15. *For every integer $n > 0$ and every positive parameter $\gamma_0 = \Omega(1/(\log n)^{O(1)})$, there is an efficient coding scheme $(\mathsf{Enc}, \mathsf{Dec})$ of block length n and rate at least $1 - \gamma_0$ such that the coding scheme is strongly non-malleable against bit-tampering adversaries, achieving error at most $\exp(-\tilde{\Omega}(n))$, Moreover, there is an efficient randomized algorithm that, given n, outputs a description of such a code with probability at least $1 - \exp(-\Omega(n))$.*

If, instead, we instantiate Theorem 12 with the construction of Theorem 14, we obtain the following strong non-malleable extractor (even though the construction of [1] is not strong).

Corollary 16. *For every integer $n > 0$ and every positive parameter $\gamma_0 = \Omega(1/(\log n)^{O(1)})$, there is an explicit and efficient coding scheme $(\mathsf{Enc}, \mathsf{Dec})$ of block length n and rate at least $1-\gamma_0$ such that the coding scheme is strongly non-malleable against bit-tampering adversaries and achieves error upper bounded by $\exp(-\tilde{\Omega}(n^{1/7}))$.* □

4.3 Proof of Theorem 12

It is clear that, given (Z', C'), the decoder can unambiguously reconstruct the message s; that is, $\mathsf{Dec}(\mathsf{Enc}(s)) = s$ with probability 1. Thus, it remains to demonstrate non-malleability of $\mathsf{Enc}(s)$ against bit-tampering adversaries.

Fix any such adversary $f \colon \{0,1\}^N \to \{0,1\}^N$. The adversary f defines the following partition of $[N]$:

- $\mathsf{Fr} \subseteq [N]$; the set of positions frozen to either zero or one by f.
- $\mathsf{Fl} \subseteq [N] \setminus \mathsf{Fr}$; the set of positions flipped by f.
- $\mathsf{Id} = [N] \setminus (\mathsf{Fr} \cup \mathsf{Fl})$; the set of positions left unchanged by f.

Since f is not the identity function (otherwise, there is nothing to prove), we know that $\mathsf{Fr} \cup \mathsf{Fl} \neq \emptyset$.

We use the notation used in the description of the encoder Enc and decoder Dec for various random variables involved in the encoding and decoding of the message s. In particular, let $(\bar{Z}', \bar{C}') = f(Z', C')$ denote the perturbation of $\mathsf{Enc}(s)$ by the adversary, and let $\bar{\Pi} := \mathsf{Perm}(\mathsf{Dec}_1(\bar{Z}'))$ be the induced perturbation of Π as viewed by the decoder Dec. In general Π and $\bar{\Pi}$ are correlated random variables, but independent of the remaining randomness used by the encoder.

We first distinguish three cases and subsequently show that the analysis of these cases suffices to guarantee non-malleability in general. The first case considers the situation where the adversary freezes too many bits of the encoding. The remaining two cases can thus assume that a sizeable fraction of the bits are not frozen to fixed values.

Case 1: Too Many Bits Are Frozen by the Adversary

First, assume that f freezes at least $n - t_2/b$ of the n bits of C'. In this case, show that the distribution of $\mathsf{Dec}(f(Z', C'))$ is always independent of the message s and thus the non-malleability condition of Definition 5 is satisfied for the chosen f. In order to achieve this goal, we rely on bounded independence property of the LECSS code C_2. We remark that a similar technique has been used in [12] for their construction of non-malleable codes (and for the case where the adversary freezes too many bits).

Observe that the joint distribution of $(\Pi, \bar{\Pi})$ is independent of the message s. Thus it suffices to show that conditioned on any realization $\Pi = \pi$ and $\bar{\Pi} = \bar{\pi}$, for any fixed permutations π and $\bar{\pi}$, the conditional distribution of $\mathsf{Dec}(f(Z', C'))$ is independent of the message s.

We wish to understand how, with respect to the particular permutations defined by π and $\bar{\pi}$, the adversary acts on the bits of the inner code blocks $C = (C_1, \ldots, C_{n_b})$.

Consider the set $T \subseteq [n_b]$ of the blocks of $C = (C_1, \ldots, C_{n_b})$ (as defined in the algorithm for Enc) that are not completely frozen by f (after permuting the action of f with respect to the fixed choice of π). We know that $|T| \leqslant t_2/b$.

Let S_T' be the string $S' = (S_1', \ldots, S_{n_b}')$ (as defined in the algorithm for Enc) restricted to the blocks defined by T; that is, $S_T' := (S_i')_{i \in T}$. Observe that the length of S_T' is at most $b|T| \leqslant t_2$. From the t_2-wise independence property of the LECSS code C_2, and the fact that the randomness of Enc_2 is independent of $(\Pi, \bar{\Pi})$, we know that S_T' is a uniform string, and in particular, independent of the original message s. Let C_T be the restriction of C to the blocks defined by T; that is, $C_T := (C_i)_{i \in T}$. Since C_T is generated from S_T (by applying the encoder Enc_0 on each block, whose randomness is independent of $(\Pi, \bar{\Pi})$), we know that the distribution of C_T is independent of the original message s as well.

Now, observe that $\mathsf{Dec}(f(Z', C'))$ is only a function of T, C_T, the tampering function f and the fixed choices of π and $\bar{\pi}$ (since the bits of C that are not picked by T are frozen to values determined by the tampering function f), which are all independent of the message s. Thus in this case, $\mathsf{Dec}(f(Z', C'))$ is independent of s as well. This suffices to prove non-malleability of the code in this case. However, in order to guarantee strong non-malleability, we need the following further claim.

Claim. Suppose $t_2 \leqslant n_2/2$. Then, regardless of the choice of the message s, $\Pr[f(Z', C') = (Z', C')] = \exp(-\Omega(\gamma_0 n)) =: \epsilon_1'$.

Proof. We upper bound the probability that the adversary leaves C' unchanged. Consider the action of f on $C = (C_1, \ldots, C_{n_b})$ (which is a permutation of how f acts on each bit according to the realization of Π). Recall that all but at most t_2/b of the bits of C (and hence, all but at most t_2/b of the n_b blocks of C) are frozen to 0 or 1 by f. Let $I \subseteq [n_b]$ denote the set of blocks of C that are completely frozen by f. We can see that $|I| \geqslant n_b/2$ by the assumption that $t_2 \leqslant n_2/2 = n_b b/2$.

In the sequel, we fix the realization of S' to any fixed string. Regardless of this conditioning, the blocks of C picked by I are independent, and each block is $\Omega(\gamma_0 B)$-wise, $\exp(-\Omega(\gamma_0 B))$-dependent by property 4 of Lemma 10. It follows that for each block $i \in I$, the probability that C_i coincides with the frozen value of the ith block as defined by f is bounded by $\exp(-\Omega(\gamma_0 B))$. Since the blocks of C picked by I are independent, we can amplify this probability and conclude that the probability that f leaves $(C_i)_{i \in I}$ (and consequently, (Z', C')) unchanged is at most

$$\exp(-\Omega(\gamma_0 B|I|)) = \exp(-\Omega(\gamma_0 B n_b/2)) = \exp(-\Omega(\gamma_0 n)) \ .$$

Consider the distribution $\mathcal{D}_{f,s}$ in Definition 5. From Claim 4.3, it follows that the probability mass assigned to <u>same</u> for this distribution is at most $\epsilon_1' = \exp(-\Omega(\gamma_0 n))$ for every s, which implies

$$\mathcal{D}_{f,s} \approx_{\epsilon_1'} \mathscr{D}(\mathsf{Dec}(f(\mathsf{Enc}(s)))),$$

since the right hand side distribution is simply obtained from $\mathcal{D}_{f,s}$ by moving the probability mass assigned to <u>same</u> to s. Since we have shown that the distribution of $\mathsf{Dec}(f(\mathsf{Enc}(s)))$ is the same for every message s, it follows that for every $s, s' \in \{0,1\}^k$,

$$\mathcal{D}_{f,s} \approx_{2\epsilon_1'} \mathcal{D}_{f,s'},$$

which proves strong non-malleability in this case.

Case 2: The Adversary Does Not Alter Π

In this case, we assume that $\Pi = \bar{\Pi}$, both distributed according to $\mathsf{Perm}(\mathcal{U}_{k_1})$ and independently of the remaining randomness used by the encoder. This situation in particular occurs if the adversary leaves the part of the encoding corresponding to Z' completely unchanged. Our goal is to upper bound the probability that Dec does not output $\perp$ under the above assumptions. We furthermore assume that Case 1 does not occur; i.e., more than $t_2/b = \gamma_2' n_2/b$ bits of C' are not frozen by the adversary.

To analyze this case, we rely on bounded independence of the permutation Π. The effect of the randomness of Π is to prevent the adversary from gaining any advantage of the fact that the inner code independently acts on the individual blocks.

Let $\mathsf{Id}' \subseteq \mathsf{Id}$ be the positions of C' that are left unchanged by f. We know that $|\mathsf{Id}' \cup \mathsf{Fl}| > t_2/b$. Moreover, the adversary freezes the bits of C corresponding to the positions in $\Pi^{-1}(\mathsf{Fr})$ and either flips or leaves the rest of the bits of C unchanged.

If $|\mathsf{Id}'| > n - \delta_2 n_b$, all but less than $\delta_2 n_b$ of the inner code blocks are decoded to the correct values by the decoder. Thus, the decoder correctly reconstructs all but less than $b(n - |\mathsf{Id}'|) \leqslant \delta_2 n_2$ bits of S'. Now, the distance property of the LECSS code $\mathcal{C}_2$ ensures that the remaining errors in S' are detected by the decoder, and thus, in this case the decoder always outputs $\perp$; a value that is independent of the original message s. Thus in the sequel we can assume that $|\mathsf{Fr} \cup \mathsf{Fl}| \geqslant \delta_2 n_2/b$. Moreover, we fix randomness of the LECSS C_2 so that S' becomes a fixed string. Recall that $C_1, \ldots, C_{n_b}$ are independent random variables, since every call of the inner encoder Enc_0 uses fresh randomness.

Since $\Pi = \bar{\Pi}$, the decoder is able to correctly identify positions of all the inner code blocks determined by C. In other words, we have

$$\bar{C} = f'(C),$$

where f' denotes the adversary obtained from f by permuting its action on the bits as defined by Π^{-1}; that is,

$$f'(x) := \Pi^{-1}(f(\Pi(x))).$$

Let $i \in [n_b]$. We consider the dependence between C_i and its tampering $\bar{C}_i$, conditioned on the knowledge of Π on the first $i-1$ blocks of C. Let $C(j)$ denote the jth bit of C, so that the ith block of C becomes $(C(1+(i-1)B), \ldots, C(iB))$. For the moment, assume that $\delta = 0$; that is, Π is exactly a ℓ-wise independent permutation.

Suppose $iB \leqslant \ell$, meaning that the restriction of Π on the ith block (i.e., $(\Pi(1+(i-1)B), \ldots, \Pi(iB))$ conditioned on any fixing of $(\Pi(1), \ldots, \Pi((i-1)B))$) exhibits the same distribution as that of a uniformly random permutation.

We define events $\mathcal{E}_1$ and $\mathcal{E}_2$ as follows. $\mathcal{E}_1$ is the event that $\Pi(1 + (i-1)B) \notin \mathsf{Id}'$, and $\mathcal{E}_2$ is the event that $\Pi(2 + (i-1)B) \notin \mathsf{Fr}$. That is, $\mathcal{E}_1$ occurs when the adversary does not leave the first bit of the ith block of C intact, and $\mathcal{E}_2$ occurs when the adversary does not freeze the second bit of the ith block. We are interested in lower bounding the probability that both $\mathcal{E}_1$ and $\mathcal{E}_2$ occur, conditioned on any particular realization of $(\Pi(1), \ldots, \Pi((i-1)B))$.

Suppose the parameters are set up so that

$$\ell \leqslant \frac{1}{2} \min\{\delta_2 n_2/b, \gamma'_2 n_2/b\}. \tag{1}$$

Under this assumption, even conditioned on any fixing of $(\Pi(1), \ldots, \Pi((i-1)B))$, we can ensure that

$$\Pr[\mathcal{E}_1] \geqslant \delta_2 n_2/(2bn),$$

and

$$\Pr[\mathcal{E}_2 | \mathcal{E}_1] \geqslant \gamma'_2 n_2/(2bn),$$

which together imply

$$\Pr[\mathcal{E}_1 \wedge \mathcal{E}_2] \geqslant \delta_2 \gamma'_2 \left(\frac{n_2}{2bn}\right)^2 =: \gamma''_2. \tag{2}$$

We let γ''_2 to be the right hand side of the above inequality.

In general, when the random permutation is ℓ-wise δ-dependent for $\delta \geqslant 0$, the above lower bound can only be affected by δ. Thus, under the assumption that

$$\delta \leqslant \gamma''_2/2, \tag{3}$$

we may still ensure that

$$\Pr[\mathcal{E}_1 \wedge \mathcal{E}_2] \geqslant \gamma_2''/2. \tag{4}$$

Let $X_i \in \{0,1\}$ indicate the event that $\mathsf{Dec}_0(\bar{C}_i) = \perp$. We can write

$$\Pr[X_i = 1] \geqslant \Pr[X_i = 1 | \mathcal{E}_1 \wedge \mathcal{E}_2] \Pr[\mathcal{E}_1 \wedge \mathcal{E}_2] \geqslant (\gamma_2''/2) \Pr[X_i = 1 | \mathcal{E}_1 \wedge \mathcal{E}_2],$$

where the last inequality follows from (4). However, by property 5 of Lemma 10 that is attained by the inner code $\mathcal{C}_0$, we also know that

$$\Pr[X_i = 1 | \mathcal{E}_1 \wedge \mathcal{E}_2] \geqslant 1/3,$$

and therefore it follows that

$$\Pr[X_i = 1] \geqslant \gamma_2''/6. \tag{5}$$

Observe that by the argument above, (5) holds even conditioned on the realization of the permutation Π on the first $i - 1$ blocks of C. By recalling that we have fixed the randomness of Enc_2, and that each inner block is independently encoded by Enc_0, we can deduce that, letting $X_0 := 0$,

$$\Pr[X_i = 1 | X_0, \ldots, X_{i-1}] \geqslant \gamma_2''/6. \tag{6}$$

Using the above result for all $i \in \{1, \ldots, \lfloor \ell/B \rfloor\}$, we conclude that

$$\Pr[\mathsf{Dec}(\bar{Z}', \bar{C}') \neq \perp] \leqslant \Pr[X_1 = X_2 = \cdots = X_{\lfloor \ell/B \rfloor} = 0] \tag{7}$$

$$\leqslant \left(1 - \gamma_2''/6\right)^{\lfloor \ell/B \rfloor}, \tag{8}$$

where (7) holds since the left hand side event is a subset of the right hand side event, and (8) follows from (6) and the chain rule.

Case 3: The Decoder Estimates an Independent Permutation

In this case, we consider the event where $\bar{\Pi}$ attains a particular value $\bar{\pi}$. Suppose it so happens that under this conditioning, the distribution of Π remains unaffected; that is, $\bar{\Pi} = \pi$ and $\Pi \sim \mathsf{Perm}(\mathcal{U}_{k_1})$. This situation may occur if the adversary completely freezes the part of the encoding corresponding to Z' to a fixed valid codeword of $\mathcal{C}_1$. Recall that the random variable Π is determined by the random string Z and that it is independent of the remaining randomness used by the encoder Enc. Similar to the previous case, our goal is to upper bound the probability that Dec does not output $\perp$. Furthermore, we can again assume that Case 1 does not occur; i.e., more than t_2/b bits of C' are not frozen by the adversary. For the analysis of this case, we can fix the randomness of Enc_2 and thus assume that S' is fixed to a particular value.

As before, our goal is to determine how each block C_i of the inner code is related to its perturbation $\bar{C}_i$ induced by the adversary. Recall that

$$\bar{C} = \bar{\pi}^{-1}(f(\Pi(C))).$$

Since f is fixed to an arbitrary choice only with restrictions on the number of frozen bits, without loss of generality we can assume that $\bar{\pi}$ is the identity permutation (if not, permute the action of f accordingly), and therefore, $\bar{C}' = \bar{C}$ (since $\bar{C}' = \bar{\pi}(\bar{C})$), and

$$\bar{C} = f(\Pi(C)).$$

For any $\tau \in [n_b]$, let $f_\tau \colon \{0,1\}^B \to \{0,1\}^B$ denote the restriction of the adversary to the positions included in the τth block of $\bar{C}$.

Assuming that $\ell \leqslant t_2$ (which is implied by (1)), let $T \subseteq [n]$ be any set of size $\lfloor \ell/B \rfloor \leqslant \lfloor t_2/B \rfloor \leqslant t_2/b$ of the coordinate positions of C' that are either left unchanged or flipped by f. Let $T' \subseteq [n_b]$ (where $|T'| \leqslant |T|$) be the set of blocks of $\bar{C}$ that contain the positions picked by T. With slight abuse of notation, for any $\tau \in T'$, denote by $\Pi^{-1}(\tau) \subseteq [n]$ the set of indices of the positions belonging to the block τ after applying the permutation Π^{-1} to each one of them. In other words, $\bar{C}_\tau$ (the τth block of $\bar{C}$) is determined by taking the restriction of C to the bits in $\Pi^{-1}(\tau)$ (in their respective order), and applying f_τ on those bits (recall that for $\tau \in T'$ we are guaranteed that f_τ does not freeze all the bits).

In the sequel, our goal is to show that with high probability, $\mathsf{Dec}(\bar{Z}, \bar{C}') = \perp$. In order to do so, we first assume that $\delta = 0$; i.e., that Π is exactly an ℓ-wise independent permutation. Suppose $T' = \{\tau_1, \ldots, \tau_{|T'|}\}$, and consider any $i \in |T'|$.

We wish to lower bound the probability that $\mathsf{Dec}_0(\bar{C}_{\tau_i}) = \perp$, conditioned on the knowledge of Π on the first $i-1$ blocks in T'. Subject to the conditioning, the values of Π becomes known on up to $(i-1)B \leqslant (|T'|-1)B \leqslant \ell - B$ points. Since Π is ℓ-wise independent, Π on the B bits belonging to the ith block remains B-wise independent. Now, assuming

$$\ell \leqslant n/2, \tag{9}$$

we know that even subject to the knowledge of Π on any ℓ positions of C, the probability that a uniformly random element within the remaining positions falls in a particular block of C is at most $B/(n-\ell) \leqslant 2B/n$.

Now, for $j \in \{2, \ldots, B\}$, consider the jth position of the block τ_i in T'. By the above argument, the probability that Π^{-1} maps this element to a block of C chosen by any of the previous $j-1$ elements is at most $2B/n$. By a union bound on the choices of j, with probability at least

$$1 - 2B^2/n,$$

the elements of the block τ_i all land in distinct blocks of C by the permutation Π^{-1}. Now we observe that if $\delta > 0$, the above probability is only affected by at most δ. Moreover, if the above distinctness property occurs, the values of C at the positions in $\Pi^{-1}(\tau)$ become independent random bits; since Enc uses fresh randomness upon each call of Enc_0 for encoding different blocks of the inner code (recall that the randomness of the first layer using Enc_2 is fixed).

Recall that by the bounded independence property of C_0 (i.e., property 4 of Lemma 10), each individual bit of C is $\exp(-\Omega(\gamma_0 B))$-close to uniform. Therefore, with probability at least $1 - 2B^2/n - \delta$ (in particular, at least $7/8$ when

$$n \geqslant 32B^2 \tag{10}$$

and assuming $\delta \leqslant 1/16$) we can ensure that the distribution of C restricted to positions picked by $\Pi^{-1}(\tau)$ is $O(B \exp(-\Omega(\gamma_0 B)))$-close to uniform, or in particular $(1/4)$-close to uniform when B is larger than a suitable constant. If this happens, we can conclude that distribution of the block τ_i of $\bar{C}$ is $(1/4)$-close to a sub-cube with at least one random bit (since we have assumed that $\tau \in T'$ and thus f does not fix all the bit of the τth block). Now, the cube property of $\mathcal{C}_0$ (i.e., property 3 of Lemma 10) implies that

$$\Pr_{\mathsf{Enc}_0} \left[\mathsf{Dec}_0(\bar{C}_{\tau_i}) \neq \perp | \Pi(\tau_1), \ldots, \Pi(\tau_{i-1}) \right] \leqslant 1/2 + 1/4 = 3/4,$$

where the extra term $1/4$ accounts for the statistical distance of $\bar{C}_{\tau_i}$ from being a perfect sub-cube.

Finally, using the above probability bound, and running i over all the blocks in T', and recalling the assumption that $\bar{C} = \bar{C}'$, we deduce that

$$\Pr[\mathsf{Dec}(\bar{Z}', \bar{C}') \neq \perp] \leqslant (7/8)^{|T'|} \leqslant \exp(-\Omega(\ell/B^2)), \tag{11}$$

where the last inequality follows from the fact that $|T'| \geqslant \lfloor \ell/b \rfloor / B$.

The General Case and Setting Up the Parameters

Recall that Case 1 eliminates the situation in which the adversary freezes too many of the bits. For the remaining cases, Cases 2 and 3 consider the special situations where the two permutations Π and $\bar{\Pi}$ used by the encoder and the decoder either completely match or are completely independent. However, in general we may not reach any of the two cases. Fortunately, the fact that the code $\mathcal{C}_1$ encoding the permutation Π is non-malleable ensure that we always end up with a *combination* of the Case 2 and 3. In other words, in order to analyze any event depending on the joint distribution of $(\Pi, \bar{\Pi})$, it suffices to consider the two special cases where Π is always the same as $\bar{\Pi}$, or when Π and $\bar{\Pi}$ are fully independent. Formal details of this argument, as well as the appropriate setting of the parameters leading to Theorem 12, appear in the full version of the paper.

5   Construction of Non-Malleable Codes Using Non-Malleable Extractors

In this section, we introduce the notion of seedless non-malleable extractors that extends the existing definition of seeded non-malleable extractors (as defined in [10]) to sources that exhibit structures of interest. This is similar to how classical seedless extractors are defined as an extension of seeded extractors to sources with different kinds of structure[7].

---

[7] For a background on standard seeded and seedless extractors, see [4, Chapter 2].

Furthermore, we obtain a reduction from the non-malleable variation of two-source extractors to non-malleable codes for the split-state model. Dziembowski et al. [11] obtain a construction of non-malleable codes encoding one-bit messages based on a variation of strong (standard) two-source extractors. This brings up the question of whether there is a natural variation of two-source extractors that directly leads to non-malleable codes for the split-state model encoding messages of arbitrary lengths (and ideally, achieving constant rate). Our notion of non-malleable two-source extractors can be regarded as a positive answer to this question.

Our reduction does not imply a characterization of non-malleable codes using extractors, and non-malleable codes for the split-state model do not necessarily correspond to non-malleable extractors (since those implied by our reduction achieve slightly sub-optimal rates). However, since seeded non-malleable extractors (as studied in the line of research starting [10]) are already subject of independent interest, we believe our characterization may be seen as a natural approach (albeit not the only possible approach) for improved constructions of non-malleable codes. Furthermore, the definition of two-source non-malleable extractors (especially the criteria described in Remark 3 below) is somewhat cleaner and easier to work with than then definition of non-malleable codes (Definition 4) that involves subtleties such as the extra care for the "same" symbol.

As discussed in Section 5.2, our reduction can be modified to obtain non-malleable codes for different classes of adversaries (by appropriately defining the family of extractors based on the tampering family being considered).

5.1   Seedless Non-malleable Extractors

First, we introduce the following notion of *non-malleable functions* that is defined with respect to a function and a distribution over its inputs. As it turns out, non-malleable "extractor" functions with respect to the uniform distribution and limited families of adversaries are of particular interest for construction of non-malleable codes.

Definition 17. A function $g\colon \Sigma \to \Gamma$ is said to be non-malleable with error ϵ with respect to a distribution $\mathcal{X}$ over Σ and a tampering function $f\colon \Sigma \to \Sigma$ if there is a distribution $\mathcal{D}$ over $\Gamma \cup \{\underline{\text{same}}\}$ such that for an independent $Y \sim \mathcal{D}$, $\mathscr{D}(g(X), g(f(X))) \approx_\epsilon \mathscr{D}(g(X), \mathsf{copy}(Y, g(X)))$.

Using the above notation, we may naturally define seedless non-malleable extractors. Roughly speaking, a seedless non-malleable extractor is a seedless extractor (in the traditional sense) that is also a non-malleable function with respect to a certain class of tampering functions. The general definition is deferred to the final version of the paper. However, for our applications we are particularly interested in the special case of two-source non-malleable extractors which is defined below.

Definition 18. A function NMExt: $\{0,1\}^n \times \{0,1\}^n \to \{0,1\}^m$ is a two-source non-malleable (k_1, k_2, ϵ)-extractor if, for every product distribution $(\mathcal{X}, \mathcal{Y})$ over $\{0,1\}^n \times \{0,1\}^n$ where $\mathcal{X}$ and $\mathcal{Y}$ have min-entropy at least k_1 and k_2, respectively, and for any arbitrary functions $f_1: \{0,1\}^n \to \{0,1\}^n$ and $f_2: \{0,1\}^n \to \{0,1\}^n$, the following hold.

1. NMExt is a two-source extractor for $(\mathcal{X}, \mathcal{Y})$; that is, $\mathsf{NMExt}(\mathcal{X}, \mathcal{Y}) \approx_\epsilon \mathcal{U}_m$.
2. NMExt is a non-malleable function with error ϵ for the distribution $(\mathcal{X}, \mathcal{Y})$ and with respect to the tampering function $(X, Y) \mapsto (f_1(X), f_2(Y))$.

The theorem below, proved in the full version, shows that non-malleable two-source extractors exist and in fact a random function is w.h.p. such an extractor.

Theorem 19. *Let* NMExt: $\{0,1\}^n \times \{0,1\}^n \to \{0,1\}^m$ *be a uniformly random function. For any* $\gamma, \epsilon > 0$ *and parameters* $k_1, k_2 \leqslant n$, *with probability at least* $1 - \gamma$, *the function* NMExt *is a two-source non-malleable* (k_1, k_2, ϵ)-*extractor provided that* $2m \leqslant k_1 + k_2 - 3\log(1/\epsilon) - \log\log(1/\gamma)$, *and* $\min\{k_1, k_2\} \geqslant \log n + \log\log(1/\gamma) + O(1)$. $\qquad\square$

In general, a tampering function may have fixed points and act as the identity function on a particular set of inputs. Definitions of non-malleable codes, functions, and extractors all handle the technicalities involved with such fixed points by introducing a special symbol "same". Nevertheless, it is more convenient to deal with adversaries that are promised to have no fixed points. For this restricted model, the definition of two-source non-malleable extractors can be modified as follows. We call extractors satisfying the less stringent requirement *relaxed* two-source non-malleable extractors. Formally, the relaxed definition is as follows.

Definition 20. A function NMExt: $\{0,1\}^n \times \{0,1\}^n \to \{0,1\}^m$ is a relaxed two-source non-malleable (k_1, k_2, ϵ)-extractor if, for every product distribution $(\mathcal{X}, \mathcal{Y})$ over $\{0,1\}^n \times \{0,1\}^n$ where $\mathcal{X}$ and $\mathcal{Y}$ have min-entropy at least k_1 and k_2, respectively, the following holds. Let $f_1: \{0,1\}^n \times \{0,1\}^n$ and $f_2: \{0,1\}^n \times \{0,1\}^n$ be functions such that for every $x \in \{0,1\}^n$, $f_1(x) \neq x$ and $f_2(x) \neq x$. Then, for $(X, Y) \sim (\mathcal{X}, \mathcal{Y})$,

1. NMExt is a two-source extractor for $(\mathcal{X}, \mathcal{Y})$; that is, $\mathsf{NMExt}(\mathcal{X}, \mathcal{Y}) \approx_\epsilon \mathcal{U}_m$.
2. NMExt is a non-malleable function with error ϵ for the distribution of (X, Y) and with respect to the following three tampering functions: $(X, Y) \mapsto (f_1(X), Y)$; $(X, Y) \mapsto (X, f_2(Y))$; and $(X, Y) \mapsto (f_1(X), f_2(Y))$.

Remark 3. In order to satisfy the requirements of Definition 20, it suffices (but not necessary) to ensure

$$(\mathsf{NMExt}(\mathcal{X}, \mathcal{Y}), \mathsf{NMExt}(f_1(\mathcal{X}), \mathcal{Y})) \approx_\epsilon (U_m, \mathsf{NMExt}(f_1(\mathcal{X}), \mathcal{Y})),$$
$$(\mathsf{NMExt}(\mathcal{X}, \mathcal{Y}), \mathsf{NMExt}(\mathcal{X}, f_2(\mathcal{Y}))) \approx_\epsilon (U_m, \mathsf{NMExt}(\mathcal{X}, f_2(\mathcal{Y}))),$$
$$(\mathsf{NMExt}(\mathcal{X}, \mathcal{Y}), \mathsf{NMExt}(f_1(\mathcal{X}), f_2(\mathcal{Y}))) \approx_\epsilon (U_m, \mathsf{NMExt}(f_1(\mathcal{X}), f_2(\mathcal{Y}))).$$

The proof of Theorem 19 shows that these stronger requirements (which are quite similar to the definition of seeded non-malleable extractor in [10]) can be satisfied with high probability by random functions.

It immediately follows from the definitions that a two-source non-malleable extractor (according to Definition 18) is a relaxed non-malleable two-source extractor (according to Definition 20) and with the same parameters. Interestingly, below we show that the two notions are equivalent up to a slight loss in the parameters (see the full version for a proof).

Lemma 21. *Let* NMExt *be a relaxed two-source non-malleable* $(k_1 - \log(1/\epsilon), k_2 - \log(1/\epsilon), \epsilon)$-*extractor. Then,* NMExt *is a two-source non-malleable* $(k_1, k_2, 4\epsilon)$-*extractor.* □

5.2   From Non-malleable Extractors to Non-malleable Codes

In this section, we present our reduction from non-malleable extractors to non-malleable codes. For concreteness, we focus on tampering functions in the split-state model. It is straightforward to extend the reduction to different families of tampering functions, for example:

1. When the adversary divides the input into $b \geqslant 2$ known parts, not necessarily of the same length, and applies an independent tampering function on each block. In this case, a similar reduction from non-malleable codes to multiple-source non-malleable extractors may be obtained.
2. When the adversary behaves as in the split-state model, but the choice of the two parts is not known in advance. In this case, the needed extractor is a non-malleable variation of the *mixed-sources extractors* studied by Raz and Yehudayoff [22].

We note that Theorem 22 below (and similar theorems that can be obtained for the other examples above) only require non-malleable extraction from the uniform distribution. However, the reduction from arbitrary tampering functions to ones without fixed points (e.g., Lemma 21) strengthens the entropy requirement of the source while imposing a structure on the source distribution which is related to the family of tampering functions being considered.

Theorem 22. *Let* NMExt: $\{0,1\}^n \times \{0,1\}^n \to \{0,1\}^k$ *be a two-source non-malleable* (n, n, ϵ)-*extractor. Define a coding scheme* (Enc, Dec) *with message length* k *and block length* $2n$ *as follows. The decoder* Dec *is defined by* $\mathsf{Dec}(x) := \mathsf{NMExt}(x)$. *The encoder, given a message* s, *outputs a uniformly random string in* $\mathsf{NMExt}^{-1}(s)$. *Then, the pair* (Enc, Dec) *is a non-malleable code with error* $\epsilon' := \epsilon(2^k + 1)$ *for the family of split-state adversaries.*

Proof. Deferred to the full version of the paper.

We can now derive the following corollary, which is the main result of this section, using Lemma 21 and Theorem 22 (see the full version for a proof).

Corollary 23. *Let* NMExt: $\{0,1\}^n \times \{0,1\}^n \to \{0,1\}^m$ *be a relaxed two-source non-malleable* (k_1, k_2, ϵ)-*extractor, where* $m = \Omega(n)$, $n - k_1 = \Omega(n)$, $n - k_2 = \Omega(n)$, *and* $\epsilon = \exp(-\Omega(m))$. *Then, there is a* $k = \Omega(n)$ *such that the following holds. Define a coding scheme* (Enc, Dec) *with message length* k *and block length* $2n$ *(thus rate* $\Omega(1)$*) as follows. The decoder* Dec, *given* $x \in \{0,1\}^{2n}$, *outputs the first* k *bits of* NMExt(x). *The encoder, given a message* x, *outputs a uniformly random string in* Dec$^{-1}(x)$. *Then, the pair* (Enc, Dec) *is a non-malleable code with error* $\exp(-\Omega(n))$ *for the family of split-state adversaries.* $\square$

Finally, using the above tools and the existence result of Theorem 19, we conclude that there are non-malleable two-source extractors defining coding schemes in the split-state model and achieving constant rates; in particular, rates arbitrarily close to $1/5$.

Corollary 24. *For every* $\alpha > 0$, *there is a choice of* NMExt *in Theorem 22 that makes* (Enc, Dec) *a non-malleable coding scheme against split-state adversaries achieving rate* $1/5 - \alpha$ *and error* $\exp(-\Omega(\alpha n))$.

Proof. First, for some α', we use Theorem 19 to show that if NMExt: $\{0,1\}^n \times \{0,1\}^n \to \{0,1\}^k$ is randomly chosen, with probability at least .99 it is a two-source non-malleable $(n, n, 2^{-k(1+\alpha')})$-extractor, provided that $k \leqslant n - (3/2)\log(1/\epsilon) - O(1) = n - (3/2)k(1 + \alpha') - O(1)$, which can be satisfied for some $k \geqslant (2/5)n - \Omega(\alpha'n)$. Now, we can choose $\alpha' = \Omega(\alpha)$ so as to ensure that $k \geqslant 2n(1 - \alpha)$ (thus, keeping the rate above $1 - \alpha$) while having $\epsilon \leqslant 2^{-k}\exp(-\Omega(\alpha n))$. We can now apply Theorem 22 to attain the desired result.

References

1. Aggarwal, D., Dodis, Y., Lovett, S.: Non-malleable codes from additive combinatorics. ECCC Technical Report TR13-081 (2013)
2. Barak, B., Rao, A., Shaltiel, R., Wigderson, A.: 2-source dispersers for subpolynomial entropy and Ramsey graphs beating the Frankl-Wilson construction. Annals of Mathematics 176(3), 1483–1544 (2012)
3. Bourgain, J.: More on the Sum-Product phenomenon in prime fields and its applications. International Journal of Number Theory 1(1), 1–32 (2005)
4. Cheraghchi, M.: Applications of Derandomization Theory in Coding. PhD thesis, Swiss Federal Institute of Technology (EPFL), Lausanne, Switzerland (2010), http://eccc.hpi-web.de/static/books/ Applications_of_Derandomization_Theory_in_Coding/
5. Cheraghchi, M., Guruswami, V.: Capacity of non-malleable codes. ECCC Technical Report TR13-118 (2013)
6. Cheraghchi, M., Guruswami, V.: Explicit optimal rate non-malleable codes for bit-tampering. IACR Technical Report 2013/565 (2013), http://eprint.iacr.org/2013/565
7. Chor, B., Goldreich, O.: Unbiased bits from sources of weak randomness and probabilistic communication complexity. SIAM Journal on Computing 2(17), 230–261 (1988)

8. Cramer, R., Dodis, Y., Fehr, S., Padró, C., Wichs, D.: Detection of algebraic manipulation with applications to robust secret sharing and fuzzy extractors. In: Smart, N.P. (ed.) EUROCRYPT 2008. LNCS, vol. 4965, pp. 471–488. Springer, Heidelberg (2008)

9. Dodis, Y., Li, X., Wooley, T.D., Zuckerman, D.: Privacy amplification and non-malleable extractors via character sums. In: Proceedings of FOCS 2011, pp. 668–677 (2011)

10. Dodis, Y., Wichs, D.: Non-malleable extractors and symmetric key cryptography from weak secrets. In: Proceedings of the 41st Annual ACM Symposium on Theory of Computing, pp. 601–610 (2009)

11. Dziembowski, S., Kazana, T., Obremski, M.: Non-malleable codes from two-source extractors. In: Canetti, R., Garay, J.A. (eds.) CRYPTO 2013, Part II. LNCS, vol. 8043, pp. 239–257. Springer, Heidelberg (2013)

12. Dziembowski, S., Pietrzak, K., Wichs, D.: Non-malleable codes. In: Proceedings of Innovations in Computer Science, ICS 2010 (2010)

13. Forney, G.D.: Concatenated Codes. MIT Press (1966)

14. Gohen, G., Raz, R., Segev, G.: Non-malleable extractors with short seeds and applications to privacy amplification. In: Proceedings of CCC 2012, pp. 298–308 (2012)

15. Guruswami, V., Smith, A.: Codes for computationally simple channels: Explicit constructions with optimal rate. In: Proceedings of FOCS 2010, pp. 723–732 (2010)

16. Justesen, J.: A class of constructive asymptotically good algebraic codes. IEEE Transactions on Information Theory 18, 652–656 (1972)

17. Kalai, Y., Li, X., Rao, A.: 2-source extractors under computational assumptions and cryptography with defective randomness. In: Proceedings of the 50th Annual IEEE Symposium on Foundations of Computer Science (FOCS), pp. 617–626 (2009)

18. Kaplan, E., Naor, M., Reingold, O.: Derandomized constructions of k-wise (almost) independent permutations. In: Chekuri, C., Jansen, K., Rolim, J.D.P., Trevisan, L. (eds.) APPROX and RANDOM 2005. LNCS, vol. 3624, pp. 354–365. Springer, Heidelberg (2005)

19. Li, X.: Non-malleable extractors, two-source extractors and privacy amplification. In: Proceedings of the 53rd Annual IEEE Symposium on Foundations of Computer Science (FOCS), pp. 688–697 (2012)

20. Rao, A.: A 2-source almost-extractor for linear entropy. In: Goel, A., Jansen, K., Rolim, J.D.P., Rubinfeld, R. (eds.) APPROX and RANDOM 2008. LNCS, vol. 5171, pp. 549–556. Springer, Heidelberg (2008)

21. Raz, R.: Extractors with weak random seeds. In: Proceedings of the 37th Annual ACM Symposium on Theory of Computing (STOC), pp. 11–20 (2005)

22. Raz, R., Yehudayoff, A.: Multilinear formulas, maximal-partition discrepancy and mixed-sources extractors. Journal of Computer and System Sciences 77(1), 167–190 (2011)

Continuous Non-malleable Codes

Sebastian Faust[2], Pratyay Mukherjee[1],
Jesper Buus Nielsen[1], and Daniele Venturi[3]

[1] Department of Computer Science, Aarhus University
[2] EPFL Lausanne
[3] Department of Computer Science, Sapienza University of Rome

Abstract. Non-malleable codes are a natural relaxation of error cor-
recting/detecting codes that have useful applications in the context of
tamper resilient cryptography. Informally, a code is non-malleable if an
adversary trying to tamper with an encoding of a given message can
only leave it unchanged or modify it to the encoding of a completely
unrelated value. This paper introduces an extension of the standard
non-malleability security notion – so-called *continuous* non-malleability
– where we allow the adversary to tamper *continuously* with an encoding.
This is in contrast to the standard notion of non-malleable codes where
the adversary only is allowed to tamper a *single* time with an encoding.
We show how to construct continuous non-malleable codes in the common split-state model where an encoding consist of two parts and the
tampering can be arbitrary but has to be independent with both parts.
Our main contributions are outlined below:
1. We propose a new *uniqueness* requirement of split-state codes which
   states that it is computationally hard to find two codewords $X = (X_0, X_1)$ and $X' = (X_0, X_1')$ such that both codewords are valid,
   but X_0 is the same in both X and X'. A simple attack shows that
   uniqueness is necessary to achieve continuous non-malleability in the
   split-state model. Moreover, we illustrate that none of the existing
   constructions satisfies our uniqueness property and hence is not secure in the continuous setting.
2. We construct a split-state code satisfying continuous non-malleability.
   Our scheme is based on the inner product function, collision-resistant
   hashing and non-interactive zero-knowledge proofs of knowledge and
   requires an untamperable common reference string.
3. We apply continuous non-malleable codes to protect arbitrary cryptographic primitives against tampering attacks. Previous applications of non-malleable codes in this setting required to *perfectly erase*
   the entire memory after each execution and required the adversary
   to be restricted in memory. We show that continuous non-malleable
   codes avoid these restrictions.

Keywords: non-malleable codes, split-state, tamper resilience.

1 Introduction

Physical attacks that target cryptographic implementations instead of breaking
the black-box security of the underlying algorithm are amongst the most severe

Y. Lindell (Ed.): TCC 2014, LNCS 8349, pp. 465–488, 2014.

threats for cryptographic systems. A particular important attack on implementations is the so-called tampering attack. In a tampering attack the adversary changes the secret key to some related value and observes the effect of such changes at the output. Traditional black-box security notions do not incorporate adversaries that change the secret key to some related value; even worse, as shown in the celebrated work of Boneh et al. [6] already minor changes to the key suffice for complete security breaches. Unfortunately, tampering attacks are also rather easy to carry out: a virus corrupting a machine can gain partial control over the state, or an adversary that penetrates the cryptographic implementation with physical equipment may induce faults into keys stored in memory.

In recent years, a growing body of work (see [21,22,17,24,1,2,19] and many more) develop new cryptographic techniques that protect against tampering attacks. Non-malleable codes introduced by Dziembowski, Pietrzak and Wichs [17] are an important approach to achieve this goal. Intuitively a code is non-malleable w.r.t. a set of tampering functions $\mathcal{T}$ if the message contained in a codeword modified via a function in $\mathcal{T}$ is either the original message, or a completely unrelated value. Non-malleable codes can be used to protect any cryptographic functionality against tampering with the memory. Instead of storing the key in memory, we store its encoding and decode it each time the functionality wants to accesses the key. As long as the adversary can only apply tampering functions from the set $\mathcal{T}$, the non-malleability property guarantees that the (possibly tampered) decoded value is not related to the original key.

The standard notion of non-malleability considers a one-shot game: the adversary is allowed to tamper a single time with the codeword and obtains the decoded output. In this work we introduce so-called *continuous non-malleable codes*, where non-malleability is guaranteed even if the adversary continuously applies functions from the set $\mathcal{T}$ to the codeword. We show that our new security notion is not only a natural extension of the standard one-shot notion, but moreover allows to protect against tampering attacks in important settings where earlier constructions fall short to achieve security.

Continuous Non-malleable Codes. A non-malleable code consists of two algorithms Code = (Encode, Decode) that satisfy the correctness property Decode(Encode(x)) = x, for all $x \in \mathcal{X}$. To define non-malleability for a function class $\mathcal{T}$, consider the random variable $\mathsf{Tamper}_{\mathsf{T},x}$ defined for every function $\mathsf{T} \in \mathcal{T}$ and any message $x \in \mathcal{X}$ in the game below:

1. Compute an encoding $X \leftarrow \mathrm{Encode}(x)$ using the encoding procedure.
2. Apply the tampering function $\mathsf{T} \in \mathcal{T}$ to obtain the tampered codeword $X' = \mathsf{T}(X)$.
3. If $X' = X$ then return the special symbol same*; otherwise, return Decode(X'). Notice that Decode(X') may return the special symbol $\bot$ in case the tampered codeword X' was invalid.

A coding scheme Code is said to be (one-shot) non-malleable with respect to functions in $\mathcal{T}$ and message space $\mathcal{X}$, if for every $\mathsf{T} \in \mathcal{T}$ and any two messages $x, y \in \mathcal{X}$ the distributions $\mathsf{Tamper}_{\mathsf{T},x}$ and $\mathsf{Tamper}_{\mathsf{T},y}$ are indistinguishable.

To define continuous non-malleable codes, we do not fix a single tampering function T a-priori.[1] Instead, we let the adversary repeat step 2 and step 3 from the above game a polynomial number of times, where in each iteration the adversary can adaptively choose a tampering function $\mathsf{T}_i \in \mathcal{T}$. We emphasize that this change of the tampering game allows the adversary to tamper continuously with the initial encoding X. As shown by Gennaro et al. [21] such a strong security notion is impossible to achieve without further assumptions. To this end, we rely on a self-destruct mechanism as used in earlier works on non-malleable codes. More precisely, when in step 3 the game detects an invalid codeword and returns $\perp$ for the first time, then it self-destructs. This is a rather mild assumption as it can, for instance, be implemented using a single public untamperable bit.

From Non-malleable Codes to Tamper Resilience. As discussed above one main application of non-malleable codes is to protect cryptographic schemes against tampering with the secret key [17,24]. Consider a reactive functionality $\mathcal{G}$ with secret state st that can be executed on input m, e.g., $\mathcal{G}$ may be the AES with key st encrypting messages m. Using a non-malleable code earlier work showed how to transform the functionality $(\mathcal{G}, st)$ into a functionality $(\mathcal{G}^{\mathsf{Code}}, X)$ that is secure against tampering with X. The transformation compiling $(\mathcal{G}, st)$ into $(\mathcal{G}^{\mathsf{Code}}, X)$ works as follows. Initially, X is set to $X \leftarrow \mathrm{Encode}(st)$. Each time $\mathcal{G}^{\mathsf{Code}}$ is executed on input m, the transformed functionality reads the encoding X from memory, decodes it to obtain $st = \mathrm{Decode}(X)$ and runs the original functionality $\mathcal{G}(st, m)$. Finally, it erases the memory and stores the new state $X \leftarrow \mathrm{Encode}(st)$. Additionally to executing evaluation queries the adversary can issue tampering queries $\mathsf{T}_i \in \mathcal{T}$. A tampering query replaces the current secret state X with a tampered state $X' = \mathsf{T}_i(X)$, and the functionality $\mathcal{G}^{\mathsf{Code}}$ continues its computation using X' as the secret state. Notice that in case of $\mathrm{Decode}(X') = \perp$ the functionality $\mathcal{G}^{\mathsf{Code}}$ sets the memory to a dummy value— resulting essentially in a self-destruct.

The above transformation guarantees continuous tamper resilience even if the underlying non-malleable code is secure only against one-shot tampering. This security "boost" is achieved by re-encoding the secret state/key after each execution of the primitive $\mathcal{G}^{\mathsf{Code}}$. As one-shot non-malleability suffices in the above cryptographic application, one may ask why we need continuous non-malleable codes. Besides being a natural extension of the standard non-malleability notion, our new notion has several important applications that we discuss in the next two paragraphs.

Tamper Resilience Without Erasures. The transformation described above necessarily requires that after each execution the entire content of the memory is erased. While such perfect erasures may be feasible in some settings, they are

---

[1] Our actual definition is slightly stronger than what is presented next (cf. Section 3).

rather problematic in the presence of tampering. To illustrate this issue consider a setting where besides the encoding of a key, the memory also contains other non-encoded data. In the tampering setting, we cannot restrict the erasure to just the part that stores the encoding of the key as a tampering adversary may copy the encoding to some different part of the memory. A simple solution to this problem is to erase the entire memory, but such an approach is not possible in most cases: for instance, think of the memory as being the hard-disk of your computer that besides the encoding of a key stores other important files that you don't want to be erased. Notice that this situation is quite different from the leakage setting, where we also require perfect erasures to achieve continuous leakage resilience. In the leakage setting, however, the adversary cannot mess around with the state of the memory by, e.g., copying an encoding of a secret key to some free space, which makes erasures significantly easier to implement.

One option to prevent the adversary from keeping permanent copies is to encode the entire state of the memory. Such an approach has, however, the following drawbacks.

1. *It is unnatural:* In many cases secret data, e.g., a cryptographic key, is stored together with non-confidential data. Each time we want to read some small part of the memory, e.g., the key, we need to decode and re-encode the entire state—including also the non-confidential data.
2. *It is inefficient:* Decoding and re-encoding the entire state of the memory for each access introduces additional overhead and would result in highly inefficient solutions. This gets even worse as most current constructions of non-malleable codes are rather inefficient.
3. *It does not work in general:* Consider a setting where we want to compute with non-malleable codes in a tamper resilient way (similar in spirit to tamper resilient circuits). Clearly, in this setting the memory will store many independent encodings of different secrets that cannot be erased. Continuous non-malleable codes are hence a first natural step towards non-malleable computation.

Using our new notion of continuous non-malleable codes we can avoid the above issues and achieve continuous tamper resilience without using *erasures* and without relying on inefficient solutions that encode the *entire* state.

Stateless Tamper Resilient Transformations. To achieve tamper resilience from one-shot non-malleability we necessarily need to re-encode the state using fresh randomness. This not only reduces the efficiency of the proposed construction, but moreover makes the transformation stateful. Using continuous non-malleable codes we get continuous tamper resilience for free, eliminating the need to refresh the encoding after each usage. This is in particular useful when the underlying primitive that we want to protect is stateless itself. Think, for instance, of any standard block-cipher construction that typically keeps the same key. Using continuous non-malleable codes the tamper resilient implementation of such stateless primitives does not need to keep any secret state. We discuss the protection of stateless primitives in further detail in Section 5.

1.1   Our Contribution

In this work, we propose the first construction of *continuous non-malleable codes* in the split-state model first introduced in the leakage setting [16,13]. Various recent works study the split-state model for non-malleable codes [24,15,1] (see more details on related work in Section 1.2). In the split-state tampering model, the codeword consists of two halves X_0 and X_1 that are stored on two different parts of the memory. The adversary is assumed to tamper with both parts independently, but otherwise can apply any efficiently computable tampering function. That is, the adversary picks two polynomial-time computable functions T_0 and T_1 and replaces the state (X_0, X_1) with the tampered state $(\mathsf{T}_0(X_0), \mathsf{T}_1(X_1))$. Similar to the earlier work of Liu and Lysyanskaya [24] our construction assumes a public untamperable CRS. Notice that this is a rather mild assumption as the CRS can be hard-wired into the functionality and is independent of any secret data.

Continuous Non-malleability of Existing Constructions. The first construction of (one-shot) split-state non-malleable codes in the standard model was given by Liu and Lysyanskaya [24]. At a high-level the construction encrypts the input x with a leakage resilient encryption scheme and generates a non-interactive zero-knowledge proof of knowledge showing that (a) the public/secret key of the PKE are valid, and (b) the ciphertext is an encryption of x under the public key. Then, X_0 is set to the secret key while X_1 holds the corresponding public key, the ciphertext and the above described NIZK proof.

Unfortunately, it is rather easy to break the non-malleable code of Liu and Lysyanskaya in the continuous setting. Recall that our security notion of continuous non-malleable codes allows the adversary to interact in the following game. First, we sample a codeword $(X_0, X_1) \leftarrow \mathsf{Encode}(x)$ and then repeat the following process a polynomial number of times:

1. The adversary submits two polynomial-time computable functions $(\mathsf{T}_0, \mathsf{T}_1)$ resulting in a tampered state $(X_0', X_1') = (\mathsf{T}_0(X_0), \mathsf{T}_1(X_1))$.
2. We consider three different cases: (1) if $(X_0', X_1') = (X_0, X_1)$ then return same*; (2) otherwise compute $x' = \mathsf{Decode}(X_0', X_1')$ and return x' if $x' \neq \bot$; (3) if $x' = \bot$ self-destruct and terminate the experiment.

The main observation that enables the attack against the scheme of [24] is as follows. For a fixed (but adversarially chosen) part X_0' it is easy to come-up with two corresponding parts X_1' and X_1'' such that both (X_0', X_1') and (X_0', X_1'') form a valid codeword that *does not* lead to a self-destruct. Suppose further that $\mathsf{Decode}(X_0', X_1') \neq \mathsf{Decode}(X_0', X_1'')$, then under continuous tampering the adversary may permanently replace the original encoding X_0 with X_0', while depending on whether the i-th bit of X_1 is 0 or 1 either replace X_1 by X_1' or X_1''. This allows to recover the entire X_1 by just $|X_1|$ tampering attacks. Once X_1 is known to the adversary it is easy to tamper with (X_0, X_1) in a way that depends on $\mathsf{Decode}(X_0, X_1)$.

Somewhat surprisingly, our attack can be generalized to break *any* non-malleable code that is secure in the information theoretic setting. Hence, also the recent breakthrough results on information theoretic non-malleability [15,1] fail to provide security under continuous attacks. Moreover, we emphasize that our attack does not only work for the code itself, but (in most cases) can be also applied to the tamper-protection application of cryptographic functionalities.

Uniqueness. The attack above exploits that for a fixed known part X_0' it is easy to come-up with two valid parts X_1', X_1''. For the encoding of [24] this is indeed easy to achieve. If the secret key X_0' is known it is easy to come-up with two valid parts X_1', X_1'': just encrypt two arbitrary messages $x_0 \neq x_1$ and generate the corresponding proofs. The above weakness motivates a new property that non-malleable codes shall satisfy in order to achieve security against continuous non-malleability. We call this property *uniqueness*, which informally guarantees that for any (adversarially chosen) valid encoding (X_0', X_1') it is computationally hard to come up with $X_b'' \neq X_b'$ such that (X_b', X_{1-b}'') forms a valid encoding. Clearly the uniqueness property prevents the above described attack, and hence is a crucial requirement for continuous non-malleability.

A New Construction. In light of the above discussion, we need to build a non-malleable code that achieves our uniqueness property. Our construction uses as building blocks a leakage resilient storage (LRS) scheme [13,14] for the split-state model (one may view this as a generalization of the leakage resilient PKE used in [24]), a collision-resistant hash function and (similar to [24]) an extractable NIZK. At a high-level we use the LRS to encode the secret message, hash the resulting shares using the hash function and generate a NIZK proof of knowledge that indeed the resulting hash values are correctly computed from the shares. While it is easy to show that collision resistance of the hash function guarantees the uniqueness property, a careful analysis is required to prove continuous non-malleability. We refer the reader to Section 4 for the details of our construction and to Section 4.1 for an outline of the proof.

Tamper Resilience for Stateless and Stateful Primitives. We can use our new construction of continuous non-malleable codes to protect arbitrary computation against continuous tampering attacks. In contrast to earlier works our construction does not need to re-encode the secret state after each usage, which besides being more efficient avoids the use of erasures. As discussed above, erasures are problematic in the tampering setting as one would essentially need to encode the entire state (possibly including large non-confidential data).

Additionally, our transformation does not need to keep any secret state. Hence, if our transformation is used for stateless primitives, then the resulting scheme remains stateless. This solves an open problem of Dziembowski, Pietrzak and Wichs [17]. Notice that while we do not need to keep any secret state, the transformed functionality requires one single bit to switch to self-destruction mode. This bit can be *public* but must be untamperable, and can for instance be implemented through one-time writable memory. As shown in the work of Gennaro et al. [21] continuous tamper resilience is impossible to achieve without such a mechanism for self-destruction.

Of course, our construction can also be used for stateful primitives, in which case our functionality will re-encode the new state during execution. Note that in this setting, as data is never erased, an adversary can always reset the functionality to a previous valid state. To avoid this, our transformation uses an untamperable *public* counter[2] that helps us to detect whenever the functionality is reset to a previous state, leading to a self-destruct. We notice that such an untamperable counter is necessary, as otherwise there is no way to protect against the above resetting attack.

Adding Leakage. As a last contribution, we show that our code is also secure against bounded leakage attacks. This is similar to the works of [24,15] who also consider bounded leakage resilience of their encoding scheme. We then show that bounded leakage resilience is also inherited by functionalities that are protected by our transformation. Notice that without perfect erasures bounded leakage resilience is the best we can achieve, as there is no hope for security if an encoding that is produced at some point in time is gradually revealed to the adversary.

1.2 Related Work

Constructions of Non-malleable Codes. Besides showing feasibility by a probabilistic argument, [17] also built non-malleable codes for bit-wise tampering and gave a construction in the split-state model using a random oracle. This result was followed by [9] which proposed non-malleable codes that are secure against block-wise tampering. The first construction of non-malleable codes in the split-state model was given by Liu and Lysyanskaya [24] assuming an untamperable CRS. Very recently two beautiful works showed how to build non-malleable codes in the split-state model without relying on a CRS [15,1] even when the adversary has unlimited computing power. Dziembowski *et al.* [15] show how to encode a single bit using the inner product function. Agrawal *et al.* [1] developed a construction that goes beyond single-bit encoding but induces a huge overhead.

See also [8,7,19] for other recent advances on the construction of non-malleable codes. We also notice that the work of Genarro *et al.* [21] proposed a generic method that allows to protect arbitrary computation against continuous tampering attacks, without requiring erasures. We refer the reader to [17] for a more detailed comparison between non-malleable codes and the solution of [21].

Other Works on Tamper Resilience. A large body of work shows how to protect specific cryptographic schemes against tampering attacks (see [4,3,23,5,25,12] and many more). While these works consider a strong tampering model (e.g., they do not require the split-state assumption), they only offer security for specific schemes. In contrast non-malleable codes are generally applicable and can provide tamper resilience of any cryptographic scheme.

In all the above works, including ours, it is assumed that the circuitry that computes the cryptographic algorithm using the potentially tampered key runs

---

[2] Note that a counter uses very small (logarithmic in the security parameter) number of bits.

correctly, and is not subject to tampering attacks. An important line of works analyze to what extent we can guarantee security when even the circuitry is prone to tampering attacks [22,20,11]. These works typically consider a restricted class of tampering attacks (e.g., individual bit tampering) and assume that large parts of the circuit (and memory) remain untampered.

2   Preliminaries

2.1   Notation

We let $\mathbb{N}$ be the set of naturals. For $n \in \mathbb{N}$, we write $[n] := \{1, \ldots, n\}$. Given a set $\mathcal{S}$, we write $s \leftarrow \mathcal{S}$ to denote that element s is sampled uniformly from $\mathcal{S}$. If S is an algorithm, $y \leftarrow \mathsf{S}(x)$ denotes an execution of S with input x and output y; if S is randomized, then y is a random variable.

Throughout the paper we denote the security parameter by $k \in \mathbb{N}$. A function $\delta(k)$ is called *negligible* in k (or simply negligible) if it vanishes faster than the inverse of any polynomial in k, i.e., $\delta(k) = k^{-\omega(1)}$. A machine S is called *probabilistic polynomial time* (PPT) if for any input $x \in \{0,1\}^*$ the computation of $\mathsf{S}(x)$ terminates in at most $poly(|x|)$ steps and S is probabilistic (i.e., it uses randomness as part of its logic).

Oracle $\mathcal{O}^\ell(s)$ is parametrized by a value s and takes as input functions L and outputs $\mathsf{L}(s)$, returning a total of at most ℓ bits.

2.2   Robust Non-interactive Zero Knowledge

Given an **NP**-relation, let $\mathcal{L} = \{x : \exists w \text{ such that } \mathcal{R}(x, w) = 1\}$ be the corresponding language. A robust non-interactive zero knowledge (NIZK) proof system for $\mathcal{L}$, is a tuple of algorithms $(\mathsf{G_{NIZK}}, \mathsf{Prove}, \mathsf{Verify}, \mathsf{Sim} = (\mathsf{Sim_1}, \mathsf{Sim_2}), \mathsf{Xtr})$ such that the following properties hold [26].

Completeness. For all $x \in \mathcal{L}$ of length k and all w such that $\mathcal{R}(x, w) = 1$, for all $\Omega \leftarrow \mathsf{G_{NIZK}}(1^k)$ we have that $\mathsf{Verify}(\Omega, x, \mathsf{Prove}(\Omega, w, x)) = \mathtt{accept}$

Multi-theorem zero knowledge. For all PPT adversaries A, we have $\mathsf{Real}(k) \approx \mathsf{Simu}(k)$, where $\mathsf{Real}(k)$ and $\mathsf{Simu}(k)$ are distributions defined via the following experiment:

$$\mathsf{Real}(k) = \left\{ \Omega \leftarrow \mathsf{G_{NIZK}}(1^k); out \leftarrow \mathsf{A}^{\mathsf{Prove}(\Omega, \cdot, \cdot)}(\Omega); \text{Output: } out. \right\}$$

$$\mathsf{Simu}(k) = \left\{ (\Omega, tk) \leftarrow \mathsf{Sim_1}(1^k); out \leftarrow \mathsf{A}^{\mathsf{Sim_2}(\Omega, \cdot, tk)}(\Omega); \text{Output: } out. \right\}.$$

Extractability. There exists a PPT algorithm Xtr such that, for all PPT adversaries A, we have

$$\mathbb{P}\left[\begin{array}{c} (\Omega, tk, ek) \leftarrow \mathsf{Sim_1}(1^k); (x, \pi) \leftarrow \mathsf{A}^{\mathsf{Sim_2}(\Omega, \cdot, tk)}(\Omega); \\ w \leftarrow \mathsf{Xtr}(\Omega, (x, \pi), ek); \\ \mathcal{R}(x, w) \neq 1 \wedge (x, \pi) \notin \mathcal{Q} \wedge \mathsf{Verify}(\Omega, x, \pi) = \mathtt{accept} \end{array} \right] \leq negl(k),$$

where the list $\mathcal{Q}$ contains the successful pairs (x_i, π_i) that A has queried to $\mathsf{Sim_2}$.

Similarly to [24], we assume that different statements have different proofs, i.e., if $\text{Verify}(\Omega, x, \pi) = \texttt{accept}$ we have that $\text{Verify}(\Omega, x', \pi) = \texttt{reject}$ for all $x' \neq x$. This property can be achieved by appending the statement to its proof.

We also require that the proof system supports labels, so that the Prove, Verify, Sim and Xtr algorithms now also take a public label λ as input, and the completeness, zero knowledge and extractability properties are updated accordingly. (This can be easily achieved by appending the label λ to the statement x.) More precisely, we write $\text{Prove}^\lambda(\Omega, w, x)$ and $\text{Verify}^\lambda(\Omega, x, \pi)$ for the prover and the verifier, and $\text{Sim}_2^\lambda(\Omega, x, tk)$ and $\text{Xtr}^\lambda(\Omega, (x, \pi), ek)$ for the simulator and the extractor.

2.3 Leakage Resilient Storage

We recall the definition of leakage resilient storage from [13,14]. A leakage resilient storage scheme $(\text{LRS}, \text{LRS}^{-1})$ is a pair of algorithms defined as follows. (1) Algorithm LRS takes as input a secret x and outputs an encoding (s_0, s_1) of x. (2) Algorithm LRS^{-1} takes as input shares (s_0, s_1) and outputs a message x'. Since the LRS that we use in this paper is secure against computationally unbounded adversaries, we state the definition below in the information theoretic setting. It is easy to extend it to also consider computationally bounded adversaries.

Definition 1 (LRS). *We call* $(\text{LRS}, \text{LRS}^{-1})$ *an ℓ-leakage resilient storage scheme (ℓ-LRS) if for all $\theta \in \{0, 1\}$, all secrets x, y and all adversaries* A *it holds that*

$$\left\{ \textsf{Leakage}_{\textsf{A}, x, \theta}(k) \right\}_{k \in \mathbb{N}} \approx_s \left\{ \textsf{Leakage}_{\textsf{A}, y, \theta}(k) \right\}_{k \in \mathbb{N}},$$

where

$$\textsf{Leakage}_{\textsf{A}, x, \theta}(k) = \left\{ \begin{array}{c} (s_0, s_1) \leftarrow \text{LRS}(x); out_\textsf{A} \leftarrow \textsf{A}^{\mathcal{O}^\ell(s_0, \cdot), \mathcal{O}^\ell(s_1, \cdot)}; \\ Output: (s_\theta, out_\textsf{A}). \end{array} \right\}.$$

We remark that Definition 1 is stronger than the standard definition of LRS, in that the adversary is allowed to see one of the two shares after he is done with leakage queries. A careful analysis of the proof, however, shows that the LRS scheme of [14, Lemma 22] satisfies the above generalized notion since the inner product function is a strong randomness extractor [10].

3 Continuous Non-malleability

We start by formally defining an encoding scheme in the common reference string (CRS) model.

Definition 2 (Split-state Encoding Scheme in the CRS Model). *A split-state encoding scheme in the common reference string (CRS) model is a tuple of algorithms* Code $= (\text{Init}, \text{Encode}, \text{Decode})$ *specified below.*

- Init *takes as input the security parameter and outputs a CRS* $\Omega \leftarrow \text{Init}(1^k)$.

- Encode *takes as input some message* $x \in \{0,1\}^k$ *and the CRS* Ω *and outputs a codeword consisting of two parts* (X_0, X_1) *such that* $X_0, X_1 \in \{0,1\}^n$.
- Decode *takes as input a codeword* $(X_0, X_1) \in \{0,1\}^{2n}$ *and the CRS and outputs either a message* $x' \in \{0,1\}^k$ *or a special symbol* $\bot$.

Consider the following oracle $\mathcal{O}_{cnm}((X_0, X_1))$, which is parametrized by an encoding (X_0, X_1) and takes as input functions $\mathsf{T}_0, \mathsf{T}_1 : \{0,1\}^n \to \{0,1\}^n$.

$\underline{\mathcal{O}_{cnm}((X_0, X_1), (\mathsf{T}_0, \mathsf{T}_1)):}$
$\quad (X'_0, X'_1) = (\mathsf{T}_0(X_0), \mathsf{T}_1(X_1))$
$\quad$ If $(X'_0, X'_1) = (X_0, X_1)$ return same*
$\quad$ If Decode$(\Omega, (X'_0, X'_1)) = \bot$, return $\bot$ and "self-destruct"
$\quad$ Else return (X'_0, X'_1).

By "self-destruct" we mean that once Decode$(\Omega, (X'_0, X'_1))$ outputs $\bot$, the oracle will answer $\bot$ to any further query.

Definition 3 (Continuous Non-Malleability). *Let* Code $=$ (Init, Encode, Decode) *be a split-state encoding scheme in the CRS model. We say that* Code *is* q-continuously non-malleable ℓ-leakage resilient $((\ell, q)$-CNMLR for short), if *for all messages* $x, y \in \{0,1\}^k$ *and all PPT adversaries* A *it holds that*

$$\left\{ \mathsf{Tamper}^{cnmlr}_{A,x}(k) \right\}_{k \in \mathbb{N}} \approx_c \left\{ \mathsf{Tamper}^{cnmlr}_{A,y}(k) \right\}_{k \in \mathbb{N}}$$

where

$$\mathsf{Tamper}^{cnmlr}_{A,x}(k) = \left\{ \begin{array}{c} \Omega \leftarrow \mathrm{Init}(1^k); (X_0, X_1) \leftarrow \mathrm{Encode}(\Omega, x); \\ out_A \leftarrow A^{\mathcal{O}^\ell(X_0), \mathcal{O}^\ell(X_1), \mathcal{O}_{cnm}((X_0, X_1))}; \mathit{Output}:\ out_A \end{array} \right\}$$

and A *asks a total of* q *queries to* $\mathcal{O}_{cnm}$.

Without loss of generality we assume that the variable out_A consists of all the bits leaked from X_0 and X_1 (in a vector Λ) and all the outcomes from oracle $\mathcal{O}_{cnm}(X_0, X_1)$ (in a vector Θ); we write this as $out_A = (\Lambda, \Theta)$ where $|\Lambda| \leq 2\ell$ and Θ has exactly q elements.

Intuitively, the above definition captures a setting where a fully adaptive adversary A tries to break non-malleability by tampering several times with a target encoding, obtaining each time some leakage from the decoding process. The only restriction is that whenever a tampering attempt decodes to $\bot$, the system "self-destructs".[3] Note that whenever the adversary mauls (X_0, X_1) to a valid encoding (X'_0, X'_1), oracle $\mathcal{O}_{cnm}$ returns (X'_0, X'_1). This is different from [17,24], where the experiment returns the output of the decoded message, i.e. Decode$(\Omega, (X'_0, X'_1))$. The recent work of Faust et al. [19] consider a similar extension where also the codeword is returned instead of the decoded message and call it super strong non-malleability. Also, we remark that Definition 3 implies strong non-malleability

---

[3] As described in [21] it is easy to see that without such a restriction non-malleability can indeed be broken, since A can simply recover the entire (X_0, X_1) after polynomially many queries.

(as defined in [17,24]) if we restrict A to ask a single query (i.e., $q = 1$) to oracle $\mathcal{O}_{cnm}$.[4] We choose the formulation above because it is stronger and at the same time achieved by our code!

3.1 Uniqueness

As we argue below, constructions that satisfy our new Definition 3 have to meet the following *uniqueness* requirement. Informally this means that for any (possibly adversarially chosen) side of an encoding X'_b it is computationally hard to find two corresponding sides X'_{1-b} and X''_{1-b} such that both (X'_b, X'_{1-b}) and (X'_b, X''_{1-b}) form a valid encoding.

Definition 4 (Uniqueness). *Let* Code $=$ (Init, Encode, Decode) *be a split-state encoding in the CRS model. We say that* Code *satisfies* uniqueness *if for all PPT adversaries* A *and for all* $b \in \{0, 1\}$ *we have:*

$$\mathbb{P}\left[\begin{array}{l} \Omega \leftarrow \text{Init}(1^k); (X'_b, X'_{1-b}, X''_{1-b}) \leftarrow \text{A}(1^k, \Omega); X'_{1-b} \neq X''_{1-b}; \\ \text{Decode}(\Omega, (X'_b, X'_{1-b})) \neq \bot; \text{Decode}(\Omega, (X'_b, X''_{1-b})) \neq \bot \end{array}\right] \leq negl(k).$$

The following attack shows that the uniqueness property is necessary to achieve Definition 3.

Lemma 1. *Let* Code *be* $(0, poly(k))$-*CNMLR. Then* Code *must satisfy uniqueness.*

Proof. For the sake of contradiction, assume that we can efficiently find a triple (X'_0, X'_1, X''_1) such that (X'_0, X'_1) and (X'_0, X''_1) are both valid and $X'_1 \neq X''_1$. For a target encoding (Y_0, Y_1), we describe an efficient algorithm recovering Y_1 with overwhelming probability, by asking $n = poly(k)$ queries to $\mathcal{O}_{cnm}((Y_0, Y_1), \cdot)$.

> For all $i \in [n]$ repeat the following:
> Prepare the i-th tampering function as follows:
> - $\mathsf{T}_0^{(i)}(Y_0)$: Replace Y_0 by X'_0;
> - $\mathsf{T}_1^{(i)}(Y_1)$: If $Y_1[i] = 0$ replace Y_1 by X'_1; otherwise replace it by X''_1.
> Submit $(\mathsf{T}_0^{(i)}, \mathsf{T}_1^{(i)})$ to $\mathcal{O}_{cnm}((Y_0, Y_1), \cdot)$ and obtain (Y'_0, Y'_1).
> If $(Y'_0, Y'_1) = (X'_0, X'_1)$, set $Z[i] \leftarrow 0$.
> Otherwise, if $(Y'_0, Y'_1) = (X'_0, X''_1)$, set $Z[i] \leftarrow 1$.
> Output Z as the guess for Y_1.

The above algorithm clearly succeeds with overwhelming probability, whenever $X'_1 \neq Y_1 \neq X''_1$.[5]

---

[4] It is easy to see that encoding from [24] satisfies the stronger variant of strong non-malleability.

[5] In case $(X'_0, X'_1) = (Y_0, Y_1)$ or $(X'_0, X''_1) = (Y_0, Y_1)$, then the entire encoding can be recovered even with more ease. In this case, whenever the oracle returns same* we know $Y_0 = X'_0$ and $Y_1 \in \{X'_1, X''_1\}$. In the next step we replace the encoding with (X'_0, X'_1); if the oracle returns same* again, then we conclude that $Y_1 = X'_1$, otherwise we conclude $Y_1 = X''_1$.

Once Y_1 is known, we ask one additional query $(T_0^{(n+1)}, T_1^{(n+1)})$ to $\mathcal{O}_{cnm}((Y_0, Y_1), \cdot)$, as follows:

- $T_0^{(n+1)}(Y_0)$ hard-wires Y_1 and computes $y \leftarrow \text{Decode}(\Omega, (Y_0, Y_1))$; if the first bit of y is 0 then T_0 behaves like the identity function, otherwise it overwrites Y_0 with 0^n.
- $T_1^{(n+1)}(Y_0)$ is the identity function.

The above clearly allows to learn one bit of the message in the target encoding and hence contradicts the fact that Code is $(0, poly(k))$-CNMLR.

Attacking existing schemes. The above procedure can be applied to show that the encoding of [24] does not satisfy our notion. Recall that in [24] a message x is encoded as $X_0 = (pk, c := \text{Enc}(pk, x), \pi)$ and $X_1 = sk$. Here, (pk, sk) is a valid key pair and π is a proof of knowledge of a pair (x, sk) such that c decrypts to x under sk *and* (pk, sk) forms a valid key-pair. Clearly, for some $X_1' = sk'$ it is easy to find two valid corresponding parts $X_0' \neq X_0''$ which violates uniqueness.

We mention two important extensions of the attack from Lemma 1, leading to even stronger security breaches:

1. In case the valid pair of encodings (X_0', X_1'), (X_0', X_1'') which violates the uniqueness property are such that $\text{Decode}(\Omega, (X_0', X_1')) \neq \text{Decode}(\Omega, (X_0', X_1''))$, one can show that Lemma 1 still holds in the weaker version of the Definition 3 in which the experiment does not output tampered encodings but only the corresponding decoded message. Note that this applies in particular to the encoding of [24].
2. In case it is possible to find both (X_0', X_1', X_1'') and (X_0', X_0'', X_1') violating uniqueness, a simple variant of the attack allows us to recover both halves of the target encoding which is a total breach of security! However, it is not clear for the scheme of [24] how to find two valid corresponding parts X_1', X_1'', because given pk' it shall of course be computationally hard to find two corresponding valid secret keys sk', sk''.

The above attack can be easily extended to the information theoretic setting to break the constructions of the non-malleable codes (in split-state) recently introduced in [15] and in [1]. In fact, in the following lemma we show that there does *not* exist any information theoretic secure CNMLR code.

Lemma 2. *It is impossible to construct information theoretically secure* $(0, poly(k))$-*CNMLR codes.*

Proof. We prove the lemma by contradiction. Assume that there exists an information theoretically secure $(0, poly(k))$-CNMLR code with $2n$ bits codewords. By Lemma 1, the code must satisfy the uniqueness property. In the information theoretic setting this means that, for all codewords $(X_0, X_1) \in \{0, 1\}^{2n}$ such that $\text{Decode}(\Omega, (X_0, X_1)) \neq \perp$, the following holds: (i) for all $X_1' \in \{0, 1\}^n$ such that $X_1' \neq X_1$, we have $\text{Decode}(\Omega, (X_0, X_1')) = \perp$; (ii) for all $X_0' \in \{0, 1\}^n$, such that $X_0' \neq X_0$, we have $\text{Decode}(\Omega, (X_0', X_1)) = \perp$.

Given a target encoding (X_0, X_1) of some secret x, an unbounded A can define the following tampering function T_b (for $b \in \{0, 1\}$): Given X_b as input, try all possible $X_{1-b} \in \{0, 1\}^n$ until $\text{Decode}(\Omega, (X_0, X_1)) \neq \bot$. By property (i)-(ii) above, we conclude that for all $X'_{1-b} \neq X_{1-b}$, the decoding algorithm $\text{Decode}(\Omega, (X_b, X'_{1-b}))$ outputs $\bot$ with overwhelming probability. Thus, T_b can recover $x = \text{Decode}(\Omega, (X_b, X_{1-b}))$ and if the first bit of the decoded value is 0 leave the target encoding unchanged, otherwise $(\mathsf{T}_0, \mathsf{T}_1)$ modifies the encoding with an invalid codeword. The above clearly allows to learn one bit of the message in the target encoding, and hence contradicts the fact that the code is $(0, poly(k))$-CNMLR.

Note that the attack of Lemma 2 requires the tampering function to be unbounded. In case when the tampering functions are computationally bounded and only the adversary is computationally unbounded we do not know how to make the above attack work.

4   The Code

Consider the following split-state encoding scheme in the CRS model (Init, Encode, Decode), based on an LRS scheme $(\text{LRS}, \text{LRS}^{-1})$, on a family of collision resistant hash functions $\mathcal{H} = \{H_t : \{0, 1\}^{poly(k)} \to \{0, 1\}^k\}_{t \in \{0, 1\}^k}$ and on a robust non-interactive zero knowledge proof system $(\mathsf{G}_{\text{NIZK}}, \text{Prove}, \text{Verify})$ which supports labels, for language $\mathcal{L}_{\mathcal{H}, t} = \{h : \exists s \text{ such that } h = H_t(s)\}$.

$\text{Init}(1^k)$. Sample $t \leftarrow \{0, 1\}^k$ and run $\Omega \leftarrow \mathsf{G}_{\text{NIZK}}(1^k)$.

$\text{Encode}(\Omega, x)$. Let $(s_0, s_1) \leftarrow \text{LRS}(x)$. Compute $h_0 = H_t(s_0)$, $h_1 = H_t(s_1)$ and $\pi_0 \leftarrow \text{Prove}^{\lambda_1}(\Omega, s_0, h_0)$, $\pi_1 \leftarrow \text{Prove}^{\lambda_0}(\Omega, s_1, h_1)$, where the labels are defined as $\lambda_0 = h_0$, $\lambda_1 = h_1$. (Note that the pre-image of h_b is s_b and the proof π_b is computed for statement h_b using label h_{1-b}.) Output $(X_0, X_1) = ((s_0, h_1, \pi_1, \pi_0), (s_1, h_0, \pi_0, \pi_1))$.

$\text{Decode}(\Omega, (X_0, X_1))$. The decoding parses X_b as $(s_b, h_{1-b}, \pi_{1-b}, \pi_b)$, computes $\lambda_b = H_t(s_b)$ and then proceeds as follows:
   (a) *Local check.* If $\text{Verify}^{\lambda_1}(\Omega, h_0, \pi_0)$ or $\text{Verify}^{\lambda_0}(\Omega, h_1, \pi_1)$ output reject in any of the two sides X_0, X_1, return $x' = \bot$.
   (b) *Cross check.* If (i) $h_0 \neq H_t(s_0)$ or $h_1 \neq H_t(s_1)$, or (ii) the proofs (π_0, π_1) in X_0 are different from the ones in X_1, then return $x' = \bot$.
   (c) *Decoding.* Otherwise, return $x' = \text{LRS}^{-1}(s_0, s_1)$.

We start by showing that the above code satisfies the uniqueness property (cf. Definition 4).

Lemma 3. *Let* Code $=$ (Init, Encode, Decode) *be as above. Then, if* $\mathcal{H}$ *is a family of collision resistant hash functions* Code *satisfies uniqueness.*

Proof. We show that Definition 4 is satisfied for $b = 0$. The proof for $b = 1$ is identical and is therefore omitted.

Assume that there exists a PPT adversary A that, given as input $\Omega \leftarrow$ Init(1^k), is able to produce (X_0', X_1', X_1'') such that both (X_0', X_1') and (X_0', X_1'') are valid, but $X_1' \neq X_1''$. Let $X_0' = (s_0', h_1', \pi_1', \pi_0')$, $X_1' = (s_1', h_0', \pi_0', \pi_1')$ and $X_1'' = (s_1'', h_0'', \pi_0'', \pi_1'')$.

Since s_0' is the same in both encodings, we must have $h_0' = h_0''$ as the hash function is deterministic. Furthermore, since both (X_0', X_1') and (X_0', X_1'') are valid, the proofs must verify successfully and therefore we must have $\pi_0' = \pi_0''$ and $\pi_1' = \pi_1''$. It follows that $X_1'' = (s_1'', h_0', \pi_0', \pi_1')$, such that $s_1'' \neq s_1'$. Clearly (s_1', s_1'') is a collision for h_1', a contradiction.

While the uniqueness property is a necessary requirement to achieve continuous non-malleability, showing that that the above code is a continuous non-malleable and leakage resilient code requires to overcome several technical challenges. We next state our main theorem and give a proof outline in the following section. The full proof of Theorem 1 is deferred to the full version of this paper.

Theorem 1. *Let* Code $=$ (Init, Encode, Decode) *be as above. Assume that* (LRS, LRS^{-1}) *is an* ℓ'*-LRS,* $\mathcal{H} = \{H_t : \{0,1\}^{poly(k)} \rightarrow \{0,1\}^k\}_{t \in \{0,1\}^k}$ *is a family of collision resistant hash functions and* (G$_{\text{NIZK}}$, Prove, Verify) *is a robust NIZK proof system for language* $\mathcal{L}_{\mathcal{H},t}$. *Then* Code *is* (ℓ, q)*-CNMLR, for any* $q = poly(k)$ *and* $\ell' \geq \max\{2\ell + (k+1)\lceil \log(q) \rceil, 2k+1\}$.

4.1   Outline of the Proof

In order to build some intuition, let us first explain why a few natural attacks do not work. Clearly, the uniqueness property (cf. Lemma 3) rules out the attack of Lemma 1. As a first attempt, the adversary could try to modify the proof π_0 to a different proof π_0', by using the fact that X_0 contains the corresponding witness s_0 and the correct label h_1. However, to ensure the validity of (X_0', X_1'), this would require to place π_0' in X_1', which should be hard without knowing a witness (by the robustness of the proof system). Alternatively, one could try to maul the two halves (s_0, s_1) of the LRS scheme, into a pair (s_0', s_1') encoding a related message.[6] This requires, for instance, to change the proof π_0 into π_0' and place π_0' in X_1', which again should be hard without knowing a witness and the correct label.

Let us now try to give a high-level overview of the proof. Given a polynomial time distinguisher D that violates continuous non-malleability of Code, we build another polynomial time distinguisher D' which breaks leakage resilience of (LRS, LRS^{-1}). Distinguisher D', which can access oracles $\mathcal{O}^{\ell'}(s_0)$ and $\mathcal{O}^{\ell'}(s_1)$, has to distinguish whether (s_0, s_1) is the encoding of message x or message y and will do so with the help of D's advantage in distinguishing Tamper$_x^{\text{cnmlr}}$ from Tamper$_y^{\text{cnmlr}}$. The main difficulty in the reduction is how D' can simulate the answers from the tampering oracle $\mathcal{O}_{\text{cnm}}$ (cf. Definition 3), without knowing the target encoding (X_0, X_1). This is the main point where our techniques diverge

---

[6] When the LRS is implemented using the inner product extractor this is indeed possible, as argued in [15].

significantly from [24] (as in [24] the reduction "knows" a complete half of the encoding). In our case, in fact, D' can only access the two halves X_0 and X_1 "inside" the oracles $\mathcal{O}^{\ell'}(s_0)$ and $\mathcal{O}^{\ell'}(s_1)$.[7] However, it is not clear how this helps answering tampering queries, as the latter requires access to *both* X_0 and X_1 for decoding the tampered message, whereas the reduction can only access X_0 and X_1 *separately*.

For ease of description, in what follows we simply assume that D' can access directly $\mathcal{O}^{\ell'}(X_0)$ and $\mathcal{O}^{\ell'}(X_1)$. Furthermore, let us assume that D can only issue tampering queries (we discuss how to additionally handle leakage briefly at the end of the outline). Like any standard reduction, D' samples some randomness r and fixes the random tape of D to r. Our novel strategy is to construct a polynomial time algorithm $\mathsf{F}(r)$ that, given access to $\mathcal{O}^{\ell'}(X_0)$, $\mathcal{O}^{\ell'}(X_1)$, outputs the smallest index j^* which indicates the round where $\mathsf{D}(r)$ provokes a self-destruct in $\mathsf{Tamper}_*^{\mathrm{cnmlr}}$. Before explaining how the actual algorithm works, let us explain how D' can complete the reduction using such a self-destruct finder F. At the beginning, it runs $\mathsf{F}(r)$ in order to leak the index j^*. At this point D' is done with leakage queries and asks to get X_0 (i.e., it chooses $\theta = 0$ in Definition 1).[8] Given X_0, distinguisher D' runs $\mathsf{D}(r)$ (with the *same* random coins r used for F). Hence, for all $1 \le j < j^*$, upon input the j-th tampering query $(\mathsf{T}_0^{(j)}, \mathsf{T}_1^{(j)})$, distinguisher D' lets $X_0' = \mathsf{T}_0^{(j)}(X_0) = (s_0', h_1', \pi_1', \pi_0')$ and answers as follows:

1. In case $X_0' = X_0$ (so called type A queries), output same^*.
2. In case $X_0' \ne X_0$ and either of the proofs in X_0' does not verify correctly (so called type B queries), output $\bot$.
3. In case $X_0' \ne X_0$ and both the proofs in X_0' verify correctly (so called type C queries), check if $\pi_1' = \pi_1$; if 'yes' (in which case there is no hope to extract from π_1') then output $\bot$.
4. Otherwise (so called type D queries), attempt to extract s_1' from π_1', define $X_1' = (s_1', h_0', \pi_0', \pi_1')$ and output (X_0', X_1').

Note that from round j^* on, all queries can be answered with $\bot$, and this is a correct simulation as $\mathsf{D}(r)$ provokes a self-destruct at round j^* in the real experiment.

In the proof of Theorem 1, we show that the above strategy is sound. with overwhelming probability over the choice of r the output produced by the above simulation is equal to the output that $\mathsf{D}(r)$ would have seen in the real experiment *until a self-destruct occurs*.[9]

Let us give some intuition why the above simulation is indeed sound. For type A queries, note that when $X_0' = X_0$ we must have $X_1' = X_1$ with overwhelming

---

[7] Looking ahead, this can be achieved by first leaking the hash values h_0, h_1 of s_0, s_1, simulating the proofs π_0, π_1, and then hard-wiring these values into all leakage queries.

[8] Recall that this is a simplification, as by choosing $\theta = 0$ the distinguisher will obtain s_0. See also footnote 7.

[9] It is crucial that both the real and simulated experiments are run with the same r.

probability, as otherwise (X_0, X_1, X_1') would violate uniqueness. In case of type B queries, the decoding process in the real experiment would output $\perp$, so D' does a perfect simulation. The case of type C queries is a bit more delicate. In this case we use the facts that (i) in the NIZK proof system we use, different statements must have different proofs and (ii) the hash function is collision resistant, to show that X_0' must be of the form $X_0' = (s_0, h_1, \pi_1, \pi_0')$ and $\pi_0' \neq \pi_1$. A careful analysis shows that the latter contradicts leakage resilience of the underlying LRS scheme. Finally, for type D queries, note that whenever D' extracts a witness from a valid proof $\pi_1' \neq \pi_1$, the witness must be valid with overwhelming probability (as the NIZK is simulation extractable).

Next, let us explain how to construct the algorithm F. Roughly, $\mathsf{F}(r)$ runs $\mathsf{D}(r)$ "inside" the oracles $\mathcal{O}^{\ell'}(X_0)$, $\mathcal{O}^{\ell'}(X_1)$ as part of the leakage functions, and simulates the answers for $\mathsf{D}(r)$'s tampering queries using only one side of the target encoding, in the exact same way as outlined in (1)-(4) above. Let $\boldsymbol{\Theta}_b$, for $b \in \{0, 1\}$, denote the output simulated by F inside $\mathcal{O}^{\ell'}(X_b)$. To locate the self-destruct index j^*, we rely on the following property: the vectors $\boldsymbol{\Theta}_0$ and $\boldsymbol{\Theta}_1$ contain identical values until coordinate $j^* - 1$, but $\boldsymbol{\Theta}_0[j^*] \neq \boldsymbol{\Theta}_1[j^*]$ with overwhelming probability (over the choice of r).

This implies that j^* can be computed as the first coordinate where $\boldsymbol{\Theta}_0$ and $\boldsymbol{\Theta}_1$ are different. Hence, F can obtain the self-destruct index by using its adaptive access to oracles $\mathcal{O}^{\ell'}(X_0)$, $\mathcal{O}^{\ell'}(X_1)$ and apply a standard binary search algorithm to $\boldsymbol{\Theta}_0$, $\boldsymbol{\Theta}_1$. Note that the latter requires at most a logarithmic number of bits of adaptive leakage.

One technical problem is that F, in order to run $\mathsf{D}(r)$ inside of, say $\mathcal{O}^{\ell'}(X_0)$, and compute $\boldsymbol{\Theta}_0$, has also to answer leakage queries from $\mathsf{D}(r)$. Clearly, all leakages from X_0 can be easily computed, however it is not clear how to simulate leakages from X_1 (as we cannot access $\mathcal{O}^{\ell'}(X_1)$ inside $\mathcal{O}^{\ell'}(X_0)$). Fortunately, the latter issue can be avoided by letting F query $\mathcal{O}^{\ell'}(X_0)$ and $\mathcal{O}^{\ell'}(X_1)$ alternately, and aborting the execution of $\mathsf{D}(r)$ whenever it is not possible to answer a leakage query. It is not hard to show that after at most ℓ steps all leakages will be known, and F can run $\mathsf{D}(r)$ inside $\mathcal{O}^{\ell'}(X_0)$ without having access to $\mathcal{O}^{\ell'}(X_1)$. (All this comes at the price of some loss in the leakage bound, but, as we show in the proof, not too much.)

5   Application to Tamper Resilient Security

In this section we apply our notion of CNMLR codes to protect arbitrary functionalities against split-state tampering and leakage attacks.

5.1   Stateless Functionalities

We start by looking at the case of *stateless* functionalities $\mathcal{G}(st, \cdot)$, which take as input a secret state $st \in \{0, 1\}^k$ and a value $x \in \{0, 1\}^u$ to produce some output $y \in \{0, 1\}^v$. The function $\mathcal{G}$ is public and can be randomized.

The main idea is to transform the original functionality $\mathcal{G}(st, \cdot)$ into some "hardened" functionality $\mathcal{G}^{\mathsf{Code}}$ via a CNMLR code Code. Previous transformations aiming to protect stateless functionalities [17,24] required to freshly re-encode the state st each time the functionality is invoked. Our approach avoids the re-encoding of the state at each invocation, leading to a stateless transformation. This solves an open question from [17]. Moreover we consider a setting where the encoded state is stored in a memory $(\mathcal{M}_0, \mathcal{M}_1)$ which is much larger than the size needed to store the encoding itself (say $|\mathcal{M}_0| = |\mathcal{M}_1| = s$ where s is polynomial in the length of the encoding). When (perfect) erasures are not possible, this feature allows the adversary to make copies of the initial encoding and tamper continuously with it, and was not considered in previous models.

Let us formally define what it means to harden a stateless functionality.

Definition 5 (Stateless hardened functionality). *Let* $\mathsf{Code} = (\mathrm{Init}, \mathrm{Encode}, \mathrm{Decode})$ *be a split-state encoding scheme in the CRS model, with* k *bits messages and* $2n$ *bits codewords. Let* $\mathcal{G} : \{0,1\}^k \times \{0,1\}^u \to \{0,1\}^v$ *be a stateless functionality with secret state* $st \in \{0,1\}^k$, *and let* $\varphi \in \{0,1\}$ *be a public value initially set to zero. We define a stateless hardened functionality* $\mathcal{G}^{\mathsf{Code}} : \{0,1\}^{2s} \times \{0,1\}^u \to \{0,1\}^v$ *with a modified state* $st' \in \{0,1\}^{2s}$ *and* $s = poly(n)$. *The hardened functionality* $\mathcal{G}^{\mathsf{Code}}$ *is a triple of algorithms* $(\mathrm{Init}, \mathrm{Setup}, \mathrm{Execute})$ *described as follows:*

- $\Omega \leftarrow \mathrm{Init}(1^k)$: *Run the initialization procedure of the coding scheme to sample* $\Omega \leftarrow \mathrm{Init}(1^k)$.
- $(\mathcal{M}_0, \mathcal{M}_1) \leftarrow \mathrm{Setup}(\Omega, st)$: *Let* $(X_0, X_1) \leftarrow \mathrm{Encode}(\Omega, st)$. *For* $b \in \{0,1\}$, *store* X_b *in the first* n *bits of* $\mathcal{M}_b$, *i.e.* $\mathcal{M}_b[1 \ldots n] \leftarrow X_b$. *(The remaining bits of* $\mathcal{M}_b$ *are set to* 0^{s-n}.) *Define* $st' := (\mathcal{M}_0, \mathcal{M}_1)$.
- $y \leftarrow \mathrm{Execute}(x)$: *Read the public value* φ. *In case* $\varphi = 1$ *output* $\perp$. *Otherwise, let* $X_b = \mathcal{M}_b[1 \ldots n]$ *for* $b \in \{0,1\}$. *Run* $st \leftarrow \mathrm{Decode}(\Omega, (X_0, X_1))$; *if* $st = \perp$, *then output* $\perp$ *and set* $\varphi = 1$. *Otherwise output* $y \leftarrow \mathcal{G}(st, x)$.

Remark 1 (On φ). The public value φ is just a way how to implement the "self-destruct" feature. An alternative approach would be to let the hardened functionality simply output a dummy value and overwrite $(\mathcal{M}_0, \mathcal{M}_1)$ with the all-zero string. As we insist on the hardened functionality being stateless, we use the first approach here.

Note that we assume that φ is untamperable. It is easy to see that this is necessary, as an adversary tampering with φ could always switch-off the self-destruct feature and apply a variant of the attack from [21] to recover the secret state.

Similarly to [17,24], security of $\mathcal{G}^{\mathsf{Code}}$ is defined via the comparison of a real and an ideal experiment. The real experiment features an adversary A interacting with $\mathcal{G}^{\mathsf{Code}}$; the adversary is allowed to honestly run the functionality on any chosen input, but also to modify the secret state and retrieve a bounded amount of information from it. The ideal experiment features a simulator S; the simulator is given black-box access to the original functionality $\mathcal{G}$ and to the adversary A, but is *not* allowed any tampering or leakage query. The two experiments are formally described below.

Experiment $\mathsf{REAL}_{\mathsf{A}}^{\mathcal{G}^{\mathsf{Code}}(st',\cdot)}(k)$. First $\Omega \leftarrow \mathsf{Init}(1^k)$ and $(\mathcal{M}_0, \mathcal{M}_1) \leftarrow \mathsf{Setup}(\Omega, st)$ are run and Ω is given to A. Then A can issue the following commands polynomially many times (in any order):

- $\langle \mathsf{Leak}, (\mathsf{L}_0^{(j)}, \mathsf{L}_1^{(j)}) \rangle$: In response to the j-th leakage query, compute $\Lambda_0^{(j)} \leftarrow \mathsf{L}_0^{(j)}(\mathcal{M}_0)$ and $\Lambda_1^{(j)} \leftarrow \mathsf{L}_1^{(j)}(\mathcal{M}_1)$ and output $(\Lambda_0^{(j)}, \Lambda_1^{(j)})$.
- $\langle \mathsf{Tamper}, (\mathsf{T}_0^{(j)}, \mathsf{T}_1^{(j)}) \rangle$: In response to the j-th tampering query, compute $\mathcal{M}_0' \leftarrow \mathsf{T}_0^{(j)}(\mathcal{M}_0)$ and $\mathcal{M}_1' \leftarrow \mathsf{T}_1^{(j)}(\mathcal{M}_1)$ and replace $(\mathcal{M}_0, \mathcal{M}_1)$ with $(\mathcal{M}_0', \mathcal{M}_1')$.
- $\langle \mathsf{Eval}, x_j \rangle$: In response to the j-th evaluation query, run $y_j \leftarrow \mathsf{Execute}(x_j)$. In case $y_j = \bot$ output $\bot$ and self-destruct; otherwise output y_j.

The output of the experiment is defined as

$$\mathsf{REAL}_{\mathsf{A}}^{\mathcal{G}^{\mathsf{Code}}(st',\cdot)}(k) = (\Omega; ((x_1, y_1), (x_2, y_2), \ldots); ((\Lambda_0^{(1)}, \Lambda_1^{(1)}), (\Lambda_0^{(2)}, \Lambda_1^{(2)}), \cdots)).$$

Experiment $\mathsf{IDEAL}_{\mathsf{S}}^{\mathcal{G}(st,\cdot)}(k)$. The simulator sets up the CRS Ω and is given black-box access to the functionality $\mathcal{G}(st, \cdot)$ and the adversary A. The output of the experiment is defined as

$$\mathsf{IDEAL}_{\mathsf{S}}^{\mathcal{G}(st,\cdot)}(k) = (\Omega; ((x_1, y_1), (x_2, y_2), \ldots); ((\Lambda_0^{(1)}, \Lambda_1^{(1)}), (\Lambda_0^{(2)}, \Lambda_1^{(2)}), \cdots)),$$

where $((x_j, y_j), ((\Lambda_0^{(j)}, \Lambda_1^{(j)})))$ are the input/output/leakage tuples simulated by S.

Definition 6 (Polyspace leak/tamper simulatability). *Let* Code *be a split-state encoding scheme in the CRS model and consider a stateless functionality* $\mathcal{G}$ *with corresponding hardened functionality* $\mathcal{G}^{\mathsf{Code}}$. *We say that* Code *is polyspace* (ℓ, q)-*leak/tamper simulatable for* $\mathcal{G}$, *if the following conditions are satisfied:*

1. *Each memory part* $\mathcal{M}_b$ *(for* $b \in \{0, 1\}$*) has size* $s = \mathrm{poly}(n)$.
2. *The adversary asks at most* q *tampering queries and leaks a total of at most* ℓ *bits from each memory part.*
3. *For all PPT adversaries* A *there exists a PPT simulator* S *such that for any initial state* st,

$$\left\{ \mathsf{REAL}_{\mathsf{A}}^{\mathcal{G}^{\mathsf{Code}}(st',\cdot)}(k) \right\}_{k \in \mathbb{N}} \approx_c \left\{ \mathsf{IDEAL}_{\mathsf{S}}^{\mathcal{G}(st,\cdot)}(k) \right\}_{k \in \mathbb{N}}.$$

We show the following result.

Theorem 2. *Let* $\mathcal{G}$ *be a stateless functionality and* Code $=$ (Init, Encode, Decode) *be any* (ℓ, q)-*CNMLR split-state encoding scheme in the CRS model. Then* Code *is polyspace* (ℓ, q)-*leak/tamper simulatable for* $\mathcal{G}$.

Proof. We discuss the overall proof approach first. We start with describing a simulator S running in experiment $\mathsf{IDEAL}_\mathsf{S}^{\mathcal{G}(st,\cdot)}(k)$ which attempts to simulate the view of adversary A running in the experiment $\mathsf{REAL}_\mathsf{A}^{\mathcal{G}^{\mathsf{Code}}(st',\cdot)}(k)$; the simulator is given black-box access to A (which can issue Tamper, Leak, and Eval queries) and to the functionality $\mathcal{G}(st,\cdot)$ for some state st. To argue that our simulator is "good" we show that if there exists a PPT distinguisher D and a PPT adversary A such that for some state st, D distinguishes the experiments $\mathsf{IDEAL}_\mathsf{S}^{\mathcal{G}(st,\cdot)}(k)$ and $\mathsf{REAL}_\mathsf{A}^{\mathcal{G}^{\mathsf{Code}}(st',\cdot)}(k)$ with non-negligible probability, then we can build another distinguisher D' and an adversary A' such that D' can distinguish $\mathsf{Tamper}_{\mathsf{A}',0^k}^{\mathsf{cnmlr}}$ and $\mathsf{Tamper}_{\mathsf{A}',st}^{\mathsf{cnmlr}}$ with non-negligible probability. In the last step essentially we reduce the CNMLR property of Code to the polyspace leak/tamper simulatability of the code itself.

The simulator starts by sampling the common reference string $\Omega \leftarrow \mathsf{Init}(1^k)$ and the public value $\varphi = 0$. Then it samples a random encoding of 0^k, namely $(Z_0, Z_1) \leftarrow \mathsf{Encode}(\Omega, 0^k)$ and sets $\mathcal{M}_b[1\ldots,n] \leftarrow Z_b$ for $b \in \{0,1\}$. The remaining bits of $(\mathcal{M}_0, \mathcal{M}_1)$ are set to 0^{s-n}. Hence, S alternates between the following two modes (starting with the normal mode in the first round):

- *Normal Mode.* Given state $(\mathcal{M}_0, \mathcal{M}_1)$, while A continues issuing queries, answer as follows:
  - $\langle \mathsf{Eval}, x_j \rangle$: Upon input the j-th evaluation query invoke $\mathcal{G}(st, \cdot)$ to get $y_j \leftarrow \mathcal{G}(st, x_j)$ and reply with y_j.
  - $\langle \mathsf{Tamper}, (\mathsf{T}_0^{(j)}, \mathsf{T}_1^{(j)}) \rangle$: Upon input the j-th tampering query, compute $\mathcal{M}_b' \leftarrow \mathsf{T}_b^{(j)}(\mathcal{M}_b)$ for $b \in \{0,1\}$. In case $(\mathcal{M}_0'[1\ldots n], \mathcal{M}_1'[1\ldots n]) = (Z_0, Z_1)$ then continue in the current mode. Otherwise go to the over-written mode defined below with state $(\mathcal{M}_0', \mathcal{M}_1')$.
  - $\langle \mathsf{Leak}, (\mathsf{L}_0^{(j)}, \mathsf{L}_1^{(j)}) \rangle$: Upon input the j-th leakage query, compute $\Lambda_b^{(j)} = \mathsf{L}_b^{(j)}(Z_b)$ for $b \in \{0,1\}$ and reply with $(\Lambda_0^{(j)}, \Lambda_1^{(j)})$.
- *Overwritten Mode.* Given state $(\mathcal{M}_0', \mathcal{M}_1')$, while A continues issuing queries, answer as follows:
  - Let $\tau = (\mathcal{M}_0', \mathcal{M}_1')$. Simulate the hardened functionality $\mathcal{G}^{\mathsf{Code}}(\tau, \cdot)$ and answer all Eval and Leak queries as the real experiment $\mathsf{REAL}_\mathsf{A}^{\mathcal{G}^{\mathsf{Code}}(\tau,\cdot)}(k)$ would do.
  - Upon input the j-th tampering query $(\mathsf{T}_0^{(j)}, \mathsf{T}_1^{(j)})$, compute $\mathcal{M}_b'' \leftarrow \mathsf{T}_b^{(j)}(\mathcal{M}_b')$ for $b \in \{0,1\}$. In case $(\mathcal{M}_0''[1\ldots n], \mathcal{M}_1''[1\ldots n]) = (Z_0, Z_1)$ then go to the normal mode with state $(\mathcal{M}_0, \mathcal{M}_1) := (\mathcal{M}_0'', \mathcal{M}_1'')$. Otherwise continue in the current mode.
- When A halts and outputs $\mathsf{view}_\mathsf{A} = (\Omega; ((x_1, y_1), (x_2, y_2), \ldots);$ $((\Lambda_0^{(1)}, \Lambda_1^{(1)}), (\Lambda_0^{(2)}, \Lambda_1^{(2)}), \cdots))$, set $\mathsf{view}_\mathsf{S} = \mathsf{view}_\mathsf{A}$ and output view_S as output of $\mathsf{IDEAL}_\mathsf{S}^{\mathcal{G}(st,\cdot)}(k)$.

Intuitively, since the coding scheme is non-malleable, the adversary can either keep the encoding unchanged or overwrite it with the encoding of some unrelated message. These two cases are captured in the above modes: The simulator starts in the normal mode and then, whenever the adversary mauls the initial encoding, it switches to the overwritten mode. However, the adversary can use the extra space to keep a copy of the original encoding and place it back at some later point in time. When this happens, the simulator switches back to the normal mode; this switching is important to maintain simulation.

To finish the proof, we have to argue that the output of experiment $\mathsf{IDEAL}_{\mathsf{S}}^{\mathcal{G}(st,\cdot)}(k)$ is computationally indistinguishable from the output of experiment $\mathsf{REAL}_{\mathsf{A}}^{\mathcal{G}^{\mathsf{Code}}(st',\cdot)}(k)$. This is done in the lemma below.

Lemma 4. *Let* S *be defined as above. Then for all PPT adversaries* A *and all* $st \in \{0,1\}^k$, *the following holds:*

$$\left\{\mathsf{REAL}_{\mathsf{A}}^{\mathcal{G}^{\mathsf{Code}}(st',\cdot)}(k)\right\}_{k\in\mathbb{N}} \approx_c \left\{\mathsf{IDEAL}_{\mathsf{S}}^{\mathcal{G}(st,\cdot)}(k)\right\}_{k\in\mathbb{N}}.$$

Proof. By contradiction, assume that there exists a PPT distinguisher D, a PPT adversary A and some state $st \in \{0,1\}^k$ such that:

$$\left|\mathbb{P}\left[\mathsf{D}(\mathsf{IDEAL}_{\mathsf{S}}^{\mathcal{G}(st,\cdot)}(k)) = 1\right] - \mathbb{P}\left[\mathsf{D}(\mathsf{REAL}_{\mathsf{A}}^{\mathcal{G}^{\mathsf{Code}}(st',\cdot)}(k)) = 1\right]\right| \geq \epsilon, \quad (1)$$

where $\epsilon(k)$ is some non-negligible function of the security parameter k.

We build a PPT distinguisher D′ and a PPT adversary A′ telling apart the experiments $\mathsf{Tamper}_{\mathsf{A}',0^k}^{\mathsf{cnmlr}}(k)$ and $\mathsf{Tamper}_{\mathsf{A}',st}^{\mathsf{cnmlr}}(k)$; this contradicts our assumption that Code is CNMLR. The distinguisher D′ is given the CRS $\Omega \leftarrow \mathsf{Init}(1^k)$ and can access $\mathcal{O}_{\mathsf{cnm}}((X_0, X_1), \cdot)$ (for at most q times) and $\mathcal{O}^\ell(X_0)$, $\mathcal{O}^\ell(X_1)$; here (X_0, X_1) is either an encoding of 0^k or an encoding of st. The distinguisher D′ keeps a flag SAME (initially set to TRUE) and a flag STOP (initially set to FALSE). After simulating the public values, D′ mimics the enviroment for D as follows:

- $\langle\mathsf{Tamper}, (\mathsf{T}_0^{(j)}, \mathsf{T}_1^{(j)})\rangle$: Upon input tampering functions $(\mathsf{T}_0^{(j)}, \mathsf{T}_1^{(j)})$, the distinguisher D′ uses the oracle $\mathcal{O}_{\mathsf{cnm}}((X_0, X_1), \cdot)$ to answer them.[10] However, it can not simply forward the queries because of the following two reasons:
  - The tampering functions $(\mathsf{T}_0^{(j)}, \mathsf{T}_1^{(j)})$ maps from s bits to s bits, whereas the oracle $\mathcal{O}_{\mathsf{cnm}}((X_0, X_1), \cdot)$ expects tampering functions mapping from n bits to n bits.
  - In both the real and the ideal experiments the tampering functions are applied to the current state (which may be different from the initial state), whereas in experiment $\mathsf{Tamper}_{\mathsf{A}',*}^{\mathsf{cnmlr}}$ the oracle $\mathcal{O}_{\mathsf{cnm}}((X_0, X_1), \cdot)$ always applies $(\mathsf{T}_0^{(j)}, \mathsf{T}_1^{(j)})$ to the target encoding (X_0, X_1).

  To take into account the above differences, D′ modifies $(\mathsf{T}_0^{(j)}, \mathsf{T}_1^{(j)})$ as follows. Define the functions $\mathsf{T}_{\mathsf{in}} : \{0,1\}^n \to \{0,1\}^s$ and $\mathsf{T}_{\mathsf{out}} : \{0,1\}^s \to \{0,1\}^n$ as $\mathsf{T}_{\mathsf{in}}(x) = (x||0^{s-n})$ and $\mathsf{T}_{\mathsf{out}}(x||x') = x$, for any $x \in \{0,1\}^n$ and $x' \in \{0,1\}^{s-n}$. The distinguisher D′ queries $\mathcal{O}_{\mathsf{cnm}}((X_0, X_1), \cdot)$ with the function

---

[10] Formally D′ has to access $\mathcal{O}_{\mathsf{cnm}}(\cdot)$ via A′. For simplicity we assume that D′ can access the oracle directly. In fact, A′ just acts as an interface between the experiment $\mathsf{Tamper}_{\mathsf{A}',*}^{\mathsf{cnmlr}}$ and D′.

pair $(\tilde{\mathsf{T}}_0^{(j)}, \tilde{\mathsf{T}}_1^{(j)})$ where each $\tilde{\mathsf{T}}_b^{(j)}$ is defined as $\tilde{\mathsf{T}}_b^{(j)} := \mathsf{T}_{\text{out}} \circ \mathsf{T}_b^{(j)} \circ \mathsf{T}_b^{(j-1)} \circ \ldots \circ \mathsf{T}_b^{(1)} \circ \mathsf{T}_{\text{in}}$ for $b \in \{0,1\}$.

In case the oracle returns $\bot$, then D′ sets STOP to TRUE. In case the oracle returns same*, then D′ sets SAME to TRUE. Otherwise, in case the oracle returns an encoding (X_0', X_1'), then D′ sets SAME to FALSE.

- $\langle$Leak, $(\mathsf{L}_0^{(j)}, \mathsf{L}_1^{(j)})\rangle$: Upon input leakage functions $(\mathsf{L}_0^{(j)}, \mathsf{L}_1^{(j)})$, the distinguisher D′ defines $(\tilde{\mathsf{L}}_0^{(j)}, \tilde{\mathsf{L}}_1^{(j)})$ (in a similar way as above), forwards those functions to $\mathcal{O}^\ell(X_0)$, $\mathcal{O}^\ell(X_1)$ and sends the answer from the oracles back to D.

- $\langle$Eval, $x_j\rangle$: Upon input an evaluation query for value x_j, the distinguisher D′ checks first that STOP equals FALSE. If this is not the case, then D′ returns $\bot$ to D. Otherwise, D′ checks that SAME equals TRUE. If this is the case, it runs $y_j \leftarrow \mathcal{G}(st, x_j)$ and gives y_j to D. Else (if SAME equals FALSE), it computes $y_j \leftarrow \mathcal{G}(st', x_j)$, where st' is the output of Decode($\Omega, (X_0', X_1')$), and gives y_j to D.

Finally, D′ outputs whatever D outputs.

For the analysis, first note that D′ runs in polynomial time. Furthermore, D′ asks exactly q queries to $\mathcal{O}_{\text{cnm}}$ and leaks at most ℓ bits from the target encoding (X_0, X_1). It is also easy to see that in case (X_0, X_1) is an encoding of $st \in \{0,1\}^k$, then D′ perfectly simulates the view of adversary D in the experiment $\mathsf{REAL}_{\mathsf{A}}^{\mathcal{G}^{\mathsf{Code}}(st', \cdot)}(k)$. On the other hand, in case (X_0, X_1) is an encoding of 0^k, we claim that D′ perfectly simulates the view of D in the experiment $\mathsf{IDEAL}_{\mathsf{S}}^{\mathcal{G}(st, \cdot)}(k)$. This is because: (i) Whenever SAME equals TRUE, then D′ answers evaluation queries by running $\mathcal{G}$ on state st and tampering/leakage queries using a pre-sampled encoding of 0^k (this corresponds to the normal mode of S); (ii) Whenever SAME equals FALSE, then D′ answers evaluation queries by running $\mathcal{G}$ on the current tampered state st' which results from applying the tampering functions to a pre-sampled encoding of 0^k (this corresponds to the overwritten mode of S).

Combining the above argument with Eq. (1) we obtain

$$\left| \mathbb{P}\left[\mathsf{D}(\mathsf{Tamper}_{\mathsf{A}',0^k}^{\text{cnmlr}}(k)) = 1\right] - \mathbb{P}\left[\mathsf{D}(\mathsf{Tamper}_{\mathsf{A}',st}^{\text{cnmlr}}(k)) = 1\right] \right| \geq \epsilon,$$

which is a contradiction to the fact that Code is (ℓ, q)-CNMLR.

5.2   Stateful Functionalities

Finally, we consider the case of primitives that update their state at each execution, i.e. functionalities of the type $(st_{\text{new}}, y) \leftarrow \mathcal{G}(st, x)$ (a.k.a. *stateful* functionalities). Note that in this case the hardened functionality re-encodes the new state at each execution.

Note that, since we do *not* assume erasure in our model, an adversary can always 'reset' the functionality to a previous valid state as follows: It could just copy the previous state to some part of the large memory and replace the current encoding by that. To avoid this, our transformation uses an untamperable *public counter* (along with the untamperable self-destruct bit) that helps us to detect whether the functionality is reset to a previous state, leading to a self-destruct. However such a counter can be implemented, for instance using $\log(k)$ bits. We notice that such a counter is necessary to protect against the above resetting attack. However, we stress that if we do not assume such a counter this "resetting" is the only harm the adversary can make in our model.

Below, we define what it means to harden a stateful functionality.

Definition 7 (Stateful hardened functionality). *Let* Code $=$ (Init, Encode, Decode) *be a split-state encoding scheme in the CRS model, with $2k$ bits messages and $2n$ bits codewords. Let $\mathcal{G} : \{0,1\}^k \times \{0,1\}^u \to \{0,1\}^k \times \{0,1\}^v$ be a stateful functionality with secret state $st \in \{0,1\}^k$, $\varphi \in \{0,1\}$ be a public value and let $\langle \gamma \rangle$ be a public $\log(k)$-bit counter both initially set to zero. We define a stateful hardened functionality $\mathcal{G}^{\mathsf{Code}} : \{0,1\}^{2s} \times \{0,1\}^u \to \{0,1\}^{2s} \times \{0,1\}^v$ with a modified state $st' \in \{0,1\}^{2s}$ and $s = \mathsf{poly}(n)$. The hardened functionality $\mathcal{G}^{\mathsf{Code}}$ is a triple of algorithms* (Init, Setup, Execute) *described as follows:*

- $\Omega \leftarrow \mathrm{Init}(1^k)$: *Run the initialization procedure of the coding scheme to sample $\Omega \leftarrow \mathrm{Init}(1^k)$.*
- $(\mathcal{M}_0, \mathcal{M}_1) \leftarrow \mathrm{Setup}(\Omega, st)$: *Let $(X_0, X_1) \leftarrow \mathrm{Encode}(\Omega, st\|\langle 1 \rangle)$ and increment $\langle \gamma \rangle \leftarrow \langle \gamma \rangle + 1$. For $b \in \{0,1\}$, store X_b in the first n bits of $\mathcal{M}_b$, i.e. $\mathcal{M}_b[1 \ldots n] \leftarrow X_b$.[11] (The remaining bits of $\mathcal{M}_b$ are set to 0^{s-n}.) Define $st' := (\mathcal{M}_0, \mathcal{M}_1)$.*
- $y \leftarrow \mathrm{Execute}(x)$: *Read the public bit φ. In case $\varphi = 1$ output $\perp$. Otherwise recover $X_b = \mathcal{M}_b[1 \ldots n]$ for $b \in \{0,1\}$ and run $(st''\|\langle \gamma' \rangle) \leftarrow \mathrm{Decode}(\Omega, (X_0, X_1))$. Read the public counter $\langle \gamma \rangle$. If $\langle \gamma \rangle \neq \langle \gamma' \rangle$ or $st'' = \perp$, set $\varphi = 1$. Else run $(st_{\mathsf{new}}, y) \leftarrow \mathcal{G}(st'', x)$ and output y. Finally, write $\mathrm{Encode}(\Omega, st_{\mathsf{new}}\|\langle \gamma + 1 \rangle)$ in $(\mathcal{M}_0[1, \ldots, n], \mathcal{M}_1[1, \ldots, n])$ and increment $\langle \gamma \rangle \leftarrow \langle \gamma \rangle + 1$.*

Remark 2 (On $\langle \gamma \rangle$). Note that the counter is incremented after each evaluation query, and the current value is encoded together with the new state. We require $\langle \gamma \rangle$ to be untamperable. This assumption is necessary, as otherwise an adversary could always use the extra space to keep a copy of a previous valid state and place it back at some later point in time. The above attack allows essentially to reset the functionality to a previous state, and cannot be simulated with black-box access to the original functionality.

In the case of stateful primitives, the hardened functionality has to re-encode the new state at each execution. Still, as the memory is large, the adversary can use the extra space to tamper continuously with a target encoding of some valid state. Security of a stateful hardened functionality is defined analogously to the stateless case (cf. Definition 6). We show the following result (for space reasons we defer the proof to the full version [18]):

---

[11] Without erasure this can be easily implemented by a stack.

Theorem 3. *Let $\mathcal{G}$ be a stateful functionality and* Code $=$ (Init, Encode, Decode) *be any (ℓ, q)-CNMLR encoding scheme in the split-state CRS model. Then* Code *is polyspace (ℓ, q)-leak/tamper simulatable for $\mathcal{G}$.*

Acknowledgments. Pratyay acknowledges support from a European Research Commission Starting Grant (no. 279447) and the CTIC and CFEM research center. Part of this work was done while this author was at the University of Warsaw and was supported by the WELCOME/2010-4/2 grant founded within the framework of the EU Innovative Economy Operational Programme.

Most of the work was done while Daniele was at Aarhus University, supported by the Danish Council for Independent Research via DFF Starting Grant 10-081612.

References

1. Aggarwal, D., Dodis, Y., Lovett, S.: Non-malleable codes from additive combinatorics. Electronic Colloquium on Computational Complexity (ECCC) 20, 81 (2013)
2. Austrin, P., Chung, K.-M., Mahmoody, M., Pass, R., Seth, K.: On the (im)possibility of tamper-resilient cryptography: Using fourier analysis in computer viruses. IACR Cryptology ePrint Archive 2013, 194 (2013)
3. Bellare, M., Cash, D., Miller, R.: Cryptography secure against related-key attacks and tampering. In: Lee, D.H., Wang, X. (eds.) ASIACRYPT 2011. LNCS, vol. 7073, pp. 486–503. Springer, Heidelberg (2011)
4. Bellare, M., Kohno, T.: A theoretical treatment of related-key attacks: RKA-PRPs, RKA-PRFs, and applications. In: Biham, E. (ed.) EUROCRYPT 2003. LNCS, vol. 2656, pp. 491–506. Springer, Heidelberg (2003)
5. Bellare, M., Paterson, K.G., Thomson, S.: RKA security beyond the linear barrier: IBE, encryption and signatures. In: Wang, X., Sako, K. (eds.) ASIACRYPT 2012. LNCS, vol. 7658, pp. 331–348. Springer, Heidelberg (2012)
6. Boneh, D., DeMillo, R.A., Lipton, R.J.: On the importance of eliminating errors in cryptographic computations. J. Cryptology 14(2), 101–119 (2001)
7. Cheraghchi, M., Guruswami, V.: Capacity of non-malleable codes. Electronic Colloquium on Computational Complexity (ECCC) 20, 118 (2013)
8. Cheraghchi, M., Guruswami, V.: Non-malleable coding against bit-wise and split-state tampering. IACR Cryptology ePrint Archive 2013, 565 (2013)
9. Choi, S.G., Kiayias, A., Malkin, T.: BiTR: Built-in tamper resilience. In: Lee, D.H., Wang, X. (eds.) ASIACRYPT 2011. LNCS, vol. 7073, pp. 740–758. Springer, Heidelberg (2011)
10. Chor, B., Goldreich, O.: Unbiased bits from sources of weak randomness and probabilistic communication complexity. SIAM J. Comput. 17(2), 230–261 (1988)
11. Dachman-Soled, D., Kalai, Y.T.: Securing circuits against constant-rate tampering. In: Safavi-Naini, R., Canetti, R. (eds.) CRYPTO 2012. LNCS, vol. 7417, pp. 533–551. Springer, Heidelberg (2012)
12. Damgård, I., Faust, S., Mukherjee, P., Venturi, D.: Bounded tamper resilience: How to go beyond the algebraic barrier. In: Sako, K., Sarkar, P. (eds.) ASIACRYPT 2013, Part II. LNCS, vol. 8270, pp. 140–160. Springer, Heidelberg (2013)
13. Davì, F., Dziembowski, S., Venturi, D.: Leakage-resilient storage. In: Garay, J.A., De Prisco, R. (eds.) SCN 2010. LNCS, vol. 6280, pp. 121–137. Springer, Heidelberg (2010)

14. Dziembowski, S., Faust, S.: Leakage-resilient cryptography from the inner-product extractor. In: Lee, D.H., Wang, X. (eds.) ASIACRYPT 2011. LNCS, vol. 7073, pp. 702–721. Springer, Heidelberg (2011)
15. Dziembowski, S., Kazana, T., Obremski, M.: Non-malleable codes from two-source extractors. In: Canetti, R., Garay, J.A. (eds.) CRYPTO 2013, Part II. LNCS, vol. 8043, pp. 239–257. Springer, Heidelberg (2013)
16. Dziembowski, S., Pietrzak, K.: Leakage-resilient cryptography. In: FOCS, pp. 293–302 (2008)
17. Dziembowski, S., Pietrzak, K., Wichs, D.: Non-malleable codes. In: ICS, pp. 434–452 (2010)
18. Faust, S., Mukherjee, P., Nielsen, J.B., Venturi, D.: Continuous non-malleable codes (2013). The full version will be available at the IACR Cryptology ePrint Archive
19. Faust, S., Mukherjee, P., Venturi, D., Wichs, D.: Efficient non-malleable codes and key-derivation for poly-size tampering circuits. IACR Cryptology ePrint Archive 2013, 702 (2013)
20. Faust, S., Pietrzak, K., Venturi, D.: Tamper-proof circuits: How to trade leakage for tamper-resilience. In: Aceto, L., Henzinger, M., Sgall, J. (eds.) ICALP 2011, Part I. LNCS, vol. 6755, pp. 391–402. Springer, Heidelberg (2011)
21. Gennaro, R., Lysyanskaya, A., Malkin, T., Micali, S., Rabin, T.: Algorithmic tamper-proof (ATP) security: Theoretical foundations for security against hardware tampering. In: Naor, M. (ed.) TCC 2004. LNCS, vol. 2951, pp. 258–277. Springer, Heidelberg (2004)
22. Ishai, Y., Prabhakaran, M., Sahai, A., Wagner, D.: Private circuits II: Keeping secrets in tamperable circuits. In: Vaudenay, S. (ed.) EUROCRYPT 2006. LNCS, vol. 4004, pp. 308–327. Springer, Heidelberg (2006)
23. Kalai, Y.T., Kanukurthi, B., Sahai, A.: Cryptography with tamperable and leaky memory. In: Rogaway, P. (ed.) CRYPTO 2011. LNCS, vol. 6841, pp. 373–390. Springer, Heidelberg (2011)
24. Liu, F.-H., Lysyanskaya, A.: Tamper and leakage resilience in the split-state model. In: Safavi-Naini, R., Canetti, R. (eds.) CRYPTO 2012. LNCS, vol. 7417, pp. 517–532. Springer, Heidelberg (2012)
25. Pietrzak, K.: Subspace lwe. In: Cramer, R. (ed.) TCC 2012. LNCS, vol. 7194, pp. 548–563. Springer, Heidelberg (2012)
26. De Santis, A., Di Crescenzo, G., Ostrovsky, R., Persiano, G., Sahai, A.: Robust non-interactive zero knowledge. In: Kilian, J. (ed.) CRYPTO 2001. LNCS, vol. 2139, pp. 566–598. Springer, Heidelberg (2001)

Locally Updatable and Locally Decodable Codes

Nishanth Chandran[1,*], Bhavana Kanukurthi[2,**], and Rafail Ostrovsky[3,***]

[1] Microsoft Research, India
[2] Department of Computer Science, UCLA
[3] Department of Computer Science and Mathematics, UCLA

Abstract. We introduce the notion of locally updatable and locally decodable codes (LULDCs). In addition to having low decode locality, such codes allow us to update a codeword (of a message) to a codeword of a different message, by rewriting just a few symbols. While, intuitively, updatability and error-correction seem to be contrasting goals, we show that for a suitable, yet meaningful, metric (which we call the Prefix Hamming metric), one can construct such codes. Informally, the Prefix Hamming metric allows the adversary to arbitrarily corrupt bits of the codeword subject to one constraint – he does not corrupt more than a δ fraction (for some constant δ) of the t "most-recently changed" bits of the codeword (for all $1 \leq t \leq n$, where n is the length of the codeword).

Our results are as follows. First, we construct binary LULDCs for messages in $\{0, 1\}^k$ with constant rate, update locality of $\mathcal{O}(\log^2 k)$, and read locality of $\mathcal{O}(k^\epsilon)$ for any constant $\epsilon < 1$. Next, we consider the case where the encoder and decoder share a secret state and the adversary is computationally bounded. Here too, we obtain local updatability and decodability for the Prefix Hamming metric. Furthermore, we also ensure that the local decoding algorithm never outputs an incorrect message – even when the adversary can corrupt an arbitrary number of bits of the codeword. We call such codes locally updatable locally decodable-detectable

* Email: nichandr@microsoft.com. Part of this work was done while this author was at AT&T Labs - Security Research Center, NY.
** Email: bhavanak@cs.bu.edu. Research supported in part by NSF grants CNS-0830803; CCF-0916574; IIS-1065276; CCF-1016540; CNS-1118126; CNS-1136174; and in part by the Defense Advanced Research Projects Agency through the U.S. Office of Naval Research under Contract N00014-11-1-0392. The views expressed are those of the author and do not reflect the official policy or position of the Department of Defense or the U.S. Government.
*** Email: rafail@cs.ucla.edu. Research supported in part by NSF grants CNS-0830803; CCF-0916574; IIS-1065276; CCF-1016540; CNS-1118126; CNS-1136174; US-Israel BSF grant 2008411, OKAWA Foundation Research Award, IBM Faculty Research Award, Xerox Faculty Research Award, B. John Garrick Foundation Award, Teradata Research Award, and Lockheed-Martin Corporation Research Award. This material is also based upon work supported by the Defense Advanced Research Projects Agency through the U.S. Office of Naval Research under Contract N00014-11-1-0392. The views expressed are those of the author and do not reflect the official policy or position of the Department of Defense or the U.S. Government.

Y. Lindell (Ed.): TCC 2014, LNCS 8349, pp. 489–514, 2014.

codes (LULDDCs) and obtain dramatic improvements in the parameters (over the information-theoretic setting). Our codes have constant rate, an update locality of $\mathcal{O}(\log^2 k)$ and a read locality of $\mathcal{O}(\lambda \log^2 k)$, where λ is the security parameter.

Finally, we show how our techniques apply to the setting of dynamic proofs of retrievability (DPoR) and present a construction of this primitive with better parameters than existing constructions. In particular, we construct a DPoR scheme with linear storage, $\mathcal{O}(\log^2 k)$ write complexity, and $\mathcal{O}(\lambda \log k)$ read and audit complexity.

1   Introduction

Standard error correcting codes (ECC) enable the recovery of a message even when a large fraction of its codeword is corrupted. One disadvantage of ECCs is that, in order to read even a single bit of the data, the entire codeword needs to be decoded. This becomes very inefficient if a user frequently needs to access specific parts of the underlying data. Locally decodable codes LDCs, introduced by Katz and Trevisan [15], overcome this problem and allow recovery of a single symbol of the message by reading only a few symbols of the potentially corrupted codeword. Another disadvantage of standard ECCs is that, in order to change even a single bit of the data, the entire codeword needs to be recomputed. A natural question to ask is: can we obtain codes which also allow us to change the underlying data by rewriting only a few symbols of the codeword? That is,

Can we build an ECC that allows you to decode and update the message by reading and/or modifying sub-linear number of symbols of the codeword?

In this work, we explore this question and its cryptographic connection.

1.1   Codes with Locality

Locally Decodable Codes. As mentioned before, locally decodable codes (LDCs), introduced by Katz and Trevisan [15] are a class of error correcting codes, where every bit of the message can be probabilistically decoded by reading only a few bits of the (possibly corrupted) codeword. In more detail, a binary locally decodable code encodes messages in $\{0,1\}^k$ into codewords in $\{0,1\}^n$. The parameters of interest in such codes are: a) the rate of the code $\rho = \frac{k}{n}$; b) the distance δ, which signifies that the decoding algorithm succeeds even when δn of the bits of the codeword are corrupted; c) the locality r which denotes the number of bits of the codeword read by the decoding algorithm; and d) the error probability ϵ that denotes that for every bit of the message, the decoding algorithm successfully decodes it with probability $1 - \epsilon$. Ideally, one would like to minimize both the length of the code as well as the locality; unfortunately, there is a trade-off between these parameters. On the one hand, we have the Hadamard code that has a locality of 2; however its length is exponential in k. (Indeed, the best code length for LDCs with constant locality are super-polynomial in k [27,8,6].). On the other hand, the best known codes with constant rate, [16,11,13], have a locality of $\mathcal{O}(n^\epsilon)$ for any constant $0 < \epsilon < 1$. For a survey on locally decodable codes, see Yekhanin's survey [28].

Locally Updatable and Locally Decodable Codes. As we mentioned before, LDCs (and error correcting codes in general) are extremely useful as they provide reliability even when many bits of the codeword may be corrupted; unfortunately, the (unavoidable) price that we pay is that even small changes to the message result in a large change to the codeword. In this work, we ask *"can we have locally decodable codes that are locally updatable?"*. That is, can we have locally decodable codes such that in order to obtain a codeword of message m' from a codeword of message m (where m and m' differ only in one bit) one only needs to modify a few bits of the codeword? We call such codes locally updatable and locally decodable codes (LULDCs); the number of bits that are modified by the update algorithm is then referred to as the *update locality* and the number of bits read by the (local) decoding algorithm is referred to as the *read locality*.

The Prefix Hamming Metric. As in the case of LDCs, our goal is to tolerate a constant fraction of errors while achieving subliniear locality (for both read and update). However, a little thought reveals that updatability and error correction are conflicting goals – if a code tolerates a δ-fraction of errors then, to change even one bit of the data, at least 2δ-fraction of the codeword symbols do need to be re-written.

In light of this, we consider a weaker, yet meaningful, adversarial model of corruption. In this model, the adversary is still allowed to corrupt constant fraction of the bits of the codeword. However, the bits of the codeword have an "age" associated with them and the adversary is allowed to corrupt fewer of the younger/newer bits and is allowed to corrupt many of the older bits. Whenever we touch (i.e., write) a particular bit i of the codeword during an update procedure, this bit becomes a young bit with an age less than every other bit in the codeword. At this point of time, the ith bit of the codeword is the youngest bit in the codeword. Now, suppose we touch the jth bit of the codeword, then this bit becomes the youngest bit, with the ith bit now becoming the second youngest bit of the codeword and so on. Note that if we were to now touch the ith bit, it would once again become the youngest bit of the codeword.

We allow the adversary to corrupt a constant fraction of the bits of the codeword subject only to one constraint – he never corrupts more than a δ fraction of the t youngest bits (for all $1 \le t \le n$). We call this metric the *Prefix Hamming Metric*. This metric models a situation where the longer the time a bit of the codeword resides in the system, the easier it is for an adversary to corrupt it. That is, stored data (codeword bits) gets "stale" unless refreshed, and hence the more time the data is untouched, the more errors it will have.

Comparison with Tree Codes. Our error model is similar to the one considered by Schulman [23],[24] in his seminal work on Tree Codes. Tree codes were specifically designed for streaming messages and allow the encoding of messages one bit at a time; the corresponding codeword symbol for every bit of the message is obtained by traversing down a tree. The codeword of the message is obtained by simply concatenating all the individual codeword symbols. Schulman's code guarantees the following: consider any two (different) paths of length t beginning at a particular node in the tree (that denote two different messages); then, the codewords corresponding to these messages have Hamming distance at least αt (for some constant α). Alternately viewed, at any given instance, as long as the adversary does not corrupt more than a α fraction of the t most recently transmitted codeword symbols, the codeword will decode to the correct message. Tree codes were designed for arbitrary (polynomial length) messages; however, we do not know of explicit constructions of tree codes with constant rate.

In our work, the message and codeword lengths are fixed in advance. But the message bits can be *updated* in a streaming fashion by rewriting certain bits of the codeword. Our adversarial error model says the following: at any given instance, as long as the adversary does not corrupt more than a particular constant fraction of the t most recently rewritten bits of the codeword (for all t), the codeword will decode to the correct message.

1.2   Our Results

Information-Theoretic Codes. We first construct an LULDC in the information-theoretic setting for the Prefix Hamming metric. We define this metric and such codes in detail in Section 2; for now, we give an overview of the result and the parameters that we achieve.

– *Result 1 (Informal):* We construct binary LULDCs for the Prefix Hamming metric for messages in $\{0,1\}^k$. Our codes have a rate of $\mathcal{O}(1)$, an amortized update locality of $\mathcal{O}(\log^2 k)$ and a worst case read locality of $\mathcal{O}(k^\epsilon)$ for any constant $\epsilon < 1$. For codes that operate on a larger alphabet Σ, with $|\Sigma| \geq \log k$, we can improve the update locality to $\mathcal{O}(\log k)$ (other parameters remaining the same).

Computational Codes. Next, we consider a scenario where the encoder and decoder share a secret state S and where the adversary is computationally bounded. In such a setting, we are able to provide the added guarantee that the (local) decoding algorithm never outputs an incorrect message, irrespective of the number of corrupted bits in the codeword. For the sake of clarity, we refer to such codes as locally updatable and locally decodable-detectable codes (LULDDCs). In addition to providing stronger guarantees, we also obtain dramatic improvements over the parameters achieved by our information-theoretic LULDC construction. In particular, we obtain the following parameters:

– *Result 2 (Informal):* We construct binary LULDDCs for messages in $\{0,1\}^k$. Our codes have constant rate, an amortized update locality of $\mathcal{O}(\log^2 k)$ and a worst case read locality of $\mathcal{O}(\lambda \log^2 k)$, where λ is the security parameter of the system.

Finally, we note that our techniques for building LULDDCs lend themselves to the construction of a Dynamic Proof of Retrievability (DPoR) scheme. Below we discuss our result on DPoR, which we believe, is of independent interest.

Dynamic Proofs of Retrievability. Informally, a proof of retrievability allows a client to store data on an untrusted server and later on, obtain a short proof from the server, that indeed all of the client's data is present on the server. In other words, the client can execute an audit protocol such that any malicious server that deletes or changes even a single bit of the client's data will fail to pass the audit protocol, except with negligible probability in the security parameter[1]. Proofs of retrievability, introduced by Juels and Kaliski [14], were initially defined on static data, building upon the closely related notion of sublinear authenticators defined by Naor and Rothblum [18]. Several works have studied the efficiency of such schemes [25,7,2,1] with the work of Cash, Küpçü, and Wichs [3] considering the notion of proofs of retrievability on dynamically changing data; in other words, they constructed a proof of retrievability scheme that allowed for efficient updates to the data. Their DPoR scheme has $\mathcal{O}(k)$ server storage, $\mathcal{O}(\lambda)$ client storage, $\mathcal{O}(\lambda \log^2 k)$ read complexity, $\mathcal{O}(\lambda^2 \log^2 k)$ write and audit complexity[2]. We improve their parameters and obtain the following result:

- *Result 3 (Informal):* We obtain a construction of a dynamic proof of retrievability with $\mathcal{O}(k)$ server storage, $\mathcal{O}(\lambda)$ client storage, $\mathcal{O}(\lambda \log k)$ read complexity, $\mathcal{O}(\log^2 k)$ write complexity and $\mathcal{O}(\lambda \log k)$ audit complexity[3].

1.3   Our Techniques

We now give a high-level overview of the techniques used to obtain our results. We shall make use of the hierarchical data structure introduced by Ostrovsky [19],[20] in the context of oblivious RAMs. Oblivious RAMs [9,19] allow efficient random access to memory without revealing the access pattern to an adversary that observes the reads and writes made to memory. ORAM protocols hide the access pattern by making use of several tools carefully put-together. Here we distill out exactly what we need for our construction. In particular, we will primarily make use of the hierarchical data structure, coupled with certain other techniques, to construct LULDCs.

Hierarchical Data Structure. At a high level, this data structure comprises of buffers $\mathsf{buff}_0, \cdots, \mathsf{buff}_\tau$ of increasing size. Buffer buff_i has 2^i elements and each

---

[1] Formally, this guarantee is provided by requiring the existence of an extractor algorithm, that given black-box rewinding access to any malicious server that passes the audit with non-negligible probability, will extract all of the client's data, except with negligible probability.

[2] The work of Cash *et al.* [3] considered the complexity without explicitly including the (storage as well as verification) complexity of the MAC; if one did this, then the parameters obtained will all be larger by a factor of $\mathcal{O}(\lambda)$.

[3] These parameters include the cost for storage and verification of the MACs.

element in the buffer is of the form (index, value). In addition, there is a special buffer, buff* which has all bits of the message in order (and hence without an index). To read a value at a particular index i, we scan the buffers in top-down manner. To write (or re-write) a value v at index i, we write it to the top buffer. Writing to buffers eventually fills them up. To handle this, buffers are periodically combined and moved to an empty buffer in some lower level in a careful manner.

LULDCs for the Prefix Hamming Metric. The first idea behind our construction in the information-theoretic setting is as follows. To achieve local decodability, we encode each buffer (including buff*) with a locally decodable code (LDC). Whenever we wish to update a bit of the message, we will write it to the topmost buffer $buff_0$ and re-encode the top buffer using an LDC to encode this latest update. Naturally, the top buffer gets full after an update operation. Whenever we encounter a full buffer, we move its contents to the buffer below it (that is, we decode the entire buffer, combine top level buffers together and re-encode them at a level below, once again using an LDC for the encoding). When we wish to (locally) decode a particular index i of the message, we scan buffers one-by-one starting with topmost buffer. Now, note that we need to check if a particular index is found in a buffer or not. In order to do this, we always ensure that buffers store (index, value) pairs that are sorted according to the index value. This will enable us to perform a binary search (decoded via the underlying LDC) to check if a buffer contains a particular index i or not. Since we are performing the binary search via the decode algorithm of the underlying LDC, we must ensure that the decode does not fail with too high a probability; hence, we repeat the decode procedure at each level some fixed number of times to ensure this and make sure that our overall local decoding algorithm succeeds except with ϵ probability. When the index is found, we stop searching lower level buffers and output the value retrieved (our construction will always ensure that if an index value was updated, then the latest value of the index will be stored at a high level buffer). If the index is not found, then we read the corresponding element from the special buffer buff*, once again using the underlying LDC.

Since we must store every updated element as a (index, value)-pair, the above described technique will decrease the rate of the code by a factor of $\mathcal{O}(\log k)$. Hence, in order to ensure that our code has constant rate, we carefully choose the total number of buffers $\tau + 1$ in our construction to ensure that we obtain constant rate codes and yet achieve good update and read locality.

Now, in the above construction, we first show that the decode and update algorithms succeed (with small locality) as long as an adversary corrupts only a constant fraction of the bits of each buffer. We then proceed to show that if an adversary corrupts bits of the codeword according to the Prefix Hamming metric, then he can only corrupt a constant fraction of the bits of each buffer (within a factor of 2). This gives us our construction of LULDCs.

Computational LULDDCs. To obtain our construction in the computational setting, at a high level, we follow our information-theoretic construction. However, there are three main differences. First, when decoding the ith bit of the codeword, we still scan each buffer to see if a "latest" copy of the ith bit is present in that buffer. However, now, because we are in the computational setting, we no longer need to store the buffer in sorted order and perform a binary search. Instead, we simply use hash functions to check if a particular index is present in a buffer or not. Furthermore, we use cuckoo hash functions to minimize our read locality in this case. Second, we store each buffer using a computational LDC that has constant rate and $\mathcal{O}(\lambda)$ locality (such codes are obtained through the construction of Hemenway et al. [12]). Third, we authenticate each bit of the codeword using a message authentication code so that we never decode incorrectly (irrespecitve of the number of errors that the adversary introduces).

The above ideas do not suffice for our construction: in particular, if we applied these techniques, we do not obtain a constant rate code as MACing each bit of the codeword would result in a $\mathcal{O}(\lambda)$ blowup in the rate of the code. One could think of MACing $\mathcal{O}(\lambda)$ bits of the codeword, block by block, but then this would result in a $\mathcal{O}(\lambda^2)$ blowup in the read locality, as we must read λ bits now in each buffer through the underlying LDC. In order to obtain our result, we MAC each bit of the codeword using a constant size MAC; this technique is similar in spirit to the use of constant size MACs when authenticating codewords in the context of optimizing privacy amplification protocols [5]. To obtain our result, we make a careful use of these constant size MACs to verify the correctness of a codeword as well as to decode correctly (except with negligible probability).

Dynamic Proofs of Retrievability. Cash et al. [3] showed how to convert any oblivious RAM (ORAM) protocol that satisfied a special property (which they define to be next-read-pattern-hiding (NRPH)) into a dynamic proof of retrievability (DPoR) scheme. We show that we do not need an ORAM scheme with this property and the techniques used to construct LULDDCs can be used to directly build a DPoR scheme. Moreover, we do not need to hide the read and write access pattern, thereby leading to significant savings in the complexity. In particular, we show, that by encoding each buffer of the ORAM *structure* using a standard error correcting code (that is also appropriately authenticated with constant size MACs), and additionally storing authenticated elements of the raw data in the clear, we can use the techniques developed for LULDDCs to construct a DPoR scheme with $\mathcal{O}(k)$ server storage, $\mathcal{O}(\lambda)$ client storage, $\mathcal{O}(\lambda \log k)$ read complexity, $\mathcal{O}(\log^2 k)$ write complexity and $\mathcal{O}(\lambda \log k)$ audit complexity. Moreover, these parameters include the cost for storage and verification of the MACs.

1.4 Organization of the Paper

In Section 2, we introduce our notion of locally updatable and locally decodable codes as well as formally define the Prefix Hamming metric. We present our construction of locally updatable and locally decodable codes for the Prefix Hamming metric in Section 3. We consider the computational setting in Section 4 and construct locally updatable and locally decodable-detectable codes.

Finally, we give our construction of a dynamic proof of retrievability scheme in Section 5. Due to the lack of space, we present further details of our schemes and proofs in the full version [4].

2   Definitions

Notation. Let k denote the length of the message. Let $\mathcal{M}$ denote a metric space with distance function $\mathsf{dis}(,)$. Let the set of all codewords corresponding to a message m be denoted by $\mathcal{C}_m$ – we will define this set shortly. Let n denote the length of all codewords. $m(i)$ denotes the ith bit of message m for $i \in [k]$, where $[k]$ denotes the set of integers $\{1, 2, \cdots, k\}$.

2.1   Codes with Locality

Locally decodable codes. We first recall the notion of locally decodable codes. Informally, locally decodable codes allow the decoding of any bit of the message by only reading a few (random) bits of the codeword. Formally:

Definition 1 (Locally decodable codes). *A binary code* $\mathcal{C}$: $\{0,1\}^k \rightarrow \{0,1\}^n$ *is* $(k, n, r_k, \delta, \epsilon)$-*locally decodable if there exists a randomized decoding algorithm* $\mathcal{D}$ *such that*

1. $\forall m \in \{0,1\}^k, \forall i \in [k], \forall c_m \in \mathcal{C}_m,$ *and for all* $\hat{c}_m \in \{0,1\}^n$ *such that* $\mathsf{dis}(c_m, \hat{c}_m) \leq \delta n$:

$$\Pr[\mathcal{D}^{\hat{c}_m}(i) = m(i)] \geq 1 - \epsilon,$$

where the probability is taken over the random coins of the algorithm $\mathcal{D}$.
2. $\mathcal{D}$ *makes at most* r_k *queries to* $\hat{c}_m$.

Locally updatable codes. We now define the notion of locally updatable and locally decodable codes. A basic property that updatable codes must have is that one can convert a codeword of message m into a codeword of message m' (where m' and m differ possibly only at the ith position), by changing only a few bits of the codeword of m. However, we will obtain codes that have a stronger property; namely, will ensure that we can convert any string that decodes to m into a string that decodes to m'. That is, let m and m' be two k-bit messages that (possibly) differ only in the ith position, where $m'(i) = b_i$. For some appropriate metric space that defines a measure of closeness, given a string $\hat{c}_m$ that is "close" to a codeword for message m, our update algorithm (that writes bit b_i at position i) must convert $\hat{c}_m$ into a new string $\hat{c}_{m'}$ that is now "close" to a codeword for message m'. Furthermore, the update algorithm must query and change only a few bits of $\hat{c}_m$. Additionally, our code should also be locally decodable.

Before we present the formal definition of a locally updatable and locally decodable code, we first need to define the set of codewords $\mathcal{C}_m$ for a message m. Conceptually, with a locally updatable code, there are two kinds of codewords that correspond to a message m – ones obtained by computing $\mathcal{E}(m)$ and those obtained by computing updating the codeword of different message m'.

We let $m^{i^{b_i}}$ denote a message that is exactly the same as m except possibly at the ith position (where it is b_i). Note that $m^{i^{b_i}}$ maybe equal to m itself.

Definition 2 (The set $\mathcal{C}_m$). *For a message m, if there exists a message $\bar{m}$, codeword $c_{\bar{m}} = \mathcal{E}(\bar{m})$ (possibly $\bar{m} = m$ and $c_{\bar{m}} = c_m$) and a (possibly empty) set of indices $\{i_1, \cdots, i_t\}$ such that $m = \bar{m}^{i_1{}^{b_1} \cdots i_t{}^{b_t}}$ and $c_m = u(....u(u(c_{\bar{m}}, i_1, b_1), i_2, b_2), ..., i_t, b_t)$, then c_m is in the set $\mathcal{C}_m$.*

It is easy to see that $\mathcal{C}_m$ contains all the codewords that decode to m. We now present the formal definition of a LULDC.

Definition 3 (Locally updatable and locally decodable codes (LULDC)). *A binary code $\mathcal{C} : \{0,1\}^k \rightarrow \{0,1\}^n$ is $(k, n, w, r, \delta, \epsilon)$-locally updatable and locally decodable if there exist (possibly) randomized algorithms $\mathcal{E}_{\text{LDC}}$, $\mathcal{U}$ and $\mathcal{D}$ such that the following conditions are satisfied:*

1. Local Updatability:
   (a) *Let $m_0 \in \{0,1\}^k$ and let $c_{m_0} = \mathcal{E}_{\text{LDC}}(m_0)$. Let m_t be a message obtained by any (potentially empty) sequence of updates. ($t = 0$ corresponds to the case where the codeword has not been updated so far.) Then $\forall m_0 \in \{0,1\}^k, \forall c_{m_0} \in \mathcal{C}_{m_0}, \forall t, \forall m_t, \forall i_{t+1} \in [k], \forall b_{t+1} \in \{0,1\}$, for all $\hat{c}_{m_t} \in \{0,1\}^n$ such that $\text{dis}(\hat{c}_{m_t}, c_{m_t}) \le \delta n$,*
      – *The actions of $\mathcal{U}^{\hat{c}_{m_t}}(i_{t+1}, b_{t+1})$, change $\hat{c}_{m_t}$ to $u(\hat{c}_{m_t}, i_{t+1}, b_{t+1}) \in \{0,1\}^n$, where $\text{dis}(u(\hat{c}_{m_t}, i_{t+1}, b_{t+1}), c_{m_{t+1}}) \le \delta n$ for some $c_{m_{t+1}} \in \mathcal{C}_{m_{t+1}}$, where m_{t+1} and m_t are identical except (possibly) at the i_{t+1}^{th} position, where $m_{t+1}(i_{t+1}) = b_{t+1}$.*
   (b) *The total number of queries and changes that $\mathcal{U}$ makes to the bits of $\hat{c}_m$ is at most w.*
2. Local Decodabilty:
   (a) *Let m_t denote the latest message. $\forall m_t \in \{0,1\}^k, \forall i \in [k], \forall c_{m_t} \in \mathcal{C}_{m_t}$, and for all $\hat{c}_{m_t} \in \{0,1\}^n$ such that $\text{dis}(c_{m_t}, \hat{c}_{m_t}) \le \delta n$:*

   $$\Pr[\mathcal{D}^{\hat{c}_{m_t}}(i) = m_t(i)] \ge 1 - \epsilon,$$

   where the probability is taken over the random coins of the algorithm $\mathcal{D}$.
   (b) *$\mathcal{D}$ makes at most r queries to $\hat{c}_{m_t}$.*

2.2   The Prefix Hamming Metric

If we want codes that are truly updatable, the update locality w needs to be $<< \delta n$. However, as mentioned earlier, we cannot hope to achieve such locality for metrics where an adversary can *arbitrarily* corrupt a constant fraction of the bits of the codeword. (Indeed, if we updated a codeword from c_m to $c_{m'}$ with a locality of w, then by corrupting those w bits of $c_{m'}$, an adversary can ensure that the decoding algorithm does not output the correct message – in particular, the decode algorithm would output m instead of m'.)

In light of this, we turn to a new, yet meaningful metric, for which we can guarantee that even if an adversary corrupts a bounded number of bits of the codeword, though not in a completely arbitrary manner, our decode algorithm still functions correctly. At a high level, bits of the codeword "age" and the adversary can corrupt a fraction of the bits as a function of their age. Our metric relies crucially on the order in which bits were written or updated during the creation of a codeword – nonetheless, we abuse notation and refer to Prefix-Hamming as a metric. We first define the "age-ordering" of a codeword.

Definition 4 (Age-ordering of a codeword). *Let $c \in \{0,1\}^n$. Let w_1 denote the index/position of the most recent bit of the codeword that was either written or updated. Let w_2 denote the unique index of the next most recent bit that was written/updated and so on, with w_n denoting the index of the earliest bit written (in comparison with the rest of the bits of the codeword). We call $w_1, \cdots, w_n$ the age-ordering of c. $c(w_i)$ denotes the bit value of the codeword at index w_i. For all $1 \le t \le n$, let $c[1, t]$ denote the bits $c(w_1), \cdots, c(w_t)$.*

We are now ready to define how the adversary in our model can corrupt bits of the codeword. That is, we define our metric space and its distance function.

Definition 5 (The Prefix Hamming Metric). *Let $c \in \{0,1\}^n$. Let $w_1, \cdots, w_n$ denote the age-ordering of c. Let $c' \in \{0,1\}^n$ and for $1 \le t \le n$, let $c'[1, t]$ denote the bits $c'(w_1), \cdots, c'(w_t)$. We say that the Prefix Hamming distance between c and c', denoted by $\mathsf{Prefix}(c, c')$ is $\le \delta n$ if for all $1 \le t \le n$, $\mathsf{Hamm}(c[1, t], c'[1, t]) \le \delta t$, where $\mathsf{Hamm}(x, y)$ denotes the Hamming Distance between any two strings x and y of equal length.*

3   LULDCs for the Prefix Hamming Metric

3.1   Our Results

In this section, we show how to construct locally updatable locally decodable error correcting codes (LULDCs) that are resilient to a constant fraction of adversarial errors for the Prefix Hamming metric that we defined in Section 2.2. Formally, we show:

Theorem 1. *Let $\tau = \log k - \log(\log k + 1) - 1$. Let $\mathcal{C}_{\mathsf{LDC}}$ be a family of $(k_i, n_i, r_i, \epsilon, \delta)$–locally decodable code for Hamming distance with algorithms $(\mathcal{E}_{\mathsf{LDC}}, \mathcal{D}_{\mathsf{LDC}})$, where $k_i = 2^i(\log k + 1)$ for all $0 \le i \le \tau$. Additionaly, let $\mathcal{C}_{\mathsf{LDC}}$ contain a $(k^*, n^*, r^*, \epsilon, \delta)$–locally decodable code for Hamming distance, where $k^* = k$. Let $\rho_i = \frac{k_i}{n_i}$ for all i and let $\rho^* = \frac{k^*}{n^*}$. Then there exists a $(k, n, w, r, \epsilon, \frac{\delta}{2}) - \mathsf{LULDC}$ code $\mathcal{C} = (\mathcal{E}, \mathcal{D}, \mathcal{U})$ for the Prefix Hamming metric with:*

- **Length of the code** (n)**:** $n = n^* + \sum\limits_{i=0}^{\tau} n_i$.

- **Update locality** (w)**:** $w = (\log k + 1) \sum\limits_{i=0}^{\tau} \frac{1}{\rho_i} + \frac{\log k + 1}{\rho^*}$, *in the amortized sense.*

- **Read locality** (r): $r = \frac{8(1-\epsilon)}{(1-2\epsilon)^2} T \log \frac{T}{\epsilon} + r^*$, where

$$T = (\log k + 1) \left(r_0 + \sum_{1 \leq j \leq \tau} j r_j \right), \text{ in the worst case.}$$

As a corollary to Theorem 1, using the LDCs from [16,11,13] we obtain:

Corollary 1. *For every* $\epsilon, \alpha > 0$, *there exists a* $(k, n, w, r, \epsilon, \delta) - \text{LULDC}$ *code* $\mathcal{C} = (\mathcal{E}, \mathcal{D}, \mathcal{U})$ *for the Prefix Hamming metric achieving the following parameters, for some constant* $0 < \delta < \frac{1}{4}$:

- **Length of the code** (n): $n = \frac{2k}{1-\alpha}$.
- **Update locality** (w): $w = \mathcal{O}(\log^2 k)$, *in the amortized sense.*
- **Read locality** (r): $r = \mathcal{O}(k^{\epsilon'})$, *for some constant* ϵ', *in the worst case.*

Large alphabet codes. We remark that for codes over larger alphabet Σ, with $|\Sigma| \geq c \log k$ for some constant c, we can modify our code to obtain a better update locality of $\mathcal{O}(\log k)$ (other parameters remaining the same).

3.2   Code Description

We will now construct the codes that will prove Theorem 1. Our codeword will have a structure similar to that of the hierarchical data-structure used by Ostrovsky [19,20] in the construction of oblivious RAMs. Let $\tau = \log k - \log(\log k + 1) - 1$. Each codeword of $\mathcal{C}$ will consist of $\tau + 1$ buffers, $\text{buff}_0, \ldots, \text{buff}_\tau$ and a special buffer buff^*. We will ensure that as updates take place, at any point of time, buff_i will be either empty or full (for all $i > 0$). A full buffer, buff_i, will contain an encoding of a set μ_i of 2^i elements. In particular, $\mu_i = [(a_i^1, v_i^1), \ldots, (a_i^{2^i}, v_i^{2^i})]$ where a_i^j is an address (between 0 and $k-1$) and v_i^j is the value corresponding to it. buff_i (when non-empty) will store $\psi_i = \mathcal{E}_{\text{LDC}}(\mu_i)$. The special buffer buff^* will contain an encoding of the bits of the entire message in order, without address values; in particular, buff^* stores $\psi^* = \mathcal{E}_{\text{LDC}}(m)$.

Encode algorithm. Our encoding algorithm works as follows:

Algorithm $\mathcal{E}(m)$:

1. Creates the $\tau + 1$ empty buffers ($\text{buff}_0, \ldots, \text{buff}_\tau$).
2. Let $\mu^* = \{m(1), \cdots, m(k)\}$, where $m(i)$ denotes the ith bit of the message. It computes $\psi^* = \mathcal{E}_{\text{LDC}}(\mu^*)$ and stores it in buff^*.

Local update algorithm. Our update algorithm updates a string $\hat{c}_m$ (such that $\text{Prefix}(\hat{c}_m, c_m) \leq \delta n$, for some $c_m \in \mathcal{C}_m$) into a string $\hat{c}'_m$, setting $m(i)$ to b_i.

Algorithm $\mathcal{U}^{\hat{c}_m}(i, b_i)$:

1. If the first buffer is empty, computes $\mathcal{E}_{\text{LDC}}(i, b_i)$ and stores it in buff_0.

2. If the first buffer is non-empty, it finds the first empty buffer. Let this be
   buff$_j$. It decodes all the buffers above it to get μ_0 to μ_{j-1} [4]. Recall that each
   μ_h is a set of (a, v) pairs where a denotes the address (of length $\log k$) and v
   denotes a value ($\in \{0, 1\}$). It merges all these pairs of values as well the pair
   (i, b_i) in a sorted manner (where the sorting is done on address) and stores
   it in μ_j. Note, there are 2^j elements and therefore μ_j is now a full buffer.
   Handling Repetitions: While merging elements from multiple buffers, we
   might encounter repetition of addresses. Instead of removing repetitions,
   we simply ensure that all values stored in the buffers until $j - 1$ store only
   the "latest value" corresponding to the repeated address. (The latest value
   is easy to determine – it is the first value corresponding to the buffer that
   you encounter when reading the buffers in a top-down manner. Of course,
   for the address being inserted, namely i, the latest value will be b_i.)
3. The update algorithm computes $\psi_j = \mathcal{E}_{\mathsf{LDC}}(\mu_j)$ and stores it in buff$_j$.
4. The buffers from $\mu_{j-1} \ldots \mu_0$, in that order, are now set to empty by writing
   special symbols into it. Looking ahead, the order in which this done is important as this ensures that buff$_h$ always has bits that are "younger" than
   the bits in buff$_{h+1}$ for all h (when considering the age-ordering of the bits).
5. If none of the buffers are empty, namely, all buffers buff$_0, \cdots,$ buff$_\tau$ are full,
   then the update algorithm simply re-computes a new encoding of the message
   using the LDC encode algorithm and stores it in buff*. In other words,
   the algorithm decodes all the buffers to obtain the latest value of each bit,
   concatenates these bits together to form $\mu^* = \{m(1), \cdots, m(k)\}$ and encodes
   these bits to compute $\psi^* = \mathcal{E}_{\mathsf{LDC}}(\mu^*)$. Once again, the buffers from buff$_\tau$ to
   buff$_0$ are set to empty in that order by writing special symbols into it.

Local decode algorithm. Recall that our buffers satisfy the following conditions:

- The buffers are always sorted (based on the address a).
- If the address a "appears" in the same buffer multiple times, then all values
  corresponding to this address are the same. (This is guaranteed by the way
  we handle repetitions during our merging procedure.)
- Finally, across multiple buffers, the most recent value corresponding to an
  address appears in the higher buffer (i.e. a lower buffer value).

Algorithm $\mathcal{D}^{\hat{c}_m}(i)$:

1. The decode algorithm starts with the top-most buffer (buff$_0$) and proceeds
   downwards until it finds the address i.
2. To search a buffer buff$_j$ for the element i, it performs a binary search on
   elements stored in that buffer. Because buff$_j$ contains an LDC encoding,
   we additionally need to use $\mathcal{D}_{\mathsf{LDC}}()$ algorithm to access these j elements.
   Since $\mathcal{D}_{\mathsf{LDC}}()$ might fail with ϵ probability to decode one coordinate of the
   underlying message, we need to repeat $\mathcal{D}_{\mathsf{LDC}}()$ multiple (i.e. λ) times to
   amplify the success probability (where λ is a carefully chosen parameter).

---

[4] Here, these buffers need not be decoded using the local decoding algorithm and one
can obtain perfect correctness by simply running the standard decoding algorithm
for the error correcting code.

3. If element i is not found in any of the buffers buff_0 through buff_τ, then the algorithm simply (locally) decodes the ith element from buff^* (which contains an LDC encoding of the message).

3.3   Proof of Theorem 1

We shall now prove Theorem 1; namely, we show that the construction described above in Section 3.2 is a locally updatable, locally encodable binary error correcting code (for the Prefix Hamming metric) with the parameters listed in Theorem 1. Instead of directly proving Theorem 1, we will instead show that the construction is a LULDC for a metric that we call the *Buffered-Hamming* metric. From this, the proof of Theorem 1 directly follows. We shall now define the Buffered-Hamming metric and its associated distance function.

Buffered-Hamming Distance. Let $c \in \{0,1\}^n$ comprise of buffers $\mathsf{buff} = \mathsf{buff}_0, \ldots, \mathsf{buff}_q$ of lengths $n_0, \ldots, n_q$ respectively. Let $c' \in \{0,1\}^n$ be another string with buffers $\mathsf{buff}' = \mathsf{buff}'_0, \ldots, \mathsf{buff}'_q$. Then we say that Buffered-Hamming Distance, $\mathsf{BHdis}(c_m, c') \leq \delta n$ if $\forall i\ \mathsf{Hamm}(\mathsf{buff}_i, \mathsf{buff}'_i) \leq \delta n_i$.

Lemma 1. *Let* $\tau = \log k - \log(\log k + 1) - 1$. *Let* $\mathcal{C}_{\mathsf{LDC}}$ *be a family of* $(k_i, n_i, r_i, \epsilon, \delta)-$*locally decodable code for Hamming distance with algorithms* $(\mathcal{E}_{\mathsf{LDC}}, \mathcal{D}_{\mathsf{LDC}})$, *where* $k_i = 2^i(\log k + 1)$ *for all* $0 \leq i \leq \tau$. *Additionaly, let* $\mathcal{C}_{\mathsf{LDC}}$ *contain a* $(k^*, n^*, r^*, \epsilon, \delta)-$*locally decodable code for Hamming distance, where* $k^* = k$. *Let* $\rho_i = \frac{k_i}{n_i}$ *for all* i *and let* $\rho^* = \frac{k^*}{n^*}$. *Then the construction described above in Section 3.2 is a* $(k, n, w, r, \epsilon, \delta) - \mathsf{LULDC}$ *code* $\mathcal{C} = (\mathcal{E}, \mathcal{D}, \mathcal{U})$ *for the Buffered-Hamming metric achieving the following parameters:*

- **Length of the code** (n): $n = n^* + \sum\limits_{i=0}^{\tau} n_i$.

- **Update locality** (w): $w = (\log k + 1) \sum\limits_{i=0}^{\tau} \frac{1}{\rho_i} + \frac{\log k + 1}{\rho^*}$, *in the worst case.*

- **Read locality** (r): $r = \frac{8(1-\epsilon)}{(1-2\epsilon)^2} T \log \frac{T}{\epsilon} + r^*$, *where*

$$T = (\log k + 1) \left(r_0 + \sum_{1 \leq j \leq \tau} j r_j \right), \text{ in the worst case.}$$

Proof. **Length of the code.** Recall that we have buffers in levels $0, 1, \ldots, \tau$. Each buffer encodes a message μ_j of length $k_j = 2^j(\log k + 1)$; the encoding is denoted ψ_j and is of length n_j. Buffer buff^* contains an LDC encoding of a message of length k. It is easy to see that the length of the code $n = n^* + \sum\limits_{i=0}^{\tau} n_i$.

Read Locality and Decode Correctness. We now analyze the read locality and the decodability of our code. Let $\hat{c}_m$ be the given (corrupted) codeword and let $\hat{c}_m$ be such that $\mathsf{BHdis}(\hat{c}_m, c_m) \leq \delta n$, where $c_m \in \mathcal{C}_m$ for the most "recent"

$m \in \{0,1\}^k$ (obtained after an encoding of a message and possible subsequent updates). We compute the read locality of our local decoding algorithm and also prove that for all $i \in [k]$, the decoding algorithm will output $m(i)$ with probability $\geq 1 - \epsilon$.

Let $\mu = \{\mu_0, \ldots, \mu_\tau\}$ and let $\psi = \{\psi_0, \ldots, \psi_\tau\}$, where $\psi_i = \mathcal{E}_{\mathsf{LDC}}(\mu_i)$. Let $\mathcal{C}_{\mathsf{LDC}}^j$ denote the locally decodable code used to encode μ_j. We use $\mu_x(y)$ to denote the yth bit of μ_x. Recall that in order to read an index i of the message $m = m_0, \ldots, m_k$, the algorithm $\mathcal{D}^\psi(i)$ does a binary-search on the buffers in a top-down manner to see if there is a value corresponding to address i. The worst case locality occurs when m_i has never been updated. In this case, the binary search needs to be done on every buffer and will then conclude by performing a (local) deocoding for the ith bit in buff* which contains $\psi^* = \mathcal{E}_{\mathsf{LDC}}(m)$.

We first calculate the number of *bits* of μ_j (for $j \geq 1$), one would need to read, if we were doing the binary search directly over μ_j. There are 2^j elements i.e.,(a, v) pairs, in level j. So the binary search would need to look at j elements (in the worst case). Each element has length $\log k + 1$. The total number of bits of μ_j we access if we did a binary search over μ_j would be $j(\log k+1)$ (for $j \geq 1$). $\mathcal{D}^\psi(i)$ learns these bits by making calls to $\mathcal{D}_{\mathsf{LDC}}^{\psi_j}$ which has locality r_j. Therefore the number of bits of ψ_j, read via calls to $\mathcal{D}_{\mathsf{LDC}}^{\psi_j}$, is at most $j(\log k + 1)r_j$ (for $1 \leq j \leq \tau$) and $(\log k + 1)r_j$ (for $j = 0$). (Recall, that in buff*, a binary search is not performed and the decode algorithm simply decodes the (single) i^{th} bit of the message via LDC decode calls to ψ^*.)

Define a set Read and add (x, y) to it if $\mu_x(y)$ was accessed; let $T = |\mathsf{Read}|$. Then,

$$T = (\log k + 1) \left(r_0 + \sum_{1 \leq j \leq \tau} jr_j \right) \text{ and} \tag{1}$$

$$\text{the total decode locality } r = T\lambda + r^* \tag{2}$$

Equation 2 follows from that fact that in order to read a bit of μ_j correctly, we must amplify the success probability of $\mathcal{D}_{\mathsf{LDC}}^{\psi_j}$, by taking the majority of λ executions (Note, that just as in standard LDCs, even though our LULDC allows a decoding error of ϵ, we cannot afford to have an error of ϵ while reading every bit of our binary search in every buffer, as this would lead to an overall worse error probability). If the element is not found in the buffers buff$_0$ through buff$_\tau$, then we only need to read 1 bit of the underlying message via a single LDC decoding call to ψ^* and hence we pay an additional r^* in our read locality.

In order to determine r, all that is left, is for us to determine λ. Let the variable #Succ(x, y) denote the number of calls such that $\mathcal{D}_{\mathsf{LDC}}^{\psi'_x}(y) = \mu(x, y)$. Let SuccRead$(x, y)$ denote that event that #Succ$(x, y) > \frac{\lambda}{2}$. First, since $\hat{c}_m$ is such that BHdis$(\hat{c}_m, c_m) \leq \delta n$, it follows that, Hamm$(\psi'_j, \psi_j) \leq \delta|\psi_j|$ for all $0 \leq j \leq \tau$

and $\mathsf{Hamm}(\psi^{*\prime}, \psi^*) \leq \delta|\psi^*|$. Now, since $C^{\psi_j^\prime}_{\mathsf{LDC}}$ has error-rate ϵ, $\mathbf{E}[\#\mathsf{Succ}(x, y)] = \lambda(1 - \epsilon)$. By the Chernoff bound[5], $\Pr[\#\mathsf{Succ}(x, y) \leq \frac{\lambda}{2}] \leq p = e^{-\frac{\lambda(1-2\epsilon)^2}{8(1-\epsilon)}}$.

In other words,

$$\Pr[\mathsf{SuccRead}(x, y) = 0] \leq p = e^{-\frac{\lambda(1-2\epsilon)^2}{8(1-\epsilon)}} \tag{3}$$

$$\text{i.e.,} \quad \sum_{(x,y)\in\mathsf{Read}} \Pr[\mathsf{SuccRead}(x, y) = 0] \leq Tp. \tag{4}$$

Our goal is to ensure that

$$\Pr\left[\bigwedge_{(\forall(x,y)\in\mathsf{Read})} \mathsf{SuccRead}(x, y) = 1 \right] (\geq 1 - Tp) \geq 1 - \epsilon.$$

In other words, we need to set λ such that $Tp \leq \epsilon$. Substituting for $p = e^{-\frac{\lambda(1-2\epsilon)^2}{8(1-\epsilon)}}$, we get that

$$\lambda \geq \frac{8(1-\epsilon)}{(1-2\epsilon)^2} \log\left(\frac{T}{\epsilon}\right).$$

By setting $\lambda = \frac{8(1-\epsilon)}{(1-2\epsilon)^2} \log\left(\frac{T}{\epsilon}\right)$ and substituting in Equation 2, we get that the decode locality,

$$r = \frac{8(1-\epsilon)}{(1-2\epsilon)^2} T \log\frac{T}{\epsilon} + r^*.$$

This proves the correctness and the read locality of our decoding algorithm.

Update Locality and Correctness. First, we count the number of coordinates accessed in order to rewrite one bit of the message m_i. This includes the total number of coordinates read and written.

It is easy to see that in algorithm $\mathcal{U}^{C_m}(x, b_x)$, buffer buff_j (for $0 \leq j \leq \tau$) is rewritten every 2^j steps. Buffer buff^* is re-written every $2^{\tau+1}$ steps. In 2^j updates (when $j < \tau + 1$), therefore, the total number of bits re-written is

$$= 2^j \frac{|\mu_0|}{\rho_0} + 2^{j-1} \frac{|\mu_1|}{\rho_1} + \ldots + 2^0 \frac{|\mu_j|}{\rho_j}$$

$$= 2^j |\mu_0| \sum_{0 \leq i \leq j} \frac{1}{\rho_i} \quad (\text{since } \mu_i = 2\mu_{i-1}, \forall i)$$

When $j \geq \tau + 1$, buff^* is re-written and hence, in this case, the total number of bits re-written is

---

[5] Recall that for a variable X with expectation $\mathbf{E}(X)$, the Chernoff bound states that for any $t > 0$, $\Pr[X \leq (1 - t)\mathbf{E}(X)] \leq e^{-\frac{t^2 \mathbf{E}(X)}{2}}$. In this case, $X = \#\mathsf{Succ}(x, y); \mathbf{E}(X) = \lambda(1 - \epsilon); t = \frac{1-2\epsilon}{2-2\epsilon}$.

$$= 2^j \frac{|\mu_0|}{\rho_0} + 2^{j-1} \frac{|\mu_1|}{\rho_1} + \ldots + 2^{j-(\tau+1)} \frac{|\mu_\tau|}{\rho_\tau} + 2^{j-(\tau+1)} \frac{|k^*|}{\rho^*}$$

$$= 2^j |\mu_0| \sum_{0 \le i \le \tau} \frac{1}{\rho_i} + 2^{j-(\tau+1)} \frac{|k^*|}{\rho^*}$$

The amortized update locality w per update is

$$|\mu_0| \sum_{0 \le i \le \tau} \frac{1}{\rho_i} + \frac{|k^*|}{2^{\tau+1} \rho^*} = (\log k + 1) \sum_{0 \le i \le \tau} \frac{1}{\rho_i} + \frac{\log k + 1}{\rho^*}.$$

Achieving a Worst-Case Guarantee. Note that, similar to the constructions of oblivious RAMs, one can convert the amortized update locality into a worst-case guarantee on the write locality, by distributing the work over many write operations. At a high level, this works by maintaining an additional "working copy" of data structure. Once levels $1, \ldots, i-1$ of the first data structure are filled in, the contents of level i are computed. This process takes place even as levels $1, \ldots, i-1$ of the second data structure are being filled in. This gives us a worst case write locality of $w = (\log k + 1) \sum_{i=0}^{\tau} \frac{1}{\rho_i} + \frac{\log k+1}{\rho^*}$ for the Buffered Hamming metric. Note, however, that a similar argument does not translate to the setting of the Prefix Hamming metric (since one would need to re-write parts of buffers at various levels at various points of time) and hence we only get an amortized bound for this metric.

To show update correctness, we must now argue, that if we begin the update algorithm with a corrupted codeword $\hat{c}_{m_t}$, such that $\mathsf{BHdis}(\hat{c}_{m_t}, c_{m_t}) \le \delta n$ and update the message m_t to m_{t+1} (where m_t and m_{t+1} differ (possibly) only at the i_t^{th} position, where $m_{t+1}(i_t) = b_{t+1}$), then we modify $\hat{c}_{m_t}$ to $\hat{c}_{m_{t+1}}$ where $\mathsf{BHdis}(\hat{c}_{m_{t+1}}, c_{m_{t+1}}) \le \delta n$ for some $c_{m_{t+1}}$ that is a codeword of m_{t+1}. To see this, observe that, the update algorithm decodes all buffers $\mathsf{buff}_0, \cdots, \mathsf{buff}_j$ for some $0 \le j \le \tau$ and possibly re-encodes these buffers into buff_{j+1}. Additionally, the update algorithm sets buffers $\mathsf{buff}_j, \cdots, \mathsf{buff}_0$ to empty. In certain cases, the update algorithm might re-write buffer buff^*. Note that if buff_{j+1} was written/re-encoded, then all buffers buff_j through buff_0 were also re-encoded. Similarly, if buff^* was re-encoded, then all buffers buff_τ through buff_0 were also re-encoded. Now, since $\mathsf{BHdis}(\hat{c}_{m_t}, c_{m_t}) \le \delta n$, it follows that all the buffers that were decoded by the update algorithm, decoded correctly and these buffers were then re-encoded without any errors. Hence, for all these buffers $0 \le h \le j+1$ in $\hat{c}_{m_{t+1}}$, $\mathsf{Hamm}(\hat{\psi}_h, \psi_h) \le \delta|\psi_h|$. For buffers that were not touched, since no change was made to these buffers, we still have that $\mathsf{Hamm}(\hat{\psi}_h, \psi_h) \le \delta|\psi_h|$ (for $h > j+1$ and for ψ^*). From these, it follows that $\mathsf{BHdis}(\hat{c}_{m_{t+1}}, c_{m_{t+1}}) \le \delta n$.

This proves the update correctness as well as the update locality of our update algorithm. This completes the proof of Lemma 1.

Lemma 2. *Let* $\mathcal{C} = (\mathcal{E}, \mathcal{D}, \mathcal{U})$ *be the above described* $(k, n, w, r, \epsilon, \delta) - \mathsf{LULDC}$ *code for the Buffered-Hamming metric. Then* $\mathcal{C}$ *is a* $(k, n, w, r, \epsilon, \frac{\delta}{2}) - \mathsf{LULDC}$ *code for the Prefix Hamming metric.*

Proof. Note that in our code construction, during a write/update operation, we never change the bits of the codeword in a buffer buff_i without changing the bits of the codeword in a buffer buff_j for any $j < i$. Furthermore, even when we change the bits of the codeword in a buffer buff_i, we then change the bits of the codeword in buffers $\mathsf{buff}_{i-1}, \cdots, \mathsf{buff}_0$ in that order. This means that if we consider the age-ordering of c_m, denoted by $w_1, \cdots, w_n$, then the indices corresponding to a buffer buff_j will always precede indices corresponding to a buffer buff_i, for any $i > j$. Now, since every buffer buff_{i+1} is twice the size of buffer buff_i, it follows that if two codewords c_m and $\hat{c}_m$ are such that $\mathsf{Prefix}(c_m, \hat{c}_m) \leq \frac{\delta n}{2}$, then $\mathsf{BHdis}(c_m, \hat{c}_m) \leq \delta n$, which gives us our result.

The proof of Theorem 1 now follows by simply combining Lemmas 1 and 2.

4   Computational Setting

4.1   Codes for Computationally Bounded Adversaries

In the previous section, we showed how to construct LULDC codes for the Prefix-Hamming metric. As noted before, we cannot construct LULDCs for metrics where the adverary can arbitrarily corrupt a constant fraction of the bits of the codeword. Since it is impossible to construct codes for the case of arbitrary adversarial errors, one could consider a setting where the decode algorithm will either decode to the correct message or *detect* if it is not able to do so; in other words, the decode algorithm will never output an incorrect message. Here too, it is easy to see that, unfortunately, one cannot have such information-theoretic error correcting codes. However, we show that by moving to the computationally-bounded adversarial setting, and by allowing the encoder/decoder to maintain a secret state S, one can construct error correcting codes with optimal rate that are locally updatable. Our code will provide the following guarantees:

- If the Prefix Hamming condition is satisfied, then every bit of the message will be locally decodable.
- Additionally, the (local) decoding algorithm will *never* output an incorrect bit of the message.

These guarantees allow us to achieve a tradeoff between *detecting* arbitrary adversarial errors and *decoding* a smaller class of errors. We will provide such a guarantee even when the adversary gets to observe the history of updates/writes made to the codeword; we denote the history of updates/writes made by hist[6].

---

[6] While this is the same guarantee that we provide even in the information-theoretic setting, we make this explicit here as we wish to endow the computationally bounded adversary with as much power as possible.

We now define such locally updatable locally decodable-detectable error correcting codes (LULDDC). As before, we provide our definition for the binary case, but this can be generalized to codes for larger alphabet Σ. Let λ be the security parameter and $\mathsf{neg}(\lambda)$ denote a function that is negligible in λ. We begin with the definition of the Prefix Hamming metric for the computational setting.

Definition 6 (The Computational Prefix Hamming Metric). *Let* $\mathsf{E} \in \{0,1\}^{r7}$. *Let* c *be of the form* $\mathsf{E}_1, \ldots, \mathsf{E}_n$. *Let* $\mathsf{w}_1, \cdots, \mathsf{w}_n$ *denote the age-ordering of* c. *For some* c' *of the form* $\mathsf{E}_1, \ldots, \mathsf{E}_n$ *and for* $1 \leq t \leq n$, *let* $c'[1,t]$ *denote the elements* $c'(\mathsf{w}_1), \cdots, c'(\mathsf{w}_t)$. *We say that the* Computational Prefix Hamming[8] *distance between* c *and* c', *denoted by* $\mathsf{Prefix}^{\mathsf{comp}}(c, c')$, *is* $\leq \delta n$ *if for all* $1 \leq t \leq n$, $\mathsf{Hamm}(c[1,t], c'[1,t]) \leq \delta t$, *where* $\mathsf{Hamm}(x,y)$ *denotes the Hamming Distance between any elements* x *and* y.

Definition 7 (Locally updatable and locally decodable-detectable codes for adversarial errors (LULDDC)). *A binary code* $\mathcal{C} : \{0,1\}^k \to \{0,1\}^n$ *is* $(k, n, w, r, \lambda, \mathsf{S})$-*locally updatable and locally decodable/detectable if there exist randomized algorithms* $\mathcal{U}$ *and* $\mathcal{D}$ *such that the following conditions are satisfied:*

1. Local Updatability:
   (a) *Let the state be initialized to* S_0. *Let* $m_0 \in \{0,1\}^k$ *and let* $c_{m_0} = \mathcal{E}(m_0, \mathsf{S}_0)$. *Let* m_t *be a message obtained by any (potentially empty) sequence of updates. (Note that the state* S *is updated everytime an update is made.) Let* hist *contain the entire history of updates made on potentially corrupted codewords. Let* $\hat{c}_{m_t}$ *be the final codeword obtained. Then* $\forall m_0 \in \{0,1\}^k, \forall t, \forall m_t, \forall i \in [k], \forall b \in \{0,1\}$, *for all probabilistic polynomial time (PPT) algorithms* $\mathcal{A}$, *for all* hist *and for all* $\hat{c}_{m_t} \in \{0,1\}^n$ *output by* $\mathcal{A}(m_t, i, b, \mathsf{hist})$, *the following condition holds with all but a negligible probability:*
     - *If* $\exists c_{m_t} \in \mathcal{C}_{m_t}$ *such that* $\mathsf{Prefix}^{\mathsf{comp}}(\hat{c}_{m_t}, c_{m_t}) \leq \delta n$, *then the actions of* $\mathcal{U}^{\hat{c}_{m_t}}(i, b, \mathsf{S}_t)$, *change* $\hat{c}_{m_t}$ *to* $u(\hat{c}_{m_t}, i, b, \mathsf{S}_t) \in \{0,1\}^n$, *where* $\mathsf{Prefix}^{\mathsf{comp}}(u(\hat{c}_{m_t}, i, b, \mathsf{S}_t), c_{m_{t+1}}) \leq \delta n$ *for some* $c_{m_{t+1}} \in \mathcal{C}_{m_{t+1}}$, *where* m_{t+1} *and* m_t *are identical except (possibly) at the* i^{th} *position, and* $m_{t+1}(i) = b$.
   (b) *The total number of queries and changes that* $\mathcal{U}$ *makes to the bits of* $\hat{c}_{m_t}$ *is at most* w.
2. Local Decodabilty-Detectability:

---

[7] We will think of E as a bit b_i followed by its constant sized authentication tag $\sigma_i = \mathsf{MAC}(b_i)$.

[8] While the definition of the distance function is not computational, we call it the computational prefix hamming distance, as this distance function is used only for the computational LULDDC construction. In our LULDDC codes, security guarantees will hold for codeword corruptions made by computationally bounded adversaries.

(a) Let $m_t \in \{0,1\}^k$ denote the latest message, as determined by hist. Then $\forall \text{hist}, \forall m_t \in \{0,1\}^k, \forall i \in [k]$, for all probabilistic polynomial time (PPT) algorithms $\mathcal{A}$ and for all $\hat{c}_{m_t} \in \{0,1\}^n$ output by $\mathcal{A}(m_t, i, \text{hist})$:

  − If $\exists c_{m_t} \in \mathcal{C}_{m_t}$ such that $\mathsf{Prefix}^{\mathsf{comp}}(\hat{c}_{m_t}, c_{m_t}) \le \delta n$, then

$$\Pr[\mathcal{D}^{\hat{c}_m}(i, \mathsf{S}) = m(i)] = 1 - \mathsf{neg}(\lambda),$$

where the probability is taken over the random coin tosses of the algorithm $\mathcal{D}$ and randomness used to generate S.

  − If $\forall c_{m_t} \in \mathcal{C}_{m_t}, \mathsf{Prefix}^{\mathsf{comp}}(\hat{c}_m, c_m) > \delta n$, then

$$\Pr[\mathcal{D}^{\hat{c}_m}(i, \mathsf{S}) = m(i) \text{ or } \bot] = 1 - \mathsf{neg}(\lambda),$$

where the probability is taken over the random coin tosses of the algorithm $\mathcal{D}$ and randomness used to generate S.

(b) $\mathcal{D}$ makes at most r queries to $\hat{c}_{m_t}$.

4.2   Our Results

In this section, we present a construction of a LULDDC in the computational setting. In particular, we show:

Theorem 2. *There exists a $(k, n, w, r, \lambda, \mathsf{S})$ locally updatable and locally decodable-detectable error correcting code $\mathcal{C} = (\mathcal{E}, \mathcal{D}, \mathcal{U})$, for the Computational Prefix Hamming metric, achieving the following parameters, for some constant $0 < \delta < \frac{1}{4}$:*

  − **Length of the code (n):** $n = \mathcal{O}(k)$.
  − **Update locality (w):** $w = \mathcal{O}(\log^2 k)$, in the amortized sense.
  − **Read locality (r):** $r = \mathcal{O}(\lambda \log^2 k)$, in the worst case.

Similar to the information-theoretic consturction, we use a *heirarchical data structure* to store our codewords. In addition, we use cuckoo hashing and private key locally decodable codes, details of which can be found in the full version.

LULDDC Overview. We start by recalling the construction of the information-theoretic LULDC code from Section 3.2. Recall that codewords had τ buffers. Each buff$_j$ encoded 2^j (address, value) pairs, stored in a sorted manner. We performed a binary search to search for a particular address, a within buff$_j$. The first difference is that we now use computational locally decodable codes to encode each buffer. (Such codes were introduced by [21]. In this work, we use the construction due to [12].) The next difference in the secret key setting is that we optimize the search performed on the buffers by using cuckoo hash functions[9]. In

---

[9] Cuckoo hash functions were first used in conjunction with the hierarchical data structure [19],[20] by Pinkas and Reinman [22] to obtain an ORAM construction. While it was shown that this construction does not hide the access pattern (i.e., which elements were read/written) [10],[17], as we will see, the underlying data structure coupled with cuckoo hashing can still be used securely to obtain a LULDDC code.

particular, an element (a, v) is inserted at location $h_{\ell,1}(a)$ or $h_{\ell,2}(a)$. To search for an address a in a particular buffer buff_ℓ, our decode algorithm only needs to read locations $h_{\ell,1}(a)$ and $h_{\ell,2}(a)$. (Of course, as in the information-theoretic case, we don't store the buffers in the clear. Rather we store an encoding of the buffers, now computed using the codes of [12] and the locations, $h_{\ell,1}(a)$ and $h_{\ell,2}(a)$, are read via calls to the underlying decode algorithm.) The second difference from the information theoretic construction is that we now use message authentication codes to *detect* a scenario where the codeword has too many errors. (To ensure local decodability, we need to authenticate each bit of the codeword separately.) This guarantees that our computational LULDDC code never decodes to an incorrect message.

Optimizing Parameters. While the above approach does give us an LULDDC construction, it doesn't give us our desired parameters. In particular, message authentication tags need to be of length at least λ, causing a blow-up of at least λ in the parameters. To avoid this, we use constant-size MACs instead.

Constant-Size Message Authentication Codes. Such message authentication codes (MAC) authenticate each bit of the message being authenticated (in this case, the codeword) with a tag of length $\mathcal{O}(1)$. While, individually, such MACs can be forged with constant probability, as we will see in our construction, they can be made secure when we are checking $\omega(\lambda)$ MAC values at a time.

At a high-level our decode algorithm will work as follows: we check the authenticity of λ randomly chosen bits of the codeword in each buffer. If most of the tags verify, we get a guarantee that less than a certain constant fraction of the bits of the codeword are corrupted. (Indeed, since each tag is computed with an independent MAC key, the odds that an adversary forges λ tags on his own, is negligible.) This, in turn, ensures that less than a constant fraction of bits of each codeword are corrupted, except with negligible probability[10], and therefore the codeword will decode correctly. (To the best of our knowledge, the idea of combining constant sized MACs with error correcting codes in such a way, was first used in the context of optimizing privacy amplification protocols in [5].) This combined with certain other ideas, give us the construction with parameters stated in Theorem 2. We now present the LULDDC construction and provide the proof of Theorem 2 in the full version.

4.3   LULDDC Construction

We now build our code (denoted $\mathcal{C}^{\text{comp}}$) in the secret key setting. The secret state S consists of a counter ctr (that is incremented everytime an update takes place), and a key to a PRF. S is used to generate the various keys used by the code. Similar to the information-theoretic case, each codeword c of $\mathcal{C}^{\text{comp}}$ consists

---

[10] This condition remains true only if all the buffers contain codewords that are at least λ-bits long. We will ensure this by starting our buffers only at a particular level.

of $\tau + 1$ buffers, $\mathsf{buff}_0, \ldots, \mathsf{buff}_\tau$, where $\tau = \log\left(\frac{k}{\log k}\right)$. In addition, there is a special buffer, buff^*, which has a structure different from the other buffers.

μ_i contains $(1 + \gamma)2^i$ cells (for some $\gamma > 1$) – each being either a "non-empty" cell containing a $(\mathsf{address}, \mathsf{value})$-pair or an "empty" cell containing a special symbol π. There are at most 2^i non-empty elements in μ_i at any point of time, and these elements are stored using cuckoo hash functions $(h_{i,1}, h_{i,2})$. The remaining locations of μ_i are filled with empty elements. We let $\psi_i = \mathcal{E}_{\mathsf{LDC}}(\mu_i)$. For each bit j of ψ_i, let $\sigma_i(j) = \mathsf{MAC}(\psi_i(j))$. Set $\eta_i = \{(\psi_i(j)||\sigma_i(j))\}$. buff_i contains η_i. μ^* contains all the bits of m in order (without the address values). $\psi^* = \mathcal{E}_{\mathsf{LDC}}(\mu^*)$ and $\eta^* = \{(\psi^*(j)||\sigma^*(j))\}$. The codeword is $c_m = [\mathsf{buff}_0, \ldots, \mathsf{buff}_\tau, \mathsf{buff}^*]$. Let α be a constant. We will pick α (as a function of δ and ζ) later on appropriately.

Encode algorithm. Our encoding algorithm works as follows:

Algorithm $\mathcal{E}(m, \mathsf{S})$:

1. Let $\mu^* = m(1), \cdots, m(k)$, where $m(i)$ denotes the ith bit of the message. Let $\psi^* = \mathcal{E}_{\mathsf{LDC}}(\mu^*)$ and $\eta^* = \{(\psi^*(j)||\sigma^*(j))\}$, where $\psi^*(j)$ is the jth bit of ψ^* and $\sigma^*(j) = \mathsf{MAC}(\psi^*(j))$.
2. Creates the $\tau + 1$ *empty* buffers $(\mathsf{buff}_\tau, \ldots, \mathsf{buff}_0)$ in that order; i.e., the underlying μ_i contains only special symbols.

Local Update Algorithm. The update algorithm takes as input a (potentially corrupted) codeword $\hat{c}$, an index i, a bit b_i, and the latest state S. Let the latest value of the message, as determined by hist, be m. Then if there exists some codeword c_m such that $c \in \mathcal{C}_m$ and $\mathsf{Prefix}^{\mathsf{comp}}(\hat{c}, c) \leq \delta n$, then the update algorithm outputs $\hat{c}'$ where $\mathsf{Prefix}^{\mathsf{comp}}(\hat{c}', c') \leq \delta n$ such that $c' \in \mathcal{C}_{m'}$ and m' and m are identical except possibly at the ith position, where $m'(i) = b_i$.

Recall that each codeword has multiple buffers of the form $\psi_i(j)||\sigma_i(j)$ where $\psi_i(j)$ is one bit of the codeword and $\sigma_i(j)$ is its constant sized message authentication tag. We refer to each of these $\psi_i(j)||\sigma_i(j)$ as an element of buff_i.

Algorithm $\mathcal{U}^{\hat{c}_m}(i, b, \mathsf{S})$:

1. If the first buffer is empty, compute $\psi = \mathcal{E}_{\mathsf{LDC}}(i||b)$; $\sigma = \mathsf{MAC}(\psi)$ and insert $\eta = (\psi||\sigma)$ into the first buffer.
2. If the first buffer is non-empty, find the first empty buffer – note this can be determined easily from ctr. Let the first empty buffer be at level j.
3. Store (i, b_i) as well as all the non-empty elements from μ_0 to μ_{j-1} into μ_j. To do this, we decode $\psi_0, \cdots, \psi_{j-1}$, insert the elements into μ_j and then compute $\mathcal{E}_{\mathsf{LDC}}(\mu_j)$ to obtain ψ_j. We compute $\eta_j(\ell) = \{\psi_j(\ell), \sigma_j(\ell)\}$. (The authentication tags $\sigma_j(\ell)$ are recomputed with the *latest key* corresponding to level j.) When decoding $\psi_0, \cdots, \psi_{j-1}$, ensure that at least $(1 - \delta)|\psi_j|$ MACs in every buffer verify; otherwise, output $\perp$.
4. Starting from buff_{j-1} up to buff_0, fill each of the buffers with empty elements in order. In other words, set the underlying μ_ℓs for each of the

buffers to contain only special symbols.

We refer the reader to the full version for further details.

Local Decode Algorithm. The algorithm for reading the ith bit works as follows:

Algorithm $\mathcal{D}^{\hat{c}m}(i, \mathsf{S})$:

1. Randomly select λ elements from each of the buffers.
2. For each of the elements, verify that $\sigma(j) = \mathsf{MAC}(\psi(j))$. (Note that this verification is done with appropriate MAC keys generated from S.)
3. If, for even one level, less than $\alpha\lambda$ of the tags verify, then output $\perp$.
4. The decode algorithm starts with the top-most buffer (buff_0) and proceeds downwards until it finds the address i.
5. For now, assume that buff_j contains μ_j instead of its encoding. Then to search a buffer buff_j for an index i, we read the locations $h_{j,1}(i)$ and $h_{j,2}(i)$. If either of these locations contains an entry (i, v) then v is the output of the algorithm. Since buff_j contains $\{\psi_j(\ell), \sigma_j(\ell)\}$, the steps we just described are implemented via calls to the underlying decoder $\mathcal{D}_{\mathsf{LDC}}$.
6. If we reach the last buffer, buff^*, we read the element v stored at address i in the buffer – once again, via calls to $\mathcal{D}_{\mathsf{LDC}}$. v is the output of the algorithm.

5 Dynamic Proof of Retrievability

A proof of retrievability scheme enables a client, storing his data on an untrusted server, to execute an audit protocol such that a malicious server that deletes or changes even a single bit of the client's data will fail to pass the audit protocol, except with negligible probability in the security parameter. Proofs of retrievability, introduced by Juels and Kaliski [14], were initially defined on static data building upon the closely related notion of sublinear authenticators defined by Naor and Rothblum [18]. The work of Cash, Küpçü, and Wichs [3] considers this notion for dynamically changing data; in other words, they constructed a proof of retrievability scheme that allowed for efficient updates to the data. We show that the techniques used to construct LULDDCs can be used to build a DPoR scheme. In addition to being conceptually simple, our construction also significantly improves the parameters achieved by [3].

A dynamic PoR scheme [3] comprises of four protocols $\mathsf{PInit}, \mathsf{PRead}, \mathsf{PWrite}$, and Audit between two stateful parties: the client $\mathcal{C}$ and a server $\mathcal{S}$ who is untrusted. The client stores some data m with the server and wishes to perform read, write, and audit operations on this data. In detail, the protocols are:

- $\mathsf{PInit}(1^\lambda, \Sigma, k)$: In this protocol, the client initializes an empty data storage on the server of length k, where each element in the data comes from an alphabet Σ. The security parameter is λ.
- $\mathsf{PRead}(i)$: In this protocol, the client reads the ith location of the data and outputs some value v_i at the end of the protocol.

- PWrite(i, v_i): In this, the client sets the ith location of the data to v_i.
- Audit(): In this protocol, the client verifies that the server is maintaining the data correctly so that they remain retrievable. The client outputs either accept or reject.

The (private) state of the client is implicitly assumed in all the above protocols and the client may also output reject during any of the protocols if it detects any malicious behavior on the part of the server. A dynamic PoR scheme must satisfy three properties: *correctness, authenticity, and retrievability*. We refer the reader to [3] for the formal definitions of these properties.

Overview of Construction. At a high-level, our construction follows the same approach as our LULDDC scheme. One main difference is that in addition to storing encoded messages in buff_0 to buff_τ and buff^*, we will store the decoded, authenticated, message of every buffer in another set of $\tau+2$ buffers (denoted by plain_0 to plain_τ and plain^*). The read algorithm works by reading these buffers (instead of the encoded buffers) and verifying their respective MACs. The write algorithm works the same as before – except that it writes to both encoded and unencoded buffers. The audit algorithm works by checking λ randomly chosen locations of each of the encoded buffers and verifying their MACs. Additionally, to obtain good write complexity, we use linear time encodable and decodable standard error correcting codes [26] to encode each buffer, as opposed to using locally decodable codes. We shall also use two types of message authentication codes: to MAC the elements of buffers buff_0 to buff_τ and buff^* (that store codewords), we shall use constant size MACs; however, to MAC the elements of buffers plain_0 to plain_τ (that store elements of the message in the clear), we shall use MACs with MAC length λ. We shall abuse notation and denote both these MACs by MAC (it will be clear from context which type of MAC we use).

- PInit($1^\lambda, \Sigma, k$): This protocol is very similar to the Encode algorithm of our LULDDC. Namely, when storing data $m = m(1), \cdots, m(k) = \mu^*$ on the server, with $m(i) \in \Sigma$, the client computes $\psi^* = \mathcal{E}_{\text{lin}}(\mu^*)$ and $\eta^* = \{(\psi^*(j)||\sigma^*(j))\}$, where $\psi^*(j)$ is the jth element of ψ^* and $\sigma^*(j) = \text{MAC}(\psi^*(j))$. The client stores η^* in buff^*. Additionally the client will also store every element of m along with its MAC in plain^*[11].
- PWrite(i, v_i): To write element v_i into position i, $\mathcal{C}$ does as follows:
  - If the first buffer is non-empty, find the first empty buffer – this can be determined using ctr, but for now, we just assume that we learn this by decoding buffers in a top-down manner and scanning them to see if they contain any non-empty element. Let the first empty buffer be at level j.
  - Update S to S' so that it now contains an incremented counter.

---

[11] In order to reduce the storage complexity, every $\frac{\lambda}{|\Sigma|}$ elements are grouped together and MACed so that the storage complexity remains at $\mathcal{O}(k)$ and does not become $\mathcal{O}(k\lambda)$.

- We store (i, b_i) as well as all the non-empty elements from μ_0 to μ_{j-1} into μ_j. To do this, we decode $\psi_0 \cdots \psi_{j-1}$, insert the elements into μ_j and then compute $\mathcal{E}_{\text{lin}}(\mu_j)$ to obtain ψ_j. We compute $\eta_j(\ell) = \{\psi_j(\ell), \sigma_j(\ell)\}$. (The authentication tags $\sigma_j(\ell)$ are recomputed with the *latest key* corresponding to level j, which in turn is computed from S').
- Additionally, we store the plain message μ_j in plain_j. Note, that whenever reading an element, we read the element along with its MAC and reject if the MAC does not verify.
- The buffers from $\mathsf{buff}_{j-1} \ldots \mathsf{buff}_0$, as well as $\mathsf{plain}_{j-1} \ldots \mathsf{plain}_0$, are now set to empty by writing special elements into it (along with appropriate MAC values).
- PRead(i): To read the ith element of the most recent message stored on the server, the client does the following:
    - The algorithm starts with the top-most buffer (plain_0) and proceeds downwards until it finds the address i.
    - Note that plain_j contains μ_j in plaintext. To search a buffer buff_j for an index i, we read the locations $h_{j,1}(i)$ and $h_{j,2}(i)$. If either of these locations contains an entry (i, v) then v is the output of the algorithm.
    - If we reach the last buffer, plain^*, we read the element v stored at address i in plain^*. If the tag σ does not verify, for any element read (in any of the buffers), then the algorithm outputs $\mathtt{reject}$, otherwise v is the output[12].
- Audit(): The audit protocol works as follows:
    - For every buffer buff_0 to buff_τ as well as buff^*, pick λ locations of the codeword ψ_j (stored in buff_j) at random and read these λ elements along with their MAC values.
    - If all the MACs verify, then output $\mathtt{accept}$, otherwise output $\mathtt{reject}$.

We defer the proof of correctness and security for construction to the full version. For now, we simply state the parameters that this construction achieves. The (worst case) complexity of the PWrite protocol is $\mathcal{O}(\log^2 k)$. The complexity of the PRead protocol is simply $\mathcal{O}(\lambda \log k)$ as we need to read a constant number of elements in each buffer (along with their MACs of length λ). Finally, the complexity of the Audit protocol is $\mathcal{O}(\lambda \log k)$ as we read λ elements of the codeword in each buffer, along with their constant-size MACs. The client storage is $\mathcal{O}(\lambda)$.

Acknowledgments. We thank the anonymous reviewers of TCC 2014 for their very valuable feedback.

References

1. Ateniese, G., Kamara, S., Katz, J.: Proofs of storage from homomorphic identification protocols. In: Matsui, M. (ed.) ASIACRYPT 2009. LNCS, vol. 5912, pp. 319–333. Springer, Heidelberg (2009)

---

[12] Note, that because of the way we MAC the plaintext values in plain buffers, when we read a single element from plain, we may have to read an additional $\frac{\lambda}{|\Sigma|}$ elements in order to verify the MAC; we ignore this in the description for ease of exposition.

2. Bowers, K.D., Juels, A., Oprea, A.: Proofs of retrievability: theory and implementation. In: Proceedings of the First ACM Cloud Computing Security Workshop, CCSW 2009, pp. 43–54 (2009)

3. Cash, D., Küpçü, A., Wichs, D.: Dynamic proofs of retrievability via oblivious RAM. In: Johansson, T., Nguyen, P.Q. (eds.) EUROCRYPT 2013. LNCS, vol. 7881, pp. 279–295. Springer, Heidelberg (2013)

4. Chandran, N., Kanukurthi, B., Ostrovsky, R.: Locally updatable and locally decodable codes. Cryptology ePrint Archive, Report 2013/520 (2013), http://eprint.iacr.org/

5. Chandran, N., Kanukurthi, B., Ostrovsky, R., Reyzin, L.: Privacy amplification with asymptotically optimal entropy loss. In: Proceedings of the 42nd ACM Symposium on Theory of Computing, STOC 2010, pp. 785–794 (2010)

6. Chee, Y.M., Feng, T., Ling, S., Wang, H., Zhang, L.F.: Query-efficient locally decodable codes of subexponential length. Computational Complexity 22(1), 159–189 (2013)

7. Dodis, Y., Vadhan, S.P., Wichs, D.: Proofs of retrievability via hardness amplification. In: Reingold, O. (ed.) TCC 2009. LNCS, vol. 5444, pp. 109–127. Springer, Heidelberg (2009)

8. Efremenko, K.: 3-query locally decodable codes of subexponential length. In: Proceedings of the 41st Annual ACM Symposium on Theory of Computing, STOC, pp. 39–44 (2009)

9. Goldreich, O.: Towards a theory of software protection and simulation by oblivious rams. In: STOC, pp. 182–194 (1987)

10. Goodrich, M.T., Mitzenmacher, M.: Privacy-preserving access of outsourced data via oblivious RAM simulation. In: Aceto, L., Henzinger, M., Sgall, J. (eds.) ICALP 2011, Part II. LNCS, vol. 6756, pp. 576–587. Springer, Heidelberg (2011)

11. Guo, A., Kopparty, S., Sudan, M.: New affine-invariant codes from lifting. In: Innovations in Theoretical Computer Science, ITCS, pp. 529–540 (2013)

12. Hemenway, B., Ostrovsky, R., Strauss, M.J., Wootters, M.: Public key locally decodable codes with short keys. In: Goldberg, L.A., Jansen, K., Ravi, R., Rolim, J.D.P. (eds.) RANDOM 2011 and APPROX 2011. LNCS, vol. 6845, pp. 605–615. Springer, Heidelberg (2011)

13. Hemenway, B., Ostrovsky, R., Wootters, M.: Local correctability of expander codes. In: Fomin, F.V., Freivalds, R., Kwiatkowska, M., Peleg, D. (eds.) ICALP 2013, Part I. LNCS, vol. 7965, pp. 540–551. Springer, Heidelberg (2013)

14. Juels, A., Kaliski, B.: Pors: proofs of retrievability for large files. In: Proceedings of the 2007 ACM Conference on Computer and Communications Security, pp. 584–597 (2007)

15. Katz, J., Trevisan, L.: On the efficiency of local decoding procedures for error-correcting codes. In: Proceedings of the 32nd Annual ACM Symposium on Theory of Computing, STOC, pp. 80–86 (2000)

16. Kopparty, S., Saraf, S., Yekhanin, S.: High-rate codes with sublinear-time decoding. In: STOC, pp. 167–176 (2011)

17. Kushilevitz, E., Lu, S., Ostrovsky, R.: On the (in)security of hash-based oblivious ram and a new balancing scheme. In: SODA, pp. 143–156 (2012)

18. Naor, M., Rothblum, G.N.: The complexity of online memory checking. In: 46th Annual IEEE Symposium on Foundations of Computer Science, FOCS 2005, pp. 573–584 (2005)

19. Ostrovsky, R.: An efficient software protection scheme. In: Brassard, G. (ed.) CRYPTO 1989. LNCS, vol. 435, pp. 610–611. Springer, Heidelberg (1990)

20. Ostrovsky, R.: Efficient computation on oblivious rams. In: Ortiz, H. (ed.) STOC, pp. 514–523. ACM (1990)
21. Ostrovsky, R., Pandey, O., Sahai, A.: Private locally decodable codes. In: Arge, L., Cachin, C., Jurdziński, T., Tarlecki, A. (eds.) ICALP 2007. LNCS, vol. 4596, pp. 387–398. Springer, Heidelberg (2007)
22. Pinkas, B., Reinman, T.: Oblivious RAM revisited. In: Rabin, T. (ed.) CRYPTO 2010. LNCS, vol. 6223, pp. 502–519. Springer, Heidelberg (2010)
23. Schulman, L.J.: Communication on noisy channels: A coding theorem for computation. In: 33rd Annual Symposium on Foundations of Computer Science, FOCS, pp. 724–733 (1992)
24. Schulman, L.J.: Deterministic coding for interactive communication. In: Proceedings of the 25th Annual ACM Symposium on Theory of Computing, STOC, pp. 747–756 (1993)
25. Shacham, H., Waters, B.: Compact proofs of retrievability. In: Pieprzyk, J. (ed.) ASIACRYPT 2008. LNCS, vol. 5350, pp. 90–107. Springer, Heidelberg (2008)
26. Spielman, D.A.: Linear-time encodable and decodable error-correcting codes. In: Proceedings of the Twenty-Seventh Annual ACM Symposium on Theory of Computing, STOC, pp. 388–397 (1995)
27. Yekhanin, S.: Towards 3-query locally decodable codes of subexponential length. In: Proceedings of the 39th Annual ACM Symposium on Theory of Computing, San Diego
28. Yekhanin, S.: Locally decodable codes. Foundations and Trends in Theoretical Computer Science 6(3), 139–255 (2012)

Leakage Resilient Fully Homomorphic Encryption

Alexandra Berkoff[1] and Feng-Hao Liu[2]

[1] Brown University
aberkoff@cs.brown.edu
[2] University of Maryland, College Park
fenghao@cs.umd.edu

Abstract. We construct the first leakage resilient variants of fully homomorphic encryption (FHE) schemes. Our leakage model is bounded adaptive leakage resilience. We first construct a leakage-resilient leveled FHE scheme, meaning the scheme is homomorphic for all circuits of depth less than some pre-established maximum set at key generation. We do so by applying ideas from recent works analyzing the leakage resilience of public key encryption schemes based on the decision learning with errors (*DLWE*) assumption to the Gentry, Sahai and Waters ([1]) leveled FHE scheme. We then move beyond simply leveled FHE, removing the need for an a priori maximum circuit depth, by presenting a novel way to combine schemes. We show that by combining leakage resilient leveled FHE with multi-key FHE, it is possible to create a leakage resilient scheme capable of homomorphically evaluating circuits of arbitrary depth, with a bounded number of distinct input ciphertexts.

1 Introduction and Related Work

Fully homomorphic encryption is a way of encrypting data that allows a user to perform arbitrary computation on that data without decrypting it first. The problem of creating a fully homomorphic encryption scheme was suggested by Rivest, Adleman, and Dertouzos in 1978 [2]. It has received renewed attention in recent years and has obvious applicability to cloud computing— If a user stores her data on someone else's servers, she may wish to store her data encrypted under a public key encryption scheme, yet still take advantage of that untrusted server's computation power to work with her data.

The first candidate for fully homomorphic encryption was proposed by Gentry in 2009 [3]. Since then, candidate schemes have been based on a variety of computational assumptions (see, for example: [4,5,6,7]) including the decision learning with errors (*DLWE*) assumption [8,9,10,1]. The latest *DLWE*-based work is due to Gentry, Sahai, and Waters (GSW) [1], and it is this work we focus most closely on in our paper.

We note that public key encryption schemes based on the *DLWE* assumption have typically been based on one of two schemes both described by Regev in the latest version of [11]. Regev originally constructed so-called "primal Regev"

Y. Lindell (Ed.): TCC 2014, LNCS 8349, pp. 515–539, 2014.

(referred to in this work as RPKE) and Gentry, Peikert, and Vaikuntanathan constructed so-called "dual Regev" [12] in 2008. The instantiations in the papers describing all the *DLWE*-based homomorphic schemes cited above use "primal Regev" as a building block. The Regev schemes have also been used as building blocks to achieve identity based encryption, attribute based encryption, and, as described in Section 1.2, leakage resilient encryption.

The term "leakage resilience" is meant to capture the security of a cryptographic algorithm when an adversary uses non-standard methods to learn about the secret key. Typically in security proofs, attackers are modeled as probabilistic polynomial time machines with only input/output access to the given cryptographic algorithm. Leakage resilience is a theoretical framework for addressing security when an attacker learns information about the secret key not obtainable through the standard interface, for example by obtaining physical access to a device, or by identifying imperfect or correlated randomness used in secret key generation.

Starting with the work of Ishai, Sahai and Wagner [13], and Micali and Reyzin [14], the cryptographic community has worked towards building general theories of security in the presence of information leakage. This has been an active topic of research over the past 15 years (see [15,16,17,18,19,20,21,22,23,24,25,26,13,14] and the references therein), resulting in many different leakage models, and cryptographic primitives such as public key encryption schemes and signature schemes secure in each model.

In our work, we, for the first time, apply the framework of leakage resilience to fully homomorphic schemes.

1.1 Non-adaptive Leakage on FHE

We start with the observation that the Decision Learning With Errors problem is, with appropriate parameter settings, leakage resilient – Goldwasser, Kalai, Peikert and Vaikuntanathan showed that the *DLWE* problem with a binary secret, and a carefully chosen bound on the size of the error term, with a leakage function applied to the secret, reduces from a *DLWE* problem with smaller dimension, modulus, and error bound, but no leakage [27]. Recently, Alwen, Krenn, Pietrzak, and Wichs extended this result to apply to a wider range of secrets and error bounds [28].

Since many FHE schemes (for example [8,9,10,1]) can be instantiated based on the *DLWE* assumption, an obvious first attempt to create leakage resilient FHE is to directly apply those results by instantiating an FHE scheme with parameters that make the underlying *DLWE* problem leakage resilient. Indeed, doing so leads immediately to non-adaptive leakage resilient FHE. We describe these results in Appendix C in the full version of our paper.

We note as well that the leakage resilience of *DLWE* leads to leakage resilient symmetric-key encryption [27], and closely related results lead to non-adaptive leakage resilience of RPKE [15].

The differentiation between adaptive and non-adaptive leakage is crucial. In the non-adaptive leakage model, an adversary can learn any arbitrary (poly-time computable, bounded output-length) function of the secret key, with the caveat that he cannot adaptively choose the function based on the scheme's public key. This leakage model is not entirely satisfactory, as typically one assumes that if a value is public, everyone, including the adversary will be able to see it at all times. In contrast, the adaptive leakage resilience model assumes that an adversary has full access to all the scheme's public parameters, and can choose its leakage function accordingly.

1.2 Adaptive Leakage on Leveled FHE

Given the gap between the non-adaptive leakage resilience model and the expected real-life powers of an adversary, in this work we primarily consider the adaptive bounded memory leakage model. The model is described in, for example, the works [15,16]. Since an adversary can choose its leakage function after seeing the public key(s), in effect we consider functions that leak on the public and secret keys together. This framework has been previously considered for non-homomorphic public key and identity based encryption schemes based on bilinear groups, lattices, and quadratic residuosity [16,26,29]. Additionally, both RPKE and "dual Regev", schemes based on *DLWE*, can be made leakage resilient; Akavia, Goldwasser, and Vaikunatanathan achieve adaptive leakage-resilient RPKE [15], and Dodis, Goldwasser, Kalai, Peikert, and Vaikuntanathan construct leakage-resilient "dual Regev" [19]. In fact, the latter scheme is secure against auxiliary input attacks—essentially, they consider a larger class of leakage functions—ones whose output length has no bound, but which no probabilistic polynomial time adversary can invert with non-negligible probability.

Unfortunately, the non-adaptive leakage resilient scheme described in Appendix C does not lead in a straightforward way to an adaptively leakage resilient scheme. The crux of the problem is that the public key is a function of the secret key, and when an adversary has leakage access to both the public and secret keys, it can choose a function which simply asks if the two are related. Existing proofs of security for *DLWE*-based FHE schemes all start by proving the public key indistinguishable from random, and such leakage functions make this impossible.

In fact, one might expect the same problem when analyzing the adaptive leakage resilience of RPKE, as the original security proof for this scheme followed the same outline [11]. Akavia, Goldwasser, and Vaikuntanathan (AGV) succeeded in constructing a leakage-resilient variant of RPKE despite this hindrance by writing a new security proof. They directly show that the *ciphertexts* are indistinguishable from random, without making any statements about the *public key* [15].

Inspired by the success of AGV, one might try to use a variation on their technique to prove prove an FHE scheme secure. We note that typically the public key of an FHE scheme consists of two parts: an "encryption key," which is used to generate new ciphertexts, and an "evaluation key," which is used to homomorphically combine the ciphertexts. A strengthening of the AGV technique leads to a secure scheme if the adversary sees the encryption key before choosing its leakage function, but unfortunately the proof fails if it also sees the evaluation key. The evaluation key is not just a *function of*, but actually an *encryption of* the secret key, and proving security when an adversary could potentially see actual decryptions of some bits of the secret key is a more complicated proposition.

Since the presence of an evaluation key is what hampers the proof, our next step is to apply this technique to a scheme *without* an evaluation key. The first leveled FHE scheme without an evaluation key was recently constructed by Gentry, Sahai, and Waters (GSW) [1]. We strengthen the results of Akavia, Goldwasser, and Vaikuntanathan to apply to a much broader range of parameters, and use this new result to construct LRGSW, a leakage-resilient variant of GSW. We present these results in sections 3 and 4.

1.3 Overcoming the "Leveled" Requirement

Note that so far, we have achieved leakage resilient leveled FHE, meaning we have a scheme where if a maximum circuit depth is provided at the time of key generation, the scheme supports homomorphic evaluation of all circuits up to that depth. In contrast, in a true, non-leveled, fully homomorphic encryption scheme, one should not need to specify a maximum circuit depth ahead of time.

The standard technique for creating a non-leveled FHE scheme, first proposed by Gentry in his original construction, is to first create a "somewhat-homomorphic" encryption scheme (all leveled schemes are automatically "somewhat homomorphic"), make it "bootstrappable" in some way, and then "bootstrap" it to achieve full homomorphism [3]. Although LRGSW is somewhat homomorphic, it needs a separate evaluation key to be bootstrappable. In fact, every known bootstrappable scheme has an evaluation key containing encryptions of the secret key, leaving us back with the same issue we sidestepped by choosing to modify the GSW scheme.

Our key insight is that while we need encryptions of the secret key to perform bootstrapping, these encryption do not need to be part of the public key. We combine a leakage resilient *leveled* FHE scheme with a N-key multi-key FHE scheme in a novel way, which allows us to store these encryptions as part of the ciphertext, letting us achieve a *non-leveled* leakage resilient FHE scheme. We provide an instantiation of this using LRGSW and the López-Alt, Tromer, and Vaikuntanathan multi-key FHE scheme [30]. We discuss these results in section 5. Our contribution is a step towards true fully homomorphic encryption, as we remove the circuit *depth* bound. An artifact of our construction is that the N from our N-key multi-key FHE scheme becomes a bound on the *arity* of our circuit instead. The problem of creating leakage resilient, true FHE is still open, and seems intimately related to the problem of creating true, non-leveled FHE without bootstrapping.

2   Preliminaries

We let bold capital letters (e.g. $\mathbf{A}$) denote matrices, and bold lower-case letters (e.g. $\mathbf{x}$) denote vectors. We denote the inner product of two vectors as either $\mathbf{x} \cdot \mathbf{y}$ or $\langle \mathbf{x}, \mathbf{y} \rangle$.

For a real number x, we let $\lfloor x \rfloor$ be the closest integer $\leq x$, and $\lfloor x \rceil$ be the closest integer to x. For an integer y, we let $[y]_q$ denote $y \mod q$. For an integer N, we let $[N]$ denote the set $\{1, 2, \ldots, N\}$.

We use $x \leftarrow \mathcal{D}$ to denote that x was drawn from a distribution $\mathcal{D}$. We use $x \xleftarrow{\$} S$ to denote that x was drawn uniformly from a set S. To denote computational indistinguishability, we write $\mathcal{X} \approx_c \mathcal{Y}$, and to denote statistical indistinguishability, we write $\mathcal{X} \approx_s \mathcal{Y}$. To denote the statistical distance between two distributions, we write $\Delta(\mathcal{X}, \mathcal{Y})$. Throughout this work, we use η to denote our security parameter.

In this work, we refer to the **ϵ-smooth average min-entropy** (first defined in [31]) of X conditioned on Y as $\tilde{H}_\infty^\epsilon(X|Y)$. We refer the reader to Appendix A where we fully define this, and other related concepts of min-entropy, and state versions of the leftover hash lemma that hold true for these concepts.

2.1   Homomorphism

We use standard definitions for fully homomorphic encryption and leveled fully homomorphic encryption, so we defer full statements of these definitions to Appendix A. We do define a new, related type of fully homomorphic encryption below:

Definition 2.1. *An encryption scheme is **bounded arity fully homomorphic** if it takes $T = poly(\eta)$ as an additional input in key generation, and is $\mathcal{T}$-homomorphic for $\mathcal{T} = \{\mathcal{T}_\eta\}_{\eta \in \mathbb{N}}$, the set of all arithmetic circuits over $\{0, 1\}$ with arity $\leq T$ and depth $poly(\eta)$.*

2.2   Leakage Resilience

Definition 2.2. *Let λ be a non-negative integer. A scheme HE is **adaptively leakage resilient** to λ bits of leakage, if for any PPT adversary $\mathcal{A}$ it holds that*

$$\mathbf{ADV}_{ALR^\lambda(b=0), ALR^\lambda(b=1)}(\mathcal{A}) = negl(\lambda)$$

where the notation $\mathbf{ADV}_{\mathcal{X}, \mathcal{Y}}(\mathcal{A}) := \mid Pr[\mathcal{A}(\mathcal{X}) = 1] - Pr[\mathcal{A}(\mathcal{Y}) = 1] \mid$

and the experiment ALR^λ is defined as follows:

1. The challenger generates $(pk, sk) \leftarrow$ HE.KeyGen(1^η) and sends pk to the adversary.
2. The adversary $\mathcal{A}$ selects a leakage function $h : \{0, 1\}^* \rightarrow \{0, 1\}^\lambda$ and sends it to the challenger.
3. The challenger replies with $h(sk)$.

4. The adversary $\mathcal{A}$ replies with (m_0, m_1)
5. The challenger chooses $b \xleftarrow{\$} \{0, 1\}$, computes $c \leftarrow$ HE.Enc(pk, m_b) and sends c to $\mathcal{A}$.
6. $\mathcal{A}$ outputs $b' \in \{0, 1\}$

In the above definition, **adaptive** refers to the fact that $\mathcal{A}$ can choose h after having seen the scheme's public parameters. In fact, an adversary could "hard-code" the scheme's public key into its leakage function, in effect seeing $h(pk, sk)$. In the remainder of this paper, we therefore consider leakage functions that leak on both the public key and the secret key together. There is a corresponding weaker notion of leakage resilience called **non-adaptive** where the adversary must choose h independently of the scheme's public key, and learns only $h(sk)$.

2.3   Learning with Errors

The learning with errors problem (*LWE*), and the related decision learning with errors problem (*DLWE*) were first introduced by Regev [11] in 2005. The problem is, given a secret $\mathbf{s} \in \mathbb{Z}_q^n$, a matrix $\mathbf{A} \xleftarrow{\$} \mathbb{Z}_q^{m \times n}$ and an error vector $x \xleftarrow{\$} \psi^m$ to distinguish $\mathbf{As} + \mathbf{x}$ from $u \xleftarrow{\$} \mathbb{Z}_q^m$. This problem is standard in the literature and we leave full definitions to Appendix A.

The following statement summarizes much of the recent work analyzing the hardness of *DLWE*.

Statement 1. *(Theorem 1 in [1], due to work of [11,32,33,34])*
Let $q = q(n) \in \mathbb{N}$ be either a prime power or a product of small (size $poly(n)$) distinct primes, and let $\beta \geq \omega(\log n) \cdot n$ Then there exists an efficiently sampleable $\beta - bounded$ distribution χ such that if there is an efficient algorithm that solves the average-case LWE problem for parameters n, q, χ, then:

- *There is an efficient quantum algorithm that solves GapSVP$_{\tilde{O}(nq/\beta)}$ on any n-dimensional lattice.*
- *If $q \geq \tilde{O}(2^{n/2})$, there is an efficient classical algorithm for GapSVP$_{\tilde{O}(nq/\beta)}$ on any n-dimensional lattice.*

In both cases, if one also considers distinguishers with sub-polynomial advantage, then we require $\beta \geq \tilde{O}(n)$ and the resulting approximation factor is slightly larger than $\tilde{O}(n^{1.5}q/\beta)$.

The *GapSVP$_\gamma$* problem is, given an arbitrary basis of an n dimensional lattice, to determine whether the shortest vector of that lattice has length less than 1 or greater than γ.

Statement 2. *(from [8])*
The best known algorithms for GapSVP$_\gamma$ [35,36] require at least $2^{\tilde{\Omega}(n/(\log \gamma))}$ time.

These hardness results guide the setting of parameters for our scheme.

3   The **LRGSW** Scheme

We now present LRGSW, an adaptively leakage resilient variant of the Gentry, Sahai, and Waters (GSW) FHE scheme [1]. We ⏐box⏐ the differences between our scheme and GSW in our description below. The scheme encrypts messages under the "approximate eigenvector" method: For a message $\mu \in \mathbb{Z}_q$, ciphertexts are matrices $\mathbf{C} = \mathsf{Enc}(pk, \mu)$ and have the property that $\mathbf{C} \cdot \mathbf{sk} \approx \mu \cdot \mathbf{sk}$, where $\mathbf{sk}$ is the secret key vector. This means that to homomorphically multiply two ciphertexts $\mathbf{C}_1 = \mathsf{Enc}(pk, \mu_1)$ and $\mathbf{C}_2 = \mathsf{Enc}(pk, \mu_2)$, one simply computes $\mathbf{C}_{mult} = \mathbf{C}_1 \cdot \mathbf{C}_2$. Crucially, this intuitive method for homomorphic evaluation removes the need for an "evaluation key" present in other fully homomorphic schemes. Note that for the error-growth reasons Gentry, Sahai, and Waters gave in Section 3.3 of their paper [1], our modification of their scheme is designed to homomorphically evaluate only binary circuits constructed of NAND gates.

3.1   Our Leveled Scheme

(note: we define PowersOfTwo, Flatten, BitDecomp and BitDecomp^{-1} in Section 3.2 below)

LRGSW.Setup($1^{\eta}, 1^{L}$): Recalling that η is security parameter of the scheme, and $L = poly(\eta)$ is the maximum circuit depth our scheme must evaluate, let $\tau = \max\{L, \eta^2\}$. Choose a lattice dimension $n = \tau^2$, modulus $\boxed{q \geq \tau \cdot 2^{2\tau \log^2 \tau}}$, and error distribution $\boxed{\chi = \overline{\Psi}_{\beta}, \text{ where } \beta = \tau \cdot \tau^{\log \tau} \text{ bounded}}$ Choose $\boxed{m = m(\eta, L) \geq 2n \log q + 3\eta.}$ Let $params = (n, q, \chi, m)$. Let $\ell = \lfloor \log q \rfloor + 1$ and $N = (n+1) \cdot \ell$.

LRGSW.SecretKeyGen($params$): Choose $\mathbf{t} \xleftarrow{\$} \mathbb{Z}_q^n$. Let $sk = \mathbf{s} = (1, -t_1, \ldots, -t_n)$. Let $\mathbf{v} = \mathsf{PowersOfTwo}(\mathbf{s})$.

LRGSW.PublicKeyGen($\mathbf{s}, params$): Let $\mathbf{A} \xleftarrow{\$} \mathbb{Z}_q^{m \times n}$. Let $\mathbf{e} \xleftarrow{\$} \chi^m$. Let $\mathbf{b} = \mathbf{At} + \mathbf{e}$. Let $pk = \mathbf{K} = [\mathbf{b} \| \mathbf{A}]$.

LRGSW.Encrypt($\mathbf{K}, \mu$): For message $\mu \in \{0, 1\}$, choose $R \xleftarrow{\$} \{0, 1\}^{N \times m}$. Let $\mathbf{I}_N$ be the $N \times N$ identity matrix.

$$\mathbf{C} = \mathsf{Flatten}(\mu \cdot \mathbf{I}_N + \mathsf{BitDecomp}(\mathbf{R} \cdot \mathbf{K})) \in \mathbb{Z}_q^{N \times N}$$

LRGSW.Decrypt($\mathbf{s}, \mathbf{C}$): Let i be the index among the first ℓ elements of $\mathbf{v}$ such that $\mathbf{v}_i = 2^i \in (\frac{q}{4}, \frac{q}{2}]$. Let $\mathbf{C}_i$ be the ith row of $\mathbf{C}$. Compute $x_i = \langle \mathbf{C}_i, \mathbf{v} \rangle$. Output $\mu' = \left\lfloor \frac{x_i}{v_i} \right\rceil$

LRGSW.NAND($\mathbf{C}_1, \mathbf{C}_2$): Output $\mathsf{Flatten}(\mathbf{I}_N - \mathbf{C}_1 \cdot \mathbf{C}_2)$

3.2   Elementary Vector Operations in **LRGSW**

The above scheme description makes use of a number of vector operations that we describe below. Let $\mathbf{a}, \mathbf{b}$ be vectors of dimension k. Let $\ell = \lfloor \log q \rfloor + 1$. Note that the operations we describe are also defined over matrices, operating row by row on the matrix, and that all arithmetic is over $\mathbb{Z}_q$.

BitDecomp($\mathbf{a}$) = the $k \cdot \ell$ dimensional vector $(a_{1,0}, \ldots, a_{1,\ell-1}, \ldots, a_{k,0}, \ldots a_{k,\ell-1})$
 where $a_{i,j}$ is the jth bit in the binary representation of a_i, with bits ordered from least significant to most significant.

BitDecomp$^{-1}(\mathbf{a}')$ For $\mathbf{a}' = (a_{1,0}, \ldots, a_{1,\ell-1}, \ldots, a_{k,0}, \ldots a_{k,\ell-1})$, let
 BitDecomp$^{-1}(\mathbf{a}') = (\sum_{j=0}^{\ell-1} 2^j a_{1,j}, \ldots, \sum_{j=0}^{\ell-1} 2^j a_{k,j})$, but defined even when $\mathbf{a}'$ isn't binary.

Flatten($\mathbf{a}'$) = BitDecomp(BitDecomp$^{-1}(\mathbf{a}')$)

PowersOfTwo($\mathbf{b}$) = $(b_1, 2b_1, 4b_1, \ldots, 2^{\ell-1}b_1, \ldots, b_k, \ldots 2^{\ell-1}b_k)$.

3.3   Correctness

Correctness of the scheme follows because: $\mathbf{C}\mathbf{v} = \mu\mathbf{v} + \mathbf{R}\mathbf{A}\mathbf{s} = \mu\mathbf{v} + \mathbf{R}\mathbf{e}$, so, $x_i = \mu \cdot \mathbf{v}_i + \langle \mathbf{R}_i, \mathbf{e} \rangle$. Since $\mathbf{v}_i > \frac{q}{4}$, if we let $B = \|\mathbf{e}\|_\infty$, since $\mathbf{R}_i$ is an N-dimensional binary vector, as long as $NB < \frac{q}{8}$, decryption will be correct.

Gentry et al. analyze the error growth of GSW and determine that if χ is β-bounded, and if $\mathbf{C}$ is the result of L levels of homomorphic evaluation, then with overwhelming probability, $B < \beta(N+1)^L$. To maintain correctness of their scheme, they set $B = \frac{q}{8}$, which gives us: $\frac{q}{\beta} > 8(N+1)^L$. This same analysis applies to LRGSW, and we set our ratio of q to β the same way.

4   Leakage Resilient Leveled FHE

Below we prove that LRGSW is leakage resilient, describe the efficiency tradeoffs we make to achieve leakage resilience, and briefly describe and why our leveled result but does not extend easily to full non-leveled homomorphism.

4.1   Adaptive Leakage Resilience of LRGSW

Theorem 4.1. *The leveled* LRGSW *scheme is resilient to adaptive bounded leakage of λ bits, where $\lambda \leq n - 2\log q - 4\eta$.*

Proof. We consider a probabilistic polynomial time adversary's advantage at playing the ALR^λ game (described in Definition 2.2). Recall that in this game, the adversary's view is $(\mathbf{K}, \mathbf{C}_b, h(\mathbf{K}, \mathbf{s}))$ where $\mathbf{C}_b$ is a correctly formed encryption of $b \in \{0, 1\}$.

Let $\mathbf{C}'_b = $ BitDecomp$^{-1}(\mathbf{C}_b) = $ BitDecomp$^{-1}(b \cdot \mathbf{I}_N) + \mathbf{R} \cdot \mathbf{K}$. Since BitDecomp^{-1} is a deterministic operation, it suffices to consider a probabilistic polynomial time adversary who plays the ALR^λ game with $\mathbf{C}'_b$.

In fact, an adversary's view after playing the ALR^λ game is $(\mathbf{K}, $ BitDecomp^{-1} $(b \cdot \mathbf{I}_N) + \mathbf{R} \cdot \mathbf{K}, h(\mathbf{K}, \mathbf{s}))$. Therefore, it is sufficient to show $(\mathbf{K}, \mathbf{R}\mathbf{K}, h(\mathbf{K}, \mathbf{s})) \approx_c$ $(\mathbf{K}, \mathbf{U} \xleftarrow{\$} \mathbb{Z}_q^{N \times n}, h(\mathbf{K}, \mathbf{s}))$.

Recall that $\mathbf{K} = [\mathbf{b}||\mathbf{A}]$ where $\mathbf{A} \xleftarrow{\$} \mathbb{Z}_q^{m \times n}$, $\mathbf{t} \xleftarrow{\$} \mathbb{Z}_q^n$, $\mathbf{e} \xleftarrow{\$} \chi^m$, $\mathbf{b} = \mathbf{At} + \mathbf{e}$, and $\mathbf{s} = (1, -\mathbf{t}_1, \ldots, -\mathbf{t}_n)$. So define:

$$\mathcal{H}_{ALR} := (\mathbf{b}, \mathbf{A}, \mathbf{Rb}, \mathbf{RA}, h(\mathbf{A}, \mathbf{t}, \mathbf{e})), \mathcal{H}_{RAND} := (\mathbf{b}, \mathbf{A}, \mathbf{u}', \mathbf{U}, h(\mathbf{A}, \mathbf{t}, \mathbf{e}))$$

Our goal is to show that $\mathcal{H}_{ALR} \approx_c \mathcal{H}_{RAND}$. We can think of the matrix $\mathbf{R}$ as a collection of N independent binary vectors $\mathbf{r}_i \xleftarrow{\$} \{0,1\}^m$. So, $\mathcal{H}_{ALR} = (\mathbf{b}, \mathbf{A}, \{\mathbf{r}_i \cdot \mathbf{b}\}_{i \in [N]}, \{\mathbf{r}_i \mathbf{A}\}_{i \in [N]}, h(\mathbf{A}, \mathbf{t}, \mathbf{e}))$

Now, define a series of hybrid games $\mathcal{H}_i$, for $0 \le i \le N$, where in game i, for $j < i$, $\mathbf{r}_j \cdot \mathbf{b}$ is replaced by $u'_j \xleftarrow{\$} \mathbb{Z}_q$, and $\mathbf{r}_j \mathbf{A}$ is replaced by $\mathbf{u} \xleftarrow{\$} \mathbb{Z}_q^n$, and for $j \ge i$, those terms are generated as they were in game $\mathcal{H}_{i-1}$.

It follows by inspection that $\mathcal{H}_0 = \mathcal{H}_{ALR}$ and $\mathcal{H}_N = \mathcal{H}_{RAND}$, so all that remains to show is that $\mathcal{H}_i \approx_c \mathcal{H}_{i+1}$.

We use Lemma 4.1, stated below, together with a simple reduction to prove this. Lemma 4.1 says that for a single $\mathbf{r} \xleftarrow{\$} \{0,1\}^m$, $\mathcal{H}_{real} := (\mathbf{b}, \mathbf{A}, \mathbf{r} \cdot \mathbf{b}, \mathbf{rA}, h(\mathbf{A}, \mathbf{t}, \mathbf{e})) \approx_c \mathcal{H}_{rand} := (\mathbf{b}, \mathbf{A}, u', \mathbf{u}, h(\mathbf{A}, \mathbf{t}, \mathbf{e}))$.

So, given an input $\mathcal{H} = (\mathbf{b}, \mathbf{A}, b', a', h(\mathbf{A}, \mathbf{t}, \mathbf{e}))$ that is equal to either $\mathcal{H}_{real}$ or $\mathcal{H}_{rand}$, if, for $j \le i$ choose $u'_j \xleftarrow{\$} \mathbb{Z}_q$, $\mathbf{u}_j \xleftarrow{\$} \mathbb{Z}_q^n$, and for $j > i + 1$, choose $r_j \xleftarrow{\$} \{0,1\}^m$, we prepare the following distribution:

$$\left(\mathbf{b}, \mathbf{A}, \{u'_j\}_{j \le i}, b', \{\mathbf{r}_j \cdot \mathbf{b}\}_{j > i+1}, \{\mathbf{u}_j\}_{j \le i}, a', \{r_j \mathbf{A}\}_{j > i}, h(\mathbf{A}, \mathbf{t}, \mathbf{e}) \right)$$

Then if $\mathcal{H} = \mathcal{H}_{real}$, this distribution is equal to $\mathcal{H}_i$, whereas if $\mathcal{H} = \mathcal{H}_{rand}$, the distribution is equal to $\mathcal{H}_{i+1}$. Since Lemma 4.1 (proven below) tells us that $\mathcal{H}_{real} \approx_c \mathcal{H}_{rand}$, we conclude that no probabilistic polynomial time adversary can distinguish $\mathcal{H}_i$ and $\mathcal{H}_{i+1}$ with non-negligible advantage.

Lemma 4.1. *Given* $\mathbf{A} \xleftarrow{\$} \mathbb{Z}_q^{m \times n}$, $\mathbf{e} \leftarrow \chi^m$, $\mathbf{t} \xleftarrow{\$} \mathbb{Z}_q^n$, $\mathbf{r} \xleftarrow{\$} \{0,1\}^m$, $\mathbf{b} = \mathbf{At} + \mathbf{e}$, *and* $\mathbf{u} \xleftarrow{\$} \mathbb{Z}_q^n$, *and* $u' \xleftarrow{\$} \mathbb{Z}_q$, *and* m, q, n *defined as in the* LRGSW *scheme*,

$$\mathcal{H}_{real} := (\mathbf{b}, \mathbf{A}, \mathbf{r} \cdot \mathbf{b}, \mathbf{rA}, h(\mathbf{A}, \mathbf{t}, \mathbf{e})) \approx_c \mathcal{H}_{rand} := (\mathbf{b}, \mathbf{A}, u', \mathbf{u}, h(\mathbf{A}, \mathbf{t}, \mathbf{e}))$$

Proof. Our proof proceeds as follows: We define a series of intermediate hybrid games, $\mathcal{H}_a, \mathcal{H}_b, \mathcal{H}_c$, and show:
$\mathcal{H}_{real} \approx_s \mathcal{H}_a \approx_c \mathcal{H}_b \approx_s \mathcal{H}_c \approx_c \mathcal{H}_{rand}$. Our hybrids are:

- $\mathcal{H}_a := (\mathbf{At} + \mathbf{e}, \mathbf{A}, \mathbf{ut} + \mathbf{r} \cdot \mathbf{e}, \mathbf{u}, h(\mathbf{A}, \mathbf{t}, \mathbf{e}))$, where $\mathbf{u} \xleftarrow{\$} \mathbb{Z}_q^N$.
- $\mathcal{H}_b := (\tilde{\mathbf{A}}\mathbf{t} + \mathbf{e}, \tilde{\mathbf{A}}, \mathbf{ut} + \mathbf{r} \cdot \mathbf{e}, \mathbf{u}, h(\tilde{\mathbf{A}}, \mathbf{t}, \mathbf{e}))$, where $\tilde{\mathbf{A}} \leftarrow$ Lossy, as defined by Lemma 4.2.
- $\mathcal{H}_c := (\tilde{\mathbf{A}}\mathbf{t} + \mathbf{e}, \tilde{\mathbf{A}}, u', \mathbf{u}, h(\tilde{\mathbf{A}}, \mathbf{t}, \mathbf{e}))$, where $u' \xleftarrow{\$} \mathbb{Z}_q$.

Lemma 4.2, stated below, immediately gives us $\mathcal{H}_a \approx_c \mathcal{H}_b$, and $\mathcal{H}_c \approx_c \mathcal{H}_{rand}$, because it tells us that $\tilde{\mathbf{A}} \approx_c \mathbf{A}$. Thus, no further work is needed for these two steps.

We use Claim 1 to show that $\mathcal{H}_{real} \approx_s \mathcal{H}_a$.

Finally, we use Claim 2 to prove $\mathcal{H}_b \approx_s \mathcal{H}_c$.

Claim 1. $\mathcal{H}_{real} \approx_s \mathcal{H}_a$

Proof. The only difference between games $\mathcal{H}_{real}$ and $\mathcal{H}_a$ is that $\mathbf{rA}$ is replaced by $\mathbf{u}$ where $\mathbf{u} \xleftarrow{\$} \mathbb{Z}_q^N$. Note that if we can show:

$$(\mathbf{At} + \mathbf{e}, \mathbf{A}, \mathbf{rAt}, \mathbf{r} \cdot \mathbf{e}, \mathbf{rA}, h(\mathbf{A}, \mathbf{t}, \mathbf{e})) \approx_s (\mathbf{At} + \mathbf{e}, \mathbf{A}, \mathbf{u} \cdot \mathbf{t}, \mathbf{r} \cdot \mathbf{e}, \mathbf{u}, h(\mathbf{A}, \mathbf{t}, \mathbf{e}))$$

this implies our claim.

To prove the above, we use the generalized form of the leftover hash lemma (Lemma A.2 in Appendix A of this paper), which tells us that for any random variable x, if $\tilde{H}_\infty(\mathbf{r}|x)$ is high enough, then $(\mathbf{A}, \mathbf{rA}, x) \approx_s (\mathbf{A}, \mathbf{u}, x)$, which in turn implies that for any $\mathbf{t}$, $(\mathbf{A}, \mathbf{rA}, \mathbf{rAt}, x) \approx_s (\mathbf{A}, \mathbf{u}, \mathbf{u} \cdot \mathbf{t}, x)$. So, set $x = (\mathbf{At} + \mathbf{e}, \mathbf{r} \cdot \mathbf{e}, h(\mathbf{A}, \mathbf{t}, \mathbf{e}))$. Since $\mathbf{r}$ is an m-dimensional binary vector chosen uniformly at random and $\mathbf{r} \cdot \mathbf{e}$ is $\ell = \lfloor \log q \rfloor + 1$ bits long, and $\mathbf{r}$ is independent of $\mathbf{e}$, we have:

$$\tilde{H}_\infty(\mathbf{r}|\mathbf{At} + \mathbf{e}, \mathbf{r} \cdot \mathbf{e}, h(\mathbf{A}, \mathbf{t}, \mathbf{e}))$$
$$\geq \tilde{H}_\infty(\mathbf{r}|\mathbf{r} \cdot \mathbf{e}, \mathbf{e}) \geq \tilde{H}_\infty(\mathbf{r}|\mathbf{e}) - \ell = m - \ell$$

For Lemma A.2 to hold, we need $n \leq \frac{m - \ell - 2\eta - O(1)}{\log q}$. Choosing $m \geq 2n \log q + 3\eta$ suffices.

Claim 2. $\mathcal{H}_b \approx_s \mathcal{H}_c$

Proof. The difference between $\mathcal{H}_b$ and $\mathcal{H}_c$ is that $\mathbf{u} \cdot \mathbf{t} + \mathbf{r} \cdot \mathbf{e}$ is replaced by $u' \xleftarrow{\$} \mathbb{Z}_q$. We employ a similar strategy to that from claim Claim 1, using the leftover hash lemma to show

$$(\tilde{\mathbf{A}}\mathbf{t} + \mathbf{e}, \tilde{\mathbf{A}}, \mathbf{ut}, \mathbf{r} \cdot \mathbf{e}, \mathbf{u}, h(\tilde{\mathbf{A}}, \mathbf{t}, \mathbf{e})) \approx_s (\tilde{\mathbf{A}}\mathbf{t} + \mathbf{e}, \tilde{\mathbf{A}}, v, \mathbf{r} \cdot \mathbf{e}, \mathbf{u}, h(\tilde{\mathbf{A}}, \mathbf{t}, \mathbf{e}))$$

where $v \xleftarrow{\$} \mathbb{Z}_q$. Note that this distribution contains both $\mathbf{ut}$ and $\mathbf{r} \cdot \mathbf{e}$, whereas the adversary only sees $\mathbf{ut} + \mathbf{r} \cdot \mathbf{e}$. Proving that $\mathbf{ut}$ can be replaced by v implies that in the adversary's actual view, $\mathbf{ut} + \mathbf{re}$ can be replaced by $u' \xleftarrow{\$} \mathbb{Z}_q$.

Now, we bound the ϵ-smooth min-entropy of $\mathbf{t}$. There exists $\epsilon = negl(\eta)$ such that

$$\tilde{H}_\infty^\epsilon(\mathbf{t}|\tilde{\mathbf{A}}\mathbf{t} + \mathbf{e}, \tilde{\mathbf{A}}, \mathbf{r} \cdot \mathbf{e}, h(\tilde{\mathbf{A}}, \mathbf{t}, \mathbf{e})))$$
$$\geq \tilde{H}_\infty^\epsilon(\mathbf{t}|\tilde{\mathbf{A}}\mathbf{t} + \mathbf{e}, \tilde{\mathbf{A}}) - BitLength(\mathbf{r} \cdot \mathbf{e}) - BitLength(h(\tilde{\mathbf{A}}, \mathbf{t}, \mathbf{e}))$$
$$\geq \tilde{H}_\infty^\epsilon(\mathbf{t}|\tilde{\mathbf{A}}\mathbf{t} + \mathbf{e}, \tilde{\mathbf{A}}) - \ell - \lambda$$

and Lemma 4.2 (stated and proven below), tells us that $\tilde{H}_\infty^\epsilon(\mathbf{t}|\tilde{\mathbf{A}}\mathbf{t} + \mathbf{e}, \tilde{\mathbf{A}}) \geq n$.

Applying the ϵ-smooth variant of the leftover hash lemma (Corollary A.1), we see that we need $n - \ell - \lambda$ to be high enough that $\log q \leq (n - \ell - \lambda) - 2\eta - O(1)$. So, if we set h to leak at most $\lambda \leq n - 2 \log q - 4\eta$ bits, the claim follows.

Since $\mathcal{H}_{real} \approx_s \mathcal{H}_a \approx_c \mathcal{H}_b \approx_s \mathcal{H}_c \approx_c \mathcal{H}_{rand}$, we know that $\mathcal{H}_{real} \approx_c \mathcal{H}_{rand}$.

We now state Lemma 4.2, used both to prove Claim 2, and to show $\mathcal{H}_a \approx_c \mathcal{H}_b$, and $\mathcal{H}_c \approx_c \mathcal{H}_{rand}$.

Lemma 4.2. *There exists a distribution* Lossy *such that* $\tilde{\mathbf{A}} \leftarrow$ Lossy $\approx_c \mathbf{U} \xleftarrow{\$}$ $\mathbb{Z}_q^{m \times n}$ *and given* $\mathbf{t} \xleftarrow{\$} \mathbb{Z}_q^n$, *and* $\mathbf{e} \leftarrow \chi$, $\tilde{H}_\infty^\epsilon(\mathbf{t}|\tilde{\mathbf{A}}, \tilde{\mathbf{A}}\mathbf{t} + \mathbf{e}) \geq n$, *where* $\epsilon = negl(\eta)$.

In our proof, we define a distribution Lossy as follows:

- Choose $\mathbf{C} \xleftarrow{\$} \mathbb{Z}_q^{m \times n'}$, $\mathbf{D} \xleftarrow{\$} \mathbb{Z}_q^{n' \times n}$, and $\mathbf{Z} \leftarrow \overline{\Psi}_\alpha^{m \times n}$, where $\frac{\alpha}{\beta} = negl(\eta)$ and $n' \log q \leq n - 2\eta + 2$.
- Let $\tilde{\mathbf{A}} = \mathbf{CD} + \mathbf{Z}$
- output $\tilde{\mathbf{A}}$.

This distribution was first defined in [27] and as our proof is closely related to proofs in their paper, we defer it to Appendix B.

4.2   The Cost of Leakage Resilience: GSW versus LRGSW

In order to make the GSW scheme leakage resilient, we needed to make a number of tradeoffs. First, there's a penalty to efficiency, as a number of the scheme's parameters need to be set higher than they are in GSW in order to maintain equivalent security in the presence of leakage. Second, our proof relies crucially on the fact that the LRGSW scheme does not have an evaluation key. The leveled version of the GSW scheme does not have an evaluation key, but the version that allows for full (non-leveled) FHE does have one. For this reason, LRGSW cannot be easily extended to a non-leveled scheme.

Parameter Setting. The hardness constraints and the correctness constraints of our scheme are in conflict. The hardness constraints tell us that the ratio of the dimension to the error bound affects the relative hardness of the *DLWE* problems, with a higher β leading to more security. However, the correctness constraint shows us that $\frac{q}{\beta}$ must grow exponentially with the depth of the circuit, which shows both that β should be set low, and since there is a limit to how low β can be set, q must grow exponentially with depth. However, the hardness constraints also tell us that if the depth is $O(n)$ or bigger, since L, the circuit depth, is in the exponent of q, the underlying *GapSVP* problems become easy. To protect against this, we must ensure that n is polynomial in L. We describe these constraints in more detail and show how to set the parameters to meet all of them in Appendix B.

Also in the appendix, we present Lemma B.1, which can replace Lemma 4.2 in our proofs above. This new lemma uses techniques from Alwen, Krenn, Pietrzak, and Wichs [28] which, as summarized in Corollary B.1, allow us to reduce the size of q and β (in particular, β is no longer super-polynomial in η), at a cost of a lower value for λ.

In Table 1 we provide sample parameter settings that simultaneously meet all correctness and security constraints. We compare these settings to those of GSW. In the table, $\tau_1 = \max\{L, \eta^2\}$, and $\tau_2 = \max\{L, \eta^3\}$.

Table 1. Sample settings of GSW v. LRGSW

Parameter	GSW	LRGSW with Lemma 4.2	LRGSW with Lemma B.1
n	$O(\eta)$	τ_1^4	τ_2^3
q	$2^{L\log n}$	$2^{\tau_1 \log^2 \tau_1}$	$2^{\tau_2 \log n}$
χ	$O(n)$ -bounded	$\overline{\Psi}_\beta,\ \beta = 2^{\log^2 \tau_1}$	$\beta = 3n^3\tau_2^3$
m	$2n\log q$	$2n\log q + 3\eta$	$2n\log q + 3\eta$
λ	0	$n - 2\log q - 4\eta$	$n - (2+\eta)\log q - \eta\log m - 4\eta$

Evaluation Keys and the Problem with Bootstrapping. Our current techniques are sufficient for proving leakage resilience of a leveled fully homomorphic encryption scheme, but do not extend to a non-leveled scheme. The bootstrapping paradigm, first defined by Gentry in [3], is to take a scheme that is capable of homomorphically evaluating its own decryption circuit and transform it into one that can evaluate functions f of arbitrary depth by performing the homomorphic-decrypt operation after each gate in f. All existing fully homomorphic schemes, including the GSW scheme, achieve full, as opposed to leveled fully homomorphic encryption through bootstrapping.

The bootstrapping paradigm tells us that given a somewhat homomorphic scheme, publishing an encryption of the scheme's secret key, together with any other data necessary to allow the scheme to homomorphically evaluate its own decryption procedure, makes the scheme fully homomorphic [3]. Thus, the scheme must be secure when an adversary sees $(pk, \mathsf{Enc}_{pk}(sk))$ (circular security). However, a scheme that is secure when the adversary sees $(pk, \mathsf{Enc}_{pk}(sk))$ or when the adversary sees $(pk, h(pk, sk))$, as is the case in the leakage resilience definition, is not necessarily secure when it sees $(pk, \mathsf{Enc}_{pk}(sk), h(pk, sk, \mathsf{Enc}_{pk}(sk)))$ all together. Formal definitions of bootstrapping and circular security are presented in Appendix A.

If we tried to make the LRGSW scheme bootstrappable, we would need not only circular security (which current FHE schemes assume rather than prove), but circular security in the presence of leakage.

If we were to create an evk that contained an encryption of the secret key under that same secret key, we would have something of the form $\mathbf{A}, \mathbf{At} + \mathbf{e} + \mathsf{BitDecompose}(\mathbf{t})$. One might try to follow the same technique outlined in the proof of Lemma 4.2, and show that the average min-entropy of $\mathbf{t}$, conditioned on seeing $\mathbf{A}, \mathbf{At}+\mathbf{e}+\mathsf{BitDecompose}(\mathbf{t})$, is still high. Unfortunately, for this technique to work, $\mathbf{t}$ needs to be only in the secret term, not in the error term as well.

To get around this, we might consider trying to "chain" our $DLWE$ secrets, so that we have two $DLWE$ secrets: $\mathbf{t}$ and $\mathbf{t}'$, but only consider our secret key to be $\mathbf{t}'$. In this case, our encryption key would be $(\mathbf{A}, \mathbf{At} + \mathbf{e})$, and our evaluation key would be $(\mathbf{A}', \mathbf{A}'\mathbf{t}' + \mathbf{e}' + \mathsf{BitDecomp}(\mathbf{t}))$. In this case, we would still need to show that $\tilde{H}_\infty(\mathbf{t}|\mathbf{A}'\mathbf{t}' + \mathbf{e}' + \mathsf{BitDecomp}(\mathbf{t}))$ was sufficiently high, and since $\mathbf{t}$

is in the error term instead of the secret term, our current techniques will not suffice.

Notice, as well, that these limitations apply to any LWE-based FHE scheme with an evaluation key. Since all other existing LWE based FHE schemes use an evaluation key, our result for the GSW scheme cannot be easily extended to these schemes either.

5   Going beyond Leveled Homomorphism

In this section we present several new ideas for achieving full (as opposed to leveled) FHE that is also leakage resilient.

5.1   Our First Approach

We observe that by definition, a leakage function h is a function of the scheme's public and secret keys. This means an adversary can see $h(pk, sk, \text{Enc}_{pk}(sk))$ only if $\text{Enc}_{pk}(sk)$ is part of the scheme's public key. If instead, we can somehow generate $\text{Enc}_{pk}(sk)$ on-the-fly as it is needed, the adversary sees only $h(pk, sk)$, instead.

More precisely, let $\mathsf{E} = (\mathsf{KeyGen}(), \mathsf{Enc}(), \mathsf{Dec}())$ be any encryption scheme (*not* necessarily homomorphic) that is also resilient to adaptive bounded leakage of λ bits, and let $\mathsf{HE} = (\mathsf{KeyGen}(), \mathsf{Enc}(), \mathsf{Dec}(), \mathsf{Eval}())$ be any (leveled) fully homomorphic encryption scheme. Then we consider the following hybrid scheme:

Scheme1.$\mathsf{KeyGen}(1^\eta)$: Run $(pk, sk) \leftarrow \mathsf{E.KeyGen}(1^\eta)$. Set the public and secret keys to be pk, sk.

Scheme1.$\mathsf{Enc}_{pk}(m)$: To encrypt a message m, first run $(pk', sk') \leftarrow \mathsf{HE.KeyGen}(1^\eta)$. Then output
$(pk', \mathsf{HE.Enc}_{pk'}(m), \mathsf{E.Enc}_{pk}(sk'))$ as the ciphertext.

Scheme1.$\mathsf{Dec}_{sk}(c)$: To decrypt a ciphertext c, first parse $c = (pk', c_1, c_2)$, and obtains $sk' = \mathsf{E.Dec}_{sk}(c_2)$. Then output $\mathsf{HE.Dec}_{sk'}(c_1)$.

Scheme1.$\mathsf{Eval}_{pk}(f, c)$: To evaluate a function f over a ciphertext c, first parse $c = (pk', c_1, c_2)$ and then output $(pk', \mathsf{HE.Eval}_{pk'}(f, c_1), c_2)$.

It is not hard to obtain the following theorem:

Theorem 5.1. *If* E *is an encryption scheme that is resilient to adaptive bounded leakage of* λ *bits and* HE *is a (leveled) fully homomorphic encryption scheme, then* Scheme1 *is a (leveled) fully homomorphic scheme that has the following properties:*

1. *It is resilient to adaptive bounded leakage of* λ *bits.*
2. *It allows unary homomorphic evaluation over any single ciphertext.*
3. *If* HE *is fully homomorphic, then* Scheme1 *has succinct ciphertexts (whose lengths do not depend on the size of circuits supported by the evaluation), while if* HE *is L-leveled homomorphic, then the size of the ciphertexts in* Scheme1 *depends on L.*

A word is in order about property 2 above. If HE is a bit-encryption scheme, then we can think of the message space as bit-strings, so a message $\mathbf{m} \in \{0,1\}^t$, and define encryption to be bit-by bit.

In this case, "unary" refers to functions over the bits of $\mathbf{m}$. Another way to think of this is that Scheme1 is (leveled) fully homomorphic for any group of bits batch-encrypted at the same time.

The proof of this theorem is simple and quite similar to that of Theorem 5.2, so we omit the proof here, and refer the reader to our proof of that theorem below.

5.2   Our Second Approach

Our next step is to extend our result so that we can homomorphically combine ciphertexts regardless of when they were created. The reason we cannot do so above is because two ciphertexts formed at different times will be encrypted under different public keys of the underlying HE scheme. To solve this issue, we consider instantiating HE with a multi-key FHE scheme, as recently defined and constructed by López-Alt, Tromer and Vaikuntanathan (LTV) [30].

A scheme $HE^{(N)}$ is a N-Key Multikey (leveled) FHE scheme if it is a (leveled) FHE scheme with the following two additional algorithms:

- $mEval(f, pk_1, \ldots, pk_t, c_1, \ldots, c_t)$ that takes as input an t-ary function f, t evaluation keys and ciphertexts, and output a combined ciphertext c^*.
- $mDec(sk_1, \ldots, sk_t, c^*)$ that takes c^*, generated by mEval and t secret keys such that sk_i corresponds to pk_i for $i \in [t]$, and outputs $f(m_1, m_2, \ldots m_t)$.

where the above holds for any $t \leq T$, with $c_1, \ldots c_t$ any ciphertexts under $pk_1, \ldots pk_t$, i.e. $c_i = Enc_{pk_i}(m_i)$ for all $i \in [t]$.

If we replace HE with $HE^{(N)}$, we get the following evaluation function:

Scheme2.$Eval_{pk}(f, c_1, \ldots, c_t)$: To evaluate a function f over ciphertexts $c_1, \ldots c_t$, first parse $c_i = (pk'_i, c_{i,1}, c_{i,2})$ for $i \in [t]$. Then, calculate $c_1^* = HE^{(N)}.Eval(pk'_1, \ldots, pk'_t, c_{1,1}, \ldots, c_{t,1})$. Finally, output $(pk'_1, \ldots pk'_t, c_1^*, c_{1,2}, \ldots, c_{t,2})$.

The problem with this approach is that the resulting ciphertext needs to include all the public keys and secret keys from $HE^{(N)}$ in order to run multikey decryption ($HE^{(N)}.mDec$). This means that outputs of the Eval function will have a different format than freshly generated ciphertexts, and no longer be compact. Thus Scheme2 cannot possibly meet the definition of fully homomorphic.

5.3   The Final Scheme

We now observe that the LTV construction actually achieves multi-key FHE with a more fine-grained definition than we provided above: one where not only *ciphertexts*, but also *keys* can be combined. As described in Section 3.4 of their paper, given $c_1 = \mathsf{LTV.Enc}(pk_1, m_1)$, $c_2 = \mathsf{LTV.Enc}(pk_2, m_2)$, one step of $\mathsf{LTV.Eval}$ is to calculate $pk^* = pk_1 \cup pk_2$. We can separate out this step and generalize it, defining $\mathsf{CombinePK}(pk_1, pk_2, \ldots, pk_t) = \bigcup_{i=1}^{t} pk_i$. Similarly, in their scheme, the secret keys are polynomials, and they show how to create a "joint secret key" by multiplying the polynomials together. We give this procedure a name, defining $\mathsf{CombineSK}(sk_1, sk_2, \ldots sk_t) = \prod_{i=1}^{t} sk_k$.

Definition 5.1. *A scheme* $\mathsf{HE}^{(N)}$ *is an* **N-Key Multikey (leveled) FHE scheme** *if it is a (leveled) FHE scheme with the following additional algorithms: For any* $t \leq N$, *let* $c_1, \ldots c_t$ *be any ciphertexts under* $pk_1, \ldots pk_t$, *i.e.* $c_i = \mathsf{Enc}_{pk_i}(m_i)$ *for all* $i \in [t]$.

- $pk^* = \mathsf{CombinePK}(pk_1, pk_2, \ldots, pk_t)$.
- *A multi-key encryption algorithm* $\mathsf{mEval}(f, pk_1, \ldots, pk_t, c_1, c_2, \ldots, c_t)$ *that first calls* $pk^* = \mathsf{CombinePK}(pk_1, pk_2, \ldots, pk_t)$, *and then produces* c^*, *and outputs* c^* *and* pk^*. *Note that this* c^* *and* pk^* *can be used as input for successive calls to* mEval.
- $sk^* = \mathsf{CombineSK}(sk_1, sk_2, \ldots, sk_t)$.
- *A multikey decryption algorithm* $\mathsf{mDec}(sk_1, \ldots, sk_t, c^*)$ *that calls* $\mathsf{CombineSK}$ *and then runs* $\mathsf{Dec}(sk^*, c^*)$ *to produce* $f(m_1, m_2, \ldots m_t)$.

As long as the outputs of $\mathsf{CombineSK}$ and $\mathsf{CombinePK}$ are succinct, we can update our scheme to make ciphertexts succinct.

Let $\mathsf{SHE} = (\mathsf{KeyGen}(), \mathsf{Enc}(), \mathsf{Dec}(), \mathsf{Eval}())$ be any somewhat[1] homomorphic encryption scheme that is also resilient to adaptive bounded leakage of λ bits, and let $\mathsf{HE}^{(N)} = (\mathsf{KeyGen}(), \mathsf{Enc}(), \mathsf{Dec}(), \mathsf{mEval}(), \mathsf{CombinePK}(), \mathsf{CombineSK}())$ be any N-key multikey fully homomorphic encryption scheme. Then we consider the following combined scheme:

$\mathsf{Scheme3.KeyGen}(1^\eta)$: Run $(pk, sk) \leftarrow \mathsf{SHE.KeyGen}(1^\eta)$. Set the public and secret keys to be pk, sk.

$\mathsf{Scheme3.Enc}(pk, m)$: First, run $(pk', sk') \leftarrow \mathsf{HE.KeyGen}(1^\eta)$.
  Then output $(pk', \mathsf{HE.Enc}(pk', m), \mathsf{SHE.Enc}(pk, sk'))$ as the ciphertext.

$\mathsf{Scheme3.Eval}(pk, f, c_1, \ldots, c_t)$: First parse $c_i = (pk'_i, c_{i,1}, c_{i,2})$ for $i \in [t]$.
  Then, calculate $c_1^* = \mathsf{HE}^{(N)}.\mathsf{Eval}(pk'_1, \ldots, pk'_t, f, c_{1,1}, \ldots, c_{t,1})$,
  $pk'^* = \mathsf{HE}^{(N)}.\mathsf{CombinePK}(pk'_1, \ldots, pk'_t)$,
  $c_2^* = \mathsf{SHE.Eval}(pk, \mathsf{HE.CombineSK}, c_{1,2}, \ldots, c_{t,2})$.
  Finally, output (pk'^*, c_1^*, c_2^*).

$\mathsf{Scheme3.Dec}(sk, c)$: To decrypt a ciphertext c, first parse $c = (pk', c_1, c_2)$, and obtain $sk' = \mathsf{SHE.Dec}(sk, c_2)$. Then output $\mathsf{HE.Dec}(sk', c_1)$.

This lets us achieve the following theorem.

Theorem 5.2. *Let* SHE *be a* $\mathcal{C}$-*homomorphic encryption scheme for some circuilt class* $\mathcal{C}$ *such that* $\mathsf{HE}^{(N)}.\mathsf{CombineSK} \in \mathcal{C}$. *Let* $\mathsf{HE}^{(N)}$ *be an* N-*Key multikey*

---

[1] SHE must support circuits large enough to evaluate CombineSK, but does not need to be fully homomorphic.

FHE scheme. If SHE *is resilient to adaptive, bounded leakage of* λ *bits, then* Scheme3 *has the following properties:*

1. *It allows homomorphic evaluation of (up to) N-ary circuits of arbitrary $(poly(\eta))$ depth.*
2. *If* SHE *is a leveled homomorphic encryption scheme, then the ciphertext size depends on N. If* SHE *is fully homomorphic, then* Scheme3 *has succinct ciphertexts (whose lengths do not depend N).*
3. *It is resilient to adaptive bounded leakage of λ bits.*

Proof. We address each statement in turn.

1. This follows immediately from the fact that by definition, $\mathsf{HE}^{(N)}$ allows homomorphic evaluation of (up to) N-ary circuits of arbitrary $(poly(\eta))$ depth.
2. If SHE is leveled, its key-size is dependent on L, the number of levels of homomorphic evaluation it can support. To instantiate Scheme3, we need SHE to homomorphically evaluate CombineSK, an N-ary circuit whose depth is a function of its arity. Thus, the key size of SHE, and by extension, of Scheme3 is a function of N. In contrast, if SHE is not leveled, its key size is independent of L, and thus of N as well.
3. A simple reduction shows that if SHE is leakage resilient, then Scheme3 will be as well. Given a probabilistic polynomial time adversary $\mathcal{A}$ who wins the ALR game with Scheme3 with non-negligible advantage, it is easy to construct a ppt $\mathcal{B}$ who wins the ALR game with SHE with the same advantage. Upon receiving the public key from SHE, $\mathcal{B}$ simply forwards this information to $\mathcal{A}$. Whenever $\mathcal{A}$ requests an encryption of a message, $\mathcal{B}$ simply runs HE.KeyGen, and then follows Scheme3.Enc(), and forwards the result to $\mathcal{A}$. When $\mathcal{A}$ decides upon a leakage function, $\mathcal{B}$ uses that same leakage function. $\mathcal{A}$'s view when interacting with $\mathcal{B}$ is exactly its view when interacting with Scheme3 so its advantage is the same. Therefore, $\mathcal{B}$ would have the same advantage when interacting with Scheme3.

5.4   Instantiation

We instantiate Scheme3 using LRGSW for SHE and LTV for $\mathsf{HE}^{(N)}$. The LTV construction can be summarized by the following theorem:

Theorem 5.3. *(from theorem 4.5 in [30]) For every $N = poly(\eta)$, under the $DSPR^2$ and $RLWE^3$ assumptions with proper parameters, there exists an N-key multi-key (leveled) Fully Homomorphic Encryption Scheme. Under the additional assumption of weak circular security, we can remove the "leveled" constraint.*

The above theorem lets us instantiate Scheme3 with LTV and LRGSW, and together with with theorem 4.1 gives us the following corollary:

---

[2] The $DSPR$ assumption is the "Decisional Small Polynomial Ratio" introduced in [30].
[3] $RLWE$ stands for "Ring Learning With Errors," first introduced in [37].

Corollary 5.1. *For every $T = poly(\eta)$ there exists an FHE scheme that supports homomorphic evaluation of all t-nary circuits for $t \leq T$, and depth $poly(\eta)$, under appropriate DSPR, RLWE, and DLWE assumptions. Under appropriate choices of n and q chosen so that certain DLWE assumptions hold, the scheme is resilient to adaptive bounded leakage of λ bits, where $\lambda \leq n - 2\log q - 4\eta$.*

Acknowledgements. This work was done under the support of NSF Grants 1012060 and 0964541. We would like to thank Anna Lysyanskaya for many useful discussions and our reviewers for their helpful comments.

References

1. Gentry, C., Sahai, A., Waters, B.: Homomorphic encryption from learning with errors: Conceptually-simpler, asymptotically-faster, attribute-based. IACR Cryptology ePrint Archive 2013, 340 (2013)
2. Rivest, R., Adleman, L., Dertouzos, M.: On data banks and privacy homomorphisms. In: Foundations on Secure Computation, pp. 169–179. Academia Press (1978)
3. Gentry, C.: Fully homomorphic encryption using ideal lattices. In: Proceedings of the 41st Annual ACM Symposium on Theory of Computing, STOC 2009, pp. 169–178. ACM, New York (2009)
4. van Dijk, M., Gentry, C., Halevi, S., Vaikuntanathan, V.: Fully homomorphic encryption over the integers. In: Gilbert, H. (ed.) EUROCRYPT 2010. LNCS, vol. 6110, pp. 24–43. Springer, Heidelberg (2010)
5. Smart, N.P., Vercauteren, F.: Fully homomorphic encryption with relatively small key and ciphertext sizes. In: Public Key Cryptography, pp. 420–443 (2010)
6. Coron, J.-S., Mandal, A., Naccache, D., Tibouchi, M.: Fully homomorphic encryption over the integers with shorter public keys. In: Rogaway, P. (ed.) CRYPTO 2011. LNCS, vol. 6841, pp. 487–504. Springer, Heidelberg (2011)
7. Brakerski, Z., Vaikuntanathan, V.: Fully homomorphic encryption from ring-LWE and security for key dependent messages. In: Rogaway, P. (ed.) CRYPTO 2011. LNCS, vol. 6841, pp. 505–524. Springer, Heidelberg (2011)
8. Brakerski, Z.: Fully homomorphic encryption without modulus switching from classical gapsvp. IACR Cryptology ePrint Archive 2012, 78 (2012)
9. Brakerski, Z., Gentry, C., Vaikuntanathan, V.: (leveled) fully homomorphic encryption without bootstrapping. In: ICTS, pp. 309–325 (2012)
10. Brakerski, Z., Vaikuntanathan, V.: Efficient fully homomorphic encryption from (standard) lwe. In: FOCS, pp. 97–106 (2011)
11. Regev, O.: On lattices, learning with errors, random linear codes, and cryptography. In: Proceedings of the Thirty-Seventh Annual ACM Symposium on Theory of Computing, STOC 2005, pp. 84–93. ACM, New York (2005)
12. Gentry, C., Peikert, C., Vaikuntanathan, V.: Trapdoors for hard lattices and new cryptographic constructions. In: STOC, pp. 197–206 (2008)
13. Ishai, Y., Sahai, A., Wagner, D.: Private circuits: Securing hardware against probing attacks. In: Boneh, D. (ed.) CRYPTO 2003. LNCS, vol. 2729, pp. 463–481. Springer, Heidelberg (2003)
14. Micali, S., Reyzin, L.: Physically observable cryptography. In: Naor, M. (ed.) TCC 2004. LNCS, vol. 2951, pp. 278–296. Springer, Heidelberg (2004)

15. Akavia, A., Goldwasser, S., Vaikuntanathan, V.: Simultaneous hardcore bits and cryptography against memory attacks. In: Reingold, O. (ed.) TCC 2009. LNCS, vol. 5444, pp. 474–495. Springer, Heidelberg (2009)

16. Alwen, J., Dodis, Y., Naor, M., Segev, G., Walfish, S., Wichs, D.: Public-key encryption in the bounded-retrieval model. In: Gilbert, H. (ed.) EUROCRYPT 2010. LNCS, vol. 6110, pp. 113–134. Springer, Heidelberg (2010)

17. Boyko, V.: On the security properties of oaep as an all-or-nothing transform. In: Wiener, M. (ed.) CRYPTO 1999. LNCS, vol. 1666, pp. 503–518. Springer, Heidelberg (1999)

18. Canetti, R., Dodis, Y., Halevi, S., Kushilevitz, E., Sahai, A.: Exposure-resilient functions and all-or-nothing transforms. In: Preneel, B. (ed.) EUROCRYPT 2000. LNCS, vol. 1807, pp. 453–469. Springer, Heidelberg (2000)

19. Dodis, Y., Goldwasser, S., Tauman Kalai, Y., Peikert, C., Vaikuntanathan, V.: Public-key encryption schemes with auxiliary inputs. In: Micciancio, D. (ed.) TCC 2010. LNCS, vol. 5978, pp. 361–381. Springer, Heidelberg (2010)

20. Dodis, Y., Kalai, Y.T., Lovett, S.: On cryptography with auxiliary input. In: STOC, pp. 621–630 (2009)

21. Dodis, Y., Ong, S.J., Prabhakaran, M., Sahai, A.: On the (im)possibility of cryptography with imperfect randomness. In: FOCS, pp. 196–205 (2004)

22. Dziembowski, S., Pietrzak, K.: Leakage-resilient cryptography. In: FOCS, pp. 293–302 (2008)

23. Goldwasser, S., Kalai, Y.T., Rothblum, G.N.: One-time programs. In: Wagner, D. (ed.) CRYPTO 2008. LNCS, vol. 5157, pp. 39–56. Springer, Heidelberg (2008)

24. Naor, M., Segev, G.: Public-key cryptosystems resilient to key leakage. In: Halevi, S. (ed.) CRYPTO 2009. LNCS, vol. 5677, pp. 18–35. Springer, Heidelberg (2009)

25. Pietrzak, K.: A leakage-resilient mode of operation. In: Joux, A. (ed.) EUROCRYPT 2009. LNCS, vol. 5479, pp. 462–482. Springer, Heidelberg (2009)

26. Rivest, R.L.: All-or-nothing encryption and the package transform. In: Biham, E. (ed.) FSE 1997. LNCS, vol. 1267, pp. 210–218. Springer, Heidelberg (1997)

27. Goldwasser, S., Kalai, Y.T., Peikert, C., Vaikuntanathan, V.: Robustness of the learning with errors assumption. In: ICS, pp. 230–240 (2010)

28. Alwen, J., Krenn, S., Pietrzak, K., Wichs, D.: Learning with rounding, revisited - new reduction, properties and applications. In: Canetti, R., Garay, J.A. (eds.) CRYPTO 2013, Part I. LNCS, vol. 8042, pp. 57–74. Springer, Heidelberg (2013)

29. Katz, J., Vaikuntanathan, V.: Signature schemes with bounded leakage resilience. In: Matsui, M. (ed.) ASIACRYPT 2009. LNCS, vol. 5912, pp. 703–720. Springer, Heidelberg (2009)

30. López-Alt, A., Tromer, E., Vaikuntanathan, V.: On-the-fly multiparty computation on the cloud via multikey fully homomorphic encryption. IACR Cryptology ePrint Archive 2013, 94 (2013)

31. Dodis, Y., Ostrovsky, R., Reyzin, L., Smith, A.: Fuzzy extractors: How to generate strong keys from biometrics and other noisy data. SIAM J. Comput. 38(1), 97–139 (2008)

32. Peikert, C.: Public-key cryptosystems from the worst-case shortest vector problem: extended abstract. In: STOC, pp. 333–342 (2009)

33. Micciancio, D., Mol, P.: Pseudorandom knapsacks and the sample complexity of lwe search-to-decision reductions. In: Rogaway, P. (ed.) CRYPTO 2011. LNCS, vol. 6841, pp. 465–484. Springer, Heidelberg (2011)

34. Micciancio, D., Peikert, C.: Hardness of sis and lwe with small parameters. IACR Cryptology ePrint Archive 2013, 69 (2013)

35. Schnorr, C.P.: A hierarchy of polynomial time lattice basis reduction algorithms. Theor. Comput. Sci. 53, 201–224 (1987)
36. Micciancio, D., Voulgaris, P.: A deterministic single exponential time algorithm for most lattice problems based on voronoi cell computations. In: STOC, pp. 351–358 (2010)
37. Lyubashevsky, V., Peikert, C., Regev, O.: On ideal lattices and learning with errors over rings. In: Gilbert, H. (ed.) EUROCRYPT 2010. LNCS, vol. 6110, pp. 1–23. Springer, Heidelberg (2010)
38. Impagliazzo, R., Levin, L.A., Luby, M.: Pseudo-random generation from one-way functions (extended abstracts). In: STOC, pp. 12–24 (1989)

A   Full Definitions

Below we provide full, formal definitions of concepts used throughout our paper.

A.1   Homomorphism

Definition A.1. *A **homomorphic (public-key) encryption scheme***

$$\mathsf{HE} = (\mathsf{HE.Keygen}, \mathsf{HE.Enc}, \mathsf{HE.Dec}, \mathsf{HE.Eval})$$

is a quadruple of probabilistic polynomial time algorithms as described below:

- **Key Generation**[4] *The algorithm* $(pk, sk) \leftarrow \mathsf{HE.Keygen}(1^\kappa)$ *takes a unary representation of the security parameter, and outputs a public key pk and a secret decryption key sk.*
- **Encryption** *The algorithm* $c \leftarrow \mathsf{HE.Enc}_{pk}(\mu)$ *takes the public key pk and a message* $\mu \in \{0, 1\}$ *and outputs a ciphertext c.*
- **Decryption** *The algorithm* $\mu^* \leftarrow \mathsf{HE.Dec}_{sk}(c)$ *takes the secret key sk, a ciphertext c, and outputs a message* $\mu^* \in \{0, 1\}$.
- **Homomorphic Evaluation** *The algorithm* $c_f \leftarrow \mathsf{HE.Eval}_{pk}(f, c_1, \ldots, c_t)$ *takes the public key, pk, a function* $f : \{0, 1\}^t \to \{0, 1\}$, *and a set of t ciphertexts* $c_1, \ldots, c_t$ *and outputs a ciphertext* c_f. *In our paper, we will represent functions f as binary circuits constructed of NAND gates.*

Definition A.2. *Let* HE *be* $\mathcal{L}$ − *homomorphic and let* f_{nand} *be the augmented decryption function defined below:*

$$f_{nand} = \mathsf{HE.Dec}(sk, c_1) \ NAND \ \mathsf{HE.Dec}(sk, c_2)$$

Then HE *is **bootstrappable** if* $f_{nand} \in \mathcal{L}$

Definition A.3. *A public key encryption scheme* (Gen, Enc, Dec) *has **weak circular security** if it is secure even against an adversary with auxiliary information containing encryptions of all secret key bits.*

---

[4] In many schemes, the public key is split into two parts, the *pk*, which is used to encrypt fresh messages, and the evaluation key (*evk*) that is used to homomorphically evaluate circuits, so the output of the algorithm is: $(pk, evk, sk) \leftarrow \mathsf{HE.Keygen}(1^\kappa)$.

Definition A.4. *For any class of circuits* $\mathcal{C} = \{\mathcal{C}_\eta\}_{\eta \in \mathbb{N}}$ *over* $\{0,1\}$. *A scheme* HE *is* $\mathcal{C}$ $-$ **homomorphic** *if for any function* $f \in \mathcal{C}$, *and respective inputs* $\mu_1, \ldots, \mu_t \in \{0,1\}$, *it holds that*

$$Pr[\text{HE.Dec}_{sk}(\text{HE.Eval}_{pk}(f, c_1, \ldots, c_t)) \neq f(\mu_1, \ldots, \mu_t)] = negl(\eta)$$

where $(pk, sk) \leftarrow \text{HE.Keygen}(1^\kappa)$ *and* $c_i \leftarrow \text{HE.Enc}_{pk}(\mu_i)$.

Definition A.5. *A homomorphic scheme* HE *is* **compact** *if there exists a poly-nomial* $p = p(\eta)$ *such that the output length of* HE.Eval$(\cdots)$ *is at most* p *bits long (regardless of* f *or the number of inputs).*

Definition A.6. *A scheme is* **leveled fully homomorphic** *if it takes* 1^L *as additional input in key generation, where* $L = poly(\eta)$, *and otherwise satisfies the definitions for a compact,* $\mathcal{L}$-*homomorphic encryption scheme, where* $\mathcal{L}$ *is the set of all circuits over* $\{0,1\}$ *of depth* $\leq L$.

Definition A.7. *A scheme* HE *is* **fully homomorphic** *if it is both compact and* $\mathcal{C}$- *homomorphic, where* $\mathcal{C} = \{\mathcal{C}_\eta\}_{\eta \in \mathbb{N}}$ *is the set of all circuits with arity and depth polynomial in* η.

A.2 Learning with Errors

Definition A.8. *The Decision Learning with Errors Problem:*
Given a secret $\mathbf{s} \leftarrow \mathbb{Z}_q^n$, $m = poly(n)$ *samples* $\mathbf{a}_i \xleftarrow{\$} \mathbb{Z}_q^n$, *and corresponding noise* $x_i \leftarrow \chi$, *Distinguish* $\{A_{s,\chi}\}_i = \{\mathbf{a}_i, \langle \mathbf{a}_i, \mathbf{s}\rangle + x_i\}_i$ *from* $\{\mathbf{a}_i, b_i\}_i \xleftarrow{\$} \mathbb{Z}_q^\ell \times \mathbb{Z}_q$.
We denote an instance of the problem as $DLWE_{n,q,\chi}$. *The* **decision learning with errors assumption** *is that no probabilistic polynomial time adversary can solve* $DLWE_{n,q,\chi}$ *with more than negligible advantage.*

Definition A.9. *A family of distributions* χ *is called* $\boldsymbol{\beta}$-**bounded** *if* $\boldsymbol{Pr}_{x \leftarrow \chi(\eta)}[||x|| > \beta] = negl(\eta)$.

Definition A.10. *The Gaussian distribution in one dimension with standard deviation* β *is* $D_\beta := \exp(-\pi(x/\beta)^2)/\beta$. *For* $\beta \in \mathbb{Z}_q$, *the* **discretized Gaussian**, $\overline{\Psi}_\beta$, *is defined by choosing* β' *such that* $\beta = \beta' \cdot q$, *then choosing* $x \xleftarrow{\$} D_{\beta'}$ *and computing* $\lfloor q \cdot x \rceil$. *Note that* $\overline{\Psi}_\beta$ *is* β-*bounded when* β *is super-polynomial in* η. *When* $\chi = \overline{\Psi}_\beta$ *we denote the DLWE instance as* $DLWE_{n,q,\beta}$.

A.3 Min-entropy and the Leftover Hash Lemma

Definition A.11. *A distribution* $\mathcal{X}$ *has* **min entropy** $\geq k$, *denoted* $H_\infty(\mathcal{X}) \geq k$, *if*

$$\forall x \in \mathcal{X}, \boldsymbol{Pr}[\mathcal{X} = x] \leq 2^{-k}$$

Definition A.12. *(From [31]) For two random variables X and Y, the **average min-entropy** of X conditioned on Y, denoted $\tilde{H}_\infty(X|Y)$ is*

$$\tilde{H}_\infty(X|Y) := -\log \mathop{E}_{y \leftarrow Y}\left[\left[\max_x \boldsymbol{Pr}[X = x|Y = y]\right] = -\log\left[\mathop{E}_{y \leftarrow Y}\left[2^{-H_\infty(X|Y=y)}\right]\right]\right.$$

Definition A.13. *(From [31]) For two random variables X and Y, the ϵ-**smooth average min-entropy** of X conditioned on Y, denoted $\tilde{H}_\infty^\epsilon(X|Y)$ is*

$$\tilde{H}_\infty^\epsilon(X|Y) = \max_{(X',Y'):\Delta((X,Y),(X',Y'))<\epsilon} \tilde{H}_\infty(X'|Y')$$

Note that in particular, for any random variable X, given distributions $\mathcal{D}_Y \approx_s \mathcal{D}_Z$ with $Y \leftarrow \mathcal{D}_Y$, $Z \leftarrow \mathcal{D}_Z$, there exists some ϵ such that $\Delta(Y,Z) < \epsilon = negl(\eta)$, and

$$\tilde{H}_\infty^\epsilon(X|Y) \geq \tilde{H}_\infty^\epsilon(X|Z)$$

We now-restate a version of the leftover hash lemma [38] relating to matrix-vector multiplication in $\mathbb{Z}_q$, as it was stated in, for example, [27].

Lemma A.1. *[Leftover Hash Lemma] For a security parameter η, let $n = poly(\eta)$, let $\mathbf{C} \xleftarrow{\$} \mathbb{Z}_q^{m \times n}$ Let $\mathbf{s} \leftarrow \mathcal{D} \in \mathbb{Z}_q^n$, and let $k = H_\infty(\mathcal{D})$. If $m \log q \leq k - 2\log(\frac{1}{\epsilon}) + 2$ then $\Delta((\mathbf{C}, \mathbf{Cs})(\mathbf{C}, \mathbf{u} \xleftarrow{\$} \mathbb{Z}_q^m)) \leq \epsilon$.*

In particular, by setting $\epsilon = 2^{-\eta}$, if $m \log q \leq k - 2\eta + 2$ then $(\mathbf{C}, \mathbf{Cs}) \approx_s (\mathbf{C}, \mathbf{u} \xleftarrow{\$} \mathbb{Z}_q^m)$

The leftover hash lemma can easily be generalized to the case where $\mathbf{s}$ has high conditional average min-entropy.

Lemma A.2. *[Generalized Leftover Hash Lemma] (from lemma 2.4 in [31]) For a security parameter η, let $n = poly(\eta)$, let $\mathbf{C} \xleftarrow{\$} \mathbb{Z}_q^{m \times n}$ Let $\mathbf{s} \leftarrow \mathcal{D} \in \mathbb{Z}_q^n$, let t be any random variable, and let $k = \tilde{H}_\infty(\mathbf{s}|t)$. If $m \log q \leq k - 2\log(\frac{1}{\epsilon}) + 2$ then $\Delta((\mathbf{C}, \mathbf{Cs}, t)(\mathbf{C}, \mathbf{u} \xleftarrow{\$} \mathbb{Z}_q^m), t) \leq \epsilon$. In particular, setting $\epsilon = 2^{-\eta}$, if $m \log q \leq k - 2\eta + 2$ then $(\mathbf{C}, \mathbf{Cs}, t) \approx_s (\mathbf{C}, \mathbf{u} \xleftarrow{\$} \mathbb{Z}_q^m, t)$*

An immediate consequence of the above lemma is the following corollary:

Corollary A.1 (Epsilon-Smooth Variant of LHL). *For a security parameter η, let $n = poly(\eta)$, let $\mathbf{C} \xleftarrow{\$} \mathbb{Z}_q^{m \times n}$ Let $\mathbf{s} \leftarrow \mathcal{D} \in \mathbb{Z}_q^n$, let t be any random variable, and let $\tilde{H}_\infty^{\epsilon_1}(\mathbf{s}|t) \geq k$. If $m \log q \leq k - 2\log(\frac{1}{\epsilon_2}) + 2$ Then $\Delta((\mathbf{C}, \mathbf{Cs}, t)(\mathbf{C}, \mathbf{u} \xleftarrow{\$} \mathbb{Z}_q^m), t) \leq 2\epsilon_1 + \epsilon_2$.*

Proof. The definition of ϵ-smooth average min-entropy means there exists a random variable $\mathbf{s}'$ over the same domain as $\mathbf{s}$ and a random variable t' over the same domain as t such that $\Delta((\mathbf{s}, t)(\mathbf{s}', t')) \leq \epsilon_1$, and $\tilde{H}_\infty(\mathbf{s}'|t') \geq k$. Lemma A.2 tell us that $\Delta((\mathbf{C}, \mathbf{Cs}', t')(\mathbf{C}, \mathbf{u} \xleftarrow{\$} \mathbb{Z}_q^m), t') \leq \epsilon_2$. Furthermore, clearly $\Delta(t, t') \leq \epsilon_1$. Finally, since statistical distance is a metric, we can conclude $\Delta((\mathbf{C}, \mathbf{Cs}, t)(\mathbf{C}, \mathbf{u}, t)) \leq 2\epsilon_1 + \epsilon_2$

B   More Details about Parameter Setting

We now describe in more detail the constraints that drive our setting of parameters. We include full proofs of Lemma 4.2 and Lemma B.1, which drive the setting of many of our parameters.

B.1   Parameter Setting Using Lemma 4.2

Below we restate and prove Lemma 4.2.

Lemma 4.2. *There exists a distribution* Lossy *such that* $\tilde{\mathbf{A}} \leftarrow$ Lossy $\approx_c \mathbf{U} \xleftarrow{\$}$ $\mathbb{Z}_q^{m \times n}$ *and given* $\mathbf{t} \xleftarrow{\$} \mathbb{Z}_q^n$, *and* $\mathbf{e} \leftarrow \chi$, $\tilde{H}_\infty^\epsilon(\mathbf{t} | \tilde{\mathbf{A}}, \tilde{\mathbf{A}}\mathbf{t} + \mathbf{e}) \geq n$, *where* $\epsilon = negl(\eta)$.

Proof. Recall that we define Lossy as follows:

> Choose $\mathbf{C} \xleftarrow{\$} \mathbb{Z}_q^{m \times n'}$, $\mathbf{D} \xleftarrow{\$} \mathbb{Z}_q^{n' \times n}$, and $\mathbf{Z} \leftarrow \overline{\Psi}_\alpha^{m \times n}$, where $\frac{\alpha}{\beta} = negl(\eta)$ and $n' \log q \leq n - 2\eta + 2$.
> Let $\tilde{\mathbf{A}} = \mathbf{C}\mathbf{D} + \mathbf{Z}$
> output $\tilde{\mathbf{A}}$.

First, observe that $\tilde{\mathbf{A}} \approx_c \mathbf{U} \xleftarrow{\$} \mathbb{Z}_q^{m \times n}$:
$\tilde{\mathbf{A}}$ is a *DLWE* instance, with $\mathbf{D}$ as the secret and $\mathbf{Z}$ as the error term, so as long as $DLWE_{n',q,\alpha}$ is hard, then $\tilde{\mathbf{A}} \approx_c \mathbb{Z}_q^{m \times n}$.

Next, observe that $\tilde{H}_\infty^\epsilon(\mathbf{t} | \tilde{\mathbf{A}}\mathbf{t} + \mathbf{e}) = n$, where $\epsilon = negl(\eta)$:
Since $\mathbf{t} \xleftarrow{\$} \mathbb{Z}_q^m$ is identically distributed to $\mathbf{t} = \mathbf{t_0} + \mathbf{t_1}$ where $\mathbf{t_0} \xleftarrow{\$} \{0,1\}^m$, and $\mathbf{t_1} \xleftarrow{\$} \mathbb{Z}_q^m$, we may consider consider $\mathbf{t} = \mathbf{t_0} + \mathbf{t_1}$.

Clearly for any ϵ, $\tilde{H}_\infty^\epsilon(\mathbf{t} | \tilde{\mathbf{A}}\mathbf{t} + \mathbf{e}) \geq \tilde{H}_\infty^\epsilon(\mathbf{t_0} | \tilde{\mathbf{A}}\mathbf{t} + \mathbf{e})$, so it suffices to bound the min-entropy of $\mathbf{t_0}$.

We can then rewrite $\tilde{\mathbf{A}}\mathbf{t} + \mathbf{e}$ as

$$= \mathbf{C}\mathbf{D}\mathbf{t_0} + \mathbf{Z}\mathbf{t_0} + \mathbf{C}\mathbf{D}\mathbf{t_1} + \mathbf{Z}\mathbf{t_1} + \mathbf{e}$$

Since $\mathbf{e}$ is drawn from a discretized Gaussian distribution, and since each element of $\mathbf{Z}\mathbf{t_0}$ is negligibly small compared to the corresponding element of $\mathbf{e}$, we know that $\mathbf{e} + \mathbf{Z}\mathbf{t_0} \approx_s \mathbf{e}$. Thus there exists some $\epsilon_1 = negl(\eta)$ such that

$$\tilde{H}_\infty^{\epsilon_1}(\mathbf{t_0} | \mathbf{C}\mathbf{D}\mathbf{t_0} + \mathbf{C}\mathbf{D}\mathbf{t_1} + \mathbf{Z}\mathbf{t_1} + \mathbf{e}) \geq \tilde{H}_\infty(\mathbf{t_0} | \mathbf{C}\mathbf{D}\mathbf{t_0} + \mathbf{C}\mathbf{D}\mathbf{t_1} + \mathbf{Z}\mathbf{t_1} + \mathbf{Z}\mathbf{t_0} + \mathbf{e})$$

Since $\tilde{H}_\infty(\mathbf{t_0} | \mathbf{C}\mathbf{D}\mathbf{t_1} + \mathbf{Z}\mathbf{t_1} + \mathbf{e}) \geq n$, Lemma A.2 tells us that for our choice of n', $(\mathbf{C}\mathbf{D}\mathbf{t_0} + \mathbf{C}\mathbf{D}\mathbf{t_1} + \mathbf{Z}\mathbf{t_1} + \mathbf{e}) \approx_s (\mathbf{C}\mathbf{u_0} + \mathbf{C}\mathbf{D}\mathbf{t_1} + \mathbf{Z}\mathbf{t_1} + \mathbf{e})$, where $\mathbf{u_0} \xleftarrow{\$} \mathbb{Z}_q^{n'}$. Since the statistical distance between these two distributions is some $\epsilon_2 = negl(\eta)$, there is some $\epsilon = \epsilon_1 + \epsilon_2 = negl(\eta)$ such that

$$\tilde{H}_\infty^\epsilon(\mathbf{t_0} | \mathbf{C}\mathbf{u_0} + \mathbf{C}\mathbf{D}\mathbf{t_1} + \mathbf{Z}\mathbf{t_1} + \mathbf{e}) \geq \tilde{H}_\infty(\mathbf{t_0} | \mathbf{C}\mathbf{D}\mathbf{t_0} + \mathbf{C}\mathbf{D}\mathbf{t_1} + \mathbf{Z}\mathbf{t_1} + \mathbf{Z}\mathbf{t_0} + \mathbf{e})$$

Since each of $\mathbf{C}, \mathbf{u_0}, \mathbf{D}, \mathbf{Z}, \mathbf{t_1}, \mathbf{e}$, is independent of $\mathbf{t_0}$, this quantity equals $H_\infty(\mathbf{t_0}) = n$. Thus, we can conclude that $\tilde{H}_\infty^\epsilon(\mathbf{t} | \tilde{\mathbf{A}}\mathbf{t} + \mathbf{e}) \geq n$ as well.

When using Lemma 4.2, the following constraints affect our parameter setting:

1. **Statistical Indistinguishability:** There are three different places in our hybrid argument where we prove that two distributions are statistically indistinguishable.
   - In Lemma 4.2, we argue that the distribution $\mathbf{Zt_0} + \mathbf{e}$ is statistically close to $\mathbf{e}$, because the magnitude of each element of $\mathbf{Zt_0}$ is small. This argument requires that $\mathbf{e}$ be a discretized Gaussian distribution, rather than just a bounded distribution, as required by the original GSW scheme.
   - In Claim 1, inside our proof of Lemma 4.1, we use the leftover hash lemma to show we can replace $\mathbf{rA}$ with $\mathbf{u} \xleftarrow{\$} \mathbb{Z}_q^n$. This step is part of the security proof of all variations on the RPKE scheme, but an artifact of our proof technique is that we consider an adversary who can see $\mathbf{r} \cdot \mathbf{e}$, which is $\ell = O(\log q)$ bits long. So for $\mathbf{r}$ of dimension m, we have $\tilde{H}_\infty(\mathbf{r}) = m - \ell$. The analogous step in the GSW security proof assumes $H_\infty(\mathbf{r}) = m$. This leads us to increase the value of m. In our scheme, m is set to $2n \log q + 3\eta$.
   - Again in Lemma 4.1, in Claim 2, we use the leftover hash lemma to show that given $\mathbf{u} \xleftarrow{\$} \mathbb{Z}_q^n$, we can replace $\mathbf{u} \cdot \mathbf{t}$ with $u' \xleftarrow{\$} \mathbb{Z}_q$. As described in our proof, the ϵ-smooth average min-entropy of $\mathbf{t}$ is $n - \ell - \lambda$, where λ is the the number of bits of leakage we can tolerate. Thus, we must set λ to a value that keeps $\tilde{H}_\infty(\mathbf{t})$ high enough for the leftover hash lemma to apply. That is how we arrive at $\lambda \leq n - 2 \log q - 4\eta$.

2. ***DLWE* Considerations:** The security of our scheme is based on the hardness of two different *DLWE* problems: $DLWE_{n',q,\alpha}$, where the n' and α come from Lemma 4.2, and $DLWE_{n,q,\beta}$. For our scheme to be secure, the following three things need to be true:
   - $\frac{\alpha}{\beta} = negl(\eta)$. This is a necessary condition in our proof of Lemma 4.2.
   - $DLWE_{n',q,\alpha}$ is hard. We refer to Statement 1, which shows that this problem is at least as hard as $GapSVP_{n'q/\alpha}$, and to Statement 2, which says the best known algorithms for solving $GapSVP_{n'q/\alpha}$ run in time $2^{\tilde{\Omega}\left(\frac{n'}{log(n'q/\alpha)}\right)}$. This quantity should be at least super-polynomial in our security parameter for the scheme to be secure.
   - $DLWE_{n,q,\beta}$ is hard. Using the same theorems, we see that we need $2^{\tilde{\Omega}\left(\frac{n}{log(nq/\beta)}\right)}$ to be super-polynomial in η as well.

3. **Correctness:** The scheme needs $8(N + 1)^L < \frac{q}{\beta}$, where L is the depth of the circuit, β is the error bound, and $N = (\log q + 1)n$, in order to ensure the noise never gets large enough to hamper accurate decryption.

Since our FHE scheme supports evaluation of circuits whose depth is polynomial in the security parameter as long as that polynomial is pre-specified, we know that there exists some constant c such that $L \leq \eta^c$. Let $\tau = \max\{L, \eta^2\}$. Setting the parameters as follows satisfies all of the hardness and correctness constraints for the scheme:

Let $n = \tau^4$. Let $q = 2^{\tau \log^2 \tau}$. Let $\beta = 2^{\log^2 \tau}$. Recall that $n' = (n - 2\eta)/\log q$, and let $\alpha = n'$.

Note that $\frac{\alpha}{\beta}$ is clearly negligible in η as required. Since the best algorithm for $GapSVP_{n'q/\alpha}$ runs in time $2^{\tilde{\Omega}\left(\frac{n'}{\log(n'q/\alpha)}\right)}$, we look more closely at the exponent $\frac{n'}{\log(n'q/\alpha)}$. We can rewrite it as $\frac{n'}{\log(q)} = \frac{n - 2\eta}{\log^2 q} = \frac{\tau^4 - 2\eta}{\tau^2 \log^4 \tau}$. Since $\tau \geq \eta^2$, we know that the above quantity is $\geq \frac{\eta^8 - 2\eta}{2\eta^4 \log^2 \eta} \geq \eta$ for $\eta \geq 16$. Thus the hardness is $2^{\tilde{\Omega}(\eta)}$.

Similarly, to bound the hardness of $GapSVP_{n,q,\beta}$ we consider the exponent of $2^{\frac{n}{\log(nq/\beta)}}$.

$$
\frac{n}{\log(nq/\beta)} = \frac{n}{\log n + \log q - \log \beta}
$$
$$
= \frac{\tau^4}{4 \log \tau + \tau \log^2 \tau - \log^2 \tau}
$$
$$
\geq \tau
$$
$$
\geq \eta
$$

This means that $DLWE_{n,q,\beta}$ is exponentially hard as well. Finally, we verify that our parameter settings maintain the correctness of the scheme: We need $8(N+1)^L < \frac{q}{\beta}$, and since we chose $\tau \geq L$, it is sufficient to show $8(N+1)^\tau < \frac{q}{\beta}$. We can upper bound the left hand side of this inequality as follows:

$$
8(N+1)^\tau \leq 8(2n \log q)^\tau
$$
$$
= 8(2\tau^4 \tau \log^2 \tau)^\tau
$$
$$
\leq 2^3 2^\tau \tau^{6\tau}
$$
$$
= 2^{3 + \tau + 6 \log^2 \tau}
$$

Meanwhile, the right hand side is equal to $2^{(\tau - 1) \log^2 \tau}$, which is clearly greater than the left hand side for sufficiently high τ.

Finally, the number of bits of leakage we can support is $n - 2 \log q - 4\eta = \tau^4 - 2\tau \log^2 \tau - 4\eta = \eta^8 - 32\eta^4 \log \eta - 4\eta$, which is positive for any $\eta \geq 3$.

B.2   Efficiency/Leakage Tradeoff

We can prove our scheme secure using the following alternate lemma, which gives us better efficiency but a lower leakage bound.

Lemma B.1. *For $n, m, n', q, \alpha, \beta$ such that $DLWE_{n',q,\alpha}$ and $DLWE_{n,q,\beta}$ are hard, if $\beta \geq \alpha nm$, there exists a distribution* Lossy' *such that $\tilde{\mathbf{A}} \leftarrow$* Lossy' $\approx_c$ $\mathbf{U} \xleftarrow{\$} \mathbb{Z}_q^{m \times n}$ *and given* $\mathbf{t} \xleftarrow{\$} \mathbb{Z}_q^n$, *and* $\mathbf{e} \leftarrow \overline{\Psi}_\beta$, $\tilde{H}_\infty^\epsilon(\mathbf{t}|\tilde{\mathbf{A}}, \tilde{\mathbf{A}}\mathbf{t} + \mathbf{e}) \geq n - \eta(\log m + 2 \log n)$, *where $\epsilon = negl(\eta)$.*

The proof of this lemma closely follows the outline of Lemma B.4 in [28], so we defer it to the full version of our paper.

The new lemma leads immediately to the following:

Corollary B.1. *The* LRGSW *scheme is resilient to* $\lambda \leq n - (2 + \eta) \log q - \eta \log m - 4\eta$ *bits of leakage when* $\overline{\Psi}_\beta$ *is chosen so that* $\frac{\beta}{m} \geq \frac{n^2}{\log q}$.

Proof. This corollary is true as long as with the new parameter settings, the scheme still maintains its correctness, so $8(N + 1)^L \leq \frac{q}{\beta}$ and its hardness: $DLWE_{n',q,\alpha}$, and $DLWE_{n,q,\beta}$, as well as the new requirement that $\frac{\beta}{m} \geq n\alpha$. If we choose an $\alpha = O(n')$, which we need for $DLWE_{n',q,\alpha}$ to be hard, then $\alpha \leq n/\log q$, so our setting of β is sufficient to meet this new requirement. Note that with these settings, β is no longer super-polynomial in η, and though q will remain superpolynomial in β, this allows for a much smaller value of q as well. For example, if we let $\tau = \max\{L, \eta^3\}$, $n = \tau^3$, $q = 2^{\tau \log n}$, $m = 2n \log q + 3\eta$, $n' = (n - 2\eta + 2)/\log q$, $\alpha = \tau^2$, $\beta = 3n^3\tau^3$, then we have:

- $DLWE_{n',q,\alpha}$ is hard:

  note that $n' = \frac{\tau^3 - 2\tau + 2}{3\tau \log \tau} \leq \tau^2 = \alpha$. So $2^{\frac{n'}{\log(n'q/\alpha)}} \geq 2^{n'/\log q}$. We can rewrite that exponent as $\frac{n - 2\eta + 2}{\log^2 q} = \frac{\eta^9 - 2\eta + 2}{81\eta^6 \log \eta}$. So for $\eta \geq 20$, we have $2^{n'/\log q} \geq 2^\eta$.

  Thus we can conclude that since $DLWE_{n',q,\alpha}$ takes time $2^{\tilde{\Omega}\left(\frac{n'}{\log(n'q/\alpha)}\right)}$ to solve, it is super-polynomially hard to solve in η.

- $DLWE_{n,q,\beta}$ is hard:

  Since $\beta > n$, we know that $\frac{n}{\log(nq/\beta)} > \frac{n}{\log q} = \frac{\tau^3}{3\tau \log \tau} \geq \tau/3 \geq \eta^6/3$. So $2^{\tilde{\Omega}\left(\frac{n}{\log nq/\beta}\right)}$ is exponential in η as well.

- The scheme is correct:

  We need to show: $8(N + 1)^L \leq \frac{q}{\beta}$. First, we rewrite $N + 1$.

$$N + 1 = n(\log q + 1) + 1$$
$$= \tau^3(\tau \log \tau^3) + \tau^3 + 1$$
$$= 3\tau^4 \log \tau + \tau^3 + 1$$
$$\leq 4\tau^4 \log \tau$$

So we have that $8(N + 1)^L \leq 2^3 2^{2L} 2^{4 \log \tau} 2^{\log \log \tau}$, and since $L \leq \tau$, we have, that this is $\leq 2^{2\tau + 4 \log \tau + 3 + \log \log \tau} \leq 2^{2\tau + 5 \log \tau}$.

Meanwhile,

$$\frac{q}{\beta} = 2^{\tau \log n}/(3n^3\tau^3)$$

$$= \frac{1}{3} 2^{\tau \log n - 3 \log n - 3 \log \tau}$$

$$\geq 2^{3\tau \log \tau - 12 \log \tau}$$

This quantity is $\geq 2^{2\tau + 5 \log \tau}$ for sufficiently high τ, (for example, if $\tau \geq 9$, meaning $\eta \geq 3$), so the scheme is secure.

Securing Circuits and Protocols against $1/\operatorname{poly}(k)$ Tampering Rate

Dana Dachman-Soled[1] and Yael Tauman Kalai[2]

[1] University of Maryland
danadach@ece.umd.edu
[2] Microsoft Research
yael@microsoft.com

Abstract. In this work we present an efficient compiler that converts any circuit C into one that is resilient to tampering with $1/\operatorname{poly}(k)$ fraction of the wires, where k is a security parameter *independent* of the size of the original circuit $|C|$. Our tampering model is similar to the one proposed by Ishai *et al.* (Eurocrypt, 2006) where a tampering adversary may tamper with any wire in the circuit (as long as the overall number of tampered wires is bounded), by setting it to 0 or 1, or by toggling with it. Our result improves upon that of Ishai *et al.* which only allowed the adversary to tamper with $1/|C|$ fraction of the wires.

Our result is built on a recent result of Dachman-Soled and Kalai (Crypto, 2012), who constructed tamper resilient circuits in this model, tolerating a *constant* tampering rate. However, their tampering adversary may learn logarithmically many bits of sensitive information. In this work, we avoid this leakage of sensitive information, while still allowing leakage rate that is *independent* of the circuit size. We mention that the result of Dachman-Soled and Kalai (Crypto, 2012) is only for Boolean circuits (that output a single bit), and for circuits that output k bits, their tampering-rate becomes $1/O(k)$. Thus for cryptographic circuits (that output k bits), our result strictly improves over (Dachman-Soled and Kalai, Crypto, 2012).

In this work, we also show how to generalize this result to the setting of two-party protocols, by constructing a general 2-party computation protocol (for any functionality) that is secure against a tampering adversary, who in addition to corrupting a party may tamper with $1/\operatorname{poly}(k)$-fraction of the wires of the computation of the honest party and the bits communicated during the protocol.

Keywords: Tamper-resilient circuits, Two-party computation.

1 Introduction

Constructing cryptographic schemes that are secure against physical attacks is a fundamental problem which has recently gained much attention in the cryptographic community. Indeed, physical attacks exploiting the implementation (rather than the functionality) of cryptographic schemes such as RSA have been known in theory for several years [41,8] and recent works have shown that these attacks can be carried out in practice [9,49]. There are many different types of physical attacks in the literature. For instance, Kocher et al. [42] demonstrated how one can possibly learn the secret

Y. Lindell (Ed.): TCC 2014, LNCS 8349, pp. 540–565, 2014.

key of an encryption scheme by measuring the power consumed during an encryption operation, or by measuring the time it takes for the operation to complete [41]. Other types of physical attacks include: inducing faults to the computation [7,8,42], using electromagnetic radiation [28,54,53], and several others [53,39,43,31].

Although these physical attacks have proven to be a significant threat to the practical security of cryptographic devices, until recently cryptographic models did not take such attacks into account. In fact, traditional cryptographic models idealize the parties interaction and implicitly assume that an adversary may only observe an honest partys input-output behavior. Over the past few years, a large and growing body of research has sought to introduce more realistic models and to secure cryptographic systems against such physical attacks. The vast majority of these works focus on securing cryptographic schemes against various leakage attacks (e.g. [10,34,47,29,33,18,50,1,48,38,15,14,22,35,30]). In these attacks an adversary plays a passive role, learning information about the honest party through side-channels but not attempting to interfere with the honest partys computation. However, as mentioned above, physical attacks are not limited to leakage, and include active tampering attacks, where an adversary may actively modify the honest partys memory or circuit. In this work, we focus on constructing schemes that are secure even in the presence of tampering.

1.1 Our Results

We present a compiler that converts any circuit into one that is resilient to (a certain form of) tampering. Then, we generalize this result, and show how to construct a general two-party computation protocol that is secure against such tampering. We consider the tampering model of Ishai et al. [33]. Specifically, we consider a tampering adversary that may tamper with any (bounded) set of wires of the computation.

We note that our compiler that converts any circuit into a "tamper resilient" one, cannot guarantee correctness of the computation in the presence of tampering. This is the case, since the adversary may always tamper with the final output wire of the circuit. Therefore, as in [33], we do not guarantee correctness, but instead ensure privacy. In particular, we consider circuits that are associated with a secret state. We model such circuits as standard circuits (with AND, OR, and NOT gates), with additional secret, persistent memory that contains the secret state. The circuit itself is public and its topology is fully known to the adversary, whereas the memory content is secret. Following the terminology of [33], we refer to such circuits as private circuits. Our notion of security guarantees that the secret state of the circuit is protected even when an adversary may run the circuit on arbitrary inputs while continuously tampering with the wires of the circuit.

There are several fundamental impossibility results for tampering, which any positive result must circumvent. In the following, we discuss some of these limitations.

Class of Tampering Functions. It is not hard to see that it is impossible to construct private circuits resilient to arbitrary tampering attacks, since an adversary may modify the circuit so that it simply outputs the entire secret state in memory. Thus, we must specify a class of allowed tampering functions. As in [33], in this we consider tampering adversaries who can tamper with individual wires [33,23,12] and individual memory gates [11,29,19]. More specifically, in each run of the circuit we allow the adversary to specify a set of tampering instructions, where each instruction is of the form: Set a wire (or a memory gate) to 0 or 1, or toggle with the value on a wire (or a memory gate). However, in contrast to [37], where the tampering rate achieved is $1/|C|$, where $|C|$ is the size of the original circuit, we allow the adversary to tamper with any $1/\operatorname{poly}(k)$-fraction of wires and memory gates in the circuit, where k is security parameter and $\operatorname{poly}(k)$ is independent of the size of the original circuit. We note that the recent work of [12] gave a construction that is resilient to constant tampering rate. However, in their construction a tampering adversary may learn logarithmically many bit on the secret state of the circuit, and their guarantee was that such an adversary learns only logarithmically many bits about the secret state. We give the guarantee that a tampering adversary does not learn anything beyond the input/output behavior.

Necessity of Feedback. As noted by [29], it is impossible to construct private circuits resilient against tampering on wires without allowing feedback into memory, i.e. without allowing the circuit to overwrite its own memory. Otherwise, an adversary may simply set to 0 or 1 one memory gate at a time and observe whether the final output is modified or not.

Even if we allow feedback, and place limitations on the type of tampering we allow, it is not a priori clear how to build tamper-resilient circuits. As pointed out in [33,12], the fundamental problem is that the part of the circuit which is supposed to detect tampering and overwrite the memory, may itself be tampered with. Indeed, this self-destruct mechanism itself needs to be resilient to tampering.

As in [33,12], we prove security using a simulation based definition, where we require that for any adversary who continually tampers with the circuit (as described above), there exists a simulator who simulates the adversarys view. Like in [33], we give the simulator only black-box access to the original private circuit with no additional leakage on the secret state. This is in contrast to the work of [12], who achieve a constant tampering rate, but where the simulator requires $O(\log k)$ bits of leakage on the secret state, where k is security parameter, in order to simulate. Thus, our result is meaningful in settings where [12] is not.

For example,[1] consider a setting where the same cryptographic key is placed on several devices, which are all obtained by an adversary. In this case, [12] does not guarantee any privacy for the cryptographic key, since $O(\log(k))$ bits leaked from each of several devices may give enough information to reconstruct the entire cryptographic key. Another example is a setting where secrecy of an algorithm is desired in order to protect intellectual property. In this case, the secret state of the device is the algorithm and the circuit is the universal circuit. Here, the same algorithm is placed on a large number of devices and is marketed. Thus, if $O(\log(k))$ bits are leaked from each device, then it may be possible to recover the entire algorithm.

Finally, we show how one can use our tamper-resilient compiler to achieve tamper-resilient secure two-party computation. We elaborate on this result in Section 1.4, but

---

[1] The following example, which was the motivating force behind this research, was brought to our attention by Shamir [55].

mention here that the results of [12] do not apply to this regime. Loosely speaking, the reason is that in this setting, the secret state of the circuit consists of the private input and randomness of each party, and (even logarithmic) leakage on the input and randomness of each party may completely compromise security of the two-party computation protocol.

Our Results More Formally. We present a general compiler T that converts a circuit C with a secret state s (denoted by C_s) into a circuit $T(C_s)$. We consider $\mathcal{PPT}$ adversaries $\mathcal{A}$ who receive access to $T(C_s)$ and behave in the following way: $\mathcal{A}$ runs the circuit many times with arbitrary and adaptively chosen inputs. In addition, during each run of the circuit the adversary $\mathcal{A}$ may specify tampering instructions of the form "set wire w to 1","set wire w to 0", "flip value of wire w", as well as "set memory gate g to 1", "set memory gate g to 0", "flip value of memory gate g", for any wire w or memory gate g. We restrict the number of tampering instructions $\mathcal{A}$ may specify per run to be at most $\lambda \cdot \sigma$, where $\lambda = \frac{1}{\mathrm{poly}(k)}$ and σ is the size of the circuit $T(C_s)$. Thus, in each run, $\mathcal{A}$ may tamper with a $1/\mathrm{poly}(k)$-fraction of wires and memory gates.

Theorem 1 (Main Theorem, Informal). *There exists an efficient transformation T which takes as input any circuit C_s with private state s, and outputs a circuit $T(C_s)$ such that the following two conditions hold:*

Correctness: For every input x, $T(C_s)(x) = C_s(x)$.
Tamper-Resilience: For every $\mathcal{PPT}$ adversary $\mathcal{A}$, which may tamper with $\lambda = 1/\mathrm{poly}(k)$-fraction of wires and memory gates in $T(C_s)$ per run, there exists an expected polynomial time simulator Sim, which can simulate the view of $\mathcal{A}$ given only black-box access to C_s.

Intuitively, the theorem asserts that adversaries who may observe the input-output behavior of the circuit while tampering with at most a λ-fraction of wires and memory gates in each run, do not get extra knowledge over what they could learn from just input-output access to the circuit.

1.2 Comparison with Ishai *et al.* [33] and Dachman-Soled *et al.* [12]

Our work follows the line of work of [33,12]. As in our work, both these works consider circuits with memory gates, and consider the same type of faults as we do. Similarly to us, they construct a general compiler that converts any private circuit into a tamper resilient one. In the following, we discuss some similarities and differences among these works.

- In our construction, as in the construction of [33], we require the use of "randomness gates", which output a fresh random bit in each run of the circuit.[2] In contrast, the construction of [12] is deterministic.

---

[2] Alternatively, [33] can get rid of these randomness gates at the cost of relying on a computational assumption.

- The constructions of [33,12] provide information-theoretic security, while our construction requires computational assumptions.
- As mentioned previously, [33] constructs tamper resilient circuits that are resilient only to local tampering: To achieve resilience to tampering with t wires per run, the circuit size blows up by a factor of at least t. In contrast, our tamper-resilient circuits are resilient to a $1/\operatorname{poly}(k)$-fraction of tampering, where k is security parameter. Thus, our tampering rate is *independent* of the original circuit size.
- The construction of [12] achieves a constant tampering rate, but requires $O(\log k)$ leakage on the secret state in order to simulate. As discussed above, in some settings the guarantees provided by [12] are too weak, while our construction still guarantees meaningful security.

  Moreover, [12] achieves constant tampering rate only for Boolean circuits that output a *single* bit. For circuits with k bit output, the resulting tampering-resilient circuit is only resilient to $1/k$-fraction of tampering.
- The tampering model of [33] allows for "persistent faults", e.g, if a value of some wire is fixed during one run, it remains set to that value in subsequent runs. We note that in our case, we allow "persistent faults" only on memory gates (and not on wires), so if a memory value is modified during one run, it remains modified for all subsequent runs.

1.3   Overview of Our Construction

Intuitively, our compiler works by first applying to the circuit C_s the leakage-resilient compiler T_{LR} of Juma and Vahlis [35]. The Juma-Vahlis compiler, T_{LR}, converts the circuit C_s into two subcomputations (or modules), $\mathsf{Mod}^{(1)}$ and $\mathsf{Mod}^{(2)}$, and provides the guarantee that (continual) leakage on the sub-computations $\mathsf{Mod}^{(1)}$ and $\mathsf{Mod}^{(2)}$ leak no information on the secret seed s. We refer the reader to [35] for the precise security guarantee. We emphasize that $T_{\mathsf{LR}}(C_s)$ has no security guarantees against a tampering adversary (rather only against a leaking adversary).

Our next idea is to use the tamper-resilient compiler T_{TR} of [12]. This compiler provides security against a (continual) *tampering* adversary, guaranteeing that the adversary learns at most $\log n$ bits about the secret s. In this work our goal is to remove this leakage from the security guarantee. To this end, we apply the tamper-resilient compiler T_{TR} to each sub-computation separately, each of which is now resilient to leakage.

We note however, that the Juma-Vahlis compiler relies on a secure hardware component. We do not want to rely on any such tamper-proof component. Therefore, we replace the tamper-proof component with a secure implementation. We describe our compiler in stages:

- First, we present a compiler (as above) that takes as input a circuit C_s and outputs a compiled circuit $T^{(1)}(C_s)$ that consists of 4 components. We prove that $T^{(1)}(C_s)$ is secure against adversaries that tamper with at most a $1/\operatorname{poly}(k)$ fraction of wires overall, but do not tamper with any of the wires in the first component, where the first component corresponds to the hardware component in the [35] construction (See Section 3.1).
- Then, we show how to get rid of the tamper-proof component and allow $1/\operatorname{poly}(k)$-fraction tampering overall (See Sections 3.2 and 5).

1.4 Extension to Tamper-Resilient Secure Two-Party Computation

We consider the two-party computation setting, where in addition to corrupting parties, an adversary may tamper with the circuits of the honest parties and the messages sent by the honest parties. In this setting, we show how to use our construction of tamper-resilient circuits to obtain a general tamper-resilient secure two-party computation protocol, where an adversary may actively corrupt parties and additionally tamper with $1/\text{poly}(k)$-fraction of wires, memory gates, and message bits overall.

To achieve our result, we start with any two-party computation (2-PC) protocol that is secure against malicious corruptions, and where the total number of bits exchanged depends only on security parameter k, and not on the size of the circuit computing the functionality. Such a 2-PC protocol can be constructed from fully homomorphic encryption and (interactive) CS-proofs. In addition we assume that each message sent in the protocol is accompanied with a signature. Then, for each party and each round of the protocol, we consider the private circuit computing the next message function, where the secret state is the party's private input and randomness and the public input is the transcript. We then run (a slight modification of) our tampering compiler on each such next message circuit to obtain a circuit that is resilient to $1/\text{poly}(k)$-fraction of tampering. Since the total number of such circuits is $\text{poly}(k)$, we achieve resilience to a $1/\text{poly}(k)$-fraction of tampering overall. We refer the reader to Section 6 for details.

1.5 Related Work

The problem of constructing error resilient circuits dates back to the work of Von Neumann from 1956 [56]. Von Neumann studied a model of *random* errors, where each gate has an (arbitrary) error independently with small fixed probability, and his goal was to obtain *correctness* (as opposed to privacy). There have been numerous follow up papers to this seminal work, including [13,52,51,25,20,32,26,21], who considered the same noise model, ultimately showing that any circuit of size σ can be encoded into a circuit of size $O(\sigma \log \sigma)$ that tolerates a fixed constant noise rate, and that any such encoding must have size $\Omega(\sigma \log \sigma)$.

There has been little work on constructing circuits resilient to *adversarial* faults, while guaranteeing correctness. The main works in this arena are those of Kalai *et al.* [37], Kleitnam *et al.* [40], and Gál and Szegedy [27]. The works of [40] and [37] consider a different model where the only type of faults allowed are short-circuiting gates. [27] consider a model that allows arbitrary faults on gates, and show how to construct tamper-resilient circuits for symmetric Boolean functions. We note that [27] allow a constant fraction δ of adversarial faults *per level* of the circuit. Moreover, if there are less than $1/\delta$ gates on some level, they allow no tampering at all on that level. [27] also give a more general construction for any circuit which relies on PCP's. However, in order for their construction to work, they require an entire PCP proof π of correctness of the output to be precomputed and handed along with the input to the tamper-resilient circuit. Thus, they assume that the input to the circuit is already encoded via an encoding which depends on the *output* value of that very circuit. We (similarly to [12]) also use the PCP methodology in our result, but do not require any precomputations or that the input be encoded in some special format.

Recently, the problem of physical attacks has come to the forefront in the cryptography community. From the viewpoint of cryptography, the main focus is no longer to ensure correctness, but to ensure *privacy*. Namely, we would like to protect the honest party's secret information from being compromised through the physical attacks of an adversary. There has been much work on protecting circuits against leakage attacks [34,47,18,50,16,24,35,30]. However, there has not been much previous work on constructing circuits resilient to tampering attacks. In this arena, there have been two categories of works. The works of [29,19,11,44,36,45,17] allow the adversary to only tamper with and/or leak on the *memory* of the circuit in between runs of the circuit, but do not allow the adversary to tamper with the circuit itself. We note that this model of allowing tampering only with memory is very similar to the problem of "related key attacks" (see [4,2] and references therein). In contrast, in our work, as well as in the works of [33,23,12], the focus is on constructing circuits resilient to tampering with both the memory as well as the wires of the circuit.

Faust *et al.* [23] consider a model that is reminiscent to the model of [33,12] and to the model we consider here. They consider adversarial faults where the adversary may actually tamper with all wires of the circuit but each tampering attack fails independently with some probability δ. As in [12], they allow the adversary to learn a logarithmic number of bits of information on the secret key. In addition, their result requires the use of small tamper-proof hardware components.

2   The Tampering Model

2.1   Circuits with Memory Gates

Similarly to [33], we consider a circuit model that includes memory gates. Namely, a circuit consists of (the usual) AND, OR, and NOT gates, connected to each other via wires, as well as input wires and output wires. In addition, a circuit may have memory gates. Each memory gate has one (or more) input wires and one (or more) output wires. Each memory gate is initialized with a bit value 0 or 1. This value can be updated during each run of the circuit.

Each time the circuit is run with some input x, all the wires obtain a $0/1$ value. The values of the input wires to the memory gates define the way the memory is updated. We allow only two types of updates: delete or unchange. Specifically, if an input wire to a memory gate has the value 0, then the memory gate is overwritten with the value 0. If an input wire to a memory gate has the value 1, then the value of the memory gate remains unchanged. We denote a circuit C initialized with memory s by C_s.

2.2   Tampering Attacks

We consider adversaries, that can carry out the following attack: The adversary has black-box access to the circuit, and thus can repeatedly run the circuit on inputs of his choice. Each time the adversary runs the circuit with some input x, he can tamper with the wires and the memory gates. We consider the following type of faults: Setting a wire (or a memory gate) to 0 or 1, or toggling with the value on a wire (or a memory gate).

More specifically, the adversary can adaptively choose an input x_i and a set of tampering instructions (as above), and he receives the output of the tampered circuit on input x_i. He can do this adaptively as many times as he wishes. We emphasize that once the memory has been updated, say from s to s', the adversary no longer has access to the original circuit C_s, and now only has access to $C_{s'}$. Namely, the memory errors are persistent, while the wire errors are not persistent.

We denote by $\mathsf{TAMP}_{\mathcal{A}}(T(C_s))$ the output distribution of an adversary $\mathcal{A}$ that carries out the above (continual) tampering attack on a compiled circuit $T(C_s)$. We note that our tampering compiler T is randomized and so the distribution is over the coins of T. We say that an adversary $\mathcal{A}$ is a λ-tampering adversary if during each run of the circuit he tampers with at most a λ-fraction of the circuit. Namely, $\mathcal{A}$ can make at most $\lambda \cdot |T(C_s)|$ tampering instructions for each run, where each instruction corresponds either to a wire tampering or to a memory gate tampering.

Remark. In this work, we define the size of a circuit C, denoted by $|C|$, as the number of wires in C plus the number of memory gates in C. Note that this is not the common definition (where usually the size includes also the gates); however, it is equivalent to the common definition up to constant factors.

To define security of a circuit against tampering attacks we use a simulation-based definition, where we compare the real world, where an adversary $\mathcal{A}$ (repeatedly) tampers with a circuit $T(C_s)$ as above, to a simulated world, where a simulator Sim tries to simulate the output of $\mathcal{A}$, while given only black-box access to the circuit C_s, and without tampering with the circuit at all. We denote the output distribution of the simulator by Sim^{C_s}.

Definition 1. *We say that a compiler T secures a circuit C_s against $\mathcal{PPT}$ λ-tampering adversaries, if for every $\mathcal{PPT}$ λ-tampering adversary $\mathcal{A}$ there exists a simulator Sim, that runs in expected polynomial time (in the runtime of $\mathcal{A}$), such that for sufficiently large k,*

$$\{\mathsf{TAMP}_{\mathcal{A}}(T(C_s))\}_{k \in \mathbb{N}} \stackrel{c}{\approx} \{\mathsf{Sim}^{C_s}\}_{k \in \mathbb{N}}.$$

In this work we construct such a compiler that takes any circuit and converts it into one that remains secure against adversaries that tamper with $\lambda = 1/\operatorname{poly}(k)$-fraction of the wires in the circuit, where k is the security parameter. Our compiler is uses both the Juma-Vahlis leakage compiler [35] and the recent tampering compiler of [12].

3 The Compiler

3.1 Overview of the First Construction

We start by presenting our first tampering compiler $T^{(1)}$ that takes as input a circuit C_s, and generates a tamper-resilient version of C_s which requires a tamper-proof component. In the case of no tampering, we show the correctness property: $T^{(1)}(C_s)(x) = C_s(x)$. Moreover, we prove that the circuit $T^{(1)}(C_s)$ is resilient to tampering with rate $1/\operatorname{poly}(k)$, where k is the security parameter.

High-Level. On a very high-level, $T^{(1)}(C_s)$ works as follows.

1. Apply the Juma-Vahlis compiler T_{LR} to the circuit C_s to obtain a hardware component and two modules $(\text{Mod}^{(1)}, \text{Mod}^{(2)})$. First, $\text{Mod}^{(1)} = \text{Mod}^{(1)}_{PK, \text{Enc}_{PK}(s)}$ is the sub-computation that takes as input a string x and outputs the homomorphic evaluation of C_s on input x. We refer to this sub-computation as Component 2 of $T^{(1)}(C_s)$ and denote the output of this component by ψ_{comp}. Then a leakage and tamper-resilient hardware is used generate a "fresh" encryption of 0, denoted by ψ_{rand}, which is used to "refresh" the ciphertext ψ_{comp}. We refer to the leakage resilient-hardware outputting encryptions of 0 as Component 1. Component 3 of $T^{(1)}(C_s)$ then takes as input ψ_{comp} and ψ_{rand} and outputs the re-randomized ciphertext $\psi_* = \psi_{\text{comp}} + \psi_{\text{rand}}$. Finally, the second sub-computation of the Juma-Vahlis compiler, $\text{Mod}^{(2)} = \text{Mod}^{(2)}_{SK}$, takes as input the refreshed ciphertext ψ_* and decrypts it to obtain $b = C_s(x)$. This sub-computation is referred to as Component 4 of $T^{(1)}(C_s)$.

2. The next idea is to apply the tampering compiler of [12], T_{TR}, to each of the components separately. We note that this tampering compiler allows a tampering adversary learn logartihmically many bits about the secret state of the circuit. However, since we apply the compiler to Components 2, 3, 4, which inherit the leakage resilient properties of the Juma-Vahlis compiler and are thus resilient to leakage of logarithmic size, this is not a concern to us.

   Unfortunately, this does not quite work. The reason is that the security definition of the tamper-resilient compiler T_{TR} allows the adversary to tamper with the input. Hence, if we simply take the components described above, then a tampering adversary may tamper with the inputs to each of the components, and may completely ruin the security guarantees of the Juma-Vahlis compiler. In particular, the refreshed ciphertext ψ_*, may no longer be distributed correctly. Instead we do the following:

3. Compute the second component, i.e. the tamper-resilient circuit $T_{TR}(\text{Mod}^{(1)})$. However, instead of outputting a single ciphertext ψ_{comp}, the circuit $T_{TR}(\text{Mod}^{(1)})$ will output M copies of ψ_{comp}, where M is a (large enough) parameter that will be specified below. We will argue that for any tampering adversary, either self-destruct occurs or a majority of the copies of ψ_{comp} are exactly correct.

4. Next apply a version of T_{TR} to the third and fourth components, with the guarantee that now an adversary cannot tamper with the input (without causing a self destruct), since the input is replicated M times, and an adversary can only tamper with a small fraction of these wires, and the compiled circuit will check for replicas. This version of T_{TR} turns out to be much simper than T_{TR} since the is size of the third and fourth components depends only on the security parameter, independent of the size of C_s, which turns out to simplify matters significantly.

We defer the details of the construction of $T^{(1)}(C_s)$ to the full version.
We are now ready to state the main theorem of this section:

Theorem 2. $T^{(1)}(C_s)$ *is secure against all* $\mathcal{PPT}$ $\lambda = 1/\text{poly}(k)$-*tampering adversaries (as defined in Definition 1) who do not tamper with Component 1, assuming semantic security of the underlying encryption scheme* $\mathcal{E}_{FH}$.

We defer the proof of Theorem 2 to the full version.

3.2   Overview of Construction of Component 1

We now show how to construct Component 1, instead of relying on tamer-resilient hardware. Recall that our goal is to compute an encryption of 0 in a robust way so that *even after tampering* the output is statistically close to a fresh encryptions of 0 (assuming the output wires were not tampered with). Unfortunately, we don't quite manage to do this. Instead, we achieve a slightly weaker goal. We construct a circuit component that computes an encryption of 0, so that *even after tampering*, if self destruct did not occur, then the output of the computation is of the form $\psi_{fresh} + \psi_{rest}$, where ψ_{fresh} is a fresh encryption of 0, and ψ_{rest} is a simulatable (not necessarily fresh) encryption of 0 with "good" randomness and which is independent of ψ_{fresh}. Moreover, one can efficiently determine when self destruct occurred. It turns out that such a component has the security guarantees needed in order to replace the hardware component in Sections 3.1.

Clearly, this component will be randomized, since ciphertexts are randomized. We note that this is the first (and only) time randomization is used by the compiled circuit. Note that the time it takes to compute a ciphertext is completely independent of the size of the underlying circuit C_s, and depends only on the security parameter k. Moreover, recall that we allow the adversary to tamper with at most $1/\operatorname{poly}(k)$ wires.

The basic idea is the following: repeat the following sub-computation M times: Compute a fresh ciphertext of 0, along with a non-interactive zero-knowledge proof that it is indeed an encryption of 0 with "good" randomness. We denote the output of the i'th sub-computation by (ψ_i, π_i), where $\psi_i \leftarrow \mathsf{Enc}(0)$ and π_i is the corresponding NIZK. The basic observation is that at least one of these sub-computations will not be tampered with at all (due to the limit on the tampering budget), and hence one of these (untampered) sub-computations can be thought of as a secure hardware component.

Next the idea would be to add all these ciphertext together, to compute the final ciphertext $\psi = \sum_{i=1}^{M} \psi_i$. Note that if we knew that this addition computation was not tampered with, then we would be done. But clearly we do not have such a guarantee. Instead we will add a proof that this sum was computed correctly. However, in order to add a proof we need to identify the underlying language (or what exactly are we proving). Note that it is insufficient to prove that there exist ciphertexts $\psi'_1, \ldots, \psi'_M$, and corresponding proofs $\pi'_1, \ldots, \pi'_M$, such that $\psi = \sum_{i=1}^{M} \psi'_i$. This is insufficient since we will need the guarantee that at least one of these ciphetexts ψ'_i was computed without any tampering, and thus can be thought of as a fresh encryption of 0. To enforce this, we need to prove that these ciphertexts $\psi'_1, \ldots, \psi'_M$ are exactly those computed previously.

To this end, we use a signature scheme, and prove that we know a bunch of signed ciphertexts and corresponding proofs $\{\psi'_i, \sigma'_i, \pi'_i\}_{i=1}^{M}$ such that all the signatures are valid, all the proofs are valid, and $\sum_{i=1}^{M} \psi'_i = \psi$, where ψ is the claimed sum. More specifically, we fix an underlying signature scheme, and store in the memory of this component a pair of keys (sksig, vksig) for this signature scheme. The M sub-computations now each compute a triplet $(\psi_i, \sigma_i, \pi_i)$, where $\psi_i \leftarrow \mathsf{Enc}(0)$, σ_i is a signature of ψ_i, and π_i is a NIZK proof that indeed ψ_i is an encryption of 0 with "good" randomness. As before the size of each computation of $(\psi_i, \sigma_i, \pi_i)$ depends only on the security parameter and hence we can assume that at least one of these computations is not tampered with.

Once all these triplets $(\psi_i, \sigma_i, \pi_i)$ were computed, we compute $\psi = \sum_{i=1}^{M} \psi_i$ together with a succinct proof-of-knowledge that we know M triplets $(\psi_i, \sigma_i, \pi_i)$ such that $\psi = \sum_{i=1}^{M} \psi_i$, each signature σ_i is a valid signature of ψ_i, and each proof π_i is a valid proof that ψ_i is an encryption of 0 with "good" randomness. We note that this part of the computation takes as input only the outputs of the previous M subcomputations, the verification key vksig, and the CRS. Intuitively, security seems to follow from the security of the signature scheme: Since the adversary is not given the secret key sksig during this computation, he cannot forge a signature on a new message, and hence must use the M ciphertexts output by the M sub-computations.

Unfortunately, this intuition is misleading, and there is a problem with this approach that complicates our construction. The problem is that some of the subcomputations that supposedly output a triplet $(\psi_i, \sigma_i, \pi_i)$ can be completely corrupted, and instead of outputting a signature σ_i may output the secret key sksig (or an arbitrary function of sksig). In such a case, during the proof that $\psi = \sum \psi_i'$, a tampering adversary, may choose the ciphertext ψ_i' arbitrarily (and in particular, depending on the *untampered* ciphertext) and forge a signature. We get around such an attack by using a very specific (one-time) *information-theoretically secure* signature scheme.

The signature scheme we use is an information-theoretical one-time (symmetric) version Lamport's signature scheme, where there is no verification key (only a secret key which is used both for verifying and computing signatures). Recall that the secret key in Lamport's scheme consists of $2k$ random strings: sksig $= (x_{1,0}, x_{1,1}, \ldots, x_{k,0}, x_{k,1})$. A valid encryption of a message $m = (m_1, \ldots, m_k)$ is the tuple $(x_{1,m_1}, \ldots, x_{k,m_k})$. The reason we use this specific signature scheme is that it has an important feature, described below.

In our M subcomputations we use M independent secret keys. Namely, we store M independently generated keys $(\text{sksig}^i)_{i=1}^{M}$ in memory, where each $\text{sksig}^i = (x_{1,0}^i, x_{1,1}^i, \ldots, x_{k,0}^i, x_{k,1}^i)$. During the i'th subcomputation, where supposedly the triplet $(\psi_i, \sigma_i, \pi_i)$ is computed, we use only sksig^i.

Our signature scheme has the following desired property: Consider a tampering adversary, who may completely tamper with the wires of subcomputation i, and thus can set σ_i to be an arbitrary function of the secret key sksig^i. Our signature scheme has the guarantee that this *arbitrary* string σ_i can (information-theoretically) be used to sign at most one message, and this message is determined by σ_i. Thus, we have the guarantee that the witness $\{(\psi_i', \sigma_i', \pi_i')\}_{i=1}^{M}$ extracted from the proof-of-knowledge has the property that if the signatures and proofs are valid and $\psi = \sum \psi_i'$, then (with overwhelming probability) the signed ciphertexts $\{\psi_i'\}$ were generated *independently* of the untampered ciphertext, and are all "good" encryptions of 0.

The proof system we use must be a *succinct* proof-of-knowledge. The reason is that we will run the verification circuit M times, and argue that most of the verification circuits cannot be tampered with. However, to argue this we use the fact that the size of each verification circuit is of size $\text{poly}(k)$, independent of the original circuit size. To ensure that each verification circuit is indeed of size $\text{poly}(k)$ (independent of M) we need to use succinctness, since the verification circuit depends on the proof length.

The actual succinct proof-of-knowledge we use is *universal arguments* [3], which is an interactive version of CS-proofs. Universal arguments consist of 4 messages, which we denote by $(\alpha, \beta, \gamma, \delta)$. The verifier's messages α and γ (which are random) are stored in the memory, and the prover's messages (β and δ) are computed during the computation of the circuit.

There are still some technical difficulties that remain. First, everything in memory must be stored in a tamper-resilient way, with the guarantee that if something in memory is corrupted then self-destruct occurs. To this end, we store M copies of the CRS and M copies of the public key of the encryption scheme. As done in previous components, we check that all the copies are the same, and if not the component self-destructs (i.e., the memory is overwritten with zeros). We also need to store the secret keys $\mathsf{sksig}^1, \ldots, \mathsf{sksig}^M$ in a robust manner, but note that since there are M such keys, simply storing M copies of each secret key is not good enough, since we allow $\mathrm{poly}(k)$ fraction of the memory gates to be tampered with, and in particular all of the repetitions of a single secret key sksig^i can be tampered with. Instead, we compute the hash value $h(\mathsf{sksig}) = h(\mathsf{sksig}^1, \ldots, \mathsf{sksig}^M)$, where h is a collision resistant hash function, and we store M copies of $h(\mathsf{sksig})$.

In the proof-of-knowledge, the statement is the tuple $(\psi, \mathsf{CRS}, \mathsf{PK}, h(\mathsf{sksig}))$, and we prove that we know a witness $\{\psi_i, \sigma_i, \pi_i, \mathsf{sksig}^i\}_{i \in [M]}$ such that $\psi = \sum_{i=1}^{M} \psi_i$, all the proofs π_i are accepted (with respect to CRS), all the signatures σ_i are valid (with respect to sksig^i), and $h(\mathsf{sksig}^1, \ldots, \mathsf{sksig}^M) = h(\mathsf{sksig})$. Unfortunately, using a symmetric (information theoretical) signature scheme, introduces a new problem: This computation now does use the secret key, and hence a new signature may be forged during this computation.

We solve this problem by adding another proof-of-knowledge before this proof-of-knowledge, which ties the hands of the adversary, and causes him to "commit" to these signatures (without knowing the secret keys). More specifically, after the initial M sub-computations, we compute $h(\sigma) \triangleq h(\sigma_1, \ldots, \sigma_M)$ and add a universal argument that we know $(\sigma_1, \ldots, \sigma_M)$ such that $h(\sigma_1, \ldots, \sigma_M) = h(\sigma)$. Note that this computation does *not* use the secret keys $(\mathsf{sksig}^1, \ldots, \mathsf{sksig}^M)$. We think of $h(\sigma)$ as a commitment to the signatures.

Then in the next proof-of-knowledge, the instance is $(\psi, \mathsf{CRS}, \mathsf{PK}, h(\mathsf{sksig}), h(\sigma))$, and we prove that we know a witness $(\psi_i, \sigma_i, \pi_i, \mathsf{sksig}^i)$ such that $\psi = \sum_{i=1}^{M} \psi_i$, all the proofs π_i are accepted (with respect to CRS), all the signatures σ_i are valid (with respect to sksig^i), $h(\mathsf{sksig}^1, \ldots, \mathsf{sksig}^M) = h(\mathsf{sksig})$, and $h(\sigma_1, \ldots, \sigma_M) = h(\sigma)$. We use the fact that h is collision resistant to argue that even if the adversary uses the secret key here to forge signatures of new messages, these new signatures cannot hash to $h(\sigma)$ assuming the adversary cannot find collisions in h.

We now present the details of the construction and security proof for Component 1.

4   Component 1:

4.1   Universal Arguments

In what follows we give the properties of the universal argument that will be useful for us. We note that the definition below slightly differs from its original form in [3]. First, we define universal arguments for any language in $\mathsf{NTIME}(T)$ (i.e., any language computable by a non-deterministic Turing machine running in time T), for any T :

$\mathbb{N} \to \mathbb{N}$, whereas Barak and Goldreich (following Micali [46]) define it for a universal non-deterministic language. Second, our proof-of-knowledge property slightly differs from the one presented in [3], but easily follows from their original formulation.

Definition 2. *Let $T : \mathbb{N} \to \mathbb{N}$, and let L be any language in $\mathsf{NTIME}(T)$. A universal argument for L is a 4-round argument system (P, V) with the following properties:*

1. **Efficiency.** *There exists a polynomial p,[3] such that for any instance $x \in \{0, 1\}^k$ the time complexity of $V(x)$ is $p(k)$, independent of T. In particular the communication complexity is at most $p(k)$ as well. Moreover, if $x \in L$ then for any valid witness w, the runtime[4] of $P(x, w)$ is at most $T(k) \cdot \mathrm{polylog}(\mathrm{T(k)})$.*
2. **Completeness.** *For every $x \in L$ and for any corresponding witness w,*

$$\Pr[(P(x, w), V(x)) = 1] = 1.$$

3. **Computational Soundness.** *For every polynomial size circuit family $\{P_k^*\}$ and for every $x \in \{0, 1\}^k \setminus L$,*

$$\Pr[(P_k^*(x), V(x)) = 1] = \mathrm{neg}(k).$$

4. **Proof-of-Knowledge Property.** *There exists a a polynomial q and a probabilistic algorithm E (an extractor) such that for every poly-size circuit family $\{P_k^*\}$ and for every $x \in \{0, 1\}^k$, if $\Pr[(P_k^*(x), V(x)) = 1] \geq \epsilon$ then*

$$\Pr[E^{P_n^*}(x) \text{ outputs a valid witness after running in time } q(1/\epsilon, T(k))] = 1 - \mathrm{neg}(n).$$

In particular, if P^* succeeds in proving that $x \in \{0, 1\}^k \cap L$ with non-negligible probability, then E can extract a corresponding witness in expected polynomial time in $T(k)$.

4.2   A Formal Description of Component 1

We first describe the cryptographic ingredients used by Component 1.

– A one-time symmetric signature scheme $\Pi_{\mathsf{Sign}} = (\mathsf{SigGen}, \mathsf{Sign}, \mathsf{Verify})$, defined as follows:

$\mathsf{SigGen}(1^k)$: SigGen outputs a random string sksig which consists of k pairs of random strings

$$(x_{1,0}, x_{1,1}), \ldots, (x_{k,0}, x_{k,1}),$$

where each $x_{\ell, b} \in_R \{0, 1\}^{2k}$ is of length $2k$.

---

[3] This polynomial is a universal polynomial that does not depend on the language L.

[4] We note that this is not the complexity guarantee given in the work of [3]. However, this complexity can be achieved by instantiating the universal argument using the recent efficient PCP construction of [5].

Sign(sksig, m), where $|m| = k$: Let $m = m[1], \dots, m[k]$ be the bit representation of m. Sign outputs $\sigma = (x_{1,m[1]}, m[1]), \dots, (x_{k,m[k]}, m[k])$.

Verify$_{\mathsf{Sign}}$(sksig, σ, m): Verify$_{\mathsf{Sign}}$ parses $\sigma = (y_1, b[1]), \dots, (y_k, b[k])$ and checks that for every $j \in [k]$ it holds that $b[j] = m[j]$ and $y_j = x_{j,m[j]}$. If yes, it outputs 1, and otherwise it outputs 0.

- A family of collision resistant hash functions $\mathcal{H} = \{h_{\mathsf{key}}\}$, where $h_{\mathsf{key}} : \{0,1\}^* \to \{0,1\}^k$.

- A non-interactive zero-knowledge (NIZK) proof system Π_{NIZK}.

- Universal arguments, which is an interactive variant of the CS proof system. Universal arguments consist of 4 messages, which we denote by $(\alpha, \beta, \gamma, \delta)$. The messages α and γ are sent by the verifier and are uniformly random strings.

We now describe Component 1. In what follows M is a parameter chosen as in Section 3.

Remark. For the sake of simplicity (and in an effort to focus on the new and interesting aspects of our component), in our formal description below, we do not formally define the notion of a ciphertext with "good" randomness. Intuitively, by "good" randomness we mean randomness r for which the error term in the ciphertext $\mathsf{Enc}_{\mathsf{PK}}(0; r)$ is not too big, so that one can perform homomorphic operations on it (that can later be decrypted using the secret key). We use the fact that a random string r is "good" with overwhelming probability.

In what follows, we use this notion of "good" randomness in a hand-wavy manner and assume that the sum of M ciphertext with "good" randomness is a ciphertext with "good" randomness (an assumption which of course does not hold inductively).

Memory: Encoding the Memory. Generate M secret keys sksig$^1, \dots,$ sksig^M $\leftarrow$ SigGen(1^k) for the signature scheme, and place in memory. Recall that for each i, the key sksig^i consists of k pairs of random values which we denote by $(x_{1,0}^i, x_{1,1}^i)$, $\dots, (x_{k,0}^i, x_{k,1}^i)$. Let sksig $=$ sksig$^1 || \cdots ||$ sksig^M.
In what follows, for any random variable x, we let $\tilde{x} = x^M$ denote M concatenated copies of x.
Compute the following encodings and place in memory:
  1. Place $\widetilde{\mathsf{PK}}$ in memory, where PK is the public-key of the underlying (homomorphic) encryption scheme.
  2. Choose a random function h_{key} from the collision resistant family $\mathcal{H}$, and place $\widetilde{\mathsf{key}}$ in memory.
  3. Compute h(sksig) $= h_{\mathsf{key}}$(sksig) and place $h(\widetilde{\mathsf{sksig}})$ in memory.
  4. Choose a common reference string CRS for the NIZK proof system Π_{NIZK} and place $\widetilde{\mathsf{CRS}}$ in memory.
  5. Choose random strings (α_1, γ_1) to be the random coins of the verifier in the first universal argument, and (α_2, γ_2) to be the random coins of the verifier in the second universal argument. Place $\tilde{\alpha}_1, \tilde{\gamma}_1, \tilde{\alpha}_2, \tilde{\gamma}_2$ in memory.

In what follows, when the circuit computation accesses one of the stored values $x \in \{\mathsf{PK}, \mathsf{key}, \mathsf{CRS}, \mathsf{h}(\mathsf{sksig}), \alpha_1, \gamma_1, \alpha_2, \gamma_2\}$, we always assume that it is accessing the first column of $\tilde{x}$.

Segment 1.

1. The first part of the computation takes randomness of length $M \cdot \mathrm{poly}(k)$ as input and performs M parallel subcomputations. We refer to each subcomputation as a block and denote the M blocks by $\mathsf{B}_1, \ldots, \mathsf{B}_M$. For $1 \leq i \leq M$, a random string $r_i = r_i^1 \| r_i^2 \in \{0, 1\}^{\mathrm{poly}(k)}$ is generated by hardware randomness gates. Each block B_i receives the corresponding $r_i^1 \| r_i^2$ as input and performs the following computation:
   - On input r_i^1, compute $\psi_i = \mathsf{Enc}_{\mathsf{PK}_{FHE}}(0; r_i^1)$. Each bit of the output $\psi_i[1], \ldots, \psi_i[k]$ is split into 4 wires which are used later on, as specified.
   - On input ψ_i, sksig^i, compute $\sigma_i = \mathsf{Sign}(\mathsf{sksig}^i, \psi_i)$. Each bit of the output $\sigma_i[1], \ldots, \sigma_i[k]$ is split into 4 wires which are used later on, as specified.
   - On input $r_i^1, r_i^2, \mathsf{CRS}$, compute a NIZK proof π_i, using proof system Π_{NIZK} with CRS and randomness r_i^2, that there exists "good" randomness r_i^1 such that $\mathsf{Enc}_{\mathsf{PK}_{FHE}}(0; r_i^1) = \psi_i$. Each bit of the output $\pi_i[1], \ldots, \pi[\mathrm{poly}(k)]$ is split into 2 wires which are used later on, as specified.

2. The next part of the computation takes as input $\psi_1, \ldots, \psi_M$ and outputs $\psi_{rand} = \sum_{i=1}^{M} \psi_i$. Each of the k output wires corresponding to the bits of $\psi_{rand} = \psi_{rand}[1], \ldots, \psi_{rand}[k]$ will be split into $M + 2$ wires, which are used later on, as specified.

3. This part of the computation takes as input $\sigma_1, \ldots, \sigma_M$ and key, and computes $\mathsf{h}(\sigma) = h_{\mathsf{key}}(\sigma_1, \ldots, \sigma_M)$. Each of the k output wires corresponding to the bits of $\mathsf{h}(\sigma) = \mathsf{h}(\sigma)[1], \ldots, \mathsf{h}(\sigma)[k]$ is split into $2M + 4$ wires which are used later on, as specified.

4. This part of the circuit computes a universal argument that proves knowledge of signatures $\sigma_1, \ldots, \sigma_M$ that hash to $\mathsf{h}(\sigma)$. More specifically, this part of the computation takes as input a witness $\sigma_1, \ldots, \sigma_M$ and the tuple $(\mathsf{key}, \mathsf{h}(\sigma), \alpha_1, \gamma_1)$, and does the following:
   - Take α_1 to be the verifier's first message. Compute the second message β_1 of the universal argument for the following language:

   $$\mathcal{L}_1 = \{(\mathsf{h}(\sigma), \mathsf{key}) \mid \exists \sigma_1', \ldots, \sigma_M' : h_{\mathsf{key}}(\sigma_1', \ldots, \sigma_M') = \mathsf{h}(\sigma)\}.$$

   Each bit of the output $\beta_1 = \beta_1[1], \ldots, \beta_1[k]$ is split into M wires which are used later on, as specified. This part of the computation also outputs a state STATE_1 which is passed to the next part of the computation, below.
   - The next part of the computation takes as input STATE_1 and γ_1, where γ_1 is the third message of the verifier. Compute the fourth message δ_1 for the language $\mathcal{L}_1$ and statement $(\mathsf{h}(\sigma), \mathsf{key})$. Each bit of the output $\delta_1 = \delta_1[1], \ldots, \delta_1[\mathrm{poly}(k)]$ is split into M wires which are used later on, as specified.

5. This part of the circuit computes a universal argument that ψ_{rand} was computed "correctly". More specifically, this part of the computation takes as input a witness

$$((\psi_1, \sigma_1, \pi_1, \mathsf{sksig}_1), \ldots, (\psi_M, \sigma_M, \pi_M, \mathsf{sksig}_M))$$

and the tuple $(\psi_{rand}, \mathsf{key}, h(\sigma), h(\mathsf{sksig}), \mathsf{CRS}, \alpha_2, \gamma_2)$ and does the following:

- Take α_2 to be the verifier's first message and compute the second message β_2 of the universal argument for the following language:

$$\mathcal{L}_2 = \{\psi_{rand}, h(\sigma), \mathsf{key}, \mathsf{CRS} \mid \exists (\psi_1', \sigma_1', \mathsf{sksig}_1', \pi_1'), \ldots, (\psi_M', \sigma_M', \mathsf{sksig}_M', \pi_M') :$$

$$\sum_{i=1}^{M} \psi_i' = \psi_{rand};$$

$$\wedge \ h_{\mathsf{key}}(\sigma_1', \ldots, \sigma_M') = h(\sigma);$$

$$\wedge \ \text{for } 1 \leq i \leq M, \mathsf{Verify}_{\mathsf{Sign}}(\mathsf{sksig}_i', \psi_i', \sigma_i') = 1$$

$$\wedge \ \text{for } 1 \leq i \leq M, \mathsf{Verify}_{\Pi}(\mathsf{CRS}, \psi_i', \pi_i') = 1$$

$$\wedge \ h_{\mathsf{key}}(\mathsf{sksig}_1', \ldots, \mathsf{sksig}_M') = h(\mathsf{sksig})\}$$

Each bit of the output $\beta_2 = \beta_2[1], \ldots, \beta_2[k]$ is split into M wires which are used later on, as specified. This part of the computation also outputs a state STATE_2 which is passed to the next part of the computation, below.

- The next part of the computation takes γ_2 to be the third message of the verifier, and uses STATE_2 to compute the fourth message δ_2 for the language $\mathcal{L}_2$ and statement $(\psi_{rand}, h(\sigma), \mathsf{key}, \mathsf{CRS})$. Each bit of the output $\delta_2 = \delta_2[1], \ldots, \delta_2[\mathrm{poly}(k)]$ is split into M wires which are used later on, as specified.

Segment 2: Universal Argument Verification. This part consists of two subcomputations:

Verification of the Computation of $h(\sigma)$. This part consists of M copies of the verifier circuit for the universal argument for language $\mathcal{L}_1$ which takes as input the statement $(h(\sigma), \mathsf{key})$, first message α_1, second message β_1 third message γ_1, and fourth message δ_1. We denote the i-th verifier circuit for $1 \leq i \leq M$ by Verify_i^1 and its output by λ_i^1.

Verification of the Computation of ψ_{rand}. This part consists of M copies of the verifier circuit for the universal argument for language $\mathcal{L}_2$ which takes as input the statement $(\psi_{rand}, h(\sigma), \mathsf{key}, \mathsf{CRS})$, first message α_2, second message β_2 third message γ_2, and fourth message δ_2. We denote the i-th verifier circuit for $1 \leq i \leq M$ by Verify_i^2 and its output by λ_i^2.

All these $2M$ output wires are inputs to the AND gate G_{cas}. This gate has $7k \cdot M$ additional input wires that belong to Segment 3 below. The gate G_{cas} has $K'' + k \cdot M$ output wires, where K'' is the size of the entire memory of of the circuit (of all components). We denote the values on these wires by $\{\mu_i\}_{i \in [K'' + k \cdot M]}$. The first K'' output wires (with values $\{\mu_i\}_{i \in [K'']}$) belong to Segment 3, and the other $k \cdot M$ output wires belong to Segment 4.

Segment 3: Error Cascade. This part is split into two subcomputations. The first subcomputation checks that all of the encodings $\tilde{x}$ that were placed in memory are uncorrupted. The second part propagates errors and overwrites memory.

- A circuit $C_{\mathsf{code}_{i,j}}^{\widetilde{\mathsf{key}}}$ of constant size σ_{code} for $1 \leq i \leq M, 1 \leq j \leq k$:

Input: $\widetilde{\mathsf{key}}$.

Output:

$$\omega_{i,j}^1 = \neg(\widetilde{\mathsf{key}}(1,j) \oplus \widetilde{\mathsf{key}}(i,j))$$

Similar subcircuits are constructed for the remaining encodings $\widetilde{\mathsf{PK}}$, $\widetilde{h(\mathsf{sksig})}$, $\widetilde{\mathsf{CRS}}$, $\tilde{\alpha}_1$, $\tilde{\gamma}_1$, $\tilde{\alpha}_2$, $\tilde{\gamma}_2$ with corresponding output wires $[\omega_{i,j}^2]_{i\in[M],j\in[k]}$, $[\omega_{i,j}^3]_{i\in[M],j\in[k]}$, $[\omega_{i,j}^4]_{i\in[M],j\in[k]}$, $[\omega_{i,j}^5]_{i\in[M],j\in[k]}$, $[\omega_{i,j}^6]_{i\in[M],j\in[k]}$, $[\omega_{i,j}^7]_{i\in[M],j\in[k]}$, $[\omega_{i,j}^8]_{i\in[M],j\in[k]}$.
All these output wires are inputs to G_{cas}. Thus, in total, G_{cas} has $8kM + 2M$ input wires (M from Segment 1, M from Segment 2 and $8k \cdot M$ from Segment 3), and outputs:

$$\left(\bigwedge_{i\in[M]} \lambda_i^1\right) \wedge \left(\bigwedge_{i\in[M]} \lambda_i^2\right) \wedge \left(\bigwedge_{i\in[M],j\in[k],\ell\in[7]} \psi_{i,j}^\ell\right)$$

- The first K'' output wires of G_{cas} are fed to all the memory gates. If the output of G_{cas} is 0, then the memory gates are set to 0. Otherwise, the memory gates remain unchanged.

Segment 4: The Output of Component 1. This segment has k AND gates $G_{\mathsf{out},1}, \ldots,$ $G_{\mathsf{out},k}$, each with fan-in $M + 1$. This segment contains all the $k \cdot M + k$ input wires to $G_{\mathsf{out},1}, \ldots, G_{\mathsf{out},k}$: The first M input wires to each gate $G_{\mathsf{out},j}$ come from the output wires of G_{cas} (with values $\{\mu_i\}_{i=K''+(j-1)\cdot M+1}^{K''+j\cdot M}$), and the other input wire of $G_{\mathsf{out},j}$ is the j-th output wire of the Circuit Computation in Segment 1, which computes the encryption ψ_{rand}. Each AND gate $G_{\mathsf{out},j}$ has fan-out M, where the M output wires of $G_{\mathsf{out},j}$ are set to:

$$\psi_{rand}^*[j] = \psi_{rand}[j] \wedge \left(\bigwedge_{K'+(j-1)\cdot M+1\leq i\leq K'+j\cdot M} \mu_i\right).$$

The final output of Component 1 is denoted by ψ_{rand}^*.

Remark. We note that the size of Component 1 is of order $M \cdot \mathrm{poly}(k) \cdot \mathrm{polylog}(M \cdot \mathrm{poly}(k))$, which can be written as $M \cdot \mathrm{poly}(k)$ (since M is poly-sized and so $\mathrm{polylog}(M)$ is smaller than k), due to the fact that we use the recent efficient PCP construction of Ben-Sasson et al. [5] to construct our universal arguments. We note that this implies that a $1/\mathrm{poly}(k)$-tampering adversary cannot tamper with each of the M subcomputations at the beginning of Component 1. An important assumption throughout the analysis will be that at least one of the M subcomputations is untampered.

Notation. In the following theorem, for any λ-tampering adversary $\mathcal{A}$ (as defined in Definition 1), we denote by t the maximum number of times $\mathcal{A}$ runs the tampered circuit. For each run $i \in [t]$, we denote by $\psi_{rand,i}$ the ciphertext in the statement of the second universal argument. For each run $i \in [t]$, we denote by $\psi_{fresh,i}$ the ciphertext outputted by the untampered subcomputation, and denote by $\psi_{rest,i} = \psi_{rand,i} - \psi_{fresh,i}$.

We denote by

$$(i^*, (\psi_{fresh,1}, \psi_{rest,1}), \dots, (\psi_{fresh,i^*-1}, \psi_{rest,i^*-1}), \psi_{fresh,i^*}) \leftarrow \mathsf{REAL}_{\mathcal{A}},$$

where i^* is the first round where self destruct occurs in the executions with the tampering instructions of $\mathcal{A}$. If self destruct does not occur (i.e. $i^* = t + 1$) then set $\psi_{fresh,i^*} = \bot$.

Theorem 3. *Assume the soundness of the underlying universal argument, the security (existential security against adaptive chosen message attacks) of the underlying signature scheme Π_{Sign}, the semantic security of the underlying encryption scheme $\mathcal{E}_{FH}$, the security of the underlying collision resistant hash family $\mathcal{H}$, and the soundness and security of the underlying NIZK proof system Π_{NIZK}. Let $\lambda = 1/\mathrm{poly}(k)$.*

Then for any $\mathcal{PPT}$ λ-tampering adversary $\mathcal{A}$ there exists a simulator $S = (S_1, S_2)$ running in expected polynomial time, such that

$$(i', \psi'_{rest,1}, \dots, \psi'_{rest,i'-1}, \psi'_{fresh}, \mathrm{STATE}) \leftarrow S_1(1^M, \mathrm{PK}),$$

and for $(\psi_{fresh,1}, \dots, \psi_{fresh,t}) \leftarrow \mathsf{Enc}_{\mathrm{PK}}(0)$ fresh encryptions of 0,

$$j' \leftarrow S_2(1^M, \mathrm{PK}, \mathrm{STATE}, \psi_{fresh,1}, \dots, \psi_{fresh,t}),$$

such that

1. *$\psi'_{rest,1}, \dots, \psi'_{rest,i'-1}$ are (simulatable) encryptions of 0 with "good" randomness.*
2. *$j' \leq i'$.*
3. *$\mathsf{REAL}_{\mathcal{A}} \equiv (j', (\psi_{fresh,1}, \psi'_{rest,1}), \dots, (\psi_{fresh,i''-1}, \psi'_{rest,j'-1}), \tilde{\psi}_{fresh})$,*
   where $\tilde{\psi}_{fresh} = \psi_{fresh,j'}$ for $j' < i'$, and $\tilde{\psi}_{fresh} = \psi'_{fresh}$ for $j' = i' \leq t$, and $\tilde{\psi}_{fresh} = \bot$ for $j' = i' = t + 1$.

We defer the proof of Theorem 3 to the full version.

5   The Final Construction

Let $T^{(2)}(C_s)$ be our original compiled circuit $T^{(1)}(C_s)$, described in Sections 3.1, where Component 1 of $T^{(1)}(C_s)$, which was implemented by tamper-resilient hardware, is replaced with the Component 1 described in Sections 3.2 and 4.

We are now ready to state our main theorem.

Theorem 4. *Assume the soundness of the underlying universal argument, the security (existential security against adaptive chosen message attacks) of the underlying signature scheme Π_{Sign}, the semantic security of the underlying encryption scheme $\mathcal{E}_{FH}$, the security of the underlying collision resistant hash family $\mathcal{H}$, and the soundness and security of the underlying NIZK proof system Π_{NIZK}. Let $\lambda = 1/\mathrm{poly}(k)$.*

Then $T^{(2)}(C_s)$ is secure against ppt λ-tampering adversaries (as defined in Definition 1). Note that the adversary may tamper with all components, including Component 1.

Overview of Proof of Theorem 4. First, we consider a Component 1 which provides weaker guarantees than the idealized hardware component described in Section 3.1. We call this hardware component WeakComp1. Next, we show that with small modifications, we can reprove Theorem 2 when the idealized hardware component is replaced with the hardware component WeakComp1. Finally, we use Theorem 3 to show that the construction of Component 1 given in Sections 3.2 and 4 is an implementation of WeakComp1, which is secure against λ-tampering adversaries.

The WeakComp1 Hardware Component. We assume the existence of a hardware component WeakComp1, which computes ciphertexts ψ_{fresh}, ψ_{rest} and outputs M copies of $\psi_{\text{rand}} = \psi_{\text{rest}} \oplus \psi_{\text{fresh}}$, where ψ_{rest} is an arbitrary "good" encryption of 0 and ψ_{fresh} is a randomly generated encryption of 0 independent of ψ_{rest}.

Plugging in WeakComp1. We state the following lemma, which uses the component WeakComp1 defined above in order to obtain a fully tamper-resilient circuit. We defer the proof to the full version.

Lemma 1. *Replace Component 1 of $T^{(1)}(C_s)$ with* WeakComp1 *described above, yielding $\tilde{T}^{(1)}(C_s)$. Then $\tilde{T}^{(1)}(C_s)$ is secure against ppt $\lambda = 1/\operatorname{poly}(k)$-tampering adversaries (as defined in Definition 1), that do not tamper with* WeakComp1, *assuming semantic security of the underlying encryption scheme $\mathcal{E}_{FH}$.*

Putting It All Together. We now argue that our construction remains secure when we replace WeakComp1 in $\tilde{T}^{(1)}(C_s)$ with Component 1 described in Section 4 to yield $T^{(2)}(C_s)$.

Fix any ppt λ-tampering adversary $\mathcal{A}$. Now, consider the adversaries $\mathcal{A}^1$, which is the adversary $\mathcal{A}$, restricted to tampering with and running only the first component (note that we simulate the final output of the circuit—assuming self-destruct does not occur— for $\mathcal{A}^1$ in order to obtain the correct tampering function in each run). By Theorem 3, we have that there exists a simulator $S = (S_1, S_2)$ for $\mathcal{A}^1$ running in expected polynomial time, such that on input $(1^M, \text{PK})$, S_1 outputs $(i', \psi'_{rest,1}, \ldots, \psi'_{rest,i'-1}, \psi'_{fresh}, \text{STATE})$ and on input $(1^M, \text{PK}, \text{STATE}, \psi_{fresh,1}, \ldots, \psi_{fresh,t})$, where $\psi_{fresh,1}, \ldots, \psi_{fresh,t}$ are fresh encryptions of 0, S_2 outputs j', where j' is the index of the first run of Component 1, where some wire to G^1_{cas} is set to 0. Note that by combining the inputs and outputs of S_1, S_2 we obtain $\psi_{\text{rand},1} = \psi_{\text{fresh},1} + \psi_{\text{rest},1}, \ldots, \psi_{\text{rand},j'-1} = \psi_{\text{fresh},j'-1} + \psi_{\text{rest},j'-1}$. Let this sequence of ciphertexts define the input-output behavior of the hardware component WeakComp1.

Now, note that by the security properties of Component 1 (See Theorem 3 in Section 4), we are guaranteed that the following two distributions are statistically close: $\text{REAL}_{\mathcal{A}} \equiv (j', (\psi_{fresh,1}, \psi'_{rest,1}), \ldots, (\psi_{fresh,j'-1}, \psi'_{rest,j'-1}), \tilde{\psi}_{fresh})$ (where, loosely speaking, a draw from $\text{REAL}_{\mathcal{A}}$ corresponds to a setting of the above random variables in a real execution).

To simulate the view of $\mathcal{A}$, we distinguish between two cases: Simulating runs of the circuit when $i < j'$ and simulating runs of the circuit when $i \geq j'$. Consider the adversaries $\mathcal{A}^{2,3,4}$, which is the adversary $\mathcal{A}$, restricted to tampering with only the second, third, and fourth component and interacts with $\tilde{T}^{(1)}(C_s)$. As noted above, for runs $i < j'$, the input-output behavior of $T^{(2)}(C_s)$ in the presence of $\mathcal{A}$ is identical to the input-output behavior of $\tilde{T}^{(1)}(C_s)$ in the presence of $\mathcal{A}^{2,3,4}$, where WeakComp1 is defined as above. Therefore, by the security of $\tilde{T}^{(1)}(C_s)$ (see Lemma 1) we have that there exists a simulator Sim for runs $i < j'$ that simulates the view of $\mathcal{A}$.

Finally, for runs $i \geq j'$, we have that "self-destruct" already occurred and so we can perfectly simulate the view of $\mathcal{A}$ as follows: Return 0 unless $\mathcal{A}$ tampers with the output wire, in which case the circuit returns b if the tamper is "set to b", and returns 1 if the tamper is "toggle". This concludes the proof of Theorem 4.

6 Extension to Tamper-Resilient Two Party Computation

In this section, we consider a two-party computation setting, where in addition to corrupting parties, an adversary may tamper with the circuits of the honest parties and the messages sent by the honest parties. As usual, we restrict the adversary to tampering with a $\lambda = 1/\operatorname{poly}(k)$-fraction of wires, memory gates, and message bits overall.

Our security definition follows the standard ideal/real paradigm, which requires that the view of the (real world) adversary, who may tamper with λ-fraction of wires, memory gates and message bits, can be simulated by a simulator in the ideal world *without tampering*. We emphasize that the ideal world we consider is the "standard" ideal world, whereas in the real world we allow the adversary tampering power.

We note that we allow both parties a tamper-free input-dependent preprocessing phase, which does not require interaction and can be done individually, offline by each party. This phase allows the parties to prepare their tamper-resilient circuits and place their private inputs in memory, while no tampering occurs.

Our approach is quite simple. We begin with any two-party computation (2-PC) protocol secure against malicious corruptions, where the communication complexity depends only on security parameter, k, and not on the size of the circuit computing the functionality. Such a 2-PC protocol can be constructed from any fully homomorphic encryption scheme and succinct argument system (such as universal arguments [46,3]).

For each party P_b, $b \in \{0, 1\}$ and each round i of the protocol, we consider the circuit $\operatorname{Next}^i_{x_b, r_b}$, which has the (secret) values x_b and r_b hardwired into it (corresponding to the input and the random coins of Party P_b). It takes as input the current transcript TRANS and it outputs the next message for party P_b. We run (a slight modification of) our tampering compiler $T^{(2)}(\operatorname{Next}^i_{x_b, r_b})$ on each such circuit to obtain a circuit which outputs the $\operatorname{poly}(k)$-bit next message for party P_b at round i. By the security guarantees of $T^{(2)}$ (see Theorem 4), the compiled circuit $T^{(2)}(\operatorname{Next}^i_{x_b, r_b})$ is resilient to $1/\operatorname{poly}(k)$-fraction of tampering. Since the total number of such circuits is $\operatorname{poly}(k)$, we are ultimately resilient to a $\lambda = 1/\operatorname{poly}(k)$-fraction of tampering.

This idea does not quite work, since the adversary may tamper with the messages sent between the two parties, which may render the resulting protocol insecure. To get around this, we add signatures to our protocol. Namely, we assume each player is associated with a verification key. This key can be transmitted via an error-correcting code in the beginning of the protocol, and we require that the length of this key be a large enough $\operatorname{poly}(k)$ so that an adversary cannot cause this message to decode to a different key (using his tampering budget). Each time a player sends a message, he will sign his message together with the entire transcript so far. Intuitively, each party must sign the entire transcript to protect against a tampering adversary who gets signatures $\sigma_1, \ldots, \sigma_z$ on z protocol messages $m_1, \ldots, m_z$, and then forwards a transcript to an honest party which is a permutation of the z messages $m_1, \ldots, m_z$.

6.1   Overview of The Model: Tamper-Resilient 2-PC

We consider the setting where two parties P_0, with input x_0, and P_1, with input x_1 interact to compute a functionality $f : \{0,1\}^* \times \{0,1\}^* \to \{0,1\}^* \times \{0,1\}^*$, where $f = (f_0, f_1)$. P_0 wishes to obtain $f_0(x_0, x_1)$ and P_1 wishes to obtain $f_1(x_0, x_1)$. In what follows, for the sake of simplicity of notation we assume that $f_0 = f_1 = f$, though our results extend trivially to the case where f_0 and f_1 differ.

Our security definition follows the ideal/real paradigm. We emphasize that our ideal model is identical to the standard ideal model, while our real model is stronger that the standard ideal model since we consider adversaries $\mathcal{A}$ who may corrupt one or more parties P_0, P_1 and may also behave as a λ-tampering adversary on the honest parties' circuits (which the honest party may prepare via input-dependent pre-processing).

The random variable $\mathsf{IDEAL}_{f,\mathsf{Sim}}(x_0, x_1)$ is defined as the output of both parties in the ideal execution computing functionality f (where Sim controls the malicious party and chooses its output). If both parties are honest, then $\mathsf{IDEAL}_{f,\mathsf{Sim}}(x_0, x_1)$ is defined as the output of both parties in the above ideal execution along with the output of Sim.

The random variable $\mathsf{REAL}_{\Pi_{\mathsf{TAMP}},\mathcal{A}}(x_0, x_1)$ is defined as the output of both parties after running Π_{TAMP} with inputs (x_0, x_1), where the honest party outputs the output of the protocol, and the malicious party controlled by $\mathcal{A}$ may output an arbitrary function of its view. If both parties are honest then $\mathsf{REAL}_{\Pi_{\mathsf{TAMP}},\mathcal{A}}(x_0, x_1)$ is defined as the output of both honest parties, together with the output of $\mathcal{A}$, which may be an arbitrary function of its view (i.e., of the transcript).

Definition 3. *(secure tamper-resilient two-party computation): Let f and Π_{TAMP} be as above. Protocol Π_{TAMP} is said to securely compute* f *(in the malicious model and in the presence of a λ-tampering adversary) if for every probabilistic polynomial-time real-world adversary $\mathcal{A}$, who may corrupt one of the parties (or both), and may also behave as a λ-tampering adversary on the honest parties' circuits, there exists an expected polynomial-time simulator Sim in the ideal-world, such that*

$$\{\mathsf{IDEAL}_{f,\mathsf{Sim}}(x_0, x_1)\} \overset{c}{\approx} \{\mathsf{REAL}_{\Pi_{\mathsf{TAMP}},\mathcal{A}}(x_0, x_1)\}.$$

6.2   Achieving Tamper-Resilient 2-PC

Fix any two-party functionality f. We assume the existence of a secure (against active corruptions) two-party protocol $\Pi_{\mathsf{MPC}}(f)$ for computing f, where the total communication complexity is $\ell(k)$, where k is security parameter, and $\ell(\cdot)$ is a fixed polynomial, independent of the size of the circuit which computes the functionality f. It is well-known that such a two-party protocol can be constructed from fully homomorphic encryption and CS-proofs.

Protocol Π_{TAMP} for computing functionality f

Public and Private keys: Party P_b generates a pair of verification and signing keys $(\text{vksig}_b, \text{sksig}_b) \leftarrow \text{SigGen}(1^k)$. It publishes the verification key vksig_b, while keeping private the corresponding signing key sksig_b.

Inputs: Party P_b has an input x_b and a random tape r_b.

Preprocessing for Party P_b: Construct $r + 1$ tamper-resilient subcircuits:

$$T^{(2)}(\text{Next}_{\text{STATE}}^{b,1}), \ldots, T^{(2)}(\text{Next}_{\text{STATE}}^{b,r}), T^{(2)}(\text{Out}_{\text{STATE}}^{b}),$$

where $\text{STATE} = (x_b, r_b, \text{vksig}_0, \text{vksig}_1, \text{sksig}_b)$. We emphasize that the compiler is run independently $r + 1$ times, each time with fresh randomness.
For each $i \in [r]$, the circuit $\text{Next}_{\text{STATE}}^{0,i}$, is defined as follows ($\text{Next}_{\text{STATE}}^{1,i}$ is defined analogously):

It takes as input a partial transcript TRANS_{i-1}, and does the following:

1. Parse TRANS_{i-1} as $2(i - 1)$ signed messages. Consider the last two messages denoted by $(m_{0,i-1}, \sigma_{0,i-1})$ and $(m_{1,i-1}, \sigma_{1,i-1})$. Let $\text{MSG}_{i-2} = (m_{b,j})_{b \in \{0,1\}, j \in [i-2]}$ be the first $i - 2$ pairs of messages sent in TRANS_{i-1}. Check that $\sigma_{0,i-1}$ is a valid signature of $(\text{MSG}_{i-2}, m_{0,i-1})$ w.r.t. verification key vksig_0, and that $\sigma_{1,i-1}$ is a valid signature of $\text{MSG}_{i-1} = (\text{MSG}_{i-2}, m_{0,i-1}, m_{1,i-1})$ w.r.t. verification key vksig_1. If either fails, send $\perp$ to P_1 and abort.
2. Otherwise, run the next message function of protocol Π with partial transcript MSG_{i-1}, to obtain the next message $m_{0,i}$.
3. Compute $\sigma_{0,i}$, a signature of $(\text{MSG}_{i-1}, m_{0,i})$ w.r.t. secret key sksig_0.
4. Output the partial (signed) transcript $(\text{TRANS}_{i-1}, m_{0,i}, \sigma_{0,i})$.

The circuit $\text{Out}_{\text{STATE}}^{0}$ takes as input a transcript TRANS_r, and does the following ($\text{Out}_{\text{STATE}}^{1}$ is defined analogously):

1. Parse TRANS_r as $2r$ signed messages. Check the validity and consistency of the last two signatures If either fails output $\perp$.
2. Let MSG_r denote the $2r$ messages in TRANS_r.
3. Compute the output y_0 of protocol Π, assuming the messages exchanged in Π were MSG_r.
4. Output y_0.

Protocol Execution:
At round $i \in [r]$, party P_0, upon receiving a message TRANS_{i-1}, runs $T^{(2)}(\text{Next}_{\text{STATE}}^{0,i})$ on input TRANS_{i-1}, and sends the output (which is of the form $(\text{TRANS}_{i-1}, m_{0,i}, \sigma_{0,i})$) to P_1.
Analogously, party P_1, upon receiving a message $(\text{TRANS}_{i-1}, m_{0,i}, \sigma_{0,i})$ from P_0, runs $T^{(2)}(\text{Next}_{\text{STATE}}^{1,i})$ on input $(\text{TRANS}_{i-1}, m_{0,i}, \sigma_{0,i})$, and sends the output to P_0.

Output:
1. Upon receiving the last message of the protocol, each party P_b runs $T^{(2)}(\text{Out}_{\text{STATE}}^{b})$ on this last message to compute the output y_b.
2. Output y_b.

Fig. 1. Tamper-resilient, secure two-party computation protocol of functionality f

To simplify our exposition, we construct our protocol in the public key model. Here, each party P_0, P_1 publishes a verification key $\mathsf{vksig}_0, \mathsf{vksig}_1$ for a digital signature scheme $\Pi_{\mathsf{Sign}} = (\mathsf{SigGen}, \mathsf{Sign}, \mathsf{Verify})$, while storing the corresponding secret key $\mathsf{sksig}_0, \mathsf{sksig}_1$. We note that such a protocol in the public key model can easily be converted to a protocol in the standard model. We defer the details to the full version.

Let r denote the number of rounds in the two-party protocol Π_{MPC} described above. For $i \in [r]$ let $\mathsf{Next}^{0,i}_{\mathsf{STATE}}$ denote the circuit that has the secret state $\mathsf{STATE} = (x_0, r_0, \mathsf{vksig}_0, \mathsf{vksig}_1, \mathsf{sksig}_0)$ hardwired into it, where x_0 and r_0 are the input and randomness of party P_0, $(\mathsf{vksig}_0, \mathsf{sksig}_0)$ are the verification and signing keys of P_0, and vksig_1 and the verification key of P_1. The circuit $\mathsf{Next}^{0,i}_{\mathsf{STATE}}$ computes the next message function of party P_0 in the i'th round of the resulting tamper-resilient protocol.

The tamper-resilient protocol emulates Π_{MPC}. Each message of the tamper-resilient protocol consists of all the messages sent so far in Π_{MPC}, along with signatures. More formally, $\mathsf{Next}^{0,i}_{\mathsf{STATE}}$ takes as input a message TRANS_{i-1}, which consists of all the $i-1$ pairs of messages sent in Π_{MPC} during the first $i-1$ rounds, where each message is accompanied by a signature of the entire transcript thus far. The circuit $\mathsf{Next}^{0,i}_{\mathsf{STATE}}$, on input TRANS_{i-1}, does the following (the circuits $\mathsf{Next}^{1,i}_{\mathsf{STATE}}$ are defined analogously):

1. Parse TRANS_{i-1}, as $2(i-1)$ message-signature pairs $(m_{b,j}, \sigma_{b,j})_{b \in \{0,1\}, j \in [i-1]}$. Check that $\sigma_{1,i-1}$ is a valid signature of the message $\mathsf{MSG}_{i-1} = (m_{0,j}, m_{1,j})_{j \in [i-1]}$ w.r.t. vksig_1, and check that $\sigma_{0,i-1}$ is a valid signature for $(\mathsf{MSG}_{i-2}, m_{0,i-1})$, where $\mathsf{MSG}_{i-2} = (m_{0,j}, m_{1,j})_{j \in [i-2]}$. If either does not verify, output 0.
2. Otherwise, compute the next message $m_{0,i}$ of party P_0 in protocol Π_{MPC} given transcript MSG_{i-1}, randomness r_0 and input x_0.
3. Compute a signature $\sigma_{0,i}$ corresponding to the message $(\mathsf{MSG}_{i-1}, m_{0,i})$.
4. Output $(\mathsf{TRANS}_{i-1}, m_{0,i}, \sigma_{0,i})$

The tamper-resilient protocol Π_{TAMP} is depicted in Figure 1.

Theorem 5. *For every two-party functionality f, Π_{TAMP} securely computes f in the malicious model and in the presence of a $1/\mathrm{poly}(k)$-tampering adversary, where k is the security parameter, and poly is a fixed polynomial independent of the size of the circuit computing the functionality f.*

We defer the proof of Theorem 5 to the full version.

References

1. Akavia, A., Goldwasser, S., Vaikuntanathan, V.: Simultaneous hardcore bits and cryptography against memory attacks. In: Reingold, O. (ed.) TCC 2009. LNCS, vol. 5444, pp. 474–495. Springer, Heidelberg (2009)
2. Applebaum, B., Harnik, D., Ishai, Y.: Semantic security under related-key attacks and applications. Cryptology ePrint Archive, Report 2010/544 (2010), http://eprint.iacr.org/
3. Barak, B., Goldreich, O.: Universal arguments and their applications. SIAM J. Comput. 38(5), 1661–1694 (2008)
4. Bellare, M., Kohno, T.: A theoretical treatment of related-key attacks: Rka-prps, rka-prfs, and applications. In: Biham, E. (ed.) EUROCRYPT 2003. LNCS, vol. 2656, pp. 491–506. Springer, Heidelberg (2003)

5. Ben-Sasson, E., Chiesa, A., Genkin, D., Tromer, E.: On the concrete efficiency of probabilistically-checkable proofs. In: STOC, pp. 585–594 (2013)
6. Ben-Sasson, E., Goldreich, O., Harsha, P., Sudan, M., Vadhan, S.P.: Robust pcps of proximity, shorter pcps, and applications to coding. SIAM J. Comput. 36(4), 889–974 (2006)
7. Biham, E., Shamir, A.: Differential fault analysis of secret key cryptosystems. In: Kaliski Jr., B.S. (ed.) CRYPTO 1997. LNCS, vol. 1294, pp. 513–525. Springer, Heidelberg (1997)
8. Boneh, D., DeMillo, R.A., Lipton, R.J.: On the importance of checking cryptographic protocols for faults. In: Fumy, W. (ed.) EUROCRYPT 1997. LNCS, vol. 1233, pp. 37–51. Springer, Heidelberg (1997)
9. Brumley, D., Boneh, D.: Remote timing attacks are practical. Computer Networks 48(5), 701–716 (2005)
10. Canetti, R., Dodis, Y., Halevi, S., Kushilevitz, E., Sahai, A.: Exposure-resilient functions and all-or-nothing transforms. In: Preneel, B. (ed.) EUROCRYPT 2000. LNCS, vol. 1807, pp. 453–469. Springer, Heidelberg (2000)
11. Choi, S.G., Kiayias, A., Malkin, T.: BiTR: Built-in tamper resilience. In: Lee, D.H., Wang, X. (eds.) ASIACRYPT 2011. LNCS, vol. 7073, pp. 740–758. Springer, Heidelberg (2011)
12. Dachman-Soled, D., Kalai, Y.T.: Securing circuits against constant-rate tampering. In: Safavi-Naini, R., Canetti, R. (eds.) CRYPTO 2012. LNCS, vol. 7417, pp. 533–551. Springer, Heidelberg (2012)
13. Dobrushin, R.L., Ortyukov, S.I.: Upper bound for the redundancy of self- correcting arrangements of unreliable functional elements 13, 203–218 (1977)
14. Dodis, Y., Goldwasser, S., Tauman Kalai, Y., Peikert, C., Vaikuntanathan, V.: Public-key encryption schemes with auxiliary inputs. In: Micciancio, D. (ed.) TCC 2010. LNCS, vol. 5978, pp. 361–381. Springer, Heidelberg (2010)
15. Dodis, Y., Kalai, Y.T., Lovett, S.: On cryptography with auxiliary input. In: STOC, pp. 621–630 (2009)
16. Dodis, Y., Pietrzak, K.: Leakage-resilient pseudorandom functions and side-channel attacks on feistel networks. In: Rabin, T. (ed.) CRYPTO 2010. LNCS, vol. 6223, pp. 21–40. Springer, Heidelberg (2010)
17. Dziembowski, S., Kazana, T., Obremski, M.: Non-malleable codes from two-source extractors. In: Canetti, R., Garay, J.A. (eds.) CRYPTO 2013, Part II. LNCS, vol. 8043, pp. 239–257. Springer, Heidelberg (2013)
18. Dziembowski, S., Pietrzak, K.: Leakage-resilient cryptography. In: FOCS, pp. 293–302 (2008)
19. Dziembowski, S., Pietrzak, K., Wichs, D.: Non-malleable codes. In: ICS, pp. 434–452 (2010)
20. Evans, W., Schulman, L.: Signal propagation and noisy circuits. IEEE Trans. Inform. Theory 45(7), 2367–2373 (1999)
21. Evans, W., Schulman, L.: On the maximum tolerable noise of k-input gates for reliable computation by formulas. IEEE Trans. Inform. Theory 49(11), 3094–3098 (2003)
22. Faust, S., Kiltz, E., Pietrzak, K., Rothblum, G.N.: Leakage-resilient signatures. In: Micciancio, D. (ed.) TCC 2010. LNCS, vol. 5978, pp. 343–360. Springer, Heidelberg (2010)
23. Faust, S., Pietrzak, K., Venturi, D.: Tamper-proof circuits: How to trade leakage for tamper-resilience. In: Aceto, L., Henzinger, M., Sgall, J. (eds.) ICALP 2011, Part I. LNCS, vol. 6755, pp. 391–402. Springer, Heidelberg (2011)
24. Faust, S., Rabin, T., Reyzin, L., Tromer, E., Vaikuntanathan, V.: Protecting circuits from leakage: the computationally-bounded and noisy cases. In: Gilbert, H. (ed.) EUROCRYPT 2010. LNCS, vol. 6110, pp. 135–156. Springer, Heidelberg (2010)
25. Feder, T.: Reliable computation by networks in the presence of noise. IEEE Trans. Inform. Theory 35(3), 569–571 (1989)
26. Gács, P., Gál, A.: Lower bounds for the complexity of reliable boolean circuits with noisy gates. IEEE Transactions on Information Theory 40(2), 579–583 (1994)

27. Gál, A., Szegedy, M.: Fault tolerant circuits and probabilistically checkable proofs. In: Structure in Complexity Theory Conference, pp. 65–73 (1995)
28. Gandolfi, K., Mourtel, C., Olivier, F.: Electromagnetic analysis: Concrete results. In: Koç, Ç.K., Naccache, D., Paar, C. (eds.) CHES 2001. LNCS, vol. 2162, pp. 251–261. Springer, Heidelberg (2001)
29. Gennaro, R., Lysyanskaya, A., Malkin, T., Micali, S., Rabin, T.: Algorithmic tamper-proof (atp) security: Theoretical foundations for security against hardware tampering. In: Naor, M. (ed.) TCC 2004. LNCS, vol. 2951, pp. 258–277. Springer, Heidelberg (2004)
30. Goldwasser, S., Rothblum, G.N.: Securing computation against continuous leakage. In: Rabin, T. (ed.) CRYPTO 2010. LNCS, vol. 6223, pp. 59–79. Springer, Heidelberg (2010)
31. Govindavajhala, S., Appel, A.W.: Using memory errors to attack a virtual machine. In: IEEE Symposium on Security and Privacy, pp. 154–165 (2003)
32. Hajek, B.E., Weller, T.: On the maximum tolerable noise for reliable computation by formulas. IEEE Transactions on Information Theory 37(2), 388 (1991)
33. Ishai, Y., Prabhakaran, M., Sahai, A., Wagner, D.: Private circuits ii: Keeping secrets in tamperable circuits. In: Vaudenay, S. (ed.) EUROCRYPT 2006. LNCS, vol. 4004, pp. 308–327. Springer, Heidelberg (2006)
34. Ishai, Y., Sahai, A., Wagner, D.: Private circuits: Securing hardware against probing attacks. In: Boneh, D. (ed.) CRYPTO 2003. LNCS, vol. 2729, pp. 463–481. Springer, Heidelberg (2003)
35. Juma, A., Vahlis, Y.: Protecting cryptographic keys against continual leakage. In: Rabin, T. (ed.) CRYPTO 2010. LNCS, vol. 6223, pp. 41–58. Springer, Heidelberg (2010)
36. Kalai, Y.T., Kanukurthi, B., Sahai, A.: Cryptography with tamperable and leaky memory. In: Rogaway, P. (ed.) CRYPTO 2011. LNCS, vol. 6841, pp. 373–390. Springer, Heidelberg (2011)
37. Kalai, Y.T., Rao, A., Lewko, A.: Formulas resilient to short-circuit errors (2011) (manuscript)
38. Katz, J., Vaikuntanathan, V.: Signature schemes with bounded leakage resilience. In: Matsui, M. (ed.) ASIACRYPT 2009. LNCS, vol. 5912, pp. 703–720. Springer, Heidelberg (2009)
39. Kelsey, J., Schneier, B., Wagner, D., Hall, C.: Side channel cryptanalysis of product ciphers. In: Quisquater, J.-J., Deswarte, Y., Meadows, C., Gollmann, D. (eds.) ESORICS 1998. LNCS, vol. 1485, pp. 97–110. Springer, Heidelberg (1998)
40. Kleitman, D.J., Leighton, F.T., Ma, Y.: On the design of reliable boolean circuits that contain partially unreliable gates. In: FOCS, pp. 332–346 (1994)
41. Kocher, P.C.: Timing attacks on implementations of diffie-hellman, RSA, DSS, and other systems. In: Koblitz, N. (ed.) CRYPTO 1996. LNCS, vol. 1109, pp. 104–113. Springer, Heidelberg (1996)
42. Kocher, P.C., Jaffe, J., Jun, B.: Differential power analysis. In: Wiener, M. (ed.) CRYPTO 1999. LNCS, vol. 1666, pp. 388–397. Springer, Heidelberg (1999)
43. Kuhn, M.G., Anderson, R.J.: Soft tempest: Hidden data transmission using electromagnetic emanations. In: Aucsmith, D. (ed.) Information Hiding 1998. LNCS, vol. 1525, pp. 124–142. Springer, Heidelberg (1998)
44. Liu, F.-H., Lysyanskaya, A.: Algorithmic tamper-proof security under probing attacks. In: Garay, J.A., De Prisco, R. (eds.) SCN 2010. SCN, vol. 6280, pp. 106–120. Springer, Heidelberg (2010)
45. Liu, F.-H., Lysyanskaya, A.: Tamper and leakage resilience in the split-state model. In: Safavi-Naini, R., Canetti, R. (eds.) CRYPTO 2012. LNCS, vol. 7417, pp. 517–532. Springer, Heidelberg (2012)
46. Micali, S.: Cs proofs (extended abstracts). In: FOCS, pp. 436–453. IEEE Computer Society (1994)
47. Micali, S., Reyzin, L.: Physically observable cryptography (extended abstract). In: Naor, M. (ed.) TCC 2004. LNCS, vol. 2951, pp. 278–296. Springer, Heidelberg (2004)

48. Naor, M., Segev, G.: Public-key cryptosystems resilient to key leakage. In: Halevi, S. (ed.) CRYPTO 2009. LNCS, vol. 5677, pp. 18–35. Springer, Heidelberg (2009)
49. Pellegrini, A., Bertacco, V., Austin, T.M.: Fault-based attack of rsa authentication. In: DATE, pp. 855–860 (2010)
50. Pietrzak, K.: A leakage-resilient mode of operation. In: Joux, A. (ed.) EUROCRYPT 2009. LNCS, vol. 5479, pp. 462–482. Springer, Heidelberg (2009)
51. Pippenger, N.: Reliable computation by formulas in the presence of noise. IEEE Trans. Inform. Theory 34(2), 194–197 (1988)
52. Pippenger, N.: On networks of noisy gates. In: FOCS, pp. 30–38. IEEE (1985)
53. Quisquater, J.-J., Samyde, D.: Electromagnetic analysis (ema): Measures and counter-measures for smart cards. In: Attali, S., Jensen, T. (eds.) E-smart 2001. LNCS, vol. 2140, pp. 200–210. Springer, Heidelberg (2001)
54. Rao, J.R., Rohatgi, P.: Empowering side-channel attacks. Cryptology ePrint Archive, Report 2001/037 (2001), http://eprint.iacr.org/
55. Shamir, A.: Personal communication (2012)
56. von Neumann, J.: Probabilistic logics and the synthesis of reliable organisms from unreliable components (1956)

How to Fake Auxiliary Input

Dimitar Jetchev and Krzysztof Pietrzak[*]

EPFL Switzerland and IST Austria

Abstract. Consider a joint distribution (X, A) on a set $\mathcal{X} \times \{0,1\}^\ell$. We show that for any family $\mathcal{F}$ of distinguishers $f \colon \mathcal{X} \times \{0,1\}^\ell \to \{0,1\}$, there exists a simulator $h \colon \mathcal{X} \to \{0,1\}^\ell$ such that
1. no function in $\mathcal{F}$ can distinguish (X, A) from $(X, h(X))$ with advantage ε,
2. h is only $O(2^{3\ell}\varepsilon^{-2})$ times less efficient than the functions in $\mathcal{F}$.

For the most interesting settings of the parameters (in particular, the cryptographic case where X has superlogarithmic min-entropy, $\varepsilon > 0$ is negligible and $\mathcal{F}$ consists of circuits of polynomial size), we can make the simulator h *deterministic*.

As an illustrative application of our theorem, we give a new security proof for the leakage-resilient stream-cipher from Eurocrypt'09. Our proof is simpler and quantitatively much better than the original proof using the dense model theorem, giving meaningful security guarantees if instantiated with a standard blockcipher like AES.

Subsequent to this work, Chung, Lui and Pass gave an interactive variant of our main theorem, and used it to investigate weak notions of Zero-Knowledge. Vadhan and Zheng give a more constructive version of our theorem using their new uniform min-max theorem.

1 Introduction

Let $\mathcal{X}$ be a set and let $\ell > 0$ be an integer. We show that for any joint distribution (X, A) over $\mathcal{X} \times \{0,1\}^\ell$ (where we think of A as a short ℓ-bit auxiliary input to X), any family $\mathcal{F}$ of functions $\mathcal{X} \times \{0,1\}^\ell \to \{0,1\}$ (thought of as distinguishers) and any $\varepsilon > 0$, there exists an efficient simulator $h \colon \mathcal{X} \to \{0,1\}^\ell$ for the auxiliary input that fools every distinguisher in $\mathcal{F}$, i.e.,

$$\forall f \in \mathcal{F} \colon |\mathbb{E}[f(X, A)] - \mathbb{E}[f(X, h(X))]| < \varepsilon.$$

Here, "efficient" means that the simulator h is $\tilde{\mathcal{O}}(2^{3\ell}\varepsilon^{-2})$ times more complex than the functions from $\mathcal{F}$ (we will formally define "more complex" in Definition 6). Without loss of generality, we can model the joint distribution (X, A) as $(X, g(X))$, where g is some arbitrarily complex and possibly probabilistic function (where $\mathbb{P}[g(x) = a] = \mathbb{P}[A = a | X = x]$ for all $(x, a) \in \mathcal{X} \times \{0,1\}^\ell$). Let us stress that, as g can be arbitrarily complex, one cannot hope to get an efficient simulator h where $(X, g(X))$ and $(X, h(X))$ are statistically close. Yet, one can still fool all functions in $\mathcal{F}$ in the sense that no function from $\mathcal{F}$ can distinguish the distribution (X, A) from $(X, g(X))$.

[*] Supported by the European Research Council under the European Union's Seventh Framework Programme (FP7/2007-2013) / ERC Starting Grant (259668-PSPC).

Y. Lindell (Ed.): TCC 2014, LNCS 8349, pp. 566–590, 2014.

Relation to [25]. Trevisan, Tulsiani and Vadhan [25, Thm. 3.1] prove a conceptually similar result, stating that if $\mathcal{Z}$ is a set then for any distribution Z over $\mathcal{Z}$, any family $\widetilde{\mathcal{F}}$ of functions $\mathcal{Z} \to [0,1]$ and any function $\widetilde{g} \colon \mathcal{Z} \to [0,1]$, there exists a simulator $\widetilde{h} \colon \mathcal{Z} \to [0,1]$ whose complexity is only $\mathcal{O}(\varepsilon^{-2})$ times larger than the complexity of the functions from $\widetilde{\mathcal{F}}$ such that

$$\forall \widetilde{f} \in \widetilde{\mathcal{F}} \;:\; |\,\mathbb{E}[\widetilde{f}(Z)\widetilde{g}(Z)] - \mathbb{E}[\widetilde{f}(Z)\widetilde{h}(Z)]\,| < \varepsilon. \tag{1}$$

In [25], this result is used to prove that every high-entropy distribution is indistinguishable from an efficiently samplable distribution of the same entropy. Moreover, it is shown that many fundamental results including the Dense Model Theorem [23,14,21,10,24], Impagliazzo's hardcore lemma [18] and a version of Szémeredi's Regularity Lemma [11] follow from this theorem. The main difference between (1) and our statement

$$\forall f \in \mathcal{F} \colon |\,\mathbb{E}[f(X,g(X))] - \mathbb{E}[f(X,h(X))]\,| < \varepsilon \tag{2}$$

is that our distinguisher f sees not only X, but also the real or fake auxiliary input $g(X)$ or $h(X)$, whereas in (1), the distinguisher $\widetilde{f}$ only sees X. In particular, the notion of indistinguishability we achieve captures indistinguishability in the standard cryptographic sense. On the other hand, (1) is more general in the sense that the range of $\widetilde{f}, \widetilde{g}, \widetilde{h}$ can be any real number in $[0,1]$, whereas our f has range $\{0,1\}$ and g, h have range $\{0,1\}^{\ell}$.

Nonetheless, it is easy to derive (1) from (2): consider the case of $\ell = 1$ bit of auxiliary input, and only allow families $\mathcal{F}$ of distinguishers where each $f \in \mathcal{F}$ is of the form $f(X,b) = \widehat{f}(X)b$ for some function $\widehat{f} \colon \mathcal{X} \to [0,1]$. For this restricted class, the absolute value in (2) becomes

$$|\,\mathbb{E}[f(X,g(X))] - \mathbb{E}[f(X,h(X))]\,| = |\,\mathbb{E}[\widehat{f}(X)g(X)] - \mathbb{E}[\widehat{f}(X)h(X)]\,| \tag{3}$$

As $\widehat{f}$ is arbitrary, this restricted class almost captures the distinguishers considered in (1). The only difference is that the function $\widetilde{g}$ has range $[0,1]$ whereas our g has range $\{0,1\}$. Yet, note that in (1), we can replace $\widetilde{g}$ having range $[0,1]$ by a (probabilistic) g with range $\{0,1\}$ defined as $\mathbb{P}[g(x) = 1] = \widetilde{g}(x)$, thus, leaving the expectation $\mathbb{E}[\widehat{f}(X)\widetilde{g}(X)] = \mathbb{E}[\widehat{f}(X)g(X)]$ unchanged.[1]

In [25], two different proofs for (1) are given. The first proof uses duality of linear programming in the form of the min-max theorem for two-player zero-sum games. This proof yields a simulator of complexity $\mathcal{O}(\varepsilon^{-4}\log^2(1/\varepsilon))$ times the complexity of the functions in $\mathcal{F}$. The second elegant proof uses boosting and gives a quantitatively much better $\mathcal{O}(\varepsilon^{-2})$ complexity.

---

[1] The simulator $\widetilde{h}$ from [25] satisfies the additional property $|\,\mathbb{E}[\widetilde{h}(X)] - \mathbb{E}[\widetilde{g}(X)]\,| = 0$. If this property is needed, we can get it by requiring that the function $f(X,b) = b$ is in $\mathcal{F}$. Then (2) for this f implies $|\,\mathbb{E}[g(X)] - \mathbb{E}[h(X)]\,| < \varepsilon$. One can make this term exactly zero by slightly biasing h towards 0 if $\mathbb{E}[h(X)] > \mathbb{E}[g(X)]$ or 1 otherwise, slightly increasing the advantage from ε to at most 2ε.

Proof outline. As it was just explained, (1) follows from (2). We do not know if one can directly prove an implication in the other direction, so we prove (2) from scratch. Similarly to [25], the core of our proof uses boosting with the same energy function as the one used in [25].

As a first step, we transform the statement (2) into a "product form" like (1) where $\mathcal{Z} = \mathcal{X} \times \{0,1\}^\ell$ (this results in a loss of a factor of 2^ℓ in the advantage ε; in addition, our distinguishers $\hat{f}$ will have range $[-1,1]$ instead of $[0,1]$). We then prove that (1) holds for some simulator $\tilde{h} \colon \mathcal{Z} \to [0,1]$ of complexity ε^{-2} relative to $\mathcal{F}$. Unfortunately, we cannot use the result from [25] in a black-box way at this point as we need the simulator $\tilde{h} \colon \mathcal{Z} \to [0,1]$ to define a probability distribution in the sense that $\tilde{h}(x,b) \geq 0$ for all (x,b) and $\sum_{b \in \{0,1\}^\ell} \tilde{h}(x,b) = 1$ for all x. Ensuring these conditions is the most delicate part of the proof. Finally, we show that the simulator h defined via $\mathbb{P}[h(x) = b] = \tilde{h}(x,b)$ satisfies (2). Note that for h to be well defined, we need $\tilde{h}$ to specify a probability distribution as outlined above.

Efficiency of h. Our simulator h is efficient in the sense that it is only $\mathcal{O}(2^{3\ell}\varepsilon^{-2})$ times more complex than the functions in $\mathcal{F}$. We do not know how tight this bounds is, but one can prove a lower bound of $\max\{2^\ell, \varepsilon^{-1}\}$ under plausible assumptions. The dependency on 2^ℓ is necessary under exponential hardness assumptions for one-way functions.[2] A dependency on ε^{-1} is also necessary. Indeed, Trevisan et al. [25, Rem. 1.6] show that such a dependency is necessary for the simulator $\tilde{h}$ in (1). Since (1) is implied by (2) with h and $\tilde{h}$ having exactly the same complexity, the ε^{-1} lower bound also applies to our h.

1.1   Subsequent Work

The original motivation for this work was to give simpler and quantitatively better proofs for leakage-resilient cryptosystems as we will discuss in Section 4. Our main theorem has subsequently been derived via two different routes.

First, Chung, Lui and Pass [4] investigate weak notions of zero-knowledge. On route, they derive an "interactive" version of our main theorem. In Section 4, we will show how to establish one of their results (with better quantitative bounds), showing that every interactive proof system satisfies a weak notion of zero-knowledge.

Second, Vadhan and Zheng [26, Thm.3.1-3.2] recently proved a version of von Neumann's min-max theorem for two-player zero sum games that does not only guarantee existence of an optimal strategy for the second player, but also constructs a nearly optimal strategy assuming knowledge of several best responses of the second player to strategies of the first player, and provide many applications of this theorem. Their argument is based on relative entropy KL projections

---

[2] More precisely, assume there exists a one-way function where inverting becomes 2^ℓ times easier given ℓ bits of leakage. It is e.g. believed that the AES block-cipher gives such a function as $(K, X) \to (\mathsf{AES}(K,X), X)$.

and a learning technique known as weight updates and resembles the the proof of the Uniform Hardcore Lemma by Barak, Hardt and Kale [2] (see also [16] for the original application of this method). They derive our main theorem [26, Thm.6.8], but with incomparable bounds. Concretely, to fool circuits of size t, their simulator runs in time $\tilde{O}(t \cdot 2^\ell/\varepsilon^2 + 2^\ell/\varepsilon^4)$ compared to ours whose run-time is $\tilde{O}(t \cdot 2^{3\ell}/\varepsilon^2)$. In particular, their bounds are better whenever $1/\varepsilon^2 \leq t \cdot 2^{2\ell}$. The additive $2^\ell/\varepsilon^4$ term in their running time appears due to the sophisticated iterative "weight update" procedure, whereas our simulator simply consists of a weighted sum of the evaluation of $\tilde{O}(2^{3\ell}/\varepsilon^2)$ circuits from the family we want to fool (here, circuits of size t).

1.2 More Applications

Apart from reproving one of [4]'s results on weak zero-knowledge mentioned above, we give two more applications of our main theorem in Section 4:

Chain Rules for Computational Entropy. Gentry and Wichs [13] show that black-box reductions cannot be used to prove the security of any succinct non-interactive argument from any falsifiable cryptographic assumption. A key technical lemma used in their proof ([13, Lem. 3.1]) states that if two distributions X and $\tilde{X}$ over $\mathcal{X}$ are computationally indistinguishable, then for any joint distribution (X, A) over $\mathcal{X} \times \{0,1\}^\ell$ (here, A is a short ℓ-bit auxiliary input) there exists a joint distribution $(\tilde{X}, \tilde{A})$ such that (X, A) and $(\tilde{X}, \tilde{A})$ are computationally indistinguishable. Our theorem immediately implies the stronger statement that not only such an $(\tilde{X}, \tilde{A})$ exists, but in fact, it is efficiently samplable, i.e., there exists an efficient simulator $h \colon \mathcal{X} \rightarrow \{0,1\}^\ell$ such that $(\tilde{X}, h(\tilde{X}))$ is indistinguishable from $(\tilde{X}, \tilde{A})$ and thus from (X, A). Reyzin [22, Thm.2] observed that the result of Gentry and Wichs implies a chain rule for conditional "relaxed" HILL entropy. We give a short and simple proof of this chain rule in Proposition 2 of this paper. We then show in Corollary 1 how to deduce a chain rule for (regular) HILL entropy from Proposition 2 using the simple fact (Lemma 1) that short (i.e., logarithmic in the size of the distinguishers) computationally indistinguishable random variables must already be statistically close. Chain rules for HILL entropy have found several applications in cryptography [10,21,7,12]. The chain rule that we get in Corollary 1 is the first one that does not suffer from a significant loss in the distinguishing advantage (we only lose a constant factor of 4). Unlike the case of relaxed HILL-entropy, here we only prove a chain rule for the "non-conditional" case, which is a necessary restriction given a recent counterexample to the (conditional) HILL chain rule by Krenn et al. [19]. We will provide more details on this negative result after the statement of Corallary 1.

Leakage Resilient Cryptography. The original motivation for this work is to simplify the security proofs of leakage-resilient [10,20,7] and other cryptosystems [12] whose security proofs rely on chain rules for computational entropy (as discussed in the previous paragraph). The main idea is to replace the chain rules with simulation-based arguments. In a nutshell, instead of arguing that a variable X must have high (pseudo)entropy in the presence of a short leakage A, one could simply use the fact that the leakage can be efficiently simulated. This not only implies that X has high (pseudo)entropy given the fake leakage $h(X)$, but if X is pseudorandom, it also implies that $(X, h(X))$ is indistinguishable from $(U, h(U))$ for a uniform random variable U on the same set as X. In the security proofs, we would now replace $(X, h(X))$ with $(U, h(U))$ and will continue with a uniformly random intermediate variable U. In contrast, the approach based on chain rules only tells us that we can replace X with some random variable Y that has high min-entropy given A. This is not only much complex to work with, but it often gives weaker quantitative bounds. In particular, in Section 4.3 we revisit the security proof of the leakage-resilient stream-cipher from [20] for which we can now give a conceptually simpler and quantitatively better security proof.

2   Notation and Basic Definitions

2.1   Notation

We use calligraphic letters such as $\mathcal{X}$ to denote sets, the corresponding capital letters X to denote random variables on these sets (equivalently, probability distributions) and lower-case letters (e.g., x) for values of the corresponding random variables. Moreover, $x \leftarrow X$ means that x is sampled according to the distribution X and $x \leftarrow \mathcal{X}$ means that x is sampled uniformly at random from $\mathcal{X}$. Let U_n denote the random variable with uniform distribution on $\{0,1\}^n$. We denote by $\Delta(X;Y) = \dfrac{1}{2} \sum_{x \in \mathcal{X}} |\mathbb{P}[X = x] - \mathbb{P}[Y = x]|$ the statistical distance between X and Y. For $\varepsilon > 0, s \in \mathbb{N}$, we use $X \sim Y$ to denote that X and Y have the same distribution, $X \sim_\varepsilon Y$ to denote that their statistical distance is less than ε and $X \sim_{\varepsilon,s} Y$ to denote that no circuit of size s can distinguish X from Y with advantage greater than ε. Note that $X \sim_{\varepsilon,\infty} Y \iff X \sim_\varepsilon Y$ and $X \sim_0 Y \iff X \sim Y$.

Finally, if $h \colon \mathcal{X} \to \{0,1\}^\ell$ is a probabilistic (randomized) function then we will use $[h]$ to denote the random coins used by h (a notation that will be used in various probabilities and expectations).

2.2   Entropy Measures

A random variable X has min-entropy k, if no (computationally unbounded) adversary can predict the outcome of X with probability greater than 2^{-k}.

Definition 1. (Min-Entropy H_∞) *A random variable X has min-entropy k, denoted $\mathsf{H}_\infty(X) \geq k$, if $\max_x \mathbb{P}[X = x] \leq 2^{-k}$.*

Dodis et al. [8] gave a notion of average-case min-entropy defined such that X has average-case min-entropy k conditioned on Z if the probability of the best adversary in predicting X given Z is 2^{-k}.

Definition 2. (Average min-Entropy [8] $\tilde{\mathsf{H}}_\infty$) *Consider a joint distribution (X, Z), then the* average min-entropy *of X conditioned on Z is*

$$\tilde{\mathsf{H}}_\infty(X|Z) = -\log(\mathop{\mathbb{E}}_{z \leftarrow Z}\left[\max_x \mathbb{P}[X = x | Z = z]\right]) = -\log(\mathop{\mathbb{E}}_{z \leftarrow Z}\left[2^{-\mathsf{H}_\infty(X|Z=z)}\right])$$

HILL-entropy is the computational analogue of min-entropy. A random variable X has HILL-entropy k if there exists a random variable Y having min-entropy k that is indistinguishable from X. HILL-entropy is further quantified by two parameters ε, s specifying this indistinguishability quantitatively.

Definition 3. (HILL-Entropy [15] $\mathsf{H}^{\mathsf{HILL}}$) *$X$ has HILL entropy k, denoted by $\mathsf{H}^{\mathsf{HILL}}_{\varepsilon,s}(X) \geq k$, if*

$$\mathsf{H}^{\mathsf{HILL}}_{\varepsilon,s}(X) \geq k \quad \Longleftrightarrow \quad \exists Y \; : \; \mathsf{H}_\infty(Y) \geq k \quad and \quad X \sim_{\varepsilon,s} Y$$

Conditional HILL-entropy has been defined by Hsiao, Lu and Reyzin [17] as follows.

Definition 4. (Conditional HILL-Entropy [17]) *X has conditional HILL entropy k (conditioned on Z), denoted $\mathsf{H}^{\mathsf{HILL}}_{\varepsilon,s}(X|Z) \geq k$, if*

$$\mathsf{H}^{\mathsf{HILL}}_{\varepsilon,s}(X|Z) \geq k \quad \Longleftrightarrow \quad \exists(Y, Z) \; : \; \tilde{\mathsf{H}}_\infty(Y|Z) \geq k \quad and \quad (X, Z) \sim_{\varepsilon,s} (Y, Z)$$

Note that in the definition above, the marginal distribution on the conditional part Z is the same in both the real distribution (X, Z) and the indistinguishable distribution (Y, Z). A "relaxed" notion of conditional HILL used implicitly in [13] and made explicit in [22] drops this requirement.

Definition 5. (Relaxed Conditional HILL-Entropy [13,22]) *X has relaxed conditional HILL entropy k, denoted $\mathsf{H}^{\mathsf{rlx\text{-}HILL}}_{\varepsilon,s}(X|Z) \geq k$, if*

$$\mathsf{H}^{\mathsf{rlx\text{-}HILL}}_{\varepsilon,s}(X|Z) \geq k \quad \Longleftrightarrow \quad \exists(Y, W) \; : \; \tilde{\mathsf{H}}_\infty(Y|W) \geq k \quad and \quad (X, Z) \sim_{\varepsilon,s} (Y, W)$$

3 The Main Theorem

Definition 6. (Complexity of a function) *Let $\mathcal{A}$ and $\mathcal{B}$ be sets and let $\mathcal{G}$ be a family of functions $h \colon \mathcal{A} \to \mathcal{B}$. A function h has* **complexity** *C relative to $\mathcal{G}$ if it can be computed by an oracle-aided circuit of size $\mathrm{poly}(C \log |\mathcal{A}|)$ with C oracle gates where each oracle gate is instantiated with a function from $\mathcal{G}$.*

Theorem 1. (Main) *Let $\ell \in \mathbb{N}$ be fixed, let $\varepsilon > 0$ and let $\mathcal{X}$ be any set. Consider a distribution X over $\mathcal{X}$ and any (possibly probabilistic and not necessarily efficient) function $g \colon \mathcal{X} \to \{0,1\}^\ell$. Let $\mathcal{F}$ be a family of deterministic (cf. remark below) distinguishers $f \colon \mathcal{X} \times \{0,1\}^\ell \to \{0,1\}$. There exists a (probabilistic) simulator $h \colon \mathcal{X} \to \{0,1\}^\ell$ with complexity[3]*

$$O(2^{3\ell} \varepsilon^{-2} \log^2(\varepsilon^{-1}))$$

---

[3] If we model h as a Turing machine (and not a circuit) and consider the *expected* complexity of h, then we can get a slightly better $O(2^{3\ell} \varepsilon^{-2})$ bound (i.e. without the $\log^2(\varepsilon^{-1})$ term).

relative to $\mathcal{F}$ which ε-fools every distinguisher in $\mathcal{F}$, i.e.

$$\forall f \in \mathcal{F} \ : \ \left| \mathop{\mathbb{E}}_{x \leftarrow X, [g]} [f(x, g(x))] - \mathop{\mathbb{E}}_{x \leftarrow X, [h]} [f(x, h(x))] \right| < \varepsilon, \tag{4}$$

Moreover, if

$$H_\infty(X) > 2 + \log\log|\mathcal{F}| + 2\log(1/\varepsilon) \tag{5}$$

then there exists a deterministic h with this property.

Remark 1 (Closed and Probabilistic $\mathcal{F}$). In the proof of Theorem 1 we assume that the class $\mathcal{F}$ of distinguishers is closed under complement, i.e., if $f \in \mathcal{F}$ then also $1 - f \in \mathcal{F}$. This is without loss of generality, as even if we are interested in the advantage of a class $\mathcal{F}$ that is not closed, we can simply apply the theorem for $\mathcal{F}' = \mathcal{F} \cup (1 - \mathcal{F})$, where $(1 - \mathcal{F}) = \{1 - f \ : \ f \in \mathcal{F}\}$. Note that if h has complexity t relative to $\mathcal{F}'$, it has the same complexity relative to $\mathcal{F}$. We also assume that all functions $f \in \mathcal{F}$ are deterministic. If we are interested in a class $\mathcal{F}$ of randomized functions, we can simply apply the theorem for the larger class of deterministic functions $\mathcal{F}''$ consisting of all pairs (f, r) where $f \in \mathcal{F}$ and r is a choice of randomness for f. This is almost without loss of generality, except that the requirement in eq.(5) on the min-entropy of X becomes slightly stronger as $\log\log|\mathcal{F}''| = \log\log(|\mathcal{F}|2^\rho)$ where ρ is an upper bound on the number of random coins used by any $f \in \mathcal{F}$.

4   Applications

4.1   Zero-Knowledge

Chung, Lui and Pass [4] consider the following relaxed notion of zero-knowledge

Definition 7 (distributional (T, t, ε)-zero-knowledge). *Let $(\mathcal{P}, \mathcal{V})$ be an interactive proof system for a language L. We say that $(\mathcal{P}, \mathcal{V})$ is distributional (T, t, ε)-zero-knowledge (where T, t, ε are all functions of n) if for every $n \in \mathbb{N}$, every joint distributions (X_n, Y_n, Z_n) over $(L \cap \{0, 1\}^n) \times \{0, 1\}^* \times \{0, 1\}^*$, and every t-size adversary $\mathcal{V}^*$, there exists a T-size simulator S such that*

$$(X_n, Z_n, \mathsf{out}_{\mathcal{V}^*}[\mathcal{P}(X_n, Y_n) \leftrightarrow \mathcal{V}^*(X_n, Z_n)]) \sim_{\varepsilon, t} (X_n, Z_n, S(X_n, Z_n))$$

where $\mathsf{out}_{\mathcal{V}^*}[\mathcal{P}(X_n, Y_n) \leftrightarrow \mathcal{V}^*(X_n, Z_n)]$ denotes the output of $\mathcal{V}^*(X_n, Z_n)$ after interacting with $\mathcal{P}(X_n, Y_n)$.

If L in an NP language, then in the definition above, Y would be a witness for $X \in L$. As a corollary of their main theorem, [4] show that every proof system satisfies this relaxed notion of zero-knowledge where the running time T of the simulator is polynomial in t, ε and 2^ℓ. We can derive their Corollary from Theorem 1 with better quantitative bounds for most ranges of parameters than [4]: we get $\tilde{O}(t2^{3\ell}\varepsilon^{-2})$ vs. $\tilde{O}(t^3 2^\ell \varepsilon^{-6})$, which is better whenever $t/\varepsilon^2 \geq 2^\ell$.

Proposition 1. *Let $(\mathcal{P}, \mathcal{V})$ be an interactive proof system for a language L, and suppose that the total length of the messages sent by $\mathcal{P}$ is $\ell = \ell(n)$ (on common inputs X of length n). Then for any $t = t(n) \geq \Omega(n)$ and $\varepsilon = \varepsilon(n)$, $(\mathcal{P}, \mathcal{V})$ is distributional (T, t, ε)-zero-knowledge, where*

$$T = O(t 2^{3\ell} \varepsilon^{-2} \log^2(\varepsilon^{-1}))$$

Proof. Let $M \in \{0, 1\}^\ell$ denote the messages send by $\mathcal{P}(X_n, Y_n)$ when talking to $\mathcal{V}^*(X_n, Z_n)$. By Theorem 1 (identifying $\mathcal{F}$ from the theorem with circuits of size t) there exists a simulator h of size $O(t \cdot 2^{3\ell} \varepsilon^{-2} \log^2(\varepsilon^{-1}))$ s.t.

$$(X_n, Z_n, M) \sim_{\varepsilon, 2t} (X_n, Z_n, h(X_n, Z_n)) \tag{6}$$

Let $S(X_n, Z_n)$ be defined as follows, first compute $M' = h(X_n, Z_n)$ (with h as above), and then compute $\mathsf{out}_\mathcal{V}^*[M' \leftrightarrow \mathcal{V}^*(X_n, Z_n)]$. We claim that

$$(X_n, Z_n, \mathsf{out}_{\mathcal{V}^*}[\mathcal{P}(X_n, Y_n) \leftrightarrow \mathcal{V}^*(X_n, Z_n)]) \sim_{\varepsilon, t} (X_n, Z_n, S(X_n, Z_n)) \tag{7}$$

To see this, note that from any distinguisher D of size t that distinguishes the distributions in (7) with advantage $\delta > \varepsilon$, we get a distinguisher D′ of size $2t$ that distinguishes the distributions in (6) with the same advantage by defining D′ as $\mathsf{D}'(X_n, Z_n, \tilde{M}) = \mathsf{D}(X_n, Z_n, \mathsf{out}_{\mathcal{V}^*}[\tilde{M} \leftrightarrow \mathcal{V}^*(X_n, Z_n)])$. □

4.2   Chain Rules for (Conditional) Pseudoentropy

The following proposition is a chain rule for relaxed conditional HILL entropy. Such a chain rule for the non-conditional case is implicit in the work of Gentry and Wichs [13], and made explicit and generalized to the conditional case by Reyzin [22].

Proposition 2. ([13,22]) *Any joint distribution $(X, Y, A) \in \mathcal{X} \times \mathcal{Y} \times \{0, 1\}^\ell$ satisfies[4]*

$$H_{\varepsilon, s}^{\mathsf{rlx\text{-}HILL}}(X|Y) \geq k \;\Rightarrow\; H_{2\varepsilon, \hat{s}}^{\mathsf{rlx\text{-}HILL}}(X|Y, A) \geq k - \ell \text{ where } \hat{s} = \Omega\left(s \cdot \frac{\varepsilon^2}{2^{3\ell} \log^2(1/\varepsilon)}\right)$$

Proof. $H_{\varepsilon, s}^{\mathsf{rlx\text{-}HILL}}(X|Y) \geq k$ means that there exists a random variable (Z, W) such that $H_\infty(Z|W) \geq k$ and $(X, Y) \sim_{\varepsilon, s} (Z, W)$. For any $\hat{\varepsilon}, \hat{s}$, by Theorem 1, there exists a simulator h of size $s_h = O\left(\hat{s} \cdot \frac{2^{3\ell} \log^2(1/\hat{\varepsilon})}{\hat{\varepsilon}^2}\right)$ such that (we explain the second step below)

$$(X, Y, A) \sim_{\hat{\varepsilon}, \hat{s}} (X, Y, h(X, Y)) \sim_{\varepsilon, s - s_h} (Z, W, h(Z, W))$$

---

[4] Using the recent bound from [26] discussed in Section 1.1, we can get $\hat{s} = \Omega\left(s \cdot \frac{\varepsilon^2 \ell}{2^\ell} + \frac{\ell^2 \log^2(1/\varepsilon)}{\varepsilon^4}\right)$

The second step follows from $(X,Y) \sim_{\varepsilon,s} (Z,W)$ and the fact that h has complexity s_h. Using the triangle inequality for computational indistinguishability[5] we get

$$(X,Y,A) \sim_{\hat{\varepsilon}+\varepsilon,\min(\hat{s},s-s_h)} (Z,W,h(Z,W))$$

To simplify this expression, we set $\hat{\varepsilon} := \varepsilon$ and $\hat{s} := \Theta(s\varepsilon^2/2^{3\ell}\log^2(1/\varepsilon))$, then $s_h = O(s)$, and choosing the hidden constant in the Θ such that $s_h \leq s/2$ (and thus $\hat{s} \leq s - s_h = s/2$), the above equation becomes

$$(X,Y,A) \sim_{2\varepsilon,\hat{s}} (Z,W,h(Z,W)) \tag{8}$$

Using the chain rule for average case min-entropy in the first, and $\mathsf{H}_\infty(Z|W) \geq k$ in the second step below we get

$$\tilde{\mathsf{H}}_\infty(Z|W,h(Z,W)) \geq \tilde{\mathsf{H}}_\infty(Z|W) - \mathsf{H}_0(h(Z,W)) \geq k - \ell . \tag{9}$$

Now equations (8) and (9) imply $\mathsf{H}^{\mathsf{rlx\text{-}HILL}}_{2\varepsilon,\hat{s}}(X|Y,A) = k - \ell$ as claimed. $\quad\square$

By the following lemma, conditional relaxed HILL implies conditional HILL if the conditional part is short (at most logarithmic in the size of the distinguishers considered.)

Lemma 1. *For a joint random variable (X,A) over $\mathcal{X} \times \{0,1\}^\ell$ and $s = \Omega(\ell 2^\ell)$ (more concretely, s should be large enough to implement a lookup table for a function $\{0,1\}^\ell \to \{0,1\}$) conditional relaxed HILL implies standard HILL entropy*

$$\mathsf{H}^{\mathsf{rlx\text{-}HILL}}_{\varepsilon,s}(X|A) \geq k \quad \Rightarrow \quad \mathsf{H}^{\mathsf{HILL}}_{2\varepsilon,s}(X|A) \geq k$$

Proof. $\mathsf{H}^{\mathsf{rlx\text{-}HILL}}_{\varepsilon,s}(X|A) \geq k$ means that there exist (Z,W) where $\tilde{\mathsf{H}}_\infty(Z|W) \geq k$ and

$$(X,A) \sim_{\varepsilon,s} (Z,W) \tag{10}$$

We claim that if $s = \Omega(\ell 2^\ell)$, then (10) implies that $W \sim_\varepsilon A$. To see this, assume the contrary, i.e., that W and A are not ε-close. There exists then a computationally unbounded distinguisher D where

$$|\,\mathbb{P}[D(W) = 1] - \mathbb{P}[D(A) = 1]| > \varepsilon.$$

Without loss of generality, we can assume that D is deterministic and thus, implement D by a circuit of size $\Theta(\ell 2^\ell)$ via a lookup table with 2^ℓ entries (where the ith entry is $D(i)$.) Clearly, D can also distinguish (X,A) from (Z,W) with advantage greater than ε by simply ignoring the first part of the input, thus, contradicting (10). As $A \sim_\varepsilon W$, we claim that there exist a distribution (Z,A) such that

$$(Z,W) \sim_\varepsilon (Z,A). \tag{11}$$

---

[5] which states that for any random variables α, β, γ we have

$$\alpha \sim_{\varepsilon_1,s_1} \beta \quad \& \quad \beta \sim_{\varepsilon_2,s_2} \gamma \quad \Rightarrow \quad \alpha \sim_{\varepsilon_1+\varepsilon_2,\min(s_1,s_2)} \gamma$$

This distribution (Z, A) can be sampled by first sampling (Z, W) and then outputting $(Z, \alpha(W))$ where α is a function that is the identity with probability at least $1 - \varepsilon$ (over the choice of W), i.e., $\alpha(w) = w$ and with probability at most ε, it changes W so that it matches A. The latter is possible since $A \sim_\varepsilon W$.

Using the triangle inequality for computational indistinguishability (cf. the proof of Proposition 2) we get with (10) and (11)

$$(X, A) \sim_{2\varepsilon, s} (Z, A) \tag{12}$$

As $\widetilde{H}_\infty(Z|W) \geq k$ (for α as defined above)

$$\widetilde{H}_\infty(Z|W) \geq k \quad \Rightarrow \quad \widetilde{H}_\infty(Z|\alpha(W)) \geq k \quad \Rightarrow \quad \widetilde{H}_\infty(Z|A) \geq k \tag{13}$$

The first implication above holds as applying a function on the conditioned part cannot decrease the min-entropy. The second holds as $(Z, A) \sim (Z, \alpha(W))$. This concludes the proof as (12) and (13) imply that $H^{\mathsf{HILL}}_{2\varepsilon, s}(X|A) \geq k$. $\quad\square$

As a corollary of Proposition 1 and Lemma 1, we get a chain rule for (non-conditional) HILL entropy. Such a chain rule has been shown by [10] and follows from the more general Dense Model Theorem (published at the same conference) of Reingold et al. [21].

Corollary 1. *For any distribution* $(X, A) \in \mathcal{X} \times \{0, 1\}^\ell$ *and* $\hat{s} = \Omega\left(\frac{s \cdot \varepsilon^2}{2^{3\ell} \log^2(\ell)}\right)$

$$H^{\mathsf{HILL}}_{\varepsilon, s}(X) \geq k \Rightarrow H^{\mathsf{rlx\text{-}HILL}}_{2\varepsilon, \hat{s}}(X|A) \geq k - \ell \Rightarrow H^{\mathsf{HILL}}_{4\varepsilon, \hat{s}}(X|A) \geq k - \ell$$

Note that unlike the chain rule for relaxed HILL given in Proposition 2, the chain rule for (standard) HILL given by the corollary above requires that we start with some non-conditional variable X. It would be preferable to have a chain rule for the conditional case, i.e., and expression of the form $H^{\mathsf{HILL}}_{\varepsilon, s}(X|Y) = k \Rightarrow H^{\mathsf{HILL}}_{\varepsilon', s'}(X|Y, A) = k - \ell$ for some $\varepsilon' = \varepsilon \cdot p(2^\ell)$, $s' = s/q(2^\ell, \varepsilon^{-1})$ (for polynomial functions $p(.), q(.)$), but as recently shown by Krenn et al. [19], such a chain rule does not hold (all we know is that such a rule holds if we also allow the security to degrade exponentially in the length $|Y|$ of the conditional part.)

4.3 Leakage-Resilient Cryptography

We now discuss how Theorem 1 can be used to simplify and quantitatively improve the security proofs for leakage-resilient cryptosystems. These proofs currently rely on chain rules for HILL entropy given in Corollary 1. As an illustrative example, we will reprove the security of the leakage-resilient stream-cipher based on any weak pseudorandom function from Eurocrypt'09 [20], but with much better bounds than the original proof.

For brevity, in this section we often write B^i to denote a sequence $B_1, \ldots, B_i$ of values. Moreover, $A \| B \in \{0, 1\}^{a+b}$ denotes the concatenation of the strings $A \in \{0, 1\}^a$ and $B \in \{0, 1\}^b$.

A function $F\colon \{0,1\}^k \times \{0,1\}^n \to \{0,1\}^m$ is an (ε, s, q)-**secure weak PRF** if its outputs on q random inputs look random to any size s distinguisher, i.e., for all D of size s

$$\left| \underset{K, X^q}{\mathbb{P}}[D(X^q, F(K, X_1), \ldots, F(K, X_q)) = 1] - \underset{X^q, R^q}{\mathbb{P}}[D(X^q, R^q) = 1] \right| \leq \varepsilon,$$

where the probability is over the choice of the random $X_i \leftarrow \{0,1\}^n$, the choice of a random key $K \leftarrow \{0,1\}^k$ and random $R_i \leftarrow \{0,1\}^m$ conditioned on $R_i = R_j$ if $X_i = X_j$ for some $j < i$.

A **stream-cipher** $SC\colon \{0,1\}^k \to \{0,1\}^k \times \{0,1\}^n$ is a function that, when initialized with a secret initial state $S_0 \in \{0,1\}^k$, produces a sequence of output blocks $X_1, X_2, \ldots$ recursively computed by

$$(S_i, X_i) := SC(S_{i-1})$$

We say that SC is (ε, s, q)-secure if for all $1 \leq i \leq q$, no distinguisher of size s can distinguish X_i from a uniformly random $U_n \leftarrow \{0,1\}^n$ with advantage greater than ε given $X_1, \ldots, X_{i-1}$ (here, the probability is over the choice of the initial random key S_0)[6], i.e.,

$$\left| \underset{S_0}{\mathbb{P}}[D(X^{i-1}, X_i) = 1] - \underset{S_0, U_n}{\mathbb{P}}[D(X^{i-1}, U_n)] \right| \leq \varepsilon$$

A **leakage-resilient** stream-cipher is $(\varepsilon, s, q, \ell)$-secure if it is (ε, s, q)-secure as just defined, but where the distinguisher in the jth round not only gets X_j, but also ℓ bits of arbitrary adaptively chosen leakage about the secret state accessed during this round. More precisely, before $(S_j, X_j) := SC(S_{j-1})$ is computed, the distinguisher can choose any leakage function f_j with range $\{0,1\}^\ell$, and then not only get X_j, but also $\Lambda_j := f_j(\hat{S}_{j-1})$, where $\hat{S}_{j-1}$ denotes the part of the secret state that was modified (i.e., read and/or overwritten) in the computation $SC(S_{j-1})$.

Figure 1 illustrates the construction of a leakage-resilient stream cipher SC^F from any weak PRF $F\colon \{0,1\}^k \times \{0,1\}^n \to \{0,1\}^{k+n}$ from [20]. The initial state is $S_0 = \{K_0, K_1, X_0\}$. Moreover, in the ith round (starting with round 0), one computes $K_{i+2} \| X_{i+1} := F(K_i, X_i)$ and outputs X_{i+1}. The state after this round is $(K_{i+1}, K_{i+2}, X_{i+1})$.[7] In this section we will sketch a proof of the following security bound on SC^F as a leakage-resilient stream cipher in terms of the security of F as a weak PRF.

---

[6] A more standard notion would require $X_1, \ldots, X_q$ to be indistinguishable from random; this notion is implied by the notion we use by a standard hybrid argument losing a multiplicative factor of q in the distinguishing advantage.

[7] Note that X_i is not explicitly given as input to f_i even though the computation depends on X_i. The reason is that the adversary can choose f_i adaptively after seeing X_i, so X_i can be hard-coded it into f_i.

Lemma 2. *If* F *is a* $(\varepsilon_F, s_F, 2)$*-secure weak PRF then* SC^F *is a* $(\varepsilon', s', q, \ell)$*-secure leakage resilient stream cipher where*

$$\varepsilon' = 4q\sqrt{\varepsilon_F 2^\ell} \qquad s' = \Theta(1) \cdot \frac{s_F \varepsilon'^2}{2^{3\ell}}$$

The bound above is quantitatively much better than the one in [20]. Setting the leakage bound $\ell = \log \varepsilon_F^{-1}/6$ as in [20], we get (for small q) $\varepsilon' \approx \varepsilon_F^{5/12}$, which is by over a power of 5 better than the $\varepsilon_F^{1/13}$ from [20], and the bound on $s' \approx s_F \varepsilon_F^{4/3}$ improves by a factor of $\varepsilon_F^{5/6}$ (from $s_F \varepsilon_F^{13/6}$ in [20] to $s_F \varepsilon_F^{8/6}$ here). This improvement makes the bound meaningful if instantiated with a standard block-cipher like AES which has a keyspace of 256 bits, making the assumption that it provides $s_F/\varepsilon_F \approx 2^{256}$ security.[8]

Besides our main Theorem 1, we need another technical result which states that if F is a weak PRF secure against two queries, then its output on a single random query is pseudorandom, even if one is given some short auxiliary information about the uniform key K. The security of weak PRFs with non-uniform keys has first been proven in [20], but we will use a more recent and elegant bound from [1]. As a corollary of [1, Thm.3.7 in eprint version], we get that for any $(\varepsilon_F, s_F, 2)$-secure weak PRF F: $\{0,1\}^k \times \{0,1\}^n \to \{0,1\}^m$, uniform and independent key and input $K \sim U_k, X \sim U_n$ and any (arbitrarily complex) function $g \colon \{0,1\}^k \to \{0,1\}^\ell$, one has[9]

$$(X, F(K,X), g(K)) \sim_{\hat\varepsilon, s_F/2} (X, U_m, g(K)) \text{ where } \hat\varepsilon = \varepsilon_F + \sqrt{\varepsilon_F 2^\ell} + 2^{-n} \approx \sqrt{\varepsilon_F 2^\ell} \tag{14}$$

Generalizing the notation of $\sim_{\varepsilon,s}$ from variables to interactive distinguishers, given two (potentially stateful) oracles G, G', we write $G \sim_{\varepsilon,s} G'$ to denote that no oracle-aided adversary A of size s can distinguish G from G', i.e.,

$$G \sim_{\varepsilon,s} G' \iff \forall A, |A| \le s \ : \ |\mathbb{P}[A^G \to 1] - \mathbb{P}[A^{G'} \to 1]| \le \varepsilon.$$

Proof (of Lemma 2 (Sketch)). We define an oracle G_0^{real} that models the standard attack on the leakage-resilient stream cipher. That is, G_0^{real} samples a random initial state S_0. When interacting with an adversary $A^{G_0^{real}}$, the oracle G_0^{real} expects as input adaptively chosen leakage functions $f_1, f_2, \ldots, f_{q-1}$. On input f_i, it computes the next output block $(X_i, K_{i+1}) := SC(K_{i-1}, X_{i-1})$ and the leakage $\Lambda_i = f_i(K_{i-1})$. It forwards X_i, Λ_i to A and deletes everything except the state $S_i = \{X_i, K_i, K_{i+1}\}$. After round $q-1$, G_0^{real} computes and forwards X_q (i.e., the next output block to be computed) to A. The game G_0^{rand} is defined in the same way, but the final block X_q is replaced with a uniformly random U_n.

---

[8] We just need security against two random queries, so the well known non-uniform upper bounds on the security of block-ciphers of De, Trevisan and Tulsiani [6,5] do not seem to contradict such an assumption even in the non-uniform setting.

[9] The theorem implies a stronger statement where one only requires that K has $k - \ell$ bits average-case min-entropy (which is implied by having K uniform and leaking ℓ bits), we state this weaker statement as it is sufficient for our application.

To prove that SC^F is an $(\varepsilon', s', \ell, q)$-secure leakage-resilient stream cipher, we need to show that

$$G_0^{real} \sim_{\varepsilon', s'} G_0^{rand}, \tag{15}$$

for ε', s' as in the statement of the lemma.

Defining games G_i^{real} and G_i^{rand} for $1 \leq i \leq q - 1$. We define a series of games $G_1^{real}, \ldots, G_{q-1}^{real}$ where G_{i+1}^{real} is derived from G_i^{real} by replacing X_i, K_{i+1} with uniformly random values $\tilde{X}_i, \tilde{K}_{i+1}$ and the leakage Λ_i with simulated fake leakage $\tilde{\Lambda}_i$ (the details are provided below). Games G_i^{rand} will be defined exactly as G_i^{real} except that (similarly to the case $i = 0$), the last block X_q is replaced with a uniformly random value.

For every i, $1 \leq i \leq q - 1$, the variables $\tilde{K}_i, \tilde{X}_i$ as defined by the oracles realizing the games G_j^{rand} and G_j^{real} where $j \geq i$ will satisfy the following properties (as the initial values (X_0, K_0, K_1) never get replaced, for notational convenience we define $(\tilde{X}_0, \tilde{K}_0, \tilde{K}_1) \stackrel{\text{def}}{=} (X_0, K_0, K_1)$)

i. $\tilde{K}_i, \tilde{X}_i$ are uniformly random.
ii. Right before the $(i - 1)$th round (i.e. the round where the oracle computes $X_i \| K_{i+1} := \mathsf{F}(\tilde{X}_{i-1}, \tilde{K}_{i-1})$), the oracle has leaked no information about $\tilde{K}_{i-1}$ except for the ℓ bits fake leakage $\tilde{\Lambda}_i$.
iii. Right before the $(i - 1)$th round $\tilde{K}_{i-1}$ and $\tilde{X}_{i-1}$ are independent given everything the oracle did output so far.

The first two properties above will follow from the definition of the games. The third point follows using Lemma 4 from [9], we will not discuss this here in detail, but only mention that the reason for the alternating structure of the cipher as illustrated in Figure 1, with an upper layer computing $K_0, K_2, \ldots$ and the lower layer computing $K_1, K_3, \ldots$, is to achieve this independence.

We now describe how the oracle G_{i+1}^{real} is derived from G_i^{real}. For concreteness, we set $i = 2$. In the third step, G_2^{real} computes $(X_3, K_4) := \mathsf{F}(\tilde{K}_2, \tilde{X}_2)$, $\Lambda_3 = f_3(\tilde{K}_2)$ and forwards X_3, Λ_3 to A. The state stored after this step is $S_3 = \{X_3, \tilde{K}_3, K_4\}$. Let $V_2 \stackrel{\text{def}}{=} \{\tilde{X}^2, \tilde{\Lambda}^2\}$ be the view (i.e. all the outputs she got from her oracle) of the adversary A after the second round.

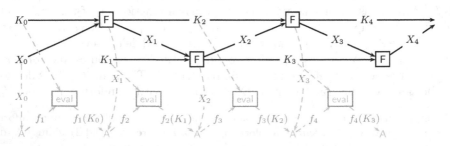

Fig. 1. Leakage resilient stream-cipher SC^F from a any weak pseudorandom function F. The regular evaluation is shown in black, the attack related part is shown in gray with dashed lines. The output of the cipher is $X_0, X_1, \ldots$.

Defining an intermediate oracle. We now define an oracle $G_{2/3}^{real}$ (which will be in-between G_2^{real} and G_3^{real}) derived from G_2^{real} by replacing $\Lambda_3 = f_3(\tilde{K}_2)$ with fake leakage $\tilde{\Lambda}_3$ computed as follows: let $h(\cdot)$ be a simulator for the leakage $\tilde{\Lambda}_3 := f_3(\tilde{K}_2)$ such that (for $\hat{\varepsilon}, \hat{s}$ to be defined)

$$(Z, h(Z)) \sim_{\hat{\varepsilon}, \hat{s}} (Z, \tilde{\Lambda}_3) \quad \text{where} \quad Z = \{V_2, X_3, K_4\} \tag{16}$$

By Theorem 1, there exists such a simulator of size $s_h \overset{\text{def}}{=} O(\hat{s} 2^{3\ell}/\hat{\varepsilon}^2)$. Note that h not only gets the pseudorandom output X_3, K_4 whose computation has leaked bits, but also the view V_2. The reason for the latter is that we need to fool an adversary who learned V_2. Equation (16) then yields

$$G_2^{real} \sim_{\hat{\varepsilon}, \hat{s} - s_0} G_{2/3}^{real}, \tag{17}$$

where s_0 is the size of a circuit required to implement the real game G_0^{real}. The reason we loose s_0 in the circuit size here is that in a reduction where we use a distinguisher for G_2^{real} and $G_{2/3}^{real}$ to distinguish $(Z, h(Z))$ and $(Z, \tilde{\Lambda}_3)$ we must still compute the remaining $q - 4$ rounds, and s_0 is an upper bound on the size of this computation.

The game G_3^{real} is derived from $G_{2/3}^{real}$ by replacing the values $X_3 \| K_4 := F(\tilde{K}_2, \tilde{X}_2)$ with uniformly random $\tilde{X}_3 \| \tilde{K}_4$ right after they have been computed (let us stress that also the fake leakage that is computed as in (16) now uses these random values, i.e., $Z = \{V_2, \tilde{X}_3, \tilde{K}_4\}$).

Proving indistinguishability. We claim that the games are indistinguishable with parameters

$$G_{2/3}^{real} \sim_{\sqrt{\varepsilon_F 2^\ell}, s_F/2 - s_h - s_0} G_3^{real} \tag{18}$$

Recall that in $G_{2/3}^{real}$, we compute $X_3 \| K_4 := F(\tilde{K}_2, \tilde{X}_2)$ where by i. $\tilde{X}_2, \tilde{K}_2$ are uniformly random, by ii. only ℓ bits of $\tilde{K}_2$ have leaked and iii. $\tilde{X}_2$ and $\tilde{K}_2$ are independent. Using these properties, equation (14) implies that the outputs are roughly $(\sqrt{\varepsilon_F 2^\ell}, s_F/2)$ pseudorandom, i.e.,

$$(\tilde{X}_2, X_3 \| K_4, \tilde{\Lambda}_1) \sim_{\sqrt{\varepsilon_F 2^\ell}, s_F/2} (\tilde{X}_2, \tilde{X}_3 \| \tilde{K}_4, \tilde{\Lambda}_1), \tag{19}$$

from which we derive (18). Note the loss of s_h in circuit size in equation (18) due to the fact that given a distinguisher for $G_{2/3}^{real}$ and G_3^{real}, we must recompute the fake leakage given only distributions as in (19).

We will assume that $s_0 \leq \hat{s}/2$, i.e., the real experiment is at most half as complex as the size of the adversaries we will consider (the setting where this is not the case is not very interesting anyway.) Then $\hat{s} - s_0 \geq \hat{s}/2$.

Up to this point, we have not yet defined what $\hat{\varepsilon}$ and $\hat{s}$ are, so we set them to

$$\hat{\varepsilon} \overset{\text{def}}{=} \sqrt{\varepsilon_F 2^\ell} \quad \text{and} \quad \hat{s} \overset{\text{def}}{=} \Theta(1) \frac{s_F \hat{\varepsilon}^2}{2^{3\ell}} \quad \text{then} \quad s_F = 8 \cdot s_h = \Theta(1) \frac{\hat{s} 2^{3\ell}}{\hat{\varepsilon}^2}.$$

With (17) and (18), we then get $G_2^{real} \sim_{2\hat{e},\hat{s}/2} G_3^{real}$. The same proof works for any $1 \leq i \leq q - 1$, i.e., we have

$$G_i^{real} \sim_{2\hat{e},\hat{s}/2} G_{i+1}^{real} \; , \; G_i^{rand} \sim_{2\hat{e},\hat{s}/2} G_{i+1}^{rand}.$$

Moreover, using i.-iii. with (14),

$$G_{q-1}^{real} \sim_{2\hat{e},\hat{s}/2} G_{q-1}^{rand}.$$

Using the triangle inequality $2q$ times, the two equations above yield

$$G_0^{real} \sim_{4q\hat{e},\hat{s}/2} G_0^{rand},$$

which which completes the proof of the lemma.

References

1. Barak, B., Dodis, Y., Krawczyk, H., Pereira, O., Pietrzak, K., Standaert, F.X., Yu, Y.: Leftover hash lemma, revisited. In: Rogaway, P. (ed.) CRYPTO 2011. LNCS, vol. 6841, pp. 1–20. Springer, Heidelberg (2011)
2. Barak, B., Hardt, M., Kale, S.: The uniform hardcore lemma via approximate bregman projections. In: Mathieu, C. (ed.) SODA, pp. 1193–1200. SIAM (2009)
3. Bellare, M., Rompel, J.: Randomness-efficient oblivious sampling. In: FOCS, pp. 276–287 (1994)
4. Chung, K.M., Lui, E., Pass, R.: From weak to strong zero-knowledge and applications. Cryptology ePrint Archive, Report 2013/260 (2013), http://eprint.iacr.org/
5. De, A., Trevisan, L., Tulsiani, M.: Non-uniform attacks against one-way functions and prgs. Electronic Colloquium on Computational Complexity (ECCC) 16, 113 (2009)
6. De, A., Trevisan, L., Tulsiani, M.: Time space tradeoffs for attacks against one-way functions and PRGs. In: Rabin, T. (ed.) CRYPTO 2010. LNCS, vol. 6223, pp. 649–665. Springer, Heidelberg (2010)
7. Dodis, Y., Pietrzak, K.: Leakage-resilient pseudorandom functions and side-channel attacks on Feistel networks. In: Rabin, T. (ed.) CRYPTO 2010. LNCS, vol. 6223, pp. 21–40. Springer, Heidelberg (2010)
8. Dodis, Y., Reyzin, L., Smith, A.: Fuzzy extractors: How to generate strong keys from biometrics and other noisy data. In: Cachin, C., Camenisch, J.L. (eds.) EUROCRYPT 2004. LNCS, vol. 3027, pp. 523–540. Springer, Heidelberg (2004)
9. Dziembowski, S., Pietrzak, K.: Intrusion-resilient secret sharing. In: 48th FOCS, pp. 227–237. IEEE Computer Society Press (October 2007)
10. Dziembowski, S., Pietrzak, K.: Leakage-resilient cryptography. In: 49th FOCS, pp. 293–302. IEEE Computer Society Press (October 2008)
11. Frieze, A.M., Kannan, R.: Quick approximation to matrices and applications. Combinatorica 19(2), 175–220 (1999)
12. Fuller, B., O'Neill, A., Reyzin, L.: A unified approach to deterministic encryption: New constructions and a connection to computational entropy. In: Cramer, R. (ed.) TCC 2012. LNCS, vol. 7194, pp. 582–599. Springer, Heidelberg (2012)
13. Gentry, C., Wichs, D.: Separating succinct non-interactive arguments from all falsifiable assumptions. In: Fortnow, L., Vadhan, S.P. (eds.) 43rd ACM STOC, pp. 99–108. ACM Press (June 2011)

14. Gowers, T.: Decompositions, approximate structure, transference, and the Hahn–Banach theorem. Bull. London Math. Soc. 42(4), 573–606 (2010)
15. Håstad, J., Impagliazzo, R., Levin, L.A., Luby, M.: A pseudorandom generator from any one-way function. SIAM Journal on Computing 28(4), 1364–1396 (1999)
16. Herbster, M., Warmuth, M.K.: Tracking the best linear predictor. Journal of Machine Learning Research 1, 281–309 (2001)
17. Hsiao, C.Y., Lu, C.J., Reyzin, L.: Conditional computational entropy, or toward separating pseudoentropy from compressibility. In: Naor, M. (ed.) EUROCRYPT 2007. LNCS, vol. 4515, pp. 169–186. Springer, Heidelberg (2007)
18. Impagliazzo, R.: Hard-core distributions for somewhat hard problems. In: FOCS, pp. 538–545 (1995)
19. Krenn, S., Pietrzak, K., Wadia, A.: A counterexample to the chain rule for conditional HILL entropy - and what deniable encryption has to do with it. In: Sahai, A. (ed.) TCC 2013. LNCS, vol. 7785, pp. 23–39. Springer, Heidelberg (2013)
20. Pietrzak, K.: A leakage-resilient mode of operation. In: Joux, A. (ed.) EUROCRYPT 2009. LNCS, vol. 5479, pp. 462–482. Springer, Heidelberg (2009)
21. Reingold, O., Trevisan, L., Tulsiani, M., Vadhan, S.P.: Dense subsets of pseudorandom sets. In: 49th FOCS, pp. 76–85. IEEE Computer Society Press (October 2008)
22. Reyzin, L.: Some notions of entropy for cryptography - (invited talk). In: Fehr, S. (ed.) ICITS 2011. LNCS, vol. 6673, pp. 138–142. Springer, Heidelberg (2011), http://www.cs.bu.edu/~reyzin/papers/entropy-survey.pdf
23. Tao, T., Ziegler, T.: The primes contain arbitrarily long polynomial progressions. Acta Math. 201, 213–305 (2008)
24. Trevisan, L.: Guest column: additive combinatorics and theoretical computer science. SIGACT News 40(2), 50–66 (2009)
25. Trevisan, L., Tulsiani, M., Vadhan, S.P.: Regularity, boosting, and efficiently simulating every high-entropy distribution. In: IEEE Conference on Computational Complexity, pp. 126–136 (2009)
26. Vadhan, S., Zheng, C.J.: A uniform min-max theorem with applications in cryptography. In: Canetti, R., Garay, J.A. (eds.) CRYPTO 2013, Part I. LNCS, vol. 8042, pp. 93–110. Springer, Heidelberg (2013)

A   Proof of Theorem 1

We will prove Theorem 1 not for the family $\mathcal{F}$ directly, but for a family $\widehat{\mathcal{F}}$ which for every $f \in \mathcal{F}$ contains the function $\widehat{f} \colon \mathcal{X} \times \{0,1\}^\ell \to [-1,1]$ defined as

$$\widehat{f}(x,b) = f(x,b) - w_f(x) \quad \text{where} \quad w_f(x) = \mathop{\mathbb{E}}_{b \leftarrow \{0,1\}^\ell}[f(x,b)] = 2^{-\ell} \sum_{b \in \{0,1\}^\ell} f(x,b)$$

Any simulator that fools $\widehat{\mathcal{F}}$ also fools $\mathcal{F}$ with the same advantage since $\forall \widehat{f} \in \widehat{\mathcal{F}}$,

$$\left| \mathop{\mathbb{E}}_{x \leftarrow X, [g]}[\widehat{f}(x, g(x))] - \mathop{\mathbb{E}}_{x \leftarrow X, [h]}[\widehat{f}(x, h(x))] \right|$$

$$= \left| \mathop{\mathbb{E}}_{x \leftarrow X, [g]}[f(x, g(x)) - w_f(x)] - \mathop{\mathbb{E}}_{x \leftarrow X, [h]}[f(x, h(x)) - w_f(x)] \right|$$

$$= \left| \mathop{\mathbb{E}}_{x \leftarrow X, [g]}[f(x, g(x))] - \mathop{\mathbb{E}}_{x \leftarrow X, [h]}[f(x, h(x))] \right|$$

Evaluating $\widehat{f}$ requires 2^ℓ evaluations of f as we need to compute $w_f(x)$. We thus lose a factor of 2^ℓ in efficiency by considering $\widehat{\mathcal{F}}$ instead of $\mathcal{F}$. The reason that we prove the theorem for $\widehat{\mathcal{F}}$ instead of for $\mathcal{F}$ is because in what follows, we will need that for any x, the expectation over a uniformly random $b \in \{0,1\}^\ell$ is 0, i.e.,

$$\forall \widehat{f} \in \widehat{\mathcal{F}}, x \in \mathcal{X}: \quad \mathop{\mathbb{E}}_{b \leftarrow \{0,1\}^\ell} [\widehat{f}(x,b)] = 0. \tag{20}$$

To prove the theorem, we must show that for any joint distribution $(X, g(X))$ over $\mathcal{X} \times \{0,1\}^\ell$, there exists an efficient simulator $h: \mathcal{X} \to \{0,1\}^\ell$ such that

$$\forall \widehat{f} \in \widehat{\mathcal{F}}: \quad \left| \mathop{\mathbb{E}}_{x \leftarrow X} [\widehat{f}(x, g(x)) - \widehat{f}(x, h(x))] \right| < \varepsilon. \tag{21}$$

Moving to product form. We define the function $\widetilde{g}: \mathcal{X} \times \{0,1\}^\ell \to [0,1]$ as $\widetilde{g}(x,a) := \mathop{\mathbb{P}}_{[g]} [g(x) = a]$. Note that for every $x \in \mathcal{X}$, we have

$$\sum_{a \in \{0,1\}^\ell} \widetilde{g}(x,a) = 1. \tag{22}$$

We can write the expected value of $\widehat{f}(X, g(X))$ as follows:

$$\mathop{\mathbb{E}}_{x \leftarrow X,[g]} [\widehat{f}(x, g(x))] = \sum_{a \in \{0,1\}^\ell} \mathop{\mathbb{E}}_{x \leftarrow X} \left[\widehat{f}(x, a) \mathop{\mathbb{P}}_{[g]}[g(x) = a] \right] =$$

$$= \sum_{a \in \{0,1\}^\ell} \mathop{\mathbb{E}}_{x \leftarrow X} \left[\widehat{f}(x, a) \widetilde{g}(x, a) \right] =$$

$$= 2^\ell \mathop{\mathbb{E}}_{x \leftarrow X, u \leftarrow \{0,1\}^\ell} [\widehat{f}(x, u) \widetilde{g}(x, u)]. \tag{23}$$

We will construct a simulator $\widetilde{h}: \mathcal{X} \times \{0,1\}^\ell \to [0,1]$ such that for $\gamma > 0$ (to be defined later),

$$\forall \widehat{f} \in \widehat{\mathcal{F}}: \quad \mathop{\mathbb{E}}_{x \leftarrow X, b \leftarrow \{0,1\}^\ell} [\widehat{f}(x, b)(\widetilde{g}(x, b) - \widetilde{h}(x, b))] < \gamma. \tag{24}$$

From this $\widetilde{h}$, we can then get a simulator $h(\cdot)$ like in (21) assuming that $\widetilde{h}(x, \cdot)$ is a probability distribution for all x, i.e., $\forall x \in \mathcal{X}$,

$$\sum_{b \in \{0,1\}^\ell} \widetilde{h}(x, b) = 1, \tag{25}$$

$$\forall b \in \{0,1\}^\ell : \widetilde{h}(x, b) \geq 0. \tag{26}$$

We will define a sequence $h_0, h_1, \ldots$ of functions where $h_0(x,b) = 2^{-\ell}$ for all x, b.[10] Define the energy function

$$\Delta_t = \mathop{\mathbb{E}}_{x \leftarrow X, b \leftarrow \{0,1\}^\ell} [(\widetilde{g}(x, b) - h_t(x, b))^2].$$

---

[10] It is not relevant how exactly h_0 is defined, but we need $\sum_{b \leftarrow \{0,1\}^\ell} [h_0(x, b)] = 1$ for all $x \in \mathcal{X}$.

Assume that after the first t steps, there exists a function $\widehat{f}_{t+1} \colon \mathcal{X} \times \{0,1\}^{\ell} \to [-1,1]$ such that

$$\mathop{\mathbb{E}}_{x \leftarrow X, b \leftarrow \{0,1\}^{\ell}} [\widehat{f}_{t+1}(x,b)(g(x,b) - h_t(x,b))] \geq \gamma,$$

and define

$$h_{t+1}(x,b) = h_t(x,b) + \gamma \widehat{f}_{t+1}(x,b) \tag{27}$$

The energy function then decreases by γ^2, i.e.,

$$
\begin{aligned}
&\Delta_{t+1} \\
&= \mathop{\mathbb{E}}_{x \leftarrow X, b \leftarrow \{0,1\}^{\ell}} [(\widetilde{g}(x,b) - h_t(x,b) - \gamma \widehat{f}_{t+1}(x,b))^2] = \\
&= \Delta_t + \underbrace{\mathop{\mathbb{E}}_{x \leftarrow X, b \leftarrow \{0,1\}^{\ell}} [\gamma^2 \widehat{f}_{t+1}(x,b)]}_{\leq \gamma^2} - \underbrace{\mathop{\mathbb{E}}_{x \leftarrow X, b \leftarrow \{0,1\}^{\ell}} [2\gamma f_{t+1}(x,b)(\widetilde{g}(x,b) - h_t(x,b))]}_{\geq 2\gamma^2} \\
&\leq \Delta_t - \gamma^2.
\end{aligned}
$$

Since $\Delta_0 \leq 1$, $\Delta_t \geq 0$ for any t (as it is a square) and $\Delta_i - \Delta_{i+1} \geq \gamma^2$, this process must terminate after at most $1/\gamma^2$ steps meaning that we have constructed $\widetilde{h} = h_t$ that satisfies (24). Note that the complexity of the constructed $\widetilde{h}$ is bounded by $2^{\ell} \gamma^{-2}$ times the complexity of the functions from $\mathcal{F}$ since, as mentioned earlier, computing $\widetilde{f}$ requires 2^{ℓ} evaluations of f. In other words, $\widetilde{h}$ has complexity $\mathcal{O}(2^{\ell} \gamma^{-2})$ relative to $\mathcal{F}$.

Moreover, since for all $x \in \mathcal{X}$ and $\widehat{f} \in \widehat{\mathcal{F}}$, we have $\sum_{b \in \{0,1\}^{\ell}} h_0(x,b) = 1$ and $\sum_{b \in \{0,1\}^{\ell}} \widehat{f}(x,b) = 0$, condition (25) holds as well. Unfortunately, (26) does not hold since it might be the case that $h_{t+1}(x,b) < 0$. We will explain later how to fix this problem by replacing $\widehat{f}_{t+1}$ in (27) with a similar function $\widehat{f}^*_{t+1}$ that satisfies $h_{t+1}(x,b) = h_t + \gamma \widehat{f}^*_{t+1} \geq 0$ for all x and b in addition to all of the properties just discussed. Assume for now that $\widetilde{h}$ satisfies (24)-(26).

Let $h \colon \mathcal{X} \to \{0,1\}^{\ell}$ be a probabilistic function defined as follows: we set $h(x) = b$ with probability $\widetilde{h}(x,b)$. Equivalently, imagine that we have a biased dice with 2^{ℓ} faces labeled by $b \in \{0,1\}^{\ell}$ such that the probability of getting the face with label b is $\widetilde{h}(x,b)$. We then define $h(x)$ by simply throwing this dice and reading off the label. It follows that $\mathop{\mathbb{P}}_{[h]}[h(x) = b] = \widetilde{h}(x,b)$. This probabilistic function satisfies

$$
\begin{aligned}
\mathop{\mathbb{E}}_{[h], x \leftarrow X} [\widehat{f}(x, h(x))] &= \mathop{\mathbb{E}}_{x \leftarrow X} \sum_{a \in \{0,1\}^{\ell}} \widehat{f}(x,a) \mathop{\mathbb{P}}_{[h]}[h(x) = a] \\
&= \mathop{\mathbb{E}}_{x \leftarrow X} \sum_{a \in \{0,1\}^{\ell}} \widehat{f}(x,a) h_t(x,a) \\
&= \mathop{\mathbb{E}}_{x \leftarrow X, u \leftarrow \{0,1\}^{\ell}} 2^{\ell} \widehat{f}(x,u) h_t(x,u). \tag{28}
\end{aligned}
$$

Plugging (28) and (23) into (24), we obtain

$$\forall \widehat{f} \in \mathcal{F} \ : \ \mathop{\mathbb{E}}_{x \leftarrow X, [h]} \left[\frac{\widehat{f}(x, g(x))}{2^\ell} - \frac{\widehat{f}(x, h(x))]}{2^\ell} \right] < \gamma.$$

Equivalently,

$$\forall \widehat{f} \in \mathcal{F} \ : \ \mathop{\mathbb{E}}_{x \leftarrow X, [h]} \left[\widehat{f}(x, g(x)) - \widehat{f}(x, h(x)) \right] < \gamma 2^\ell \qquad (29)$$

We get (4) from the statement of the theorem by setting $\gamma := \varepsilon/2^\ell$. The simulator $\widetilde{h}$ is thus of complexity $\mathcal{O}(2^{3\ell}(1/\varepsilon)^2)$ relative to $\mathcal{F}$.

Enforcing $h_t(x, b) \geq 0$ for $\ell = 1$. We now fix the problem with the positivity of $h_t(x, b)$. Consider the case $\ell = 1$. Consider the following properties:

i. $\displaystyle\sum_{b \in \{0,1\}} h_t(x, b) = 1$ for $x \in \mathcal{X}$,

ii. $\forall b \in \{0, 1\}, h_t(x, b) \geq 0$ for $x \in \mathcal{X}$,

iii. $\displaystyle\mathop{\mathbb{E}}_{x \leftarrow X, b \leftarrow \{0,1\}} [\widehat{f}_{t+1}(x, b)(g(x, b) - h_t(x, b))] \geq \gamma$ for $\gamma > 0$.

Assume that $h_t \colon \mathcal{X} \to \{0, 1\}$ and $\widehat{f}_{t+1} \colon \mathcal{X} \times \{0, 1\} \to [-1, 1]$ satisfy *i)* and *ii)* for all $x \in \mathcal{X}$ and *iii)* for some $\gamma > 0$. Recall that $\Delta_t = \mathop{\mathbb{E}}_{x \leftarrow X, b \leftarrow \{0,1\}} [(\widetilde{g}(x, b) - h_t(x, b))^2]$. We have shown that $h_{t+1} = h_t + \gamma \widehat{f}_{t+1}$ satisfies

$$\Delta_{t+1} \leq \Delta_t - \gamma^2. \qquad (30)$$

Moreover, for all $x \in \mathcal{X}$, h_{t+1} will still satisfy *i)* but not necessarily *ii)*. We define a function $\widehat{f}_{t+1}^*$ such that setting $h_{t+1} = h_t + \gamma \widehat{f}_{t+1}^*$ will satisfy *i)* and *ii)* for all $x \in \mathcal{X}$ and an inequality similar to (30).

First, for any $x \in \mathcal{X}$ for which condition *ii)* is satisfied, let $f_{t+1}^* = \widehat{f}_{t+1}$. Consider now $x \in \mathcal{X}$ for which *ii)* fails for some $b \in \{0, 1\}$, i.e., for which $h_t(x, b) + \gamma \widehat{f}_{t+1}(x, b) < 0$. Let $\gamma' = -h_t(x, b)/f_{t+1}(x, b)$. Note that $0 \leq \gamma' \leq \gamma$ and $h_t(x, b) + \gamma' \widehat{f}_{t+1}(x, b) = 0$. Let

$$\widehat{f}_{t+1}^*(x, b) = \frac{\gamma'}{\gamma} \widehat{f}_{t+1}(x, b) \qquad \widehat{f}_{t+1}^*(x, 1 - b) = \widehat{f}_{t+1}(x, 1 - b) + \frac{1 - \gamma'}{\gamma} \widehat{f}_{t+1}(x, b).$$

Let $h_{t+1}(x, \cdot) = h_t(x, \cdot) + \gamma \widehat{f}_{t+1}^*(x, \cdot)$ and note that

$$\sum_{b \in \{0,1\}} \widehat{f}_{t+1}^*(x, b) = \sum_{b \in \{0,1\}} \widehat{f}_{t+1}(x, b) = 0.$$

Condition *i)* is then satisfied for h_{t+1} for any $x \in \mathcal{X}$. By the definition of γ', condition *ii)* is satisfied for any $x \in \mathcal{X}$ as well. Condition *iii)* is more delicate and in fact need not hold. Yet, we will prove the following:

Lemma 3. *If $\widehat{f}_{t+1}$ and h_t satisfy i) and ii) for every $x \in \mathcal{X}$, and iii) then*

$$\mathop{\mathbb{E}}_{x \leftarrow X, b \leftarrow \{0,1\}} [\widehat{f}_{t+1}(x,b)(g(x,b) - h_t(x,b))]$$

$$- \mathop{\mathbb{E}}_{x \leftarrow X, b \leftarrow \{0,1\}} [\widehat{f}^*_{t+1}(x,b)(g(x,b) - h_t(x,b))] \leq \frac{\gamma}{4}. \tag{31}$$

Proof. To prove (31), it suffices to show that for every $x \in \mathcal{X}$,

$$\sum_{b \in \{0,1\}} \widehat{f}_{t+1}(x,b)(g(x,b) - h_t(x,b)) - \sum_{b \in \{0,1\}} \widehat{f}^*_{t+1}(x,b)(g(x,b) - h_t(x,b)) \leq \frac{\gamma}{2}. \tag{32}$$

If $x \in \mathcal{X}$ is such that $ii)$ is satisfied for h_{t+1} then there is nothing to prove. Suppose that $ii)$ fails for some $x \in \mathcal{X}$ and $b \in \{0,1\}$. For brevity, let $f := \widehat{f}_{t+1}(x,b)$, $g := g(x,b)$, $h = h_t(x,b)$. We have $-1 \leq f < 0$, $h + \gamma f < 0$, $0 \leq g \leq 1$ and $h = -\gamma f^*$. Using $g - h \geq -h$, the left-hand side of (32) then satisfies

$$2(f + h/\gamma)(g - h) \leq 2(f + h/\gamma)(-h)$$

$$= \frac{2}{\gamma}(-f\gamma - h)h \leq \frac{2}{\gamma} \left(\frac{-f\gamma - h + h}{2} \right)^2 = \frac{\gamma f^2}{2} \leq \frac{\gamma}{2}, \tag{33}$$

where we have used the inequality $uv \leq \left(\dfrac{u+v}{2} \right)^2$.

If $iii)$ holds then Lemma 3 implies $\gamma - \mathop{\mathbb{E}}_{x \leftarrow X, b \leftarrow \{0,1\}} [\widehat{f}^*_{t+1}(x,b)(g(x,b) - h_t(x,b))] \leq \frac{\gamma}{4}$. Equivalently,

$$\mathop{\mathbb{E}}_{x \leftarrow X, b \leftarrow \{0,1\}} [\widehat{f}^*_{t+1}(x,b)(g(x,b) - h_t(x,b))] \geq \frac{3\gamma}{4}. \tag{34}$$

Defining $h_{t+1} = h_t + \gamma \widehat{f}^*_{t+1}$, we still get

$$\Delta_{t+1} \leq \Delta_t - \left(\frac{3\gamma}{4} \right)^2 = \Delta_t - \frac{9\gamma^2}{16}. \tag{35}$$

Remark 2. In this case, the slightly worse inequality (35) will increase the complexity of $\widetilde{h}$, but only by a constant factor of $16/9$, i.e., $\widetilde{h}$ will still have complexity $\mathcal{O}(2^\ell \gamma^{-2})$ relative to $\mathcal{F}$.

Enforcing $h_t(x,b) \geq 0$ for general ℓ. Let $\widehat{f}_{t+1}(x,b)$ be as before and suppose that there exists $x \in \mathcal{X}$ such that $h_t(x,b) + \gamma \widehat{f}_{t+1}(x,b) < 0$ for at least one $b \in \{0,1\}^\ell$. We will show how to replace $\widehat{f}_{t+1}$ with another function $\widehat{f}^*_{t+1}$ such that it satisfies an inequality of type (34) and such that $h_{t+1}(x,b) = h_t(x,b) + \gamma \widehat{f}^*_{t+1}(x,b) \geq 0$. Let S be the set of all elements $b \in \{0,1\}^\ell$ for which $h_t(x,b) + \gamma \widehat{f}_{t+1}(x,b) < 0$. For $b \in S$, it follows that $\widehat{f}_{t+1}(x,b) < 0$. As before, for $b \in S$, define $\widehat{f}^*_{t+1}(x,b) = -\dfrac{h_t(x,b)}{\gamma}$. Note that for each such b, we have added a positive

mass $-\dfrac{h_t(x,b) + \gamma\widehat{f}_{t+1}(x,b)}{\gamma}$ to modify each $\widehat{f}_{t+1}(x,b)$. Let

$$M = \sum_{b \in S} -\left(\widehat{f}_{t+1}(x,b) + \frac{h_t(x,b)}{\gamma}\right) \tag{36}$$

be the total mass. For $b \notin S$, define $\widehat{f}^*_{t+1}(x,b) = \widehat{f}_{t+1}(x,b) - \dfrac{M}{2^\ell - s}$. Clearly,

$\underset{b \leftarrow \{0,1\}^\ell}{\mathbb{E}} \widehat{f}^*_{t+1}(x,b) = 0$. We will now show the following

Lemma 4. *For every $x \in \mathcal{X}$, the function $\widehat{f}^*_{t+1}$ satisfies*

$$\sum_{b \in \{0,1\}^\ell} (\widehat{f}_{t+1}(x,b) - \widehat{f}^*_{t+1}(x,b))(g(x,b) - h_t(x,b)) < 2^{\ell-1}\gamma.$$

Proof. Let $s = |S|$ and $h_S = \displaystyle\sum_{i=1}^{s} h_t(x,b_i)$. First, note that (as in the case $\ell = 1$)

$$\forall b \in S: \left(\widehat{f}_{t+1}(x,b) + \frac{h_t(x,b)}{\gamma}\right)(g(x,b) - h_t(x,b))$$

$$\leq -\left(\widehat{f}_{t+1}(x,b) + \frac{h_t(x,b)}{\gamma}\right)h_t(x,b). \tag{37}$$

Moreover,

$$\sum_{b \notin S} g(x,b) \leq \sum_{b \in \{0,1\}^\ell} g(x,b) = 1. \tag{38}$$

The difference that we want to estimate is then

$$\Delta = \sum_{b \in \{0,1\}^\ell} (\widehat{f}_{t+1}(x,b) - \widehat{f}^*_{t+1}(x,b))(g(x,b) - h_t(x,b))$$

$$= \sum_{b \in S} \left(\widehat{f}_{t+1}(x,b) + \frac{h_t(x,b)}{\gamma}\right)(g(x,b) - h_t(x,b)) + \frac{M}{2^\ell - s}\sum_{b \notin S}(g(x,b) - h_t(x,b))$$

$$\overset{(37),(38)}{\leq} \sum_{b \in S} -\left(\widehat{f}_{t+1}(x,b) + \frac{h_t(x,b)}{\gamma}\right)h_t(x,b) + \frac{M}{2^\ell - s}\underbrace{\left(1 - \sum_{b \notin S} h_t(x,b)\right)}_{=h_S}$$

$$\overset{(36)}{=} \underbrace{\sum_{b \in S} -\left(\widehat{f}_{t+1}(x,b) + \frac{h_t(x,b)}{\gamma}\right)h_t(x,b)}_{\leq \gamma/4} + \frac{h_S}{2^\ell - s}\sum_{b \in S} -\left(\widehat{f}_{t+1}(x,b) + \frac{h_t(x,b)}{\gamma}\right)$$

$$\overset{(33)}{\leq} \frac{s\gamma}{4} - \frac{h_S}{2^\ell - s}\sum_{b \in S}\widehat{f}_{t+1}(x,b) - \frac{h_S^2}{\gamma(2^\ell - s)} = \frac{s\gamma}{4} + \frac{h_S f_S}{2^\ell - s} - \frac{h_S^2}{\gamma(2^\ell - s)},$$

where $f_S = -\sum_{b \in S} \widehat{f}_{t+1}(x, b)$. Note that $\sum_{b \in S} -\widehat{f}_{t+1}(x, b) \leq s$ and (using (20))

$$\sum_{b \in S} -\widehat{f}_{t+1}(x, b) = \sum_{b \notin S} \widehat{f}_{t+1}(x, b) \leq 2^\ell - s, \text{ i.e., } f_S \leq \min\{s, 2^\ell - s\}. \text{ Since}$$

$$\frac{h_S f_S}{2^\ell - s} - \frac{h_S^2}{\gamma(2^\ell - s)} = \frac{1}{\gamma(2^\ell - s)} h_S(\gamma f_S - h_S)$$

$$\leq \frac{1}{\gamma(2^\ell - s)} \left(\frac{h_S + (\gamma f_S - h_S)}{2} \right)^2 \leq \frac{s\gamma}{4},$$

where we have used that $f_S^2 \leq s(2^\ell - s)$. Since $s < 2^\ell$, we obtain $\Delta \leq \frac{s\gamma}{2} < 2^{\ell-1}\gamma$ which proves the lemma.

To complete the proof, note that the above lemma implies that

$$\mathop{\mathbb{E}}_{x \leftarrow X, b \leftarrow \{0,1\}^\ell} [\widehat{f}_{t+1}(x, b)(g(x, b) - h_t(x, b))]$$

$$- \mathop{\mathbb{E}}_{x \leftarrow X, b \leftarrow \{0,1\}^\ell} [\widehat{f}^*_{t+1}(x, b)(g(x, b) - h_t(x, b))] < \frac{\gamma}{2},$$

and hence,

$$\mathop{\mathbb{E}}_{x \leftarrow X, b \leftarrow \{0,1\}^\ell} [\widehat{f}^*_{t+1}(x, b)(g(x, b) - h_t(x, b))] > \frac{\gamma}{2}. \tag{39}$$

Remark 3. Similarly, the slightly worse inequality (39) will increase the complexity of $\widetilde{h}$ by a constant factor of 4, i.e., $\widetilde{h}$ will still have complexity $\mathcal{O}(2^\ell \gamma^{-2})$ relative to $\mathcal{F}$.

A.1   Derandomizing $\widetilde{h}$

Next, we discuss how to derandomize $\widetilde{h}$. We can think of the probabilistic function $\widetilde{h}$ as a deterministic function $\widetilde{h}'$ taking two inputs where the second input represents the random coins used by $\widetilde{h}$. More precisely, for $R \leftarrow \{0,1\}^\rho$ (ρ is an upper bound on the number of random bits used by $\widetilde{h}$) and for any x in the support of X, we have $\widetilde{h}'(x, R) \sim \widetilde{h}(x)$.

To get our derandomized $\widehat{h}$, we replace the randomness R with the output of a function ϕ chosen from a family of t-wise independent functions for some large t, i.e., we set $\widehat{h}(x) = \widetilde{h}'(x, \phi(x))$. Recall that a family Φ of functions $\mathcal{A} \to \mathcal{B}$ is t-wise independent if for any t distinct inputs $a_1, \ldots, a_t \in \mathcal{A}$ and a randomly chosen $\phi \leftarrow \Phi$, the outputs $\phi(a_1), \ldots, \phi(a_t)$ are uniformly random in $\mathcal{B}^t$. In the proof, we use the following tail inequality for variables with bounded independence:

Lemma 5 (Lemma 2.2 from [3]). *Let $t \geq 6$ be an even integer and let $Z_1, \ldots, Z_n$ be t-wise independent variables taking values in $[0, 1]$. Let $Z = \sum_{i=1}^n Z_i$, then for any $A > 0$*

$$\mathbb{P}[|Z - \mathbb{E}[Z]| \geq A] \leq \left(\frac{nt}{A^2} \right)^{t/2}$$

Recall that the min-entropy of X is $H_\infty(X) = -\log\left(\max_x \mathbb{P}[X = x]\right)$, or equivalently, X has min-entropy k if $\mathbb{P}[X = x] \leq 2^{-k}$ for all $x \in \mathcal{X}$.

Lemma 6. (Deterministic Simulation) *Let $\varepsilon > 0$ and assume that*

$$H_\infty(X) > 2 + \log\log|\mathcal{F}| + 2\log(1/\varepsilon). \tag{40}$$

For any (probabilistic) $\widetilde{h} \colon \mathcal{X} \to \{0,1\}^\ell$, there exists a deterministic $\widehat{h}$ *of the same complexity relative to $\mathcal{F}$ as $\widetilde{h}$ such that*

$$\forall f \in \mathcal{F}: \left| \mathop{\mathbb{E}}_{x \leftarrow X, [\widetilde{h}]} [f(x, \widetilde{h}(x))] - \mathop{\mathbb{E}}_{x \leftarrow X} [f(x, \widehat{h}(x))] \right| < \varepsilon \tag{41}$$

Remark 4. **About the condition (40).** A lower bound on the min-entropy of X in terms of $\log\log|\mathcal{F}|$ and $\log(1/\varepsilon)$ as in (40) is necessary. For example one can show that for $\varepsilon < 1/2$, (41) implies $H_\infty(X) \geq \log\log|\mathcal{F}|$. To see this, consider the case when X is uniform over $\{0,1\}^m$ (so $H_\infty(X) = m$), $\mathcal{F}$ contains all 2^{2^m} functions $f \colon \{0,1\}^m \times \{0,1\} \to \{0,1\}$ satisfying $f(x, 1-b) = 1 - f(x,b)$ for all $x, b \in \{0,1\}^{m+1}$, and $\widetilde{h}(x) \sim U_1$ is uniformly random for all x (so it ignores its input). Now, given any deterministic $\widehat{h}$, we can choose $f \in \mathcal{F}$ where $f(x, \widehat{h}(x)) = 1$ for all $x \in \{0,1\}^m$ (such an f exists by definition of $\mathcal{F}$). For this f,

$$\left| \underbrace{\mathop{\mathbb{E}}_{x \leftarrow X, [\widetilde{h}]} [f(x, \widetilde{h}(x))]}_{=1/2} - \underbrace{\mathop{\mathbb{E}}_{x \leftarrow X} [f(x, \widehat{h}(x))]}_{=1} \right| = 1/2.$$

In terms of $\log(1/\varepsilon)$, one can show that (41) implies $H_\infty(X) \geq \log(1/\varepsilon) - 1$ (even if $|\mathcal{F}| = 1$). For this, let $\widetilde{h}$ and X be as above, $\mathcal{F} = \{f\}$ is defined as $f(x,b) = b$ if $x = 0^m$ and $f(x,b) = 0$ otherwise. For any deterministic $\widehat{h}$, we get

$$\left| \underbrace{\mathop{\mathbb{E}}_{x \leftarrow X, [\widetilde{h}]} [f(x, \widetilde{h}(x))]}_{1/2^{m+1}} - \underbrace{\mathop{\mathbb{E}}_{x \leftarrow X} [f(x, \widehat{h}(x))]}_{1/2^m \text{ or } 0} \right| = 1/2^{m+1}$$

and thus, $\varepsilon = 1/2^{m+1}$. Equivalently $H_\infty(X) = m = \log(1/\varepsilon) - 1$. The condition (40) is mild and in particular, it covers the cryptographically interesting case where $\mathcal{F}$ is the family of polynomial-size circuits (i.e., for a security parameter n and a constant c, $|\mathcal{F}| \leq 2^{n^c}$), X has superlogarithmic min-entropy $H_\infty(X) = \omega(\log n)$ and $\varepsilon > 0$ is negligible in n. Here, (40) becomes

$$\omega(\log n) > 2 + c\log n + 2\log\varepsilon^{-1}$$

which holds for a negligible $\varepsilon = 2^{-\omega(\log n)}$.

Proof (Proof of Lemma 5). Let $m = H_\infty(X)$. We will only prove the lemma for the restricted case where X is flat, i.e., it is uniform on a subset $\mathcal{X}' \subseteq \mathcal{X}$ of size 2^m. [11] Consider any fixed $f \in \mathcal{F}$ and the 2^m random variables $Z_x \in \{0,1\}$ indexed by $x \in \mathcal{X}'$ sampled as follows: first, sample $\phi \leftarrow \Phi$ from a family of t-wise independent functions $\mathcal{X} \to \{0,1\}^\rho$ (recall that ρ is a upper bound on the number of random bits used by $\widetilde{h}$). Now, Z_x is defined as

$$Z_x = f(x, \widetilde{h}'(x, \phi(x))) = f(x, \widehat{h}(x))$$

and $Z = \sum_{x \in \mathcal{X}'} Z_x$. Note that the same ϕ is used for all Z_x.

1. The variables Z_x for $x \in \mathcal{X}'$ are t-wise independent, i.e., for any t distinct $x_1, \ldots, x_t$, the variables $Z_{x_1}, \ldots, Z_{x_t}$ have the same distribution as $Z'_{x_1}, \ldots, Z'_{x_t}$ sampled as $Z'_{x_i} \leftarrow f(x_i, \widetilde{h}'(x_i, R))$. The reason is that the randomness $\phi(x_1), \ldots, \phi(x_t)$ used to sample the $Z_{x_1}, \ldots, Z_{x_t}$ is uniform in $\{0,1\}^\rho$ as ϕ is t-wise independent.

2. $\mathbb{E}[Z_x] = \underset{\phi \leftarrow \Phi}{\mathbb{E}}[f(x, \widetilde{h}'(x, \phi(x)))] = \underset{[\widetilde{h}]}{\mathbb{E}}[f(x, \widetilde{h}(x))]$.

3. $\underset{x \leftarrow X, [\widetilde{h}]}{\mathbb{P}}[f(x, \widetilde{h}(x)) = 1] = \underset{\phi \leftarrow \Phi}{\mathbb{E}}[Z/2^m]$.

Let $\mu = \underset{\phi \leftarrow \Phi}{\mathbb{E}}[Z] = \underset{\phi \leftarrow \Phi}{\mathbb{E}}\left[\sum_{x \in \mathcal{X}'} Z_x\right]$. By Lemma 5, we have

$$\mathbb{P}[|Z - \mu| \geq \varepsilon 2^m] \leq \left(\frac{t}{\varepsilon^2 2^m}\right)^{t/2}.$$

Let us call ϕ bad for f if $|Z - \mu| \geq \varepsilon 2^m$ (or equivalently, using iii)),

$$\left| \underset{x \leftarrow X, [\widetilde{h}]}{\mathbb{E}}[f(x, \widetilde{h}(x))] - \underset{x \leftarrow X}{\mathbb{E}}[f(x, \phi(x))] \right| \geq \varepsilon$$

We want to choose t such that the probability of ϕ being bad for any particular $f \in \mathcal{F}$ is less than $1/|\mathcal{F}|$, i.e.

$$(t/\varepsilon^2 2^m)^{t/2} < |\mathcal{F}|^{-1}. \tag{42}$$

We postpone for a second how to choose t and discussing when this is even possible. Assuming (42),

$$\underset{\phi \leftarrow \Phi}{\mathbb{P}}[|Z - \mu| \geq \varepsilon 2^m] \leq \left(\frac{t}{\varepsilon^2 2^m}\right)^{t/2} < |\mathcal{F}|^{-1},$$

---

[11] Any distribution satisfying $H_\infty(X) = m$ can be written as a convex combination of flat distributions with min-entropy m. Often, this fact is sufficient to conclude that a result proven for flat distributions with min-entropy m implies the result for any distribution with the same min-entropy. Here, this is not quite the case, because we might end up using a different ϕ for every flat distribution. But as the only property we actually require from X is $\mathbb{P}[X = x] \leq 2^{-m}$, the proof goes through for general X, but becomes somewhat more technical.

and by taking a union bound over all $f \in \mathcal{F}$, we get

$$\Pr_{\phi \leftarrow \varPhi}[\exists f \in \mathcal{F} \; : \; |Z - \mu| \geq \varepsilon 2^m] < 1,$$

which implies that there exits $\phi \in \varPhi$ such that

$$\forall f \in \mathcal{F} \; : \; |Z - \mu| < \varepsilon 2^m.$$

Equivalently, using how Z and μ were defined,

$$\forall f \in \mathcal{F} \; : \; \left| \sum_{x \in \mathcal{X}'} f(x, \widehat{h}(x)) - \sum_{x \in \mathcal{X}'} f(x, \widetilde{h}(x)) \right| < \varepsilon 2^m.$$

Finally, using that X is uniform over $\mathcal{X}'$, we get (for the above choice of ϕ) the statement of the lemma

$$\forall f \in \mathcal{F} \; : \; \left| \underset{x \leftarrow X}{\mathbb{E}}[f(x, \widehat{h}(x))] - \underset{x \leftarrow X, [\widetilde{h}]}{\mathbb{E}}[f(x, \widetilde{h}(x))] \right| < \varepsilon.$$

We still have to determine when t can be chosen so that (42) holds. By taking logarithm and rearranging the terms, (42) becomes

$$mt/2 > \log|\mathcal{F}| + (t/2)\log(t) + t\log(1/\varepsilon),$$

i.e.,

$$m > 2\log|\mathcal{F}|/t + \log(t) + 2\log(1/\varepsilon).$$

Setting $t = \log|\mathcal{F}|$, we get

$$m > 2 + \log\log|\mathcal{F}| + 2\log(1/\varepsilon).$$

which holds as it is the condition (5) we made on the min-entropy $m = H_\infty(X)$.

Standard versus Selective Opening Security: Separation and Equivalence Results*

Dennis Hofheinz and Andy Rupp

Karlsruhe Institute of Technology, Germany
{dennis.hofheinz,andy.rupp}@kit.edu

Abstract. Suppose many messages are encrypted using a public-key encryption scheme. Imagine an adversary that may adaptively ask for openings of some of the ciphertexts. Selective opening (SO) security requires that the *unopened* ciphertexts remain secure, in the sense that this adversary cannot derive any nontrivial information about the messages in the unopened ciphertexts.

Surprisingly, the question whether SO security is already implied by standard security notions has proved highly nontrivial. Only recently, Bellare, Dowsley, Waters, and Yilek (Eurocrypt 2012) could show that a strong form of SO security, *simulation-based* SO security, is not implied by standard security notions. It remains wide open, though, whether the potentially weaker (and in fact comparatively easily achievable) form of *indistinguishability-based* SO (i.e., IND-SO) security is implied by standard security. Here, we give (full and partial) answers to this question, depending on whether active or passive attacks are considered. Concretely, we show that:

(a) For active (i.e., chosen-ciphertext) security, standard security does *not* imply IND-SO security. Concretely, we give a scheme that is IND-CCA, but not IND-SO-CCA secure.

(b) In the case of passive (i.e., chosen-plaintext) security, standard security *does* imply IND-SO security, at least in a generic model of computation and for a large class of encryption schemes. (Our separating scheme from (a) falls into this class of schemes.)

Our results show that the answer to the question whether standard security implies SO security highly depends on the concrete setting.

Keywords: security definitions, public-key encryption, selective opening security.

1   Introduction

Motivation. It is a challenging task to find a useful and achievable definition of security for encryption schemes. There seems to be no "one size fits all" security notion; for instance, certain settings involve key-dependent messages

---

* Supported by DFG grant GZ HO 4534/2-1.

Y. Lindell (Ed.): TCC 2014, LNCS 8349, pp. 591–615, 2014.

(e.g., [7, 9, 2]) or leakage of key material (e.g., [18, 13, 1]). In most of these specific settings, it is easily seen that standard encryption security notions (such as IND-CPA or IND-CCA security) do not provide any reasonable security guarantees. However, one particularly challenging setting is the setting of *selective opening attacks*, which models a specific (and realistic) form of adaptive corruptions. The topic of this paper is the connection of standard and selective opening security.

Selective Opening Attacks. The premise of a selective opening (SO) attack is as follows: suppose an adversary observes many ciphertexts c_i, and then gets to request openings of some of them. (Here, an opening corresponds to an adaptive corruption of the sender, and yields not only the plaintext m_i but also the random coins used during encryption.) The question is: can the adversary learn anything about the *unopened* m_i? Of course, if the encrypted messages are related, then the opened messages may already reveal information about the unopened messages. (In fact, this is the main source of trouble when trying to define selective opening security.) However, we would like to express that the unopened messages remain "as secure as possible", given the opened messages.

Selective Opening Security Notions... Dwork et al. [12] were the first to propose a formal SO security notion; their notion is simulation-based and was formulated for commitments. Bellare et al. [5] gave a public-key encryption (PKE) version of the definition of [12] (SIM-SO-CPA[1]), along with a weaker, indistinguishability-based notion (weak IND-SO-CPA).[2] Most relations among SO security notions (and between SO and standard security notions) have already been investigated (see also Figure 1). Specifically, [8] provided separations[3] between SO notions, and Bellare et al. [4] have separated SIM-SO-CPA security from IND-CPA security. The *only* remaining open question (that we approach in this paper) is thus

> Does standard security already imply *indistinguishability-based* selective opening security?

...and Constructions. Bellare et al. [5] proved lossy encryption [22, 21, 20] weakly IND-SO-CPA secure, and the scheme of Goldwasser and Micali [15] SIM-SO-CPA secure. Subsequently, several works have developed chosen-ciphertext secure (i.e., weakly IND-SO-CCA and SIM-SO-CCA secure) PKE schemes [14, 16, 17]. However, it seems safe to say that (weak) indistinguishability-based SO security is significantly easier to achieve than simulation-based SO security. In

---

[1] The naming of SO notions is not quite consistent in the literature. We follow the naming of Böhl et al. [8].

[2] There is also a stronger indistinguishability-based SO notion called *full* IND-SO-CPA. Weak and full IND-SO-CPA security differ in the sense that the considered (joint) message distributions are arbitrary in full IND-SO-CPA, but restricted in weak IND-SO-CPA security. No fully IND-SO-CPA secure schemes are known.

[3] Here, with a "separation" between two security notions X and Y, we mean that there is a scheme that achieves X but not Y (or vice versa). We do *not* mean that a Y-secure scheme cannot be constructed from an X-secure one (or vice versa).

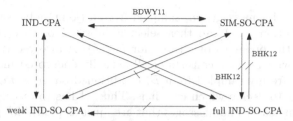

Fig. 1. Relations among notions of selective opening security and IND-CPA security. Solid arrows denote implications, crossed arrows denote concrete counterexamples, and the dashed arrow stands for the remaining open question investigated in this paper.

particular, the most efficient SO secure PKE schemes are not known to be SIM-SO secure. This makes the question whether standard security implies weak IND-SO security even more interesting.

Our Contribution. We tackle this last remaining question both in the chosen-plaintext (CPA) and in the chosen-ciphertext (CCA) case. We give a definite answer in the CCA case and a partial answer in the CPA case. First, we separate IND-CCA and IND-SO-CCA security: we give an IND-CCA secure but IND-SO-CCA insecure PKE scheme. Our result utilizes the standard model of computation and works under the minimal assumption that IND-CCA secure PKE schemes exist. Nonetheless, the IND-SO-CCA attack on our scheme is completely generic and does not make use of, e.g., non-black-box techniques (such as using the internal structure of the IND-CCA secure scheme). Our second result shows that IND-CPA and IND-SO-CPA security are equivalent in a generic model of computation and with respect to a restricted class of PKE schemes. We stress that the generic model considered for the CPA equivalence is realistic: it covers, e.g., ElGamal, Cramer-Shoup and similar encryption schemes, and in fact also concrete instantiations (e.g., based on Cramer-Shoup) of our separating example for the CCA case (including our attack on its weak IND-SO-CCA security). Interestingly, [4] shows that there is no such equivalence in the case of SIM-SO-CPA for the class of committing encryption schemes which also includes ElGamal and Cramer-Shoup. The adversary for which they can show that no simulator exists is a simple generic algorithm.

Another interesting point of view on our results is the following: For a broad class of encryption schemes (including instances of our separating scheme), it holds that any generic IND-SO-CPA adversary can be turned into a generic IND-CPA adversary, while this does not hold in the CCA case. For instance, there exists an efficient generic IND-SO-CCA adversary against our separating scheme, while there are no generic (or even non-generic) IND-CCA adversaries.

Details on Our IND-CCA/IND-SO-CCA Separation. To construct our separating scheme, we take an arbitrary IND-CCA secure PKE scheme and modify it such that a weak IND-SO-CCA attack becomes possible. To understand the basic idea behind our modification, recall that in the weak IND-SO-CCA experiment, an adversary A first receives a ciphertext vector $\mathbf{c} = (c_i)_i$ with

$c_i \leftarrow \mathsf{Enc}(pk, m_i)$ for messages m_i sampled from a (joint) adversarially selected message distribution $\mathcal{D}$. A can then select a subset $\mathcal{I}$ of all c_i to be opened. In addition to the openings of all m_i (for $i \in \mathcal{I}$), A also receives a full message vector $\mathbf{m}$ which *either* consists of all actually encrypted messages m_i, *or* of messages m_i' freshly sampled from $\mathcal{D}$, conditioned on $m_i' = m_i$ for all $i \in \mathcal{I}$. As usual, A has to decide which case it is. Thus, A has to distinguish between the encrypted messages and messages that are "just as plausible" given only the opened messages.

To obtain our separating scheme, we take an IND-CCA secure scheme and modify its decryption algorithm. Namely, we now allow a special type of decryption queries (soa, Z) in which Z contains a whole ciphertext vector $\mathbf{c}$, along with openings of a subset of these ciphertexts. (For now, it is easiest to imagine that this subset is selected externally and randomly.) If the openings are valid, then decryption will return an error-corrected version of the message vector from $\mathbf{c}$. (Hence, the scheme itself actually helps an adversary that can prove that it is taking part in an SO attack.)

This immediately gives rise to a weak IND-SO-CCA attack: a suitable adversary A essentially only has to relay between its decryption oracle and the SO experiment to obtain the decryption of all challenge ciphertexts. The message distribution considered in the attack will only select codewords, so that the mentioned error correction will not disturb the decryption. Moreover, the underlying code has the property that a codeword is not fixed by the openings that occur during the attack. (Hence, a re-sampling will lead to a different message vector and can thus be detected.)

It is more challenging to prove that our modification does not harm the scheme's IND-CCA security. Intuitively, an IND-CCA adversary B could try to embed its own (IND-CCA) challenge c^* into a ciphertext vector $\mathbf{c}$ and obtain the decryption of c^* through a suitable (soa, Z) query. (With a little luck, B will not have to open c^*, so decryption will return the full message vector, including the decryption of c^*.)

To cope with such an IND-CCA adversary B, we will answer (soa, Z) only with the error-corrected message vector. Decryption will ensure (by the random choice of $\mathcal{I}$ and by ensuring suitably valid openings) that most of the encrypted messages m_i *and all opened messages* are consistent with a unique single codeword. (If this is not the case, then the query is rejected. Of course, we will have to make sure that B also learns nothing from the fact that the query was rejected.) Decryption then returns this unique codeword, and not simply the decryption of all individual ciphertexts. This procedure makes sure that a single ciphertext c^* embedded into $\mathbf{c}$ alone has no significant influence on the returned value.

Our strategy is somewhat reminiscent of the strategy of Bellare et al. [5], who show a black-box impossibility for IND-SO secure commitments. Our approach can be seen as a refinement and adaptation of their ideas to the PKE setting and to the standard model.

Note that our attack only uses two decryption queries; furthermore, one of these queries can be substituted by a random oracle query when adapting the scheme to the random oracle model. Thus, our scheme also gives rise to a separation between IND and weak IND-SO security in a bounded CCA setting [10]. Moreover, since CCA settings with only 1 decryption query and non-malleability are tightly related [6], our counterexample has also implications for non-malleability notions of security. (See Section D for details.)

Details on Our Generic Group IND-CPA/IND-SO-CPA Equivalence. Our equivalence result applies to a broad class of encryption schemes over prime order groups for which public keys as well as ciphertexts can be described by (low-degree) polynomials "in the exponent". We model the underlying group as generic (following Shoup's formalization) with respect to the IND-SO-CPA adversary and the adversarial message sampling algorithm. That means that the only basic group operations such algorithms may perform are equality testing, application of the group law, and computation of inverse elements. However, note that this is already sufficient, e.g., for realizing our efficient IND-SO-CCA adversary (see also Section C). A potential hash function utilized by the encryption scheme is modeled as a Random Oracle. Although the model we consider for our equivalence result may appear highly idealized, proving the equivalence is anything but trivial. There are several novel and challenging aspects about this proof; we only highlight a few here.

The common strategy of a proof in the generic group model is to show that, with overwhelming probability, an adversary does not obtain any information about the underlying secrets (e.g., secret keys, the challenge bit in indistinguishability games, etc.) of the considered game (IND-SO-CPA in our case). Thus, it can only win by mere guessing. To this end, one shows by means of a simulation game (where all secrets are replaced by formal variables) that a generic algorithm may only obtain information about secrets from nontrivial equations that hold between low-degree combinations of these secrets. (An equation is called trivial if it holds for all possible choices of the secrets.) If the secret values are chosen uniformly at random then by applying standard techniques (e.g., the Schwartz-Zippel Lemma in the case of prime power order groups) one can see that such equations may occur only very rarely. However, in our setting also the adversarial messages, which are chosen according to an arbitrary (efficiently re-samplable) distribution, belong to the secrets for which we want to argue that they are hidden information-theoretically. Moreover, in the opening phase, parts of the secrets are even disclosed to the adversary. We cope with these issues by modifying the way we usually simulate in the generic model and, hence, how non-trivial equations are defined. In particular, we need to adapt the simulation when the opening phase starts and show that a non-trivial equation and "bad" message distribution can be leveraged to win the IND-CPA game.

In a nutshell, our proof is split into two parts: First, we show that in order to win the IND-CPA game, it suffices that for all possible public keys and encryptions of a message vector, we can efficiently compute a non-trivial representation of the neutral group element in terms of the public key and (at least one of) the corresponding ciphertexts. The idea is to replace one of the messages with a different one for which this equation does not hold anymore and use the two messages in the IND-CPA game. Second, we show that from any generic IND-SO-CPA adversary, such a representation and message vector can be extracted.

Outline. After recalling some definitions in Section 2, we describe our separation in Section 3. The generic equivalence in the passive case can be found in Section 4. Sections A, B, and C discuss the restrictions we make for the CPA case, and in particular show that our separating scheme from the CCA case (and its analysis) is generic in our sense. Finally, Section D briefly describes extensions to our CCA separation.

2   Preliminaries

Notation. For $n \in \mathbb{N}$, let $[n] := \{1, \ldots, n\}$. Throughout the paper, $k \in \mathbb{N}$ denotes the security parameter. For a finite set $\mathcal{S}$, we denote by $s \leftarrow \mathcal{S}$ the process of sampling s uniformly from $\mathcal{S}$. For a probabilistic algorithm A, we denote with $\mathcal{R}_A$ the space of A's random coins. $y \leftarrow A(x; R)$ denotes the process of running A on input x and with randomness $R \leftarrow \mathcal{R}_A$, and assigning y the result. We write $y \leftarrow A(x)$ for $y \leftarrow A(x; R)$ with uniform R. If A's running time is polynomial in k, then A is called probabilistic polynomial-time (PPT).

PRFs. A pseudorandom function (PRF) is a function $\mathsf{PRF} : \mathcal{K} \times \mathcal{D} \to \mathcal{R}$ for finite $\mathcal{K}, \mathcal{R}$, such that oracle access to $\mathsf{PRF}_K(\cdot)$ (for $K \leftarrow \mathcal{K}$) is indistinguishable from oracle access to a truly random function $RF : \mathcal{D} \to \mathcal{R}$. Concretely, for a distinguisher D, let $\mathsf{Adv}^{\mathsf{prf}}_{\mathsf{PRF},D}(k) := \Pr\left[D^{\mathsf{PRF}_K(\cdot)} = 1\right] - \Pr\left[D^{RF(\cdot)} = 1\right]$. We require that $\mathsf{Adv}^{\mathsf{prf}}_{\mathsf{PRF},D}$ is negligible for all PPT D.

PKE Schemes. A public-key encryption (PKE) scheme PKE with message space $\mathcal{M}$ consists of three PPT algorithms $\mathsf{Gen}, \mathsf{Enc}, \mathsf{Dec}$. Key generation $\mathsf{Gen}(1^k)$ outputs a public key pk and a secret key sk. Encryption $\mathsf{Enc}(pk, m)$ takes pk and a message $m \in \mathcal{M}$, and outputs a ciphertext c. Decryption $\mathsf{Dec}(sk, c)$ takes sk and a ciphertext c, and outputs a message m. For correctness, we want $\mathsf{Dec}(sk, c) = m$ for all $m \in \mathcal{M}$, all $(pk, sk) \leftarrow \mathsf{Gen}(1^k)$, and all $c \leftarrow \mathsf{Enc}(pk, m)$.

Standard Security Notions. Let PKE be a PKE scheme as above. For an adversary A, consider the following experiment: first, the experiment samples $(pk, sk) \leftarrow \mathsf{Gen}(1^k)$ and runs A on input pk. Once A outputs two messages m_0, m_1, the experiment flips a coin $b \leftarrow \{0, 1\}$ and runs A on input $c^* \leftarrow \mathsf{Enc}(pk, m_b)$. We say that A wins the experiment iff $b' = b$ for A's final output b'. We denote A's advantage with $\mathsf{Adv}^{\mathsf{ind\text{-}cpa}}_{\mathsf{PKE},A}(k) := \Pr\left[A \text{ wins}\right] - 1/2$ and say that PKE is IND-CPA secure iff $\mathsf{Adv}^{\mathsf{ind\text{-}cpa}}_{\mathsf{PKE},A}$ is negligible for all PPT A. Similarly, write $\mathsf{Adv}^{\mathsf{ind\text{-}cca}}_{\mathsf{PKE},A}(k) := \Pr\left[A \text{ wins}\right] - 1/2$ for A's winning probability when A additionally gets access to a decryption oracle $\mathsf{Dec}(sk, \cdot)$ at all times. (To avoid trivialities, A may not query Dec on c^*, though.) PKE is IND-CCA secure iff $\mathsf{Adv}^{\mathsf{ind\text{-}cca}}_{\mathsf{PKE},A}$ is negligible for all PPT A.

Security under Selective Openings. Following [5, 16, 8], we present an indistinguishability-based definition for security under selective openings that captures security of an encryption scheme under adaptive attacks.

Intuitively, an adversary A that receives a vector of ciphertexts, along with openings of a subset of these ciphertexts, should not be able to distinguish the messages in the unopened ciphertexts from independently selected messages. The encrypted message vector is selected according to a (joint) message distribution selected by A. A also selects the set of ciphertexts to be opened, in a way possibly depending on the ciphertexts themselves. Since we currently do not know how to achieve this security notion for arbitrary (efficiently samplable) message distributions, we further restrict to efficiently re-samplable message distributions:

$$\boxed{\begin{aligned}
&\textbf{Experiment } \mathsf{Exp}_{\mathsf{PKE},A}^{\mathsf{weak\text{-}ind\text{-}so\text{-}cpa}} \\
&b \leftarrow \{0,1\} \\
&(pk, sk) \leftarrow \mathsf{Gen}(1^k) \\
&\mathsf{samp}(\cdot) \leftarrow A(pk) \\
&\mathbf{m}_0 := (m_i)_{i \in [n]} \leftarrow \mathsf{samp}() \\
&\mathbf{R} := (R_i)_{i \in [n]} \leftarrow (\mathcal{R}_{\mathsf{Enc}})^n \\
&\mathbf{c} := (c_i)_{i \in [n]} := (\mathsf{Enc}(pk, m_i; R_i))_{i \in [n]} \\
&\mathcal{I} \leftarrow A(\mathsf{sel}, \mathbf{c}) \\
&\mathbf{m}_1 \leftarrow \mathsf{samp}(\mathbf{m}_{\mathcal{I}}) \\
&out_A \leftarrow A(\mathsf{out}, (R_i)_{i \in \mathcal{I}}, \mathbf{m}_b) \\
&\text{return } 1 \text{ if } out_A = b, \text{ and } 0 \text{ otherwise}
\end{aligned}}$$

Fig. 2. Weak IND-SO-CPA experiment

Definition 1 (Efficiently re-samplable). *Let $n = n(k) > 0$, and let $\mathcal{D}$ be a joint distribution over $\mathcal{M}^n$. We say that $\mathcal{D}$ is efficiently re-samplable if there is a PPT algorithm samp such that for any $\mathcal{I} \subseteq [n]$ and any partial vector $\mathbf{m}'_{\mathcal{I}} := (m'_i)_{i \in \mathcal{I}} \in \mathcal{M}^{|\mathcal{I}|}$, $\mathsf{samp}(\mathbf{m}'_{\mathcal{I}})$ samples from $\mathcal{D} \mid \mathbf{m}_{\mathcal{I}}$, i.e., from the distribution $\mathcal{D}$, conditioned on $m_i = m'_i$ for all $i \in \mathcal{I}$. Note that in particular, $\mathsf{samp}()$ samples from $\mathcal{D}$.*

Definition 2 (Weak indistinguishability-based selective opening security). *For a PKE scheme $\mathsf{PKE} = (\mathsf{Gen}, \mathsf{Enc}, \mathsf{Dec})$, a polynomially bounded function $n = n(k) > 0$, and a stateful PPT adversary A, consider the experiment in Figure 2. We only allow A that always output re-sampling algorithms as in Definition 1. We call PKE weakly IND-SO-CPA secure if*

$$\mathsf{Adv}_{\mathsf{PKE},A}^{\mathsf{ind\text{-}so\text{-}cpa}}(k) := \Pr\left[\mathsf{Exp}_{\mathsf{PKE},A}^{\mathsf{weak\text{-}ind\text{-}so\text{-}cpa}}(k) = 1\right] - \frac{1}{2}$$

is negligible for all PPT A. Similarly, we define an experiment $\mathsf{Exp}_{\mathsf{PKE},A}^{\mathsf{weak\text{-}ind\text{-}so\text{-}cca}}$ (with advantage $\mathsf{Adv}_{\mathsf{PKE},A}^{\mathsf{ind\text{-}so\text{-}cca}}$) that is identical to $\mathsf{Exp}_{\mathsf{PKE},A}^{\mathsf{weak\text{-}ind\text{-}so\text{-}cpa}}$, except that A gets access to a decryption oracle $\mathsf{Dec}(sk, \cdot)$ at all times. To avoid trivialities, we only allow A that never query their decryption oracle with any ciphertext from $\mathbf{c}$. We say that PKE is weakly IND-SO-CCA secure if $\mathsf{Adv}_{\mathsf{PKE},A}^{\mathsf{ind\text{-}so\text{-}cca}}(k)$ is negligible.

There are some minor technical differences between Definition 2 and the IND-SO-ENC definition from [5]: IND-SO-ENC security universally quantifies over all (efficiently re-samplable) message distributions. We let A choose samp instead, e.g., to allow a message distribution that depends on the public key pk. (In fact, otherwise it is not even clear that the resulting definition implies IND-CPA security.) Besides, unlike Böhl et al. [8], we model only one round of openings for simplicity. (However, our results hold also for multiple rounds of openings.)

3 Our Separating Encryption Scheme

In this section, we describe a PKE scheme that is IND-CCA secure, but not weakly IND-SO-CCA secure. So our scheme separates standard security from even the weakest considered form of selective opening security.

3.1 The Scheme

Specific Notation and Assumptions. In the following, let $\mathbb{F} = \mathbb{Z}_p$ be the finite field of size p for a prime p. (We will later choose a $(k+1)$-bit p as part of the public key of our scheme.) By $\mathsf{ipol}((X_i, Y_i)_{i=0}^d)$ (for pairwise different X_i), we denote the unique degree-$\le d$ polynomial $F \in \mathbb{F}[X]$ with $F(X_i) = Y_i$ for all i. Let $\mathcal{S}_\ell^S$ denote the set of all ℓ-sized subsets of S. We will assume a PRF $\mathsf{PRF} : \{0,1\}^k \times \{0,1\}^* \to \mathcal{S}_k^{[3k]}$ (such that oracle access to $\mathsf{PRF}_K(\cdot)$ for uniform $K \in \{0,1\}^k$ cannot be distinguished from access to a truly random function that maps arbitrary bitstrings to uniform k-sized subsets of $[3k]$). We will also assume an IND-CCA secure PKE scheme $\mathsf{PKE}' = (\mathsf{Gen}', \mathsf{Enc}', \mathsf{Dec}')$ with message space $\mathbb{F}$. (The requirement about the message space is without loss of generality [19]; see also Section B for a scheme with a group as message space.)

Construction. $\mathsf{PKE} = (\mathsf{Gen}, \mathsf{Enc}, \mathsf{Dec})$ is constructed from PKE':

Key generation adds a PRF key to sk: $\mathsf{Gen}(1^k)$ outputs $(pk, sk) = ((pk', p), (sk', K))$ for $(pk', sk') \leftarrow \mathsf{Gen}'(1^k)$, a uniformly chosen $(k+1)$-bit prime p, and $K \leftarrow \{0,1\}^k$.

Encryption marks ciphertexts as "regular": $\mathsf{Enc}(pk, m)$ runs $c' \leftarrow \mathsf{Enc}'(pk', m)$ and outputs $c = (\mathbf{reg}, c')$.

Decryption decrypts "regular" ciphertexts as PKE', but also offers possibilities to evaluate PRF_K and perform a special type of attack by decrypting "non-regular" ciphertexts:

$$\mathsf{Dec}(sk, c) = \begin{cases} \mathsf{Dec}'(sk', c') & \text{if } c = (\mathbf{reg}, c') \text{ for some } c', \\ \mathsf{PRF}_K(Z) & \text{if } c = (\mathbf{sel}, Z) \text{ for some } Z, \\ \mathsf{SOA}(sk, Z) & \text{if } c = (\mathbf{soa}, Z) \text{ for some } Z, \\ \bot & \text{else.} \end{cases}$$

Here, the function $\mathsf{SOA}(sk, Z)$ operates as follows:

1. Parse Z as $Z = ((c_i')_{i \in [3k]}, (m_i, R_i)_{i \in \mathcal{I}})$, where $\mathcal{I} = \mathsf{PRF}_K((c_i')_{i \in [3k]})$.
2. If there are indices $i \ne j$ with $c_i' = c_j'$, then return $\bot$.
3. If there is an $i \in \mathcal{I}$ with $\mathsf{Enc}'(pk', m_i; R_i) \ne c_i'$, then return $\bot$.
4. Decrypt the unopened ciphertexts by $m_i = \mathsf{Dec}'(sk', c_i')$ for $i \in [3k] \setminus \mathcal{I}$.
5. First, determine if there is a degree-$\le k$ polynomial $F \in \mathbb{F}[X]$ with $F = \mathsf{ipol}((i, m_i)_{i \in \mathcal{I} \cup \{j\}})$ for more than k values $j \in [3k] \setminus \mathcal{I}$. Note that there are only $2k$ candidates $F_\ell = \mathsf{ipol}((i, m_i)_{i \in \mathcal{I} \cup \{\ell\}})$ (for $\ell \in [3k] \setminus \mathcal{I}$) for F; hence, if such an F exists, it can be found efficiently (and in fact is unique). Return F, or $\bot$ if no such F exists.

Intuitively, $\mathsf{SOA}(sk, Z)$ returns a polynomial F that is consistent with *all* opened values m_i (for $i \in \mathcal{I}$), and *most* unopened values m_i (for $i \in [3k] \setminus \mathcal{I}$). (This slight distinction will be crucial to ensure that access to SOA does not enable IND-CCA attacks.)

Rationale and Intuition for Security Analysis. The rationale of our modifications to $\mathsf{PKE'}$ is to enable a specific attack that only a weak IND-SO-CCA adversary is able to perform. Concretely, once an adversary supplies $3k$ ciphertexts along with openings of k of them (in a suitable $\mathsf{Dec}(sk, (\mathsf{soa}, Z))$ query), the scheme itself helps to decrypt all ciphertexts. Indeed, PKE is weakly IND-SO-CCA insecure with respect to the message distribution $\mathcal{D} = (F(i))_{i \in [3k]}$ with a uniform degree-$\leq k$ polynomial F: by relaying between the experiment and its Dec oracle, an adversary can obtain *all* (i.e., even unopened) challenge messages.

The difficult part will be to prove that our modification preserves the IND-CCA security of $\mathsf{PKE'}$. That is, we will have to prove that (sel, Z) and (soa, Z) decryption queries do not help an IND-CCA adversary A on PKE. For (sel, Z) queries, this is intuitively clear, as they are answered independently of the "actual" secret key sk'. For (soa, Z) queries, we will argue that the answer can already be deduced by "regular" decryption queries (reg, c'). Concretely, if the $\mathsf{PKE'}$ ciphertext c^* from A's own challenge (reg, c^*) does not appear as ciphertext in Z, A can itself use Dec queries to emulate $\mathsf{SOA}(sk, Z)$. And even if Z contains c^*, A can still use Dec to decrypt all ciphertexts in Z except for c^*. We will show that $\mathsf{SOA}(sk, Z)$ can be reasonably well approximated when knowing all plaintexts encrypted in Z except for at most one. Namely, in order not to be rejected by $\mathsf{SOA}(sk, Z)$, almost all of the ciphertexts in Z must already decrypt to a value $F(i)$ that is consistent with one F. Knowing all but one plaintext allows a simulation to compute this F, and thus $\mathsf{SOA}(sk, Z)$'s answer.

Variations. Section D gives variations for bounded CCA security and non-malleability.

3.2 Why PKE Is Not Weakly IND-SO-CCA Secure

We now formally show that PKE allows for a simple weak IND-SO-CCA attack.

Theorem 1. *The PKE scheme PKE from Section 3.1 is not weakly IND-SO-CCA secure.*

Proof. We construct a weak IND-SO-CCA adversary A on PKE. On input pk, A outputs the $3k$-message distribution

$$\mathcal{D} = \left\{ (F(1), \ldots, F(3k)) \mid F \in \mathbb{F}[X] \text{ uniformly chosen degree-} \leq k \text{ polynomial} \right\}$$

along with a suitable (re-)sampling algorithm samp. (For instance, samp can randomly extend its input $(F(i))_{i \in \mathcal{I}}$ to $k+1$ evaluation points as necessary and then use polynomial interpolation to retrieve F and thus all $F(i)$.) Note that k messages $m_i = F(i)$ from a $\mathcal{D}$-sample do not fully determine F and thus the whole message vector.

Once A receives a ciphertext vector $\mathbf{c} := (\mathbf{reg}, c_i')_{i \in [3k]}$, it queries its decryption oracle on $(\mathbf{sel}, (c_i')_{i \in [3k]})$ to receive a k-sized subset $\mathcal{I} \subset [3k]$. This $\mathcal{I}$ is the subset that A submits to its weak IND-SO-CCA experiment. Let $\mathsf{bad_{coll}}$ be the event that $c_i' = c_j'$ for some $i \neq j$. By the correctness of PKE', this can only happen if $m_i = m_j$ for these i, j. By definition of $\mathcal{D}$, we have $\Pr[m_i = m_j] = 1/|\mathbb{F}| < 1/2^k$ for any fixed i, j. Hence, a union bound over all i, j shows that $\Pr[\mathsf{bad_{coll}}] < \frac{3k(3k-1)}{2} \cdot \frac{1}{2^k} < \frac{5k^2}{2^k}$. We will thus assume $\neg\mathsf{bad_{coll}}$ hereafter.

Upon receiving openings $(m_i, R_i)_{i \in \mathcal{I}}$ and a message vector $\mathbf{m}^* = (m_i^*)_{i \in [3k]}$, A queries its decryption oracle on $(\mathbf{soa}, ((c_i')_{i \in [3k]}, (m_i, R_i)_{i \in \mathcal{I}}))$. By definition of Dec (and the function SOA), A will thus receive a polynomial F with $m_i = F(i)$ for all $i \in [3k]$ and can thus obtain the actually encrypted messages m_i. Finally, A will output $out_A = 0$ iff $m_i^* = F(i)$ for all $i \in [3k]$.

Still assuming $\neg\mathsf{bad_{coll}}$, it is clear that A will output $out_A = 0$ when $b = 0$, i.e., when $\mathbf{m}^* = \mathbf{m}_0$. On the other hand, if $b = 1$, then $\mathbf{m}^* = \mathbf{m}_1$ has been re-sampled subject to $m_i^* = m_i$ for all $i \in \mathcal{I}$. However, since a message vector $\mathbf{m}$ from $\mathcal{D}$ is not fixed by only $k = |\mathcal{I}|$ values m_i, we have that $\mathbf{m}^* \neq \mathbf{m}_0$ (so that A outputs $out_A = 1$) except with probability at most $1/|\mathbb{F}| < 1/2^k$. Summarizing, we get

$$\mathsf{Adv}_{\mathsf{PKE},A}^{\mathsf{ind\text{-}so\text{-}cca}}(k) = \Pr[out_A = b] - \frac{1}{2} \geq \Pr[out_A = b \mid \neg\mathsf{bad_{coll}}] - \Pr[\mathsf{bad_{coll}}] - \frac{1}{2}$$

$$> \frac{1}{2}\left(1 + \left(1 - \frac{1}{2^k}\right)\right) - \frac{5k^2}{2^k} - \frac{1}{2} = \frac{1}{2} - \frac{5k^2 + 2}{2^k},$$

which is non-negligible (and in fact negligibly close to the maximal advantage).

3.3 Why PKE Is Still IND-CCA Secure

We show that PKE inherits PKE''s IND-CCA security.

Theorem 2. *The PKE scheme* PKE *from Section 3.1 is IND-CCA secure, assuming that* PKE' *is IND-CCA secure, and* PRF *is pseudorandom.*

Proof. Let A be a PPT adversary on PKE that makes exactly q decryption queries. We proceed in games, and let out_i denote the output of Game i.

Game 1 is the original IND-CCA game with A. Consequently,

$$\Pr[out_1 = 1] - 1/2 = \mathsf{Adv}_{\mathsf{PKE},A}^{\mathsf{ind\text{-}cca}}(k).$$

In **Game** 2, we answer decryption queries of the form $(\mathbf{sel}, Z)$ with $RF(Z)$ instead of $\mathsf{PRF}_K(Z)$ for a truly random function $RF : \{0,1\}^* \to \mathcal{S}_k^{[3k]}$. (We will assume that RF is efficiently implemented, e.g., using lazy sampling.) A straightforward reduction to PRF's pseudorandomness yields

$$\Pr[out_1 = 1] - \Pr[out_2 = 1] = \mathsf{Adv}_{\mathsf{PRF},D}^{\mathsf{prf}}(k)$$

for a suitable PRF distinguisher D.

In **Game** 3, we slightly change the way decryption queries of the form (soa, Z) are answered. Our goal is to avoid a decryption of c^*, where (reg, c^*) is A's own challenge ciphertext. Informally, we simply skip decrypting c'_i if $c'_i = c^*$ in Step 4 of the function $\mathsf{SOA}(sk, Z)$. In Step 5, we skip any comparison of m_i for $c'_i = c^*$. Formally, we change Steps 4 and 5 into

4. Let $\mathcal{I}^*$ be the set of all $i \in [3k] \setminus \mathcal{I}$ with $c'_i = c^*$. (Note that $|\mathcal{I}^*| \leq 1$.) Decrypt the unopened ciphertexts not equal to c^* by $m_i = \mathsf{Dec}'(sk', c'_i)$ for $i \in [3k] \setminus (\mathcal{I} \cup \mathcal{I}^*)$.

5. If there is a degree-$\leq k$ polynomial $F \in \mathbb{F}[X]$ with $F = \mathsf{ipol}((i, m_i)_{i \in \mathcal{I} \cup \{j\}})$ for more than k values $j \in [3k] \setminus (\mathcal{I} \cup \mathcal{I}^*)$, then return F. Else return $\perp$.

This modified version SOA' only yields different values from that of Game 2 if

(a) $Z = ((c'_i)_{i \in [3k]}, (m_i, R_i)_{i \in \mathcal{I}})$ with pairwise different c'_i and $\mathcal{I} = RF((c'_i)_{i \in [3k]})$,

(b) all openings are valid in the sense $\mathsf{Enc}'(m_i; R_i) = c'_i$ for $i \in \mathcal{I}$, and

(c) there are *exactly* $k + 1$ indices $i \in [3k] \setminus \mathcal{I}$ with $m_i = F(i)$ for a degree-$\leq k$ polynomial F.

In this case, $\mathsf{SOA}(sk, Z)$ will return F, while $\mathsf{SOA}'(sk, Z)$ might return $\perp$ (in case there is an unopened $c'_i = c^*$ with $m_i = F(i)$). Let us call a query (soa, Z) satisfying (a)-(c) *implausible*. Denote by $\mathsf{bad_{impl}}$ the event that A ever submits an implausible decryption query. Unless $\mathsf{bad_{impl}}$ occurs, Game 2 and Game 3 are identical, so that $\Pr[\mathsf{bad_{impl}}]$ is the same in these games.

Intuitively, $\mathsf{bad_{impl}}$ is unlikely, because it necessitates that the (randomly chosen) subset $\mathcal{I} = RF((c'_i)_{i \in [3k]})$ happens to contain only indices i with $m_i = F(i)$ for the uniquely determined polynomial F. However, requirement (c) states that $k - 1$ indices i are *not* compatible with F, meaning $m_i \neq F(i)$. The probability that any such i is contained in $\mathcal{I}$ is overwhelming. We prove the following lemma after the main proof.

Lemma 1. $\Pr[\mathsf{bad_{impl}}] \leq q \cdot \left(\frac{5}{6}\right)^k$ *for* $k \geq 2$.

Using Lemma 1, we thus get for $k \geq 2$:

$$|\Pr[out_3 = 1] - \Pr[out_2 = 1]| \leq \Pr[\mathsf{bad_{impl}}] \leq q \cdot \left(\frac{5}{6}\right)^k.$$

Finally, we have

$$\Pr[out_3 = 1] = \mathsf{Adv}^{\mathsf{ind\text{-}cca}}_{\mathsf{PKE}, B} + 1/2 \tag{1}$$

for a suitable IND-CCA adversary B on PKE'. Concretely, observe that the whole Game 3 only uses sk' to decrypt PKE' ciphertexts different from c^*. Hence, B can simulate A, using its own challenge public key pk' and ciphertext c^* as A's public key and challenge. A's choice of challenge messages m_0, m_1 is also used by B. A's decryption queries (reg, c') are relayed (as c') to B's decryption oracle; (sel, Z) and (soa, Z) queries are answered by B for A, using B's own decryption oracle as necessary for (soa, Z) queries. (Note that B's challenge c^* will never have to be decrypted by our change from Game 3.) This adversary B thus perfectly simulates Game 3 for A, so we get (1).

Taking things together yields

$$\left|\mathsf{Adv}_{\mathsf{PKE},A}^{\mathsf{ind\text{-}cca}} - \mathsf{Adv}_{\mathsf{PKE},B}^{\mathsf{ind\text{-}cca}}\right| = |\Pr\left[out_1\right] - \Pr\left[out_3\right]| \leq \left|\mathsf{Adv}_{\mathsf{PRF},D}^{\mathsf{prf}}(k)\right| + q \cdot \left(\frac{5}{6}\right)^k$$

for $k \geq 2$, which shows the theorem.

It remains to prove Lemma 1:

Proof (Proof of Lemma 1). Given a ciphertext vector $\mathbf{c} = (c_i')_{i \in [3k]}$, define $m_i = \mathsf{Dec}(sk', c_i')$ for all i. (Correctness implies that these m_i are the same that will be recovered by SOA, either using openings given by A, or by decryption.) Say that there is a (unique) degree-$\leq k$ polynomial F and a $(2k+1)$-sized subset $\overline{\mathcal{I}} \subset [3k]$ with $m_i = F(i) \Leftrightarrow i \in \overline{\mathcal{I}}$. (Note that this is a prerequisite for $\mathsf{bad}_{\mathsf{impl}}$.)

The crucial observation is that the set $\mathcal{I} = RF(\mathbf{c})$ that determines which ciphertexts A must open is chosen independently and uniformly from the set of all k-sized subsets of $[3n]$. Furthermore, $\mathcal{I}$ is only chosen once A makes a (sel, Z) or (soa, Z) query that involves $\mathbf{c}$. If $\mathcal{I} \not\subset \overline{\mathcal{I}}$, then there can be no implausible query with this $\mathbf{c}$. (Condition (b) would require that some $i^* \notin \overline{\mathcal{I}}$ is opened, so that Condition (c) cannot be met, as $m_{i^*} \neq F(i^*)$.) Hence $\mathcal{I} \subset \overline{\mathcal{I}}$ is a necessary requirement for an implausible query with this $\mathbf{c}$. But $\mathcal{I} \subset \overline{\mathcal{I}}$ means that a random k-sized subset $\mathcal{I}$ of $[3k]$ is a subset of a fixed $(2k+1)$-sized subset $\overline{\mathcal{I}} \subset [3k]$. Hence,

$$\Pr\left[\mathcal{I} \subset \overline{\mathcal{I}}\right] = \frac{\binom{2k+1}{k}}{\binom{3k}{k}} = \frac{(2k+1)!(2k)!}{(3k)!(k+1)!} = \frac{(2k+1)\cdots(k+2)}{(3k)\cdots(2k+1)} \overset{k \geq 2}{\leq} \left(\frac{5}{6}\right)^k.$$

Since A makes only q decryption queries, it can only submit at most q different $\mathbf{c}$. For each $\mathbf{c}$, the probability is at most $(5/6)^k$ that an implausible query with this $\mathbf{c}$ exists. Hence, a union bound shows that

$$\Pr\left[\mathsf{bad}_{\mathsf{impl}}\right] \leq q \cdot \left(\frac{5}{6}\right)^k.$$

4 Equivalence of IND-SO-CPA and IND-CPA in the GGM

We give evidence towards the equivalence of IND-SO-CPA and IND-CPA by showing that for a broad class of encryption schemes any efficient generic IND-SO-CPA adversary can be turned into an efficient IND-CPA adversary.

In the following, some additional notation is needed: For a vector of variables $\mathbf{X}$ or polynomials $\mathbf{P}$, let $|\mathbf{X}|$ and $|\mathbf{P}|$ denote the size of the corresponding vector. For a polynomial P, let $|P|$ denote the number of non-zero monomials.

4.1   The Class of $(\mathbf{P}, \mathbf{E}, \mathbf{H}, \mathcal{H})$-CS-Type Encryption Schemes

The following definition covers a broad class of public-key encryption schemes over prime order groups where messages are group elements. This includes El-Gamal, Cramer-Shoup (CS), and also a slight variation of the separating scheme from Section 3.1, e.g., instantiated with Cramer-Shoup (see Section B). Note that the restrictions on the polynomials in Definition 3 are reasonable for meaningful encryption (see Section A).

Definition 3. *Let G be a group of prime order p with generator g and $\mathbb{F} = \mathbb{Z}_p$. Furthermore, let u_1, u_2, u_3, v_1, $v_2 \in \mathbb{N}$,*

- $\mathbf{P} = (P_1 = 1, P_2, \ldots, P_{u_1})$ *be public key polynomials in* $\mathbb{F}[X_1, \ldots, X_{v_1}]$,
- $\mathbf{E} = (E_1, \ldots, E_{u_2})$ *be polynomials in* $\mathbb{F}[X_1, \ldots, X_{v_1}, Y_1, \ldots, Y_{v_2}, Z, M]$, *called encryption polynomials, where all monomials have the form*

$$\alpha P^{e_1} Z^{e_2} M^{e_3} \prod_{i=1}^{v_2} Y_i^{d_i}$$

   with $P \in \mathbf{P}$, $e_1, e_3 \in \{0, 1\}$, $e_1 + e_3 \le 1$, and $e_2, d_i \in \mathbb{N}_0$,
- $\mathbf{H} = (H_1, \ldots, H_{u_3})$ *be tuple of hash input polynomials, where $H_i \in \mathbf{E}$, $\deg_Z(H_i) = 0$, and for at least one H_i it holds that $\deg_M(H_i) > 0$ or $\max_j(\deg_{Y_j}(H_i)) > 0$,*
- $\mathcal{H} : G^{u_3} \mapsto \mathbb{F}$ *be a hash function.*

Then we call an encryption scheme over G a $(\mathbf{P}, \mathbf{E}, \mathbf{H}, \mathcal{H})$-CS-type encryption scheme if the following conditions are satisfied:

- *The public key is of the form $(g^{P(\mathbf{x})})_{P \in \mathbf{P}}$, where $\mathbf{x} \leftarrow \mathbb{F}^{v_1}$.*
- *The ciphertext of a message $m = g^{m'}$ is of the form $\mathbf{c} = (g^{E(\mathbf{x}, \mathbf{y}, z, m')})_{E \in \mathbf{E}}$, where $\mathbf{y} \leftarrow \mathbb{F}^{v_2}$ and z is the output of $\mathcal{H}$ given $g^{H_1(\mathbf{x}, \mathbf{y}, m')}, \ldots, g^{H_{u_3}(\mathbf{x}, \mathbf{y}, m')}$.*

Example 1. Cramer-Shoup encryption scheme can be viewed as $(\mathbf{P}, \mathbf{E}, \mathbf{H}, \mathcal{H})$-CS-type encryption scheme, where we assume that generator g from Definition 3 has been chosen randomly and

- $P_1 = 1$, $P_2 = X_1$, $P_3 = X_2 + X_1 X_3$, $P_4 = X_4 + X_1 X_5$, $P_5 = X_6$
- $E_1 = P_1 Y_1$, $E_2 = P_2 Y_1$, $E_3 = P_3 Y_1 + P_4 Y_1 Z$, $E_4 = P_5 Y_1 + M$
- $\mathcal{H} : G^3 \mapsto \mathbb{F}$ is a collision resistant hash function computed over group elements with exponents of the form $H_1 = E_1$, $H_2 = E_2$, $H_3 = E_4$.

4.2   IND-SO-CPA in the Generic Group Model

We base the formalization of the IND-SO-CPA game for generic adversaries on the generic group model (GGM) introduced by Shoup [23]. In Shoup's GGM elements are encoded as unique random bit strings, ensuring that no special property of a group's representation can be exploited. More precisely, let $\mathbb{E} \subset \{0,1\}^{\lceil \log_2(p) \rceil}$, where $|\mathbb{E}| = p$, denote the set of possible element encodings of a cyclic group G of order p. Since any such group G is isomorphic to $(\mathbb{F}, +)$, we will always use $\mathbb{F}$ for the internal representation of G. A *generic group oracle* defines the random map between group elements and encodings and allows A to perform operations from $\Omega = \{+, -\}$ on encoded group elements. Equality testing can be done without the help of $\mathcal{O}$ since encodings are unique.

Internal State of $\mathcal{O}$. The oracle maintains two lists $\mathcal{L}$ and $\mathcal{E}$ which are used to define the random mapping between $\mathbb{F}$ and $\mathbb{E}$ in a lazy manner: $\mathcal{L} \subset \mathbb{F}$ will be initially populated with the elements comprising the public key of the considered encryption scheme. While A interacts with $\mathcal{O}$, additional elements are added to $\mathcal{L}$. The list $\mathcal{E} \subset \mathbb{E}$ contains the random encodings corresponding to the elements in $\mathcal{L}$. More precisely, the i-th encoding $\mathcal{E}_i$ represents the i-th element $\mathcal{L}_i$. We will denote the encoding of an element $a \in \mathcal{L}$ by $[\![a]\!]$.

Encoding of Elements. Each time an element a should be added to $\mathcal{L}$, $\mathcal{O}$ checks if a is already contained in $\mathcal{L}$. If this is the case, $[\![a]\!]$ is already defined and will be appended to $\mathcal{E}$ again. Otherwise, a fresh encoding $\sigma \leftarrow \mathbb{E} \setminus \mathcal{E}$ is sampled and appended to $\mathcal{E}$. The encoding $[\![a]\!]$ is sent to A. We may assume that a generic algorithm A always outputs encodings it has previously received by the oracle: A fresh encoding not contained in $\mathcal{E}$ is associated with a random $a \in \mathbb{F} \setminus \mathcal{L}$. Assuming $|\mathcal{L}|$ is polynomial in $\log(p)$, such an element can be efficiently generated by A itself with overwhelming probability $1 - \frac{|\mathcal{L}|}{p}$ by sampling a random $a \in \mathbb{F}$. The corresponding encoding $[\![a]\!]$ can be computed from $[\![1]\!]$ using double-and-add. Similarly, in our upcoming IND-SO-CPA setting, A will be able to output an encoding that has been computed by another generic algorithm, but has not explicitly given to A, only with negligible probability of at most $\frac{|\mathcal{E}|}{p}$.

Query Operations. A may ask $\mathcal{O}$ to perform an operation $\circ \in \Omega$ on encoded elements $[\![a]\!], [\![b]\!] \in \mathcal{E}$. Then $a \circ b$ is added to $\mathcal{L}$ and $[\![a \circ b]\!]$ is sent to A.

The GGM IND-SO-CPA Game. The IND-SO-CPA game for generic adversaries against an $(\mathbf{P}, \mathbf{E}, \mathbf{H}, \mathcal{H})$-CS-type encryption scheme is shown in Figure 3a. By abuse of notation we assume that for a new input $[\![a]\!]$ given to a generic algorithm (in the sense that $a \notin \mathcal{L}$), a is first added to $\mathcal{L}$ and then an encoding is determined. The hash function $\mathcal{H} : G^{u_4} \mapsto \mathbb{F}$ is modeled as a Random Oracle, which on input of u_4 concatenated encodings, outputs a fresh hash value $z \leftarrow \mathbb{F}$ if it has not received this input before. Otherwise, the hash value which has been chosen previously is returned. Furthermore, as can be seen from Figure 3a, both the adversary A and samp are modeled as generic algorithms. The algorithm samp is stateless but its output may depend on the public key $[\![P(\mathbf{x})]\!]_{P \in \mathbf{P}}$ since this was given as input to A before samp was created.

4.3   Equivalence for $(\mathbf{P}, \mathbf{E}, \mathbf{H}, \mathcal{H})$-CS-Type Encryption

As a warm-up, consider the IND-SO-CPA game for ElGamal, viewed as a CS-type encryption scheme where $\mathbf{P} = (1, X)$ and $\mathbf{E} = (Y, XY + M)$, in the GGM. In this model, it is not hard to show (by means of a simulation game) that the only source of information for the adversary about the challenge bit b are non-trivial equations between elements that are linear combinations of the secret key x, the unopened random coins for encryption $(y_j)_{j \notin \mathcal{I}}$, and the unopened encrypted messages $(m'_{0,j})_{j \notin \mathcal{I}}$. More precisely, an adversary may only obtain information about b if the difference $\Delta(x, y_1, \ldots, y_n, m'_{0,1}, \ldots, m'_{0,n})$ of two computed elements is zero, where Δ is a non-zero polynomial of the form

$$\Delta = \alpha_0 + \alpha_1 X + \sum_{j=1}^{n} \beta_j Y_j + \sum_{j=1}^{n} \gamma_j (XY_j + M_j)$$

and $\beta_j = \gamma_j = 0$ for $j \in \mathcal{I}$ ($\mathcal{I} = \emptyset$ if we are not yet in the opening phase). What is the probability that this happens? Note that in contrast to the secret key and the random coins, the messages $m'_{0,j}$ are not necessarily uniformly chosen. So the well-known Schwartz-Zippel Lemma cannot immediately be applied. However, since samp is also assumed to be generic, $m'_{0,j}$ will be of the form $m'_{0,j} = R_j(x)$ for some polynomial R_j of the form $R_j = \alpha_0 + \alpha_1 X$. Let us consider the polynomial $\Delta' = \Delta(R_1, \ldots, R_n)$ which results from replacing any occurrence of M_j by R_j. It is easy to see that $\Delta' \neq 0$ if $\Delta \neq 0$. Finally, we can apply Schwartz-Zippel to Δ' to upper bound the probability that $\Delta(x, y_1, \ldots, y_n, m'_{0,1}, \ldots, m'_{0,n}) = \Delta'(x_1, y_1, \ldots, y_n)$ is zero, yielding the bound $\frac{2}{p}$.

Note that for more general public key and encryption polynomials as considered in Definition 3, Δ' is not guaranteed to be non-zero anymore: For instance, consider the slightly modified encryption polynomials $\mathbf{E} = (Y, XY + YM)$ and the difference polynomial $\Delta = -Y_1 + (XY_1 + Y_1 M_1)$. Here Δ' becomes zero for $R_1 = 1 - X$. Fortunately, it turns out that in this case the corresponding encryption scheme is already IND-CPA insecure. In our example, this is obvious: An IND-CPA adversary could choose $m_0 = g(g^x)^{-1}$ and a random message m_1 and check whether for the challenge ciphertext $c = (c_1, c_2)$ holds that $c_1^{-1} c_2 = g^{\Delta(X=x, Y_1=y_1, M_1=m'_b)}$ is equal to 1, where $m'_0 = 1 - x$ and $m'_1 = \log_g(m_1)$. With overwhelming probability this will not hold for $b = 1$.

More generally, we can show that any Δ and R_j's can be used to build an IND-CPA adversary that works similarly. This is done in the first part (Theorem 3) of our proof which is actually independent of the generic model. It essentially says that if for all possible public keys and all possible encryptions of certain messages, we can efficiently compute a non-trivial representation of $1 \in G$ in terms of the public key and the ciphertexts, then we can win the IND-CPA game with overwhelming probability. The idea is to replace one of the messages with a different one for which the equation does not hold anymore and use these two messages in the scope of the IND-CPA game. In the second part (Theorem 4), we show that any efficient generic adversary who wins the IND-SO-CPA game with non-negligible probability gives rise to such a representation (in form of a polynomial) and corresponding messages.

Theorem 3. *Let a* $(\mathbf{P}, \mathbf{E}, \mathbf{H}, \mathcal{H})$*-CS-type encryption scheme* PKE *over a group* G *of prime order* p *be given. Furthermore, let a polynomial* Δ *of the form*

$$\Delta = \sum_{P \in \mathbf{P}} \alpha_P P(\mathbf{X}) + \sum_{j=1}^{n} \sum_{E \in \mathbf{E}} \beta_{j,E} E(\mathbf{X}, \mathbf{Y}_j, Z_j, M_j)$$

and polynomials $R_1, \ldots, R_n$ of the form $R_i = \sum_{P \in \mathbf{P}} \alpha_P P(\mathbf{X})$ over $\mathbb{F}$ be given (where the coefficients α, β in the above representation are known) such that $\Delta \neq 0$ but $\Delta(M_1 = R_1, \ldots, M_n = R_n) = 0$. Then we can build a generic adversary B who wins the IND-CPA game for PKE, modeling $\mathcal{H}$ as a Random Oracle, with probability at least $1 - \frac{\deg(\Delta)}{2p}$ using $O((\max_{E \in \mathbf{E}}(|E|)|\mathbf{E}||\mathbf{Y}| + |\mathbf{P}|) \log(p)n)$ multiplications over G and $\mathbb{F}$.

Proof. First, observe that there is some $1 \leq i \leq n$ such that $\Delta(R_1, \ldots, R_{i-1}) \neq 0$ but $\Delta(R_1, \ldots, R_i) = 0$. Note that in this case, we know that $\deg_{M_j}(\Delta) = \deg_{Z_j}(\Delta) = \deg_Y(\Delta) = 0$, for all $Y \in \mathbf{Y}_j$ and $j > i$. Clearly, also for uniform $\mathbf{x}, \mathbf{y}_1, \ldots, \mathbf{y}_i, z_1, \ldots, z_i$, it holds that $\Delta(\mathbf{x}, \mathbf{y}_1, \ldots, \mathbf{y}_i, z_1, \ldots, z_i, m_1', \ldots, m_i') = 0$, where $m_1' = R_1(\mathbf{x}), \ldots, m_i' = R_i(\mathbf{x})$. Furthermore, if we additionally choose $m_i'' \in \mathbb{F}$ at random, the probability that

$$\Delta(\mathbf{x}, \mathbf{y}_1, \ldots, \mathbf{y}_i, z_1, \ldots, z_i, m_1', \ldots, m_{i-1}', m_i'') = 0$$

is upper bounded by $\frac{\deg(\Delta)}{p}$. This follows from the Schwartz-Zippel Lemma observing that $\Delta(R_1, \ldots, R_{i-1})$ is a non-zero and of degree at most $\deg(\Delta)$.

Now, we are prepared to describe the IND-CPA adversary. First, B receives the public key $(g^{P(\mathbf{x}^*)})_{P \in \mathbf{P}}$ of the $(\mathbf{P}, \mathbf{E}, \mathbf{H}, \mathcal{H})$-CS-type encryption scheme from the challenger. Using this key it creates the message $g^{m_0^*} = g^{R_i(\mathbf{x}^*)} = \prod_{P \in \mathbf{P}}(g^{P(\mathbf{x}^*)})^{\alpha_P}$, where $R_i = \sum_{P \in \mathbf{P}} \alpha_P P(\mathbf{X})$, and $g^{m_1^*}$, where $m_1^* \leftarrow \mathbb{F}$. Then it sends them to the challenger who responds with the ciphertext

$$(g^{E(\mathbf{x}^*, \mathbf{y}^*, z^*, m_b^*)})_{E \in \mathbf{E}}, \tag{2}$$

where $b \leftarrow \{0, 1\}$ and z^* is the hash value associated with the message $g^{m_b^*}$. Since we consider the IND-CPA game in the Random Oracle Model z^* has been chosen uniformly at random from $\mathbb{F}$. Next, B creates the remaining values in order to evaluate Δ as exponent: It computes the messages $g^{m_j'} = g^{R_j(\mathbf{x}^*)}$ and chooses $\mathbf{y}_j \leftarrow \mathbb{F}^{v_2}$, $z_j \leftarrow \mathbb{F}$, for $j < i$. Finally, it computes

$$g^{\Delta(\mathbf{x}^*, \mathbf{y}_1, \ldots, \mathbf{y}_{i-1}, \mathbf{y}^*, z_1, \ldots, z_{i-1}, z^*, m_1', \ldots, m_{i-1}', m^*)}, \tag{3}$$

where it is easy to see that B is in fact able to evaluate this polynomial in the exponent (cf. paragraph on runtime). If the resulting element equals 1, B outputs $out_B = 0$ and otherwise 1.

As we know from the previous analysis, the element in Equation 3 happens to be 1 for both messages with probability at most $\frac{\deg(\Delta)}{p}$. So in this case B's guess is correct with probability $\frac{1}{2}$. If this failure does not happen B's guess is correct with probability 1. In total, we have a probability of at least $\frac{1}{2}\frac{\deg(\Delta)}{p} + 1 - \frac{\deg(\Delta)}{p}$.

Let us briefly consider the runtime of B. Note that elements involving $\mathbf{y}_i = \mathbf{y}^*$, and $m_i' = m^*$ or $z_i = z^*$, are given and do not need to be computed (cf. Equation 2). First, constructing the messages $g^{m_j'}$ requires $O(\log(p)|\mathbf{P}|n)$ group operations. To compute the group element from Equation 3, B uses the known representation in $\mathbf{P}$ and $\mathbf{E}$. For the first part

$\prod_{P \in \mathbf{P}}(g^{P(\mathbf{x}^*)})^{\alpha_P}$ about $O(\log(p)|\mathbf{P}|)$ group operations are required. To compute the second at most $n - 1$ encryptions are needed. More precisely, the second part $\prod_{j=1}^{n} \prod_{E \in \mathbf{E}} g^{\beta_{j,E} E(\mathbf{x}, \mathbf{y}_j, z_j, m'_j)}$ can be computed (for $j \neq i$) as a multi-exponentiation (the exponents are the monomials of each E) with elements of the form

$$a^{\prod_{k=1}^{v_2} y_{j,k}^{d_{j,k}} z_j^{e_2} \beta_{j,E}},$$

where $a = g^{(P(x^*))^{e_1}}$ or $a = g^{m'_j{}^{e_3}}$ which requires $O(\log(p)|\mathbf{Y}| \max_{E \in \mathbf{E}}(|E|)|\mathbf{E}|n)$ multiplications.

Theorem 4 says that from any generic IND-SO-CPA adversary A certain polynomials as required for Theorem 3 can be extracted using "white-box access" to A. Here the extraction algorithm B does not only play the role of the IND-SO-CPA challenger and restricts itself to considering the in- and output of A (in this case we would be in the standard model) but closely observes the operations A performs on its inputs, i.e., B substitutes (and modifies) the generic oracle. More precisely, B's strategy is as follows: It turns the real IND-SO-CPA game in the generic model into a simulation game which does not reveal any information about the secret bit b chosen by the challenger. So A has no better chance than mere guessing to win the simulation game. Since the simulation game and the real game are equivalent unless a certain failure event occurs, an adversary who has a non-negligible advantage in winning the real game must cause this simulation failure with non-negligible probability. The crucial point is that a failure event is defined in a way such that it gives rise to the polynomials from Theorem 3.

Theorem 4. *Let a* $(\mathbf{P}, \mathbf{E}, \mathbf{H}, \mathcal{H})$-*CS-type encryption scheme* PKE *over a group* G *of prime order* p *be given. Furthermore, let* $d = \max_{S \in \mathbf{P} \cup \mathbf{E}}(\deg(S))$, $d' = \max_{S \in \mathbf{P}}(\deg(S))$, *and* $r = \max_{S \in \mathbf{P} \cup \mathbf{E}}(|S|)$, *and* $s = \max(|\mathbf{X}|, |\mathbf{Y}|)$. *Suppose there is a generic group adversary* A *that wins the IND-SO-CPA game for* PKE, *where we model* $\mathcal{H}$ *as Random Oracle, with advantage* $\mathsf{Adv}_{\mathsf{PKE},A}^{\text{ind-so-cpa}}$, *and by using* n *challenge messages. Let* $O(t)$ *and* $O(t')$ *denote the runtime of* A *and* samp, *respectively. Then there is a generic algorithm* B *which, by white-box access to* A, *extracts a polynomial* Δ *of degree at most* d *as well as polynomials* $R_1, \ldots, R_n$ *satisfying the conditions of Theorem 3 with a probability of at least* $\mathsf{Adv}_{\mathsf{PKE},A}^{\text{ind-so-cpa}} - \frac{dd'}{p}$ *and by performing at most* $O(r(|\mathbf{P}| + |\mathbf{E}|n)((t + t' + |\mathbf{P}| + |\mathbf{E}|n)^2 + \log(p)s))$ $\mathbb{F}$-*operations.*

Proof. **Game** 1 is the real IND-SO-CPA game as shown in Figure 3a.
Game 2 is the transition game shown in Figure 3b, which is actually equivalent to the real IND-SO-CPA game. Here, we define a new oracle $\mathcal{O}_1$ as follows: $\mathcal{O}_1$ uses polynomials to internally represent elements from $\mathbb{F}$. More precisely, we have $\mathcal{L} \subset \mathbb{F}[\mathbf{X}, \mathbf{Y}_1, \ldots, \mathbf{Y}_n, \mathbf{Z}, \mathbf{M}]$. Initially, the list is populated with the polynomials $\mathbf{P}$ describing the public key. Later, for each message, polynomials $\mathbf{E}$ describing its ciphertext are added. Applying the group operation in this polynomial representation translates to polynomial addition over $\mathbb{F}$. Moreover, the oracle receives certain elements $\mathbf{x}, \mathbf{y}_1, \ldots, \mathbf{y}_n, \mathbf{z}, \mathbf{m}'_0$ which are used to evaluate

Initialization & Challenge
$b \leftarrow \{0,1\}$
$\mathbf{x} \leftarrow \mathbb{F}^{v_1}$
$\mathsf{samp}(\cdot) \leftarrow A^{\mathcal{O}_0, \mathcal{H}}(\llbracket P(\mathbf{x}) \rrbracket_{P \in \mathbf{P}})$
$\llbracket m'_{0,i} \rrbracket_{i \in [n]} \leftarrow \mathsf{samp}^{\mathcal{O}_0, \mathcal{H}}()$
$(\mathbf{y}_1, \dots, \mathbf{y}_n) \leftarrow (\mathbb{F}^{v_2})^n$
$z_i \leftarrow \mathcal{H}(\llbracket H(\mathbf{x}, \mathbf{y}_i, m'_{0,i}) \rrbracket_{H \in \mathbf{H}}), 1 \leq i \leq n$
$\mathcal{I} \leftarrow A^{\mathcal{O}_0, \mathcal{H}}(\mathsf{sel}, \llbracket E(\mathbf{x}, \mathbf{y}_i, z_i, m'_{0,i}) \rrbracket_{E \in \mathbf{E}, i \in [n]})$

Opening
$\llbracket m'_{1,i} \rrbracket_{i \in [n]} \leftarrow \mathsf{samp}^{\mathcal{O}_0, \mathcal{H}}(\llbracket m'_{0,i} \rrbracket_{i \in \mathcal{I}})$
$out_A \leftarrow A^{\mathcal{O}_0, \mathcal{H}}(\mathsf{out}, (\mathbf{y}_i)_{i \in \mathcal{I}}, \llbracket m'_{b,i} \rrbracket_{i \in [n]})$

(a) Real Game

Initialization & Challenge
$b \leftarrow \{0,1\}$
$\mathbf{x} \leftarrow \mathbb{F}^{v_1}$
$\mathsf{samp}(\cdot) \leftarrow A^{\mathcal{O}_1(\mathbf{x}), \mathcal{H}}(\llbracket P(\mathbf{X}) \rrbracket_{P \in \mathbf{P}})$
$\llbracket m'_{0,i} \rrbracket_{i \in [n]} \leftarrow \mathsf{samp}^{\mathcal{O}_1(\mathbf{x}), \mathcal{H}}()$
$(\mathbf{y}_1, \dots, \mathbf{y}_n) \leftarrow (\mathbb{F}^{v_2})^n$
$z_i \leftarrow \mathcal{H}(\llbracket H(\mathbf{X}, \mathbf{Y}_i, M_i) \rrbracket_{H \in \mathbf{H}}), 1 \leq i \leq n$
$\mathcal{I} \leftarrow A^{\mathcal{O}_1(\mathbf{x}, \mathbf{y}_1, \dots, \mathbf{y}_n, \mathbf{z}, \mathbf{m}'_0), \mathcal{H}}(\mathsf{sel}, \llbracket E(\mathbf{X}, \mathbf{Y}_i, Z_i, M_i) \rrbracket_{E \in \mathbf{E}, i \in [n]})$

Opening
$\llbracket m'_{1,i} \rrbracket_{i \in [n]} \leftarrow \mathsf{samp}^{\mathcal{O}_2(\mathbf{x}, \mathbf{y}_1, \dots, \mathbf{y}_n, \mathbf{z}, \mathbf{m}'_0), \mathcal{H}}(\llbracket m'_{0,i} \rrbracket_{i \in \mathcal{I}})$
$out_A \leftarrow A^{\mathcal{O}_2(\mathbf{x}, \mathbf{y}_1, \dots, \mathbf{y}_n, \mathbf{z}, \mathbf{m}'_0), \mathcal{H}}(\mathsf{out}, (\mathbf{y}_i)_{i \in \mathcal{I}}, \llbracket m'_{b,i} \rrbracket_{i \in [n]})$

(b) Transition Game

Initialization & Challenge
$b \leftarrow \{0,1\}$
$\mathbf{x} \leftarrow \mathbb{F}^{v_1}$
$\mathsf{samp}(\cdot) \leftarrow A^{\mathcal{O}_2(), \mathcal{H}}(\llbracket P(\mathbf{X}) \rrbracket_{P \in \mathbf{P}})$
$\llbracket m'_{0,i} \rrbracket_{i \in [n]} \leftarrow \mathsf{samp}^{\mathcal{O}_2(), \mathcal{H}}()$
$(\mathbf{y}_1, \dots, \mathbf{y}_n) \leftarrow (\mathbb{F}^{v_2})^n$
$z_i \leftarrow \mathcal{H}(\llbracket H(\mathbf{X}, \mathbf{Y}_i, M_i) \rrbracket_{H \in \mathbf{H}}), 1 \leq i \leq n$
$\mathcal{I} \leftarrow A^{\mathcal{O}_2(), \mathcal{H}}(\mathsf{sel}, \llbracket E(\mathbf{X}, \mathbf{Y}_i, Z_i, M_i) \rrbracket_{E \in \mathbf{E}, i \in [n]})$

Opening
$\llbracket m'_{1,i} \rrbracket_{i \in [n]} \leftarrow \mathsf{samp}^{\mathcal{O}_2((\mathbf{y}_i, z_i, m'_{0,i})_{i \in \mathcal{I}}), \mathcal{H}}(\llbracket m'_{0,i} \rrbracket_{i \in \mathcal{I}})$
$out_A \leftarrow A^{\mathcal{O}_2((\mathbf{y}_i, z_i, m'_{0,i})_{i \in \mathcal{I}}), \mathcal{H}}(\mathsf{out}, (\mathbf{y}_i)_{i \in \mathcal{I}}, \llbracket m'_{b,i} \rrbracket_{i \in [n]})$

(c) Simulation Game

Fig. 3. IND-SO-CPA Games: From the Real Game to the Simulation Game

the polynomials in order to determine encodings: Two elements $R_1, R_2 \in \mathcal{L}$ are assigned the same encoding if

$$((R_1 - R_2)(\mathbf{M} = \mathbf{m}'_0))(\mathbf{X} = \mathbf{x}, \mathbf{Y}_1 = \mathbf{y}_1, \dots, \mathbf{Y}_n = \mathbf{y}_n, \mathbf{Z} = \mathbf{z}) \equiv 0 \bmod p. \quad (4)$$

Note that a message $m'_{0,j}, 1 \leq j \leq n$, might be a non-constant polynomial of the form $\sum_{P \in \mathbf{P}} \alpha_P P(\mathbf{X})$ in which case we assume that it is also evaluated with $\mathbf{x}$. Now, each time a polynomial R_1 is added to $\mathcal{L}$, the list is searched for a polynomial R_2 satisfying Equation 4. If such an element is found, the corresponding encoding is returned, otherwise a fresh, unused encoding is sampled.

There is only a minor technical difference between the two oracles: $\mathcal{O}_0$ immediately evaluates polynomials and calculates with $\mathbb{F}$-elements, whereas $\mathcal{O}_1$ does the calculation with polynomials and delays the evaluation to the point when encodings are determined. However, this is equivalent and so A has the same success probability in the real and the transition game.

Game 3 is the simulation game, as shown in Figure 3c, in which the computation is independent of the bit b. More precisely, we make the computation independent of all (unopened) secrets and messages. Thus, A has no better chance than guessing b. The simulation game is equivalent to the transition game unless a simulation failure occurs yielding polynomials which can be used to build an IND-CPA adversary.

For the simulation game, we slightly modify $\mathcal{O}_1$ resulting in a oracle $\mathcal{O}_2$:

- During Initialization & Challenge, $\mathcal{O}_2$ assigns two elements $R_1, R_2 \in \mathcal{L}$ the same encoding if they are equal as polynomials over $\mathbb{F}$, i.e., $(R_1 - R_2) \equiv 0$.
- In the Opening Phase the oracle receives the choices $\{\mathbf{y}_i, z_i, m'_{0,i}\}_{i \in \mathcal{I}}$ revealed to A and assigns the same encoding if $(R_1 - R_2)(\mathbf{y}_i, z_i, m'_{0,i})_{i \in \mathcal{I}} \equiv 0$.

The reason why we need to simulate differently in the Opening Phase is that the adversary obtains additional information about part of the secrets. For instance, he now can compute the encryption of $m'_{0,i}$, for $i \in \mathcal{I}$, on his own. So we need to make sure that he receives the same encodings for the ciphertext that the oracle has assigned in the previous phase.

Now, the crucial observation is that in the simulation game given $[\![m'_{b,1}]\!], \ldots,$ $[\![m'_{b,n}]\!]$ the only source of information about b would be encodings given to A in previous steps that depend on $m'_{0,i}$ or $m'_{1,i}$ for $i \notin \mathcal{I}$ since $m'_{0,i} = m'_{1,i}$ for $i \in \mathcal{I}$. However, encodings representing (combinations of) encryptions are independent of $m'_{0,i}$ (and $m'_{1,i}$) for $i \notin \mathcal{I}$, since we never evaluate the variables M_i. Hence, the probability that out_A equals b in the simulation game is $\frac{1}{2}$.

Clearly, due to the modification of $\mathcal{O}_1$ we changed the mapping between encodings and group elements. This might lead to a different behavior of generic algorithms when interacting with $\mathcal{O}_2$ in comparison to $\mathcal{O}_1$. More precisely, a simulation failure occurs during the

- Initialization & Challenge Phase (bad_1) if there exists $R_1, R_2 \in \mathcal{L}$ such that

$$(R_1 - R_2) \not\equiv 0 \text{ but } ((R_1 - R_2)(\mathbf{m}'_0))(\mathbf{x}, \mathbf{y}_1, \ldots, \mathbf{y}_n, \mathbf{z}) \equiv 0 \quad (5)$$

- Opening Phase (bad_2) if there exists $R_1, R_2 \in \mathcal{L}$ such that

$$((R_1 - R_2)(m'_{0,i})_{i \in \mathcal{I}})(z_i, \mathbf{y}_i)_{i \in \mathcal{I}} \not\equiv 0$$
$$\text{but} \quad (6)$$
$$((R_1 - R_2)(\mathbf{m}'_0))(\mathbf{x}, \mathbf{y}_1, \ldots, \mathbf{y}_n, \mathbf{z}) \equiv 0$$

Note that if failure event bad_1 did not happen (during Initialization & Challenge) then bad_2 may only be caused by a new polynomial computed during the Opening Phase. So the Initialization & Challenge Phases of the simulation and transition game are equivalent unless bad_1 happens and the Opening Phases are equivalent unless bad_2 occurs. Hence, A's probability in winning the IND-SO-CPA game is upper bounded by $\frac{1}{2} + \Pr[\mathsf{bad}_1 \vee \mathsf{bad}_2]$. In other words, A causes a simulation failure with probability at least $\mathsf{Adv}_{\mathsf{PKE},A}^{\mathsf{ind\text{-}so\text{-}cpa}}$.

It remains to show that we can extract polynomials as required for Theorem 3 in case bad_1 or bad_2 occurs. The extraction algorithm B plays the IND-SO-CPA simulation game with A and takes over the role of the simulation oracle $\mathcal{O}_2$. B checks if bad_1 happens during the Initialization & Challenge Phase. If this is the case, it considers the corresponding polynomials which have caused the failure. Otherwise, it executes the Opening Phase and checks whether bad_2 occurs.

Let us now assume that bad_1 happens for some $\Delta := R_1 - R_2$ and $m'_{0,1}, \ldots, m'_{0,n}$ as well as $\mathbf{x}, \mathbf{y}_1, \ldots, \mathbf{y}_n, \mathbf{z}$ chosen uniformly at random by B.[4] Note that since generic algorithms are only able to add polynomials whose encodings they receive as input, Δ is of the form

$$\Delta = \sum_{P \in \mathbf{P}} \alpha_P P(\mathbf{X}) + \sum_{j=1}^{n} \sum_{E \in \mathbf{E}} \beta_{j,E} E(\mathbf{X}, \mathbf{Y}_j, Z_j, M_j) \tag{7}$$

and $m'_{0,1}, \ldots, m'_{0,n}$ are of the form $m'_{0,i} = \sum_{P \in \mathbf{P}} \alpha_P P(\mathbf{X})$. The degree of Δ is upper bounded by $d = \max_{S \in \mathbf{P} \cup \mathbf{E}}(\deg(S))$ and the degree of $m'_{0,i}$ is upper bounded by $d' = \max_{S \in \mathbf{P}}(\deg(S))$.

In case bad_2 occurs, we consider the partially evaluated polynomial $\Delta := ((R_1 - R_2)((m'_{0,i})_{i \in \mathcal{I}}))((z_i, \mathbf{y}_i)_{i \in \mathcal{I}})$ and the polynomials $m'_{0,1}, \ldots, m'_{0,n}$ as before. Due to the form of the monomials of E, evaluation of E with $m'_{0,i}$, z_i, and $\mathbf{y}_i$ results in polynomials of the form $\sum_{P \in \mathbf{P}} \alpha_P P(\mathbf{X})$. Hence, also Δ can be viewed as a polynomial of the form in Equation 7, where the $\beta_{i,E}$ coefficients are zero for $i \in \mathcal{I}$. The upper bounds d and d' specified above also hold in this case.

To summarize, with probability at least $\Pr[\mathsf{bad}_1 \vee \mathsf{bad}_2]$, B can extract a non-zero polynomial Δ as in Equation 7 and polynomials $m'_{0,1}, \ldots, m'_{0,n}$. Δ becomes zero when evaluated with $m'_{0,1}, \ldots, m'_{0,n}$ and uniformly and independently chosen values $\mathbf{x}, \mathbf{y}_j, z_j$, where $j \in \{1, \ldots, n\}$ for the case bad_1 and $j \notin \mathcal{I}$ for the case bad_2. Applying Lemma 2 stated below yields that Δ already becomes zero when evaluated with the messages with probability at least $\Pr[\mathsf{bad}_1 \vee \mathsf{bad}_2] - \frac{dd'}{p}$. In this case B has found polynomials as required in Theorem 3.

Lemma 2. Let $d, d' \in \mathbb{N}_0$, $k, i \in \mathbb{N}$ with $1 \leq i \leq k$. Let dist be a distribution over $(i+1)$-tuples $(P, x_1, \ldots, x_i)$ of polynomials from $\mathbb{F}[X_1, \ldots, X_k]$ where $P \neq 0$, $\deg(P) \leq d$, and $\deg(x_j) \leq d'$ for $1 \leq j \leq i$. Then it holds that

$$\Pr_{(P, x_1, \ldots, x_i) \leftarrow \mathsf{dist}}[P(X_1 = x_1, \ldots, X_i = x_i) = 0] \geq$$

$$\Pr_{\substack{(P, x_1, \ldots, x_i) \leftarrow \mathsf{dist} \\ x_{i+1}, \ldots, x_k \leftarrow \mathbb{F}}}[(P(X_1 = x_1, \ldots, X_i = x_i))(X_{i+1} = x_{i+1}, \ldots, X_k = x_k) = 0] - \frac{dd'}{p}$$

---

[4] Note that the hash values z_j are indeed uniformly chosen since the input to the Random Oracle is guaranteed to be different for the n encryptions made: For $1 \leq j \leq n$, the variable M_j or $Y_{j,i} \in \mathbf{Y}_j$ appear in at least one of the encryption polynomials ensuring that the corresponding encoding, input to the hash function, is fresh.

Proof.

$$\Pr[(P(x_1,\ldots,x_i))(x_{i+1},\ldots,x_k) = 0]$$
$$= \Pr[(P(x_1,\ldots,x_i))(x_{i+1},\ldots,x_k) = 0 \wedge P(x_1,\ldots,x_i) = 0]$$
$$+ \Pr[(P(x_1,\ldots,x_i))(x_{i+1},\ldots,x_k) = 0 \wedge P(x_1,\ldots,x_i) \neq 0]$$
$$\leq \Pr[P(x_1,\ldots,x_i) = 0]$$
$$+ \Pr[(P(x_1,\ldots,x_i))(x_{i+1},\ldots,x_k) = 0 \mid P(x_1,\ldots,x_i) \neq 0]$$
$$\leq \Pr[P(x_1,\ldots,x_i) = 0] + \frac{dd'}{p}$$

The last inequality follows from the Schwartz-Zippel Lemma.

Let us briefly estimate the runtime of B. The algorithm runs A once, samp twice, plays the role of the IND-SO-CPA challenger, the generic oracle $\mathcal{O}_2$, and checks for a simulation failure. We will count the number of operations on polynomials and group elements: B maintains the list $\mathcal{L}$ of polynomials on behalf of $\mathcal{O}_2$. This requires at most $O(t+t')$ additions of polynomials. Additionally, to determine encodings, B needs to compute at most $O(|\mathcal{L}|^2) = O((t+t'+|\mathbf{P}|+|\mathbf{E}|n)^2)$ difference polynomials Δ. Note that the monomials of all these polynomials come from a set of at most at most $r(|\mathbf{P}| + |\mathbf{E}|n)$ different monomials. Thus, one polynomial addition results in at most $r(|\mathbf{P}| + |\mathbf{E}|n)$ operations over $\mathbb{F}$.

To check for simulation failures, B needs to evaluate the difference polynomials Δ. To do so, B maintains a second list $\mathcal{L}' \subset \mathbb{F}$ just like the real $\mathcal{O}_0$ would do and computes the corresponding differences. Evaluating $\mathbf{P}$ and $\mathbf{E}$ when added to $\mathcal{L}'$ requires $O(\log(p)(|\mathbf{X}| \max_{P \in \mathbf{E}}(|P|)|\mathbf{P}| + |\mathbf{Y}| \max_{E \in \mathbf{E}}(|E|)|\mathbf{E}|n)$ $\mathbb{F}$-operations and computing the differences $O(|\mathcal{L}'|^2) = O((t + t' + |\mathbf{P}| + |\mathbf{E}|n)^2)$.

To check for a failure during the Opening Phase, B evaluates all polynomials in $\mathcal{L}$ with $\mathbf{y}_i, z_i, m'_{0,i}$, for $i \in \mathcal{I}$, when the Opening Phase starts. This requires $O(\log(p)|\mathbf{Y}|n \max_{E \in \mathbf{Q}}(|E|)|\mathbf{E}|)$ $\mathbb{F}$-operations. These evaluations do not increase the size of the set of monomials polynomials in $\mathcal{L}$ may consist of.

Note that the success probability of the IND-CPA adversary B from Theorem 3 is non-negligible if the degrees of the public key and encryption polynomials of PKE are small, i.e., polynomial in $\log(p)$. Moreover, B is efficient if the representation of these polynomials is polynomial in $\log(p)$ (always the case for an efficient encryption scheme) as well as the number n of involved message polynomials R_i. The same statement holds for the polynomial extraction algorithm from Theorem 4, where we additionally need to assume that the runtime of the IND-SO-CPA adversary is polynomial and its advantage is non-negligible.

Acknowledgements. We would like to thank the anonymous reviewers for very helpful and constructive comments.

References

[1] Alwen, J., Dodis, Y., Wichs, D.: Leakage-resilient public-key cryptography in the bounded-retrieval model. In: CRYPTO 2009. LNCS, vol. 5677, pp. 36–54. Springer, Heidelberg (2009)

[2] Barak, B., Haitner, I., Hofheinz, D., Ishai, Y.: Bounded key-dependent message security. In: Gilbert, H. (ed.) EUROCRYPT 2010. LNCS, vol. 6110, pp. 423–444. Springer, Heidelberg (2010)

[3] Bellare, M., Desai, A., Pointcheval, D., Rogaway, P.: Relations among notions of security for public-key encryption schemes. In: Krawczyk, H. (ed.) CRYPTO 1998. LNCS, vol. 1462, pp. 26–45. Springer, Heidelberg (1998)

[4] Bellare, M., Dowsley, R., Waters, B., Yilek, S.: Standard security does not imply security against selective-opening. In: Pointcheval, D., Johansson, T. (eds.) EUROCRYPT 2012. LNCS, vol. 7237, pp. 645–662. Springer, Heidelberg (2012)

[5] Bellare, M., Hofheinz, D., Yilek, S.: Possibility and impossibility results for encryption and commitment secure under selective opening. In: Joux, A. (ed.) EUROCRYPT 2009. LNCS, vol. 5479, pp. 1–35. Springer, Heidelberg (2009)

[6] Bellare, M., Sahai, A.: Non-malleable encryption: Equivalence between two notions, and an indistinguishability-based characterization. In: Wiener, M. (ed.) CRYPTO 1999. LNCS, vol. 1666, pp. 519–536. Springer, Heidelberg (1999)

[7] Black, J., Rogaway, P., Shrimpton, T.: Encryption-scheme security in the presence of key-dependent messages. In: Nyberg, K., Heys, H.M. (eds.) SAC 2002. LNCS, vol. 2595, pp. 62–75. Springer, Heidelberg (2003)

[8] Böhl, F., Hofheinz, D., Kraschewski, D.: On definitions of selective opening security. In: Fischlin, M., Buchmann, J., Manulis, M. (eds.) PKC 2012. LNCS, vol. 7293, pp. 522–539. Springer, Heidelberg (2012)

[9] Boneh, D., Halevi, S., Hamburg, M., Ostrovsky, R.: Circular-secure encryption from decision Diffie-Hellman. In: Wagner, D. (ed.) CRYPTO 2008. LNCS, vol. 5157, pp. 108–125. Springer, Heidelberg (2008)

[10] Cramer, R., Hanaoka, G., Hofheinz, D., Imai, H., Kiltz, E., Pass, R., Shelat, A., Vaikuntanathan, V.: Bounded CCA2-secure encryption. In: Kurosawa, K. (ed.) ASIACRYPT 2007. LNCS, vol. 4833, pp. 502–518. Springer, Heidelberg (2007)

[11] Dolev, D., Dwork, C., Naor, M.: Non-malleable cryptography (1998) (manuscript)

[12] Dwork, C., Naor, M., Reingold, O., Stockmeyer, L.J.: Magic functions. In: 40th FOCS, pp. 523–534. IEEE Computer Society Press (October 1999)

[13] Dziembowski, S., Pietrzak, K.: Leakage-resilient cryptography. In: 49th FOCS, pp. 293–302. IEEE Computer Society Press (October 2008)

[14] Fehr, S., Hofheinz, D., Kiltz, E., Wee, H.: Encryption schemes secure against chosen-ciphertext selective opening attacks. In: Gilbert, H. (ed.) EUROCRYPT 2010. LNCS, vol. 6110, pp. 381–402. Springer, Heidelberg (2010)

[15] Goldwasser, S., Micali, S.: Probabilistic encryption. Journal of Computer and System Sciences 28(2), 270–299 (1984)

[16] Hemenway, B., Libert, B., Ostrovsky, R., Vergnaud, D.: Lossy encryption: Constructions from general assumptions and efficient selective opening chosen ciphertext security. In: Lee, D.H., Wang, X. (eds.) ASIACRYPT 2011. LNCS, vol. 7073, pp. 70–88. Springer, Heidelberg (2011)

[17] Hofheinz, D.: All-but-many lossy trapdoor functions. In: Pointcheval, D., Johansson, T. (eds.) EUROCRYPT 2012. LNCS, vol. 7237, pp. 209–227. Springer, Heidelberg (2012)

[18] Micali, S., Reyzin, L.: Physically observable cryptography (extended abstract). In: Naor, M. (ed.) TCC 2004. LNCS, vol. 2951, pp. 278–296. Springer, Heidelberg (2004)

[19] Myers, S., Shelat, A.: Bit encryption is complete. In: 50th FOCS, pp. 607–616. IEEE Computer Society Press (October 2009)

[20] Naor, M., Pinkas, B.: Efficient oblivious transfer protocols. In: Kosaraju, S.R. (ed.) 12th SODA, pp. 448–457. ACM-SIAM (January 2001)

[21] Peikert, C., Vaikuntanathan, V., Waters, B.: A framework for efficient and composable oblivious transfer. In: Wagner, D. (ed.) CRYPTO 2008. LNCS, vol. 5157, pp. 554–571. Springer, Heidelberg (2008)

[22] Peikert, C., Waters, B.: Lossy trapdoor functions and their applications. In: Ladner, R.E., Dwork, C. (eds.) 40th ACM STOC, pp. 187–196. ACM Press (May 2008)

[23] Shoup, V.: Lower bounds for discrete logarithms and related problems. In: Fumy, W. (ed.) EUROCRYPT 1997. LNCS, vol. 1233, pp. 256–266. Springer, Heidelberg (1997)

A   Some Remarks on Definition 3

We would like to note that the restrictions on the form and degrees of polynomials made in Definition 3 are not of artificial nature and just derived from the proofs but need to be satisfied by a meaningful encryption scheme. On the other hand, we would like to stress that they are not sufficient for such a scheme as, e.g., no conditions on the nature of the decryption algorithm are made.

In particular, the encryption polynomials E may not be "arbitrary" polynomials in $\mathbf{X}$ since during encryption we are usually only given $P(\mathbf{x})$ and do not know how to evaluate encryption polynomials not being "combinations" of public key polynomials. Furthermore, any public key polynomial may only appear linearly in any encryption polynomial. Otherwise, in absence of a pairing we do not know how to compute, e.g., P^2 efficiently. In fact, in the case of a single group this translates to solving the Square-DH problem. For this reason, also any monomial of an encryption polynomial might only contain at most one public key polynomial. Moreover, assume an encryption polynomial E would include a monomial of the form $\alpha P^{e_1} Z^{e_2} M^{e_3} \prod_{i=1}^{v_2} Y_i^{d_i}$ with $e_1 + e_3 > 1$. This would mean that we have to solve a variant of the DH problem to encrypt a message unless we know the DL of the message.

Finally, the condition on the input of the hash function ensures that the input is not constant for different m'. The use of a hash function for constant input would be meaningless. Note that for an encryption scheme without hash function like ElGamal e_2 is simply set to zero in all encryption polynomials.

B   A Separating CS-type Encryption Scheme

Interestingly, we again obtain a CS-type scheme if we instantiate (a slight variation of) the separating scheme from Section 3.1 with a CS-type scheme according to Definition 3. Some details can be found in the following.

- Compared to the original definition of the separating scheme, the message space of a CS-type scheme is a group G of order p and not $\mathbb{F}$. But as already mentioned, this is no real issue. What we can do in our particular case here is the following: We only need a means such as the degree-$\leq k$ polynomial F before that allows to generate and reconstruct the data points $(i, m_i)_{i \in [3k]}$, where $m_i = g^{m_i'} \in G$. This can easily be done "using F in the exponent".

Especially, Lagrange Interpolation works in this case. To evaluate a degree-$\leq k$ interpolation polynomial F defined by $k+1$ data points (i, m_i) in the exponent with some $\ell \in [3k]$, one would compute $\prod_i m_i^{Q_i(\ell)}$, where $Q_i(x) = \prod_{t \neq i} \frac{x-t}{i-t}$ is a Lagrange basis polynomial. To determine if there exists a unique F in Step 5 of $\mathsf{SOA}(sk, Z)$, one could check if there exists some $j \in [3k] \setminus \mathcal{I}$ such that

$$\prod_{i \in \mathcal{I} \cup \{j\}} m_i^{Q_i(\ell)} = m_\ell$$

for k values $\ell \in [3k] \setminus (\mathcal{I} \cup \{j\})$. If this is the case, it suffices to return m_j.

- The secret key of the CS-type scheme needs to be extended by a key K for the PRF. Note that this is not forbidden by Definition 3. Moreover, due to our slight modification above, we do not need an additional prime p in the public key of the CS-type scheme.

- In order to mark a ciphertext as regular, Enc' simply adds a fixed group element to each regular CS-type ciphertext. For instance, this could translate to adding polynomials $P_6 = 0$ and $E_5 = P_6$ to the specification of Cramer-Shoup as CS-type encryption scheme shown in Section 4.1. The other types of inputs we allow to Dec' can be marked similarly using other fixed elements.

- Note that apart from these markers, we do not need to care about how the decryption function looks like since Definition 3 only specifies the form of public keys and (regular) ciphertexts.

C   Our CCA-Separation Works in the GGM

In this section, we briefly argue why our CCA-separation also holds in the GGM, i.e., there exists a IND-CCA secure generic group encryption scheme that can be efficiently broken by a generic group IND-SO-CCA adversary.

First, it is easy to see that our separating scheme works over any (prime order) group G when it is instantiated with a generic group encryption scheme like Cramer-Shoup: The original IND-CCA secure scheme (CS in our case) is treated as a black box and also the modifications applied work for any group. In particular, in Section B we show how the message space can be switched to G and how $\mathsf{SOA}(sk, Z)$ can be implemented in this case. It is clear, that the resulting separating scheme is IND-CCA secure despite this switch of the message space. To summarize, it perfectly makes sense and is meaningful to consider the IND-SO-CCA game for our separating scheme in the GGM.

Second, the IND-SO-CCA adversary A and its sampling algorithm only apply generic group operations: The message distribution A outputs will be

$$\mathcal{D} = \left\{ (g^{F(1)}, \ldots, g^{F(3k)}) \mid F \in \mathbb{F}[X] \text{ uniformly chosen degree-}\leq k \text{ polynomial} \right\},$$

where $\langle g \rangle = G$, and can be implemented by a generic group algorithm. Moreover, the polynomial interpolation A's resampling algorithm uses can be realized over generic groups as shown in Section B. As also shown there, the interpolation polynomial returned by the decryption oracle on a soa-query can be evaluated using only multiplications with given group elements.

D   Variations of Our CCA-Separation

An Observation. We remark that the attack from Theorem 1 actually only uses two decryption queries. Moreover, one of these queries is a query (sel, Z) to a (pseudo)random function. Our proof would work also in the random oracle model, if we defined $\mathcal{I} = \mathcal{RO}((c_i')_{i \in [3k]})$ (instead of $\mathcal{I} = \mathsf{PRF}_K((c_i')_{i \in [3k]})$). With this change, we would get the same separation in the random oracle model, but with a weak IND-SO-CCA attack that requires only one decryption query.

Bounded CCA Security. Cramer et al. [10] define a bounded notion (called IND-q-CCA security) of IND-CCA security, in which an adversary only gets an a-priori bounded number q of decryption queries. If we define weak IND-SO-q-CCA security in the obvious way, our observation above immediately yields a separation between IND-2-CCA and weak IND-SO-2-CCA security. Furthermore, we get a separation between IND-1-CCA and weak IND-SO-1-CCA security in the random oracle model.

Non-malleability. IND-1-CCA security is known to be tightly related to non-malleability [11, 3]. Concretely, Bellare and Sahai [6] show that non-malleability under chosen-plaintext attacks (NM-CPA) is equivalent to a mild form of IND-CCA security, which in turn implies IND-1-CCA security. Since our results yield a separation between IND-1-CCA and IND-SO-1-CCA security in the random oracle model, we can expect a similar separation between between NM-CPA and NM-SO-CPA security. Here, NM-SO-CPA stands for "non-malleability under chosen-plaintext selective opening attacks," a notion which has not yet been formally defined. (We leave such a definition for future work; however, if one opts to simply equip an NM-CPA adversary with an "opening oracle" for NM-SO-CPA, the random oracle variation of our result seems to directly apply.)

Dual System Encryption via Predicate Encodings

Hoeteck Wee*

ENS, Paris, France

Abstract. We introduce the notion of *predicate encodings*, an information-theoretic primitive reminiscent of linear secret-sharing that in addition, satisfies a novel notion of reusability. Using this notion, we obtain a unifying framework for adaptively-secure public-index predicate encryption schemes for a large class of predicates. Our framework relies on Waters' dual system encryption methodology (Crypto '09), and encompass the identity-based encryption scheme of Lewko and Waters (TCC '10), and the attribute-based encryption scheme of Lewko et al. (Eurocrypt '10). In addition, we obtain obtain several concrete improvements over prior works. Our work offers a novel interpretation of dual system encryption as a methodology for amplifying a one-time private-key primitive (i.e. predicate encodings) into a many-time public-key primitive (i.e. predicate encryption).

1 Introduction

Predicate encryption [42, 10, 32] is a new paradigm for public-key encryption that enables fine-grained access control for encrypted data. In predicate encryption, ciphertexts are associated with descriptive values x in addition to a plaintext, secret keys are associated with functions f, and a secret key decrypts the ciphertext if and only if $f(x) = 1$. Here, f may express an arbitrarily complex access policy, which is in stark contrast to traditional public-key encryption, where access is all or nothing. Predicate encryption generalizes both identity-based encryption (IBE) [43, 8, 19] where f checks for equality, and attribute-based encryption (ABE) [42, 27], where f encodes a boolean formula. The security requirement for predicate encryption enforces resilience to collusion attacks, namely any group of users holding secret keys for different functions learns nothing about the plaintext if none of them is individually authorized to decrypt the ciphertext. This should hold even if the adversary *adaptively* decides which secret keys to ask for.

Terminology. Throughout this work, we use *predicate encryption* to refer to public-index predicate encryption, and reserve attribute-based encryption for the special case where the predicate is computed by a boolean formula.

Dual System Encryption. In [45], Waters introduced the powerful *dual system encryption* methodology for building adaptively secure predicate encryption. In a dual system encryption scheme, there are two types of keys and ciphertexts: normal and

* wee@di.ens.fr. CNRS (UMR 8548) and INRIA. Part of this work was done at George Washington University, supported by NSF Awards CNS-1237429 and CNS-1319021.

semi-functional. Normal keys and ciphertexts are used in the real system, while the semi-functional objects are gradually introduced in the hybrid security proof. The proofs are often quite complex and delicate. In spite of the large body of work relying on the dual system encryption methodology (e.g. [34, 37, 40, 35, 38, 33, 41]), there seems to be no concrete, overarching framework explaining these schemes. In particular, even in the simplest information-theoretic setting in composite-order groups, we do not have a clear understanding of why the Lewko-Waters heuristic [34] for deriving dual system encryption schemes via "embedding" works for the IBE scheme in [7] but not the ABE scheme in [27] (even under the "one use" restriction). We also do not have a formal, systematic approach for deriving the semi-functional objects used in the security proof: for instance, the semi-functional keys in the dual system IBE in [34] have independent random semi-functional components, whereas those in the ABE scheme in [37] are carefully designed to have certain correlations.

Decoupling Functionalities? A recurring trend in predicate encryption, which arose with both the introduction of dual system encryption and lattice-based techniques [24, 13], is a systematic adaption of prior selectively secure in bilinear groups to achieve either improved parameters (e.g. shorter ciphertexts for HIBE) or larger classes of functionalities (e.g. from IBE to ABE). Moreover, the new schemes often bear a structural resemblance to prior schemes. The phenomenon suggests that we should aim to decouple the way we encode a predicate/functionality in an encryption scheme from the design and analysis of the scheme.

1.1   Our Contributions

We present a framework for the design and analysis of dual system encryption schemes in composite-order bilinear groups, which allows us to also decouple the predicate from the security proof. The crux of our framework is a notion of *predicate encodings*. Roughly speaking, predicate encodings are an information-theoretic primitive reminiscent of secret-sharing schemes that in addition, satisfies a novel notion of reusability. Using predicate encodings, we obtain new insights into the dual system encryption methodology and new concrete predicate encryption schemes. Before we describe our results, we present an overview of predicate encodings.

Predicate Encodings. A predicate encoding for a Boolean predicate $P(\cdot, \cdot)$, is specified by a pair of algorithms (sE, rE) with a common private input w and in addition,

- sender encoding sE takes as input (x, w) and outputs $sE(x, w)$.
- receiver encoding rE takes as input (α, y, w) and randomness r, and outputs $rE(\alpha, y, w; r)$.

The basic requirements for α are the same as that for secret-sharing:

(reconstruction.) if $P(x, y) = 1$, we can recover α from the encodings;

(privacy.) if $P(x, y) = 0$, the encodings hide α perfectly.

The key conceptual novelty in predicate encoding (over other existing notions e.g. [46, 4, 21, 29, 25, 30, 1]) which enables us to handle collusions in predicate encryption is w-**hiding**. Informally, w-hiding stipulates that we can hide all information about w in the receiver encoding by setting the randomness r to some fixed value (e.g. we can hide w in the expression rw by setting r to 0). Note that the definition of w-hiding treat w and r differently. Finally, we impose some algebraic structure in the encodings similar to that for linear secret-sharing, in order to carry out encoding and reconstruction "in the exponent" in the encryption scheme.

We stress that the requirements for predicate encodings are fairly basic and indeed, we readily obtain predicate encodings for a large class of predicates like HIBE, doubly spatial encryption and ABE, many of which are implicit in prior selectively secure schemes [7, 11, 9]. Moreover, privacy for these encodings follows readily from linear algebra, as is typically the case for information-theoretic primitives and constructions. On the other hand, the encodings in [27, 3] do not satisfying our requirements (c.f. Section 5.5); this provides a partial explanation as to why the Lewko-Waters heuristic [34] cannot be applied to these schemes.

Predicate Encryption from Predicate Encodings. Starting from a predicate encoding for P, we construct a predicate encryption scheme in composite-order bilinear groups whose order is the product of three primes p_1, p_2, p_3, and establish adaptive security in a *modular* manner via Waters' dual system encryption methodology. Here, a secret key sk_y can decrypt a ciphertext ct_x iff $P(x, y) = 1$. We associate ciphertext with sender encoding and secret keys with receiver encodings. Correctness will rely on the reconstruction property modulo p_1, whereas security against collusions will rely on privacy and w-hiding modulo p_2. The third subgroup corresponding to p_3 is used for additional randomization which we ignore in this overview. Roughly speaking, the master public key, secret key and ciphertext are of the form:

$$\mathsf{mpk} := (g_1, g_1^w, e(g_1, g_1)^\alpha), \qquad \mathsf{sk}_y := g_1^{\mathsf{rE}(\alpha, y, w; r)}, \qquad \mathsf{ct}_x := ((g_1^{\mathsf{sE}(x, w)})^s, e(g_1, g_1)^{\alpha s} \cdot m)$$

where g_1 is a generator of order p_1. Observe that the lengths of w, sE and rE correspond naturally to the sizes of the public parameters, ciphertexts and secret keys. If $P(x, y) = 1$, decryption works by reconstructing α from $\mathsf{sE}(x, w)$ and $\mathsf{rE}(\alpha, y, w; r)$ in the exponent via a pairing.

Proof Strategy. We outline the key challenges in establishing adaptive security of the predicate encryption scheme, which yields new insights into dual system encryption methodology:

- First, predicate encoding is essentially a private-key primitive, in that α-privacy against an adversary that does not see the shared randomness w, whereas w must be made public in order that encryption uses the same w as that used for decryption. The scheme overcomes this conundrum by publishing only g_1^w in the public parameters. This leaks information about w (mod p_1) so that we can exploit α-reconstruction modulo p_1, but completely hides w (mod p_2) so that α-privacy holds modulo p_2. In the final step in the hybrid security proof, the message is masked by α modulo p_2 whereas the public parameters and all secret keys reveal no information about α modulo p_2. Security then follows via a simple information-theoretic argument.

- Second, predicate encoding only provides one-time security, that is, α-privacy no longer holds if we use w across more than one receiver encoding, as will be the case when an adversary requests multiple secret keys. We overcome this difficulty by ensuring that in each step in the proof of security, at most one secret key leaks information about w (mod p_2). In particular, both normal and semi-functional keys reveals no information about w (mod p_2). We only leak information about w (mod p_2) when transitioning from a normal to a semi-functional key, one key at a time. During the transition, we rely on w-hiding to "erase" information about w (mod p_2) from all remaining keys (see Fig 2 and Lemma 3).

- Finally, predicate encoding only provides non-adaptive security, namely α-privacy only holds if x, y are fixed in advance. On the other hand, an adversary may choose a key query y after seeing the challenge ciphertext for x, which leaks $rE(x, w, r)$. This is where we rely crucially on the fact that the encoding achieves *perfect* α-privacy, for which non-adaptive implies adaptive privacy. To the best of our knowledge, this is the first time this requirement is explicitly pointed out for use in dual system encryption. (A recent work [6] highlights several subtleties in defining and achieving adaptive privacy in the related setting of garbled circuits.)

In short, dual system encryption allows us to boost security in a private-key, one-time, non-adaptive setting to a full-fledged public-key, many-time, adaptive setting! Along the way, we introduce a conceptual simplification where we define the semi-functional entities via auxiliary algorithms, reminiscent of Cramer-Shoup projective hashing [20].

Instantiations. Our final predicate encryption scheme is adaptively secure under the standard Subgroup Decision Assumptions in composite order bilinear groups. We note that our implementation of the dual system encryption methodology differs in subtle ways from prior composite-order instantiations in [34, 37] (see e.g. Remark 2). In addition to a unifying proof of security for a large class of predicates, we obtain the several concrete improvements over prior works:

- We eliminate the need for an additional computational assumption which refers to the target group, as used in the prior composite-order HIBE and ABE [34, 37]. In particular, we show how to execute the final transition in the proof of security with an information-theoretic argument instead of a computational one.

- We reduce the key size of the (key-policy) ABE in [37] by half. The improvement comes from eliminating some redundant randomization in the associated encoding.

- We obtain novel (to the best of our knowledge) and simple constructions of adaptively-secure non-zero inner product encryption and doubly spatial encryption in composite-order bilinear groups.

1.2   Discussion

Predicate encodings decouple and modularize the essential information-theoretic properties from the broader mechanics of a dual system cryptosystem and its analysis.

Previous dual-system proofs are often monolithic and hard to follow, and the core new ideas are sometimes buried underneath lots of algebraic notation that is repeated (or only slightly tweaked) from one scheme to another. Our framework allows us to distill the core argument that is common to dual system cryptosystems from a separate information-theoretic argument which is tailored to the underlying predicate.

Open Problems. This work raises a number of open problems.

- Do bilinear (or multi-linear) predicate encodings exist for all polynomial-time computable predicates? An affirmative answer would yield adaptively secure ABE for circuits [26, 22], without relying on complexity leveraging. However, even achieving perfect α-hiding without the bilinear requirements would likely require overcoming long-standing barriers.

- Can we prove lower bounds on the length of $\mathbf{w}$, rE or sE for predicate encodings (corresponding to public parameters, secret keys and ciphertext sizes respectively)? In particular, the encodings for ABE require that rE grows with the size of the formula (c.f. Section 5.5) and we conjecture that such a dependency is in fact necessary for perfect α-privacy.

- Finally, we note that our work does not cover more recent applications of dual system encryption in the computational setting for ABE with short ciphertexts [36]. There, α-privacy is computational, for which we no longer get adaptive from non-adaptive security "for free". We leave these extensions for future work.

Subsequent Work. In subsequent works [16, 17], we built upon the ideas introduced here in several ways. In [16], we introduced *dual system groups*, a step towards abstracting the underlying group structure needed to support the dual system encryption methodology. This is orthogonal and complementary in this work, which is about abstracting how we encode the predicate/functionality. In [17], we presented the first adaptively secure IBE where the security loss does not depend on the number of secret key queries, partially resolving an open problem in [44, 23]. The crucial insight lies in replacing the one-time predicate encoding for IBE (a randomized MAC) with a reusable one (a pseudorandom function). Specifically, we rely on dual system encryption methodology to "compile" the Naor-Reingold PRF [39] which is a private-key primitive into a fully secure IBE.

Organization. We formalize predicate encodings in Section 3. We present the generic construction of a predicate encryption scheme in Section 4. We describe instantiations of predicate encodings in Section 5. Preliminaries are given in Section 2.

2   Preliminaries

Notation. We denote by $s \leftarrow_R S$ the fact that s is picked uniformly at random from a finite set S. By PPT, we denote a probabilistic polynomial-time algorithm. Throughout, we use 1^λ as the security parameter. We use $\cdot$ to denote multiplication as well as component-wise multiplication. We use lower case boldface to denote (column) vectors over scalars and upper case boldcase to denote vectors of group elements as well as matrices. Given two vectors $\mathbf{x} = (x_1, x_2, \ldots), \mathbf{y} = (y_1, y_2, \ldots)$ over scalars, we use $\langle \mathbf{x}, \mathbf{y} \rangle$ to denote the standard dot product $\mathbf{x}^\top \mathbf{y}$. Given a group element g, we write $g^{\mathbf{x}}$ to denote $(g^{x_1}, g^{x_2}, \ldots)$.

2.1 Composite Order Bilinear Groups and Cryptographic Assumptions

We instantiate our system in composite order bilinear groups, which were introduced in [12] and used in [32, 34, 37]. A generator $\mathcal{G}$ takes as input a security parameter λ and outputs a description $\mathbb{G} := (N, G, G_T, e)$, where N is product of distinct primes of $\Theta(\lambda)$ bits, G and G_T are cyclic groups of order N, and $e : G \times G \to G_T$ is a non-degenerate bilinear map. We require that the group operations in G and G_T as well the bilinear map e are computable in deterministic polynomial time. We consider bilinear groups G whose orders N are products of three distinct primes p_1, p_2, p_3 (that is, $N = p_1 p_2 p_3$). We can write $G = G_{p_1} G_{p_2} G_{p_3}$ where $G_{p_1}, G_{p_2}, G_{p_3}$ are subgroups of G of order p_1, p_2 and p_3 respectively. In addition, we use $G_{p_i}^*$ to denote $G_{p_i} \setminus \{1\}$. We will often write g_1, g_2, g_3 to denote random generators for the subgroups $G_{p_1}, G_{p_2}, G_{p_3}$ of order p_1, p_2 and p_3 respectively.

Cryptographic Assumptions. Our construction relies on the following two assumptions which are essentially the first two of three assumptions used in [34, 37] and are instances of the General Subgroup Decision Assumption in composite-order groups [5]. We define the following two advantage functions:

$$\mathrm{Adv}_{\mathcal{G},\mathcal{A}}^{\mathrm{SD1}}(\lambda) := \left| \Pr[\mathcal{A}(\mathbb{G}, D, T_0) = 1] - \Pr[\mathcal{A}(\mathbb{G}, D, T_1) = 1] \right|$$

where $\mathbb{G} \leftarrow \mathcal{G}, T_0 \leftarrow \boxed{G_{p_1}}, T_1 \leftarrow_{\mathrm{R}} \boxed{G_{p_1} G_{p_2}}$

and $D := (g_1, g_3, g_{\{1,2\}}), g_1 \leftarrow_{\mathrm{R}} G_{p_1}^*, g_3 \leftarrow_{\mathrm{R}} G_{p_3}^*, g_{\{1,2\}} \leftarrow_{\mathrm{R}} G_{p_1} G_{p_2}$

$$\mathrm{Adv}_{\mathcal{G},\mathcal{A}}^{\mathrm{SD2}}(\lambda) := \left| \Pr[\mathcal{A}(\mathbb{G}, D, T_0) = 1] - \Pr[\mathcal{A}(\mathbb{G}, D, T_1) = 1] \right|$$

where $\mathbb{G} \leftarrow \mathcal{G}, T_0 \leftarrow \boxed{G_{p_1}^* G_{p_3}}, T_1 \leftarrow_{\mathrm{R}} \boxed{G_{p_1}^* G_{p_2}^* G_{p_3}}$

and $D := (g_1, g_3, g_{\{1,2\}}, g_{\{2,3\}}), g_1 \leftarrow_{\mathrm{R}} G_{p_1}^*, g_3 \leftarrow_{\mathrm{R}} G_{p_3}^*, g_{\{1,2\}} \leftarrow_{\mathrm{R}} G_{p_1} G_{p_2}, g_{\{2,3\}} \leftarrow_{\mathrm{R}} G_{p_2} G_{p_3}$

Assumption 1 (resp. 2) asserts that for all PPT adversaries $\mathcal{A}$, the advantage $\mathrm{Adv}_{\mathcal{G},\mathcal{A}}^{\mathrm{SD1}}(\lambda)$ (resp. $\mathrm{Adv}_{\mathcal{G},\mathcal{A}}^{\mathrm{SD2}}(\lambda)$) is a negligible function in λ.

2.2 Predicate Encryption

We define predicate encryption in the framework of key encapsulation. A predicate encryption scheme for a predicate $\mathsf{P}(\cdot, \cdot)$ consists of four algorithms (Setup, Enc, KeyGen, Dec):

Setup$(1^\lambda, \mathcal{X}, \mathcal{Y}) \to (\mathsf{pp}, \mathsf{mpk}, \mathsf{msk})$. The setup algorithm gets as input the security parameter λ, the attribute universe $\mathcal{X}$, the predicate universe $\mathcal{Y}$ and outputs the public parameter (pp, mpk), and the master key msk. All the other algorithms get pp as part of its input.

$\mathsf{Enc}(\mathsf{mpk}, x) \to (\mathsf{ct}_x, \kappa)$. The encryption algorithm gets as input mpk and an attribute $x \in \mathcal{X}$. It outputs a ciphertext ct_x and a symmetric key $\kappa \in \{0, 1\}^\lambda$. Note that x is public given ct_x.

$\mathsf{KeyGen}(\mathsf{msk}, y) \to \mathsf{sk}_y$. The key generation algorithm gets as input msk and a value $y \in \mathcal{Y}$. It outputs a secret key sk_y. Note that y is public given sk_y.

$\mathsf{Dec}(\mathsf{sk}_y, \mathsf{ct}_x) \to \kappa$. The decryption algorithm gets as input sk_y and ct_x such that $\mathsf{P}(x, y) = 1$. It outputs a symmetric key κ.

Correctness. We require that for all $(x, y) \in \mathcal{X} \times \mathcal{Y}$ such that $\mathsf{P}(x, y) = 1$,

$$\Pr[(\mathsf{ct}_x, \kappa) \leftarrow \mathsf{Enc}(\mathsf{mpk}, x); \mathsf{Dec}(\mathsf{sk}_y, \mathsf{ct}_x) = \kappa)] = 1,$$

where the probability is taken over $(\mathsf{mpk}, \mathsf{msk}) \leftarrow \mathsf{Setup}(1^\lambda, \mathcal{X}, \mathcal{Y})$ and the coins of Enc.

Security Definition. For a stateful adversary $\mathcal{A}$, we define the advantage function

$$\mathsf{Adv}^{\mathsf{PE}}_{\mathcal{A}}(\lambda) := \Pr \left[b = b' : \begin{array}{l} (\mathsf{mpk}, \mathsf{msk}) \leftarrow \mathsf{Setup}(1^\lambda, \mathcal{X}, \mathcal{Y}); \\ x \leftarrow \mathcal{A}^{\mathsf{KeyGen}(\mathsf{msk}, \cdot)}(\mathsf{mpk}); \\ b \leftarrow_{\mathsf{R}} \{0, 1\}; \kappa_1 \leftarrow_{\mathsf{R}} \{0, 1\}^\lambda \\ (\mathsf{ct}_x, \kappa_0) \leftarrow \mathsf{Enc}(\mathsf{mpk}, x); \\ b' \leftarrow \mathcal{A}^{\mathsf{KeyGen}(\mathsf{msk}, \cdot)}(\mathsf{ct}_x, \kappa_b) \end{array} \right] - \frac{1}{2}$$

with the restriction that all queries y that $\mathcal{A}$ makes to $\mathsf{KeyGen}(\mathsf{msk}, \cdot)$ satisfies $\mathsf{P}(x, y) = 0$ (that is, sk_y does not decrypt ct_x). A predicate encryption scheme is *adaptively secure* if for all PPT adversaries $\mathcal{A}$, the advantage $\mathsf{Adv}^{\mathsf{PE}}_{\mathcal{A}}(\lambda)$ is a negligible function in λ.

3   Bilinear Predicate Encodings

In this section, we describe *predicate encodings* more formally. Then, we discuss several examples, before describing the *bilinear* requirement.

3.1   Predicate Encodings

Fix a predicate $\mathsf{P} : \mathcal{X} \times \mathcal{Y} \to \{0, 1\}$. A *predicate encoding* for P is a pair of algorithms $(\mathsf{sE}, \mathsf{rE})$, where sE is deterministic and takes as input $(x, w) \in \mathcal{X} \times \mathcal{W}$; and rE is randomized and takes as input $(\alpha, y, w) \in \mathcal{D} \times \mathcal{Y} \times \mathcal{W}$ and randomness $r \in \mathcal{R}$. (We stress that $\mathcal{W}$ and $\mathcal{R}$ play very different roles, as evident in the w-hiding property.) In addition, we require that $(\mathsf{sE}, \mathsf{rE})$ satisfy the following three properties:

(α-**reconstruction.**) For all $(x, y) \in X \times Y$ such that $P(x, y) = 1$ and for all r, we can (efficiently) recover α given $x, y, sE(x, w), rE(\alpha, y, w; r)$.

(α-**privacy.**) For all $(x, y) \in X \times Y$ such that $P(x, y) = 0$, and for all $\alpha \in D$, the joint distribution $sE(x, w), rE(\alpha, y, w; r)$ *perfectly* hides α. That is, for all $\alpha, \alpha' \in D$, the following joint distributions are *identically* distributed:

$$\{x, y, \alpha, sE(x, w), rE(\alpha, y, w; r)\} \quad \text{and} \quad \{x, y, \alpha, sE(x, w), rE(\alpha', y, w; r)\}$$

where the randomness is taken over $(w, r) \leftarrow_R W \times R$.

(w-**hiding.**) There exists some element $0 \in R$ such that for all $(\alpha, y, w) \in D \times \in Y \times W$, $rE(\alpha, y, w; 0)$ is statistically independent of w, that is, for all $w' \in W$:

$$rE(\alpha, y, w; 0) = rE(\alpha, y, w'; 0)$$

Remark 1. We rely crucially on the fact that α is *perfectly* hidden in the proof of security, so that non-adaptive indistinguishability implies adaptive indistinguishability. (This is not true in the statistical or computational setting.) Concretely, we claim that α-privacy implies that even if y is chosen *adaptively* after seeing $(x, \alpha, rE(x, w))$, the distributions

$$rE(\alpha, y, w; r) \quad \text{and} \quad rE(0, y, w; r)$$

are perfectly indistinguishable. This simply follows from the fact that an adaptive distinguisher with advantage ϵ can be converted into a non-adaptive distinguisher with advantage $\epsilon / |Y|$ via random guessing. Since any non-adaptive distinguisher has advantage 0, we must have $\epsilon = 0$ to begin with. The same argument applies to the setting where x is chosen *adaptively* after seeing $(y, \alpha, rE(\cdot, y, w; r))$.

Remark 2. We note that w-hiding as defined is not the only way to achieve "w-reusability". For instance, for the equality predicate as in IBE, the Lewko-Waters scheme [34] achieves reusability by essentially masking $rE(\alpha, y, w; r)$ with a fresh one-time pad for each secret key query. This works for IBE and HIBE because $rE(\alpha, y, w; r)$ has the uniform distribution for every y. However, this approach does not work for the ABE predicate. Indeed, by using w-hiding, we obtain a different proof of security of the Lewko-Waters HIBE.

Example 1: Equality. Fix an integer N to be the product of three λ-bit primes. Consider the equality predicate where $X = Y = [N]$ and $P(x, y) = 1$ iff $x = y$. The following is a predicate encoding for equality used in [7, 34]:

- $D := \mathbb{Z}_N; W := \mathbb{Z}_N^2; R := \mathbb{Z}_N^*.$
- $sE(x, (w_1, w_2)) := (1, w_1 + w_2 x)$
- $rE(\alpha, y, (w_1, w_2); r) := (\alpha + r(w_1 + w_2 y), -r)$

For α-reconstruction when $x = y$, simply take the dot product of the two vectors. For α-privacy when $x \neq y$,[1] we exploit the fact that $(w_1 + w_2 x, w_1 + w_2 y)$ are pairwise

---

[1] Here, we will even assume $\gcd(x - y, N) = 1$; otherwise, we can find a non-trivial factor of N.

independent and $r \in \mathbb{Z}_N^*$. (Note that perfect α-privacy does not hold if we set $\mathcal{R}$ to be $\mathbb{Z}_N$ instead of $\mathbb{Z}_N^*$.) To achieve w-hiding, we simply set $r = 0$.[2]

Example 2: Equality. Consider the same predicate and construction as before, but replace rE by

$$\mathsf{rE}(\alpha, y, (w_1, w_2)) := (\alpha + (w_1 + w_2 y), -1).$$

This still satisfies α-reconstruction and α-privacy, but not w-hiding nor linear receiver encoding (the latter property is defined in the next Section).

3.2 Bilinearity

Fix a prime p. Let $(\mathsf{sE}, \mathsf{rE})$ be a predicate encoding for $\mathsf{P} : \mathcal{X} \times \mathcal{Y} \to \{0, 1\}$, where $\mathcal{X}$ and $\mathcal{Y}$ may depend on p. We say that $(\mathsf{sE}, \mathsf{rE})$ is *p-bilinear* if it satisfies the following properties:

(input domains.) $\mathcal{D} = \mathbb{Z}_p$, $\mathcal{W} = \mathbb{Z}_p^{\ell_{\mathcal{W}}}$ and $\mathcal{R} = \mathbb{Z}_p^{\ell_{\mathcal{R}}} \times (\mathbb{Z}_p^*)^{\ell_{\mathcal{R}}'}$ for some integers $\ell_{\mathcal{W}}, \ell_{\mathcal{R}}, \ell_{\mathcal{R}}'$.[3]

(output domains.) The output of sE and rE are vectors over $\mathbb{Z}_p$.

(affine sender encoding.) For all $x \in \mathcal{X}$, $\mathsf{sE}(x, \cdot)$ is affine in $\mathbf{w}$.

(linear receiver encoding.) For all $(\alpha, y, \mathbf{w}) \in \mathcal{D} \times \mathcal{Y} \times \mathcal{W}$, $\mathsf{rE}(\cdot, y, \mathbf{w}; \cdot)$ is linear in $\alpha, \mathbf{r}$.

(bilinear α-reconstruction.) For all (x, y) such that $\mathsf{P}(x, y) = 1$, we can efficiently compute a linear map $\mathbf{M}_{xy}$ (a matrix over $\mathbb{Z}_p$) such that for all $\mathbf{r} \in \mathcal{R}$,

$$\langle \mathsf{sE}(x, \mathbf{w}), \mathbf{M}_{xy} \mathsf{rE}(\alpha, y, \mathbf{w}; \mathbf{r}) \rangle = \alpha$$

(w-hiding.) For all $(\alpha, y, \mathbf{w}) \in \mathcal{D} \times \in \mathcal{Y} \times \mathcal{W}$, we have

$$\mathsf{rE}(\alpha, y, \mathbf{w}; \mathbf{0}) = \mathsf{rE}(\alpha, y, \mathbf{0}; \mathbf{0})$$

where we use $\mathbf{0}$ to refer to the all zeroes vector in $\mathbb{Z}_p^{\ell_{\mathcal{W}}}$ and in $\mathbb{Z}_p^{\ell_{\mathcal{R}} + \ell_{\mathcal{R}}'}$.[4]

The above definition extends to any integer N by replacing $\mathbb{Z}_p, \mathbb{Z}_p^*$ with $\mathbb{Z}_N, \mathbb{Z}_N^*$ respectively.

Remark 3. We will exploit the affine sender encoding and linear receiver encoding to compute sE and rE "in the exponent". Fix $g \in G_N$.

---

[2] This does not actually work since $0 \notin \mathcal{R}$, but we will consider a slight weakening of w-hiding in the next section.

[3] The distinction between $\mathbb{Z}_p$ and $\mathbb{Z}_p^*$ is significant because we require *perfect* α-privacy.

[4] This is in fact a slight relaxation of the general w-hiding property since $\mathbf{0}$ does not lie in $\mathcal{R}$ whenever $\ell_R' > 0$.

Property	Where it is used
bilinear α-reconstruction	Dec and correctness
affine sender encoding	Enc
linear receiver encoding	KeyGen, $\widetilde{\text{KeyGen}}$
α-privacy	pseudo-normal to pseudo-SF secret keys, Lemma 3
$\mathbf{w}$-hiding	pseudo-normal to pseudo-SF secret keys, Lemma 3

Fig. 1. Properties of predicate encodings and where they are used

- Affine sender encoding implies that given $x \in \mathcal{X}$ along with $g, g^{\mathbf{w}}$, we can compute $g^{\mathsf{sE}(x,\mathbf{w})}$; indeed, we will slightly abuse notation and write this as $\mathsf{sE}(x, g^{\mathbf{w}})$.
- Similarly, linear receiver encoding implies that given $(y, \mathbf{w}) \in \mathcal{Y} \times \mathcal{W}$ along with $g^{\alpha}, g^{\mathbf{r}}$ (but not g), we can compute $g^{\mathsf{rE}(\alpha,y,\mathbf{w};\mathbf{r})}$; again, we will write this as $\mathsf{rE}(g^{\alpha}, y, \mathbf{w}; g^{\mathbf{r}})$.

Extensions. We also consider two extensions, first to handle randomized sender's encoding in Section 5.5 and second to support delegation in Section 5.3.

4 Predicate Encryption from Bilinear Encoding

We present a predicate encryption scheme in composite-order bilinear groups whose order is the product of three primes (c.f. Section 2.1), for any predicate $\mathsf{P}(\cdot, \cdot)$ which admits a bilinear predicate encoding. In addition, we show that the scheme is adaptively secure under the General Subgroup Decision Assumption. We refer to Section 1.1 for an overview of the construction and the proof.

4.1 Construction

Fix a predicate $\mathsf{P} : \mathcal{X} \times \mathcal{Y} \to \{0, 1\}$. Given a N-bilinear predicate encoding $(\mathsf{sE}, \mathsf{rE})$ for P, we may construct a predicate encryption scheme for P as follows:

Setup$(1^{\lambda}, \mathcal{X}, \mathcal{Y})$: On input $(1^{\lambda}, \mathcal{X}, \mathcal{Y})$, first generate $\mathbb{G} \leftarrow \mathcal{G}(1^{\lambda})$, then sample $\mathsf{H} : G_T \to \{0, 1\}^{\lambda}$ from a family of pairwise-independent hash functions. In addition, sample $\alpha \leftarrow_{\mathsf{R}} \mathbb{Z}_N, \mathbf{w} \leftarrow_{\mathsf{R}} \mathcal{W}$, and output[5]

$$\mathsf{pp} := (\mathbb{G}, \mathsf{H}, g_1, g_3) \quad \text{and} \quad \mathsf{mpk} := (g_1^{\mathbf{w}}, e(g_1, g_1)^{\alpha}) \quad \text{and} \quad \mathsf{msk} := (g_1^{\alpha} g_2^{\alpha}, \mathbf{w})$$

---

[5] If we want to be able to derive multiple $(\mathsf{mpk}, \mathsf{msk})$ from the same pp, we will need to append a random generator of $G_{\{1,2\}}$ to pp, which we can then use to sample msk. Note that this will not affect the proof of security, since such a generator is provided to the distinguisher in both Assumption 1 and 2.

KeyGen(msk, y): On input msk $= (g_1^\alpha g_2^\alpha, \mathbf{w})$ and a predicate y, sample $\mathbf{r} \leftarrow_R \mathcal{R}$ and output[6]

$$\mathsf{sk}_y := \mathrm{rand3}(\mathrm{rE}(g_1^\alpha g_2^\alpha, y, \mathbf{w}; g_1^\mathbf{r})) = \mathrm{rand3}(g_1^{\mathrm{rE}(\alpha, y, \mathbf{w}; \mathbf{r})} \cdot g_2^{\mathrm{rE}(\alpha, y, \mathbf{w}; 0)})$$

Here, $\mathrm{rand3}$ is an algorithm that randomizes the G_{p_3}-components, namely on input a vector $\mathbf{C} \in G_N^\ell$, outputs $\mathbf{C} \cdot g_3^{\mathbf{r}'}$ where $\mathbf{r}' \leftarrow_R \mathbb{Z}_N^\ell$.

Enc(mpk, x): On input an attribute $x \in \mathcal{X}$, sample $s \leftarrow_R \mathbb{Z}_N$ and output the ciphertext and symmetric key

$$\mathsf{ct}_x := (\mathsf{sE}(x, g_1^\mathbf{w}))^s = g_1^{\mathsf{sE}(x, \mathbf{w})s} \quad \text{and} \quad \kappa := \mathsf{H}((e(g_1, g_1)^\alpha)^s)$$

Dec($\mathsf{sk}_y, \mathsf{ct}_x$): On input sk_y and ct_x where $\mathsf{P}(x, y) = 1$, output

$$\mathsf{H}(e(\mathsf{ct}_x, \mathsf{sk}_y^{\mathbf{M}_{xy}}))$$

where $\mathbf{M}_{xy}$ is the matrix for bilinear reconstruction and $e(\mathsf{ct}_x, \mathsf{sk}_y^{\mathbf{M}_{xy}}) := \sum_i e((\mathsf{ct}_x)_i, \sum_j (\mathsf{sk}_y)_j^{(\mathbf{M}_{xy})_{i,j}})$.

Correctness. For all $(x, y) \in \mathcal{X} \times \mathcal{Y}$ such that $\mathsf{P}(x, y) = 1$, we have

$$
\begin{aligned}
\mathsf{Dec}(\mathsf{sk}_y, \mathsf{ct}_x) &= \mathsf{Dec}(g_1^{\mathrm{rE}(\alpha, y, \mathbf{w}; \mathbf{r})} g_2^{\mathrm{rE}(\alpha, y, \mathbf{w}; 0)} \mathbf{Z}_3, g_1^{\mathsf{sE}(x, \mathbf{w})s}) \\
&= \mathsf{H}(e(g_1^{\mathsf{sE}(x, \mathbf{w})s}, (g_1^{\mathrm{rE}(\alpha, y, \mathbf{w}; \mathbf{r})} g_2^{\mathrm{rE}(\alpha, y, \mathbf{w}; 0)} \mathbf{Z}_3)^{\mathbf{M}_{xy}})) \\
&= \mathsf{H}(e(g_1^{\mathsf{sE}(x, \mathbf{w})s}, (g_1^{\mathrm{rE}(\alpha, y, \mathbf{w}; \mathbf{r})})^{\mathbf{M}_{xy}})) \\
&= \mathsf{H}(e(g_1, g_1)^{\langle \mathsf{sE}(x, \mathbf{w})s, \mathbf{M}_{xy} \mathrm{rE}(\alpha, y, \mathbf{w}; \mathbf{r}) \rangle}) \\
&= \mathsf{H}((e(g_1, g_1)^\alpha)^s)
\end{aligned}
$$

4.2 Proof of Security

We prove the following theorem:

Theorem 1. *Under Assumptions 1 and 2 (c.f. Section 2.1), the predicate encryption scheme described in Section 4.1 is adaptively secure (c.f. Section 2.2). More precisely, for any adversary $\mathcal{A}$ that makes at most q queries against the predicate encryption scheme, there exist adversaries $\mathcal{A}_1, \mathcal{A}_2, \mathcal{A}_3$ whose running times are essentially the same as that of $\mathcal{A}$, such that*

$$\mathsf{Adv}_{\mathcal{A}}^{\mathrm{PE}}(\lambda) \le \mathsf{Adv}_{\mathcal{G}, \mathcal{A}_1}^{\mathrm{SD1}}(\lambda) + q \cdot \mathsf{Adv}_{\mathcal{G}, \mathcal{A}_2}^{\mathrm{SD2}}(\lambda) + q \cdot \mathsf{Adv}_{\mathcal{G}, \mathcal{A}_3}^{\mathrm{SD2}}(\lambda) + 2^{-\Omega(\lambda)}$$

The proof follows via a series of games, outlined in Section 1.1 and summarized in Fig 2. Following Waters' dual system encryption methodology [45], there are two types of keys and ciphertexts: normal and semi-functional. We first describe two auxiliary algorithms (analogous to "private evaluation" algorithms in Cramer-Shoup projective

---

[6] Refer to Remark 3 for the notation $\mathrm{rE}(\cdots)$ as used here.

hashing [20]), and then defining the semi-functional distributions via these auxiliary algorithms.

Auxiliary Algorithms. We consider the following algorithms: a deterministic algorithm $\widehat{\mathsf{Enc}}$ for computing ciphertexts and a randomized algorithm $\widehat{\mathsf{KeyGen}}$ for computing secret keys.

$\widehat{\mathsf{Enc}}(\mathsf{pp}, x; \mathsf{msk}', C)$: On input $x \in \mathcal{X}$, along with $\mathsf{msk}' = (h, \mathbf{w}) \in G_N \times \mathcal{W}$ and $C \in G_N$, output:

$$(\mathsf{ct}_x, \kappa) := (C^{\mathsf{sE}(x,\mathbf{w})}, \mathsf{H}(e(C, h)))$$

Observe that for all $(\mathsf{pp}, \mathsf{mpk}, \mathsf{msk})$ output by Setup and for all $s \in \mathbb{Z}_N$, we have

$$\mathsf{Enc}(\mathsf{mpk}, x; s) = \left(g_1^{\mathsf{sE}(x,\mathbf{w})s}, \mathsf{H}(e(g_1, g_1)^{\alpha s}) \right) = \widehat{\mathsf{Enc}}(\mathsf{pp}, x; \mathsf{msk}, g_1^s)$$

$\widehat{\mathsf{KeyGen}}(\mathsf{msk}', y; R)$: On input $\mathsf{msk}' = (h, \mathbf{w}) \in G_N \times \mathcal{W}$, $y \in \mathcal{Y}$ and $R \in G_N$, sample $\mathbf{r} \leftarrow_{\mathsf{R}} \mathcal{R}$ and output

$$\mathsf{sk}_y := \mathsf{rand3}(\mathsf{rE}(h, y, \mathbf{w}; R^{\mathbf{r}}))$$

Observe that, for any msk', y and any $R \in G_{p_1}^* G_{p_3}$, the following three distributions are identical:

$$\mathsf{KeyGen}(\mathsf{msk}', y) \quad \text{and} \quad \widehat{\mathsf{KeyGen}}(\mathsf{msk}', y; g_1) \quad \text{and} \quad \widehat{\mathsf{KeyGen}}(\mathsf{msk}', y; R)$$

That is, we have three different but equivalent ways to generate real secret keys. The equivalence of the first two distributions is straight-forward. For the equivalence of the second and the third, we use the fact that $\mathcal{R}$ is of the form $\mathbb{Z}_N^{\ell_{\mathcal{R}}} \times (\mathbb{Z}_N^*)^{\ell_{\mathcal{R}}'}$ and that we randomize using $\mathsf{rand3}$.

Auxiliary Distributions. We consider the following auxiliary distributions for ciphertext and secret keys, where $(\mathsf{pp}, \mathsf{mpk}, \mathsf{msk}, \alpha, \mathbf{w})$ are sampled as in Setup.

- semi-functional (SF) master secret key: $\widehat{\mathsf{msk}} = (\boxed{g_1^\alpha}, \mathbf{w})$.

- normal ciphertexts:

$$\widehat{\mathsf{Enc}}(\mathsf{pp}, x; \mathsf{msk}, C), \quad C \leftarrow_{\mathsf{R}} \boxed{G_{p_1}}$$

this is identically distributed to real ciphertexts as computed using $\mathsf{Enc}(\mathsf{mpk}, x)$.

- semi-functional (SF) ciphertexts:

$$\widehat{\mathsf{Enc}}(\mathsf{pp}, x; \mathsf{msk}, \widehat{C}), \quad \widehat{C} \leftarrow_{\mathsf{R}} \boxed{G_{p_1} G_{p_2}}$$

- normal secret keys:

$$\widehat{\mathsf{KeyGen}}(\boxed{\mathsf{msk}}, y; R), \quad R \leftarrow_{\mathsf{R}} \boxed{G_{p_1}^* G_{p_3}}$$

this is identically distributed to real secret keys as computed using $\mathsf{KeyGen}(\mathsf{msk}, y)$.

Game	Ciphertext / Key (ct_x, κ)	Secret Key sk_y	Justification	Remark
0	$(0,0)$	$rE(\alpha, y, \mathbf{w}; 0)$		actual scheme
1	$\boxed{sE(x,\mathbf{w})s, \alpha s}$	$rE(\alpha, y, \mathbf{w}; 0)$	Assumption 1	normal to SF (ct_x, κ)
2.i.1	$sE(x,\mathbf{w})s, \alpha s$	$rE(\alpha, y, \mathbf{w}; \boxed{\mathbf{r}})$	Assumption 2	normal to pseudo-normal sk_y
3.i.2	$sE(x,\mathbf{w})s, \alpha s$	$rE(\boxed{0}, y, \mathbf{w}; \mathbf{r})$	α-privacy & $\mathbf{w}$-hiding	pseudo-normal to pseudo-SF sk_y
2.i.3	$sE(x,\mathbf{w})s, \alpha s$	$rE(0, y, \mathbf{w}; \boxed{0})$	Assumption 2	pseudo-SF to SF sk_y
3	$sE(x,\mathbf{w})s, \boxed{\text{random}}$	$rE(0, y, \mathbf{w}; 0)$		

Fig. 2. Sequence of games in the semi-functional space (the G_{p_2}-subgroup), where we drew a box to highlight the differences between each game and the preceding one, and games $2.i.xx$ refer to the i'th secret key

- pseudo-normal secret keys:

$$\widehat{\text{KeyGen}}(\boxed{\text{msk}}, y; R), \quad R \leftarrow_R \boxed{G_{p_1}^* G_{p_2}^* G_{p_3}}$$

- pseudo-semi-functional (pseudo-SF) secret keys:

$$\widehat{\text{KeyGen}}(\boxed{\widehat{\text{msk}}}, y; R), \quad R \leftarrow_R \boxed{G_{p_1}^* G_{p_2}^* G_{p_3}}$$

- semi-functional (SF) secret keys:

$$\widehat{\text{KeyGen}}(\boxed{\widehat{\text{msk}}}, y; R), \quad R \leftarrow_R \boxed{G_{p_1}^* G_{p_3}}$$

Remark 4 (decryption capabilities). Observe that all types of secret keys can decrypt a normal ciphertext. In addition, only normal and pseudo-normal secret keys can decrypt a semi-functional ciphertext, whereas pseudo-SF and SF keys cannot. The latter is consistent with the fact that we exploit α-hiding and $P(x,y) = 0$ when we switch from pseudo-normal to pseudo-SF keys, which is precisely why we lose decryption capabilities in the G_{p_2}-components.

Game Sequence. We present a series of games. We write Adv_{xx} to denote the advantage of $\mathcal{A}$ in Game_{xx}.

- Game_0: is the real security game (c.f. Section 2.2).
- Game_1: is the same as Game_0 except that the challenge ciphertext is semi-functional. We also modify the distribution of κ_0 accordingly.
- $\text{Game}_{2,i}$ for $i = 1, \ldots, q$: is the same as Game_1, except the first $i - 1$ keys are semi-functional, and the last $q - i$ keys are normal. There are 4 sub-games, where the i'th key transitions from normal in $\text{Game}_{2.i.0}$, to pseudo-normal in $\text{Game}_{2.i.1}$, to pseudo-SF in $\text{Game}_{2.i.2}$, to SF in $\text{Game}_{2.i.3}$.
- Game_3: is the same as $\text{Game}_{2,q,3}$, except that $\kappa_0 \leftarrow_R \{0,1\}^\lambda$.

In Game_3, the view of the adversary $\mathcal{A}$ is statistically independent of the challenge bit β. Hence, $\text{Adv}_3 = 0$. We complete the proof by establishing the following sequence of lemmas.

Lemma 1 (normal to semi-functional ciphertexts). *There exists $\mathcal{A}_1$ whose running time is roughly that of $\mathcal{A}$ such that*

$$|\mathsf{Adv}_0 - \mathsf{Adv}_1| \leq \mathsf{Adv}^{\mathsf{SD1}}_{\mathbb{G},\mathcal{A}_1}(\lambda)$$

Proof. We will rely on Assumption 1. On input $D = (\mathbb{G}, g_1, g_3, g_1^\alpha g_2^\alpha)$ and $T \in \{T_0, T_1\}$ where $T_0 \leftarrow_{\mathrm{R}} G_{p_1}, T_1 \leftarrow_{\mathrm{R}} G_{p_1} G_{p_2}$, the adversary $\mathcal{A}_1$ simulates $\mathcal{A}$ as follows:

Setup. Sample $\mathsf{H}, \mathbf{w}$ as in Setup, set $\mathsf{msk} := (g_1^\alpha g_2^\alpha, \mathbf{w})$ and output

$$\mathsf{pp} := (\mathbb{G}, \mathsf{H}, g_1, g_3) \quad \text{and} \quad \mathsf{mpk} := (g_1^{\mathbf{w}}, e(g_1, g_1^\alpha g_2^\alpha)).$$

Ciphertext. Compute $(\mathsf{ct}_x, \kappa_0) \leftarrow \widehat{\mathsf{Enc}}(\mathsf{pp}, x; \mathsf{msk}, T)$.

Key Queries. On input the j'th key query y_j, output

$$\mathsf{sk}_j \leftarrow \widehat{\mathsf{KeyGen}}(\mathsf{msk}, y; g_1)$$

Output. Output whatever $\mathcal{A}$ outputs.

Observe that when $T = T_0 \leftarrow_{\mathrm{R}} G_{p_1}$, the output is identical to that in Game 0, and when $T = T_1 \leftarrow_{\mathrm{R}} G_{p_1} G_{p_2}$, the output is identical to that in Game 1.   □

Lemma 2 (normal to pseudo-normal secret keys). *There exists $\mathcal{A}_2$ whose running time is roughly that of $\mathcal{A}$ such that for all $i = 1, 2, \ldots, q$,*

$$|\mathsf{Adv}_{2.i.0} - \mathsf{Adv}_{2.i.1}| \leq \mathsf{Adv}^{\mathsf{SD2}}_{\mathbb{G},\mathcal{A}_2}(\lambda)$$

Proof. We will rely on Assumption 2. On input $D = (\mathbb{G}, g_1, g_3, g_{\{1,2\}}, g_{\{2,3\}})$ and $T \in \{T_0, T_1\}$ where $T_0 \leftarrow_{\mathrm{R}} G^*_{p_1} G_{p_3}, T_1 \leftarrow_{\mathrm{R}} G^*_{p_1} G^*_{p_2} G_{p_3}$, the adversary $\mathcal{A}_2$ simulates $\mathcal{A}$ as follows:

Setup. Sample $\alpha, \mathsf{H}, \mathbf{w}$ as in Setup, set

$$\mathsf{msk} := (g_1^\alpha \cdot g_{\{2,3\}}, \mathbf{w}) \quad \text{and} \quad \widehat{\mathsf{msk}} := (g_1^\alpha, \mathbf{w})$$

and output

$$\mathsf{pp} := (\mathbb{G}, \mathsf{H}, g_1, g_3) \quad \text{and} \quad \mathsf{mpk} := (g_1^{\mathbf{w}}, e(g_1, g_1^\alpha)).$$

Ciphertext. Compute $(\mathsf{ct}_x, \kappa_0) \leftarrow \widehat{\mathsf{Enc}}(\mathsf{pp}, x; \mathsf{msk}, g_{\{1,2\}})$.

Key Queries. On input the j'th key query y_j, output

$$\mathsf{sk}_{y_j} \leftarrow \begin{cases} \widehat{\mathsf{KeyGen}}(\widehat{\mathsf{msk}}, y_j; g_1) & \text{if } j < i \quad \text{(semi-functional key)} \\ \widehat{\mathsf{KeyGen}}(\mathsf{msk}, y_j; T) & \text{if } j = i \quad \text{(normal vs pseudo-normal key)} \\ \widehat{\mathsf{KeyGen}}(\mathsf{msk}, y_j; g_1) & \text{if } j > i \quad \text{(normal key)} \end{cases}$$

Output. Output whatever $\mathcal{A}$ outputs.

Observe that when $T = T_0 \leftarrow_{\mathrm{R}} G^*_{p_1} G_{p_3}$, the output is identical to that in Game 2.i.0, and when $T = T_1 \leftarrow_{\mathrm{R}} G^*_{p_1} G^*_{p_2} G_{p_3}$, the output is identical to that in Game 2.i.1.   □

Lemma 3 (pseudo-normal to pseudo-SF secret keys). *For all* $i = 1, 2, \ldots, q,$

$$|\mathsf{Adv}_{2.i.1} - \mathsf{Adv}_{2.i.2}| = 0$$

Proof. Observe that the only difference between Game 2.i.1 and Game 2.i.2 lies in the distribution of sk_{y_i}, which we sample using $\mathsf{msk} = g_1^\alpha g_2^\alpha$ and $\widehat{\mathsf{msk}} = g_1^\alpha$ respectively. This means the only difference between Game 2.i.1 and Game 2.i.2 lies in the G_{p_2}-component of sk_{y_i}, which are given by

$$g_2^{\mathsf{rE}(\alpha, y_i, \mathbf{w}; \mathbf{r})} \quad \text{and} \quad g_2^{\mathsf{rE}(0, y_i, \mathbf{w}; \mathbf{r})} \quad (*)$$

respectively, where $\mathbf{r} \leftarrow_{\mathsf{R}} \mathcal{R}$. By the Chinese Remainder Theorem, it suffices to focus on the G_{p_2}-components of challenge ciphertext and secret keys, which are independent of the corresponding G_{p_1}-components. Observe that for all $j \neq i$, the G_{p_2}-component of sk_{y_j} is given by:

$$\begin{cases} \mathsf{rE}(0, y_j, \mathbf{w}; \mathbf{0}) = \mathsf{rE}(0, y_j, \mathbf{0}; \mathbf{0}) & \text{if } j < i \text{ (semi-functional key)} \\ \mathsf{rE}(\alpha, y_j, \mathbf{w}; \mathbf{0}) = \mathsf{rE}(\alpha, y_j, \mathbf{0}; \mathbf{0}) & \text{if } j > i \text{ (normal key)} \end{cases}$$

where the equality above follows by $\mathbf{w}$-hiding. This means that only the challenge ciphertext and the sk_{y_i} leaks any information about $\mathbf{w} \pmod{p_2}$. It now follows from the α-privacy property (modulo p_2) and $\mathsf{P}(x, y_i) = 0$ that

$$\mathsf{rE}(\alpha, y_i, \mathbf{w}; \mathbf{r}) \pmod{p_2} \quad \text{and} \quad \mathsf{rE}(0, y_i, \mathbf{w}; \mathbf{r}) \pmod{p_2}$$

are identically distributed from the view-point of the adversary. (Here, we also use secrecy of $\mathbf{r} \pmod{p_2}$.) This holds even if the adversary chooses y_i adaptively after seeing the challenge ciphertext ct_x, or if the challenge x is chosen after the adversary sees sk_{y_i} (c.f. Remark 1). $\qquad\square$

Lemma 4 (pseudo-SF to SF secret keys). *There exists* $\mathcal{A}_3$ *whose running time is roughly that of* $\mathcal{A}$ *such that for all* $i = 1, 2, \ldots, q,$

$$|\mathsf{Adv}_{2.i.2} - \mathsf{Adv}_{2.i.3}| \leq \mathsf{Adv}_{\mathcal{G}, \mathcal{A}_3}^{\mathsf{SD2}}(\lambda)$$

Proof. We will again rely on Assumption 2. The proof is completely analogous to Lemma 2, except $\mathcal{A}_3$ uses $\widehat{\mathsf{msk}}$ instead of msk to sample sk_{y_j}. That is, $\mathcal{A}_3$ outputs

$$\mathsf{sk}_{y_j} \leftarrow \begin{cases} \widehat{\mathsf{KeyGen}}(\widehat{\mathsf{msk}}, y_j; g_1) & \text{if } j < i \text{ (semi-functional key)} \\ \widehat{\mathsf{KeyGen}}(\widehat{\mathsf{msk}}, y_j; T) & \text{if } j = i \text{ (pseudo-SF vs SF key)} \\ \widehat{\mathsf{KeyGen}}(\mathsf{msk}, y_j; g_1) & \text{if } j > i \text{ (normal key)} \end{cases}$$

Observe that when $T = T_1 \leftarrow_{\mathsf{R}} G_{p_1}^* G_{p_2}^* G_{p_3}$, the output is identical to that in Game 2.i.2, and when $T = T_0 \leftarrow_{\mathsf{R}} G_{p_1}^* G_{p_3}$, the output is identical to that in Game 2.i.3. $\qquad\square$

Lemma 5 (final transition).

$$|\mathsf{Adv}_{3.q.3} - \mathsf{Adv}_4| \leq 2^{-\Omega(\lambda)}$$

Proof. In Game 3.q.3, all the secret keys are semi-functional, which means they leak no information whatsoever about $\alpha \pmod{p_2}$. Next, let us examine the (semi-functional) challenge ciphertext. Observe that the quantity (from which the symmetric key κ_0 is derived)

$$e(\widehat{C}, g_1^\alpha) \cdot e(\widehat{C}, g_2^\alpha)$$

has $\log p_2 = \Theta(\lambda)$ bits of min-entropy as long as $\widehat{C} \in G_{p_1} G_{p_2}^*$, which occurs with probability $1 - 1/p_2$. Then, by the left-over hash lemma, $\kappa_0 = H(e(\widehat{C}, g_1^\alpha) \cdot e(\widehat{C}, g_2^\alpha))$ is $2^{-\Omega(\lambda)}$-close to the uniform distribution over $\{0,1\}^\lambda$. The claim follows readily.   □

5 Instantiations of Predicate Encodings

We present N-bilinear predicate encodings for a large class of predicates that have been considered in the literature. For concreteness, think of N as the order of the composite-order bilinear group. Note that in the proof of α-privacy, whenever we compute some value $v \neq 0 \in \mathbb{Z}_N$, we will simply assume that $\gcd(v, N) = 1$; otherwise, we will be able to compute a non-trivial factor of N. Instantiated via our framework, we obtain the adaptively-secure composite-order (H)IBE, ABE and spatial encryption schemes in [34, 37, 14]. In addition, we obtain novel (to the best of our knowledge) and simple constructions of adaptively-secure NIPE and doubly spatial encryption.

5.1 Inner Product (IPE)

Predicate [32]. Here, $\mathcal{X} = \mathcal{Y} := \mathbb{Z}_N^d$ and

$$P(\mathbf{x}, \mathbf{y}) = 1 \text{ iff } \langle \mathbf{x}, \mathbf{y} \rangle = 0$$

First Encoding (short secret keys) [7].

- $\mathcal{W} := \mathbb{Z}_N \times \mathbb{Z}_N^d; \mathcal{R} := \mathbb{Z}_N^*$.
- $\mathsf{sE}(\mathbf{x}, (u_0, \mathbf{u})) := (u_0 \mathbf{x} + \mathbf{u}, 1)$
- $\mathsf{rE}(\alpha, \mathbf{y}, (u_0, \mathbf{u}); r) := (r, \alpha - r \langle \mathbf{u}, \mathbf{y} \rangle)$

Second Encoding (short ciphertext) [11].

- $\mathcal{W} := \mathbb{Z}_N \times \mathbb{Z}_N^d; \mathcal{R} := \mathbb{Z}_N \times \mathbb{Z}_N^*$.
- $\mathsf{sE}(\mathbf{x}, (u_0, \mathbf{u})) := (1, u_0 + \langle \mathbf{x}, \mathbf{u} \rangle)$
- $\mathsf{rE}(\alpha, \mathbf{y}, (u_0, \mathbf{u}); (r', r)) := (r\mathbf{u} - r'\mathbf{y}, r, \alpha - u_0 r)$

5.2 Non-Zero Inner Product (NIPE)

Predicate [2]. Here, $\mathcal{X} = \mathcal{Y} := \mathbb{Z}_N^d$ and

$$P(\mathbf{x}, \mathbf{y}) = 1 \text{ iff } \langle \mathbf{x}, \mathbf{y} \rangle \neq 0$$

The constructions exploit the following simple algebraic fact: given $\mathbf{x}, \mathbf{y}, u_0 \mathbf{x} + \mathbf{u}, \langle \mathbf{y}, \mathbf{w} \rangle$,

- if $\langle \mathbf{x}, \mathbf{y} \rangle \neq 0$, then we can recover u_0.
- if $\langle \mathbf{x}, \mathbf{y} \rangle = 0$, then u_0 is perfectly random.

First Encoding (short ciphertext).

- $\mathcal{W} := \mathbb{Z}_N^d; \mathcal{R} := \mathbb{Z}_N^*.$
- $sE(\mathbf{x}, \mathbf{w}) := (\langle \mathbf{w}, \mathbf{x} \rangle, 1)$
- $rE(\alpha, \mathbf{y}, \mathbf{w}; r) := (r, \alpha \mathbf{y} - r\mathbf{w})$

Second Encoding (short secret keys).

- $\mathcal{W} := \mathbb{Z}_N \times \mathbb{Z}_N^d; \mathcal{R} := \mathbb{Z}_N^*.$
- $sE(\mathbf{x}, (u_0, \mathbf{u})) := (u_0 \mathbf{x} + \mathbf{u}, 1)$
- $rE(\alpha, \mathbf{y}, (u_0, \mathbf{u}); r) := (r, \alpha + u_0 r, r \langle \mathbf{u}, \mathbf{y} \rangle)$

5.3   Spatial Encryption

Predicate [9]. Here, $\mathcal{X} := \mathbb{Z}_N^d, \mathcal{Y} := \mathbb{Z}_N^{d \times \ell}$ and

$$P(\mathbf{x}, \mathbf{Y}) = 1 \text{ iff } \mathbf{x} \in \text{span}(\mathbf{Y})$$

Recall from [9] that spatial encryption generalizes HIBE.

Supporting Delegation. Consider a predicate P that supports delegation, namely, there is a partial ordering $\leq$ on $\mathcal{Y}$ such that for all $x \in \mathcal{X}$, the predicate $P(x, \cdot)$ is monotone, i.e.

$$(y \leq y') \wedge P(x, y) = 1 \implies P(x, y') = 1.$$

For instance, in HIBE, $y \leq y'$ iff y' is a prefix of y. A bilinear encoding (sE, rE) for such a predicate supports delegation if given y, y' such that $y \leq y'$, we can efficiently compute a linear map L such that for all $(\alpha, \mathbf{w}, \mathbf{r}) \in \mathcal{D} \times \mathcal{W} \times \mathcal{R}$, L maps $(\mathbf{w}, rE(\alpha, y', \mathbf{w}; \mathbf{r}))$ to $rE(\alpha, y, \mathbf{w}; \mathbf{r})$. Note that we can always rerandomize the output due to linearity of receiver encoding.

Encoding (short ciphertext) [9, 11, 34, 14].

- $\mathcal{W} = \mathbb{Z}_N \times \mathbb{Z}_N^d; \mathcal{R} = \mathbb{Z}_N^*.$
- $sE(\mathbf{x}, (u_0, \mathbf{u})) := (u_0 + \mathbf{u}^\top \mathbf{x}, 1)$
- $rE(\alpha, \mathbf{Y}, (u_0, \mathbf{u}); r) := (r\mathbf{u}^\top \mathbf{Y}, -r, \alpha + r u_0)$

α-privacy holds for all $r \in \mathbb{Z}_N^*$, and relies on the fact that if $\mathbf{x} \notin \text{span}(\mathbf{Y})$, then $\mathbf{u}^\top \mathbf{x}$ is statistically independent of $\mathbf{u}^\top \mathbf{Y}$ for a random $\mathbf{u} \leftarrow_R \mathbb{Z}_N^d$.

5.4   Doubly Spatial Encryption

Predicate [28]. Here, $\mathcal{X} := \mathbb{Z}_N \times \mathbb{Z}_N^{d \times \ell}, \mathcal{Y} := \mathbb{Z}_N^{d \times \ell'}$ and

$$P((\mathbf{x}_0, \mathbf{X}), \mathbf{Y}) = 1 \text{ iff } (\mathbf{x}_0 + \text{span}(\mathbf{X})) \cap \text{span}(\mathbf{Y}) \neq \emptyset$$

Encoding [28].

- $\mathcal{W} = \mathbb{Z}_N \times \mathbb{Z}_N^d; \mathcal{R} = \mathbb{Z}_N^*$.
- $\mathsf{sE}((x_0, X), (u_0, \mathbf{u})) := (u_0 + \mathbf{u}^\top x_0, \mathbf{u}^\top X, 1)$
- $\mathsf{rE}(\alpha, Y, (u_0, \mathbf{u}); r) := (r\mathbf{u}^\top Y, -r, \alpha + r u_0)$

α-privacy holds for all $r \in \mathbb{Z}_N^*$, and relies on the fact that if $(x_0 + \mathrm{span}(X)) \cap \mathrm{span}(Y) = \varnothing$ then $\mathbf{u}^\top x_0$ is statistically independent of $\mathbf{u}^\top X, \mathbf{u}^\top Y$ for a random $\mathbf{u} \leftarrow_R \mathbb{Z}_N^d$.

5.5   Attribute-Based Encryption (ABE)

We define (monotone) access structures using the language of (monotone) span programs [31].

Definition 1 (access structure [4, 31]). *A* (monotone) access structure *for attribute universe* $[n]$ *is a pair* $(\mathbf{M}, \rho)$ *where* $\mathbf{M}$ *is a* $\ell \times \ell'$ *matrix over* $\mathbb{Z}_N$ *and* $\rho : [\ell] \to [n]$. *Given* $\mathbf{x} = (x_1, \ldots, x_n) \in \{0, 1\}^n$, *we say that*

$$\mathbf{x} \text{ satisfies } (\mathbf{M}, \rho) \text{ iff } \mathbf{1} \in span\langle \mathbf{M_x}\rangle,$$

Here, $\mathbf{1} := (1, 0, \ldots, 0) \in \mathbb{Z}^{\ell'}$ *is a row vector;* $\mathbf{M_x}$ *denotes the collection of vectors* $\{\mathbf{M}_j : x_{\rho(j)} = 1\}$ *where* $\mathbf{M}_j$ *denotes the* j'th row of $\mathbf{M}$; *and span refers to linear span of collection of (row) vectors over* $\mathbb{Z}_N$.

That is, $\mathbf{x}$ satisfies $(\mathbf{M}, \rho)$ iff there exists constants $\omega_1, \ldots, \omega_\ell \in \mathbb{Z}_N$ such that

$$\sum_{j:x_{\rho(j)}=1} \omega_j \mathbf{M}_j = \mathbf{1}.$$

Observe that the constants $\{\omega_j\}$ can be computed in time polynomial in the size of the matrix $\mathbf{M}$ via Gaussian elimination.

KP-ABE Predicate [27, 42].   Here, $\mathcal{X} := \mathbb{Z}_N^\ell, \mathcal{Y} := \{(\mathbf{M}, \rho) : \mathbf{M} \in \mathbb{Z}_N^{\ell \times \ell'}, \rho : [\ell] \to [\ell] \text{ is a permutation}\}$ (that is, $\ell = n$) and

$$\mathsf{P}(\mathbf{x}, (\mathbf{M}, \rho)) = 1 \text{ iff } \mathbf{x} \text{ satisfies } (\mathbf{M}, \rho)$$

Encoding.   Our encoding improves upon that in [37] by reducing the length of rE (and thus the secret key size) from 2ℓ to $\ell + 1$ elements.

- $\mathcal{W} = \mathbb{Z}_N^\ell; \mathcal{R} = \mathbb{Z}_N^{\ell'-1} \times \mathbb{Z}_N^*$.
- $\mathsf{sE}(\mathbf{x}, \mathbf{w}) := (x_1 w_1, \ldots, x_\ell w_\ell, 1)$
- $\mathsf{rE}(\alpha, (\mathbf{M}, \rho), \mathbf{w}; (\mathbf{u}, r)) := (\alpha_1 - r w_{\rho(1)}, \ldots, \alpha_\ell - r w_{\rho(\ell)}, r)$ where $\alpha_i := \mathbf{M}_i \binom{\alpha}{\mathbf{u}}$ is the i'th share of α and $\binom{\alpha}{\mathbf{u}}$ denotes the column vector in $\mathbb{Z}_N^{\ell'}$ formed by concatenating $\alpha \in \mathbb{Z}_N$ and $\mathbf{u} \in \mathbb{Z}_N^{\ell'-1}$.

In the prior construction [37], rE is given by

$$(\alpha_1 - r_1 w_{\rho(1)}, \ldots, \alpha_\ell - r_\ell w_{\rho(\ell)}, r_1, \ldots, r_\ell).$$

Here, α-privacy holds for all $r \in \mathbb{Z}_N^*$, and relies crucially on the fact that ρ is injective.

Remark 5 (GPSW encoding [27]). It is instructive here to revisit the encoding used in the selective ABE in [27] where rE is given by

$$(\alpha_1 / w_{\rho(1)}, \ldots, \alpha_\ell / w_{\rho(\ell)}).$$

This implies α-privacy but only in a statistical sense (the encoding only hides non-zero shares). Moreover, it does not satisfy **w**-hiding.

CP-ABE Predicate [27, 18]. As before with $\mathcal{X}$ and $\mathcal{Y}$ switched, so that

$$P((\mathbf{M}, \rho), \mathbf{y}) = 1 \text{ iff } \mathbf{y} \text{ satisfies } (\mathbf{M}, \rho)$$

Encoding. In the following encoding, we allow sE to be randomized:

- $\mathcal{W} = \mathbb{Z}_N^\ell \times \mathbb{Z}_N; \mathcal{R} = \mathbb{Z}_N^*$.
- $\mathsf{sE}((\mathbf{M}, \rho), (\mathbf{w}, v); \mathbf{u}) := (1, w_{\rho(1)} + v_1, \ldots, w_{\rho(\ell)} + v_\ell)$ where $v_i := \mathbf{M}_i \binom{v}{\mathbf{u}}$ is the i'th share of v.
- $\mathsf{rE}(\alpha, \mathbf{y}, (\mathbf{w}, v); r) := (\alpha + r v, r, \{w_j r\}_{j : y_j = 1})$

Randomized Sender Encodings. We may handle the extension to randomized sender encodings where sE takes additional randomness **u** as follows:

- the requirement for α-privacy holds over random coin tosses of sE;
- affine sending encoding says that we can compute $g^{\mathsf{sE}(x, \mathbf{w})}$ given $g^{\mathbf{w}}, x$ and the coin tosses used in sE;
- we extend the definition of Enc and $\widehat{\mathsf{Enc}}$ to use randomized sE in a straight-forward manner;
- the proof remains largely unchanged except for accounting for sender randomness when invoking α-privacy in the proof of Lemma 3.

Acknowledgments. I would like to thank Jie Chen, Kai-Min Chung, Yuval Ishai, Allison Lewko and Vinod Vaikuntanathan for insightful discussions.

References

[1] Applebaum, B., Ishai, Y., Kushilevitz, E.: Cryptography in NC^0. SIAM J. Comput. 36(4), 845–888 (2006)

[2] Attrapadung, N., Libert, B.: Functional encryption for inner product: Achieving constant-size ciphertexts with adaptive security or support for negation. In: Nguyen, P.Q., Pointcheval, D. (eds.) PKC 2010. LNCS, vol. 6056, pp. 384–402. Springer, Heidelberg (2010)

[3] Attrapadung, N., Libert, B., de Panafieu, E.: Expressive key-policy attribute-based encryption with constant-size ciphertexts. In: Catalano, D., Fazio, N., Gennaro, R., Nicolosi, A. (eds.) PKC 2011. LNCS, vol. 6571, pp. 90–108. Springer, Heidelberg (2011)

[4] Beimel, A.: Secure Schemes for Secret Sharing and Key Distribution. Ph.D., Technion - Israel Institute of Technology (1996)

[5] Bellare, M., Waters, B., Yilek, S.: Identity-based encryption secure against selective opening attack. In: Ishai, Y. (ed.) TCC 2011. LNCS, vol. 6597, pp. 235–252. Springer, Heidelberg (2011)

[6] Bellare, M., Hoang, V.T., Rogaway, P.: Adaptively secure garbling with applications to one-time programs and secure outsourcing. In: Wang, X., Sako, K. (eds.) ASIACRYPT 2012. LNCS, vol. 7658, pp. 134–153. Springer, Heidelberg (2012)

[7] Boneh, D., Boyen, X.: Efficient selective-ID secure identity-based encryption without random oracles. In: Cachin, C., Camenisch, J.L. (eds.) EUROCRYPT 2004. LNCS, vol. 3027, pp. 223–238. Springer, Heidelberg (2004)

[8] Boneh, D., Franklin, M.K.: Identity-based encryption from the Weil pairing. SIAM J. Comput. 32(3), 586–615 (2003)

[9] Boneh, D., Hamburg, M.: Generalized identity based and broadcast encryption schemes. In: Pieprzyk, J. (ed.) ASIACRYPT 2008. LNCS, vol. 5350, pp. 455–470. Springer, Heidelberg (2008)

[10] Boneh, D., Waters, B.: Conjunctive, subset, and range queries on encrypted data. In: Vadhan, S.P. (ed.) TCC 2007. LNCS, vol. 4392, pp. 535–554. Springer, Heidelberg (2007)

[11] Boneh, D., Boyen, X., Goh, E.-J.: Hierarchical identity based encryption with constant size ciphertext. In: Cramer, R. (ed.) EUROCRYPT 2005. LNCS, vol. 3494, pp. 440–456. Springer, Heidelberg (2005a)

[12] Boneh, D., Goh, E.-J., Nissim, K.: Evaluating 2-DNF formulas on ciphertexts. In: Kilian, J. (ed.) TCC 2005. LNCS, vol. 3378, pp. 325–341. Springer, Heidelberg (2005b)

[13] Cash, D., Hofheinz, D., Kiltz, E., Peikert, C.: Bonsai trees, or how to delegate a lattice basis. In: Gilbert, H. (ed.) EUROCRYPT 2010. LNCS, vol. 6110, pp. 523–552. Springer, Heidelberg (2010)

[14] Chen, C., Zhang, Z., Feng, D.: Fully secure doubly-spatial encryption under simple assumptions. In: Takagi, T., Wang, G., Qin, Z., Jiang, S., Yu, Y. (eds.) ProvSec 2012. LNCS, vol. 7496, pp. 253–263. Springer, Heidelberg (2012)

[15] Chen, J., Wee, H.: Fully (almost) tightly secure IBE and dual system groups. In: Canetti, R., Garay, J.A. (eds.) CRYPTO 2013, Part II. LNCS, vol. 8043, pp. 435–460. Springer, Heidelberg (2013)

[16] Chen, J., Wee, H.: Dual system groups and its applications — compact HIBE and more. Full version in preparation (2013); Preliminary version in [15]

[17] Chen, J., Wee, H.: Fully (almost) tightly secure IBE from standard assumptions. IACR Cryptology ePrint Archive, Report 2013/803 (2013); Preliminary version in [15]

[18] Cheung, L., Newport, C.C.: Provably secure ciphertext policy ABE. In: ACM Conference on Computer and Communications Security, pp. 456–465 (2007)

[19] Cocks, C.: An identity based encryption scheme based on quadratic residues. In: Honary, B. (ed.) Cryptography and Coding 2001. LNCS, vol. 2260, pp. 360–363. Springer, Heidelberg (2001)

[20] Cramer, R., Shoup, V.: Universal hash proofs and a paradigm for adaptive chosen ciphertext secure public-key encryption. In: Knudsen, L.R. (ed.) EUROCRYPT 2002. LNCS, vol. 2332, pp. 45–64. Springer, Heidelberg (2002)

[21] Feige, U., Kilian, J., Naor, M.: A minimal model for secure computation. In: STOC, pp. 554–563 (1994)

[22] Garg, S., Gentry, C., Halevi, S., Sahai, A., Waters, B.: Attribute-based encryption for circuits from multilinear maps. In: Canetti, R., Garay, J.A. (eds.) CRYPTO 2013, Part II. LNCS, vol. 8043, pp. 479–499. Springer, Heidelberg (2013)

[23] Gentry, C.: Practical identity-based encryption without random oracles. In: Vaudenay, S. (ed.) EUROCRYPT 2006. LNCS, vol. 4004, pp. 445–464. Springer, Heidelberg (2006)

[24] Gentry, C., Peikert, C., Vaikuntanathan, V.: Trapdoors for hard lattices and new cryptographic constructions. In: STOC, pp. 197–206 (2008)

[25] Gertner, Y., Ishai, Y., Kushilevitz, E., Malkin, T.: Protecting data privacy in private information retrieval schemes. J. Comput. Syst. Sci. 60(3), 592–629 (2000)

[26] Gorbunov, S., Vaikuntanathan, V., Wee, H.: Attribute-based encryption for circuits. In: STOC, pp. 545–554 (2013)

[27] Goyal, V., Pandey, O., Sahai, A., Waters, B.: Attribute-based encryption for fine-grained access control of encrypted data. In: ACM Conference on Computer and Communications Security, pp. 89–98 (2006)

[28] Hamburg, M.: Spatial Encryption. Ph.D., Stanford University, Also, Cryptology ePrint Archive, Report 2011/389 (2011)

[29] Ishai, Y., Kushilevitz, E.: Private simultaneous messages protocols with applications. In: ISTCS, pp. 174–184 (1997)

[30] Ishai, Y., Kushilevitz, E.: Randomizing polynomials: A new representation with applications to round-efficient secure computation. In: FOCS, pp. 294–304 (2000)

[31] Karchmer, M., Wigderson, A.: On span programs. In: Structure in Complexity Theory Conference, pp. 102–111 (1993)

[32] Katz, J., Sahai, A., Waters, B.: Predicate encryption supporting disjunctions, polynomial equations, and inner products. In: Smart, N. (ed.) EUROCRYPT 2008. LNCS, vol. 4965, pp. 146–162. Springer, Heidelberg (2008)

[33] Lewko, A.: Tools for simulating features of composite order bilinear groups in the prime order setting. In: Pointcheval, D., Johansson, T. (eds.) EUROCRYPT 2012. LNCS, vol. 7237, pp. 318–335. Springer, Heidelberg (2012)

[34] Lewko, A., Waters, B.: New techniques for dual system encryption and fully secure HIBE with short ciphertexts. In: Micciancio, D. (ed.) TCC 2010. LNCS, vol. 5978, pp. 455–479. Springer, Heidelberg (2010)

[35] Lewko, A., Waters, B.: Decentralizing attribute-based encryption. In: Paterson, K.G. (ed.) EUROCRYPT 2011. LNCS, vol. 6632, pp. 568–588. Springer, Heidelberg (2011)

[36] Lewko, A., Waters, B.: New proof methods for attribute-based encryption: Achieving full security through selective techniques. In: Safavi-Naini, R., Canetti, R. (eds.) CRYPTO 2012. LNCS, vol. 7417, pp. 180–198. Springer, Heidelberg (2012)

[37] Lewko, A., Okamoto, T., Sahai, A., Takashima, K., Waters, B.: Fully secure functional encryption: Attribute-based encryption and (hierarchical) inner product encryption. In: Gilbert, H. (ed.) EUROCRYPT 2010. LNCS, vol. 6110, pp. 62–91. Springer, Heidelberg (2010)

[38] Lewko, A., Rouselakis, Y., Waters, B.: Achieving leakage resilience through dual system encryption. In: Ishai, Y. (ed.) TCC 2011. LNCS, vol. 6597, pp. 70–88. Springer, Heidelberg (2011)

[39] Naor, M., Reingold, O.: Number-theoretic constructions of efficient pseudo-random functions. J. ACM 51(2), 231–262 (2004)

[40] Okamoto, T., Takashima, K.: Fully secure functional encryption with general relations from the decisional linear assumption. In: Rabin, T. (ed.) CRYPTO 2010. LNCS, vol. 6223, pp. 191–208. Springer, Heidelberg (2010)

[41] Okamoto, T., Takashima, K.: Adaptively attribute-hiding (hierarchical) inner product encryption. In: Pointcheval, D., Johansson, T. (eds.) EUROCRYPT 2012. LNCS, vol. 7237, pp. 591–608. Springer, Heidelberg (2012)

[42] Sahai, A., Waters, B.: Fuzzy identity-based encryption. In: Cramer, R. (ed.) EUROCRYPT 2005. LNCS, vol. 3494, pp. 457–473. Springer, Heidelberg (2005)

[43] Shamir, A.: Identity-based cryptosystems and signature schemes. In: Blakely, G.R., Chaum, D. (eds.) CRYPTO 1984. LNCS, vol. 196, pp. 47–53. Springer, Heidelberg (1985)

[44] Waters, B.: Efficient identity-based encryption without random oracles. In: Cramer, R. (ed.) EUROCRYPT 2005. LNCS, vol. 3494, pp. 114–127. Springer, Heidelberg (2005)

[45] Waters, B.: Dual system encryption: Realizing fully secure IBE and HIBE under simple assumptions. In: Halevi, S. (ed.) CRYPTO 2009. LNCS, vol. 5677, pp. 619–636. Springer, Heidelberg (2009)

[46] Yao, A.C.-C.: Theory and applications of trapdoor functions. In: FOCS, pp. 80–91 (1982)

(Efficient) Universally Composable Oblivious Transfer Using a Minimal Number of Stateless Tokens

Seung Geol Choi[1,*], Jonathan Katz[2,**], Dominique Schröder[3,***],
Arkady Yerukhimovich[4,†], and Hong-Sheng Zhou[5,‡]

[1] United States Naval Academy
choi@usna.edu
[2] University of Maryland
jkatz@cs.umd.edu
[3] Saarland University
ds@ca.cs.uni-saarland.de
[4] MIT Lincoln Laboratory
arkady@cs.umd.edu
[5] Virginia Commonwealth University
hszhou@vcu.edu

Abstract. We continue the line of work initiated by Katz (Eurocrypt 2007) on using *tamper-proof hardware* for universally composable secure computation. As our main result, we show an efficient oblivious-transfer (OT) protocol in which two parties each create and exchange a single, stateless token and can then run an unbounded number of OTs. Our result yields what we believe is the most *practical* and *efficient* known approach for oblivious transfer based on tamper-proof tokens, and implies that the parties can perform (repeated) secure computation of arbitrary functions without exchanging additional tokens.

Motivated by this result, we investigate the minimal number of stateless tokens needed for universally composable OT/secure computation. We prove that our protocol is *optimal* in this regard for constructions making *black-box* use of the tokens (in a sense we define). We also show that nonblack-box techniques can be used to obtain a construction using only a single stateless token.

* Work done in part at the University of Maryland and Columbia University.
** Work supported in part by NSF awards #1111599 and #1223623.
*** Work done in part at the University of Maryland, and supported by the German Ministry for Education and Research (BMBF) through funding for the Center for IT-Security, Privacy, and Accountability (CISPA www.cispa-security.org) and also by an Intel Early Career Award.
† Work done in part at the University of Maryland.
‡ Work done at the University of Maryland, and supported by an NSF CI postdoctoral fellowship.

Y. Lindell (Ed.): TCC 2014, LNCS 8349, pp. 638–662, 2014.

1   Introduction

The universal composability (UC) framework [6] provides a way of analyzing protocols while ensuring strong security guarantees. In particular, protocols proven secure in this framework remain secure when run concurrently with arbitrary other protocols in a larger networked environment. Unfortunately, most interesting cryptographic tasks are impossible to realize in the "plain" UC framework when an honest majority cannot be assumed and, in particular, in the setting of two-party secure computation [8,9,36]. This stark negative result has motivated researchers to explore various extensions/variants of the UC framework in which secure computation can be achieved [7], with notable examples being the assumption of a common reference string (CRS) [6,8,10] or a public-key infrastructure [6,3]. In the real world, implementing either of these approaches seems to require the existence of some *trusted entity* that parties agree to use (though see [11] for some ideas on using a naturally occurring high-entropy source in place of a CRS).

Katz [32] suggested using *tamper-proof hardware tokens* for UC computation. That is, Katz proposed a model where parties can construct hardware tokens to compute functions of their choice such that an adversary given a token $\mathcal{T}_F$ for a function F can do no more than observe the input/output characteristics of this token. The motivation for this being that the existence of tamper-proof hardware can be viewed, in principle, as a *physical* assumption rather than an assumption of trust in some external entity. (In fact, in Katz's model the parties may create the tamper-proof tokens themselves—rather than obtain them from a trusted provider [28]—and a malicious party can put any algorithm on a token it creates.) In addition, secure hardware may also potentially result in more efficient protocols; indeed, it has been suggested for improving efficiency in other settings (e.g., [17,13,14,5,27,31,34,22]). In addition to introducing the model, Katz showed that tamper-proof hardware tokens can be used for universally composable computation of arbitrary functions. His work motivated an extensive amount of follow-up work [12,37,23,15,24,25,19] that we discuss in detail later.

As our main result, we show here a new protocol for universally composable 1-out-of-2 string oblivious transfer (OT) based on tamper-proof hardware tokens, secure against a static, malicious adversary. Our work yields what we believe to be the most *practical* and *efficient* known protocol since it simultaneously achieves all the following (which are not achieved all at once by any other solution; see Table 1 for a detailed comparison):

– Our protocol is based on *stateless* tokens, which seem easier/cheaper to create in practice and are (automatically) resistant to resetting attacks.
– Our protocol requires the parties to exchange a *single* pair of tokens. This can be done in advance, before the parties' inputs are known. Furthermore, the tokens can be used to implement an *unbounded* number of oblivious transfers, rather than requiring the parties to exchange a fresh pair of tokens for every oblivious transfer they want to compute. Thus, by relying on known completeness results [33,30], the parties can use the same tokens to perform

an unlimited number of secure computations (of possibly different functions, and on different inputs).

– Our protocol is efficient. It is *black-box*, and each OT needs mostly standard symmetric-key operations along with only a few (unique) digital signatures. Moreover, any desired number of OTs can be obtained (in parallel) in *constant* rounds.

– If the total number of OTs is a priori bounded, then a variant of our protocol can realize any bounded number of OTs in constant rounds based only on the existence of collision-resistant hash functions.

Inspired by our result, we investigate the *minimal* number of stateless tokens needed for universally composable OT/secure computation. We show that two tokens—one created by each party—are needed even to obtain a single universally composable OT as long as only "black-box techniques" are used. (We explain what we mean by "black-box techniques" in the relevant section of our paper.) Our protocol, above, is thus optimal in this regard. Our impossibility result is somewhat surprising, since a single *stateful* token suffices for OT [19]. Our results thus demonstrate an inherent difference between stateful and stateless tokens.

Since protocols based on nonblack-box techniques tend to be impractical, our work pins down the minimal number of stateless tokens needed as far as practical protocols are concerned. From a theoretical point of view, however, it is still interesting to completely resolve the question. In this vein, we show a protocol for carrying out an unbounded number of secure computations using only a *single* (stateless) token. Our construction uses a variant of the nonblack-box simulation technique introduced by Barak [2].

In summary, our work shows that efficient, universally composable oblivious transfer can be realized from two stateless tokens without any additional setup assumptions, and is unlikely using a single stateless token. On the other hand, using (inefficient) nonblack-box techniques, a single stateless token serves as a sufficient setup for general universally-composable two-party computation.

1.1   Related Work

Katz's original protocol for secure computation using tamper-proof tokens [32] required stateful tokens and relied on number-theoretic assumptions (specifically, the DDH assumption). Subsequent work has mainly focused on improving one or both of these aspects of his work.

Several researchers have explored constructions using *stateless* tokens. Stateless tokens are presumably easier and/or cheaper to build, and are resistant to resetting attacks whereby an adversary cuts off the power supply and thus effectively "rewinds" the token. Chandran et al. [12] were the first to eliminate the requirement of stateful tokens. They construct UC commitments based on the existence of one-way functions, and oblivious transfer based on any enhanced trapdoor permutation (eTDP). They also introduce a variant security model in which an adversary need not know the code of the tokens he produces, thus capturing scenarios where an adversary may pass along tokens whose code he doesn't know, e.g., via token replication. (We do not consider this model here.[1]) From a practical perspective, however, their work has several drawbacks. Their OT protocol makes nonblack-box use of the underlying primitives, runs in $\Theta(\lambda)$ rounds (where λ is the security parameter), and uses the heavy machinery of concurrent non-malleable zero-knowledge. Improving upon their work, Goyal et al. [25] show a black-box construction of oblivious transfer; their protocol runs in constant rounds assuming a collision-resistant hash function, or $\Theta(\lambda/\log\lambda)$ rounds based on one-way functions. However, their protocol requires the parties to exchange $\Theta(\lambda)$ tokens for *every* oblivious transfer the parties wish to execute. Compared with these results, our protocol is much more efficient at the expense of the stronger assumption of existence of unique signature schemes.

A second direction has explored the possibility of eliminating computational assumptions altogether. This line of work was initiated by Moran and Segev [37], who showed how to realize statistically secure UC commitments using a single stateful token. (Note that commitment does not imply OT, or general secure computation, in the unconditional setting.) Their construction can be used for any bounded number of commitments, still using only one token, and the authors note that they can achieve an unbounded number of commitments (with computational security) based on one-way functions. Goyal et al. [25] show an unconditional construction of oblivious transfer (and hence general secure computation) using $\Theta(\lambda)$ stateful tokens. Recently, Döttling et al. [19] show how to construct unconditionally secure OT using only a single stateful token. Goyal et al. [24] showed that unconditional security from stateless tokens is impossible (unless the token model is extended to allow tokens to encapsulate each other). If such encapsulation is allowed, then they show how to realize statistically secure OT in constant rounds using $\Theta(\lambda)$ stateless tokens.

Kolesnikov [34] showed an efficient construction of oblivious transfer from stateless tokens. However, his work is not in the UC setting, and he achieves only covert security [1] rather than security against malicious parties. Dubovitskaya et al. [21] constructed an OT protocol from two stateful tokens. Their work assumes tokens are not reused (it is not clear how this is enforced), and is also not in the UC setting.

Our Work in Relation to Prior Work. For our main result, we carefully combine the techniques of [25] and [19] to achieve the most *practical* and *efficient* known protocol for universally composable OT based on tamper-proof hardware tokens. Our protocol uses two *stateless* tokens (one per party), and can be used for either a bounded number of OTs (assuming the existence of collision-resistant hash functions) or for an unbounded number of OTs (additionally assuming the existence of unique signatures or, equivalently, verifiable random

---

[1] Although we do not formally consider this model, it appears that our efficient, two-token protocol would remain secure in that model since the simulator in our security proof does not refer to the code of the tokens.

Table 1. Universally composable OT based on tamper-proof hardware tokens. The security parameter is denoted by λ and TE means token encapsulation.

	stateless tokens					stateful tokens		
	Here 1	Here 2	[12]	[25]	[24]	[32]	[25]	[19]
Tokens:	2	2	2	$\Theta(\lambda)$	$\Theta(\lambda)$	2	$\Theta(\lambda)$	1
Rounds:	$\Theta(1)$	$\Theta(1)$	$\Theta(\lambda)$	$\Theta(1)$	$\Theta(1)$	$\Theta(1)$	$\Theta(1)$	$\Theta(1)$
Asmpt.:	CRHF, VRF	CRHF	eTDP	CRHF	TE	DDH	none	none
# OTs:	unbounded	bounded	unbounded	1	1	unbounded	1	bounded

functions (VRFs)). Both instantiations run in constant rounds.[2] A detailed comparison of this protocol to relevant prior work is given in Table 1.

In addition to the above, we show two other results: there is no "black-box" construction of universally composable OT using fewer than two stateless tokens, but universally composable coin tossing (and hence OT) *can* be based on a single stateless token using nonblack-box techniques.

Concurrent Work. Independently, Döttling et al. [20] show a different nonblack-box construction of UC coin-tossing from a single stateless token, and argue (without proof) that nonblack-box techniques are needed. Here, we provide a rigorous version of their argument. Our efficient, black-box OT protocol using two stateless tokens—which we consider our primary contribution—has no analog in their work.

2   Preliminaries

Let λ be the security parameter. For a set S, we let $x \leftarrow S$ denote choosing x uniformly at random from S. We assume readers are familiar with pseudorandom functions, collision-resistant hash functions, strong extractors, commitment schemes, digital signature schemes, message-authentication codes (MACs), and witness-indistinguishable arguments of knowledge (WI-AoKs). Due to space restrictions, we omit the formal definitions; here, we mainly introduce notation.

Throughout the paper, a pseudorandom function is denoted by PRF, and it is assumed that the output length is sufficiently long so that it can be truncated to the appropriate length. We let MAC = (Sig, Vrfy) be a deterministic message-authentication code, where Vrfy is a canonical verification procedure that checks the validity of a tag τ by recomputing it. We slightly abuse the notation to let (Kg, Sig, Vrfy) also denote a digital signature scheme (the context should make it obvious whether the notation indicates a MAC or a signature). A digital signature scheme is called *unique* if for every possible verification key vk and

---

[2] The CRHF assumption is only used to make our protocols constant round by instantiating constant round statistically-hiding commitments. Thus, we can instead instantiate our protocols in $\Theta(\lambda/\log \lambda)$ rounds using only one-way functions.

every message m, there is a unique signature σ such that $\mathsf{Vrfy}_{vk}(m,\sigma) = 1$. Dodis and Yampolskiy [18] give an efficient construction for unique signatures based on a certain number-theoretic assumption. Let $\mathsf{Ext}: \{0,1\}^{2\lambda} \times \{0,1\}^d \to \{0,1\}^\lambda$ denote a strong randomness extractor where the source has length 2λ and the seed has length d. If the min-entropy of the source is at least $2\lambda - O(\log \lambda)$, the output is statistically close to uniform.

We use SCom to denote a (possibly interactive) statistically-hiding and computationally-binding commitment scheme [16,26] and Com to denote a (possibly interactive) computationally-hiding and strongly-binding commitment scheme. We let $\mathsf{com}_m \leftarrow \mathsf{Com}(m; r_m, \breve{r}_m)$ denote a commitment to a message m, where the sender (resp., receiver) uses uniform random coins r_m (resp., $\breve{r}_m$) and the final transcript is com_m; sometimes we omit $\breve{r}_m$ when it is clear from the context. (We also use similar notation for SCom.) In a strongly binding commitment scheme [29], with overwhelming probability over the receiver's coins $\breve{r}_m$, for any commitment com there is at most one (m, r_m) such that $\mathsf{com} = \mathsf{Com}(m; r_m, \breve{r}_m)$. Although this definition is stronger than usual, the Naor scheme [38] satisfies it. Without loss of generality, we assume a canonical decommit phase in which the sender sends m together with the randomness r_m (i.e., $\mathsf{decom}_m = (m, r_m)$). Then the receiver runs the algorithm $\mathsf{Open}(\mathsf{com}_m, m, r_m)$ which checks if (m, r_m) is consistent with the transcript com_m. If so, Open outputs m; otherwise it outputs $\perp$.

Linear Algebra. By $\mathbb{F}_2$ we denote the finite field with two elements. If $a \in \mathbb{F}_2^\lambda$ and $b \in \mathbb{F}_2^k$ are two column-vectors, then $(ab^T) = (a_i b_j)_{ij} \in \mathbb{F}_2^{\lambda \times k}$ is the outer product (or tensor product) of a and b, and $a^T b = \sum_{i=1}^\lambda a_i b_i \in \mathbb{F}_2$ the inner product.

Let $C \in \mathbb{F}_2^{\lambda \times 2\lambda}$. Then $\dim(\ker(C)) \geq \lambda$. Let $B = \{b_1, \ldots, b_\lambda\} \subseteq \ker(C)$ be a linearly independent set. One can choose a set $B^* = \{b_{\lambda+1}, \ldots, b_{2\lambda}\}$ such that $B \cup B^*$ is a basis of $\mathbb{F}_2^{2\lambda}$. Let $e_i \in \mathbb{F}_2^\lambda$ be the ith unit-vector. Then, there exists a matrix $G \in \mathbb{F}_2^{\lambda \times 2\lambda}$ such that $Gb_i = e_i$ for $i = 1, \ldots, \lambda$ and $Gb_i = 0$ for $i = \lambda+1, \ldots, 2\lambda$. This matrix is called the *complementary matrix* of C and we denote by $G \leftarrow \mathsf{Comp}(C)$ its computation. It holds that $\mathrm{rank}(G) = \lambda$ and $B^* \subseteq \ker(G)$. For such C and G, we can always solve the linear system $Cx = r$, $Gx = s$ by solving $Cx_{B^*} = r$ and $Gx_B = s$ independently with $x_B \in \mathrm{span}(B) \subseteq \ker(C)$ and $x_{B^*} \in \mathrm{span}(B^*) \subseteq \ker(G)$, and then setting $x := x_B + x_{B^*}$.

Token Functionality. We model tamper-proof hardware tokens as an ideal functionality in the UC framework, following Katz [32]; see Figure 1. Our ideal functionality models stateful tokens: although all our protocols use stateless tokens, an adversarially generated token may be stateful.

Oblivious-Transfer Functionality. The OT functionality is standard, but we wish here to model a multi-session variant where the sender and receiver repeatedly (in different sub-sessions) execute any agreed-upon number m of parallel OTs (in a given sub-session). We refer to this functionality as $\mathcal{F}_{\mathsf{multi\text{-}OT}}$, and describe it in Figure 2. We note that the sub-sessions are executed sequentially. Additionally, as highlighted in [25], the sender is notified each time

---

Functionality $\mathcal{F}_{\mathsf{wrap}}$

The functionality is parameterized by a polynomial $p(\cdot)$ and an implicit security parameter λ.

Create: Upon receiving an input (CREATE, $\langle \mathsf{sid}, \mathsf{C}, \mathsf{U} \rangle, \mathcal{M}$) from a party C (i.e., the token creator), where U is another party (i.e., the token user) in the system and $\mathcal{M}$ is an interactive Turing machine, do:

> If there is no tuple of the form $\langle \mathsf{C}, \mathsf{U}, \star, \star, \star \rangle$ stored, store $\langle \mathsf{C}, \mathsf{U}, \mathcal{M}, 0, \emptyset \rangle$. Reveal (CREATE, $\langle \mathsf{sid}, \mathsf{C}, \mathsf{U} \rangle$) to the adversary.

Ready: Upon receiving a message (READY, $\langle \mathsf{sid}, \mathsf{C}, \mathsf{U} \rangle$) from the adversary, send (READY, $\langle \mathsf{sid}, \mathsf{C}, \mathsf{U} \rangle$) to U.

Execute: Upon receiving an input (RUN, $\langle \mathsf{sid}, \mathsf{C}, \mathsf{U} \rangle, \mathsf{msg}$) from the user U, find the unique stored tuple $\langle \mathsf{C}, \mathsf{U}, \mathcal{M}, i, \mathsf{state} \rangle$. If no such tuple exists, do nothing. Otherwise, do:

> If the Turing machine $\mathcal{M}$ has never been used yet, i.e., $i = 0$, then choose ω uniformly at random from $\{0,1\}^{p(\lambda)}$ and set $\mathsf{state} := \omega$ before running the Turing machine. Run $(\mathsf{out}, \mathsf{state}') \leftarrow \mathcal{M}(\mathsf{msg}; \mathsf{state})$ for at most $p(\lambda)$ steps where out is the response and state' is the new state of $\mathcal{M}$ (set $\mathsf{out} :=\perp$ and $\mathsf{state}' := \mathsf{state}$ if $\mathcal{M}$ does not respond in the allotted time). Send (RESPONSE, $\langle \mathsf{sid}, \mathsf{C}, \mathsf{U} \rangle, \mathsf{out}$) to U. Erase $\langle \mathsf{C}, \mathsf{U}, \mathcal{M}, i, \mathsf{state} \rangle$ and store $\langle \mathsf{C}, \mathsf{U}, \mathcal{M}, i+1, \mathsf{state}' \rangle$.

Fig. 1. The ideal $\mathcal{F}_{\mathsf{wrap}}$ functionality for stateful tokens

the receiver obtains output. We define the bounded OT functionality similarly except that the sender and receiver only execute a single sub-session of m parallel OTs. This allows the sender and receiver to execute any bounded number of OTs.

3   Efficient Oblivious Transfer Using Two Stateless Tokens

In this section, we first give the details of our unbounded OT protocol. Then, in Section 3.3 we briefly sketch how this protocol can be modified to achieve a bounded OT protocol using only CRHFs. We provide some intuition and background before giving the details of our protocol. Our starting point is the unconditionally secure OT protocol from [19], which uses a single *stateful* token. We sketch a simplified version of their protocol for the case of a single OT carried out between the sender S with input $(x_0, x_1) \in \{0,1\}^\lambda \times \{0,1\}^\lambda$ and the receiver R with input $b \in \{0,1\}$. We canonically identify the vector space $\mathbb{F}_2^\lambda$ with the set $\{0,1\}^\lambda$ of strings of length λ. The main steps of the protocol are as follows:

1. S creates a token $\mathcal{T}_\mathsf{S}$ holding random vector $a \in \mathbb{F}_2^{2\lambda}$ and matrix $B \in \mathbb{F}_2^{2\lambda \times 2\lambda}$ and gives it to R.

Functionality $\mathcal{F}_{\mathsf{multi\text{-}OT}}$

$\mathcal{F}_{\mathsf{multi\text{-}OT}}$ interacts with sender S, receiver R, and the adversary. The functionality is parameterized by a security parameter λ. It also maintains a variable curr-id initialized to $\perp$.

Upon receiving $(\textsc{send}, \langle sid, S, R \rangle, ssid, \langle m, \{(x_i^0, x_i^1)\}_{i=1}^m \rangle)$ from S with $x_i^0, x_i^1 \in \{0,1\}^\lambda$, if curr-id $\notin \{\perp, ssid\}$ then ignore it. Otherwise, set curr-id $:= ssid$ and record $\langle ssid, m, \{(x_i^0, x_i^1)\}_{i=1}^m \rangle$, and reveal $(\textsc{send}, \langle sid, S, R \rangle, ssid)$ to the adversary. Ignore further $(\textsc{send}, \langle sid, S, R \rangle, ssid, \ldots)$ inputs with this ssid.

Upon receiving $(\textsc{receive}, \langle sid, S, R \rangle, ssid, \langle m, \{b_i\}_{i=1}^m \rangle)$ from R with $b_i \in \{0,1\}$, if curr-id $\notin \{\perp, ssid\}$ then ignore it. Otherwise, set curr-id $:= ssid$ and record the tuple $\langle ssid, m, \{b_i\}_{i=1}^m \rangle$, and reveal $(\textsc{receive}, \langle sid, S, R \rangle, ssid)$ to the adversary. Ignore further $(\textsc{receive}, \langle sid, S, R \rangle, ssid, \ldots)$ inputs with this ssid.

Upon receiving $(\textsc{go}, \langle sid, S, R \rangle, ssid)$ from the adversary, ignore it if curr-id $\neq ssid$, or either $\langle ssid, m, \{(x_i^0, x_i^1)\}_{i=1}^m \rangle$ or $\langle ssid, m, \{b_i\}_{i=1}^m \rangle$ is not recorded. Otherwise, do the following: set curr-id $:= \perp$, return $(\textsc{received}, \langle sid, S, R \rangle, ssid)$ to S, and return $(\textsc{received}, \langle sid, S, R \rangle, ssid, \{x_i^{b_i}\}_{i=1}^m)$ to R. Ignore further $(\textsc{go}, \langle sid, S, R \rangle, ssid)$ messages with this ssid from the adversary.

Fig. 2. The $\mathcal{F}_{\mathsf{multi\text{-}OT}}$ functionality

2. R chooses a random matrix $C \in \mathbb{F}_2^{\lambda \times 2\lambda}$ and sends it to S. In turn, S computes $\tilde{a} := Ca$, $\tilde{B} := CB$, and a complementary matrix $G \in \mathbb{F}_2^{\lambda \times 2\lambda}$ to C and sends these to R.
3. R sends random $h \in \mathbb{F}_2^{2\lambda}$ to S. Then, S sends $\tilde{x}_0 := x_0 + GBh$ and $\tilde{x}_1 := x_1 + GBh + Ga$ to R.
4. R queries the token with a random $z \in \mathbb{F}_2^{2\lambda}$ such that $z^T h = b$. The token will in turn output $V := az^T + B$. Then R checks that $CV = \tilde{a}z^T + \tilde{B}$ and, if this is the case, outputs $x_b := \tilde{x}_b - GVh$. (Otherwise it detects that S was cheating.)

The basic idea of their protocol is as follows. The receiver R performs a secret sharing of its input b into shares z and h; by using only h with the sender S and only z with the token, R maintains the privacy of b. In order to obtain the output x_b, the receiver R has to compute the mask (i.e., GBh or $GBh + Ga$). This is achieved by querying the token with z, since $GVh = G(az^T + B)h = b(Ga) + GBh$. To ensure that the token outputs the correct value V, the receiver checks that $CV = \tilde{a}z^T + \tilde{B}$. Since the token does not know C, incorrect behavior is detected with overwhelming probability. To achieve security against a malicious receiver, it is crucial that the receiver is only able to query the token once, and this is enforced by making the token *stateful*. In particular, the token "self destructs" after the first query by the receiver.

Using Stateless Tokens. We carefully combine the techniques of [19] with ideas from [25] so that we can use two *stateless* tokens instead of one stateful token. This entails several difficulties:

Multiple queries. A stateless token can be executed multiple times while the token remains oblivious about it (in contrast, stateful tokens "self destruct" after their use [23]), so we need to ensure that a malicious party queries the token only once. Motivated by similar techniques in [25], we handle this issue by modifying the token so that it only replies to *authenticated* inputs. That is, instead of querying z directly to the token, R queries $(\mathsf{com}_z, z, r_z, \sigma_z)$ where com_z is a commitment to z, the value σ_z is a digital signature on com_z (received from S) with respect to a verification key vk_S, and z, r_z is the decommitment of com_z (see the discussion below about why this technique is useful).

Extracting the inputs. Using a stateless token additionally introduces the difficulty of extracting the sender's inputs during the simulation. Extraction from a stateful token is possible by having the simulator rewind the token and query it multiple times. (The fact that the simulator can query the token multiple times is an advantage of the simulator over the real-world parties.) Once we move to a stateless token in the real world, and authenticate the queries as described above, even the simulator is no longer able to rewind and send multiple (authenticated) queries to the token.

To resolve this issue we introduce a second stateless token $\mathcal{T}_\mathsf{R}$ sent from the receiver to the sender which allows directly extracting the sender's inputs.

At a high level, the above results in the following changes to the protocol:

1. The receiver sends a token $\mathcal{T}_\mathsf{R}$ to the sender. (The behavior of this token will become clear below.)
2(a). S generates a statistically hiding[3] commitment $\mathsf{com} \leftarrow \mathsf{SCom}(a\|B)$ and sends com to R. Then, R chooses a matrix C and authenticates com by computing $\sigma := \mathsf{Sig}_{sk_\mathsf{R}}(\mathsf{com}\|C)$ (where sk_R is a secret key also embedded in $\mathcal{T}_\mathsf{R}$), and sends C and σ to S.
2(b). S queries $\mathcal{T}_\mathsf{R}$ on input $(C, \mathsf{com}, \mathsf{decom}, \sigma)$. The token checks that the signature σ is valid and that decom is a valid opening of com; otherwise it aborts. $\mathcal{T}_\mathsf{R}$ returns $\tilde{a} := Ca$ and $\tilde{B} := CB$ together with a signature $\sigma' := \mathsf{Sig}_{sk_\mathsf{R}}(\tilde{a}\|\tilde{B})$. Then S sends $(\tilde{a}, \tilde{B}, \sigma')$ to R.

Intuitively, the value (a, B) remains hidden from a corrupted R as before. For a corrupted sender, the simulator will emulate $\mathcal{F}_\mathsf{wrap}$ and observe the inputs to $\mathcal{T}_\mathsf{R}$. To guarantee that the simulator extracts the sender's inputs correctly, it is necessary that the sender queries the token exactly once with a value (a, B). The binding property of the commitment scheme, together with the unforgeability of the signature scheme, guarantees that S can make at most one valid query to the token. The unforgeability of the signature scheme assures that S must query the token at least once to generate a valid next message of the protocol. A similar argument holds for the inputs $(\mathsf{com}_z, z, r_z, \sigma_z)$ to the sender's token $\mathcal{T}_\mathsf{S}$.

---

[3] We were not able to prove security using a computationally hiding commitment scheme. Indeed, it is an interesting open problem to achieve a constant-round OT protocol with two stateless tokens based only on one-way functions.

On input (key):
ouput vk_S.

On input $(ssid, i, \mathsf{com}_z, z, r_z, \sigma_z)$:
$v := \mathsf{Vrfy}_{vk_S}(ssid\|i\|\mathsf{com}_z, \sigma_z)$
if $z = \mathsf{Open}(\mathsf{com}_z, z, r_z)$ and $v = 1$
 $a := \mathsf{PRF}_{k_a}(ssid\|i)$
 $B := \mathsf{PRF}_{k_B}(ssid\|i)$
 $V := az^\mathsf{T} + B$
 $w := \mathsf{Sig}_{sk_S}(ssid\|i)$
 output (V, w)
else output $(\bot, \bot)$

Fig. 3. The Turing machine $\mathcal{M}_S$ to be embedded in the sender-created token $\mathcal{T}_S$. It is initialized with (sk_S, vk_S, k_a, k_B) given by the creator.

On input (key):
ouput vk_R.

On input $(ssid, i, \mathsf{com}_{a\|B}, a, B, r_{a\|B}, \sigma_{a\|B})$:
$v := \mathsf{Vrfy}_{vk_R}(ssid\|i\|0\|\mathsf{com}_{a\|B}, \sigma_{a\|B})$
if $a\|B = \mathsf{Open}(\mathsf{com}_{a\|B}, a\|B, r_{a\|B})$
 and $v = 1$
 $\tilde{a} := Ca;\ \tilde{B} := CB;$
 $\sigma_{\tilde{a}\|\tilde{B}} := \mathsf{Sig}_{sk_R}(ssid\|i\|1\|\tilde{a}\|\tilde{B})$
 output $(\tilde{a}, \tilde{B}, \sigma_{\tilde{a}\|\tilde{B}})$
else output $(\bot, \bot, \bot)$

Fig. 4. The Turing machine $\mathcal{M}_R$ to be embedded in the receiver-created token $\mathcal{T}_R$. It is initialized with (sk_R, vk_R, C) given by the creator.

Malicious tokens. There is one additional issue to be taken care of. The above protocol is secure assuming the tokens are generated as specified by the protocol. However, a malicious token may try to leak some information about the other party's queries to the token creator. For example, a malicious token $\mathcal{T}_{R^*}$ may output $(\tilde{a}, \tilde{B}, \sigma')$ where the bits of σ' leak information about (a, B). To protect against this, we require the underlying signature scheme to be *unique*; then, σ' carries no more information about (a, B) than $(\tilde{a}, \tilde{B})$ does. This ensures that the only information leakage that can occur is from a *token abort*. For $\mathcal{T}_S$, this type of leakage is already handled by the protocol of [19]. For $\mathcal{T}_R$, we handle the leakage using strong extractors; see the formal description of the protocol below.

The above suffices for a single OT from each pair of tokens. To achieve an unbounded number of OTs we replace all the secret keys and secret inputs with pseudorandom values output by a pseudorandom function.

3.1 The Protocol

The protocol π between a sender S and a receiver R consists of an initial token-exchange phase after which the parties can carry out an unlimited number of oblivious transfers. We describe π now; see also Figure 8.

Token-Exchange Phase. Each party generates a single token and sends it to the other party. The sender's token $\mathcal{T}_S$ encapsulates the code described in Figure 3, where $k_a, k_B, \leftarrow \{0, 1\}^\lambda$ and $(sk_S, vk_S) \leftarrow \mathsf{Kg}()$. The receiver's token $\mathcal{T}_R$ encapsulates the code described in Figure 4, where $(sk_R, vk_R) \leftarrow \mathsf{Kg}()$ and $C \leftarrow \mathbb{F}_2^{2\lambda \times 4\lambda}$. Formally, each party sends the relevant code to $\mathcal{F}_{\mathsf{wrap}}$ using an appropriate (CREATE, ...) message. Then, S runs $\mathcal{T}_R(key)$ to obtain vk_R; likewise, R obtains vk_S by running $\mathcal{T}_S(key)$. Finally, R sends C to S, and in turn S computes the complementary matrix $G \leftarrow \mathsf{Comp}(C)$ and sends it to R.

Oblivious Transfer Phase. Following the token-exchange phase, the parties can sequentially run an unbounded number of sub-sessions where in each sub-session they carry out any desired number m of oblivious transfers. (Each sub-session uses only a constant number of rounds.) In each sub-session, S gets (SEND, $\langle \text{sid}, \text{S}, \text{R}\rangle$, ssid, $\langle m, \{(x_i^0, x_i^1)\}_{i=1}^m\rangle$) and R gets (RECEIVE, $\langle \text{sid}, \text{S}, \text{R}\rangle$, ssid, $\langle m, \{b_i\}_{i=1}^m\rangle$) from the environment, and they execute the following protocol:

S → R: For $i \in [m]$, the sender computes $a_i := \text{PRF}_{k_a}(\text{ssid}\|i)$ and $B_i := \text{PRF}_{k_B}(\text{ssid}\|i)$. Here, we have $a_i \in \mathbb{F}_2^{4\lambda}$, $B_i \in \mathbb{F}_2^{4\lambda \times 4\lambda}$. Then, the sender commits to (a_i, B_i) by executing $\text{com}_{a_i\|B_i} \leftarrow \text{SCom}(a_i\|B_i; r_{a_i\|B_i})$, and sends $\{\text{com}_{a_i\|B_i}\}_{i=1}^m$ to R.

R → S: R chooses $h_i \leftarrow \mathbb{F}_2^{4\lambda}$ and $z_i \leftarrow \mathbb{F}_2^{4\lambda}$ subject to the constraint that $b_i = z_i^{\text{T}} h_i$, and commits to z_i by executing $\text{com}_{z_i} \leftarrow \text{SCom}(z_i; r_{z_i})$. It next authenticates the commitment $\text{com}_{a_i\|B_i}$ by computing $\sigma_{a_i\|B_i} := \text{Sig}_{sk_R}(\text{ssid}\|i\|0\|\text{com}_{a_i\|B_i})$ for $i \in [m]$. Finally, it sends $\{(\sigma_{a_i\|B_i}, \text{com}_{z_i})\}_{i=1}^m$ to S.

S → R: The sender checks if $\text{Vrfy}_{vk_R}(\text{ssid}\|i\|0\|\text{com}_{a_i\|B_i}, \sigma_{a_i\|B_i}) = 1$ for $i \in [m]$; if the check fails, then S aborts the protocol. For $i \in [m]$, the sender runs the token $\mathcal{T}_R$ with (ssid, i, $\text{com}_{a_i\|B_i}$, a_i, B_i, $r_{a_i\|B_i}$, $\sigma_{a_i\|B_i}$) and obtains in return $(\tilde{a}_i, \tilde{B}_i, \sigma_{\tilde{a}_i\|\tilde{B}_i})$. Then S checks that $\tilde{a}_i = Ca_i$, $\tilde{B}_i = CB_i$, $\text{Vrfy}_{vk_R}(\text{ssid}\|i\|1\|\tilde{a}_i\|\tilde{B}_i, \sigma_{\tilde{a}_i\|\tilde{B}_i}) = 1$ for $i \in [m]$; if the check fails, S aborts the protocol. Otherwise, for $i \in [m]$ it authenticates the commitment com_{z_i} by computing $\sigma_{z_i} := \text{Sig}_{sk_S}(\text{ssid}\|i\|\text{com}_{z_i})$. Finally, it sends $\{(\tilde{a}, \tilde{B}, \sigma_{\tilde{a}_i\|\tilde{B}_i}, \sigma_{z_i})\}_{i=1}^m$ to R.

R → S: The receiver checks that $\text{Vrfy}_{vk_R}(\text{ssid}\|i\|1\|\tilde{a}_i\|\tilde{B}_i, \sigma_{\tilde{a}_i\|\tilde{B}_i}) = 1$, and $\text{Vrfy}_{vk_S}(\text{ssid}\|i\|\text{com}_{z_i}, \sigma_{z_i}) = 1$ for $i \in [m]$. If this check fails, it aborts the protocol. Otherwise, R runs the token $\mathcal{T}_S$ with (ssid, i, com_{z_i}, z_i, r_{z_i}, σ_{z_i}) and obtains in return (V_i, w_i), for $i \in [m]$. The receiver checks that $\text{Vrfy}_{vk_S}(\text{ssid}\|i, w_i) = 1$ and $CV_i = \tilde{a}_i z_i^{\text{T}} + \tilde{B}_i$ for $i \in [m]$. If the check fails, then it aborts the protocol. Otherwise, it sends $\{(h_i, w_i)\}_{i=1}^m$ to S.

S → R: The sender checks that $\text{Vrfy}_{vk_S}(\text{ssid}\|i, w_i) = 1$ for $i \in [m]$; if not, it aborts the protocol. Otherwise, for $i \in [m]$ it chooses $v_i^0, v_i^1 \leftarrow \{0,1\}^d$ and computes $\tilde{x}_i^0 := \text{Ext}(GB_i h_i, v_i^0) \oplus x_i^0$ and $\tilde{x}_i^1 := \text{Ext}(GB_i h_i + Ga_i, v_i^1) \oplus x_i^1$. Here d is an appropriate seed length for the extractor. Finally it sends $\{(v_i^0, v_i^1, \tilde{x}_i^0, \tilde{x}_i^1)\}_{i=1}^m$ to the receiver.

R → S: For $i \in [m]$, the receiver computes $x_i^{b_i} := \tilde{x}_i^{b_i} \oplus \text{Ext}(GV_i h_i, v_i^{b_i})$ and outputs it.

Theorem 1. *Assume* PRF *is a pseudorandom function,* SCom *is statistically hiding and computationally binding, and* (Kg, Sig, Vrfy) *is a unique signature scheme. Then protocol* π *securely realizes* $\mathcal{F}_{\text{multi-OT}}$ *in the* $\mathcal{F}_{\text{wrap}}$*-hybrid model.*

3.2   Proof Idea

In this section, we briefly sketch the main ideas behind the proof of Theorem 1. A complete proof is deferred to the full version.

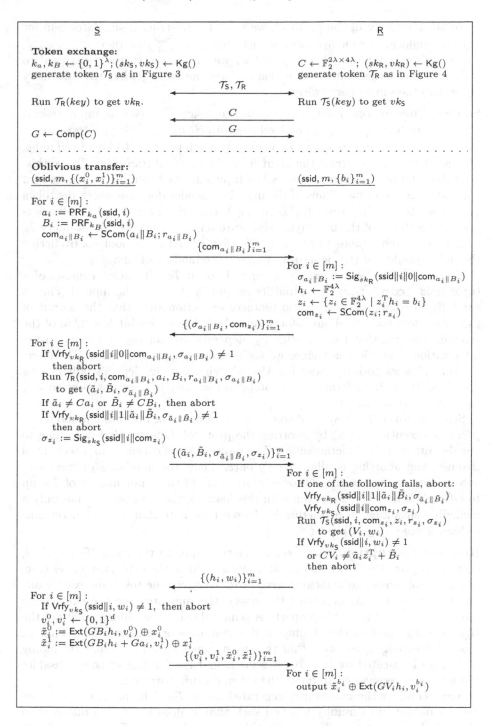

Fig. 5. An OT protocol π from two stateless tokens

To show security of the protocol, we need to construct a simulator Sim for any non-uniform PPT environment $\mathcal{Z}$ such that $\text{EXEC}^{\mathcal{F}_{\text{wrap}}}_{\pi,\mathcal{A},\mathcal{Z}} \approx \text{IDEAL}_{\mathcal{F}_{\text{multi-OT}},\text{Sim},\mathcal{Z}}$, where $\mathcal{A}$ is the dummy adversary. Below, we briefly sketch the ideas used to construct such a simulator. Note that we assume w.l.o.g. that the adversary never asks the same query twice.

Sender Corruption. First, recall that the token $\mathcal{T}_R$ takes as input (ssid, i, $\text{com}_{a_i \| B_i}, a_i, B_i, r_{a_i \| B_i}, \sigma_{a_i \| B_i}$) and returns $(\tilde{a}_i, \tilde{B}_i, \sigma_{\tilde{a}_i \| \tilde{B}_i})$. Our protocol forces the malicious sender S^* to query $\mathcal{T}_R$ on exactly one input $(a_i \| B_i)$ for a fixed index i. This allows Sim to extract the input from the protocol transcript. To see this, note that the receiver authenticates the inputs to its token $\mathcal{T}_R$, which in turn authenticates its output. Thus, if the malicious sender does not query the token $\mathcal{T}_R$ but sends a "valid" tuple $(\tilde{a}_i \| \tilde{B}_i, \sigma_{\tilde{a}_i \| \tilde{B}_i})$, then this immediately contradicts the unforgeability of the underlying signature scheme. Also, S^* cannot query $\mathcal{T}_R$ with two different values $(a_i \| B_i), (a_i' \| B_i')$ (for a fixed i) without contradicting the unforgeability of the signature $\sigma_{a_i \| B_i}$ or the binding of $\text{com}_{a_i \| B_i}$.

Next, consider the maliciously generated token $\mathcal{T}_{S^*}$. Its input consists of a tuple (ssid, i, $\text{com}_{z_i}, z_i, r_{z_i}, \sigma_{z_i}$), and its output is (V_i, w_i). The input is chosen carefully in combination with the protocol execution such that the output of the token does not reveal any information about the choice bit $b_i = z_i^T h_i$ of the receiver. Observe that the signature σ_{z_i} depends only on vk_S and com_{z_i} in the information thoeretic sense, since we use a unique signature scheme. Therefore, the token knows nothing about h_i. Also, the signature w_i depends only on vk_R and (ssid, i) in the information theoretic sense, and it does not contain any information about z_i.

Still, the token $\mathcal{T}_{S^*}$ may send some limited amount of information about the *previous* executions to S^* by aborting the protocol. Observe that according to our definition of the functionality, OT sub-sessions are executed in a sequential manner, and aborting is allowed *only once*. Thus, the number ℓ of successful sub-sessions so far can encode some information of the input history of $\mathcal{T}_{S^*}$ up to $O(\log \lambda)$ bits. Therefore, even with this leakage, the adversary S^* has only a negligible advantage in predicting b_i. To see this, note that $b_i = z_i^T h_i$ remains unfixed given $O(\log \lambda)$ bits of z_i.

Receiver Corruption. First, recall that the input to the token $\mathcal{T}_S$ is (ssid, i, $\text{com}_{z_i}, z_i, r_{z_i}, \sigma_{z_i}$) and that its output is (V_i, w_i). As in the malicious sender case, our protocol forces the malicious receiver R^* to query the token on exactly one input z_i for each i. By observing this query, the simulator Sim can extract the input $\hat{b}_i = z_i^T h_i$ of R^*. This property is achieved using the unforgeability of the signature scheme and the binding of the commitment scheme SCom. That is, the unforgeability guarantees that the token runs only on the input (including com_{z_i}) authenticated by the honest sender, and the binding of com_{z_i} disables the malicious receiver R^* to query the token on two distinct z_is.

Next, consider the maliciously generated token $\mathcal{T}_{R^*}$. The input to the token and its output are carefully handled such that it does not reveal information about (x_i^0, x_i^1) of the sender. To see this, recall the token $\mathcal{T}_{R^*}$ takes as input (ssid, i, $\text{com}_{a_i \| B_i}, a_i, B_i, r_{a_i \| B_i}, \sigma_{a_i \| B_i}$) and returns $(\tilde{a}_i, \tilde{B}_i, \sigma_{\tilde{a}_i \| \tilde{B}_i})$. In particular, since we use a unique signature scheme, the signature $\sigma_{\tilde{a}_i\|\tilde{B}_i}$ information-theoretically depends only on $(vk_\mathsf{R}, \tilde{a}_i, \tilde{B}_i)$, so any malicious attempt by R^* and $\mathcal{T}_{\mathsf{R}^*}$ to gain information about (a_i, B_i) beyond $(\mathsf{com}_{a_i\|B_i}, \tilde{a}_i, \tilde{B}_i)$ is not possible.

Still, the token $\mathcal{T}_{\mathsf{R}^*}$ may reveal to the environment some limited amount of information about the previous executions to R^* by aborting the protocol (only allowed one-time), that is, the number ℓ of successful sub-sessions so far can encode some information of the input history of $\mathcal{T}_{\mathsf{R}^*}$. This value ℓ can encode at most $O(\log \lambda)$ bits. Using appropriate extractors in computing the masks for $\tilde{x}_i^0$ and $\tilde{x}_i^1$, we can handle this amount of leakage.

3.3   Bounded Oblivious Transfer from Collision Resistant Hash Functions

The bounded OT protocol between a sender S and a receiver R consists of an initial token-exchange phase after which the parties can carry out a bounded number of oblivious transfers in a single sub-session. We describe π now; see also Figure 8. Here, let $(\mathsf{Sig}, \mathsf{Vrfy})$ be a MAC scheme.

The main intuition for this protocol is that if we allow *only one sub-session* to be executed, then we only need to worry about covert channels that transmit information observed by each party/token so far up to the covert communication. We do not need to worry about a later sub-session sending information about a prior sub-session. This allows us to eliminate the need for unique signature and use MACs instead, thus giving a protocol based only on the existence of CRHFs.

We eliminate the unique signatures in two ways. First, some of the checks in the unbounded protocol can be removed. In particular, in the protocol, the receiver doesn't check the authentication on com_{z_i} before forwarding it to the sender-created token $\mathcal{T}_\mathsf{S}$. This is because it cannot contain any useful information about z_i beyond com_{z_i}. For other checks, unique signatures are replaced with (roughly) the following technique: a party A commits to a MAC key, authenticates the message using the MAC key, and later sends the decommitment. Due to the unforgeablity of MAC, the party A makes sure that B runs the token only once before receiving messages of B derived from the token execution result. Given the decommitment, the other party B can check that all the messages from A so far have been legitimate.

Token-Exchange Phase. Each party generates a single token and sends it to the other party. The sender's token $\mathcal{T}_\mathsf{S}$ encapsulates the code described in Figure 6, where $k_a, k_B, k_w, k_W, s_2 \leftarrow \{0,1\}^\lambda$. The receiver's token $\mathcal{T}_\mathsf{R}$ encapsulates the code described in Figure 7, where $s \leftarrow \{0,1\}^\lambda$ and $C \leftarrow \mathbb{F}_2^{2\lambda \times 4\lambda}$. Formally, each party sends the relevant code to $\mathcal{F}_{\mathsf{wrap}}$ using an appropriate $(\mathrm{CREATE}, \ldots)$ message. Finally, R sends C to S, and in turn S computes the complementary matrix $G \leftarrow \mathsf{Comp}(C)$ and sends it to R.

Oblivious-Transfer Phase. Following the token-exchange phase, the parties enter the oblivious transfer phase, where m OT instances are executed in parallel. In this phase, S gets $(\mathrm{SEND}, \langle sid, \mathsf{S}, \mathsf{R}\rangle, \langle m, \{(x_i^0, x_i^1)\}_{i=1}^m\rangle)$ and R gets

On input $(i, \text{com}_z, z, r_z, \tau_z)$:
$\quad v := \text{Vrfy}_{s_2}(i \| \text{com}_z, \tau_z)$
$\quad$ if $z = \text{Open}(\text{com}_z, z, r_z)$ and $v = 1$
$\quad\quad a := \text{PRF}_{k_a}(i); \ B := \text{PRF}_{k_B}(i)$
$\quad\quad w := \text{PRF}_{k_w}(i); \ r_w := \text{PRF}_{k_W}(i)$
$\quad\quad V := az^{\mathsf{T}} + B$
$\quad\quad$ output (V, w, r_w)
$\quad$ else output $(\bot, \bot, \bot)$

Fig. 6. The Turing machine $\mathcal{M}_\mathsf{S}$ to be embedded in the sender-created token $\mathcal{T}_\mathsf{S}$. It is initialized with $(k_a, k_B, k_w, k_W, s_2)$ given by the creator.

On input $(i, \text{com}_{a\|B}, a, B, r_{a\|B}, \tau_{a\|B})$:
$\quad v := \text{Vrfy}_s(i \| 0 \| \text{com}_{a\|B}, \tau_{a\|B})$
$\quad$ if $a\|B = \text{Open}(\text{com}_{a\|B}, a\|B, r_{a\|B})$
$\quad\quad$ and $v = 1$
$\quad\quad \tilde{a} := Ca; \ \tilde{B} := CB;$
$\quad\quad \tau_{\tilde{a}\|\tilde{B}} := \text{Sig}_s(i \| 1 \| \tilde{a} \| \tilde{B})$
$\quad\quad$ output $(\tilde{a}, \tilde{B}, \tau_{\tilde{a}\|\tilde{B}})$
$\quad$ else output $(\bot, \bot, \bot)$

Fig. 7. The Turing machine $\mathcal{M}_\mathsf{R}$ to be embedded in the receiver-created token $\mathcal{T}_\mathsf{R}$. It is initialized with (s, C) given by the creator.

(RECEIVE, $\langle \text{sid}, \mathsf{S}, \mathsf{R} \rangle, \langle m, \{b_i\}_{i=1}^m \rangle$) from the environment, and they execute the following protocol:

$\mathsf{S} \to \mathsf{R}$: For $i \in [m]$, the sender computes $a_i := \text{PRF}_{k_a}(i)$, $B_i := \text{PRF}_{k_B}(i)$, $w_i := \text{PRF}_{k_w}(i)$, and $r_{w_i} := \text{PRF}_{k_W}(i)$, and chooses $u_i \leftarrow \{0,1\}^{8\lambda^2 + 3\lambda}$. Here, we have $a_i \in \mathbb{F}_2^{4\lambda}$, $B_i \in \mathbb{F}_2^{4\lambda \times 4\lambda}$, and $w_i \in \{0,1\}^\lambda$. Then, the sender commits to (a_i, B_i), w_i, and u_i by executing $\text{com}_{a_i\|B_i} \leftarrow \text{SCom}(a_i\|B_i; r_{a_i\|B_i})$, and $\text{com}_{w_i} \leftarrow \text{Com}(w_i; r_{w_i})$, and $\text{com}_{u_i} \leftarrow \text{Com}(u_i; r_{u_i})$, respectively. The sender sends $\{(\text{com}_{a_i\|B_i}, \text{com}_{w_i}, \text{com}_{u_i})\}_{i=1}^m$ to R.

$\mathsf{R} \to \mathsf{S}$: The receiver commits to s by executing $\text{com}_s \leftarrow \text{Com}(s; r_s)$. It also chooses $z_i \leftarrow \mathbb{F}_2^{4\lambda}$, and commits to z_i by executing $\text{com}_{z_i} \leftarrow \text{SCom}(z_i; r_{z_i})$. It next authenticates the commitment $\text{com}_{a_i\|B_i}$ by computing

$$\tau_{a_i\|B_i} := \text{Sig}_s(i \| 0 \| \text{com}_{a_i\|B_i}) \text{ for } i = 1, \ldots, m.$$

Finally, it sends $(\text{com}_s, \{(\tau_{a_i\|B_i}, \text{com}_{z_i})\}_{i=1}^m)$ to S.

$\mathsf{S} \to \mathsf{R}$: For $i \in [m]$, the sender runs the token $\mathcal{T}_\mathsf{R}$ with $(i, \text{com}_{a_i\|B_i}, a_i, B_i, r_{a_i\|B_i}, \tau_{a_i\|B_i})$ and obtains in return $(\tilde{a}_i, \tilde{B}_i, \tau_{\tilde{a}_i\|\tilde{B}_i})$. Then S checks that $\tilde{a}_i = Ca_i$ and $\tilde{B}_i = CB_i$ for $i \in [m]$. If the check fails, S aborts the protocol. Otherwise, S computes $U_i := u_i \oplus (\tilde{a}_i\|\tilde{B}_i\|\tau_{\tilde{a}_i\|\tilde{B}_i})$ for $i \in [m]$ and sends $\{U_i\}_{i=1}^m$ to R.

$\mathsf{R} \to \mathsf{S}$: R sends the decommitment (s, r_s) of com_s to S.

$\mathsf{S} \to \mathsf{R}$: The sender validates all values transmitted so far as follows: It checks that $s = \text{Open}(\text{com}_s, s, r_s)$ and that $\text{Vrfy}_s(i \| 0 \| \text{com}_{a_i\|B_i}, \tau_{a_i\|B_i}) = 1$, as well as $\text{Vrfy}_s(i \| 1 \| \tilde{a}_i \| \tilde{B}_i, \tau_{\tilde{a}_i\|\tilde{B}_i}) = 1$ for $i \in [m]$. If the check fails, then S aborts the protocol. Otherwise, for $i \in [m]$ it authenticates the commitment com_{z_i} by computing $\tau_{z_i} := \text{Sig}_{s_2}(i \| \text{com}_{z_i})$. Finally, it sends $(\{(u_i, r_{u_i}, \tau_{z_i})\}_{i=1}^m)$ to R.

$\mathsf{R} \to \mathsf{S}$: For $i \in [m]$, the receiver checks that $u_i = \text{Open}(\text{com}_{u_i}, u_i, r_{u_i})$; if not, it aborts. Then R sets $\tilde{a}_i\|\tilde{B}_i\|\tau_{\tilde{a}_i\|\tilde{B}_i} := U_i \oplus u_i$. It next checks that

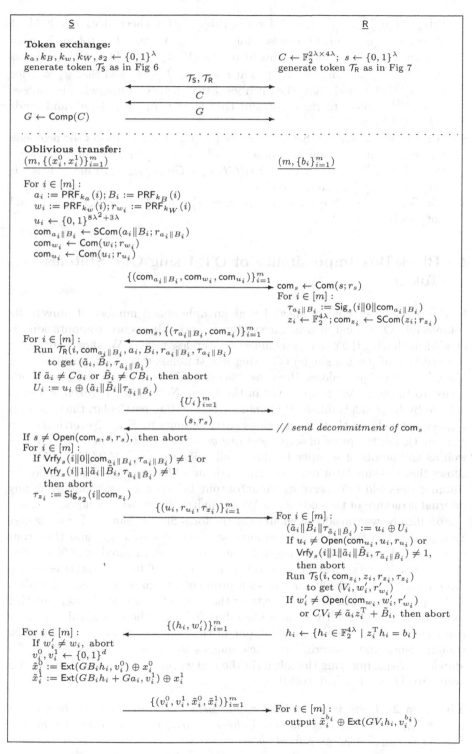

Fig. 8. A bounded OT protocol π from two stateless tokens

$\mathsf{Vrfy}_s(i\|1\|\tilde{a}_i\|\tilde{B}_i, \tau_{\tilde{a}_i\|\tilde{B}_i}) = 1$ for $i \in [m]$. If this check does not hold, it aborts the protocol. Otherwise, for $i \in [m]$, it runs the token $\mathcal{T}_S$ with $(i, \mathsf{com}_{z_i}, z_i, r_{z_i}, \tau_{z_i})$ and obtains in return (V_i, w'_i, r'_{w_i}). The receiver checks that $w'_i = \mathsf{Open}(\mathsf{com}_{w_i}, w'_i, r'_{w_i})$, that $CV_i = \tilde{a}_i z_i^{\mathrm{T}} + \tilde{B}_i$, and that $w'_i \neq \bot$ for $i \in [m]$. If the check fails, then it aborts the protocol. Otherwise, it chooses $h_i \leftarrow \mathbb{F}_2^{4\lambda}$ subject to the constraint that $b_i = z_i^{\mathrm{T}} h_i$ for $i \in [m]$, and sends $\{(h_i, w'_i)\}_{i=1}^m$ to S.

S $\to$ R: The sender checks that $w_i = w'_i$ for $i \in [m]$; if not, it aborts the protocol. Otherwise, for $i \in [m]$ it chooses $v_i^0, v_i^1 \leftarrow \{0,1\}^d$ and computes $\tilde{x}_i^0 := \mathsf{Ext}(GB_i h_i, v_i^0) \oplus x_i^0$ and $\tilde{x}_i^1 := \mathsf{Ext}(GB_i h_i + Ga_i, v_i^1) \oplus x_i^1$. Finally it sends $\{(v_i^0, v_i^1, \tilde{x}_i^0, \tilde{x}_i^1)\}_{i=1}^m$ to the receiver.

R $\to$ S: For $i \in [m]$, the receiver computes $x_i^{b_i} := \tilde{x}_i^{b_i} \oplus \mathsf{Ext}(GV_i h_i, v_i^{b_i})$ and outputs it.

4 Black-Box Impossibility of OT Using One Stateless Token

In the previous section we showed that an unbounded number of universally composable OTs (and hence universally composable secure computation) is possible by having the parties exchange two stateless tokens. We show here that a construction of (even a single) OT using *one* stateless token is impossible using black-box techniques alone. Here, by "black-box" we refer to the simulator's access to the code $\mathcal{M}$ encapsulated in the token. Note that the token model as defined by Katz [32] is inherently *non*black-box and, in particular, the simulator is given the code $\mathcal{M}$ that the malicious party submits to $\mathcal{F}_{\mathsf{wrap}}$. Nevertheless, an examination of the proof of security of our two-token protocol in Section 3— as well as the proofs of security in almost all prior work [12,37,23,15,24,25,19]— shows that the simulator only uses this code in a black-box fashion, namely, by running the code to observe its input/output behavior but without using any internal structure of the code itself. We formalize this in what follows.

Specifically, we consider simulators of the form $\mathsf{Sim} = (\mathsf{Sim}_{code}, \mathsf{Sim}_{bb})$, where Sim_{code} gets the code $\mathcal{M}$ that the adversary submits to $\mathcal{F}_{\mathsf{wrap}}$, and then runs Sim_{bb} as a subroutine while giving it oracle access to $\mathcal{M}$. Inspired by [8,9] we show that, restricting to such simulators, constructions of OT from one stateless token are impossible. Intuitively, for any such protocol proven secure using black-box techniques, Sim_{bb} must be able to extract the input of a corrupted token-creating party by interacting with $\mathcal{M}$ in a black-box fashion. The real-world adversary can then use Sim_{bb} to extract the input of the honest token-creating party by running Sim_{bb} and answering its oracle queries by *querying the token* itself; for *stateless* tokens, querying the token (in the real world) is equivalent to black-box access to $\mathcal{M}$ (in the ideal world).

Theorem 2. *There is no protocol π that uses one stateless token to securely realize $\mathcal{F}_{\mathsf{OT}}$ in the $\mathcal{F}_{\mathsf{wrap}}$-hybrid model whose security is proven using a simulator $\mathsf{Sim} = (\mathsf{Sim}_{code}, \mathsf{Sim}_{bb})$ as defined above.*

Proof (Sketch).   For completeness, the (single-session) OT functionality $\mathcal{F}_{\mathsf{OT}}$ is given in Figure 9. Let π be a protocol between a sender S and a receiver R, in which a single token is sent from the sender to the receiver. (The other case is handled analogously.) Consider the following environment $\mathcal{Z}'$ and dummy adversary $\mathcal{A}'$ corrupting the sender: $\mathcal{Z}'$ chooses random bits s_0, s_1, and b, and instructs the sender to run π honestly on input (s_0, s_1), including submitting some code to $\mathcal{F}_{\mathsf{wrap}}$ to create a token. Once the honest receiver outputs s, then $\mathcal{Z}'$ outputs 1 if $s = s_b$ and 0 otherwise.

Suppose that π securely realizes $\mathcal{F}_{\mathsf{OT}}$, where security is proved via a simulator $\mathsf{Sim} = (\mathsf{Sim}_{code}, \mathsf{Sim}_{bb})$ as previously described. In the course of the proof, Sim_{bb} *plays the role of a receiver* while interacting with $\mathcal{A}'$, and Sim_{code} provides Sim_{bb} with black-box access to whatever code $\mathcal{A}'$ submits to $\mathcal{F}_{\mathsf{wrap}}$. At some point during its execution, Sim_{bb} must send some inputs $(\tilde{s}_0, \tilde{s}_1)$ to the ideal functionality $\mathcal{F}_{\mathsf{OT}}$. It is not hard to see that we must have $(\tilde{s}_0, \tilde{s}_1) = (s_0, s_1)$ with all but negligible probability.

We now consider a different environment $\mathcal{Z}$ and an adversary $\mathcal{A}$ *corrupting the receiver*. $\mathcal{Z}$ chooses random bits s_0, s_1, b and provides (s_0, s_1) as input to the honest sender; it outputs 1 iff $\mathcal{A}$ outputs (s_0, s_1). Note that $\mathcal{A}$ receives a token from the honest sender as specified by π. Adversary $\mathcal{A}$ works as follows:

> Run Sim_{bb}, relaying messages from the honest sender to this internal copy of Sim_{bb}. Whenever Sim_{bb} makes a query to Sim_{code} to run $\mathcal{M}$ (the code created by the honest sender) on some input q, adversary $\mathcal{A}$ runs the token with input q (formally, $\mathcal{A}$ sends $(\mathrm{RUN}, \langle sid, \mathsf{S}, \mathsf{R} \rangle, q)$ to $\mathcal{F}_{\mathsf{wrap}}$ and gives the response to Sim_{bb}). At some point, Sim_{bb} sends $(\tilde{s}_0, \tilde{s}_1)$ to $\mathcal{F}_{\mathsf{OT}}$, at which point $\mathcal{A}$ outputs $(\tilde{s}_0, \tilde{s}_1)$ and halts.

The key point is that *because the token is stateless*, there is no difference between Sim_{code} running the code $\mathcal{M}$, and $\mathcal{A}$ querying the token via $\mathcal{F}_{\mathsf{wrap}}$. Thus, $\mathcal{A}$ provides a perfect simulation for Sim_{bb}, and we conclude that $(\tilde{s}_0, \tilde{s}_1) = (s_0, s_1)$ with all but negligible probability in an execution of π in the $\mathcal{F}_{\mathsf{wrap}}$-hybrid world. But this occurs with probability at most $1/2$ in an ideal-world evaluation of $\mathcal{F}_{\mathsf{OT}}$. Thus, $\mathcal{Z}$ can distinguish between the real- and ideal-world executions, contradicting the claimed security of π.   □

5   Coin Tossing Using One Stateless Token

In the previous section we showed that universally composable OT cannot be realized from one stateless token if only "black-box techniques" are used. We complement that result by showing that UC secure computation from one stateless token *is* feasible via *non*black-box techniques. Here, we find it somewhat simpler to construct a protocol for universally composable coin tossing rather than OT. (The coin-tossing functionality $\mathcal{F}_{\mathsf{coin}}$ is defined in the natural way; see Figure 10 .) Note that coin tossing suffices for general secure computation under a variety of cryptographic assumptions [10,35].

Functionality $\mathcal{F}_{OT}$

$\mathcal{F}_{OT}$ interacts with sender S, receiver R, and the adversary.

Upon receiving an input (SEND, $\langle sid, S, R \rangle, \langle x_0, x_1 \rangle$) from S with $x_0, x_1 \in \{0, 1\}$, record the tuple $\langle x_0, x_1 \rangle$, and reveal (SEND, $\langle sid, S, R \rangle$) to the adversary. Ignore further (SEND, ...) inputs.

Upon receiving an input (RECEIVE, $\langle sid, S, R \rangle, b$) from R with $b \in \{0, 1\}$, record the bit b, and reveal (RECEIVE, $\langle sid, S, R \rangle$) to the adversary. Ignore further (RECEIVE, ...) inputs.

Upon receiving a message (GO, $\langle sid, S, R \rangle$) from the adversary, ignore the message if $\langle x_0, x_1 \rangle$ or b is not recorded. Otherwise, do the following: return (RECEIVED, $\langle sid, S, R \rangle$) to S, and return (RECEIVED, $\langle sid, S, R \rangle, x_b$) to R. Ignore further (GO, $\langle sid, S, R \rangle$) messages from the adversary.

Fig. 9. The ideal $\mathcal{F}_{OT}$ functionality for bit-OT

Functionality $\mathcal{F}_{coin}$

$\mathcal{F}_{coin}$ interacts with two parties, Alice A and Bob B, and the adversary. The functionality is parameterized by a security parameter λ. It also maintains variables $(b_A, b_B, coins)$ initialized to (false, false, $\perp$).

Upon receiving an input (TOSS, $\langle sid, A, B \rangle$) from party $P \in \{A, B\}$, then set $b_P :=$ true, and reveal (TOSS, $\langle sid, A, B \rangle, P$) to the adversary. Ignore further (TOSS, $\langle sid, A, B \rangle$) inputs from the party P.

Upon receiving a message (GO, $\langle sid, A, B \rangle, P$) from the adversary for $P \in \{A, B\}$, ignore the message if $b_A =$ false or $b_B =$ false. Otherwise, do the following: if coins $= \perp$, i.e, coins has not been set yet, then randomly choose $u \leftarrow \{0, 1\}^\lambda$ and set coins $:= u$; return (COIN, $\langle sid, A, B \rangle, coins$) to party P. Ignore further (GO, $\langle sid, A, B \rangle, P$) messages for the party P from the adversary.

Fig. 10. The ideal $\mathcal{F}_{coin}$ functionality for coin tossing

At a high level, our protocol follows the general structure of Blum's coin-tossing protocol [4]. This protocol consists of three moves. In the first move (B1), Alice commits to a random x and sends com_x to Bob, who in return chooses a random value y and sends it to Alice (B2). In the last move (B3), Alice sends the decommitment to x and both parties output $x \oplus y$.

To obtain a UC coin-tossing protocol that follows this basic approach, we need to be able to *simulate* each party. In particular, if Alice is malicious then the simulator needs to extract the message contained in com_x (extractability), whereas when Bob is corrupted the simulator needs to open the commitment in an arbitrary way (equivocation). We achieve both goals by having Bob send a single stateless token $\mathcal{T}_B$ to Alice. This token behaves in two different ways, depending on its input. The first task of the token is to generate a random value

e upon seeing com_x (we discuss the details later); the second task is to generate a notification t for Bob that Alice knows the decommitment x. The value e is used for equivocation, and the notification t gives the simulator the ability to extract x. We give further details next. In the high-level description here, we assume the token is created honestly; we deal with a potentially malicious token when we formally define the protocol, below.

Achieving Extractability. Similar to Section 3, the token only works on authenticated inputs. That is, whenever Alice wants to query the token on some inputs, she needs to first ask Bob to compute a MAC on those inputs. In order to achieve extractability, we let Alice query the token on input $(\mathsf{com}_x, x, r_x, \tau_x)$, where $\mathsf{com}_x \leftarrow \mathsf{SCom}(x; r_x)$ comes from Alice and τ_x is a MAC tag on com_x. The output of the token is a random value t that can be seen as a notification to Bob that Alice knows the decommitment x. As in the previous OT protocol, the authentication of the inputs guarantees that Alice makes exactly one valid query to the token: more than one valid query would imply that Alice has violated security of the MAC or binding of com_x; if Alice makes no query, then she cannot guess t. By forcing Alice to query the token exactly once, we can now easily construct a simulator that extracts the value x while emulating $\mathcal{F}_{\mathsf{wrap}}$. Modifying Blum's coin-tossing protocol, we have the following step:

(B1) Alice commits to x by executing $\mathsf{com}_x \leftarrow \mathsf{SCom}(x; r_x)$. She sends the commitment com_x to Bob, who in turn computes a tag τ_x on com_x and sends τ_x to Alice. Alice runs the token with $(\mathsf{com}_x, x, r_x, \tau_x)$ to obtain output t, and sends t to Bob. Bob checks if t is correct.

Achieving Equivocation. Toward this goal, we further modify the protocol and the token. Alice sends com_x and a dummy commitment com_M before getting the tag τ_x on $\mathsf{com}_x \| \mathsf{com}_M$ from Bob. The token also gets as an additional input this commitment com_M; it outputs a random value e on input $(\mathsf{com}_x, \mathsf{com}_M, \tau_x)$ and a random value t on input $(\mathsf{com}_x, \mathsf{com}_M, x, r_x, \tau_x)$. In step (B3), instead of sending the decommitment (x, r_x) of com_x to Bob, Alice sends x together with a witness-indistinguishable (WI) proof that either x is a valid decommitment of com_x, or com_M contains code that outputs the actual output e of the token $\mathcal{T}_{\mathsf{B}}$. (This is where we use the nonblack-box techniques of Barak [2].)

Due to the binding property of com_M, and because the notification e is unpredictable, Alice cannot commit to such code in com_M. The simulator, however, takes advantage of the fact that it obtains the code of the token generated by Bob while emulating $\mathcal{F}_{\mathsf{wrap}}$. Then, as in [2], the simulator's ability to predict the output of the token beforehand can be used to achieve equivocation. We remark that in contrast to Barak's work, we do not need to use universal arguments; this is because $\mathcal{F}_{\mathsf{wrap}}$ is parameterized with a fixed polynomial bounding the running time of the token (whereas Barak had to handle any polynomial running time).

More formally, we change the protocol as follows:

Token. The token input is either $(\mathsf{com}_x, \mathsf{com}_M, \tau_x)$ or $(\mathsf{com}_x, x, r_x, \mathsf{com}_M, \tau_x)$ and in both cases it checks the validity of τ_x before responding. In the first

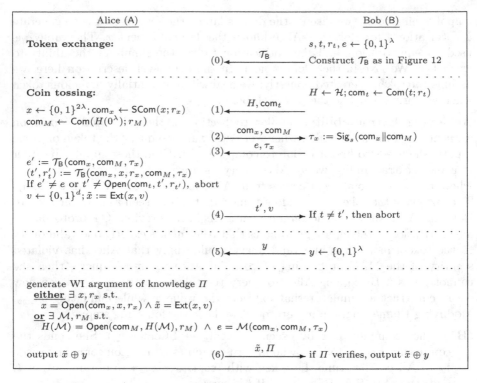

Fig. 11. A coin-tossing protocol ψ from a single stateless token

case, the token outputs a random value e. In the second case, it returns a random value t.

Protocol. Once Alice obtains the token, she runs the following protocol with Bob:

(B1). Alice commits to x and sends com_x together with a dummy commitment com_M to Bob. In turn, Bob authenticates $\mathrm{com}_x\|\mathrm{com}_M$ using a MAC and sends (τ_x, e) to Alice. Alice invokes the token twice, first with $(\mathrm{com}_x, \mathrm{com}_M, \tau_x)$ and then with $(\mathrm{com}_x, x, r_x, \mathrm{com}_M, \tau_x)$, obtaining the values e' and t', respectively. She checks if $e = e'$ and, if so, sends t' to Bob. Finally, Bob checks if $t' = t$.

(B2). Bob sends Alice a random value y.

(B3). Alice sends x together with a WI proof that either (i) (x, r_x) is the decommitment of com_x, or (ii) com_M is a commitment to a Turing machine $\mathcal{M}$ such that $\mathcal{M}(\mathrm{com}_x, \mathrm{com}_M, \tau_x) = e$.

5.1 Formal Description of the Protocol

The protocol ψ between Alice A and Bob B consists of an initial token-exchange phase, followed by a coin-tossing phase for generating a random λ-bit string. We now describe ψ formally; see also Figure 11.

On input $(\text{com}_x, \text{com}_M, \tau_x)$ do:
  if $\text{Vrfy}_s(\text{com}_x \| \text{com}_M, \tau_x) = 1$
    output e
  else output $\bot$

On input $(\text{com}_x, x, r_x, \text{com}_M, \tau_x)$ do:
  if $\text{Vrfy}_s(\text{com}_x \| \text{com}_M, \tau_x) = 1$
    and $x = \text{Open}(\text{com}_x, x, r_x)$
    output (t, r_t)
  else output $(\bot, \bot)$

Fig. 12. The Turing machine $\mathcal{M}$ embedded in the sender-created token $\mathcal{T}_B$. Here, $e, s, t, r_t \in \{0,1\}^\lambda$ are chosen uniformly at random and embedded in the token.

Token-Exchange Phase. Bob generates a single token $\mathcal{T}_B$ and sends it to Alice. Bob's token $\mathcal{T}_B$ encapsulates the code $\mathcal{M}$ described in Figure 12, where s, e, t, r_t are each chosen uniformly from $\{0,1\}^\lambda$.

Coin-Tossing Phase. In this phase, Alice and Bob proceed as follows.

B → A: Bob chooses a collision-resistant hash function $H \leftarrow \mathcal{H}$. He also commits to the value t (that he used when creating the token) by executing $\text{com}_t \leftarrow \text{Com}(t; r_t)$. He sends H, com_t to Alice.

A → B: Alice chooses a value $x \leftarrow \{0,1\}^{2\lambda}$ and commits to x and $H(0^\lambda)$ by executing $\text{com}_x \leftarrow \text{SCom}(x; r_x)$ and $\text{com}_M \leftarrow \text{Com}(H(0^\lambda); r_M)$. She sends com_x and com_M to Bob.

B → A: Bob generates a MAC tag $\tau_x := \text{Sig}_s(\text{com}_x \| \text{com}_M)$, and sends (e, τ_x) to Alice. Recall that Bob has already chosen e and embedded the value in the token $\mathcal{T}_B$ in the token exchange phase.

A → B: Alice runs the token $\mathcal{T}_B$ with $(\text{com}_x, \text{com}_M, \tau_x)$ and obtains e' in response. Then she runs the token with $(\text{com}_x, x, r_x, \text{com}_M, \tau_x)$ and obtains (t', r_t'). Alice checks if $e' = e$ and $t' = \text{Open}(\text{com}_t, t', r_t')$. If not, she aborts the protocol. Otherwise, she chooses $v \leftarrow \{0,1\}^d$ and sends (t', v) to Bob, where d is an appropriate seed length for the extractor.

B → A: Bob checks that $t = t'$, and aborts if not. Otherwise he chooses $y \leftarrow \{0,1\}^\lambda$ and sends it to Alice.

A → B: Alice sends $\tilde{x} := \text{Ext}(x, v)$ and gives a WI argument of knowledge that $(H, \text{com}_x, \text{com}_M, e, \tau_x, v, \tilde{x})$ belongs to the $\mathcal{NP}$ language L defined by the following relation R_L:

$R_L((H, \text{com}_x, \text{com}_M, e, \tau_x, v, \tilde{x}), (\alpha, \beta)) = 1$ if either one of the following holds:

(i) $\alpha = \text{Open}(\text{com}_x, \alpha, \beta)$ and $\tilde{x} = \text{Ext}(\alpha, v)$.

(ii) $H(\alpha) = \text{Open}(\text{com}_M, H(\alpha), \beta)$. In addition, treating α as the description of a Turing machine, the execution of $\alpha(\text{com}_x, \text{com}_M, \tau_x)$ outputs e in time at most $p(\lambda)$, where p is the polynomially bounded running time defined by $\mathcal{F}_{\text{wrap}}$.

If the proof succeeds, both parties output $\tilde{x} \oplus y$.

Theorem 3. *Assume* Com *is computationally hiding and strongly binding,* SCom *is statistically hiding and computationally binding,* MAC *is a deterministic, unforgeable message-authentication code,* $\mathcal{H}$ *is a family of collision-resistant hash functions, and the proof system is a witness-indistinguishable argument of knowledge. Then* ψ *securely realizes* $\mathcal{F}_{\text{coin}}$ *in the* $\mathcal{F}_{\text{wrap}}$*-hybrid model.*

5.2   Proof Idea

In this section, we briefly sketch the main ideas behind the proof of Theorem 3. A complete proof is deferred to the full version.

To show the security of the protocol, we need to construct a simulator Sim for any PPT environment $\mathcal{Z}$ such that $\text{EXEC}^{\mathcal{F}_{\text{wrap}}}_{\mathcal{A},\pi,\mathcal{Z}} \approx \text{IDEAL}_{\mathcal{F}_{\text{coin}},\text{Sim},\mathcal{Z}}$, where $\mathcal{A}$ is the dummy adversary. Below, we briefly provide ideas for simulation.

Corrupting Alice. Note that Alice cannot commit to a TM $\mathcal{M}$ in com_M that will output e, since e is chosen at random independently. Therefore, from the collision-resistance of H, Alice has to show that $\tilde{x} = \text{Ext}(x, v)$ in the WI proof. Since v appears in the communication transcript, by forcing malicious Alice A^* to query $\mathcal{T}_B$ with x exactly once, the simulator Sim can extract the value $\tilde{x}$ from binding of com_x. Then, upon receiving the random string coins from $\mathcal{F}_{\text{coin}}$, the simulator can send $y = \tilde{x} \oplus \text{coins}$ to A^*. To see how the protocol forces exactly one query to the token, note that malicious Alice is not able to generate a valid t without querying $\mathcal{T}_B$, since t is random and com_t is hiding. Also, note that A^* cannot query $\mathcal{T}_B$ with different xs without contradicting the unforgeability of the tag τ_x and/or the binding of com_x. This allows the simulator to extract the value of x.

Corrupting Bob. The simulator Sim needs to equivocate the value $\tilde{x}$ so that it may hold that $\tilde{x} = \text{coins} \oplus y$, where coins is the random string from $\mathcal{F}_{\text{coin}}$. This is achieved as follows. While emulating $\mathcal{F}_{\text{wrap}}$, the simulator obtains the token code $\mathcal{M}$ generated by Bob, then it generates $\text{com}_M \leftarrow \text{Com}(H(\mathcal{M}); R_M)$. Given the hiding property of com_M, the simulated transcript is indistinguishable from that in the $\mathcal{F}_{\text{wrap}}$-hybrid world. Now, with the witness $(\mathcal{M}, R_M)$ in the WI proof, the simulator can send any value $\tilde{x}$.

References

1. Aumann, Y., Lindell, Y.: Security against covert adversaries: Efficient protocols for realistic adversaries. In: TCC 2007. LNCS, vol. 4392, pp. 137–156. Springer, Heidelberg (2007)
2. Barak, B.: How to go beyond the black-box simulation barrier. In: 42nd Annual Symposium on Foundations of Computer Science (FOCS), pp. 106–115. IEEE (2001)
3. Barak, B., Canetti, R., Nielsen, J.B., Pass, R.: Universally composable protocols with relaxed set-up assumptions. In: 45th Annual Symposium on Foundations of Computer Science (FOCS), pp. 186–195. IEEE (2004)
4. Blum, M.: Coin flipping by telephone. In: Proc. IEEE COMPCOM, pp. 133–137 (1982)
5. Brands, S.: Untraceable off-line cash in wallets with observers (extended abstract). In: Stinson, D.R. (ed.) CRYPTO 1993. LNCS, vol. 773, pp. 302–318. Springer, Heidelberg (1994)
6. Canetti, R.: Universally composable security: A new paradigm for cryptographic protocols. In: 42nd Annual Symposium on Foundations of Computer Science (FOCS), pp. 136–145. IEEE (2001) (December 2005),
   http://eprint.iacr.org/2000/067

7. Canetti, R.: Obtaining universally compoable security: Towards the bare bones of trust (invited talk). In: Kurosawa, K. (ed.) ASIACRYPT 2007. LNCS, vol. 4833, pp. 88–112. Springer, Heidelberg (2007)
8. Canetti, R., Fischlin, M.: Universally composable commitments. In: Kilian, J. (ed.) CRYPTO 2001. LNCS, vol. 2139, pp. 19–40. Springer, Heidelberg (2001)
9. Canetti, R., Kushilevitz, E., Lindell, Y.: On the limitations of universally composable two-party computation without set-up assumptions. Journal of Cryptology 19(2), 135–167 (2006)
10. Canetti, R., Lindell, Y., Ostrovsky, R., Sahai, A.: Universally composable two-party and multi-party secure computation. In: 34th Annual ACM Symposium on Theory of Computing (STOC), pp. 494–503. ACM Press (2002)
11. Canetti, R., Pass, R., Shelat, A.: Cryptography from sunspots: How to use an imperfect reference string. In: 48th Annual Symposium on Foundations of Computer Science (FOCS), pp. 249–259. IEEE (2007)
12. Chandran, N., Goyal, V., Sahai, A.: New constructions for UC secure computation using tamper-proof hardware. In: Smart, N. (ed.) EUROCRYPT 2008. LNCS, vol. 4965, pp. 545–562. Springer, Heidelberg (2008)
13. Chaum, D., Pedersen, T.P.: Wallet databases with observers. In: Brickell, E.F. (ed.) CRYPTO 1992. LNCS, vol. 740, pp. 89–105. Springer, Heidelberg (1993)
14. Cramer, R.J.F., Pedersen, T.P.: Improved privacy in wallets with observers (extended abstract). In: Helleseth, T. (ed.) EUROCRYPT1993. LNCS, vol. 765, pp. 329–343. Springer, Heidelberg (1994)
15. Damgård, I., Nielsen, J.B., Wichs, D.: Universally composable multiparty computation with partially isolated parties. In: Reingold, O. (ed.) TCC 2009. LNCS, vol. 5444, pp. 315–331. Springer, Heidelberg (2009)
16. Damgård, I., Pedersen, T.P., Pfitzmann, B.: On the existence of statistically hiding bit commitment schemes and fail-stop signatures. Journal of Cryptology 10(3), 163–194 (1997)
17. Desmedt, Y., Quisquater, J.-J.: Public-key systems based on the difficulty of tampering (is there a difference between DES and RSA?). In: Odlyzko, A.M. (ed.) CRYPTO 1986. LNCS, vol. 263, pp. 111–117. Springer, Heidelberg (1987)
18. Dodis, Y., Yampolskiy, A.: A verifiable random function with short proofs and keys. In: Vaudenay, S. (ed.) PKC 2005. LNCS, vol. 3386, pp. 416–431. Springer, Heidelberg (2005)
19. Döttling, N., Kraschewski, D., Müller-Quade, J.: Unconditional and composable security using a single stateful tamper-proof hardware token. In: Ishai, Y. (ed.) TCC 2011. LNCS, vol. 6597, pp. 164–181. Springer, Heidelberg (2011)
20. Döttling, N., Mie, T., Müller-Quade, J., Nilges, T.: Basing obfuscation on simple tamper-proof hardware assumptions. Cryptology ePrint Archive, Report 2011/675 (2011), http://eprint.iacr.org/
21. Dubovitskaya, M., Scafuro, A., Visconti, I.: On efficient non-interactive oblivious transfer with tamper-proof hardware. Cryptology ePrint Archive, Report 2010/509 (2010)
22. Fischlin, M., Pinkas, B., Sadeghi, A.-R., Schneider, T., Visconti, I.: Secure set intersection with untrusted hardware tokens. In: Kiayias, A. (ed.) CT-RSA 2011. LNCS, vol. 6558, pp. 1–16. Springer, Heidelberg (2011)
23. Goldwasser, S., Kalai, Y.T., Rothblum, G.N.: One-time programs. In: Wagner, D. (ed.) CRYPTO 2008. LNCS, vol. 5157, pp. 39–56. Springer, Heidelberg (2008)
24. Goyal, V., Ishai, Y., Mahmoody, M., Sahai, A.: Interactive locking, zero-knowledge PCPs, and unconditional cryptography. In: Rabin, T. (ed.) CRYPTO 2010. LNCS, vol. 6223, pp. 173–190. Springer, Heidelberg (2010)

25. Goyal, V., Ishai, Y., Sahai, A., Venkatesan, R., Wadia, A.: Founding cryptography on tamper-proof hardware tokens. In: Micciancio, D. (ed.) TCC 2010. LNCS, vol. 5978, pp. 308–326. Springer, Heidelberg (2010)
26. Haitner, I., Reingold, O.: Statistically-hiding commitment from any one-way function. In: 39th Annual ACM Symposium on Theory of Computing (STOC), pp. 1–10. ACM Press (2007)
27. Hazay, C., Lindell, Y.: Constructions of truly practical secure protocols using standard smartcards. In: ACM CCS 2008: 15th ACM Conf. on Computer and Communications Security, pp. 491–500. ACM Press (2008)
28. Hofheinz, D., Müller-Quade, J., Unruh, D.: Universally composable zero-knowledge arguments and commitments from signature cards. In: Proc. 5th Central European Conference on Cryptology, MoraviaCrypt (2005)
29. Horvitz, O., Katz, J.: Bounds on the efficiency of "black-box" commitment schemes. In: Caires, L., Italiano, G.F., Monteiro, L., Palamidessi, C., Yung, M. (eds.) ICALP 2005. LNCS, vol. 3580, pp. 128–139. Springer, Heidelberg (2005)
30. Ishai, Y., Prabhakaran, M., Sahai, A.: Founding cryptography on oblivious transfer - efficiently. In: Wagner, D. (ed.) CRYPTO 2008. LNCS, vol. 5157, pp. 572–591. Springer, Heidelberg (2008)
31. Järvinen, K., Kolesnikov, V., Sadeghi, A.-R., Schneider, T.: Embedded SFE: Offloading server and network using hardware tokens. In: Sion, R. (ed.) FC 2010. LNCS, vol. 6052, pp. 207–221. Springer, Heidelberg (2010)
32. Katz, J.: Universally composable multi-party computation using tamper-proof hardware. In: Naor, M. (ed.) EUROCRYPT 2007. LNCS, vol. 4515, pp. 115–128. Springer, Heidelberg (2007)
33. Kilian, J.: Founding cryptography on oblivious transfer. In: STOC, pp. 20–31 (1988)
34. Kolesnikov, V.: Truly efficient string oblivious transfer using resettable tamper-proof tokens. In: Micciancio, D. (ed.) TCC 2010. LNCS, vol. 5978, pp. 327–342. Springer, Heidelberg (2010)
35. Lin, H., Pass, R., Venkitasubramaniam, M.: A unified framework for concurrent security: Universal composability from stand-alone non-malleability. In: 41st Annual ACM Symposium on Theory of Computing (STOC), pp. 179–188. ACM Press (2009)
36. Lindell, Y.: General composition and universal composability in secure multiparty computation. Journal of Cryptology 22(3), 395–428 (2009)
37. Moran, T., Segev, G.: David and Goliath commitments: UC computation for asymmetric parties using tamper-proof hardware. In: Smart, N. (ed.) EUROCRYPT 2008. LNCS, vol. 4965, pp. 527–544. Springer, Heidelberg (2008)
38. Naor, M.: Bit commitment using pseudorandomness. Journal of Cryptology 4(2), 151–158 (1991)

Lower Bounds in the Hardware Token Model

Shashank Agrawal[1,*], Prabhanjan Ananth[2,**], Vipul Goyal[3],
Manoj Prabhakaran[1,***], and Alon Rosen[4,†]

[1] University of Illinois Urbana-Champaign
{sagrawl2,mmp}@illinois.edu
[2] University of California Los Angeles
prabhanjan@cs.ucla.edu
[3] Microsoft Research India
vipul@microsoft.com
[4] IDC Herzliya
alon.rosen@idc.ac.il

Abstract. We study the complexity of secure computation in the tamper-proof hardware token model. Our main focus is on non-interactive unconditional two-party computation using bit-OT tokens, but we also study computational security with stateless tokens that have more complex functionality. Our results can be summarized as follows:

- There exists a class of functions such that the number of bit-OT tokens required to securely implement them is at least the size of the sender's input. The same applies for receiver's input size (with a different class of functionalities).
- Non-adaptive protocols in the hardware token model imply efficient (decomposable) randomized encodings. This can be interpreted as evidence to the impossibility of non-adaptive protocols for a large class of functions.
- There exists a functionality for which there is no protocol in the stateless hardware token model accessing the tokens at most a constant number of times, even when the adversary is computationally bounded.

En route to proving our results, we make interesting connections between the hardware token model and well studied notions such as *OT hybrid model*, *randomized encodings* and *obfuscation*.

1 Introduction

A protocol for secure two-party computation allows two mutually distrustful parties to jointly compute a function f of their respective inputs, x and y, in

* Research supported in part by NSF grants 1228856 and 0747027.
** Part of this work done while visiting IDC Herzliya. Supported by the ERC under the EU's Seventh Framework Programme (FP/2007-2013) ERC Grant Agreement no. 307952.
*** Research supported in part by NSF grants 1228856 and 0747027.
† Supported by ISF grant no. 1255/12 and by the ERC under the EU's Seventh Framework Programme (FP/2007-2013) ERC Grant Agreement no. 307952.

Y. Lindell (Ed.): TCC 2014, LNCS 8349, pp. 663–687, 2014.

a way that does not reveal anything beyond the value $f(x, y)$ being computed. Soon after the introduction of this powerful notion [41,18], it was realized that most functions $f(x, y)$ do not admit an unconditionally-secure protocol that satisfies it, in the sense that any such protocol implicitly implies the existence (and in some case requires extensive use [2]) of a protocol for Oblivious Transfer (OT) [7,3,29,23,35]. Moreover, even if one was willing to settle for computational security, secure two-party computation has been shown to suffer from severe limitations in the context of protocol composition [15,5,32,33].

The above realizations have motivated the search for alternative models of computation and communication, with the hope that such models would enable bypassing the above limitations, and as a byproduct perhaps also give rise to more efficient protocols. One notable example is the so called *hardware token model*, introduced by Katz [28]. In this model, it is assumed that one party can generate hardware tokens that implement some efficient functionality in a way that allows the other party only black-box access to the functionality.

The literature on hardware tokens (sometimes referred to as *tamper proof tokens*[1]) discusses a variety of models, ranging from the use of *stateful* tokens (that are destroyed after being queried for some fixed number of times) to *stateless* ones (that can be queried for an arbitrary number of times), with either *non-adaptive* access (in which the queries to the tokens are fixed in advance) or *adaptive* access (in which queries can depend on answers to previous queries). Tokens with varying levels of complexity have also been considered, starting with simple functions such as bit-OT, and ranging all the way to extremely complex functionalities (ones that enable the construction of UC-secure protocols given only a single call to the token).

The use of hardware tokens opened up the possibility of realizing information-theoretically and/or composable secure two-party protocols even in cases where this was shown to be impossible in "plain" models of communication. Two early examples of such constructions are protocols for UC-secure computation [28], and one-time programs [20]. More recently, a line of research initiated by Goyal et al. [22] has focused on obtaining unconditionally-secure two-party computation using stateful tokens that implement the bit-OT functionality. In [21], Goyal et al. went on to show how to achieve UC-secure two party computation using stateless tokens under the condition that tokens can be encapsulated: namely, the receiver of a token A can construct a token B that can invoke A internally. Finally, Dottling et al. [12] have shown that it is possible to obtain information-theoretically secure UC two-party protocols using a single token, assuming it can compute some complex functionality.

Generally speaking, the bit-OT token model has many advantages over a model that allows more complex tokens. First of all, the OT functionality is simple thus facilitating hardware design and implementation. Secondly, in many cases [22], the bit-OT tokens do not depend on the functionality that is being computed. Hence, a large number of bit-OT tokens can be produced "offline"

---

[1] There are papers which deal with the leakage of tokens' contents. We do not consider such a setting in this work.

and subsequently used for any functionality. The main apparent shortcoming of bit-OT tokens in comparison to their complex counterparts is that in all previous works the number of tokens used is proportional to the size of the circuit being evaluated, rendering the resulting protocols impractical. This state of affairs calls for the investigation of the minimal number of bit-OT token invocations in a secure two-party computation protocol.

In this work we aim to study the complexity of constructing secure protocols with respect to different measures in the hardware token model. Our main focus is on non-interactive information-theoretic two-party computation using bit-OT tokens, but we also study computational security with stateless tokens that compute more complex functionalities. En route to proving our results, we make interesting connections between protocols in the hardware token model and well studied notions such as *randomized encodings*, *obfuscation* and the *OT hybrid model*. Such connections have been explored before mainly in the context of obtaining feasibility results [22,13].

The first question we address is concerned with the number of bit-OT tokens required to securely achieve information-theoretic secure two-party computation. The work on one-time programs makes use of bit-OT tokens in order to achieve secure two party computation in the computational setting, and the number of tokens required in that construction is proportional to the receiver's input size. On the other hand, the only known construction in the information-theoretic setting [22] uses a number of tokens that is proportional to the size of the circuit. This leads us to the following question: is it possible to construct information theoretic two party computation protocols in the token model, where the number of tokens is proportional to the size of the functionality's input? Problems of similar nature have been also studied in the (closely related) OT-hybrid model [11,2,40,36,37,39].

The second question we address is concerned with the number of levels of adaptivity required to achieve unconditional two party computation. The known constructions [22] using bit-OT tokens are highly adaptive in nature: the number of adaptive calls required is proportional to the depth of the circuit being computed. The only existing protocols which are non-adaptive are either for specific complexity classes ([26] for NC1) or in the computational setting [20]. An interesting question, therefore, is whether there exist information-theoretic non adaptive protocols for all efficient functionalities.

The works of [21,34] give negative results on the feasibility of using stateless tokens in the information-theoretic setting. Goyal et al. [22] have shown that it is feasible to construct protocols using stateless tokens under computational assumptions. So, a natural question would be to determine the minimum number of calls to the (stateless) token required in a computational setting.

1.1   Our Results

We exploit the relation between protocols in the hardware token model and cryptographic notions such as randomized encodings and obfuscation to obtain lower bounds in the hardware token model. We focus on non-interactive two-party protocols, where only one party (the sender) sends messages and tokens to the other party (the receiver). Our results are summarized below.

Number of Bit-OT Tokens in the Information-Theoretic Setting. Our first set of results establishes lower bounds on the number of bit-OT tokens as a function of the parties' input sizes. Specifically:

- We show that there exists a class of functionalities such that the number of tokens required to securely implement them is at least the size of the sender's input. To obtain this result, we translate a similar result in the correlated distributed randomness model by Winkler et al. [39] to this setting.
- We provide another set of functionalities such that the number of tokens required to securely implement them is at least the size of the receiver's input.

While this still leaves a huge gap between the positive result (which uses number of tokens proportional to the size of the circuit) and our lower bound, we note that before this result, even such lower bounds were not known to exist. Even in the case of OT-hybrid model, which is very much related to the hardware token model (and more deeply studied), only lower bounds known are in terms of the sender's input size.

Non-adaptive Protocols and Randomized Encodings. In our second main result we show that non-adaptive protocols in the hardware token model imply efficient randomized encodings. Even though currently known protocols [22] are highly adaptive, it was still not clear that non adaptive protocols for all functionalities were not possible. In fact, all functions in NC1 admit non adaptive protocols in the hardware token model [26]. To study this question, we relate the existence of non-adaptive protocols to the existence of a "weaker" notion of randomized encodings, called *decomposable randomized encodings*. Specifically, we show that if a function has a non adaptive protocol then correspondingly, the function has an efficient decomposable randomized encoding. The existence of efficient decomposable randomized encodings has far-reaching implications in MPC, providing strong evidence to the impossibility of non-adaptive protocols for a large class of functions.

Constant Number of Calls to Stateless Tokens. In our last result we show that there exists a functionality for which there does not exist any protocol in the *stateless* hardware token model making at most a constant number of calls. To this end, we introduce the notion of an *obfuscation complete oracle scheme*, a variant of obfuscation tailored to the setting of hardware tokens. Goyal et al. [22] have shown such a scheme can be realized under computational assumptions (refer to Section 6.2.2 in the full version). We derive a lower bound stating that a constant number of calls to the obfuscation oracle does not suffice. This result can then be translated to a corresponding result in the hardware token model. This result holds even if the hardware is a complex stateless token (and hence still relevant even in light of our previous results) and (more importantly) against computational adversaries. Previous known lower bounds on complex tokens were either for the case of stateful hardware [19,20,14] or in the information theoretic setting [21,34].

Our hope is that the above results will inspire future work on lower bounds in more general settings in the hardware token model and to further explore the connection with randomized encodings, obfuscation and the OT-hybrid model.

2 Preliminaries

2.1 Model of Computation

Hardware Tokens: Hardware tokens can be divided into two broad categories – stateful and stateless. As the name implies, stateful tokens can maintain some form of state, which might restrict the extent to which they can be used. On the other hand, stateless tokens cannot maintain any state, and could potentially be used an unbounded number of times. The first formal study of hardware tokens modeled them as stateful entities [28], so that they can engage in a two-round protocol with the receiver. Later on, starting with the work of Chandran et al. [6], stateless tokens were also widely studied.

The token functionality models the following sequence of events: (1) a player (the creator) 'seals' a piece of software inside a tamper-proof token; (2) it then sends the token to another player (the receiver), who can run the software in a black-box manner. Once the token has been sent, the creator cannot communicate with it, unlike the setting considered in [9,10]. We also do not allow token *encapsulation* [21], a setting in which tokens can be *placed inside* other tokens.

Stateless Tokens: The $\mathcal{F}_{wrap}^{stateless}$ functionality models the behavior of a stateless token. It is parameterized by a polynomial $p(.)$ and an implicit security parameter k. Its behavior is described as follows:

- **Create**: Upon receiving (create, sid, P_i, P_j, mid, M) from P_i, where M is a Turing machine, do the following: (a) Send (create, sid, P_i, P_j, mid) to P_j, and (b) Store $(P_i, P_j, \text{mid}, M)$.
- **Execute**: Upon receiving (run, sid, P_i, mid, msg) from P_j, find the unique stored tuple $(P_i, P_j, \text{mid}, M)$. If no such tuple exist, do nothing. Run $M(msg)$ for at most $p(k)$ steps, and let out be the response (out $= \perp$ if M does not halt in $p(k)$ steps). Send (sid, P_i, mid, out) to P_j.

Here sid and mid denote the session and machine identifier respectively.

Stateful Tokens: In the class of stateful tokens, our primary interest is the One Time Memory (OTM) token, studied first in [20]. This token implements a single Oblivious Transfer (OT) call, and hence is also referred to as OT token. Oblivious transfer, as we know, is one of the most widely studied primitives in secure multi-party computation. In the $\binom{n}{t}$-OT^k variant, sender has n strings of k bits each, out of which a receiver can pick any t. The sender does not learn anything in this process, and the receiver does not know what the remaining $n - t$ strings were. The behavior of an OTM token is similar to $\binom{2}{1}$-OT^k.

The primary difference between the OT functionality and an OTM token is that while the functionality forwards an acknowledgment to the sender when the receiver obtains the strings of its choice, there is no such feedback provided by the token. Hence, one has to put extra checks in a protocol (in the token model) to ensure that the receiver opens the tokens when it is supposed to (see, for example, Section 3.1 in [22]). Formal definitions of $\mathcal{F}^{OT}$ and $\mathcal{F}^{OTM}$ are given below. We would be dealing with OTMs where both inputs are single bits. We will refer to them as bit-OT tokens.

Oblivious Transfer (OT): The functionality $\mathcal{F}^{OT}$ is parameterized by three positive integers n, t and k, and behaves as follows.

- On input $(P_i, P_j, \mathsf{sid}, \mathsf{id}, (s_1, s_2, \ldots, s_n))$ from party P_i, send $(P_i, P_j, \mathsf{sid}, \mathsf{id})$ to P_j and store the tuple $(P_i, P_j, \mathsf{sid}, \mathsf{id}, (s_1, s_2, \ldots, s_n))$. Here each s_i is a k-bit string.
- On receiving $(P_i, \mathsf{sid}, \mathsf{id}, l_1, l_2, \ldots, l_t)$ from party P_j, if a tuple $(P_i, P_j, \mathsf{sid}, \mathsf{id}, (s_1, s_2, \ldots, s_n))$ exists, return $(P_i, \mathsf{sid}, \mathsf{id}, s_{l_1}, s_{l_2}, \ldots, s_{l_t})$ to P_j, **send an acknowledgment** $(P_j, \mathsf{sid}, \mathsf{id})$ to P_i, and delete the tuple $(P_i, P_j, \mathsf{sid}, \mathsf{id}, (s_1, s_2, \ldots, s_n))$. Else, do nothing. Here each l_j is an integer between 1 and n.

One Time Memory (OTM): The functionality $\mathcal{F}^{OTM}$ which captures the behavior an OTM is described as follows:

- On input $(P_i, P_j, \mathsf{sid}, \mathsf{id}, (s_0, s_1))$ from party P_i, send $(P_i, P_j, \mathsf{sid}, \mathsf{id})$ to P_j and store the tuple $(P_i, P_j, \mathsf{sid}, \mathsf{id}, (s_0, s_1))$.
- On receiving $(P_i, \mathsf{sid}, \mathsf{id}, c)$ from party P_j, if a tuple $(P_i, P_j, \mathsf{sid}, \mathsf{id}, (s_0, s_1))$ exists, return $(P_i, \mathsf{sid}, \mathsf{id}, s_c)$ to P_j and delete the tuple $(P_i, P_j, \mathsf{sid}, \mathsf{id}, (s_0, s_1))$. Else, do nothing.

Non-interactivity: In this paper, we are interested in non-interactive two-party protocols (i.e., where only one party sends messages and tokens to the other). Some of our results, however, hold for an interactive setting as well (whenever this is the case, we point it out). The usual setting is as follows: Alice and Bob have inputs $x \in \mathcal{X}_k$ and $y \in \mathcal{Y}_k$ respectively, and they wish to securely compute a function $f : \mathcal{X}_k \times \mathcal{Y}_k \to \mathcal{Z}_k$, such that only Bob receives the output $f(x, y) \in \mathcal{Z}_k$ of the computation (here, k is the security parameter). Only Alice is allowed to send messages and tokens to Bob.

Circuit Families. In this work, we assume that parties are represented by circuit families instead of Turing machines. A circuit is an acyclic directed graph, with the gates of the circuit representing the nodes of the graph, and the wires representing the edges in the graph. We assume that a circuit can be broken down into layers of gates such that the first layer of gates takes the input of the circuit and outputs to the second layer of the gates which in turn outputs to the third layer and so on. The output of the last layer is the output of the circuit.

A circuit is typically characterized by its size and its depth. The size of the circuit is the sum of the number of gates and the number of wires in the circuit. We define the depth of a circuit C, denoted by $\mathsf{Depth}(C)$ to be the number of layers in the circuit. There are several complexity classes defined in terms of depth and size of circuits. One important complexity class that we will refer in this work is the NC1 complexity class. This comprises of circuits which have depth $O(\log(n))$ and size $\mathsf{poly}(n)$, where n is the input size of the circuit. Languages in P can be represented by a circuit family whose size is polynomial in the size of the input.

2.2 Security

Definition 1 (Indistinguishability). *A function $f : \mathbb{N} \to \mathbb{R}$ is negligible in n if for every polynomial $p(.)$ and all sufficiently large n's, it holds that $f(n) < \frac{1}{p(n)}$. Consider two probability ensembles $X := \{X_n\}_{n \in \mathbb{N}}$ and $Y := \{Y_n\}_{n \in \mathbb{N}}$. These ensembles are computationally indistinguishable if for every PPT algorithm $\mathcal{A}$, $|\Pr[\mathcal{A}(X_n, 1^n) = 1] - \Pr[\mathcal{A}(Y_n, 1^n) = 1]|$ is negligible in n. On the other hand, these ensembles are statistically indistinguishable if $\Delta(X_n, Y_n) = \frac{1}{2} \sum_{\alpha \in \mathcal{S}} |\Pr[X_n = \alpha] - \Pr[Y_n = \alpha]|$ is negligible in n, where $\mathcal{S}$ is the support of the ensembles. The quantity $\Delta(X_n, Y_n)$ is known as the statistical difference between X_n and Y_n.*

Statistical Security: A protocol π for computing a two-input function $f : \mathcal{X}_k \times \mathcal{Y}_k \to \mathcal{Z}_k$ in the hardware-token model involves Alice and Bob exchanging messages and tokens. In the (static) semi-honest model, an adversary could corrupt one of the parties at the beginning of an execution of π. Though the corrupted party does not deviate from the protocol, the adversary could use the information it obtains through this party to learn more about the input of the other party. At an intuitive level, a protocol is secure if any information the adversary could learn from the execution can also be obtained just from the input and output (if any) of the corrupted party. Defining security formally though requires that we introduce some notation, which we do below.

Let the random variables $\mathsf{view}_A^\pi(x, y) = (x, R_A, M, U)$ and $\mathsf{view}_B^\pi(x, y) = (y, R_B, M, V)$ denote the views of Alice and Bob respectively in the protocol π, when Alice has input $x \in \mathcal{X}_k$ and Bob has input $y \in \mathcal{Y}_k$. Here R_A (resp. R_B) denotes the coin tosses of Alice (resp. Bob), M denotes the messages exchanged between Alice and Bob, and U (resp. V) denotes the messages exchanged between Alice (resp. Bob) and the token functionality. Also, let $\mathsf{out}_B^\pi(x, y)$ denote the output produced by Bob. We can now formally define security as follows.

Definition 2 (ϵ-secure protocol [39]). *A two-party protocol π computes a function $f : \mathcal{X}_k \times \mathcal{Y}_k \to \mathcal{Z}_k$ with ϵ − security in the semi-honest model if there exists two randomized functions $\mathcal{S}_A$ and $\mathcal{S}_B$ such that for all sufficiently large values of k, the following two properties hold for all $x \in \mathcal{X}_k$ and $y \in \mathcal{Y}_k$:*

− $\Delta((\mathcal{S}_A(x), f(x, y)), (\mathsf{view}_A^\pi(x, y), \mathsf{out}_B^\pi(x, y))) \leq \epsilon(k)$,

$$- \Delta(\mathcal{S}_B(y, f(x,y)), \mathsf{view}_B^\pi(x,y)) \leq \epsilon(k).$$

If π computes f with ϵ-security for a negligible function $\epsilon(k)$, then we simply say that π securely computes f. Further if $\epsilon(k) = 0$, π is a prefectly secure protocol for f.

Information Theory: We define some information-theoretic notions which will be useful in proving unconditional lower bounds. *Entropy* is a measure of the uncertainty in a random variable. The entropy of X given Y is defined as:

$$H(X|Y) = -\sum_{x \in \mathcal{X}} \sum_{y \in \mathcal{Y}} \Pr[X = x \wedge Y = y] \log \Pr[X = x \mid Y = y].$$

For the sake of convenience, we sometimes use $h(p) = -p \log p - (1-p) \log(1-p)$ to denote the entropy of a binary random variable which takes value 1 with probability p $(0 \leq p \leq 1)$.

Mutual information is a measure of the amount of information one random variable contains about another. The mutual information between X and Y given Z is defined as follows:

$$I(X; Y|Z) = H(X|Z) - H(X|YZ).$$

See [8] for a detailed discussion of the notions above.

3 Lower Bounds in Input Size for Unconditional Security

In this section, we show that the number of simple tokens required to be exchanged in a two-party unconditionally secure function evaluation protocol could depend on the input size of the parties. We obtain two bounds discussed in detail in the sections below. Our first bound relates the number of hardware tokens required to compute a function with the input size of the sender. (This bound holds even when the protocol is interactive.) In particular, we show that the number of bit-OT tokens required for oblivious transfer is at least the sender's input size (minus one). Our second result provides a class of functions where the number of bit-OT tokens required is at least the input size of the receiver.

3.1 Lower Bound in Sender's Input Size

In this subsection we consider k to be fixed, and thus omit k from $\mathcal{X}_k$, $\mathcal{Y}_k$ and $\epsilon(k)$ for clarity. In [39], Winkler and Wullschleger study unconditionally secure two-party computation in the semi-honest model. They consider two parties Alice and Bob, with inputs $x \in \mathcal{X}$ and $y \in \mathcal{Y}$ respectively, who wish to compute a function $f : \mathcal{X} \times \mathcal{Y} \to \mathcal{Z}$ such that only Bob obtains the output $f(x,y) \in \mathcal{Z}$ (but Alice and Bob can exchange messages back and forth). The parties have access to a functionality $\mathcal{G}$ which *does not take any input*, but outputs a sample (u, v) from a distribution p_{UV}. Winkler and Wullschleger obtain several lower bounds on the information-theoretic quantities relating U and V for a secure implementation of the function f.

Here, we would like to obtain the minimum number of bit-OT tokens required for a secure realization of a function. The functionality which models the token behavior $\mathcal{F}^{OTM}$ is an interactive functionality: not only does $\mathcal{F}^{OTM}$ give output to the parties, but also take inputs from them. Therefore, as such the results of [39] are not applicable to our setting. However, if we let U denote all the messages exchanged between Alice and $\mathcal{G}$, and similarly let V denote the entire message transcript between Bob and $\mathcal{G}$, we claim that the following lower bound (obtained for a non-interactive $\mathcal{G}$ in [39]) holds even when the functionality $\mathcal{G}$ is interactive. This will allow us to apply this bound on protocols where hardware tokens are exchanged.

Theorem 1. *Let* $f : \mathcal{X} \times \mathcal{Y} \to \mathcal{Z}$ *be a function such that*

$$\forall x \neq x' \in \mathcal{X} \ \exists y \in \mathcal{Y} : f(x, y) \neq f(x', y).$$

If there exists a protocol that implements f *from a functionality* $\mathcal{G}$ *with* ϵ *security in the semi-honest model, then*

$$H(U|V) \geq \max_{y \in \mathcal{Y}} H(X|f(X, y)) - (3|\mathcal{Y}| - 1)(\epsilon \log |\mathcal{Z}| + h(\epsilon)) - \epsilon \log |\mathcal{X}|,$$

where $H(U|V)$ *is the entropy of* U *given* V.

In order to prove that Theorem 1 holds with an interactive $\mathcal{G}$, we observe that the proof provided by Winkler and Wullschleger for a non-interactive $\mathcal{G}$ *almost* goes through for an interactive one. An important fact they use in their proof is that for any protocol π, with access to a non-interactive $\mathcal{G}$, the following mutual information relation holds: $I(X; VY|UM) = 0$, where M denotes the messages exchanged in the protocol. (In other words, $X - UM - VY$ is a Markov chain.) If one can show that the aforementioned relation holds even when $\mathcal{G}$ can take inputs from the parties (and U and V are redefined as discussed above), the rest of the proof goes through, as can be verified by inspection. Hence, all that is left to do is to prove that $I(X; VY|UM) = 0$ is true in the more general setting, where U and V stand for the *transcripts* of interactions with $\mathcal{G}$. This follows from a simple inductive argument; for the sake of completeness, we provide a proof in full version.

Theorem 1 lets us bound the number of tokens required to securely evaluate a function, as follows. Suppose Alice and Bob exchange ℓ bit-OT tokens during a protocol. If Bob is the recipient of a token, there is at most one bit of information that is hidden from Bob after he has queried the token. On the other hand, if Bob sends a token, he does not know what Alice queried for. Therefore given V, entropy of U can be at most ℓ (or $H(U|V) \leq \ell$). We can use this observation along with Corollary 3 in [39] (full version) to obtain the following result.

Theorem 2. *If a protocol* ϵ-*securely realizes* m *independent instances of* $\binom{n}{t}$-OT^k, *then the number of bit-OT tokens* ℓ *exchanged between Alice and Bob must satisfy the following lower bound:*

$$\ell \geq ((1 - \epsilon)n - t)km - (3\lceil n/t \rceil - 1)(\epsilon m t k + h(\epsilon)).$$

We conclude this section with a particular case of the above theorem which gives a better sense of the bound. Let us say that Alice has a string of n bits, and Bob wants to pick one of them. In other words, Alice and Bob wish to realize an instance of $\binom{n}{1}$-OT^1. Also, assume that they want to do this with perfect security, i.e., $\epsilon = 0$. In this case, the input size of Alice is n, but Bob's input size is only $\lceil \log n \rceil$. Now, we have the following corollary.

Corollary 1. *In order to realize the functionality $\binom{n}{1}$-OT^1 with perfect security, Alice and Bob must exchange at least $n - 1$ tokens.*

Suppose Alice is the only party who can send tokens. Then, we can understand the above result intuitively in the following way. Alice has n bits, but she wants Bob to learn exactly one of them. However, since she does not know which bit Bob needs, she must send her entire input (encoded in some manner) to Bob. Suppose Alice sends ℓ bit-OT tokens to Bob. Since Bob accesses every token, the ℓ bits it obtains from the tokens should give only one bit of information about Alice's input. The remaining ℓ positions in the tokens, which remain hidden from Bob, must contain information about the remaining $n - 1$ bits of Alice's input. Hence, ℓ must be at least $n - 1$.

One can use Protocol 1.2 in [4] to show that the bound in Corollary 1 is tight.

3.2   Lower Bound in Receiver's Input Size

In this section, we show that the number of bit-OT tokens required could depend on the receiver's input size. We begin by defining a non-replayable function family, for which we shall show that the number of tokens required is at least the input size of the receiver.

Definition 3. *Consider a function family $f : \mathcal{X}_k \times \mathcal{Y}_k \to \mathcal{Z}_k$, $k \in \mathbb{I}^+$. We say that f is replayable if for every distribution $\mathcal{D}_k$ over $\mathcal{X}_k$, there exists a randomized algorithm $\mathcal{S}_B$ and a negligible function ν, such that on input $(k, y, f(x, y))$ where $(x, y) \leftarrow \mathcal{D}_k \times \mathcal{Y}_k$, $\mathcal{S}_B$ outputs $\perp$ with probability at most $3/4$, and otherwise outputs (y', z) such that (conditioned on not outputting $\perp$) with probability at least $1 - \nu(k)$, $y' \neq y$ and $z = f(x, y')$.*

Theorem 3. *Let $f : \mathcal{X}_k \times \mathcal{Y}_k \to \mathcal{Z}_k$ be a function that is not replayable. Then, in any non-interactive protocol π that securely realizes f in the semi-honest model using bit-OT tokens, Alice must send at least $n(k) = \lfloor \log |\mathcal{Y}_k| \rfloor$ tokens to Bob.*

Proof. For simplicity, we omit the parameter k in the following. Suppose Alice sends only $n - 1$ bit-OT tokens to Bob in the protocol π. We shall show that f is in fact replayable, by constructing an algorithm $\mathcal{S}_B$ as in Definition 3, from a semi-honest adversary $\mathcal{A}$ that corrupts Bob in an execution of π.

Let the input of Alice and Bob be denoted by x and y respectively, where x is chosen from $\mathcal{X}$ according to the distribution $\mathcal{D}$, and y is chosen uniformly at random over $\mathcal{Y}$. On input x, Alice sends tokens $(\Upsilon_1, \cdots, \Upsilon_{n-1})$ and a message m to Bob. Bob runs his part of the protocol with inputs y, m, a random tape

r, and (one-time) oracle access to the tokens. Without loss of generality, we assume that Bob queries all the $n-1$ tokens. Bob's view consists of y, m, r, and the bits $b = (b_1, \ldots, b_{n-1})$ received from the $n-1$ tokens $(\Upsilon_1, \ldots, \Upsilon_{n-1})$. Let $q = (q_1, \ldots, q_{n-1})$ denote the query bits that Bob uses for the $n-1$ tokens. For convenience, we shall denote the view of Bob as (y, m, r, q, b) (even though q is fully determined by the rest of the view).

We define $\mathcal{S}_B$ as follows: on input (y_1, z_1), it samples a view (y, r, m, q, b) for Bob in an execution of π conditioned on $y = y_1$ and Bob's output being z_1. Next, it samples a second view (y', r', m', q', b') conditioned on $(m', q', b') = (m, q, b)$. If $y' = y$, it outputs $\perp$. Else, it computes Bob's output z' in this execution and outputs (y', z').

To argue that $\mathcal{S}_B$ meets the requirements in Definition 3, it is enough to prove that when $x \in \mathcal{X}$ is sampled from any distribution $\mathcal{D}$, $y \leftarrow \mathcal{Y}$ is chosen uniformly, and $z = f(x, y)$: (1) (y', r', m', q', b') sampled by $\mathcal{S}_B(y, z)$ is distributed close (up to a negligible distance) to Bob's view in an actual execution with inputs (x, y'), and (2) with probability at least $\frac{1}{4}$, $y' \neq y$. Then, by the correctness of π, with overwhelming probability, whenever $\mathcal{S}_B$ outputs (y', z'), it will be the case that $z' = f(x, y')$, and this will happen with probability at least $1/4$.

The first claim follows by the security guarantee and the nature of a token-based protocol. Consider the experiment of sampling (x, y) and then sampling Bob's view (y, r, m, q, b) conditioned on input being y and output being $z = f(x, y)$. Firstly, this is only negligibly different from sampling Bob's view from an actual execution of π with inputs x and y, since by the correctness guarantee, the output of Bob will indeed be $f(x, y)$ with high probability. Now, sampling (x, y, r, m, q, b) in the actual execution can be reinterpreted as follows: first sample (m, q, b), and then conditioned on (m, q, b), sample x and (y, r) independent of each other. This is because, by the nature of the protocol, conditioned on (m, q, b), Bob's view in this experiment is independent of x. Now, (y', r') is also sampled conditioned on (m, q, b) in the same manner (without resampling x), and hence (x, y', r', m, q, b) is distributed as in an execution of π with inputs (x, y').

To show that $\mathcal{S}_B$ outputs $\perp$ with probability at most $\frac{3}{4}$, we rely on the fact that the number of distinct inputs y for Bob is 2^n, but the number of distinct queries the Bob can make to the tokens q is at most 2^{n-1}. Below, we fix an (m, b) pair sampled by $\mathcal{S}_B$, and argue that $\Pr[y = y'] \leq \frac{3}{4}$ (where the probabilities are all conditioned on (m, b)).

For each value of $q \in \{0, 1\}^{n-1}$ that has a non-zero probability of being sampled by $\mathcal{S}_B$, we associate a value $Y(q) \in \{0, 1\}^n$ as $Y(q) = \operatorname{argmax}_y \Pr[y|q]$, where the probability is over the choice of $y \leftarrow \mathcal{Y}$ and the random tape r for Bob. If more than one value of y attains the maximum, $Y(q)$ is taken as the lexicographically smallest one. Let $\mathcal{Y}^* = \{y | \exists q \text{ s.t. } y = Y(q)\}$. Then, $|\mathcal{Y}^*| \leq |\mathcal{Y}|/2$, or equivalently (since the distribution over $\mathcal{Y}$ is uniform), $\Pr[y \notin \mathcal{Y}^*] \geq \frac{1}{2}$.

Let $Q^* = \{q | \Pr[Y(q)|q] > \frac{1}{2}\}$. Further, let $\beta = \min\{\Pr[Y(q)|q]|q \in Q^*\}$. Note that $\beta > \frac{1}{2}$. We claim that $\alpha := \Pr[q \in Q^*] \leq \frac{1}{2}$. This is because

$$\frac{1}{2} \leq \Pr[y \notin \mathcal{Y}^*] = \sum_{y \notin \mathcal{Y}^*, q \in Q^*} \Pr[y, q] + \sum_{y \notin \mathcal{Y}^*, q \notin Q^*} \Pr[y, q]$$

$$\leq \sum_{q \in Q^*} (1 - \beta) \Pr[q] + \sum_{q \notin Q^*} \beta \Pr[q] = \alpha(1 - \beta) + \beta(1 - \alpha)$$

Since $\beta > \frac{1}{2}$, if $\alpha > \frac{1}{2}$ then $\alpha(1 - \beta) + \beta(1 - \alpha) < \frac{1}{2}$, which is a contradiction. Hence $\alpha \leq \frac{1}{2}$. Now,

$$\Pr[y = y'] \leq \alpha \Pr[y = y'|q \in Q^*] + (1 - \alpha) \Pr[y = y'|q \notin Q^*]$$

$$\leq \alpha + (1 - \alpha)\frac{1}{2} \leq \frac{3}{4}.$$

$\square$

We give a concrete example of a function family that is not replayable. Let $\mathcal{X}_k = \{1, 2, \ldots, k\}$ be the set of first k positive integers. Let $\mathcal{Y}_k = \{S \subseteq \mathcal{X}_k : |S| = k/2 \wedge 1 \in S\}$. Define $f : \mathcal{X}_k \times \mathcal{Y}_k \rightarrow \{0, 1\}$ as follows: for all k, $x \in \mathcal{X}_k$ and $y \in \mathcal{Y}_k$, $f(x, S) = 1$ if $x \in S$, and 0 otherwise.

Fix a value of k. Suppose a simulator $\mathcal{S}_B$ is given S and $f(X, S)$ as inputs, where X, S denote random variables uniformly distributed over $\mathcal{X}_k$ and $\mathcal{Y}_k$ respectively. From this input, $\mathcal{S}_B$ knows that X could take one of $k/2$ possible values. Any $S' \neq S$ intersects S' or its complement in at most $k/2 - 1$ positions. Hence, $\mathcal{S}_B$ can guess the value of $f(X, S')$ with probability at most $1 - 2/k$. This implies that if $\mathcal{S}_B$ outputs (S', Z) with probability $1/4$, with a non-negligible probability $Z \neq f(X, S')$.

Note that the number of bits required to represent an element of $\mathcal{X}_k$ is only $\lceil \log k \rceil$, but that required to represent an element of $\mathcal{Y}_k$ is $n(k) = \lceil \log \frac{1}{2}\binom{k}{k/2} \rceil$, which is at least a polynomial in k. Since f is not replayable, it follows from Theorem 3 that in any protocol that realizes f, Alice must send at least $n(k)$ tokens to Bob.

4   Negative Result for Non-adaptive Protocols

4.1   Setting

In this section, we explore the connection between the randomized encodings of functions and the protocols for the corresponding functionalities [2] in the bit-OT (oblivious transfer) token model. We deal with only protocols which are non-adaptive, non-interactive and are perfectly secure. The notions of non-interactivity (Section 2.1) and perfect security (Definition 2 in Section 2.2) have already been dealt with in the preliminaries. We will only explain the notion of

---

[2] Here, we abuse the notation and interchangeably use functions and functionalities.

non-adaptivity. A protocol in the bit-OT token model is said to be *non-adaptive* if the queries to the tokens are fixed in advance. This is in contrast with the adaptive case where the answers from one token can used to generate the query to the next token.

Such (non-adaptive and non-interactive) protocols have been considered in the literature and one-time programs [20,22] is one such example, although one-time programs deal with malicious receivers. Henceforth, when the context is clear we will refer to "perfectly secure non-adaptive non-interactive protocols" as just "non-adaptive protocols".

We show that the existence of non-adaptive protocols for a function in the bit-OT token model implies an efficient (polynomial sized) decomposable randomized encoding for that function. This is done by establishing an equivalence relation between decomposable randomized encodings and a specific type of non-adaptive protocols in the bit-OT token model. Then, we show that a functionality having any non-adaptive protocol also has this specific type of protocol thereby showing the existence of a DRE for this functionality. Since decomposable randomized encodings are believed to not exist for all functions in P [17,38,16,27], this gives a strong evidence to the fact that there cannot exist non-adaptive protocols in the bit-OT token model for all functions in P.

4.2   Randomized Encodings

We begin this section by describing the necessary background required to understand randomized encodings [25]. A randomized encoding for a function f consists of two procedures - encode and decode. The encode procedure takes an input circuit for f, x which is to be input to f along with randomness r and outputs $\hat{f}(x; r)$. The decode procedure takes as input $\hat{f}(x; r)$ and outputs $f(x)$. There are two properties that the encode and decode procedures need to satisfy for them to qualify to be a valid randomized encoding. The first property is (perfect) correctness which says that the decode algorithm always outputs $f(x)$ when input $\hat{f}(x; r)$. The second property, namely (perfect) privacy, says that there exists a simulator such that the output distribution of the encode algorithm on input x is identical to the output distribution of the simulator on input $f(x)$.

We deal with a specific type of randomized encodings termed as decomposable randomized encodings [30,17,24,26,31] which are defined as follows.

Definition 4. *An (efficient) Decomposable Randomized Encoding, denoted by DRE, consists of a tuple of PPT algorithms* (RE.Encode, RE.ChooseInpWires, RE.Decode)*:*

1. RE.Encode: *takes as input a circuit* C *and outputs* $(\tilde{C}, \text{state})$ *,where* state $= ((s_1^0, s_1^1), \ldots, (s_m^0, s_m^1))$ *and* m *is the input size of the circuit.*
2. RE.ChooseInpWires: *takes as input* (state, x) *and outputs* $\tilde{x}$*, where* x *is of length* m *and* $\tilde{x} = (s_1^{x_1}, \ldots, s_m^{x_m})$ *and* x_i *is the* i^{th} *bit of* x.

3. RE.Decode: *takes as input* $(\tilde{C}, \tilde{x})$ *and outputs* out.

A decomposable randomized encoding needs to satisfy the following properties.

(Correctness):- Let RE.Encode on input C output $(\tilde{C}, \text{state})$. Let RE.ChooseInpWires on input (state, x) output $\tilde{x}$. Then, RE.Decode$(\tilde{C}, \tilde{x})$ always outputs $C(x)$.

(Perfect privacy):- There exists a PPT simulator Sim such that the following two distributions are identical.

- $\left\{ (\tilde{C}, \tilde{x}) \right\}$, where $(\tilde{C}, \text{state})$ is the output of RE.Encode on input C and $\tilde{x}$ is the output of ChooseInpWires on input (state, x).
- $\left\{ (\tilde{C}_{\text{Sim}}, \tilde{x}_{\text{Sim}}) \right\}$, where $(\tilde{C}_{\text{Sim}}, \tilde{x}_{\text{Sim}})$ is the output of the simulator Sim on input $C(x)$.

In the above definition, ϵ-privacy can also be considered instead of perfect privacy where the distributions are ϵ far from each other for some negligible ϵ. In this section, we only deal with DRE with perfect privacy. It can be verified that a decomposable randomized encoding is also a randomized encoding. There are efficient decomposable randomized encodings known for all functions in NC^1 [30,26]. However, it is believed that there does not exist efficient decomposable randomized encodings for all functions in P. The existence of efficient decomposable randomized encodings for all efficiently computable functions has interesting implications, namely, multiparty computation protocols in the PSM (Private Simultaneous Message) model [17], constant-round two-party computation protocol in the OT-hybrid model [38,16] and multiparty computation with correlated randomness [27].

We now proceed to relate the existence of non-adaptive protocols for a functionality to the existence of randomized encodings, and more specifically DRE, for the corresponding function. But first, we give an overview of our approach and then we describe the technical details.

4.3   Overview

We first make a simple observation which is the starting point to establish the connection between randomized encodings and non-adaptive protocols in the bit-OT token model. Consider the answer obtained by the receiver of the non-adaptive protocol after querying the tokens. This answer can be viewed as a decomposable randomized encoding. The message contained in the bit-OT tokens along with the software sent by the sender corresponds to the output of the encode procedure. The choose-input-wires procedure corresponds to the algorithm the receiver executes before querying the bit-OT tokens. The decode procedure corresponds to the decoding of the answer from the tokens done by the receiver to obtain the output of the functionality. Further, these procedures satisfy the correctness and the privacy properties. The correctness of the decoding of the output follows directly from the correctness of the protocol. The privacy of the decomposable randomized encoding follows from the fact that the answer obtained from the tokens can be simulated which in turn follows from the privacy of the protocol. At this point it may seem that this observation directly gives us a decomposable randomized encoding from a non-adaptive protocol. However, there are two main issues. Firstly, the output of the encode procedure given by the protocol can depend on the input of the function while in the case of DRE, the encode procedure is independent of the input of the function. Secondly, the choose-inputs-procedure given by the protocol might involve a complex preprocessing on the receiver's input before it queries the tokens. This is in contrast to the choose-inputs-procedure of a DRE where no preprocessing is done on the input of the function.

We overcome these issues in a series of steps to obtain a DRE for a function from a non-adaptive protocol for that function. In the first step, we split the sender's algorithm into two parts - the first part does computation solely on the randomness and independent of the input while the second part does preprocessing on both its input as well as the randomness. We call protocols which have the sender defined this way to be SplitState protocols. We observe that every function that has a non-adaptive protocol also has a SplitState protocol. In the next step, we try to reduce the complexity of the preprocessing done on both the sender's as well as the receiver's inputs. The preprocessing refers to the computation done on the inputs before the hardware tokens are evaluated. We call protocols which have no preprocessing on its inputs to be *simplified* protocols. Our goal is then to show that if a protocol has a SplitState protocol then it also has a simplified protcol. At the heart of this result lies the observation that all NC^1 protocols have simplified protocols. We use the simplified protocols for NC^1 to recursively reduce the complexity of the preprocessing algorithm in a SplitState protocol to finally obtain a simplified protocol. Finally, by using an equivalence relation established between simplified protocols and efficient DRE, we establish the result that a function having a non-adaptive protocol also has an efficient DRE. We now proceed to the technical details.

4.4   Equivalence of RE and Simplified Protocols

We now show the equivalence of randomized encodings and simplified protocols in the bit-OT token model.

SplitState **Protocols.** Consider the protocol Π in the bit-OT token model. We say that Π is a SplitState protocol if the sender and the receiver algorithms in SplitState protocol are defined as follows. The sender in Π consists of the tuple of algorithms (Π.InpFreePP, Π.Preproc$_{sen}$, Π.EvalHT$_{sen}$). It takes as input x with randomness R_{sen} and executes the following steps.

- It first executes Π.InpFreePP on input R_{sen} to obtain the tokens (htokens$_{sen}$, htokens$_{rec}$) and Software.
- It then executes Π.Preproc$_{sen}$ on input (x, R_{sen}) to obtain x'.

- It then executes $\Pi.\mathsf{EvalHT}_{\mathsf{sen}}$ on input $(x', \mathsf{htokens}_{\mathsf{sen}})$. The procedure $\Pi.\mathsf{EvalHT}_{\mathsf{sen}}$ evaluates the ith token in $\mathsf{htokens}_{\mathsf{sen}}$ with the ith bit of x' to obtain $\tilde{x}_i$. The value $\tilde{x}$ is basically the concatenation of all $\tilde{x}_i$.
- The sender then outputs $(\mathsf{htokens}_{\mathsf{rec}}, \mathsf{Software}, \tilde{x})$.

Notice that the third step in the above sender's procedure involves the sender evaluating the tokens $\mathsf{htokens}_{\mathsf{sen}}$. This seems to be an unnecessary step since the sender himself generates the tokens. Later we will see that modeling the sender this way simplifies our presentation of the proof significantly.

The receiver, on the other hand, consists of the algorithms $(\Pi.\mathsf{Preproc}_{\mathsf{rec}}, \Pi.\mathsf{EvalHT}_{\mathsf{rec}}, \Pi.\mathsf{Output})$. It takes as input y, randomness R_{rec} along with $(\mathsf{htokens}_{\mathsf{rec}}, \mathsf{Software}, \tilde{x})$ which it receives from the sender and does the following.

- It executes $\Pi.\mathsf{Preproc}_{\mathsf{rec}}$ on input $(y, R_{\mathsf{rec}}, \mathsf{Software}, \tilde{x})$ to obtain (q, state).
- It then executes $\Pi.\mathsf{EvalHT}_{\mathsf{rec}}$ by querying the tokens $\mathsf{htokens}_{\mathsf{rec}}$ on input q to obtain $\tilde{y}$. The ith token in $\mathsf{htokens}_{\mathsf{rec}}$ is queried by the ith bit of q to obtain the ith bit of $\tilde{y}$.
- Finally, $\Pi.\mathsf{Output}$ is run on input $(\mathsf{state}, \tilde{y})$ to obtain z which is output by the receiver.

This completes the description of Π. The following lemma shows that there exists a SplitState protocol for a functionality if the functionality has a non-adaptive protocol. The proof of the below lemma is provided in the full version.

Lemma 1. *Suppose a functionality f has a non-interactive and a non-adaptive protocol in the bit-OT token model. Then, there exists a SplitState protocol for the functionality f.*

Whenever we say that a functionality has a protocol in the bit-OT token model we assume that it is a SplitState protocol. In the class of SplitState protocols, we further consider a special class of protocols which we term as *simplified protocols*.

Simplified Protocols. These are SplitState protocols which have a trivial preprocessing algorithm on the sender's as well as receiver's inputs. In more detail, a protocol is said to be a *simplified protocol* if it is a SplitState protocol, and the sender's preprocessing algorithm $\mathsf{Preproc}_{\mathsf{sen}}$ as well as the receiver's preprocessing algorithm $\mathsf{Preproc}_{\mathsf{rec}}$ can be implemented by depth-0 circuits. Recall that depth-0 circuits which solely consists of wires and no gates. We now explore the relation between the simplified protocols and decomposable randomized encodings. We show, for every functionality, the equivalence of DRE and simplified protocols in the bit-OT token model. The proof can be found in full version.

Theorem 4. *There exists an efficient decomposable randomized encoding for a functionality f iff there exists a simplified protocol for f in the bit-OT token model.*

Ishai et al. [26] show that there exists decomposable randomized encodings for all functions in NC^1. From this result and Theorem 4, the following corollary is immediate.

Corollary 2. *There exists a simplified protocol for all functions in NC^1.*

4.5   Main Theorem

We now state the following theorem that shows that every function that has a non-adaptive protocol in the bit-OT token model also has a simplified protocol. Essentially this theorem says the following. Let there be a non-adaptive protocol in the bit-OT token model for a function. Then, no matter how complex the preprocessing algorithm is in this protocol, we can transform this into another protocol which has a trivial preprocessing on its inputs. Since a function having a non-adaptive protocol also has a SplitState protocol from Lemma 1, we will instead consider SplitState protocols in the below theorem.

Theorem 5. *Suppose there exists a* SplitState *protocol for f in the bit-OT, token model having $p(k)$ number of tokens, for some polynomial p. Then, there exists a simplified protocol for f in the bit-OT token model having $O(p(k))$ number of tokens.*

Proof. Consider the set S of all SplitState protocols for f each having $O(p(k))$ number of tokens. In this set S, consider the protocol Π'_{sen} having the least depth complexity of $\mathsf{Preproc}_{sen}$. That is, protocol Π'_{sen} is such that the following quantity is satisfied.

$$\mathsf{Depth}(\Pi'_{sen}.\mathsf{Preproc}_{sen}) = \min_{\Pi \in S} \left\{ \mathsf{Depth}(\Pi.\mathsf{Preproc}_{sen}) \right\}$$

We claim that the $\Pi'_{sen}.\mathsf{Preproc}_{sen}$ is a depth-0 circuit. If it is not a depth-0 circuit, then we arrive at a contradiction. We transform Π'_{sen} into Π''_{sen}, and show that $\mathsf{Depth}(\Pi'_{sen}.\mathsf{Preproc}_{sen}) < \mathsf{Depth}(\Pi''_{sen}.\mathsf{Preproc}_{sen})$. This would contradict the fact that the depth of $\Pi'_{sen}.\mathsf{Preproc}_{sen}$ is the least among all the protocols in S. To acheive the transformation, we first break $\Pi'_{sen}.\mathsf{Preproc}_{sen}$ into two circuits $\Pi'_{sen}.\mathsf{Preproc}^{up}_{sen}$ and $\Pi'_{sen}.\mathsf{Preproc}^{low}_{sen}$ such that, $\Pi'_{sen}.\mathsf{Preproc}_{sen}$ will first execute $\Pi'_{sen}.\mathsf{Preproc}^{low}_{sen}$ and its output is fed into $\Pi'_{sen}.\mathsf{Preproc}^{up}_{sen}$ whose output determines the output of $\Pi'_{sen}.\mathsf{Preproc}_{sen}$. Further, $\Pi'_{sen}.\mathsf{Preproc}^{up}_{sen}$ consists of a single layer of the circuit and hence has depth 1 (If $\Pi'_{sen}.\mathsf{Preproc}_{sen}$ was just one layer to begin with then $\Pi'_{sen}.\mathsf{Preproc}^{low}_{sen}$ would be a depth-0 circuit.). Then we define a functionality which executes the algorithms $\Pi'_{sen}.\mathsf{Preproc}^{up}_{sen}$ and $\Pi'_{sen}.\mathsf{EvalHT}_{sen}$. We observe that this functionality can be realized by an NC^1 circuit. Then, we proceed to replace the procedures $\Pi'_{sen}.\mathsf{EvalHT}_{sen}$ and $\Pi'_{sen}.\mathsf{Preproc}^{up}_{sen}$ by the sender algorithm of a simplified protocol defined for this functionality, the existence of which follows from Corollary 2. The $\mathsf{Preproc}_{sen}$ of the resulting protocol just consists of $\Pi'_{sen}.\mathsf{Preproc}^{low}_{sen}$ and this would contradict the choice of Π'_{sen}. We now proceed to the technical details.

The sender algorithm of Π'_{sen} can be written as $(\Pi'_{sen}.\mathsf{InpFreePP}, \Pi'_{sen}.\mathsf{Preproc}_{sen}, \Pi'_{sen}.\mathsf{EvalHT}_{sen})$ and the receiver of Π'_{sen} can be written as $(\Pi'_{sen}.\mathsf{Preproc}_{rec}, \Pi'_{sen}.\mathsf{EvalHT}_{rec}, \Pi'_{sen}.\mathsf{Output})$. The description of these algorithms are given in Section 4. Consider the following functionality, denoted by $f^{sen}_{\mathsf{NC}^1}$.

$f_{\mathsf{NC}^1}^{\mathsf{sen}}(s, \mathsf{temp}_x; \bot)$:- On input (s, temp_x) from the sender, it first executes $\Pi'_{\mathsf{sen}}.\mathsf{Preproc}_{\mathsf{sen}}^{\mathsf{up}}(\mathsf{temp}_x)$ to obtain x'. It then parses s as $((s_1^0, s_1^1), \ldots, (s_m^0, s_m^1))$, where the size of x' is m. It then computes $\tilde{x} = (s_1^{x_1'}, \ldots, s_m^{x_m'})$, where x_i' is the ith bit of x'. Finally, output $\tilde{x}$. This functionality does not take any input from the receiver.

Observe that $f_{\mathsf{NC}^1}^{\mathsf{sen}}$ is a NC^1 circuit and has a simplified protocol from Corollary 2. Let us call this protocol $\Pi_{\mathsf{NC}^1}^{\mathsf{sen}}$. Since, the receiver's input is $\bot$, the sender algorithm in this protocol does not output any tokens [3]. We use Π'_{sen} and $\Pi_{\mathsf{NC}^1}^{\mathsf{sen}}$ to obtain Π''_{sen}. The protocol Π''_{sen} is described as follows.

Before we describe the sender algorithm of Π''_{sen}, we modify the sender of Π'_{sen} such that, the algorithm $\Pi'_{\mathsf{sen}}.\mathsf{InpFreePP}$ instead of outputting $\mathsf{htokens}_{\mathsf{rec}}$ just outputs s, which is nothing but the string contained in $\mathsf{htokens}_{\mathsf{sen}}$. The sender algorithm of Π''_{sen} on input (x, R_{sen}), does the following.

- It first executes $\Pi'_{\mathsf{sen}}.\mathsf{InpFreePP}(R_{\mathsf{sen}})$ to obtain $(\mathsf{Software}, s, \mathsf{htokens}_{\mathsf{rec}})$, where s, as described before is the string obtained by concatenating all the bits in $\mathsf{htokens}_{\mathsf{sen}}$.
- It then executes $\Pi'_{\mathsf{sen}}.\mathsf{Preproc}_{\mathsf{sen}}^{\mathsf{low}}$ on input (x, R_{sen}) to obtain temp_x.
- It then executes the sender algorithm of $\Pi_{\mathsf{NC}^1}^{\mathsf{sen}}$ with input (s, temp_x). Let the output of this algorithm be $\mathsf{Software}^{\mathsf{NC}^1}$.
- Send $(\mathsf{Software}, \mathsf{Software}^{\mathsf{NC}^1}, \mathsf{htokens}_{\mathsf{rec}})$ across to the receiver (*recall that the sender of $\Pi_{\mathsf{NC}^1}^{\mathsf{sen}}$ does not output any tokens.*).

The receiver on input (y, R_{rec}) along with $(\mathsf{Software}, \mathsf{Software}^{\mathsf{NC}^1}, \mathsf{htokens}_{\mathsf{rec}})$ which it receives from the sender, does the following.

- It executes the receiver algorithm of $\Pi_{\mathsf{NC}^1}^{\mathsf{sen}}$ on input $\mathsf{Software}^{\mathsf{NC}^1}$ as well as its internal randomness to obtain $\tilde{x}$. Note that the receiver of $\Pi_{\mathsf{NC}^1}^{\mathsf{sen}}$ does not have its own input.
- It then executes the receiver algorithm of Π'_{sen} on input $(y, R_{\mathsf{rec}}, \mathsf{Software}, \tilde{x}, \mathsf{htokens}_{\mathsf{rec}})$. Let the output of this algorithm be out.
- Output out.

We first claim that the protocol Π''_{sen} satisfies the correctness property. This follows directly from the correctness of the protocols Π'_{sen} and $\Pi_{\mathsf{NC}^1}^{\mathsf{sen}}$. The security of the above protocol is proved in the following lemma.

Lemma 2. *Assuming that the protocol Π'_{sen} and $\Pi_{\mathsf{NC}^1}^{\mathsf{sen}}$ is secure, the protocol Π''_{sen} is secure.*

Proof Sketch. To prove this, we need to construct a simulator $\mathsf{Sim}_{\Pi''_{\mathsf{sen}}}$, such that the output of the simulator is indistinguishable from the output of the sender of

---

[3] From the Corollary 2 and Ishai et al. [26], the simplified protocols defined for NC^1 functionalities are such that the sender does not send any tokens to the receiver if the receiver does not have any input.

Π''_{sen}. To do this we use the simulators of the protocols Π'_{sen} and $\Pi^{\mathsf{sen}}_{\mathsf{NC}^1}$ which are denoted by $\mathsf{Sim}_{\Pi'_{\mathsf{sen}}}$ and $\mathsf{Sim}_{\Pi^{\mathsf{sen}}_{\mathsf{NC}^1}}$ respectively.

The simulator $\mathsf{Sim}_{\Pi''_{\mathsf{sen}}}$ on input out, which is the output of the functionality f, along with y' which is the query made by the receiver to the OT tokens does the following. It first executes $\mathsf{Sim}_{\Pi'_{\mathsf{sen}}}(\mathsf{out}, y')$ to obtain $(\mathsf{Software}, \tilde{x}, \tilde{y})$. Then, $\mathsf{Sim}_{\Pi^{\mathsf{sen}}_{\mathsf{NC}^1}}$ on input $\tilde{x}$ is executed to obtain $\mathsf{Software}^{\mathsf{NC}^1}$. The output of $\mathsf{Sim}_{\Pi''_{\mathsf{sen}}}$ is $(\mathsf{Software}, \mathsf{Software}^{\mathsf{NC}^1}, \tilde{y})$. By standard hybrid arguments, it can be shown that the output of the simulator $\mathsf{Sim}_{\Pi''_{\mathsf{sen}}}$ is indistinguishable from the output of the sender of Π''_{sen}.

The above lemma proves that Π''_{sen} is a secure protocol for f. We claim that the number of tokens in Π''_{sen} is $O(p(k))$. This follows directly from the fact that the number of tokens output by the sender of Π''_{sen} is the same as the number of tokens output by Π'_{sen}. And hence, the number of tokens output by the sender of Π''_{sen} is $O(p(k))$. Further, the the depth of $\mathsf{Preproc}_{\mathsf{sen}}$ of Π''_{sen} is strictly smaller than the depth of $\Pi'_{\mathsf{sen}}.\mathsf{Preproc}_{\mathsf{sen}}$. This contradicts the choice of Π'_{sen} and so, the $\mathsf{Preproc}_{\mathsf{sen}}$ algorithm of Π'_{sen} is a depth-0 circuit.

Now, consider a set of protocols, $S' \subset S$ such that the $\mathsf{Preproc}_{\mathsf{sen}}$ algorithms of all the protocols in S' are implementable by depth-0 circuits. From the above arguments, we know that there is at least one such protocol in this set. We claim that there exists one protocol in S' such that its $\mathsf{Preproc}_{\mathsf{rec}}$ algorithm is implementable by a depth-0 circuit. Now, the $\mathsf{Preproc}_{\mathsf{sen}}$ algorithm of this protocol is also implementable by a depth-0 circuit since this protocol is in the set S. From this, it follows that there exists a simplified protocol for f having $O(p(k))$ tokens. The argument for this is similar to the previous case and due to lack of space, we present this part in the full version.   □

We now show that the existence of a non-adaptive protocol for a function implies the existence of a decomposable randomized encoding for that function. Suppose there exists a non-interactive and a non-adaptive protocol for f in the bit-OT token model. Then, from Theorem 5 it follows that there exists a simplified protocol for f. Further, from Theorem 4, it follows that there exists a DRE, and hence an efficient randomized encoding for f. Summarising, we have the following.

Theorem 6. *If there exists a non-interactive and a non-adaptive protocol in the bit-OT token model for a function f then there exists an efficient decomposable randomized encoding for f.*

5   Lower Bound for Obfuscation Complete Oracle Schemes

In this section, we study the notion of an obfuscation complete oracle scheme. Roughly speaking, an obfuscation complete oracle scheme consists of an oracle generation algorithm whose execution results in: (a) a secret obfuscation complete circuit (whose size is only dependent on the security parameter), and, (b) a public obfuscation function. We call an oracle implementing the functionality of the secret obfuscation complete circuit an obfuscation complete (OC) oracle. The public obfuscation function can be applied on any desired (polynomial size) circuit to produce an *obfuscated oracle circuit*. This oracle circuit would make calls to the OC oracle during its execution. The OC oracle implements a fixed functionality and cannot keep any state specific to the execution of any obfuscated program. Informally, our security requirement is that for every polynomial size circuit C, whatever can be computed given access to the obfuscated oracle circuit for C and the OC oracle, can also be computed just given access to an oracle implementing the functionality of C. An obfuscation complete oracle scheme is formally defined as follows.

Definition 5. *A secure obfuscation complete oracle scheme consists of a randomized algorithm OracleGen called the oracle generation algorithm such that an execution OracleGen(1^κ) (where κ denotes the security parameter) results in a tuple $(T, \mathcal{O}^T)$. The string T is the description of the circuit called the secret obfuscation complete circuit while $\mathcal{O}^T$ is a function (or the description of a Turing machine) called the public obfuscation function.[4] The tuple $(T, \mathcal{O}^T)$ has the following properties:*

1. **Preserve Functionality.** *The application of the function $\mathcal{O}^T(\cdot)$ to a circuit C results in an obfuscated oracle circuit $\mathcal{O}^T(C)$ (which during execution might make calls to the oracle T implementing the functionality T). We require the obfuscated oracle circuit $\mathcal{O}^T(C)$ to have the same functionality as the circuit C. In other words, $\forall C, \forall x$, we must have:*

$$\mathcal{O}^T(C) = C(x)$$

2. **Polynomial Slowdown.** *There exist polynomials $p(\cdot, \cdot)$ and $q(\cdot)$ such that for sufficiently large κ and $|C|$, we have:*

$$|\mathcal{O}^T(C)| \le p(|C|, \kappa), \quad and, \quad |T| \le q(\kappa)$$

Observe that the size of the circuit T is dependent only on the security parameter.

3. **Virtual Black Box.** *For every PPT adversary A, there exists a PPT simulator Sim and a negligible function negl($\cdot$) such that for every PPT distinguisher D, for every circuit C and for every polynomial size auxiliary input z:*

$$Pr[D(A^T(\mathcal{O}^T(C), z), z) = 1] - Pr[D(Sim^C(1^{|C|}, T, z), z) = 1] \le negl(\kappa)$$

---

[4] The modeling of T as a circuit rather than a Turing machine is to reflect the fact that given the security parameter, the size and the running time of T is fixed and it handles inputs of fixed size (so that T can, for example, be implemented in a small tamper proof hardware token).

In other words, we require the output distribution of the adversary A and that of the simulator Sim to be computationally indistinguishable.

By replacing the above virtual black box definition by the "predicate" virtual black box definition used by Barak et al. (see [1] for more details), we obtain a relaxed security notion for obfuscation complete oracles schemes. This relaxed version will be used for our lower bounds.

5.1   Lower Bounds

In Section 6.2.2 [22] (full version), Goyal et al. construct an obfuscation complete oracle scheme in the $\mathcal{F}_{wrap}^{stateless}$-hybrid model[5]. In their scheme, if the size of original circuit is $|C|$, then the obfuscated oracle circuit makes $O(|C| \cdot \log(|C|))$ calls to the OC oracle, which is embedded inside a stateless token. Thus, a natural question is: "Do there exist obfuscation complete oracles schemes for which the above query complexity is lower?" Towards that end, we show a lower bound which rules out obfuscation complete oracles schemes where this query complexity is a constant.

Turing Machines. We start by proving the lower bound result for the case of Turing machines. While this case is significantly simpler, it would already illustrate the fundamental limitations of OC Oracle schemes with low query complexity. For an OC scheme, denote by $Q(|M|)$ the number of queries the obfuscated Oracle Turing machine $\mathcal{O}^{\mathcal{T}}(M)$ makes to the Oracle $\mathcal{T}$. We now have the following theorem.

Theorem 7. *For every constant q, there does not exist any obfuscation complete oracle scheme such that for every Turing machine M, query complexity $Q(|M|) \leq q$.*

Proof. We prove the above theorem by contradiction. Assume that there exists such an OC Oracle scheme would query complexity $Q(|M|) \leq q$. Let the size of response to a query to the Oracle $\mathcal{T}$ be bounded by $p(k)$. Hence, observe that the information "flowing" from the Oracle $\mathcal{T}$ to the obfuscated Oracle TM $\mathcal{O}^{\mathcal{T}}(M)$ is bounded by $q \cdot p(k)$. We will show that this communication between the Oracle and the obfuscated TM is not sufficient for successful simulation. Let $f_1 : \{0,1\}^{\leq poly(k)} \to \{0,1\}^{q \cdot p(k)+k}$ and $f_2 : \{0,1\}^{\leq poly(k)} \to \{0,1\}^{poly(k)}$ denote functions drawn from a pseudorandom function ensemble. Now define a functionality $F_{f_1,f_2,s}(.,.)$ as follows. For $b \in \{1,2\}$, we have $F_{f_1,f_2,s}(b,x) = f_b(x)$. For $b = 3$ (referred to as mode 3), we interpret the input x as the description of an Oracle TM M and a sequence of q strings $a_1, \ldots, a_q$. The function outputs $\perp$ if there exists an i s.t. $|a_i| > p(k)$. Otherwise, run the machine $M(1, f_2(M))$. When the machine makes the ith Oracle query, supply a_i as the response (irrespective

---

[5] They actually construct a secure protocol for stateless oblivious reactive functionalities. However, it is easy to see that the same protocol gives us an obfuscation complete oracle scheme.

of what the query is). Now, if $M(1, f_2(M)) = f_1(f_2(M))$, output s, else output $\perp$. To summarize, check if the Oracle TM behaves like the PRF f_1 on a random point (determined by applying PRF f_2 on the description of the machine) and if so, output the secret s. (The function $F_{f_1, f_2, s}(., .)$ is actually uncomputable. However, similar to [1], we can truncate the execution after poly(k) steps and output 0 if M does not halt.) Denote the obfuscated Oracle TM for this function as $\mathcal{O}^{\mathcal{T}}(F_{f_1, f_2, s})$.

Consider the real world when the adversary is given access to description of the Oracle TM $M' = \mathcal{O}^{\mathcal{T}}(F_{f_1, f_2, s})$ and is allowed to query the Oracle $\mathcal{T}$. In this case, the adversary can recover s as follows. First recover $d = M'(2, M')$ (by simply running the obfuscated Oracle TM M' on its own description string with the help of the Oracle $\mathcal{T}$). Now the adversary executes $M'(1, d)$ and stores responses of $\mathcal{T}$ to all the queries made by $M'(1, d)$. Call the responses $a_1, \ldots, a_q$. Finally, prepare a string x containing the description of M' along with the strings $a_1, \ldots, a_q$ and execute $M'(3, x)$. M' will in turn execute $M'(1, d)$ using $a_1, \ldots, a_q$ and, by construction, will get $f_1(f_2(M'))$. Thus, the adversary will receive s as output. Hence, we have constructed a real word adversary A such that:

$$\Pr[A^{\mathcal{T}}(\mathcal{O}^{\mathcal{T}}(F_{f_1, f_2, 0})) = 1] - \Pr[A^{\mathcal{T}}(\mathcal{O}^{\mathcal{T}}(F_{f_1, f_2, 1})) = 1] = 1 \qquad (1)$$

Now consider the ideal world where the adversary S only has Oracle access to the functionality $F_{f_1, f_2, s}$. For simplicity, we first consider the hybrid ideal world where the functions f_1 and f_2 are truly random (that is, for each input, there exists a truly random string which is given as the output). Without loss of generality, we assume that S does not query $F_{f_1, f_2, s}$ multiple times with the same input. Consider a query $(2, M)$ to the functionality $F_{f_1, f_2, s}$. Then it is easy to see that, except with negligible probability, S has not issued the query $(1, f_2(M))$ so far (where the probability is taken over the choice of truly random function f_2). Now when $M(1, f_2(M))$ is executed, depending upon how the Oracle queries are answered, the total number of possible outputs is $2^{q \cdot p(k)}$. Lets call this output set S_o. The probability (taken over the choice of f_1) that $f_1(f_2(M)) \in S_o$ can be bounded by $\frac{1}{2^k}$ ($= \frac{|S_o|}{2^{|f_1(f_2(M))|}}$) which is negligible. Thus, when S queries with $(3, M||a_1|| \ldots ||a_q)$, except with negligible probability, it will get $\perp$ as the output no matter what $a_1, \ldots, a_q$ are. By a straightforward union bound, it can be seen that except with negligible probability, all the queries of S in mode 3 will result in $\perp$ as the output (as opposed to s). By relying on the pseudorandomness of f_1 and f_2, this will also be true not only in the hybrid ideal world but also in the actual ideal world. Hence we have shown that for all ideal world adversaries S,

$$\Pr[S^{F_{f_1, f_2, 0}}(1^k) = 1] - \Pr[S^{F_{f_1, f_2, 1}}(1^k) = 1] \le \text{negl}(k) \qquad (2)$$

Combining equations 1 and 2, we get a contradiction with the relaxed virtual black box property (see the predicate based virtual black box property in [1]) of the OC Oracle scheme.   □

Circuits. In extending the impossibility result to the case of circuits, the basic problem is that since the input length of the circuit is fixed, it may not be possible to execute a circuit on its own description. To overcome this problem, [1] suggested a functionality "implementing homomorphic encryption". This allowed the functionality to let the user (or adversary) evaluate a circuit "gate by gate" (as opposed to feeding the entire circuit at once) and still test certain properties of the user circuit. These techniques do not directly generalize to our setting. This is because in our setting, the Oracle queries made by the adversary's circuit will have to be seen and answered by the adversary. This might leak the input on which the circuit is being "tested" by the functionality. Thus, once the adversary knows the input and hence the "right output", he might, for example, try to change the circuit or tamper with intermediate encrypted wire values to convince the functionality that the circuit is giving the right output. We use the techniques developed in Section 6.2.2 [22] to overcome these problems. Note that these problems do not arise in the setting of Barak et al [1] since there the adversary never gets to see the input on which his circuit is being tested (and hence cannot pass the test even if he can freely force the circuit to give any output of his choice at any time). We now state our impossibility results for circuits.

Theorem 8. *For every constant q, there does not exist any obfuscation complete oracle scheme such that for every circuit C, query complexity $Q(|C|) \leq q$.*

A proof can be found in the full version.

References

1. Barak, B., Goldreich, O., Impagliazzo, R., Rudich, S., Sahai, A., Vadhan, S., Yang, K.: On the (im)possibility of obfuscating programs. J. ACM 59(2), 6:1–6:48 (2012)
2. Beimel, A., Malkin, T.: A quantitative approach to reductions in secure computation. In: Naor, M. (ed.) TCC 2004. LNCS, vol. 2951, pp. 238–257. Springer, Heidelberg (2004)
3. Beimel, A., Malkin, T., Micali, S.: The all-or-nothing nature of two-party secure computation. In: Wiener, M. (ed.) CRYPTO 1999. LNCS, vol. 1666, pp. 80–97. Springer, Heidelberg (1999)
4. Brassard, G., Crepeau, C., Santha, M.: Oblivious transfers and intersecting codes. IEEE Transactions on Information Theory 42(6), 1769–1780 (1996)
5. Canetti, R., Kilian, J., Petrank, E., Rosen, A.: Black-box concurrent zero-knowledge requires (almost) logarithmically many rounds. SIAM J. Comput. 32(1), 1–47 (2002)
6. Chandran, N., Goyal, V., Sahai, A.: New constructions for UC secure computation using tamper-proof hardware. In: Smart, N. (ed.) EUROCRYPT 2008. LNCS, vol. 4965, pp. 545–562. Springer, Heidelberg (2008)
7. Chor, B., Kushilevitz, E.: A zero-one law for boolean privacy. SIAM J. Discrete Math. 4(1), 36–47 (1991)
8. Cover, T.M., Thomas, J.A.: Elements of information theory, vol. 2. Wiley (2006)
9. Damgård, I., Nielsen, J.B., Wichs, D.: Isolated proofs of knowledge and isolated zero knowledge. In: Smart, N. (ed.) EUROCRYPT 2008. LNCS, vol. 4965, pp. 509–526. Springer, Heidelberg (2008)

10. Damgård, I., Nielsen, J.B., Wichs, D.: Universally composable multiparty computation with partially isolated parties. In: Reingold, O. (ed.) TCC 2009. LNCS, vol. 5444, pp. 315–331. Springer, Heidelberg (2009)
11. Dodis, Y., Micali, S.: Lower bounds for oblivious transfer reductions. In: Stern, J. (ed.) EUROCRYPT 1999. LNCS, vol. 1592, pp. 42–55. Springer, Heidelberg (1999)
12. Döttling, N., Kraschewski, D., Müller-Quade, J.: Unconditional and composable security using a single stateful tamper-proof hardware token. In: Ishai, Y. (ed.) TCC 2011. LNCS, vol. 6597, pp. 164–181. Springer, Heidelberg (2011)
13. Döttling, N., Mie, T., Müller-Quade, J., Nilges, T.: Basing obfuscation on simple tamper-proof hardware assumptions. Technical Report 675 (2011)
14. Döttling, N., Mie, T., Müller-Quade, J., Nilges, T.: Implementing resettable UC-Functionalities with untrusted tamper-proof hardware-tokens. In: Sahai, A. (ed.) TCC 2013. LNCS, vol. 7785, pp. 642–661. Springer, Heidelberg (2013)
15. Dwork, C., Naor, M., Sahai, A.: Concurrent zero knowledge, pp. 409–418 (1998)
16. Even, S., Goldreich, O., Lempel, A.: A randomized protocol for signing contracts. Communications of the ACM 28(6), 637–647 (1985)
17. Feige, U., Killian, J., Naor, M.: A minimal model for secure computation. In: Proceedings of the Twenty-Sixth Annual ACM Symposium on Theory of Computing, pp. 554–563. ACM (1994)
18. Goldreich, O., Micali, S., Wigderson, A.: How to play ANY mental game, pp. 218–229 (1987)
19. Goldreich, O., Ostrovsky, R.: Software protection and simulation on oblivious RAMs. J. ACM 43(3), 431–473 (1996)
20. Goldwasser, S., Kalai, Y.T., Rothblum, G.N.: One-time programs. In: Wagner, D. (ed.) CRYPTO 2008. LNCS, vol. 5157, pp. 39–56. Springer, Heidelberg (2008)
21. Goyal, V., Ishai, Y., Mahmoody, M., Sahai, A.: Interactive locking, zero-knowledge PCPs, and unconditional cryptography. In: Rabin, T. (ed.) CRYPTO 2010. LNCS, vol. 6223, pp. 173–190. Springer, Heidelberg (2010)
22. Goyal, V., Ishai, Y., Sahai, A., Venkatesan, R., Wadia, A.: Founding cryptography on tamper-proof hardware tokens. In: Micciancio, D. (ed.) TCC 2010. LNCS, vol. 5978, pp. 308–326. Springer, Heidelberg (2010)
23. Harnik, D., Naor, M., Reingold, O., Rosen, A.: Completeness in two-party secure computation: A computational view. J. Cryptology 19(4), 521–552 (2006)
24. Ishai, Y., Kushilevitz, E.: Private simultaneous messages protocols with applications. In: Proceedings of the Fifth Israeli Symposium on Theory of Computing and Systems, 1997, pp. 174–183. IEEE (1997)
25. Ishai, Y., Kushilevitz, E.: Randomizing polynomials: A new representation with applications to round-efficient secure computation. In: Proceedings of the 41st Annual Symposium on Foundations of Computer Science, 2000, pp. 294–304. IEEE (2000)
26. Ishai, Y., Kushilevitz, E.: Perfect constant-round secure computation via perfect randomizing polynomials. In: Widmayer, P., Eidenbenz, S., Triguero, F., Morales, R., Conejo, R., Hennessy, M. (eds.) ICALP 2002. LNCS, vol. 2380, pp. 244–256. Springer, Heidelberg (2002)
27. Ishai, Y., Kushilevitz, E., Ostrovsky, R., Sahai, A.: Extracting correlations. In: 50th Annual IEEE Symposium on Foundations of Computer Science, FOCS 2009, pp. 261–270. IEEE (2009)
28. Katz, J.: Universally composable multi-party computation using tamper-proof hardware. In: Naor, M. (ed.) EUROCRYPT 2007. LNCS, vol. 4515, pp. 115–128. Springer, Heidelberg (2007)

29. Kilian, J.: Founding cryptography on oblivious transfer. In: STOC, pp. 20–31 (1988)
30. Kilian, J.: Founding crytpography on oblivious transfer. In: Proceedings of the Twentieth Annual ACM Symposium on Theory of Computing, pp. 20–31. ACM (1988)
31. Kolesnikov, V.: Gate evaluation secret sharing and secure one-round two-party computation. In: Roy, B. (ed.) ASIACRYPT 2005. LNCS, vol. 3788, pp. 136–155. Springer, Heidelberg (2005)
32. Lindell, Y.: General composition and universal composability in secure multi-party computation (2003)
33. Lindell, Y.: Lower bounds for concurrent self composition. In: Naor, M. (ed.) TCC 2004. LNCS, vol. 2951, pp. 203–222. Springer, Heidelberg (2004)
34. Mahmoody, M., Xiao, D.: Languages with efficient zero-knowledge pcps are in szk. In: Sahai, A. (ed.) TCC 2013. LNCS, vol. 7785, pp. 297–314. Springer, Heidelberg (2013)
35. Maji, H.K., Prabhakaran, M., Rosulek, M.: Complexity of multi-party computation problems: The case of 2-party symmetric secure function evaluation. In: Reingold, O. (ed.) TCC 2009. LNCS, vol. 5444, pp. 256–273. Springer, Heidelberg (2009)
36. Prabhakaran, V., Prabhakaran, M.: Assisted common information. In: 2010 IEEE International Symposium on Information Theory Proceedings (ISIT), pp. 2602–2606 (2010)
37. Prabhakaran, V., Prabhakaran, M.: Assisted common information: Further results. In: 2011 IEEE International Symposium on Information Theory Proceedings (ISIT), pp. 2861–2865 (2011)
38. Rabin, M.O.: How to exchange secrets with oblivious transfer. IACR Cryptology ePrint Archive, 2005:187 (2005)
39. Winkler, S., Wullschleger, J.: On the efficiency of classical and quantum oblivious transfer reductions. In: Rabin, T. (ed.) CRYPTO 2010. LNCS, vol. 6223, pp. 707–723. Springer, Heidelberg (2010)
40. Wolf, S., Wullschleger, J.: New monotones and lower bounds in unconditional two-party computation. IEEE Transactions on Information Theory 54(6), 2792–2797 (2008)
41. Yao, A.C.-C.: Protocols for secure computations (extended abstract). In: FOCS, pp. 160–164 (1982)

Unified, Minimal and Selectively Randomizable Structure-Preserving Signatures

Masayuki Abe[1], Jens Groth[2,*], Miyako Ohkubo[3], and Mehdi Tibouchi[1]

[1] Secure Platform Laboratories, NTT Corporation, Japan
{abe.masayuki,tibouchi.mehdi}@lab.ntt.co.jp
[2] University College London, UK
j.groth@ucl.ac.uk
[3] Security Architecture Lab, NSRI, NICT, Japan
m.ohkubo@nict.go.jp

Abstract. We construct a structure-preserving signature scheme that is selectively randomizable and works in all types of bilinear groups. We give matching lower bounds showing that our structure-preserving signature scheme is optimal with respect to both signature size and public verification key size.

State of the art structure-preserving signatures in the asymmetric setting consist of 3 group elements, which is known to be optimal. Our construction preserves the signature size of 3 group elements and also at the same time minimizes the verification key size to 1 group element.

Depending on the application, it is sometimes desirable to have strong unforgeability and in other situations desirable to have randomizable signatures. To get the best of both worlds, we introduce the notion of selective randomizability where the signer may for specific signatures provide randomization tokens that enable randomization.

Our structure-preserving signature scheme unifies the different pairing-based settings since it can be instantiated in both symmetric and asymmetric groups. Since previously optimal structure-preserving signatures had only been constructed in asymmetric bilinear groups this closes an important gap in our knowledge. Having a unified signature scheme that works in all types of bilinear groups is not just conceptually nice but also gives a hedge against future cryptanalytic attacks. An instantiation of our signature scheme in an asymmetric bilinear group may remain secure even if cryptanalysts later discover an efficiently computable homomorphism between the source groups.

Keywords: Structure-preserving signatures, automorphic signatures, selective randomizability.

* The research leading to these results has received funding from the Engineering and Physical Sciences Research Council grant EP/G013829/1 and the European Research Council under the European Union's Seventh Framework Programme (FP/2007-2013) / ERC Grant Agreement n. 307937.

Y. Lindell (Ed.): TCC 2014, LNCS 8349, pp. 688–712, 2014.

1   Introduction

Structure-preserving signatures [3] (SPS) are signatures defined over groups with a bilinear pairing where messages, signatures and public verification keys all consist of group elements and the verification algorithm evaluates verification equations consisting of products of pairings of these group elements. Based on such signatures, one can easily design modular cryptographic protocols with reasonable efficiency, in particular in combination with non-interactive zero-knowledge (NIZK) proofs of knowledge about group elements [21]. Numerous applications of SPS, including blind signatures [3,17], group signatures [3,17,25], homomorphic signatures [24,9], delegatable anonymous credentials [16], compact verifiable shuffles [14], network encoding [8], oblivious transfer [19,12], tightly secure encryption [22,2], anonymous e-cash [26], etc., have been presented in the literature.

1.1   Symmetric and Asymmetric Bilinear Pairings

Bilinear pairing groups are usually instantiated as groups of points of certain restricted families of elliptic curves (or more rarely, other abelian varieties), and can be broadly classified into several types [18] according to the efficient morphisms that exist between the cyclic groups of prime order $\mathbb{G}_1$, $\mathbb{G}_2$ associated with the bilinear pairing $e \colon \mathbb{G}_1 \times \mathbb{G}_2 \to \mathbb{G}_T$. The two most important ones are Type I pairings, where $\mathbb{G}_1 = \mathbb{G}_2$, and Type III pairings defined as the ones that do not have an efficiently computable isomorphism between $\mathbb{G}_1$ and $\mathbb{G}_2$ in either direction. Type II parings, like Type III pairings, have $\mathbb{G}_1 \neq \mathbb{G}_2$ but with an efficiently computable isomorphism from one group to the other. We will also refer to Type I pairings as *symmetric* bilinear groups because $\mathbb{G}_1 = \mathbb{G}_2$ and refer to other types types of pairings where $\mathbb{G}_1 \neq \mathbb{G}_2$ as *asymmetric* bilinear groups.

Type I, or symmetric, pairings are obtained from supersingular curves, and have traditionally had an efficiency edge in implementations on resource-constrained devices, although recent advances on the discrete logarithm problem over finite fields of small characteristic [10] call this into question (large characteristic Type I pairings remain secure, but they are not as efficient). Pairing-based protocol designers often present their schemes in the symmetric setting, as protocol descriptions and security arguments tend to be simpler.

Type III pairings, which are the more efficient kind of asymmetric pairings, are obtained from special families of ordinary curves, and tend to be more compact, faster at least in software, and support stronger and more compact hardness assumptions such as the DDH assumption in their source groups. Thus, Type III pairings are often preferred for practical purposes. However, certain protocol descriptions given in the symmetric setting do not easily translate to the Type III setting.

1.2   Unified Structure-Preserving Signatures

Since SPS are a relatively low-level building block, their efficiency is of crucial importance. That efficiency is usually measured in terms of the number of group elements in signatures and the number of verification equations, and a significant amount of research has been devoted to obtaining lower and upper bounds with respect to these measures. Abe et al. [4] gave a construction of an SPS with 3 group element signatures and 2 verification equations using Type III bilinear groups. They also gave a matching lower bound in Type III bilinear groups of 3 group elements for signatures and 2 verification equations, which showed that the construction is optimal with respect to signature size and verification complexity.

In contrast to the work on Type III pairings, very little is known about SPS over symmetric bilinear groups. The best known construction for Type I pairings has signatures with 7 group elements, and no non-trivial lower bounds or more efficient constructions have been proposed to date. One could hope that symmetric functions such as $e(X, X)$ that are only possible in Type I pairings would make more efficient designs possible. Besides, it seems plausible that having only one group in the symmetric case admits lower complexity than separately handling two groups as must be done in the asymmetric case. On the other hand, the ability to use elements as the input in either side of the pairing may give the adversary additional flexibility and cause additional vulnerabilities. So it is not *a priori* clear whether symmetric pairings are advantageous for designers or for attackers.

We answer this question in a strong sense by providing a unified structure-preserving signature scheme that works in all types of bilinear groups. The design of the scheme does not exploit any symmetry or maps between the source groups and can therefore be instantiated in any type of bilinear group. At the same time though, it is resistant to adversaries that are allowed to exploit symmetry. Our signature scheme has 3 group element signatures and 2 verification equations and is therefore optimal with respect to Type III pairings. We will also show similar lower bounds hold for Type I pairings and the scheme is therefore also optimal in the symmetric setting.

Designing unified structure-preserving schemes that can be used in either type of bilinear group is of course conceptually appealing since it is simpler than having separate schemes for each setting. Unified signature may also be more resistant to cryptanalysis. Currently Type III pairings are the most efficient but building cryptographic schemes in this setting may leave us vulnerable if cryptanalysts find an efficiently computable homomorphism between $\mathbb{G}_1$ and $\mathbb{G}_2$. However, if we use a unified structure-preserving signature we can even resist attacks where the adversary has an isomorphism between $\mathbb{G}_1$ and $\mathbb{G}_2$ that is efficiently computable in both directions. It is a fascinating question whether there are other cryptographic tasks for which we can construct unified structure-preserving schemes without sacrificing efficiency.

1.3 Minimal Verification Keys

An important efficiency measure that has not received much attention in the literature on structure-preserving signatures is the size of the public verification key. For applications that involve certification chains the public key size is of high importance. If the size of the public key exceeds the size of the messages the signature scheme can handle, it becomes difficult and cumbersome to build certification chains and in the world of structure-preserving signatures it is not possible to use collision-resistant hash-functions to reduce the size of the messages since such hash-functions destroy the structure we are trying to preserve.

Abe et al. [4] considered only the size of the signatures and the number of verification equations but did not try to minimize the size of the public key. To sign a single group element they have a structure-preserving signature scheme with strong existential unforgeability where the public key consists of 3 group elements and a randomizable signature scheme where the public key consists of 2 group elements. This means that their schemes cannot easily be used to sign public keys. There are generic methods to extend the message space of SPS [7] but they incur a significant overhead so it is preferable to have an atomic scheme that can be used to sign verification keys.

Fuchsbauer [15,3] defined an automorphic signature scheme as a structure-preserving signature scheme where the verification keys belong to the message space itself. This makes certification chains easy and cheap to construct and indeed automorphic signatures have been used in the construction of anonymous delegatable credentials [15,16]. Current automorphic signatures, however, are more expensive than the most efficient structure-preserving signatures. The scheme in [3] has verification keys that consist of 2 group elements, the signatures consist of 5 group elements and the scheme uses 3 verification equations.

In contrast to these works, we also minimize the size of the verification key. As in the first construction of automorphic signatures [15,3], we allow the setup to include some random group elements in the public parameters describing the bilinear group to help shortening the verification key. In our case, we assume that a bilinear group has been generated and a random group element X is included in these parameters. With this type of setup, it is possible to get a public verification key that consists of just one single group element.

If the signer runs the setup algorithm, she is ensured that the setup is correct. However, even if the signer uses a pre-existing setup it is a moderate trust assumption since we do not need to trust anybody to store any secret trapdoors associated with the setup; it is for instance not necessary for the signer to know the discrete logarithm of X in our scheme. If the setup is generated by a trusted third party, we therefore only need to assume the trusted third party is honest at the particular time it is generating the setup without storing a secret trapdoor at that point in time. Alternatively, we may sample the setup in an oblivious manner from a trusted source of random bits such as a multi-party coin-flipping protocol or extract it from a physical source of randomness, e.g., solar activity in a given time interval.

With a single group element as verification key, it becomes easy to build certification chains. In the symmetric setting, we get an automorphic signature scheme where the verification key space is identical to the message space. In Type III pairings, our construction is not automorphic because the message and the verification key belong to different groups. However, it is easy to create a certification ladder where we use a verification key in $\mathbb{G}_2$ to certify a verification key in $\mathbb{G}_1$, which can then be used to certify a verification key in $\mathbb{G}_2$, etc.

1.4   Selective Randomizability

We introduce a new feature called *selective randomizability* that allows a strongly unforgeable signature to be randomized with the help of a randomization token. Selective randomizability reconciles the notions of strong unforgeability, where it is impossible to create new signatures on signed messages, and randomizability, where it is possible to randomize signatures. Depending on the application, different parties may hold randomization tokens corresponding to certain signatures and they may randomize the signatures, while other parties cannot randomize the signatures.

Randomizability is useful in reducing the size of the proofs when the SPS is combined with the Groth-Sahai proof system since a part of a randomized signature can be shown in the clear. There are other applications and theoretical results on (not selectively) randomizable signatures in the literature, e.g. [27,23]. Selective randomizability may also have uses on its own; a selectively randomizable signature can for instance be used as a service token. Fee paying users get a signature on the time period they have paid for and a randomization token and can in each use reveal a fresh randomized signature. Fraudsters on the other hand do not know the randomization tokens and cannot modify the signatures and can therefore only copy previous signatures.

We show that our structure-preserving signature scheme is selectively randomizable. Our randomization tokens consist of a single group element, so also here we achieve minimal size.

1.5   Related Work

Abe et al. [3] first used the term structure-preserving signatures but there are earlier works in the area. Groth [20] proposed the first structure-preserving signature but the construction involves hundreds of group elements and is not practical. Green and Hohenberger [19] gave a structure-preserving signature scheme, which is secure against random message attack, but is not known to be secure against adaptive chosen message attack. Cathalo, Libert and Yung [13] constructed a signature scheme that structure-preserving in a relaxed sense that permits the verification key to include target group elements.

Abe et al. [4] showed that structure-preserving signatures in Type III bilinear groups require at least 3 group elements and 2 verification equations. They also gave structure-preserving signatures matching those bounds that are secure in the generic group model. Abe et al. [5] later showed 3 element signatures cannot be proven secure under a non-interactive assumption using black-box reductions, so strong assumptions are needed to get optimal efficiency.

Hofheinz and Jager [22] and Abe et al. [1,2] investigated the possibility of basing structure-preserving signatures on standard assumptions. They give structure-preserving signatures based on the decision linear (DLIN) assumption. The use of a nice security assumption, however, comes at the price of reducing efficiency.

Table 1. Comparison of structure-preserving signatures on a single group element. *1: Strongly unforgeable. *2: Randomizable. *3: Automorphic. *4: Selectively Randomizable.

Scheme	Signature	Ver. key	Equations	Type	Assumption	Notes
[20]	Many	Many	Many	I	DLIN	
[13]	11	11	9	I	HSDH, FlexDH, S2D	*1
[3]	7	13	2	Any	SFP	*2
[15,3]	5	2	3	Any	ADH-SDH, AWF-CDH	*3
[4]	4	4	2	III	Non-interactive	*1
[4]	3	3	2	III	Interactive	*1
[4]	3	2	2	III	Interactive	*2
[1]	17	27	9	I	DLIN	
[1]	11	21	5	III	SXDH, XDLIN	
[2]	14	22	7	I	DLIN	
Ours	3	1	2	Any	Interactive	*4

1.6 Our Contributions

We construct a selectively randomizable structure-preserving signature scheme with message space $\mathcal{M} = \mathbb{G}_1$, where a verification key is 1 group element, a signature is 3 group elements and the verifier uses 2 verification equations to verify the signature. The setup for the signature scheme consists of the description of a bilinear group and a single random group element. Our signature scheme is unified, i.e., it can be used in both symmetric and asymmetric bilinear groups.

We prove our signature scheme secure in the generic group model. The security of the signature scheme can therefore be viewed as an interactive security assumption. However, as shown by Abe et al. [7] it is impossible to base the security of structure-preserving signature schemes with 3 group element signatures on non-interactive intractability assumptions using black-box reductions, so at least in the Type III setting we could not hope to base security on a non-interactive assumption. On the positive side, being unified provides a hedge against cryptanalytic attacks. Even if cryptanalysts uncover efficiently computable homomorphisms between $\mathbb{G}_1$ and $\mathbb{G}_2$, our structure-preserving signature scheme may remain secure.

Table 1 compares our results to previous work on structure-preserving signatures in the symmetric and asymmetric settings. We only consider the case of a single group element and in the table we therefore compare all schemes on the same terms, i.e., the cost for signing a single group element, with the exception of Fuchsbauer's automorphic signature scheme, which is tailored to sign Diffie-Hellman pairs of group elements.

To complement our signature scheme, we provide the first analysis of lower bounds in the symmetric setting. We demonstrate that in the symmetric setting a signature must be at least 3 group elements and the verifier must use at least 2 verification equations. This matches the Type III setting previously analyzed in [4] and shows that our signature scheme is optimal also in symmetric bilinear groups.

Interestingly it turns out that in the case of one-time signatures there is actually a difference between Type I and Type III pairings. While it is known that Type III pairings admit one-time signatures with a single verification equation, we show this is not the case for Type I pairings. The lower bound of 2 verification equations also applies to one-time signatures.

The lower bound of 3 group elements for the size of signatures does not apply though. We demonstrate this by constructing a one-time signature scheme in the symmetric setting with 2 group element signatures. We also analyze one-time signatures with respect to the size of the verification key. We show that both Type I and Type III pairings have structure-preserving one-time signature schemes with 1 group element verification keys.

2   Preliminaries

2.1   Bilinear Groups

Let $\mathcal{G}$ be a bilinear group generator that returns $(p, \mathbb{G}_1, \mathbb{G}_2, \mathbb{G}_T, e, G, H) \leftarrow \mathcal{G}(1^k)$ given security parameter k with the following properties:

- $\mathbb{G}_1, \mathbb{G}_2, \mathbb{G}_T$ are groups of prime order p
- $e : \mathbb{G}_1 \times \mathbb{G}_2 \to \mathbb{G}_T$ is a bilinear map s.t. $e(G^a, H^b) = e(G, H)^{ab}$ for all $a, b \in \mathbb{Z}$
- G generates $\mathbb{G}_1$, H generates $\mathbb{G}_2$, and $e(G, H)$ generates $\mathbb{G}_T$
- There are efficient algorithms for computing group operations, evaluating the bilinear map, comparing group elements and deciding membership of the groups

Bilinear groups can be classified in the three types according to the efficient morphisms that exist between the source groups $\mathbb{G}_1$ and $\mathbb{G}_2$. Type I pairings have $\mathbb{G}_1 = \mathbb{G}_2$ and $G = H$. Type II pairings have an efficiently computable isomorphism from one source group to the other but none in the reverse direction. Type III pairings have no efficiently computable isomorphism from either source group to the other.

2.2   Generic Algorithms

In a bilinear group $(p, \mathbb{G}_1, \mathbb{G}_2, \mathbb{G}_T, e, G, H)$ generated by $\mathcal{G}$ we refer to deciding group membership, computing group operations in $\mathbb{G}_1$, $\mathbb{G}_2$ or $\mathbb{G}_T$, comparing group elements and evaluating the bilinear map as the generic group operations. The signature schemes we construct only use generic group operations.

As a matter of notation, we will use capital letters $G, H, M, R, S, T, U, V, W$ for group elements in $\mathbb{G}_1$ and $\mathbb{G}_2$. We will use small letters $1, m, r, s, t, u, v, w$ for the corresponding discrete logarithms of group elements with respect to base G or H.

2.3 Setup

Our signature schemes work over a bilinear group generated by $\mathcal{G}$. This group may be generated by the signer and included in the public verification key. In many cryptographic schemes it is convenient for the signer to work on top of a pre-existing bilinear group though. We will therefore in the description of our signatures explicitly distinguish between a setup algorithm $\mathcal{P}$ that produces a public parameter PP and a key generation algorithm the signer uses to generate her own keys.

The setup algorithms we use generate a bilinear group $(p, \mathbb{G}_1, \mathbb{G}_2, \mathbb{G}_T, e, G, H)$ $\leftarrow \mathcal{G}(1^k)$. They may then extend the description of the bilinear group with additional random group elements. As discussed in Sect. 1.3 this is a moderate setup assumption since the signer does not need to know the discrete logarithms of the random group elements. The group elements may therefore be sampled obliviously without learning the discrete logarithms or the discrete logarithms may be erased immediately upon generation.

2.4 Secure Signature Schemes

A digital signature scheme (with setup algorithm $\mathcal{P}$) is a quadruple of efficient algorithms $(\mathcal{P}, \mathcal{K}, \mathcal{S}, \mathcal{V})$. The setup algorithm $\mathcal{P}$ takes the security parameter and outputs a public parameter PP. The key generation algorithm $\mathcal{K}$ takes PP as input and returns a public verification key VK and a secret signing key SK. We will always assume that VK includes PP and that SK includes VK. The signing algorithm $\mathcal{S}$ takes a signing key SK and a message M in the message space $\mathcal{M}$ defined by PP and VK as input and returns a signature Σ. The verification algorithm $\mathcal{V}$ takes the verification key VK, a message M and the signature Σ and returns either 1 (accept) or 0 (reject).

Definition 1 (Correctness). *We say the signature scheme $(\mathcal{P}, \mathcal{K}, \mathcal{S}, \mathcal{V})$ is (perfectly) correct if for all security parameters $k \in \mathbb{N}$*

$$\Pr \left[\begin{array}{l} PP \leftarrow \mathcal{P}(1^k) \\ (VK, SK) \leftarrow \mathcal{K}(PP) \\ M \leftarrow \mathcal{M} \\ \Sigma \leftarrow \mathcal{S}_{SK}(M) \end{array} : \mathcal{V}_{VK}(M, \Sigma) = 1 \right] = 1.$$

A signature scheme is said to be existentially unforgeable if it is hard to forge a signature on a new message that has not been signed before. The adversary may see signatures on other messages before making the forgery. We distinguish between a random message attack (RMA), where the adversary gets pairs of random messages and corresponding signatures, and an adaptive chosen message attack (CMA) where the adversary can choose arbitrary messages and receive signatures on them. Our signatures will be existentially unforgeable against the strong adaptive chosen message attack, but our lower bounds on the complexity of signature schemes will hold even for the weaker random message attacks.

We now formally define existential unforgeability under adaptive chosen message attack.

Definition 2 (EUF-CMA). *A signature scheme* $(\mathcal{P}, \mathcal{K}, \mathcal{S}, \mathcal{V})$ *is existentially unforgeable under adaptive chosen message attack if for all non-uniform polynomial time* $\mathcal{A}$

$$\Pr\left[\begin{array}{l} PP \leftarrow \mathcal{P}(1^k) \\ (VK, SK) \leftarrow \mathcal{K}(PP) \\ (M, \Sigma) \leftarrow \mathcal{A}^{\mathcal{S}_{SK}(\cdot)}(VK) \end{array} : M \notin Q \ \wedge \ \mathcal{V}_{VK}(M, \Sigma) = 1 \right] = \mathrm{negl}(k),$$

where Q *is the set of queries made by* $\mathcal{A}$ *to the signing oracle.*

Sometimes it is also useful to prevent the adversary from issuing a new signature for a message that has already been signed. A signature scheme is strongly existentially unforgeable if it is hard to find a signature on a message that has not been signed before and also hard to find a new signature for a message that has already been signed. This notion, denoted by sEUF-CMA, is formally captured in the same way as the definition of EUF-CMA except for additionally requiring $(M, \Sigma) \notin Q$ where Q is the set of message-signature pairs from $\mathcal{A}$'s queries to the signing oracle.

We get the definition for existential unforgeability against random message attack (EUF-RMA) by modifying the signing oracle to picking $M \leftarrow \mathcal{M}$ at random, computing $\Sigma \leftarrow \mathcal{S}_{SK}(M)$ and returning (M, Σ) to the adversary whenever the signing oracle is queried.

Corresponding security notions for one-time signature schemes can be obtained by restricting the adversary to only calling the signing oracle once in the above definitions.

2.5   Selectively Randomizable Signatures

Some applications require signatures to be strongly unforgeable, while in other applications it is desirable that a signature on a message can be randomized into a new random signature on the same message. A randomizable signature scheme can only be EUF-CMA secure though since a randomized signature would violate sEUF-CMA security. In order to reconcile the two notions and get the best of both worlds, we define the notion of selective randomizability where the signer can select to make specific signatures randomizable by providing randomization tokens for them.

In a selectively randomizable signature scheme the signing algorithm returns both a signature and a randomization token. Furthermore, there is a randomization algorithm $\mathcal{R}$ that given a message, signature and randomization token returns a random signature on the message. We require that the randomization algorithm $\mathcal{R}$ given a message M, signature Σ and corresponding randomization token W computes a signature $\Sigma' \leftarrow \mathcal{R}_{VK}(M, \Sigma, W)$ such that for all correctly generated inputs $R_{VK}(M, \Sigma, W)$ and $\mathcal{S}_{SK}(M)$ have identical probability distributions.

Since the signatures are randomizable it is not possible to have *strong* existential unforgeability if the randomization tokens are given to the adversary. However, we can get strong existential unforgeability for signatures on messages for which the adversary does not have randomization tokens. Formally we define security against a chosen message and token attack (CMA-TA) as follows.

Definition 3 (sEUF-CMA-TA). *A selectively randomizable signature scheme* $(\mathcal{P}, \mathcal{K}, \mathcal{S}, \mathcal{R}, \mathcal{V})$ *is strongly existentially unforgeable under chosen message and token attack if for all non-uniform polynomial time* $\mathcal{A}$

$$\Pr \left[\begin{array}{ll} PP \leftarrow \mathcal{P}(1^k) & (M, \Sigma) \notin Q \\ (VK, SK) \leftarrow \mathcal{K}(PP) & : \quad M \notin Qt \\ (M, \Sigma) \leftarrow \mathcal{A}^{\mathcal{S}_{SK}(\cdot), \mathcal{S}t_{SK}(\cdot)}(VK) & \mathcal{V}_{VK}(M, \Sigma) = 1 \end{array} \right] = \mathrm{negl}(k),$$

where $\mathcal{S}$ *is a signing oracle that is given a message and returns a signature on the message,* $\mathcal{S}t$ *is a token-signing oracle that is given a message and returns both a signature and a randomization token,* Q *is the set of messages and signatures observed by the signing oracle, and* Qt *is the set of messages observed by the token-signing oracle.*

Please observe that $\mathcal{A}$ can send M to $\mathcal{S}$ and $\mathcal{S}t$ an arbitrary number of times to get (random) signatures on M so we do not need to provide the adversary with a randomization oracle in the security definition.

2.6 Structure-Preserving Signature Schemes

We study structure-preserving signature schemes [3] on bilinear groups generated by group generator $\mathcal{G}$. In a structure preserving signature scheme the verification key, the messages and the signatures consist only of group elements from $\mathbb{G}_1$ and $\mathbb{G}_2$ and the verification algorithm evaluates the signature by deciding group membership of elements in the signature and by evaluating pairing product equations, which are equations of the form

$$\prod_i \prod_j e(X_i, Y_j)^{a_{ij}} = 1,$$

where $X_1, X_2, \ldots \in \mathbb{G}_1$, $Y_1, Y_2, \ldots \in \mathbb{G}_2$ are group elements appearing in PP, VK, M and Σ and $a_{11}, a_{12}, \ldots \in \mathbb{Z}_p$ are constants stored in PP. Structure-preserving signatures are extremely versatile because they mix well with other pairing-based protocols. Groth-Sahai proofs [21] are for instance designed with pairing product equations in mind and can therefore easily be applied to structure-preserving signatures.

Definition 4 (Structure-preserving signatures). *A digital signature scheme* $(\mathcal{P}, \mathcal{K}, \mathcal{S}, \mathcal{V})$ *is said to be structure preserving over bilinear group generator* $\mathcal{G}$ *if*

- *PP includes a bilinear group* $(p, \mathbb{G}_1, \mathbb{G}_2, \mathbb{G}_T, e, G, H)$ *generated by* $\mathcal{G}$*, and constants in* $\mathbb{Z}_p$*,*
- *the verification key consists of PP and group elements in* $\mathbb{G}_1$ *and* $\mathbb{G}_2$*,*
- *the messages consist of group elements in* $\mathbb{G}_1$ *and* $\mathbb{G}_2$*,*
- *the signatures consist of group elements in* $\mathbb{G}_1$ *and* $\mathbb{G}_2$*, and*
- *the verification algorithm only needs to decide membership in* $\mathbb{G}_1$ *and* $\mathbb{G}_2$ *and evaluate pairing product equations.*

When proving our lower bounds, we will relax the above definition to allow arbitrary target group elements $Z \in \mathbb{G}_T$ to be included in the verification key and to appear in the verification equations. This gives the strongest possible results: our lower bounds hold in a relaxed model of structure-preserving signatures and our constructions of signatures satisfy the strict model of structure-preserving signatures.

Generic signer. Abe et al. [3] did not explicitly require the signing algorithm to only use generic group operations when they defined structure-preserving signatures. However, all existing structure-preserving signatures in the literature have generic signing algorithms and we believe it would be a surprising result in itself to construct a structure-preserving signature with a non-generic signer. Our constructions have generic signer algorithms and some of our lower bounds will assume the signer is generic.

3   Selectively Randomizable Structure-Preserving Signatures

Fig. 1 gives a selectively randomizable structure-preserving signature scheme with 1 element verification keys, 3 group element signatures and 2 verification equations. The signature scheme is sEUF-CMA-TA secure. The lower bounds in [4] and Sect. 5 show that this construction is optimal with respect to size and verification complexity in both Type I and Type III bilinear groups.

Setup $\mathcal{P}(1^k)$: Run $(p, \mathbb{G}_1, \mathbb{G}_2, \mathbb{G}_T, e, G, H) \leftarrow \mathcal{G}(1^k)$, pick $X \leftarrow \mathbb{G}_1$, and return $PP = (p, \mathbb{G}_1, \mathbb{G}_2, \mathbb{G}_T, e, G, X, H)$.

Key generation $\mathcal{K}(PP)$: Choose $v \leftarrow \mathbb{Z}_p$, compute $V \leftarrow H^v$, and return $VK = (PP, V)$ and $SK = (PP, v)$.

Signing $\mathcal{S}_{SK}(M)$: On $M \in \mathbb{G}_1$ choose $r \leftarrow \mathbb{Z}_p^*$ and compute signature $\Sigma = (R, S, T)$ and randomization token W as:

$$R \leftarrow H^r, \qquad S \leftarrow M^{\frac{v}{r}} X^{\frac{1}{r}}, \qquad T \leftarrow S^{\frac{v}{r}} G^{\frac{1}{r}}, \qquad W \leftarrow G^{\frac{1}{r}}.$$

Randomization $\mathcal{R}_{VK}(M, (R, S, T), W)$: Pick $\alpha \leftarrow \mathbb{Z}_p^*$ and compute the randomized signature $\Sigma' = (R', S', T')$ given by:

$$R' \leftarrow R^{\frac{1}{\alpha}}, \qquad S' \leftarrow S^{\alpha}, \qquad T' \leftarrow T^{\alpha^2} W^{\alpha(1-\alpha)}.$$

Verification $\mathcal{V}_{VK}(M, (R, S, T))$: Accept if and only if $M, S, T \in \mathbb{G}_1$, $R \in \mathbb{G}_2$ and

$$e(S, R) = e(M, V)e(X, H) \quad \text{and} \quad e(T, R) = e(S, V)e(G, H).$$

Fig. 1. Minimal structure-preserving signature scheme

Randomized signatures are perfectly indistinguishable from real signatures since both types of signatures are uniquely determined by the uniformly random non-trivial group element R. Somebody who has a signature on a particular message and a corresponding randomization token can create as many uniformly random signatures on the message as she wants. An additional feature is that the randomization token can also be randomized together with the signature by computing $W' \leftarrow W^\alpha$, so the power to randomize can be delegated to others.

The signature scheme is designed with Groth-Sahai proofs in mind. If we have a secret randomization token and use it to randomize a signature, we may reveal the random group element R without this leaking any information about the message or the original signature from which the randomized signature was derived. When R is public both verification equations become linear, which makes Groth-Sahai proofs very efficient.

We will now prove that the signature scheme with selective randomization is sEUF-CMA-TA secure. This implies as two special cases that the signature in Fig. 1 is EUF-CMA secure even when all randomization tokens are revealed and sEUF-CMA secure if no randomization tokens are revealed.

Theorem 1. *The signature scheme in Fig. 1 is* sEUF-CMA-TA *secure in the generic group model.*

Proof. We will without loss of generality show that the signature scheme is sEUF-CMA-TA secure in the symmetric setting where $\mathbb{G}_1 = \mathbb{G}_2$ since this setting gives the adversary the most degrees of freedom and hence the best chance of breaking the scheme. Moreover, the scheme is secure in the generic group model even if the discrete logarithm $\log_G(H)$ is known to the adversary and we will therefore without loss of generality assume $H = G$.

A generic adversary only uses generic group operations. This means that in $\mathbb{G}_1$ it can only compute linear combinations of group elements from the verification key and the signatures it has seen. Linear combinations on verification key elements and signature elements correspond to formal Laurent polynomials (of degree ranging from $-2q$ to $2q + 1$ after q signature queries) in the discrete logarithms of the group elements. We will show that no linear combinations produce formal Laurent polynomials corresponding to a forgery. By the master theorem in [11] this means that the signature scheme is secure in the generic group model.

The group elements in VK are G, X, V with corresponding discrete logarithms $1, x, v$. On a query M_i with discrete logarithm m_i from the adversary, the signature oracle responds with a signature (R_i, S_i, T_i) and possibly a rerandomization token W_i with discrete logarithms

$$r_i \leftarrow \mathbb{Z}_p^* \qquad s_i = \frac{m_i v}{r_i} + \frac{x}{r_i} \qquad t_i = \frac{m_i v^2}{r_i^2} + \frac{xv}{r_i^2} + \frac{1}{r_i} \qquad w_i = \frac{1}{r_i}.$$

Suppose the adversary after q queries constructs (M, R, S, T). Since the adversary is generic it can only construct m, r, s, t that are linear combinations of $1, x, v, r_1, s_1, t_1, w_1, \ldots, r_q, s_q, t_q, w_q$, i.e.,

$$m = \mu + \mu_x x + \mu_v v + \sum_{i=1}^{q} \left[\mu_{r_i} r_i + \mu_{s_i} \left(\frac{m_i v}{r_i} + \frac{x}{r_i} \right) + \mu_{t_i} \left(\frac{m_i v^2}{r_i^2} + \frac{xv}{r_i^2} + \frac{1}{r_i} \right) + \mu_{w_i} \frac{1}{r_i} \right]$$

$$r = \rho + \rho_x x + \rho_v v + \sum_{i=1}^{q} \left[\rho_{r_i} r_i + \rho_{s_i} \left(\frac{m_i v}{r_i} + \frac{x}{r_i} \right) + \rho_{t_i} \left(\frac{m_i v^2}{r_i^2} + \frac{xv}{r_i^2} + \frac{1}{r_i} \right) + \rho_{w_i} \frac{1}{r_i} \right]$$

$$s = \sigma + \sigma_x x + \sigma_v v + \sum_{i=1}^{q} \left[\sigma_{r_i} r_i + \sigma_{s_i} \left(\frac{m_i v}{r_i} + \frac{x}{r_i} \right) + \sigma_{t_i} \left(\frac{m_i v^2}{r_i^2} + \frac{xv}{r_i^2} + \frac{1}{r_i} \right) + \sigma_{w_i} \frac{1}{r_i} \right]$$

$$t = \tau + \tau_x x + \tau_v v + \sum_{i=1}^{q} \left[\tau_{r_i} r_i + \tau_{s_i} \left(\frac{m_i v}{r_i} + \frac{x}{r_i} \right) + \tau_{t_i} \left(\frac{m_i v^2}{r_i^2} + \frac{xv}{r_i^2} + \frac{1}{r_i} \right) + \tau_{w_i} \frac{1}{r_i} \right]$$

Similarly, each query m_i is a linear combination of $1, x, v, r_1, s_1, t_1, w_1, \ldots, r_{i-1}, s_{i-1}, t_{i-1}, w_{i-1}$.

We will show that the signature scheme is EUF-CMA secure, i.e., an adversary cannot construct a valid signature (R, S, T) on M where the discrete logarithms m, r, s, t satisfy the verification equations

$$sr = mv + x \qquad tr = sv + 1 = mv^2 + xv + 1$$

unless it reuses $M = M_j$ from a previous query.

If a randomization token has not been given for a particular message the attacker must use $\tau_{w_i} = 0$ for all indices i where this message was queried. We will show that the adversary can only randomize a signature by using some $\tau_{w_j} \neq 0$. This means the signature scheme is strong for those messages where no randomization token has been given, which gives us sEUF-CMA-TA security.

Our proof strategy is to use the first verification equation $sr = mv + x$ to simplify the descriptions of s and r by demonstrating that many of the coefficients σ_* and ρ_* are 0. After narrowing the solution space down to four distinct cases, we use the second verification equation $tr = sv + 1$ to rule out three cases and determine a single type of possible solutions. These solutions correspond exactly to randomization of signatures and if no randomization token is given then the solution must be an exact copy of a previous signature.

In order to get to the core of our proof, we delay the proof of the following claim.

Claim. The first verification equation $sr = mv + x$ can only be satisfied if the adversary picks $\sigma_{t_i} = 0, \sigma_{w_i} = 0, \rho_{t_i} = 0$ and $\rho_{w_i} = 0$ for all $i = 1, \ldots, q$.

We now know that $\sigma_{t_i}, \sigma_{w_i}, \rho_{t_i}$ and ρ_{w_i} are zero for all $i = 1, \ldots, q$. Obviously we cannot rule out the existence of some j for which $\sigma_{s_j} \neq 0$ since the adversary could simply copy a previous signature $s = s_j$ by setting $\sigma_{s_j} = 1$. We will now analyze the structure of s and r when there exists a j such that $\sigma_{s_j} \neq 0$.

We can write $sr = mv + x$ as

$$\left(\sigma + \sigma_x x + \sigma_v v + \sum_{i=1}^{q} \sigma_{r_i} r_i + \sigma_{s_i}\left(\frac{m_i v}{r_i} + \frac{x}{r_i}\right)\right) \cdot \left(\rho + \rho_x x + \rho_v v + \sum_{i=1}^{q} \rho_{r_i} r_i + \rho_{s_i}\left(\frac{m_i v}{r_i} + \frac{x}{r_i}\right)\right)$$

$$= \left(\mu + \mu_x x + \mu_v v + \sum_{i=1}^{q} \mu_{r_i} r_i + \mu_{s_i}\left(\frac{m_i v}{r_i} + \frac{x}{r_i}\right) + \mu_{t_i}\left(\frac{m_i v^2}{r_i^2} + \frac{xv}{r_i^2} + \frac{1}{r_i}\right) + \mu_{w_i}\frac{1}{r_i}\right)v + x.$$

We first look at the term $\frac{x^2}{r_j^2}$. Observe that all verification key elements and signatures are linear in x and therefore all elements $m, r, s, t, m_1, \ldots, m_q$ constructed using generic group operations must also be linear in x. This shows that the term $\frac{x^2}{r_j^2}$ has coefficient 0 in $mv + x$.

Let us now determine the coefficient of $\frac{x^2}{r_j^2}$ in the product sr. Whenever the adversary makes a query m_i to get a signature (r_i, s_i, t_i) the message m_i is multiplied by v or v^2 by the signing oracle. It is not possible to decrease the degree of v, so these queries cannot contribute to the $\frac{x^2}{r_j^2}$ term. Looking at the terms in s and r we then see that the coefficient of $\frac{x^2}{r_j^2}$ in sr is $\sigma_{s_j}\rho_{s_j}$.

Comparing the coefficients of $\frac{x^2}{r_j^2}$ from the two sides of the verification equation we get $\sigma_{s_j}\rho_{r_j} = 0$. Since we assumed $\sigma_{s_j} \neq 0$ this implies $\rho_{s_j} = 0$. Using a similar analysis of the terms $\frac{x^2}{r_i r_j}$ give us $\sigma_{s_i}\rho_{s_j} + \sigma_{s_j}\rho_{s_i} = \sigma_{s_j}\rho_{s_i} = 0$ and therefore $\rho_{s_i} = 0$ for all i.

The term $\frac{x^2}{r_j}$ gives us $\sigma_{s_j}\rho_x = 0$ and therefore $\rho_x = 0$. The term $\frac{x}{r_j}$ gives us $\sigma_{s_j}\rho = 0$ and therefore $\rho = 0$. The terms $\frac{x r_i}{r_j}$ give us $\sigma_{s_j}\rho_{r_i} = 0$ and therefore $\rho_{r_i} = 0$ for all $i \neq j$. Finally, the term x gives us $\sigma_{s_j}\rho_{r_j} = 1$ and therefore $\rho_{r_j} = \frac{1}{\sigma_{s_j}}$.

We now have $r = \rho_v v + \rho_{r_j} r_j$ with $\rho_{r_j} = \frac{1}{\sigma_{s_j}} \neq 0$. Let us proceed to analyze the structure of s. The terms $\frac{x r_j}{r_i}$ give us $\sigma_{s_i}\rho_{r_j}$ and therefore $\sigma_{s_i} = 0$ for all $i \neq j$. The term r_j^2 gives us $\sigma_{r_j}\rho_{r_j} = 0$ and therefore $\sigma_{r_j} = 0$. The terms $r_i r_j$ give us $\sigma_{r_i}\rho_{r_j} + \sigma_{r_j}\rho_{r_i} = \sigma_{r_i}\rho_{r_j} = 0$ and therefore $\sigma_{r_i} = 0$ for all i. The term r_j gives us $\sigma\rho_{r_j} = 0$ and therefore $\sigma = 0$. The term $x r_j$ gives us $\sigma_x\rho_{r_j} = 0$ and therefore $\sigma_x = 0$. We conclude that $s = \sigma_v v + \sigma_{s_j}(\frac{m_j v}{r_j} + \frac{x}{r_j})$.

By symmetry we now have two possible cases:

Case	s	r
1: $\sigma_{s_j} \neq 0$	$s = \sigma_v v + \sigma_{s_j}(\frac{m_j v}{r_j} + \frac{x}{r_j})$	$r = \rho_v v + \frac{1}{\sigma_{s_j}} r_j$
2: $\rho_{s_j} \neq 0$	$s = \sigma_v v + \frac{1}{\rho_{s_j}} r_j$	$r = \rho_v v + \rho_{s_j}(\frac{m_j v}{r_j} + \frac{x}{r_j})$

There still remains the possibility that $\sigma_{s_i} = 0$ and $\rho_{s_i} = 0$ for all i. We can then write $sr = mv + x$ as

$$\left(\sigma + \sigma_x x + \sigma_v v + \sum_{i=1}^{q} \sigma_{r_i} r_i\right) \cdot \left(\rho + \rho_x x + \rho_v v + \sum_{i=1}^{q} \rho_{r_i} r_i\right)$$

$$= \left(\mu + \mu_x x + \mu_v v + \sum_{i=1}^{q} \mu_{r_i} r_i + \mu_{s_i}\left(\frac{m_i v}{r_i} + \frac{x}{r_i}\right) + \mu_{t_i}\left(\frac{m_i v^2}{r_i^2} + \frac{xv}{r_i^2} + \frac{1}{r_i}\right) + \mu_{w_i}\frac{1}{r_i}\right)v + x.$$

The term x^2 shows that $\rho_x \sigma_x = 0$, so they cannot both be non-zero. The term x on the other hand shows $\sigma \rho_x + \sigma_x \rho = 1$ so at least one of ρ_x or σ_x is non-zero. Let us in the following assume $\sigma_x \neq 0$ and therefore $\rho = \frac{1}{\sigma_x} \neq 0$. The constant term gives us $\sigma \rho = 0$ and therefore $\sigma = 0$. The terms r_i give us $\sigma_{r_i} = 0$ for all i and the terms xr_i give us $\rho_{r_i} = 0$ for all i. This means we have $s = \sigma_x x + \sigma_v v$ and $r = \frac{1}{\sigma_x} x + \rho_v v$. By symmetry we now have two additional cases

Case	s	r
3 : $\sigma_x \neq 0$	$s = \sigma_x x + \sigma_v v$	$r = \frac{1}{\sigma_x} + \rho_v v$
4 : $\rho_x \neq 0$	$s = \frac{1}{\rho_x} + \sigma_v v$	$r = \rho_x x + \rho_v v$

We will now analyze the four cases we have identified with the help of the second verification equation $tr = sv + 1$. In case 4 where $r = \rho_x x + \rho_v v$ we see that in tr all terms involve x or v. This means we do not have a constant term in either tr or sv, which makes it impossible to get $tr = sv + 1$.

A similar argument can be used in case 2 where $r = \rho_v v + \rho_{s_j}\left(\frac{m_j v}{r_j} + \frac{x}{r_j}\right)$, since both in tr and in sv all terms involve x or v and therefore it is impossible to get $tr = sv + 1$.

Let us now analyze case 3 where $r = \rho + \rho_v v$ and $s = \frac{1}{\rho} x + \sigma_v v$. We get

$$\left(\tau + \tau_x x + \tau_v v + \sum_{i=1}^{q} \tau_{r_i} r_i + \tau_{s_i}\left(\frac{m_i v}{r_i} + \frac{x}{r_i}\right) + \tau_{t_i}\left(\frac{m_i v^2}{r_i^2} + \frac{xv}{r_i^2} + \frac{1}{r_i}\right) + \tau_{w_i}\frac{1}{r_i}\right) \cdot (\rho + \rho_v v)$$

$$= \frac{1}{\rho} xv + \sigma_v v^2 + 1.$$

The constant term gives us $\tau \rho = 1$ and therefore $\rho = \frac{1}{\tau} \neq 0$. The term x gives us $\tau_x \rho = 0$ and therefore $\tau_x = 0$. But now the xv term yields a contradiction since it gives us $0 = \frac{1}{\rho} \neq 0$.

The only remaining possibility is case 1 where $r = \rho_v v + \rho_{r_j} r_j$ with $\rho_{r_j} = \frac{1}{\sigma_{s_j}}$ and $s = \sigma_v v + \sigma_{s_j}\left(\frac{m_j v}{r_j} + \frac{1}{r_j}\right)$. Inserting it in the second verification equation we get

$$\left(\tau + \tau_x x + \tau_v v + \sum_{i=1}^{q} \tau_{r_i} r_i + \tau_{s_i}\left(\frac{m_i v}{r_i} + \frac{x}{r_i}\right) + \tau_{t_i}\left(\frac{m_i v^2}{r_i^2} + \frac{xv}{r_i^2} + \frac{1}{r_i}\right) + \tau_{w_i}\frac{1}{r_i}\right) \cdot \left(\rho_v v + \rho_{r_j} r_j\right)$$

$$= \sigma_v v^2 + \sigma_{s_j}\left(\frac{m_j v^2}{r_j} + \frac{xv}{r_j}\right) + 1.$$

The terms $\frac{xvr_j}{r_i^2}$ give us $\tau_{t_i}\rho_{r_j} = 0$ and therefore $\tau_{t_i} = 0$ for $i \neq j$. The terms $\frac{1}{r_i r_j}$ give us $\tau_{w_i}\rho_{r_j} = 0$ and therefore $\tau_{w_i} = 0$ for $i \neq j$. The terms $\frac{xr_j}{r_i}$ give us $\tau_{s_i}\rho_{r_j} = 0$ and therefore $\tau_{s_i} = 0$ for all $i \neq j$. The term x gives us $\tau_{s_j}\rho_{r_j} = 0$ and therefore $\tau_{s_j} = 0$. The terms r_j^2 and $r_i r_j$ give us $\tau_{r_i} = 0$ for all i. The term xr_j gives us $\tau_x = 0$ and the term r_j gives us $\tau = 0$.

Since $\rho_{r_j} = \frac{1}{\sigma_{s_j}}$ we have now simplified the second verification equation to

$$\left(\tau_v v + \tau_{t_j}\left(\frac{m_j v^2}{r_j^2} + \frac{xv}{r_j^2} + \frac{1}{r_j}\right) + \tau_{w_j}\frac{1}{r_j}\right) \cdot \left(\rho_v v + \frac{1}{\sigma_{s_j}}r_j\right) = \sigma_v v^2 + \sigma_{s_j}\left(\frac{m_j v^2}{r_j} + \frac{xv}{r_j}\right) + 1.$$

The term $\frac{xv}{r_j}$ gives us $\tau_{t_j} \cdot \frac{1}{\sigma_{s_j}} = \sigma_{s_j}$ giving us $\tau_{t_j} = \sigma_{s_j}^2$. The constant term gives us $(\tau_{t_j} + \tau_{w_j}) \cdot \frac{1}{\sigma_{s_j}} = 1$ giving us $\tau_{w_j} = \sigma_{s_j}(1 - \sigma_{s_j})$. The vr_j term gives us $\tau_v = 0$. The $\frac{xv^2}{r_j^2}$ term gives us $\rho_v = 0$. Finally, the v^2 term gives us $\sigma_v = 0$.

The adversary can therefore only compute a valid signature by using

$$r = \frac{1}{\sigma_{s_j}}r_j \qquad s = \sigma_{s_j}s_j \qquad t = \sigma_{s_j}^2 t_j + \sigma_{s_j}(1 - \sigma_{s_j})w_j.$$

The first verification equation then gives us $mv + x = sr = s_j r_j = m_j v + x$, showing $m = m_j$ and therefore the signature scheme is EUF-CMA secure even in the presence of randomization tokens. Furthermore, if no randomization token w_j has been provided for the message then $\tau_{w_j} = 0$. Since $\tau_{w_j} = \sigma_{s_j}(1 - \sigma_{s_j})$ and $\sigma_{s_j} \neq 0$ this shows $\sigma_{s_j} = 1$. This implies $r = r_j, s = s_j$ and $t = t_j$, which shows that the signature scheme is sEUF-CMA-TA secure.

Let us now prove Claim 3.

Proof. Starting with the first verification equation $sr = mv + x$ we have

$$\left(\sigma + \sigma_x x + \sigma_v v + \sum_{i=1}^{q}\sigma_{r_i}r_i + \sigma_{s_i}\left(\frac{m_i v}{r_i} + \frac{x}{r_i}\right) + \sigma_{t_i}\left(\frac{m_i v^2}{r_i^2} + \frac{xv}{r_i^2} + \frac{1}{r_i}\right) + \sigma_{w_i}\frac{1}{r_i}\right)$$

$$\cdot\left(\rho + \rho_x x + \rho_v v + \sum_{i=1}^{q}\rho_{r_i}r_i + \rho_{s_i}\left(\frac{m_i v}{r_i} + \frac{x}{r_i}\right) + \rho_{t_i}\left(\frac{m_i v^2}{r_i^2} + \frac{xv}{r_i^2} + \frac{1}{r_i}\right) + \rho_{w_i}\frac{1}{r_i}\right)$$

$$=\left(\mu + \mu_x x + \mu_v v + \sum_{i=1}^{q}\mu_{r_i}r_i + \mu_{s_i}\left(\frac{m_i v}{r_i} + \frac{x}{r_i}\right) + \mu_{t_i}\left(\frac{m_i v^2}{r_i^2} + \frac{xv}{r_i^2} + \frac{1}{r_i}\right) + \mu_{w_i}\frac{1}{r_i}\right)v + x$$

We first show that $\sigma_{t_i} = 0$ for all i. Assume for contradiction that there exists a j such that $\sigma_{t_j} \neq 0$. We start by looking at the coefficients of $\frac{x^2 v^2}{r_j^4}$. The Laurent polynomials corresponding to r, s, m and $m_1, \ldots, m_q$ are all linear in x. Terms involving x^2 can therefore only arise in the product sr. This shows that the coefficient of $\frac{x^2 v^2}{r_j^4}$ is 0 in $mv + x$. We will in the following argue the coefficient of $\frac{x^2 v^2}{r_j^4}$ in sr is $\sigma_{t_j}\rho_{t_j}$, which by our assumption $\sigma_{t_j} \neq 0$ implies $\rho_{t_j} = 0$.

To see that indeed the coefficient of $\frac{x^2 v^2}{r_j^4}$ in sr is $\sigma_{t_j}\rho_{t_j}$, we need to rule out that other cross terms in sr can be $\frac{x^2 v^2}{r_j^4}$. Observe that in all terms of s and r the degree of r_j ranges from -2 to 1 and only has degree -2 in the term $t_j = \frac{m_j v^2}{r_j^2} + \frac{xv}{r_j^2} + \frac{1}{r_j}$ and in subsequent signatures on queries m_i that include a t_j term. However, if a term m_i involves t_j then the resulting signature terms s_i and t_i multiply the t_j by v or v^2. By using the fact that the degree of v never decreases, we see that all other cross terms involving $\frac{x^2}{r_j^4}$ have degree 3 or higher in v. The coefficients of $\frac{x^2 v^2}{r_j^4}$ therefore do indeed give us $\sigma_{t_j}\rho_{t_j} = 0$ and therefore $\rho_{t_j} = 0$.

Next we look at the term $\frac{x^2 v^2}{r_j^2 r_i^2}$ for $i \neq j$. A similar analysis shows that the coefficients satisfy $\sigma_{t_j}\rho_{t_i} + \sigma_{t_i}\rho_{t_j} = 0$. Since $\rho_{t_j} = 0$ and $\sigma_{t_j} \neq 0$ this implies $\rho_{t_i} = 0$ for all i.

We proceed to the term $\frac{x^2 v}{r_j^3}$ and will show the coefficient in sr of this term is $\sigma_{t_j}\rho_{s_j}$. Since the degree of r_j is -3 in the term, we see that t_j must be used either directly, or indirectly through a signature on a subsequent query m_i involving t_j. However, whenever m_i involves t_j the degree of v is increased to at least 2 and such subsequent queries cannot contribute to the term. An inspection of the different cross terms now shows that indeed $\sigma_{t_j}\rho_{s_j}$ is the coefficient in sr for the term $\frac{x^2 v}{r_j^3}$. Since cross terms involving x^2 can only arise in sr and not in $mv + x$ we then have $\sigma_{t_j}\rho_{s_j} = 0$ and since we assumed $\sigma_{t_j} \neq 0$ this means $\rho_{s_j} = 0$.

A similar analysis shows that for $i \neq j$ the terms $\frac{x^2 v}{r_i r_j^2}$ have coefficient $\sigma_{t_i}\rho_{s_j} + \sigma_{t_j}\rho_{s_i} = \sigma_{t_j}\rho_{s_i} = 0$ and therefore $\rho_{s_i} = 0$ for all i.

We now look at the term $\frac{x^2 v}{r_j^2}$. Again looking at the degrees of v in subsequent queries with m_i using t_j we see that they cannot contribute to the coefficient of $\frac{x^2 v}{r_j^2}$ in sr and therefore the coefficient is $\sigma_{t_j}\rho_x$. Since there are no terms involving x^2 in $mv + 1$ this means $\sigma_{t_j}\rho_x = 0$ and therefore $\rho_x = 0$.

Using the term $\frac{xv}{r_j^3}$ we see that $\sigma_{t_j}\rho_{w_j} = 0$ and therefore $\rho_{w_j} = 0$. The terms $\frac{xv}{r_i r_j^2}$ give us $\sigma_{t_j}\rho_{w_i} + \sigma_{w_i}\rho_{t_j} = \sigma_{t_j}\rho_{w_i} = 0$, which implies $\rho_{w_i} = 0$ for all i.

The term $\frac{xv}{r_j^2}$ gives us $\sigma_{t_j}\rho = 0$ and therefore $\rho = 0$.

The terms $\frac{xvr_i}{r_j^2}$ give us $\sigma_{t_j}\rho_{r_i} = 0$ and therefore $\rho_{r_i} = 0$ for all $i \neq j$. The term $\frac{xv}{r_j}$ gives us $\sigma_{t_j}\rho_{r_j} = 0$ and therefore $\rho_{r_j} = 0$.

We now have $r = \rho_v v$, which means all terms in sr and mv have at least degree 1 in v. It is therefore impossible to get $sr = mv + x$ when there exists a j such that $\sigma_{t_j} \neq 0$.

By symmetry we can also rule out the existence of $\rho_{t_j} \neq 0$. We conclude that both r and s must have $\rho_{t_i} = 0$ and $\sigma_{t_i} = 0$ for all $i = 1, \ldots, q$ and will use that simplification in the rest of our proof.

Next, we will show that for all i we have $\sigma_{w_i} = 0$. Assume for contradiction $\sigma_{w_j} \neq 0$ for some j. We can write $sr = mv + x$ as

$$\left(\sigma + \sigma_x x + \sigma_v v + \sum_{i=1}^q \sigma_{r_i} r_i + \sigma_{s_i} \left(\frac{m_i v}{r_i} + \frac{x}{r_i} \right) + \sigma_{w_i} \frac{1}{r_i} \right)$$

$$\cdot \left(\rho + \rho_x x + \rho_v v + \sum_{i=1}^q \rho_{r_i} r_i + \rho_{s_i} \left(\frac{m_i v}{r_i} + \frac{x}{r_i} \right) + \rho_{w_i} \frac{1}{r_i} \right)$$

$$= \left(\mu + \mu_x x + \mu_v v + \sum_{i=1}^q \mu_{r_i} r_i + \mu_{s_i} \left(\frac{m_i v}{r_i} + \frac{x}{r_i} \right) + \mu_{t_i} \left(\frac{m_i v^2}{r_i^2} + \frac{xv}{r_i^2} + \frac{1}{r_i} \right) + \mu_{w_i} \frac{1}{r_i} \right) v + x.$$

The term $\frac{1}{r_j^2}$ gives us $\sigma_{w_j} \rho_{w_j} = 0$ since all other terms involving r_j^{-2} are multiplied by powers of x or v. With $\sigma_{w_j} \neq 0$ this means $\rho_{w_j} = 0$. Similarly, the terms $\frac{1}{r_i r_j}$ give us $\sigma_{w_i} \rho_{w_j} + \sigma_{w_j} \rho_{w_i} = \sigma_{w_j} \rho_{w_i} = 0$ yielding $\rho_{w_i} = 0$ for all i.

The term $\frac{x}{r_j^2}$ now gives us $\sigma_{w_j} \rho_{s_j} = 0$ and therefore $\rho_{s_j} = 0$. The terms $\frac{x}{r_i r_j}$ give us $\sigma_{w_i} \rho_{s_j} + \sigma_{w_j} \rho_{s_i} = \sigma_{w_j} \rho_{s_i} = 0$ and therefore $\rho_{s_i} = 0$ for all i.

The term $\frac{1}{r_j}$ gives us $\sigma_{w_j} \rho = 0$ and therefore $\rho = 0$. The term $\frac{x}{r_j}$ now gives us $\sigma_{w_j} \rho_x = 0$ and therefore $\rho_x = 0$.

The constant term gives us $\sigma_{w_j} \rho_{r_j} = 0$ and therefore $\rho_{r_j} = 0$. The terms $\frac{r_i}{r_j}$ give us $\sigma_{w_j} \rho_{r_i} = 0$ and therefore $\rho_{r_i} = 0$ for all i.

We now have $r = \rho_v v$ giving us $sr = \rho_v sv = mv + x$. Since signing queries only increase the degree of v this equation cannot be satisfied because of the x. The contradiction leads us to conclude $\sigma_{w_i} = 0$ for $i = 1, \ldots, q$. By symmetry this also shows $\rho_{w_i} = 0$ for all $i = 1, \ldots, q$.

4   Optimal One-Time Signatures

The construction of a 3-element structure-preserving signature scheme in Sect. 3 leaves open the question whether 2-element one-time signatures exist. (A one-time signature scheme with 3-element signatures already exists in the symmetric setting [6]. It is sEUF-CMA under the simultaneous double-pairing assumption.) We will now give a candidate for an sEUF-CMA secure one-time structure-preserving signature scheme in the symmetric setting, which matches the 2-element lower bound from Sect. 5. This one-time signature beats the 3-element lower bound for general structure-preserving signatures in Theorem 5. Moreover, the scheme is deterministic, so it also demonstrates that Lemma 1 requiring general structure-preserving signatures to be randomized does not apply to one-time signatures.

The case of one-time signatures also indicates a difference between the symmetric and the asymmetric Type III setting. Abe et al. [4] constructed a one-time signature scheme with a single verification equation for messages belonging exclusively to one of the groups $\mathbb{G}_1$ or $\mathbb{G}_2$ and in Sect. 4.1 we show that it is even possible to make 1 element signatures in Type III groups. On the other hand, there is no known structure-preserving (one-time) signature scheme in the asymmetric setting for messages that contain groups elements in both $\mathbb{G}_1$ and $\mathbb{G}_2$ with signature size less than 3.

The construction of our one-time signature is given in Fig. 2. We observe that the verification key has two group elements V, W and the signer needs to know the discrete logarithm of both of these elements. It is an interesting question whether a 2-element structure-preserving one-time signature scheme can be constructed with just a single variable verification key element like we did for 3-element signatures in Fig. 1, but we have some initial indications (not included in this paper) that for some classes of one-time signature schemes this may not be possible and that the signer needs to know at least two discrete logarithms.

Setup $\mathcal{P}(1^k)$: Run $(p, \mathbb{G}, \mathbb{G}_T, e, G) \leftarrow \mathcal{G}(1^k)$ and return $PP = (p, \mathbb{G}, \mathbb{G}_T, e, G)$.
Key generation $\mathcal{K}(PP)$: Choose $v, w \leftarrow \mathbb{Z}_p$ and compute

$$V \leftarrow G^v, \qquad W \leftarrow G^w.$$

Return $(VK, SK) = ((PP, V, W), (PP, v, w))$.
Signing $\mathcal{S}_{SK}(M)$: Given $M \in \mathbb{G}$, return the signature $\Sigma = (S, T)$ given by:

$$S = M^v G^{w^2}, \qquad T = S^v.$$

Verification $\mathcal{V}_{VK}(M, (S, T))$: Accept if all the input elements are in $\mathbb{G}$ and if:

$$e(S, G) = e(M, V)e(W, W) \quad \text{and} \quad e(T, G) = e(S, V).$$

Fig. 2. One-time structure-preserving signature scheme in the symmetric setting

Theorem 2. *The scheme given in Fig. 2 is an* sEUF-CMA *secure one-time signature scheme in the generic group model.*

Proof. A generic adversary only uses generic group operations, which means that in $\mathbb{G}$ it can only compute linear combinations on group elements from the verification key or the signature from the one-time chosen message attack. We will show that linear combinations of verification key elements and signature elements correspond to formal polynomials (of degree 3 or less) in the corresponding discrete logarithms of these elements and that no linear combinations will produce formal polynomials corresponding to a forgery. By the master theorem in [11] this means that the signature scheme is secure in the generic group model.

Suppose the adversary gets a one-time signature (S, T) on a query M and then outputs a valid signature (S^*, T^*) on M^*. Since the adversary is generic it computes M^*, S^*, T^* as linear combinations of G, V, W, S, T. This means the discrete logarithms are of the form

$$m^* = \mu + \mu_v v + \mu_w w + \mu_s(mv + w^2) + \mu_t(mv^2 + w^2 v)$$
$$s^* = \sigma + \sigma_v v + \sigma_w w + \sigma_s(mv + w^2) + \sigma_t(mv^2 + w^2 v)$$
$$t^* = \tau + \tau_v v + \tau_w w + \tau_s(mv + w^2) + \tau_t(mv^2 + w^2 v)$$

where m itself is a linear combination of $1, v, w$.

The second verification equation $t^* = s^*v = m^*v^2 + w^2v$ gives us

$$\tau + \tau_v v + \tau_w w + \tau_s(mv + w^2) + \tau_t(mv^2 + w^2v)$$
$$= \mu v^2 + \mu_v v^3 + \mu_w wv^2 + \mu_s(mv^3 + w^2v^2) + \mu_t(mv^4 + w^2v^3) + w^2v$$

The coefficients of w^2v^3 give us $\mu_t = 0$. The coefficients of w^2v give us $\tau_t = 1$. The coefficients of w^2 give us $\tau_s = 0$. The coefficients of w^2v^2 give us $\mu_s = 0$. The coefficients of $1, v, w$ give us $\tau = 0, \tau_v = 0, \tau_w = 0$. This means $mv^2 = \mu v^2 + \mu_v v^3 + \mu_w wv^2$, which implies $m = \mu + \mu_v v + \mu_w w = m^*$. Since the verification equations uniquely determined the signature once the message is fixed, $m^* = m$ implies $s^* = s$ and $t^* = t$. This means $(M^*, S^*, T^*) = (M, S, T)$, which was the message and signature pair from the query.   □

4.1   Optimal One-Time Signatures in the Type III Setting

In Fig. 3, we present a one-time signature scheme over asymmetric bilinear groups with single element signatures. It can be used to sign vectors of n group elements in the second base group $\mathbb{G}_2$.

Setup $\mathcal{P}(1^k)$: Return $PP= (p, \mathbb{G}_1, \mathbb{G}_2, \mathbb{G}_T, e, G, H)$ generated by asymmetric bilinear group generator $\mathcal{G}(1^k)$.

Key generation $\mathcal{K}(PP)$: Choose $v, a_1, \ldots, a_n \leftarrow \mathbb{Z}_p$ and compute:

$$V = G^v, \quad A_1 = G^{a_1}, \quad \ldots, \quad A_n = G^{a_n}.$$

Return $(VK, SK) = ((PP, V, A_1, \ldots, A_n), (PP, v, a_1, \ldots, a_n))$.

Signing $\mathcal{S}_{SK}(M)$: On input $M = (M_1, \ldots, M_n) \in \mathbb{G}_2^n$, return the signature:

$$S \leftarrow H^v \prod_{i=1}^{n} M_i^{a_i}.$$

Verification $\mathcal{V}_{VK}((M_1, \ldots, M_n), S)$: Accept if $M_1, \ldots, M_n, S \in \mathbb{G}_2$ and if:

$$e(G, S) = e(V, H) \prod_{i=1}^{n} e(A_i, M_i).$$

Fig. 3. One-time structure-preserving signature with 1 element signatures in the Type III setting

Theorem 3. *The scheme given in Fig. 3 is an* sEUF-CMA *secure one-time signature in the generic group model.*

Proof. A generic adversary can only compute linear combinations of group elements in the base groups, which means its signing query must be $(M_1, \ldots, M_n) = (H^{m_1}, \ldots, H^{m_n})$ with known discrete logarithms $m_1, \ldots, m_n$. The generic adversary gets a signature $S = H^{v + \sum_{i=1}^{n} a_i m_i}$ as response.

Suppose now the generic adversary computes a message $(M_1^*, \ldots, M_n^*) = (H^{m_1^*}, \ldots, H^{m_n^*})$ and a valid signature $S^* = H^{s^*}$. Since the adversary only uses linear combinations of existing group elements it knows $\mu_1, \ldots, \sigma_s \in \mathbb{Z}_p$ such that

$$m_j^* = \mu_j + \mu_{s,j}(v + \sum_{i=1}^{n} m_i a_i) \qquad \text{for } j \in \{1, \ldots, n\}$$

$$s^* = \sigma + \sigma_s(v + \sum_{i=1}^{n} m_i a_i).$$

The verification equation gives us $s^* = v + \sum_{i=1}^{n} a_i m_i^*$. This means:

$$(\sigma_s - 1)v = -\sigma - \sigma_s \sum_{j=1}^{n} m_j a_j + \sum_{j=1}^{n} \mu_j a_j + \sum_{j=1}^{n} \mu_{s,j} a_j (v + \sum_{i=1}^{n} m_i a_i).$$

It then holds that $\sigma_s = 1$, $\sigma = 0$, $\mu_j = m_j$ and $\mu_{s,j} = 0$ for all j. This means $m_j^* = m_j$ and $s^* = s$, so $((M_1^*, \ldots, M_n^*), S^*) = ((M_1, \ldots, M_n), S)$, which is not a valid forgery. $\qquad \square$

5   Lower Bounds in the Symmetric Setting

We will show that in the Type I setting structure-preserving signatures must have at least two verification equations and consist of at least three group elements. This matches the lower bounds in the Type III setting [4]. One-time signature can be just two group elements but still require two verification equations. Our lower bounds hold even when the verification key may also include target group elements $Z \in \mathbb{G}_T$, and the security is relaxed to random message attacks.

Theorem 4. (No one-equation signatures) *The verification algorithm $\mathcal{V}$ of a (one-time) EUF-RMA secure structure-preserving signature scheme over a symmetric pairing group must evaluate at least two pairing product equations.*

Proof. By diagonalizing the corresponding quadratic form, we may assume without loss of generality that the single verification equation for a signature $\Sigma = (S_1, \ldots, S_n)$ on a one-element message M has the following form:

$$e(M, M)^a \cdot e(M, U \prod_{i=1}^{n} S_i^{b_i}) \cdot \prod_{i=1}^{n} e(S_i, S_i)^{c_i} \cdot e(S_i, V_i) = Z. \qquad (1)$$

Let us fix an arbitrary message $M \in \mathcal{G}$ and a signature $\Sigma = (S_1, \ldots, S_n)$ on M which is valid with respect to the verification equation (1). We will construct an explicity forgery (M^*, Σ^*) such that Σ^* coincides with Σ on all components except one. We distinguish between two cases: either all the coefficients c_i in the verification equation (1) are nonzero or at least one of the c_i's is zero.

Case 1: $c_i \neq 0$ for all i. We first assume that all the c_i's are nonzero, and fix an arbitrary index $i \in \{1, \ldots, n\}$. Let M be any message and $\Sigma = (S_1, \ldots, S_n)$ a valid signature on M. We concentrate on the component $S = S_i$ of Σ, and claim that we can find a pair $(M^*, S^*) \neq (M, S)$ such that $\Sigma^* = (S_1, \ldots, S_{i-1}, S^*, S_{i+1}, \ldots, S_n)$ is a valid signature on M^*. In terms of discrete logarithms, this is equivalent to finding $(m^*, s^*) \neq (m, s)$ such that:

$$am^2 + m(u + bs + k) + cs^2 + sv = am^{*2} + m^*(u + bs + k) + cs^{*2} + s^*v \quad (2)$$

where we let $b = b_i$, $c = c_i$, $V = V_i$, and $K = \prod_{j \neq i} S_j^{b_j}$. To find such a pair, we look for m^*, s^* of the form:

$$m^* = \mu_0 u + \mu_1 v + (1 + \mu_2)m + \mu_3 s + \mu_4 k,$$
$$s^* = \sigma_0 u + \sigma_1 v + \sigma_2 m + (1 + \sigma_3)s + \sigma_4 k.$$

such that equation (2) is satisfied regardless of the discrete logarithms, i.e. such that the corresponding coefficients of the left-hand side and right-hand side of equation (2), when regarded as polynomials in $\mathbb{Z}_p[u, v, m, s, k]$, are pairwise equal.

This gives a quadratic system of 15 equations in the 10 unknowns $\mu_0, \ldots, \mu_4$, $\sigma_0, \ldots, \sigma_4$, which we solve by computing a Gröbner basis of the corresponding ideal. We obtain, in particular, a rational one-parameter family of solutions. Let ω be any element in $\mathbb{Z}_p$ such that $\tau = b^2 - 4ac - \omega^2 \neq 0$. Then the following is a solution:

$$(\mu_0, \mu_1, \mu_2, \mu_3, \mu_4) = 2/\tau \cdot (2c, \omega - b, b\omega - \delta, 2c\omega, 2c)$$
$$(\sigma_0, \sigma_1, \sigma_2, \sigma_3, \sigma_4) = (\omega - b)/(c\tau) \cdot (2c, \omega - b, b\omega - \delta, 2c\omega, 2c)$$

(where $\delta = b^2 - 4ac$) and defines corresponding group elements (M_ω^*, S_ω^*).

This is a successful forgery provided that we can find some ω such that $M \neq M_\omega^*$. Suppose that this is not the case. Then for all ω such that $\tau \neq 0$, we must have:

$$M_\omega^* \cdot M^{-1} = U^{\mu_0} \cdot V^{\mu_1} \cdot M^{\mu_2} \cdot S^{\mu_3} \cdot K^{\mu_4} = 1.$$

By raising to the power $\tau/2$, this gives:

$$U^{2c} V^{\omega-b} M^{b\omega-b^2+4ac} S^{2c\omega} K^{2c} = \left(VM^bS^{2c}\right)^\omega \cdot \left(U^{2c}V^{-b}M^{-b^2+4ac}K^{2c}\right) = 1,$$

and since this relation is verified for all $\omega \in \mathbb{Z}_p$ except at most two values, this implies in particular that $VM^bS^{2c} = 1$, or in other words $S = V^{-1/2c} \cdot M^{-b/2c}$. Now recall that all the c_i's are nonzero. By the previous argument, we can either carry out the previous attack for at least one index i, or the signature on a message M must be given, with overwhelming probability, by $\Sigma = (S_1, \ldots, S_n)$ where $S_i = V_i^{-1/2c_i} \cdot M^{-b_i/2c_i}$ for all i, which is obviously insecure.

Case 2: $c_i = 0$ for some i Suppose $c_i = 0$ for some i. We concentrate on that index like before, and look again for a forgery (M^*, S^*) given a signature Σ on an arbitrary message M. With the same notation as before, we find a one-parameter family (M_ω^*, S_ω^*) of solutions, given by:

$$(\mu_0, \mu_1, \mu_2, \mu_3, \mu_4) = -\omega/(b\omega + 1) \cdot (0, 1, b, 0, 0)$$
$$(\sigma_0, \sigma_1, \sigma_2, \sigma_3, \sigma_4) = \omega \cdot (1, -a\omega/(b\omega + 1), a(b\omega + 2)/(b\omega + 1), b, 1)$$

for all ω such that $b\omega + 1 \neq 0$. This gives a forgery unless $M_\omega^* = M$ for all such ω, namely $(VM^b)^{-\omega/(b\omega+1)} = 1$. As a result, we get a forgery on any message except $V^{-1/b}$ (or any message if $b = 0$). This completes the proof.   □

Corollary 1. (Two group elements required for one-time signatures.)
A structure-preserving one-time signature scheme that is existentially unforgeable against a one-time random message attack must have at least 2 group elements.

Proof. Suppose there is a scheme where a signature is a single group element S. If a linear combination of the verification equations give us a non-trivial equation that is linear in S, then this equation uniquely determines S and we can just use this equation as the verification equation instead of all the other verification equations. If there is no linear combination of the verification equations that yield a non-trivial linear equation in S then they must all be linearly dependent and we can again reduce to the case where there is a single verification equation.   □

For structure-preserving signatures where the adversary can ask multiple signature queries there is a stronger lower bound of 3 group elements.[1]

Theorem 5. (Three group elements required for structure-preserving signatures.) *A structure-preserving signature scheme with a generic signer that is existentially unforgeable against random message attacks must have at least 3 group elements.*

Proof. We begin by proving the following lemma.

Lemma 1. *A structure-preserving signature scheme with a generic signer that is existentially unforgeable against random message attacks must for each message have a superpolynomial number of potential signatures.*

Proof. Suppose that for a message M there are only polynomially many signature vectors Σ. Since the signer is generic this means there is a polynomial set $\{(\vec{\alpha}, \vec{\beta})\}_{i=1}^{\text{poly}(k)}$ of vectors in $\mathbb{Z}_p^n$ creating signature vectors $\Sigma = G^{\vec{\alpha}} M^{\vec{\beta}}$ by entrywise exponentiation. Given signatures Σ_0 and Σ_1 on random messages M_0 and

---

[1] Our proof of the lower bound is much simpler than the proof for the similar lower bound of 3 group elements in [4] in the asymmetric Type III setting and can with minor modifications be adapted to Type III groups. More generally, the proof of Theorem 5 indicates that in general if there are m verification equations, then the signature size needs to be $m + 1$.

M_1 we have $\frac{1}{\text{poly}(k)^2}$ probability that they are constructed with the same $(\vec{\alpha}, \vec{\beta})$ pair. In that case

$$\Sigma^* = \Sigma_0^r \Sigma_1^{1-r} = G^{\vec{\alpha}} (M_0^r M_1^{1-r})^{\vec{\beta}}$$

is a signature on $M^* = M_0^r M_1^{1-r}$ for all $r \in \mathbb{Z}_p$. $\qquad\square$

Now suppose that we have an SPS with just two group elements (S, T) and a minimal number of verification equations. We know there must be at least two verification equations. This means the discrete logarithms s, t of the signature elements must satisfy two quadratic equations. By using a linear combination of the two verification equations, we can without loss of generality ensure the first equation is linear in t, i.e., $t = as^2 + bs + c$ for some $a, b, c \in \mathbb{Z}_p$ determined by the message and the verification key. We can then substitute this into the second verification equation to get a quartic equation in s. If the equation is non-trivial, then there are at most 4 solutions for s and therefore at most 4 signatures in total contradicting Lemma 1. On the other hand if the equation is trivial, then the second verification equation was redundant and could be eliminated, which contradicts our initial assumption that we had a minimal number of verification equations. $\qquad\square$

References

1. Abe, M., Chase, M., David, B., Kohlweiss, M., Nishimaki, R., Ohkubo, M.: Constant-size structure-preserving signatures: Generic constructions and simple assumptions. In: Wang, X., Sako, K. (eds.) ASIACRYPT 2012. LNCS, vol. 7658, pp. 4–24. Springer, Heidelberg (2012)
2. Abe, M., David, B., Kohlweiss, M., Nishimaki, R., Ohkubo, M.: Tagged one-time signatures: Tight security and optimal tag size. In: Kurosawa, K., Hanaoka, G. (eds.) PKC 2013. LNCS, vol. 7778, pp. 312–331. Springer, Heidelberg (2013)
3. Abe, M., Fuchsbauer, G., Groth, J., Haralambiev, K., Ohkubo, M.: Structure-preserving signatures and commitments to group elements. In: Rabin, T. (ed.) CRYPTO 2010. LNCS, vol. 6223, pp. 209–236. Springer, Heidelberg (2010)
4. Abe, M., Groth, J., Haralambiev, K., Ohkubo, M.: Optimal structure-preserving signatures in asymmetric bilinear groups. In: Rogaway, P. (ed.) CRYPTO 2011. LNCS, vol. 6841, pp. 649–666. Springer, Heidelberg (2011)
5. Abe, M., Groth, J., Ohkubo, M.: Separating short structure-preserving signatures from non-interactive assumptions. In: Lee, D.H., Wang, X. (eds.) ASIACRYPT 2011. LNCS, vol. 7073, pp. 628–646. Springer, Heidelberg (2011)
6. Abe, M., Haralambiev, K., Ohkubo, M.: Signing on group elements for modular protocol designs. IACR ePrint Archive, Report 2010/133 (2010), http://eprint.iacr.org
7. Abe, M., Haralambiev, K., Ohkubo, M.: Efficient message space extension for automorphic signatures. In: Burmester, M., Tsudik, G., Magliveras, S., Ilić, I. (eds.) ISC 2010. LNCS, vol. 6531, pp. 319–330. Springer, Heidelberg (2011)
8. Attrapadung, N., Libert, B., Peters, T.: Computing on authenticated data: New privacy definitions and constructions. In: Wang, X., Sako, K. (eds.) ASIACRYPT 2012. LNCS, vol. 7658, pp. 367–385. Springer, Heidelberg (2012)
9. Attrapadung, N., Libert, B., Peters, T.: Efficient completely context-hiding quotable and linearly homomorphic signatures. In: Kurosawa, K., Hanaoka, G. (eds.) PKC 2013. LNCS, vol. 7778, pp. 386–404. Springer, Heidelberg (2013)

10. Barbulescu, R., Gaudry, P., Joux, A., Thomé, E.: A quasi-polynomial algorithm for discrete logarithm in finite fields of small characteristic. IACR ePrint Archive, Report 2013/400 (2013), http://eprint.iacr.org/

11. Boneh, D., Boyen, X.: Short signatures without random oracles and the SDH assumption in bilinear groups. J. Cryptology 21(2), 149–177 (2008)

12. Camenisch, J., Dubovitskaya, M., Enderlein, R.R., Neven, G.: Oblivious transfer with hidden access control from attribute-based encryption. In: Visconti, I., De Prisco, R. (eds.) SCN 2012. LNCS, vol. 7485, pp. 559–579. Springer, Heidelberg (2012)

13. Cathalo, J., Libert, B., Yung, M.: Group encryption: Non-interactive realization in the standard model. In: Matsui, M. (ed.) ASIACRYPT 2009. LNCS, vol. 5912, pp. 179–196. Springer, Heidelberg (2009)

14. Chase, M., Kohlweiss, M., Lysyanskaya, A., Meiklejohn, S.: Malleable proof systems and applications. In: Pointcheval, D., Johansson, T. (eds.) EUROCRYPT 2012. LNCS, vol. 7237, pp. 281–300. Springer, Heidelberg (2012)

15. Fuchsbauer, G.: Automorphic signatures in bilinear groups. IACR ePrint Archive, Report 2009/320 (2009), http://eprint.iacr.org

16. Fuchsbauer, G.: Commuting signatures and verifiable encryption. In: Paterson, K.G. (ed.) EUROCRYPT 2011. LNCS, vol. 6632, pp. 224–245. Springer, Heidelberg (2011)

17. Fuchsbauer, G., Vergnaud, D.: Fair blind signatures without random oracles. In: Bernstein, D.J., Lange, T. (eds.) AFRICACRYPT 2010. LNCS, vol. 6055, pp. 16–33. Springer, Heidelberg (2010)

18. Galbraith, S.D., Paterson, K.G., Smart, N.P.: Pairings for cryptographers. Discrete Applied Mathematics 156(16), 3113–3121 (2008)

19. Green, M., Hohenberger, S.: Universally composable adaptive oblivious transfer. In: Pieprzyk, J. (ed.) ASIACRYPT 2008. LNCS, vol. 5350, pp. 179–197. Springer, Heidelberg (2008)

20. Groth, J.: Simulation-sound NIZK proofs for a practical language and constant size group signatures. In: Lai, X., Chen, K. (eds.) ASIACRYPT 2006. LNCS, vol. 4284, pp. 444–459. Springer, Heidelberg (2006)

21. Groth, J., Sahai, A.: Efficient noninteractive proof systems for bilinear groups. SIAM J. Comput. 41(5), 1193–1232 (2012)

22. Hofheinz, D., Jager, T.: Tightly secure signatures and public-key encryption. In: Safavi-Naini, R., Canetti, R. (eds.) CRYPTO 2012. LNCS, vol. 7417, pp. 590–607. Springer, Heidelberg (2012)

23. Hofheinz, D., Jager, T., Knapp, E.: Waters signatures with optimal security reduction. In: Fischlin, M., Buchmann, J., Manulis, M. (eds.) PKC 2012. LNCS, vol. 7293, pp. 66–83. Springer, Heidelberg (2012)

24. Libert, B., Peters, T., Joye, M., Yung, M.: Linearly homomorphic structure-preserving signatures and their applications. In: Canetti, R., Garay, J.A. (eds.) CRYPTO 2013, Part II. LNCS, vol. 8043, pp. 289–307. Springer, Heidelberg (2013)

25. Libert, B., Peters, T., Yung, M.: Group signatures with almost-for-free revocation. In: Safavi-Naini, R., Canetti, R. (eds.) CRYPTO 2012. LNCS, vol. 7417, pp. 571–589. Springer, Heidelberg (2012)

26. Zhang, J., Li, Z., Guo, H.: Anonymous transferable conditional e-cash. In: Keromytis, A.D., Di Pietro, R. (eds.) SecureComm 2012. LNICST, vol. 106, pp. 45–60. Springer, Heidelberg (2013)

27. Zhou, S., Lin, D.: Unlinkable randomizable signature and its application in group signature. In: Pei, D., Yung, M., Lin, D., Wu, C. (eds.) Inscrypt 2007. LNCS, vol. 4990, pp. 328–342. Springer, Heidelberg (2008)

On the Impossibility of Structure-Preserving Deterministic Primitives

Masayuki Abe[1], Jan Camenisch[2], Rafael Dowsley[3], and Maria Dubovitskaya[2,4]

[1] NTT Corporation, Japan
abe.masayuki@lab.ntt.co.jp
[2] IBM Research - Zurich, Switzerland
{jca,mdu}@zurich.ibm.com
[3] Karlsruhe Institute of Technology, Germany
rafael.dowsley@kit.edu
[4] ETH Zurich, Switzerland
dumaria@inf.ethz.ch

Abstract. Complex cryptographic protocols are often constructed in a modular way from primitives such as signatures, commitments, and encryption schemes, verifiable random functions, etc. together with zero-knowledge proofs ensuring that these primitives are properly orchestrated by the protocol participants. Over the past decades a whole framework of discrete logarithm based primitives has evolved. This framework, together with so-called generalized Schnorr proofs, gave rise to the construction of many efficient cryptographic protocols.

Unfortunately, the non-interactive versions of Schnorr proofs are secure only in the random oracle model, often resulting in protocols with unsatisfactory security guarantees. Groth and Sahai have provided an alternative non-interactive proof system (GS-proofs) that is secure in the standard model and allows for the "straight line" extraction of witnesses. Both these properties are very attractive, in particular if one wants to achieve composable security. However, GS-proofs require bilinear maps and, more severely, they are proofs of knowledge only for witnesses that are group elements. Thus, researchers have set out to construct efficient cryptographic primitives that are compatible with GS-proofs, in particular, primitives that are structure-preserving, meaning that their inputs, outputs, and public keys consist only of source group elements. Indeed, structure-preserving signatures, commitments, and encryption schemes have been proposed. Although deterministic primitives such as (verifiable) pseudo-random functions or verifiable unpredictable functions play an important role in the construction of many cryptographic protocols, no structure-preserving realizations of them are known so far.

As it turns out, this is no coincidence: in this paper we show that it is impossible to construct *algebraic* structure-preserving deterministic primitives that provide provability, uniqueness, and unpredictability. This includes verifiable random functions, unique signatures, and verifiable unpredictable functions as special cases. The restriction of structure-preserving primitives to be algebraic is natural, in particular as otherwise it is not possible to prove with GS-proofs that an algorithm has been run correctly. We further extend our negative result to pseudorandom functions and deterministic public key encryption as well as

Y. Lindell (Ed.): TCC 2014, LNCS 8349, pp. 713–738, 2014.

non-strictly structure-preserving primitives, where target group elements are also allowed in their ranges and public keys.

Keywords: Verifiable random functions, unique signatures, structure-preserving primitives, Groth-Sahai proofs.

1 Introduction

Most practical cryptographic protocols are built from cryptographic primitives such as signature, encryption, and commitments schemes, pseudorandom functions, and zero-knowledge (ZK) proofs. Thereby the ZK proofs often "glue" different building blocks together by proving relations among their inputs and outputs. The literature provides a fair number of different cryptographic primitives (e.g., CL-signatures [20,21], Pedersen Commitments [49], ElGamal and Cramer-Shoup encryption [30,27], verifiable encryption of discrete logarithms [23], verifiable pseudo-random functions [29]) that are based on the discrete logarithm problem and that together with so-called generalized Schnorr protocols [50,18] provide a whole framework for the construction of practical protocols. Examples of such constructions include anonymous credential systems [19,6], oblivious transfer with access control [15], group signatures [10,43], or e-cash [17]. The non-interactive versions of generalized Schnorr protocols are secure only in the random oracle model as they are obtained via the Fiat-Shamir heuristic [31] and it is well known the random oracles cannot be securely instantiated [25]. Consequently, many protocols constructed from this framework unfortunately are secure only in the random oracle model.

A seminal step towards a framework allowing for security proofs in the *standard model* was therefore the introduction of the so-called GS-proofs by Groth and Sahai [36]. These are efficient non-interactive proofs of knowledge or language membership and are secure in the standard model. They make use of bilinear maps to verify statements and therefore are limited to languages of certain types of equations, including systems of pairing product and multi exponentiation equations. In particular, GS-proofs are proofs of *knowledge* only for witnesses that are group elements but not for exponents. Thus, it is unfortunately not possible to use GS-proofs as a replacement for generalized Schnorr proofs in the "discrete logarithm based framework of cryptographic primitives" described earlier. To alleviate this, the research community has engaged on a quest for alternative cryptographic primitives that are *structure-preserving*, i.e., for which the public keys, inputs, and output consist of (source) group elements and the verification predicate is a conjunction of pairing product equations, thus making the primitives "GS-proof compatible" and enabling a similar, GS-proof-based, framework for the construction of complex cryptographic protocols. Such a framework is especially attractive because GS-proofs are "on-line" extractable, a property that is essential for the construction of UC-secure [24] protocols.

Structure-preserving realizations exist for primitives such as signature schemes [3,4,38,13,2], commitment schemes [3,5], and encryption schemes [16]. However, so far no *structure-preserving* constructions are known for important primitives including pseudorandom functions (PRF) [34,28], verifiable unpredictable functions (VUF)

[46], verifiable random functions (VRF) [46,40], simulatable verifiable random functions (VRF) [26], unique signatures (USig) [35,46,45], and deterministic encryption (DE) [9,12] despite the fact that these primitives are widely employed in the literature. Examples include efficient search on encrypted data [12] from deterministic encryption; micropayments [48] from unique signatures; resettable zero-knowledge proofs [47], updatable zero-knowledge databases [44], and verifiable transaction escrow schemes [42] from verifiable random functions. PRFs together with a proof of correct evaluation have been used to construct compact e-cash [17], keyword search [32], set intersection protocols [37], and adaptive oblivious transfer protocols [22,14,41]. We further refer to Abdalla et al. [1] and Hohenberger and Waters [40] for a good overview of applications of VRFs .

Our Results. In this paper we show that it is no coincidence that no structure-preserving constructions of PRF, VRF, VUF, USig, and DE are known: it is in fact impossible to construct them with *algebraic* algorithms. To this end, we provide a generic definition of a secure Structure-Preserving Deterministic Primitive (SPDP) and show that such a primitive cannot be built using algebraic operations only. The latter is a very reasonable restriction, indeed all constructions of structure-preserving primitives known to date are algebraic. We then show that PRF, VRF, VUF, and USig are special cases of a SPDP. We further extend our results to deterministic encryption and to "non-strictly" structure-preserving primitives which are allowed to have target group elements in their public keys and ranges. Regarding the latter, we show that such primitives cannot be constructed for asymmetric bilinear maps and that the possible constructions for symmetric maps are severely restricted in the operations they can use.

Let us point out that of course our results do not rule out the possibility of constructing efficient protocols from GS-proofs and non-structure-preserving primitives. Indeed a couple of such protocols are known where although some of the inputs include exponents (e.g., x) it turned out to be sufficient if only knowledge of a group elements (e.g., g^x) is proved. Examples here include the construction of a compact e-cash scheme [7] from the Dodis-Yampolskiy VRF [29] and of a so-called F-unforgeable signature scheme [6] and its use in the construction of anonymous credentials.

Related Work. Some impossibility results and lower bounds for structure-preserving primitives are known already. Abe et al. [4] show that a signature from a structure-preserving signature scheme must consist of at least three group elements when the signature algorithm is algebraic. They also give constructions meeting this bound. Lower bounds for structure-preserving commitment schemes are presented by Abe, Haralambiev and Ohkubo [5]. They show that a commitment cannot be shorter than the message and that verifying the opening of a commitment in a symmetric bilinear group setting requires evaluating at least two independent pairing product equations. They also provide optimal constructions that match these lower bounds.

To the best of our knowledge, there are no results about the (im)possibility of structure-preserving *deterministic* primitives.

Paper Organization. In Section 2 we specify our notation, define the syntax and security properties of an algebraic structure-preserving deterministic primitive, and show

that such primitives are impossible to construct. In Section 3 we present some general-
izations to the non-strictly structure-preserving case. Then, in Section 4, we show how
our result can be applied to structure-preserving PRF, VRF, VUF, and unique signatures.
Section 5 is devoted to the impossibility results for structure-preserving deterministic
encryption. Finally, Section 6 concludes the paper and points to open problems and
possible future research directions.

2 Definitions and Impossibility Results for Algebraic Structure-Preserving Deterministic Primitives

2.1 Preliminaries

Notation. We say that a function is *negligible* in the security parameter λ if it is asymptotically smaller than the inverse of any fixed polynomial in λ. Otherwise, the function
is said to be *non-negligible* in λ. We say that an event happens with *overwhelming* probability if it happens with probability $p(\lambda) \geq 1 - \mathsf{negl}(\lambda)$, where $\mathsf{negl}(\lambda)$ is a negligible
function of λ.

We denote by $Y \xleftarrow{\$} \mathsf{F}(X)$ a *probabilistic* algorithm that on input X outputs Y. A
similar notation $Y \leftarrow \mathsf{F}(X)$ is used for a *deterministic* algorithm with input X and
output Y. We abbreviate polynomial time as PT.

We use an upper-case, multiplicative notation for group elements. By $\mathbb{G}_1$ and $\mathbb{G}_2$ we
denote source groups and by $\mathbb{G}_T$ a target group. Let $\mathcal{G}$ be a bilinear group generator
that takes as input a security parameter 1^λ and outputs the description of a bilinear
group $\Lambda = (p, \mathbb{G}_1, \mathbb{G}_2, \mathbb{G}_T, e, G_1, G_2)$ where $\mathbb{G}_1$, $\mathbb{G}_2$, and $\mathbb{G}_T$ are groups of prime
order p, e is an efficient, non-degenerated bilinear map $e : \mathbb{G}_1 \times \mathbb{G}_2 \to \mathbb{G}_T$, and
G_1 and G_2 are generators of the groups $\mathbb{G}_1$ and $\mathbb{G}_2$, respectively. We denote by $\Lambda^* = (p, \mathbb{G}_1, \mathbb{G}_2, \mathbb{G}_T, e)$ the description Λ without the group generators. By Λ_{sym} we denote
the symmetric setting where $\mathbb{G}_1 = \mathbb{G}_2$ and $G_1 = G_2$. In the symmetric setting we
simply write $\mathbb{G}$ for both $\mathbb{G}_1$ and $\mathbb{G}_2$, and G for G_1 and G_2.

We also denote the set of all possible vectors of group elements from both $\mathbb{G}_1$ and
$\mathbb{G}_2$ as $\{\mathbb{G}_1, \mathbb{G}_2\}^*$, and from $\mathbb{G}_1$, $\mathbb{G}_2$ and $\mathbb{G}_T$ as $\{\mathbb{G}_1, \mathbb{G}_2, \mathbb{G}_T\}^*$. For example, if $H_1 \in \mathbb{G}_1, H_2 \in \mathbb{G}_2$ then $(H_2, H_1) \in \{\mathbb{G}_1, \mathbb{G}_2\}^*$ and $(H_1^a, H_2^b, H_2^c, H_1^d) \in \{\mathbb{G}_1, \mathbb{G}_2\}^*$ for
$a, b, c, d \in \mathbb{Z}_p$.

Algebraic Algorithms. For a bilinear group Λ generated by $\mathcal{G}$, an algorithm Alg that
takes group elements $(X_1, \ldots, X_n)$ as input and outputs a group element Y is called
algebraic if Alg always "knows" a representation of Y with respect to $(X_1, \ldots, X_n)$,
i.e., if there is a corresponding extractor algorithm Ext that outputs $(c_1, \ldots, c_n)$ such
that $Y = \prod X_i^{c_i}$ holds for all inputs and outputs of Alg. We consider this property with
respect to the source groups only. A formal definition for the minimal case where Alg
takes group elements from only one group $\mathbb{G}$ and outputs one element of this group is
provided below.

Definition 1 (Algebraic Algorithm). *Let Alg be a probabilistic PT algorithm that
takes as an input a bilinear group description Λ generated by $\mathcal{G}$, a tuple of group elements $(X_1, \ldots, X_n) \in \mathbb{G}^n$ for some $n \in \mathbb{N}$, and some auxiliary string $aux \in \{0, 1\}^*$*

and outputs a group element Y and a string ext. The algorithm Alg *is algebraic with respect to* $\mathcal{G}$ *if there is a probabilistic PT extractor algorithm* Ext *that takes the same input as* Alg *(including the random coins) and generates output* $(c_1, \ldots, c_n, ext)$ *such that for all* $\Lambda \xleftarrow{\$} \mathcal{G}(1^\lambda)$, *all polynomial sizes* n, *all* $(X_1, \ldots, X_n) \in \mathbb{G}^n$ *and all auxiliary strings* aux *the following inequality holds:*

$$\Pr\left[\begin{array}{l} (Y, ext) \leftarrow \mathsf{Alg}(\Lambda^*, X_1, \ldots, X_n, aux; r) ; \\ (c_1, \ldots, c_n, ext) \leftarrow \mathsf{Ext}(\Lambda^*, X_1, \ldots, X_n, aux; r) \end{array} \middle| Y \neq \prod X_i^{c_i} \right] \leq \mathsf{negl}(\lambda),$$

where the probability is taken over the choice of the coins r.

It is straightforward to extend this definition to algorithms that output multiple elements of the groups $\mathbb{G}_1$ and $\mathbb{G}_2$ of Λ. We note that all known constructions of structure-preserving primitives are algebraic in the sense defined here. Indeed if the considered algorithms were non-algebraic one could no longer prove their correct execution with GS-proofs.

One may see a similarity between the above definition and the knowledge of exponent assumption (KEA) [11] as both involve an extractor. We, however, emphasize that the algebraic algorithm definition characterizes honest algorithms, whereas the KEA is an assumption on adversaries.

2.2 Definitions of Structure-Preserving Deterministic Primitives

We define the syntax of a structure-preserving deterministic primitive (SPDP). An SPDP consists of the tuple of the following algorithms: (Setup, KeyGen, Comp, Prove, Verify). Besides the parameters generation (Setup), key generation (KeyGen), and main computation function (Comp), it includes proving (Prove) and verification (Verify) algorithms that guarantee that the output value was computed correctly using Comp. We call it provability property. It captures the verifiability notion of some deterministic primitives such as verifiable random functions, unique signatures, and verifiable unpredictable functions. Furthermore, for the deterministic primitives that do not have an inherent verification property such as pseudorandom functions and deterministic encryption, it covers their widely used combination with non-interactive proof systems. Indeed, one of the main advantages of the structure-preserving primitives and one of the reasons to construct those is their compatibility with the existing non-interactive zero-knowledge proof systems.

Definition 2 (Provable Structure-Preserving Deterministic Primitive). *Let* $\mathcal{G}$ *be a bilinear group generator that takes as an input a security parameter* 1^λ *and outputs a description of a bilinear group* $\Lambda = (p, \mathbb{G}_1, \mathbb{G}_2, \mathbb{G}_T, e, G_1, G_2)$. *Let* $\mathcal{SK}, \mathcal{PK}, \mathcal{X}, \mathcal{Y}, \mathcal{P}$ *be a secret key space, a public key space, a domain, a range, and a proof space, respectively. Let* $\mathsf{F} : \mathcal{SK} \times \mathcal{X} \to \mathcal{Y}$ *be a family of deterministic PT computable functions. A primitive* $\mathfrak{P} = $ (Setup, KeyGen, Comp, Prove, Verify) *that realizes* F *is called a Structure-Preserving Deterministic Primitive with respect to* $\Lambda, \mathcal{SK}, \mathcal{PK}, \mathcal{X}, \mathcal{Y}$, *and* $\mathcal{P}$ *if:*

- $\mathcal{PK}, \mathcal{X}, \mathcal{Y}, \mathcal{P} \subset \{\mathbb{G}_1, \mathbb{G}_2\}^*$. *Namely, the public key space, the domain, range and the proof space consist only of the source group elements.*

- $CP \xleftarrow{\$} \mathsf{Setup}(\Lambda)$ is a probabilistic algorithm that takes as input the group description Λ and outputs the common parameters CP. Without loss of generality we assume $\Lambda \in CP$.
- $(PK, SK) \xleftarrow{\$} \mathsf{KeyGen}(CP)$ is a probabilistic key generation algorithm that takes as input the common parameters and outputs a public key $PK \in \mathcal{PK}$ and a secret key $SK \in \mathcal{SK}$. It is assumed without loss of generality that PK includes CP, and SK includes PK.
- $Y \leftarrow \mathsf{Comp}(X, SK)$ is a deterministic algorithm that takes $X \in \mathcal{X}$ and a secret key SK as input and outputs $Y \in \mathcal{Y}$.
- $\Pi \xleftarrow{\$} \mathsf{Prove}(X, SK)$ is a probabilistic algorithm that takes X, SK as input and outputs a proof $\Pi \in \mathcal{P}$ for relation $Y = \mathsf{Comp}(X, SK)$.
- $0/1 \leftarrow \mathsf{Verify}(X, Y, \Pi, PK)$ is a deterministic verification algorithm that takes $(X \in \mathcal{X}, Y \in \mathcal{Y}, \Pi \in \mathcal{P}, PK \in \mathcal{PK})$ as input and accepts or rejects the proof that Y was computed correctly. The verification operations are restricted to the group operations and evaluation of pairing product equations (PPE), which for a bilinear group Λ and for group elements $A_1, A_2, \ldots \in \mathbb{G}_1, B_1, B_2, \ldots \in \mathbb{G}_2$ contained in X, Y, Π, PK and constants $c_{11}, c_{12}, \ldots \in \mathbb{Z}_p$ are equations of the form:

$$\prod_i \prod_j e(A_i, B_j)^{c_{ij}} = 1.$$

The following properties are required from a provable *structure-preserving deterministic primitive*:

1. **Uniqueness:** For all $\lambda, \Lambda, CP \xleftarrow{\$} \mathsf{Setup}(\Lambda)$ there are no values $(PK, X, Y, Y', \Pi, \Pi')$ such that $Y \neq Y'$ and $\mathsf{Verify}(X, Y, \Pi, PK) = \mathsf{Verify}(X, Y', \Pi', PK) = 1$.

2. **Provability:** For all $\lambda, \Lambda, CP \xleftarrow{\$} \mathsf{Setup}(\Lambda)$; $(PK, SK) \xleftarrow{\$} \mathsf{KeyGen}(CP)$; $X \in \mathcal{X}$; $Y \leftarrow \mathsf{Comp}(X, SK)$; $\Pi \xleftarrow{\$} \mathsf{Prove}(X, SK)$ it holds that $\mathsf{Verify}(X, Y, \Pi, PK) = 1$.

Now, we define two security properties. The *unpredictability* property states that no PT adversary can predict the output value Y for a fresh input X after having called the Comp and Prove oracles with inputs that are different from X. The *pseudorandomness* property states that no PT adversary can distinguish the output value Y from a random value.

Definition 3 (Unpredictability). *A Structure-Preserving Deterministic Primitive* $\mathfrak{P}$ *is unpredictable if for all probabilistic PT algorithms* $\mathcal{A}$

$$\Pr\left[\begin{array}{l} CP \xleftarrow{\$} \mathsf{Setup}(\Lambda) ; \\ (PK, SK) \xleftarrow{\$} \mathsf{KeyGen}(CP) ; \\ (X, Y) \leftarrow \mathcal{A}^{\mathsf{Comp}(\cdot, SK), \mathsf{Prove}(\cdot, SK)}(PK) \end{array} \middle| \begin{array}{l} Y = \mathsf{Comp}(X, SK) \wedge \\ X \notin S \end{array} \right] \leq \mathsf{negl}(\lambda)$$

where S is the set of inputs queried to the oracles Comp *and* Prove.

Definition 4 (Pseudorandomness). *A Structure-Preserving Deterministic Primitive* $\mathfrak{P}$ *is pseudorandom if for all probabilistic PT distinguishers* $\mathcal{D} = (\mathcal{D}_1, \mathcal{D}_2)$

$$
\Pr\left[
\begin{array}{l}
CP \xleftarrow{\$} \mathsf{Setup}(\Lambda)\,;\, (PK, SK) \xleftarrow{\$} \mathsf{KeyGen}(CP)\,; \\
(X, st) \leftarrow \mathcal{D}_1{}^{\mathsf{Comp}(\cdot, SK), \mathsf{Prove}(\cdot, SK)}(PK)\,; \\
Y_{(0)} \leftarrow F_{SK}(X)\,;\, Y_{(1)} \xleftarrow{\$} \mathcal{Y}\,;\, b \xleftarrow{\$} \{0,1\}\,; \\
b' \xleftarrow{\$} \mathcal{D}_2{}^{\mathsf{Comp}(\cdot, SK), \mathsf{Prove}(\cdot, SK)}(Y_{(b)}, st)
\end{array}
\,\middle|\,
\begin{array}{l}
b = b' \wedge \\
X \notin S
\end{array}
\right] \leq \frac{1}{2} + \mathsf{negl}(\lambda),
$$

where S is the set of queries to the oracles Comp *and* Prove.

One can see that a provable SPDP having the unpredictability property is a structure-preserving verifiable unpredictable function (VUF), and a provable SPDP with the pseudorandomness property is a structure-preserving verifiable random function (VRF).

2.3 Inexistence of Structure-Preserving Verifiable Unpredictable Functions

Now, we prove that a structure-preserving VUF as defined in the previous section cannot exist. Namely, we show that a provable SPDP cannot unpredictable according to Definition 3 because of its uniqueness property.

Theorem 1. *Let $\mathcal{G}$ be a bilinear group generator that takes as an input a security parameter 1^λ and outputs a description of bilinear groups $\Lambda = (p, \mathbb{G}_1, \mathbb{G}_2, \mathbb{G}_T, e, G_1, G_2)$. Let $\mathfrak{P}$ = (Setup, KeyGen, Comp, Prove, Verify) be a Provable Structure-Preserving Deterministic Primitive as in Definition 2. Suppose that the discrete logarithm problem is hard in the groups $\mathbb{G}_1, \mathbb{G}_2$ of Λ and let KeyGen, Comp, and Prove be restricted to the class of algebraic algorithms over Λ. Then $\mathfrak{P}$ is not unpredictable according to Definition 3.*

Proof. For simplicity, we first consider a symmetric bilinear setting ($\Lambda = \Lambda_{\mathsf{sym}}$), where $\mathcal{PK}, \mathcal{X}, \mathcal{Y}, \mathcal{P} \subset \{\mathbb{G}\}^*$. Furthermore, we consider the input X to consist only of a single group element. We then show that the same result holds for the input being a tuple of group elements from $\mathbb{G}$ and also in the asymmetric setting, for both Type 2 pairings (where an efficiently computable homomorphism from $\mathbb{G}_2$ to $\mathbb{G}_1$ exists and there is no efficiently computable homomorphism from $\mathbb{G}_1$ to $\mathbb{G}_2$), and Type 3 pairings (where there are no efficiently computable homomorphisms between $\mathbb{G}_1$ and $\mathbb{G}_2$) [33].

The outline of the proof is as follows. First, in Lemma 1 we show that because of the provability and uniqueness properties of $\mathfrak{P}$ as specified in Definition 2, the output of Comp must have a particular format, namely $\mathsf{Comp}(X, SK) = (G^{a_1} X^{b_1}, \ldots, G^{a_\ell} X^{b_\ell})$ for (secret) constants $a_1, \ldots, a_\ell, b_1, \ldots, b_\ell \in \mathbb{Z}_p$. Then, in Lemma 2, we prove that if the output of Comp has this format then the unpredictability property from Definition 3 does not hold for $\mathfrak{P}$. This means that a structure-preserving VUF cannot exist.

Lemma 1. *Let $\mathfrak{P}$ = (Setup, KeyGen, Comp, Prove, Verify) be a Structure-Preserving Deterministic Primitive such that KeyGen, Comp, and Prove are algebraic algorithms over Λ. If the discrete-logarithm problem is hard in the base group of Λ and $\mathfrak{P}$ meets the provability and uniqueness property as defined in Definition 2, then with an overwhelming probability it holds that $\mathsf{Comp}(X, SK) = (Y_1, \ldots, Y_\ell) = (G^{a_1} X^{b_1}, \ldots, G^{a_\ell} X^{b_\ell})$ for constants $a_1, \ldots, a_\ell, b_1, \ldots, b_\ell \in \mathbb{Z}_p$.*

Proof. Fix $(PK, SK) \xleftarrow{\$} \mathsf{KeyGen}(CP)$, where $PK \subset \{\mathbb{G}\}^*$. Let $x \xleftarrow{\$} \mathbb{Z}_p, X = G^x$.

First, notice that because Comp, Prove and KeyGen are algebraic algorithms, their outputs can be expressed as

$$\mathsf{Comp}(X, SK) = Y = (Y_1, \ldots, Y_\ell) \text{ with } Y_i = G^{a_i} X^{b_i},$$

$$\mathsf{Prove}(X, SK) = \Pi = (\Pi_1, \ldots, \Pi_n) \text{ with } \Pi_j = G^{u_j} X^{v_j}, \text{ and}$$

$$PK = (S_1, \ldots, S_m) \text{ with } S_f = G^{s_f},$$

where $a_i = H_{1,i}(X, SK), b_i = H_{2,i}(X, SK), v_j = H_{3,j}(X, SK; r), u_j = H_{4,j}(X, SK; r)$, and $H_{\ell,m}$ are arbitrary functions, and r is the randomness used by the Prove algorithm. We note that a_i, b_i, u_j, and v_j can depend on X in an arbitrary manner, but, as Comp and Prove are algebraic, one can extract a_i, b_i, u_j, and v_j as values from $\mathbb{Z}_p$ using the extractors of algorithms Comp and Prove.

Second, we recall that according to Definition 2 the verification algorithm consists of pairing product equations (PPE). Let the k-th PPE used in the verification algorithm be

$$\prod_{f=1}^{m} e(S_f, X^{c_{k,1,f}}) \prod_{t=1}^{m} S_t^{c_{k,2,f,t}} \prod_{i=1}^{\ell} Y_i^{c_{k,3,f,i}} \prod_{j=1}^{q} \Pi_j^{c_{k,4,f,j}}) \prod_{q=1}^{n} e(\Pi_q, \prod_{j=1}^{n} \Pi_j^{c_{k,5,q,j}}) \cdot$$

$$\cdot e(X, X^{c_{k,6}} \prod_{i=1}^{\ell} Y_i^{c_{k,7,i}} \prod_{j=1}^{q} \Pi_j^{c_{k,8,j}}) \prod_{w=1}^{\ell} e(Y_w, \prod_{i=1}^{\ell} Y_i^{c_{k,9,w,i}} \prod_{j=1}^{n} \Pi_j^{c_{k,10,w,j}}) = 1.$$

The intuition behind the proof is the following. We note that Comp should perform the computation without necessarily knowing the discrete logarithm of the input – otherwise one can use Comp to solve the discrete logarithm for the input X. Now, one can see that the relation in the exponents of the k-th PPE for the tuple (X, Y, Π, PK) induce a polynomial $Q_k(x)$ in the discrete logarithm $x = \log_G X$. Basically, we can re-write the k-th PPE as $e(G, G)^{Q_k(x)} = 1$. So, first, we prove that $Q_k(x)$ is a trivial function, otherwise it is possible to solve the discrete logarithm problem for the given X by solving Q_k. Second, we show that if Q_k is trivial then, by the uniqueness property, a_i and b_i are constants. Let a_i, b_i, u_j, and v_j be the values computed for one specific $X : Y_i = G^{a_i} X^{b_i}$, $\Pi_j = G^{u_j} X^{v_j}$, and $\mathsf{Verify}(PK, X, Y, \Pi) = 1$. Proposition 2 shows that these values can be reused to compute a correct $\tilde{Y}$ for any other $\tilde{X} \in \mathcal{X}$. So, if $\tilde{Y}_i$ is computed as $G^{a_i} \tilde{X}^{b_i}$ and $\tilde{\Pi}_j$ as $G^{u_j} \tilde{X}^{v_j}$, instead of using the normal computation procedures, then $\tilde{X}, \tilde{Y}, \tilde{\Pi}, PK$ is also accepted by the verification algorithm due to the triviality of Q_k. Then, from the uniqueness property, it follows that these a_i, b_i are the only valid values, i.e., constants.

Now we provide the proof in detail. First, we prove that all polynomials Q_k induced by the verification PPEs, as described above, are constants with overwhelming probability.

Proposition 1. *If the discrete logarithm problem in the base group of Λ is hard, then Q_k is a trivial function.*

Proof. The proof is done by constructing a reduction algorithm R that takes as an input a group description $\Lambda_{\mathsf{sym}} = (p, \mathbb{G}, \mathbb{G}_T, e, G)$ generated by a group generator $\mathcal{G}(1^\lambda)$ and a random element $X \in \mathbb{G}$ and outputs $x \in \mathbb{Z}_p$ that satisfies $X = G^x$ with a high probability.

The reduction algorithm R works as follows. It first takes Λ as an input and sets the common parameters $CP = \Lambda$. R then runs KeyGen(CP), Comp(X, SK), and Prove(X, SK) for the given X. It also runs the corresponding extractors for KeyGen, Comp, and Prove. The extractor for KeyGen outputs representations s_f that satisfy $S_f = G^{s_f}$ with overwhelming probability. Similarly, the extractor for Comp outputs representations a_i and b_i such that $Y_i = G^{a_i} X^{b_i}$, and the extractor for Prove outputs u_j and v_j such that $\Pi_j = G^{u_j} X^{v_j}$ as concrete values in $\mathbb{Z}_p$.

This set of extracted exponents s_f, a_i, b_i, u_j, and v_j induce a quadratic equation Q_k in the exponents of the k-th pairing product verification equation (PPE). Let us call the variable of this exponent equation $\tilde{x}$, then we can write the k-th PPE as $e(G, G)^{Q_k(\tilde{x})} = 1$. Given the representations, R can compute $Q_k(\tilde{x}) : d_2\tilde{x}^2 + d_1\tilde{x} + d_0 = 0$ in $\mathbb{Z}_p$. The condition that $Q_k(\tilde{x})$ is non-trivial guarantees that $d_2 \neq 0$ or $d_1 \neq 0$. But then R can solve $Q_k(\tilde{x})$ for $\tilde{x}$ with standard algebra. Due to the provability property, x is one of the possible solutions to $\tilde{x}$. So if the equation is non-trivial, then we can solve this equation for $\tilde{x}$ and obtain the discrete logarithm of $X : \tilde{x} = x$. Therefore, if the discrete logarithm problem is hard in the base group of Λ, Q_k must be trivial. $\square$

Now we show that if Q_k is trivial then by the provability and uniqueness properties a_i and b_i are constants.

Proposition 2. *Fix (PK, SK, X) and let $a_i \leftarrow H_{1,i}(X, SK)$, $b_i \leftarrow H_{2,i}(X, SK)$, $u_j \leftarrow H_{3,j}(X, SK, r)$ and $v_j \leftarrow H_{4,j}(X, SK, r)$. If all the relations in the exponents of the PPEs are trivial, then, for any $\tilde{X} \in \mathbb{G}$, $\tilde{Y} = (\tilde{Y}_1, \ldots, \tilde{Y}_\ell)$ with $\tilde{Y}_i = G^{a_i}\tilde{X}^{b_i}$ and $\tilde{\Pi} = (\tilde{\Pi}_1, \ldots, \tilde{\Pi}_n)$ with $\tilde{\Pi}_j = G^{u_j}\tilde{X}^{v_j}$, it holds that $(\tilde{X}, \tilde{Y}, \tilde{\Pi}, PK)$ will be accepted by the verification algorithm.*

Proof. Consider fixed (PK, SK, X), any $\tilde{X} \in \mathbb{G}$, and $\tilde{Y}$ and $\tilde{\Pi}$ computed from $\tilde{X}$ as specified in the proposition. Note that the verification algorithm only evaluates PPEs and performs group memberships tests. First, all group memberships tests are clearly successful for the above tuple $(\tilde{X}, \tilde{Y}, \tilde{\Pi}, PK)$. Since all polynomials Q_k are trivial and due to the way in which $\tilde{Y}$ and $tilde\Pi$ are defined, it holds that the result of evaluating the k-th PPE will be the same for any tuple $(\tilde{X}, \tilde{Y}, \tilde{\Pi}, PK)$. Therefore, Verify$(\tilde{X}, \tilde{Y}, \tilde{\Pi}, PK)$ should output the same value for every $\tilde{X} \in \mathbb{G}$. Now, considering the case where $\tilde{X} = X$ we have that $\tilde{Y} = (\tilde{Y}_1, \ldots, \tilde{Y}_\ell)$ with $\tilde{Y}_i = G^{a_i}X^{b_i}$ and $\tilde{\Pi} = (\tilde{\Pi}_1, \ldots, \tilde{\Pi}_n)$ with $\tilde{\Pi}_j = G^{u_j}X^{v_j}$. But due to the correctness of the extractors of Comp and Prove, these $\tilde{Y}$ and $\tilde{\Pi}$ are exactly the outputs of Comp(X, SK) and Prove(X, SK). Therefore, by the provability property, it holds that Verify$(X, \tilde{Y}, \tilde{\Pi}, PK) = 1$ for $\tilde{X} = X$; and thus, for any $\tilde{X} \in \mathbb{G}$, Verify$(\tilde{X}, \tilde{Y}, \tilde{\Pi}, PK) = 1$ also. $\square$

Now, for an arbitrary $\tilde{X} \in \mathbb{G}$, consider the tuple $(PK, SK, \tilde{X}, \tilde{Y}, \tilde{\Pi}, a_1, \ldots, a_\ell, b_1, \ldots, b_\ell)$ of values as defined above. $\tilde{\Pi}$ is valid proof for $(\tilde{X}, \tilde{Y})$ and thus the uniqueness property guarantees that there is no other $\hat{Y} \neq \tilde{Y}$ for which there is a valid proof that $\hat{Y}$ is the output corresponding to $\tilde{X}$. But the provability property guarantees that for

$(\tilde{X}, \text{Comp}(\tilde{X}, SK))$ there is a valid proof of correctness. Hence, for any $\tilde{X} \in \mathbb{G}$, it holds that

$$\text{Comp}(\tilde{X}, SK) = \tilde{Y} = (\tilde{Y}_1, \ldots, \tilde{Y}_\ell) = (G^{a_1}\tilde{X}^{b_1}, \ldots, G^{a_\ell}\tilde{X}^{b_\ell}). \qquad \square$$

Lemma 2. *Suppose that* $\mathfrak{P} = (\text{Setup}, \text{KeyGen}, \text{Comp}, \text{Prove}, \text{Verify})$ *is a provable Structure-Preserving Deterministic Primitive such that* $\text{Comp}(X, SK) = (Y_1, \ldots, Y_\ell)$ $= (G^{a_1}X^{b_1}, \ldots, G^{a_\ell}X^{b_\ell})$ *for some constants* $a_1, \ldots, a_\ell, b_1, \ldots, b_\ell \in \mathbb{Z}_p$. *Then* $\mathfrak{P}$ *does not satisfy the unpredictability requirement from Definition 3.*

Proof. Pick $\hat{X}, \tilde{X}$ and define $\overline{X}$, such that $\overline{X} = \hat{X}^2/\tilde{X} \notin \{\hat{X}, \tilde{X}\}$. Then an adversary that learns

$$\text{Comp}(\hat{X}, SK) = (\hat{Y}_1, \ldots, \hat{Y}_\ell) = (G^{a_1}\hat{X}^{b_1}, \ldots, G^{a_\ell}\hat{X}^{b_\ell}) \text{ and}$$

$$\text{Comp}(\tilde{X}, SK) = (\tilde{Y}_1, \ldots, \tilde{Y}_\ell) = (G^{a_1}\tilde{X}^{b_1}, \ldots, G^{a_\ell}\tilde{X}^{b_\ell})$$

can compute the value of $\text{Comp}(\overline{X}, SK)$ as:

$$\left(\frac{\hat{Y}_1^2}{\tilde{Y}_1}, \ldots, \frac{\hat{Y}_\ell^2}{\tilde{Y}_\ell}\right) = \left(\frac{G^{2a_1}\hat{X}^{2b_1}}{G^{a_1}\tilde{X}^{b_1}}, \ldots, \frac{G^{2a_\ell}\hat{X}^{2b_\ell}}{G^{a_\ell}\tilde{X}^{b_\ell}}\right) =$$

$$\left(G^{a_1}\left(\frac{\hat{X}^2}{\tilde{X}}\right)^{b_1}, \ldots, G^{a_\ell}\left(\frac{\hat{X}^2}{\tilde{X}}\right)^{b_\ell}\right) = \left(G^{a_1}\overline{X}^{b_1}, \ldots, G^{a_\ell}\overline{X}^{b_\ell}\right) = \text{Comp}(\overline{X}, SK),$$

and therefore $\mathfrak{P}$ is not unpredictable. $\square$

Now, we show that the same result holds for the input being a tuple of group elements from $\mathbb{G}$.

X Is a Tuple of Group Elements. Both Lemmas 1 and 2 can be easily modified to the case where X consists of more than one (say t) group element as follows. The reduction algorithm, after receiving the discrete logarithm challenge X_1, will choose $t-1$ random exponents $x_2, \ldots, x_t$ and fix X_i as G^{x_i} for $i = 2, \ldots, t$. Then the lemmas use the first group element X_1 in the place of the original X. Note that in the computation of the Y_j and Π_j the exponents corresponding to $X_2, \ldots, X_t$ can essentially be incorporated into $H_{1,j}(X, SK)$ and $H_{3,j}(X, SK)$ since the prover knows $x_2, \ldots, x_t$. If the quadratic equations $Q_k(\tilde{x}_1)$ in the exponents of the PPEs are not trivial, then the first element of the input X can be used to solve the discrete logarithm problem; otherwise, supposing that the uniqueness and provability properties hold, the elements of the output will be of the form $\tilde{Y}_i = G^{a_i}\tilde{X}_1^{b_i}$ (for the fixed values $x_2, \ldots, x_t$) and this can be used to break the unpredictability by asking two queries in which only the first elements of the inputs are different (i.e., $\hat{X}_1$ and $\tilde{X}_1$) and then learning the output corresponding to a third input which has $\overline{X}_1 = \hat{X}_1^2/\tilde{X}_1$ and the remaining elements equal to the ones of the oracle queries.

Asymmetric Bilinear Groups Setting. Lemmas 1 and 2 can be generalized to the asymmetric setting as well. We consider both Type 2 and Type 3 pairings. The case where there are efficiently computable homomorphisms in both directions can be reinterpreted as a symmetric setting [33]. If $\mathcal{X}$ consists of t group elements, we choose $t-1$ random exponents $x_2, \ldots, x_t$ and fix X_i as $G_1^{x_i}$ if the i-th input element is in group $\mathbb{G}_1$, or $G_2^{x_i}$ if the i-th input element is in group $\mathbb{G}_2$. Then either some quadratic equation $Q_k(\tilde{x}_1)$ in the exponents of the PPEs is not trivial in x_1 and this can be used to solve the discrete logarithm problem in the base group in which X_1 is contained, or one of the three security properties (provability, uniqueness and unpredictability) does not hold.

In the case of Type 3 pairings, where there are no efficiently computable homomorphisms between the groups, each Y_j (let $\mathbb{G}_c$ denote the group in which it is and G_c its generator) is of the form

$$Y_j = G_c^{H_{1,j}(X,SK)} X_1^{H_{2,j}(X,SK)}$$

(where $H_{2,j}(X, SK) = 0$ if X_1 and Y_j are not in the same group) and each Π_j (that is in the group $\mathbb{G}_c$) is of the form $\Pi_j = G_c^{H_{3,j}(X,SK)} X_1^{H_{4,j}(X,SK)}$ (where $H_{4,j}(X, SK) = 0$ if X_1 and Π_j are not in the same group), in both cases with the exponents corresponding to $X_2, \ldots, X_t$ incorporated into $H_{1,j}(X, SK)$ and $H_{3,j}(X, SK)$. Then the argument continues as in the previous cases.

In the case of Type 2 pairings, there is an efficiently computable homomorphism $\phi : \mathbb{G}_2 \to \mathbb{G}_1$. Then an element Y_j of the output (or an element Π_j of the proof) that is in the group $\mathbb{G}_1$ can depend on both group generators and on X_1 or its mapping $\phi(X_1)$ into $\mathbb{G}_1$.

I.e., if $X_1 \in \mathbb{G}_1$, Y_j and Π_j have the form:

$$Y_j = G_1^{H_{1,j}(X,SK)} X_1^{H_{2,j}(X,SK)} \phi(G_2)^{H_{5,j}(X,SK)};$$

$$\Pi_j = G_1^{H_{3,j}(X,SK)} X_1^{H_{4,j}(X,SK)} \cdot \phi(G_2)^{H_{6,j}(X,SK)};$$

or if $X_1 \in \mathbb{G}_2$, Y_j and Π_j have the form:

$$Y_j = G_1^{H_{1,j}(X,SK)} \phi(X_1)^{H_{2,j}(X,SK)} \phi(G_2)^{H_{5,j}(X,SK)};$$

$$\Pi_j = G_1^{H_{3,j}(X,SK)} \phi(X_1)^{H_{4,j}(X,SK)} \phi(G_2)^{H_{6,j}(X,SK)}.$$

Then we should have $H_{1,j}(X, SK) = a_j$, $H_{2,j}(X, SK) = b_j$ and $H_{5,j}(X, SK) = z_j$ for constants a_j, b_j, and z_j if the provability and uniqueness hold. But in this case the unpredictability does not hold for the same reasons as before.

Putting Lemmas 1 and 2 together completes the proof of Theorem 1.   □

3   Impossibility Results for "Non-strictly" Structure-Preserving Primitives

One can see that the definition above only captures so-called "strictly" structure-preserving primitives, i.e., $\mathcal{PK}$ and $\mathcal{Y}$ can contain only source group elements. Let us discuss the case of structure-preserving primitives that also have target group elements in their public key space and/or their range. A target group element can be represented by 2 source group elements using pairing randomization techniques [3] or even deterministically, by fixing the "randomization" exponents. By this the provability property can be preserved. Now, the question is: if the uniqueness property holds is the output unpredictable according to the definition above? In this section, we show that our impossibility result can be extended to some cases of "non-strictly" structure-preserving primitives, formally defined below:

Definition 5 ("Non-strictly" Structure-Preserving Deterministic Primitive). *Let $\mathcal{G}$ be a bilinear group generator that takes as an input a security parameter 1^λ and outputs a description of bilinear groups $\Lambda = (p, \mathbb{G}_1, \mathbb{G}_2, \mathbb{G}_T, e, G_1, G_2)$. Let $\mathcal{SK}, \mathcal{PK}, \mathcal{X}, \mathcal{Y}, \mathcal{P}$ be the secret key space, public key space, domain, range, and the proof space, respectively. Let $\mathfrak{P} =$ (Setup, KeyGen, Comp, Prove, Verify) be a Structure-Preserving Deterministic Primitive as defined in Definition 2, except that the range of Comp and KeyGen can contain also target group elements ($\mathcal{Y}, \mathcal{PK} \subset \{\mathbb{G}_1, \mathbb{G}_2, \mathbb{G}_T\}^*$). Then the primitive $\mathfrak{P}$ is called a "non-strictly" structure-preserving deterministic primitive.*

First, we extend the notion of the algebraic algorithms from Definition 1 to operate in all groups of Λ. We provide a formal definition as follows.

Definition 6 (Algebraic Algorithms over Λ). *Let Alg be a probabilistic PT algorithm that takes as an input a bilinear group description Λ generated by $\mathcal{G}$, two tuples of group elements $(X_{1,1}, \ldots, X_{1,n}) \in \{\mathbb{G}_1\}^n$ and $(X_{2,1}, \ldots, X_{2,m}) \in \{\mathbb{G}_2\}^m$ for some $n, m \in \mathbb{N}$, and some auxiliary string $aux \in \{0,1\}^*$ and outputs group elements $Y \in \mathbb{G}_1, W \in \mathbb{G}_2$, and $Z \in \mathbb{G}_T$ and a string ext. The algorithm Alg is algebraic with respect to $\mathcal{G}$ if there is a probabilistic PT extractor algorithm Ext that takes the same input as Alg (including the random coins) and generates output $(c = (c_1, \ldots, c_n), d = (d_1, \ldots, d_m), f = (f_1, \ldots, f_{nm}), ext)$ such that for all $\Lambda \xleftarrow{\$} \mathcal{G}(1^\lambda)$, all polynomial sized n, m, all $(X_{1,1}, \ldots, X_{1,n}) \in \{\mathbb{G}_1\}^n, (X_{2,1}, \ldots, X_{2,m}) \in \{\mathbb{G}_2\}^m$ and all auxiliary strings aux the following inequality holds over the choice of the coins r and for $X = (X_{1,1}, \ldots, X_{1,n}, X_{2,1}, \ldots, X_{2,m})$:*

$$\Pr\left[\begin{array}{l}(Y, W, Z, ext) \leftarrow \mathsf{Alg}(\Lambda^*, X, aux; r) \,; \\ (c, d, f, ext) \leftarrow \mathsf{Ext}(\Lambda^*, X, aux; r)\end{array} \middle| \begin{array}{l} Y \neq \prod_{i=1}^n X_{1,i}^{c_i} \vee \\ W \neq \prod_{j=1}^m X_{2,j}^{d_j} \vee \\ Z \neq \prod_{i=1}^n \prod_{j=1}^m \\ \quad e(X_{1,i}, X_{2,j})^{f_{(j-1)n+i}} \end{array}\right] \leq \mathsf{negl}(\lambda),$$

where the probability is taken over the choice of the coins r.

Similarly to Definition 1, this definition can be extended to algorithms that output multiple elements of the groups of Λ. Then one can use the extractors of KeyGen and Comp to also extract representations for target group elements output by them.

Now, as we discussed in Section 2, pairing randomization techniques allow us to preserve the provability property. Here we show that if the uniqueness property (according to Definition 2) holds, then the unpredictability property does not hold in the asymmetric setting and also in some cases for the symmetric setting.

Theorem 2. *Let* $\mathfrak{P}$ = (Setup, KeyGen, Comp, Prove, Verify) *be a "non-strictly" structure-preserving deterministic primitive as defined in Definition 5. Suppose that the discrete logarithm problem is hard in the source groups of the asymmetric bilinear groups* Λ *and let* KeyGen, Comp, *and* Prove *be restricted to the class of algebraic algorithms over* Λ. *Then* $\mathfrak{P}$ *is not unpredictable according to Definition 3.*

Proof. The outline of the proof is the same as the one for Theorem 1. First, in Lemma 3 we show that for any $\mathfrak{P}$ that is provable and has the uniqueness property as specified in Definition 2, the output of Comp must have a particular format, namely Comp(X, SK) = $(Y_1, \ldots, Y_\ell, W_1, \ldots, W_n, Z_1, \ldots, Z_m)$ with $Y_i = G_1^{a_{1,i}} X_1^{b_{1,i}}, W_i = G_2^{a_{2,i}}$ $X_2^{b_{2,i}}$, and $Z_i = e(G_1, G_2)^{a'_i} e(X_1, G_2)^{b'_i} e(G_1, X_2)^{c'_i} e(X_1, X_2)^{d'_i}$ for $X_1, G_1 \in \mathbb{G}_1$, and $X_2, G_2 \in \mathbb{G}_2$ and for constants $a_{j,i}, b_{j,i}, a'_i, b'_i, c'_i, d'_i \in \mathbb{Z}_p$. Then in Lemma 4 we prove that if the output of Comp has this format then the unpredictability property from Definition 3 does not hold for $\mathfrak{P}$, and thus the latter cannot exist.

We note that Lemma 3 holds for both symmetric and asymmetric settings. Lemma 4, however, holds only for the asymmetric setting, thus the result of this theorem holds only for the asymmetric setting.

Lemma 3. *If the discrete-logarithm problem is hard in the source groups of* Λ *and* $\mathfrak{P}$ *has the provability and the uniqueness properties (Definition 2), then with overwhelming probability it holds that* Comp(X, SK) = $(Y_1, \ldots, Y_\ell, W_1, \ldots, W_n, Z_1, \ldots, Z_m)$ *with* Y_i = $G_1^{a_{1,i}} X_1^{b_{1,i}}, W_i$ = $G_2^{a_{2,i}} X_2^{b_{2,i}}$, *and* Z_i = $e(G_1, G_2)^{a'_i} e(X_1, G_2)^{b'_i} e(G_1, X_2)^{c'_i} \cdot e(X_1, X_2)^{d'_i}$ *for* $X_1, G_1 \in \mathbb{G}_1; X_2, G_2 \in \mathbb{G}_2$; *and for constants* $a_{j,i}, b_{j,i}, a'_i, b'_i, c'_i, d'_i \in \mathbb{Z}_p$.

Proof. Similarly to Lemma 1, we start with the symmetric setting ($\mathbb{G}_1 = \mathbb{G}_2 = \mathbb{G}$) and a single group element as an input for simplicity. Fix $(PK, SK) \xleftarrow{\$}$ KeyGen(CP), where a public key consists of both source and target group elements: $PK \subset \{\mathbb{G}, \mathbb{G}_T\}^*$. Let $x \xleftarrow{\$} \mathbb{Z}_p, X = G^x$.

First, since Comp is deterministic and KeyGen, Comp and Prove are all algebraic algorithms over Λ, without loss of generality, their outputs can be expressed as

$$\mathsf{Comp}(X, SK) = Y = (Y_1, \ldots, Y_\ell, Z_1, \ldots, Z_{\ell'}) \text{ with } Y_i = G^{a_{1,i}} X^{b_{1,i}},$$

$$Z_i = e(G^{a_i} X^{b_i}, G^{c_i} X^{d_i}) = e(G, G)^{a'_i} e(X, G)^{b'_i} e(G, X)^{c'_i} e(X, X)^{d'_i};$$

$$\mathsf{Prove}(X, SK) = \Pi = (\Pi_1, \ldots, \Pi_n) \text{ with } \Pi_j = G^{u_j} X^{v_j},$$

$$PK = (S_1, \ldots S_m, T_1, \ldots T_{m'}) \text{ with } S_f = G^{s_f}, T_{f'} = G^{t_{f'}};$$

where $a_i = H_{a,i}(X, SK), b_i = H_{b,i}(X, SK), a'_i = H_{a',i}(X, SK), b'_i = H_{b',i}(X, SK), c'_i = H_{c',i}(X, SK), d'_i = H_{d',i}(X, SK), v_j = H_{v,j}(X, SK, r), u_j = H_{u,j}(X, SK, r)$, and $H_{*,*}$ are arbitrary functions. We note that $a_i, b_i, a'_i, b'_i, c'_i, d'_i, u_j, v_j$ can depend on X in an arbitrary manner, but since Comp and Prove are algebraic, one can extract $a_i, b_i, a'_i, b'_i, c'_i, d'_i, u_j$, and v_j as values from $\mathbb{Z}_p$ using the extractors of algorithms KeyGen, Comp and Prove.

Second, we recall that, according to Definition 2, the verification algorithm consists of pairing product equations (PPE). Let the k-th PPE from the verification algorithm be

$$e(X, X^{c_{k,6}} \prod_{i=1}^{\ell} Y_i^{c_{k,7,i}} \prod_{j=1}^{q} \Pi_j^{c_{k,8,j}}) \prod_{w=1}^{\ell} e(Y_w, \prod_{i=1}^{\ell} Y_i^{c_{k,9,w,i}} \prod_{j=1}^{n} \Pi_j^{c_{k,10,w,j}}) \prod_{i=1}^{\ell'} Z_i^{c_{k,i}} \cdot$$

$$\cdot \prod_{f=1}^{m} e(S_f, X^{c_{k,1,f}} \prod_{t=1}^{m} S_t^{c_{k,2,f,t}} \prod_{i=1}^{\ell} Y_i^{c_{k,3,f,i}} \prod_{j=1}^{q} \Pi_j^{c_{k,4,f,j}}) \prod_{q=1}^{n} e(\Pi_q, \prod_{j=1}^{n} \Pi_j^{c_{k,5,q,j}}) = T_k.$$

The proof works very similarly to the one of Lemma 1. One can see that the relation in the exponents of the k-th PPE for the tuple (X, Y, Π, PK) induce a polynomial $Q_k(x)$ in the discrete logarithm $x = \log_G X$. Basically, we can re-write the k-th PPE as $e(G, G)^{Q_k(x)} = 1$. So, first, we prove that $Q_k(x)$ is a trivial function, otherwise it is possible to solve the discrete logarithm problem for the given X by solving Q_k. Second, if Q_k is trivial, then by the uniqueness property $a_i, b_i, a_i', b_i', c_i'$, and d_i' are constants. Let $a_i, b_i, a_i', b_i', c_i', d_i', u_j$, and v_j be the correct values computed for one specific X. Then we have that $Y_i = G^{a_{1,i}} X^{b_{1,i}}$, $Z_i = e(G, G)^{a_i'} \cdot e(X, G)^{b_i'} e(G, X)^{c_i'} e(X, X)^{d_i'}$, $\Pi_j = G^{u_j} X^{v_j}$, and $\text{Verify}(PK, X, Y, \Pi) = 1$. Proposition 2 shows that these values can be reused to compute a correct $\tilde{Y}$ for any other $\tilde{X} \in \mathcal{X}$. So if $\tilde{Y}_i, \tilde{Z}_i$, and $\tilde{\Pi}_i$ are computed as $\tilde{Y}_i = G^{a_{1,i}} \tilde{X}^{b_{1,i}}$, $\tilde{Z}_i = e(G, G)^{a_i'} e(\tilde{X}, G)^{b_i'} e(G, \tilde{X})^{c_i'} e(\tilde{X}, \tilde{X})^{d_i'}$, and $\tilde{\Pi}_j = G^{u_j} \tilde{X}^{v_j}$, respectively, instead of using the normal computation procedures, then $(\tilde{X}, \tilde{Y}, \tilde{\Pi})$ are also accepted by the verification algorithm due to the triviality of Q_k. Then, from the uniqueness property, it follows that these $a_i, b_i, a_i', b_i', c_i'$, and d_i' are the only valid values, i.e., constants.

Similarly to the proof of Lemma 1 the proof above can be extended to the asymmetric setting and to the case when the input consists of a tuple of group elements. □

Lemma 4. *Let Λ be a description of asymmetric bilinear groups. Let $\mathfrak{P} =$ (Setup, KeyGen, Comp, Prove, Verify) be a "non-strictly" structure-preserving deterministic primitive as defined in Definition 5. If $\mathfrak{P}$ is provable and unique according to Definition 2 then $\mathfrak{P}$ does not satisfy the unpredictability requirement of Definition 3.*

Proof. We consider an input $X = (X_1, X_2)$ consisting of a single element from the first source group $X_1 \in \mathbb{G}_1$ and a single element from the second source group $X_2 \in \mathbb{G}_2$. Applying Lemma 3, the output of Comp looks as follows, without loss of generality:

$$\text{Comp}(X, SK) = (Y_1, \ldots, Y_\ell, W_1, \ldots, W_{\ell''}, Z_1, \ldots, Z_{\ell'}) \text{ with } Y_i = G_1^{a_{1,i}} X_1^{b_{1,i}},$$

$$W_i = G_2^{a_{2,i}} X_2^{b_{2,i}}, \text{ and } Z_i = e(G_1^{a_i} X_1^{b_i}, G_2^{c_i} X_2^{d_i}) =$$

$$e(G_1, G_2)^{a_i'} e(X_1, G_2)^{b_i'} e(G_1, X_2)^{c_i'} e(X_1, X_2)^{d_i'}.$$

Pick $\hat{X}_1, \tilde{X}_1 \in \mathbb{G}_1$ and $X_2 \in \mathbb{G}_2$, set $\hat{X} = (\hat{X}_1, X_2)$ and $\tilde{X} = (\tilde{X}_1, X_2)$ and define $\overline{X}$, such that $\overline{X}_1 = \hat{X}_1^2/\tilde{X}_1 \notin \{\hat{X}_1, \tilde{X}_1\}$, and $\overline{X}_2 = X_2$. As we proved in Lemma 2, the unpredictability does not hold for source group elements. Now we show that with this choice of the input the adversary can also compute the target group elements of the output. For simplicity, let us now assume the output to consist only of the target group elements.

An adversary that learns $\mathsf{Comp}(\hat{X}, SK) = (\hat{Z}_1, \ldots, \hat{Z}_{\ell'})$ with $\hat{Z}_i = e(G_1^{a_i} \hat{X}_1^{b_i}, G_2^{c_i} \hat{X}_2^{d_i}) = e(G_1, G_2)^{a_i} e(\hat{X}_1, G_2)^{b_i} e(G_1, \hat{X}_2)^{c_i} e(\hat{X}_1, \hat{X}_2)^{d_i}$ and $\mathsf{Comp}(\tilde{X}, SK) = (\tilde{Z}_1, \ldots, \tilde{Z}_{\ell'})$ with $\tilde{Z}_i = e(G_1^{a_i} \tilde{X}_1^{b_i}, G_2^{c_i} \tilde{X}_2^{d_i}) = e(G_1, G_2)^{a_i} e(\tilde{X}_1, G_2)^{b_i} e(G_1, \tilde{X}_2)^{c_i} e(\tilde{X}_1, \tilde{X}_2)^{d_i}$ can already compute the value of $\mathsf{Comp}(\overline{X}, SK) = (\overline{Z}_1, \ldots, \overline{Z}_{\ell'})$ as $\left(\frac{\hat{Z}_1^2}{\tilde{Z}_1}, \ldots, \frac{\hat{Z}_{\ell'}^2}{\tilde{Z}_{\ell'}} \right)$, because we have that

$$\frac{\hat{Z}_i^2}{\tilde{Z}_i} = \frac{e(G_1, G_2)^{2a_i'} e(\hat{X}_1, G_2)^{2b_i'} e(G_1, \hat{X}_2)^{2c_i'} e(\hat{X}_1, \hat{X}_2)^{2d_i'}}{e(G_1, G_2)^{a_i'} e(\tilde{X}_1, G_2)^{b_i'} e(G_1, \tilde{X}_2)^{c_i'} e(\tilde{X}_1, \tilde{X}_2)^{d_i'}} =$$

$$e(G_1, G_2)^{a_i'} \cdot e \left(\frac{\hat{X}_1^2}{\tilde{X}}, G_2 \right)^{b_1'} \cdot e(G_1, \overline{X}_2)^{c_1'} \cdot e \left(\frac{\hat{X}_1^2}{\tilde{X}_1}, \overline{X}_2 \right)^{d_1'} = \overline{Z}_i.$$

Therefore, $\mathfrak{P}$ is not unpredictable for the target group elements either. $\qquad\square$

One can see that for the symmetric setting the result above holds only if there is no element $e(X_1, X_2)^{d_i}$ in the output. I.e., since $X_1 = X_2 = X$, when X appears on both sides of the pairing, the relation will not be linear – X^2 will induce the power of 4 in the output: $e(X, X)^4$.

Corollary 1. *Let $\mathfrak{P} = (\mathsf{Setup}, \mathsf{KeyGen}, \mathsf{Comp}, \mathsf{Prove}, \mathsf{Verify})$ be "non-strictly" structure-preserving deterministic primitive as defined in Definition 5. Suppose that the discrete logarithm problem is hard in the source groups of the symmetric bilinear groups Λ_{sym} and let KeyGen, Comp, and Prove be restricted to the class of algebraic algorithms over Λ. If the output of Comp algorithm is not of the form $e(X, X)^{d_i'}$, then $\mathfrak{P}$ is not unpredictable according to Definition 3.*

Finally, allowing just a public key to contain target group elements would also induce the impossibility result in both symmetric and asymmetric settings (see the following corollary).

Corollary 2. *Let $\mathcal{G}$ be a bilinear group generator that takes as an input a security parameter 1^λ and outputs a description of bilinear groups $\Lambda = (p, \mathbb{G}_1, \mathbb{G}_2, \mathbb{G}_T, e, G_1, G_2)$. Let $\mathcal{SK}, \mathcal{PK}, \mathcal{X}, \mathcal{Y}$, and $\mathcal{P}$ be a secret key space, public key space, domain, range and a proof space, respectively. Let $\mathfrak{P} = (\mathsf{Setup}, \mathsf{KeyGen}, \mathsf{Comp}, \mathsf{Prove}, \mathsf{Verify})$ be a Structure-Preserving Deterministic Primitive as defined in Definition 2, except that the public key can contain also target group elements ($\mathcal{PK} \in \{\mathbb{G}_1, \mathbb{G}_2, \mathbb{G}_T\}^*$). Then $\mathfrak{P}$ is not unpredictable according to Definition 3.*

One can see that if Comp does not contain target group elements, but the public key does, then the result follows from Lemmas 1 and 2 with a slight modification of Lemma 1. Namely, in this case KeyGen is algebraic according to Definition 6 and one can use its extractor to compute the exponents for both source and target group elements of the public key.

4 Impossibility Results for Structure-Preserving PRF, VRF and Unique Signatures

In this section, we show how the definition of an abstract provable structure-preserving deterministic primitive (SPDP) given in Section 2 relates to the definitions of structure-preserving verifiable random function (VRF), and unique signatures (USig). We show that the security properties of an SPDP are necessary conditions for any VRF or USig to be secure.[1] We also discuss how the SPDP definition relates to structure-preserving PRF.

We recall the standard definitions of PRF and USig with a slight adaptation to our notation in Appendix A. Here we only explain how the requirements for structure-preserving variants of these primitives are captured by our SPDP definition.

4.1 Impossibility of Structure-Preserving Unique Signatures

A unique signature scheme consists of the setup Setup, key generation KeyGen, signing Sign and verification Verify algorithms as formally defined below:

Definition 7 (Unique Signatures [45]). *A function family $\sigma : \mathcal{SK} \times \mathcal{X} \to \mathcal{Y}$ is an unique signature scheme (USig) if there exists probabilistic PT algorithms* Setup *and* KeyGen, *and deterministic PT algorithms* Sign *and* Verify *(in case* Verify *is probabilistic, the adjustment to the definition is straightforward) such that:*

- *$CP \xleftarrow{\$}$ Setup(Λ) is a common parameter generation algorithm that takes as input a group description Λ and outputs the common parameters CP.*
- *$(PK, SK) \xleftarrow{\$}$ KeyGen(CP) is a key generation algorithm that takes as input the common parameters CP and outputs a public key PK and the corresponding secret key SK.*
- *$Y \leftarrow$ Sign(X, SK) is a deterministic algorithm that takes as input $X \in \mathcal{X}, SK \in \mathcal{SK}$ and outputs the signature $Y = \sigma_{SK}(X) \in \mathcal{Y}$.*
- *$0/1 \leftarrow$ Verify(X, Y, PK) is a verification algorithm that takes as input a public key PK, $X \in \mathcal{X}, Y \in \mathcal{Y}$ and verifies whether $Y = \sigma_{SK}(X)$.*

The following properties are required from an unique signature scheme:

1. **Uniqueness of the signature:** *There are no values (PK, X, Y, Y') such that $Y \neq Y'$ and* Verify$(X, Y, PK) =$ Verify$(X, Y', PK) = 1$.
2. **Security:** *For all probabilistic PT adversaries $\mathcal{A}$:*

$$\Pr\left[\begin{array}{l} CP \xleftarrow{\$} \text{Setup}(\Lambda)\,; \\ (PK, SK) \xleftarrow{\$} \text{KeyGen}(CP)\,; \\ (X, Y) \xleftarrow{\$} \mathcal{A}^{\text{Sign}(\cdot, SK)}(PK) \end{array} \middle| \text{Verify}(X, Y, PK) = 1 \wedge X \notin S \right] \leq \text{negl}(\lambda),$$

where S is the set of queries to the oracle Sign.

---

[1] Note that the requirements are necessary conditions, but maybe not sufficient conditions, e.g., in the case of VRF pseudorandomness is a stronger requirement than unpredictability.

Goldwasser and Ostrovsky [35] proposed a relaxed definition for USig. Namely, they require a proof that the signature is correct as an additional input to the verification algorithm. This proof is also an output of the signing algorithm together with the signature, but it might not be unique. This definition is sufficient to construct a VRF from USig [46].

Applying the definition of an SPDP to the context of unique signatures, one can see that Comp is the signing algorithm, and that the Prove algorithm does not exist, which is equivalent to a Prove algorithm that always returns an empty string. From the security point of view, the uniqueness of the unique signatures according to Definition 7 is the same as in Definition 2. Now we see the match for the unpredictability property.

One can see that in the security game from Definition 7 an adversary can output a forgery Y that passes the verification equation, but it can be computed in an arbitrary manner. However, in the unpredictability game from Definition 3 a forgery must be computed using the Comp algorithm. But because of the provability and uniqueness properties the former condition (Verify$(X, Y, PK) = 1$) actually implies the latter one ($Y = $ Comp(X, SK)). Therefore, the unpredictability property from Definition 3 is equivalent to the Security property of USig described above. Thus, the following corollary holds:

Corollary 3. *Assuming the hardness of the discrete logarithm problem in the base groups of Λ, there is no unique signature that is algebraic and secure.*

4.2 Impossibility of Structure-Preserving Verifiable Random Functions

The syntax of a verifiable random function (VRF) follows our generic definition of an SPDP: VRF consists of Setup, KeyGen, Comp, Prove, and Verify algorithms. Structure-preserving VRF has the same restriction on the public key space, domain, range and proof space as an SPDP, namely, they consist only of source group elements. VRF is also provable and unique, but instead of unpredictability, VRF has a pseudorandomness property (see Definition 4).

Lemma 5. *If a verifiable random function is pseudorandom according to Definition 4 then its translation to a generic deterministic primitive satisfies unpredictability as defined in Definition 3.*

Proof. The distinguisher $\mathcal{D} = (\mathcal{D}_1, \mathcal{D}_2)$ of the pseudorandomness from Definition 4 can use the adversary $\mathcal{A}$ that breaks the unpredictability according to Definition 3. $\mathcal{D}_1$ executes a copy of $\mathcal{A}$ internally and forwards the oracle queries/answers appropriately. If $\mathcal{A}$ produces an output pair (X, Y) where Y is an output value for a fresh input X that was not queried to the oracle before, then $\mathcal{D}_1$ uses X as his output and forwards Y to $\mathcal{D}_2$ who uses Y to distinguish if the returned challenge $Y_{(b)}$ is a random value or the output of the real function. If no such pair (X, Y) is produced by $\mathcal{A}$, then $\mathcal{D}$ makes a random guess. $\square$

Given the above, the impossibility of an SPDP that provides unpredictability implies the impossibility of a VRF that provides pseudorandomness:

Corollary 4. *Assuming the hardness of the discrete logarithm problem in the base groups of Λ, there is no verifiable random function that is algebraic and secure.*

One can see that this result also rules out the construction of a structure-preserving simulatable VRF (sVRF) [26], which is a special case of a VRF with the public parameters (see Definition 1 from [26]) and is a key building block in some e-cash schemes [7].

4.3 Impossibility of Structure-Preserving Pseudorandom Functions

The standard definition of a PRF (Definition 11 in Appendix A) does not feature Prove and Verify algorithms. However, the reason one wants a PRF to be structure-preserving is that one can use GS-proofs so that one party can prove to another party that the Comp algorithm was followed as prescribed. As we mentioned before, one of the examples of using PRF coupled with non-interactive zero-knowledge (NIZK) proofs is in e-cash systems [17,7].

This approach essentially adds Prove and Verify algorithms to the definition of a PRF. We formalize it later in this section. First, note that adding a proof that a PRF was computed correctly does not result in a VRF as in the latter case there is a public key and one wants to verify that the VRF was really computed with a specific public key. Whereas here one is interested in proving that the PRF was correctly computed w.r.t. any secret key, which is a weaker requirement. We are thus interested in the question of whether it is possible to construct a PRF for which one can prove the correctness of computation with GS-proofs. Or, more generally, with NIZK proofs that use only PPE for verification. With this in mind, we define a variant of Definition 2 and Definition 4, which are extensions of PRF definitions with Prove and Verify algorithms and further have Setup generate parameters for NIZK proofs. Note that one can of course always trivially prove that a PRF was computed correctly without using NIZK proofs by just revealing the secret key (as is for instance done in the Hohenberger-Waters signature scheme [39]). Formally, this proof method follows the definition we give, nevertheless, it is easy to see that a straightforward adaptation of our impossibility proof rules out the existence of PRFs (or even functions which instead of being pseudorandom are only unpredictable) that are algebraic and secure according to the suitable modification of Definition 2 and Definition 4 to allow the verification algorithm to use SK.

Non-interactive Zero-Knowledge Proof System. First, we define a Non-Interactive Zero-Know-ledge Proof System. Let R be an efficiently computable binary relation. For pairs $(W, S) \in R$ we call S the statement and W the witness. Let $\mathcal{L}$ be the language consisting of statements in R.

Definition 8 (Non-Interactive Zero-Knowledge Proof System (NIZK)). *Let $\mathcal{G}$ be a bilinear group generator that takes as an input a security parameter 1^λ and outputs a description of a bilinear group $\Lambda = (p, \mathbb{G}_1, \mathbb{G}_2, \mathbb{G}_T, e, G_1, G_2)$. The non-interactive* *zero-knowledge proof system for a language $\mathcal{L}$ consists of the following algorithms and protocols:*

- $CP \xleftarrow{\$} \mathsf{Setup}_{nizk}(\Lambda)$: *On input Λ, it outputs the common parameters (CP) for the proof system.*
- $\Pi \xleftarrow{\$} \mathsf{Prove}_{nizk}(CP, W, S)$: *On input the common parameters CP, a statement S, and a witness W, it generates a zero-knowledge proof that the witness W satisfies the statement S.*
- $0/1 \leftarrow \mathsf{Verify}_{nizk}(CP, \Pi, S)$: *On input a statement S and a proof Π, it outputs 1 if Π is valid, and 0 otherwise.*

In this work we refer to Groth-Sahai proofs [36] as the instantiation of the NIZK proof system.

Theorem 3. *[36] The Groth-Sahai ZK proof system is a non-interactive zero-knowledge (NIZK) proof system with perfect correctness, perfect soundness and composable zero-know-ledge for satisfiability of a set of equations over a bilinear group where the K-linear assumption holds.*

We refer to [36] for detailed security definitions and proofs. We also note that the results from this section and Section 5 hold for non-interactive witness-indistinguishable proofs as well that can be also instantiated with NIWI proofs by Groth and Sahai ([36]).

Combining Structure-Preserving PRF with a NIZK Proof System. Below we provide a formal definition for the construction of a PRF coupled with NIZK proofs, where verification operations are restricted to checking group membership and evaluating pairing product equations. Note that Prove_{nizk} and Verify_{nizk} algorithms take NIZK parameters and a proof statement as an input as well. Since in our case the statement is always a correctness of Comp algorithm, for consistency of notation with Definition 2 we omit the statement input.

In order to distinguish functions and variables with the same name among different primitives, we may give subscripts that represents the primitive in obvious manner. For instance Comp_{prf} denote Comp of the PRF in mind.

Definition 9 (Structure-Preserving Pseudorandom Function with a Proof of Computation Correctness). *Let $\mathcal{G}$ be a bilinear group generator that takes as an input a security parameter 1^λ and outputs a description of a bilinear group $\Lambda = (p, \mathbb{G}_1, \mathbb{G}_2, \mathbb{G}_T, e, G_1, G_2)$. A structure-preserving pseudorandom function with a proof of computation correctness with respect to Λ is a set of the following algorithms:*

- $CP \xleftarrow{\$} \mathsf{Setup}(\Lambda)$: *Run $CP_{prf} \xleftarrow{\$} \mathsf{Setup}_{prf}(\Lambda)$ and $CP_{nizk} \xleftarrow{\$} \mathsf{Setup}_{nizk}(\Lambda)$, and return $CP = (CP_{prf}, CP_{nizk})$.*
- $(PK, SK) \xleftarrow{\$} \mathsf{KeyGen}(CP)$: *Run $SK \xleftarrow{\$} \mathsf{KeyGen}_{prf}(CP_{prf})$. Set an empty string to PK. Return (PK, SK).*
- $Y \leftarrow \mathsf{Comp}(X, SK)$: *Run $Y \leftarrow \mathsf{Comp}_{prf}(X, SK)$. Return Y.*
- $\Pi \xleftarrow{\$} \mathsf{Prove}(X, SK)$: *Run $Y \leftarrow \mathsf{Comp}_{prf}(X, SK)$ and $\Pi \xleftarrow{\$} \mathsf{Prove}_{nizk}(CP_{nizk}, SK, (X, Y))$. (We consider Prove_{nizk} for the relation $R = \{(SK, (X, Y)) : Y = F_{SK}(X)\}$, where (X, Y) is the proof statement and SK is the witness.) Return Π.*
- $0/1 \leftarrow \mathsf{Verify}(X, Y, \Pi)$: *Run $b \leftarrow \mathsf{Verify}_{nizk}(CP_{nizk}, \Pi, (X, Y))$ and return b.*

We now show that the above primitive provides provability, uniqueness and pseudo-randomness based on the security of underlying PRF and NIZK.

Lemma 6. *The above pseudorandom function with a proof of computation correctness is a provable structure-preserving deterministic primitive defined in Definition 2 and provides unpredictability according to Definition 3 if the underlying PRF is pseudorandom and NIZK is correct and sound.*

Proof. Syntactical consistency can be verified by inspection. We focus on the security properties. First of all, provability holds from correctness of NIZK as Verify is identical to Verify$_{nizk}$. Uniqueness holds due to the soundness of NIZK and the fact that a PRF is deterministic. Namely, if (SK, X, Y) satisfies the relation defined by the PRF, (SK, X, Y') for $Y \neq Y'$ does not satisfy the relation since a PRF is deterministic. Thus, by the soundness of NIZK, there is no Π' that is accepted by the verification algorithm for (SK, X, Y'). The pseudorandomness holds due to the definition of a PRF (see Definition 11). The unpredictability follows from it due to the same reason as stated in Lemma 5.   □

We observe, that similarly to a pair VRF-VUF, where the pseudorandomness implies unpredictability, one can follow Lemma 6 and show that any unpredictable function can be coupled with NIZK to get SPDP with required properties.

Corollary 5. *Assuming the hardness of the discrete logarithm problem in the base groups of Λ, there is no triple of pseudorandom function (or even functions that are only unpredictable), and prove and verification algorithms that is algebraic and satisfies Definition 9.*

The Corollary follows by a trivial adaptation of the proof in the context of Definition 9. As discussed in Section 4.2, if the adversary can break the unpredictability property and compute the output value himself, then he can obviously break the pseudorandomness property.

5   Impossibility Results for Structure-Preserving Deterministic Encryption

Since deterministic encryption (DE) does not fit into Definition 2 both from the syntax and security perspective, we discuss it separately here. DE consists of the following algorithms: Setup, KeyGen, Enc, Dec (see Definition 12 from Appendix A). A structure-preserving encryption scheme has public keys, messages, and ciphertexts that consist entirely of source group elements. Moreover, the encryption and decryption algorithms perform only group and bilinear map operations.

Following Definition 2, one can view Comp as the encryption algorithm Enc from Definition 12. If we add Prove and Verify algorithms, then Prove will output a proof of the correct computation of the encryption algorithm. Below we provide a formal definition for the DE with this in mind, similarly as we did for the PRF in Section 4.3. We note that even though we assume the instantiation of Prove and Verify with GS-proofs our result holds in general for proofs that require only PPE for verification.

Definition 10 (Structure-Preserving Deterministic Encryption with a Proof of Encryption Correctness). *Let $\mathcal{G}$ be a bilinear group generator that takes as an input a security parameter 1^λ and outputs a description of a bilinear group $\Lambda = (p, \mathbb{G}_1, \mathbb{G}_2, \mathbb{G}_T, e, G_1, G_2)$. Structure-Preserving Deterministic Encryption with a NIZK proof of encryption correctness with respect to Λ is a tuple of the following PT algorithms:*

- *$CP \xleftarrow{\$} \mathsf{Setup}(\Lambda)$: Run $CP_{de} \xleftarrow{\$} \mathsf{Setup}_{de}(\Lambda)$ and $CP_{nizk} \xleftarrow{\$} \mathsf{Setup}_{nizk}(\Lambda)$, and return $CP = (CP_{de}, CP_{nizk})$.*
- *$(PK, SK) \xleftarrow{\$} \mathsf{KeyGen}(CP)$: Run $(SK, PK) \xleftarrow{\$} \mathsf{KeyGen}_{de}(CP_{de})$. Return (PK, SK).*
- *$Y \leftarrow \mathsf{Comp}(X, SK)$: Compute PK from SK. Run $Y_{de} \leftarrow \mathsf{Enc}(X, PK)$. Return $Y = (Y_{de}, PK)$.*
- *$\Pi \xleftarrow{\$} \mathsf{Prove}(X, SK)$: Compute PK from SK, run $Y_{de} \leftarrow \mathsf{Enc}(X, PK)$ and $\Pi \xleftarrow{\$} \mathsf{Prove}_{nizk}(CP_{nizk}, X, (Y, PK))$.*
  (We consider Prove_{nizk} for the relation $R = \{(X, (Y_{de}, PK)) : Y_{de} = \mathsf{Enc}(X, PK)\}$, where (Y_{de}, PK) is the proof statement and X is the witness.) Return Π.
- *$0/1 \leftarrow \mathsf{Verify}(X, Y, \Pi)$: Run $b \leftarrow \mathsf{Verify}_{nizk}(CP_{nizk}, \Pi, (Y, PK))$ and return b.*

Note that X is not really referred in Verify but it is anyway consistent to the syntax of an SPDP.

We would like to point out that since Comp, as an encryption algorithm, takes a public key as an input, no unpredictability-style property can hold in this case. Thus, we cannot require both uniqueness and unpredictability properties (see Definitions 2 and 3) to hold for the primitive from the Definition 10 to be secure. Nevertheless, we show that the latter primitive, which is provable and unique, cannot exist.

First, we show that two security properties, provability and uniqueness, of an SPDP hold for the above DE coupled with (GS) NIZK-proofs. Formally:

Lemma 7. *The above deterministic encryption scheme with a proof of encryption correctness has the provability and uniqueness properties defined in Definition 2, if NIZK is correct and sound.*

The proof of the above lemma is the same as that for Lemma 6 with obvious modification and thus omitted.

Theorem 4. *Assuming the hardness of the discrete logarithm problem in the base groups of Λ, there is no algebraic structure-preserving deterministic encryption scheme, which is secure and where encryption can be verified by an algorithm, which takes X, Y, Π, PK as input and only performs group operations and PPE evaluations.*

Proof. As we mentioned before, by the definition of DE and the correctness of decryption, the uniqueness property holds if we consider Enc to be the Comp algorithm. According to Lemma 1 the ciphertext that encrypts a group element X looks as follows: $\mathsf{Comp}(X, PK) = Y = (G^{a_1} X^{b_1}, \ldots, G^{a_\ell} X^{b_\ell})$, where $a_1, \ldots, a_\ell, b_1, \ldots, b_\ell$ are constants in $\mathbb{Z}_p$, and G is a group generator. To encrypt X, it is obvious that G^{a_i} and $b_i, i = 1, \ldots, \ell$ should be efficiently derivable from the public key. This means that the ciphertext can be decrypted using the public key. $\qquad\square$

6   Conclusion

In this paper we proved that it is impossible to construct algebraic structure-preserving VRF, VUF and USig. It is also shown that PRF and DE coupled with non-interactive proof system cannot be structure-preserving, either. We further extend our results to "non-strictly" structure preserving primitives, which are allowed to have target group elements in their public keys and ranges. Regarding the latter, we show that such primitives cannot be constructed for asymmetric bilinear maps and that the possible constructions for symmetric maps are severely restricted on the operation they can use.

Although our results are restricted to the class of algebraic algorithms, all known constructions of structure-preserving primitives consist of algebraic algorithms. Finding constructions of secure structure-preserving algorithms that allow non-algebraic operations but whose correctness of computation still can be verified using a system of PPE is an interesting problem. We also would like to point out that it might be possible to extend our impossibility result to the quasi-deterministic case where the uniqueness condition can be relaxed to have at most $poly(\lambda)$ output values corresponding to each input value.

Finally, we note that the deterministic primitives might exist in a restricted form, where only one query to the oracle is allowed. Namely, one-time deterministic primitives might still be possible in the world of structure-preserving cryptography.

Acknowledgments. The authors would like to thank Kristiyan Haralambiev for the useful discussions and the anonymous reviewers for their helpful comments and suggestions. The research leading to these results was supported in part by the European Community's Seventh Framework Programme for the projects ABC4Trust (grant agreement no. 257782) and PERCY (grant agreement no. 321310).

References

1. Abdalla, M., Catalano, D., Fiore, D.: Verifiable random functions from identity-based key encapsulation. In: EUROCRYPT 2009. LNCS, vol. 5479, pp. 554–571. Springer, Heidelberg (2009)
2. Abe, M., Chase, M., David, B., Kohlweiss, M., Nishimaki, R., Ohkubo, M.: Constant-size structure-preserving signatures: Generic constructions and simple assumptions. In: Wang, X., Sako, K. (eds.) ASIACRYPT 2012. LNCS, vol. 7658, pp. 4–24. Springer, Heidelberg (2012)
3. Abe, M., Fuchsbauer, G., Groth, J., Haralambiev, K., Ohkubo, M.: Structure-preserving signatures and commitments to group elements. In: Rabin, T. (ed.) CRYPTO 2010. LNCS, vol. 6223, pp. 209–236. Springer, Heidelberg (2010)
4. Abe, M., Groth, J., Haralambiev, K., Ohkubo, M.: Optimal structure-preserving signatures in asymmetric bilinear groups. In: Rogaway, P. (ed.) CRYPTO 2011. LNCS, vol. 6841, pp. 649–666. Springer, Heidelberg (2011)
5. Abe, M., Haralambiev, K., Ohkubo, M.: Group to group commitments do not shrink. In: Pointcheval, D., Johansson, T. (eds.) EUROCRYPT 2012. LNCS, vol. 7237, pp. 301–317. Springer, Heidelberg (2012)
6. Belenkiy, M., Chase, M., Kohlweiss, M., Lysyanskaya, A.: P-signatures and noninteractive anonymous credentials. In: Canetti, R. (ed.) TCC 2008. LNCS, vol. 4948, pp. 356–374. Springer, Heidelberg (2008)

7. Belenkiy, M., Chase, M., Kohlweiss, M., Lysyanskaya, A.: Compact e-cash and simulatable VRFs revisited. In: Shacham, H., Waters, B. (eds.) Pairing 2009. LNCS, vol. 5671, pp. 114–131. Springer, Heidelberg (2009)
8. Bellare, M., Boldyreva, A., O'Neill, A.: Deterministic and efficiently searchable encryption. In: Menezes, A. (ed.) CRYPTO 2007. LNCS, vol. 4622, pp. 535–552. Springer, Heidelberg (2007)
9. Bellare, M., Fischlin, M., O'Neill, A., Ristenpart, T.: Deterministic encryption: Definitional equivalences and constructions without random oracles. In: Wagner, D. (ed.) CRYPTO 2008. LNCS, vol. 5157, pp. 360–378. Springer, Heidelberg (2008)
10. Bellare, M., Micciancio, D., Warinschi, B.: Foundations of group signatures: Formal definitions, simplified requirements, and a construction based on general assumptions. In: Biham, E. (ed.) EUROCRYPT 2003. LNCS, vol. 2656, pp. 614–629. Springer, Heidelberg (2003)
11. Bellare, M., Palacio, A.: The knowledge-of-exponent assumptions and 3-round zero-knowledge protocols. In: Franklin, M. (ed.) CRYPTO 2004. LNCS, vol. 3152, pp. 273–289. Springer, Heidelberg (2004)
12. Boldyreva, A., Fehr, S., O'Neill, A.: On notions of security for deterministic encryption, and efficient constructions without random oracles. In: Wagner, D. (ed.) CRYPTO 2008. LNCS, vol. 5157, pp. 335–359. Springer, Heidelberg (2008)
13. Camenisch, J., Dubovitskaya, M., Haralambiev, K.: Efficient structure-preserving signature scheme from standard assumptions. In: Visconti, I., De Prisco, R. (eds.) SCN 2012. LNCS, vol. 7485, pp. 76–94. Springer, Heidelberg (2012)
14. Camenisch, J., Dubovitskaya, M., Neven, G.: Oblivious transfer with access control. In: ACM CCS 2009. ACM Press (November 2009)
15. Camenisch, J., Dubovitskaya, M., Neven, G., Zaverucha, G.M.: Oblivious transfer with hidden access control policies. In: Catalano, D., Fazio, N., Gennaro, R., Nicolosi, A. (eds.) PKC 2011. LNCS, vol. 6571, pp. 192–209. Springer, Heidelberg (2011)
16. Camenisch, J., Haralambiev, K., Kohlweiss, M., Lapon, J., Naessens, V.: Structure preserving CCA secure encryption and applications. In: Lee, D.H., Wang, X. (eds.) ASIACRYPT 2011. LNCS, vol. 7073, pp. 89–106. Springer, Heidelberg (2011)
17. Camenisch, J., Hohenberger, S., Lysyanskaya, A.: Compact e-cash. In: Cramer, R. (ed.) EUROCRYPT 2005. LNCS, vol. 3494, pp. 302–321. Springer, Heidelberg (2005)
18. Camenisch, J., Kiayias, A., Yung, M.: On the portability of generalized schnorr proofs. In: Joux, A. (ed.) EUROCRYPT 2009. LNCS, vol. 5479, pp. 425–442. Springer, Heidelberg (2009)
19. Camenisch, J., Lysyanskaya, A.: An efficient system for non-transferable anonymous credentials with optional anonymity revocation. In: Pfitzmann, B. (ed.) EUROCRYPT 2001. LNCS, vol. 2045, pp. 93–118. Springer, Heidelberg (2001)
20. Camenisch, J., Lysyanskaya, A.: A signature scheme with efficient protocols. In: Cimato, S., Galdi, C., Persiano, G. (eds.) SCN 2002. LNCS, vol. 2576, pp. 268–289. Springer, Heidelberg (2003)
21. Camenisch, J., Lysyanskaya, A.: Signature schemes and anonymous credentials from bilinear maps. In: Franklin, M. (ed.) CRYPTO 2004. LNCS, vol. 3152, pp. 56–72. Springer, Heidelberg (2004)
22. Camenisch, J., Neven, G., Shelat, A.: Simulatable adaptive oblivious transfer. In: Naor, M. (ed.) EUROCRYPT 2007. LNCS, vol. 4515, pp. 573–590. Springer, Heidelberg (2007)
23. Camenisch, J., Shoup, V.: Practical verifiable encryption and decryption of discrete logarithms. In: Boneh, D. (ed.) CRYPTO 2003. LNCS, vol. 2729, pp. 126–144. Springer, Heidelberg (2003)
24. Canetti, R.: Universally composable security: A new paradigm for cryptographic protocols. In: 42nd FOCS. IEEE Computer Society Press (October 2001)
25. Canetti, R., Goldreich, O., Halevi, S.: The random oracle methodology, revisited (preliminary version). In: 30th ACM STOC, pp. 209–218. ACM Press (May 1998)

26. Chase, M., Lysyanskaya, A.: Simulatable VRFs with applications to multi-theorem NIZK. In: Menezes, A. (ed.) CRYPTO 2007. LNCS, vol. 4622, pp. 303–322. Springer, Heidelberg (2007)

27. Cramer, R., Shoup, V.: Design and analysis of practical public-key encryption schemes secure against adaptive chosen ciphertext attack. SIAM Journal on Computing 33(1), 167–226 (2003)

28. Dodis, Y.: Efficient construction of (distributed) verifiable random functions. In: Desmedt, Y.G. (ed.) PKC 2003. LNCS, vol. 2567, pp. 1–17. Springer, Heidelberg (2002)

29. Dodis, Y., Yampolskiy, A.: A verifiable random function with short proofs and keys. In: Vaudenay, S. (ed.) PKC 2005. LNCS, vol. 3386, pp. 416–431. Springer, Heidelberg (2005)

30. El Gamal, T.: A public key cryptosystem and a signature scheme based on discrete logarithms. In: Blakely, G.R., Chaum, D. (eds.) CRYPTO 1984. LNCS, vol. 196, pp. 10–18. Springer, Heidelberg (1985)

31. Fiat, A., Shamir, A.: How to prove yourself: Practical solutions to identification and signature problems. In: Odlyzko, A.M. (ed.) CRYPTO 1986. LNCS, vol. 263, pp. 186–194. Springer, Heidelberg (1987)

32. Freedman, M.J., Ishai, Y., Pinkas, B., Reingold, O.: Keyword search and oblivious pseudorandom functions. In: Kilian, J. (ed.) TCC 2005. LNCS, vol. 3378, pp. 303–324. Springer, Heidelberg (2005)

33. Galbraith, S.D., Paterson, K.G., Smart, N.P.: Pairings for cryptographers. Discrete Applied Mathematics 156(16), 3113–3121 (2008)

34. Goldreich, O., Goldwasser, S., Micali, S.: How to construct random functions. In: 25th FOCS. IEEE Computer Society Press (October 1984)

35. Goldwasser, S., Ostrovsky, R.: Invariant signatures and non-interactive zero-knowledge proofs are equivalent (extended abstract). In: Brickell, E.F. (ed.) CRYPTO 1992. LNCS, vol. 740, pp. 228–245. Springer, Heidelberg (1993)

36. Groth, J., Sahai, A.: Efficient non-interactive proof systems for bilinear groups. In: Smart, N. (ed.) EUROCRYPT 2008. LNCS, vol. 4965, pp. 415–432. Springer, Heidelberg (2008)

37. Hazay, C., Lindell, Y.: Efficient protocols for set intersection and pattern matching with security against malicious and covert adversaries. In: Canetti, R. (ed.) TCC 2008. LNCS, vol. 4948, pp. 155–175. Springer, Heidelberg (2008)

38. Hofheinz, D., Jager, T.: Tightly secure signatures and public-key encryption. In: Safavi-Naini, R., Canetti, R. (eds.) CRYPTO 2012. LNCS, vol. 7417, pp. 590–607. Springer, Heidelberg (2012)

39. Hohenberger, S., Waters, B.: Short and stateless signatures from the RSA assumption. In: Halevi, S. (ed.) CRYPTO 2009. LNCS, vol. 5677, pp. 654–670. Springer, Heidelberg (2009)

40. Hohenberger, S., Waters, B.: Constructing verifiable random functions with large input spaces. In: Gilbert, H. (ed.) EUROCRYPT 2010. LNCS, vol. 6110, pp. 656–672. Springer, Heidelberg (2010)

41. Jarecki, S., Liu, X.: Efficient oblivious pseudorandom function with applications to adaptive OT and secure computation of set intersection. In: Reingold, O. (ed.) TCC 2009. LNCS, vol. 5444, pp. 577–594. Springer, Heidelberg (2009)

42. Jarecki, S.: Handcuffing big brother: an abuse-resilient transaction escrow scheme. In: Cachin, C., Camenisch, J.L. (eds.) EUROCRYPT 2004. LNCS, vol. 3027, pp. 590–608. Springer, Heidelberg (2004)

43. Kiayias, A., Yung, M.: Group signatures with efficient concurrent join. In: Cramer, R. (ed.) EUROCRYPT 2005. LNCS, vol. 3494, pp. 198–214. Springer, Heidelberg (2005)

44. Liskov, M.: Updatable zero-knowledge databases. In: Roy, B. (ed.) ASIACRYPT 2005. LNCS, vol. 3788, pp. 174–198. Springer, Heidelberg (2005)

45. Lysyanskaya, A.: Unique signatures and verifiable random functions from the DH-DDH separation. In: Yung, M. (ed.) CRYPTO 2002. LNCS, vol. 2442, pp. 597–612. Springer, Heidelberg (2002)

46. Micali, S., Rabin, M.O., Vadhan, S.P.: Verifiable random functions. In: 40th FOCS. IEEE Computer Society Press (October 1999)
47. Micali, S., Reyzin, L.: Soundness in the public-key model. In: Kilian, J. (ed.) CRYPTO 2001. LNCS, vol. 2139, pp. 542–565. Springer, Heidelberg (2001)
48. Micali, S., Rivest, R.L.: Micropayments revisited. In: Preneel, B. (ed.) CT-RSA 2002. LNCS, vol. 2271, pp. 149–163. Springer, Heidelberg (2002)
49. Pedersen, T.P.: Non-interactive and information-theoretic secure verifiable secret sharing. In: Feigenbaum, J. (ed.) CRYPTO 1991. LNCS, vol. 576, pp. 129–140. Springer, Heidelberg (1992)
50. Schnorr, C.-P.: Efficient signature generation by smart cards. Journal of Cryptology 4(3) (1991)

A   Definitions of PRF and DE

In this section we recall the definitions of the cryptographic primitives that we are concerned with, i.e., pseudorandom functions and deterministic encryption. To make the notation consistent throughout the paper we slightly adjust the original definitions of the primitives described in this section. Also, for all primitives we assume that a group description is a public parameter and is given as input to a setup algorithm.

For all definitions we consider the following. Let $\mathcal{G}$ be a bilinear group generator that takes as input a security parameter 1^λ and outputs a description of bilinear groups $\Lambda = (p, \mathbb{G}_1, \mathbb{G}_2, \mathbb{G}_T, e, G_1, G_2)$. Let $\mathcal{SK}, \mathcal{X}, \mathcal{Y}, \mathcal{P}$ be the secret key space, public key space, domain, range and a proof space, respectively.

A.1   Pseudorandom Functions

Pseudorandom functions (PRF) were introduced in [34]. Below we give an adaptation of the original definition to the notation used in this paper.

Definition 11 (Pseudorandom Function). *A function family $F : \mathcal{SK} \times \mathcal{X} \to \mathcal{Y}$ is called a pseudorandom function (PRF) if there are probabilistic PT algorithms* Setup *and* KeyGen *and a deterministic PT algorithm* Comp *such that:*

- *$CP \xleftarrow{\$} \mathsf{Setup}(\Lambda)$ is an algorithm that takes as input a group description Λ and outputs the common parameters CP.*
- *$SK \xleftarrow{\$} \mathsf{KeyGen}(CP)$ is an algorithm that takes as input the common parameters CP and outputs a (secret) key $SK \in \mathcal{SK}$.*
- *$Y \leftarrow \mathsf{Comp}(X, SK)$ is a deterministic algorithm that takes as input $X \in \mathcal{X}$ and $SK \in \mathcal{SK}$ and outputs the function value $Y = F_{SK}(X) \in \mathcal{Y}$.*

The following property is required from a PRF:
Pseudorandomness: *For all probabilistic PT distinguishers $\mathcal{D} = (\mathcal{D}_1, \mathcal{D}_2)$ we have*

$$\Pr\left[\begin{array}{l} CP \xleftarrow{\$} \mathsf{Setup}(\Lambda)\,;\ SK \xleftarrow{\$} \mathsf{KeyGen}(CP)\,; \\ (X, st) \leftarrow \mathcal{D}_1{}^{\mathsf{Comp}(\cdot, SK)}(CP)\,; \\ Y_{(0)} \leftarrow \mathsf{F}_{SK}(X)\,;\ Y_{(1)} \xleftarrow{\$} \mathcal{Y}\,;\ b \xleftarrow{\$} \{0,1\}\,; \\ b' \xleftarrow{\$} \mathcal{D}_2{}^{\mathsf{Comp}(\cdot, SK)}(Y_{(b)}, st) \end{array} \middle| \ b = b' \wedge X \notin S \right] \leq \frac{1}{2} + \mathsf{negl}(\lambda),$$

where S is the set of queries to the oracle Comp.

A.2 Deterministic Encryption

Deterministic encryption was introduced by was introduced by Bellare, Boldyreva, and ONeill in [8]. Here we provide a slightly adapted definition of DE.

Definition 12 (Deterministic Encryption). *A function family* $F : \mathcal{PK} \times \mathcal{X} \rightarrow \mathcal{Y}$ *is called a structure-preserving deterministic encryption (SPDE) if there are probabilistic PT algorithms* Setup *and* KeyGen, *and a deterministic PT algorithms* Enc *amd* Dec *such that:*

- *$CP \xleftarrow{\$}$ Setup(Λ) is a probabilistic algorithm that takes as input the security parameter and outputs the common parameters CP that consists of the group description $\Lambda = (p, \mathbb{G}_1, \mathbb{G}_2, \mathbb{G}_T, e, G_1, G_2)$ generated by $\mathcal{G}(1^\lambda)$ and possibly also constants in $\mathbb{Z}_p$.*
- *$(PK, SK) \xleftarrow{\$}$ KeyGen(CP) is a probabilistic key generation algorithm that takes as input the common parameters and outputs a public key PK and a secret key SK. It is assumed without loss of generality that SK includes PK.*
- *$Y \leftarrow$ Enc(X, PK) is a deterministic algorithm that takes as input $X \in \mathcal{X}$ and a public key PK and outputs a ciphertext $Y \in \mathcal{Y}$.*
- *$X \leftarrow$ Dec(Y, SK) is a deterministic algorithm that takes as input a ciphertext $Y \in \mathcal{Y}$ and a secret key SK and outputs a plaintext $X \in \mathcal{X}$.*

Intuitively, the security notion for deterministic encryption, called a PRIV game, that was introduced in [8], states that it should be hard to guess any public key independent information of a list of messages given their encryptions, as long as the list has component-wise high min-entropy. Or, in other words, the adversary should not be able to distinguish ciphertexts that correspond to messages that come from two message distributions with high min-entropy. We refer the reader to [8] for the formal definition of the game.

Author Index